saur

International Encyclopedia of Abbreviations and Acronyms in Science and Technology

Compiled by Michael Peschke

Internationale Enzyklopädie der Abkürzungen und Akronyme in Wissenschaft und Technik

Bearbeitet von Michael Peschke

Vol 10 / Band 10 · SUPPLEMENT

M–Z

K · G · Saur München 1999

Die Deutsche Bibliothek – CIP-Einheitsaufnahme

Peschke, Michael:
International encyclopedia of abbreviations and acronyms in science and technology = Internationale Enzyklopädie der Abkürzungen und Akronyme in Wissenschaft und Technik / comp. by Michael Peschke. – München : Saur.
ISBN 3-598-22970-4
Vol.10 – Suppl. M–Z. – 1999
ISBN 3-598-22988-7

Gedruckt auf säurefreiem Papier/
Printed on acid-free paper

Part of Reed Elsevier
Printed in the Federal Republic of Germany

Datenerfassung und Satz / Computer-controlled keyboarding, data preparation and automatic data processing by
Michael Peschke, Berlin

Druck/Printed by Strauss Offsetdruck, Mörlenbach
Binden/Bound by Buchbinderei Schaumann, Darmstadt

ISBN 3-598-22970-4 (17 Volume-Set)
ISBN 3-598-22988-7 (Vol 10)

Contents / Inhalt

Supplement 2

M - Z

Preface

Abbreviations, Acronyms, and Initials in all their various forms confront us today more than ever. We encounter them in all areas of daily life, both at home and at work: they are often the basis of documentation and communication and frequently themselves contain the central information in a given context. Decoding them is, however, often difficult, sometimes even impossible because adequate reference sources are either not available or not readily accessible.

As the use and the invention of abbreviations is increasing world-wide, no one reference work can hope to answer all queries concerning them correctly. The present *International Encyclopedia of Abbreviations in Science and Technology* with appr. 680,000 entries in 8 volumes attempts to encompass the most important acronyms from the most diverse general and scientific fields. It does, however, limit itself to languages which use the Roman alphabet.

The International Encyclopedia also lists common vernacular abbreviations in general use. Thus it goes beyond official terminology. Historical abbreviations have also been included.

The flood of abbreviations is growing constantly and was reason to add two supplements to the edition. The supplements have appr. 170,000 entries again.

Where the use of punctuation within the abbreviation is erratic, the most commonly-used form has been given preference. Further information of this nature can be found in the "Notes for the User".

Notes for the User

1. The abbreviations are presented in alphabetical order from a to z. Ä, æ, ç, ö, ö, , ü are arranged as ae, c, oe, ue. Symbols and (diacritic) signs such as . - () [] ^ ^° & + are of no importance in regard to the alphabetical order.
1. 1. Alphanumeric sequences are listed in the alphabetical place of the respective letter(s).
1.2. Roman numerals are arranged in their value as literal character. Arabs numerals in alphanumeric sequences are arranged in their value as number. Abbreviations beginning with a number are arranged after the character z.
2. Styles of type are used as follows:
2.1. boldface for the abbreviation
2.2. plain for the remaining text
3. Within each entry (beginning with the abbreviation protruding to the left), the meanings are likewise listed a) in alphabetical order of the label indicating the language in which the abbreviation is used, b) in alphabetical order.
3.1. Each abbreviation is followed by the label indicating the area or sphere in which the abbreviation is used and/or the pertinent language.
3.2. The sign of equality (=) normally separates explanatory remarks (usually abbreviated) from each other as well as from the remaining text. In connection with the grammalogue of a language the colon (:) refers to a different expression in another language of the respective meaning.
3.3. Any additional explanations are enclosed in parentheses ().
3.4. Brackets [] are components of abbreviations as well as of meanings. However, parentheses are also used for marking a word that can be substituted for the preceding word, and for marking that part of a word, which, if omitted, will result in a different meaning or spelling.

Language Codes

A	Afrikaans	kr	Croatian
a	Danish	l	Latin
b	Bulgarian	ma	Malayan
c	Czech	N	Norwegian
ch	Chinese	n	Dutch
d	German	P	Polish
e	English	p	Portuguese
F	Finnish	ph	Philippine
f	French	R	Russian
g	Greek	r	Swedish
H	Hungarian	ru	Romanian
h	Hebrew	S	Slovenian
i	Italian	s	Spanish
ir	Irish	se	Serbian
in	Indonesian	U	Ukrainian
j	Japanese	z	Turkish
ko	Korean		

Berlin, October 1998

Michael Peschke

Vorwort

Abkürzungen, Akronyme und Initialen in ihren mannigfachen Erscheinungsformen begegnen uns heute überall und mehr denn je. Sie treten in allen Bereichen des täglichen und beruflichen Lebens auf; sie sind oft Grundlage von Dokumentation und Kommunikation und vielfach die entscheidenden Informationsträger. Ihre Bedeutung zu entschlüsseln, ist gleichwohl oft schwierig, manchmal unmöglich, weil geeignete Hilfsmittel fehlen oder diese nur schwer greifbar sind.

Weil der Gebrauch und die Erfindung von Abkürzungen und Abbreviaturen in der ganzen Welt ständig zunehmen, vermag niemand allein auf diesem Gebiet alle Fragen zutreffend zu beantworten. Die vorliegende *Internationale Enzyklopädie der Abkürzungen und Akronyme in Wissenschaft und Technik*, die ca. 680.000 Einträge in 8 Bänden verzeichnet, versucht, möglichst umfassend die wichtigsten Akronyme aus den unterschiedlichsten Fach- und Lebensbereichen zusammenzustellen. Dabei beschränkt sie sich freilich auf die "lateinisch" schreibende Welt.

Die Internationale Enzyklopädie berücksichtigt auch alltagssprachlich übliche und gebräuchliche Formen. Sie enthält also nicht nur die offiziellen Versionen. Auch Abkürzungen, die der Vergangenheit angehören, sind aufgeführt.

Die Flut neuer Abkürzungen gab Veranlassung zwei Supplementbände an die Hauptausgabe anzufügen. Diese verzeichnen nochmals ca. 170.000 Einträge.

Abkürzungen, die teils mit, teils ohne Punkt geschrieben werden, sind nach der jeweils überwiegenden Verwendung verzeichnet. Im übrigen geben die "Hinweise für den Benutzer" alle Erläuterungen über die Anlage der Einträge wie des gesamten Werkes.

Hinweise für den Benutzer

1. Die Abkürzungen werden in der alphabetischen Reihenfolge a...z gebracht. Ä, æ, ç, ö, ö, , ü werden wie ae, c, oe, ue eingereiht. Zeichen wie . - ( ) [ ] ^ ^ ° & + sind für die alphabetische Reihe ohne Bedeutung.

1.1. Alphanumerische Folgen stehen an der von dem (den) Buchstaben bestimmten alphabetischen Stelle.

1.2. Römische Zahlzeichen werden nach ihrem Buchstabenwert eingereiht. Arabische Zahlzeichen innerhalb alphanumerischer Folgen werden nach ihrem Zahlenwert geordnet. Abkürzungen, die mit einer arabischen Ziffer beginnen, stehen nach dem Buchstaben Z.

2. Die Schriftarten werden wie folgt verwendet:

2.1. halbfett für die Abkürzung

2.2. Grundschrift für den übrigen Text

3. Innerhalb einer (mit einer links herausgerückten Abkürzung beginnenden) Wortstelle sind die Bedeutungen a) nach ihrem Bedeutungsbereich (Sortierung Sprache), b) in ihrer alphabetischen Reihenfolge aufgeführt.

3.1. Der Abkürzung folgt die Angabe der Sprache.

3.2. Das Gleichheitszeichen (=) trennt im allgemeinen abgekürzte erläuternde Hinweise voneinander und vom übrigen Text. Der Doppelpunkt (:) in Verbindung mit einem Sprachkürzel verweist auf anderssprachige Bezeichnungen der entsprechenden Bedeutung.

3.3. Sonstige Erläuterungen stehen in runden Klammem ().

3.4. Eckige Klammern [] schließen Bestandteile von Abkürzungen und von Bedeutungen ein. Bei den Bedeutungen werden diese Klammern mitunter auch dazu benutzt, ein Wort zu kennzeichnen, das an die Stelle des vorangehenden gesetzt werden kann, bzw. Wortteile zu markieren, die, läßt man sie weg, eine andere Bedeutung oder eine andere Schreibung ergeben.

Sprachenkürzel

A	afrikaans	kr	kroatisch
a	dänisch	l	lateinisch
b	bulgarisch	ma	malaiisch
c	tschechisch	N	norwegisch
ch	chinesisch	n	holländisch
d	deutsch	P	polnisch
e	englisch	p	portugiesich
F	finnisch	ph	philippinisch
f	französisch	R	russisch
g	griechisch	r	schwedisch
H	ungarisch	ru	rumänisch
h	hebräisch	S	slowenisch
i	italienisch	s	spanisch
ir	irisch	se	serbisch
in	indonesisch	U	ukrainisch
j	japanisch	z	türkisch
ko	koreanisch		

Berlin, Oktober 1998

Michael Peschke

Supplement

M - Z

M

M a: Centrum Demokraterne
- d: Menschen machen Medien
- e: Manganese
- e: Mathematics
- e: Mega
- e: Million
- e: Miniatures
- e: Monatsmittel
- e: Monthly
- e: More
- e: Thousand
- f: Magasin
- f: Maire
- f: Maître
- f: Majoritaire
- f: Majorité
- f: Marchand
- f: Marchandise
- f: Marché
- f: Mardi
- f: Marié
- f: Masque
- f: Matin
- f: Médecine
- f: Médical
- f: Médicament
- f: Médicinal
- f: Membre
- f: Mémoires
- f: Mensuel
- f: Mercredi
- f: Midi
- f: Mille
- f: Mixte
- f: Modèle
- f: Moderne
- f: Monsieur
- f: Mort
- f: Mot
- N: Med
- r: Mach-tal
- r: Marin
- r: Matros
- r: Minnesregler
- r: Missvisning
- R: Modifikatsirovanny
- R: Modifikatsiya
- r: Motståndskraft
- s: Máquina
- s: Marina
- s: Militar
- s: Minador
- s: Minuto
- s: Mortero

m e: beliebige ganze Zahl
- e: micrometer
- e: milli

m. a: med
- e: married
- e: masculine
- e: meter
- e: meters

M. e: Mark
- e: Marquess
- N: Marine

m/ a: model

M0 e: Macrophagelike

M1 e: The Crab Nebula
- e: The standard infantry gun for US soldiers

3M e: Minnesota Mining and Manufacturing

m³ e: cubic meter

M-4 e: Bison [Soviet]

M-12 e: Mail [Soviet]

M-14 e: MAC Airlift Simulation Model

M16 e: The Eagle Nebula
- e: standard gun for US infantry

M31 e: Andromeda Galaxy

M-42 e: Duster [antiaircraft weapon]

M45 e: The Pleiades Nebula

M-47 e: Dragon

M 60 r: Modell 1960

M-60 e: Tank, combat

M548 e: cargo carrier

MA a: Medlem af Amtsrådet
- a: Mentalalder
- d: Mathematik
- d: Mediaananalyse
- e: Commonwealth of Massachusetts
- e: Deputy Commander for Maintenance
- e: Division of Military Application
- e: Maintenance [Administrator]
- e: Maintenance Agency
- e: Maintenance and Adaptation
- e: Maleic Anhydride
- e: Management Agent
- e: Marijuana Anonymous
- e: Mask Aligner
- e: Maximum Averages
- e: Mean Anomaly
- e: Mean Motion
- e: Meteorological Applications
- e: Milliamperes
- e: Mission Abort
- e: Mission Application
- e: Morocco
- f: Machine à Affranchir
- f: Machine Agricole
- f: Magistrature Assise
- f: Maison Assurée
- f: Maison d'Arrêt
- f: Maître-Artisan
- f: Majorité Absolue
- f: Mandat d'Arrêt
- f: Marchand Ambulant
- f: Matériel Agricole
- f: Mécanique Analytique
- f: Mesures Agraires
- f: Ministère de l'Agriculture
- f: Monnaie d'Aloi
- f: Monnaie d'Appoint
- f: Montant des Allocations
- f: Mort Accidentelle
- f: Moyen Âge
- N: Marinens Artilleri
- N: Meierienes Arbeidsgiverforening
- N: Motorførernes Avholdsforbund
- p: Ministério de Agricultura
- r: Materialadministration
- r: Militärapoteket
- s: Mecánico Ajustador

mA e: Milliamp

Ma d: Studiengang Mathematik [Diplom]
- e: Millions of years before the present
- f: Mardi
- N: Maskin

ma. a: mandag

Ma. s: Martes

Mä d: Männer

M.A. e: Master of Arts
- l: Magister Artium

M&A e: Mergers and Acquisitions

M48A3 e: Tank, combat

M88A1 e: Recovery vehicle

MAA a: Medlem af Akademisk Arkitektforening
- e: Material Access Area
- e: Medical Artists Association
- e: Mission Area Analysis
- e: Moist-Adiabatic Adjustment
- e: Mutual Assistance Association
- e: Myositis Association of America

- s: Mecánico Ajustador de Armas
MAAC e: Metro Atlantic Athletic Conference
- e: Mid-Atlantic Athletic Conference
MAAG e: Military Assistance Advisory Group
MAAMA e: Middleton Air Material Agency
MAAP e: Maintenance And Administration Panel
- e: More Able Autistic Persons
MAAR e: Monthly Associate Administrator's Review
MAAS e: Manpower Allocations and Accounting System
Maast. e: Maastrichtian
MAB d: Marienberg, Sachsen
- d: Der Mensch und die Biosphäre
- e: Man and [the] Biosphere [Program[me]]
- e: Marine Amphibious Brigade
- e: Material Advisory Board
- e: Mauritius Advertising Bureau
- f: Mise À Bord
- f: Programme sur l'homme et la biosphère
- s: Programa sobre el Hombre y la Biosfera
MABE e: MA Beamforming Equipment
MABIE e: Mauna Loa Aerosol Backscatter Intercomparison Experiment
MABL e: Marine Atmospheric Boundary Layer
MABS e: Marine Air Base Squadron
- e: Methyl Methacrylate/ABS copolymer
- e: Muscovite-Almandine-Biotite-sillimanite
MaBV d: Makler- und Bauträgerverordnung
MAC e: Machine Aided Cognition
- e: Macintosh
- e: MacPaint
- e: Macro
- e: Management Action Center
- e: Management Advisory Committee
- e: Mandatory Access Control[ler]
- e: Manufacturers Address Code
- e: Maricopa Agricultural Center
- e: Material Accounting Center
- e: Material Accounting system
- e: Maximum Allowable Cost
- e: Media Access Control[ler]
- e: Medium Access [Control[ler]]
- e: Membrane Attack Complex
- e: Memory Access Control[ler]
- e: Message Authentication Check
- e: Message Authentification Code
- e: Middle Atlantic Conference
- e: Middle Atmosphere Cooperation
- e: Military Air Command
- e: Minimal Antibiotic Concentration
- e: Ministry of Agriculture and Livestock
- e: Mobile Air Conditioner
- e: Move, Add, Change
- e: Multi Access Computer
- e: Multiphase Atmospheric Chemistry
- e: Multiple Access Controller
- e: Multiple Analogue Component
- e: Multiply Accumulate Cycle
- e: Multiply Accumulator Unit
- e: Multisensor Aircraft Campaign
- e: Multispecialty Advisory Committee
- e: Mycobacterium Avium Complex
- f: Mouvement Anti-Colonialiste
- f: Mouvement d'Action Civique
Mac e: Apple Macintosh computer
MACA e: Military Air Clearance Authority
- e: Modular Automatic Conferencing Arranger
MACAF e: MAC Numbered Air Force
MACALT e: MAC Alternate Headquarters
MacApp d: Von Apple entwickeltes Objektgitter
MACARMS e: MAC Aircrew Resources Management System
MACARNET e: MAC Airlift Recovery Network
MACAWS e: Multi-center Airborne Coherent Atmospheric Wind Sensor
MACBS e: Multi Access Cable Billing System
MACC f: Manufacture d'Armes et de Cartouches Congolaises
MACCONNET e: MAC Contingency Network
MACCSU e: MAC Accelerated Command Center System Upgrade
MACCT e: Multiple Assembly Cooling Cask Test
MACDD e: Marine and Coastal Data Directory of Australia
MacDFT e: Macintosh Distributed-Function Terminal
MACDIF e: Map[ping] and Chart[ing] Data Interchange Format
MACE e: Military Airlift Capability Estimator
- e: Military Airlift Center Europe
- e: Multi-spectral Atmospheric Column Extinction
Mace e: version of the MGM-1C Matador missile
MACH e: Multilayer Actuator Head
Mach e: Speed of Sound
MACHIS e: Michigan Association of Consumer Health Information Specialists
MACHO e: Massive Compact Halo Object
- e: Massive Condensed Halo Object
MACHT e: Maximum Achievable Control Technology
MacInfos d: Macintosh-Zeitschrift für Wirtschaft, Kunst, Kultur und Freizeit
MACIPS e: MAC Information Processing System
MACISIN e: MAC Information Systems Internetting
MACLO e: Military Airlift Command Liaison Officer
MACLOGFOR e: MAC Logistics Force Packing System
MACMA e: MAC Minicomputer Acquisition
MACMIP e: MAC Mainframe Internetting Project
MACNW e: MAC Northwest
MACOM e: Major [Army] Command
MACOPS e: MAC Operational Phone Systems
MACOS e: Macintosh Operating System
MacOS e: Macintosh Operation System
MACP e: Motion Adaptive Color Plus
MacPPP e: Macintosh Point to Point Protocol
Macq. e: Macquarie
MACR e: MAC Regulation
- e: Material Acquisition Control Record
MACRES e: Malaysian Centre for Remote Sensing
MACRIIM e: Marine and Coastal Resources Integrated Information and Management System
MACRO e: MAC Resource Optimization
- e: Monopoles, Astrophysics and Cosmic Ray Observatory
MACS e: Marine Air Control Squadron
- e: Middle Atmosphere in the Climate System
- e: Mixed Aloha Carrier Sense
- e: Modem Access Control System
- e: Modular Attitude Control System
- e: Multicenter AIDS Cohort Study
- e: Multiple Access Communication System
MACSAM e: MAC System Architecture Modernization
MAC/SINE e: Middle Atmosphere Cooperation/Summer in Northern Europe
MACSO e: Military Airlift Command Support Office
MACSYMA e: MACs Symbolic Manipulation System
MACT e: Maximum Achievable Control Technology
Mac/TCP d: Macintosh-Adaptation des TCP-Protokolls
MACTEC e: A private company
- e: MAC Technical Services Company
MACTIS e: Marine and Coastal Technology Information Service
- f: Service d'Information sur les Techniques Marines et Côtières
- s: Servicio de Información sobre Tecnología Marina y Costera
MACV e: Military Assistance Command in Vietnam
MACWACC e: MAC Washington Area Computer Center
MACWS e: Machine Area Cooling Water System
MAD d: Militärischer Abschirmdienst
- e: Magnetic Anomaly Detection Technology
- e: Maintenance Assembly/Disassembly
- e: Mass Air Delivery
- e: Maximum Applicable Dose
- e: Message Address Directory
- e: Multispectral Analyzer Display
- e: Mutual Assured Destruction
Mad f: Madame
MADA e: Multiple Access Demand Assignment
MADAR e: Malfunction Analysis, Detection and Reporting
MADCAP e: Model of Advection, Diffusion and Chemistry for Air Pollution
Mad Cow Disease e: Bovine Spongiform Encephalopathy
MADD e: Mirror Added Dual Doppler
- e: Mothers Against Drunk Driving
MADE e: Manufacturing and Automated (or Automation) and Design Engineering

Madelle f: mademoiselle
MADER e: Management of Atmospheric Data for Evaluation and Research
MADIMPORT e: Malagasy Import Agency
- f: Comptoir Malgache d'Importation
MADN e: Multiple Access Directory Numbers
MADP e: Material Acquisition Decision Process
Madr. e: Madras
Madrid e: WWII IR-guidance system for missiles
MADRS e: Montgomery-Asberg Depression Rating Scale
MADYMO e: Mathematical Dynamic Modeling
MAE e: Macintosh Application Environment
- e: Magnetocrystalline Anisotropy Energy
- e: Master of Automotive Engineering
- e: Mechanical and Aerospace Engineer[ing]
- e: Metropolitan Area Ethernet
- e: Metropolitan Area Exchange
- e: Ministry of Atomic Energy
- e: Missions-Activities-End Products
- f: Machine à Écrire
- f: Ministère des Affaires Économiques
- f: Ministère des Affaires Étrangères
- s: Ministerio de Asuntos Exteriores
MAED e: Master of Arts in Education
MAEDS e: Meteosat Argos Extended Dissemination Service
- e: Multisatellite Applications Extended Dissemination Service
MAER e: Maximum Allowable Emission Rate
MAERC e: Minority Access to Energy Related Research Centers
M-AERI e: Marine-Atmospheric Emitted Radiance Interferometer
MAERI e: Marine AERI
MAES e: Management Assistant Expert System
- e: Mexican American Engineers and Scientists
MAESA e: Measurements for Assessing [the] Effects of Stratospheric Aircraft
MAESTRO e: Multiple Airborne Experiments Towards Radar Observation
- e: Multiple-array, Assimilation, Evaporation, Stand, Tree, Radiation Orgy
MAF d: Marine Abkürzungs Fimmel
- e: Management Application Function
- e: Marine Amphibious Force
- e: Mass Air Flow
- e: Mineral-Ash Free
- e: Multiple Access Forward
- r: Maskeringsreglemente för Flygvapnet
- s: Muy Alta Frecuencia
MA-F e: Multiple Access-Forward
MAFAC e: Marine Fisheries Advisory Committee
MAFALDA f: Modélisation et Analyse Fonctionnelle des Applications des Liaisons de Données Air-sol
MAFET e: Microwave and Analog Front End Technology
MAFF e: Ministry of Agriculture, Fisheries and Food
- e: Ministry of Agriculture, Fisheries and Food
- f: Ministère de l'Agriculture, des pêches et de l'alimentation
- s: Ministerio de Agricultura, Pesca y Alimentación
M. af I. a: Medlem af Ingeniør-sammenslutningen
Mafic e: Mg- and Fe-rich
MAFIS e: Malaysian Aquatic Sciences and Fisheries Information System
- e: Mobile Area Field Instrumentation System
- f: Système Malaisien d'Information sur les Sciences Aquatiques et la pêche
- s: Sistema Malayo de Información sobre las Ciencias Acuáticas y la Pesca
M. af K.E. a: Medlem af Københavns Ejedomsmæglerforening
MAFMA e: Midwest Advanced Food Manufacturers Alliance
MAFREMO e: Malawi Freedom Movement
MAG d: Mehrgeräteanschluss
- e: Magnetometer/Gradiometer
- e: Management Advisory Group
- e: Marine Air Group
- e: Maritime Air Group
- e: Master of Agriculture
- e: Military Airlift Group
- e: Monochrome Graphic Adapter
- e: Myelin Associated Glycoprotein
- p: Missão Geográfica de Angola
- r: Målangivare
Mag N: Magasin
Mag. e: Magazine
MAGA s: Ministerio de Agricultura, Ganado y Alimentación
mag. avd. N: magasinavdeling
MAGE e: Marine Aerosol and Gas Exchange
- e: Marine Gas Emissions, Atmospheric Chemistry and Climate
Mage f: Message
MAGEBT e: Tigray Peoples Progressive Association
MAGED e: Master of Education
MAGI e: Maryland Automated Geographic Information System
MAGIC e: Maritime Air-Ground Intelligence Center
- e: Mode A Goes Image Converter
- e: Model of Acidification of Groundwater in Catchment
- e: Mozambique, Angola, Guinea-Bissau Information Center (or Centre)
MAGIIC e: Mobile Army Ground Imagery Interpretation Center
Magin f: Magasin
MAGISIAC e: Marine Air Ground Intelligence System Intelligence Analysis Center
mag. lag N: magasinlag
MAGMA p: Empresa Nacional de Minas
magn. a: magnetisk
MAGNA 8 e: Fourth-Generation Software
Magnaflux e: trade name for electro-magnetic inspection given to BHA
MAGNAS3 e: Multiphase Analysis of Groundwater, Nonaqueous Phase Liquid, Chemical and Soluble Components in Three Dimensions
Magnes Res e: Magnesium Research
Magn Reson Imaging e: Magnetic Resonance Imaging
Magn Reson Imaging Clin N Am e: Magnetic Resonance Imaging Clinics of North America
Magn Reson Med e: Magnetic Resonance in Medicine
Magn Reson Q e: Magnetic Resonance Quarterly
MAGNUS e: Danish CD-ROM bibliography
MAGPIE e: Mega Ampere Generator for Plasma Implosion Experiments
Magplut r: Magasinspluton
MAGRET f: Modèle Agrométéorologique d'Évaporation et de Temperature
MAGS e: Mackenzie GEWEX Study
- e: Midwestern Association of Graduate Schools
MAGSAT e: Magnetometer Satellite
MAGT e: Magnetic Tape
MAGTF e: Marine Air-Ground Task Force
MAh e: Milliampere-Hour
MAH f: Médecin Assistant des Hôpitaux
Mahedco e: Maharish Heaven on Earth Development Corporation
MAHLOVS e: Middle and High Latitudes Oceanic Variability Study
MAI d: Mai
- e: Mail
- e: Maintenance Administrative Instruction
- e: Management Action Indicator
- e: Mean Annual Increment
- e: Miscarriage After Infertility
- e: Multilateral Assistance Initiative
- e: Multiple Access Interference
- e: Multiple Applications Interface
- e: Mycobacterium Avium Intracellulare
- N: Marinens Artilleriinspeksjon
Mai f: Maison
MAID e: Magnetic Anti-Intrusion Detector
- e: Market Analysis and Information Database
- e: Modular AUTODIN Interface Device
Mail d: private Nachricht
Mailing d: Verschicken von Rundschreiben
Mainframe d: Grossrechner bzw. Hauptrechner
MAINSAIL e: Machine Independent SAIL
MAIRS e: Military Airlift Integrated Reporting System
MAIS e: Maintenance Analysis Information Sheets
- e: Major Automated Information System[s]
- e: Master of Arts in Interdisciplinary Studies
- e: Military Airlift Intelligence System
- e: Mobile Automated Instrumentation Suite
- e: Module Airborne Imaging Spectrometer
- e: Multiresource Analysis and Information System
MAISRC e: Major Automated Information System[s] Review Council

MAJ e: Major
- f: Mise À Jour
Maj f: Majoration
MAJO e: International Monitoring Systems
Majp f: Majorité présidentielle
MAK r: Materielanskaffningskommittén
Mak. e: Makerere
Makro d: Einheitsbefehl
MAKS e: Martin and Kitzis Sampler
maks. a: maksimal[t]
- N: maksimalt
- N: Maksimum
MAL d: Mittelarmlehne
- e: Malayalam
- r: Markförsvarsledare
Mal f: Médical
Mal. e: Malaya[n]
Malays J Pathol e: Malaysian Journal of Pathology
MALDI e: Matrix-Assisted Laser Desorption [and] Ionization
MALE e: Military Airlift Estimator
MALS e: Miniature Air Launched System
MALT e: Mobile Analytical Laboratory Truck
- e: Mucosa-Associated Lymphoid Tissue
MALVINE e: Manuscripts and Letters via Integrated Networks in Europe
MAM e: Management Analysis Model
- e: Mirror Attenuation Mosaic
- e: Munitions Assessment Model
- s: Mecánico Ajustador de Máquinas y herramientas
- s: Movimiento Autónomo de Masas
Màm f: Mot à mot
MAMA e: Moist Soil Management Advisor
- e: Multi-Anode Micrchannel [plate] Array
MAMBO e: Mediterranean Association of Marine Biological Oceanography
MA Merc s: Marina Mercante
Mamm Genome e: Mammalian Genome
MAMP e: Mission Area Materiel Plan
M&AMPO e: Maintenance and Operations
MAMS e: Medical Administration Management System
- e: Multispectral Atmospheric Mapping Sensor
MA/MSc e: Master of Arts/Master of Science
MAMSER e: Mass Mobilization for Self-Reliance, Social Justice and Economic Recovery
MAMS-II e: Maintenance Activity Management System II
MAMS-R e: Medical Administrative Management System-Revision
MAN A: Magaksie-Afrikaner Nasionalisme
- d: Maschinenfabrik Augsburg-Nürnberg AG
- e: Manchester
- e: Manual [Area Network]
- e: Metropolitan Area Network
- e: Metropolitan-Area Network
- e: Muslim Association of Nigeria
man N: manøver[ing]
Man s: Maniobra
Man. e: Manual
Mån r: Månad
MANA e: Malawi News Agency
- e: Michigan Association of Nurse Anesthetists
Managem. e: Management
Manc. e: Manchester [Victoria University]
Manch. e: Manchester
MANCOVA e: Multiple Analysis of Covariance
MANDOP e: Multiple Analytical Doppler
MANFOR e: Manpower Force Packaging System
MANIAC e: Mathematical Analyzer Numerical Integrator and Computer
MANIET e: Manifest Update
ManO r: Manöverofficer
MANO s: Movimiento de Antocomunista Nacional Organizado
Manp s: Manipulación
MANPADS e: Man-Portable Air-Defense System
MANPAGE e: Manual Page
MANPER e: Manpower/Personnel Module
MANPER-B e: Manpower and Personnel Module-Base Level
MANPER-M e: Manpower and Personnel Module-Major Command Level
MANPRINT e: Manpower and Personnel Integration
MANREQ e: Manpower Requirements System
MANS e: Mission Analysis for Missile and NUDET Surveillance
MANSEV e: Market Authorization by Network Submission and Evaluation
MAN T&EC e: Manchester and North HPC Training and Education Centre
MANTECH/MT e: Manufacturing Technology
MANTIS e: Manual, Alternative and Natural Therapy Index System
Mantj r: Manövertjänst
Mant[o] s: Mantenimiento
MANU e: Mozambique African National Union
Manuf. e: Manufacture
m.a.o. N: med andre ord
- r: med andra ord
MAO e: Major Attack Option
- e: Muslims Against Oppression
- f: Musique Assistée par Ordinateur
- s: Mecánico Ajustador de Óptica
MAOA e: Panel on Meteorological Aspects of Ocean Affairs
- f: Groupe d'experts sur les Aspects Météorologiques de l'Océan
MAOPR e: Minimum Acceptable Operational Performance Requirement
MAOS e: Metal-Al_2O_3-SiO_2-Semiconductor
- e: Microbial Analysis Of Sediments
- e: Minimum Aircraft Operating Strips
MAP e: Linker Map
- e: Maintenance Administration Panel
- e: Maintenance Analysis Procedure[s]
- e: Maintenance and Administration Position
- e: Management Application Protocol
- e: Management Assesment Program
- e: Management Assistance Program
- e: Manufactoring Automation Protocol
- e: Manufacturing Package
- e: Map Analysis Package
- e: Margins of Active Plates
- e: Market Applications
- e: Marketing Assistance Program
- e: Master Attack Plan
- e: Master Automation Plan
- e: Medical Assistance Program International
- e: Mediterranean Action Plan
- e: Memory Allocation Map
- e: Message Access Protocol
- e: Methamphetamine
- e: Micro-tubule Associated Protein
- e: Microwave Anisotropy Probe
- e: Middle Atmosphere Program
- e: Middle Atmosphere Programme
- e: Mission Automation Plan
- e: Mitigation Action Plan
- e: Moa Anbesa [Conquering Lion] Party
- e: Mobile Application Part
- e: Modern Aids to Planning
- e: Modular Array Processor; Hughes Aircraft Co.
- e: Moral Alliance for Peace
- e: Mouse Antibody Production
- e: Mutual Assistance Program
- f: Marges des Plaques Actives
- s: Márgenes de las Placas Activas
- s: Mina Antipersonal
Map s: Mapa
MAPAD e: Military Assistance Program Address Directory
MAPAQ f: Ministère de l'Agriculture, des Pêcheries et de l'Alimentation du Québec
MAPC e: Manpower Model for Control Agencies
MAPD p: Movimento Amplo para a Democracia
MAPEP e: Mixed Analyte Performance Evaluation Program
MAPES e: Management of Personnel Records
MAPEX e: Mapping Experiment From Space
MAPI e: Mail API [Application[s] Programmer (or Programming) Interface]
- e: Message (or Messaging) API [Application[s] Program[mer] (or Programming) Interface]
- e: Microsoft Messaging Application Programming Interface
MAPICS e: Manufacturing, Accounting and Production Information Control System
MAPIPS e: McCaulay Automated Photogrammetric and Interactive Image Processing System
Maple Flag e: Exercise held in Canada
MAPMIS e: Manpower and Personnel Management Information System

MAPMOPP e: Marine Pollution [Petroleum] Monitoring Pilot Project
- f: Projet pilote de surveillance continue de la pollution des mers [hydrocarbures]
- s: Proyecto Piloto de Vigilancia de la Contaminación del Mar [Petróleo]
Mapo e: Moscow Aircraft Production Organization
MAP[P] s: Mina Antipersonal de Presión
MAPP e: Modern Aids to Planning Program
MAPPA p: Movimento Amplo para a Paz e Pluralismo em Angola
MAPPER e: Maintain and Prepare Professional Executive Reports
- e: Maintaining, Preparing and Processing Executive Reports
MAPR e: ARM Multiple Antenna Profiler
- e: Marine Airborne Patrol Radar
- e: Miniaturization and Automation of Personnel Record
MAPRP e: Mesoscale Atmospheric Processes Research Program
MAP3S e: Multistate Atmospheric Power Production Pollution Study
MAPS e: MAC Automated Planning System
- e: MaCaulay Automated Photogrammetric System
- e: Manpower and Personnel System
- e: Map Analysis Package System
- e: Map Assisted Photo Interpretation System
- e: Measurement of Air Pollution from Satellite
- e: Measurement [of] Air Pollution from Space
- e: Measurement of Atmospheric Pollution from Satellites
- e: Mechanism of AIDS Pathogenesis
- e: Mesoscale Analysis and Prediction System
- e: Microwave Active and Passive Spectrometer
- e: Migrant Access Projects Scheme
- e: Mitsubishi Array Processing System
- e: Mobile Aerial Port Squadron
- e: Mobile Atmospheric Profiling System
- e: Mobility Analysis and Planning System
- e: Modular Antenna Pointing System
- e: Multiple Aircraft Positioning System
- e: Multispectral Active and Passive System
- e: Multispectral Active/Passive Scanner
- e: Region 2 Environmental Map Catalog System
MAPSAT e: Mapping Satellite
MAPSE e: Minimal APSE
MAPSIM e: Mesoscale Air Pollution Simulation Model
MAP3S/PCN e: Multistate Atmospheric Power Production Pollution Study/ Precipitation
MAPSS e: Maintenance and Analysis Plan for Special Services
MAP[T] s: Mina Antipersonal de Tracción
MAP/TOP e: MAP Technical Office Protocol
MAPTOP e: Manufacturing Automation Protocol/Technical Office Protection
MAPW e: Medical Association for Prevention of War
MAP/WINE e: Middle Atmosphere Program/Winter in Northern Europe
MAQ e: Macquarie Island
- e: Minimum Accountable Quantity
Maq s: máquinas
MAR e: Management Assessment Review; Monthly Activity Report
- e: Management Assistance Review
- e: March
- e: Marocco
- e: Microprogram Address Register
- e: Minimally Attended Radar
- e: Mississippi-Atchafalaya River
- e: Multicast Address Registers
- e: Multiple Access Return
- f: Dorsale médio-Atlantique
- r: Maskinrum
- s: Cordillera Central Atlántica
Mar f: Mardi
- f: Maréchal [ferrant]
- f: Marignane
- f: Marne
- f: Marseille
mar N: marine
Mar. s: Marina
mar. s: marzo
MÄR d: März
MA-R e: Multiple Access-Return
MARA e: Multimode Airborne Radar Altimeter
MARAD e: Maritime Administration
Maragra e: Marracuene Agricultural-Sugar Estates
MARAIRMED e: Maritime Air Force Mediterranean
Mar. Behav. Physiol. e: Marine Behaviour and Physiology
MARBI e: Machine Readable form of Bibliographic Information
Mar. Biol. Lett. e: Marine Biology Letters
Mar. Biol. [NY]/[Berl.] e: Marine Biology [New York]/[Berlin]
MARC e: MAC ALCE Reactions Communications
- e: Machine Readable card Catalog
- e: Machine-Readable Catalog[[u]ing]
- e: Market Analysis of Revenue and Customers system
- e: Mid-America Remote Sensing Center
- e: Middle Atmosphere Responses to Change
- e: Mining and Reclamation Council
- e: Minority Access to Research Careers Program
- e: Monitoring and Assessment Research Centre
- e: The Middle Atmosphere Responses to Changes
- f: Centre de recherche pour la surveillance et l'évaluation
- f: Mouvement d'Action pour la Résurrection du Congo
- s: Centro de Investigaciones sobre Vigilancia y Evaluación
MARCCO e: Mans Role in Command and Control
MARCH e: Master of Architecture
March f: Marchandise
MARC II e: Machine-Readable Catalogue
MARCO e: Microelectronics Advanced Research Corporation
MARCORPS e: Marine Corps
MARCORSYSCOM e: Marine Corps Systems Command
MARCS e: Multispectral Airborne Reference-Aided Calibrated Scanner
MARD e: Marine Assessment Research Division
- e: Modernization and Associated Restructuring Demonstration
MARDI e: Malaysian Agricultural Research & Development Inst.
MARDIS e: Modernized Army Research and Development Information System
MARDIV e: Marine Division
MARDOS e: Sources of Radioactivity in the Marine Environment and Their Relative Contributors to Overall Dose Assessment from Marine Radioactivity
MARE e: Mare Island Naval Shipyard Island
- e: Master of Architectural Engineering
MARECO e: Marine Radioecology Working Group
Mar. Ecol. Prog. Ser. e: Marine Ecology Progress Series
MAREMICS e: Maintenance, Repair and Minor Construction
MAREMP e: Marine Resources and Environment Management Programme
MAREP e: Maritime Reporting System
MARF e: Medical Acupuncture Research Foundation
- e: Metadata Archive Retrieval Facility
MARFOR e: Marine Forces
MARGS e: Marine Research Grant Scheme
- f: Programme de Subventions pour les Recherches Marines
- s: Programa de Subsidios para las Investigaciones Marinas
MARI e: Mexico City Air Quality Research Initiative
MARIA e: Methods for Assessing the Radiological Impact of Accidents
MARIC e: Marine Resources Information Center
MARICA f: Mariage de l'Image et de la Carte
MARIN e: Netherlands Maritime Research Institute
MARINF e: Marine Information Document
- f: Document d'Information Marine
- s: Documento de Información Marina
Marininfreg r: Marininfanteriregemente
Marintoff r: Marinintendenturofficer
MARIS e: Mississippi Automated Resource Information System
MARISAT e: Marine Satellite System
MARK e: Mid-Atlantic Ridge Kane Fracture Zone
- e: Middle Arkansas River Basin
MARKLEY e: Markley Hall
Markrbbrig r: Markrobotbrigad
Markstr r: Markstrid
Markup d: Formatierungsmerkmal
MarLK r: Marinläkarkurs

MARMAP e: Marine Resources Monitoring, Assessment and Prediction Program[me]
- f: Programme de contrôle, d'évaluation et de prévision des ressources marines
- s: Programa de Vigilancia, Evaluación y Predicción de los Recursos Marinos

MARMIC e: Command Level Maintenance Repair and Minor Construction Program Reporting System

MARNET e: Military Area Radio Network

MARNR e: Ministry of the Environment and Natural Resources

MARP e: Manpower Requirements Plan
- e: Mating and Ranging Program

MARPOL d: Internationales Übereinkommen zur Verhütung der Meeresverschmutzung durch Schiffe
- e: International Convention for the Marine Prevention of Pollution from Ships

MARPOL73/78 e: International Convention for the Prevention of Pollution from Ships, 1973, as modified by the Protocol of 1978 relating thereto
- f: Convention internationale de 1973 pour la prévention de la pollution par les navires, telle que modifiée par le Protocole de 1978 y relatif
- s: Convenio Internacional para Prevenir la Contaminación por los Buques, 1973, en su forma modificada por el correspondiente Protocolo de 1978

MARPOLMON e: Marine Pollution Monitoring System
- f: Système de surveillance continue de la pollution marine
- s: Sistema de Vigilancia de la Contaminación del Mar

MARPRO e: Country Marine Profile Database
- f: Base de données sur les profils maritimes des pays
- s: Base de Datos de Perfiles Marinos por Países

MARR e: Manpower Authorization Requirement Review

MARRES e: Manual Radar Reconnaissance Exploitation Systems

MARS e: New series of Russian Mars spacecraft
- e: Management Analysis and Reporting System
- e: Marine Corps Ammunition Reporting System
- e: Marine Research Stations Network
- e: Message Archiving and Retrieval Service
- e: Mid-Air Recovery/Retrieval System
- e: Military Affiliate Radio System
- e: Military Amateur Radio Service
- e: Mirror Advanced Reactor Study
- e: Mitigation and Adaption Research Strategies
- e: Mobile Agricultural Radar Scatterometer-Truck Mounted Platform
- e: Mobile Analytical Reconnaissance System
- e: Mobile Microwave Atmospheric Remote Sensing System
- e: Monitoring Agriculture by Remote Sensing
- e: Monitoring Agro-Ecological Resources with Remote Sensing and Simulation
- e: Monograph Acquisitions and Record System
- e: Multicast Address Resolution Server
- e: Multiuser Archival and Retrieval System
- e: Multivariate Adaptive Regression Splines
- e: nickname for a corporate planning board initiative to develop rapid methods for analyzing tank wastes
- f: Réseau de stations de recherches marines
- s: Red de Estaciones de Investigaciones Marinas

MARSEN e: Marine Remote Sensing Experiment

MARSIS e: Marine Remote Sensing Information System

MARSS e: Microwave Airborne Radiometer Scanning System

MARSSS e: Modular Airborne Remote Sampling and Sensing System

MARSTO e: North American Research Strategy on Tropospheric Ozone

MART e: A coded mask X-ray telescope on Spectrum-X-Gamma
- e: Martin Field
- e: Modular Automated Remote Terminal

MARTD e: Marine Air Reserve Training Detachment

MARVEL e: Machine-Assisted Realization [of] [a/the] Virtual Electronic Library

MARVS e: Material Acquisition Requirements Validation System

Mar. Zool. e: Marine Zoologist

MAS e: Astronomical Observatory of Cordoba
- e: Magic-Angle Spinning
- e: Main Store
- e: Man-day Account System
- e: Manifest Anxiety Scale
- e: Mass Announcement System
- e: May Aerial Survey
- e: Mexican American Studies
- e: Microbeam Analysis Society
- e: Microwave Active Spectrometer System
- e: Military Agency for Standardization
- e: Military Airlift Squadron
- e: Millimeter-Wave Atmospheric Sounder
- e: Mobile Atmospheric Spectrometer
- e: Modis Airborne Simulator
- e: Multiplex Applications Software
- s: Movimiento de Acción Solidaria

mas e: milli-arcseconds

MAS90 d: Mittelstands-Anwendungs-System 90

MASA e: Medical Association of South Africa

MASAR e: Mid-Atlantic Slope and Rise

MASB e: MAS Bus

MASBAL e: RCRA Mass Balance System

MASC e: MAGTF Automated Services Center
- e: MAS Controller
- e: Mass Communication
- e: Mountain Administration (or Administrative) Support Center
- e: Multiple Award Schedule Contract

masc. e: masculine

MASCDC e: Military Aircraft Storage and Disposition Center

MaschEl A d: Institut für Maschinenelemente

MaschEl B d: Institut für Maschinenkonstruktion und Getriebebau

Ma [school] r: Maskinskola

MASD e: Marine and Atmospheric Sciences Directorate

MASDC e: Military Aircraft Storage and Disposition Center

MASDCMIS e: MASDC Management Information System

MASER e: Microwave Amplification by/through Stimulated Emission of Radiation
- e: Multispectral Airborne System for Environmental Research

MASEX e: Mesoscale Air-Sea Exchange

MASF e: Maintenance and Storage Facility
- e: Mobile Aeromedical Staging Facilities

MASFC e: MAGTF All Source Fusion Center

MASFLEX e: Marginal Sea Flux Experiment in the West Pacific

MASH e: MgO-Al2O3-SiO2-H2O system

MASI f: [Laboratoire de] Méthodologie et d'Architecture des Systèmes Informatiques
- N: Marinens asdic-inspeksjon

MASIIS e: Maintenance Analysis Structural Integrity Information System

MASINT e: Measurement and Signal Intelligence

Mask N: Maskering

MASM e: Macro Assembler
- e: MAS Memory

MASMAP e: Mapping and Analysis With Small Format Aerial Photography

MASNET e: Management Support Network

MASO e: Midlands AIDS Support Organisation

MASP e: Master Alarm Setpoint Panel
- e: Minimum Aviation System Performance Standards
- e: Multi Angle Scattering Probe

MASR e: Microwave Atmospheric Sounding Radiometer

MASRO e: Mid-Atlantic Society of Radiation Oncologists

MASS e: Marine Survival of Salmon
- e: Massachusetts Bay
- e: Maximum Availability and Support Subsystem
- e: Mesoscale Analysis and Space Sensor
- e: Mesoscale Atmospheric Simulation Sensor
- e: MICAP Asset Sourcing System
- e: Military Airlift Support Squadron
- e: Mobilization Automated Support System

MassivB d: Institut für Konstruktion und Entwurf II
MassPIRG e: Massachusetts Public Interest Research Group
MAS-SS e: Millimeter-Wave Atmospheric Sounder-Space Shuttle
MAST e: Marine Science and Technology Programme
- e: Montery Area Ship Track
- e: Multi-Application Sonar Trainer
- e: Multiple Antigen Stimultaneous Test
- e: Specific Research and Technological Development Programme in the field of Marine Science and Technology
- f: Programme spécifique de recherche et de développement technologique en matière de science et technologie marines
- s: Programa Específico de Investigación y Desarrollo Tecnológico de Ciencia y Tecnología Marinas
MASTER e: Main Access to Storage and Tracking Engineering Records
MASTEX e: Mediterranean Aircraft and Ship Transmission Experiment
MASTIR e: Handbook on Marine Scientific and Technological Information Resources
- f: Manuel sur les ressources en information dans les sciences et techniques de la mer
- s: Manual de Recursos de Información Científica y Técnica Marinas
MaSTS e: Marine Strategy for the Torres Strait
Masui j: Masui. Japanese Journal of Anesthesiology
MASURCA e: Pin-type shielding facility
MASW e: Mission Application Software
MAT e: Maintenance Access Terminal
- e: Marine Air Temperature
- e: Master Archive Tape
- e: Materiel
- e: Miller Analogy Test
- e: Mobilization Assistance Team
- s: Material
- s: Mina Antitanque
Mat f: Maternelle
- f: Maternité
mat. a: matemati[s]k
Mat. d: Matrikel
MATA s: Matadero
- s: Mina Antitanque Activada
Mataph. e: Metaphysics
MATC e: Maximum Acceptable Toxic Concentration
- e: Maximum Allowable Toxicant Concentration
MAT-CALS e: Marine Air Traffic-Control And Landing System
MATCALS e: Marine Air Traffic and Landing Systems
MATCO e: Military Air Traffic Coordinating Office
MATCU e: Military Air Traffic Coordinating Unit
MATE e: Mobile Analysis and Training Extension
- e: Modular AUTODIN Terminal Equipment
- e: Modular Automated Test Equipment
MATELT f: Ministère de l'Aménagement du Territoire, de l'Equipement, du Logement et du Tourisme
Mater Med Pol e: Materia Medica Polona
Matern Child Nurs J e: Maternal-Child Nursing Journal
MATES e: Mobilization and Training Equipment Site
MATFAP e: Metropolitan Area Transmission Facility Analysys Program
MATH d: Mathematik
- e: Astronomy/Math/Statistics Library
- e: Mathematics and Computer Laboratory
Math. e: Mathematical
- e: Mathematics
Math Biosci e: Mathematical Biosciences
Mat Hid s: Material Hidrográfico
MATHLAC e: Mathematical Sciences Library Advisory Committee
MATIF f: marché à terms des instruments financières
MATLID e: Multi-Frequency Atmospheric Lidar
Mätlysptr r: Mätlyspatron
MATMAN e: ODP [materials management] inventory database system matrices
MATMS e: Marine Aviation Training Management System
MATOC e: Mobility Air Terminal Operations Center
Matr N: Materiell
MATRA f: [Société générale de] Mécanique-Aviation-Traction
Matrgrp HFT N: Materiellgruppen av Hærens Forsyningstropper
Matrix Biol e: Matrix Biology
Matrix Suppl e: Matrix. Supplement
matr.nr. a: matrikelnummer
MATS e: Management (or Managers) Action Tracking System
- e: Mesoscale Atmospheric Transport Studies
Mat Top s: Material Topográfico
MATV e: Multi-Axis Thrust Vectoring
MAU e: Marine Amphibious Unit
- e: Math Acceleration Unit
- e: Media Access Unit
- e: Media Adapter Unit
- e: Medium Access Unit
- e: Medium Adaptor Unit
- e: Medium Attachement Unit
- e: Million Accounting Units
- e: Multistation Access Unit
MAUU e: Massachusetts Alliance of Utility Unions
MAV e: Municipal Association of Victoria
- s: Ministerios de Agua Viva & Cía
MAVDM e: Multiple Application VDM
MAVMS e: MACOM Automated Vehicle Management System
MAVSS e: Monostatic Acoustic Vortex Sensing System
MAVT e: Mobile Audio Visual Terminal
MAW e: International Monitoring System
- e: Management Application Workshop
- e: Marine Aircraft Wing
- e: Microsoft At Work
- e: Ministry of Agriculture and Water
- e: Mission Adaptive Wing
m.a.W. D: mit anderen Worten
MAWS e: Minimum Additive Waste Stabilization System
- e: Missile Approach Warning System
MAWTS e: Marine Aviation Weapons & Tactics Squadron
MAWU e: Mechanical and Allied Workers Union
- e: Metal and Allied Workers Union
MAX e: Massively parallel Unix
- e: Message Automatic Exchange
max. a: maksimal[t]
- a: maksimum
MAXC e: Maximum prime service length
MAXDOSE e: Maximum Individual Dose Model
MAXEL e: Maximum Elevation
Maxhast r: Maximalhastighet
MAXI e: Modular Architecture for the Exchange of Information
- e: Modular Architecture for the Exchange of Intelligence
MAXIS e: Multi-tasking Acquisition and Imaging System
MAXS e: Maximum count or time per schedule period
MAXSEP e: Maximum Separation time
MAXX e: Maximum Access to Diagnosis and Therapy: The Electronic Library of Medicine
MAY e: Mad About You
- e: May
- s: Mayo[r]
May f: Mayenne
MAYO e: Mamelodi Youth Organization
Mayo Clin Proc e: Mayo Clinic Proceedings
MAYPOLE e: May Polarization Experiment
MAZ d: Deutscher Serviceprovider
- d: Magnet[bild]aufzeichnung
- e: Mikroelektronik Anwendungszentrum Hamburg GmbH
Maz f: Mazout
MB a: Medlem af Borgerrepræsentationen
- a: Mellenbøge
- e: Machine Batch
- e: Mailbox
- e: Management Board
- e: Manipulator Bridge
- e: Manitoba
- e: Manned Base
- e: Marine Board
- e: Maxwell-Boltzmann
- e: MBit
- e: MByte
- e: Medical Battalion
- e: Mega Byte
- e: Megabyte[s]
- e: Miamisburg Area Office
- e: Middle Button
- e: Moored Buoy
- e: MotherBoard
- e: Motor Brake
- e: Music Building
- f: Marge Bénéficiaire
- f: Monnaie de Billon
- N: Morgenbladet
- r: Materielbyrå
- r: Militärbefälhavare
- r: Monopolutredningsbyrån

Mb d: Meßbereich
- e: Megabit
- r: Medelbar[t]
- r: Motorbåt
mb e: millibar[n]
M.B. e: Bachelor of Medicine
MBA e: Main Battle Area
- e: Management Business Analyst
- e: Master Business Administration
- e: Master of Business Administration
- e: Masters in Business Administration
- e: Material Balance Accounting
- e: Material Balance Areas
- e: MONJU Blanket Assembly
MBa a: Midtjydsk Byggeadministration A/S
- e: Medulloblastoma
Mba e: Megabar
MBA/ICA e: Material Balance Area/Item Control Area
MBAS e: Mail Based Archive Server
- e: Methylene-Blue-Active Substances
- e: Molecular Beam Atomic Scattering
MBASIC e: Microsoft BASIC
MbauMstrV d: Modellbaumeisterverordnung vom 19.3.1998
MBB d: Messerschmitt-Bölkow-Blohm
- e: Messerschmitt-Bölkow-Blohm
- e: Modular Building Blocks
- f: Médecin du Bureau de Bienfaisance
- N: møteplass for bortkomne biler
MB B r: Militärbefälhavare-Bergen
MBBA e: N-[p-methoxybenzylidene]-p-butylaniline
MBBI e: Mulitple-Bit Binary Input
MBBL e: Mounted Battlespace Battle Lab
MBBLS e: Thousands of Barrels
MBBO e: Multiple-Bit Binary Output
MBC e: International Monitoring System
- e: Machine Bath Collection
- e: Malawi Broadcasting Corporation
- e: Marine Biological Centre
- e: Mass Bias Correction
- e: Mauritius Broadcasting Corporation
- e: Memory Control Block
- e: Merchant Banking Company
- e: Meteor Burst Communications
- e: Military Budget Committee
- e: Minimal Bactericidal Concentration
- e: Museum of Broadcast Communications
- f: Centre de Biologie Marine
- s: Centro de Biología Marina
- s: Programa sobre Biodiversidad Marina
M.B.C.I. e: Medical Books for China International
MBCP e: Missile Base Communications Processor
MBCS e: Multi-Byte Character Set
MBD e: Manual Branch Driver
- e: Programme on Marine Biodiversity
- f: Programme sur la Biodiversité Marine
mbd e: million barrels per day
MBDA e: Minority Business Development Agency
MBDC e: Minority Business Development Center
MBDL e: Missile Battery Data Link
M.B.E. e: Member of the Order of the British Empire
MBE e: Master of Business Education
- e: Mega-Buck Epitaxy
- e: Member of the [Order of the] British Empire
- e: Minority Business Enterprise[s]
- e: Molecular-Beam-Epitaxy
- e: Money Buys Everything
MBEP e: Minority Business Enterprise Program
- e: Region 4 Minority Business Tracking
MBER e: Minority Business Enterprise Representative
MBE/WBE e: Certification Questionnaire-Minority/Women Business Enterprise
MBF e: Thousand Board Feet
MBFR e: Mutual[ly] [and] Balanced Force Reduction
MBFT e: Multipoint Binary File Transfer
MBG e: Missouri Botanical Garden
MBHO e: Managed Behavioral Healthcare Organization
MBI d: Max-Born-Institut für Nichtlineare Optik und Kurzzeitspektroskopie, Berlin
- e: Management Buy-In
- e: Max Born Institute
MBIS e: MacKenzie Basin Impact Study
Mbit[s] e: Megabit[s]
MBK e: Medications and Bandage Kit
MBL e: Marine Boundary Layer
- r: Medbestämmandeavtal
MBLV e: Multiple-Beam Laser Velocimeter
M.b.m. e: Thousand board measure
MBM r: Massabunden Makadam
MBMS e: Mocecular Beam Mass Spectrometry
MB NN r: Militärbefälhavare-Nordnorrland
MBO e: Management Buy-Out
- e: Management By Object[ive]s
- e: Million Barrels of Oil
- e: Secondary Vocational Education
MBÖ r: Militärbefälhavare-Öst
MB ÖN r: Militärbefälhavare-Östnorrland
MBONE e: Multicast Backbone
MBP e: Major Basic Protein
- e: Materials and Bulk Processes
- e: Myelin Basic Protein
MBPC e: Model-Based Process Control
Mb per mn e: Megabytes per minute
MBPlG d: Magnetschwebebahnplanungsgesetz vom 23.11.1994
MBPS e: Million Bits Per Second
Mbps e: Megabits per second
MBps e: Megabytes per second
MBq e: Megabequerel
MBR e: Master Beneficiary Record
- e: Master Boot Record
- e: Mechanical Bit Release
- e: MILA Bermuda Re-engineering
- R: Mezhkontenentalnaya Ballisticheskaya Raketa-ICBM
M/BR N: Med Brannrør
Mbra s: Maniobra
Mbrf e: Meters below rig floor
MBRRGChIN R: MBR s Razdelyayuchimi Golovnyym Chastyami Indiviualnogo Navedeniya-MIRV
MBRS e: Minority Biomedical Research Support Program
MB S r: Militärbefälhavare-Söder
MBS d: Mobiles Breitband-System
- e: m-Maleimidobezoyl-N-hydroxysuccinimide ester
- e: Malawi Bureau of Standards
- e: Maximum Burst Size
- e: Methyl methacrylate Butadiene Styrene terpolymer
- e: Mission Bit Stream
- e: Mobile Base System
- e: Mobile Broadband Systems
- e: Multiple-Blade Slurry
Mbsf e: Meters below seafloor
MBSGM e: Multi-Base Sortie Generation Model
MBSHC e: Mediterranean and Black Seas Hydrographic Commission
- s: Comisión Hidrográfica de los Mares Mediterráneo y Negro
Mbsl e: Meters below sea level
MBSTA e: Mobilization Station
MBT e: Molecular Beam Technique[s]
- f: Bathythermographe Mécanique
- s: Batitermógrafo Mecánico
MBTC e: model-based temperature control
MBTE e: Methane Butane Tertiary Ethane
MBUILD e: Mental BUILD of conceptualization
MB V r: Militärbefälhavare för V Militäromrade
- r: Militärbefälhavare-Väst
MBWA e: Management By Walking Around
MBX e: [Electronic] Mailbox
- e: Message Bus Exchange
MByte e: Megabyte
MC e: Main Cross-Connect
- e: Management Committee
- e: Manipulator Crane
- e: Manual Chapter
- e: Marginal Cost
- e: Marine Corps
- e: Maritime Commission
- e: Maritime Committee
- e: Mast Cell
- e: MasterCard
- e: Materiel Control
- e: Medical Corps
- e: Medium Coeli
- e: Mekhong Committee
- e: Melter Cell
- e: Meteorological Codes
- e: Methoxychlor
- e: Micro Channel
- e: Military Committee
- e: Minges Coliseum
- e: Miniature Card
- e: Miscarriage
- e: Mission Capable
- e: Mixing Component
- e: Mobility Control Center
- e: Monitor and Control CSCI
- e: Monitoring Center
- e: Monte Carlo [simulation]
- e: Motorola Chip
- f: Comité Maritime
- f: Magasin de Couchage
- f: Maison de Campagne
- f: Maison de Commerce
- f: Maître de Conférences

- f: Maître-Clerc
- f: Maladie Contagieuse
- f: Mandat Commercial
- f: Mandat de Comparution
- f: Mandat-Carte
- f: Mandat-Contributions
- f: Mappemonde Céleste
- f: Marché au Comptant
- f: Mariage Civil
- f: Marine Commerciale
- f: Marque Commerciale
- f: Marque de Contrôle
- f: Matrice Cadastrale
- f: Médecin-Chef
- f: Médecin-Conseil
- f: Mois Courant
- f: Mon Compte
- f: Mort Civile
- f: Motoculture

Mc a: Motorcykel
- f: Monnaie de Compte
- r: Maskinchef
- r: Motståndscentrum
- s: Médico Cirujano

M.C. e: Military Cross
m/c e: miscarriage
- s: mi cuenta

M3C e: MAJCOM Manpower Management Computer
MC-130 e: Combat Talon
MC540 e: Merocyanine 540
MCA e: Major Constituent Analyzer
- e: Manufacturing Chemists Association
- e: Measurement Capability Analysis
- e: Mechanical Computer Aided Design
- e: Medicine Control Agency
- e: Micro Channel Adapter
- e: Micro Channel Architecture
- e: Military Construction, Army
- e: Mission Critical Application[s]
- e: Motion Capture and Analysis
- e: Movement Control Agency
- e: Multi-Channel Access
- e: Multi-Channel Analyzer
- e: Multi-Criteria Analysis
- e: Multiprocessor Communication Adapter
- f: Mouvement Corse pour l'Autodétermination

MC&A e: Material Control and Accountability
MCAD e: Mechanical [Engineering] Computer-Aided Design
MCAE e: Mechanical Computer Aided Engineering
MCAF e: Monocyte Chemotactic and Activating Factor
MCAIERS e: Microcomputer Aerial Image Enhancement Reconnaissance System
Mcal s: Mariscal
MC/A/N/AF/M e: Military Construction, Army/Navy/Air Force/USMC
MCAP e: Major Command ADP Plan
- e: Material Control and Accountability Plan
- e: Maximum Calling Area Procedure
- e: Mobile Consolidated Aerial Port Subsystem

MCAR e: Military Construction Army Reserve
- e: Multi-Channel Acoustic Relay

MCAS e: Marine Corps Air Station
- e: Material Control Accounting System
- e: MgO-CaO-Al2O3-SiO2 system

MCATF e: Mechanized Combined Arms Task Force
MCATS e: Medium Capacity Automated Telecommunications System
MCAV e: Modified Constant Angular Velocity
MCAWN e: Microcomputer Connectivity to AWN
MCB e: Main Control Building
- e: Management of Marine Catchment Basins
- e: Memory Control Block
- e: Message Control Block
- e: Mobile Construction Battalion
- e: Mode Control Box
- e: Molecular and Cellular Biology
- e: Multi-Channel Buffer
- f: Gestion des bassins versants marins
- s: Gestión de las cuencas receptoras marinas
- f: Mouvement Culturel Berbère

MCBA e: Mean Cycles Between Assists
MCBF e: Mean Cycle Between Failure
MCBI e: Mean Cycles Between Interrupts
MCBIAS e: Microcomputer Based Image Analysis System
MCBS e: Masonry Core-Barrel Sampling
MCBSE e: Mean Cycles Between Scrap Event
MCBSF e: Mixed Commission for Black Sea Fisheries
- f: Commission mixte des pêcheries de la mer Noire
- s: Comisión Mixta para las Pesquerías del Mar Negro

MCC d: Mitteldeutsches Communication Center
- e: Main Combustion Chamber
- e: Manchester Code Converter
- e: Manchester Computing Centre
- e: Master Control Center
- e: Materials Characterization Center
- e: Materials Control Center
- e: Medicines Control Council
- e: Megacaryocyte
- e: Melter Cell Crane
- e: Member Conduct Committee
- e: Memory Cache Controller
- e: Mesoscale Convective Complex
- e: Micro Compact Car AG
- e: Microelectronics and Computer [Consortia]
- e: Microelectronics and Computer technology Corporation
- e: Midwest Climate Center
- e: Mission and Control Centre
- e: Mobile Command Center
- e: Mobile Communications Center
- e: Mobility Control Center
- e: Motor Control Center
- f: Ministère de la Culture et de la Communication
- s: Mantenimiento y Custodia de Carreteras
- s: Misil Contra Carro

MC C e: McCoy Hall
m/c.c. s: mi cuenta corriente
MCCA-E e: Manual Cloud Cover Assessment by EOSAT
MCCC e: Maragone Convection on Closed Containers
- e: Marine Corps Command Center

MCCDC e: Marine Corps Combat Development Command
MCCISWG e: Military Command, Control Information Systems Working Group
MCCMF e: Mixed Commission for Cooperation in Marine Fishing
- f: Commission mixte de coopération pour la pêche en mer
- s: Comisión Mixta para la Cooperación en la Pesca Marítima

MCCN e: Marine and Coastal Community Network
MCCOI e: Multimedia Communications Community of Interest
MCCP e: Military Consolidated Command Post
- e: Mountain Cloud Chemistry Project

MCCR e: Mission Critical Computer Resources
MCCRES e: Marine Corps Combat Readiness Evaluation System
MCCS e: Mechanized Calling Card Service
- e: Mission Critical Computer System

MCCSS e: Mobile Command Center Strategic System
M&CD e: Metals and Ceramics Division
M.C.D. e: Melbourne College of Divinity
MCD e: Caliper tool; Measures hole Diameter
- e: Mad Computer Disease
- e: Magneto CD-Disk
- e: Marketing Comunication Design
- e: Medical Crew Director
- e: Mini Client Driver
- e: Molecular Column Density
- e: Multichannel Detector
- e: Municipal Construction Division

McD e: McDonald Observatory Library
Mcd e: Meters composite depth
MCDA e: Micro Channel Developers Association
MCDD e: Marine and Coastal Data Directory
MCDDI f: Mouvement Congolais pour la Démocratie et le Développement Intégral
MCDF e: Master Code Descriptor File
MCDN e: Marine Corps Data Network
MCDS e: Migration Control Data Set
MCDW e: Monthly Climate Data for the World
MCE e: Mapping and Charting Establishment
- e: Mission Control Element
- e: Modular Control Equipment
- e: Motor Carrier Evaluation
- f: Marché commun européen

MCEAMS e: Marine Corps Expeditionary Aircraft Maintenance Shelter

MCEB e: Military Communications Electronic Board
- e: Military Communications-Electronics Board
MCEF e: Mixed Cellulose Ester Filter
MCEP e: Motor Carrier Evaluation Plan
- e: Motor Carrier Evaluation Program
MCES e: Modular Command and Control Evaluation Structure
MCF e: Maintenance Control Facility
- e: Malignant Catarrhal Fever
- e: Message Communication Function
- e: Meta Content Format
- e: Meta Content Framework
- e: Methane Conversion Factor
- e: Microcode Fix
- e: Mission Control Facility
- e: Mission Control Facility within the MCS
- e: Mission Control Forecast
- e: Modulation Contrast Function
Mcf e: thousand cubic feet
MCFD e: Malta College of Family Doctors
MCFFA e: Ministerial Council of/on Forestry, Fisheries and Aquaculture
MCFMIS e: Marine Corps Food Management Information System
MCG e: Movement Control Group
- f: Modèle de Circulation Générale
- s: Movimiento Cooperativista Guatemalteco
McG. e: McGill
MC&G e: Mapping, Charting and Geodesy
MCGA e: Multi Colo[u]r Graphics Adapter (or Adaptor)
- e: Multi Colo[u]r Graphics Array
MCG&I e: Mapping, Charting, Geodetic Data and Imagery
MCGS e: Manual Cover Gas System
MCH e: Greater Manchester
- e: Maintenance Channel
- e: Mother-Child Health
- e: Moveable Cultural Heritage
- f: Médecin-Chef de l'Hôpital
Mch s: Marcha
M.Ch. e: Master of Surgery
MCHA e: Moveable Cultural Heritage Act
MCHB e: Maintenance Channel Buffer
- e: Maternal and Child Health Bureau
MCHJ e: Maternal and Child Health Journal
MCHPRC e: Maternal and Child Health Policy Research Center
MCHQ e: Marine Corps Headquarters
MCHSF e: Medicare Continuous History Sample File
MC/I e: Microphone Included
MCI d: Amerikanischer Serviceprovider
- e: Data-Media Communications company
- e: Management Cost Improvement
- e: MCI Communications Corp.
- e: Measurement Layer Interface
- e: Media Control Interface
- e: Microwave Communications, Inc.
- e: MIDAS Component Index
- e: Mission Capability Inspection
- e: Monsoon Climatological Index
- f: Ministère du Commerce et de l'Industrie
- s: Movimiento Campesino Independiente
mCi e: millicurie
MCIA e: Methyl Chloride Industry Alliance
- e: Micro Channel Interface Adapter
MCIAS e: Multi Channel Intelligent Announcement System
- e: Multi Channel Intercept Announcement System
MCIC e: Managed Care Information Center
McIDAS e: Man computer Interactive Data Access System
MCINTOSH e: Julian & Vera McIntosh Theatre, School of Music
MCIS e: Materials Compatibility In Sodium
MCL e: Mauritius Confederation of Labor
- e: Maximum Clearance Level
- e: Maximum Contaminant (or Contamination) Level[s]
- e: Microcode Level
- e: Microsoft Compatibility Laboratories
- e: Mobilization Cross-Leveling
- e: Monoclinic
MCLA e: Monetary Centre for Latin America
M-CLASS e: Mobile CLASS
MCLBA e: Marine Corps Logistics Base Albany, Georgia
MCLD e: Milestone Control Log Document
MCLG e: Maximum Contaminant (or Contamination) Level Goal[s]
- e: Maximum Contaminant Level Guidelines
MCLI5 e: Manual of Clinical Laboratory Immunology, Fifth Edition
MCLN f: Mouvement Centrafricain de Libération Nationale
MCM d: Multichipmodul
- e: Manufacturing Cycle Management
- e: Mechanical Current Meter
- e: Media and Communication Management
- e: Military Committee Memorandum
- e: Mine Countermeasures [ship]
- e: Minimum Critical Mass
- e: Ministerial Council Meeting
- e: Molecular Crystal Model
- e: Multichip Module
- e: Multichipmodul and Communication Management
- f: Montant Compensatoire Monétaire
- s: Medidas Contraminas
mcm e: million cubic metre
MCM6 e: Manual of Clinical Microbiology, 6th Edition
MCMCC e: Marine Corps Manpower Control Center
MCM-C/D/L e: Multi-Chip Module-Ceramic/Deposited/Laminate
MCMIS e: Motor Carrier Management Information System
MCMST e: Montana College of Mineral Science and Technology
MCN e: American Journal of Maternal/ Child Health Nursing
- e: MACIMS Communication Network
- e: Metropolitan Campus Network
- e: Minimum Critical Number
- e: Molecular and Cellular Neuroscience
MCN Am J Matern Child Nurs e: MCN; American Journal of Maternal Child Nursing
MCNC e: Microelectronics Center of North Carolina
MCNE e: Master Certified Novell Engineer
MCNF e: Mean Cumulative Number of Failures
MCNP e: Monte Carlo [computer code for] Neutron Photon [transport]
- e: Monte Carlo N-Particle Transport Code
MCNS e: Multimedia Cable Network System Partners Ltd.
MCO e: Magnetic-field-Controlled Oscillator
- e: Magnetically Controlled Oscillator
- e: Main Contractor
- e: Managed Care Organization
- e: Manual Change Order
- e: Mill Cuts Out
MCODB e: Marine Climatological and related Ocean Data Base
- f: Base de données de climatologie maritime et de données océaniques connexes
- s: Base de Datos Marinos Climatológicos y Oceánicos Conexos
MCOM e: Marketing Communications
MCom f: Ministère du Commerce
MCOTEA e: Marine Corps Operational Test and Evaluation Agency
MCP e: Machine Control Panel
- e: Macintosh Coprocessor Platform
- e: Main Caution Panel
- e: Malawi Congress Party
- e: Management and Control of Provisioning
- e: Marine Corps Capabilities Plan
- e: Master COMSEC Plan
- e: Master Control Programm
- e: Meteorological Communications Package
- e: Micro-Channel Plate[s]
- e: Microsoft Certified Professional
- e: Military Construction Program
- e: Military Construction Project
- e: Mission Coordinating Paper
- e: Molten Carbonate Process
- e: Multi-Chip Package
- e: Multiport Communication Processor
- e: Multiprotocol Communication Processor
- e: Municipal Compliance Plan
MCP[/]AS e: Master Control Program/ Advanced System
MCPC e: Multi Channel Per Carrier
MC-PGA e: Metallized Ceramic-Pin Grid Array
MCPPM e: Materiel Control Policies and Procedure Manual
MCPS e: Microsoft Certified Product Specialist
MCPW e: Microstrip Co-Planar Waveguide

MCQ e: Multiple Choice Question
MC-QFP e: Metallized Ceramic-Quad Flat Pack
MCR e: Manpower Change Request
- e: Map Compilation and Reproduction Division
- e: Memory Controller
- e: Minimum Cell Rate
- e: Modem Control Register
- e: Modification Completion Report
- e: Modified Community Rating
- e: Multichannel Radiometer
- e: Multispectral Cloud Radiometer
MCRC e: Master Control and Reporting Center
- e: Master Control Relay Cabinets
MCRFP e: Monitoring of Chemical Residues in Food Products
MCRL e: Master Cross Reference Library
MCRP e: Master of City and Regional Planning
MCR[PP] e: Military Compact Reactor Program
MCRS e: Malaysian Centre for Remote Sensing
MCRSCMS e: Marine Corps Reserve Support Center Management System
MCS e: Mailroom Control System
- e: Maintenance Control System
- e: Maintenance Cost System
- e: Management [and] Control System
- e: Maritime Communications System
- e: Material Control Station
- e: Material Control System
- e: Meeting Communications Service
- e: Mesoscale Cloud System
- e: Mesoscale Convective System[s]
- e: Message, Command, Status
- e: Method of Corresponding Solutions
- e: Method of Corresponding States
- e: Microsoft Consulting Services
- e: Mission Control System
- e: Monitor [and] Control [Sub]system
- e: Multicast Server
- e: Multichannel Communications System
- e: Multichannel Scaling
- e: Multichannel Seismic
- e: Multichannel Sonic
- e: Multichannel Spectrometer
- e: Multiple Chemical Sensitivities (or Sensitivity) .
- e: Multipoint Communication Services
- f: Movement des Combatants Socialistes
MCSA e: Motor Carrier Safety Act
MCSAP e: Motor Carrier Safety Assistance Program
MCSCF e: Multiple Configuration Self-Consistent Field
MCSE e: Microsoft Certified Systems Engineer
MCSF e: Marine Corps Security Force
- e: Mobile Cryptologic Support Facility
MCSM e: Master of Construction Science and Management
MCS/P e: Maneuver Control System/ PHOENIX
MCSR e: Materiel Condition Status Reporting System
MCSS e: Marine Climatological Summaries Scheme
- f: Programme de résumés de climatologie maritime
- s: Programa de Resúmenes de Climatología Marina
MCSSCC e: Multicolor Spin-Scan Cloud Camera
MCSST e: Multi-Channel SST [Sea Surface Temperature]
MCST e: Ministers Committee for Science and Technology
- e: Ministers Council on Science and Technology
MCSV e: Mars Crew Sortie Vehicle
MCT e: Market Cattle Testing
- e: Maximum Corrective Time
- e: Mean Corrective Time
- e: Mercury Cadmium Telluride [detector]
- e: Mobile Communications Terminal
- e: MOS-Controlled Thyristor
- e: Motor-Control Transformer
- e: Movement Control Team
MCt f: Mois Courant
MCT p: Ministério da Ciência e Tecnologia
MCTD e: Mean Cell Transfer Delay
MCTEX e: Marine Continent Thunderstorm Experiment
- e: Maritime Continent Thunderstorm Experiment
MCTSSA e: Maritime Corps Tactical Systems Support Activity
MCTU e: Mauritius Confederation of Trade Unions
MCU e: Memory Control[ler] Unit
- e: Micro-Controller Unit
- e: Minimum Coded Unit
- e: Mobile Calibration Unit
- e: Multi-Chip Unit
- e: Multipoint Conferencing Unit
- e: Multipoint Control Unit
MCV e: Mean Cell Volume
- e: Mesoscale Convectively-generated Vortices
- e: Movable Closure Valve
- f: Mandat-Carte de Versement
MCVA e: Modified Constant Angular Velocity
MCVD e: Metal Chemical Vapor Deposition
MCWG e: Marine Chemistry Working Group
- f: Groupe de travail sur la chimie marine
- s: Grupo de Trabajo sobre Química Marina
MCWH e: Maternal, Child and Womens Health
MCWO e: Miscellaneous Capital Work Order
MCX e: Marine Corps Exchange
MCY e: Media-City
MCZ e: Magnetic Czochralski
- e: Mesoscale Compressible Community Model
- e: Museum of Comparitive Zoology, Harvard University
MD a: Mejeriselskabet Danmark
- d: Meteorologischer Dienst der DDR
- e: Fissile Materials Disposition, Office of
- e: Mail Drop
- e: Make Directory
- e: Management Division
- e: Maritime Defense
- e: Marshaled Deployability posture
- e: Master Directory
- e: Material Developer
- e: Mediation Device
- e: Message Digest
- e: Metal Detector surveys
- e: Middle Distillate
- e: Military Department
- e: Mini Disc
- e: Misfit Dislocation[s]
- e: Missile Defense
- e: Molecular Deposition
- e: Molecular Diversity
- e: Monochrome Display
- e: Motor Drive
- f: Magistrature Debout
- f: Mandat de Dépôt
- f: Mandat de Député
- f: Marque Déposée
- f: Mineur Délinquant
- f: Mission Diplomatique
- f: Modèle Déposé
- f: Mois Dernier
- N: Miljøverndepartementet
- r: Marindistrikt
- r: Motoriserad skyttedivision
- s: Módulo [de municiones]
- s: Motor Diesel
- s: Movimiento de los Descamisados
Md e: Milliard
- f: Marchand
md e: Moldova [Internet]
md. a: mand
- N: måned
M.D. e: Doctor of Medicine
M-D e: Manic Depression
- e: Martin-Decker rig-weight indicator
MD5 e: Message Digest algorithm 5
MDA e: Macdonald, Dettwiler & Association Ltd.
- e: Mail Delivery Agent
- e: McDonnell Douglas Astronautics
- e: Mediated Digest Authentication
- e: Medicated Digest Authentication
- e: Mesocyclone Detection Algorithm
- e: Methodology of Decision Analysis
- e: Methylenedianilline
- e: Milestone Decision Authority
- e: Minimum Detectable Activity
- e: Muscular Dystrophy Association
- e: Muscular Dystrophy Association Australia
- s: Mando de la Defensa Aérea
MdA d: Mitglied des Abgeordnetenhauses [von Berlin]
M.D.A. a: Medlem af Dansk Arkitektforening
MDAAQS e: Miscellaneous Data Analysis and Air Quality Simulation Studies
MDAC/ACDM e: Muscular Dystrophy Association of Canada/Association Canadienne de la Dystrophie Musculaire
MDACC e: MD Anderson Cancer Center

MDACS e: Modular Digital Access Control System
MDAD e: Monitoring and Data Analysis Division
m-dag N: mobiliseringsdag
MDAP e: Major Defense Acquisition Program
- e: Momentum Distribution of Annihilation Pairs
MDARNG e: Maryland Army National Guard
MDAS e: Magnetic Drum Auxiliary Sender
MDASER e: Multispectral Data Analysis System for Earth Resources
MDB e: Management Database
- e: Microsoft Database
- e: Murray-Darling Basin
MDBC e: Murray-Darling Basin Commission
MDBMC e: Murray-Darling Basin Ministerial Council
MDBMS e: Multidimensional DBMS
MDBS e: Microcomputer Data Base System
- e: Mobile Data Base Station
MDBs e: Multilateral Development Banks
MDC d: Max-Delbrück-Centrum für Molekulare Medizin, Berlin
- e: Major Diagnostic Category
- e: Malawi Development Corporation
- e: Management Development Centre
- e: Manufacturing Development Center
- e: Marker Distributor Control
- e: Maximum Dependable Capacity
- e: Mead Data Central
- e: Meridian Digital Centre
- e: Message Display Console
- e: Message Distribution Center
- e: Miniature Detonating Cord
- e: Minimum Detectable Concentration
- e: Missile Display Conference
- e: More Developed Country
- f: Médecin en Chef
- f: Mouvement de Défense des Contribuables
- f: Mouvement Démocratique Chrétien
- r: Marinens Dykericentrum
MDCB e: Motor-Driven Core Barrel
MDCCS e: Mourning Dove Call Count Survey
MDCI e: Multidiscipline Counter-Intelligence
MDCM f: Mouvement des Démocrates Chrétiens Malgaches
MD Comput e: MD Computing
MDD e: Major Depressive Disorder
- e: Management Division Director
- e: Medium-Doped Drain
- e: Meteorological Data Dissemination
- e: Meteorological Data Distribution
- f: Mouvement pour la Démocratie et le Développement
- f: Transmission de Données Météorologiques
- s: Distribución de Datos Meteorológicos
mdd. s: millones de dólares
MDDBMS e: Multidimensional Data Base Management System
MDDS e: Media Documentation Distribution Set
MDE d: Maschinendatenerfassung
- e: Manufacturing Development Engineering
- e: Maryland Department of the Environment
- e: Message Distribution Element
- e: Middle East
- e: Modular Design Environment
- s: Memorandum de Entendimiento
MDEC e: MAINDEC maintenance on DEC
- e: Marine Corps Development and Education Command
MDEP e: Management Decision Package
MDEQ e: Massachusetts Department of Environmental Quality
MDF e: Macular Degeneration Foundation
- e: Manpower Data File
- e: Medium Density Fibre
- e: Medium-Density Fiberboard
- e: Menu Definition File
MdF f: Milliard[s] de Francs
MDFP f: Mouvement pour la Démocratie des Forces du Progrès
MDFRC e: Murray-Darling Freshwater Research Centre
MDG e: Major Donors Group
- e: Mission Data Generator
MD[H] e: Homeopathic Medical Doctor
MDH e: Multimedia Data Handling
MDHS e: Meteorological Data Handling System
MDI e: Management Development I
- e: Medium-Dependent Interface
- e: Memory Display Interface
- e: Metered Dose Inhaler
- e: Methylene Diisocyanate
- e: Multiple Document Interface
- e: Multiplex Device Interface
- f: Mouvement pour la Démocratie et l'Indépendance
MDIA p: Movimento de Defesa dos Interesses de Angola
MDIC e: Manchester Decoder and Interface Chip
MDIF e: Map Data Interchange Format
MDII e: Management Development II
MDIII e: Management Development III
MDIO e: Multi-Dimensional Input and Output
MDIP e: Molded Dual-In-Line Package
MDIS e: McDonnell Information Systems Group
- e: Mobile Data Intermediate System
MDIV e: Master of Divinity
MDJCL e: Maryland Junior Classical League
MDK d: Medizinischer Dienst der Krankenversicherung
- e: Multimedia Developers (or Development) Kit
MDL e: Maintenance Deficiency List
- e: Master Deficiency List
- e: Master Deliverables List
- e: Maximum Detectable Limit
- e: Mechanical Development Laboratory
- e: Method Detection Limit
- e: Microstation Development Language
- e: Minimum Description Length
- e: Minimum Detection Limit
Mdl a: Mådelig [karakter]
mdl. a: mandlig[t]
- a: månedlig[t]
- a: medlem[mer]
MDLP e: Modile Data Link Protocol
MDM e: Mass Democratic Movement
- e: Medical Decision Making
- e: Mission Data Message
- e: Working Group on Marine Data Management
- f: Groupe de travail sur la gestion des Données Marines
- f: Médecins du Monde
- f: Mont-de-Marsan
- s: Grupo de Trabajo sobre Gestión de Datos Marinos
- s: Medicos del Mundo
Md Med J e: Maryland Medical Journal
MDMMS e: Multidimensional Microscopes and Maize Structures Research Group
MD-MOS e: Multi-Drain Metal-Oxide Semiconductor
MDMS e: Maintenance Data Management System
MDN s: Movimiento Democratico Nacionalista
mDNA e: mitochondrial DNA
Mdo s: Mando
MDOC e: Michigan Department of Commerce
MDOT e: Department of Transportation
MDP e: Malagasy Press Agency
- e: Master Data Processor
- e: Methylene Diphosphonate
- e: Multi-Divisional Program
- f: Mouvement Démocratique Populaire
- f: Mouvement des Démocrates Progressistes
- f: Mouvement pour la Démocratie et le Progrès
MDPA f: Mines Domaniales de Potasse de l'Alsace
MDPI e: Management Development and Productivity Institute
- e: Manual Dial Position Indicator
MDPS e: Mission Data Preparations Systems
- e: MOPS Data Processing System
MDQ e: Market Driven Quality
MDQS e: Management Data Query System
- e: Multi Device Queueing System

MDR d: Monatsschrift für Deutsches Recht
- e: Master Deficiency Report
- e: Materiel Deficiency Report
- e: Medium Data Rate
- e: Mini D-Ribbon
- e: Minimum Design Requirement
- e: Minimum Detectable count Rate
- e: Minimum Detectable Radiance
- f: Mouvement Démocratique Républicain
- f: Mouvement pour la Démocratie et la République

mdr. N: måneder

MDRA e: Material Deficiency Report Analysis

MDRAM e: Multibank Dynamic Random Access Memory

MDRC e: Manufacturing Design Rule Checker

MDRE e: Mass Driver Reaction Engine

MDRF e: Mark Diamond Research Fund

MDRM f: Mouvement Démocratique de Rénovation Malgache

MDRO e: Mission Disaster Relief Officer

MDRRA f: Ministère du Développement Rural et de la Réforme Agraire

MDS e: Manpower Data System
- e: Manufacturing Design System
- e: Meteorological Distribution System
- e: Milestone Data Sheet
- e: Milestone Description Sheet
- e: Mission, Design and Series
- e: Mobilization, Deployment and Sustainment
- e: Model Design Series
- e: Modify Device Status
- e: Modular Display System
- e: Multi Dimensional Scaling
- e: Multichannel Distribution Service
- e: Multipoint Distribution Service[s]
- e: Multipoint Distribution System
- f: Mouvement Démocratique et Social

Mds f: Mesdames

MDSD e: Monitoring and Data Support Division

Mdse f: Marchandise

MDSL e: Moderate Speed Digital Subscriber Line

MDS-MU e: Multiple Domain Support-Message Unit

MDSRS e: Morphology-Dependent Stimulated Raman Scattering

MDSS e: MPS [Maritime Pre-Positioned Ships] Decision Support System
- e: Multiple Distribution Switching Subsystem

MDSSC e: McDonnell Douglas Space Systems Corp.

MDST f: Mouvement Démocratique pour le Socialisme au Tchad

MDStV d: Mediendienste-Staatsvertrag

MDT d: Mittlere Datentechnik
- e: Mean Downtime
- e: Mechanical Desktop
- e: Mobile Data Terminal
- e: Modified Distance Transform
- e: Multidisciplinary Team
- f: Mandat-Télégramme

Mdt f: Mandat

MDT. e: Modified Data Tag

MDTA/D e: Micro-Distillation Tube Assembler/Disassembler

Mdt-Ct f: Mandat-Contributions

mdtl. a: mundtlig

MDTS e: Mission Data Transfer System
- e: Modular Data Transport System-part of Needs

MDU e: Marker Decoder Unit
- e: Modular Dispensing Unit
- f: Mouvement Démocratique Uni

MDu. e: Middle Dutch

MDUC e: Meteorological Data Utilization Center
- e: Meteorological Data Utilization Centre
- f: Centre d'Utilisation des Données Météorologiques
- s: Centro de Utilización de Datos Meteorológicos

MDUS e: Medium-Scale Data Utilization Station
- e: Meteorological Data Utilization Station

MDV e: Medieval Studies
- f: Mouvement Démocratique Voltaïque

MDW e: Military District of Washington

MDWS e: Missile Detection & Warning System

MDX e: Modular Digital Exchange
- e: Multiplexer/Demultiplexer

MD/XMP e: Message Distribution/Xerox Message Processor

MDY e: Month Day Year

MDZ e: Maritime Defense Zone

MDZL e: Maritime Defense Zone Atlantic

MDZP e: Maritime Defense Zone Pacific

ME e: Main Engine
- e: Mapping Entity
- e: Materials Engineering
- e: Mechanical Engineer[ing]
- e: Medium Energy
- e: Ministry of Education
- e: Mission Effectiveness
- e: Moisture Element
- e: Mutation Engine
- e: Myalgic Encephalomyelitis
- e: Opening Information
- f: Machine à Écrire
- f: Maison d'Éducation
- f: Maître d'École
- f: Manoeuvre Électorale
- f: Manufacture de l'Etat
- f: Médecin-Expert
- f: Mineur Émancipé
- f: Mission Étrangère
- f: Mode d'Emploi
- f: Mortalité Enfantine
- f: Mouvement Européen
- f: Moyenne Entreprise
- p: Ministério de Educação
- r: Medicinalkårexpedition
- s: Mecánico Electricista
- s: Memorándum de Entendimiento

Me e: Mercury
- f: Maître
- r: Motoreldare

M E e: Mechanical Engineering
- e: Monitoring and Enforcement

ME. e: Middle English

M&E e: Material and Equipment

MEA e: Maintainability Engineering Analysis
- e: Major Emergency Actions
- e: Maryland Energy Administration
- e: Monoethanolamine
- s: Mecánico Electricista de Armas
- s: Muerto En Acción

ME-a e: Melanoma

MEAT e: Meat Lab

MEB e: Main Electronics Box
- e: Materials-Energy Balance
- e: Memory Expansion Board

MEBES e: Manufacturing Electron-Beam Exposure System

MEBS e: Management Evaluation Guides
- e: Medium Energy Backscattering Spectrometry

MEBU e: Minimum Essential Backup
- e: Mission Essential Backup

MEC e: Member of the Executive Council
- e: Members of the Executive Council
- e: Minimal Effective plasma Concentration
- e: Ministry of Education and Culture
- e: Missile Event Conference
- e: Mobile Equipment Console
- e: Model Energy Code
- p: Ministério de Educação e Cultura
- s: Movimiento Emergent de Concordancia

Mec f: Mécanicien
- e: Mécanique
- f: Mécanisé
- s: Mecánico
- s: Mecanizado

MECA e: Manufacturers of Emission Controls Association
- e: Missile Electronic Computer Assembly
- f: Machines et Equipements de Conception Avancée

Meca s: Mecanógrafo

MECANO e: Mechanism of Automatic Comparison of Answers with OPACs

MecApp f: Mécanique Appliquée

MECB p: Missao Espacial Completa Brasileira

MECCA e: Model Evaluation Consortium for Climate Assessment

MECH e: Mechanical Engineering
- e: Mechanical Shop
- e: Mechanized

Mech. e: Mechanical
- e: Mechanics

Mech A e: Institut A für Mechanik

Mech Ageing Dev e: Mechanisms of Ageing and Development

Mech B e: Institut B für Mechanik

MechBau e: Institut für Mechanik [Bauwesen]

Mech Dev e: Mechanisms of Development

MechE e: Mechanical Engineering

MECL e: Families of ECL logic
- e: Minimum Essential Circuit List

MECN e: Mobile Extended Corporate Network

MECO e: Main Engine Cutoff

MECo e: Massachusetts Electric Company

MECU e: Million European Currency Units

MED d: Medizin und Veterinärmedizin
- e: Manhattan Engineer[ing] District
- e: Manipulative Electronic Deception
- e: Master of Education
- e: Median Effective Dose
- e: Mediterranean Region
- e: Message Element Dictionary
- e: Ministry of Education
- e: Modeling for Equipment Design
- e: Modem Equivalent Device
- e: Office of Medical Services
Med r: Medicin[sk]
- s: Médico
- s: Médio
Med. e: Medical
med. N: meddelelse
- N: medisin[sk]
Méd f: Médecin[e]
- f: Médical
- f: Médicament
- f: Médiocre
MEDA e: Maintenance Error Decision Aid
- e: Military Emergency Diversion Airfield
MEDAC e: Meteor Echo Detection and Collection System
MEDAD e: Mean Effective Depth of the Atmospere Duct
MEDALPEX e: Mediterranean Alpine Experiment
- f: Expérience Alpine en Méditerranée
- s: Experimento Alpino en el Mediterráneo
MEDALS e: Military Engineering Data Asset Locator
MEDALUS e: Mediterranean Desertification and Land Use Impacts
Med Anthropol e: Medical Anthropology
Med Anthropol Q e: Medical Anthropology Quarterly
Med Arh R: Medicinski Arhiv
MEDAS e: Automated Meteorological Data Acquisition System
MedAu e: Medicine Australia-The Online Journal of Medicine
Med Biol Eng Comput e: Medical and Biological Engineering and Computing
MEDBLOOD e: Medical Blood
Med Care e: Medical Care
MedCh f: Médecin-Chef
Med Clin [Barc] e: Medicina Clinica
Med Clin North Am e: Medical Clinics of North America
MEDCOAST e: International Conference on the Mediterranean Coastal Environment
- f: Conférence internationale sur l'environnement Côtier Méditerranéen
- s: Conferencia Internacional sobre el Medio Ambiente de las Zonas Costeras Mediterráneas
MEDCOM e: Mediterranean Communication System
- e: Mediterranean Planning Committee
- e: Mediterranean Regional Committee [for START]
Medd r: Meddelande
medd. a: meddelelse[r]
Med Decis Making e: Medical Decision Making
Med Dosim e: Medical Dosimetry
Med Dosw Mikrobiol R: Medycyna Doswiadczalna i Mikrobiologia
MEDEA e: Measurements of Earth Data for Environmental Analysis
Med Educ e: Medical Education
Med Eng Phys e: Medical Engineering and Physics
MEDFAD e: Medical Field Assistance Branch
medflg. a: medfølgende
MEDFLUX e: Quantification of pollutant dry deposition fluxes over Mediterranean ecosystems
MEDGLOSS e: Mediterranean Global Sea-Level Observing System
- e: Sistema Mundial de Observación del Nivel del Mar Mediterráneo
- f: Système mondial d'observation du niveau de la mer Méditerranée
Med Hist e: Medical History
Med Hist Suppl e: Medical History. Supplement
MED-HYCOS e: Mediterranean Hydrological Cycle Observing System
- f: Système d'observation du cycle hydrologique méditerranéen
- s: Sistema de Observación del Ciclo Hidrológico en el Mediterráneo
Med Hypotheses e: Medical Hypotheses
MEDI e: Sistema de Reenvío de Datos e Información relativos al Medio Marino
- e: Système d'accès aux données et informations sur le milieu marin
Media e: Mauritius Export Development Authority
MEDIAS e: Mediterranean [Basin] and Subtropical Africa [Regional Research Network]
- f: Reseau de Recherche Regionale pour le Bassin Mediterraneen et l'Afrique Subtropicale
MEDIF a: Medicinimportørernes Forening
Medigap e: Medicare Supplement Insurance
Med Inf [Lond] e: Medical Informatics
MEDINTEL f: Médecine Informatique Télématique
Med Interna e: Medicina Interna
MEDIS e: Medical Dietary Information System
MEDISLE e: Network on Small Islands Marine and Coastal Research in the Mediterranean
Med J Aust e: Medical Journal of Australia
Med J Malaysia e: Medical Journal of Malaysia
Med J Osaka Univ e: Medical Journal of Osaka University
Med Klin d: Medizinische Klinik
med.L. e: medieval Latin
medl. a: medlem[mer]
- r: medlem
Med Lav s: Medicina del Lavoro
medlb. a: medlemsbidrag
Médlég f: Médecine légale
Med Leg J e: Medico-Legal Journal
Med Lett Drugs Ther e: Medical Letter on Drugs and Therapeutics
MEDLINE e: Medlars Online [Medical database with bibliography-Online database equivalent of Index Medicus]
MEDLOG e: Medical Logistics
Med Microbiol Immunol [Berl] e: Medical Microbiology and Immunology [Berlin]
Med Monatsschr Pharm d: Medizinische Monatsschrift für Pharmazeuten
MEDNET e: Medical Systems Network
MEDO e: Mediterranean Development and Environment Observatory
- s: Observatorio del Desarrollo y el Medio Ambiente del Mediterráneo
Med Oncol e: Medical Oncology
MEDP e: Mission Essential Data Processing
MédP f: Médecin Principal
MEDPAR e: Medical Patient Accounting and Reporting
Med Parazitol [Mosk] R: Meditsinskaia Parazitologiia i Parazitarnye Bolezni
MED-PED e: Make Early Diagnoses-Prevent Early Deaths
Med Pediatr Oncol e: Medical and Pediatric Oncology
Med Pediatr Oncol Suppl e: Medical and Pediatric Oncology. Supplement
MEDPERFAC e: Medical and Personnel Planning Factors Report
MEDPOL e: Mediterranean Action Plan Pollution Monitoring and Research Program[me]
- f: Programme coordonné d'étude et de surveillance de la pollution en Méditerranée
- s: Programa Coordinado de Vigilancia e Investigación de la Contaminación en el Mediterráneo
MED POL-I e: Pilot Project on Baseline Studies and Monitoring of Oil and Petroleum Hydrocarbons in Marine Waters
- f: Projet pilote sur les études de base et la surveillance continue du pétrole et des hydrocarbures pétroliers dans les eaux marines
- s: Proyecto Piloto sobre Estudios Básicos y Vigilancia del Petróleo e Hidrocarburos del Petróleo en las Aguas del Mar
Med Pr P: Medycyna Pracy
Med Pregl P: Medicinski Pregled
Med Prog Technol e: Medical Progress Through Technology
medr. a: medregnet
MEDRAMS e: Medical Readiness Assemblage Materiel System
MEDRED e: Medical Readiness Report Capability
MEDREG e: Medical Regulation
MEDREQ e: Medical Requirements Model
Med Res Rev e: Medicinal Research Reviews
MEDRTS e: Medical Requirements Model

MEDS e: Maintenance Engineering Data System
- e: Marine Ecological Database System
- e: Medicine
- e: Meteorological and Environmental Data Services
- e: Meteorological Data Systems
- e: Modular Environment for Data Systems
- s: Servicio de Datos sobre el Medio Marino

MEDS 70 e: Management Engineering Data System 70s
Med Sci Law e: Medicine, Science and the Law
Med Sci Sports Exerc e: Medicine and Science in Sports and Exercise
MEDSOM e: Medical Supply, Optical and Maintenance Unit
MEDSTAT e: Medical Statistical System
MEDSTOCK e: Medical Stock Control System
MedStyr r: Medicinalstyrelsen
Medsupp e: Medicare Supplement Insurance
Med Tekh R: Meditsinskaia Tekhnika
Med Trop f: Medecine Tropicale
Med Tr Prom Ekol R: Meditsina Truda I Promyshlennaia Ekologiia
MEDU e: Mediterranean Action Plan Coordinating Unit
- f: Unité de coordination du Plan d'action pour la Méditerranée
- s: Dependencia del Plan de Acción para el Mediterráneo

MEDUNSA e: Medical University of Southern Africa
Médvét f: Médecine vétérinaire
Med Vet Entomol e: Medical and Veterinary Entomology
MED-VI e: Pilot Project on Problems of Coastal Transport of Pollutants
- f: Projet pilote sur les mouvements des polluants le long des côtes
- s: Proyecto Piloto sobre Problemas del Transporte Costero de Contaminantes

Med War e: Medicine and War
MED-WRAP e: Medical War Reserve Automated Process
MEE e: Migration-Enhanced Epitaxy
- f: Machine à Écrire Électrique

MEEA e: Mobil Exploration Equatorial Africa, Inc.
MEEC e: Massachusetts Energy Efficiency Council
MEECI f: Mouvement des Etudiants et des Elèves de Côte d'Ivoire
MEECN e: Minimum Essential Emergency Communication Network
MEEK e: International Monitoring System
ME&ES e: Maintenance Environmental and Engineering Services
MEETS e: Minimum Engine Tracking System
MEF d: Medizinische Fakultät
- e: Marine Expeditionary Force
- e: Mauritian Employers Federation
- e: Minimum Essential Facilities
- e: Minimum Essential Functions
- e: Ministry of Economics and Finance
- f: Mouvement des Étudiants de France
- N: Maskinentreprenørenes Forening
- p: Ministério de Economia e Finanças

MEFA a: Foreningen af Danske Medicinfabrikker
MEFPAK e: Manpower and Equipment Force Packaging System
MEFS e: Midterm Energy Forecasting System
MEFT e: Minimum Essential Functional Task
MEFV e: Maintenance Equipment Floor Valve
MEG e: Mega Evil Grin
MEGAS e: Multienergy Gamma Assay System
MEGO e: My Eyes Glaze Over
MEGW e: Megawatt[s]
MEHALICE e: Methane and Halocarbons Intercalibration[s] Experiment
MEH-PPV e: Poly[2-Methoxy-5-[2'-Ethyl-Hexyloxy]-p-Phenylenevinylene]
MEI d: Meißen, Sachsen
- e: Major End Item
- e: Management Effectiveness Inspection
- e: Maximally (or Maximum) Exposed Individual
- e: Most Exposed Individual

MEIC e: Multichannel Electronic Image Classifier
MEIDA e: Metals and Engineering Industries Development Association
MEIDS e: Military Electronic Information Delivery System
- e: Miniaturized Electronic Information Delivery System

MEIR e: Minimum Essential Information Requirement
MEIS e: Mineral Exploration Information System
- e: Multispectral Electro-Optical Imaging Scanner
- e: Multispectral Electro-Optical Imaging Spectrometer [forestry and mapping]
- e: Multispectral Electro-Optical Imaging System

Meison e: All-Ethiopia Socialist League
MEISR e: Minimum Essential Improvement in System Reliability
MEITS e: Mission Effective Information Transmission System
- e: Mission Essential Information Transmission System

Mej s: Mensaje
Meje s: mensajero
Mek r: Mekaniker
- r: Mekaniserad

mek. N: mekanikk
- N: mekanisert
- N: mekanisk

MekB r: Mekaniserad Brigad
MEL e: Material and Equipment Log
- e: Medium Energy Laser
- e: Middle Eastern Languages and Cultures
- e: National Herbarium of Victoria International Acronym

Mel f: Mélangeur
MELAB e: Michigan English Language Assessment Battery
Melanoma Res e: Melanoma Research
MELAS e: Mitochondrial Encephalo-myopathy Lactic Acidosis Stroke[-like episodes]
MELD e: Mechanized Engineering and Layout for Distributing frames
MELF e: Metal Electrical Face
MEL-FS e: Multiprogram Energy Laboratories (or Laboratory) Facilities Support
MELISR e: National Herbarium of Victoria Specimen Information Registrer
mell. a: mellem
- r: mellersta

Melle f: Mademoiselle
MELU e: Melbourne University [Herbarium International Acronym]
MELV e: Medium Expendable Launch Vehicle
MEM e: Manufacturing Enterprise Model
- e: Manufacturing Equipment Monitor
- e: Memory Management
- e: Micro-Electro Mechanical
- e: Micro-Erosion Meter
- e: Ministry for Energy and Mines
- e: Modal Emission Model
- f: Mémoire
- s: Mecánico Electricista Montador

Mem. e: Memoir[s]
Mém f: Mémorandum
MEMAC e: Marine Emergency Mutual Aid Centre
- f: Centre d'assistance mutuelle en cas d'urgence dans le milieu marin
- s: Centro de Ayuda Mutua para Emergencias Marinas

MEMACO e: Metal Marketing Corporation
MEMCO e: Miller Electric Manufacturing Company
MEMDT e: Message Execution Matrix Display Task
Mem Inst Oswaldo Cruz s: Memorias do Instituto Oswaldo Cruz
Memo s: Memorándum
MEMO/MAINT e: Medical Equipment Management Office/Maintenance
MEMS e: Master of Emergency Medical Services
- e: Master of Engineering-Manufacturing Systems
- e: Microelectromechanical System[s]

MEN e: Menu
- e: Multiple Endocrine Neoplasia
- f: Ministère de l'Éducation Nationale

Men f: Menuiserie
men. N: menig
Mén f: Ménage
MENA e: Middle East and North Africa
MEND e: Mine Environment Neutralization Draining
M.Eng e: Master of Engineering

MENJS f: Ministère de l'Education Nationale, de la Jeunesse et des Sports
MeNNG e: N-methyl-N'-nitroso-N-nitrosoguanidine
MENS e: Mission Element Need Statement
Mens f: mensuel
Ment Health Stat Note e: Mental Health Statistical Note
MENTOR e: Meteor Echoes-No Transmitter-Only Receivers
Ment Retard e: Mental Retardation
MEO e: Medium Earth Orbit
- e: Message Exchange Occurrence
MEOCAM f: Mouvement d'Etudiants de l'Organisation Commune Africaine, Malgache et Mauricienne
MEOS e: Medium Earth Orbit Satellites
- e: Multidisciplinary Earth Observation Satellite System
MEOSS e: Monocular Electro-Optical Stereo Scanner
- e: Multispectral Electro-Optical Stereo Scanner
MEP e: Maintenance Engineering Program
- e: Management Engineering Plan
- e: Member of the European
- e: Member of the European Parliament
- e: Meteosat Exploitation Project
- e: Microcircuit Emulation Program
- e: Minimum Entry Point
- e: Mission Equipment Package
- e: Multiple Extraction Procedure
- f: Mise En Place
- s: Ministerio de Educación Pública
MEP-91 e: Mesoscale Evolution Project-1991
MEPA e: Marine and Estuarine Protected Area
- e: Massachusetts Environmental Protection Act
- e: Meteorological (or Meteorology) and Environmental Protection Administration
- f: Administration de météorologie et de protection de l'environnement
- s: Organismo de Meteorología y Protección del Medio Ambiente
MEPAI f: Mouvement Etudiant du Parti Africain de l'Indépendance
MEPAS e: Multimedia Environmental Pollutant (or Pollution) Assessment System
MEPC e: Malawi Export Promotion Council
- e: Marine Environment Protection Committee
MEPCOM e: Military Entrance Processing Command
MEPED e: Medium Energy Proton and Electron Detector
MEPHISTO f: Matériel pour l'Etude des Phénomènes Interéssant la Solidification sur Terre et en Espace
MEPIS e: Management Engineering Program Information System
MEPP e: Middle East Peace Process
MEPR e: Medical Expense and Performance Report
MEPROBA f: Mouvement des Etudiants Progressistes Bahutu
MEPRS e: Medical Expense and Performance Reporting System
MEPS e: Medium-Energy Particle Spectrometer
- e: Message Editing and Preparation Service
- e: Message Editing and Processing Station
- e: Military Entrance Processing Stations
MEPU e: Ministry of Pre-University Education
MEQC e: Marine Environmental Quality Committee
- f: Comité d'étude de la qualité du milieu marin
- s: Comité para la Calidad del Medio Marino
MEQPT e: Major Equipment ID Code
MER d: Merseburg, Sachsen-Anhalt
- e: Mechanical Equipment Room
- e: Merchant Ship
- e: Meridian
- f: Mise En Route
mer. r: meridian
MERASEX e: Mesoscale Rain and Snowfall Experiment
MERC e: Middle East Regional Cooperation Program
- e: Morgantown Energy Research Center
Merc f: Mercredi
- f: Mercuriale
MERCAST e: Merchant Shop Broadcast System
MERCI e: Multimedia European Research Conferencing Integration
MERCOSUR f: Mercado Común del Sur
- s: Mercado Comén del Sur
Mercosur s: Mercado Comun del Cono Sur
MERCURY e: Multisource Evidential Reasoning Classification of Klut Probabilities by Orthogonal Summation
MEREP e: Merchant Ship Report
MERES e: Mineral and Energy Resources Exploration Satellite
MERG e: integrated energy-economic-environmental model
MERGE e: Model for Evaluating Regional and Global Effects of GHG reduction policies
MERIE e: Magnetically Enhanced Reactive Ion Etching
MERIS e: [European] Medium Resolution Imaging Spectrometer
MERIT e: Maastricht Economic Research Institute on Innovation and Technology
- e: Method to Extend Research in Time Award
- e: Michigan Education Research Insruction Triad
- e: Military Exploitation of Reconnaissance and Intelligence Technology
- e: Monitoring Earths Rotation and Intercomparison of Techniques
MERL e: Municipal Environmental Research Laboratory
MERLIN e: Multi-Element Radio Linked Interferometer Network
MERP e: Marine Ecosystem Response Project
- e: Middle Earth Role Playing
MERRF e: Myoclonic Epilepsy; Ragged Red Fiber
MERS e: Medium Resolution Imaging Spectrometer
- e: Mobile Emergency Response System
- e: Most Economic Route Selection
MERSHIP e: Merchant Shipping system
MERTU e: Medical Entomology Research and Training Unit, University de Valle
MES d: Mein erstes Semester [Erstsemesterinfo des AStA FU Berlin]
- e: Management Engineering Squadron
- e: Managing Editor Software, Inc.
- e: Manufacturing Execution System[s]
- e: Marine Environmental Services
- e: Message Entry Subsystem
- e: Middle Eastern Studies
- e: Military Engineering Services
- e: Missing End Surface
- e: Mobile End System
- e: Modal Emission Model
- e: Molecular Environmental Science
Mes f: Mesdames
- f: Message
M-ES e: Mobile End System
ME&S e: Management Evaluation & Support
MESA e: Marine Education Society of Australia
- e: Mathematics, Engineering and Science Achievement
- e: Mathematics, Engineering, Science Achievement
- e: Microsurgical Epididymal Sperm Aspiration
- e: Middle East Studies Association
- e: Model Experimental Systems Analysis
- e: Modular Experimental platform for Science and Applications
- e: Myoepithelial Sialadenitis
- f: Analyse des Écosystèmes Marins
- s: Análisis de Ecosistemas Marinos
MESAMAC e: Engine Status Accounting System
MESAN f: Mouvement de l'Evolution Sociale de l'Afrique Noire
MESC e: Midcontinent Ecological Science Center
- e: Modular Equipment Standards Committee
- e: Modular Equipment Sub-Committee for Communications
MESCAL e: United States Biological Cruise off Baja California
- f: Croisière biologique américaine au large de la Basse-Californie
- s: Crucero Biológico Norteamericano Frente a las Costas de la Baja California
MESCAM e: Microwave Emission and Scattering Model
Mesd f: Mesdames
MESDIS e: Message Distribution System
MESE e: Maximum Entropy Spectral Estimation
MESEC e: Multi-Cultural Environmental Science Education Centers
MESEEC e: Methodologies to Estimate Social, Environmental and Economic Consequences

MESFET e: Metal-Semiconductor Field-Effect Transistor
MESH e: Managing Engineering Safety Health
MESHA e: Manager ES&H Awareness
MESI e: Modified, Exclusive, Shared [and] Invalid
- N: Maskinentreprenørenes Service- og Innkjøpslag
MESL e: Marine Environment Studies Laboratory
- e: Minimum Essential Subsystem List
- f: Laboratoire d'étude du milieu marin
- s: Laboratorio de Estudios del Medio Ambiente Marino
MESN f: Mouvement pour l'Edification de la Société Nouvelle
MESOPAC e: Mesoscale Meteorological Preprocessor Program
- e: Mesoscale meteorological reprocesser program
MESOPLUME e: Mesoscale "Bent Plume" Model
MESOPUFF e: Mesoscale Puff Model
MESOSOFT e: Mesoscale Data Management Facility
Mesoz. e: Mesozoic
MESP e: Management Engineering Scheduling Program
MESPREP e: Message Preparation
MESS e: Military Equipment Support Subsystem
- e: Model Evaluation Support System
MES-SR e: Molecular Environmental Science Synchrotron Radiation
MESSR e: Multispectral Electronic Self-Scanning Radiometer
Messrs f: Messieurs
MET a: Mellemeuropæisk Tid
- e: Management Engineering Team[s]
- e: Memory Enhancement Technology
- e: Metadata
- e: Metafile
- e: Metallurgy
- e: Metropolitan Trasport Authority
- e: Mid[dle] European Time
- e: Ministry of Education and Training
- e: Mobile Environmental Teams
- e: Multibutton Electronic Telephone
- e: Sensor Meteorological Satellite Sensor
- f: Météo
- N: Mellomeuropeisk Tid
- s: Mecánico Electricista de Transmisiones
Met f: Métier
- N: Meteorolog[i][sk]
- s: Meteorólogo
Mét f: Métal[lique]
- f: Métallurgie
- f: Métallurgique
- f: Météorologie
- f: Météorologique
META e: Management Enhancement Team Approach
Metab Brain Dis e: Metabolic Brain Disease
Metabolism e: Metabolism: Clinical and Experimental
Metab Pediatr Syst Ophthalmol e: Metabolic, Pediatric and Systemic Ophthalmology
MetaCenter e: National MetaCenter for Computational Science and Engineering
META-DATA e: Ancillary data characterising a data set
METAL a: Dansk Metalarbejderforbundet
Métall f: Métallique
- f: Métallurgie
Metallk d: Institut für Metallkunde, Lehrstuhl
METAP e: Mediterranean Environment Technical Assistance Programme
- f: Programme d'assistance technique pour l'environnement méditerranéen
- s: Programa de Asistencia Técnica Ambiental del Mediterráneo
METAR e: Aviation Rout[in]e Weather Report
- e: Meteorological Aviation Report
Meta Tags d: Versteckter Text auf einer WWW-Seite
Meta-TS e: Meta Trouble Shooter
Meta-XP e: Meta-Explanation Pattern
METCAM e: Metric Camera
METCON e: Metropolitan Consortium for Minorities in Science and Engineering
METDAS e: Meteorological Data Analysis System
METEO e: Meteorologie Consult GmbH, Glashütten
METEOR e: Meteorological Event Oriented Reporting System
- e: Operational Weather Satellite
- e: Russian meteorological satellite system
Meteor e: Research vessel
Météor f: Météorologie
Meteo[r] s: Meteorología
Meteorol. e: Meteorological
METEOSAT e: ESA [European Space Agency] Geostationary Meteorological Satellite [operated by EUMETSAT]
- f: Satellite météorologique géostationnaire de l'ASE
- s: Satélite Meteorológico Geoestacionario del ESA
METEOSAT-3 e: European Space Agency Geostationary Meteorological Satellite
Méth f: Méthode
- f: Méthodique
Methods Biochem Anal e: Methods of Biochemical Analysis
Methods Cell Biol e: Methods in Cell Biology
Methods Enzymol e: Methods in Enzymology
Methods Find Exp Clin Pharmacol e: Methods and Findings in Experimental and Clinical Pharmacology
Methods Inf Med e: Methods of Information in Medicine
Methods Mol Biol e: Methods in Molecular Biology
Métie f: Métairie
Met Ions Biol Syst e: Metal Ions in Biological Systems
METIS d: Meteorologisches Informations-system [Rheinland-Pfalz]
M-et-L f: Maine-et-Loire
METL e: Metallurgical Engineering
- e: Metals Data Base
- e: Mission Essential Task List
- e: Multielement Two-Layer
- e: Region 4 Metals System
METLA e: Finnish Forest Research Institute
MET LAB e: Metallurgical laboratory
M-et-M f: Meurthe-et-Moselle
METMF e: Marine Corps Meteorological Mobile Facility
METNO e: Advance telegraphic notification relating to the operation of WWW
- f: Notification télégraphique préalable de modifications concernant le fonctionnement de la VMM
- s: Notificación telegráfica previa de modificaciones relativas al funcionamiento de la VMM
METOC e: Bureau of Meteorology and Oceanographic Services
- e: Meteorological and Oceanographic
METOP e: Meteorological Operation[al] [Weather Satellite]
METOX e: Metal Toxicity in humans
METP f: Marché d'Entreprises de Travaux Publiques
METPAK e: Meteorological Analysis Software Package
Metr f: Mission à l'étranger
METRACOM e: Meteorologist Tracking Computer
METRAN e: Managed European Transmission Network
METRASAT e: Meteorological Transmission Satellite Service
METRIC e: Multi-Echelon Technique Recoverable Item Control
METRO e: Meteorological
Metro Dade e: Metro Dade County [Florida]
METS e: Maintenance Equipment Transport System
- e: Materials and Equipment Trading Service
- e: Mission Events Transformation System
- e: Mobile Electronic Test Set
- e: Multi-Engined Training Squadron
METSAAT e: Meteorological Satellite
METSII/ACI e: Military Export Traffic System II-Enhanced/Automated Carrier Interface
METSIS e: Meteorological Satellite Information System
METT e: Mission, Enemy, Terrain and Troops
METTT e: Mission, Enemy, Terrain, Troops and Time
MET Watch e: Meteorological Watch
MEU e: Mission Essential Unit
Meu f: Meuse
MEU[SOC] e: Marine Expeditionary Unit [Special Operations Capable]
MEV e: Mars Excursion Vehicle
MeV e: Mega Electro[n] Volt[s]
- e: Million Electron Volts
MeVerfT d: Institut für Mechanische Verfahrenstechnik
MEW e: Maintenance Equipment Wing

MEWSG e: Maritime Electronic Warfare Support Group
- e: Multi-Service Electronic Warfare Support Group
MEWSS e: Mobile Electronic Warfare Support System
MEWUSA e: Metal and Electrical Workers Union of South Africa
MEXAMS e: Metals Exposure Analysis Modeling System
MEY e: Maximum Economic Yield
MEZ d: Mitteleuropäische Zeit
- e: Maritime Exclusion Zone
- e: Missile Engagement Zone
MF a: Medlem af Folketinget
- a: Møbelsnedkernes Fagforening
- d: Marinefunker
- d: Multifunktion[stastatur]
- e: Management Function
- e: Manipulator Field
- e: Media Filter
- e: Medium Frequency
- e: Middle French
- e: Million Francs
- e: Molecular Formula
- e: More Fragments
- e: Multi Frequency [telephony]
- e: Multi-Function[al]
- e: Multiple Frequency signaling
- f: Maison Familiale
- f: Manoeuvres Frauduleuses
- f: Marché Ferme
- f: Marché Forfaitaire
- f: Marque de Fabrique
- f: Médecin de Famille
- f: Menus Frais
- f: Mère de Famille
- f: Million[s] de Francs
- f: Ministère des Finances
- f: Monnaie Fiduciaire
- f: Mouvements de Fonds
- N: Mellomfrekvens
- N: Menighetsfakultetet
- r: Marinförvaltning
- r: Mellanfrekvens
mf. a: midtfor
M/F N: Motorferge
m/f r: motorfartyg
- s: mi favor
M[/]F a: Motorfærge
MFA e: Master Fine Arts
- e: Materiel Fielding Agreement
- e: Mississippi Forestry Association
- e: MONJU Fuel Assembly
- e: Moving Filter Assembly
- f: Mission Française d'Assistance
- f: Mouvement des Forces Armées
- N: Middeltung Feltartilleriet
MFAbn N: Middeltung Feltartilleri-bataljon
MFASYS e: Management, Financial and Accounting Information System
MFB e: Management Function Block
- e: Monochrome Frame Buffer
- r: Målfaktorberäkningsmetod
MFBI e: Major Fuel Burning Installation
MFBP f: Mouvement des Femmes et du Bas Peuple pour la Paix
MFBS e: Multifrequency Binary Sequencing (or Sequence)
M1FC e: Model 1 Full Capacity
MFC d: Multifunktionale Chipkarte
- e: Maintenance of Facilities Capability
- e: Mass Flow Controller
- e: Message Formatting Committee
- e: Metal Finishing Category
- e: Meteorological and oceanographic Forecast Center
- e: Microsoft Foundation Class[es]
- e: Millard Fillmore College
- e: Missile Fire Control
MFCA e: Maryland Food Center Authority
MFCC e: Marriage, Family and Child Counselor
- e: McFarland Cascade Company
MFCI e: Molten Fuel Coolant Interaction
MFCMA e: Magnuson Fishery Conservation and Management Act
MFCS e: Mortar Fire Control System
- s: Medidas Destinadas a Fomentar la Confianza y la Seguridad
MFD e: Multi-Function Display
- e: Municipal Facilities Division
MFDC f: Mouvement des Forces Démocratiques de Casamance
MFDP e: Mesoscale Frontal Dynamics Project
MFDS e: Maxey Flats Disposal Site
MFE e: Magnetic Fusion Energy
- e: Manpower Force Element
- f: Mouvement Fédéraliste Européen
MFEL e: Manpower Force Element Listing
MFE/MAGNOLIA e: Magnetic Field Experiment
MFENET e: Defense data network
- e: Magnetic Fusion Energy Network
MFF e: Metal fuel
MFFS e: Microsoft Flash File System
MFG a: Ministeriet For Grønland
- e: Manufacturing [Systems Engineering]
- e: More Friendly Garbage
MfG d: Mit freundlichem Gruß
- d: Mit freundlichen Grüßen
MFHBF e: Mean Flight Hours Between Failure
MFI e: Mainframe Interactive
- e: Metafile Input
- N: Minefartøyinspeksjon
MFIA e: Magtape Formatter Interface Adapter
MFI/MFO d: Mathematisches Forschungsinstitut in Oberwolfach
MFIMR e: Multi-Frequency Imaging Microwave Radiometer
MFIR e: Moderate Field-of-View Imaging Radiometer
MFJ e: Modification of Final Judgement
- e: Modified Final Judgment
- e: Movement for Freedom and Justice
MFKP signaling e: Multifrequency Key Pulse signaling
MFKS d: Multifunktionales Konferenzsystem
MFKY e: Maxey Flats, Kentucky
MFL e: Master Fishermens Licence
- e: Master Force List
- f: Machines Françaises Lourdes
Mfl. a: Med flere
m.fl.[d.] r: med flera
MFLOP e: Million Floating Point Operations
MFLOPS e: Million [of] Floating Point Operations Per Second
m.fl.st. a: med flere steder
MFM e: Magnetic Flowmeter
- e: Magnetic Force Microscope
- e: Magnetic Force Microscopy
- e: Mass Flow Meter
- e: Movable Fine-mesh Model
- e: Municipal Finance and Management
- f: Modèle à maille Fine Mobile
- f: Mouvement Français des Mères
- s: Modelo de Malla Fina Movible
M2FM e: Modified Double Frequency Modulation
MFMA e: Maple Flooring Manufacturers Association
MF/MAGNOLIA e: Magnetic Field Experiment
MFMR e: Multifrequency Microwave Radiometer
MFN e: Most Favored Nation
MFO e: Market Facing Organization
- e: Metafile Output
- e: Mixed-Function Oxidase
MFOM f: Ministère de la France d'Outre-Mer
MFOV e: Multiple Field Of View
MFP e: Major Force Program
- e: Marematlou Freedom Party
- e: Master Personnel File
- e: Material (or Materiel) Fielding Plan
- e: Mixed Fission Products
- e: Mobile Flux Platform
- e: Module Field Panel
- e: Monofluorophosphate
- e: Mosaic Focal Plane
- e: Multi-Function Peripheral
- e: Multi-Function Polis
- e: Multi-Function Product
- e: Multi-Purpose Facility
MFPA e: Mosaic Focal Plane Array
- e: Multi Function Peripheral Association
MFPE e: Misson From Planet Earth
MFPI e: Multi Function Peripheral Interface
MFR e: Magnetic Fusion Reactor
- e: Master of Forest Resources
- e: Memorandum for the Record
- e: Multi Frequency Receivers
- e: Multi Filter Radiometer
- f: Mouvement Familial Rural
Mfr f: Manufacture
mfr f: mille francs
MFRAM e: Moving Filter Radioactive Aerosol Monitor
- e: Moving Filter Radiological Aerosol Monitor
MFRP a: Det Mongolske Folkerevolutionære Parti
- e: Midwest Fuel Recovery Plant, Morris, Illinois
MFRSR e: Multifilter Rotating Shadowband Radiometer

MFS e: Macintosh File System
- e: Magnetic tape Field Search
- e: Magnetic-Field Sensor
- e: Management Function Set
- e: Maximum Frame Size
- e: Mediterranean Forecasting System
- e: Memory File System
- e: Message Format Service
- e: Message Formatting System
- e: Metro Fiber Service
- e: Metropolitan Fiber Systems
- e: Minor Field Study
- e: Modified Filing System
- f: Etude mineure sur le terrain
- f: Système méditerranéen de prévision
- N: Mobil Forsyningsstasjon
- s: Estudio Circunscrito sobre el Terreno
- s: Sistema de Previsiones para el Mediterráneo

Mfs f: Menus frais
MFSG e: Microbiological Food Surveillance Group
MFSK e: Multilevel Frequency Shift Keying
MFT e: Master File Table
- e: Master of Family Therapy
- e: Melter Feed Tank
- e: Minimum Flexible Targeting
- e: Multiprogramming with a Fixed number of Tasks
- f: Militants pour la Concrétisation de la Révolution

MFTA e: Multiduct Fuel Test Assembly
MFTL e: My Favorite Toy Language
MFTP e: Multicast File Transfer Protocol
MFV d: Mehrfrequenz-Wahlverfahren
- e: Main Fuel Valve
- e: Maintenance Floor Valve
- e: Minimum Fluidization Velocity

MFZ e: Microsoft Free Zone
MG a: Menig
- d: Magisterstudiengang
- d: Maschinengewehr
- d: Modulationsgenerator
- e: Madagascar
- e: Magnetometer surveys
- e: Management Guide
- e: Manufactured Goods
- e: Master Group
- e: Measurements Group
- e: Modulation Generator
- e: Monoglycerides
- e: Motor-Generator
- f: Main Gauche
- f: Marque de Garantie
- f: Meubles Gagés
- f: Minimum Garanti
- f: Mutilé de Guerre
- N: Maskingevær
- r: Marin Gräns
- s: Material de Guerra

mg e: milligram
Mg a: Meget godt [karakter]
Mg2+ e: Magnesium-Kation
MGA d: Mehrgeräteanschluß
- e: Matrox Graphics Adapter
- e: Meteorological and Geoastrophysical Abstracts
- e: Monochrome Graphic[s] Adapter
- e: Multimode Graphic Adapter

MGAP f: Mouvement Gabonais d'Action Populaire
MGB e: Maintenance Glovebox
MGBNVCP e: Murray Geological Basin Native Vegetation Clearance Policy
MGC d: Metallgesellschaft Chemiehandel GmbH
- e: Manufactured Goods Collection

MGCG s: Movimiento Gnóstico Cristiano Guatemalteco
MGCS e: Manual Gas Control Station
- e: Meteosat Ground Computer System

MGD e: Millions of Gallons per Day
MGD77 e: Marine Geophysical Data Exchange Format
- f: Format pour l'échange des données géophysiques marines
- s: Formato de Intercambio de Datos Geofísicos Marinos

MGDS e: Mined Geologic Disposal System
MGDSR e: Mined Geologic Disposal System Requirements
MGE e: Microstation foundation and modular GIS Environment
- e: Modular Gis Environment

M GEN e: Major General
MGET e: Multiple Get
MGF f: Mutuelles Générales de France
MgF_2 d: Magnesium Fluoride
MgF2 e: Magnesium Fluoride
mg-gjeng N: maskingeværgjeng
MGI e: Marine Geological Institute, Indonesia
- e: Marine Hydrophysical Institute, Ukraine Academy of Sciences Sevastopol
- f: Institut de Géologie Marine
- s: Instituto de Geología Marina

Mgía s: Meteorología
MGIS e: Military Geographic Information System
MGL e: Management Guidance Letters
- f: Minière des Grands Lacs Africains

mgl. a: manglende
- a: mangler

mg/l e: milligrams per liter
MG-lag N: Maskingeværlag
MGM d: Media Gruppe München
- e: Medical Group Management Journal
- e: Metro-Goldwyn Mayer
- e: Mobile, guided Ground-to-ground Missile

M-gM e: Multi-gas Monitor
MGMA e: Medical Group Management Association
MGN d: Meiningen, Thüringen
- r: Minsta Gemensamma Nämnare

MGO e: Main Geophysical Observatory
MGPA p: Marinha de Guerra Popular de Angola
MGPF e: Multiprogram General Purpose Facilities
MGPGP e: Master of Group Process and Group Psychotherapy
MGR e: Mobile, Ground-targeted Rocket
- e: Multi Gigabit Router
- f: Mouvement des Radicaux de Gauche

Mgr f: Monseigneur
- r: Marschgruppering

MG/R e: Message Generator/Recorder
MGRA e: Master Geographical Reference Area
mg-rede N: maskingeværrede
MGRIDS e: Machine-Generated Record Inventory and Disposal Schedule
MGR N e: Manager Notice
Mgrp r: Marschgruppering
MGRR s: Movimiento Guatemalteco de Reconstrucción Rural
MGS e: Master Ground Station
- e: McMurdo Ground Station
- e: Metallurgical-Grade Silicon
- e: Mission Ground Station
- e: Mobile Ground System

MGS2 e: Message Generation System 2
MGSE e: Mechanical GSE
Mgsf e: Million gross square feet
mg-skytter N: maskingeværskytter
MG&SR e: Mission Generic and Specific Requirements
mg-stilling N: maskingeværstilling
MGT e: Master Group Translator
- e: Mobile Ground Terminal
- e: Mustang GT
- s: Megatonelada

MGU e: Mid-course Guidance Unit
- e: Moscow State University

MGUS e: Mathematical Geologists of the United States
- e: Monoclonal Gammopathy of Unknown Significance

MH a: Medlem af Højesteret
- d: Musikhochschule
- e: Fe3O4-Fe2O3 buffer
- e: Malignant Hyperthermia
- e: Man Hour[s]
- e: Marshall Islands
- e: Modified Huffman
- f: Malade Hospitalisé
- f: Médecin des Hôpitaux
- f: Médecine Homéopathique
- f: Motif d'Hospitalisation

M H e: Mason Hall
m/h r: med hustru
MHA e: Canadian Malignant Hyperthermia Association
- e: Master of Health Administration
- e: Maximum Hypothetical Accident
- e: Mental Health Association
- e: Microangiopathic Hemolytic Anemia
- e: Mobile High power plant

Mha e: Million hectare
m.h.a. r: med hjälp av
MHAC e: Midwest Hispanic AIDS Coalition
MHAUS e: Malignant Hyperthermia Association of the United States
MHC e: Coastal Minehunter
- e: Managed Health Care
- e: Mauritian Housing Corporation
- e: Mental Health Commission

MH/CD e: Mental Health/Chemical Dependency
MHD e: Magnethydrodynamics
- e: Magneto-HydroDynamic[s] [simulation]

MHDDV e: Medium Heavy-Duty Diesel Vehicle
MHDEMP e: Magentohydrodynamic EMP
MHDL e: MIMIC Hardware Description Language

MHE e: Message Handling Element
MHEC e: Maryland Higher Education Commission
MHEG e: Multimedia [and] Hypermedia [information coding] Expert Group
MHF e: Magnetic Hyperfine Field
- e: Medium High Frequency
- e: Message Handling Facility
MHFARS e: Minuteman High Frequency Antenna Replacement System
MHFEP e: Multi-Host Front End Processor
MHFES e: Materials Handbook for Fusion Energy Systems
MHG d: Gesetz zur Regelung der Miethöhe
- e: Meteorological Hydrogen Generator
MHG. e: Middle High German
MHHWL e: Mean Higher High Water Line
MHI e: Manufactured Housing Institute
- e: Material Hazard Index
mhi e: misc health infertility newsgroup
MHKP r: Medeltung Helikopter
MHL d: Mühlhausen, Thüringen
MHMS e: Modular Hydrologic Modeling System
MHN e: Material hold notice
- e: Mental Health Net
MHO e: Millstone Hill Observatory
- e: My Humble Opinion [Internet]
MHOTY e: My Hats Off To You [Internet]
mhp. a: med henblik på
MHPCC e: Maui High Performance Computing Center
MHPG e: 3-Methoxy-4-Hydroxyphenyl-glycol Psychoactive hormone
MHR e: Microwave Hologram Radar
MHRA e: Medical and Health Research Association of New York City
MHRD e: Master of Human Research Development
M H R I e: Mental Health Research Institute
MHRI e: Miami Heart Research Institute
MHRS e: Modified Hazard Ranking System
MHS e: Mail Handling Service
- e: Mail Handling System
- e: Message Handling System[s]
- e: Microwave Humidity Sensor
- e: Microwave Humidity Sounder
- e: Military Health System
- f: Matra Harris Semiconductor
- r: Militärhögskolan
MH/SA e: Mental Health/Substance Abuse
MHS AAK r: Militärhögskolans Allmänna Kurs
MHS/AMSU-B e: Microwave Humidity Sounder/Advanced Microwave Sounding Unit
MHSMHS e: Message Handling Service
- e: Message Handling System
MHS ROK r: Militärhögskolans reservofficerskurs
MHSS e: Military Health Services System
m. h. t. r: med hänsyn till
m.h.t. N: med hensyn til
mht. a: med hensyn til
MHTGR e: Modular High-Temperature Gas-cooled Reactor
MHTS e: Main Heat Transport System
MHV e: Miniature High Voltage
MHWL e: Mean High Water Line
MHWN e: Mean High Water Neap
MHWS e: Mean High Water Spring
MHZ e: Mega Hertz
MHz. d: Megahertz
MI d: Medizinische Informatik
- e: Machine Intelligence
- e: Machine Interface
- e: Mafic Index
- e: Management Indicator
- e: Management Interface
- e: 3-Methylindole
- e: Minority Institution
- e: Mode Indicate (or Indicator)
- e: Module interface
- e: Monsoon Index
- e: Mutual Interference
- f: Maison Individuelle
- f: Majoration d'Impôts
- f: Maladie Infantile
- f: Marque [de fabrique] Internationale
- f: Matière Imposable
- f: Ministère de l'Information
- f: Ministère de l'Intérieur
- f: Mortalité Infantile
- r: Maskiningenjör
- s: Manual de Infantería
- s: Manual de Instrucción
- s: Mapa Itinerario
mi e: mile
Mi d: Mittwoch
- f: Minéralogie
- f: Minéralogique
- f: Minéralogiste
- f: Mineur
mi² e: square mile[s]
MIA e: Magazine Indexer Assembly
- e: Management Information Analysis
- e: Maryland Insurance Administration
- e: Miami, FL
- e: Multivariate Image Analysis
- e: Murrumbidgee Irrigation Area
- e: Music Industries Association
- s: Movimiento Islámico Armado
mia a: milliard[er]
MIADMB e: Murrumbidgee Irrigation Area and Districts Management Board
MIAL e: Maine Image Analysis Laboratory
MIAP f: Mouvement International d'Action pour la Paix
MIAR e: International Monitoring System
MIAS d: Meßinstitut für Immissions-, Arbeits- und Strahlenschutz, Mainz
- e: Marine Information [and] Advisory Service
- e: Muhyiddin Ibn 'Arabi Society
MIB e: Management Information Base
- e: Management Information Briefing
- e: Men In Black
- e: Military Industrial Base
- e: Military Intelligence Board
- e: Military Intelligence Branch
- e: Missile Interceptor Base
MIBA e: Master of International Business Administration
MIBG e: [131-I]-Metaiodobenzyl-guanidine
MI [BN][CO] e: Military Intelligence [Battalion] [Company]
MIBS e: Master of International Business Studies
M3IC e: Monolithic Microwave and Millimeter Wave Integrated Circuit
MIC e: Macro Interpretative Command
- e: Management Interface Connector
- e: Marine Information Centre
- e: Master Item Code
- e: Media Information Control
- e: Media Interface Connector
- e: Medium Interface Connector
- e: Memory In Cassette
- e: Message Integrity Check
- e: Meteorologist-In-Charge
- e: Methyl Isocyanate
- e: Microbiology
- e: Minimal Invasive Chirugie
- e: Ministy of Information and Communication
- e: Mobile Intelligence Center
- e: Mode Indicate Common
- e: Motorcycle Industry Council
- e: Multimedia Interactive Control
- e: Music Information Centre
- f: Centre d'Information des sciences de la Mer
- f: Ministère de l'Industrie et du Commerce
- f: Modulation par Impulsions Codées
- s: Centro de Información de las Ciencias del Mar
MICA e: Modem ISDN Channel Aggregation
- e: Modular Interactive Classification Analyzer
MICAD e: Micro-Installation, Inc.
- e: Multipurpose Integrated Chemical Agent Detector
MICADO e: Michaelson Interferometer for Coordinated Auroral Doppler Observations
- f: Mission pour la Conception Assistée et le Dessin par Ordinateur
MICAF e: Measuring Improved Capabilities of Army Forces
MICAP e: Mission Capability [analysis system]
- e: Mission Impaired Capability Awaiting Parts
MICA-SYSTEM e: Modular Interactive Classification Analyzer
MICB e: Mission Integration Control Board
MICE e: Mans Impact on Coastal and Estuarine Ecosystems
- e: Management Information Capability for Enforcement
- e: Map Image Congruency Evaluation
- e: Modular Integrated Communications Environment
- e: Multimedia Integrated Conferencing for Europe
- f: Impact des activités humaines sur les écosystèmes côtiers et estuariens
- s: Repercusiones de la Acción Humana en los Ecosistemas Costeros y los Estuarios
MICH LEAGUE e: Michigan League

Mich LR e: Michigan Law Review
Mich Med e: Michigan Medicine
MICH RM e: Michigan Room, School of Bus
MICIS e: Midwestern Climate Information System
MICLP e: Medical Informatics Cultural Literacy Project
MICNS e: Modular Integrated Communications and Navigation System
MICo e: Military Intelligence Company
MICOM e: Missle Command
MICR e: Microbiology
MICRA e: Motion, Imaging Contouring and Roughness Amplitude
MICRAD e: Microwave Radiometric System
MICRO e: One-millionth
MICRO-ASEAN e: Micro-Computer Based Advanced System for Environmental Analysis With Remote Sensing Data
Microb f: Microbiologie
Microbiol Immunol e: Microbiology and Immunology
Microbiol Mol Biol Rev e: Microbiology and Molecular Biology Reviews
Microbiol Res e: Microbiological Research
Microbiol Rev e: Microbiological Reviews
Microb Pathog e: Microbial Pathogenesis
Microcirc Endothelium Lymphatics e: Microcirculation, Endothelium and Lymphatics
MICROFX e: Intelligence Data Processing Set
Micrgr f: Micrographie
Microl f: Micrologie
MICROLYNX e: Serials control system
MICROMORT e: A One in a Million Chance of Death from [an] Environmental Hazard[s]
MICRON e: One millionth of a meter
MICRONET e: Network of Centres of Excellence in Microelectronics
MICROSAT e: Microwave Satellite
Microsc Res Tech e: Microscopy Research and Technique
Microvasc Res e: Microvascular Research
MICS e: Macro Interpretive Commands
- e: Maintenance Instrumentation Calibration System
- e: Materiel Intransit Control System
Mics. e: Miscellaneous
- e: Miscellany
MICSIS e: Micro Spatial Information Processing System
MICSLP e: Maintenance Inventory Control Stock Levels Projection
MICWU e: Motor Industry Combined Workers Union
MID e: Material ID
- e: Message Identification
- e: Message Identifier [Digit]
- e: Military Intelligence Detachment
- e: Multiplexing Identifier
MIDAP e: Multichannel Image Display and Analysis Program
MIDAS e: Maintenance Information Data Automation System
- e: Management In a Distributed Application and Service
- e: Map Information and Data Analysis System
- e: Marketmaker Information And Dealing System
- e: Master Information and Data Acquisition System
- e: Microcomputer-Based
- e: Microprocessor-Based
- e: Model for Intertheater Deployment And Scheduling
- e: Multi-discipline Interactive Display and Analysis System
- e: Multi-task Ice Data Analysis System
- e: Multivariate Interactive Digital Analysis System
- e: Munich Data Analysis System
Midas e: Mythological Idiotic Digital-Aurum Sovereign
MIDB e: Major Item Data Base
Middle East J Anesthesiol e: Middle East Journal of Anesthesiology
MIDDS e: Meteorological Interactive Data Display System
MIDEH s: Indigenous Humanist Democratic Movement
MIDEMA f: Minoterie de Matadi
MIDI e: A standard for optically isolated serial linking
- e: Musical Instrument[s] Digital Interface
MID-INFRARED e: Infrared radiation extending approximately from 1.3 to 3.0 micrometer and being part of the reflective infrared
MIDIST f: Mission Interministérielle de l'Information Scientifique et Technique
MIDL e: Microsoft Interface Definition Language
midl. e: midland [dialect]
midl. ud N: midlertidig udyktig
MIDP e: Major Item Distribution Plan
MIDRC e: Microbiology and Infectious Diseases Research Committee
MIDS e: Matrix Information and Directory Services
- e: MPAC Information Data System
- e: Multifunction[al] Information Distribution System[s]
MIDSD e: Management Information and Data Systems Division
MIDSS e: Methods of Investigation of the Dead Sea Scrolls and the Khirbet Qumran Site
MIE e: Magnetron Ion Etching
- e: Major Item of Equipment
- e: Military Information Environment
- e: Most Intense Earthquake
Mie f: Mairie
M&IE e: Meals and Incidental Expenses
MIEC e: Meteorological Information Extraction Centre
MIEMSS e: Maryland Institute Emergency Medical Services
MIES e: Manufacturing Informations and Execution System[s]
MIF e: Maker Interchange Format
- e: Management Information File[s]
- e: Management Information Format
- e: Management Integration Forum
- e: Modernization and Implementation Facility
- e: Module Interconnection Facility
MIF. A r: Manskapsinstruktion för flottan. Allmän del
MIFASS e: Marine Integrated Fire And Support System
MIFED f: Mouvement Ivoirien des Femmes Démocratiques
MIFERGUI f: Mines de Fer de Guinée
MIFERSO e: Eastern Senegal Iron Ore Mining Company
MIFU a: Margarine-Industriens Fælles-Udvalg
MIG e: Metal In Gap
- f: Médecin Inspecteur Général
MiG R: Mikoyan-Gurevich
MIGA e: Multilateral Investment Guarantee Agency
MiGCAP e: MiG Combat Air Patrol
MIGRI r: Miljövårdsgruppernas Riksförbund
MIGS e: Metal-Induced Gap States
MIH f: Médecin Interne des Hôpitaux
MIHMA e: Model-based Integration Heterogeneous Multimedia Archives
MII e: Media Independent Interface
- e: Microsoft, IBM, Intel
- e: Movement Indépendantiste Ivoirien
MIIA e: Medical Intelligence Information Agency
MIIDS e: Military Intelligence Integrated Data System
MIIST f: Mission Interministérielle de l'Information Scientifique et Technique
MIIT e: Materials Interface Interaction Test
MIJI e: Meaconing, Interference, Jamming and Intrusion
MIJID f: Marché International des Jeux Vidéo de l'Informatique Individuelle et Domestique
MIKIN e: Michelson Interferometer
Mikrobiol d: Institut für Mikrobiologie
Mikrobiol Z R: Mikrobiolohichnyi Zhurnal
MIL d: Miltenberg, Bayern
- e: Machine Interface Layer
- e: Marine Instrumentation Laboratory
- e: Merritt Island Tracking Station
- e: Mila Station
- e: Mother In-Law [Internet]
Mil r: Militär
- s: Miliciano
Mil. e: Military usage
mil. N: militær
MILAP e: Maintenance Information Logically Analyzed and Presented
Milassist r: Militärassistent
Milbank Q e: Milbank Quarterly
Milbef r: Militärbefälhavare
milbef f 1 milo r: militärbefälhavare för 1 militärområdet
milbefstab r: militärbefälhavarens stab
MILCAP e: Military Standard Contract Administration Procedures

MILCOMSAT e: Military Communications Satellites
MILCON e: Military Construction Program
MILDEP e: Military Department
- e: Military Deputy
MILDEX e: Mixed Layer Dynamics Experiment
MILES e: Magnetic Intrusion Line Sensor
MILFAC e: Military Facilities File
MILFAX r: Militär faximilöverföring
Milik, Books e: J. T. Milik, The Books of Enoch: Aramaic Fragments of Qumrân Cave 4 [Oxford: Clarendon Press, 1976].
Milik, MS f: J. T. Milik, 'Milki-sedeq et Milki-resha dans les anciens écrits juifs et chrétiens,'Journal of Jewish Studies 23 [1972]
Milik, Years e: J. T. Milik, Ten Years of Discovery in the Wilderness of Judaea [London: SCM, 1959].
mill. a: million
- N: million[er]
Milled r: Militärled
mill[s] e: 0.001 U.S. dollars
Mil Med e: Military Medicine
MILNET e: Military Network
Milo r: Militärområde
milo B r: Bergslagens militärområde
milobef r: militärområdebefälhavare
MILOGS e: Marine Integrated Logistic System
milolvoff r: militärområdesluftvärnsofficer
milo NN r: militärområde Nordnorrland
milo Ö r: militärområde Öst
milo ÖN r: militärområde Östnorrland
Milorg N: Hjemmefrontens Hemmelige Militære Organisasjon
Milorg. N: Militærorganisasjon
milo S r: militärområde Syd
milosambandsoff r: militärområde-sambandsofficer
milotygoff r: militärområdestygofficer
milo V r: militärområde Väst
MILOX e: Mid-Latitude Ecosystems and Photochemical Oxidants
- e: Mid-Latitude Ecosystems as Sources and Sinks for Atmospheric Oxidants
MILPAY e: Military Pay
MILPERCEN e: U.S. Army Military Personnel Center
MILS e: Military Standard Logistic System
MILSATCOM e: Military Satellite Communications
MILSCAP e: Military Standard Contract Administration Procedure
MIL SPEC e: Military Specifications
MILSPETS e: Military Standard Petroleum System
MILSTAMP e: Military Standard Movement Procedures
MILSTAR e: Military Strategic and Tactical Relay Satellite
MIL STD e: Military Standards
MILSTEP e: Military Supply and Transportation Evaluation Procedures
MILSTRAP e: Military Standard Transaction Reports and Accounting Procedures
MILSTRIP e: Military Standard Requisitioning and Issuing Procedures
mil. strl. N: militær straffelov
MILSVC e: Military Services
MILTRACS e: Military Air Traffic Control System
MILVAN e: Military-owned demountable container
MIL VIG e: Military Vigilance
MIM d: Metall-Isolator-Metall
- d: Meteorologisches Institut der Universität München
- e: Maroc Informations Médicales
- e: Mechanical Interaction Model
- e: Media Interface Module
- e: Metal-Insulator-Metal
- e: Mobile SAM. Examples
- e: Morality In Media
MIMA e: Mighty Image Analyzer
MIMAC e: Measurement and Improvement of Manufacturing Capacity
MIMD e: Management Information of Metrology Data
- e: Multiple Instruction [stream] Multiple Data [stream] [machine]
MIME e: Multimedia Internet Mail Extensions
- e: Multipart Internet Mail Extensions
- e: Multipurpose Internet Mail Extension[s]
MIMI e: Millimetric Ice Mapping Instrument
- e: Missile Intensive Maintenance Item
MI/MIC e: Mode Indicate/Mode Indicate Common
MIMICS e: Michigan Microwave Canopy Scattering Model
MIMIM e: Metal-Insulator-Metal-Insulator-Metal
MIMIS e: Major Item Management Information System
MIMO e: Multi[ple]-Input/Multi[ple]-Output
MIMOS e: Malaysian Institute of Microelectronic Systems
MIMP e: Motorola Impedance Matching Program
MIMPT e: Multiresource Inventory Methods Pilot Test
MIMR e: Multiband Imaging Microwave Radiometer
- e: Multifrequency Imaging Microwave Radiometer
- e: Multispectral Imaging Microwave Radiometer
MIMS e: Magnetic Ion Mass Spectrometer
- e: Manifest Information Management System
- e: Meteosat Infrared and Microwave Sounder
- e: Mobile Information System
MIMSE e: Microwave-Infrared Mesopheric-Stratospheric Experiment
MIN e: Minimum [Identification Number]
- e: Mobile ID [Identification Number]
- e: Multistage Interconnection Network[s]
min a: minut
- e: minute[s]
- r: ministerium
Min d: Institut für Mineralogie und Kristallchemie
- f: Médecin
- f: Minéral
- f: Ministère
- f: Ministre
- f: Moulin
- s: Mina
- s: Minador
- s: Minuto
min. a: minister[ium]
- e: minimum
- e: minute
- N: minister
- N: minutt
Min. e: Mineralogical
- s: Ministerio
MINC e: Minimum prime service length
MINCE e: MINCE Is Not Complete Emacs
- p: Ministério do Comércio Externo
MINCI p: Ministério de Comércio Interno
MINDAT e: Minerals Database
- f: Base de Données Minérales
- s: Base de Datos sobre Minerales
MINDECO e: Mining Development Corporation
mindiv r: mindivision
MINDM e: Miniature Dyna-Metric Model
MINE e: Mines Building
- e: Mining Engineering
mineavd. N: mineavdeling
MINED e: Master of Industrial Education
- p: Ministério de Educação
Minér f: Minéral[ogie]
- f: Minéralogique
Mineral. e: Mineralogical
Miner Electrolyte Metab e: Mineral and Electrolyte Metabolism
Minerva Anestesiol s: Minerva Anestesiologica
Minerva Cardioangiol s: Minerva Cardioangiologica
Minerva Chir s: Minerva Chirurgica
Minerva Endocrinol s: Minerva Endocrinologica
Minerva Gastroenterol Dietol s: Minerva Gastroenterologica e Dietologica
Minerva Ginecol s: Minerva Ginecologica
Minerva Med s: Minerva Medica
Minerva Pediatr s: Minerva Pediatrica
Minerva Psichiatr s: Minerva Psichiatrica
Minerva Stomatol s: Minerva Stomatologica
Minerva Urol Nefrol s: Minerva Urologica e Nefrologica
MINESEB f: Ministère de l'Enseignement Secondaire et de l'Education de Base
MINET e: Movement Information Network
MinFlj r: Minutläggarflottilj
Minförr r: Minförråd
Minftg r: Minfartyg
M.Ing.F. a: Medlem af Dansk Ingeniørforening
MIng K r: Mariningenjörkår
MINIBIB e: Danish government libraries catalogue
Minim Invasive Neurosurg e: Minimally Invasive Neurosurgery

MINIS e: Mini Intraport Network Information System
- e: Multipurpose Interactive NASA Information System
MINISINS e: Miniature Ships Inertial Navigation System
MINI-TAC e: Mini-Terminal Access Controller
MINK e: Missouri-Iowa-Nebraska-Kansas
MinklargL r: Minklargöringsledare
MIN, MAX d: Minimum, bzw. Maximum oder Minimierungs-, bzw. Maximierungsforderung
Minn Med e: Minnesota Medicine
MinNUL S-H d: Minister für Natur, Umwelt und Landesentwicklung, Schleswig-Holstein
MinO r: Minofficer
MinÖlStG d: Mineralölsteuergesetz
Minpbv r: Minpansarbandvagn
MinPlen f: Ministre Plénipotentiaire
Minröjavdelning r: Minröjningsavdelning
Minröjflottilj r: Minröjningsflottilj
Minröjftg r: Minröjningsfartyg
Minröjpbv r: Minröjningspansarbandvagn
MINS e: Mare Island Naval Shipyards
- e: Minimum count or time per schedule period
MINSA e: Ministry of Health
MINSE p: Ministério da Segurança do Estado
MINSEP e: Minimum Separation time
Minspärrto r: Minspärrtrupp
Minstn r: Minstation
M.Inst.P.I. e: Member of the Institute of Patentees [Inc.]
MINSTREL e: Management Information Software Tool-Research in Libraries
Minsvep r: Minsvepare
Minsvepdiv r: Minsvepardivision
MINT e: Malaysian Institute for Nuclear Technology Research
- e: Management Information Network for Training
- e: Mint is Not TOS
MINTEK e: Council for Mineral Technology
- e: Mineral Technology
MINTEQ e: Geochemical Model
MIntK r: Marinintendenturkår
MINTSS e: MAC Intelligence Support System
MINUAR f: Mission des Nations Unies pour l'Assistance au Rwanda
MINUET e: Minnesota Internet Users Essential Tool[s]
MINUGUA s: Misión de Naciones Unidas para la Verificación de los Derechos Humanos en Guatemala
MINURSO e: United Nations Mission for the Referendum in Western Sahara
MINX e: Multimedia Information Network Exchange
MIO e: Management Integration Office
- e: Materials Integration Office
- e: Memory Input/Output
- e: Michigan International Office
Mio. e: Million[en]
- e: Miocene
MIOG d: Mitteilungen des Instituts für Österreichische Geschichte
MIOS e: Metal-Insulator-Oxide-Semiconductor
MIP e: Mainframe Internetting Project
- e: Maintenance Implementation Plan
- e: Management [and] Integration Plan
- e: Maximum Integration Phone
- e: Meeting In Person [Internet]
- e: Message Interface Processor
- e: Missile Impact Prediction
- e: Mission Integration Plan
- e: Model Installation Program
- e: Mooring Implementation Panel
- e: Multi-Isotope Production
- e: Multiple Insertions Performances
- e: Museum Informatics Project
- e: Museums Informatics Project
- f: Ministère de l'Instruction Publique
MIPALS e: Management Integration of Program and Launch Site
MIP/A/N/AF e: Missile Procurement, Army/Navy/Air Force
MIPAS e: Michelson Interferometer [for] Passive Atmospheric Sounder (or Sounding)
- e: Michelson Interferometric Passive Atmosphere Sounder
- f: Interféromètre Michelson pour sondage passif de l'atmosphère
- s: Interferómetro Michelson para Sondeo Pasivo de la Atmósfera
MIPB e: Material Improvement Project Board
MIPL e: Multimission Image Processing Laboratory
MIPR e: Military Interagency Procurement Requisition
- e: Military Interagency Purchase Request
- e: Military Interdepartmental Procurement Request
MIPRA f: Mouvement Indépendant PRA
MIPS e: Macaulay Image Processing System
- e: Map and Image Processing System
- e: Marine Integrated Personnel system
- e: Microprocessor without Interlocked Piped Stages
- e: Million[s] [of] Instruction[s] Per Second
- e: Mini Image Processing System
- e: Modified Integrated Program Summary
- e: MSS Image Processing System
- e: Multi-Image Processing System
- e: Multipurpose Image Processing System
- f: Million d'Instructions Par Seconde
MIP-TV f: Marché International des Programmes de Télévision
MIR e: Management Implementation Review
- e: Maximum Information Rate
- e: Micro-Instruction Register
- e: Millimeter-wave Imaging Radiometer
- e: orbital system
- e: Russian manned space station
- r: Marininfanteriregemente
MIRAS e: Mir Infrared Spectrometer
MIRCEN e: Microbial Resources Centre
- e: Microbiological Resources Centre
- f: Centre de ressources microbiennes
- s: Centro de Recursos Microbiológicos
MIRI e: Midinfrared Imager
- e: Multichannel Infra Red Interferometer
MIRIAM d: Mikroprozessorgesteuertes Registriersystem des Instrumenten
MIRIAM-U d: Mikroprozessorgesteuertes Registriersystem des Instrumenten universal [alle Komponenten]
MIRIAM-W d: Mikroprozessorgesteuertes Registriersystem des Instrumenten nur für Windmessung
MIRIS e: Michigan Inventory and Resource Information System
MIRN p: Movimento Independente por a Reconstrução Nacional
MIROC e: Mining Industry Research Organization of Canada
MIROS e: Microwave Remote Sensor for the Ocean Surface
MIRP e: Manipulated Information Rate Processor
MIRS e: Management Information Retrival System
- e: Mid-America Interventional Radiological Society
- e: Motorola Integrated Radio System
- e: Musical Information Retieval System
MirSIP e: [Belgium] Mirage Safety Improvement Programme
MIRSL e: Microwave Remote Sensing Laboratory
MIRV e: Multiple Independently Targetable (or Targeted) Reentry Vehicle[s]
MIS d: Meteorologisches Informationssystem
- e: Maintenance Information System
- e: Management Information System[s]
e: Master of Information Systems
- e: Master of Interdisciplinary Studies
- e: Membrane Inlet System
- e: Metal Insulator Silicon
- e: Mineral Industry Surveys
- e: Missile Squadron
- e: Multimedia Information Sources
- e: Multispectral Imaging Spectrometer
Mis f: Maquis
MISA e: Media Institute of Southern Africa
MISAWI d: Meteorologisches Informationssystem
MISC e: Miscellaneous
misc. e: miscarriage
MISCAP e: Mission Capability Statement
MISCF e: Miscellaneous Frame
MISD e: Multiple Instruction Single Data
MISE e: Mississippi Embayment
Mise f: Marchandise
MISICC e: Management Information System Input to Command and Control
MIS/INAS e: MIS for Industrial Naval Air Stations
MISIS e: Micro Integrated Storm Information System
MISM e: MAJCOM Information System Manager
MISMO e: MAJCOM Information Systems Management Office
MISNAV s: Misión Naval
MISP e: MAJCOM Information Systems Plan

MI2SPM e: Micom Infrared Imaging Systems Performance Model
MISR e: Multiangle Imaging Spectro[radio]meter
- e: Multidirectional Imaging Spectroradiometer
MISREP e: Joint Tactical Air Reconnaissance/Surveillance Mission Report
MISS e: Manpower Information Systems Support
- e: Metal-Insulator-Semiconductor Switch
MISSEC f: Mission Économique
MISSI e: Multi-level Information System Security Initiative
MISS-IDA e: Meteorological Interactive Software System for Image Data Analysis
MIST e: Microburst and Severe Thunderstorm Experiment
- e: Multispectral Image Storage Tape Format
MISTC e: Multiple Information Set Tracking Correlator System
MISTEL e: Delft monograph document delivery and catalogue
MISTER e: Model-Based Intelligent Sar Target Exploitation and Recognition
MISTR e: Management of Items Subject to Repairs
MISU e: Meteorological Institute Stockholm University
MIS-User e: Management Information Service User
MIT e: Management Information Tree
- e: Massachusetts Institute of/for Technology
- e: Material[s] Interaction Test
- e: Metal-Insulator Transition
- e: Minimal Inversive Therapie
- e: MIPS Technologies Inc.
- e: Most Intense Tornado
- e: Multi-function Instrument Tree
- f: Médecin Inspecteur du Travail
MITAS e: Martin Marietta Interactive Thermal Analysis System
MitbestErgG d: Gesetz zur Ergänzung des Gesetzes über die Mitbestimmung der Arbeitnehmer in den Aufsichtsräten und Vorständen der Unternehmen des Bergbaus und der Eisenerzindustrie
MitbestG d: Gesetz über die Mitbestimmung der Arbeitnehmer [Mitbestimmungsgesetz]
MITEC e: Machine Intelligent Technical Controller
MITES e: Model-Driven Iterative Texture Segmentation
MITI e: Ministry of International Trade and Industry
- e: Multilingual Intelligent Interface
MIT/LL ETS e: Massachusetts Institute of Technology Lincoln Laboratory Experimental Test System
MITO e: Microphyte Toxins
mitr N: mitraljøse
MITRE e: Massachusetts Institute of Technology Research Establishment
- e: Miniature Individual Transmitter Receiver Equipment
mitrgjeng N: mitraljøsegjeng
mitr. kp. N: mitraljøsekompani
mitr tropp N: mitraljøsetropp
MITS e: MAC Intra-theater Transmission System
- e: Management Information Tracking System
- e: Microcomputer Interactive Test System
- e: Model Instrumentation Telemetry System
MITT e: Mobile Imagery Tactical Terminal
MittHV d: Mitteilungen des Hochschulverbandes
MIU e: Multiplexing Interface Unit
MiU e: University of Michigan, Physics-Astronomy Library
Mius r: Minutkiksstation
MIUU e: Meteorological Institute of the University of Uppsala, Sweden
MIVIS e: Multispectral Infrared and Visible Imaging Spectrometer
MIW e: Missile Wing
MIWG e: MAC Interoperability Working Group
MIX e: Member Information Exchange
MIZ d: Materialien und Informationen zur Zeit
- e: Marginal Ice Zone
MIZEX e: Marginal Ice Zone Experiment
- f: Expérience sur la zone marginale de glaces
- s: Experimento sobre la Zona de Hielos Marginales
MJ e: Modular Jack
- f: Ministère de la Justice
- f: Mouvement de Jeunesse
Mj r: Major
MJA e: Medical Journal of Australia
MJAR e: Matsushiro Array in Japan
MJC e: Muslim Judicial Council
MJCOM e: Major Command
MJCS e: Joint Chiefs of Staff Memorandum
MJE e: Manual Journal Entry
MJED e: Master of Jewish Education
MJEMS e: Modular Jet Engine Management Simulator
MJNPE e: Mobile Joint Nuclear Planning Element
MJO e: Madden-Julian Oscillation
MJPD f: Mouvement des Jeunes Patriotes pour le Développement
M-JPEG e: Motion JPEG
MJT e: Maintenance Job Tracking System
MJUPS f: Mouvement des Jeunes de l'Union Progressiste Sénégalaise
MK a: Manufaktur-Kompagniet A/S
- A: Umkhonto we Sizwe
- e: Member of Knesset
- e: Missionary Kid
- N: Marinekommando
- r: Marinkommando
mk e: Macedonia [Internet]
m. k. N: maskinkanon
M&K a: Mund- og Klovesyge
m/k a: mand[lig]/kvinde[lig]
M/K N: Motorkutter
MK18 e: Mark 18 fuel assembly
MK22 e: Mark 22 fuel assembly
MKA r: Medelkvadratavvikelse
Mkan N: Maskinkanon
MKB r: Marinens Kustbevakning
Mkb r: Motorkanonbåt
MKDIR e: Make Directory
MKG d: Klinik und Poliklinik für Mund-Kiefer-Gesichtschirurgie
- r: Gotlands Militärkommando
MKN N: Marinekommando Nord-Norge
- r: Marinkommando Nord
MK/NGO N: Militærkontoret ved Norges Geografiske Oppmåling
MK/NSB N: Militærkontoret ved Norges Statsbaner
MKØ N: Marinekommando Østlandet
MKÖ r: Marinens Krigsövningar
- r: Marinkommando Öst
MKS d: Maul- und Klauenseuche
- d: Ministerium für Kultur und Sport
- e: Meter, Kilogram, Second
- e: Mortice Kern SystemsGmbH
- N: Marinekommando Sørlandet
- r: Marinkommando Syd
- r: Marinkompaniskola
m. kp. r: med kapp
mkp e: misc kids pregnancy newsgroup
mksa e: meter-kilogram-second-ampere
- N: meter-kilogram-sekund-ampere
MKSAP e: Medical Knowledge Self-Assessment Program
MKSF r: Marinkommando Syds Förvaltningar
MK-sjef N: Marinekommando-sjef
MKT e: Marketing
- N: Marinekommando Trøndelag
Mkt r: Mycket
MK/T N: Militærkontoret ved Telegrafstyret
MKTP e: Market Planning
MKV e: Miniature Kill Vehicle
- N: Marinekommando Vestlandet
- r: Marinkommando Väst
MK/V N: Militærkontoret ved Vegdirektorat
MKW d: Mineralöl Kohlen Wasserstoffe
ML e: Mali
- e: Maryland Lottery Agency
- e: Master list of data submittals
- e: Maximum Leader
- e: Medical Logistics
- e: Meta Language
- e: Meteorology Laboratory
- e: Micro Linear
- e: Microlog
- e: Military Leave
- e: Milliliter
- e: Mixed Layer
- e: Modern Languages
- e: Monolayer
- e: Music Library
- f: Mandat Légal
- f: Mariage Légitime
- f: Médecin Légiste
- f: Médecine Légale
- f: Médico-Légal
- f: Mesures Légales
- f: Motif Légitime
- f: Musée du Louvre
- N: Minelegger
- s: Mortero Ligero
- s: Motolancha

Ml f: Municipal
ml. a: mellem
m.l. N: marxist-leninist[isk]
m/l s: mi letra
ML1 e: Migration Level 1
ML-1 e: Mobile Low-Power Plant, No. 1
ML2 e: Migration Level 2
MLA e: Medical Library Association
- e: Member of [Canadian] Legislative Assembly
- e: Military Liaison Assistant
- e: Modern Language Association
- e: Multiple Letter Acronym
- e: Multispectral Linear Array
- e: Music Library Association
- e: My Local Address
ML-1A e: Mobile Low-Power Plant, No. 1A
Mläk r: Marinläkare
MLäk K r: Marinläkarkår
MLAN e: Mobile LAN
MLAP e: Migrant Legal Action Program
MLAPI e: Multilingual Application Programming Interface
MLAR e: Multi-Layer Anti-Reflection
MLAS e: Master of Laboratory Animal Sciences
MLAT e: Mutual Legal Assistance Treaty
M L B e: Modern Languages Building
MLB e: Message Link Buffer
- e: Multilayer Board
MLC e: Mauritius Labor Congress
- e: Military Liaison Committee
- e: Miniline Card
- e: Mixed Lymphocyte Culture
- e: Multilayer Capacitor
- e: Multilayer Ceramic
- e: Multilevel Cell
- e: Multiline Communications Controller
- e: Multilink Control Field
- f: Mandat-Lettre de Crédit
MLCAD e: Maintenance and Logistics Factors in Computer Aided Design
MLCC e: Multilayer Ceramic Capacitors
MLCD e: Multi Line Call Detail
MLCMC e: Multilayer Ceramic Multichip Module
MLCR e: Module Load Completion Rate
MLD e: MAJCOM Level Data
- e: Mixed Layer Depth
- e: MOTA Lift Device
- e: Movement for the Liberation of Djibouti
- e: Multiloop Device
- f: Maladie de Longue Durée
- f: Mouvement de Libération de Djibouti
- f: Profondeur de la couche de mélange océanique
- n: Marine Luchtvaart Dienst
- s: Profundidad de la capa de mezcla oceánica
MLDD e: Matched-Line Directional Divider
MLDR e: Maximum likelihood decision rule
MLDT e: Mean Logistics Delay Time
MLE a: Marxistisk-Leninistisk Enhedsfront
- e: Maximum Likelihood Estimation
- e: Mid-Latitude Ecosystem
- e: Millenium Language Extension
- e: Multi-Line Editor
- e: Multi-Line Entry field
Mle f: Matricule
- f: Modèle
MLEC f: Mouvement pour la Libération de l'Enclave de Cabinda
MLEM e: Multi Language Environment
MLF e: Mauritius Labor Federation
- f: Mesures Légales Françaises
- f: Mouvement de Libération des Femmes
- N: Mobelhandlernes Landsforbund
MLG. e: Middle Low German
MLI e: Machine Level Interface
- e: Multi-Layer Insulation
- e: Multiple Link Interface
MLID e: Multi[ple] Link Interface Driver
m-linje r: motståndslinje
MLIS e: Multilingual Information Society
MLK r: Militärläkarkurs
MLL e: Mars Logistics Lander
- e: Minimum Level of Living
- e: Modify Lot Location
Mlle f: Mademoiselle
- f: Marseille
MLLW e: Mean Lower Low Water
- e: Mixed Low-Level Waste
MLM e: Mailing List Manager
- e: Mixed Layer Model
- e: Multi-Level Metal
- e: Multilongitudinal Mode
Mlm r: Mer eller mindre
MLMA e: Multi Level Multi Access
MLMS e: Multipurpose Lightweight Missile System
MLN a: Maskinfabriken Lindholm
- e: Message Ledger Number
- f: Mouvement de [la] Libération Nationale
- s: Movimiento de Liberación Nacional
MLNC f: Mouvement de Libération National des Comores
MLO e: Material, Labor, Overhead
- e: Mauna Loa Observatory
MLOPEX e: MLO [Mauna Loa Observatory] Photochemical (or Photochemistry) Experiment
MLOPEX II e: Mauna Loa Observatory Photochemical Experiment II
MLP d: Marschollek, Lautenschläger und Partner
- e: Message Link Protocol
- e: Mobile Launch Platform
- e: Multi-layer Link Protocol
- e: Multi-Layer Perceptron
- e: Multi-Layer Protocol
- e: Multi-Level Precedence
- e: Multi-Link Protocol
- f: Mouvement de Libération du Peuple
MLPC f: Mouvement de Libération du Peuple Centrafricain
MLPP e: Multi-Level Precedence and Preemption
- e: Multi-Level Priority and Preemption
ML-PPP e: Multi Link PPP [Point-to-Point Protocol]
MLR e: International Monitoring System
- e: Message Log Report
- e: Multi-Channel Linear Recording
- e: Multiple Linear Regression
- s: Movimiento Laboral Registrado
MLRS e: Multiple Launch Rocket System
MLS e: Master Line State
- e: Master of Liberal Studies
- e: Master of Library Studies
- e: Masters of Library Science
- e: Microwave Landing System
- e: Microwave Limb Sounder
- e: Mid-Latitude Summer
- e: Moored Listening Station
- e: Movement for the Liberation of the Sanwi
- e: Multi-Layered Structure
- e: Multi-Level Secure (or Security)
- e: Multiple Link Support
- e: Multiple Listing Service
- f: Mouvement de Libération du Sanwi
Mls r: Militär lotsstation
MLSA e: Mid-Latitude Standard Atmosphere
MLS DDN e: Multi-Level Secure Defense Data Network
MLSF e: Mobile Laboratory Storage Facility
- e: Mobile Logistics Support Force
MLSS e: Mixed Liquor Suspended Solids
MLSTP e: Movement for the Liberation of Sao Tome and Principe
- f: Mouvement pour la Libération de São Tomé e Príncipe
MLSTP-PSD p: Movimento de Libertação de São Tome e Príncipe-Partido Social Democrata
MLT e: Mechanized Loop Testing
- e: Median Lethal Time
- e: Melatonin
- e: Multi-Level Transition
- e: Multi-Level Transmission
- r: Marknadslönetillägg
MLTF e: Multi-loading Test Facility
MLTG e: Multi-Link Transmission Group
MLTT e: Magazine Loader Transfer Table
MLU a: Mejeriernes og Landbrugets Ulykkesforsikring
- e: Mid-Life Upgrade
MLV e: Medium Lift Vehicle
- e: Memory Loader/Verifier
- e: Modified Live Virus
- e: Modify Logging Versions
MLVSS e: Mixed Liquor Volatile Suspended Solids
MLW e: Mid-Latitude Winter
MLWN e: Mean Low-Water Neap
MLWS e: Mean Low-Water Spring
MLX e: Mustang LX

MM d: Maschinenlesbares Merkmal
- e: Machine Model
- e: Manipulator Main
- e: Manufacturing Methods
- e: Marriage-Minded
- e: Maryknoll Missioners
- e: Material Manufacturer
- e: Materiel Management
- e: Mesoscale Model
- e: Methadone Maintenance
- e: Millimole/liter
- e: Minuteman [missile]
- e: Minutes and Month
- e: Mission Model
- e: Mixer Module
- e: Mobile (or Mobility) Management
- e: Modified Mercalli [seismic measurement]
- e: Month
- e: Multimedia
- e: Multiple Master
- e: Multiple Mode
- e: Multiple Myeloma
- e: Myanmar
- e: Program Maintenance Manual
- f: Maison Maternelle
- f: Maître Mineur
- f: Maître-Maçon
- f: Marine Marchande
- f: Médecine Mentale
- f: Médicaments Magistraux
- f: Menue Monnaie
- f: Mer Méditerranée
- f: Messieurs
- f: Métier Manuel
- f: Meubles Meublants
- f: Monnaie-Matière
- f: Mur Mitoyen
- N: Marinens Minevesen
- s: Mapa de Mando
- s: Medalla Militar
- s: Medios Móviles
- s: Mortero Medio

mm d: millimeter
- e: millimeter
- r: minmatroser

mm. e: millimeters

m.m. a: med mere
- r: med mera

m/m f: même mois

M-M f: Meurthe-et-Moselle

M2M e: Manager-to-Manager

MM2 f: Marié Moins de 2 ans

MM4 e: Mesoscale Meteorological model-version 4

MM5 e: Mesoscale Model version 5

MMA d: Multi-Megabit-Anpassung
- e: Mastitis-Metritis-Agalactia complex
- e: Microcomputer Managers Association
- e: MIDI Manufacturer[s] Association
- e: Millimeter Array project

mm[a] r: motormaskinist

MMAC e: Multi Media Access Center

MMAI N: Marinens Maskininspeksjon

MM-Al e: total Monomeric Aluminum

MMAP e: Marine Mammal[s] Action Plan
- f: Plan d'Action pour les Mammifères Marins
- s: Plan de Acción para los Mamíferos Marinos

MMAS e: Medical Materiel Accounting System

MMA [S] e: Material Measurement and Accounting [System]

MMAXCT e: Maximum Corrective Time

MMBDC e: Michigan Minority Business Development Center

MMBF e: Million Board Feet

MMBR e: Microbiology and Molecular Biology Reviews

MMC e: Man Machine Communication
- e: Manufacturing Methods Council
- e: Marine Mammal Commission
- e: Matched Memory Cycle
- e: Material Management Center
- e: Metal Matrix Composite[s]
- e: Microcomputer Marketing Council
- e: Microsoft Management Console
- e: Midi Machine Control
- e: Minerals Marketing Corporation
- e: Minicomputer Maintenance Center
- e: Money Market Certificate
- e: Movements Monitoring Center
- e: Multi Media Card
- e: Multi Media Commands
- e: Multi Media Conference

MMCC e: Medicare Managed Care Contract
- e: Mission Management and Control Centre

MM[-]CD e: Multi Media Compact Disc (or Disk)

MMCDE e: Multi Media Compact Disk-Erasable

MMCF e: Multimedia Communications Forum

MMcf/d e: Million cubic feet per day

MMCH4 e: Methane emitted from animal manure

MMCR e: Millimeter Wave Cloud Radar

MMCS e: Mauritius Marine Conservation Society
- f: Association de Maurice pour la protection et la sauvegarde de la mer
- s: Asociación de Mauricio para la Protección y la Salvaguardia del Mar

MMCSRS e: MACOM Materiel Condition Status Reporting System

MMCX e: Multimedia Communication Exchange

MMD e: Median Mass Diameter
- e: Micro Miniature Diode
- e: Microlithographic Mask Development program
- e: Movement for Multiparty Democracy

MMDA e: Money Market Deposit Account

MMDB e: Molecular Modeling DataBase

MMDC e: Master Message Display Console

Mmde f: Marine marchande

MMDF e: Multi channel Memo[randum] Distribution Facility

MMDS e: Multichannel Memorandum Distribution Facility
- e: Multichannel, Multipoint Distribution System

MME e: Master Mode Entry
- e: Microsoft Multimedia Extension
- e: Military Message Experiment
- e: Mobile Meteorological Equipment
- e: Multi Media Extension[s]

Mme f: Madame

MMEC e: Material Management Executive Committee

MMES e: Martin Marietta Energy Systems, Inc.

Mmes f: Mesdames

MMF e: Magnetomotive Force
- e: Master Message File
- e: Multi-Mode Fiber[-optic cable]
- s: Modelo de Malla Fina para áreas Restringidas

MMFAB e: Metals and Machine Fabricators

MMFM f: Front des Journalistes

MMFT f: Front des Enseignants

MMG e: Motor-Motor Generator

MMGT e: Multi Master Group Translator

MMH e: Mono-Methyl Hydrazine

MMH/FH e: Maintenance Man Hours per Flying Hour

mmHg e: millimeter of Mercury

MMH/OH e: Maintenance Man Hours per Operating Hour

MMI e: Man-Machine Interaction
- e: Man-Machine Interface
- e: Modified Mercalli Intensities (or Intensity)
- s: Mapa Militar Itinerario

MMIA e: Methods and Models for Integrated Assessment

MMIC e: Millimeter/Microwave Integrated Circuit
- e: Monolithic Microwave IC [Integrated Circuit[s]]

MMIF e: My Mouth Is Full [Internet]

MMIS e: Master of Management Information Studies
- e: Materials Manager Information System

MMJ e: Modified Modular Jack

MMK d: Multimediale Kommunikation
- d: Museum für Moderne Kunst Frankfurt

mmk r: magnetomotorisk kraft

MML e: Man Machine Language
- e: Math Markup Language

MMLP e: Massachusetts Municipal Light Plants

MMLS e: Mobile Microwave Landing System

MMM e: Material Movement Management
- e: Mauritian Militant Movement
- e: Meaningful Measures of Merit
- e: Memory Mapper Modul
- e: Mesoscale and Micro-scale Meterology
- e: Mode Management Module
- e: Multi Media Mediocrity
- e: Multi Media Modeling
- f: Ministère de la Marine Marchande
- f: Mouvement Militant Mauricien
- s: mortero medio en montaje

MMMC e: Machine Material Movement Component

MMMDCS e: Multi-National Maintenance Management Data Collection System

Mmmes f: Mesdames

MMMLTG e: Mixed Media Multi-Link Transmission Group
MMMS e: Material Movement Management Standard
- e: Medical Materiel Management System
MMMS-OL e: Medical Materiel Management System On-Line
MMMSP f: Mouvement Militant Mauricien-Socialiste Progressif
Mmo e: Maximum operating mach number
MMO e: Mission Management Office
- e: Mission Management Office within the MCS
- e: Multimodel Optimization
MMOA e: Martin Marietta Orlando Aerospace
MMOC e: Minicomputer Maintenance Operations Center
Mmol e: Millimoles
MMOS e: Modified MOS
MMP e: Maintenance Management Program
- e: Marine Meteorology Programme
- e: Massive Multi Processing
- e: Master Mobilization Plan
- e: Materials Management Plan
- e: Matrix Metalloproteinase
- e: Medical Mission Planner
- e: MEECN Master Plan
- e: MEECN Message Processing Mode
- e: Mixed Media Publishing
- e: Modified Modular Plug
MMPA e: Magnetic Material Producers Association
MMPI-D e: Depression scale of the Minnesota Multiphasic Personality Inventory
MMPM e: Marine Management Programming Model
- e: Multi Media Presentation Manager
MMPM2 e: Multi Media Presentation Manager/2
MMPO e: Mission Management and Planning Office
M M POOL e: Matt Mann Pool
MMPS e: Manpower Mobilization Planning System
- e: MEECN Message Processing System
- e: Million Multiplies Per Second
MMR d: Multimedia und Recht. Zeitschrift für Informations- Telekommunikations- und Medienrecht
- e: Maintenance Management Report
- e: Marketing and Media Research
- e: Maternal Mortality Ratio
- e: Measles-Mumps-Rubella Vaccine
- e: Modified Modified Read [compression]
MMrs f: Messieurs
MMS e: Main Memory Status
- e: Manufacturing Message Specification
- e: Manufacturing Message Standards
- e: Manufacturing Multimedia Survey
- e: Marine Meteorological Services
- e: Mast-Mounted Sight
- e: Master Milestone Schedule
- e: Maximum Message Size
- e: Meteorological Measuring System
- e: Mineral[s] Management Service
- e: Movement Monitoring System
- e: Multimedia Survey
- e: Multimission Modular Spacecraft
- e: Munitions Maintenance Squadron
- f: Servicios de Meteorología Marina
MMSA e: Multi-Mission Surveillance Aircraft
mmscmd e: million metric standard cubic metre per day
MMSE e: Minimum Mean Square Error
- e: Multiuse Mission Support Equipment
MMSG e: Molecular Manufacturing Shortcut Group
MMSI e: Maritime Mobile Service Identity
MMSM e: National Front for the Defense of Malagasy Socialism
MMSS e: Master Milestone Summary Schedule
MMST e: Microelectronics Manufacturing Science and Technology
MMT e: Manufacturing Methods Technology
- e: Master Message Terminal
- e: Micro Miniature Transistor
- e: Million Metric Tons
MMTC e: Midwest Manufacturing Technology Council
MMTCA e: Multimedia Toolbox for Cooperative Applications
MMTO e: Multiple Mirror Telescope Observatory
MMTPS e: Maintenance Materiel Transaction Processing System
MMTS e: Maximum-Minimum Temperature System
MMU e: Main Memory Unit
- e: Manned Maneuvering Unit
- e: Master of Music
- e: Milli-Mass-Unit
- s: Módulo de Municiones
MMUC r: Militärmedicinsk Undersökningscentral
MMVF e: Multimedia Video File
MMW d: Münchener Medizinische Wochenschrift
- e: Millimeter Wave
- e: Multi Mega Watt
MMWEC e: Massachusetts Municipal Wholesale Electric Company
MMWR e: Millimeter-Wave Radar
- e: Morbidity and Mortality Weekly Report
MMWR CDC Surveill Summ e: MMWR CDC Surveillance Summaries
MMWR Morb Mortal Wkly Rep e: MMWR. Morbidity and Mortality Weekly Report
MMX e: Mastergroup Multiplex
- e: Matrix Manipulation Extensions
- e: Microprocessor from Intel
- e: Multi Media Extension[s]
- e: Multi Media X [pentium-family processor]
MN f: Magasin de Nouveautés
- f: Marché Noir
- f: Mer Noire
- f: Montant Nominal
- f: Mort Naturelle
- f: Mort-Né
- f: Musée National
- N: Marinens Navigasjonsvesen
- s: Mapa Nacional
- s: Minador
- s: Motoniveladora
mn e: Mongolia [Internet]
Mn e: Manganese
- Mean range [between mean high and mean low water]
- s: Manual
- s: Mina[do]
Mn-54 e: Manganese-54
MNA e: Malawi News Agency
- e: Massachusetts Nurses Association
- e: Member of National Assembly
- e: Michigan Nurses Association
- e: Minnesota Nurses Association
- e: Mississippi Nurses Association
- f: Mouvement National Arménien
- N: Medlem Norske Advokatsforening
Mña s: Montaña
MNADREG e: Movement of National Alliance for Democratic Restoration of Equatorial Guinea
MNAL N: Medlem Norske Arkitekters Landsforbund
m-näst r: motståndsnäst
MNB f: Mouvement National Belge
MNC e: Major NATO Command[er]
- e: Mononuclear Cells
- e: Multinational Corporation
- p: Movimento Nacionalista Cabindes
Mnch f: Manche
MNC L[umumba] f: Mouvement National Congolais Lumumba
MNC[/]R f: Mouvement National Congolais/Renové
MNCS e: Master Net Control Station
MND e: Mission Need Determination
- f: Mouvement National Djiboutien
mnd. N: måned
MNDA e: Motor Neurone Disease Association
MNDD f: Mouvement National pour la Démocratie et le Développement
MNDEP e: Mineral Deposits Data Base
MNDLD f: Mouvement National Djiboutien pour l'Instauration de la Démocratie
MNDS f: Mouvement National pour le Développement des Sciences
Mne e: Mach number Never to be Exceeded
MNE f: Mouvement National de l'Épargne
- f: Mouvement Nationaliste des Étudiants
MNF d: Mathematisch-Naturwissenschaftliche Fakultät
MNG e: Multi-Net Gateway

MNGIE e: Myopathy and external ophthalmoplegia Neuropathy Gastro-Intestinal Encephalopathy
MNHN f: Muséum National d'Histoire Naturelle, Paris
MNIF N: Medlem av den Norske Ingeniørforening
MNIND e: Mineral Industry Data Base
MNIT e: Mobile Networks Integration Technology
MNJ f: Mouvement National Judiciaire
MNJD f: Mouvement National de la Jeunesse Démocratique
MNJTS f: Mouvement National des Jeunes Travailleurs du Sénégal
MNLC f: Mouvement National de Libération Congolaise
MNLC/L-R f: Mouvement National de Libération Congolais/Lumumba-Renové
Mnm f: Minimum
MNMHS e: Metal and Nonmetal Mine Health and Safety
MNNP e: Master of Nursing, Nurse Practitioner
MNNR e: Marine National Nature Reserve
Mno r: Marinorder
MNOS e: Metal-[silicon] Nitride-[silicon] Oxide-Semiconductor
MNP e: Medical Necessity Program
- e: Microcom Network Protocol
- e: Multiple Networking Program
- e: Multiple Networking Protocol
- f: Mouvement National des Pionniers
MNPVC e: Multinetwork PVC
MNPWR e: Manpower Standard Application Program
MNR A: Mediese Navorsingsraad
- e: Ministry of Natural Resources
- e: Mozambique National Resistance
- f: Mouvement National de la Révolution
- f: Mouvement National de Résistance
- f: Mouvement National pour le Renouveau
MNRE e: Manual Not Read Error
MNRG f: Mouvement National de la Révolution Gabonaise
MNR [RENAMO] e: Mozambique National Resistence
MNRS e: Midwest Nursing Research Society
MNRT e: Minister of State for Research and Technology
- f: Mouvement National des Renovateurs Tchadiens
MNS e: Malayan Nature Society
- e: Marine Science
- e: Metal-Nitride Semiconductor
- e: Mission Need Statement
- e: Mission Needs Statement
- f: Manufacture Nationale de Sèvres
- N: Medlem av den Norske Sakførerforening
MNSD-Nassara f: Mouvement National pour la Société de Développement-Nassara
M/N/T e: Main/Satellite/Tributary Network
Mnt f: Minuit
- s: Monitor
MNTD e: Maximum Nontoxic Dose
MNTF N: Medlem av den Norske Tannlegeforening
MNU e: Menu
MNUR f: Mouvement National d'Union et de Réconciliation
MNYC e: Mauritius National Youth Council
MO a: Moralsk Oprustning
- e: Magneto-Optic[al]
- e: Managed Object
- e: Managment Object
- e: Metal-Organic
- e: Microelectronic Outline
- e: Microscope Objective
- e: Missouri
- e: Mobile Originated
- e: Molecular Orbital
- e: Monocyte
- e: Monocytic
- f: huile Minérale Ordinaire
- f: Main-d'Œuvre
- f: Maintien de l'Ordre
- f: Marchandise d'Occasion
- f: Matières Organiques
- f: Médecin-Opthalmologiste
- f: Médecine Opératoire
- f: Médicaments Officinaux
- f: Moyen-Orient
- l: Modus Operandi
- N: Militær Orientering
- N: Moralsk Opprustning
- r: Militärområde
- r: Militieombudsmannens Ämbetsberättelse
- s: Microonda
Mo d: Montag
- e: Megaoctet[s]
- f: Maître-ouvrier
- N: Motorordon[n]ans
mo e: Macau [Internet]
mo. e: month
- N: motorsykkelordon[n]ans
M&O e: Maintenance and Operations
- e: Management and Operations
M/O e: Meteorological and Oceanographic Sensing System
m/o f: mon ordre
- s: mi orden
Mo-99 e: Molybdenum-99
MOA e: Manual Off Auto
- e: Maryland Office on Aging
- e: Measure[s] Of Adequacy
- e: Memoranda of Agreement
- e: Military Operations Area
- e: Ministry of Aviation
MoA e: Museum of Australia
MOAS e: Merit Oriented Advancement Scheme
MOATRS e: MAJCOM On-Line Aerospace Vehicle Training Report System
MOB e: Main Operating Base; Missile Order-of-Battle
- e: Main Operations Base
- e: Memory-Order Buffer
- e: Military Order of Battle
- f: Mouvement pour l'Organisation de la Bretagne
mob N: mobilisasjon
- N: mobilisere
- N: mobilisering
- N: mobilisert
MOBA e: Museum of Bad Art
Mobavd r: Mobiliseringsavdelning
MOBC2 e: Mobilization Command and Control
MOBCON e: Mobilization Construction Plan
MOBDET e: Mobility Deployment Planning System
MOBERS e: Mobilization Equipment Redistribution System
Mobförberedelser r: Mobiliserings-förberedelser
Mobförråd r: Mobiliseringsförråd
mobhær N: mobiliseringshær
Mobil f: Mobilier
MOBILE e: Mobile Source Emission[s] Model
MOBILE5A e: Mobile Source Emission Factor Model
MOBILHY f: Mobilisation du Bilan Hydrique
- f: Modélisation du Bilan Hydrique
Mobkalender r: Mobiliseringskalender
MOBL e: Mobile River And Tributaries
MOBO e: Mother Board
MobÖ r: Mobiliseringsövningar
moboff[icer] r: mobiliseringsofficer
moborder r: mobiliseringsorder
mobort r: mobiliseringsort
MOBPERS e: Mobilization Personnel Processing System
mobplan r: mobiliseringsplan
MOBPLANS e: Mobility Plans
MOBRAP f: Mouvement Burkinabe de Lutte contre le Racisme, l'Apartheid et pour l'Amitié entre les Peuples
MOBREM e: Mobilization Base Requirements Model
MOBS e: Multiple Orbital Bombardment System
MOBSCOPE e: Mobilization Shipments Configured for Operations, Planning and Execution
MOBSS e: Mobile Support Squadron
mobtab r: mobiliseringstabell
MOBTDA e: Mobilization Table of Distribution and Allowances
mobtid r: mobiliseringstid
MOBTRAN e: Mobilization and Deployment Transportability
MOC e: Managed Object Class
- e: Management and Operating Contractor
- e: Mars Observer Camera
- e: Massachusetts Oilheat Council
- e: Materials Operation Center
- e: Memorandum of Cooperation
- e: Minimum Operating Capability
- e: Mission Operations Center
- f: Main-d'Œuvre Civile
- f: Mystérieux Objets Célestes
MOCA e: Minimum Obstacle Clearance Altitude
- e: Missouri Classical Association
MOCAS e: Mechanization of Contract Administration Sources

MOCD e: Mission Operations Concept Document
MOCERT e: Motorcycle Certification Data
MOCIL p: Moçambique Importadora
MOCNESS e: Multiple Opening-Closing Net Environmental Sampling System
- e: Multiple Opening/Closing Net and Environmental Sensing System
MOCOCO e: Movement Coordinating Committee
MOCP e: Mission Operational Computer Program
MOCS e: MacPhail Operant Chambers
- e: Managed Object Conformance Statement
- e: Multichannel Ocean Colour Sensor
MOCVD e: Metal-Organic CVD [Chemical Vapor Deposition]
MOD e: MAC Operational Directive
- e: Magneto-Optical Disk
- e: Management and Organization Division
- e: Manufacturers Operations Division
- e: March Of Dimes
- e: Marine Operations Division
- e: Mesio Occlusal Distal
- e: Ministry of Defense
- e: Mission Operating Directive
- e: Modem
- e: Module
- s: Módulo [de municiones]
MoD e: Ministry of Defence
Mod f: Modèle
- f: Modéré
- f: Moderne
- f: Modifié
- s: Modelo
mod. e: modern
- N: moderne
Mod. s: modificación
MÖD r: Marinöverdirektör
MO&DA e: Mission Operations and Data Analysis
ModArch d: Institut für Grundlagen moderner Architektur und Entwerfen
MODAS e: Maintenance and Operational Data Access Subsystem
- e: Model for Intertheater Deployment by Air and Sea
MO-DCA e: Mixed Object Document Content Architecture
MO:DCA e: Mixed Object Document Content Architecture
MODCA e: Mixed Object Document Content Architecture
MODE e: Mid-Ocean Dynamics Experiment
- e: Middeck Zero-Gravity Dynamics Experiment
- e: Music on Demand
- f: Expérience sur la dynamique en haute mer
- s: Experimento de Dinámica Mesoceánica
MODE 1 e: Power operation
MODE 2 e: Startup and low-power operation
MODE 3 e: Shutdown
MODE 4 e: Refueling
MODECS e: Molecular Design of Chemical Systems
MODEF f: Mouvement de Défense [et de coordination] des Exploitations agricoles Familiales
MODELS e: Modernization of the Defense Logistics Standard System
MODEM e: Modulator/Demodulator
- f: Modulateur-Demodulateur
Modem d: Kunstwort aus Modulator und Demodulator
ModEnG d: Gesetz zur Förderung der Modernisierung von Wohnungen und von Maßnahmen zur Einsparungen von Heizenergie [Modernisierungs- und Energieeinsparungsgesetz]
MODES e: Mode Optimization and Delivery Estimation System
MODEST f: Movement pour la Démocratie Sociale et la Tolérance
MODF e: Modification
MODFET e: Modulation-doped FET
MODHIWAY e: Modified Highway Program
MODIS e: Moderate-resolution Imaging Spectro[radio]meter
MODIS-N e: Moderate-resolution Imaging Spectrometer-Nadir
MODIS-T e: Moderate-resolution Imaging Spectrometer-Tilt
mod.L. e: modern Latin
Mod LR e: Modern Law Review
MODMAP e: Modernization of Mapping techniques
MODMAST e: Modify Master
Mod Pathol e: Modern Pathology
Mod Probl Pharmacopsychiatry e: Modern Problems of Pharmacopsychiatry
MO&DS e: Mission Operations and Data System
MODS e: Mobilization Planning Data System
MODs e: Modifications
mods. a: modsat
- a: modsætning
MO&DSD e: Mission Operations and Data Systems Directorate
MODSIN e: Mission Operations and Data Systems Information Network
ModSt d: Institut für Modellstatik
modsv. a: modsvarende
Modt. a: Modtagelse
- a: Modtager
- a: Modtaget
MODTRAN e: Moderate Resolution Atmospheric Radiance and Transmittance
MODU e: Mobile Offshore Drilling Unit
Modula d: Höhere modulare Programmiersprache
MODWORS e: Modification Work Order/Status System
MODY e: Maturity Onset Diabetes of the Young
MOE e: Magnitude Of Effort
- e: Margin Of Exposure
- e: Measure[s] Of Effectiveness
- e: Minimum Operable Equipment
- f: Main-d'Œuvre Étrangère
MOEE e: Ministry of Environment and Energy
MOEM e: Micro-Opto-Electro-Mechanical
MOET e: Multiple Ovulation and Embryo Transfer
MOF e: Matter Of Fact
- e: Minister Of Finance
- e: Ministry Of Finance of Japan
MOFA e: Ministry Of Foreign Affairs of Japan
- e: Ministry Of Foreign Affairs, Thailand
MOFFF e: Mixed-Oxide Fuel Fabrication Facility
MOFW e: Men Of Few Words
MOG e: Machinery of Government
- e: Managed Object Group
- e: Maximum [Aircraft] On the Ground
- e: Melter Off Gas
- e: Minicomputer Operations Group
MOGUNTIA e: Global atmospheric transport model from Mainz
- e: Model of the General (or Global) Universal Tracer Transport in the Atmosphere
m. o. h. N: meter over havet
MOH d: Meteorologisches Observatorium Hamburg
- e: Master of Occupational Health
- e: Ministry Of Health
Möh r: Meter över havet
MOHAVE e: Measurement Of Haze And Visual Effects
MOHLL e: Machine Oriented High Level Language
Moho e: Mohorovii discontinuity
Mohole e: a hypothetical drill hole that was to penetrate the Moho
MOHp d: Meteorologisches Observatorium Hohenpeißenberg
MOI e: Memorandum Of Information
- e: Memorandum Of Intent
- e: Mission Oriented Items
MÖI r: Marinöverintendent
MOIA e: Mission Oriented Item Activity
MOIL e: Marine Operations and Instrumentation Laboratory
MOIRA e: Model of International Relations in Agriculture
MOJA e: Movement for Justice in Africa
möjl. r: möjlig[en]
MOK N: Marinens Overkommando
MOKE e: Magneto-Optical Kerr Effect
MOL d: Märkisch-Oderland, Brandenburg
- e: Mathematics Of Language
- e: Maximum Order Limitation
- e: Meteorologisches Observatorium Lindenberg
- e: Molecular Biology
mol% e: mole percent
MÖL r: Marinöverläkare
MOLA e: Mars Observer Laser Altimeter
MOLAP e: Multidimensional On-Line Analytical Processing
Mol Aspects Med e: Molecular Aspects of Medicine
Mol Biochem Parasitol e: Molecular and Biochemical Parasitology
Mol Biol Cell e: Molecular Biology of the Cell
Mol Biol Evol e: Molecular Biology and Evolution

Mol Biol [Mosk] R: Molekuliarnaia Biologiia
Mol Biol Rep e: Molecular Biology Reports
Mol Biotechnol e: Molecular Biotechnology
Mol Carcinog e: Molecular Carcinogenesis
Mol Cell Biochem e: Molecular and Cellular Biochemistry
Mol Cell Biol e: Molecular and Cellular Biology
Mol Cell Biol Hum Dis Ser e: Molecular and Cell Biology of Human Diseases Series
Mol Cell Endocrinol e: Molecular and Cellular Endocrinology
Mol Cell Neurosci e: Molecular and Cellular Neurosciences
Mol Cell Probes e: Molecular and Cellular Probes
Mol Chem Neuropathol e: Molecular and Chemical Neuropathology
Mol Ecol e: Molecular Ecology
Mol Endocrinol e: Molecular Endocrinology
Mol Gen Genet e: Molecular and General Genetics
Mol Gen Mikrobiol Virusol R: Molekuliarnaia Genetika, Mikrobiologia, i Virusologa
MOLI e: Microsoft Online Institute
MOLICA p: Movimento de Libertação da Cabinda
MOLIDE f: Mouvement pour la Liberté et la Démocratie
MOLIFUGE f: Mouvement pour la Liberté et le Futur de la Guinée Equatoriale
Mol Immunol e: Molecular Immunology
MOLINACO f: Mouvement de Libération Nationale des Comores
MOLINK e: Moscow-Washington Direct Communications Link
MOLIS e: Minority On-Line Information System
Moll f: Mollusque
Mol Mar Biol Biotechnol e: Molecular Marine Biology and Biotechnology
Mol Med e: Molecular Medicine
Mol Membr Biol e: Molecular Membrane Biology
Mol Microbiol e: Molecular Microbiology
Mol Neurobiol e: Molecular Neurobiology
MOLP e: Microsoft Open License Pak
Mol Pharmacol e: Molecular Pharmacology
Mol Phylogenet Evol e: Molecular Phylogenetics and Evolution
Mol Plant Microbe Interact e: Molecular Plant-Microbe Interactions
Mol Reprod Dev e: Molecular Reproduction and Development
MOLS e: Machine Oriented Language Lines
mol.v. a: molekylarvægt
MOM e: Manager Of Managers
- e: Message-Oriented Middleware
- e: Metal-Oxide-Metal
- e: Microsoft Office Manager
- e: MIS Modification Module
- e: Mission Operations Manager
- e: Modular Ocean Model
MoM e: Map oriented Machine
MOMBE e: Metal-Organic Molecular Beam Epitaxy
Mo Med e: Missouri Medicine
MOMPLS e: Moment Please
m-område r: motståndsområde
MOMS a: Merværdiomsætningsafgift
- d: Modularer Optoelektronischer Multispectral-Scanner
- e: Modular Optoelectronic Multispectral Scanner
- r: Mervärdesskatt
MOM SEDCOs e: Marine Operations Manual
MOMspider e: Multi-Owner Maintenance Spider
MoMuC e: Mobile Multimedia Communication
MOMV e: Manned Orbital Maneuvering Vehicle
MON e: Motor Octane Number
Mon f: Maison
MONA e: Missouri Nurses Association
Monaldi Arch Chest Dis e: Monaldi Archives for Chest Disease
MONAMO p: Movimento Nacionalista Moçambicano
MONAP s: Movimiento Nacional de Pobladores
MONAPOL f: Mouvement National Populiste Liquidateur
Monare e: Movement for National Redemption
MONAS e: Movement on National Affairs
Monash U. e: Monash University
MonAwV d: Verordnung zur Einreichung von Monatsausweisen nach dem Gesetz über das Kreditwesen [Monatsausweisverordnung] vom 29.12.1997
MONBUSHO e: Ministry of Education, Science and Culture
Mond f: Mondial
MONEG e: Monsoon Numerical Experimental (or Experimentation) Group
- f: Groupe chargé de l'expérimentation numérique relative à la mousson
- s: Grupo de Experimentación Numérica sobre los Monzones
MONESTO f: Mouvement National des Etudiants et Stagiaires Togolais
MONET e: [high data rate] Mobile Internet
MONEX f: Expérience de la mousson
- s: Experimento sobre los Monzones
Mon For f: Maison Forestière
MONICA e: Multilateral Monitoring of Trends and Determinants in Cardiovascular Disease
MONIMA f: Mouvement National pour l'Indépendance de Madagascar
Monit f: Moniteur
Mon LR e: Monash University Law Review
Monn f: Monnaie
Monogr f: Monographie
Monogr Allergy e: Monographs in Allergy
Monogr Clin Cytol e: Monographs in Clinical Cytology
Monogr Dev Biol e: Monographs in Developmental Biology
Monogr Endocrinol e: Monographs on Endocrinology
Monogr Natl Cancer Inst e: Monographs/ National Cancer Institute
Monogr Oral Sci e: Monographs in Oral Science
Monogr Pathol e: Monographs in Pathology
Monogr Soc Res Child Dev e: Monographs of the Society for Research in Child Development
Monop f: Monopole
Monr s: Monitor
Mons. s: Monseñor
MONSEE e: Monitoring of the Sun-Earth Environment
MONSU e: Movement for National Student Union
Mont f: Monument
MontanMitbestG d: Gesetz über die Mitbestimmung der Arbeitnehmer in den Aufsichtsräten und Vorständen der Unternehmen des Bergbaus und der Eisen und Stahl erzeugenden Industrie
MOO e: Miami MOO
- e: MUD [Multi-User Device] Object Orient[at]ed [environment]
MOOSE e: Major Object-Oriented SQL Extensions
MOP d: Meteorologisches Observatorium Potsdam
- d: Übertragungsprotokoll im Netzwerk
- e: Maintenance Operation[s] Protocol
- e: Management Oversight Practices
- e: Management Overview Program
- e: Maximum Operating Pressure
- e: Measure[s] Of Performance
- e: Memorandum Of Participation
- e: Memorandum Of Policy
- e: Memory Organization Packet
- e: Message Output Processor
- e: Meteorological Operational Programme
- e: Meteosat Operational Programme
- e: Modify Operating Procedures
- e: Monthly Obligation Plan
- e: Multiple Original Prints
MOPAD e: Movement for Peace and Democracy
MOPH e: Ministry of Public Health
- p: Ministério das Obras Públicas e Habitação
MOPITT e: Measurement[s] of Pollution in the Troposphere
MoPL e: Map oriented Programming Language
MopÖ r: Marin operativ Övning
MOPP e: Master Operational Program Plan
- e: Mission Objective Protective Posture
MOPS e: Marine Oil Pollution System
- e: Message Output Processing System

MOR e: Mid-Ocean Ridge
- e: Monthly Operating Review
- f: Mandat Ordonnancé
MORACC e: Modeling Radiation Clouds And Convection
MORB e: Mid-Ocean-Ridge Basalt
Morb f: Morbihan
MORC e: Manually Operated Rail Cart
MORE e: MARC Optical Recognition
- e: Minority Outreach Research and Education
Møre Fdi N: Møre Forsvarsdistrikt
MORENA f: Mouvement de Redressement National
- f: Mouvement de Rénovation Nationale
MORF e: Male Or Female? [Internet]
MORFLOT e: Ministry of the Merchant Marine of the Soviet Union
MORIE e: Metal-Organic Reactive Ion Etching
MORL e: Morrill Hall
MORR e: Morrill Hall
- e: Morrison Library
MORS e: Group of Experts Monitoring of Radioactive Substances in the Baltic Sea
- e: Military Operations Research Society
MORSA e: Movement Requirements for Staff Planning and Special Studies Applications
Mor[t] s: Mortero
MORT e: HERL-RTP Mortality Data Base
- e: Management Oversight Risk Tree
- e: Ministry of Research and Technology
MOS d: Neckar-Odenwald-Kreis in Mosbach, Baden-Württemberg
- e: Magneto-Optic Storage
- e: Management Operations Staff
- e: Margin Of Safety
- e: Marine Observation[al] (or Observations) Satellite
- e: Maritime Observation Satellite
- e: Mass Online Storage
- e: Mean Opinion Score
- e: Measure Of Suitability
- e: Medical Outcomes Study
- e: Memorandum Of Support
- e: Metal-Oxide-Semiconductor
- e: Military Occupational Skill
- e: Military Occupational Specialty
- e: Military Oceanography Subcommittee
- e: Minimum Operating Strip
- e: Ministry Of Supply
- e: Monolithic Oxide Silicon
Mos f: Moselle
MOS-1 e: Marine Observation[s] Satellite-1
- f: Satellite d'observation marine
- s: Satélite de Observación Marina
MOSAIC e: Guide for multimedia within Internet
- e: Multi-User, On-Line System for Automated Information Communication
MOSAICS e: Multi Observation Satellite Image Correction System
MOS-C e: MOS [Metal Oxide Semiconductor] Capacitor
MOSCAMED s: Programa de Erradicación de la Mosca del Mediterraneo
MOSCAP e: MOS Capacitor
MOSES e: Major Open Systems Environment Standards
- e: Mission, Objectives and Strategies Execution plan and Success indicators
MOSFET e: Metal-Oxide Semiconductor Field-Effect Transistor
MOSHER e: Mosher Jordan Hall
MOSI e: Minimal Open Systems Interconnection
MOSIS e: MOS Implementation System
MOSIX e: Multicomputer Operating System for UNIX
mOsm e: milliosmol[s]
MOSOPESCA p: Empresa Moçambicana-Soviética de Pescas
MOSP e: Mission Operations Support Plan
MOSPF e: Multicast Extensions to OSPF
MOSPOR e: Movement for the Struggle for Political Rights
Mosq s: Mosquetón
MOSS e: Maintenance and Operation Subsystem
- e: Maintenance Operations Support Set
- e: Map Overlay and Statistics System
- e: Map Overlay Statistical System
- e: MIME Object Security Services
- e: Mobile Oceanography Support System
- e: Moving Object Support System
MOSS-E e: Maintenance and Operation Subsystem-Extended
Mossgas e: Mossel Bay Gas Project
MOST e: Ministry of Science and Technology
- e: Molonglo Observatory Synthesis Telescope
- e: MOS Transistor
MOSTT e: Mosaic Optical Sensor Technical Testbed
MOT e: Means Of Testing
- e: Ministry Of Transport [of Japan]
- e: Ministry Of Transportation
- e: Multimedia Object Transfer
- r: Oktantal [Motormetoden]
Mot N: Motorisert
-s: Motocicleta
mot- r: motoriserad-
MOTA e: Materials open test assembly
MOTA-IC e: Alloy test
MOTAS e: Member Of The Appropriate Sex [Internet]
MOTD e: Message Of The Day
MOT&E e: Multinational Operational Test and Evaluation
MOTEL e: Motorola and Intel compatible
MOTH e: Humorous Collectibles Forum variation on Member of the Month
MOTI e: Management Of Technological Innovation
MOTIC e: Management Of Technology Information Center
MOTIS e: Message-Orientated Text Interchange System[s]
MOTM e: Member Of The Month
Motm r: Motmedel
Moto e: Motorola
MOTO s: Motorizado
MOTORAGRI f: Société pour le Développement de la Motorisation de l'Agriculture
MotorS r: Motorskola
MOTOS e: Member Of The Opposite Sex [Internet]
MOTRON e: MacPhail Motron System
MOTS e: Multiple OPLAN Tasking Summary
Motskreg r: Motoriserad skytteregemente
MOTSS e: Member Of The Same Sex [Internet]
Motsv r: Motsvarande
Mott r: Mottagning
MOTT HOSP e: C. S. Mott Childrens Hospital
MotU e: Mark of the Unicorn
MOTV e: Manned Orbital Transfer Vehicle
MOU e: Memoranda (or Memorandum) Of Understanding
- f: Mémorandum d'accord
- s: Memorando de Acuerdo
MOUND e: Mound Plant, Miamisburg, Ohio
MOURAD f: Mouvement pour la Renovation de l'Action Démocratique
MOUT e: Military Operations in/on Urban[ized] Terrain
Mouv f: Mouvement
MOV e: Manual Override Valve
- e: Metal-Oxide Varistor
- e: Motor-Operated Valve
- e: Mov[i]e
Mov s: Móvil
- s: Movilización
- s: Movimiento
Mov Disord e: Movement Disorders
MOVE e: Mobility Opportunities Via Education
MOVEET e: Ministers of Vocational Education, Employment and Training
MoVEM e: Motor Vehicle Emissions Model
MOVPE e: Metal-Organic Vapor-Phase Epitaxy
MOVREP e: Movement Reporting
MOVS e: Move String
MOWORM e: Magneto-Optical Write Once Read Many
MOX e: Mixed Oxide
MOXE e: All sky monitor on Spectrum-X-Gamma
- e: Monitoring X-ray Experiment
Moy f: Moyen[s]
MOZ d: Mittlere Ortszeit
- d: Motor Oktan Zahl
MOZAIC e: Measurement of Ozone on Airbus In-Service Aircraft

MP d: Meßpunkt[e]
- d: Musikpädagogik
- e: Main Process
- e: Maintenance Planning
- e: Major Procurement
- e: Major Project
- e: Management Policy
- e: Management Process
- e: Manifold Pressure
- e: Manipulator Power
- e: Manpower and Personnel
- e: Manpower Plan
- e: Manufacturing-use Product
- e: Massively Parallel [processing]
- e: Material Process
- e: Melamine Phenolic
- e: Melchizedek Priesthood
- e: Melting Point
- e: Message Processing
- e: Metal Particle
- e: Metropolitan Police
- e: Micro Processor
- e: Midplane
- e: Military Police[man]
- e: Missed Packet
- e: Mixed Projection
- e: Mobilization Plan
- e: Modular Plug
- e: Mounting Plate
- e: Multi Point
- e: Multilink PPP
- e: Multi[link] Protocol
- e: Multiple Processors
- e: Multiprocessing
- e: Northern Mariana Islands
- f: Maître Primaire
- f: Majorité Parlementaire
- f: Maladie Professionnelle
- f: Mandat-Poste ordinaire
- f: Marché Parallèle
- f: Mariage Putatif
- f: Médecin Principal
- f: Médecine Préventive
- f: Menus Plaisirs
- f: Mesures Préventives
- f: Ministère Public
- f: Ministre Plénipotentiaire
- f: Mise à Prix
- f: Moins-Perçu
- f: Mois Prochain
- f: Monnaie-Papier
- f: Montant d'une Pension
- f: Mouvement Politique
- f: Mouvement Populaire
- f: Mouvement Progressiste
- f: Municipalité parisienne
- f: Mutation de Propriété
- N: Meldeplass
- N: Militærpoliti
- s: Ministerio Público
- s: Mortero Pesado
Mp f: [en] Mains propres
mp. N: maskinpistol
M.P. e: Member of Parliament
M&P e: Maintenance and Process engineering
- e: Materials and Producibility
m/P s: mi Pagaré
MP2 f: Marié depuis Plus de 2 ans
MPA d: Staatliche Materialprüfungsanstalt
- e: Main Political Administration
- e: Man Powered Aircraft
- e: Marine Protected Area
- e: Massively Parallel Architecture
- e: Master of Public Administration
- e: Minimum Perceptable Action
- e: Multiple-use Planning Area
- f: Magasins Populaires d'Arrondissement
- f: Mouvement Panafricain Anticommuniste
- f: Mouvement Populaire Africain
- f: Mouvement Populaire de l'Azawad
MPa e: Megapascal
m.p.a. r: med personlig ansvarighet
MPAC e: Materials and Parts Availability Control
MPACC e: Master of Professional Accountancy
MP/A/N/AF/M e: Military Personnel, Army/Navy/Air Force/USMC
MPAP e: Mar del Plata Action Plan
- f: Plan d'action de Mar del Plata
- s: Plan de Acción de Mar del Plata
MPB e: Monobutyl Phosphate
MP-befal N: Militærpolitiets befal
MPBRS e: Maintenance Production/ Backlog Reporting System
MPBX e: Multiple Position Borehole Extensometer
MPC e: Main Process Cell
- e: Massively Parallel Computer
- e: Master of Pastoral Counseling
- e: Metal Properties Council
- e: Microsoft Certified Professional
- e: Minor Planets Circular
- e: Missile Practice Camp
- e: Mobile Processing Center
- e: Monitoring Proportional Counter
- e: MPOA [Multi-Protocol over ATM [Asynchronous Transfer Mode]] Client
- e: Multimedia PC [Personal Computer] [marketing Council]
- e: Multiparty Conference
- e: Multipath Channel
- e: Multiprocess Communications
- e: Multmedia Personal Computer
- p: Movimento do Povo de Cabinda
- s: Medidas de Protección
MPCA e: Microbial Pest Control Agent
- e: Multi-Protocol Communications Adapter
MPCAG e: Military Parts Control Advisory Group
MPCASS e: Modernization Parts Control Automated Support System
MPCC e: Microprocessor Common Control
MPCFM f: Ministério de Portos e Caminhos de Ferro
MPCH e: Main Parallel Channel
MPCI f: Mouvement Progressiste de la Côte d'Ivoire
MPCMC e: Multimedia PC Marketing Council
MP/CS e: Manufacturing Planning and Control System
MPCS e: Mission Planning and Control Station
MPCVIL s: Mando de la Protección Civil
MPD e: Metropolitan Police Department
- e: Multi-Purpose Display
- e: Multiple Personality Disorder
- p: Movimento Para a Democracia
MPDD f: Mouvement Populaire pour la Démocratie et le Développement
MPD [IMPD] e: Institute for Multi-Party Democracy
MPDM p: Movimento para Paz e Democracia em Moçambique
MPDR e: Microwave Plasma Deposition Reactor
MPDT e: Message Processing Data Terminal
- e: Microwave Polarization Difference Temperature
MPE e: Maximum Permissible Exposition
- e: Ministry of Planning and Environment
- e: Mission to Planet Earth
- e: Most Probable Earthquake
- e: Multiple Programming Executive
Mpe a: Mælkeproduktionsenhed
MPEC e: Multi-Purpose Experiment Canister
MPEG e: Motion Picture[s] Expert[s] Group[-compression standard]
- e: Moving Picture[s] Expert[s] Group
MPEG 1 e: MPEG [Motion picture Experts Group] video compression standard
MPEG 2 e: Standard for MPEG that supports interlaced video, multiple-resolution scaling and bit rates of 3-15 Mbps
MPEG 4 e: MPEG standard for very low bit rate images
MPER e: Missing Plant and Equipment Report
MPersD r: Marinens Personaldelegation
MPERSS e: Marine Pollution Emergency Response Support System
- s: Sistema de Apoyo a la Respuesta de Emergencia en caso de Contaminación Marina
MPES e: Management Planning and Evaluation Staff
MPESS e: Multiple Purpose Experiment Support Structure
MPF e: Mapping Field
- e: Marematlou Freedom Party
- e: Maximum Probable Fire
- e: Maximum Probably Flood
- e: Military Personnel Flight
- e: Mission Planning Facility
- e: Mission Planning Forecast
- e: Momentary Power Failure
MPG d: Max Planck Gesellschaft zur Förderung der Wissenschaften e.V.
- d: Medizinprodukte-Gesetz vom 2.8.1994
- e: Mobile Protected Gun
- e: Motion Picture Guide
- f: Milice Populaire Guinéenne
MPGA d: Archiv der Max Planck-Gesellschaft
MPG Algorithm e: Marr-Poggio-Grinson Algorithm
MPH e: Master of Public Health
mph e: miles per hour

MPI d: Max Planck Institut für Meteorologie, Hamburg, Germany
- e: Machine Performance Integration
- e: Maintenance Programs Integration
- e: Management Process Integrators
- e: Marine Pollution Incident
- e: Max-Planck Institute for Meteorology
- e: Maximum Permitted Intake
- e: Message Passing Interface
- e: Multi-Protocol Interchange
- e: Multiprecision Integer
- e: Multiprocessor Interface
- e: Multiprocessor Interconnect Bus
- f: Moyennes et Petites Industries
- f: Pollution accidentelle de la mer
- r: Militärpsykologiska Institutet
- s: Incidente de Contaminación Marina

MPIE d: Max-Planck-Institut für Extraterrestriche Physik
MPIF e: Message Passing Interface Forum
- e: Multiprocessor Interface

MPI/Festk d: Max-Planck-Institut für Festkörperforschung, Max-Planck-Institut für Metallforschung, Institut für Physik Zentralbibliothek Büsnau
MPI-HH d: Max-Planck-Institut für Meteorologie, Hamburg
MPIK d: Max-Planck-Institut für Kernphysik
MPIM d: Max-Planck-Institut für Meteorologie
MPIP e: Maintenance Posture Improve Program
MPIPh d: Max Planck-Institut für Physik in München
MPIR e: Multispectral Pushbroom Imaging Radiometer
Mpist N: Maskinpistol
MPI/Werkst de: Max-Planck-Institut für Metallforschung, Institut für Werkstoff-wissenschaft-Bibliothek
MPJ f: Mouvement Panafricain de la Jeunesse
MPK d: Ministerpräsident[innen]-konferenz
- R: Maliy Protivolodochniy Korabl

MPL e: Mandatory Parts List
- e: Mechanical Properties Loop
- e: Member of the Provincial Legislature
- e: Micropulse Lidar
- f: Mouvement Populaire de Libération
- N: Møbelprodusentenes Landsforening

Mpl f: Montpellier
mpl. N: meldeplass
MPLA p: Movimento Popular de Libertação de Angola
MPlag N: Militærpolitilag
MPLC p: Movimento Popular de Libertação da Cabinda
MPLD f: Mouvement Populaire pour la Liberté et le Développement
MPLM e: Mini-Pressurized Logistics Module
MPLR e: Movement for the Liberation of Reunion
- f: Mouvement pour la Libération de la Réunion

MPLS e: Multi Protocol Label Switching
Mpls. e: Minneapolis
MPLT f: Mouvement Populaire pour la Libération du Tchad
MPLTT f: Mouvement Populaire pour la Libération Totale du Tchad
MPM e: Mahorais Peoples Movement
- e: Mauritian Patriotic Movement
- e: Medical Planning Module
- e: Message Passing Library
- e: Movement Planning Module
- e: Multiprocessor Module
- f: Mouvement Patriotique Mauricien
- f: Mouvement Populaire Mahorais

MP/M e: Multiprogramming control Program for Microprocessors
MPMD e: Multiple Processor/Multiple Data
MPMIS e: Military Personnel Management Information System
MPMP e: Master Program Management Plan
MPMR e: Multifrequency Passive Microwave Radiometer
MPMS e: Missile Performance Measurement System
MPM-TIRF e: Multifrequency Phase and Modulation Total Internal Reflection Fluorescence
MPN e: Maximum Possible Number
- e: Mobility Position Number
- e: Mouvement Patriotique National

MPNA e: Working Group on Pollution Baseline and Monitoring Studies in the North Atlantic
- f: Groupe de travail sur les études de base et la surveillance continue de la pollution dans l'Atlantique Nord
- s: Grupo de Trabajo sobre Estudios Básicos y Vigilancia de la Contaminación en el Atlántico Norte

MPN-XX e: Mobile Precision Number-XX
- e: Next Generation Mobile Ground Radar

MPO e: Manufacturers Point of Origin
- e: Manufacturing Production Order
- e: Memorandum Purchase Order
- e: Metropolitan Planning Organization
- e: Mission Priority Order
- f: Ministère des Pêches et des Océans
- s: Ministerio de Pesquerías y Océanos

MPOA e: Multi-Protocol Over ATM [Asynchronous-Transfer-Mode]
MP-OES e: Multi-Point Optical Emission Spectroscopy
Mpol r: Marinpolis
Mpolgrp r: Marinpolisgrupp
Mpolplut r: Marinpolispluton
MPOW e: Multiple Purpose Operator Workstation
MPP e: Market Promotion Program
- e: Massive Paralell Processing
- e: Massively Parallel Processing
- e: Massively Parallel Processor[s]
- e: Medical Planning Program
- e: Member of [Canadian] Provincial Parliament
- e: Merit Promotion Plan
- e: Message Posting Program
- e: Message Posting Protocol
- e: Message Processing Program
- e: Message Processing Protocol
- e: Microcom network Protocol
- e: Microprocessor Pascal
- e: Monthly Payment Plan
- e: Multiphase Printing
- e: Multiple Parallel Processing

MPPA e: Michigan Press Photographers Association
MPPCS e: Manpower and Personnel Contingency Support System
MPPD e: Multi-Purpose Peripheral Device
MPPEAS e: Regional Programme for the Prevention and Management of Marine Pollution in the East Asian Seas
- f: Programme régional pour la prévention et la gestion de la pollution marine dans les mers d'Asie orientale
- s: Programa Regional de Prevención y Gestión de la Contaminación Marina en los Mares de Asia Oriental

MPPI p: Movimento Para a Paz Imediata
MPPL e: Mobilization Procurement Planning List
MPPSC e: Multipurpose Process Storage Complex
MPPT e: Message Preparation Processing Task
MPQP e: Multi-Protocol Quad Port
MPR a: Minipopulationsregisteret
- e: keyboard standard
- e: Management Policies and Requirements
- e: Mars Pressurized Rover
- e: Master Personnel Records
- e: Material Procurement Rate
- e: Medium Power Radar
- e: Minnesota Public Radio
- e: Monthly Program Review
- e: Multi-Protocol Router
- e: Multipart Repeater
- e: Multiple Provider Router
- e: National Board for Metrology and Testing
- f: Mouvement Populaire de la Révolution
- r: Mät- och Provningsrådet

MPR-2 d: Strahlenschutz-Norm für Bildschirme
MPRES e: Modular Plasma Reactor Simulator
MPRI e: Multi-state Professional Responsibility Exam
MPRM e: Marine Pollution Research and Monitoring
- f: Etude et surveillance continue de la pollution marine
- s: Investigación y Vigilancia de la Contaminación en el Medio Marino

MPRR e: Mobility Personnel Resource Roster

MPRS e: Mission Planning and Rehearsal System
MPRSA e: Marine Protection, Research and Sanctuaries Act
- e: Marine Protection, Research and Sanctuaries Act
MPRT e: Multipurpose Rail Transporter
MPRTM e: Master of Parks, Recreation and Tourism Management
MPS e: Management of Parts Shortages
- e: Manpower Planning System
- e: Maritime Prepositioning Ships
- e: Master Production Schedule
- e: Master Program Schedule
- e: Medical Products Sales
- e: Medical Protection Society
- e: Medium-Powered Satellite
- e: Megabytes Per Second
- e: Message Processing Subsystem
- e: Meteorological Profiler System
- e: Microwave Pressure Sounder
- e: Mission Planning and Scheduling
- e: Mission Planning System
- e: Mission Processing Schedule
- e: MPOA [Multi-Protocol over ATM [Asynchronous Transfer Mode]] Server
- e: Multipage Signal
- e: Multiple Protective Structure
- e: Multiprocessor Specification
- e: Multipurpose Sampler
- f: Mouvement Patriotique de Salut
- N: Sjøforsvarets Militærpolitiets skole
MPS2 e: Region 2 Merit Pay System
MPSC e: Multi-Purpose Processing and Storage Complex
MPSK e: Multilevel Phase Shift Keying
MP-skole N: Militærpolitiets skole
MPSØ N: Militærpolitiets skole- og øvingsavdeling
MPSR e: Material/Parts Substitution Request
- e: Military Petroleum Stocks Report
MPSS f: Mouvement Populaire pour la Sauvegarde du Sénégal
MPST d: Mikroprogrammsteuerwerk
MPT a: Militærpsykologisk Tjeneste
- e: Magnetic Particle Testing
- e: Manpower, Personnel and Training
- e: Maryland Public Broadcasting
- e: Master of Physical Therapy
- e: Message Processing Time
- e: Minimum Pressure-Temperature
- e: Ministry of Post[s] and Telecommunications
- e: Missed Packet Tally Counter register
- e: MultiPort Transceiver
- e: Multipurpose Transfer
- f: Mouvement Populaire Togolais
- f: Mouvement pour le Parti du Travail
- s: Movimiento Pro-Tierra
MPT 1327 e: European hybrid digital trunked radio
MPTA e: Main Propulsion Test Article
- e: Multi Protocol Transport Architecture
MPTDS e: MPTER Model with Deposition and Settling of Pollutants
MPTER e: Multiple Point Source Model with Terrain
MPTF e: Multiple Protocol Transport Feature
MPTM f: Mouvement Travailliste et Patriotique Malgache
MPTN e: Multi[ple] Protocol Transport Network[ing]
MPTP e: Micro Payment Transfer Protocol
MP-tropp N: Militærpolitiets tropp
MPTS e: Multi[ple] Protocol Transport Service[s]
MPTS2 e: Multi Protocol Transport Services/2
MPTT f: Ministère des Postes, Télégraphes et Téléphones
MPU e: Marine Pollution Research and Monitoring Unit
- e: Math Processing Unit
- e: Message Passing Unit
- e: Message Processing Unit
- e: Microprocessor Unit
- e: MIDI Processing Unit
- e: Mobility Processing Unit
- f: Unité d'étude et de surveillance continue de la pollution marine
- s: Unidad de Investigación y Vigilancia de la Contaminación del Mar
MP-utvalg N: Militærpolitiets utvalg
MPV e: Mars Piloted Vehicle
- e: Methane-Powered Vehicle
MPW e: Macintosh Programmers Workbench
- e: Macintosh Programmers Workshop
- f: Mouvement Populaire Wallon
MPWC e: Multiprocess Wet Cleaning
MPX e: Multiplex
- e: Multiprocessor Extension
Mp 1x1,y1,z1 e: Meßpunkt auf OG-Probeplättchen
Mp2x2,y2,z2 e: Meßpunkt auf OG-Probeplättchen
Mp3x3,y3,z3 e: Meßpunkt auf OG-Probeplättchen
Mp4x4,y4,z4 e: Meßpunkt auf OG-Probeplättchen
mpy e: mils per year
MQ e: Martinique [French]
- e: Message Queuing
- f: Médecin de Quartier
- s: Mina Química
Mq f: Manque
MQC or STJD 11 e: The Madrid Qumran Congress: Proceedings of the International Congress on the Dead Sea Scrolls, Madrid 18-21 March, 1991
MQE e: Managed Query Environment
MQH e: Memory Queue Handler
MQI e: Message Queue[ing] Interface
- e: Messaging and Queuing Interface
MQS e: Mail Questionnaire Survey
MQT e: Mission Qualification Training
MQWR e: Multiple Quantum Wire
MR d: Beginn der Milchreife
- e: Magnetic Resonance
- e: Magneto-Resistance
- e: Magneto-Resistive
- e: Magneto-Resistivity
- e: Management Reserve
- e: Materials Research
- e: Mauretania
- e: Mental Retardation
- e: Methane Recovery
- e: Microwave Radiometer
- e: Miniatures Rules
- e: Mission Ready
- e: Modem Ready
- e: Modified Read
- e: Movement Region
- f: Maison de Rapport
- f: Maison de Repos
- f: Maison de Retraite
- f: Maître des Requêtes
- f: Majorité Relative
- f: Maximum des Ressources
- r: Manöverrum
- r: Motoriserad Regemente
- r: Motorrum
- s: Marca Registrada
mR e: milliroentgen
Mr f: Monsieur
M&R e: Maintenance and Repair
- e: Meter and Relay
- e: Mines and Resources
m-r s: marxista-revolucionario
MR-13 s: Movimiento Revolucionario 13 de Noviembre
MRA d: Motorrestwärmeausnutzung
- e: Magnetic Resonance Angiography
- e: Market Research Africa
- e: Minimum Retirement Age
- e: Mission Ready and Available
- f: Mission pour la Réforme Administrative
MRAA e: Mental Retardation Association of America
MRAAM e: Medium-Range Air-to-Air Missile
MRAC f: Musée Royal de l'Afrique Centrale
mrad e: millirad
- e: millirad [1 x 10-3 rad]
MRAI N: Marinens Radarinspeksjon
MRA&L e: Manpower Accounting System
- e: Manpower, Reserve Affairs and Logistics
MRAL e: Mobile Radionuclide Analysis Laboratory
MRAM e: MacPhail Radial Arm Maze
- e: Magnetoresistive Random Access Memory
MRAO e: Mullard Radio Astronomy Observatory
MRAP f: Mouvement contre el Racisme, l'Antisémitisme et pour la Paix
MRAR e: Material Receiving Acceptance Report
MRASM e: Medium Air-to-Surface Missile
- e: Medium Range ASM
MRAX f: Mouvement contre le Racisme, l'Antisémitisme et la Xénophobie

MRB e: Management Reserve Budget
- e: Management Review Board
- e: Materials Review Board
- e: Method Request Broker
- e: Multi-Role Bomber

Mrbb r: Motorrobotbåt
MRBF e: Main Rounds Before Failure
MRBM e: Medium Range Ballistic Missile
MRC e: Machine-Readable Code
- e: Maintenance Requirement Cards
- e: Marine Research Corporation
- e: Massachusetts Rehabilitation Commission
- e: Material[s] Research Center
- e: Media Resources Center
- e: Medical Research Council
- e: Medical Research Council of Canada
- e: Middle Roll Change
- e: Mission Research Corporation
- e: Monthly Recurring Charge
- f: Matrice du Rôle des Contributions

Mrc f: Marocain
MRCA e: Multi[ple] Role Combat Aircraft
MRCB f: Musée Royal du Congo Belge
MRCC e: Maritime Rescue Coordination Center
- e: Movement Reports Control Center

MRC/CRM e: Medical Research Council of Canada/Conseil de Recherches Médicales du Canada
MRCD e: Movement for Responsible Coastal Development
MRCE f: Maître des Requêtes au Conseil d'Etat
MRCF e: Microsoft Realtime Compression Format
MRCI e: Microsoft Realtime Compression Interface
- e: Multireference Configuration Interaction

MRCMC e: Murrumbidgee Regional Catchment Management Committee
MRCMCCR e: Medical Responsibility Center Manager Cost Center Report
MRCO e: Manufacturing Research Corporation of Ontario
MRCP e: Master of Regional and Community Planning
MRCS e: Multi-Rate Circuit Switching
MRD d: Milliarde
- e: Marketing Requirements Document
- e: Material Requirement Document
- e: Materials Required Data
- e: Mesoscale Research Division
- e: Mesozoic Radiolaria Database
- e: Minimal Residual Disease

MRDA e: Mission Requirements Definition and Analysis
MRDB e: Materiel Returns Data Base
MRDC e: Microelectronic Research and Development Center
MRDFS e: Manpack Radio Direction Finding System
MRDS e: Maintenance Requirements Data Systems
- e: Man-carried Radiological Data System

MRDT f: Mouvement Démocratique de la Rénovation Tchadienne
MRE e: Magnetoresistive Element
- e: Meal, Ready-to-Eat
- e: Morally Reprehensible Elite
- e: Multiple cytokine and second-messenger Responsive Element

M-region e: Maunder region
mrem e: millirem
Mrep e: Millirep
MrepL r: Maskinreparationsledare
MRF e: Maintenance Reset Function
- e: Markov Random Fields
- e: Materials Recovery Facility
- e: Medium Range Forecast
- e: Medium-Range Forecast
- e: Motorcycle Riders Foundation
- e: Multi-Role Fighter

MRFL e: Mandatory File and Record Locking
MRFSS e: Marine Recreational Fishing Statistics Survey
MRG e: Management Review Group
- e: Movement Requirements Generator
- f: Mouvement des Radicaux de Gauche

MRH e: Monitoring River Health
- s: Manual de Referencia del HOMS

MRHI e: Monitoring River Health Initiative
MRHT e: Marked Rated Holding Time
MRI e: Magnetic Resonance Imaging
- e: Management Recruiters International
- e: Manual Revision Instruction
- e: Master Record Index
- e: Material Receiving Instruction
- e: Materials Research Institute
- e: Medical Records Institute
- e: Meteorology Research, Inc.

MRID e: Master Record Identification number
MRIF e: Materials Research Instrument Facility
MRIPS e: Multi-modality Radiological Image Processing System
MRIR e: Medium Resolution Infra-red Radiometer
- f: Radiomètre Infrarouge à Résolution Moyenne
- s: Radiómetro de Infrarrojo de Resolución Media

MRIT e: Mobile Remote Intelligence Terminal
MRJ e: Macintosh Runtime Java
MRK R: Maliy Raketniy Korabl
mrk. a: mærket
MRL e: Main Ring Length
- e: Marine Research Laboratory
- e: Master Requirements List
- e: Materials Reference Library
- e: Maximum-Residue Limit
- e: Meteorological Research Laboratory
- e: Minimal (or Minimum) Risk Level
- f: Ministère de la Reconstruction et du Logement
- f: Mouvement Républicain de la Libération

MRLC e: Interagency Multi-Resolution Land Cover Characterization
MRLDC e: Milton Roy Lab Data Control
MRM e: Magnetometric Resistivity Method
- e: Maintenance Resource Management
- e: Management Review Meeting
- e: Materiel Readiness and Modernization
- e: Miscellaneous Radioactive Material
- e: Most Recently-used Master
- e: Motif Resource Manager
- e: Movement for the Redemption of Muslims in Liberia

MRMC e: Medical Research Modernization Committee
MRMF e: Module Reduction and Monitoring Facility
mRNA e: messenger RNA [Ribonucleic Acid]
MRND f: Mouvement pour la Révolution Nationale Démocratique
- f: Mouvement Révolutionnaire National pour le Développement

MRNET e: Minnesota Regional Network
MRNI f: Mouvement de la Résistance Nationale Iranienne
MRO e: Maintenance, Repair and Operational
- e: Material Release Order
- e: Medical Regulating Office
- e: Medical Review Officer
- e: Memory Read-Out
- e: Movement Report Office
- e: Multi-Region Operation

MROC e: Multiple-Command Required Operational Capability
MROCC e: Medical Review Officer Certification Council
Mröj- r: Minröjnings-
MröjA r: Minröjningsavdelning
Mröjdiv r: Minröjningsdivision
Mröjflj r: Minröjningsflottilj
Mröjlag r: Minröjningslag
MROS e: Material Requirements and Operating Standards
- e: MIDI Realtime Operating System

MRP e: Management Requirements and Practices
- e: Management Requirements and Procedures
- e: Manpower Requirements and Personnel
- e: Manufacturing Requirements Planning
- e: Manufacturing Resource Planning
- e: Material Resource Planning
- e: Material[s] Requirement[s] Planning
- e: Mid Range Plan
- e: Militarized Reconfigurable Processor
- e: Mississippi River Plume
- e: Mobile Repair Party
- e: Multi-Roller Press
- e: Multiple Rate Processor
- f: Mouvement pour la Réconciliation et la Paix
- f: Mouvement Républicain Populaire
- f: Mouvement Révolutionnaire du Peuple

MRP-II e: Manufacturing Resource Planning II
MRPL e: Main Ring Path Length
MRR e: Material Receiving Report
- e: Material Release Point
- e: Minimum Risk Route
- e: Mission Requirements Request
- f: Mouvement Républicain Réformiste

MRRD e: Marine Resources Research Division
MRRG d: Melderechtsrahmengesetz
MRRL e: Midlands Regional Research Laboratory
MRRM e: Medical Resupply Requirements Module
MRRR e: Mobility Requirements/Resource Roster
MRS e: Magnetic Resonance Spectroscopy
- e: Material Reject Stock
- e: Material Requirements Schedule
- e: Material Research Society Meeting
- e: Materials Research Society
- e: Media Recognition System
- e: Medical Readiness System
- e: Medical Resupply Set Model
- e: Microcontroller Readable Sector
- e: Miniature Rawinsonde System
- e: Minority Research Supplement
- e: Mission Radio System
- e: Moderator Recovery System
- e: Monitored Retrievable Storage [program]
- e: Multispectral Resource Sampler
- f: Manuel de Référence du SHOFM
- f: Mouvement Républicain Sénégalais
Mrs f: messieurs
mrs. a: marts
MRSA e: Materiel Readiness Support Activity
- e: Methicillin Resistant Staphylococcus Aureus
MRSEC e: Materials Research Science and Engineering Center
MRSF e: Monitored Retrievable Storage Facility
MRS/IS e: Monitored Retrievable Storage/Interim Storage
MRSL e: Microwave Remote Sensing Laboratory
MRSP e: Mobility Readiness Spares Package
MRSQ e: Medical Reference Services Quarterly
MRSR e: Mars Rover [and] Sample Return
MRSRM e: Mars Rover and Sample Return Mission
MRSS e: Monitored Retrievable Surface Storage
MRT d: Meß- und Regelungstechnik
- e: Maintenance Recovery Team
- e: Management Review Team
- e: Maximum-Reading Thermometer
- e: Miniature Receive Terminal
- f: Ministère de la Recherche et de la Technologie
- r: Motorreparationstjänst för Totalförsvaret
MRTB e: Medical Radiation Technologists Board
MRTFB e: Major Range and Test Facility Base
M_2RTMS e: Multispectral Radiative Transfer Models
MRTS e: Modification Request Tracking System
MRTT e: Modular Record Traffic Terminal
MRU e: Mano River Union
- e: Maximum Receive Unit
- e: Military Radar Unit
- e: Most Recently Used
- f: Ministère de la Reconstruction et de l'Urbanisme
MRV e: Machine Readable Visa
- e: Multiple Re-entry Vehicle
MRW e: Maximum Ramp Weight
MS a: Mellemfolkeligt Samvirke
- a: Minestryger
- d: Massenspektrometer
- d: Massenspektrometrie
- e: Brody Medical Science Building
- e: Mail Stop
- e: Maintenance State
- e: Management Service[s]
- e: Management Staff
- e: Management System[s]
- e: Manipulator Signal
- e: Manual Switch
- e: Manuscript
- e: Marine Stratus
- e: Mass Spectrometer (or Spectrometry)
- e: Mass Spectroscopy
- e: Material Supplier
- e: Matrix Spike
- e: McIntire-Stennis
- e: Medical School
- e: Member State
- e: Meta Signaling
- e: Metal-Semiconductor
- e: Microsecond
- e: Microsoft Corp.[, Inc.]
- e: Microwave Scanner
- e: Middle School
- e: Military Science
- e: Military Specifications
- e: Millisecond[s]
- e: Mission Support
- e: Mobile Station
- e: Mobile Systems
- e: Mobilization Station
- e: Moisture Separator
- e: Monitoring the Stratosphere
- e: Montserrat
- e: Mössbauer Spectroscopy
- e: Multilateral Staff
- e: Multiple Sclerosis
- e: Multiple Sclerosis Society of Canada
- f: Maison de Santé
- f: Maison Sinistrée
- f: Majoration Semestrielle
- f: Maladie Sporadique
- f: Médecin Spécialiste
- f: Médico-Social
- f: Mission Scientifique
- f: Mission Spéciale
- f: Monnaie Scripturale
- f: Morte-Saison
- f: Scanneur en hyperfréquence
- l: Memoriae Sacrum
- N: Målskjema
- r: [Arméns] Motorskola
- r: Marinstab
- s: Escáner de Microondas
- s: Mando Superior
Ms f: Manuscrit
- f: Même sens
- N: Megasykle
- N: Motorsykkel
- r: Maskinsikte
- s: Misión
ms. a: Manus[kript]
- e: maschinenschriftlich
M.S. a: Motorskib
- r: Medicinalstyrelsen
M&S e: Materials and services
- e: Modeling and Simulation
M-S e: McIntire-Stennis
M/S e: Milestone
- e: Modeling and Simulation
- N: Motorskip
M3S e: MAJCOM Manpower Management System
- e: Marine Corps Standard Supply System
MSA e: Maintenance Support Activities
- e: Major Statistical Areas
- e: Major System Acquisition
- e: Management Sciences of America
- e: Management System
- e: Management System Audits
- e: Maritime Safety Agency
- e: Maryland State Archives
- e: Master of Science in Accounting
- e: Material Surveillance Assemblies
- e: Methane Sulfonic Acid
- e: Methane Sulphonic Acid
- e: Methanesulfonate (or -sulphonate)
- e: Method of Standard Addition
- e: Metropolitan Statistical Area[s]
- e: Microscopy Society of America
- e: Minnesota Software Association
- e: Mobile Subscriber Access
- e: Muslim Students Association
- e: Mutual Security Agency
- f: Maître ès Sciences Agricoles
MSAA e: Master Safeguards and Security Agreement
MSACM e: Microsoft Audio Compression Manager
MS/Adm r: Chef för Marinstabadministration
MSAEP e: Mediterranean Scientific Association for Environmental Protection
- f: Association Scientifique Méditerranéenne de Protection de l'Environnement
- s: Asociación Científica del Mediterráneo para la Protección del Medio Ambiente
MSAGE e: Master of Science in Agricultural Engineering
MSAIS e: Multiplexer Section Alarm Indication Signal
MSAM e: Multikeyed indexed Sequential file Access Method
MSAP e: Management Service Access Point
- e: Management Skills Assessment Program
- e: Military Security Assistance Program
- e: Minislotted Alternating Priorities
MSA[R] e: Mine Safety Appliance
MSAR e: Manpower Standards Application Routine
- e: Mine Safety Appliance Report
MSARC e: Marine Systems Acquisition Review Council

MSAS e: Manpower Standards Application System
MSAT e: Master of Accounting and Taxation
- e: Mobile Telecommunications Satellite
MSAU e: Multi-Station Access Unit
MSAV e: Microsoft Anti Virus
MSAW e: Minimum Safe Altitude Warning
MSB e: Main Support Battalion
- e: Medical School Building
- e: Medical Sciences Bulletin
- e: Monetary Stabilization Bond
- e: Most Significant Bit
- e: Most Significant Byte
MSBE e: Molten Salt Breeder Experiment
MSBF e: Mean Swaps Between Failures
MSBI N: Marinens Sambandsinspeksjon
ms bsf e: milliseconds below seafloor
MSBVW e: MagnetoStatic Backward Volume Wave[s]
MSC e: Major Subordinate Command
- e: Management Science
- e: Management Service Center
- e: Management Service Component
- e: Management Systems Council
- e: Marie Sklodowska Curie fund
- e: Marine Sciences Council
- e: Maritime Safety Committee
- e: Master of Science
- e: Material Support Concept
- e: Media Stimulated Calling
- e: Medical Service Corps
- e: Message Sequence Charts
- e: Meteorological Systhesising Centre
- e: Microsoft C
- e: Midwest Science Center
- e: Mission Support Committee
- e: Mobile Satellite Communication
- e: Mobile Switching Center
- e: Motorcycle Safety Center
- e: Muslim Supreme Council
- f: Mission Scientifique et Technique
M SC e: Medical Science
M.Sc e: Master of Science
MSCA e: Modified Simulated Core Assembly
- f: Mouvement Socialiste Centrafricain
MSCD e: Military Support to Civil Defense
MSCDEX e: Microsoft CD [Compact Disk] Extension[s]
MSCE e: Master of Science on Civil Engineering
MSCI e: Molten Steel Coolant Interaction
MSC-IMIS e: Military Sealift Command Integrated Management Information System
MSCIS e: Master of Science in Computer Information Science
MSCIT e: Medium Speed Communications Interface Terminal
MSCM e: Mobile Surface Contamination Monitor
- e: Multiple Slots on Continuation Mechanism
MSCN e: Master of Science in Clinical Nutrition
MSCP e: Mass Storage Control Protocol
- e: Meteorological and Climate Services Project
- s: Perfiles Nacionales en las Ciencias del Mar
MSCPs e: Marine Science Country Profiles
MSCS e: Management Scheduling Control System
- e: Manned Spaceflight Control Squadron
- e: Manual Staff Consultative Sub-Committee
- e: Multi Service Communications System
MSCT e: Marine Science Contents Tables
- f: Actualités des sciences de la mer
- s: Indices de Revistas sobre Ciencias Marinas
MSD e: Management Systems Designers
- e: Management Systems Division
- e: Marble Spleen Disease
- e: Mass Selective Detector
- e: Mass Spectral Detector
- e: Material Support Date
- e: Materials Science Division
- e: Metal Shipment Documentation
- e: Meteorological Services Department
- e: Microsoft System Diagnostics
- e: Modified Signed Digit
- e: Molten Salt Destruction
- e: Movement for Socialism and Democracy
- e: Movement of Social-Democrats
- e: Multisertdrum
- f: Mouvement des Socio-Démocrates
- f: Mouvement pour le Socialisme et la Démocratie
MSDB e: Mission Schedule Data Base
MSDC e: Mass Spectrometry Data Centre
MSDE e: Department of Education
MSDE-BSNT e: Department of Education-Business Services
MSDE-CA e: Department of Education-C and A
MSDE-DORS e: Department of Education-Rehabilitation
MSDE-LBRY e: Department of Education-Libraries
MSDE-PRIM e: Department of Education-Planning, Results
MSDE-SCO e: Department of Education-Community Outreach
MSDE-SIS e: Department of Education-Improvement
MSDN e: International Microbial Strain Data Network
- e: Macintosh Software Distribution Network
- e: Message-Switched Data Network
- e: Microsoft Developer[s] Network
- f: Réseau international de données sur les souches microbiennes
- s: Red Internacional de Datos sobre Cepas Microbianas
MS[-]DOS e: Microsoft DOS [Disk Operating System]
MSDP e: Maintenance Systems Development Program
MSDR e: Multiplexed Streaming Data Request
MSDS e: Manpower Standards Development System
- e: Marine Science and Development Shop
- e: Material Safety Data Sheet
- e: Material Safety Data Sheets
- e: Microsoft Developer Support
MSDT f: Mouvement Social pour la Démocratie au Tchad
MSE e: Master of Science in Engineering
- e: Mobile Subscriber Equipment
- e: Modernization Support Environment
- e: Molten-Salt Extraction
- e: Motional Stark Effect
Ms[ec] e: Millisecond
MSED e: Micro and Small Enterprise Development
- e: Mobile Subscriber Equipment Division
MSEE e: Major Source Enforcement Effort
m/sek. a: meter per sekund
MSEL e: Master Scenario Events List
MSEM e: Metrology Specific Equipment Model
- e: Mobile Subscriber Equipment
MSEMECH e: Master of Science in Engineering Mechanics
MSEN e: Multiple Sclerosis Education Network
MSENG e: Master of Science in Engineering
MSES e: Master of Environmental Studies
MSE/TPN e: Mobile Subscriber Equipment/Tactical Packet Network
MSF e: Managerial, Scientific and Financial
- e: Master of Science in Forestry
- e: MIDI Standard File
- e: Mission Support Facility
- e: Mobile Strike Force
- e: Motorcycle Safety Foundation
- e: Multiple Sclerosis Foundation
- e: Munitions Storage Facility
- f: Médecin[s] Sans Frontière[s]
- s: Médicos Sin Fronteras
MSFC e: [George C.] Marshall Space Flight Center
- e: Medical Students For Choice
MSFDC e: Microsoft and First Data Corporation
MSFP e: Managemaent Service Focal Point
MSFR e: Minimum Security Function[ality] Requirements
MSFS e: Maintenance Shop Floor System
MSFVW e: Magnetostatic Forward Volume Wave[s]
MSG e: Makerere Students Guild
- e: Management Steering Group
- e: Message [Internet]
- e: Meteosat Second Generation
- e: Modular Steam Generator
- e: Mycoses Study Group
- e: Program Message
MSGC e: Microstrip Gas Counters
Msg-ID e: Message-Identification
MSGID e: Message Identifier [Set]
MSGL e: Thousand square feet of Single Glueline
MSGT e: Multi-pillar Surrounding-Gate Transistor

MSH e: MAC Standard Header
- e: Main Steam Header
- e: Management Sciences for Health
- e: Marvel Super Heroes
- e: Melanocyte-Stimulating Hormone
- e: Minesweeper, coastal

MSHA e: Mine Safety and Health Administration

MSHJ e: Medical Staff Hospital Joint Venture

MSHP e: Main steam Header Pressure
- e: Master of Science in Health Physics

MSI e: Management System Indicator
- e: Manager Staff Instruction
- e: Manufacturing Support Item
- e: Maritime Safety Information
- e: Medium Scale Integration
- e: Mental Sciences Institute
- e: Metal-Semiconductor Interface
- e: Multi-Source Integration
- e: Multi-Spec Imaging
- e: Multimedia Service International GmbH
- e: Multispectral Imagery

MSIB e: Master of Science in International Business

MSIC e: Missile and Space Intelligence Center
- e: Mixed-Signal Integrated Circuit

MSIE e: Microsoft Internet Explorer

MSIN e: Mail Stop Identification Number

MS/Info r: Chef för Marinstabinformation

MSIO e: Master of Industrial Operations

MSIP e: Multi-Spectral Imagery Processor
- e: Multi{national[Stage[d] Improvement Program

MSIS e: Mass Spectrometer Incoherent Scatter
- e: Master of Computer-Based Information Systems
- e: Master of Science in Information Systems
- e: Model State Information System

MSISDN e: Mobile Station ISDN Number

M Sjv r: Minnesregler för Sjukvårdare

MSK e: Minimum Shift Keying
- e: Minimum Spares Kit

Msk N: Mannskap

MSKCC e: Memorial Sloan-Kettering Cancer Center

MSL e: Mail Services Log
- e: Map Specification Library
- e: Marine Sciences Laboratory
- e: Master Submittal List
- e: Maximum Segment Lifetime
- e: Mean Sea Level
- e: Member Specification Library
- e: Microsoft Software Library
- e: Microwave Landing System
- e: Mirrored Server Link
- e: Modify System Logging
- e: Motor Simulation Laboratory

Msl r: Motorslup

MSL-91 e: Multistrategy Learning workshop 1991

MSLD e: Mass Spectrometer Lead Detector
- e: Mass Spectromenter Leak Detector

MSLF e: Mountain States Legal Foundation

MSLOC e: Million Source Lines of Code

MSLP e: Medium Speed Line Printer
- e: Mixed Signal and Logic Product

MSM e: Magnetospheric Specification Model
- e: Malagasy Socialist Movement
- e: Master-Slave Manipulator
- e: Mauritius Socialist Movement
- e: Media Support Module
- e: Medical Sourcing Model
- e: MEECN System Model
- e: Mission Support Manager
- e: Mission Support Model
- e: Mode Select Mechanism
- e: Mouvement Social Muhutu
- e: Multispectral Mapper
- e: Museum Course
- f: Mouvement Socialiste Malgache
- f: Mouvement Socialiste Mauricien

MSMA e: Monosodium Methanearsonates

MSMINE e: Master of Mining Engineering

MSMLC e: Medium Speed Multiline Controller

MSMPD e: Metal-Semiconductor-Metal Photodetectors

MS/MS e: Mass Spectrometer/Mass Spectrometer
- e: tandem Mass Spectrometry

MSMS e: Master of Materials Science

MS/MSD e: Matrix Spike/Matrix Spike Duplicate

MSMU e: Moscow State Mining University

MSN e: Maintenance and Support Network
- e: Manhattan Street Network
- e: Manufacturing Sequence Number
- e: Master of Nursing
- e: Microsoft Network
- e: Monitoring cell Sequence Number
- e: Multiple Subscriber Number
- f: [Groupe] Matériaux et Structures Navals

MSNA e: Maine State Nurses Association

MSNF e: Multiple Systems Networking Facility

MSO e: Management Service Organization
- e: Mandatory Scramble Order
- e: Manufacturers Statement of Ownership
- e: Mees Solar Observatory
- e: Metarie Site Office
- e: Military Satellite Communications Systems Office
- e: Multiple System[s] Operator

MS&O e: Mission Systems and Operations directorate of the RADARSAT project

MSOA e: Mines Surface Officials Association

MSOCC e: Multi-Satellite Operations Control Center

MSOGS e: Molecular Sieve Oxygen Generating System

MSOH e: Multiplexer Section Overhead

Mson f: Maison

MSORP e: Michael Scott Oral Records Project Windhoek

MSOS d: Mitteilungen des Seminars für orientalische Sprachen

MSOW e: Modular Stand Off Weapon[s]

MSP d: Main-Spessart-Kreis in Karlstadt, Bayern
- e: Mach Sweep Programmer
- e: Manpower Support Plan
- e: Marketing and Sales Productivity
- e: Maryland State Police
- e: Mass Storage Pedestal
- e: Master Sampling Plan
- e: Materials and Structures Program
- e: Message Security Protocol
- e: Message Send Protocol
- e: Microsoft Paint
- e: Mission Support Plan
- e: Mobile Sediment Processor
- e: Motion-Sensing Platform
- e: Multiplexer Section Protection
- f: Ministère de la Santé Publique
- f: Ministre Sans Portefeuille
- f: Mouvement pour le Socialisme par la Participation
- r: Moderata Samlingspartiet

Msp r: Minspärrning

MSPA e: Mauritius Sugar Producers Association
- e: Migrant and Seasonal worker Protection Act

MSPAS s: Ministerio de Salud Pública y Asistencia Social

MSPB e: Merit System[s] Protection Board

MSPE e: Mercenaries, Spies and Private Eyes

MSPF e: Maritime Special Purpose Force

MSPP f: Ministère de la Santé Publique et de la Population

MSPS e: Million Samples Per Second
- e: Mobilization Stationing and Planning System

MSPTA e: Mission Schedule Period Totals Array

Mspto r: Minspärrtrupp

MSR d: Meß-, Steuer- und Regelungssysteme
- e: Individual Motor Starter
- e: Machine Stress Rated
- e: Management Summary Report
- e: Marine Scientific Research
- e: Medical Software Reviews
- e: Message Service Routine
- e: Microwave Scanning Radiometer
- e: Microwave Sounding Radiometer
- e: Mission Status Report
- e: Mission Support Requirements
- e: Model Specific Register
- e: Molten Salt Reactor
- e: Monthly Status Review
- e: Movement for Salvation and Reconstruction
- e: Multitrack Serpentine Recording
- f: Mouvement pour le Salut et la Reconstruction
- f: Recherche Scientifique Marine
- s: Investigación Científica Marina

MSR. e: Magnetic Stripe Reader

MSRB e: Managerial Selection Review Board

MSRC e: Marine Sciences Research Center
- e: Molecular Science Research Center

MSRD e: Marine Services Research Division
MSRDP e: Medical Service Research and Development Plan
MSRDPS e: Missile Site Radar Data Processing System
MSRE e: Molten Salt Reactor Experiment
MSRL e: Molecular Sciences Research Center
MSRM e: Mixture and Systemic Toxicant Risk Model
MSRN e: Mobile Station Roaming Number
MSRP e: Manufacturer Suggested Retail Price
MSRPS e: State Retirement and Pension System
Msrr r: Minspaningsradar
MSRT e: Mobile Subscriber Radio[-telephone] Terminal
MSS d: Multispektraler Scanner
- e: Maintenance Support System
- e: MAN Switching System
- e: Management Support Staff
- e: Management Support System
- e: Management Systems Staff
- e: Manufacturing Standardization Society
- e: Manuscripts
- e: Mass Storage Subsystem
- e: Maximum Segment Size
- e: Medical Social Services
- e: Medical Support Services
- e: MEECN System Simulator
- e: Message Support Subsystem
- e: Minority Satellite Supplement
- e: Mission Support System
- e: Mobile Satellite Systems
- e: Mobile Servicing System
- e: Mobility Support Squadron
- e: Mobilization Stations
- e: Modify System State
- e: Multiple Satellite System
- e: Multiprotocol Switched Services
- e: Multispectral Scanner [Landsat]
- e: Muslim Students Society
- s: Escáner Multiespectral
Mss f: Manuscrits
MSSA e: Master Safeguards and Security Agreement
MSSC e: MAJCOM Spare Support List
MSSD e: Model Secondary School for the Deaf
MSSG e: MAU Service Support Group
MSSM e: Minimal Supersymmetric [extension of the] Standard Model
MSSQ e: Mission Support Squadron
MSSR e: Milestone Schedule and Status Report
- e: Moldavian Soviet Socialist Republic
MSSs e: Metropolitan Substructures
MSSS e: Multisatellite Support System
- f: Ministère de la Santé et des Services Sociaux
MSSTM e: Military Space Systems Technology Model
MSSW e: Mission Support Software
MST d: Mikrosystemtechnik
- e: Material Services Team
- e: Materials Science and Technology
- e: Mathematical Statistics
- e: Mesosphere-Stratosphere-Troposphere
- e: Microsystem Technology
- e: Ministry of Environment and Science
- e: Mission Support Team
- e: Mountain Standard Time
- e: Multi-Stage Tapped
- e: Multisensor Track
- e: Multislotted Token
- f: Maladie Sexuellement Transmissible
- f: Mission Scientifique et Technique
- N: Marinens Skipstilsyn
- s: Mando Secreto de las Tropas
Mst r: Minstyrman
MSTAB e: Manufacturing Systems Technical Advisory Board
MST&E e: Multiservice Test and Evaluation
MSTE e: Multi-Source Thermal Evaporation
- e: Multiplexer Section Terminating Equipment
MSTF e: Molten Salt Test Facility
MSTGS e: Marine Sciences and Technologies Grants Scheme
M. St. J. e: Ordinary Member, styled Serving Brother or Sister of the Order of St. John of Jerusalem
MST3K e: Mystery Science Theater 3000
MSTM e: Military Standard Transportation and Movement
MSTMC e: Materials and Structures Technology Management Center ORNL
MST-RADAR e: Mesosphere-Stratosphere-Troposphere Radar
MSTS e: Military Sea Transport Service
e: Multiphase Subsurface Transport Simulator
- N: Marinens Stabsskole
MSU e: Management Services Unit
- e: Medical Support Unit
- e: Message Switching Unit
- e: Message Switching Upgrade
- e: Microwave Sounder Unit
- e: Mission Service Utilization
- e: Morgan State University
- e: Moscow State University
- f: Université d'Etat de Moscou
- s: Universidad Estatal de Moscú
MSUD f: Mouvement pour un Syndicalisme Uni Et Démocratique
MS/und r: Marinstabens underrättelseavdelning
MSV N: Minesveiper
Msv r: Medelsvår
mSV e: milli Sievert
MSVA e: Mechanical Shock and Vibration Arresters
msv art r: medelsvårt artilleri
MSVC e: Meta-Signaling Virtual Channel
- e: Microsoft Visual C
MSVD e: Message Status Visual Display
MSVDP e: Message Status Visual Display Panel
msvp r: minsvepare
msvpavd r: minsvepningsavdelning
msvpdiv r: minsvepningsdivision
msvpflj r: minsvepningsflottilj
msvpk r: kustminsvepare
msvpm r: min- och svepmatroser
MSW e: Machine Status Word
- e: Master [of] Social Work
- e: Masters in Social Work
MSWG e: Message Standards Working Group
MSWL[F] e: Municipal Solid Waste Landfill
MSX e: Microsoft Super Extended BASIC
- e: Mid-Course Space Experiment
MSY e: Merseyside
MT e: Magnetic [particle] Test
- e: Management Threshold
- e: Manual Transmission
- e: Manufacturing Technology
- e: Marine Technician
- e: Marine Technology
- e: Massoretic Text
- e: Material Transfer
- e: McGinnis Auditorium
- e: Measurement Ton
- e: Mechanical Technologies division
- e: Medical Transcription
- e: Medium Terminal
- e: Megaton
- e: Message Type
- e: Metric Ton
- e: Metric Tons
- e: Micro-Tubules
- e: Million Tonne
- e: Millitesla
- e: Miscellaneous Treatment
- e: Mobile Terminated
- e: Mobile Transporter
- e: Montana
- f: Mandat Télégraphique
- f: Manufacture de Tabac
- f: Mappemonde Terrestre
- f: Marchandise en Transit
- f: Marché à Terme
- f: Médecin du Travail
- f: Médecin Traitant
- f: Même Texte
- f: Message Télégraphique
- f: Message Téléphonique
- f: Modification Tarifaire
- f: Mutilé du Travail
- N: Marinens Telekommunikasjonsvesen
- r: Maskinteknisk
- r: Medeltid
- R: Morskoy Tralshchik
- s: Mecánico auto [de Transportes]
- s: Megatonelada
mt e: Malta [Internet]
Mt d: Megatonne
- f: Moment
- f: Mont[ant]
- N: Motortransport
- r: Medeltung
- s: Metro
- s: Monitor
- s: Moto
- s: Motorizado
Mt. e: Mount[ain]
M&T e: Maintenance and Transportation
M/T e: Midterm
- N: Motortankskip
- r: Motortankfartyg

M[/]T a: Motortankskib
MTA e: Hungarian Academy of Sciences
- e: MAC Transportation Authorization
- e: Mail Transfer Agent
- e: Major Trading Area
- e: Material Transfer Agreement
- e: Message Transfer Agent
- e: Message Transfer Architecture
- e: Metropolitan Transit Authority
- e: Military Training Area
- N: Møbel- og Trebearbeidingsindustriens Arbeidsgiverforening
MTAA e: Motor Trades Association of Australia
MTAC e: Management Training and Advisory Center
- e: Michigan Technical Assistance Center
MTACC e: Marine Tactical Command and Control System
MTACCS e: Marine Corps Tactical C2 System
Mtar s: Militar
MTAS e: Millimetric Target Acquisition System
- e: Modular Target Acquisition System
- e: Multi-Sensor Target Acquisition System
MTASS e: Machine Transferable AN 10 YK 20 Support Software
MTB e: Materials Transportation Bureau
- e: Mountain Bike
- N: Missiltorpedobåt
- N: Motortorpedobåt
- r: Motortorpedbåt
MTBA e: Mean Time Between Assists
MTBAp e: Mean [productive] Time Between Assists
MTBB e: Mean Time Between Breakdowns
MTBCF e: Mean Time Between Critical Failure
MTBCMA e: Mean Time Between Corrective Maintenance Actions
Mtbdiv r: Motortorpedbåtsdivision
MTBDL e: Mean Time Between Data Loss
MTBE d: Methyl-Tertiär-Butylether
- e: Methyl Ter[tiary] Butyl Ether
MTBF e: Mean Time Before Failure
- e: Mean Time Before Fix
- e: Mean Time Between Failure[s]
MTBFp e: Mean [productive] Time Between Failures
Mtbgr r: Motortorpedbåtsgrupp
MTBI e: Mean Time Between Incident[s]
- e: Mean Time Between Interrupt
- N: Motortorpedobåtinspeksjon
MTBJ e: Mean Time Between Jams
Mtbm r: Mindre motortorpedbåt
MTBMA e: Mean Time Between Maintenance Actions
MTBO e: Mean Time Between Outage
MTBR e: Mean Time Between Removal[s]
- e: Mean Time Between Repairs
MTBRP e: Mean Time Between Parts Replacement
Mtbs r: större Motortorpedbåt
MTBSP e: Mobilization Troop Basic Stationing Plan
MTC e: Man Tended Capability
- e: Maneuver Training Command
- e: Manufacturing Technology Centre[s]
- e: Maryland Tax Court
- e: Mass Transfer Coefficient
- e: Mass Transfer Condition
- e: Maximum Tolerated plasma Concentration
- e: MIDI Time Code
- e: Missile Threat Conference
- e: Missionary Training Center
- e: Monitor Test Console
MTCA e: Model Toxic Control Act
MTCC e: Modular Tactical Communications Center
- e: Modular Tactical Control Center
MTCE e: Million Tonnes of Carbon Equivalent
MTCIP e: Manual Technical Control Improvement Program
Mtcl s: Motocicleta
MT-CLIM e: Mountain Climate Model
MT-CMOS e: Multi-Threshold CMOS
MTCN e: Management Task Control Number
MTCR e: Million Tonnes of Coal Replacement
- e: Missile Technology Control Regime
MTD e: Maximum Tolerated Dose
- e: Median Toxic Dose
- e: Memory Technology Drivers
- e: Month To Date
- e: Moscow Time Daylight
- e: Movement of Democratic Laborites
- e: Moving Target Detection
- f: Mouvement Togolais pour la Démocratie
MTDA e: Mean Time Data Availability
MTDC e: Modified Total Direct Costs
MTDD e: Master Time Division Demultiplexer
MTDDIS e: Mesoscale Transport Diffusion and Deposition Model for Industrial Sciences
- e: Mesoscale Transport Diffusion and Deposition Model for Industrial Sources
MTDM e: Master Time Division Multiplexer
mtDNA e: mitochondrial Deoxyribonucleic Acid
MTE e: Maintenance Training Equipment
- e: Measurement and Test Equipment
- e: Mutating Engine
Mte f: mixte
M&TE e: Measuring and Test Equipment
MT&E e: Materials Testing and Evaluation
MtEC f: Millions de tonnes Équivalent-Charbon
MTER e: Multitest Evaluation Report
Mtère f: Ministère
MTE/STE e: Maintenance Test Equipment/Specified Test Equipment
MTF e: Master Test Frame
- e: Materials Test Facility
- e: Medical Treatment Facility
- e: Memorandums-To-File
- e: Message Test Formatting
- e: Message Text Format[s]
- e: Message Transfer Facility
- e: Microsoft Tape Format
- e: Model Test Facility
- e: Modernization Task Force
MTFBWY e: May The Force Be With You [Internet]
MTFF e: [Columbus] Man-Tended Free-Flyer
MTFS e: Medical Treatment Facility System
MTG d: Malteser Trägergesellschaft GmbH
- e: Media Task Group
MTH e: Magnetic Tape Holder
- e: Mathematics
- e: Methamphetamine
MTHM e: Metric Ton of Heavy Metal
MTI e: Material Technology International
- e: Moving-Target-Indication
- e: Multi Terminal Interface
MTIA e: Many Thanks In Advance
- e: Metal Trades Industry Association
MTIF e: Michigan Training Incentive Fund
MTIHM e: Metric Tons Initial Heavy Metal
MTINA e: Miscellaneous and Technical Immigration and Naturalization Amendments
Mtj C r: Maskintjänstchef
Mtj L r: Maskintjänstledare
MTK d: Main-Taunus-Kreis in Hofheim, Hessen
- d: Medizinaltarif-Kommission
MTL e: Master Target List
- e: Master Tracking List
- e: Materials
- e: Mean Tide Level
- e: Mobiltherm Light
MTLCCS e: Mediterranean Tactical Logistic Command and Control System
MTM e: Mandatory Training Matrix
- e: Mechanical and Thermal Model
- e: Method[s] Time Measurement[s]
- e: Metric Ton of Metal
- e: Mission Table Management
- f: Mouvement des Travailleurs Mauriciens
- r: Metodtidmatning
- s: Mecánico auto Montador Eléctrico
MTMA e: Military Traffic Management Agency
MT/MANTECH e: Manufacturing Technology
MTMC e: Military Traffic Management Command
- e: Military Transportation Management Command
MTMCEA e: Military Traffic Management Command, Eastern Area
MTMCTEA e: Military Traffic Management Command Transportation Engineering Agency
MTMC-TTCE e: MTMC Transportation Terminal Command Europe

MTMCWA e: Military Traffic Management Command, Western Area
MTM/D e: Million Ton-Miles-Per-Day
MTMI N: Marinens Torpedo- og Mineinspeksjon
MTM/RVN f: Madagasikara Tany Maitso-Rassemblement des Verts de Madagascar
MTN e: Main Telecommunication[s] Network
Mtn f: Matin
Mtña s: Montaña
MTNG e: Meeting
MTO e: Material Take-Off
- e: Ministry of Transportation of Ontario
- e: Multilateral Trade Organization
Mto f: Météorologique
MTOC e: Monitoring Transport of Ocean Currents
MTOE e: Million Tonnes of Oil Equivalent
- e: Modified Table of Organization and Equipment
MTOGW e: Maximum Take-Off Gross Weight
MTOL e: Mean Time Off Line
- e: Mean Time On Line
MTOW e: Maximum Take-Off Weight
MTP e: Magnetic Tape Processors
- e: Mail Transfer Part
- e: Mail Transfer Protocol
- e: Marema Tlou Party
- e: Mark Twain Project
- e: Materiel Transfer Plan
- e: Melbourne Trading Post
- e: Metatarsophalangeal
- e: MIDI Time Piece
- e: MITRE Technical Paper
- e: Movement for Tolerance and Progress
- e: Multi-Megawatt Terrestrial Power Plant
- f: Ministère des Travaux Publics
- f: Mouvement pour la Tolérance et le Progrès
- r: Medeltung Transportflygplan
MTPA e: Mega Tonnes Per Annum
MTPE e: Mean Time Prior to Error
- e: Mission to Planet Earth [program]
MTPR e: Miniature Temperature Pressure Recorder
MTPTT f: Ministère des Travaux Publics, Transports et Tourisme
MTR e: Magnetic Tape Recording
- e: Marine Transportation-Related
- e: Material Test Report
- e: Material Transaction Report
- e: Material Transfer Request[s]
- e: Materials Test Reactor
- e: Maximum Report Time
- e: Mechanized Time Reporting
- e: Membrane Technology [and] Research, Inc.
- e: Miniature Temperature Recorder
- e: Minimum Techn[olog]ical Requirement
- e: Missile Tracking Radar
- e: MITRE Technical Report
M/Tr. N: Motortråler
MTRANS e: Mental Transfer of information
Mtre f: maître
MTRF e: Mean Time to Repair Failures
Mtrltpkomp r: Materiel transportkompani
Mtrp r: Medeltreffpunkt
MTRR e: Memory Type Range Register
MTRS e: McMurdo TDRSS Relay System
MTRUW e: Mixed Transuranic Waste
MTS e: Management Tracking System
- e: Material Tracking Standard
- e: Mauritius Telecommunication Services
- e: Message Telecommunications Service
- e: Message Transfer Service[s]
- e: Message Transfer System
- e: Message Transport System
- e: Microsoft Transaction Server
- e: Microwave Temperature Sounder
- e: Modular Training System
- e: Moscow Time Standard
- e: Multichannel Television Sound
- e: Multimedia Teleschool
- e: Musculoskeletal Tumor Society
- N: Måletroppsjef
- s: Mecánico auto chapista Soldador
Mts s: Metros
M&TS e: Mines and Technical Surveys
MTsD e: All Union Scientific Institute on Hydrometeorological Information - World Data Center, State Committee on Hydrometeorology
MTSE e: Multitool Support Enclosure
Mt Sinai J Med e: Mount Sinai Journal of Medicine
MTSL e: Monitoring and Technical Support Laboratory
MTSS f: Ministère du Travail et de la Sécurité Sociale
MTT e: Main Treatment Train
- e: Ministerial Trade Talks
- e: Mobile Training Team
- e: Model Test Train
- e: Multi-Transaction Timer
- e: Multiple Target Tracking
MTTA e: Mean Time To Assist
MTTC e: Mean Time To Correct
MTTD e: Mean Time To Diagnose
M/Tte f: Ma Traite
MTTF e: Mean Time To Failure
- e: Median Time To Fail[ure]
MTTFF e: Mean Time To First Failure
MTTR e: Mean Time to React
- e: Mean Time To Recovery
- e: Mean Time Trouble Repair
MTTS e: Magnetic Tape Terminal Station
MTTY e: Maintenance Module Teletype
MTU e: Magnetic Tape Unit
- e: Maintenance Termination Unit
- e: Maximum Transfer Unit
- e: Maximum Transmission Unit[s]
- e: Media Tech Unit
- e: Message Transfer Unit
- e: Metric Ton[s] [of] Uranium
- e: Michigan Technical University
- e: Mobile Treatment Unit
MTUC e: Mauritius Trade Union Congress
MTUF e: Magnetic Tape Unit Formatter
MTV d: Main-Taunus-Verkehrs-gesellschaft
- e: Monitoring of Tropical Vegetation
- e: Music Television
MTX e: Methotrexate
- e: Motor Transport
Mtz s: Motorizado
MU d: Medizinische Universität
- e: Master Unit
- e: Media Union
- e: Melbourne University
- e: Microprocessor Unit
- e: Microwave Unit
- f: Mouvement Universitaire
mu e: Mauritius [Internet]
- e: murine
MUA e: Mail User Agent
MUAC e: Mid-Upper-Arm Circumference
MUBEF f: Mouvement Unifié Belge des Etudiants Francophones
Muc r: Medicinska Undersökningscentral
MUCEU e: Makarere University College Employees Union
MUCG e: Management-Union Consultative Group
µCi e: microcurie
MUCIA e: Midwest Universities Consortium for International Activities, Inc.
MUCV e: Monash University, Melbourne
MUD e: Multi-User Device
- e: Multi-User Domain
- e: Multi[ple] User Dialog[ue][s]
- e: Multi[ple] User Dimension
- e: Multi[ple] User Dungeon
MUDAB d: Meeresumweltdatenbank
MUDAR p: Movimento de Unidade Democrática Angolana para a Reconstrução
MUDE p: Movimento para a Democracia
- s: Mujeres en Desarollo
MUDEBV s: Mujeres Unidas para el Desarollo de Baja Verapaz
MUDI e: Module User Debug Interface
MUDS e: Man-hour Use Data System
MUeW d: Monopolübertragungsweg
MUF e: Material Unaccounted For
- e: Maximal Usable Frequency
- r: Moderata Ungdomsförbundet
MUFF d: Maus und Fenster-Firlefanz
µg e: microgram
MUG d: Mietenüberleitungsgesetz
- e: Multi User Game
MUI e: Multimedia User Interface
- e: Multiple User Interference
MUID e: Map Unit Identification Data
MUIR e: Map Unit Interpretation Record
MUJ e: Macromedia User Journal
MUL e: Maximum Up Limit
- e: Mobile User Link
- e: Modify User Login
Mul r: Minutläggare
mul. a: mulig[vis]
MULDEM e: Multiplexer-Demultiplexer
Muldiv r: Minutläggningsdivision
MULPOC e: Multinational Programming and Operational Center
MULR e: Melbourne University Law Review
MULSE e: Modified Universal Soil Loss Equation
MULTA e: Multiple-Use Land Alliance
MULTAM e: Multiple Target Assessment Model
Multicast d: Sonderform des Broadcast
MULTICS e: Multiplex Information Computer Services

MULTILIS e: Library system from SOBECO
Multimedia Archiving/Retrieval d: Dokumente in einer gemeinsamen Online-Datenbasis zu speichern, zu verwalten und auf Abruf bereitzustellen
Multimedia Conferencing d: Ton-, Bild und Datenkommunikation in einer gemeinsamen Sitzung [Konferenz] von zwei oder mehr Teilnehmern mit entsprechenden Endgeräten
Multimedia-Datenbanken d: bildliche und tonliche ortsunabhängige Datenbank-Erweiterung
Multimedia Messaging d: Offline-Pendant zu Videokonferenzen mit gemeinsamer Dokumentenbearbeitung
MULTIPLEX e: Combine multiple circuits into one data stream
MULTOTS e: Multiple Unit Link Test and Operational Training System
µm e: micrometer
MUM e: Multiple User Mode
MUMLIB e: Multimedia [Technology] in Libraries
MUMMS e: Marine Corps Unified Material Management System
MUMPS e: Massachusetts general hospital Utility Multi[-]Programming System
- e: Multi-User MEMS Processes
Mu[n] s: Munición
MUN e: Mine Workers Union of Namibia
- e: Model United Nations
- s: Movimiento de Unificación Nacional
Mün. d: Münster
MUN CT e: Municipal Court
MUNDI e: Multiplexed Network for Distributive and Interactive Services
MUNEX e: Munitions Allocation and Expenditure Tracking System
MUNG e: Mung Until No Good
MUNGE f: Movimiento de Unión Nacional de Guinea Ecuatorial
Munic f: Municipal
MUNIREP e: Air Munitions Report
MUNSCS e: MUNSS Small Computer System
MUNSS e: Munition Storage Site
- e: Munition Support Squadron
MUNSTOR e: Munitions Storage Facilities
Munto s: Municionamiento
MUOS r: Marinens Underofficersskola
MUP e: Manufacturing-Use Product
- e: Master of Urban Planning
- e: Melbourne University Press
- e: Modify User Password
- e: Multi-[User] Project
Mup s: Municipio
MUPP e: Master of Urban Planning and Policy
µR e: microroentgen
MUR e: Management Update and Retrieval [System]
MURA e: Midwestern Universities Research Association
MURC e: Murrumbidgee Users Rehabilitation Committee
Murd. e: Murdoch
MUREX e: Multiple Reflection of X-rays
MURIEL e: Multimedia Remote Interactive Electronic Documents
MURP e: Master of Urban and Regional Planning
MURR e: Murrow Communication Center
µs. e: Microsecond
MUS e: A. G. Ruthven Museums Building
Mus. e: Museum
- e: Music[al]
MUSA p: Movimento de Unidade Socialista de Angola
MUSAIOS e: Software for searching and viewing Classical text CD-ROMs from TLG and PHI
MUSARC e: Major U.S. Army Reserve Command
MUSC e: Music
MuSchG d: Gesetz zum Schutz der erwerbstätigen Mutter [Mutterschutzgesetz]
Muscle Nerve e: Muscle and Nerve
Muscle Nerve Suppl e: Muscle and Nerve. Supplement
Mus.Doc. e: Doctor of Music
MUSE e: Esprit4 project concerned with music delivery for recording companies
- e: Multi-User Shared Environment
MUSH e: Multi-User Shared Hallucination[s]
MUSI e: Music Library
MUSIC e: Multi-Spectral Infrared Camera
MuSIC e: Multi-System Integrated Control
MUSICAM e: Masking Universal Subband Integrated Coding And Multiplex
MUST e: MAC UHF Satellite Terminal
- e: Melton Valley Storage and Evaporator Concentrate Storage Tank
- r: Militär Underrättelse- och Säkerhetstjänst
µSv e: microsievert
MUSWTCH e: Mussel Watch
MUT e: Monitor Under Test
Mut f: Mutualiste
MUTA e: Mutagenicity
Mutat Res e: Mutation Research
MUTE e: Multiplied Unit For Transmission Elimination
MUTES e: Multiple Threat Emitter System
MUTOS e: MultiUser [or multitasking] Operating System
MUTSU e: Japanese prototype nuclear-powered ship
MüW d: Monopolübertragungsweg
MUX e: Multiplexer
MUX/DEMUX e: Multiplexer/ Demultiplexer
MUZ e: Mineworkers Union of Zambia
MV d: Müllverbrennung
- e: MicroVAX
- e: Millivolt[s]
- e: Motor operated Valve
- e: Productive infix and suffix for IBM-3000 series machine names
- f: Marchandises-Voyageurs
- f: Médecine Vétérinaire
- f: Minimum Vital
- f: Mise en Vente
- f: Moins-Value
- f: Mort Violente
- f: Moulin à Vent
- f: Mouvement des Valeurs
- r: Militär Väderlekstjänst
mv. a: med videre
- e: Millivolt
m.v. N: med videre
- r: med vidare
m/v s: meses vista
MVA e: Market Value Adjustment
- e: Multi-domain Vertical Alignment
MVAA d: Mitteilungen aus dem Verein zur Abwehr des Antisemitismus
MVB e: Multimedia Viewer Book
Mvbef r: Maskinvaktsbefäl
MVC d: Micro-Bit-Virus-Zentrum, Karlsruhe
- e: Model View Controller
- e: Multimedia Viewer Compiler
- r: Militär Väderlekscentral
MVCP e: Management Visibility and Control of Provisioning
MVD e: Mean Volume Diameter
- e: Ministry of Internal Affairs
- e: Multi Valued Dependency
- s: Magnitud de Variación de la Distancia
MVDM e: Multiple Virtual DOS Machines
MVDS e: Microwave Video Distribution System
MVEL e: Motor Vehicle Emission[s] Laboratory
MVF d: Magazinverlag am Fleetrund, Hamburg
MVG d: Marktverantwortungsgebiet
MVGA e: Monochrome Video Graphics Array
MVI e: Microgravity Vestibular Investigation
MVICSA e: Motor Vehicle Information and Cost Savings Act
MVIF e: Multi-Vendor Interacting Forum
MVIP e: Multi Vendor Integration Protocol
- e: Multi Vendor Interaction Protocol
MVIRI e: Meteosat Visible and Infrared Instrument
MVISIP e: Murray Valley Irrigation and Salinity Investigation Program
MVISR e: Multispectral Visible and IR Scan Radiometer
MVL d: Mecklenburg-Vorpommern Landesregierung und Landtag
- N: Mekaniske Verksteders Landsforening
MVLDC e: Murray Valley League for Development and Conservation
MVM s: Magnitud de Variación de la Marcación [verdadera]

MVMA e: Motor Vehicles Manufacturers Association
M.V.O. e: Member of the Royal Victorian Order
MVO e: Montserrat Volcano Observatory
MVP e: Mitral-Valve Prolapse
- e: Most Valuable Professional
- e: Multiline Variety Package
- e: Multimedia for ps/Value Point pc
- e: Multimedia Video Processor
MVR e: Monthly Variance Report
MVRS e: Marine Vapor Recovery System
MVS e: Minimum-Variance-Simultaneous
- e: Multiple Virtual Storage [operating system]
MVS-ESA e: Multiple Virtual Storage/ Extended System Architecture
MVS/SA e: Multiple Virtual Storage/ System Architecture
MVSSP e: Multiple Virtual Storage System Product
MVS[/]XA e: Multiple Virtual Storage/ Extended Architecture
MVT d: Mechanische Verfahrenstechnik
- e: Multiple Variable Tasks
- f: Mouvement
MVTS e: Motor Vehicle Tampering Survey
MVV d: Münchener Verkehrsverbund
MW d: Mehrzweckwaffe
- e: Main Window
- e: Missile Warning
- e: Missile Wing
- e: Mixed Waste
- e: Modified wire Wrap
- e: Multiwink
- f: Marchandises Warrantées
m.w. e: mixed widths
e: molecular weigth
MWA e: Metropolitan Waterworks Authority of Thailand
- e: Minimum Warning Attack
MWAP e: Minimum Warning Attack Plan
MWARS e: MACOM Worldwide Ammunition Reporting System
MWASA e: Media Workers Association of South Africa
MWBC e: Mean Wafers Between Cleans
MWBD e: Mean Wafers Between Defects
MWBP e: Missile Warning Bypass
MWC e: Machine Wet Cleaning
- e: Measurement While Coring
MWD d: Mehrwertdienst
MWDR e: Mid-Western Development Region
MWDS e: Missile Warning and Display System
MWe e: Megawatt electrical [power]
MWF d: Ministerium für Wissenschaft und Forschung [Baden-Württemberg]
MWG d: Marianne Weber Gymnasium
MWh e: Megawatt hours
MWI e: Ministry of Water and Irrigation
MWIP e: Mixed Waste Integrated Program
MWIR e: Medium Wave Infrared
- e: Mixed Waste Inventory Report
MWK d: Ministerium für Wissenschaft, Forschung und Kunst
MWL e: Municipal Waste Leachate
- f: Mouvement de la Wallonie Libre
MWLID e: Mixed Waste Landfill Integrated Demonstration
MWM e: Motif Window Manager
MWN d: meines Wissens nach
- e: Managed WAN Networks
MWnT e: Multi-Walled NanoTub[ul]e
MWO e: Missile Warning Office
MWP e: Meteorologists Weather Processor
- e: Mixed Waste Project
- e: numerical Weather Prediction
MWR e: Microwave Radiometer
- e: Morale, Welfare and Recreation
MWRF e: Multi-Function Waste Remediation Facility
MWRMQAS e: Medical War Reserve Materiel Quality Assurance Subsystem
MWS e: Matsushita White Skipping
MWs e: Megawatt-second
MWS e: Microwave Spectrometer
- e: Missile Warning Squadron
- e: Multi-Wire Slurry
- f: Spectromètre à hyperfréquences
- s: Espectrómetro de Microondas
MWSC e: Missile Warning Systems Conference
MWSCAS e: Midwest Symposium on Circuits and Systems
MWSOC e: Missile Warning and Space Operations Center
MWSR e: Microwave Water Substance Radiometer
MWSS e: Marine Wing Support Squadron
MwSt d: Mehrwertsteuer ausweisbar
MWT e: Marxist Workers Tendency
MWt e: Megawatt-thermal
MWT e: Monitor Wafer Turner
MWTA e: Medical Waste Tracking Act
MWTF e: Multi-Function Waste Tank Facility
MWTIP e: Mixed Waste Treatment Program
MWTTY e: Missile Warning Teletypewriter
MWU A: Mynwerkersunie
- e: Mine Workers Union
MWWP e: Mine Warfare Working Party
MWyr e: Megawatt-year
MX e: Experimental Missile
- e: Mail Exchange[r]
- e: New generation ICBM
- e: Peacekeeper Missile
MXCC e: Module X-bus Cache Controller
MXS e: Microsoft Exchange Server
MXSP e: Maintenance Support Plan
MY e: Model Year
- f: Mutilés des Yeux
My N: Mikron
m.y. e: million years
M/Y N: Motoryacht
MYCL e: Mauritius Young Communist League
Mycol f: Mycologie
MYDP e: Multi-Year Development Plans
MYE e: Model Year Emissions
MYK d: Mayen-Koblenz, Rheinland-Pfalz
MYM e: Muslim Youth Movement
MYNIC e: Malaysian Network Information Center
MYOB e: Mind Your Own Business [Internet]
MyoD1 e: Myogenic regulatory gene
MYP e: Malawi Young Pioneers
- e: Multi-Year Procurement
MYPP e: Multi-Year Program Plan
My[r] s: Mayor
MYRIADE e: Interface for accessing CCN
MYSD e: Maybe You Should Drive [Internet]
MYSTIC STAR e: Presidential Support Airground Communications System
Mythol. e: Mythology
MZ e: Mark Zbikowski
- s: Mecanizado
MZG d: Merzig-Wadern, Saarland
MZI e: Mach-Zehnder Interferometer
MZR e: Multiple Zone Recording
MZT e: Monozygotic Twins

N

N a: Nit
- a: Numerisk faktor
- d: Nebenfach
- d: Niederschlagssumme
- d: Nord[en]
- d: Zahl der Einzelmessungen d: pH-Wert des Niederschlags und Leitfähigkeit des Niederschlags
- d: Stickstoff
- e: Avogadros number
- e: Nonmetal
- e: Normalized
- e: Notice
- e: Number of neutrons in a nucleus
- e: Nursing
- f: Néerlandais
- f: Nom
- f: Numéro
- N: Nord-Norge
- r: Ned[gång]
- r: Nedslags-
- r: Nordlig
- r: Norr
- s: [buque] Nodriza
- s: Nacional
- s: Norte

n e: Anzahl der beteiligten Farbfilter bei einer Vollfilterung
- d: Anzahl der Unbekannten
- d: Anzahl der Zielfunktionsvariablen
- d: Brechzahl
- d: Potenz einer Exponentbildung
- s: nuestro

n. e: neuter
N. e: North
N0 e: sea level refractivity
- e: spectral noise density

N_2 e: Nitrogen
NA e: Narcotics Anonymous
- e: National Assembly
- e: Network Access
- e: Non-Australian
- e: Nonattainment
- e: North America[n]
- e: Not Applicable
- e: Not Available
- e: Nuclear Assessment
- f: Narcotiques Anonymes
- f: Nord-Africain
- f: Nord-Atlantique
- f: Normalisation de l'Automobile
- f: Nouvel An
- f: Nouvel Avis
- f: Numéro de l'Arrondissement
- s: Niebla Artificial

Na e: Sodium
- f: Nota

n-a r: norra
N/A e: Not Applicable
Na+ e: Natrium-Kation
Na-22 e: sodium-22
NA48 e: The CP Violation experiment
NAA e: N-Acetyl Aspartate
- e: National Academy on Aging
- e: Nonattainment Area
- e: Not At All [Internet]

NAACCR e: North American Association of Central Cancer Registries
NAADM e: North American Defense Modernization
NAAEC e: North American Agreement on Environmental Cooperation
NAAFETEE e: North American Association For Exports To Eastern Europe
NAAFI e: Navy, Army and Air Force Institutes
NAAG-DPG e: North Atlantic-Arctic Gateways Detailed Planning Group
NAAIDC e: National Advisory Allergy and Infectious Disease Council
NAAO e: North America[n] Aircraft Operations
NAAP e: National Association of Activity Professionals
NAAQ e: National Ambient Air Quality
NAAQS e: National Ambient Air Quality Standards [program]
NAAR e: National Alliance for Autism Research
- e: Night Air-to-Air Refueling

NAAS e: North American Automated Systems co.
NAAWU e: National Automobile and Allied Workers Union
NAAWUL e: National Agriculture and Allied Workers Union of Liberia
NABE e: North Atlantic Bloom Experiment
NABEL f: National d'observation des polluants atmosphériques
NABIR e: Natural and Accelerated Biomediation Research
NABIS e: Northern Alberta Brain Injury Society
NABN e: EMSL-RTP National Atmospheric Background Network
NABS e: NATO Air Base SATCOM
- e: NATO Airborne SATCOM [Satellite Communications]
- e: Nuclear-Armed Bombardment Satellites

NABSO e: Namibian Black Students Organization
NAC e: CDC National AIDS Clearinghouse
- e: National Adoption Center
- e: National Agency Check
- e: Naval Administrative Command
- e: Network Access Component/Centers
- e: Network Adapter Card
- e: Network Administration Center
- e: Network Analysis Corportion
- e: Network Applications Consortium
- e: Nevada Administrative Code
- e: NMCS Automated Control
- e: Null Attachment Concentrator

NACA e: National Agricultural Chemicals (or Chemists) Association
- e: Network of Aquaculture Centres in Asia and the Pacific
- f: Réseau des centres d'aquaculture d'Asie et du Pacifique
- s: Red de Centros de Acuicultura en Asia y el Pacífico

NACB e: National Academy of Clinical Biochemistry
- e: National Accreditation Certification Body

NACC e: National Air Communications Council
- e: North American Calibration Cooperation
- e: North Atlantic Cooperation Council

NACCB e: National Accreditation Council for Certification Bodies
NACCCSA e: NATO Air Command and Control System Agency
NACCHO e: National Association of County and City Health Officials
NACCIMA e: Nigerian Association of Chambers of Commerce, Industry, Mines and Agriculture

NACCP e: North American Cambridge Classics Project
NACCSM e: National Advisory Committee on Control Surveys and Mapping
NACD e: National Association of Computer Dealers
NAC/DPC e: North Atlantic Council/ Defense Planning Committee
NACE e: National Association of Corrosion Engineers
- e: NMCS Automated Control Executive
- e: North American Commission on the Environment
NACEPT e: National Advisory Council for Environmental Policy and Technology
NACES e: Naval Aircrew Common Ejection Seat
NACF e: Naval Air Combat Fighter
NACHRI e: National Association of Childrens Hospitals and Related Institutions
Nachrichtenblatt d: Nachrichtenblatt. Jüdische Wochenzeitung. Amtliches Organ für die Synagogen-Gemeinden Hannover und Braunschweig. Jg. 1-15 [1924-1938]
NachrT d: Institut für Elektrische Nachrichtentechnik
NachrÜ d: Institut für Nachrichtenübertragung
NachrV d: Institut für Nachrichten-vermittlung und Datenverarbeitung
NachwG d: Gesetz über den Nachweis der für ein Arbeitsverhältnis geltenden wesentlichen Bedingungen [Nachweis-gesetz] vom 20.7.1995
NachwV d: VO über Verwertungs- und Beseitigungsnachweise [Nachweis-verordnung] vom 10.9.1996
NACI e: National Advisory Council on Innovation
- e: National Agency Check and Inquiry
NACISC e: NATO Communications and Information Systems Committee
NACK e: National Advisory Committee on Kangaroos
- e: Negative Acknowledge
NaCl e: Sodium Chloride
NACLANG e: Nacelle Angle
NACM e: Network Authentication Control Module
NACNEMS e: North American Cooperative Network of Enhanced Measurement Sites
NACOM e: National Alert Communications
NACOSA e: National AIDS Co-ordinating Committee of South Africa
NACOST e: National Advisory Council on Science and Technology
NACOTU e: National Congress of Trade Unions
NACP e: National AIDS Control Program
NACR e: National Association of Chemical Recyclers
NACS e: Netware Asynchronous Communciation Services
- e: Netware Asynchronous Communication Server
- e: Network Access Control System
- e: Network Asynchronous Communications Services
NACSEM e: National Communications Security Emanation Memorandum
NACSI e: National COMSEC [Communications Security] Instruction
NACSIM e: National Communications Security Information Memorandum
- e: National Communications Security Instruction Memorandum
- e: National COMSEC/EMSEC Information Memorandum
NACSIS e: National Center for Science Information Service
- e: National Center for Science Information Systems [of the Japanese Ministry of Education, Science and Culture]
NACSM e: National Advisory Committee on Surveys and Mapping
- e: National Communications Security Information Memoranda
NACSN e: North American Commission on Stratigraphic Nomenclature
NACTU e: National Council of Trade Unions
NAD e: Naval Air Department
- e: Naval Aviation Depot
- e: Nuclear Accident Dosimeter
NAD+ e: Oxidized form of Nicotinamide Adenine Dinucleotide
NAD[-] 27 e: North American Datum [of] 1927
NAD 83 e: North American Datum [of] 1983
NADA e: National Acupuncture Detoxification Association
- e: National Democratic Alliance
- e: Native Affairs Department Annual
- e: Numerical Analysis and Computing Science
N-ADAMS e: National Agricultural Drought Assessment and Monitoring System
NADAMS e: National Agricultural Drought Assessment and Monitoring System
NADAR e: NADIR Altimeter for Rain
NADB e: EPA National Air Data Branch
- e: National Atmospheric Databank
- e: Network Accessible Data Base
NADC e: NATO Air Defense Command
- e: North American Digital Cellular system
NADDO e: NATO Design and Development Objective
NADECO e: National Democratic Coalition
NADEPA e: National Democratic Party of Tanzania
NADF e: National Adrenal Diseases Foundation
- e: North American Directory Forum
NADH e: Nicotinamide Adenine Dinucleotide
- e: Reduced form of Nicotinamide Adenine Dinucleotide
NADI e: National Disengagement Intention
NADIS e: NATO Defense Information System
NADL e: National Association of Democratic Lawyers
NADOC e: Naval Aviation Depot Operations Center
NADONA/LTC e: National Association of Directors of Nursing Administration in Long Term Care
NADP e: National Association of Dental Plans
NADPH e: Nicotinamide Adenine Dinucleotide Phosphate
NADP/NTN e: National Atmospheric Deposition Program/National Trends Network
NADPSC e: National ADP Service Center
NADS e: NATO Air Defense System
NADUC e: Nimbus/ATS Data Utilization Center
NADW f: Eau profonde de l'Atlantique Nord
- s: Agua Profunda del Atlántico Norte
NAE e: North American Environmental, Inc.
- e: Not Above or Equal
Nae f: Non autrement énuméré
NAEC e: Hungarys National Atomic Energy Commission
- e: National Aboriginal Education Committee
- e: Naval Air (or Aviation) Engineering Center
- e: Novell Authorized Education Center
NAEG e: Nevada Applied Ecology Group
NAEGIS e: NATO Airborne Early Warning and Ground Control Integration Segment
NAEHCA e: National Association of Employers on Health Care Action
NAEI e: Naval Aviation Executive Institute
NAEMSE e: National Association of EMS Educators
NAEMSP e: National Association of EMS Physicians
NAEPP e: National Asthma Education and Prevention Program
NAER e: National Association for Emergency Response
Nær a: Nærum
NAESU e: Naval Air Electronics Support Unit
NAEWF e: NATO Airborne Early Warning Force
NAF d: Nutzungsanforderung
- e: National Anxiety Foundation
- e: Non-Appropriated Funds
- e: Northern Africa [region]
- N: Nordisk Andelsforbund
- N: Norge-Amerika Foreningen
- N: Norges Apotekerforening
- N: Norges Automobil Forbund
- N: Norsk Arbeidsgiverforening
- N: Norske Annonsørers Forening
NAFA a: Nordjyske Andelsslagteriers Fabrikker, Andelsselskab
- e: Nordic Association for Andrology

NAFAU e: Namibian Food and Allied Workers Union
NAFAX e: National Facsimile
NAFB e: Norton Air Force Base
NAFC e: National Association for Continence
NAFCO e: National Food Corporation
NAFCOC e: National African Federated (or Federation) Chamber of Commerce
NAFCOM e: Northern Africa [regional] Committee [for START]
NAFCR e: NAF Computer Replacement
NAFE e: National Association of Female Executives
NAFEN e: National Foundation of Indian Engineers
NAFIS e: NASA Accounting and Financial Information System
NAFISS e: Non-Appropriated Funds Information Standard System
NAFMIS e: Non-Appropriated Funds Management Information System
NAFO e: Northwest Atlantic Fisheries Organization
- e: Organización de Pesca del Atlántico Noroeste
NaFöG d: Nachwuchsförderungsgesetz
NAFSEP f: Nouvelle Association Française des Sclérosés En Plaques
NAFTA e: North America Free Trade Agreement
NAFTI e: National Film and Television Institute
NAG e: National Algorithms Group
- e: Network Architecture Group
- e: Numerical Algorithms Group, Ltd., Oxford
NAGIS e: Natal/KwaZulu Association for Geographical Information Systems
NAGP e: National Assessment Governing Board
NAGPS e: National Association of Graduate and Professional Students
NAGRA e: National Association of Government Archives and Records Administration
- e: National Cooperative for the Disposal of Radioactive Waste
NAGUA e: Numerical Algorithms Group Users Association
NAHAT e: National Association of Health Authorities and Trusts
NAHMOR e: National Association of HMO Regulators
Nahosa e: National Hostel Residents Association
NAHS e: North American Hyperthermia Society
NAI d: Netzwerk Arbeitswelt Informatik
- e: The Nordic Africa Institute Uppsala, Sweden
NAIC e: National Academic Information Centre
- e: National Aging Information Center
- e: National Air Intelligence Center
- e: National Asian Information Centre
- e: Network Applications [and] Information Center
- e: Newly Agro-Industrialized Country
- e: Nigerian Army Intelligence Corps
- e: Nuclear Accident/Incident Control
NAICO e: Nuclear Accident/Incident Control Officer
NAIF e: Navigation and Ancillary Information Facility
NAIL e: New African Investments, Ltd
- N: Norges Assuranse-Inspektørlag
NAILDD e: North American Interlibrary Loan and Document Delivery
NAILS e: NCAR Airborne Infrared Lidar System
NAIPS e: NMCS Automated Input Processing System
NAIS e: National Atlas Information Service
- e: Neutral Administrative Inspection Scheme
- e: Neutral Administrative Inspection System
NAJIT e: National Association of Judiciary Interpreters and Translators
NAK e: Negative Acknowledge Character
- e: Negative Acknowledgement
- e: Non-Acknowledgement
- e: Not Acknowledged
NaK e: Sodium-potassium alloy
Näk r: Näken
NAL e: National Aerospace Laboratory
- e: National Alliance of Liberals
- e: Network Adaptation Layer
- e: Novell Application Launcher
- N: Den Norske Amerikalinje
- N: Norske Agenters Landsforbund
- N: Norske Arkitekters Landsforbund
- N: Norske Avisers Landsforbund
Nal f: National
NALC e: Naval Aviation Logistics Command
- e: North American Landscape Characterization
NALCO e: National Association of Local Councils
NALCOMIS e: Naval Aviation Logistics Command Management Information System
NALCOP e: NORAD Alternate Command Post
NALD e: Nonattainment Areas Lacking Demonstrations
NALDA e: Naval Aviation Logistic Data Analysis System
NALE e: Naval [and] Amphibious Liaison Element
NALF N: Norges Arbeidslederforbund
NALICON e: National Liberation Council of Nigeria
NALIN e: National Library Information Network
NALLA e: National Long Lines Agency
NALOPKT e: Not A Lot Of People Know That [Internet]
NALT e: Nasal Associated Lymphoid Tissue
- e: Nasopharyngeal Associated Lymphoid Tissue
NALTOACS e: Naval Laboratories Technical Office Automation and Communications System
NAM e: Name and Address Module
- e: National Account Manager
- e: Naval Armaments Model
- e: Netware Application Manager
- e: Network Access Mode
- e: Network Advisory Message
- e: Network Asset Management
NAMA e: National Air Monitoring Audits
- e: National Alliance of Methadone Advocates
NAMBOARD e: National Agricultural Marketing Board
NAMCP e: National Association of Managed Care Physicians
NAMDA e: National Medical and Dental Association
NAMDI f: Inventaire National des Données Marines
- s: Inventario Nacional de Datos Marinos
NA&ME e: Naval Architecture and Marine Engineering Building
N.Amer. e: North America[n]
NAMES e: National Association for Medical Equipment Services
NAMIE e: National AIDS Minority Information and Education Program
N. Am. J. Fish. Manage. e: North American Journal of Fisheries Management
näml. r: nämligen
NAMLS e: Naval Automated Medical Logistics System
NAMM e: North American Music Merchants
NAMMA e: NATO MRCA development and procurement Management Agency
NAMMCO e: North Atlantic Marine Mammal Commission
- s: Comisión de Mamíferos Marinos del Atlántico Septentrional
NAMO e: Naval Air Maintenance Office
NAMOC e: Northwest Atlantic Mid-Ocean Canyon
NAMP e: Netware Asynchronous Messaging Protocol
NAMPA e: Namibian Press Agency
NAMPS e: Narrow [band] Advanced Mobile Phone Service
- e: Narrow-Band Analog Mobile Phone Service
N-AMPS e: Narrowband Advanced Mobile Phone System
NAMRIA e: National Mapping and Resource[s] Information Authority
- f: Agence nationale de cartographie et d'information sur les ressources
- s: Servicio Nacional de Información sobre Cartografía y Recursos

NAMS e: National Air Monitoring Station[s]
- e: National Association of Multimedia Shareware
- e: NATO AWACS Maintenance System
- e: North American Micropaleontology Section
NAMS/MIS e: National Air Monitoring Stations Management Information System
NAMSO e: Namibian Students Organization
NAMSS e: National Association Medical Staff Services
NAN e: News Agency of Nigeria
Nan. ma: Nanyang [now incorporated in National University of Singapore]
NANAD e: National Association of Anorexia Nervosa and Associated Disorders
Nancy Rae e: Boeing JKC-135A TELINT aircraft
NAND e: Not AND [gate]
NANHPH e: National Association of Nursing Homes and Private Hospitals [Australia]
NANMT e: National Association of Nurse Massage Therapists
NANN e: National Association of Neonatal Nurses
NANO e: One thousand-millionth
NANOS e: North American Neuro-Ophthalmology Society
Nanosecond e: One billionth of a second
NANP e: North American Numbering Plan
NANPRH e: National Association of Nurse Practitioners in Reproductive Health
NANS e: National Association of Nigerian Students
NANSO e: Namibian National Students Organization
NANTU e: Namibia National Teachers Union
NANU e: Namibia African National Union
- e: Notice Advisory to NAVSTAR User
NAO e: NOAA Administrative Order
NAOC e: National Airborne Operations Center
NAOMA e: National Acupuncture and Oriental Medicine Alliance
NAOMI e: NADIN/AUTODIN Mode-1 Interface
NAOMIS e: Navy Materiel Transportation Office Operations and Management Information System
NAOS e: North American Atmospheric Observing System
- e: North Atlantic Ocean Station
- f: Station Océanique de l'Atlantique Nord
- s: Estación Oceánica del Atlántico Norte
NAP e: National Action Party
- e: National Action Programme on electronic highways
- e: National Afforestation Program
- e: National AIDS Programme
- e: National Aim Point
- e: National Awareness Partner[s]
- e: Nigeria Advance Party
- e: Nonarborescent Pollen
- e: Not Authorized POMCUS
- e: Null Amsterdam Pegel
- f: Nouvelle Agence de Presse
NAPA e: National Association of the Partners of the Americas
- e: Naval Association of Physician Assistants
NAPAFASA e: National Asian Pacific American Families Against Substance Abuse
NAPAG e: National Academies Policy Advisory Group
NAPAL e: National Patriotic Association of Liberia
NAPAP e: National Acid[ic] Precipitation Assessment Program [emission inventories]
NAPAW e: North American Process Algebra Workshop
NAPBC e: National Action Plan on Breast Cancer
NAPBTAC e: National Air Pollution Control Technical Advisory Committee
NAPC e: North American Paleontology Convention
NAPCAN e: National Association for Prevention of Child Abuse and Neglect
NAPCO e: National Poultry Company
NAPCP e: National AIDS Prevention and Control Program
NAPE e: National Association of Physicians for the Environment
NAPEC e: National Association of Professional Environmental Communicators
NAPEM e: National Alliance for Photonics Education in Manufacturing
NAPF e: North Atlantic Polar Front
NAPFSC e: National Association of Professional Forestry Schools and Colleges
NAPH e: National Association of Public Hospitals and Health Systems
NAPL e: NAPL Reproduction Centre
- e: National Airphoto Library
- e: Non-Aqueous Phase Liquid
NAPLDS e: North American Presentation Level Autocol Syntax
NAPLPS e: North American Presentation Layer Protocol Suite
- e: North American Presentation Layer Protocol Syntax
NAPLS e: North American Presentation Layer Protocol Syntax
NAPM e: National Association of Purchasing Management
NAPMA e: NATO AWACS Program Management Agency
NAPMO e: NATO AWACS Program Management Office
NAPO e: Namibia Peoples Organization
- e: NATO Production Objective
NAPOL e: Namibian Police
Naposu e: National Police Service Union
NAPP f: Noyau Armé Pour l'autonomie Populaire
NAPPS e: Navy Automated Publishing and Printing System
NAPRALERT e: Natural Product Data Base
NAPS e: National Air Pollution Surveillance
- e: Navy Automated Publishing System
- e: Negative-Acting Proofing System
NAPSA e: North American Pediatric Subspecialty Association
Naptosa e: National Professional Teachers Organization of South Africa
NAPWA e: National Association of People Living With HIV/AIDS
NAPWU e: Namibian Public Workers Union
NAQ d: Nationale Arbeitsgemeinschaft für Qualitätsförderung im Gesundheitswesen
NAR e: National Asbestos Registry
- e: Nominal Acceleration Radar
- e: Non-Advocacy Review
- e: Not At Risk
- e: Nucleic Acids Research
NARA e: National Agrichemical Retailers Association
- e: National Aquatic Resources Agency
- e: National Archive and Records Administration
- f: Agence Nationale des Ressources Aquatiques
- s: Organismo Nacional de Recursos Acuáticos
NARAL e: National Abortion and Reproductive rights Action League
NARC e: National Agricultural Research Committee of the Joint Council
- e: National Atmospheric Radiation Centre
NARCISSE e: IMPACT study on optical coding of art treasures
NARCO e: National Ranching Company
NARCOMS e: North American Research Consortium On MS [Multiple Sclerosis]
NARD e: National Association of Resident Doctors
NARDAC e: Naval Regional Data Automation Center
NARE e: North Atlantic Regional Experiment
NAREL e: National Air and Radiation Environmental Laboratory
NARF e: Network Access Request Form
Nar F N: Narvik Festning
närfplut r: närförsvarspluton
NARHP e: National Adolescent Reproductive Health Partnership
NARIC e: National Academic Recognition Information Centres
Narko a: Narkotika
Narkomaviaprom e: State Commissariat for the Aviation Industry [USSR]
NARL e: National Arctic Research Laboratory
- e: Naval Arctic Research Laboratory
närlv r: närluftvärn

NARM e: Naturally occurring or Accelerator-generated (or produced) Radioactive Material[s]
- e: Navy Resource[s] Model
- e: Nuclear Accelerator-generated Radioactive Material

NARMAL p: Núcleo de Apoio aos Refugiados e Movimentos de Libertação
NARMC e: North Atlantic Regional Medical Center
NARM-DPG e: North Atlantic Rifted Margins Detailed Planning Group
NARP e: National Association of Rail Passengers
- e: Neuropathy, Ataxia, Retinitis Pigmentosa

NARPPS e: National Association of Rehabilitation Professionals in the Private Sector
Narr. e: Narrative
NARS e: National Archives and Records Services
- e: National Asbestos-Contractor Registry System
- e: Sodium [Na] Removal System

NARSAD e: National Alliance on Schizophrenia and Depression
NARSD e: National Alliance for Research on Schizophrenia and Depression
NARSHA e: NATO Acoustic and Remote Sensing Shallow Water Experiment
NARSNDF e: North Atlantic Regional Study Narrative Data File
NARSTO e: North American Research Strategy for Tropospheric Ozone [charter]
NART e: National Adult Reading Test
NARU e: North Australia Research Unit
NAS e: National Academy of Science[s]
- e: National Aerospace Standard
- e: Naval Air Squadron
- e: Naval Air Station
- e: Naval Air transport Service
- e: Nerve Attenuation Syndrome
- e: Netware Access Services
- e: Network Access Server
- e: Network Application Services
- e: Network Application Support
- e: Network Attached Storage
- e: Non-indigenous Aquatic Species
- e: North Arabian Sea
- e: Numerical Aerodynamics Simulator
- N: Norsk Akademikersamband

NaS e: Sodium-Sulfur [battery]
NASA e: National Aeronautical (or Aeronautics) [and] Space Administration
- e: National Aeronautics and Space Agency
- e: National Air and Space Administration
- e: National Security Agency
- e: Nitrogen Atmosphere Sampling Analysis
- s: Administración Nacional de Aeronáutica y del Espacio de los EE.UU.

NASA-4 e: large vacuum chamber at Ball for testing
NASAA e: National Association for Sustainable Agriculture in Australia
NASACO e: National Shipping Company Agencies, Ltd.
NASA/EDC e: NASA/Environmental Data Centre
NASA/GSFC e: NASA [National Aeronautics and Space Administration]/ Goddard Space Flight Center
NASA-MSFC e: NASA Marshall Space Flight Center
NASAO e: National Aeronautics and Space Administration Offices
NASAP e: Nonproliferation Alternate Systems Assessment Program
NASA-TR e: Plum Brook Reactor Facility
NASC e: North American Stratigraphic Code
- e: North Atlantic Salmon Convention

NASCAR e: National Association of Stock Car Auto Racing
NASCAT e: NASA Scatterometer
NASCI e: North American Society for Cardiac Imaging
NASCOM e: NASA Communication[s] [link (or network, or systems)]
- e: Naval Air Station Communication

NASCOM II e: upgraded version of NASCOM
NASD e: NMCS Applications Software Description
NASDA e: National [Aero]space Development Agency of Japan
- f: Agence nationale de développement spatial du Japon
- s: Organismo Nacional de Investigaciones Espaciales del Japón

NASDAQ e: National Association of Securities Dealers Automated (or Automatic) Quotation [system]
NASE e: National Agency for Science and Engineering
n-ASER e: neutron-Accelerated Soft-Error Rate
NASG e: New-Age Sensitive Guy
NASI e: Netware Asynchronous Service[s] Interface
NASIG e: North American Serials Interest Group
NASIGNET e: NASIG Network
NASIRC e: Nasa Automated Systems Incident Response Capability
- e: NASA Automated Systems Internet Response Capability

NASIRE e: National Association of State Information Resource Executives
NASIS e: Northeast Australian Satellite Imagery System
NASK e: Polish academic and research network
- P: Naukowa i Akademicka Siec Komputerowa

NASKER e: NASA Ames Kernel
NASL e: Native American Studies Library
NASM e: National Academy of Sports Medicine
NASMO e: NATO Starfighter Management Office
NASMSA e: National Association of State Motorcycle Safety Administrators
NASN e: National Air Surveillance Network
NAS/NAE-SECAN e: NAS/NAE Science and Engineering Committee Advisory to NOAA
NASO e: Native American Student Organization
- e: North American Space Operations

NAS/OAB e: National Academy of Sciences Ocean Affairs Board
NASOG e: National Association of Specialist Obstetricians and Gynaecologists
NASP e: National Aero-Space Plane [project]
- e: National Airspace System Plan
- e: NATO Atomic Supply Point

NASQUAN e: National Stream Water Quality Accounting Network
Nasrec e: National Exhibition Center
NASRS e: Non-Available Status Reporting System
NAST e: National Academy of Science and Technology
NASTRAN e: National Aeronautics and Space Administration Structural Analysis Program
NAT d: Mathematik und Naturwissenschaften
- e: Network Address Translation (or Translator)
- e: Not Air Transportable
- N: Norsk Artikkel-Tjeneste

Nat f: Natation
- f: Nationalistes
- f: Nationalité
- f: Naturel

nat. a: national[t]
- a: naturvidenskabelig

Nat. e: Natural
NATAC e: North Atlantic Chemistry Experiment
NATAS e: National Appropriate Technology Assistance Service
- e: NOAA AVHRR Transcription and Archive System

NATD d: Normaler Alter Telefondienst
- e: Nuclear and Advanced Technology Division

NATED e: National Education Department
NATF N: Norsk Apotekteknikerforbund
NATFAX e: National Weather Facsimile Network
Nat Genet e: Nature Genetics
NATHERS e: National House Energy Rating Scheme
Nat. Hist. e: Natural History
NATI e: National Association of Translators and Interpreters
NATICH e: National Air Toxics Information Clearinghouse
Nat Immun e: Natural Immunity
National Register e: National Register of Historic Places
NATLFED e: National Labor Federation
Natl Med J India e: National Medical Journal of India
NATMAP e: Division of National Mapping
Nat Med e: Nature Medicine

NATMIRC e: National Marine Information and Research Centre
- f: Centre national d'information et de recherche marines
- s: Centro Nacional de Información e Investigación Marinas
NatMIS e: National Marine Information System
NATO e: North Atlantic Treaty Organization
NATOA e: National Association of Telecommunications Officers and Advisors
NATO ASI e: NATO-subsidized Advanced Study Institute
NATOPS e: Naval Air Training and Operating (or Operational) Procedures Standardization
NATO-ZABK d: Zusatzabkommen zu dem Abkommen der Parteien des Nordatlantikvertrags über die Rechtstellung ihrer Truppen [NATO-Truppenstatut]
NATRE e: North Atlantic Tracer Release Experiment
NATS e: NMCS AUTODIN Terminal Subsystem
- e: North Atlantic Track System
NATSF e: Naval Air Technical Support Facility
nat.-soc. a: nationalsocialistisk
Nat Struct Biol e: Nature Structural Biology
NATT e: Northern Australia Tropical Transect
- e: Northern Australian Tropical Transect
nattjfpl r: nattjaktflygplan
Nat Toxins e: Natural Toxins
NAU d: Nauen, Brandenburg
- e: Natal Agricultural Union
- e: Network Accessible Unit
- e: Network Analysis Unit
- e: Network Attachment Unit
- e: Network-Addressable Unit
NAU. e: Network Addressable Unit
NAUHF [Green Pine] e: Northern Area Ultra High Frequency Radio System
NAUN e: Nearest Active Upstream Neighbour
Naunyn Schmiedebergs Arch Pharmacol e: Naunyn-Schmiedebergs Archives of Pharmacology
naut. a: nautisk
- e: nautical
Naut. e: Nautical language
NAUTS e: National Union of Tanganyika Students
NAV e: Navigation [top level computer software component]
Nav N: Navigasjon
- N: Navigatør
- s: Navegación
- s: Navío
NAVADS e: Navy Automated Transportation Documentation System
NAVAF e: Navy-Air Force
NAVAID f: Aide à la Navigation
- s: Ayuda a la Navegación
NAVAIDS e: Navigation Aids
- e: Navigational Airborne Integrated Data Systems
NAVAIR e: Naval Air Systems command headquarters
NAVAIRINST e: NAVAIR Instruction
NAVAIRLANT e: Naval Air Force Atlantic Fleet
NAVAIRMIS e: Naval Aviation LOGCEN MIS: Industrial NAIR Station Workload Control
NAVAIRPAC e: Naval Air Force Pacific Fleet
NAVCAMS e: Naval Communications Area Master Station
NAVCERT e: Naval Computer Incident Response Team
NAVCIPS e: Navy Standard Civilian Pay System
NAV/COM[M] e: Navigation/ Communication
NAVCOMPARS e: Naval (or Navy) Communications Processing and Routing System
NAVCOMSTA e: Naval Communications Station
NAVCOMTELSTA e: Naval Computer and Telecommunications Station
NAVCO-SSACT e: Naval Command System Support Activity
NAVD88 e: North American Vertical Datum of 1988
NAVDAF e: Navy Data Automation Facilities
NAVDF e: Naval Direction Finding Network
NAVELEC f: [Société] Navale d'Électricité
NAVELEX e: Naval Electronic Systems command headquarters
NAVEODTECHCEN e: Naval Explosive Ordnance Disposal Technology Center
NAVEUR e: Naval Forces, Europe
NAVF N: Norges Almenvitenskapelige Forskningsråd
NAVFAC e: Naval Facilities [engineering] Command [headquarters]
NAVFACSYSCOM e: Naval Facilities Systems Command
NAVFCO f: [Société] Navale de Formation et Conseil
NAVICECEN e: Naval Ice Center
NAVIQUE p: Empresa Moçambicana de Navegação
NAV/LOC e: Navigation/Localizer
NAVMACS e: Navy Modular Automated Communications System
NAVMAT e: Navy Material Command
NAVMATCOM e: Naval Material Command
NAVMAT INST e: NAVMAT Instruction
NAVMEDCOM e: Naval Medical Command
NAVMIC e: Naval Military Intelligence Center
NAVMOD e: Navy Model
NAVO e: Naval Oceanographic Office
NAVOCEANO s: Servicio Naval Oceanográfico de EE.UU.
NAVORD e: Naval Ordnance Systems command headquarters
NAVPLANTREP e: Naval Plant Representative
NAVPRO e: Naval (or Navy) Plant Representative[s] Office
NAVSAT e: Navigational Satellite
- e: Navy Satellite System
- e: Specific space-based radionavigation system
- f: Satellite de Navigation
- s: Satélite de Navegación
NAVSEA INST e: NAVSEA Instruction
NAVSEASYSCOM e: Naval Sea Systems Command
NAVSEC e: Naval Ship Engineering Center
NAVSECGRU e: Navy Security Group
NAVSHIPS e: Naval Ship Systems command headquarters
NAVSOC e: Naval Special Warfare Special Operations Component
NAVSOF e: Naval Special Warfare Forces
NAVSOUTH e: Naval Forces, Southern Region
NAVSPACECOM e: Naval (or Navy) Space Command
NAVSPASUR e: Naval Space Surveillance System
NAVSTAR e: Navigation Satellite Timing and Ranging
- e: Navigation System [using] Timing And Range [Global Positioning Satellite]
- e: Navigation System with Time and Ranging [satellite]
- e: Navigational Satellite Timing and Ranging
NAVSTAR-GPS e: Navigational Satellite Timing and Ranging-Global Positioning System
NAVSUPFAC e: Naval Supply Facility
NAVSWC e: NAV Surface Warfare Center
- e: NAV Surface Weapons Center
NAVTA e: North American Veterinary Technician[s] Association
NAVTACSTANS e: Naval Standards for Tactical Data Systems
NAVTAG e: Naval Tactical Gaming
NAVTEX e: Navigational Warnings by Telex
NAVTIP e: Navy Training Improvement Program
NavtjL r: Navigationstjänstledare
NAVWASS e: Navigation and Weapon Aiming Sub System
NAVY e: Navy Building
- e: Never Again Volunteer Yourself
NAW e: National Association of Wholesaler-Distributors
NAWAC e: National Aviation Weather Advisory Committee
NAWAU e: National Aviation Weather Advisory Unit
N-AWB A: Nuwe Afrikaner Weerstandsbeweging
NAWC e: Naval Air Warfare Center
NAWC-WD e: Naval Air Warfare Center Wespons Division
NAWHO e: National Asian Womens Health Organization
NAWHP e: National Association of Womens Health Professionals

NAWPC e: National Aviation Weather Program Council
NAWPP e: National Aviation Weather Program Plan
NAWQC e: National Ambient Water Quality Criteria
NAWS e: Negotiate About Window Size
NAWU e: Ovambo Namibia Workers Union
NAX d: Nachrichten auf Fax
NB a: Nordre Birk
- d: Nebenbedingung[en]
- e: Narrowband
- e: Narrowbeam
- e: Neuroblastoma
- e: Neutral Buoyancy
- e: New Brunswick [laboratory]
- e: Nonbargaining
- e: Nursing Building
- f: Nom de Baptême
- n: Het Nieuwsblad
- r: Norrlands brigad
- s: Niebla[s]
Nb r: Nederbörd
n. b. N: nordlig bredde
N/B e: Narrowband
Nb-95 e: niobium-95
NBA d: Niels Bohr-Archiv in Kopenhagen
- e: Niger Basin Authority
NBAC e: National Bioethics Advisory Commission
NbAl3 e: Niobium Aluminide
NBAR e: Non-Binding Allocation of Responsibility
NBB e: Nike/Black Brant
NBBL N: Norges Boligbyggelags Landsforbund
NBC e: Namibian Broadcasting Corporation
- e: Narrow-Body CRAF
- e: National Bank of Commerce
- e: Not Backward-Compatible
- e: Nuclear, Biological [and] Chemical [warfare]
- e: Nuclear, Biological, Chemical [defense]
- e: Nuclear-Biochemical
- N: Nordens Bondeorganisationers Centralråd
NBCC e: National Board for Certified Counselors, Inc.
- e: Nuclear, Biological and Chemical Contamination
NBCD e: Network Based Collaborative Development
NBC defense e: Nuclear defense, Biological defense and Chemical defense
NBCDI e: National Black Child Develpment Institute
NBCHN e: National Board for Certification of Hospice Nurses
NBCWRS e: Nuclear, Biological, Chemical Warfare Reporting System
- e: Nuclear/Biological/Chemical Warning and Reporting System
NBD e: National Bibliographic Database
- e: No Big Deal [Internet]
- n: Nederlandse Bibliotheek Dienst
NBDP e: Narrow-Band Direct Printing
NBE e: National Bank of Ethiopia
- e: Not Below or Equal
NBEW e: National Business Employment Weekly
NBF e: NetBEUI Frame
- e: Netbios Frame Transport Protocol
- N: Norges Bilbransjeforbund
- N: Norsk Blindeforbund
- N: Norsk Blomsterhandlerforbund
- N: Norske Bankfunksjonærers Forbund
- N: Norske Byggvareprodusenters Forening
NBFCP e: Netbios Frames Control Protocol
NBFD e: No Big Fucking Deal [Internet]
NBFI e: Non-Bank Financial Institution
NBG N: Norsk Bilforsikring Gjensidige
NBHU e: Nitrogen Blower Heater Unit
NBI e: Non-Battle-Injury
- N: Norges Byggforskningsinstitutt
NBIC e: National Board Inspection Code
- e: National Building and Investment Corporation
NBIFTZ e: National Board of Industrial Free Trade Zones
NBII e: National Biodiversity Information Infrastructure
- e: National Biological Information Infrastructure
NBIN e: National Biodiversity Information Network
NBIOME e: Northern Biosphere Observation and Model[l]ing Experiment
NBIS e: Neutral Beam Injection System
- e: Northern Biosphere Information System
NBK f: Nouvelle Banque de Kinshasa
NBL d: Neue Bundesländer
- e: Neutral Beam Line
- e: Nocturnal Boundary Layer
- N: Norges Befalslag
NBMA e: Non-Broadcast Multiple Access
NBMO e: NonBonding Molecular Orbital
NBNA e: National Benchmark Network for Agrometeorology
NB1 ... NBm e: Aktuelle Nebenbedingungen
NBNSC e: National Board of Nutrition Support Certification
NBO e: Network Build Out
- e: Network Business Opportunity
- r: Nya Befälsordning
NBP e: Name-Binding Protocol
- e: Near-real-time and Batch Processing
- e: Netbios Protocol
- N: Norges Bondelags Pressekontor
NBPC e: Neutral Beam Power Conversion
NBQ s: Nuclear, Biológico y Químico
NBR N: Norske Bedriftsavisers Redaktørforening
n.br. N: nordlig bredde
- r: nordlig bredd
Nb[r] f: Nombre
Nbre f: Novembre
NBRI e: National Botanical Research Institute
NBRVF e: Narrowband Radio Voice Frequency
NBS e: Natal Building Society Holdings
- e: National Biological Service
- e: National Biological Survey
- e: National Bureau of Standards
- N: Norges Bonde- og Småbrukerlag
nBSA e: normalized Body Surface Area
NBS-DATA e: Nordic Building Research Cooperation Group, Working Group for Information Technology [DATA]
NBSL e: National Biological Standards Laboratory
NBST e: Narrow-Band Subscriber Terminal
- e: Narrowband Secure Terminal
NBT e: Near-Ballistic Transport
- e: NETBIOS on TCP/IP
NBTF e: National Biomedical Tracer Facility
- e: Neutral Beam Test Facility
NBTI N: Norges Betongtekniske Institutt
NBTS e: Neutral Beam Test Stand
NBTT e: Nigerian Bible Translation Trust
NBTU e: Narrowband Trunking Unit
NBU a: Nordisk Benzinhandler Union
- e: National Biodiversity Unit
NBVM e: Narrow Band Voice Modulation
NBVS e: Neutral Beam Vacuum System
NBz d: Nitrobenzene
NC e: Narrowbeam Coverage
- e: National Committee
- e: Network Computer[s]
- e: Network Control
- e: Network Coordinator
- e: New Caledonia
- e: Newcastle Connection
- e: No Carry
- e: NOAA Corps [operations]
- e: Noncarcinogen
- e: Norton Commander
- e: Number of unallocated Channels at node
- e: Numeric[al] Control/25
- e: Old Cafeteria Building
- f: Nombre Cardinal
- N: Norsk Cementforening
- s: Nudo de Comunicaciones
nc r: nitrocellulosa
N-C e: Nitrogen-Carbon ratio
n/c f: non compté
- s: nuestra cuenta
NCA e: National Climatological Center
- e: National Command Authorities
- e: National Committee on Agrometeorology
- e: National Communication Association
- e: National Computerization Agency
- e: National Contact Address
- e: Network Communications Adapter
- e: Network Computing Architecture
- e: Network Control Analysis
- e: Neutralized Current Acid
NCAA e: NATO Civil Air Augmentation
- e: Non-nuclear Consumable Annual Analysis System
NC-AAPSO e: Nigerian Committee of the Afro-Asian Peoples Solidarity Organization
NCADI e: National Clearing House for Alcohl and Drug Information
NCAL e: New Caledonia
NCAM e: Neural Cell Adhesion Molecule
NCAQ e: National Commission of/on Air Quality

NCAR e: National Center [for] Atmospheric Research
- e: National Committee Against Removals
- f: Centre National de Recherche Atmosphérique
- s: Centro Nacional de Investigaciones Atmosféricas

NCAS e: National Centre for Supercomputing Applications

NCASC e: National Capital Administrative Support Center

NCASI e: National Council for/of/on the Paper Industry [for] Air and Stream Improvement[s], Inc.

NCAT e: National Center for Asphalt Technology

NCATA e: National Coalition of Arts Therapies Associations

NCAW e: Neutralized Current Acid Waste

NCB e: National Classification Board American Trucking Association
- e: National Cooperative Bank
- e: Navidrill Core Barrel
- e: NETBIOS Control Block
- e: Net Change of Biomass
- e: Nuclear Contingency Branch
- r: Norrlands Skogsägares Cellulosa AB

NCBA e: National Cattlemens Beef Association

NCBHR e: National Council on Bioethics in Human Research [Canada]

NCBP e: National Contaminant[s] Biomonitoring Program

NCBR e: Near-Commercial Breeder Reactor

NCC e: Dutch National Computer Catalogue
- e: Namibia Chiefs Council
- e: Namibia Council of Churches
- e: National Cancer Center [Japan]
- e: National Classification Commission
- e: National Climate Center (or Centre)
- e: National College of Chiropractic
- e: National Command Center
- e: National Communications System
- e: National Computer Center
- e: National Consultative Committee
- e: National Consultative Council
- e: National Coordinating Center
- e: National Coordinating Committee
- e: National Council on Codependence
- e: Nature Conservancy Council
- e: Negotiated Contract Cost
- e: Nepera Chemical Company
- e: Network Communications Central
- e: Network Computer Client
- e: Noncancellable Commitments
- e: NORAD Command Center
- f: Notre Cause Commune
- n: Nederlandse Centrale Catalogus

n/c.c. s: nuestra cuenta corriente

NCCA e: National Christian Counselors Association
- e: National Commission for the Certification of Acupuncturists
- e: North Carolina Classical Association

NCCAOM e: National Certification Commission for Acupuncture and Oriental Medicine

NCCAS e: National Committee for Climate Change and Atmospheric Sciences

NCCAW e: National Consultative Committee on Animal Welfare

NCCC e: National Climate Climatological Center
- e: National Council of Churches of Christ

NCCDPC e: NATO Command, Control and Information System and ADP Committee

NCCDPHP e: National Center for Chronic Disease Prevention and Health Promotion

NCCDS e: NCC Data System
- e: Network Control Center Data System[s]

NCCE e: National Center for Computational Electronics

NCCF e: National Childhood Cancer Foundation
- e: Network Communication[s] Control Facility
- e: NOAA Central Computer Facility

NCCH e: National Centre for Classification in Health

NCCHS e: Nonirradiated Core Component Handling System

NCCI e: Namibian Chamber of Commerce and Industry

NCCIDSA e: North Central Chapter Infectious Diseases Society of America

NCCIS e: NATO Command, Control and Information System

NCCK e: National Consultative Committee on Kangaroos
- e: National Council of Churches of Kenya

NCCM e: Nchanga Consolidated Copper Mines

NCCMAG e: National Computer Center Management Advisory Group

NCCN e: National Comprehensive Cancer Network
- e: National Coordinating Committee of Nkrumahists

NCCNSW e: Nature Conservation Council of New South Wales

NCCPA e: National Commission on Certification of Physician Assistants

NCCPM e: Navy Command and Control Process Model

NCCR e: National Convention for Construction and Reform
- e: National Coordinating Committee for the Repatriation of South African Exiles

NCC-RIPE e: RIPE Network Coordination Center

NCCS e: National Childrens Cancer Society
- e: National Credit Counseling Services
- e: Naval (or Navy) Command and Control System
- e: NORAD Command and Control System
- e: Nuclear Command and Control System
- e: Numerical Control Computer Sciences

NC curves e: Noise Criterion curves

NCCWS e: Nuclear Component Cooling Water System

NCD e: Nascom Circuit Directory
- e: National Christian Democratic Party
- e: National Commission for Democracy
- e: Net Control Device
- e: Network Computing Device[s]
- e: New Chemicals Data base
- e: Nonlinear Circular Dichroism

NCDB e: National Cancer Data Base
- e: National Cartographic Data Base

NCDC e: National Climate (or Climatic) Data Center (or Centre) [of NOAA]
- f: Centre National de Données Climatiques
- s: Centro Nacional de Datos Climáticos

NCDCDS e: National Committee-Digital Cartographic Data Standards

NCDCS e: Narrow band Coherent Data Collection System

NCDDG e: National Cooperative Drug Discovery Group

NCDF e: Non-Coded Digital Facsmile
- f: Fac-similé numérique non codé
- s: Facsímil numérico no codificado

NCDMDP e: National Cladding/Duct Materials Development Program

NCDP e: National Christian Democratic Party

NCDRS e: NMCS Crisis Data Retrieval System

NCDS e: NASA Climate Data System
- e: National Centre for Development Studies
- e: National Cooperative Dialysis Study

NCDT e: Non-Chargeable Down Time

NCE e: National Centre of Excellence
- e: National Committee for the Environment
- e: Net Carbon Emission
- e: Network Control Element

NCEB e: NATO Communications-Electronics Board

NCEC e: North Coast Environment Centre

NCEE e: National Center for Earthquake Engineering Research
- e: Northeast Consortium for Engineering Education

NCEES e: National Council of Examiners for Engineering and Surveying

NCEG e: Numerical C Extensions Group

NCEH e: National Center for Environmental Health

NCEHIC e: National Center for Environmental Health and Injury Control

NCEMCH e: National Center for Education in Maternal and Child Health

NCEMI e: National Center for Emergency Medicine Informatics

NCEP e: National Center[s] (or Centre) for/of Environmental Prediction

NCEPA e: National Civil Emergency Planning Authorities

NCEPH e: National Centre for Epidemiology and Population Health

NCEPI e: National Center for Environmental Publications and Information

NCEPOD e: National Confidential Enquiry into Perioperative Deaths

NCEUR e: National Security Agency Europe

NCF e: Netware Command File
- e: Network Control Facility
- e: No Clean Flux
NCFA e: North Carolina Forestry Association
NCFAE e: National Council of Forest Association Executives
NCFB N: Norsk Centralforening for Boktrykkere
NCFST e: National Center for Food Safety and Technology
NCG e: NCSS to CSS Guard
NCGA e: Northern California Golf Association
NCGCP e: National Climate and Global Change Program
NCGEN f: Nouvelle Compagnie Générale Électrique de Nancy
NCGIA e: National Center (or Centre) for Geographic Information and Analysis
ncg/krut r: nitroglycerin krut
ncgl r: nitroglycerin
ncgl-krut r: nitroglycerinkrut
NCGR e: National Center for Genome Research
- c: National Center for Genome Resources
NCHC e: National Coalition on Health Care
NCHE e: National Commission on Higher Education
NCHEC e: National Commission for Health Education Credentialing
NCHF e: Northern California Hemophilia Foundation
NCHGR e: National Center for Human Genome Research
NCHSE e: National Centre for Human Settlements and Environment
NCHSR e: National Center for Healtlh Services Research and Health Care Technology Assessment
NCHSTE e: National Consortium of Health Science and Technology Education
NCHSTP e: Natinal Center for HIV, STD, and TB Prevention
NCI e: Network Control Information
- e: Non Coded Information
nCi e: nanocurie
NCIA e: Native Client Interface Architecture
NCICAS e: National Cooperative Inner-City Asthma Study
NCID e: National Center for Infectious Diseases
NCIFR e: National Committee to Improve Financial Reporting
NCIGBP e: Nationaa Committee for IGBP
NCIHE e: National Committee of Inquiry into Higher Education
NCIPC e: National Center for Injury Prevention and Control
NCIS e: NADIR Climate Interferometer/ Spectrometer
- e: NADIR Cryogenic Interferometer/ Spectrometer
nc-krut r: nitrocellulosakrut
Ncl s: Nuclear
NCLAN e: National Crop Loan Assessment Network
NCLANG e: Nacelle Angle
NCLEHA e: Natural Conference of Local Environmental Health Administrators
Ncle[NSW] e: Newcastle [New South Wales]
Ncle[UK] e: Newcastle upon Tyne [UK]
NCLEX e: National Council Licensure Examination
NCLF e: National Childrens Leukemia Foundation
NCLP e: National Contract Laboratory Program
NCLS e: NEIC Library System
NCM e: National Coastal Monitoring
- e: National Convention Movement
- e: Network Configuration Management
- e: Network Control and Management
- e: Notice of Commencement of Manufacture
NCMA e: National Concrete Masonry Association
NCMB e: NATO Configuration Management Board
NCMC e: NORAD [North American Air Defense [Command]] Cheyenne Mountain Complex
N C Med J e: North Carolina Medical Journal
NCMO e: NAVAIDS Communications Management Office
- e: Navigational Aids/Communications Management Officer
NCMOS e: N-Channel [Silicon Gate Reversed] CMOS
NCMP e: NATO Configuration Management Plan
- e: Navy Capabilities and Mobilization Plan
NCMRED e: National Council of Marine Resources and Engineering
NCMRWF e: National Centre for Medium Range Weather Forecasting
NCMS e: National Center for Manufacturing and Science
- e: National Classification Management Society, Inc.
- e: Navy Configuration Management System
NCMT e: Numerical Control for Machine Tools
NCN e: New Century Network
NCNA e: National Council Nonprofit Associations
- e: North Carolina Nurses Association
NCNC e: National Convention of Nigerian Citizens
- e: National Council of Nigeria and Cameroon
NCNCNC e: No Coffee, No Chocolate, No Computer [Internet]
NCND e: National Center for Nutrition and Dietetics
NCNIA e: Northern California Nursing Informatics Association
NCO e: NCEP Central Operations
- e: Negotiated Consent Order
- e: Net Control Operator
- e: Non Commissioned Officer
- f: Nom Commercial
NCOA e: National Council on the Aging
- e: Non Commissioned Officers Association
NCOC e: NORAD [North American Air Defense [Command]] Combat Operations Center
NCOFF e: National Center On Fathers and Families
NCO-HPCC e: National Coordination Office for HPCC
NCOIC e: Non-Commissioned Officer-in-Charge
NCOL f: Nouveau Courant Opportuniste et Liquidateur
NCOQ e: Non-Cost Of Quality
NCOS e: National Committee for Oceanic Sciences
- e: Network Computer Operating System
NCP e: National Climate Programme
- e: National Convention Party
- e: National oil and hazardous substances [pollution] Contingency Plan
- e: Netware Core Protocol
- e: Network Configuration Process
- e: Network Control Processor
- e: Non-Carbon Paper
- e: Non-Consultative Party
- e: Noncompliance Penalties
- e: Nonconformance Penalty
- e: Nonconformance Pickup
- e: NORAD Command Post
- e: Not Copy Protected
NCPA e: National Center for Policy Analysis
NCPCA e: National Committee to Prevent Child Abuse
NCPD e: National Council for Population and Development
NCPDI e: National Coastal Pollutant Discharge Inventory
NCPDP e: National Council of Prescription Drug Programs
NCPEG e: Navy Contractor Performance Evaluation Group
NCPIASI e: National Council of the Paper Industry for Air and Stream Improvements
NCPIE e: National Council on Patient Information and Education
NCPOA e: National Chronic Pain Outreach Association
NCPP e: National Colored Peoples Party
NCPS e: National Crime Prevention Strategy
- e: Network Command Processing System
NCQA e: National Committee for/on Quality Assurance
NCR e: National Capital Region
- e: National Cash Register Co.
- e: Non-Compliance Report
- e: Non-Compliance Request
- e: Non-Conformance Report
- e: Number of Collisions Register
NCRA e: National Cancer Registrars Association
- e: National Centre for Radio Astrophysics
- e: National Court Reporters Association
N C R B e: North Campus Recreation Building
NCRC e: NAVAIDS [Navigational Aids] Communications Readiness Center
NCRDP e: Nimba County Rural Development Plan

NCRF e: National Court Reporters Foundation
NCRP e: National Committee on Radiation Protection and Measurement
NCRR e: National Center for Research Resources
NCRW e: Neutralized Cladding Removal Waste
- e: Nuclear Cladding Removal Waste
NCS e: INMARSAT Network Coordination Station
- e: NASI Connection Service
- e: National Calibration Service
- e: National Communications Service
- e: National Communication[s] Systems
- e: National Compliance Strategy
- e: National Conservation Strategy
- e: Network Communication Standard
- e: Network Computer Server
- e: Network Control Station
- e: Network Control Subsystem
- e: NORAD Computer System
- f: Station de Coordination du Réseau INMARSAT
NCSA e: National Center for Super computer (or computing) Applications
- e: National Conservation Strategy for Australia
- e: Nebraska Council of School Administrators
NCSAC e: National Conservation Strategy Advisory Council
NCSC e: National Communications Security Committee
- e: National Computer (or Computing) Security Center
- e: Nuclear Criticality Safety Committee
NCSD e: National Communications Security Directive
NCSEB e: National Center for Synthesis in Experimental Biology
NCSI e: Network Communication Services Interface
NCSL e: National Computer System[s] Laboratory
- e: National Council of State Legislatures
NCSNDR e: Network Computing System Network Data Representation
NCSORG e: Navy Control of Shipping Organization
NCSP e: National Certified School Psychologist
- e: National Soil Conservation Program
NCS-R e: NCS Replacement
NCSR e: National Centre for Scientific Research
- e: National Committee for/on Space Research
NCSRP e: North Central Soybean Research Program
NCSS e: Naval Control Shipping System
- e: NGT Command and Status System
NCS&T e: National Council for Science and Technology
NCST e: National Council of Science and Technology
NC SU CAPPS e: North Carolina State University Center for Aseptic Processing and Packaging Studies
NCT e: Network Control Terminal[s]
- f: Niveau de Culture
NCTAF e: Nuclear Communications Task Force
NCTAMS e: Naval Computer and Telecommunications Area Master Station
NCTC e: Naval Computer and Telecommunications Command
NCTE e: National Council of Teachers of English
- e: Network Channel-Terminating Equipment
NCTH e: National Clearinghouse on Tobacco and Health
NCTI/R e: Non-Cooperative Target Identification/Recognition
NCTL e: National Computer and Telecommunications Laboratory
NCTN e: NASA Commercial Technology Network
NCTR e: Noncooperative Target Recognition
NCTTA e: National Coompetitive Technology Transfer Act
NCU e: University of North Carolina, Math-Physics Library
NCV e: Nerve Conduction Velocities
- e: Noze Control Vehicle
NCVDG e: National Cooperative Vaccine Development Group
NCVQ e: National Council for Vocational Qualifications
NCW e: Naval Coastal Warfare
NCWA e: Neutralized Current Acid Waste
NCWC e: Naval Coastal Warfare Commander
NCWCD e: National Commission for Wildlife Conservation and Development
NCWM e: National Conference of Weights and Measures
NCWP e: Near Coastal Waters Program
NCWS e: National Council of Womens Societies
- e: Non-Community Water System
Ncy f: Nancy
NCYD e: National Center for Youth with Disabilities
ND d: Natürliche Dummheit
- e: Network Disk
- e: Networks Division
- e: No Data
- e: No Degree Objective
- e: Nondetect
- e: Normal Deployability posture
- e: Normalized Difference
- e: Not Determined
- e: Not Documented
- f: Non[-]Disponible
- f: Note Diplomatique
- f: Numération Décimale
Nd f: Nord
N/D e: N-Dimensional
NDA d: Norddeutsche Datenautobahn
- e: Nigerian Defense Academy
- e: Non-Destructive Analysis
- e: Non-Disclosure Agreement
- e: Not Diagnosed with Anything
Nda f: Non dénommé ailleurs
NDAA e: National Defense Authorization Act
- e: Non-Developmental Airlift Aircraft
NDACAN e: National Data Archive on Child Abuse and Neglect
NDADS e: NSSDC Data Archive and Distribution System
NDAIS e: Nondestructive Assay Isotopic System
NDAP e: National Democratic Action Party
NDAR e: National Directory of Australian Resources
NDATS e: Naval Data Automation Technical Standard
NDB e: Non Directional [radio] Beacon
NDBMS e: Nonstandard Database Management System
NDBO e: National Data Buoy Office
NDC e: National Data Center (or Centre)
- e: National Defense Committee
- e: National Democratic Congress
- e: Negative Differential Conductance
- e: Non-Destructive Cleaning
- e: Nondairy Cattle
- e: Normalized Device Coordinate
- e: Nuclear Detector Circuit
Ndca f: Non dénommé ni compris ailleurs
NDCC e: National Defense Command Center Canada
- e: Network Data Collection Center
NDCDB e: National Digital Cartographic Data Base
NDCMP e: North Dakota Cloud Modification Project
NDCS e: Network Data Control System
NDD e: Negotiation Decision Document
- e: Network Delivery Device
NDDF e: National Drug Data File
NDDK e: Network Device Development Kid
NDDL e: Neutral Data Definition Language
NDDN e: National Dry Deposition Network
NDDP e: National Decontamination and Decommissioning Program
NDDRP e: Natural Disaster and Drought Relief Policies
NDE e: National Directorate of Employment
- e: News Development Environment
- e: Non-Destructive Evaluation
NDEF e: Not [to be] Defined
NDEI e: National Diabetes Education Initiative
Nde lA f: Note de l'Auteur
Nde lE f: Note de l'Éditeur
N de Obst s: Nudo de Obstáculos
NDEP e: National Diabetes Education Program
- e: Nevada Division of Environmental Protection
NDEW e: Nuclear Directed Energy Weapon[s]
NDF e: Namibian Defense Force
- e: Narrative Data File
- e: Network Definition Facility
- e: Network Definition File
- e: No Defect[s] Found

Ndf f: Nécessité de fermer
ndf. a: nedenfor
NDGA e: Nordihydroguaiarectic Acid
$NdGaO_3$ e: Neodymium Gallate
NdGaO3 e: Neodymium Gadolate
NDHC e: National Defense Headquarters, Canada
NDHX e: NaK-to-air Dump Heat Exchanger
- e: Natural Draft Heat Exchanger
NDI e: National Democratic Institute for International Affairs
- e: Non-Development[al] Item[s]
NDIC e: National Defense Intelligence Center
NDIR d: Nichtdispersive Infrarot-absorption
- e: Non-Dispersive Infra-Red analysis
- e: Non-Dispersive Infra-Red [gas analyzer]
- e: Non-Dispersive Infra-Red spectroscopy
NDIS e: Network Device Interface Specification
- e: Network Driver Interface Specifications
NDL e: Logical Network Disk
- e: National Digital Library
- e: Network Data[base] Language
NDLA f: Note de l'Auteur
NDLE f: Note de l'Éditeur
NDLEA e: National Drug Law Enforcement Agency
NDLR f: Note de la Rédaction
NDLT f: Note de la Traductrice
NDM d: Netzwerk-Datenmodell
- e: Netlaboratories Dual Manager
- e: Netview Distribution Manager
- e: Nigerian Democratic Movement
NDM. e: Normal Disconnect Mode
NDMAC e: Nonprescription Drug Manufacturers Association of Canada
NDMC e: National Diamond Mining Company
NDMCS e: Nellis Digital Microwave Communications System
NDMDA e: National Depressive and Manic-Depressive Association
NDMS e: Netware Distributed Management Services
NDN e: Nodal Data Network
- e: Non-Delivery Notice
ndn r: nämndeman
NDOC e: National Defense Operations Center
NDOW e: Nevada Division of Wildlife
NDP e: National Drought Policy
- e: Neutron Depth Profiling
- e: New Democratic Party
- e: Numeric Data Package
NDPF e: NASA Data Processing Facility
NDPL e: National Democratic Party of Liberia
NDPS e: Netware Distributed Printing Services
- e: Non-Disruptive Path Switching
NDR e: Network Data Representation
- e: New Document Reference
Ndr. a: Nordre
NDRA e: Natural Disaster Relief Arrangements
NDRF e: National Defense Reserve Fleet
NDRO e: Non-Destructive Read Out
NDS e: National Design Specification for wood construction
- e: Netware Directory Service[s]
- e: NPIC Data System
- e: Nuclear Detonation Detection System
- e: NUDET [Nuclear Detection] System
- f: Néo-Démocrates Sénégalais
- f: Note De Service
- N: Det Nordlandske Dampskipsselskap A/S
Nds. D: Niedersachsen
NDSC e: National Defense and Security Council
- e: Network for [the] Detection of Stratospheric Change
NDSM e: Notre Dame Saint Marys [College]
NDST e: Non-Dimensional Special Tooling
NDT e: Network Design Tool
- e: Non-Destructive Testing
- e: Normal Device Termination
NDTA e: Non-Destructive Testing Association
NDTE e: North Dakota Tracer Experiment
NDTM e: National Defense Transportation Model
NDTS e: Nonlinear Dynamics Time Series
NDTT e: Nil Ductility Transitions Temperature
NDU e: National Democratic Union
- e: National Development Unit
NDUV e: Non-Dispersive Ultraviolet Spectroscopy
NDV e: NASP [National Aerospace Plane] Derived Vehicle
NDVI e: Normalized Difference Vegetation Index
NDX e: Index
NDY e: Not Diagnosed Yet
Nd-YAG e: Neodymium [[3+]-doped] Yttrium Aluminum Garnet [laser]
NE e: National Estate
- e: Network Element
- e: Non Exempt
- e: Office of Nuclear Energy
- e: University of New England Herbarium, Armidale International Acronym
- f: Nationalité Étrangère
- f: Nord-Est
- f: Note Explicative
- f: Nouvelle Édition
- f: Numéro d'Édition
- s: Nordeste
Ne f: Naissance
N.E. e: New England
- e: North East
ne1 e: anyone [Internet]
NEA d: Neustadt-Bad Windsheim/Aisch, Bayern
- d: Nichtdeterministischer Endlicher Automat
- e: National Educators Association
- e: Net Expenditure Accrued
- e: Nuclear Energy Agency of the OECD
- f: Nouvelles Editions Africaines
- f: Numérotation Européenne des Articles
NE-a e: Neuroepithelioma
NEAA e: North-Eastern Anthropological Association
NEACP TTY e: NEACP Teletypewriter Network
NEACRP e: Nuclear Energy Agency Committee on Reactor Physics
NEADS e: Northeast Air Defense Sector
NEAG e: Nevada Environmental Advisory Group
NEAP e: National Environmental Action Plan
NEAPACC e: North East Atlantic Palaeooceanography and Climate Change
NEAR e: National Electronic Accounting and Reporting system
- e: Near East Asian Region
NEAR GOOS e: North-East Asian Regional GOOS
- f: Composante régionale du GOOS pour l'Asie du Nord-Est
- s: GOOS Regional para el Nordeste de Asia
NEARGOOS e: North East Asian Regional GOOS
NEARnet e: New England Academic and Research Network
NEARSS e: Northeast Area Remote Sensing System Association
NEAT e: National Environment Awareness Trust
- e: Network Expert Advisory Tool
- e: Next Advanced Technology
Neau f: Nouveau
NEAWS e: NORAD/ADCOM Emergency Airborne Warning System
NEB d: Nebra/Unstrut, Sachsen-Anhalt
- e: National Environment Board, Thailand
- e: National Environmental Board
- e: Net Environmental Benefits
- e: Nonisothermal Energy Balance
NEBIOS e: Network Basic Input/Output System
NEBS e: New Equipment Building System
NEC e: National Electoral Commission
- e: National Electric Conference
- e: National Exhibition Centre, Birmingham
- e: Network Equipment Configuration
- e: Nippon Electronic Corporation
- e: Nuclear Energy Commission
NEC 9801 e: A series of 386/486 machines popular in Japan
NECB e: New England Coastal Basins
NECC e: National Education Computing Conference
- e: North Equatorial Counter Current
- e: NTB [National Test Bed] Experiment Control Center
NECJ e: New England Classical Journal
NECL e: Nonepitheliotropic Cutaneous Lymphosarcomas
NECON e: National Electoral Commission of Nigeria
NECOP e: Nutrient-Enhanced Coastal Ocean Productivity
NECOR e: North East Consortium
NECOS e: Network Coordination Station
NECPRC e: National Economic Committee of the Peoples Republic of China
NECS e: Navy Embedded Computer System
- e: Normal Environmental Control System

NECT e: National Research Council of Thailand
- e: North East China Transect

NECTEC e: National Electronics and Computer Technology Center, Thailand

NECZ e: Near-Equatorial Convergence Zone

NECZAM e: National Educational Company of Zambia

NED e: NASA Extragalactic Database
- e: Netview-DECnet Interface Option
- e: Network of Egghead Divers
- e: Noise Emitting Diode

N.E.D. e: New English Dictionary

ned. a: nederst

NEDALS e: Navy Engineering Data Asset Locator System

NEDARC e: National EMSC Data Analysis Resource Center

NEDBAG e: NATO Emitter Database Advisory Group

NEDBAS e: NATO Emitter Database

NEDCOM e: National Education Committee

nedenn. a: nedennævnte

nedenst. a: nedenstænde

NEDER f: Norme d'Echange de Données en Environnement RTC

nederl. a: nederlandsk

nedild N: nedholdende ild

NEDIPS e: NEC Data-Flow Image Processing System

NEDLAC e: National Economic Development and Labo[u]r Council

NEDLIB e: Networked European Deposit Library

NEDN e: Navy Environmental Data Network

NEDR e: Noise Equivalent Delta Reflectivity

NEDRES e: National Environmental Data Referential (or Referral) Services
- e: National Environmental Data Referral System

NEDSA e: NAVSEA Engineering Drawing Support Activity

NEEC e: National Environmental Enforcement Council

NEEDHA e: National Electrical Engineering Department Heads Association

NEEDS e: NASA End-to-End Data System[s] [programs]
- e: National Engineering Education Delivery System
- e: Needs Survey

NEEE e: Near Eastern Equine Encephalomyelitis

NEEJ e: National Environmental Enforcement Journal

NEETS e: Naval Electronic Environment Training System

NEF d: Notarzteinsatzfahrzeug
- e: NASA Extended FORTRAN
- e: Network Element Function Block
- e: Northern Elders Forum
- N: Norges Eiendomsmeglerforbund
- N: Norges Elektrohandlerforbund
- N: Norsk Elektroteknisk Forening
- N: Norske Eksportørers Forening

NEFA e: North East Forest Alliance

NEFC e: North-East Fisheries Center

NEFMA e: NATO EFA [Eurofighter Management Agency]

NEFS e: Network Extensible File System

NEG e: Negate
- e: Numerical Experimentation Group
- e: WOCE Numerical Experimentation Group

Nég f: Négociant

NEGL e: Negligible

NEGP e: National Estate Grants Program

NEHAP e: National Environmental Health Action Plan

NEHAWU e: National Education, Health and Allied Workers Union

NEHP e: National Register of Historic Places

NEHRP e: National Earthquake Hazards Reduction Program

NEI e: National Eye Institute

NEIC e: National Enforcement Investigation Center
- e: Netherlands East India Company
- f: Centre national d'information sur les tremblements de terre
- f: Service national d'information sur les tremblements de terre
- s: Centro Nacional de Información Sísmica
- s: Servicio Nacional de Información Sísmica

NEIDA e: National Estuarine Inventory Data Atlas

NEIL e: Neil Hall

NEK N: Norsk Elektroteknisk Komité

NEKF N: Norsk Elektriker og Kraftstasjonforbund

NELA e: National Employment Lawyers Association

NELC e: Naval Environmental Protection Act

NELIMO p: Núcleo de Estudo de Línguas Moçambicanas Faculdade de Letras, Universidade Eduardo Mondlane, Moçambique

NELS e: Nuclear Effects Link Simulator

NEMA e: National Electric[al] Manufacturers Association[-Medical]
- e: National Electronic Manufacturers Association
- e: Network for Environmental Management and Auditing

NEMD e: Non-Equilibrium Molecular Dynamics

NEMF N: Norske Elektromontørers Forening

NEMFB e: New England Motor Freight Bureau

NEMI e: National Electrical Manufacturers Initiative

NEMKO N: Norsk Elektrisk Material Kontroll

neml. a: nemlig

NEMO e: Nano-Electronic Modeling project

NEMOC e: Naval European Meteorology and Oceanography Center

NEMRL e: New England Marine Research Laboratory

NEMS e: NASA Equipment Management System
- e: National Emergency Management System
- e: National Environment Management Strategies (or Strategy) [program]
- e: Nimbus-E Microwave Spectrometer
- s: Estrategia Nacional de Gestión del Medio Ambiente

NEMSA e: National EMS Alliance

NEO e: Network Extension Option
- e: New Employee Orientation
- e: Non-Combatant Evacuation Operation
- N: Norsk Elevorganisasjon

NEOC e: National Earth Observation Center

NEODAT e: Inter-Institutional Database of Fish Biodiversity in the Neotropics

Neog. e: Neogene

NEON e: NBLC host facility

NEONS e: Navy Environmental Operational Nowcasting System

NEOP e: New Employees Orientation Program

NEOS e: Networked Educational On-line System

NEP e: National Earthquake loss reduction Program
- e: National Environmental Policy
- e: National Estuary Programs
- e: National Extension Programs
- e: Net Ecosystem Productivity
- e: Network Entry Point
- e: New Economic Plan
- e: Noise Equivalent Power
- e: Nominal Entry Point
- e: Nuclear Environment Protection

NEPA e: National Environment Protection Authority
- e: National Environmental Policy Act
- e: National Environmental Policy Administration
- e: National Environmental Protection Act
- e: National Environmental Protection Agency
- e: Nuclear Energy [for the] Propulsion [of] Aircraft
- e: Nuclear Environmental Protection Agency
- f: Agence Nationale de Protection de l'Environnement
- s: Organismo Nacional de Protección del Medio Ambiente

NEPA-CX e: National Environmental Policy Act/categorical exclusion

NEPADTS e: NEPA Document Tracking System

NEPATS e: NEPA Tracking System

NEPC e: National Environment Protection Council
- e: Nigerian Export Promotions Council

NEPCG s: National Presbyterian Church of Guatemala

Nephrol Dial Transplant e: Nephrology, Dialysis, Transplantation

NEPI e: National Environmental Policy Institute

NEPM e: National Environment Protection Measure

NEPMs e: National Environment Protection Measures
NEPMU5 e: Navy Environmental and Preventive Medicine Unit No.5
NEPP e: National Early Psychosis Project
- e: National Environment Policy Plan
- e: National Environment Protection Plan
NEPPS e: National Environmental Performance Partnership System
NEPRU e: Namibian Economic Policy Research Unit
- e: Namibian Economic Policy Review Unit
NER e: Nondestructive Examination Report
NERA e: National EMSC Resource Alliance
NERC e: National Environmental Research Center (or Centre)
- e: National Environmental Research Council
NERC-EPFS e: NERC Equipment Pool for Field Spectroscopy
NERD&D e: National Energy Research Development and Demonstration
NERDDC e: National Energy Research, Development and Demonstration Council
NEREIS e: European deep-sea drilling program ship
NEREP e: Nuclear Execution and Reporting
NERF e: Navy Emitter Reference File
NERFC e: North East River Forecast Center
NERI e: National Employee Rights Institute
- e: National Environmental Research Institute
NERL e: National Exposure Research Laboratory
NEROS e: ASRL-RTP Northeast Regional Oxidant Study
NERP e: National Environmental Research Park
NERRS e: National Estuarine Research Reserve System
NERRT e: Nuclear Energy Reactor Review Team
NERSA e: European Fast Reactor Power Station
- e: Builder and future operator of SUPERPHENIX
NERSC e: National Energy Research Scientific Computing
- e: National Energy Research Supercomputer (or Supercomputing) Center [at LLNL]
NES d: Rhön-Grabfeld in Bad Neustadt/Saale, Bayern
- e: National Education Supercomputer
- e: National Enuresis Society
- e: National Environment Secretariat
- e: Network Encryption System
- e: Network of Emerging Scientists
- e: Nuclear Energy System
- e: Nuclear Estate with Small holdings
NESA a: Nordsjællands Elektricitets og Sporvejs A/S
- e: Nigerian English Studies Association
NESAC e: National Electronic Switching Assistance Center
NESB e: Non-English Speaking Background
NESBU e: Nuclear Energy Systems Business Unit Westinghouse
NESC e: National Environmental Supercomputing Center
- e: New England Science Center
- e: Non-English Speaking Country
- e: Nuclear and Environmental Safety Council
NESCAUM e: Northeast States for Coordinated Air Use Management
NesCom e: IEEE-SA Standards Board New Standards Committee
NESDB e: National Economic and Social Development Board, Thailand
NESDD e: NOAA Earth Science Data Directory
- e: NOAA Earth System Data Directory
NESDIP e: NESTOR Shielding and Dosimetry Improvement Program
NESDIS e: National Environment, Satellite, Data and Info[rmation] Service
- e: National Environmental Satellite Data Information System
- f: Service national de satellites, de données et d'information en matière d'environnement
- s: Servicio Nacional de Satélites, de Datos e Informaciones Ambientales
NESDP e: National Economic and Social Development Plan
NESDRES e: National Environmental Data Referral Service
NESEA e: Northeast Sustainable Energy Association
NESEC e: Naval Electronic Systems Engineering Center
NESFD e: Noise Equivalent Spectral Flux Density
NESHAP[s] e: National Emission[s] Standard[s] for Hazardous Air Pollutants
NESIP e: Naval Explosive Safety Improvement Program
NESO e: NAVAIR Engineering Support Office
- e: New Employee Safety Orientation
NESP e: Navy EHF [Extremely High Frequency] Satellite Program
- e: Navy EHF SATCOM Program
NES PHYTO e: National Phytoplankton Data Base
NESR e: Noise Equivalent Spectral Radiance
NESS e: National Earth Satellite Service
- e: Navy EHF SATCOM System
- e: NMCS Environmental Support System
- f: Service national des satellites pour l'étude de l'environnement
- s: Servicio Nacional de Satélites para el Estudio del Medio Ambiente
NEST e: Netware Embedded Systems Technology
- e: Non-surgical Embryonic Selective Thinning
- e: Novell Embedded Systems Technology
- e: Nuclear Energy Search Team
nestk. N: nestkommanderende
NESTOR e: Netherlands Educational and Scientific Titles for Online Retrieval
- e: Nuclear Reactor Winfrith
- e: Older family of tactical secure voice equipment
NESW e: National Earthquake Strategy Working Group
NET e: Nerve Excitability Test
- e: Network Entity Title
- e: Networks
- e: New England Telephone
- e: No Earlier Than
- f: Norme[s] Européennes de[s] Télécommunication
Net e: Network
- f: Neutralisation
net e: network; Internet-Domain
NETA e: Nigerian English Teaching Association
N et B f: Noir et Blanc
NetBEUI e: NetBIOS Extend[ed] User Interface
NetBIOS e: Network Basic Input and Output System
NETBLT e: Network Block Transfer
NETCAP e: Network Capability
NETCDF e: Network Common Data Form
netCDF e: network Common Data Format
NETEV e: Network Experimental Version
NETI N: Norsk Forening for Teknisk Informasjon
NETID e: Network Identifier
Netiquette d: Network und Etiquette bezeichnet Regeln über das Verhalten im Netzwerk
NETLIB e: A collection of mathematical software, papers and databases
NETMON e: Network Monitor
NET[P] e: New Equipment Training [Plan]
NETP e: New Equipment Training Plan
NETPDC e: Naval Education and Training Program Development Center
NETPMSA e: Naval Education and Training Program Management Support Activity
NetRPC e: Network Remote Procedure Call
NETS e: National (or Nationwide) Emergency Telecommunications Service
- N: Norsk Elevsamband for Teknisk Studerende
NETSC e: Naval Education and Training Support Center
NetSP e: Network Security Program
NETTING e: Emission trading used to avoid PSD/NSR permit review requirements
Netzw d: Institut für Netzwerk- und Systemtheorie
NEU e: Neuroscience
Neu-Epi-a e: Neuroepithelioma
Neural Comput e: Neural Computation
Neurobiol Learn Mem e: Neurobiology of Learning and Memory
Neurogastroenterol Motil e: Neuro-gastroenterology and Motility
Neuroimaging Clin N Am e: Neuro-imaging Clinics of North America

Neurol Med Chir [Tokyo] e: Neurologia Medico-Chirurgica
Neurol Neurochir Pol P: Neurologia i Neurochirurgia Polska
Neuromuscul Disord e: Neuromuscular Disorders
Neuropathol Appl Neurobiol e: Neuropathology and Applied Neurobiology
Neurophysiol Clin f: Neurophysiologie Clinique
Neuropsychol Rev e: Neuropsychology Review
NEURO SC e: Neuroscience Building
Neurosci Biobehav Rev e: Neuroscience and Biobehavioral Reviews
Neurosci Lett Suppl e: Neuroscience Letters. Supplement
Neurosci Res e: Neuroscience Research
Neurosurg Clin N Am e: Neurosurgery Clinics of North America
Neurosurg Rev e: Neurosurgical Review
Neurotoxicol Teratol e: Neurotoxicology and Teratology
Neurourol Urodyn e: Neurourology and Urodynamics
NEUSA e: National Education Union of South Africa
Neut s: Neutralización
NEV e: Network Experimental Version
NEW d: Neustadt/Waldnaab, Bayern
- e: New Information
- e: Newark, NJ
- e: NOAA Environmental Watch
- e: North, East and West
Newbie d: Neuling
New Biol. e: New Biologist
New Horiz e: New Horizons
New Look e: Adaptive High Frequency Test Program Test
NEWMAG e: NATO Electronic Warfare Management Assessment Group
New Microbiol e: New Microbiologica
News d: öffentliche Nachricht in einer Newsgroup
NEWS e: Netware Early Warning System
- e: Networked Extensible Window[ing] System
- e: Newspaper/Microcopy Library
NEWS. e: Current events [USENET Newsgroup Category]
NewsAgent e: Project under eLib [electronic Libraries Programme]
Newsgroup d: Themenbezogene Diskusionsgruppe zum Austausch von Nachrichten
NEWT e: News Terminal
NEX e: Navy Exchange
- e: Non-Cyclic Executive Lover Level Computer Software Component
- e: Nonfueled Experiment
NexGen e: Next Generation
NEXPRI n: Nederlands Expertise Centruum voor Ruimtelijke Informatiererwerkig
NEXRAD e: Next-generation [Doppler] Radar for weather surveillance
NEXT e: Near-End Crosstalk
- e: New Extended Technology
NEYAL e: North East and Yorkshire Academic Libraries Consortium
NF a: Nationernes Forbund
- a: Nordisk Forfatterfond
- e: National Facility
- e: National Forum
- d: Nebenfach
- e: Neurofibromatosis, Inc.
- e: Not Releasable to Foreign Nationals
- f: Nationalité Française
- f: Naturalisé Français
- f: Nom de Famille
- f: Normes Françaises
- N: Nasjonenes Forbund
- N: Norges Fiskerlag
- N: Norges Funksjonærforbund
- N: Norsk Film
- N: Norsk Forlagsselskap
- N: Norsk Front
n.f. a: nord for
- N: nord for
NFA e: No Further Action
- e: No-Flow Assembly
- e: Non-Food Agriculture
- N: Norsk Forening for Akvakulturforskning
NFAA e: National Food Assistance Agency
NFADB e: Naval Facilities Automated Data Base
NFAP e: National Fisheries Adjustment Program
- e: Nationwide Forestry Applications Program
NFAR e: Network File Access Routine
NFATF N: Norsk Forbund for Arbeidsledere og Tekniske Funksjonærer
NFAZ e: National Farmers Association of Zimbabwe
NFB e: National Federation for the Blind
NFBB N: Norsk Forening for Bolig- og Byggeplanlegging
NFC e: National Finance Center
- e: National Forum Committee
- e: Northern Fisheries Committee
NFCP e: Nuclear Fuel Cycle and Production Division
NFCS e: Nuclear Force Communications Satellite
NFCVP e: National Funding Collaborative on Violence Prevention
NFD e: Northeast Frontier District
- e: Nuclear Energy and Surplus Facilities DOE-RL Management Division
- e: Nuclear Fuels Division
NFDLF e: Northern Frontier District Liberation Front
NFDR e: National Front for the Defense of the Revolution
NFDRS e: National Fire Danger Rating System
NFDS N: Nordenfjeldske Dampskibsselskab
nf-EBM e: noise forced Energy Balance Model
NFEPWM e: National Fund for Environmental Protection and Water Management
NFERC e: National Fertilizer and Environmental Research Center
NFF e: National Farmers Federation
- e: No Failures Found
- f: Nouveau Franc Français
- N: Norges Farmaceutiske Forening
- N: Norges Farvehandlerforbund
- N: Norges Fiskehandlerforbund
NFFR N: Norges Fiskeriforskningsråd
NFG e: No Freeping Good
- e: Not Functioning Good
NFGF e: Naval Flying Grading Flight
NFH e: NATO Frigate Helicopter
NFHE e: Nonirradiated Fuel Handling Equipment
NFI e: National Fisheries Center
- e: National Forest Inventory
- e: No Frigging Idea [Internet]
NFIA e: National Families In Action
- e: National Flood Insurance Act
NFIC e: National Fisheries Industry Council
- e: National Foreign Intelligence Community
NFID e: National Foundation for Infectious Diseases
NFIEC e: Niagara Frontier Industry Education Council
NFIPA e: National Fire Protection Association
NFj N: Norsk Fjørfeavlslag
NFL e: New Foreign Launch
NFLC e: Northern Forest Lands Council
Nfld e: Memorial University of Newfoundland
NF&M e: Nuclear Fuels and Materials Department
NFM e: Network File Manager
- e: Northern Fowl Mite
NFMA e: National Forest Management Act
NFMC e: Not Fully Mission Capable
NFMCS e: Not Fully Mission Capable Supply
NFNT e: New Font Numbering Table
NFO e: Navy Finance Office
- e: Navy Finance Officer
NFOM e: Near-Field Optical Microscopy
NFOVR e: Narrow Field Of View
NFP d: Nationales Forschungsprojekt
- d: Nur für Positionierung
- e: National Focal Point
- e: Navy Flag Plot
- s: Centro Nacional de Coordinación
- s: Point Focal National
NFPA e: National Fire Protection Agency
NFPC e: National Forest Planning Committee
NFPDI e: National Flat-Panel Display Initiative
nfplut r: närförsvarspluton
NFPP e: National Family Planning Program
NFPPE e: Nuclear Force Policy, Planning and Execution
NFPS e: National Forest Policy Statement
NF[r]. e: Northern French
NFR e: Naval Reactor Facility
- e: Not a Functional Requirement
NFRAP e: No Further Remedial Action Planned

NFRC e: National Forest Research Council
- e: National Freight Rail Corporation
- e: Northwest First Regional Consultants

NFRR N: Norsk Forening av Rådgivende Rasjonaliseringsfirmaer

NFS a: Den Nordiske Faglige Samarbejds-organisation
- e: National Fertility Survey
- e: National Forests Strategy
- e: Network File Service
- e: Not Fully Successful
- e: Nuclear Facility Safety
- e: Nuclear Fuel Services
- f: Ne pas Faire Suivre
- N: Nordens Faglige Samorganisasjon
- r: Nordens Fackliga Samorganisation

NFSC e: Nuclear Fuel Services Corporation

NFSNET e: National Science Foundation Network

NFSS e: New Font Selection Scheme

NFST e: National Fleet Surgical Team

NFT e: Network File Transfer
- N: Norsk Forsvars Teknologi

NFU e: Nihon Fukushi University

NFV a: Naturvidenskabeligt Forskningsråd
- d: Niedersächsische Forstliche Versuchsanstalt, Göttingen

NFW e: No Freeping Way [Internet]
- e: No Frigging Way [Internet]
- e: No Fucking Way [Internet]

NFWF e: National Fish and Wildlife Foundation

NFWS e: Naval Fighter Weapons School

NFZ e: No-Fly Zone

NG e: National Guard
- e: Newsgroup
- e: Next Generation
- e: Nose Guard
- s: Norte Geográfico

ng e: nanogram

N&G e: Navigation and Guidance

NGA e: Natural Gas Analyzer
- e: Natural Gas Association
- e: New Generation Alliance
- f: Navire Géographique Auxiliaire

NGAC e: National Greenhouse Advisory Committee

NGAP f: Nomenclature Générale des Actes Professionnels des médecins, des chirurgiens-dentistes, des sages-femmes et des auxiliaires médicaux

NGAS e: North General Area Services

NGATM e: New Generation Air Traffic Manager

NGB e: Nordic Gene Bank

NGC e: New General Catalog[ue] [of nebulae and clusters of stars]
- e: Numismatic Guarantee Corporation

NGCADMM e: Next Generation Computer-Aided Design

NgCAM e: Neural-Glial Cell Adhesion Molecule

NGCR e: New Generation Computer Resources

NGDB e: National Geodetic Data Base

NGDC e: National Geographic Data Center

Ngdo s: Negociado

NGE e: National Greek Exam
- e: Not Greater or Equal

n. gen. e: new genus

NGF e: Nerve GF
- N: Norges Grossistforbund
- N: Norsk Geodetisk Forening
- N: Norsk Geofysisk Forening

NGG d: Gewerkschaft Nahrung, Genuß, Gaststätten

NGI a: Nordsjællands Galvano Industri
- e: Next Generation Internet
- e: Non-Government Institution
- N: Norges Geotekniske Institutt

NGIC e: National Geomagnetic Information Center

NGK A: Nederduits[e] Gereformeerde Kerk

Ngkr r: Nitroglycerinkrut

NGL N: Norsk Galvano-Teknisk Landsforening
- N: Norsk Grafisk Leverandørforening

NGLDM e: Neighboring Grey Level Dependence Matrix

NGM e: Nested Grid Model
- e: Netware Global Messaging

NGMA e: National Geoscience Mapping Accord

NGMC e: National Grid Management Council

NGMF e: Netview Graphic Monitor Facility

NGMH e: New Generation Military Hospital

NGMHS e: Netware Global Message Handling System

NGMNV e: Novgorod Museum

NGNA e: National Gerontological Nursing Association

NGnnn e: Nutating Grating <mode>

NGO e: Non-Government[al] Organisation[s] (or Organization[s])
- N: Norges Geografiske Oppmåling

NGPF e: North General Purpose Facilities

NGPSA e: National Gas Pipeline Safety Act

NGR e: National Guard Regulation
- e: Natural Gamma Ray
- e: Nigeria
- e: Nuclear Gamma Resonance

NGRS e: National Greenhouse Response Strategy

NGS e: National Genealogical Society
- e: Non-Government Standard[s]
- N: Norges Gymnasistsamband

NGSB e: Non-Government Standards Bodies

NGSC e: National Greenhouse Steering Committee

NGSF N: Norges Glass- og Stentøihandleres Forbund

NGSK A: Nederduitse Gereformeerde Sendingkerk

NGT e: Natural-Gamma spectrometry Tool

Ngt f: Négociant

NGU e: Non-Gonococcal Urethritis
- N: Norges Geologiske Undersøkelser

ngu-krut r: nitroguanidinkrut

NGV e: Natural Gas for Vehicles

NGVD e: National Geodetic Vertical Datum

NGVD 29 e: National Geodetic Vertical Datum of 1929

NGVO d: Baden-Württemberg: VO des Ministeriums für Kultur und Sport zur Änderung der VO über die Jahrgangsstufen 12 und 13 sowie die Abiturprüfung an Gymnasien

NGVTP e: Natural Gas Vehicle Technology Partnership

NGW e: Newly Generated Waste

Ngwasorep e: Ngwane Socialist Revolutionary Party

NGWLM e: Next Generation Water Level Measurement system
- f: Système de mesure du niveau de la mer de conception nouvelle
- s: Sistema de Medición del Nivel del Mar de Nueva Generación

NH a: Nordisk Handels-Hus A/S
- e: Ammonia
- e: National Highway
- f: Normes Homologuées
- f: Notes d'Honoraires

N H e: North Hall

n.h. r: nitton hundrade

N_2H_4 e: Hydrazine

NH_3 d: Ammoniak
- e: Ammonia

NH_{4+} e: Ammonium-ion

NHA e: National Housing Authority, Thailand

NHAAP e: National Heart Attack Alert Program

N Hål Fdi N: Nord-Hålogaland Forsvarsdistrikt

NHANES e: National Health and Nutrition Examination Study

NHAP e: National High-Altitude Photography [program]

NHATS e: National Human Adipose Tissue Survey

NHB e: NASA Hand Book
- e: National Housing Board

NHBIA e: New Hampshire Brain Injury Association

NHBU e: Networking Hardware Business Unit

NHC e: National Housing Corporation
- e: New Zealand National Health Committee
- e: Next Hop Client
- e: Northwest Hydraulic Consultants
- e: Numatec Hanford Corp. [Inc.]
- e: United States National Hurricane Center

NHCAA e: National Health Care Anti-Fraud Association

NHCSSP e: National Health Care Skill Standards Project

NHDA e: National Hydrometeorological Development System-FOA

NHDP e: National Committee on Human Dimensions of Global Environmental Change
- f: Comité national sur les dimensions humaines des changements de l'environnement planétaire
- s: Comité Nacional sobre las Dimensiones Humanas del Cambio Ambiental Mundial

NHDR e: National Hydrometeorological Development System-RFO

NHDW e: National Hydrometeorological Development System-WFO
NHEFS e: NHANES I Epidemiologic Followup Study
NHEML e: National Hurricane Experimental Meteorological Laboratory
NHF a: Nordens Husmoder-Forbund
- e: National Housing Forum
- N: Norges Handelsstands Forbund
- N: Norges Håndverkerforbund
- N: Norske Havforskeres Forening
NHFC e: National Housing Finance Corporation
NHFFO N: Norsk Handelsstands Forbunds Forsknings- og Opplysningsfond
NHGRI e: National Human Genome Research Insitute
NHGS e: Non-Hydrogenous Gas Delivery System
NHH N: Norges Handelshøyskole
NHHCS e: National Home and Hospice Survey
NHIC e: National Health Information Center
NHIF N: Norske Håndverks- og Industribedrifters Forbund
NHIK e: National Health Information Knowledgebase
nh. ild N: nedholdende ild
NHIRC e: National Health Information Resource Center
NHIS/SA e: National Health Information System of South Africa
NHK A: Nederduits Hervormde Kerk van Afrika
- e: National Host Germany
- j: Nippon Hosou Kyokai
NHKF N: Norges Handels- og Kontorfunksjonærers Forbund
NHL e: Non-Hodgkin[s] Lymphoma
- N: Norges Hydrodynamiske Laboratorier
NHLBI e: National Heart, Lung and Blood Institute
NHLBI OEI e: NHLBI Obesity Education Initiative
NHLF N: Nord-Hålogaland Landforsvar
NHLW e: Neutralized High-Level Waste
NHM e: Northamptonshire
NHMP e: National Human Milk Monitoring Program
NHMR e: National Health and Medical Research Council
NH[&]MRC e: National Health and Medical Research Council
NHMT e: Nuclear-Hardened Mosaic Technology
NHNA e: New Hampshire Nurses Association
NHO N: Næringslivets Hovedorganisasjon
NHOP e: National Hurricane Operations Plan
NHPA e: National Historic Preservation Act
NHPC e: Natural Holistic Pet Care
NHPI e: National Health Provider Inventory
NHPP e: Non-Homogeneous Poisson Process
NHR e: National Handwriting Recognition
- e: Non Hierarchial Routing
NHRC e: National Hydrology Research Centre
- e: Nigerian Human Rights Community
NHRE e: National Hail Research Experiment
NH&RF N: Norsk Hotell- og Restaurantforbund
NHRP e: Next Hop Resolution Protocol
- e: Next Hop Routing Protocol
NH&S e: Nuclear Hardening (or Hardness) and Survivability
NHS e: International Society for the Prevention and Mitigation of Natural Hazards
- e: National Highway System
- e: National Housing Strategy
- e: Next Hop Server
- f: Société internationale pour la prévention et l'atténuation des risques naturels
- s: Asociación Internacional para la Prevención y la Atenuación de los Riesgos Naturales
NHSA e: National Healthcare Staffing Association
NHSDA e: National Household Survey on Drug Abuse
NHSE e: National HPCC Software Exchange
NHTF e: National Hover Test Facility
NHTSA e: National Highway Traffic [and] Safety Administration
- e: National Highway Traffic Safety Act
NHW e: Neutralized High-Level Waste
NHWRDDC e: NW Hazardous Waste Research, Development and Demonstration Center
NI e: Natural Intelligence
- e: Navigation Initialization lover level computer software component
- e: Network Interconnect
- e: Network Interface
- f: Non-Inscrit
- f: Normalisation Internationale
- f: Numéro d'Immatriculation
- f: Numéro d'Impression
- s: No Identificado
N/I e: Non-Interlaced
Ni-63 e: Nickel-63
NIA e: Natal Indian Association
- e: National Intelligence Agency
- e: Network Integration and Analysis
- e: Network Interoperability Alliance
- e: Next Instruction Address
NIAA e: Nursery Industry Association of Australia
NIAC e: Naval Intelligence Automation Center
NIAE e: National Institute of Agricultural Engineering
NIAES e: National Institute of Agro-Environmental Studies
NIAF N: Norges Ingeniør- og Arkitektforening
NIAF/2 e: Netview Installation and Administration Facility/2
NIAID e: National Institute of Allergy and Infectious Disease
Ni3Al e: Nickel Aluminide
NIAM e: Natural Language Information Analysis Model
NIAST e: National Institute of Aeronautical Scientific Technology
NIB e: Noise Investigation Bureau
N I B e: North Ingalls Building
NIBGF e: Nested Interpolation Based Grid File
NIBS e: Nuffield Interactive Book System
NIBT e: Net Income Before Taxes
NIC e: Natal Indian Congress
- e: National Ice Center
- e: National Illumination Committee
- e: National Insurance Corporation
- e: National Interim Council
- e: National Investigations Committee
- e: National ISDN Council
- e: Network Information Center (or Centre)
- e: Newly Industrialized Countries (or Country)
- e: Numeric Intensive Computing
NICAD e: Nickel Cadmium
NICD e: Nickel Cadmium [batterie]
NICE e: National Institute For Consumer Education
- e: National Irradiation Creep Effort
NICEIC e: National Inspection Council Electrical Installation Contracting
NICHCY e: National Information Center for Children and Youth with Disabilities
NICHSR e: National Information Center on Health Services Research and Health Care Technology
NICK d: Netzweites Informationssystem im Campus der Universität Karlsruhe
NICKA e: Codeword, Nickname and Exercise Term System
NICMIS e: Navy Integrated Command/ Management Information System
NICMOS e: Near Infrared Camera/Multi Object Spectrometer
NICNAS e: National Industrial Chemicals Notification and Assessment Scheme
NICOLAS e: Network Information Center Online Aid System
NICON e: National Insurance Corporation of Nigeria
NICSOI e: NICS Operating Instruction
NICT e: National Incident Coordination Team
NICU e: Neonate Intensive Care Unit
NICWA e: National Indian Child Welfare Association
NID d: Nichtraucher Initiative Deutschland
- e: Network Information Database
- e: Network Inward-Dialing
- e: Next ID
- e: Non-Interactive Display
NIDA e: n-Nitrosoiminodiacetic Acid
NIDA Res Monogr e: NIDA Research Monograph
NIDCD e: National Institute of Deafness and Other Communication Disorders
NIDDESC e: Navy-Industry Digital Data Exchange Standards Committee
NIDDK e: National Institute of Diabetes and Digestive and Kidney Disease
NIDL e: National Information Display Laboratory

NIDM e: National Institute of Disaster Mobilization
NIDMS e: NMCS Information for Decision Makers System
NIDR e: National Institute of Dental Research
- e: Network[ed] Information Discovery and Retrieval
NIDRR e: National Institute on Disability and Rehabilitation Research
NIDS e: NMCS Information [and] Display System
NIE e: National Index of Ecosystems
- e: National Intelligence Estimate[s]
- e: Newly Industrializing Economy
- e: Non-Interference Experiment
NIEC e: National Import and Export Corporation
NIED e: National Research Institute for Earth Science and Disaster Prevention
Niederschl. D: Niederschlag
NIEE e: National Institute for Engineering Ethics
NIEHS e: National Institute for/of Environmental Health Sciences
- e: National Institute of Environmental Health Services
NIEI e: National Indoor Environmental Institute
NIEL e: Non-Ionizing Energy Loss
NIEM e: Network for Industrial Environmental Management
NIES e: National Industry Extension Service
- e: National Institute for Environmental Studies
- e: National Institute of Environmental Science
NIF e: National Ignition Facility
- e: Neighbor Information Frame
- e: Notice of Intent to Fine
- N: Norges Industriforbund
- N: Norsk Ingeniørforening
NIFC e: National Interagency Fire Center
NIFFTE e: Non-cooperative Identification Friend or Foe Technical Evaluation
NIFM e: Northwest Independent Forest Manufacturers
NIFMA e: Nigerian Furniture Manufacturers Association
NIFTP e: File transfer network in the United Kingdom
NIFV N: Nærforsvarsinspektøren for Flyvåpnet
NIG e: National Institute of Genetics
- e: Non Isolated Gate
NIGA e: Nuclear Indirect Gamma Activity
NIH e: Nonirradiated Handling
- e: Not Invented Here [Internet]
NIHCL e: NIH [National Institute of Health] Class Library
NIH Consens Statement e: NIH Consensus Statement
NIHEC e: Northern Ireland Higher Education Council
NIHECS e: National Institutes of Health Electronic Computer Store
NIH Guide Grants Contracts e: NIH Guide for Grants and Contracts
Nihon Rinsho Meneki Gakkai Kaishi j: Nihon Rinsho Meneki Gakkai Kaishi
Nihon Shinkei Seishin Yakurigaku Zasshi j: Nihon Shinkei Seishin Yakurigaku Zasshi
NII e: National Information Infrastructure
NIIA e: Nigerian Institute of International Affairs
NIIG e: National ISDN Interface Group
NIIIP e: National Industrial Information Infrastructure Protocols
NIIM e: National Integrated Information Management Program
NIIR e: Non-Imaging Infrared Sensor
NIIT e: National Information Infrastructure Testbed
NII V-VS R: Nauchno-Issledovatelsky Institut V-VS
NIJ e: Nigerian Institute of Journalism
NIKO e: Nigerian Korean Company
NIKT d: Neue Informations- und Kommunikationstechnologien
NIL e: Noise Immunity Level
- N: Nærforsvarsinspektøren for Luftforsvaret
NILE e: Network Integration Lab Environment
NILFO N: Norsk Innkjøpslederforbund
NILU e: Norwegian Institute for Air Research
- N: Norsk Institutt for Luftforskning
NIM e: National Impact Model
- e: National Institute for Maritime Research
- e: Network Installation Management (or Manager)
- e: Nuclear Instrument Modules
- e: Neutron Interactive Materials program
NIMA e: National Imagery Mapping Agency
- e: National Infomercial Marketing Association
NiMAnorgM d: Institut für Nicht-metallische Anorganische Materialien
NIMBUS e: A series of American polar orbiting meteorological satellites
- e: Network Information Management Client-Based User Service
- f: Satellite météorologique des Etats-Unis
- s: Satélite Meteorológico de los Estados Unidos
NIMBUS-7 e: Meteorological satellite
NIMBY e: Not In My Back Yard [Internet]
NIMH e: Nickel Metal Hybrid [batterie]
NiMH e: Nickel Metal-Hybride [battery]
NIMHANS e: National Institute of Mental Health and Neuro Sciences
NIMMS e: NAVAIR Industrial Material Management System
NiMo e: Niagara-Mohawk
NiMoV e: Nickel-Molybdenum-Vanadium
NIMP e: Netware Internetwork Messaging Protocol
NIMR e: National Institute for Materials Research
NIMS e: Near-Infrared Mapping Spectrometer
NINDS e: National Institute of Neurological Disease and Stroke
- e: National Institute of Neurological Disorders and Stroke
NINE e: National Institute of Nutritional Education
NINF N: Norsk Institutt for Næringsmiddelforskning
NINO e: No Input, No Output [Internet]
- e: Nothing In, Nothing Out [Internet]
NINP e: Norfolk Island National Park
NINR e: National Institute for/of Nursing Research
NIO e: Native Input/Output
- f: Institut national d'océanographie
- s: Instituto Nacional de Oceanografía
NIOC e: National Iron Ore Company
NIOD e: Network Inward/Outward Dialing
NIOF e: National Institute of Oceanography and Fisheries
- f: Institut national d'océanographie et de pêche
- s: Instituto Nacional de Oceanográphia y Pesca
NIOMR e: Nigerian Institute for Oceanography and Marine Research
- f: Institut nigérian d'océanographie et de recherche marine
- s: Instituto Nigeriano de Oceanografía e Investigación Marina
NIOZ e: Netherlands Institute for Sea Research
- e: Netherlands Institute of Ocean Sciences
- f: Institut néerlandais de recherche marine
- s: Instituto Neerlandés de Investigaciones Marinas
NIP e: National Immunization Program
- e: National Independence Party
- e: National Integration Party
- e: NATO Interoperability Plan
- e: Network Implementation Plan
- e: Neuroleptic-Induced Parkinsonism
- e: Normal Incidence Pyreheliometer
- e: Numismatic Indexes Project
NIPA e: Non-local Independent Pixel Approximation
- N: Norsk Institutt for Personaladministrasjon
NIPC e: Nacelle Ice Protector Controller
- e: National Inhalant Prevention Coalition
NIPD e: NATO Interoperability Planning Document
- e: Nightly Intermittent Peritoneal Dialysis
NIPDE e: National Initiative for Product Data Exchange
NIPF e: National Incontinentia Pigmenti Foundation
- e: Nonindustrial Private Forest Owners
NIPG e: National ISDN Parameter Group
NIPHLE e: National Institute of Packaging, Handling and Logistics Engineers
NIPHYS e: Nitrogen Physiology of Forest Plants and Soils
Nippon Eiseigaku Zasshi j: Nippon Eiseigaku Zasshi. Japanese Journal of Hygiene
Nippon Ganka Gakkai Zasshi j: Nippon Ganka Gakkai Zasshi. Acta Societatis Ophthalmologicae Japonicae

Nippon Geka Gakkai Zasshi j: Nippon Geka Gakkai Zasshi. Journal of Japan Surgical Society
Nippon Geka Hokan j: Nippon Geka Hokan. Archiv für Japanische Chirurgie
Nippon Hinyokika Gakkai Zasshi j: Nippon Hinyokika Gakkai Zasshi. Japanese Journal of Urology
Nippon Hoigaku Zasshi j: Nippon Hoigaku Zasshi. Japanese Journal of Legal Medicine
Nippon Igaku Hoshasen Gakkai Zasshi j: Nippon Igaku Hoshasen Gakkai Zasshi. Nippon Acta Radiologica
Nippon Ika Daigaku Zasshi j: Nippon Ika Daigaku Zasshi. Journal of the Nippon Medical School
Nippon Jibiinkoka Gakkai Kaiho j: Nippon Jibiinkoka Gakkai Kaiho [Journal of the Oto-Rhino-Laryngological Society of Japan]
Nippon Jinzo Gakkai Shi j: Nippon Jinzo Gakkai Shi. Japanese Journal of Nephrology
Nippon Koshu Eisei Zasshi j: Nippon Koshu Eisei Zasshi [Japanese Journal of Public Health]
Nippon Kyobu Geka Gakkai Zasshi j: Nippon Kyobu Geka Gakkai Zasshi. Journal of the Japanese Association for Thoracic Surgery
Nippon Kyobu Shikkan Gakkai Zasshi j: Nippon Kyobu Shikkan Gakkai Zasshi. Japanese Journal of Thoracic Diseases
Nippon Naika Gakkai Zasshi j: Nippon Naika Gakkai Zasshi. Journal of Japanese Society of Internal Medicine
Nippon Rai Gakkai Zasshi j: Nippon Rai Gakkai Zasshi. Japanese Journal of Leprosy
Nippon Rinsho j: Nippon Rinsho. Japanese Journal of Clinical Medicine
Nippon Ronen Igakkai Zasshi j: Nippon Ronen Igakkai Zasshi. Japanese Journal of Geriatrics
Nippon Saikingaku Zasshi j: Nippon Saikingaku Zasshi. Japanese Journal of Bacteriology
Nippon Sanka Fujinka Gakkai Zasshi j: Nippon Sanka Fujinka Gakkai Zasshi. Acta Obstetrica et Gynaecologica Japonica
Nippon Seikeigeka Gakkai Zasshi j: Nippon Seikeigeka Gakkai Zasshi. Journal of the Japanese Orthopaedic Association
Nippon Seirigaku Zasshi j: Nippon Seirigaku Zasshi. Journal of the Physiological Society of Japan
Nippon Shokakibyo Gakkai Zasshi j: Nippon Shokakibyo Gakkai Zasshi. Japanese Journal of Gastroenterology
Nippon Yakurigaku Zasshi j: Nippon Yakurigaku Zasshi. Folia Pharmacologica Japonica
NIPR e: National Institute for/of Polar Research
NIPS e: National Institute for Physiological Sciences
- e: Network I/O[s] Per Second
- e: NMCS Information Processing System
- e: Nottingham Image Processing System
NIPSS e: National Institute for Policy and Strategic Studies
NIR e: National Institute of Development Research and Documentation University of Botswana
- e: Near Infra-Red
- e: Network Information Registry
- e: Network Information Retrieval
- e: Networked Information Resource
- e: Networked Information Retrieval
- e: Nighttime Infrared
NIRA e: Designer of SUPERPHENIX with Novatome
- e: Italian Fast Nuclear Reactor Co., Genova
- e: Near Infrared Reflection Analysis
- e: Near-Infrared Reflectance Analysis
NIRF e: National Irrigation Research Fund
NIRP e: National Institute for Radiation Protection
NIRRL e: Northern Ireland Regional Research Laboratory
NIS d: Nachbarschaftsinformationssystem
- e: National Institute of Standards and Technology
- e: Nationwide Inpatient Sample
- e: NATO Infrastructure System
- e: Near Infrared Source
- e: Network Information Services
- e: New[ly] Independent States of the former Soviet Union
- e: NMCS Information System
- e: Not In Scope
NISA e: New Information Services Architecture
NISAC e: National Industrial Security Advisory Committee
NISC e: Network Information and Support Center
N[-]ISDN e: Narrowband ISDN [Integrated Services Digital Network]
- e: National ISDN [Integrated Services Digital Network]
NISEG e: NATO In-Service Goals
NISER e: National Institute for Social and Economic Research University of Namibia
NISI e: Northwest Instrument Systems, Inc.
NISIE e: NATO Integrated System for Information Exchange
NISMAWU e: National Iron, Steel, Metal and Allied Workers Union
NISO e: National Information Standards Organisation
NISORS e: Nigerian Society of Remote Sensing
NISP e: Networked Information Service[s] Project
NISR e: National Institute for Scientific Research
NISR-INPE e: National Institute for Space Research-Instituto de Pesquisas Espacias, Brazil
NISS e: National Information [on] Software and Services
- e: National Information Services and Systems
NIST e: National Institute for/of Standards and Technology
- e: National Institute[s] of Science and Technology
- e: National Intelligence Support Team
- e: North-American Institute for Standards and Telecommunications
- f: Institut national des normes et des technologies
- s: Instituto Nacional de Normas y Tecnologías
NISTAV d: Niederschlags-Verzeichnis der Niederschlagsstationen
NISTEP e: National Institute of Science and Technology Policy
NISU e: National Injury Surveillance Unit
NISYP e: Network Information System/Yellow Pages
NIT e: Institute of Technology
- e: Network Integrity Test
- e: Network Interface Tap
NITADS e: NATO Interoperability of Tactical Automated Data Systems
NITC e: National Information Technology Center
NITE e: N-Interval Trajectory Estimation
NITE-2 e: NTH Internal Trajectory Estimation Program
NITEC e: National Information Technology in Education Centre
NITE-K e: N Interval Trajectory Estimation-K Program
NITEL e: Nigeria Telecommunications Limited
NITE-OP e: Night Imaging Trough Electro-Optics
NITES e: Naval Integrated Tactical Environmental Subsystem
NITF e: National Imagery Transmission Format
- e: Nuclear Instrument Test Facilities
NITFS e: National Imagery Transmission Format Standard
NITMP e: Navy Integrated Technical Manual Processing
NITO N: Norges Ingeniør- og Teknikerorganisasjon
NITOL e: Norway-net with IT for Open Learning
NITREX e: Nitrogen Saturation Experiments
nitric e: anhydride
NITS e: National Institute of Test Standards
- e: Network Independent Transport Service
NIU e: Nacelle Interface Unit
- e: National Intelligence Unit
- N: Nasjonalkomitéen for Internasjonalt Ungdomsarbeid
NIUF e: National ISDN Users Forum
- e: North-American ISDN User[s] Forum
NIV e: Non-Immigrant Visa
NIVA e: Norwegian Institute for Water Research
- N: Norsk Institutt for Vannforskning

NIVR e: Netherlands Agency for Aerospace Programmes
NIWA e: Namibian Information Workers Association Windhoek
- e: National Institute of Water and Atmospheric Research Ltd
NIWARS e: Netherlands Interdepartmental Working Community for the Application of Remote Sensing Techniques
NIWC e: Northern Ireland Womens Coalition
NJ N: Norges Juristforbund
- N: Norsk Journalistlag
nj r: nattjakt
NJC e: Not Just Cows
NJCA e: New Jersey Classical Association
NJCL e: National Junior Classical League
NJDEPE e: New Jersey Department of Environmental Protection and Energy
NJF N: Norges Jernvarehandleres Forbund
- N: Norsk Jernbaneforbund
NJFF N: Norges Jeger- og Fiskerforening
njfpl r: nattjaktflygplan
NJHA e: New Jersey Hospital Association
NJLS N: Norges Jordskiftedommer- og Landmålersamband
N J Med e: New Jersey Medicine
NJSNA e: New Jersey State Nurses Associaton
NJUG e: National Joint Utilities Group
NJW d: Neue Juristische Wochenschrift
NJWE-HV d: NJW-Entscheidungsdienst Haftpflicht- und Versicherungsrecht
NJWE-MietR d: NJW-Entscheidungsdienst Mietrecht
NJWE-WettbR d: NJW-Entscheidungsdienst Wettbewerbsrecht
NK a: Nordisk Kontakt
- e: Natural killer [cells]
- N: Norsk Kvinneparti
- r: Nitroglycerinkrut
- r: Nobelkrut
- r: Nordiska Kompaniet AB
- r: Norrlands Kustartillerifôrsvar
nk d: Anzahl der Vollfilter zugeordnet zur k-ten Partialfläche
Nk. N: Nestkommanderende
NKA e: Nordic Liaison Committee for Atomic Energy
NKAL N: Norske Kvinnelige Akademikeres Landsforbund
NKCF e: National Keratoconus Foundation
NKE e: Nuclear Technology Standards Committee
NKEW e: Nuclear Kinetic Energy Weapon
NKF N: Norges Kjøtt og Fleskescentral
- N: Norsk Kommunalteknisk Forening
- N: Norsk Kommuneforbund
- N: Norsk Kvinneforbund
- N: Norsk Kvinnesaksforening
NKIF N: Norsk Kjemisk Industriarbeiderforbund
- N: Norske Kommunale Ingeniørveseners Forening
NKK N: Nordiske Kulturkommisjon
NKL d: Norddeutsche Klassenlotterie
NKLF N: Norges Kolonial- og Landhandlerforbund
NKMU e: National Kangaroo Monitoring Unit
NKN N: Norske Kvinners Nasjonalråd
NKO n: Nederlandse Kalibratie Organisatie
NKP N: Norges Kommunistiske Parti
- N: Norsk Kollektiv Pensjonskasse A/S
nkr. a: norske kroner
NKS e: Network Knowledge Server
- e: Nordic Nuclear Safety Project
- N: Nordiske Kvinners Samarbeid
- N: Norges Kommunistiske Student-Forbund
- N: Norges Kristelige Studentbevegelse
- N: Norsk Korrespondanseskole
- N: Norske Kvinners Sanitetsforening
NKSS N: Norges Kristelige Student- og Skoleungdomslag
NKSU N: Nordens Konservative Studentunion
NKT a: Nordiske Kabel- og Traadfabriker A/S
NKTF N: Nordisk Komité for Transportøkonomisk Forskning
NKU N: Nasjonalt Koordinerende Utvalg
- N: Norges Kommunistiske Ungdomsforbund
NKUF N: Norges Kristelige Ungdomsforbund
NKVD R: Narodny Komitet Vnutrennih Del
NL e: Netherlands
- e: NeuroLeptic
- e: Number of unallocated channels on Link
- f: Naissance Légitime
- f: Nations Latines
- f: Nombre Limité
- f: Non-Lieu
- r: Normallinjen
nl. a: nemlig
n.l. r: nordlig latitud
n/l s: nuesta letra
NLA e: National Liberal Alliance
- e: Not Long Ago
- N: Norske Landskapsarkitekters Forening
NLAC e: National Landcare Advisory Committee
NLANR e: National Laboratory for Applied Networking Research
NLAP e: National Laboratory Audit Program
NLB d: Niedersächsische Landesbibliothek
NLBMDA e: National Lumber and Building Materials Dealers Association
NLC e: National Laboratory Consortium
- e: National Liberation Council
- e: Nigerian Labor Congress
- e: Noctilucent Cloud
NLCC e: NORTHAG Logistics Command Center
NLD e: Netherland
- e: Northumberland
NLDN e: National Lightning Detection Network
NLDP e: National Launch Development Program
NLE e: National Latin Exam
- e: National Library for the Environment
- e: Nonlinear Editing
- e: Not Less or Equal
- r: Navigationslärarkurs
NLEA e: Nutrition Labeling Act
NLETS e: National Law Enforcement Teletype System[s]
NLF e: National Labor Federation
- N: Norges Lastebileier-Forbund
- N: Norges Lotteforbund
- N: Norsk Landbruksakademikerforbund
- N: Norsk Lokomotivmannsforbund
- N: Norsk Losforbund
NLGDB e: National Local Government Boundary Database
NLGERN e: National Local Government Environmental Resource Network
NLGHA e: National Lesbian and Gay Health Association
NLH N: Norges Landbrukshøyskole
NLI e: National Localities Index
NLIN e: NOAA Library and Information Network
NLIS e: Niedersächsisches Landesamt für Immissionsschutz, Hannover
Nlj r: Normalljus
NLJ/DCM e: National Library Jerusalem: Department of Collections and Manuscripts
NLK a: Nordisk Lito Kemi A/S
NLL e: Night Low Level
Nlle f: Nouvelle
NLM e: Netware Loadable Modul
NLMOC e: Naval Atlantic Meteorology and Oceanography Center
NLM SIS e: National Library of Medicine Specialized Information Services
NLMWT f: Mouvement de Libération Nationale du Togo Occidental
NLNB d: Nichtlineare Nebenbedingung[en]
NLO e: National Liaison Officer
NLÖ d: Niedersächsisches Landesamt für Ökologie, Hannover
NLOS e: Non-Line of Sight
NLP d: Neurolinguistische Programmierung
- e: National Landcare Plan
- e: National Landcare Program
- e: National Liberal Party
- e: National Liberation Party
- e: Natural Language Processing
- e: Nigerian Labor Party
NLPBB e: Navy Leadership Policy Bulletin Board
NLPID e: Network Layer Protocol Identifier
NLQ e: Near Letter Quality
NLR e: National Aerospace Laboratory
- n: Nationaal Lucht- en Ruimtevaartlaboratorium
NLRC e: NASA Lewis Research Center
NLRP e: NMCS Long-Range Plan
NLS e: Native Language Support
- e: Native Language System
- e: Network License Server
- e: Noise Line State
NLSC e: National Logistics Supply Center
NLSF N: Norges Liberale Studentforbund

NLSFUNC e: National Language Support Function
NLSP e: Netware Link Service[s] Protocol
NLSST e: Nonlinear Sea Surface Temperature
NLSUS e: National Launch System Upper Stage
NLSYM e: National Launch System Model
NLT e: No Later Than [Internet]
- e: Non-Linear Time sequence
- e: Nucleus Load Table
NLTE e: Non-Local Thermodynamic[s] Equilibrium
NLTN e: National Laboratory Training Network
NLUUG d: Niederländische UNIX User Group
NLV e: National Language Version
NLVF e: North Las Vegas Facility
- e: Norway Agricultural Research Council
- N: Norges Landbruksvitenskapelige Forskningsråd
NM a: Nordisk Maskinfabrik
- a: Nordisk Medicin
- d: Netzwerk Management
- e: Nanometre
- e: Nautical Mile[s]
- e: Network Manager
- e: Network Mangement
- e: Network Module
- e: No[t] Mark[ed]
- e: Nuclear materials
- f: Note Ministérielle
- f: Nouveaux Mariés
- s: Norte Magnético
nm e: nano-meters
Nm r: nautisk mil
n.m. N: nautisk mil
Nm³ e: Normal meter³
nm³ e: natural metre cube
NMA e: National Museum of Australia
- e: Network Management Architecture
- e: Nigerian Medical Association
- e: Nuclear Materials Accountability
NMAB e: National Materials Advisory Board
- e: Network Monitoring and Analysis Branch
NMACS e: Nuclear Materials Accounting Cost System
NMANG e: New Mexico ANG
NMAQCR e: New Mexico Air Quality Control Region
NMAQD e: New Mexico Air Quality District
NMAS e: Network Management Alert System
NMB d: Naumburg/Saale, Sachsen-Anhalt
NMC e: National Manpower Commission
- e: National Marine Center
- e: National Meteorological Center[s] (or Centre)
- e: National Milling Corporation
- e: National Mission Centre
- e: National Mobilization Committee
- e: National Movement of the Congo
- e: Naval Medical Command
- e: Network Management and Control
- e: Network Management Center
- e: Network Monitoring Center
- e: New Muon Collaboration
- e: NSSDC Master Catalog
- e: Nuclear Material Control
NMC&A e: Nuclear Material Control and Accountability
NMCA e: National Muscle Car Association
NM CAMP e: University Division, National Music Camp, Interlochen
NMCB e: Not Mission Capable-Both maintenance and supply
NMCC/MC e: NMCC/Message Center
NMCCS e: Navy Manpower Configuration Control System
- e: Nuclear Materials Control Computer System
NMCM e: Noble Metals Compatibility Melters
- e: Not Mission Capable-Maintenance
NMCMS e: Not Mission Capable Maintenance Scheduled
NMCMU e: Not Mission Capable Maintenance Unscheduled
NMCP e: Network Management Control Protocol
NMCS e: Not Mission Capable-Supply
NMC SEB e: New Media Center, School of Education Building
NMD e: Doctor of Naturopathy
- e: NASA Master Directory
- e: National Missile Defense
- e: Neuromotor Disease
- e: Nuclear Medicine Department
- N: Norsk Medisinaldepot
NMDA e: N-Methyl-D-Aspartate
NMDAS e: Navy Manpower Data Accounting System
NMDC e: National Medical Device Coalition
- e: Nonmagnetic Drill Collar
NMDF e: National Material Development Facility
- e: NATO Main Distribution Frame
NMDIS e: National Marine Data and Information Service
- f: Service national de données et d'information marines
- s: Servicio Nacional de Información y Datos Marinos
NMDS e: National Missile Defense System
- e: Network and Mission Operations Support
NME e: Navy Materiel Establishment
NMEFC e: National Marine Environmental Forecasting Center (or Centre)
- f: Centre national des prévisions relatives au milieu marin
- s: Centro Nacional de Previsiones relativas al Medio Marino
NMEIB e: New Mexico Environmental Improvement Board
NMEN e: National Materials Exchange Network
NMEP e: NMCS Management Engineering Plan
NMET e: Naval Mobile Environmental Team
NMF e: Network Management Forum
- e: Neutron Multiplier Facility
- N: Norges Markedsføringsforbund
- N: Norges Merkantile Funksjonærforbund
- N: Norsk Mensendieckforbund
- N: Norsk Meteorologforening
NMFA e: National Master Freight Agreement
NMFC e: National Motor Freight Classification
NMFECC e: National Magnet Fusion Energy Computer Center
NMFS e: National Marine Fisheries Service
- e: Nuclear Materials Facilities Stabilization
- f: Service national des pêches maritimes
- s: Servicio Nacional de Pesquerías Marinas
NMFWF e: National Fish and Wildlife Foundation
NMG e: NCC Message
NMHC e: Non-Methane Hydrocarbons
NMHF e: National Mental Health Foundation
NMI d: Nicht Maskierbarer Interrupt
- e: New Model Introduction
- e: Non-Maskable Interrupt
- e: Northeast Midwest Institute
- e: Nuclear Materials Information
NMIB e: Network Management Information Base
NMIC e: National Military Intelligence Center]
- e: Navy Management Information System
NMIC-SS e: National Military Intelligence Center Support System
Nmil N: Nautisk mil
NMIMC e: Naval Medical Information Management Center
NMIM[&]T e: New Mexico Institute of Mining and Technology
NMIS e: Network Medical Information Services
NMIST e: National Military Intelligence Support Team
NMIT e: Nuclear Material Item Transfer
NMITC e: Navy and Marine Corps Intelligence Training Center
NMJIC e: National Military Joint Intelligence Center
NML e: Network Management Layer
- e: Nuclear-Magnetism Log
- N: Norske Melkeprodusenters Landsforbund
- N: Norske Murmesteres Landsforening

NMLCF e: Nuclear Measurements and Logging Calibration
NMM e: National Motorsports Marketing
- e: Netware Management Map
- e: Network Management Modul
- f: Niveau Moyen de la Mer
- s: Nivel Medio del Mar
NMMIS e: Naval Materiel Management Information System
NMMP e: Network Management Manager Process
NMMPS e: National Military Message Processor System
NMMSA e: National Minority Medical Suppliers Association
NMN d: Niedersächsischer Monographiennachweis
- N: Norsk Nærings- og Nydelsesmiddelarbeiderforbund
NMNA e: New Mexico Nurses Association
NMnnn e: Nutating Mirror <mode>
NMO e: Network Management Option
- f: Navire Météorologique Océanique
NMOC e: National Meteorological Operartions Centre
- e: Naval Meteorology and Oceanography Command
- e: Non-Methane Organic Component
- e: Non-Methane Organic Compounds
NMOS e: N[egative]-channel MOS [Metal-Oxide-Semiconductor]
- e: N[egative]-channel MOSFET
- e: Network and Mission[s] Operation[s] Support [contract]
NMP e: National Mobilization Program
- e: National Municipal Policy
- e: Network Management Protocol
- e: New Millenium Program
- e: NMCS Master Plan
NMPC e: Naval Military Personnel Command
NMPE e: Network Management Protocol Entry
NMPIS e: National Marine Pollution Information Systems
NMPPO e: National Marine Pollution Program Office
NMR e: NATO Military Representative
- e: Natural Remanent Magnetization
- e: Neonatal Mortality Rate
- e: Network Managment Responder
- e: New Mexico Regulations
- e: New Mobile Radar
- e: No Maintenance Required
NMRA e: Not Mission Ready and Available
NMR Biomed e: NMR in Biomedicine
NMRS e: Navy Manpower Requirement System
NMS d: Neumünster, Schleswig-Holstein
- d: Neutrale Markierungssprache
- e: National Meteorologicial Service
- e: National Military Strategy
- e: National Mortality Survey
- e: Neonatal Maladjustment Syndrome
- e: Netware Management Station
- e: Netware Management System
- e: Network Management Station
- e: New Management System
- e: Non-Member State
- N: Norske Meieriers Salgssentral
N&MS e: Networks and Mission Services
NMSAE e: Netherlands Ministry for Social Affairs and Employment
N&MSP e: Networks and Mission Services Project
NMSR e: New Mexico State Road
NMT r: Ny MediaTeknik
NMTN e: National Telecommunications Network
NMTR e: National Weather Service Modernization, Test and Integration System River Forecast Office
NMTW e: National Weather Service Modernization, Test and Integration System Weather Forecast Office
NMUP e: Network Management User Process
NMV e: National Museum of Victoria, Melbourne
NMVOC e: Non-Methane Volatile Organic Chemicals
NMVT e: Network Management Vector Transport [protocol]
NMWA e: National Mine Workers Association
NMWG e: Numerical Modeling Working Group
- e: Numerical Modelling Working Group
- f: Groupe de travail sur la modélisation numérique
- s: Grupo de Trabajo sobre Modelización Numérica
NMWP e: National Mixed Waste Program
NMWU e: National Mine Workers Union
NMX e: Network Magement Extensions
NN e: Nearest Neighbo[u]r
- e: Network Node
- e: Neural Net[work]
- e: Next Node
- e: No News
- e: Office of Nonproliferation and National Security
- s: No Nombrado
nn e: A unix program for browsing newsgroups
n.n. d: nicht nachgewiesen
- d: nicht nachweisbar
N.N. e: noch nicht besetzt
- l: nomen nescio
N/N e: Noise-to-Noise
n1, n2 e: Brechzahl der Luft, bzw. des Farbfilterglases
NNA e: Nevada Nurses Association
NNC e: Notice of Noncompliance
NNCCI e: Namibia National Chamber of Commerce and Industry
NNCS e: NICS Network Control System
NNDC e: Namibia National Democratic Coalition
NNDP e: Nigerian National Democratic Party
NNE e: North by North East
- f: Nord-Nord-Est
- s: Nornordeste
NNEDDC e: N-[naphthyl[1]]-ethylenediamine
NNEEDD e: NOAA N-ROSS/ERS-1 Environmental Data Development
NNEMS e: National Network for Environmental Management Studies
NNF a: Nærings- og Nydelsesmiddelforbund
- a: Nyt Nordisk Forlag
- e: Namibia National Front
NNHSF e: National Nursing Home Survey Followup
NNI e: Network Node Interface
- e: Network-[to-]Network Interface
- e: Next Node Indicator
- n: Nederlands Normalisatie-Instituut
NNK e: Non-Nuclear Kill
NNLC e: Ngwane National Liberation Congress
NN/LM e: National Network of Libraries of Medicine
NN[/]LM e: National Network of Libraries of Medicine
NNLP e: National Natural Landmarks Program
NNMF e: Nearest Neighbour Median Filter
NNMI e: Nonlinear Normal Mode Initialization
NNMS e: National Network Management System
NNMSB e: Non-Nuclear Munitions Safety Board
NNN e: Next-Nearest Neighbor
NNNA e: Non-Native Network Attachment
NNO d: Nordnordosten
- e: Ni-NiO buffer
- s: Nornoroeste
Nno N: Nordnordost
NNØ a: Nordnordøst
NNOC e: Nigerian National Oil Corp.
NNOR e: Non-Nuclear Ordnance Requirements Model
NNOT e: Not Necessarily On Topic
NNP a: Nettonationalprodukt
- e: Network Node Processor
NNPC e: Nigerian National Petroleum Company
NNPSPP e: National Nonpoint Source Pollution Program
NNR e: Network News Reader for IBM CMS
- e: Novell Network Registry
NNRIS e: National Natural Resources Information System
NNRMS e: National Natural Resource Management System
NNRTI e: Non-Nucleoside Reverse-Transcriptase Inhibitors
NNS e: Netware Name Service
- e: Newport News Shipbuilding
NNSA e: National Nuclear Safety Administration

NNSC e: NSF Network Service Center
NNSDC e: NASA National Space Science Data Center
NNSDO e: National Nursing Staff Development Organization
NNSDP e: National Nutrition and Social Development Programme
NNSF e: National Natural Science Foundation
NNSS e: National Navigation Satellite System
- e: Naval (or Navy) Navigation Satellite System
NNT e: Nearest Neighbor Tool
NNTP e: Net[work] News Transfer Protocol
- e: Network News Transmission Protocol
- e: Network News Transport Protocol
NNU e: Namibia National Union
NNV a: Nordnordvest
- r: Nordnordväst
NNW d: Nordnordwesten
- e: North by North West
NNWDB e: National Nuclear Weapons Data Base
NNWS e: Non-Nuclear Weapons State[s]
NNWSI e: Nevada Nuclear Waste Storage Investigations
NNX e: Network Numbering Exchange
NO d: Nordost[en]
- d: Stickstoffmonoxid
- e: Nitric Oxide
- e: Nitrogen Monoxide
- e: Not Ours
- f: Navire Océanographique
- f: Nord-Ouest
- f: Nouvel Ordre
- s: Noroeste
No e: Nobelium
- f: Nombre
- f: Numéro
- N: Nordost
- s: Número
no. a: norsk
- a: numero
- a: nummer
NØ a: Nordøst
Nö r: Nordöst[lig]
n/o s: nuestra orden
n.ø. N: nordøst
NO_2 d: Stickstoffdioxid
- e: Nitrogen Dioxide
- e: Nitrogen Peroxide
NO_{2-} e: Nitrite
N_2O d: Distickstoffoxid
- e: Nitrous Oxide
N_2O_4 e: Nitrogen Tetroxide
N_2O_5 e: Nitrogen Pentoxide, nitric anhydride
NO_3 e: Nitrate Radical
NO_{3-} e: Nitrate
NOA e: New Obligation Authority
- e: Notice of Arrival
- e: Notice of Availability
- e: NSWC Office Automation
- e: Scientific Test Aerodrome [Russia]
- s: Nueva Organización Anticomunista
NOAA e: National Ocean[ograph]ic and Atmospheric Administration (or Agency)
- e: NOAA Satellite
- s: Administración Nacional Oceánica y Atmosférica
NOAADIR e: NOAA Earth System Data Directory
- e: NOAA Environmental Services Data Directory
NOAAPORT e: access to NOAA real-time database system
NOAA Tech. Rep. e: NOAA Technical Report
NOAC e: Nature of Action Code
NOAEL e: No Observable Adverse Effect Level
NOAH e: National Organization for Albinism and Hypopigmentation
- e: Network Of Animal Health
- e: Network Oriented Application Harmonization
- e: New York Online Access to Health
- e: Nitrous Oxide and Halocarbons Division
NOAL s: No Alineados
NOAMI e: National Oceanographic (or Oceanography) and Maritime Institute
NOARL e: Naval Ocean[ograph][ic] and Atmosphere (or Atmospheric) Research Laboratory
N O B e: North Outpatient Building
NOBC e: Navy Officer Billet Classification
N O B LIB e: North Outpatient Building Library
NOBREX e: North Brazil[ian] Current Experiment
NOBS e: Naval Operations Bases System
NOC e: National Oceanographic Centre
- e: National Oceanographic Committee
- e: Network Operations Centre
- e: Notice of Commencement
- e: Number Of Children
- f: Centre Océanographique National
- f: Comité Océanographique National
- s: Centro Oceanográfico Nacional
- s: Comité Oceanográfico Nacional
NOCIRC e: National Organization of Circumcision Information Resource Centers
NOCMIS e: NOC Management Information System
NOCN e: National Ocean Communications Network
NOCO e: Norwegian Oil Consortium
NOCR e: Network For Oncology Communication and Research
NOCS e: Network Operations Center System
NOCT e: Nominal Operating Cell Temperature
NOD e: Network Outward-Dialing
- e: Not Otherwise Defined
- e: Notice Of Deficiency
NODC e: National Oceanographic Data Center (or Centre)
- e: National Ozone Data Center
- s: Centro Nacional de Datos Oceanográficos
NODD e: NORD Orphan Drug Designation Database
NODDES e: Naval Oceanographic Data Distribution and Expansion System
NODDS e: Naval (or Navy) Oceanographic Data Distribution System
- e: NOAA Oceanographic Data Distribution System
NODIS e: NSSDC Online Data and Information System
NODS e: NASA Ocean Data Service
- e: NASA Ocean Data System
- e: National Ocean Data System
nødv. a: nødvendig[t]
NOE e: Nap[e]-Of-the-Earth
- e: Nuclear Overhauser Effect
- e: Nuclear Overhauser Enhancement
- e: Number Of Employees
- e: Number Of Errors
NoE e: Network of Excellence
- f: Numéro d'Édition
NOEA e: National Outdoor Events Association
NOEI f: Nouvel Ordre Economique International
NOEL e: No[n] Observable Effect[s] Level
- e: Number Of Errors Left
NOEMFO N: Norsk Olje- og Energimedarbeideres Fellesorganisasjon
NOESY e: Nuclear Overhauser enhancement and Exchange Spectroscopy
NOF e: National Oceanographic Facility
- e: Node Operator Facility
- e: Notice of Findings
No-Fa-Ki N: Norges Farmasøytisk-Kjemisk Industriforening
NOFAS e: National Organization on Fetal Alcohol Syndrome
NOFORN e: No Foreign Nationals
NOH d: Grafschaft Bentheim in Nordhorn, Niedersachsen
NOHALICE e: Nitrous Oxide and Halocarbon Intercalibration Experiment
NOHARMM e: National Organization to Halt the Abuse and Routine Mutilation of Males
NOHSC e: National Occupational Health and Safety Commission
NOHSCP e: National Oil and Hazardous Substances Contingency Plan
NOI e: Notice of Inquiry
- e: Notice of Intent
NoI f: Numéro d'Impression
NOIC e: National Oceanographic Instrumental Center
- e: Naval Operational Information Center
- e: Notice of Intent to Cancel
NOIK n: Nederlandse Oos-Indiese Kompanjie
NOIS e: Notice of Intent to Suspend
NOIWAD e: National Operations and Intelligence Watch Officers Advisory
NOIWON e: National Operations Intelligence Watch Officers Network
NOK A: Die Nywerheidsontwikkelings-Korporasie
- d: Nationales Olympisches Komitee
- N: Norges Olympiske Komité
- N: Norsk Oseanografisk Komité
NØK a: Nordisk Økonomisk Kvægavl

NOLF e: Nursing Organization Liaison Forum
NOLS e: Nuclear Ordnance Logistics Systems
NOLTP e: National Organization of Lung Transplant Patients
NOM d: Northeim, Niedersachsen
- e: Individual with Personal Site
- e: National Orientation Movement
- e: Ninth October Movement
- e: No Offense Meant [Internet]
- e: Nomenclature
nom. e: nominative
NOMA e: National Online Media Association
NOMAC e: Noise Modulation and Correlation
NOMAD e: Navy Oceanographic Meteorological Association
- e: Neutrino Oscillation Magnetic Detector
Nombr f: Nombreux
NOMHICE e: Non-Methane Hydrocarbon Intercomparison Experiment
Nomi f: Nominal
NOMIC e: National Organization for Mentally Ill Children
NOMOFO N: Norske Motoroverhalingsverksteders Forbund
NOMS e: Nuclear Operations Monitoring Systems
NoMSG e: National organization Mobilized to Stop Glutamate
NOMSS e: National Operational Meteorological Satellite System
- e: Naval Oceanographic and Meteorological Support System
NOMTS e: Naval Ordnance Missile Test Station
N_2O-N e: Nitrous Oxide Nitrogen trace gas emission ratio
NON e: Notice of Noncompliance
NO3=NO32 e: Nitrat-Ion
Non-par e: Non-participating Provider
NONS d: Nordostniedersachsen
NOO e: National Oceanographic Office
- e: National Organization Order
NOP e: National Oceanographic Programme
- e: Nuclear Operating Plan
- s: Programa Oceanográfico Nacional
nop e: not operative
NoP f: Numéro de Publication
NÖP d: Niederösterreichisches Pressehaus
NOPAC e: Network OPAC
NOPECO e: Northern Transvaal Peoples Congress
NOPEF N: Norsk Olje og Petro-Kjemiforbund
NOPEX e: Northern Hemisphere Climate Process [land-surface] Experiment
- f: Expérience surface terrestre processus climatiques dans l'hémisphère Nord
- s: Experimento en la Superficie Terrestre sobre los Procesos Climáticos en el Hemisferio Norte
NOPLAN e: No Operational Plan Published
- e: No Plan
NOPP e: National Oil Palms Limited
NOR e: New Orleans, LA
- e: Norfluoxetine
- e: Norfolk
- e: Notice[s] of Revision
- N: Norske Oljeraffinerier
Nor N: Norsk
- s: Norte
NORA e: National Oil Recyclers Association
- e: Notice of Recruitment Activity
NORAD e: North American aerospace Defense
- e: North American [Air] Defense [command]
- e: North Atlantic Aerospace Defense Command
- e: Norwegian Agency for Development cooperation
- e: Norwegian Agency for international Development
- f: Agence norvégienne pour le développement international
- N: Direktoratet for Utviklingshjelp
- s: Organismo Noruego para el Desarrollo Internacional
NOR ADMIN e: NORAD Administrative Telephone Network
NORAD, Norad e: Norwegian Agency for International Development
NORAGRIC e: Norwegian Centre for International Agricultural Development
NOR ALERT e: NORAD Voice Alert Network
NORATS e: Naval Operational Radio-Telephone System
NOR COMNET e: NORAD Teletypewriter Network
NORCSEX e: Norwegian Continental Shelf Experiment
NORD f: Nord-Aviation
nord. a: nordisk
NORDA e: Naval Ocean Research and Development Activity
- e: Naval Oceanographic Research and Development Administration
NORDCO e: Newfoundland Ocean Research and Development Corp.
NORDEL r: Nordiskt Samarbete på Elkraftområde
NORDIHS e: Nordic Integrated Hydrographic System
Nordkdo N: Nordkommandoen
NORDMAP e: European Nordic Pollen Data Mapping Project
NORDQUA e: Nordic Association for Quaternary Research
NORIF e: Non-stimulated Oocyte Retrieval In Fertilization
NORM e: Naturally Occurring Radioactive Material
- e: Not Operational Ready-Maintenance
- s: Normalización
Norm e: Normative analysis
- f: Normand
NORMA e: No Remote Memory Access
NORMAC e: Northern Prawn Fishery Management Committee
Normal e: Town of Normal, Illinois
NORMARSEN e: Norwegian Maritime Remote Sensing [programme]
NORMED f: Chantiers du nord et de la Méditerranée
NOROL e: State-dominated refinery and oil products marketing group
NORPAX e: North Pacific Expedition
- f: Expérience dans le Pacifique Nord
- s: Experimento del Pacífico Norte
norr. N: norrøn
NORS e: Not Operational Ready-Supply
NORSE e: Nuclear Optical and Radar System Effects
NORSEX e: Norwegian Remote Sensing Experiment Programme
NORSMAP e: Norwegian Remote Sensing Spectroscopy for Mapping and Monitoring of Algal Blooms and Pollution
NOR TAC e: NORAD HQ ADC Tactical Telephone Network
north. e: northern dialect
NORTHAG e: Northern Army Group, Central Europe
Norv f: Norvège
- f: Norvégien
NOR WARN e: NORAD Warning Network
NOS e: National Ocean Service
- e: Network Operating System
- e: Network Operation Center
- e: Night Observation Surveillance
- e: Nirtous Oxide Systems
- e: Nodal Operating System
e: Northern Tropospheric Oxidants Study
- e: Not of Specific Origin
- e: Not Otherwise Specified
- N: Norges Offisielle Statistikk
Nos f: Numéros
NOSC e: NATO Operations Support Center
NosCom e: IEEE-SA Standards Board New Opportunities in Standards Committee
No Shinkei Geka j: No Shinkei Geka. Neurological Surgery
NOSIC e: Navy Ocean Surveillance Information Center
NOSIE e: NODC Ocean Science Information Exchange
NOSL e: Night-time Optical Survey of Lightning
- e: Night/day Optical Survey of Lightning
NOSP e: NASA Orbital Support Plan
NOSPI e: Newsletter on Serials Pricing Issues
NOSS f: Système National de Satellites Océaniques
- s: Sistema Nacional de Satélites Oceánicos
NOSWA e: National Organization of South-West Africa
NOT e: Nottinghamshire
Not f: Notamment
- f: Notice
- f: Notification
- s: Noticia
NOTAMS e: Notice to Airmen and Sailors
NOTAR e: No Tail Rotor
NotBZ d: Zeitschrift für die notarielle Beratungs- und Beurkundungspaxis
NOTCC e: Network Operational/ Technical Coordination Center

NOTEMPS e: Nontemporary Storage of Household Goods System
NOTIS e: Network Operator Trouble Information System
- e: North-western Online Totally Integrated System
No To Hattatsu j: No To Hattatsu [Brain and Development]
No To Shinkei j: No To Shinkei. Brain and Nerve
NOTS e: NMCS Operations Team Support Subsystem
NoTT e: Nordic Theatre Technicians
NOTU e: National Organization of Trade Unions
NOU N: Norges Offentlige Utredninger
Nouv f: Nouveau
Nouv Rev Fr Hematol f: Nouvelle Revue Francaise d'Hematologie
NOV e: November
- s: Noviembre
Nov f: Novembre
Nova e: laser facility at Lawrence Livermore National Laboratory
NOVA e: NOAA Operational VAS Assessment [program]
- e: Northern Ohio Valley Area
NOVAM e: Naval Oceanic Vertical Aerosol Model
nov[br.] a: november
NOV/CD e: Notice of Violation/ Compliance Demand
NOVEM e: Netherlands Development Corporation for Energy and the Environment
Novette e: laser system at Lawrence Livermore National Laboratory
NOVRAM e: Nonvolatile Random Access Memory
NOW e: Network Of Workstations
- e: NOAA Ocean Wave [model]
NOW-account e: Negotiable Order of Withdrawal account
NOWES e: Northern Wetlands Study
NOWESP e: North-West European Shelf Programme
NOWPAP e: North-West Pacific Action Plan
- f: Plan d'action pour le Pacifique du Nord-Ouest
- s: Plan de Acción para el Pacífico Noroccidental
NOWrad e: trademarked name of WSI, Inc. product
NOWS e: NVG Operations Weather Software
NOWT e: Netherlands Observatory of Science and Technology
No_x d: Stickstoff-Sauerstoffverbindungen
- e: [generi]c Nitrogen oxides
NO_y e: Menge des reaktiven Stickstoffs
- e: total [re]active Nitrogen
NOYB e: None Of Your Business
NOZE e: National Ozone Expedition
NP a: Nautisk Prøvekammer
- A: Nasionale Party
- e: National Party
- e: National Priority
- e: Network Performance
- e: Neutron Porosity
- e: No Protest
- e: No-Phonon
- e: Noise Parameter
- e: Non-deterministic-Polynomial
- f: Neuropsychiatre
- f: Nom Patronimique
- f: Nom Propre
- f: Non-Paiement
- f: Notoriété Publique
- f: Nouveau Prix
- N: Norsk Presseforbund
Np e: Neper
- f: Notes payables
np e: no problem [Internet]
n/P s: nuestro Pagaré
n/p s: nuestro pago
Np-239 e: Neptunium-239
NPA e: National Prescription Audit
- e: National Prostate Association
- e: National Provincial Administration
- e: Network Performance Analyzer
- e: Network Printer Alliance
- e: Network Printing Alliance
- e: Nigel Press Association Ltd.
- e: No Power Alarm
- e: Non-Par Approved
- e: Non-Parallel Application
- e: Number Plan Area
- f: Nouvelle Politique Agricole
NP&AA e: Noise Pollution and Abatement Act
NPAA e: National Oceanic and Atmospheric Administration
NPAC e: National Plantations Advisory Committee
NPACE e: Nurse Practitioner Association for Continuing Education
NPAD News e: National Panic/Anxiety Disorder Newsletter
NPAFC e: North Pacific Anadromous Fish Commission
- f: Commission des poissons anadromes du Pacifique Nord
- s: Comisión de Peces Anádromos del Pacífico Septentrional
NPAN e: Newspapers Proprietors Association of Nigeria
NPANSW e: National Parks Association of New South Wales
NPAP e: Network Printing Alliance Protocol
NPAQ e: National Parks Association of Queensland
NPAR e: Nuclear Plant Aging Research
NPAS e: Network Planning and Analysis System
NPB e: Neutral Particle Beam
- e: Neutron Particle Beam
- e: No Problem [Internet]
NPBSV e: Neutral Particle Beam Space Vehicle
NPC e: National Paralympic Committee
- e: National Pertroleum Council
- e: National Population Commission
- e: National Population Council
- e: Network Performance Control
- e: Network Personal Computer
- e: Non-Player Character
NPCMCH e: National Parent Consortium on Maternal and Child Health
NPCMW e: North Pacific Central Mode Water
NPCPL e: Normal Process Complementary Pass transistor Logic
NPCR e: National Program of Cancer Registries
- e: Normalized Protein Catabolic Rate
NPCS e: Narrow-band Personal Communication Service
NPD d: Nationaldemokratische Partei Deutschlands
- e: Narcissistic Personality Disorder
- e: Navy Procurement Directive
- e: Network Products Division
NPDES e: Laboratory Performance Evaluation
NPDES FILE e: National Pollutant Discharge Elimination System Compliance Files
NPDF e: N-Dimensional Probability Density Function
NPDP e: National Professional Development Program
NPDS e: National Military Command System Processing and Display System
- e: NMCC Processing and Display System
NPDU e: Network Layer Protocol Data Unit
NPDWR e: National Primary Drinking Water Regulation
NPE e: Naval Preliminary Examination
- e: Nuclear Planning and Execution
NPES e: Nuclear Planning and Execution System
NPF e: National Patriotic Front
- e: National Provident Fund
- e: NAVSTAR Processing Facility
- N: Norsk Plastforening
- N: Norsk Postforbund
NPFL e: National Patriotic Front of Liberia
NPFVAS e: Northern Prawn Fishery Voluntary Assistance Scheme
NPG e: Negative Population Growth, Inc.
- e: Network Protocol Gateway
- e: New Power Generation
- e: New Product Group
- e: NIOSH Pocket Guide to Chemical Hazards
- e: Non-Unit Personnel Generator
NPH e: No Parse Headers
- e: Non Processed Header
- e: Normal Paraffin Hydrocarbons
NPHAP e: National Pesticide Hazard Assessment Program

NPI e: National Pollution Inventory
- e: Network Printer Interface
- e: Neuropsychiatric Institute
- e: New Partnership Initiative
- e: New Product Introduction
- e: NEXRAD Product Interface
- e: Nuclear Propulsion Initiative
- e: Numbering Plan Identification
- f: Neuropsychiatrie Infantile
- f: Nouvelle Politique Industrielle
- N: Norsk Produktivitetsinstitutt

NPIAP e: National Pesticide Impact Assessment Program
NPIC e: National Photographic Interpretation Center
- e: National Programme on Improved Chulhas

NPIW e: North Pacific Intermediate Water
NPJ e: Number of projects in the RTD-projects
NPK a: kvælstof-fosfor-kalium [om gødning]
NPL e: National Physical Laboratory
- e: National Priorities (or Priority) List[ing]
- e: NOAAPORT Liaison
- e: Non-Procedural Language
- N: Norske Papirhandleres Landsforbund

NPLE e: National Patriotic Front of Liberia
NPLF e: Namibia Peoples Liberation Front
NPLT e: North Platte Basin
NPM e: NetView Performance Monitor
NPMA e: NATO Project Management Agency
NPMK e: National Plan of Management of Kangaroos
NPMOC e: Naval Pacific Meteorology and Oceanography Center
NPMP e: National Pesticide Monitoring Program
NPMR e: Native Pasture Management Research
NPMS e: Named Pipes/Mail Slots
NPN e: National Particulate Network
- e: National Party of Nigeria
- e: Negative Positive Negative
- e: Non-par Not Approved
- e: Non-Physician Practitioner
- N: Norsk Polarnavigasjon A/S

nPNA e: normalized Protein equivalent of total Nitrogen Appearance
NPND e: National Parent Network on Disabilities
NPO e: Naval Petroleum and Oil Shale Reserves in Colorado, Utah and Wyoming
- e: Nothing Per Os [Internet]
- R: Naucnyj-Proisvodstvennyj Organizacii

NPO"Vega" e: Scientific-Industrial Corporation "Vega"
NPOC e: National Point Of Contact
- e: Naval Polar Oceanography Center

NPODS e: Navy Print on Demand System
NPOE R: Naucnyj-Proisvodstvennyj Organizacii Energiya
NPOESS e: National Polar-orbiting Operational Environmental Satellite System
NPOP e: NASA Polar Orbiting Platform
NPO-PM R: Naucnyj-Proisvodstvennyj Organizacii Prikladnoi Mekhaniki
NPP e: National Peoples Party
- e: National Pesticides Policy
- e: Net Primary Production
- e: New Patriotic Party
- e: New Policy Proposal
- e: Nigerian Peoples Party

NPP 435 e: Namibian Peace Plan Group 435
NPPC e: Northwest Power Planning Council
NPPHCN e: National Progressive Primary Health Care Network
NPPRT e: Line Trace Report
NPPS e: Navy Printing and Publishing Service
- e: Navy Publication and Printing Service

NPPSBO e: Navy Printing and Publishing Service Branch Office
NPPSIS e: National Parent to Parent Support and Information System
NPPSO e: Naval Publications Printing Support Office
NPR e: National Performance Review
- e: Naval Petroleum Reserves in California
- e: Near-Pristine Rivers
- e: Non Persistant Reference
- e: Notice of Proposed Rulemaking
- e: Nuclear Posture Review
- e: Office of New Production Reactors

Npr f: Nom propre
NPRA e: National Patriotic Reconstruction Assembly
NPRC e: National Provisional Ruling Council
NPRF e: Noise Power Ratio Floor
NPRGCC e: National Program for Responding to Greenhouse Climate Change
NPRH e: Nurse Practitioners in Reproductive Health
NPRI e: National Pollution Release Inventory
NPRIS e: Nuclear Planning and Resource Information System
NPRM e: Notice of Proposed Rulemaking
NPRPM e: National Program for River Process and Management
NPRS e: National Programme for Remote Sensing
NPS e: NAPS [North Atlantic Publishing Systems] Publishing System
- e: National Park Service
- e: National Permit Strategy
- e: National Provider System
- e: Naval Postgraduate School, Monterey
- e: Navy Postgraduate School
- e: Network Print Server
- e: Newspaper Press Union
- e: Nominal Pipe Size
- e: Non-Parallel Share
- e: Non-Point Source
- e: Novell Productivity Specialist

$NPSH_a$ e: Net Positive Suction Head Available
$NPSH_r$ e: Net Positive Suction Head Required
NPSI e: NCP Packet Switching Interface
- e: Network Protocol Service Interface

NPSMS e: Nonpoint Source Management System
NPSN e: NAS Processing System Network
NPSU e: Nuclear Plant Safety Unit
NPT a: Nordisk Post Tidsskrift
- e: National PDES [Product Data Exchange using STEP [Standard for the Exchange of Product model data]] Testbed]
- e: Non-Programmable Terminal
- e: Non-Proliferation Treaty
- e: Nuclear Nonproliferation Treaty

NPTN e: National Pesticide[s] Telecommunications Network
- e: National Public Telecomputing Network

NPTZ e: North Pacific Transition Zone
NPU e: National Peoples Union
- e: Natural Processing Unit
- e: Newspaper Press Union

NPW e: National Parks and Wildlife
NPWC Act e: National Parks and Wildlife Conservation Act
NPX e: Numeric Processor Extension
NQ e: No Quote
n.q. e: no quotations
N.&Q. e: Notes and Queries
NQA e: No Questions Asked
NQCC e: Network Quality Control Center
NQCRRP e: North Queensland Community Rainforest Reforestation Program
NQE e: Network Queuing Environment
NQF e: National Qualification[s] Framework
NQL e: Network Query Language
NQNC e: North Queensland Naturalists Club
NQS e: Network Queu[e]ing System
NR a: Nordisk Råd
- a: Nordisk Rationalisering A/S
- e: National Range [operations directorate]
- e: Nature Reserve
- e: Naval Reactors
- e: No Return
- e: Not Ranked
- f: Nombre
- f: Notes Recevables
- N: Norges Rederforbund
- r: Nordiska Rådet
- s: Nuestra Referencia

Nr f: Numéro
Nr. a: Norre
N[R] e: Receive sequence Number
NRA e: National Reconstruction Alliance
- e: National Recovery Administration
- e: National Registration Authority for agricultural and veterinary chemicals
- e: Natural Resource Accounting

n.r.a. f: non repris ailleurs
NR-A e: National Range-Data Sciences Division
NRAC e: NATO [North Atlantic Treaty Organization] Requirements And Criteria
- e: Natural Resources Audit Council

NRAJ e: Natural Resource Authority of Jordan
NRAL e: National Radio Astronomy Laboratory, Jodrell Bank
NRALCC e: Northern Region Airlift Control Center
NRALD e: Northern Region Airlift Division

NRAO e: National Radio Astronomy Observatory [library]
NRB e: Normalized Relative Backscatter
nRBC e: nucleated Red Blood Cells
NRBP e: Natural Resource-Based Products
NRC e: National Rail Corporation
- e: National Reception Committee
- e: National Reconditioning Order
- e: National Referral Center
- e: National Reformation Council
- e: National Regulatory Commission
- e: National Republican Convention
- e: National Research Council
- e: National Resistance Council
- e: National Resources College
- e: Network Resource Controller
- e: Nichols Research Corporation
- e: Nigerian Railway Corporation
- e: Non-Recurring Charges
- e: Non-Recurring Cost[s]
- e: Non-Unit Related Cargo
- e: Nonreusable Container
- e: NORAD Readiness Center
- e: Nuclear Regulatory Commission
- s: Consejo Nacional de Investigación
NR-C e: National Range-Operations Control Division
NRCA e: Nonconformance Reporting and Corrective Action
NRCan e: Natural Resources Canada
NRCC e: Naval Regional Contracting Center
- e: NORAD Region Control Center
NRCCSA e: National Resource Center on Child Sexual Abuse
NRCDES e: National Research Council Division of Earth Sciences
NR-CE e: National Range-Safety Branch
NR-CF e: National Range-Flight Safety Branch
NRCL e: Natural Resources Conservatiopn League
NRCLV e: Natural Resources Conservation League of Victoria
NRCP e: National Rainforest Conservation Program
- e: National Research Council for the Philippines
NR-CS e: National Range-Range Services Branch
NRCS e: National Resource Conservation Service
- e: Natural Resource[s] Conservation Service
NRCSM e: Natural Resources Conservation Service
NRCT e: National Research Council of Thailand
- f: Conseil National de la Recherche de Thaîlande
- s: Consejo Nacional de Investigación de Tailandia
NR-D e: National Range-Data Collection Division
NRD e: Naural Resource Damage
NRDA s: Nube Radi[o]activa
NRDC e: National Revolutionary Development Campaign
- e: Natural Resource Defense Council
NRDC-CPSE e: National Revolutionary Development Campaign and Central Planning Supreme Council
NRDP e: National Rural Development Partnership
- e: National Rural Development Program
NRE e: Non-Recurring Engineering
N Reactor e: 105-N Reactor
N/REF f: Notre Référence
NRef[a] s: Nuestra Referencia
NREL e: National Renewable Energy Laboratory
NREM e: National Research and Engineering Network
NREN e: NASA Research and Educational Network
- e: National Research and Education Network
NRER e: Non-Rejected Earth Radiance
NRES s: Núcleo de Reserva
NRF e: Naval Reactor[s] Facility
- e: Naval Reserve Force
- N: Norges Rutebileierforbund
- N: Norsk Rasjonaliserings Forbund
- N: Norske Radiofabrikanters Forbund
- N: Norske Rørleggerbedrifters Landsforening
NRFC e: National Rail Freight Corporation
NRFF N: Norsk Forening for Rådgivende Rasjonaliseringsfirmaer
NRHP e: National Register of Historic Places
NRI e: National Agricultural Research Initiative
NRIC e: National Resource Information Centre
NRK N: Norges Røde Kors
NRL e: Naval (or Navy) Research Laboratories (or Laboratory)
- e: Nuclear Referral List
- e: NUWEP Reconnaissance List
NRLF e: Northern Regional Library Facility
NRM e: National Resistance Movement
- e: National Resources Management project
- e: Network Resource Management (or Manager)
- s: Niños Refugiados del Mundo
NRM. e: Normal Response Mode
NRM/A e: National Resistance Movement/Army
NRMA e: National Roads and Motoring Association
NRMC e: National Resources Management Centre
- e: Navy Regional Medical Center
NRMCA e: National Ready-Mixed Concrete
NRMDP e: Natural Resource[s] Management Development Project
NRMP e: National Resident Match Program
NRMS e: Natural Resource Management Strategy
NRMSS e: National Resource Management System
NRMT e: Nuclear Resonance Magnetometer Tool
NRN e: Netware Remote Node
- e: No Reply Necessary [Internet]
NRO e: Nation River Observatory
- e: National Reconnaissance Office
- e: National Reconnaissance Organization
- e: National Research Office
NROD e: National Range Operations Directorate
NROFF e: New ROFF [Run-OFF]
NROK e: Northern Rockies Intermontane Basins
NROMM e: Netherlands Register of Microform Masters
N-ROSS e: Navy Remote Ocean Sensing System
- e: Navy Remote Sensing Satellite
NROTC e: Naval Reserve Officer Training Corps
NR-P e: National Range-Range Programs Division
NRP e: NASA Resource Protection
- e: National Recovery Party
- e: National Religious Party
- e: National Research Program
- e: Navigation Reference Point
- e: New Republic Party
- e: Non-unit Related Personnel
NRPB e: National Radiation Protection Board
NRPE e: National Range Product Engineer
NRPF e: National Retinitis Pigmentosa Foundation
NRPGC e: National Research Program for Global Change
NRPMP e: National Rivers Processes and Management Program
NRR e: Navy Research Requirements
NRRC e: Nuclear Risk Reduction Center
NRRDC e: National Resources Research and Development Corporation
NRRF e: Naval Radio Receiving Facility
NRS e: National Residues Survey
- e: Navy Reporting Structure
NRSA e: National Remote Sensing Agency
- e: National Research Service Award
NRSA-NIOD DRG e: National Institute of Health, Division of Research Grants
NRSC e: Nansen Remote Sensing Centre
NRSCP e: National Reserves System[s] Cooperative Program
NRSP e: National Remote Sensing Program[me]
NRT e: Near Real-Time
- e: Non-Real-Time
- N: Nettoregistertonn
- s: Nudo Radiotécnico
NRTC e: National Road Transport Commission
NRTF e: National Rural Tourism Foundation
NRTL e: Nationally Recognized Testing Laboratory
NRTMR e: Near Real Time Target Motion Resolution
NRTS e: Not Repairable This Station
NRT-VBR e: Non Real-Time Variable Bit Rate
NRV e: Natural Resource Valuation
- e: Nigerian Revolutionary Vanguard
NRW d: Nordrhein-Westfalen

NRZ e: No[n]-Return-to-Zero [level (or recording)]
- e: Not Return to Zero
NRZ1 e: Non-Return-to-Zero, change on 1s
NRZI e: Non-Return-to-Zero Indicates (or Invert[ed]) [on ones]
NRZ[-]L e: Non-Return to Zero-Level
NRZ-M e: Non-Return-to-Zero-Mark
NRZ-S e: Non-Return-to-Zero-Space
NS d: Niederschlagstaub
- e: Name Server
- e: Nano-Seconds
- e: Navier-Stokes
- e: Network Services
- e: Network Supervisor
- e: Non Standard
- e: Non Stop
- e: Non-Smoker
- e: Noonans Syndrome
- f: Niveau Social
- f: Non Salariés
- f: Non Standardisé
- f: Nord-Sud
- f: Note de Service
- N: Norges Standardiseringsforbunds-kontrollmerke
- N: Norges Statspoliti
- r: Numerisk Styrning
- s: Nodriza de Submarinos
N S e: Edward Henry Kraus Natural Science Building
- e: Naval Science
ns d: nanosekunde
Ns e: Nielsbohrium
- f: Nous
NS. e: Nonsequenced
N[S] e: Send Sequence Number
NSA d: Nichtschlüsselattribut
- e: National Security Administration
- e: National Security Association
- e: Next Station Addressing
- e: No Suitable Applicant
- e: Non-Self-Averaging
- e: Non-surgical Sperm Aspiration
- e: Northern Slope of Alaska
- e: Not Seasonally Adjusted
NSA/AAO e: North Slope of Alaska and Adjacent Arctic Ocean
- e: NSA [National Security Agency]/Adjacent Arctic Ocean
NSA/CSS e: National Security Agency/Central Security Service
NSACSS Circular 80-17 e: System Test, Evaluation and Transition
NSAI e: National Standards Authority of Ireland
NSAID e: Non-Steroidal Anti-Inflamatory Drugs
NSAP e: Navy Scientific Assistance Program
NSAPA e: Network Service Access Point Address
NSAPI e: Netscape Server API
NSAPOLY e: National Security Agency Polygraph
NSAU e: National Space Agency of Ukraine
N S AUD e: Kraus Auditorium
NSB e: National Savings Bank
- e: National Sorghum Breweries
- e: North Slope Borough
NSBE e: National Society for Black Engineers
NSBMK N: Norges Statsbanenes militærkontor
NSC a: Nordisk Solar Compagni A/S
- e: National Security Council
- e: National Semiconductor
- e: National Simulations Center
- e: National Skin Centre, Singapore
- e: National Sports Congress
- e: National Statutory Council
- e: National Support Center
- e: Natural Sciences
- e: Network Service Center
- e: Network Signon Coordinator
- e: Network Support Center
- e: Neuroscience Center
- e: Norwegian Space Center
- e: Notice of Schedule Change
- e: Nuclear Services Corporation
NSC1 e: Network Systems Cypher One
NSCACE e: NSC Auxiliary Control Element
NSCAT e: N-ROSS Scatterometer
- e: NASA [advanced] Scatterometer
- f: Diffusiomètre perfectionné de la NASA
- s: Difusímetro perfeccionado de la NASA
NSCC e: N-SITE Control Center
e: New Sudan Council of Churches
NSCCA e: Nuclear Safety Cross-Check Analysis
NSCD e: National Spatial Cadastral Database
NSCN e: Navy Satellite Control Network
NSCO e: Navy Supply Center, Oakland
NSCORT e: NASA-sponsored Specialized Center of Research and Training
NSCS e: National Soil Conservation Strategy
- e: Network Service Center System
NSD e: Nairobi Sheep Disease
- e: National Security and Defense
- e: Navy Support Data
- e: NMCS Support Division
Nsd f: Non spécialement dénommé
NSDA e: Non-Self Deployment Aircraft
NSDD e: National Security Decision Directive
- e: National Security Defense Directive
NSDI e: National Spatial Data Infrastructure
NS/DOS e: Networking Services/DOS
NSDS e: Navy Source Data System
NSDSSO e: NASA Science Data Systems Standards Office
NSDWR e: National Secondary Drinking Water Regulations
NSE e: Nairobi Stock Exchange
- e: Namibian Stock Exchange
- e: Natural Sciences and Engineering
- e: Network Software Environment
- e: Neuron Specific Enolase
- e: Neutral Stream Etch
NSEC e: National System for Emergency Coordination
- e: Network Switching Engineering Center
NSEI N: Norsk Selskap for Elektronisk Informasjonsbehandling
NSEL e: Network Service Selector
NSEM e: Network Software Environment
NS/EP e: National Security Emergency Preparedness
NSEP e: National Security Emergency Preparedness
- e: National System for Emergency Preparedness
NSERC e: National Sciences and Engineering Research Council
- e: Natural Science[s] and Engineering Research Council
NSERCC e: Natural Sciences and Engineering Research Council of Canada
NSESCC e: NASA Space and Earth Sciences Computing Center
NSESD e: National Strategy and Ecologically Sustainable Development
NSF e: N-ethylmaleimide-Sensitive Fusion protein
- e: National Schizophrenia Fellowship
- e: National Science Foundation
- e: National Sleep Foundation
- e: Navy Stock Fund
- e: Non Standard Format
- e: Non-Sufficient Funds
- e: Notes Storage File
- N: Norges Sivil Forsvar
- N: Norges Skibsførerforbund
- N: Norges Standardiserings-Forbund
- N: Norsk Sjømannsforbund
- N: Norsk Speidergutt-Forbund
- N: Norsk Sykepleierforbund
- N: Norske Siviløkonomers Forening
NSFA e: Naval Support Force Antarctica
NSFI N: Norges Skipsforskningsinstitutt
NSFIP d: Nextstep Für Intelprozessoren
NSFNET e: National Science Foundation Network
NSFNet e: NSF Network
NSFO e: Navy Standard Fuel Oil
NSFS e: Naval Surface Fire Support
NSG N: Norsk Sau- og Geiteavlslag
NSGC e: National Society of Genetic Counselors, Inc.
N S H e: Northville State Hospital
NSH e: Nutritional Secondary Hyperparathyroidism
- f: Nouvelle Société Helvétique
NSHC s: Comisión Hidrográfica del Mar del Norte
NSI e: NASA Science Internet
- e: National System of Innovation
- e: Network Solutions, Inc.
- e: Non-SNA Interconnect
- e: Not Seriously Injured
NSIC e: National Storage Industry Consortium
NSIDC e: National Snow and Ice Data Center (or Centre)
NSI/DECNET e: DEC Network
NSIP e: Nonlinear Signal and Image Processing
NSIPS e: NRL Satellite Image Processing System
NSIS e: National Schengen Information System

N-SITE e: Near-term System Integration Technology Experiment
- e: Near-Term System Integration Test and Evaluation [experiment]

NSIU N: Norges Svalbard og Ishavs Undersøkelsen

NSIWG e: NATO SATCOM Interoperability Working Group

NSJI e: North-South Joint Implementation

NSKF N: Norske Studenters Kristelige Forbund

NSKV N: Norges Sjøkartsverk

NSL a: Nordsjællands Spær- & Limtræindustri A/S
- e: Network Service Layer
- e: Network Software Laboratory
- e: Network Space Load
- N: Norske Skofabrikkers Landssammenslutning

Nsl r: Nedslags-

NSLF N: Norsk Skog- og Landarbeiderforbund

NSLOOKUP e: Name Server Lookup

NSLRSDA e: National Satellite Land Remote Sensing Data Archives

NSLSI e: Nova Scotia Land Survey Institute

NSM e: Negative-Stiffness-Mechanism
- e: Netscape Server Manager
- e: Netz-Service-Modul
- e: NMCS Simulation Model
- f: [Banque de] Neuflize, Schlumberger et Mallet

NS&MF N: Norske Sykkel- og Mopedfabrikanters Forening

NS&MP e: Navy Support and Mobilization Plan

NSMS e: National Security Management System

NSMTFS e: NATO Standard Message Text Formatting System

NSN e: NASA Science Network
- e: NASA Space Network
- e: Non-Strategic Nuclear Forces

NSNA e: No Stock Number Assigned

NSNF e: Non-Strategic Nuclear Forces

NSNS e: National Spallation Neutron Source

NSO e: NASA Safety Officer
- e: National Security Organization
- e: National Solar Observatory [library]
- e: National Sports Organization [general term]
- e: Native Signal Processing
- e: Network Security Officer
- e: Nitrogen, Sulfur, Oxygen
- e: Non-SIOP Option
- e: Nuclear Safety Office

NSOC e: National Signals Intelligence Operations Center

NSOF e: Navy Status of Forces File

NSOM e: Near-field Scanning Optical Microscopy

NSOR e: NATO Statement of Operational Requirement
- e: NORAD Statement of Operational Requirement

NSOW e: Norwegian Sea Overflow Water

NSP e: Name Service Protocol
- e: Native Signal Processing
- e: Network Automation Protocol
- e: Network Service Part
- e: Network Service Protocol
- e: Network Service Provider
- e: Neurological Shellfish Poisoning
- e: Numeric Space
- N: Norges Speditørforbund

n. sp. e: new species

NSpC e: National Space Council

NSPD e: National Space Policy Directive

NSPE e: National Society for Professional Engineers

NSPHPC e: National Survey of Personal Health Practices and Consequences

NSPMP e: Network Switching Performance Measurement Plan

NSPPO e: Navy Signal Processing Program Office

NSPRS e: Nigerian Society for Photogrammetry and Remote Sensing

NSPS[R] e: New Source Performance Standard [Review]

NSQAC e: National Specialist Qualification Advisory Committee

NSR d: Niederspannungsrichtlinie
- e: NATO Staff Requirement
- e: New Source Review
- e: Non Sequential Recording
- e: Non-Shared Resources
- e: Non-Source Routed

NSRA r: Nordvästra Skånes Radioamatörer

NSRF N: Norges Salgs- og Reklameforbund

NSRI A: Nasionale Seereddings-Institut

NSRL e: National SIGINT Requirements List
- e: National Strategic Reconnaissance List

NSRM e: National Strategy for Rangeland Management

Nsrp f: Ne se représent pas

NSR/PSD e: New Source Review and Prevention of Significant Deterioration Permitting

Nsrr r: Närspaningsradar

NSRS e: National Spatial Reference System

NSS e: Namibia Scientific Society
- e: National Security Service
- e: National Service Scheme
- e: National Strategic Stockpile
- e: National Stream Survey
- e: Natural Security-System
- e: No Shit, Shylock [Internet]
- e: No Study Section [in house]
- e: Nodal Switching Subsystem
- N: Norsk Studentsamband

nss e: non-sea-salt

NSSA e: National Sjogrens Syndrome Association
- e: National Social Security Authority

NSSC e: NASA Standard Spacecraft Computer
- e: National Space Surveillance Center
- e: NORAD Space Surveillance Center
- e: Nordic Symposium on Super-Conductivity

NSSC-1 e: NASA Standard Spacecraft Computer

NSSD e: National Security Study Directive

NSSDC e: NASA Space Science Data Center
- e: National Space Science Data Center

NSSF e: Near Surface Storage Facility
- N: Nordiske Sosialdemokratiske Studentforbund

NSSFC e: National Severe Storms Forcast Center

NSSN e: National Standards System Network

NSSR N: Norsk Selskab til Skibbrudnes Redning

NST a: Nordisk Systemtryk A/S
- e: National Security Technology
- e: NATO Staff Target

NSTAC e: National Security Telecommunications Advisory Committee

NSTAG e: National Science and Technology Analysis Group

NSTC e: National Science [and] Technology Council

NSTF e: National Science and Technology Forum
- e: Near Surface Test Facility
- e: North Sea Task Force
- f: Equipe spéciale sur la mer du Nord
- N: Norges Skogteknikerforbund
- s: Equipo de Trabajo sobre el Mar del Norte

NSTIS e: Navy Standard Technical Information System[s]

NSTL e: Nation Space Technology Laboratory
- e: National Software Testing Laboratories
- e: National Strategic Targets List

NSTP e: National Security Telecommunication Policy

NSTSPO e: National Space Transportation System Program Office

NSTVA d: Nebenstellen-Vermittlungsanlage

NSTX e: National Spherical Tokamak Experiment

NStZ-RR d: NStZ-Rechtsprechungsreport

NSU e: Network Service/Interface Unit
- e: Networking Support Utilities
- N: Norske Studentersamfund
- N: Norsk Studentunion

NSÜ d: Netzsteuerungs- und -überwachungssystem

NSUB d: Niedersächsische Staats- und Universitätsbibliothek Göttingen

NSUe d: Netzsteuerungs- und -überwachungssystem

NSUF N: Nasjonal Samlings Ungdomsfylking

NSUG e: Nihon Sun Users Group

NSV e: National Strategic Vision

NSVN e: NATO Secure Voice Network

NSVS e: NATO Secure Voice System

NSW d: Nebelscheinwerfer
- e: Herbarium of New South Wales International Acronym
- e: Naval Special Warfare

NSWAIP e: Australian Institute of Physics, NSW Branch

NSWC e: Naval Surface Warfare Center

NSWCDD e: Naval Surface Warfare Center Dahlgren Division
N.S.W.C.M. e: New South Wales State Conservatorium of Music
NSWEPA e: New South Wales Environment Protection Authority
NSWFC e: New South Wales Forestry Commission
NSWIT e: New South Wales Institute of Technology
NSWNPWS e: New South Wales National Parks and Wildlife Service
NSWP e: National Space Weather Program
NSWS e: National Surface Water Survey
NSWTE e: Naval Special Warfare Task Element
NSWTG/TU e: Naval Special Warfare Task Group/Task Unit
NSWTU e: Naval Special Warfare Task Unit
NSWU e: Naval Special Warfare Unit
NT a: Nordisk Textiltryk A/S
- a: Nye Testamente
- a: Nørresundby Transportkompagni
- d: Netzwerk Terminator
- e: Microsoft Windows NT [New Technology]
- e: Neurologically Typical
- e: New Trouble
- e: Nitrogen Tetroxide
- e: Node Type
- e: Noise Thermometer
- e: Normal Threat
- e: Northern Tablelands
- e: Northern Territory
- e: Northern Territory herbarium International Acronym
- e: Saudiarab [Irak]
- f: Nouveau Tarif
- N: Det Nye Testamente
- r: Norrlands Tygstation
Nt f: Nantes
- f: Négociant
N.T. e: New Testament
NT[-]1 e: Network Terminator Type 1
NT2 e: Network Termination Type 2
NT-3 e: Neurotrophin-3
NTA e: New Transatlantic Agenda
- e: Nigerian Television Authority
NTAPS e: Near Term Active Protection System
NTAS e: New Technology Advanced Server
- e: NORAD Tactical Automatic Voice Network System
- e: NORAD Tactical AUTOVON System
- e: NT Advamced Server
- e: Windows NT Advanced Server
NTB e: Network Test Bed
- N: Norsk Telegrambyrå
NTBA d: ISDN-Netzbasisanschluss
NTBIC e: National Test Bed Integrating (or Integration) Contractor
NTB/JPO e: National Test Bed/Joint Program Office
NTBJPO e: National Test Bed/Joint Program Office
NTC e: National Training Contact
- e: National Transportation Corporation
- e: Network Terminal Concentrator
- f: Correspondant national pour la formation
- s: Contacto Nacional de Formación
NTCA e: National Weather Service Training Center System 1, River Forecast Center
NTCC e: Naval Telecommunications Center
NTCF e: NATO Technical Control Facility
NTCOC e: Naval Telecommunications Command Operations Center
NTC-OIS e: National Training Center Objective Instrumentation System
NTCS e: Naval Telecommunications and Computer System
NTCSA e: Navy Tactical Command System Afloat
NTD s: Núcleo de Tropas Divisionarias
NTDB e: National Topographic Data Base
- e: National Toponymic Data Base
NTDLM d: Neue Telefondienstleistungsmerkmale
NTDO e: Naval Technical Data Office
NTDP e: New Technology Demonstration Program
NTDR e: Near-Term Data Radio
NTDRP e: Northern Territory Drought Releif Policy
NTDS e: Naval (or Navy) Tactical Data System
- e: Naval Tactical Distribution System
- e: Navy Technical Data System
NTDS-B e: Navy Tactical Data System Type B
NTDS-E e: Navy Tactical Data System Type E
NTE e: National Teaching Examination
- e: Non-Traditional Exports
Nte. s: Norte
NTEC e: Network Technical Equipment Center
NTEU e: National Tertiary Education Union
NTF e: National Tidal Facility
- e: National Transfer Format
- e: Nigeria Trust Fund
- e: Nuclear Nonproliferation Treaty
- N: Norges Tekstilkjøpmenns Forbund
- N: Norsk Tannlægeforening
- N: Norsk Tolltjenestemannsforbund
NTF JPO e: National Test Facility Joint Program Office
NTFMRS e: Navy Total Force Manpower Requirements System
NTFS e: Network of Tropical Fisheries Scientists
- e: New Technology File System
- e: NT File System
- f: Réseau de spécialistes des pêches tropicales
- s: Red de Científicos de Pesquerías Tropicales
NTG d: Nachrichtentechnische Gesellschaft e. V.
- d: Netzwerk und Telematic GmbH
NTGDS e: Non-Tritium Gas Delivery System
NTGS e: National Technical Guidance Studies
NTH e: Norwegian Institute of Technology
- N: Norges Tekniske Høyskole
NTHG e: Nothing
NTI e: National Technology Initiative
- e: National Toxics Inventory
- e: Near-Term Improvements
- e: Network Terminating Interface
- f: Nouvelles Technologies de l'Information
- N: Norsk Treteknisk Institutt
NTIA e: National Telecommunications [and] Information Agency (or Administration)
NTIM e: Not That It Matters [Internet]
NTIMM e: Not That It Matters Much [Internet]
NTIPS e: NAC Total Image Processing System
- e: Navy Technical Information Presentation System
NTIS e: National Technical Information Service[s]
- e: NMOS Technical Information Services
- f: Service National d'Information Technique
- s: Servicio Nacional de Información Técnica
NTISA e: Naval Tactical Interoperability Support Activity
- e: Naval Tactical Interoperability Support Agency
NTISSC e: National Telecommunications and Information Systems Security Committee
NTK d: Niedertemperaturkessel
- e: Net Tonne Kilometre
- e: Newton Tool Kit
NTL e: Nevertheless
- N: Norsk Tjenestemannslag
- N: Norske Trevarefabrikkers Landsforbund
- N: Norske Tønneprodusenters Landsforening
NTLF e: National Tritium Labeling Facility
NTLIS e: Northern Territory Land Information System
NTLPSS e: Non-real-Time Launch Processing Software System
NTM e: Network Model
- e: Network Terminal Manager
- e: Northern Territory Art Gallery and Museum, Darwin
NTMP e: NMCS Technical Management Plan
NTMS e: National Topographic Mapping Series
NTN e: National Trends Network
NTNCWS e: Non-Transient Non-Community Water System
NTNF N: Norges Teknisk-Naturvitenskapelig Forskningsråd
NTNU N: Norges Teknisk-Naturvitenskapelige Universitet
nto. a: netto

NTP e: Navy Telecommunications Publications
- e: Near-Term Penetrator
- e: Network Time Protocol
- e: Nontree Pollen

NTPCC e: National Tree Program Coordinating Committee
- e: Northern Transvaal People's Coordinating Committee

NTPF e: Near-Term Pre-positioning Force
NTPK N: Norsk Teknisk Pressekontor
NTPO e: National Transuranic Waste Program Office
NTPS e: Naval Test Pilots School
NTR e: Nine Thousand Remote
- e: Non Tactical Radio
- e: Nuclear Thermal Rocket[ry]
- e: Nutrition

NTRAS e: NT Remote Access Services
NTRD e: Nascom Traffic Requirements Database
NTRI e: NCP Token-Ring Interconnect
- e: NCP Token-Ring Interface

NTRID e: Netted Telephone Radio Interface
NTrl Fdi N: Nord-Trøndelag Forsvarsdistrikt
NTRM e: NIST Traceable Reference Materials
NTRS e: National Technology Roadmap for Semiconductors
NTS e: National Technical Systems
- e: NCC Test System
- e: Network Technical Support
- e: Network Test System
- e: Network Transport Services
- e: Nevada Test Site
- e: New Testament Studies

NTS[/]2 e: Network Transport Services/2
NTSA e: Netware Telephony Services Architecture
- e: Networking Technical Support Alliance

NTSC e: National Television Standard Coding
- e: National Television Standards Committee
- e: National Television Systems Committee
- e: Navy Training Systems Center
- e: Never The Same Color
- e: TV standard in Japan and most of the Americas

NTSK N: Nordisk Telesatellittkomité
NTSL e: Nonintegrated Two-Stage Liquid
NTSMP e: Naval Telecommunications System Master Plan
NTSN e: National Threatened Species Network
NTT e: National Tree Trust
- e: Nippon Telegraph and Telephone Corp.

NTTAWWT e: Not That There's Anything Wrong With That
NTTC e: National Technology Transfer Center
NTTF e: Network Technical Training Facility
- e: Networks Test and Training Facility
- e: Netwrok and Telecommunications Taskforce
- N: Norges Tekstil Teknisk Forbund
- N: Norsk Telegraf- og Telefonforbund

NTT-IT e: Nippon Telegraph and Telephone-Intelligent Technology
NTTP e: Network News Transfer Protocol
NTTR e: Non-Tactical Telecommunications Requirements
NTTS e: NATO Terrestrial Transmission System
NTU e: Nanyang Technological University
- e: Nephelometric Turbidity Unit[s]

NTUF e: Nigerian Trade-Union Federation
NTVES e: Northern Territory Voluntary Euthanasia Society
NTW e: Not To Worry
NTYMI e: Now That You Mention it
NTZ e: Nitazoxanide
NU e: Astronomy Department Library, Nanjing University
- e: Network Utility
- N: Norges Ungdomslag
- r: Näringsutskottet

Nu r: Normalutrustning
N U B e: North University Building
NuBus d: Interner 32-Bit-Datenbus mit hoher Leistungsfähigkeit
NUC e: National Underseas research Center
- e: Naval Undersea research and Development Center

NUCAP e: Nuclear Capabilities Data Base
NUCEWA e: Nuclear Weapons Availability
Nucleic Acids Res e: Nucleic Acids Research
Nucleic Acids Symp Ser e: Nucleic Acids Symposium Series
Nucl Med Biol e: Nuclear Medicine and Biology
Nucl Med Commun e: Nuclear Medicine Communications
NUCMATTS e: Nuclear Materials Transportation Tracking System
NUCOR e: Nuclear Development Corporation of South Africa
NUCW e: National Union of Cameroon Workers
NUCWAL e: Nuclear Weapons Allocation Logistics
NUCWEP e: Nuclear Weapons
NUD e: Non-Update
NUDA e: Nigerian Unions of Dispensing Assistants
NUDAP e: Nuclear Detonation Data Point
NUDETS e: Nuclear Detonation Detection and Reporting System
NUDIAS e: National University Digital Image Analysis System
NUDIS e: NUDET and Damage Information Summary
NUDO e: National United (or Unity) Democratic Organization
NUDP e: National Urban Development Program
NUE d: Nürnberg, Germany [Airport]
NUES e: National Union of Ethiopian Students
NUEUS e: National Union of Ethiopian University Students

NUF e: National Unifying Force
- e: National Union of Farmworkers
- N: Nasjonal Ungdomsfylking
- N: Norges Urmakerforbund

NUFATS e: Nebraska University Fortran Algorithm for Terms
NUFCOR e: Nuclear Fuels Corporation
NUFED s: Núcleos Familiares y Educativos para el Desarollo
NUFFIC e: Netherlands Universities Foundation for International Cooperation
NUFU N: Nasjonalt utvalg for Utviklings-relatert Forskning og Utdanning
NUG e: National Union of Ghanaian Students
NUGs e: Non-Utility Generators
NUH N: Norges Ungdomsherberger
NUI e: Network User Indentification
- e: Network User Interface
- N: Norsk Undervannsinstitutt

N.U.I. e: National University of Ireland
NUI. e: Network user identifier
NUICCS e: NORAD-USSPACECOM Integrated Command and Control System
NUJ e: Nigerian Union of Journalists
NUKS e: National Union of Kenya Students
NUL e: Dummy Device
- e: National University of Lesotho
- e: No Device

NULGE e: Nigerian Union of Local Government Employees
NULT e: National Union of Liberian Teachers
N[u]m e: Book of Numbers
NUM e: Nagoya University Museum
- s: Número

Num f: Numéral
- f: Numéro

NUM. e: Numeric
NUMA e: Non-Uniform Memory Access
- e: Non-Uniform Memory Address
- e: Non-Uniform Memory Architecture

NUMEC e: Nuclear Uranium Materials and Equipment Corporation
Numér f: Numération
NUMERIS e: French ISDN Network
NUMSA e: National Union of Metalworkers of South Africa
NUNA-TEK a: Greenland Technical Organization
NUNBE e: National Union of Nigerian Bank Employees
NUNS e: National Union of Nigerian Students
NUNW e: National Union of Namibian Workers
NUO e: North Up
NUOR e: Nuclear Ordnance
NUPENG e: National Union of Petroleum and Natural Gas Workers
NU[P]I N: Norsk Utenrikspolitisk Institutt
NUPI e: Norwegian Institute for International Affairs
NUPTW e: National Union of Postal and Telecommunications Workers
NuR d: Natur und Recht
Nurb d: Bezierkurventyp
NURB e: Non-Uniform Rational B-spline

NURBS e: Non-Uniform Rational B-Spline
NURC e: Non-Unit Related Cargo
NUREG e: Criteria for Preparation and Evaluation of Radiological Regulations and Guides
- e: Nuclear Regulatory Commission
NUREP e: Nuclear Reporting
NURF e: NAPAP Utility Reference File
NURO e: National Undergraduate Research Observatory
NURP e: NOAA Undersea Research Program
- e: Non-Unit Related Personnel
Nurs Clin North Am e: Nursing Clinics of North America
Nurse Pract e: Nurse Practitioner
Nurs Hist Rev e: Nursing History Review
Nurs Outlook e: Nursing Outlook
Nurs Res e: Nursing Research
NURTW e: National Union of Road Transport Workers
NURW e: National Union of Railway Workers
NURWEC e: Netherlands U.K. Radar Wavebuoy Experimental Comparisons
NUS e: No Upper Stage
- e: NUS Corporation, Gaithersburg, Maryland
NUSAS A: Nasionale Unie van Suid-Afrikaanse Studente
NUSC e: Naval Underwater Systems Center
- e: Naval Underwater Systems Command
NUSDE e: National Union of Shop and Distributive Employees
NUSS e: Nuclear Safety Standards
NUSU e: National Union of Students of Uganda
NUSWAS e: National Union of South-West African Students
NUT e: Network Utility File Transfer
- e: Nutrition
NUTA e: National Union of Tanganyika Workers
NUTEK e: Swedish National Board for Industrial and Technical Development
NUTIS e: NERC Unit for Thematic Information Systems
- f: Unit for Thematic Information Systems du NERC
- s: Unidad de Sistemas de Información Temática del NERC
Nutr Cancer e: Nutrition and Cancer
Nutr Health e: Nutrition and Health
Nutr Hosp e: Nutricion Hospitalaria
Nutr Rev e: Nutrition Reviews
NUTS f: Nomenclature des Unités Territoriales Statistiques
NutzEV d: Nutzungsentgeltverordnung
NUUS e: National Union of Uganda Students
NUV N: Norges Unge Venstre
nuv. a: nuværende
- r: nuvarande
NUWAX e: Nuclear Weapon Accident
NUWC e: Naval Underwater Warfare Center
NUWEP e: Nuclear Weapons Employment Policy
- e: Nuclear Weapons Employment Procedures
NUYO e: National Union of Youth Organizations
NUZS e: National Union of Zambia Students
NV a: Nordvest
- e: No Overflow
- f: Non-Valeur
- r: Nordväst
nV e: normalized Volume
Nv f: Nièvre
n.v. N: nordvest
- r: nuvarande
NVA e: North-Vietnamese Army
NVAP e: NOAA water Vapor Project
NVB n: Nederlandse Vereniging van Bibliothecarissen, documentalisten en literatuuronderzoekers
NVBF e: Nordic federation of Research Libraries
- N: Nordisk Videnskabeligt Biblioteker-Forbund
NVBR e: Nevada Basin And Range
NVB-SG e: Dutch Serials Group
NVC e: National Visa Center
NVCJD e: New Variant of Creutzfeldt-Jakob Disease
NVDET e: Network Virtual Data Entry Terminal
NVDM2 e: NetView Distribution Manager/2
NVDS e: NOAA Virtual Data System
NVE d: Nummer der Versandeinheit
- e: Network-Visible Entities
- N: Norges Vassdrags- og Elektrisitetsvesen
NVG d: Rheinland-Pfalz: Landesgesetz über den öffentlichen Personennahverkehr [Nahverkehrsgesetz] vom 17.11.1995
- e: Night Vision Goggle
NVGC e: Night Vision Goggle Compatibiity
NVH e: Noise, Vibration, Harshness
NVI e: Normalized difference Vegetation Index
NVIS e: Near Vertical Incident Skywave
NVLAP e: National Validation Laboratory Program
NVLSPM e: Night Vision Laboratories Static Performance Model
NVM e: Non Volatile Memory
NVMS e: Noise and Vibration Monitoring System
NVN e: National Videotext Network
NVO e: Non-Visual User Object
NVoD e: Near Video on Demand
NVOD e: Near-Video On Demand
NVP e: Nominal Velocity of Propagation
NVPP e: National Vehicle Population Poll
NVPS e: Night Vision Pilotage System
NVR[AM] e: Non Volatile RAM [Random Access Memory]
NVROM e: Non-Volatile ROM [Read-Only Memory]
NVS e: NonVolatile Storage
NVSIMM e: Non-Volatile Single Inline Memory Module
nvSRAM e: nonvolatile Static Random Access Memory
NVT e: Netware Virtual Terminal
- e: Novell Virtual Terminal
NVwZ d: Neue Zeitschrift für Verwaltungsrecht
NW e: Narrow Well
- e: North West
- e: Science-Engineering Library, Northwestern University
NWA e: Northwest Airline
NWAFC e: Northwest and Alaska Fishery Center
NWC e: Naval (or Navy) War College
- e: Nuclear Weapons Center
- e: Nuclear Weapons Council
NWCC e: National Womens Consultative Council
NWCF e: Northwest Citizens Forum
NWCP e: Noxious Weeds Control Program
NWDA e: National Wholesale Drug Association
- e: National Wholesale Druggists Association
NWDC e: National Wildlife Defence Council
NWDS e: Networkwide Directory System
NWE e: Nuclear Weapon[s] Effect[s]
n-well e: A region of n-doped silicon for the fabrication of a p-channel MOSFET
NWF e: Nuclear Waste Fund
NWG d: Nachweisgrenze
- d: Namibische Wissenschaftliche Gesellschaft
- e: Network Working Group
NWGCN e: National Working Group on Coastal Management
NWHC e: National Wildlife Health Center
NWHIC e: National Womens Health Information Center
NWHRC e: National Womens Health Resource Center
NWI e: National Wilderness Inventory
NWIS e: Navy WWMCCS Information System
NWL e: National Weapons Laboratory
NWLISN e: Northwest Land Information System Network
NWLON e: National Water-Level Observation Network
NWMP e: Nuclear Weapons Master Plan
NWnet e: North West Net
NWNET e: Northwestern states Network
NWO e: Netherlands Organisation for Scientific Research
- e: No World Order
NWOC e: New Way Of Computing
NWP e: Nationwide Permit Program
- e: Navy Warfare Publication
- f: Prévision numérique du temps
NWPA e: National Wildlife Protection Act
- e: Nuclear Waste Policy Act
NWPC e: National Womens Political Caucus
NWPM e: Numerical Weather Prediction Model
NWPU e: National Wildlife Protection Unit

NWQMS e: National Water Quality Management Strategy
NWR e: News Relationships
- e: Niwot Ridge
- e: NOAA Weather Radio
NWRC e: National Wetlands Research Center
- e: National Womens Resource Center
NWRI s: Instituto Nacional de Investigación Hidrológica
NWRP e: Nuclear Weapons Release Procedures
NWS e: National Weather Services
- e: Netware Web Server
- e: Northern Wetland Study
- e: Nuclear Weapons Storage
- f: Service météorologique national des Etats-Unis
- s: Servicio Meteorológico Nacional de los EE.UU.
NWS. e: Nonprogrammable Workstation
NWSA e: Nose Wheel Steering Actuator
NWSFO e: NEXRAD Weather Service Forecast Office
- e: NWS Forecast Office
NWSM e: Nuclear Weapon Stockpile Memorandum
NWSP e: Nuclear Weapon Stockpile Plan
NWSRFS e: NWS River Forecast[ing] System
NWSS e: Navy WWMCCS Software Standardization
NWSSG e: Nuclear Weapons Systems Safety Group
NWSTG e: NWS Telecommunications Gateway
NWT e: North West Territories [Australia]
NWTDB e: Naval Warfare Tactical Data Base
NWTP e: NMCS WWMCCS Transition Plan
NWV A: Namibië Wetenskaplike Vereniging
NWWS e: NOAA Weather Wire Service
NXP e: Norberts XML Parser
NXX e: NANP syntax for a three digit string
Ny f: Nancy
N.Y. e: New York [University]
NYC e: New York City, NY
NYCLU e: New York Civil Liberties Union
NYCOP e: National Youth Council of Ogoni People
NYCRC e: New York Civil Rights Coalition
N. Y. Fish Game J. e: New York Fish and Game Journal
NYI e: Not Yet Impacted
NYJM e: New York Journal of Mathematics
NYKL N: Norges Yrkeskvinners Landsforbund
NYL N: Norges Yrkeslærerlag
NYLJ e: New York Law Journal
NYM e: Anonymous
- e: National Youth Movement
NYMOL e: National Youth Movement of Liberia
NYNEX e: New York, New England and the unknown [X]
NYOC e: National Youth Organizing Commission
NYR e: New York Review of Books
nys. a: nysynet
NYSC e: National Youth Service Corps
NYSERDA e: New York State Energy Research and Development Authority, Albany, New York
NYSERNET e: New York State Education and Research NETwork
NYSERNet e: New York State Educational and Research Network
NYSHESC e: New York State Higher Education Service
NYSNA e: New York State Nurses Association
NYSSA e: New York State Student Assembly
N Y State Dent J e: New York State Dental Journal
NYT e: New York Times
NYTBR e: New York Times Book Review
NYU/Ultra e: New York University's Ultracomputer
NZAP e: New Zealand Antarctic Program[me]
NZA-RR d: NZA-Rechtsprechungsreport
NZAV A: Nederlandsch Zuid-Afrikaansche Vereeniging
NZBA e: New Zealand Biotechnology Association
N Z Dent J e: New Zealand Dental Journal
NZFP e: New Zealand Family Physician
NZM d: Neue Zeitschrift für Miet- und Wohnungsrecht
N Z Med J e: New Zealand Medical Journal
NZN d: Niedersächsischer Zeitschriften-nachweis
NZULR e: New Zealand Universities Law Review
NZUSUGI e: New Zealand Unix System User Group, Inc.

O

O a: Observatør
- d: obligatorische Lehrveranstaltung
- e: Angel
- e: for those innocent souls
- e: Immediate Precedence
- e: Mittelpunkt des Koordinatensystems
- e: Ohms
- e: Omega
- e: Ozone
- f: [Journal] Officiel
- f: Océan[ique]
- f: Officiel
- f: Offre
- f: Omission
- f: Ordonnance
- f: Origine
- f: Ouvert[ure]
- f: Ouvrier
- r: Olycka
- r: Original [upplaga]
- s: Objetivo
- s: Obús
- s: Oficial
- s: Origen

o e: obligatorisch
- r: och

o. a: omkring
- N: ordre

Ø a: Ørdre
- N: Øst[en]
- N: øst[lig]
- N: Østlandet
- N: østre

Ø. a: Øster

Ö r: Ögonblicklig[t]
- r: Östlig

ö. r: över

o- r: observations-

o= e: A burning candle for flames

O_2 e: [normal] molecular Oxygen

O_3 d: Ozon
- e: Ozone

O_{18} e: Oxygen-18

OA a: Organismealder
- d: Oberbayerisches Archiv
- e: Office Automation
- e: Office of Audits
- e: Open Access
- e: Operational Architecture
- e: Operational Area
- e: Operational Assessment
- e: Optic Axis
- e: Order Administration
- e: Osteoarthritis, degenerative joint disease
- e: Outside Air
- f: Officier d'Administration
- f: Ordonnance d'Application
- f: Ordre Additif
- f: Ordre Alphabétique
- f: Ordre des Avocats
- f: Outre-Atlantique
- f: Ouvrier Agricole
- r: Operationsanalys
- s: Operación Anfibia
- s: Orden de Ataque

ÖÄ r: Överståthållarämbetet

o.a. a: og andet
- a: og andre
- a: oversætterens anmærkning
- N: og andre
- N: oversetterens anmerkning
- r: och andra
- r: och annan

o.ä. d: oder ähnlich

ö.a. r: översättarens anmärkning

O/A e: Operations/Administration

OAA e: Obstetric Anaesthetists Association
- e: Office of Assessment and Assurance
- e: Open Agent Architecture
- f: Organisation pour l'Alimentation et l'Agriculture

OAACE f: Organisation Afro-Asiatique de Coopération Economique

OAAH f: Organisation Afro-Asiatique pour l'Habitation

OAARR f: Organisation Afro-Asiatique pour la Reconstruction Rurale

OAAV f: Organisation Autonome de l'Allocation Vieillesse

OAB e: One-to-All Broadcast

OA-37B e: Dragonfly

OAC e: Army Ordnance Ammunition Command
- e: Oceanographic Advisory Committee
- e: Official Acceptance of Construction
- e: Operator Access Console

OACES e: Ocean-Atmosphere Carbon Exchange Study

OACI f: Organisation de l'Aviation Civile Internationale

OACIS e: Ocean Atmosphere Climate Interaction Studies
- f: Etudes sur l'Interaction Océan-Atmosphère-Climat
- s: Estudios de la Interacción Océano-Atmósfera-Clima

OACT f: Organisation Africaine de Cartographie et [de] Télédétection

o.a.d. a: og andet deslige

OAD e: Open Architecture Driver
- e: Organization Address

OADEMQA e: Office of Acid Deposition, Environmental Monitoring and Quality Assurance

OADPM e: Office of ADP Management

OADR e: Originating Agencys Determination Required

OADSS e: Office Assignment Decision Support System

OAE A: Organisasie van/vir Afrika-Eenheid
- d: Organisation für Afrikanische Einheit
- e: Office of Analysis and Evaluation
- e: Otoacoustic Emission[s]
- f: Office d'Approvisionnement de l'Etat
- s: Organisación de los Estados Americanos

OAF e: Optimum Array Filter
- N: Oljeselskapenes Arbeidsgiverforening
- N: Oslo-Avisenes Forening

OAG e: Official Airline Guide
- e: Online Air[line] Guide
- e: Open Application Group

OAH e: Office of Administrative Hearings

OAI d: Offene Anwendungs- und Interkommunikationssysteme
- e: Open Application[s] Interface
- e: Or-And-Invert
- e: Organization of African Immigrants

OAIC f: Organisation Autonome de l'Industrie et du Commerce

OAII e: Ocean-Atmosphere-Ice Interactions Program

OAISN e: OJCS Automated Information System Network

OAK e: Oakland [Operations Office]

OAL d: Ostallgäu in Marktoberdorf, Bayern
- f: Organisation pour l'Afrique Libre

OALF e: Oromo Abo Liberation Front

OALOS e: Office for Ocean Affairs and the Law of the Sea
- f: Bureau des affaires maritimes et du droit de la mer
- s: Oficina de Asuntos Oceánicos y del Derecho del Mar

OAM e: Object Access Method
- e: Office Administration Manual
- e: Operations And Management
- e: Operation[s], Administration [and] Maintenance
- e: Operator Assistance Menu
- e: Optical Absorption Monitor

OA&M e: Operations Administration and Management

OAMCAF f: Organisation Africaine et Malgache du Café

OAMCE s: Organisation Africaine y Malgache de Coopération Economique

OAML e: Oceanographic and Atmospheric Master Library

OAM&P e: Operations, Administration, Maintenance and Provisioning

OAMPI f: Office Africain et Malgache de la Propriété Industrielle

OAMRT e: Ontario Association of Medical Radiation Technologists

OAnt f: Océan Antarctique

OAO e: Over And Out

OA/OÄ e: Oberarzt/Oberärztin

OAP e: Old-Age Pension[er]
- e: Operating and Assurance Program
- e: Optical Axial Plane

OAPEC e: Organization of Arab Oil Exporting Countries

OAPI f: Organisation Africaine pour la Propriété Intellectuelle

OAPO e: Eastern Pacific Tuna Fishing Organization
- f: Organisation de la Pêche au thon du Pacifique Est
- s: Organización de Pesca del Atún en el Pacífico Oriental

OAPQ e: Office of Assistance Program Quality

OAQPS e: Office of Air Quality Planning and Standards

OA&R e: Office of Air and Radiation

OAR e: Office of Atmospheric Research
- s: Obstáculo Artificial

OArct f: Océan Arctique

OARE e: Orbital Acceleration Research Experiment

OARNET e: Ohio Academic Resources Network

OARS e: Ontario Association for Remote Sensing

OAS d: Organisation der Amerikanischen Staaten
- e: Offensive Air Support
- e: One-to-All Scatter
- e: Ophthalmic Anesthesia Society
- e: Optoacoustic Spectroscopy
- f: Organisation Armée Secrète
- s: Operaciones de apoyo aéreo ofensivo

OA/S e: Operator control station

OASAS e: Office of Alcoholism and Substance Abuse Services

OASC e: Office Automation Steering Committee
- e: Office Automation Support Center

OASDI e: Old-Age Survivors and Disability Insurance

OASD[PA] e: Office of the Assistant Secretary of Defense

OASF e: Office Automation System Facility

OASG d: Gesetz zur Sicherung der zivilrechtlichen Ansprüche der Opfer von Straftaten [Opferanspruchssicherungsgesetz] vom 8.5.1998

OASH e: Office of the Assistant Secretary for Health

OASIS e: A PC package from Dawson Technology
- e: Observation[s] At Several Interacting Scales
- e: Obstetric Anesthesia Safety Improvement Study
- e: Oceanic and Atmospheric Satellite Imaging System
- e: Oceanographic and Atmospheric Support and Information System
- e: OERR Info System
- e: On-Line Asperger Syndrome Information and Support
- e: Online Application System Interactive Software
- e: Open Access [to] Student Information System[s]
- e: Operational Analysis Strategic Interaction Simulator
- e: Operational And Supportability Implementation System
- e: Operational Application of Special Intelligence Systems
- e: Orbiter experiment Autonomous Supporting Instrumentation System
- e: Organisation for Article Standards In Science
- e: Organization for the Advancement of Structured Information Standards
- e: Overweight And Seeking Infertility Support

OASSO e: Operational Applications of Satellite Snowcover Operations

OAST e: Office of Aeronautics [and] Space Technology

OASTP e: Office of the Assistant Secretary for Technology Policy

OASYS e: Obstacle Avoidance System

OATRA f: Office d'Assistance Technique et de Réorganisation Administrative

OATS e: Office Automation Technology and Services
- e: Orbit and Attitude Tracking

OATUS e: On A Totally Unrelated Subject [Internet]

OAU a: Offentligt Ansattes Udvalg
- e: Organisation of African Unity

OAUS e: On An Unrelated Subject [Internet]

OAV e: Original Animation Video
- f: Organization Autonome de la Vallée

OAW e: Optically Assisted Winchester

OAWC e: Overseas Air Weapons Control

OAXTC e: Ocean Atmosphere Exchange of Trace Compounds

OAY e: Outstanding Airmen of the Year

OB A: Onafhanklike Beweging
- A: Orde Boerestaat
- A: Orde Boerevolk
- A: Ossewa Brandwag
- e: Obligatory
- e: Obstetrician
- e: Order of the Boer people
- e: Order of the Boer state
- f: Œuvre de Bienfaisance
- f: Office du Blé
- f: Opérations Bancaires
- f: Ordre de Bourse
- s: Obús
- s: Operación Bendición
- s: Orden de Batalla

ÖB r: Örlogsbas
- r: Överbefälhavare

Öb r: Ögonblicksbrisad

ö.b. r: överbord

Ob- e: Obligatory

OBA e: Online Banking Association
- e: Oxygen Breathing Apparatus

OBADI s: Operación Bendición de Ayuda y Desarollo Integral

OBAE F: Office des Bois d'Afrique Equatoriale

OBAK d: Objektabbildungskatalog

OBAP f: Office Belge pour l'Accroissement de la Productivité

O-bas r: Operationsbas

Öbåt r: Överskeppningsbåt

OBC e: On-Board Computer
- e: Optical Bar Camera
- e: Outside Back Cover

Obc r: Ordbehandlingscentral

OBCE f: Office Belge du Commerce Extérieur

OBD e: On-Board Diagnostics
- e: Online Bugs Database

OBDH e: On-Board Data Handling

OBDM e: On-Board Data Management

OBE e: Officer of the Order of the British Empire
- e: Online Banking Excellence

OBECI f: Office Béninois de Cinéma

OBEM e: Object-Based Equipment Model

OBEMAP f: Office Béninois des Manutentions Portuaires

OBER e: Office of Biological and Environmental Research

oberstltn. N: oberstløytnant

Obes Res e: Obesity Research

obet. r: obetalt

OBEX e: Object Exchange

ØBF a: Østjysk Biologisk Forening

OBG d: Osterburg, Sachsen-Anhalt

OB/GYN e: Obstetrician/Gynecologist
- e: Obstetrics and Gynecology

OBI e: Online Book Initiative
- e: Optical-Beam-Induced
- f: Outillage pour le Bâtiment et l'Industrie

OBIA e: Ontario Brain Injury Association

OBIC e: Optical-Beam-Induced Current
- f: Office Belge d'Investigations Commerciales

OBID e: Object Identifier

OBIGGS e: On-Board Inert Gas Generating System

OBIP f: Office des Biens et Intérêts Privés

OBIRCH e: Optical-Beam-Induced Resistance Change
OBJ e: Object [File]
- e: Objectives
Obj d: Objekt
- s: Objetivo
obj. e: object
OBK f: Office des Bauxites de Kindia
- f: Organisation pour l'aménagement de Bassin de la Kagera
OB1KB e: Order of Battle Version 1 Knowledge Based
OBL e: Object-Based Language
Obl f: Obligatoire
obl. a: obligatorisk
- e: oblique
oblast e: osteoblast
Ob Lat s: Observatorio Lateral
OBLI d: obligatorischer On-Topic Anteil einer Message
oblt N: oberstløytnant
OBM e: Office of Management and Budget
ØBM a: Økonomi- og Budgetministeriet
OBN e: Open Bibliothek Network
- e: Open Library Network
- n: Open Bibliothek Netwerk
OBO e: Oboe
- e: Output Back-Off
Oboe e: Navigation system
OBOGS e: On Board Oxygen Generating System
OBOS N: Oslo Bolig- og Sparelag
OBP e: Offshore Biological Programme
- f: Programme biologique au large
- s: Programa Biológico de Alta Mar
OBPC e: Optical Bar Panoramic Camera
öbrisad r: ögonblicksbrisad
OBS d: Objektbanksystem
- e: Ocean-Bottom Seismometer
- e: Omni Bearing Selector
- e: Organizational Breakdown Structure
- e: OSIS Baseline System
Obs N: Observasjon
- N: Observatør
- N: Observere
- s: Observación
- s: Observador
- s: Observatorio
obs r: observation[s]
Obs. e: Obsolete
obs. r: observera!
obs- a: observations-
obs. 2 N: observasjonspost nr. 2
Obs Avdo s: Observatorio Avanzado
obs. btt. N: observasjonsbatteri
Obsdr s: Observador
OBSET e: Observation Set
Obsio s: Observatorio
OBSLAT s: Observatorio Lateral
Obsn s: Observación
obs. p. N: observasjonspost
obs. pl. N: observasjonsplass
Obs Res s: Observatorio de Reserva
OBSS f: Office Béninois de Sécurité Sociale
OBST e: Object Management System of STONE
Obst s: Observatorio
- s: Obstáculo
Obst CD s: Obstáculo Contra Desembarco
Obstet Gynecol e: Obstetrics and Gynecology
Obstet Gynecol Clin North Am e: Obstetrics and Gynecology Clinics of North America
Obstet Gynecol Surv e: Obstetrical and Gynecological Survey
Obstetr. e: Obstetric
Obst Nat s: Obstáculo Natural
OBT e: Office of Building Technologies
- e: One Billion Trees program
- s: Orden Básica para las Transmisiones
Obto s: Observatorio
OBTP f: Office du Bâtiment et des Travaux Publics
OBTW e: Oh, By The Way [Internet]
OBU e: Ocean information system Baseline Upgrade
- e: Off-shore Banking Unit
- e: OSIS Baseline Upgrade
OC e: Open Commitment
- e: Open Control
- e: Operating Contractor
- e: Operational Concept
- e: Operational Condition
- e: Operations Communications
- e: Operations Contract
- e: Operator Centralization
- e: Optical Carrier
- e: Optical Coupler
- e: Order of Canada
- e: Over-Current
- e: volatile Organic Compound
- f: Obligations Communales
- f: Office Central
- f: Office des Changes
- f: Opérations Commerciales
- f: Ordre Chronologique
- r: Områdescentral
- s: Ondas Cortas
- s: Orden Circular
Oc f: Océan[ique]
O&C e: Operations and Checkout
O/C e: Observer/Controller
O[/]C e: Open Collector
OC[-]3 e: Optical Carrier level 3
OCA e: Obsessive Compulsive Anonymous
- e: Offensive Counter Air
- e: Office of Consumer Affairs
- e: Open Component Architecture
- e: Original Classification Authority
- e: Other Competitive Action
- f: Office Commercial de l'Automobile
- f: Office de Commercialisation Agricole
- s: Operaciones Contra Aéreas ofensivas
OCABE f: Office de Coordination Atomique de Belgique
OCAC e: Operations Control and Analysis Center
OCACHAR f: Office Central d'Approvisionnement des Charbonnages belges
OCADO f: Organe Central d'Achat des Denrées d'Ordinaire
OCAHO e: Office of the Chief Administrative Hearing Officer
OCALA e: Open Common Annotation Language
OC[-]ALC e: Oklahoma City Air Logistics Center
OCAM e: Afro-Malagasy Common Organization
OCAP e: Oceans and Coastal Areas Programme
- e: Operating Criteria and Procedures
OCA/PAC e: Oceans and Coastal Areas Programme Activity Centre
OCARINA e: Ocean, Atmosphere, Research and Investigation with Acoustic techniques
OCAS f: Office de Commercialisation Agricole de Sénégal
- s: Organización Campesina de Acción Social
OCATE e: Oregon Center for Advanced Technology Education
OCAW e: Oil, Chemical and Atomic Workers
OCB e: Olivocochlear Bundle
- e: Operations Center Building
- e: Outer Core Barrel
- f: Office Congolais des Bois
ÖCB r: Överstyrelse för Civil Beredskap
OCBN f: Organisation Commune Bénin-Niger des chemins de fer et des transports
OCBV f: Office Communautaire du Bétail et de la Viande
OCC e: Office Communications Cabinet
- e: Officer Candidate Class
- e: Onsite Construction Contractor
- e: Open, Cooperative Computing
- e: Operational Control Center
- e: Opposition Coordinating Committee
- e: Orange Coast College in California
- e: Organization Chart and Charter
- e: Owens Community College in Ohio
- s: Obstáculo Contracarro
Occ f: Occidental
Occ. e: Occupation[al]
OCCA e: Office of Compliance and Consumer Assistance
- e: Open, Cooperative Computing Architecture
OCCAM e: Ocean Circulation [and] Climate Advanced Modelling [project]
occas. e: occasionally
Occas Pap R Coll Gen Pract e: Occasional Paper/Royal College of General Practitioners
OCCC e: Obfuscated C Code Contest
OCCETS/CCOHTA f: Office Canadien de Coordination de l'Évaluation des Technologies de la Santé/Canadian Coordinating Office for Health Technology Assessment
OCCF f: Office Commun des Consommateurs de Ferrailles
OCCGE f: Organe (ou Organisation) de Coordination et de Coopération pour la lutte contre les Grandes Endémies en Afrique de l'Ouest
OCCL f: Office Central des Contingents et Licences
OCCM e: Ocean Carbon Cycle Model
OCC MED e: Occupational Medicine
OCC/NIBS e: Ohio Consultive Council of the National Institute of Building Sciences

OCCR f: Office Central de Chauffage Rationnel
- f: Office Central de Chimie Rationnelle
- f: Organisation-Conception-Contrôle-Réalisation

OCCULT e: Orser Complete Conversational User-Language Translator
Occup Environ Med e: Occupational and Environmental Medicine
Occup Health Saf e: Occupational Health and Safety
Occup Med [Oxf] e: Occupational Medicine
OCD e: Obsessive-Compulsive Disorder
- e: Ocean Chemistry Division
- e: Off-Chip Driver
- e: Offshore and Coastal Dispersion [model]
- e: Operations Concept Document
- e: Out-of-Cell Delineation
- f: Office de Coopération au Développement

OCDE f: Office du Crédit pour le Développement Économique
- f: Organisation de Coopération et de Développement Économique[s]
- s: Organización de/para la Cooperación y Desarrollo Económico[s]

OCDM e: Office of Civil [and] Defense [and] Mobilization
OCDS e: Offline Control Data Set
OCDW e: Ocean Climate Data Workshop
- f: Atelier sur les données relatives au système océan/climat
- s: Taller sobre Datos Climáticos Oceánicos

OCE e: Ocean Colour Experiment
- e: Ocean[ia]
- e: Office Chief of Engineers
- e: Office of Criminal Enforcement
- e: Open Collaborative Environment
- e: Other Common carrier channel Equipment

OCEAC f: Organisation de Coordination pour la lutte contre les Endémies en Afrique Centrale
OCEAN e: Ocean Color Environment Archive Network
- e: Ocean Colour European Archive Network

OCEANIC e: Ocean Information Center
Oceanogr. Mar. Biol. Annu. Rev. e: Oceanography and Marine Biology Annual Review
OCEAN-PC e: Ocean Personal Computer Project
- f: Projet d'utilisation de l'ordinateur personnel en océanographie
- s: Proyecto de Microcomputadoras para los Océanos

OCEANSAT e: Ocean Studies Satellite
OCEAN STORMS e: JISAO field experiment for the assessment of weather fronts
OCED e: Organization for Economic Co-operation and Development
OCEE f: Organisation de coopération économique européenne
OCEKA a: Olivarius & Christensens Efterfølger A/S
OCELOT e: PC-based library system
OCEM e: Office of Cooperative Environmental Management
OCEO e: Open Change Engineering Orders
OCESL e: Office of Criminal Enforcement and Special Litigation
OCETI f: Organisation Centrale Européenne pour les Transports dans l'Intérieur
OCF e: Objects Components Framework
- e: Obsessive-Compulsive Foundation
- e: Osteopathy in the Cranial Field
- f: Office Central des Fournitures
- f: Organisation des Femmes Centrafricaines

OCFT f: Office du Chemin de Fer Transcamerounais
OCGF e: Organic Crystal Growth Facility
OCH f: Office Congolais de l'Habitat
OCHAMPUS e: Office of Civilian Health and Medical Program of the Uniformed Services
OCI e: Object Code Insertion
- e: Ocean Colour Imager
- e: Office of Criminal Investigation
- e: Organization City
- e: Organizational Conflicts of Interest
- e: Other Cooperating Institutions
- e: Out of City Indicator
- e: Oxide Control and Indicating
- f: Organisation de la Conférence Islamique
- f: Organisme de Coopération Industrielle

OCIBU f: Office des Cultures Industrielles du Burundi
OCIL f: Office Central Interprofessionnel du Logement
OCIMF s: Foro Marítimo Internacional de Compañías Petroleras
OCIR e: Office of Community and Intergovernmental Relations
- f: Office des Cultures Industrielles du Rwanda

OCIS e: On-Line Chemical Information System
OCL e: Operater Control Language
- e: OS/2 inside Class Library

OCLALAV f: Organisation Commune de Lutte Antiacridienne et Antiaviaire
OCLC e: An organization that offers the CRIS technical database on CD ROM
- e: Ohio College Library Catalog
- e: Online Computer Library Center (or Centre)

OCLD f: Organization Camerounaise de Lutte pour la Démocratie
OCLEAC f: Organisation de Coordination pour la Lutte contre les Endémies en Afrique Centrale
OCLI e: Optical Coating Labs Inc.
OCLIPS e: Operational Climate Prediction and Services
OCLL e: Office, Chief of Legislative Liaison
OClO e: Chlorine Dioxide
OCM e: Ocean Circulation Model
- e: Ocean Colour Monitor
- e: Oxidative Coupling of Methane
- f: Instrument d'étude de la couleur des eaux marines
- s: Instrumento de observación del color de los mares

OCMLR f: Organisation Communiste Marxiste-Léniniste de la Réunion
OCMR e: Ontario Centre for Materials Research
OCMS e: On-Chip test and Maintenance System
OCN e: Oceania
- e: Office of the Commissioner of Namibia
- e: Operational Carrier Number

OCNMAP e: Ocean Map
OCO e: Object Code Only
- e: Operating Capital Outlay

OC-1, OC-3 e: Optical Carrier
O COMB s: Orden de Combate
OCONUS e: Outside the Continental United States
OCOS f: Ordre des Conseils en Organisation Scientifique
OCOSS f: Office Central des Œuvres et Services Sociaux
OCP e: Ocean Surveillance Product
- e: Onchocerciasis Control Program
- e: Open-Circuit Potential
- e: Optical Storage Processor
- e: Orbital Combustion Process
- f: Office des Comptes Postaux
- f: Organisation Commerciale de la Production
- s: Obstáculo Antipersonal
- s: Obstáculo Contrapersonal
- s: Oficina Conjunta de Planificación del PMIC

OCPCSB e: Operational Computer Program Configuration Sub-Board
OCPD e: Officer Commanding Police Division
OCPE f: Office Communautaire de Promotion des Echanges
OCPS e: Orbiter Camera Payload System
OCR e: Ocean Colour Radiometer
- e: Office of Community Relations
- e: Operational safety Requirements
- e: Optical Character Recocnition
- e: Optical Curve Recognition
- e: Organization of work Camps of the Revolution
- f: Organisation des Chantiers de la Révolution

OCRB f: Office Central de Répression du Banditisme
OCRC e: Ocean Climate Research Committee
OCRD e: Ocean Climate Research Division
- e: Office Chief of Research and Development United States Army
- e: Office of Crystalline Repository Development

OCREN s: Oficina de Control de Reservas de la Nación, MAGA
OCR-Leser d: Belegleser, die codierte Zeichen optisch erkennen können

OCRM e: Office of Coastal Resource Management
- e: Outer Core Restraint Module
OCRO e: Office of Central Records Operations
OCRPI f: Office Central de Répartition des Produits Industriels
OCRS e: Operational Change Reporting System
- f: Organisation Commune des Régions Sahariennes
OCRTIS f: Office Central de Répression du Trafic Illicite des Stupéfiants
OCS e: Object Compatibility Standard
- e: Ocean Colour Scanner
- e: Office of the Chief Scientist
- e: Old Church Slavonic
- e: On Call System
- e: On-Card Sequencer
- e: Operator Communications Software
- e: Operator Control Station
- f: Office de la Canne et du Sucre
- f: Organe de Contrôle des Stupéfiants
OCSA e: Office [of the] Chief of Staff, Army
OCSAN f: Organisation pour la Conservation du Saumon de l'Atlantique Nord
- s: Organización para la Conservación del Salmón del Atlántico Norte
OCSEAP e: Outer Continental Shelf Environment Assessment Program
OCSS f: Office Centrafricain de Sécurité Sociale
OCT e: Object Code Translator
- e: October
- e: Optimized Compensation Transactions
- f: Office Central des Transports internationaux par chemins de fer
- s: Octubre
Oct f: Octobre
- f: Octroi
OC&T e: Office of Certification and Training
OCTC e: Operators Console Transfer Channel
OCTD e: Observation, Conclusion, Temporary Data
OCTH e: Occupational Therapy
OCTL e: One-Channel-To-Line
OCTOPUS e: Ocean Colour Techniques for Observation, Processing and Utilization Systems
OCTRA f: Office du Chemin de Fer Transgabonais
OCTS e: Ocean Colo[u]r [and] Temperature Scanner
- e: Ocean Colo[u]r and Temperature Sensor
- e: Ocean Colo[u]r and Temperature Sounder
- f: Office Central des Traitements et Salaires
OCTW e: Optical Communications Through the shuttle Window
OCU e: Operational Capability Update
O-cut e: c-axis parallel to one edge
OCV e: Consultative Committee for Exploratory Studies
- e: Operational Compliance Value
OCVD e: Open Circuit Decay Voltage
OCW e: Ministry of Education, Culture and Science
- e: Organizing Committee World Championship[s]
OCWG e: Ocean Color Working Group
OCX e: OLE Control Extension[s]
- e: OLE Custom Control
OCY e: Organization Country
OCYF e: Office for Children, Youth and Families
OD A: Orde van die Dood
- a: Overskudsdeling
- e: Oculus Dexter
- e: Office of the Director
- e: Oktal Dump
- e: Open Doc[ument]
- e: Optical Density
- e: Optical Detection
- e: Optical Disk
- e: Order of Death
- e: Outside Diameter
- e: Oxford Dictionary
- f: Office Départemental
- f: Ordonnance de Délégation
- f: Organismes de Documentation
- s: Orden de Defensa
ØD a: Økonomisk Demokrati
o.d. a: og dylikt
- e: outside diameter
O-D e: zero Dimensional
OD3 e: Operational Data Dictionary/Directory
ODA e: Office Document Architecture
- e: Office of the Deputy Administrator
- e: Open Document Architecture
- e: Overseas Development Administration
ODAC e: Open Document Architecture Consortium
- e: Operations Distribution Administration Center
ODAI e: Origin/Destination Address Assignor Indicator
ODAI. e: Origin-Destination Assignor Identifier
ODAM e: Open Distributed Application Model
ODA/ODIF e: Office Document Architecture/Office Document Interchange Format
ODAP e: Office of Drug and Alcohol Programs
ODAPI e: Open Database Application Programming Interface
ODAR e: Omnidirectional Airborne Radar
ODAS e: Ocean Data Acquisition System
- e: Ocean[ic] Data Assimilation System
- s: Sistemas, Medios y Dispositivos de Adquisición de Datos Oceánicos
ODATS e: Office Director Assignment Tracking System
ODB d: Objektorientierte Datenbank
- e: Object Oriented Data Base
- e: Orbit Determination Beacon
ODBC e: Object-Oriented Database Connectivity
- e: Open Data Base Connectivity [driver]
ODBMS e: Object[-oriented] Database Management System
ODC e: Ocean Dynamics and Climate
- e: Other Data Center
- f: Dynamique des Océans et Climat
- f: Office de Direction et de Contrôle
- s: Dinámica de los Océanos y el Clima
ODCP e: Office of the Deputy Commissioner for Programs
ODCR e: Operations Deputies Conference Room
ODCS e: Open Distributed Computing Structure
ODCSINT e: Office of the Deputy Chief of Staff for Intelligence
ODCSLOG e: Offices Deputy Chief of Staff for Logistics
ODCSOPS e: Office of the Deputy Chief of Staff for Operations [and Plans]
ODCSPER e: Office, Deputy Chief of Staff for Personnel
ODD e: Open Data Desktop
- e: Oppositional Defiant Disorder
- e: Oppositional Developmental Disorder
- e: Order and Dispatch Desk for SAR data products
- e: Overdetermined Dual-Doppler
ODDD e: Operator Direct Distance Dialing
ODDS i: Organizzazione delle Donne Democratiche Somale
ODE e: Object Database [and] Environment
- e: Old-Dog Encephalitis
- e: Ordinary Differential Equation[s]
- e: OSF Development Environment
- e: Oxford Dictionary of the English language
- f: Office de Documentation Économique
ODECO f: Organisation de Diffusion et d'Expansion Commerciale
ODEF f: Office National de Développement et d'Exploitation des Ressources Forestières
ODEL f: Office de Développement de l'Elevage
ODEMO f: Organisation de Développement du Moyen Ouest
ODES e: Ocean Data Evaluation System
ODESSA e: Oceanographic Database and Environmental Satellite System Application
ODETTE e: Organisation for Data Exchange by Tele-Transmission in Europe
ODEX e: Optical Dynamics Experiment
ODF e: Oceanographic Data Facility
- e: Open-doc Development Framework
- e: Optical Distribution Frame[s]
- e: Optoelectronic Data Filter
ODHA s: Oficina de Derechos Humanos del Arzobispado
ODHAGUA s: Oficina de Derechos Humanos del Arzobispado
ODHI s: Oficina de Derechos Humanos del Ixcan
ODI e: Open Data[-link] Interface
- e: Open Device Interconnect
- e: Open Driver Interface
- e: Optical Digital Image
ODIC e: Oceanographic Data and Information Centre

ODID e: Operational Display and Input Development
ODIF e: Office Document Interchange Format
- e: Open Document Interchange Format
ODIG s: Organización para el Desarollo Indígena
ODIK f: Opération de Développement Intégré du Kaarta
ødild N: ødeleggelsesild
ODIMS e: Open Distributed Information Management System
ODIN e: Ocean Data and Information Network
- f: Réseau d'échange de données et d'information sur l'océan
- n: Oosterhouts Digitaal Informatie Netwerk
- n: Oosterhouts Digitaal Informatie Netwerk - a local network
- s: Red de Datos e Información Oceanográficos
ODINEA e: Ocean Data and Information Network for Eastern Africa
- f: Réseau d'échange de données et d'information sur l'océan en Afrique de l'Est
- s: Red de Datos e Información Oceanográficos para Africa Oriental
Odinsup e: ODI-NDIS Supplementary Driver
ODIPAC f: Office de Développement Intégré des Productions Arachidières et Céréalières
ODIS e: Online Data Information System
- e: Onsite Discharge Information System
ODISS e: Optical Digital Image Storage System
ODJR f: Organisation Démocratique de la Jeunesse Réunionnaise
ODL d: Ortsdosisleistung
- e: Object Definition Language
- e: Object Description Language
- e: Object Design Language
- e: Open and Distance Learning
- e: Optical Disc Library
- e: ORACLE Data Load
- e: Overseas Drilling Limited
- f: Office Départemental du Logement
OdL f: Observatoire de Lyon
ODLA s: Autonomous Labor Defense Office
ODLC e: Outboard Data Link Control
ODM e: Object Data[base] Manager
- e: Optimized Distribution Model
- p: Organizações Democráticas de Massas
ODMA e: Open Document Management API
ODMG e: Object Database Management Group
ODMR e: Optically Deflected Magnetic Resonance
- e: Optically-Detected Magnetic Resonance
ODMS e: Odesta Document Management System
ODN e: Obligation Document Number
- e: Optical Data Network
ODO e: Offensive Duty Officer
ODP e: Ocean Drilling Program
- e: Ocean Drilling Programme
- e: Office of Disclosure Policy
- e: Open Distributed Processing
- e: Organization department
- e: Overdrive Processor
- e: Ozone-Depleting Potential
- f: Office Départemental des Pensions
- f: Office Départemental du Placement
- f: Programme de forages océaniques
- p: Organização da Defesa Popular
- s: Programa de Perforaciones Oceánicas
ODPC e: ODP Council
ODP-MT f: Organization pour la Démocratie Populaire-Mouvement Travailliste
ODPRF e: Operating Documents Preliminary Review Form
ODQ f: Ordre des Dentistes du Québec
ODR e: Officer of Defense Resources
- e: Omnidirection Range
- e: Optical Digital Reference
- e: Optical Double Resonance
- e: Optically-Detected Resonance
- e: Optimized Dynamic Routing
ODS a: Ordbog over det Danske Sprog
- d: Offene Datenschnittstelle
- d: Offenes Deutsches Schulnetz
- e: Office of Disability Services
- e: Open Data Services
- e: Operating Differential Subsidy
- e: Optical Disk Storage
- e: Overhead Data Stream
- e: Ozone-Depleting Substance[s]
- f: Ordre de Service
ODSI e: Open Directory Service Interface[s]
ODT e: On Demand Technologies
- e: Open Desktop
- e: Operator Display Terminal
ODTA f: Organisation pour le Développement du Tourisme Africain
ODVA e: Open Device-Net Vendor Association
ODW e: Office of Drinking Water
ODWG e: Offset Drilling Working Group
o.dyl. a: og dylikt
o.d[yl]. r: och dylikt
OE e: Office of Emergency Planning and Operations
- e: Office of Enforcement
- e: Open Edition
- e: Operating Engineering
- e: Operating Expense[s]
- e: Operations Engineer
- e: Opportunity Evaluation
- e: Opto-Electronic[s]
- e: Order Entry
- e: Output Enable
- f: Offre d'Emploi
- f: Omissions Excepté
- f: Organisation Économique
- f: Organismes d'Enseignement
- f: Ouest-Est
- f: Ouvriers d'Etat
OE. e: Old English
O&E e: Operations and Engineering
O/E e: Optical-to-Electrical
OEA d: Ozean-Eis-Atmosphäre
- f: Ordre des Experts en Automobiles
- f: Organisation des Etats Américains
- s: Organización de Estados Americanos
OEB f: Office Européen du Brevet[s]
OEBR e: Optical Edge Bead Removal
OEC e: Office of Energy Conservation
- e: Office of Environmental Compliance
- e: Open Environment Corporation
- e: Operations and Engineering Contractor
- e: Output Edge Control
- f: Officier d'État Civil
- f: Organisation Européenne du Charbon
O&EC e: Operations and Engineering Contractor
OECD d: Organisation für wirtschaftliche Entwicklung und Zusammenarbeit
- e: Organisation (or Organization) for Economic[al] Cooperation and Development
- e: Organization for European Community Development
OECDC e: Organization for Economic Cooperation among Developing Countries
OECE f: Organisation Européenne de Coopération Économique
OECF e: Overseas Economic Cooperation Fund
OECI e: Organization of European Cancer Institutes
OECM e: Office of Enforcement and Compliance Monitoring
OECQ f: Organisation Européenne pour le Contrôle de la Qualité
OECS e: Optics of Excitors in Confined Systems
- e: Organization of Eastern Caribbean States
OED e: Operational Effectiveness Demonstration
- e: Oxford Dictionary of the English language
OEE e: Overall Equipment Effectiveness
OEEPE f: Organisation Européene d'Études Photogrammétriques Experimentales
OEET e: Office of Environmental Engineering and Technology
OEF e: Operational Efficiency Factor
ÖEF r: Överstyrelsen för Ekonomisk Försvarsberedskap
OEG e: Office of Environmental Guidance
- s: Olimpiadas Especiales de Guatemala
OEH e: Occupational and Environmental Health Library
OEI d: Osteuropa Institut [FU Berlin]
OEIC e: Opto-Electronic Integrated Circuit[s]
OELS e: Operationally Efficient Launch Site Study
- f: Organisation Européenne de Lancement de Satellites
OEM e: Odd Even Merge
- e: Opto-Electronic
- e: Original Equipment Manufacturer
- e: Original Equipment Market
OEMI e: Original Equipment Manufacture Information
- e: Original Equipment Manufacture Interface

OEMM e: Operational Experiment on Mesoscale Meteorology
OEMP e: Office of Environmental Monitoring and Prediction
- e: Operational Environmental Monitoring Program
OEMT e: Operational Emergency Management Team
OEN e: Operating Environment
OEP e: Office of Economic Planning
- e: Office of Economic Policy
- e: Office of Enforcement Policy
- e: Office of Environmental Policy
- e: Operand Execution Pipeline[s]
O&EP e: Occupational and Environmental Protection
OEPA e: Ohio Enviromenatl Protection Agency
OEPL f: Organisation de l'Enseignement Privé Laïc
OEPP e: Office of Environmental Policy and Planning
OEPSS e: Operationally Efficient Propulsion System Study
OERD e: Ocean Environment Research Division
- e: Office of Economic and Regional Development, SIUC
OERN f: Organisation Européenne pour la Recherche Nucléaire
OERR e: Office of Emergency and Remedial Response
- e: Office of Environmental Regulatory Research
OERS e: Operational Earth Resources System
- f: Organisation des Etats Riverains du Fleuve Sénégal
OES d: Olivetti Entsorgungsservice
- e: Bureau of Oceans [and] International Environment[al] and Scientific Affairs
- e: Odd Even Sort
- e: Office Evaluation System
- e: Office of Earthquake Studies
- f: Ecole nouvelle d'Organisation Économique et Sociale
OESC f: Organisation pour l'Éducation, la Science et la Culture intellectuelle
OESS e: Organizational Effectiveness Survey System
OET e: Office of Economic Transition
- e: Office of Employment Training
- s: Oficial de Enlace con el ejército de Tierra
O.E.T.A. e: Occupied Enemy Territory Administration
Oet D f: Offre et Demande
Oet M f: Organisation et Méthodes
OEU e: Operational Evaluation Unit
OEVA f: Office de l'Expérimentation et de la Vulgarisation Agricoles
OF a: Oplysningsforbundenes Forberedelses-kursus
- e: Operational Functionality
- e: Overflow Flag
- f: Obligation Foncière
- f: Opérations Financières
- f: Organisation Féminine
- f: Organisation Française
- f: Organisation Fraternelle
- f: Organisme Familial
- N: Organisasjonenes Fellesråd
- s: Oficial
ØF a: [Det Europæiske] Økonomiske Fællesskab
Of f: Officiel
öf r: ögonblickligt eller fördröjt
- r: överfurir
o.f. a: og følgende
- r: och följande
ø.f. a: øst for
- N: øst for
OF/2 e: Operator Facility/2
OFA e: Office of Flight Assurance
- e: Oklahoma Forestry Association
- e: Optimal Flexible Architecture
- e: Orthopedic Foundation for Animals
- e: Other Federal Agencies
OFAB f: Office d'Aménagement de Boke
OFACI f: Office Franco-Américain pour le Commerce et l'Industrie
OFASA s: Obra Filantropica de Asistencia Social Adventista
OFAZ d: Oberfeldarzt
OFB e: Output Feedback [mode]
OFC e: Outside Front Cover
- f: Officiel Chiffré
- s: Oficina
OFCAPPDE e: Office Application Descriptons
OFCAPPE e: Office Application Entries
Of Car s: Oficial A Cargo
OFCC e: Office of Federal Contract Compliance
Ofc Cor s: Oficina de Correos
OFCE f: Observatoire Français des Conjonctures Économiques
- f: Office Français du Commerce Extérieur
OFCM e: Office of the Federal Coordinator for Meteorology
OFCO e: Office of the Federal Coordinating Officer
- s: Oficina Colectora de Correo[s]
OFD e: Optical Frequency Division Demultiplexer (or Demultiplexing)
OFDA e: Office of Foreign Disaster Assistance
OFDE f: Organisation de Fraternité et de Développement Economique
OFDM e: Orthogonal Frequency Division Modulation
- e: Orthogonal Frequency Division Multiplex[ing]
OFE e: Open Finance Exchange
- e: Overall Factory Effectiveness
- e: Oxfordshire
OFEDES f: Office des Eaux du Sous-Sol
OFEFP f: Office Fédéral de l'Environnement, des Forêts et du Paysage
OFEMA f: Office Français d'Exportation du Matériel Aéronautique
Ofen s: Ofensiva
OFF f: [télégramme] Officiel
- N: Ombudsmannen for Forsvaret
Off f: Officiel
- f: Offre
off. a: offentlig[t]
- a: officiel[t]
- N: offentlig
- N: offiser
o.ff. a: og flere følgende
OFFA f: Office des Fers, Fontes, Aciers
OffCh f: Office des Changes
Offic f: Officiel
OFFIS d: Oldenburger Forschungs- und Entwicklungsinstitut für Informatik-Werkzeuge und -Systeme
OFHEO e: Office of Federal Housing Enterprise Oversight
OFID e: Optical Free Induction Decay
Ø. Finn avsn N: Øst-Finnmark avsnitt
Ofj F N: Oslofjord festning
o. fl. N: og flere
ofl. a: og flere
Öfl r: Örlogsflotta
o.fl.a. a: og flere andre
o.fl.st a: og flere steder
OFM e: Optical Frequency Division Multiplexer (or Multiplexing)
- e: Order of Friars Minor
- e: Ordinis Fratrum Minorum
- e: Outlet Feature Model
OFME f: Organisation Française du Mouvement Européen
OF/MET/AUX s: Oficina Meteorológica Auxiliar
OF/MET/P s: Oficina Meteorológica Principal
OFMT e: Output Format for Numbers
OFNACER f: Office National des Céréales
OFNACOM f: Office National de Commerce
OFO f: Officiel Ordinaire
OFP e: Office of Family Planning
- e: Operations Funded Project
ÖFP d: Öffentlich finanzierte Preise
OFPI f: Office Fédéral de la Propriété Intellectuelle
OFPPA e: Office of Federal Procurement Policy Act
OFPPL e: Office of Federal Procurement Policy Letters
OFPS e: Office of Field Project Support
OFPSD e: Office of Financial Policy and Systems Design
OF[r]. e: Old French
OFR e: Ocean Freight Reimbursement
- e: Order of the Federal Republic
- e: Oxygen Free Radical
OFRB f: Organisation des Femmes Révolutionnaires du Bénin
OFREIN s: Oficina Regional Iglesia del Nazareno
OFris. e: Old Frisian
OFRS f: Office Français des Recherches Sous-marines

OFS e: Object File System
- e: Office of oceanographic Facilities and Support
- e: Office of the Foreign Secretary
- e: Order of Free State
- e: Output Field Separator
Öfsar r: Ögonblicks- eller fördröjnings-spetsanslagsrör
OFSH e: Ovine Follicle Stimulation Hormone
OFSP f: Office Fédéral de la Santé Publique
OFT e: Orbital (or Orbiter) Flight Test
OFTA f: Observatoire Français des Techniques Avancées
OFTP e: Odette File Transfer Protocol
OFU e: Organization for Free Ugandans
ofullb. r: ofullbordad
OFUNC f: Organisation des Femmes de l'Union Nationale Camerounaise
OFUNIR f: Organisation des Femmes de l'UNIR
OFW d: Organisationsforum Wirtschaftskongreß
OFX e: Open Finance (or Financial) Exchange
OG e: Old Greek
- e: Operations Group
- e: Organic Geochemistry
- f: Organisations Gouvernementales
- s: Oficial de Guardia
og. r: ogift
o.g. a: over gården
O&G e: Oil and Gas
OGA e: Option Generation Aid
- f: Office Général de l'Air
OGB f: Office du Gros Bétail
- s: Observatorio de la Geosfera y la Biosfera
ÖGB d: Tätigkeitsbericht und stenographisches Protokoll des ersten Kongresses des Österreichischen Gewerkschaftsbundes
OGC e: Office of [the] General Council (or Counsel)
- e: Oregon Graduate Centre
OGCM e: Ocean[ic] General Circulation Model
- f: Modèle de Circulation Océanique Générale
- s: Modelo de Circulación Oceánica General
OGDC e: Office of Geographic Data Coordination
ÖGDG d: Baden-Württemberg: Gesetz über den öffentlichen Gesundheitsdienst vom 12.12.1994
OGDH f: Organization Guinéenne des Droits de l'Homme
OGE p: Orçamento Geral de Estado
OGEDEP f: Office de Gestion de la Dette Publique
OGI e: Oregon Graduate Institute
- e: Oregon Graduate Institute of Science and Technology
OGICSE e: Oregon Graduate Institute Computer Science and Engineering
OGIL f: Office Général d'Information pour le Logement
OGIP e: Office of Guest Investigator Programs
OGIRS e: Oklahoma Geographic Information Retrieval System
ÖGL d: Österreich in Geschichte und Literatur
og lign. a: og lignende
OGM f: Ordre Général Ministériel
- f: Organisme Génétiquement Modifié
OGMA p: Officinas Gerais de Material Aeronautica
OGN f: Océan Glacial du Nord
OGO s: Oficial de Guardia Operativo
OGP e: Office of Global Programs
- f: Office Général de la Propriété
OGR e: Office of Geologic Repositories
OGRA e: Russian thermonuclear device
ogrp r: observationsgrupp
O-gruppe N: Ordregruppe
OGS e: Other Government Securities
- f: Océan Glacial du Sud
- i: Osservatorio Geofisico Sperimentale
ogs. a: også
Ögu[ll] r: Gustavsviks örlogsdepå
OGWMN e: Operational Groundwater Monitoring Network
OGWP e: Office of Ground Water Protection
OH e: Hydroxyl
- e: Hydroxyl ion of water
- e: Off-Hook
- e: Ohio field Office
- e: Osteopathic Hospitals
ÖH r: Hårsfjärdens örlogsdepå
Öh r: Över havet
- r: Över havsytan
o.h. a: over havoverfladen
- a: over havet[s overflade]
- N: over havet
- r: och hustru
O/H e: On Hand
OH-58 e: Kiowa helicopter
OHA d: Osterode/Harz, Niedersachsen
OHC e: Other Hanford Contractor
- e: Overhead Cams
- e: Oxygen Hole Centers
OHDH e: Old Habits Die Hard [Internet]
OHDMI e: 2-Hydroxydesipramine
OHE e: Office of Hanford Environment
OHEA e: Office of Health and Environmental Assessment
- e: Office of Health Effects Assessment
OHER e: Office of Health and Environmental Research
OHF e: Old Hydrofracture Facility
OHFS e: Optimized Hartree-Fock-Slater
OHG e: Old High German
OHI f: Organisation Hydrographique Internationale
- s: Organización Hidrográfica Internacional
OHIF N: Oslo Håndverks- og Industriforening
OHIMA e: Ohio Health Information Management Association
Ohio e: College Library Center
ö hk sar r: ögonblickligt högkänsligt spetsanslagsrör
OHLM f: Office des Habitations à Loyer Modéré
OHMS e: Office of Hazardous Materials Safety
OHMTADS e: Oil and Hazardous Materials Technical Assistance Data System
OHMVR e: Off-Highway Motor Vehicle Recreation
OHOC e: Oregon Hanford Oversight Committee
OHP d: Overhead-Projektor
- e: Occupational Health Physician
- e: Ocean History Panel
- e: Operational Health Physics
- e: Operational Hydrology Program[me]
- e: Overhead Projector
- e: Overhead Transparency Panel
- f: Observatoire de Haute Provence
OHPSS e: Operational Health Physics Site Surveillance
OHQ e: Originating Headquarters
OHRDP e: Ontario Health Resources Development Program
OHRE e: Department of Energy, Office of Human Radiation Experiments
OH&S e: Occupational Health and Safety
OHS e: Occupational Health and Safety
- e: Occupational Health Services
- r: Officershögskolan
OHSPC e: Oil and Hazardous Substance Pollution Contingency
OHSS e: Occupation Health and Safety Staff
- e: Ovarian Hyperstimulation Syndrome
OHST e: Occupational Health and Safety Technicians
OHTEX e: Ocean Heat Transport Experiment
- f: Expérience sur le transport de chaleur dans l'océan
- s: Experimento sobre el Transporte del Calor Oceánico
OHV d: Oberhavel, Brandenburg
- e: Overhead Valve[s]
OHWM e: Office of Hazardous Waste Management
- e: Ordinary High Water Mark
OHZ d: Osterholz-Scharmbeck, Niedersachsen
OHZORA e: Japanese scientific satellite
OI e: Object Interface
- e: Operations Intelligence
- e: Optimum Interpolation
- e: Oxygen Index
- f: Océan Indien
- f: Office d'Identification
- f: Organisation Internationale
Ø. I N: Økonomistabens Budsjettavdeling
OIA e: Operations Intelligence Automation
- e: Operator Information Area
- s: Oficina de Información del Arzobispado
OIAB f: Office d'Identification de l'Agglomération Bruxelloise
OIAC f: Organisation Interafricaine du Café
- f: Organisation Internationale de l'Aviation Civile
OIAO e: Ogaden Islamic Alliance Organization
OIB e: Ocean-Island Basalt
- e: Operations Intelligence Branch

OIBT f: Organisation Internationale des Bois Tropicaux
OIC e: Oh, I See [Internet]
- e: On-line Instrument and Control
- e: Organisation (or Organization) of [the] Islamic Conference
- f: Office central des transports Internationaux par Chemin de fer
- f: Office International de Chimie
- f: Organisation Internationale du Commerce
OICA e: OPLAN Implementation Capabilities Report
OICC e: Officer in Charge of Construction
- e: Operational Intelligence Crisis Center
- f: Office International du Cacao et du Chocolat
- s: Oficina de Información sobre el Cambio Climático
OICR e: Operation, Implementation, Capabilities Report
OICRS f: Organisation Interentreprises sur la Coordination et la Recherche Sociale
OID e: Object Identification (or Identifier)
- e: Object Interaction Diagram
- e: Ovine Interdigital Dermatitis
- f: Organisation Internationale de Désarmement
OIDC e: Object Identifier Component
OIDL e: Object Interface Definition Language
OIE e: International Office of Epizootics
- f: Office International des Épizooties
OIEA e: Office of Integrated Environmental Analysis
- s: Organismo Internacional de/para la Energía Atómica
OIEAU f: Office International de l'Eau
- s: Oficina Internacional del Agua
OIES e: Office of Interdisciplinary Earth Studies
OIETA f: Office Inter-Etats du Tourisme Africain
OIF e: Optimum Index Factor
OIG e: Office of Inspector General
- e: Operations Interface Group
- f: Organisations Intergouvernementales
OII e: Occupational Injury/Illness
- e: Open Information Interchange
- e: Optical Imaging Instrument
Ø. II N: Økonomistabens Pengeregnskapsavdeling
Ø. III N: Økonomistabens Statistikkavdeling
OIIQ f: Ordre des Infirmières et Infirmiers du Québec
OIL e: Oil refining, storage and distribution
- e: Operator Identification Language
OILB f: Organisation Internationale de Lutte Biologique
OILCOM e: Oil Company of Malawi
OILF e: Oromo Islamic Liberation Front
OILHM e: Oil and Hazardous Material Information System
OILPOL e: International Convention for the Prevention of Pollution of the Sea by Oil
- f: Convention internationale pour la prévention de la pollution des eaux de la mer par les hydrocarbures
- s: Convenio Internacional para Prevenir la Contaminación de las Aguas del Mar por Hidrocarburos
OILSAR e: Ocean-Ice-Land Synthetic Aperture Radar
OIM e: Open systems interconnections Internet Management
- e: Orientation-Imaging Microscopy
OIMA s: Oficina para la Industria y el Medio Ambiente
OIML f: Organisation Internationale de Métrologie Légale
OIMT s: Organización Internacional de las Maderas Tropicales
OIN f: Organisation Internationale de la Normalisation
OINAC s: Organización Internacional Nueva Acrópolis de Guatemala
Öing r: Överingenjör
OIP e: Operations Improvement Program
- e: Operations Interface Procedure
OIPC f: Organisation Internationale de Police Criminelle
OIPT e: Overarching Integrated Product Team
OIQ f: L'Ordre des Ingénieurs du Québec
OIR e: Operational and Information Requirements
OIr. e: Old Irish
OIRA e: Office of Information and Regulatory Affairs
OIRM e: Office of Information Resource Management
OIS e: Office Information System[s]
- e: Osteopathic Information Service
- e: WWW Operational Information Service
- f: Service d'information sur le fonctionnement de la VMM
OISCA e: Organization for Industrial, Spiritual and Cultural Advancement
OISE e: Office Information System Equipment
OISF e: Oxidation-Induced Stacking Fault
OIS/IGP e: Office Information System Intelligent Gateway Processor
OISTAT f: Organisation Internationale des Scenographers, Techniciens et Architects de Theatre
OISTV f: Organisation Internationale pour la Science et la Technique du Vide
OIT e: Object Identifier Tree
- e: Office of Industrial Technologies
- e: Office software development and Information Technology
OITC e: Officer-In-Tactical Command
OIU e: Office Interface Unit
- e: Ogaden Islamic Union
OIV e: Object Idenitfier Value
- f: Organisation Internationale de la Vigne et du vin
OIW e: Open systems environment Implementors Workshop
- e: OSI Implementors Workshop
OIWG e: Operations-Intelligence Working Group
OJ e: Official Journal of the European Community
- f: Ordre Judiciaire
- f: Organisation de la Jeunesse
- s: Organismo Judicial
o.J. e: ohne Jahr
OJC f: Organisation Jeunesse Comorienne
OJCCT e: Online Journal of Current Clinical Trends
- e: Online Journal of Current Clinical Trials
OJD e: The Date of the edition of the Official Journal in which the decision was published
OJE e: On the Job Evaluator
OJI f: Office Juridique International
OJIB e: Ojibwe
OJIN e: Online Journal of Issues in Nursing
OJKSN e: Online Journal of Knowledge Synthesis for Nursing
OJM p: Organização da Juventude Moçambicana
OJN p: Organização Nacional de Jornalistas
OJNI e: Online Journal of Nursing Informatics
OJR e: The edition of the Official Journal in which the decision was published
OJRB f: Organisation de la Jeunesse Révolutionnaire du Bénin
OJS e: Organization Jointly Shared
OJT e: On The Job Resource Management Behaviors
- e: On the Job Training
OK a: Olympiske Komité
- d: Oberer Kanal
Ok r: Officerskurs
ØK N: øverstkommanderende
Ök r: Ögonblicklig känslig
O-K e: Pre-Storm Oklahoma-Kansas Preliminary Regional Experiment-Stormscale Operational Research and Meteorology
Okajimas Folia Anat Jpn j: Okajimas Folia Anatomica Japonica
OKB e: Design Bureau [Russia]
OKCS e: Oklahoma Climatological Survey
OKEAN e: A Soviet ocean radar satellite
OKL e: On Key Label
OKLAJUJ s: Centro de Estudios y Desarollo Integral de la Cultura Maya Kekchi
OKMA s: Oxlajuuj Keej Maya' Ajtz'iib'
ØKN N: Øverstkommanderende I Nord-Norge
Økol a: Økologisk
økol. N: økologi[sk]
Økon a: Øconomisk
økon. N: økonomi[sk]
ØKS a: Øernes Konsummælks Selskab
- N: Øverstkommanderende I Sør-Norge
Öksar r: Ögonblickligt känsligt spetsanslagsrör
okt. a: oktantal
OKU a: Odense Klædefabriks Udsalg A/S

OL a: Olympiske Lege
- e: Open-Loop
- e: Operation Limits
- e: Optical Limiter
- f: Objet de Luxe
- N: Olympiske Leker
- s: Onda Larga
- s: Orden Logística
Ol f: Olympiade
- f: Olympique
o.l. N: og lignende
ø.l. N: østlig lengde
Öl r: överläkare
OL-A e: Operating Location-A
OLA e: Ogaden Liberation Army
- e: Oklahoma Lumbermen's Association
- e: On-Line AUTODIN
- e: Original language
ØLA a: Østjyske Landboauktioner
OLADE s: Organización Latino-Americana de Energía
OLAM e: Online Alpha Monitor
OLANI f: Office du Lait du Niger
OLAO-CISM f: Office de Liaison de l'Ouest Afrique-Conseil International du Sport Militaire
OLAP e: On-Line Analytical Processing
OLC e: Operating Location Clerk
- e: Optical Loop Carrier
OLCA e: Office of Legislation and Congressional Affairs
OLCC e: On-Line Card Catalog
OLCMS e: On-Line Cargo Movement System
OLCP-EA f: Organisation de Lutte contre le Criquet Pèlerin dans l'Est Africain
OLCS e: On-line Cover [gas] System
OLD d: Ortsdosisleistung
- e: Old version
- e: Onsite Licensing Office
ØLD a: Østre Landsrets Dom
OLDEPESCA e: Latin American Fisheries Development Organization
- f: Organisation latino-américaine de développement de la pêche
- s: Organización Latinoamericana de Desarrollo Pesquero
OLDU e: On-Line Distillation Unit
OLE e: Object-Linked Environment
- e: Object Linking and Embedding
OLEC e: Other Local Exchange Carrier
OLEIS f: Œuvre pour l'Éducation de l'Image et du Son
OLETC e: Office of Law Enforcement Technology Commercialization
OLF e: Oromo Liberation Front
OLG d: Oberlandesgericht
- e: Office of Local Government
OLGA e: On-Line Guitar Archive
ø.lgd. a: østlig længde
ö. lgd. r: östlig längd
OLI e: Optical Line Interface
- e: Optical Phone Line Interface
- e: Originating Line Information
o.lign. a: og lignende
Oligo. e: Oligocene
OLI/HMD e: Online Images from the History of Medicine Division
OLIN e: Oklahoma Library Information Network
olinje r: observationslinje
OLIPaC e: Office of Land Information Policy and Coordination
OLIPAC e: Oligotrophic Pacific
OLIS e: OECDs document transmission system
- e: Oxford Library Integrated System
OLIT e: Open Look Interface Toolkit
- e: Open Look Intrinsic Toolkit
OLIX d: On-Line Informationssystem UNIX
OLLS e: Operational Logistics Support Summary
OLM e: Optical Link Modul
OLO e: On-Line Operation[s]
OLOS e: Operational Land Observation System
OLP f: Organisation de Libération de [la] Palestine
- s: Organización para la Liberación de Palestina
OLR e: Outgoing Longwave Radiation
ØLR a: Østre Landsret
Ol-r f: Olivier
OLRP e: On-Line Report Processor
OLS e: Oceanographic Lidar System
- e: On-Line Service[s]
- e: Online Library System
- e: Operation Line Scanner
- e: Operational License Stage
- e: Operational Line Scanner
- e: Operational Linescan System
- e: Optical Line Scanner
OLSChVO d: Verordnung über Orderlagerscheine
OLSNA e: Orthopaedic Laser Society of North America
OLSP e: On-Line Service Provider
OLSS e: On-Line Sodium Sampling
OLT e: On-Line Test[ing]
- e: Optical Line Termination
OLTIF f: Organisation pour la Libération du Tchad de l'Impérialisme et du Fascisme
OLTM e: Optical Line Terminating Multiplexer
OLTP e: On Line Transaction Processing
- e: On-Line Transaction Processor
OLTS e: On Line Tracking System
OLUIT e: Object Oriented Librarian User Interface Tool
OLUS e: On-Line User Services
OLVIMS e: On-Line Vehicle Interactive Management System
OLWI e: Organic Liquid Waste Incinerator
OLWM e: Open Look Window Manager
- e: Ordinary Low Water Mark
O/M e: Operations and Maintenance
OM a: Officielle Meddelelser
- e: Object Manager
- e: Occupational Medicine
- e: Ocean Mapping
- e: Odyssey of the Mind
- e: Office Management
- e: Operational Modeling
- e: Operations Module
- e: Operators Manual
- e: Optimal Mismatch
- e: Organic Matter
- f: Cartographie des Océans
- f: Officiers Ministériels
- f: Officiers Municipaux
- f: Ordonnance Médicale
- f: Ordonnance Ministérielle
- f: Organisme de Mutualité
- f: Outre-Mer
- f: Ouvrier Mécanicien
- f: Ouvrier Mineur
- s: Cartografía Oceánica
- s: Oficina Militar
- s: Onda Media
- s: Orden Ministerial
om e: Oman [Internet]
øm a: omdrejninger per minut
o.m. a: og mere
O.M. e: Order of Merit
O&M e: Operating (or Operations) and Maintenance
- e: Region 4 O&M Municipal Inventory
OMA e: Object Management Architecture
- e: Oceanography and Marine Assessment
- e: Open Management Architecture
- e: Operation and Maintenance, Army
- p: Organização das Mulheres Angolanas
o.m.a. a: og mange andre
- a: og meget andet
om[a] r: omyndig
OMAG R: Osobaya Morskaya Aviatsionna Grupa
OM-Al e: Organic Monomeric Aluminum
OM/A/N/AF/M e: Operation and Maintenance, Army/Navy/Air Force/USMC
OMAP e: Operations, Maintenance and Administration Part
omarb. a: omarbejdet
- r: omarbetad
- r: omarbetning
OMB e: Office of Management and Budget
- e: Office of Money and Banking
- e: Ombudsperson-Faculty
- e: Ombudsperson-Student
omb N: ombord
omb. r: ombud
OMC e: Occupational Measurement Center
- e: Office of Motor Carriers
- e: Operation[s] and Maintenance Center
- e: Optimum Moisture Content
- e: TOGA XBT Operations and Management Committee
- f: Comité des Opérations et de la gestion des XBT de TOGA
- f: Organisation Mondiale du Commerce
- s: Comité de Operaciones y Gestión de los XBT de TOGA
- s: Organización Mundial del Comercio

OMCA e: Organic Moderated Critical Experiment
OMCFP e: Optimized MAC Computer Flight Plan
OMCI f: Organisation intergouvernementale consultative de la navigation maritime
OMC-R e: Operation and Maintenance Center Radio
OMC-S e: Operation and Maintenance Center Switch
OMCS e: Office of Motor Carrier Standards
- e: Ozone Monitor Comparison System
OMCV p: Organização das Mulheres de Cabo Verde
OMCVD e: Organometallic CVD
Omd a: Omdrejning
OMD e: On-line Medical Dictionary
OMDP e: Orbiter Maintenance Down Period
OMDR e: Optical Memory Disk Recorder Animation System
OME e: On-board Modul Extension
- e: Open Messaging Environment
- e: Oregon Microcomputer Engineering
OMEC e: Online Medical Employment Center
OMED f: Observatoire Méditerranéen de l'Environnement et du Développement
Omega e: Long-range navigation aid
Omega Rho e: Operations Research Honor Society
OMEI e: Other Major Equipment Items
OMEP e: Office of Marine and Estuarine Protection
OMERA f: Société d'Optique, de Mécanique, d'Électricité et de Radio
OME-RESA e: Ohios Mid-Eastern Regional Education Services Agency
OMEX e: Ocean Margin[s] Exchange[s]
OMF e: Object Management Facility
- e: Object Management Framework
- e: Object Module Format
- e: Open Media Framework
- e: Open Message Format
- e: Open Modeling Forum
- e: Organization Master File
OMFB e: National Technical Development Committee [Hungary]
OMFCU e: Outboard Message Format Conventional Unit
OMFI e: Open Media Framework Interchange
omf. rum r: omformarrum
OMG e: Object Management Group
- e: Open Management Group
- e: Operational Maneuver Group
- f: Organisation Mondiale du commerce
omg. a: omgående
- a: omgang
OMGE f: Organisation Mondiale de Gastroentérologie
OMH e: Office of Minority Health
OMI e: Open Messaging Interface
- e: Open Microprocessor [systems] [Initiative]
- e: Open Modelling Interface
- e: Operations and Maintenance Instruction
- e: Operator Message Input
- e: Optical Mapping Instrument
- e: Optical Mode Interference
- e: Ozone Monitoring Instrument
- f: Office des Migrations Internationales
- s: Organización Marítima Internacional
OMIA e: Online Mendelian Inheritance in Animals
OMIPRA f: Opération Militaire de Production Agricole
OMIS e: Office of Management and Information Systems
- e: Office of Media and Information Services
OMISTN e: Optical Mode Interface Supertwisted Nematic
omkr. a: omkreds
- N: omkring
OML e: Object Manipulation Language
- e: Ocean Mixed Layer
- f: Office Médico-Legal
- N: Oslo Militære Legevakt
oml r: omladdad
OMLET e: Ocean Mixed Layer Experiment
- f: Expérience sur la couche de mélange océanique
- s: Experimento sobre la Capa de Mezcla Oceánica
OMLF e: Oromo Muslim Liberation Front
OMM e: Output Message Manual
- f: Organisation Meteorologique Mondiale
- p: Organização das Mulheres Moçambicanas
- s: Organización Meteorológica Mundial
o/m/m f: à l'ordre de moi-même
om./min N: omdreininger i minutt
OMMSQA e: Office of Modeling, Monitoring Systems and Quality Assurance
OMN e: Open Network Management
OMNI e: Mobilization and Deployment Data Base
- e: Organising Medical Networked Information
OMNIS f: Office Militaire National pour les Industries Stratégiques
OMNS e: Open Network Management System
OMO e: Occupational Medicine Office
- e: Office of Marine Operations
- e: Operator Message Output
OMP e: Ocean Margins Program
- s: Operaciones para el Mantenimiento de la Paz
OMPC e: Office of Municipal Pollution Control
OMPE e: Office of Management Planning and Evaluation
OMPF e: Operation and Maintenance Processor Frame
OMPI f: Organisation Mondiale de/pour la Propriété Intellectuelle
- s: Organización Mundial de la Propiedad Intelectual
OMPR e: Optimal Multiple Point Reassignment
OMR e: Offsite Methods Retrieval
- e: Output Message Report
omr N: område
omr. a: område
OMRB e: Outstanding national resource waters
OMRE e: Organic Moderated Reactor Experiment
OMRI e: Open Media Research Institute
OMRSD e: Operations and Maintenance Requirements Specifications Document
OMS d: Offenes Musiksystem
- e: Object Management Services
- e: Object Management System
- e: Office of Management Support
- e: Office of Mobile Sources
- e: Opcode-MIDI-System
- e: Open Music System
- e: Optical Mass Spectroscopy
- e: Orbital Mapping System
- e: Ordnance Management System
- f: Organisation Mondiale de la Santé
- N: Oslo Militære Samfund
- N: Oslo Militære Sykehus
- r: Omsättningsskatt
o.m.s. r: och mera sådant
oms. Energi a: omsættelig energi
OMT e: Object Model[l]ing Technique
- e: Oceanography and Marine Technology
- e: Other Military Targets
- f: Organisation Militante des Travailleurs
OMT-FNAS f: Organisation Militante des Travailleurs-Front National Anti-Souffrance
omtr. a: omtrent
OMU d: Original mit Untertitel
OMUP e: Operating and Maintenance User Part
OMVC f: Organisation de Mise en Valeur du Fleuve Gambie
OMVG e: Gambia River Development Organization
OMVPE e: Organo-Metallic Vapor-Phase Epitaxy
OMVS e: Senegal River Development Organization
- f: Organisation pour la Mise en Valeur du fleuve Sénégal
- s: Organización de Desarrollo del Río Senegal
OMWOG e: Ocean-based Measurements Working Group
ÖMZ d: Österreichische Militärische Zeitschrift
O/N e: Oxygen-to-Nitrogen ratio
ON e: Occupational Nursing
- f: Obligation Nominative
- f: Office de la Navigation
- f: Office National
- f: Ordre de la Nation
- f: Ordre Numérique
- f: Outillage National
- s: Octanaje Numérico
on. a: onsdag
ON. e: Old Norse

ONA e: Ohio Nurses Association
- e: Oklahoma Nurses Association
- e: Oregon Nurses Association
ONAC e: Operations Network Administration Center
ONADEC f: Organisation Nationale des Entreprises Congolaises
ONADEF f: Office National pour le Développement des Forêts
ONADER f: Opération Nationale pour le Développement de la Riziculture
ONAF f: Office National des Allocations Familiales
- f: Office National des Forêts
ONAGI f: Office National de Gestion des Biens Immobiliers de l'état
ONAH f: Office National d'Hydrocarbures
ONAM s: Oficina Nacional de la Mujer
ONAREM f: Office National des Ressources Minières
ONAREST f: Office National de la Recherche Scientifique et Technique
ONARP e: Occurrence Notification and Report Program
ONARS f: Office d'Assistance aux Réfugiés et Sinistrés
ONAS e: Office of Naval Acquisition Support
ONATEL f: Office National des Télécommunications
ONATHO f: Office National du Tourisme et de l'Hôtellerie
ONATRA f: Office National des Transports
ONB e: Office of Management and Budget
- f: Office National du Bois
ONBDT f: Œuvre Nationale Belge de Défense Contre la Tuberculose
ONBE f: Œuvre Nationale Belge
ONC e: Occurrence Notification Center
- e: Open Network Computing
- e: Open Networking Consortium
- f: Office National de Commerce
ONCA f: Office National de Commercialisation Agricole
ONCAD f: Office National de Coopération d'Assistance pour le Développement
ONCB e: Office of the Narcotics Control Board, Thailand
ONCE f: Office National du Commerce Extérieur
ONCL f: Office National des Combustibles Liquides
ONCO e: Office of NOAA Corps Operations
Oncol Nurs Forum e: Oncology Nursing Forum
Oncol Res e: Oncology Research
ONCPA f: Office National de Commercialisation des Produits Agricoles
ONCPB f: Office National de Commercialisation des Produits de Base
ONCXDR e: Open Network Computing External Data Representation
ONDAH f: Office National des Débouchés Agricoles et Horticoles
ONDE f: Office National de Développement de l'Elevage
Onderstepoort J Vet Res n: Onderstepoort Journal of Veterinary Research
ONDO e: ODP Nankai Downhole Observatory
ONDR f: Office National Tchadien de Développement Rural
ONDS e: Open Network Distribution Services
ONDT f: Œuvre Nationale de Défense contre la Tuberculose
ONE e: One number
- e: OPAC Network in Europe
- e: Open Network Environment
- e: Organization Number of Employees
- f: Œuvre Nationale de l'Enfance
ONEB e: Office of National Environmental Board
ONEGUA s: Organization of Guatemalan Blacks
ONEM f: Office National de l'Emploi
ONERA f: Office National d'Études et de Recherches Aérospatiales
ONERN s: Oficina Nacional de Evaluacion de Recursos Naturales
ONERSOL f: Office Nigérien de l'Energie Solaire
ONF e: Oncology Nursing Forum
- f: Office National des Forêts
ONF. e: Old Northern French
ONG f: Organisation[s] Non-Gouvernementale[s]
- s: Organización No Gubernamental
ONGT f: Office National Gabonais du Tourisme
ØNH N: Øre, Nese, Hals
ONI e: Office [of] Naval Intelligence
- e: Office of National waste terminal storage Integration
- f: Office des Nouvelles Internationales
- f: Office National d'Immigration
ONIC f: Office National Interprofessionnel des Céréales
ONIPPAM e: Office National Interprofessionnel des Plantes à Parfum, Aromatiques et Médicinales
ONIT f: Organisation Nationale Indépendante des Travailleurs
ONJ p: Organização Nacional de Jornalistas
ONL f: Office National de Logement
- f: Ordonnance de Non-Lieu
ONLEXCRT e: Online Exercise Critique System
ONLF e: Ogaden National Liberation Front
On-Line d: Betriebszustand, in dem ein Datenendgerät mit einem anderen Computer verbunden ist
Online-Dienste d: Kommerzielle Datennetzwerke
Online J Curr Clin Trials e: Online Journal of Current Clinical Trials
ONM e: Open Network Management
- e: Organization Name
- f: Office National de Modernisation
- f: Ordre National des Médecins
ONMOE f: Office National de la Main d'Œuvre
ONMP e: Office of Nuclear Materials Processing
ONMS e: Open Network Management System
ONN e: Open Network Node
- f: Office National de la Navigation
ONNA e: Oh No, Not Again [Internet]
ONNTA e: Oh No, Not This Again [Internet]
ONNYA e: Oh No, Not You Again [Internet]
ONO d: Ostnordosten
- e: Oxide-Nitride-Oxide
- f: Ouest-Nord-Ouest
- r: Ostnordost
ØNØ a: Østnordøst
ONP e: Open Network Provision
- e: Open Networking Platform
- f: Office National de Pêche
- p: Organização Nacional dos Professores
ONPC f: Office National du Placement et du Chômage
ONPE f: Office National de la Promotion de l'Emploi
ONPG e: Operations Nuclear Planning Group
ONPI f: Office National de la Propriété Industrielle
ONPPC f: Office National des Produits Pharmaceutiques et Chimiques
ONPT f: Office National des Postes et Télécommunications
ONPTZ[a] f: Office National des Postes et Télécommunications du Zaire
ONPV f: Office National des Produits Vivriers
ONRA f: Office National de la Recherche Aéronautique
ONRD f: Office National de Recherches et de Développement
ONRR e: Office of Nuclear Regulatory Research
ONRSD f: Office National de la Recherche Scientifique et du Développement
ONRSI f: Office National des Recherches Scientifiques et Industrielles
ONRSII f: Office national des Recherches Scientifiques et Industrielles et des Inventions
ONS e: Office of National Statistics
- e: Overseas News Service
- f: Œuvre Nationale de Secours
- f: Ouvriers Non Syndiqués
ONSAM e: Onsala Atmospheric Measurements
ONSEC s: Oficina Nacional de Servicio Civil
ONSER f: Organisation Nationale de la Sécurité Routière
ONSIL f: Organisation Nationale des Syndicats d'Infirmiers Libéraux
ONSPS e: Office of Nuclear Safety Policy and Standards
ONSS f: Œuvre Nationale de Sécurité Sociale
- f: Office National de Sécurité Sociale
ONSSU f: Office National des Sports Scolaires et Universitaires
ONST e: Office of National Security Technology
ONT f: Office National de Télécommunications

ONU e: Optical Network Unit
- f: Organisation des Nations Unies
- s: Organisación de las Naciones Unidas
ONUDI f: Organisation des Nations Unies pour le Développement Industriel
ONUEF f: Office National des Universités et Écoles Françaises
ONUESC f: Organisation des Nations Unies pour l'Éducation, la Science et la Culture
ONUMOZ e: United Nations Operation in Mozambique
ONURS s: Oficina para Combatir la Desertificación y la Sequía
ONUSAL e: United Nations Observer Mission in El Salvador
ONUSOM f: Opération des Nations Unies en Somalie
ON-VA e: Operator Network Validation Area
ONWB e: Oregon Nuclear Waste Board
ONWI e: Office of Nuclear Waste Isolation
OO d: Objektorientierung
- e: Object Oriented
- e: Operation Office
- e: Over [and] Out
- e: Skywest Airlines
o.O. e: ohne Ort
O^O e: a FEMALE moon
O_O e: a MALE moon
OOA d: Objektorientierte Analyse
- e: Object Orient[at]ed [system] Analysis
- e: Out-Of-Area
OOAD e: Object Orientated Analysis and Design
- e: Object Oriented Application Design
- e: Object Oriented Application Development
OO-ALC e: Ogden Air Logistics Center
OOB e: Offensive Operations Branch
OOC e: Operational Oceanography Center
OOCTG e: Object Orientated COBOL Task Group
OOD e: Object Oriented Design
- e: Officer Of the Deck
OODB e: Object Oriented Data Base
OODBMS e: Object-Oriented DBMS [Data Base Management System]
OODBS e: Object-Oriented DataBase System
OODL e: Object Orientated Dynamic Language
OODMS e: Object-Oriented Database Management System
OOE d: Objektorientierter Entwurf
OOF e: Other Official Flows
- e: Out Of Frame
Ø Ofj Fdi N: Østre Oslofjord Forsvarsdistrikt
OOI d: Objektorientierte Implementierung
- e: Object Oriented Implementation
OOIDA e: Owner-Operators Independent Drivers Association of America
OOK A: Orandee Ontwikkelingskorps
- A: Ovambo Ontwikkelingskorporasie
- e: On Off Keying
OOL e: Object Oriented Language
OOLP e: Object-Oriented Literate Programming
OOM e: Outside Office Memo
OOM[P]S e: Open Ocean Monitoring [Payload] Studies
OON e: Object-Oriented Nonsens
OOOS e: Object-Oriented Operating System
OOP d: Objektorientierte Programmierung
- e: Object-Orient[at]ed Programming
- e: Object-Oriented Pleasure
- f: Office d'Orientation Professionnelle
- s: Orden de Operaciones
OOPAC f: Organisme Officiel de Paiement des Allocations de Chômage
OOPC e: Ocean[s] Observation[s] Panel for Climate
- f: Groupe sur les observations océaniques pour l'étude du climat
- s: Panel de Observación del Océano en relación con el Clima
OOPL e: Object-Oriented Programming Language
O Opnes s: Orden de Operaciones
Ø Oppl Fdi N: Øst-Oppland Forsvarsdistrikt
OOPS d: Objektorientierte Programmiersprache
- e: Object-Oriented Program Support
- e: Object-Oriented Programming [and] System[s]
- e: Object-Oriented Programming Software
- e: Operational Ozone Processing System
- e: Operational Ozone Product System
OOPs e: Out-Of-Pocket costs/expenses
OOPSLA e: Object-Orient[at]ed Programming Systems, Languages and Applications
OOPSTAD e: Object-Orient[at]ed Programming for Smalltalk Application Developers (or Development) association
OORASS e: Object-Oriented Role Analysis, Synthesis and Structuring
OORDBMS e: Object-Oriented Relational DataBase Management System
OOS e: Object-Oriented Systems
- e: Ocean Observing System
- e: Off-line Operating Simulator
- e: Out-Of-Sync
- f: Système d'Observation des Océans
- s: Sistema de Observación de los Océanos
OOSA e: Object-Orientated System Analysis
OOSDP e: Ocean Observing System Development Panel
- e: Ocean Observing Systems Development Panel
- f: Groupe chargé de la mise en place du Système d'observation de l'océan
- s: Panel para el Desarrollo del Sistema de Observación de los Océanos
OOSE e: Object-Oriented Software Engineering
OOSH e: Object-Oriented Shell
OOT e: Object-Oriented Technology
OOTB e: Out Of The Box
OOTC e: Obligatory On Topic Comment [Internet]
OOTW e: Operations Other Than War
OOUI e: Object Oriented User Interface
OP a: Operationsafdeling
- e: Office of Planning
- e: Office of the President
- e: Office Power
- e: Operational Performance
- e: Order of Preachers
- e: Order Parameter
- e: Ordering Possibility
- e: Organophosphates
- e: Outside Plant
- e: Overhead Projector
- f: Obligation au Porteur
- f: Observatoire de Paris
- f: Océan Pacifique
- f: Œuvres de Placement
- f: Office Professionnel
- f: Office Public
- f: Officier de Paix
- f: Officier de Police
- f: Officier de Presse
- f: Opinion Publique
- f: Ordonnance de Police
- f: Ordre de Paiement
- f: Organisation Professionnelle
- f: Organisme Payeur
- f: Orientation Professionnelle
- f: Ouvrier Professionnel
- l: Ordo Praedicatorum
- N: Observasjonspost
- s: Obras Publicas
- s: Operación
Op f: Opération[nel]
- f: Opinion
- f: Opposition
- N: Operasjon
- r: Observationspost
- r: Operation[s]-
O&P e: Organization and Procedures
O/P e: Operating Procedure
OPA e: Office of Performance Assessment
- e: Office of Planning and Analysis
- e: Office of Policy Analysis
- e: Office of Public Affairs
- e: Office of Public Assistance
- e: Oil Pollution Act
- e: One-Photon Absorption
- e: Ontario Paramedic Association
- e: Ontario Physiotherapists Association
- e: Opaque
- e: Open Protocol Architecture
- e: Open Publishing Architecture
- e: Other Procurement Army
- e: Overview Pac Attribute
- f: Offre Publique d'Achat
- f: Organisation du Pacte Atlantique
- f: Organisation Politico-Administrative
- f: Organisations Professionnelles Agricoles
- p: Organização de Pioneiros Angolanos
OPA 90 e: Oil Pollution Control Act of 1990
OPAC e: Online Public Access Catalog[ue]
- e: Operations Planning Advicory Committee
- e: Overall Performance Appraisal Certification
- e: The Online Public Access Catalog
OPAEP f: Organisation des Pays Arabes Exportateurs de Pétrole

OPAG e: Operations and Planning Group
OPAL d: Optische Anschlußleitung
- e: Omni Purpose Apparatus for LEP
OPALE e: French bibliographic database
OPALINE e: Database management system
OPALS e: Optical Parallel Array Logic System
OPAM f: Office des Produits Agricoles du Mali
OPAN f: Organisation des Pêches de l'Atlantique du Nord-Ouest
OP/A/N/AF e: Other Procurement, Army/Navy/Air Force
OPANAL e: Agency for the Prohibition of Nuclear Weapons in Latin America and the Caribbean
- s: Organismo para la Proscripción de las Armas Nucleares en América Latina y el Caribe
- s: Organismo para la Proscripcion de las Armas Nucleares en la America Latina y el Caribe
OPAQUE e: Optical Atmospheric Quantities in Europe
OPARS e: Optimum Path Aircraft Routing System
OPAT f: Office des Produits Agricoles du Togo
OPB e: Observation Preparation Branch
- e: Office of Plans and Budget
OPBTP f: Organisme de Prévention du Bâtiment et des Travaux Publics
OPC e: Ocean Products Center
- e: Office of Peoples Counsel
- e: Office of the Parliamentary Counsel
- e: Optical Particle Counter
- e: Optical Photoconductor
- e: Optical Proximity Correction
- e: Organic Photoconductor
- e: Organization Postcode
- e: Organizational Point of Contact
- e: Originating Point Codes
- e: Origination Point Code
- e: Other Project Costs
- e: Output Control
- e: Ovamboland People's Congress
- f: Office de la Protection du Consommateur
OP&C e: Operations Planning and Control
OPCA e: Ornamental Plant Collectors Association
OPCC e: Offutt Processing and Correlation Center
OPC/COIP e: Ontario Prevention Clearinghouse/Centre Ontarien d'Information en Prévention
OPC/ESA e: Operations, Planning and Control/ESA
OPCIA f: Omnium de Produits Chimiques pour l'Industrie et l'Agriculture
OPCODE e: Operational Code
OPCONCTR e: Operations Control Center
OPCR e: Original Program Clock Reference
OPCVM f: Organisme de Placement Collectif en Valeurs Mobilières
OPD e: Office of Program Development
- e: Office of Public Defender
- e: Open Distributed Processing
- e: Original Point of Distribution
- e: Output Driver
OPDAT e: Office of Professional Development and Training
OPDD e: Overall Plant Design Description
OPDEP e: Operations Deputy
OPDF e: Omo Peoples Democratic Front
OPDIN e: Ocean Pollution Data Center
OPDN f: Organization Pro-Démocratique pour la Défense de la Nature
OPDO e: Oromo Peoples Democratic Organization
OPDOC e: Operational Documentation
Opdor s: Operador
OPDQ f: Ordre Professionnel des Diététistes du Québec
OPDS e: Officer Professional Development Seminar
- e: Overall Plant Design Specification
OPDU e: Operation Protocol Data Units
OPE e: Office of Personnel
- e: Office of Planning and Environment
- e: Operational Proficiency Examination
Op-Ed e: Opposite the Editorials
OPEEPE f: Organisation Européenne d'Études Photogrammetiques Expérimentales
OPEI f: Office National de Promotion de l'Entreprise Ivoirienne
OPEMA f: Office des Pêches Maritimes
OPEN f: Office de Promotion de l'Entreprise Nigérienne
OPEOS e: Outside Plant planning, Engineering and construction Operations Systems
OPEP f: Organisation des Pays Exportateurs de Pétrole
- s: Organización de Países Exportadores de Petróleo
OPERA e: Ordnance-Pulsed Experimental Research Assembly
OPET e: Oriented Polyester
OPEX e: Operating Expenses
- e: Operational and Executive Personnel for Civil Service
OPEZ f: Office pour la Promotion des Entreprises Zairoises
OPF e: Official Personnel File
- f: Organisation Panafricaine des Femmes
OPFI e: Office of Program and Fiscal Integrity
OPFM e: Outlet Plenum Feature Module
OPFT e: Other than Permanent Full Time
OPFTE e: Other than Permanent Full Time Equivalent
OPG e: Office of Global Programs
- e: Operations Planners Group
- s: Ordenación de Pagos
opg. a: opgang
OPGEN e: Operation Plan Generation
- e: Operations General Message
oph. a: ophørt
OPHBM f: Office Public des Habitations à Bon Marché
OPHLM f: Office Public des Habitations à Loyer Modéré
OPHQ f: Office des Personnes Handicapées du Québec
OPHS f: Office Public d'Hygiène Sociale
Ophthalm. e: Ophthalmology
Ophthalmic Epidemiol e: Ophthalmic Epidemiology
Ophthalmic Genet e: Ophthalmic Genetics
Ophthalmic Physiol Opt e: Ophthalmic and Physiological Optics
Ophthalmic Res e: Ophthalmic Research
Ophthalmic Surg Lasers e: Ophthalmic Surgery and Lasers
Ophthal Plast Reconstr Surg e: Ophthalmic Plastic and Reconstructive Surgery
OPI e: Office of Planning and Integration
- e: Open Information interchange
- e: Open Prepress Interface
- e: Open Prepress Standard
- e: Other Party Identifier
- f: Office de la Propriété Industrielle
OPIC e: Overseas Processing and Interpretation Center
- f: Office de Propagande Industrielle et Commerciale
- f: Offices Publics d'aménagement et de Construction
OPIE e: Outdoor Power Equipment Institute, Inc.
OPIM e: Operations and Information Management
OPIO e: Office of Policy, Integration and Outreach
OPIR e: Office of Program and Integrity Reviews
OPIS s: Organismo Público de Investigación
OPIT f: Opération Pilote Interministérielle de Télédétection
OPIU e: Operator Interface Unit
OPJ f: Officier[s] de [la] Police Judiciaire
- f: Organization Populaire de la Jeunesse
OPK e: Ovulation Predictor Kit
Opkl N: Oppklare
- N: Oppklaring
- N: Oppklart
Opklesk N: Oppklaringseskadron
Opk. Pt N: Oppklaringspatrulje
OPL e: Optical Path Length
Opl N: Observasjonsplass
opl. a: oplag
- a: oplysning
- a: Opløse
- a: opløsning
OPLAN e: Operations Plan
OPLAN REV e: Operation Plan Review
Oplats r: Observationsplats
OPLF e: Oromo Peoples Liberation Front
OP/LIM e: Operational Limitation
OPLO e: Oromo Peoples Liberation Organization
Opl. Regt. N: Opplandske Regiment

OPM e: Organization and Procedures Manual
- e: Original Program Model
- e: Others People Money
- e: Outside Plant Module
- f: Observatoire de Paris-Meudon
- f: Office des Pêches Maritimes
- f: Organisation Politico-Militaire
OPMC e: Office of Professional Medical Conduct
OPMEM e: Operand Memory
OPMS e: Officer Personnel Management Systems
OPMT e: Office of Program Management and Technology
OPN e: Organ Pipe National monument special upper-air observing site
ÖPN d: Öffentlicher Personennahverkehr
OPNAVINST e: Navy Operational Instruction
- e: OPNAV Instruction
Opn[es] s: Operación[es]
OPNET e: WWMCCS Operational Network
OPNL RPTS e: Operational Reports
OPNO e: Originating Public Network Operator
OPO e: Operational Performance Objectives
- e: Ovamboland Peoples Organization
- p: Ofensiva Política e Organizacional
Opo r: Operativ order
Op Ö r: Operativ Övning
op-område r: operationsområde
OPORD e: Operation Order
OPOs e: Organ Procurement Organizations
opost r: observationspost
OPP e: Off-Peak Power
- e: Other Programme Participation
- f: Organisation du Peuple Progressiste
- f: Organisme Professionnel de Prévention
opp. e: opposed
- e: opposite
OPPAS e: OPP Administrative Support Systems
OPPBTP f: Organisme Professionnel de Prévention du Bâtiment et des Travaux Publics
OPPE e: Office of Policy, Planning and Evaluation
opph. N: opphørt
OPPI e: Office of Policy, Planning and Information
oppkl N: oppklaring
oppklavd N: oppklaringsavdeling
oppkl. tr. N: oppklaringstropp
oppl. N: opplag
- N: opplysning
OPPLAN e: Operation Plan
OPPM e: Outside Principal Period of Maintenance
OPPME f: Office de Promotion des Petites et Moyennes Entreprises
oppr. N: opprettet
- N: opprinnelig
- N: opprinnelse
OPPSD e: Organic Peroxide Producers Safety Division
OPPTS e: EPA Office of Prevention, Pesticides and Toxic Substances
OPQ f: Ouvrier Professionnel Qualifié
OPQC e: Office Professional Quality Council
OPR d: Ostprignitz-Ruppin, Brandenburg
- e: Office of Primary Responsibility
- e: Officer Performance Report
- e: Opera
- e: Operation Planning Report
- f: Objets à Prix Réduits
- f: Objets à remettre par Poste Recommandée
- f: Offre Publique de Retrait
- f: Organisation du Peuple Rodriguais
- f: Organisation Politico-Religieuse
Opr f: Opérateur
- s: Operaciones
opr N: operasjon
opr. a: oprettet
- a: oprindelig
- a: oprindelse
Opr. s: Operador
OPRAG f: Office des Ports et Rades du Gabon
oprap N: operasjonsrapport
OPRC e: Oil Pollution Preparedness, Response and Cooperation Convention
- f: Convention internationale sur la préparation, la lutte et la coopération en matière de pollution par les hydrocarbures
- s: Convenio Internacional sobre Cooperación, Preparación y Lucha contra la Contaminación por Hidrocarburos
OPREP e: Operational Event/Incident
- e: Operational Report
OPRI f: Office de Protection contre les Rayonnements Ionisants
oprum r: operationsrum
OPS e: Off Premises Station
- e: Open Profiling Standard
- e: Operations Division
- e: Operations Group
- e: Operations Officer
- e: Operations Processing System
- e: Ophthalmic Photographers Society
- e: Optical Sensor
- e: Oracle Parallel Server
- e: Outlet Pipe Space
- f: Opérations
- f: Organisation Panaméricaine de la Santé
OPSAN e: Operations Analysis System
OPSCOM e: Operational Secure Communications
OPSCOMM e: Operational Communications
OPSD e: O Products and Services Division
OPSDEPS e: Operations Deputies
OPSEC e: Operations (or Operational) Security
OPSEC M&A e: Operations Security Management and Analysis Section
OPSG e: Operation Planning Steering Group
opsk. a: opskåret
OPSM e: Outside Plant Subscriber Module
OPSMOD e: Operational Validation
- e: Operations Planning Module
OPSNOTE e: Operations Note
OPSORD e: Operations Order
OPSS e: Office of Principal Staff Support
- e: Overweight and Pregnant Support
OPSTAT e: Operations Status Report
OPT e: Open Protocol Technology
- e: Optimization under uncertainty Project
- e: Options
- e: Organization Parent Company
- e: Ovulation Predictor Test
- f: Office des Postes et des Télécommunications
Opt. e: Optics
- e: Optimal
- e: Optimum
- e: Optional
OPTINOM d: Forschungsprogramm zur Optimierung des Nordsee-Monitorings
OPTIONS e: SWAGMAN program on water tables and their management
OPTIPARES e: Optical Processing of Airborne Remote Sensing
OPTO e: Optometry Library
OPTOMA e: Ocean Prediction Through Observation, Modeling and Analysis
Optom Clin e: Optometry Clinics
Optom Vis Sci e: Optometry and Vision Science
OPTR f: Office Professionnel des Transports Routiers
OPTRA f: Office Professionnel des Transports
OPTRAN e: Operational Transit
OPTS f: Office des Postes et des Télécommunications du Sénégal
OPTS RTS e: OPTS Regulation Tracking System
OP/UCOP e: Office of the President/UC Office of the President
OPUL A. Organisasie van Petroluitvoerlande
OPUS e: Officer Planning Utilization System
- e: Organization of Persistant Upwelling Structure
OPV e: Organic PhotoVoltaic
- s: Oferta Pública de Venta
OPVN f: Office des Produits Vivriers du Niger
OPWSA e: Ontario Prader-Willi Syndrome Association
OPX e: Off-Premises Extension
- e: Online Performance monitor destination
OPY e: Organization Parent Country
OQ f: Ouvrier Qualifié
O[-]QAR e: Optical Quick Access Recorder
OQL e: Object Query Language
OQPSK e: Offset Quadrature Phase Shift Keying
OQS e: Offsite Quality Surveillance

OR a: Ordrup
- e: Oak Ridge complex: Oak Ridge National Laboratory, K-25 Site and Y-12 Plant, Oak Ridge, Tennessee
- e: Oak Ridge operations office
- e: Occurrence Report
- e: Offner Relay
- e: Omega Rho
- e: Operating Revenue
- e: Operation Research
- e: Operational Report
- e: Organizational Representative
- e: Originating Register
- e: Originator/Recipient
- e: Own[ers] Risk
- f: Occasions Renault
- f: Office de Réparation
- f: Offre Réelle
- f: Ordonnance de Référé
- f: Ordre Révocable
- f: Organisme de Rééducation
- r: Operationsrum
- r: Orientering

Or f: Orne

O/R e: Originator/Recipient

OR2000 e: Ocean Rescue 2000

ORA e: Occultation Radiometer
- e: Office of Research Administration
- e: Office of Research and Applications
- e: Operational Requirements Analysis
- e: Orthopaedic Rehabilitation Association
- f: Organisation de Résistance Armée

ORACLE e: On-Line Inquiry and Report Generator
- e: Oversight of Resources and Capability for Logistics Effectiveness Relational Data
- e: Relational DBMS [Data Base Management System]

Oral Surg Oral Med Oral Pathol Oral Radiol Endod e: Oral Surgery, Oral Medicine, Oral Pathology, Oral Radiology and Endodontics

ORALTOX e: Acute Oral Toxicity for Birds, Mice, Rats

ORAM s: Oficina Regional para Africa Meridional

ORANA f: Office de Recherches pour l'Alimentation et la Nutrition Africaine

ORANI e: macroeconometric model for the Australian economy used by various Commonwealth government agancies

ORAU e: Oak Ridge Associated Universities

ORB d: Ostdeutscher Rundfunk Brandenburg
- e: Object Request Broker
- e: Office Repeater Bay
- e: Online Reference Book for medieval studies
- e: Orbital [data systems special upper-air observing site]

ORBATTOA e: Order of Battle-Transfer of Authority

ORBD-NRC e: Osteoporosis and Related Bone Diseases-National Resource Center

ORBI e: Stock ticker symbol for OSC

ORC e: Online Reactivity Computers

ORCA e: Operations Requirements Continuity Assessment
- e: Regional Office for Central America
- f: Bureau régional pour l'Amérique centrale
- f: Organisme européen de coordination des recherches sur le fluor et la prophylaxie de la carie dentaire
- s: Oficina Regional para Centroamérica

ORCON e: Originator Control of Information Release

ORCS e: Online Remote Compile System

ORD e: O'HaRe Airport at Chicago
- e: Office of Rare Diseases, National Institutes of Health
- e: Office of Research and Development
- e: Operational Requirements Document[s]
- e: Operational Suitability
- e: Oxidation-Reduced Diffusion
- f: Office Régional de Développement
- f: Organisation Rurale de Développement

Ord f: Ordinaire
- f: Ordonnance
- f: Ordre
- r: Ordonnans
- s: Ordenanza

ord. a: ordinær[t]
- a: ordineret
- N: ordon[n]ans

ORDA e: Office of Research Development Administration, SIUC

ordb. a: ordbog

ORDBMS e: Object-Relational Database Mangement System

ordf. r: ordförande

ORD/HWERL e: Office of Research and Development/Hazardous Waste Environmental Research Laboratory

ORDIS e: Office of Research and Development Information Systems

ORDLIX e: Organized Design for Line and Crew System

Ordn s: Ordinario

ordn. r: ordning

ORE e: Operations Readiness Exercise
- e: Organization Region
- f: Organisation Régionale Européenne

Oreg. Fish Comm. Contrib. e: Oregon Fish Commission Contributions

Oreg. Fish Comm. Res. Briefs e: Oregon Fish Commission Research Briefs

OREN s: Orden para el Enlace

OREO e: Operations Response Engineering Order

Orer f: Oranger

ORES f: Organisation Européenne de Recherches Spatiales

ORESTES e: Intrabattle group tactical HF circuit

ORF d: Österreichischer Rundfunk
- e: Operational Readiness Fleet

OR&F e: Operations, Research and Facilities

ORFCOM e: Oak Ridge Facility Comparison Study

ORFEUS e: Orbiting [and] Retrievable Far and Extreme Ultraviolet Spectrometer

ORG e: Non-Profit Organizations
- e: Organization [organization Domain name] [Internet]
- s: Organización

org e: organizational
- e: organizational; Internet-Domain
- N: organisasjon
- N: organisere

Org f: Organisation

org. a: organiseret
- a: organisk

ORGALIME f: Organisme de Liaison des Industries Métallurgiques Européennes

ORGANIC f: Organisation Autonome Nationale d'Assurance-vieillesse de l'Industrie et du Commerce

OrgCh d: Institut für Organische Chemie und Isotopenforschung

ORGECO f: Organisation Générale des Consommateurs

Orger f: Oranger

OrgInt f: Organisation Internationale

OrgKG d: Gesetz zur Bekämpfung der organisierten Kriminalität

ORHP e: Office of Rural Health Policy

ORI e: Inclusive Or Immediate
- e: Online Retrieval Interface
- e: Operations Research and Industrial engineering
- e: Operations Research and Information systems
- e: Original
- e: Originating body
- f: Institut de Recherche Océanographique
- s: Instituto de Investigación Oceánica

ORIAC e: CERN Accounting System

Orig f: Origine

orig. e: origin[al][ly]

ORIGEN2 e: Oak Ridge Isotope Generation and Depletion Code [Version 2]

Orig Life Evol Biosph e: Origins of Life and Evolution of the Biosphere

ORINFOR f: Office Rwandaise d'Information

ORINS e: Oak Ridge Institute of Nuclear Studies, Inc.

ORIO e: Oak Ridge Regional Inspections Office

ORION f: Office de Recherche Interdisciplinaire sur les Organisations Neurophysiologiques

ORISE e: Oak Ridge Institute for Science and Education

ORJETS e: Online Remote Job Entry Terminal System

ORK e: Orkney

ÖRK d: Pressespiegel der österreichischen Rektorenkonferenz

Orl f: Orléans

orl N: orlogs-

ORLA e: Optical Remote Sensing of the Land

ÖrlB r: Örlogsbas

ÖrlBavd r: Örlogsbasavdelning

ÖrlBÖ r: Östkustens Örlogsbas

ÖrlBS r: Örlogsbas Söder

ORL J Otorhinolaryngol Relat Spec e: ORL; Journal of Oto-Rhino-Laryngology and Its Related Specialties

ORLRAPS e: Online Remedial Action Progress System
ORM e: Object Role Modeling
- e: Office of Regional Management
- e: Office of Resource Management
- e: Optical[ly] Remote [switching] Module
- e: Other Regulated Material
ORMA e: Regional Office for Meso-America
- f: Bureau régional pour la Méso-Amérique
- s: Oficina Regional para Mesoamérica
ORMAS e: Operational Resource Management Assessment Plan
ORMH e: Office of Research on Minority Health
ORMS e: Office Resources Management System
ORMSS e: Ohio River Main Stem Study
ORMVAG f: Office Régional de Mise en Valeur Agricole eu Gharb
ORMWH e: Office of Research on Minority and Womens Health
Orn f: Ornemental
- f: Ornementation
ORNI f: Office Régulateur de la Navigation Intérieure
ORNL e: Oak Ridge National Laboratories (or Laboratory)
ORNL/ACL e: ORNL/Advanced Computing Laboratory
ORNL-CF e: Oak Ridge National Laboratory Critical Facility
ORNL-PCA e: Pool Critical Assembly, BSF Pool
ORNOGE p: Organización Nacional de la Oposición de Guinea Ecuatorial en el Exilio
ORO e. Oak Ridge Operations Office
- e: Office of Regional Operations
OR/OD e: Operations Requirements/ Operations Directive
O[-]ROM e: Optical Read Only Memory
ORP e: Occurrence Reporting Program
- e: Office of Radiation Programs
- e: Optional Retirement Program
ORPA s: Organización del Pueblo en Armas
ORPADE f: Organisation Panafricaine des Juristes et Economistes pour l'Assistance et le Développement
ORPM e: Office of Research Program Management
ORPS e: Occurrence Reporting and Processing System
ORPV f: Office Régional de Produits Vivriers
ORR e: Oak Ridge Research [reactor]
- e: Oak Ridge Reservation
- e: Office of Refugee Relief
- e: Office of Refugee Resettlement
- e: Operational Radar Replacement
- e: Operational Readiness Review
ÖrrB r: Örregrundsbas
ORRIO e: Oak Ridge Regional Investigation Office
ORRS e: Outer Radial Reflector Surveillance
ORRSR e: Onsite Routine Radioactive Shipment Record
ORS e: Office of Regulatory Support
- e: Office of Research and Statistics
- e: On-line Retrieveal System
- e: Operational Reports Section
- e: Operational Research Section
- e: Oral Rehydration Salts
- e: Oral Rehydration Solution
- e: Orthopaedic Research Society
- e: Output Record Separator
ORSA e: Oceanic Remote Sensing Assembly
- e: Operations Research Systems Analysis
ORS[agf] a: Overretssagfører
ORSEC f: Organisation des Secours
ORSER e: Office of Remote Sensing for Earth Resources
ORSI e: Office of Retirement and Survivors Insurance
ORSR e: Offsite Radioactive Shipment Record
- e: Onsite Radioactive Shipment Record
ORSSA e: Office of Regulatory Support and Scientific Analysis
ORSTOM f: Institut français de recherche scientifique pour le développement en coopération
- f: Office de la Recherche Scientifique et Technique d'Outre-mer
- s: Instituto Francés de Investigación Científica para el Desarrollo y la Cooperación
ORT e: Ongoing Reliability Test
- e: Ooty RadioTelescope
- e: Operational Reliability Test joint
- e: Organization For Rehabilitation and Training
- f: Obligation Renouvelable du Trésor
- f: Office Régional des Transports
- f: Office Régional du Travail
- f: Organisation Rationnelle du Travail
- R: Obschtschestwo Rasprostranenija Truda sredi Jewrejew w Rossji
ORTA e: Office of Research and Technology Applications
- e: Office of Research and Technology Assessment
ORTB f: Office de Radiodiffusion et Télévision du Bénin
ORTF f: Office de [la] Radio[diffusion] [et de la] Télévision Française
ORTHO e: American Orthopsychiatric Association
Orthod Fr f: Orthodontie Francaise
Orthop Clin North Am e: Orthopedic Clinics of North America
ORTIBA f: Organisation Technique et Industrielle du Bâtiment
Ortpta s: Ortópedo
ORTRP f: Office Régional des Transports de la Région Parisienne
ORTS e: Occurrence Reporting and Tracking System
- f: Office de Radio-Télévision Sénégalaise
or. u. r: originalupplaga
ORU e: Orbit Replaceable Unit
- e: Organized Research Unit
Oru s: [de] Oruga
ORUS e: Orbital Replacement Units
ORV e: Off Road Vehicle
Orv Hetil e: Orvosi Hetilap
ORWA DDR d: Ordnung über die Bildung und Arbeitsweise der Richterwahlausschüsse vom 22.7.1990
ORWH e: Office of Research on Womens Health, NIH
OS d: Oberschulamt
Os e: Oberseminar
OS e: Oculus Sinister
- e: Office of Systems
- e: Operations Superintendent
- e: Operations Support
- e: Operator Service
- e: Optical Spectroscopy
- e: Organizational Support
- e: Output Specification
- e: Outstate
- e: Oxygen Saturation
- f: Obligation Solidaire
- f: Œuvre Sociale
- f: Ordre de Service
- f: Ordre de Succession
- f: Ordre Spécial
- f: Organisation Scientifique
- f: Organisation Secrète
- f: Organisation Sociale
- f: Organisation Syndicale
- f: Ouverture d'une Succession
- f: Ouvrier Spécialisé
- f: Ouvriers Syndiqués
- r: Olympiska Spelen
- r: Områdessektion
- s: Servicio de Oscurecimiento y central eléctrica
ÖS r: Örlogsmannasällskapet
- r: Örlogsstation
- r: Överstyrelse
OS. e: Old Saxon
O&S e: Operations and Support
O/S e: Operations and Support
- e: Out of State
OS/2 d: Multitasking Betriebssystem
- e: IBMs Operating System/2
- e: Operation System 2
OS[-]9 e: Operating System-9
OS[/]360 e: Operating System /360
OS/390 e: Open Server/390
OS/400 e: Optimized Server for AS/400 computers
OS[/]400 e: Operating System /400
OSA e: Obstructive Sleep Apnea
- e: Office of Safety Assessment
- e: Office of Special Applications
- e: Office of Student Aid
- e: Office Systems Architecture
- e: Olefin modified Styrene Acrylonitrile
- e: Open Scripting Architecture
- e: Open System Adapter
- e: Open System Architecture
- e: Open Systems Environment
- e: Operational Safety Analysis
- e: Operational Safety Assessment
- e: Operational Support Aircraft
- e: Operational Support Airlift
- e: Organisation for Strategic Labour Markets
OSa e: Osteosarcoma
o.s.a. r: om svar anhålles
OSAA e: Operational Satellite Active Archive

OSAC e: Operator Services Assistance Center
- e: Oxford System of Automated Cartography

OSACA e: Open System Architecture for Controls within Automation Systems

OSACTT s: Órgano Subsidiario de Asesoramiento Científico, Técnico y Tecnológico

OSAD e: Outer Space Affairs Division

OSAF e: Origin Subarea Address Field

Osaka City Med J e: Osaka City Medical Journal

OSAP e: Office of Scientific and Academic Publications
- e: Office of Vocational Rehabilitation
- R: Otdelnii Smeshannii Aviatsonii Polk

OSAR e: Operational Safety Analysis Report

OSASA e: Obstructive Sleep Apnea Sydrome

OSATRMS e: Operational Resource Tracking and Management System

OSB e: Ocean Studies Board
- e: Operation Support Building
- e: Operations and Services Building
- e: Operations Software Branch of SESD
- e: Order of St. Benedict
- e: Oriented-Strand Board
- l: Ordinis Sancti Benedicti

OSBRB e: Overhead/Support Budget Review Board

OSC e: Ocean Science Committee
- e: Office of Scientific Computing
- e: Office of Special Counsel for employment-related immigration discrimination
- e: Ohio Soybean Council
- e: On-Scene Coordinator
- e: Open System Center
- e: Operating System Command
- e: Operational System Control
- e: Operations Support Center
- e: Operator Services Center
- e: Options Selection Committee
- e: Orbital Sciences Corporation
- e: Order to Show Cause and notice of hearing
- e: Osciallator
- e: Outer Space Committee
- f: Office Suisse de Compensation

OSCA e: Operations Systems Computing Architecture
- e: Optical Sensor Collaborative Association

OSCAR e: Oil Spill Computer Aided Response
- e: Orbiting Satellite[s] Carry[ing] Amateur Radio
- e: OSC [Operations Support Center] Automated Reports System

OSCE e: Office of Child Support Enforcement
- e: Organization for/on Security and Cooperation in Europe

Osc F N: Oscarsborg Festning

OS/CMP e: Operational Support/ Configuration Management Plan

OSCOM e: Oslo Commission for the Prevention of Marine Pollution by Dumping from Ships and Aircraft
- f: Commission d'Oslo pour la prévention de la pollution marine par les opérations d'immersion effectuées par les navires et aéronefs
- s: Comisión de Oslo para la Prevención de la Contaminación de los Mares Causada por el Vertimiento de Desechos de los Buques y Aeronaves

OSCP e: On-Site Computer Programmer

OSCR e: Ocean Surface Current Radar
- e: Off Standard Condition Report

OSCRL e: Operating System Command Response Language

OSD e: Office of the Secretary of Defense
- e: Office Secretary of Defense
- e: On-Screen-Display
- e: Open Software Distribution
- e: Open System Direction
- e: Operating Safety Document
- e: Operating Specification Document
- e: Operational Specification Document
- e: Organization Subject Details
- e: Outer Sleeve and Dome

OS[&]D e: Over, Short and Damaged

OSDA e: Onsite Sewage Disposal Act

OSDC e: Oceanographic Sub-programme [for GATE] Data Centre
- f: Centre de rassemblement des données océanographiques issues du sous-programme [pour ETGA]
- s: Centro de Datos del Subprograma Oceanográfico [para el GATE]

OSDH e: Oklahoma State Department of Health

OSDIT e: Office of Software Development and Information Technology

OSDP e: Operations System Development Program

OSDPD e: Office of Satellite Data Processing and Distribution

OSE e: Oblique Seismic Experiment
- e: Observing System Experiment[s]
- e: Open Software Environment
- e: Open Systems Environment
- e: Operation Support Equipment
- f: Expérience concernant les Systèmes d'Observation
- f: Œuvre Sociale pour Enfants
- s: Experimento del Sistema de Observación

OS/E e: Operating System/Environment

OS/2 EE e: Operating System/2 Extended Edition

OSEOS e: Operational SEOS

OSEP e: Office of Special Education Programs

OSERS e: Office of Special Education and Rehabilitative Services

o/sersj N: oversersjant

OSF e: Office of Space Flight
- e: Open Software Foundation
- e: Open System[s] Foundation
- e: Operation[al] Support Facility
- e: Operations Systems Function [block]
- e: Oppose Sun Forever
- e: Order of St. Francis
- e: Organic Storage Facility
- e: Oxidation-induced Stacking Fault

OSF[/]1 e: Open Software Foundation [system]/1

OSF/DME e: Operation System Function/ Distributed Management Environment

OSFM e: Office of Surplus Facility Management

osfr. a: og så fremdeles

ÖSG r: Göteborgs Örlogsstation

OSGP e: Operations Support Group Prototype

OSHA e: Occupational Safety and Health Act
- e: Occupational Safety and Health Administration

OSI d: Otto-Suhr-Institut [FB Politische Wissenschaft der FU Berlin]
- e: Office [of] Scientific Investigation
- e: Office [of] Special Investigations
- e: Open System Interface
- e: Open System[s] Interconnect[ion][s]
- f: Interconnexion de Systèmes Ouverts
- f: Organisation de Standardisation Internationale
- s: Interconexión Abierta de Sistemas

OSIA e: Open Systems Interconnection Association

OSI/CS e: Open Systems Interconnection/ Communications Subsystem

OSID e: Origination Signaling Identifier

OSIFA e: Operational Sequence and Information Flow Analysis

OSI/FS e: OSI File Services

OSINET e: Open System[s] Interconnection Network

OSI/NMF e: OSI Network Management Forum

OSINT e: Open Source Intelligence

OSIP e: Operational [Meteorological] Satellite Improvement Program

OSIRIS e: Online Serials Information Registration and Inquiry System
- e: Optical Simulation for Imaging Reconnaissance and Intelligence Sensors

OSI[/]RM e: Open System[s] Interconnection/Reference Model

OSIS e: Operating Systems Installation Support

o/sjt N: oversersjant

OS-JTF e: Open Systems Joint Task Force

ÖSK r: Karlskrona Örlogsstation

OSKA d: Objektschluesselkatalog

OSL d: Oberspreewald-Lausitz, Brandenburg
- e: Object Script Language
- e: Operating (or Operational) Safety Limit
- e: Optically Stimulated Luminescence

OSl. e: Old [Church] Slavonic

O-SLM e: Optically addressed Spatial Light Modulator

OSLR e: [Intergovernmental Committee] for Ocean Science[s] and Living Resources
- e: Ocean Science in relation to Living Resources
- f: Comité intergouvernemental pour l'océanologie et les ressources vivantes
- s: Comité Intergubernamental sobre las Ciencias Oceánicas y los Recursos Biológicos

OS&M e: Operational Support and Maintenance

OSM e: Off-Screen Model
- e: Office of Sample Management
- e: Office of Surface Mining Reclamation and Enforcement
- e: On-Screen Menue
- e: Oncostatin M
- e: Operating System Service Module
- e: Operating System Specific Module
- e: Operations Sampling Management
- e: Operations Systems Manager
- e: Optical Storage Manager
- f: Office de Sécurité Mutuelle
- f: Organisation Syndicale Mondiale
- s: Obra Social el Martinico
- s: Órgano Superior de la Movilización

Osm. e: Osmania

OSMC e: Object Storage Management Component

OSME e: Open Systems Message Exchange

OSMRE e: Office of Surface Mining Reclamation and Enforcement

OS/MVS e: Operating System with Multiple Virtual Storage

OSN e: Ocean Seismic Network
- e: Ocular Surgery News
- e: Office System[s] Node
- e: Open Systems Network

OSNI e: Ordinance Survey of Northern Island

OSNLR e: Ocean Science in Relation to Non-Living Resources
- f: L'océanologie et les ressources non vivantes
- s: Ciencias Oceánicas en relación con los Recursos no Biológicos

OSO d: Ostsüdosten
- e: Objective and Strategies Overview
- e: Office of Satellite Operations
- e: Office of Systems Operations
- e: Onsite Office
- e: Originating Screening Office
- e: Originating Signaling Office
- r: Ostsydost

ØSØ a: Østsydøst

O-S-O f: Ouest-Sud-Ouest

OSO-8 e: Orbiting Solar Observatory -8

OSOCC e: On-Site Operations Coordination Center

OSP d: Organisationsstrukturplan
- e: Objective and Strategies Plan
- e: Occupational Safety Programs Branch
- e: Office of Sponsored Projects
- e: OLE-DB Simple Provider
- e: Online Service Provider
- e: On-Site Programmer
- e: Operations Special Procedures
- e: Operator Station Processor
- e: Optical Storage Processor
- p: Organizações Socio-Profissionais

OSPA f: Organisation Sanitaire Panaméricaine

OSPAA f: Organisation de [la] Solidarité des Peuples Afro-Asiatiques

OSPAAAL f: Organisation de Solidarité des Peuples Afro-Asiatiques et d'Amérique Latine

OSPC e: Off-Site Property Control

OSPF d: Internes Routingprotokoll
- e: Open Shortest Path First

OSPI e: Office of Strategic Planning and Integration
- e: Operating System Programming Interface

OSPM e: Operating System directed Power Management

OSPS e: Operator Service Position System

OSQL e: Object Structured Query Language

OSR e: OEM Service Release
- e: Office of Sponsored Research
- e: Office of Standards and Regulations
- e: Office of Systems Requirements
- e: Operating (or Operational, or Operations) Safety Requirement[s]

OSR2 e: OEM Service Release 2

OSRA e: Operational Software Release Approval

OSRI e: Originating Station Routing Indicator

OSRM e: Office of Standard Reference Materials

OSRMS e: Ocean Surface Roughness Measurement System

OSRP e: Occupant Safety Research Project

OSRS e: [Subgroup on] Ocean[ic] Satellites and Remote Sensing
- f: Sous-groupe des satellites d'observation de l'océan et de la télédétection
- s: Subgrupo sobre Satélites Oceánicos y Teledetección

OSS e: Object Services Standard
- e: Observatory of the Sahara and the Sahel
- e: Observing Simulation System
- e: Office of Safeguards and Security
- e: Office of Secret Services
- e: Office of the Space Station
- e: Office of the Supervising Scientist
- e: Office Support System
- e: Online Service System
- e: Open Source Solution
- e: Open Source Spectrometer
- e: Operational Storage Site
- e: Operational Support Squadron
- e: Operations Support Services
- e: Operations Support Squadron
- e: Operation[s] Support System[s]
- e: Operator Service System
- e: Operator Support System
- f: Œuvres Sanitaires et Sociales
- f: Organismes de Sécurité Sociale
- r: Oberoende Staternas Samvälde

ÖSS r: Stockholms Örlogsstation

OSSA e: Office of Space Science and Application

OSSAG s: Oficina Servicio Social del Arzobispado de Guatemala

OSSD e: Office of Space Systems Development

OSSE e: Observation (or Observing) System Simulation Experiment[s]
- e: Oriented Scintillation Spectrometer Experiment
- f: Expérience de Simulation des Systèmes d'Observation
- s: Experimento de Simulación del Sistema de Observación

OSSHE e: Oregon State System of Higher Education

OSSI e: Office of Supplemental Security Income

OSSN e: Originating Station Serial Number

OSSV R: Ogrnichenie i Sokrashchenie Stratigicheskykh Vooruzhenii

OST e: Ocean Surface Temperature
- e: Order and Shipping Time
- f: Organisation Scientifique du Travail
- f: Organisation Socialiste des Travailleurs
- s: Temperatura de la Superficie del Océano

OSTA e: Office of Space and Terrestrial Applications
- e: Optical Storage Technology Association

Ø-stab N: Økonomistaben

O-staben N: Operasjonsstab

OSTC e: Open System Testing Consortium
- e: Open Systems Testing Consortium

Osteoarthritis Cartilage e: Osteoarthritis and Cartilage

Osteoporos Int e: Osteoporosis International

OSTF e: Off-Site Test Facility
- e: Ordnance Survey Transfer Format

OSTI e: Office of Scientific and Technological Information

O.St.J. e: Officer of the Order of St. John of Jerusalem

OSTM e: Open System Transaction Management

OST[P] e: Office of Science and Technology Policy

OSU e: Office of Systems Operations
- e: Oklahoma State University
- f: Œuvre de Secours d'Urgence
- r: Obundet Slumpmässigt Urval
OSUS e: Ocean Surveillance
OSV e: Offset Scan Voting
- f: Organisation Sociale des Volontaires
- R: Ogranichenie Stratigicheskykh Vooruzhenii
osv. a: og så videre
o.s.v. r: och så vidare
OSVC e: Outgoing Switched Virtual Circuit
- f: Ordre de Service Valant Commande
OSW e: Office of Solid Waste
- e: Office of the Status of Women
OSWER e: Office of Solid Waste and Emergency Response
- e: Office of Solid Waste and Energy Response
OSW[/]L2 e: OS-9 Windows/Level 2
OSWR e: Office of Solid Waste Research
OT e: Object Technology
- e: Occupational Therapist
- e: Off Topic
- e: Open Transport
- e: Operational TIROS
- e: Operations Team
- e: Original Title
- f: Objets Trouvés
- f: Obligation du Trésor
- f: Office des Transports
- f: Organisation du Travail
ÖT r: Östtyskland
O.T. e: Old Testament
OT&A e: Operational Test and Acceptance
OTA e: Congressional Office of Technology Assessment
- e: Office of Tax Analysis
- e: Office of Technical Assistance
- e: Office of Technology Assessment
- e: Operation-Triggered Architecture
- e: Operational Test Agency
- e: Orthopaedic Trauma Association
- e: Orthopaedic Trauma Association Newsletter
OTAA e: Office of Trade Adjustment Assistance
OTAG e: Ozone Transport Assessment Group
OTAN f: Organisation du Traite de l'Atlantique Nord
- s: Organización del Tratado del Atlántico Norte
OTAO f: Organisation des Travailleurs de l'Afrique de l'Ouest
OTAR e: Over-the-Air Rekey
OTASE f: Organisation du Traité de défense collective pour l'Asie du Sud-Est
- s: Organización del Tratado de Asia del Sureste
OTB e: Off To Bed [Internet]
- e: Off-Track-Betting
- f: Office du Thé du Burundi
OTBS e: One True Bracketing Style
OTC e: Output Technology Corporation
- e: Over-The-Counter
- e: Over-The-Counter securities market
- e: Ownership Transfer Corporation
- e: Ozone Transport Commission
- f: Organisation du Traité Central
OTCIXS e: Office-in-Tactical Command Information Exchange Subsystem
OTCRP f: Office des Transports en Commun de la Région Parisienne
OTCZ f: Office des Transports en Commun du Zaire
OTD e: Office of Technology Development
- e: On-Time Delivery
- e: Optical Time Division
- e: Optical Transient Detector
OTDB e: Operations Tasking Data Base
OTDM e: Optical Time Division Multiplexer (or Multiplexing)
OTDR e: Optical TDR [Test Data Receiver]
- e: Optical Time-Domain Reflectometer (or Reflectometry)
OTE e: Organization Turnover in Mecu
- f: Organisation des Transports Européens
Ote. s: Oeste
OT&E e: Operational Test[ing] and Evaluation
OTEA e: Operational Test and Evaluation Agency
OTEC e: Operational Test and Evaluation Center
OTECH e: Oceaneering Technology Integration
OTER f: Office Technique d'Études et de Recherches
OTES e: Operational Test and Evaluation Squadron
OTF e: On The Floor [Internet]
- e: Open Token Foundation
OTF3 e: Octamer-binding protein 3
OTFA e: Office of Technical and Financial Assistance
OTFE e: Oscillatory Thermocapillary Flow Experiment
OTG e: OPTEVFOR Tactics Guide
OTGR f: Office Technique du Génie Rural
OTH e: Over-The-Horizon [Backscatter] Radar
- f: Omnium Technique de l'Habitation
OTH[-]B e: Over The Horizon Backscatter
OTH-B WCRS e: Over-the-Horizon Backscatter West Coast Radar System
OTH-RADAR e: Over-The-Horizon Radar
OTH-T e: Over-The-Horizon-Targeting
OTI d: Odenwald Touristik Information
- e: Office of Technology Integration
- e: Office of Transition Initiative
- e: Original Title
- e: Original Transmitter Identification
- f: Office de Transactions Immobilières et commerciales
OTIS e: Optimum Thermal Interpolation System
OTK A: Oostelik Transvaalse Kooperasie
OTKA e: National Scientific Research Fund
OTL e: Out To Lunch
OTLAMR e: Ocean Temperature Large Antenna Microwave Radiometer
OTLF e: One-Time License Fee
OT&LF e: Open Taxi and Limousine Foundation
OTM e: Office of Transportation Materials
- e: On Time Marker
- e: On-The-Move
- e: Operational Test Models
- p: Organização dos Trabalhadores de Moçambique
- s: Orden de Transmisiones
- s: Organização dos Trabalhadores de Moçambique, Maputo
OTMIR e: Office of Tropical Medicine and International Research
OTOH e: On The Other Hand [Internet]
Otolaryngol Clin North Am e: Otolaryngologic Clinics of North America
Otolaryngol Head Neck Surg e: Otolaryngology - Head and Neck Surgery
Otolaryngol Pol P: Otolaryngologia Polska
OTOOH e: On The Other Other Hand [Internet]
OTP e: Office of Territorial Programs
- e: One-Time Password
- e: Open Trade Protocol
- e: Operability Test Procedures
- e: Operational Test Plan
- e: Ozone Trends Panel
- f: Office Togolais des Phosphates
OTPP e: Operational Transfer Point Plan
OTPROM e: One-Time Programable ROM [Read Only Memory]
Ot.prp. N: Odelstingsproposisjon
OTPS e: Operational Test Program Sets
OTPV2 e: Operational Test Perspective View and Visualization
OTR e: On The Rag
- e: One Touch Recording
- e: Over The Road
- e: Oxygen Transmission Rate
- e: Ozone Transport Region
OTRAG d: Orbital Transport- und Raketen-Aktiengesellschaft
OTRQ f: Ordre des Technologues en Radiologie du Québec
OTRR e: Organisation for the Total Redemption of Rwandese
OTS e: Object Transaction Service
- e: Off The Shelf
- e: Office of Technical Services
- e: Office of Toxic Substances
- e: Officer Training School
- e: Overseas Telecommunications Services
OTSG e: Office, The Surgeon General
OTS MTS e: OTS Milestone Tracking System
OTSS e: Operational Telecommunications Switching System
OTT e: Office of Transportation Technologies
OTTER e: Oregon Transect Ecosystem Research
OTTH e: On The Third Hand [Internet]
OTTOMH e: Off The Top Of My Head [Internet]

OTTOMHAROOB e: Off The Top Of My Head And Rolling Out Of Bounds [Internet]
OTU e: Organization Turnover
OTUA f: Office Technique de l'Utilisation de l'Acier
ÖTV d: Gewerkschaft Öffentliche Dienste, Transport und Verkehr
OTW e: On The Whole
OTWG e: Operational Test Working Group
OTY e: Organization Type
OU e: Operable Unit
- e: Organizational Unit
- e: Orthodox [Jewish] Union
- e: University of Oklahoma, Norman
OUA f: Organisation de l'Unité Africaine
- f: Organisation de l'Unité Africaine, Uganda
- s: Organización de la Unidad Africana
OUAA e: Ontario Universities Athletic Association
OUADP e: Operational Utility of ADP
OUC s: Ondas Ultracortas
OUCH e: Organised Unitary Content Hypothesis
OUCL e: Oxford University Computing Laboratory
OUE e: Operational Utility Evaluation
OUG e: Occam User Group
OUI e: Organizational Unit Identifier
- f: Organization Unique Identifier
OUIC e: Operational Unit Identification Code
OUK d: Orthopädische Klinik und Poliklinik
OULG f: Organisation Unifiée pour la Libération de la Guinée
OUNSPI e: Ontario Universities Non Salary Price Index
OUO e: Official Use Only
OUP e: Oxford University
OUPID e: Ontario Universities Program for Instructional Development
OUQ e: Ocean University of Qingdao
- f: Université Océanique de Qingdao
- s: Universidad Oceánica de Qindgao
OUR e: Office of the University Registrar
- e: oxygen utilization rate
OURS e: Open Users Recommended Solutions
OUSA f: Organisation de l'Unité Syndicale Africaine
OUSD[A&T] e: Office of the Under Secretary of Defense [Acquisition and Technology]
OUT e: Outlines
- e: Outsize
Outl. e: Outline
OUTP e: Output
OUTS e: Output String
Outsourcing d: Auslagern von Tätigkeitsfeldern eines Unternehmens zu Dienstleistern
OUTWATS e: OUTward Wide Area Telephone Service
Ouv f: ouvert[ure]
Ouvr f: Ouvrage
OV e: Operational Verification
- e: Orientation Visit
- f: Objet de Valeur
ÖV r: Örlogsvarv
ØV N: Øvelse
Öv r: Överste
- r: Övertrycksventil
øv. a: øverst
OV/400 e: OfficeVision/400
OVA e: Optical Visual Analysis
- e: Organic Vapor Analysis (or Analyzer)
- e: Original Video Animation
OVAIM f: Omnium des Valeurs Agricoles, Industrielles et Minières
OVAL e: Object-Based Virtual Application Language
OVAR e: Off Vertical Axis Rotation
Ovärn r: Observationsplatsvärn
øv. avd. N: øvingsavdeling
OVC e: Office for Victims of Crime
OVDP e: Outside Vapor-Deposition Process
OVE R: Otdelnii Vertoletnii Eskadrilya
ovenn. a: ovennævnte
ovenst. a: ovenstænde
OVER e: Oversize
overf. a: overført
overs. a: oversættelse
- a: oversætter
- a: oversat
- N: oversatt
Övers r: Översättare
- r: Översikt
OVERVIEW e: POM Spares Forecasting System
Övet r: Överfältveterinär[en]
ovf. a: ovenfor
- a: overført
ÖVG r: Göteborgs Örlogsvarv
OVGWU e: Orange Vaal General Workers Union
Övhgashgr r: Övningsgashandgranat
OVI e: Office of Voluntarism Initiatives
- e: Open Verilog International
OVID e: Optical Visible and near Infrared Detector
ÖVK r: Karlskrona Örlogsvarv
OVL d: Obervogtland in Klingenthal und Ölsnitz, Sachsen
- e: Program Overlay
Övl r: Överstelöjtnant
ÖvlBS r: Örlogsbas Söder
Övlöjtn r: Överstelöjtnant
Övlt r: Överstelöjtnant
OVM e: Department of Computational Mathematics, Academy of Sciences, Moscow
OV/MVS e: OfficeVision/MVS
Övn r: Övning
Övngr r: Övningsgranat
Övngrkt r: Övningsgranatkartesch
Övnhgr r: Övningshandgranat
OVNI f: Objet Volant Non-Identifié
- p: Objecto Voador Não-Identificado
Övnle r: Övningsledare
Övnlng r: Övningsladdning
Övnprj r: Övningsprojektil
Övnrörlng r: Övningsrörladdning
Övnsptr r: Övningssprängpatron
Övnstrrmina r: Övningsstridsvagnsmina
Övnvinggr r: Övningsvinggranat
OVP e: Office of the Vice President
- R: Otdelnii Vertoletnii Polk
ÖVP d: Österreichische Volkspartei
OVR e: Program Overlay
øvr. a: øvrige
övr. r: övrig
OVS A: Oranje-Vrystaat
- e: OSHA Versatile Sampler
ÖVS r: Stockholms Örlogsvarv
OVSL f: Organisation Voltaïque des Syndicats Libres
OVT e: Operational Validation Test
OV/VM e: OfficeVision/VM
Ovw e: OpenView Windows
OVW e: Overflow bit
OvWa d: Offizier vom Wachdienst
OW e: Office of Water
- e: Over Write
OWA e: Oracle Web Application
OWBO e: Office of Womens Business Ownership
OWC e: Out-of-Window Collision
OWCP e: Office of Workers Compensation Programs
OWDC e: Office of Water Data Coordination
OWDM e: Original Water Depth Mine
OWEN e: Science and Engineering Library
OWEP e: Office of Water Enforcement and Permits
- e: Oily Waste Extraction Program
OWH e: CDCs Office of Womens Health
OWHN e: One Word Host Name
OWIAA e: Ontario Womens Intercollegiate Athletic Association
OWiG d: Gesetz über Ordnungswidrigkeiten
OWJ e: Orthodontics Web Journal
OWL d: Ostwestfalen-Lippe
e: Object Window[s] Library
- e: Open Windows Library
- e: Optimal Waste Loading
e: Orthopaedic Web Links
O[-]WORM e: Optical Write Once Read Many
OWOW e: Office of Wetlands, Oceans and Watersheds
OWP e: Office of Wetlands Protection
OWPE e: Office of Waste Programs Enforcement
OWPO e: Office of Water Program Operations
OWR e: Old World Realm
OWRM e: Other War Reserve Material
OWRMR e: Other War Reserve Materiel Requirements
OWRS e: Office of Water Regulations and Standards
OWS e: Ocean Weather Ship
- e: Operator Work Station
- f: Station météorologique océanique
- s: Buque Meteorológico Oceánico
- s: Estación Meteorológica Oceánica
OWSE e: Operational WWW [World Weather Watch] Systems Evaluation
- s: Evaluación Operativa de los Sistemas de la VMM

OWSE-AF e: Operational WWW [World Weather Watch] Systems Evaluation for Africa
- s: Evaluación Operativa de los Sistemas de la VMM para Africa
OWSE-NA e: Operational WWW [World Weather Watch] Systems Evaluation-North Atlantic
- e: Operational WWW [World Weather Watch] Systems Evaluation-North America
- s: Evaluación Operativa de los Sistemas de la VMM para el Atlántico Norte
OWTD e: Office of Waste Technology Development
OWTTE e: Or Words To That Effect [Internet]
OWW e: Organic Wash Waste
OX e: Ottowa Exchange
- e: Overnight Express
Ox e: Total Oxidants
O_x e: Odd Oxygen
Oxa r: Oexploderade ammunitionseffekter
OXF e: Open Exchange Format
Oxf. e: Oxford
Oxf Surv Eukaryot Genes e: Oxford Surveys on Eukaryotic Genes
OXS e: Objective X-ray Crystal Spectrometer
OXSOME e: Oxford County Soil Moisture Experiment
OY e: Operating Year
OYG e: Operating Year Guidance
OYOC e: One-Year-On-Campus
OZ e: Australia
- f: Onze
O-Z e: Ornstein-Zernike [equation]
OZAC f: Office Zairois de Contrôle
OZACAF f: Office Zairois du Café
ÖZE d: Österreichische Zeitschrift für Elektrizitätswirtschaft
ÖZG d: Österreichische Zeitschrift für Geschichtswissenschaften
OZIPP e: Kinetics Model and Ozone Isopleth Plotting Package
OZIPPM e: Modified Ozone Isopleth Plotting Package
ÖZPW d: Österreichische Zeitschrift für Politikwissenschaft
OZRK e: Ozark Plateaus
OZRT f: Office Zairois de Radiodiffusion et de Télévision
ÖZS d: Österreichische Zeitschrift für Soziologie

P

P a: Parkering[splads]
- a: Præventivpille
- d: Präsident
- d: Prüfungsleistung
- e: Symbol einer Produktbildung
- e: Momentum
- e: Peta
- e: Phenolic
- e: Phosphate
- e: Praktikum
- e: Predicted Pictures
- e: Priority Precedence
- e: significance level
- e: variable
- f: [chèque au] Porteur
- f: [nombre] Provisoire
- f: [ouvert en] Permanence
- f: Paires
- f: Par
- f: Parc
- f: Parti
- f: Passage
- f: Patente
- f: Payé
- f: Paysans
- f: Père
- f: Pièce
- f: Pied
- f: Poids
- f: Polaire
- f: Pôle
- f: Police[r]
- f: Polytechnique
- f: Positif
- f: Postal
- f: Poste
- f: Pouce
- f: Pour
- f: Pourvoyeur
- f: Presse
- f: Printemps
- f: Priorité
- f: Privé
- f: Prix
- f: Production
- f: Productivité
- f: Professionnel
- f: Protesté
- f: Protêt
- N: Panser
- N: Panserregiment
- N: Parkeringsplass
- r: av Pressgjutgods-
- r: Pansar
- r: Pansarregemente
- R: Perekhvatchuk
- r: Persontåg
- r: Pistol
- r: Pjäschef
- r: Provflygplan
- r: Punkt
- R: Pushechny
- s: Pantalla
- s: Patrulla
- s: Petrolero
- s: Pistola
- s: Proyectil
- s: Puesto

p e: Anzahl der Partialflächen, bzw. Partialflächenfaktoren
- e: Druck
- e: pre-cell

p. e: page

P. e: Printer
- e: Protected
- l: Publius
- N: Post

P1 e: Form to arrange payment to an outside body

P2 e: Interpersonal Messaging Protocol
- e: Pollution Prevention

P 2 r: Skånska Pansarregementet

P3 e: Platform for Privacy Preferences
- e: Port Truss Section 3
- e: Protocol used by UA Entities and MTA Entities

P-3 e: Fixed wing ASW aircraft

P4 e: The video interface bus on the 3/80

P 4 r: Skaraborgs regemente

P 5 r: Norrbottens Pansarbataljon

P 6 r: Norra skånska regementet

P 7 r: Södra skånska regementet

P 10 r: Södermanlands regemente

P 18 r: Gotlands regemente

PA a: Politiassistent
- d: Präsidialamt
- e: Palaeontology Association
- e: Palestinian [Authority]
- e: Performance Analysts
- e: Performance Assessment
- e: Phosphatidic Acid
- e: Physician Assistant
- e: Planning and Analysis
- e: Play Aid
- e: Policy Analyst
- e: Position Angle
- e: Power Alarm
- e: Preamble
- e: Precision Architecture
- e: Preliminary Assessment
- e: Press Association
- e: Pressure Alarm
- e: Principal Applicant
- e: Probability of Arrival
- e: Procedure Administrator
- e: Process Alert
- e: Program Address
- e: Program Administrator
- e: Program Announcement
- e: Programmatic Agreement
- e: Project Authorization/modification
- e: Property Administrator
- e: Provisional Acceptance
- e: Public Address
- e: Public Address System
- f: Pacte Atlantique
- f: Paiement Anticipé
- f: Paiement des Allocations
- f: Par Ampliation
- f: Partage Amiable
- f: Partie Adverse
- f: Peine Afflictive
- f: Pension Alimentaire
- f: Pension d'Attente
- f: Petit Artisan
- f: Plantes Alimentaires
- f: Plantes Annuelles
- f: Police Administrative
- f: Police d'Assurance
- f: Population Active
- f: Port d'Armes
- f: Pour Acquit
- f: Pour Ampliation
- f: Pouvoir d'Achat
- f: Prairie Artificielle
- f: Premier Avis
- f: Prêt d'Argent
- f: Prime Abandonnée
- f: Prime d'Aménagement
- f: Prime d'Ancienneté
- f: Prime d'Assurance
- f: Prix Autorisé
- f: Prix d'Adjudication

- f: Produits Alimentaires
- f: Professeur Agrégé
- f: Professions Agricoles
- f: Professions Artisanales
- f: Projet d'Acte
- f: Propriété Agricole
- f: Provision Alimentaire
- r: Personaladministrativa rådet
- s: Patronato Antialcohólico
- s: Piloto Aviador
- s: Punto de Apoyo

pA d: maximales Alter der Individuen
Pa e: Pascal
- f: Par an
- f: Par analogie
- f: Paris

P A e: Public Affairs
PA. e: Program Access
P&A e: Planning and Analysis
- e: Precision Accuracy
- e: Price and Availability
- E: Professional and Administrative
- e: Programming and Analysis
- e: Protection and Advocacy

P-A e: Posterior-Anterior
p.a. a: pro anno
P[/]A s: Portaaviones
p.å N: poståpneri
PAA e: p-Azoxyanisole
- e: Pan American World Airways system
- e: Poly Acetic Acid
- e: Primary Aromatic Amine[s]
- e: Primary Authorized Aircraft
- e: Priority Abatement Areas
- e: Program Assistant Administrator
- f: Prix Autorisé par l'Arrêté
- P: Panstwowa Agencja Atomistyki

P/AA s: Protección Antiaérea
PAAA e: Price Anderson Amendments Act
PAAD e: Program Assistance Approval Document
PAAJR e: Proceedings of the American Academy of Jewish Research
PA/A/N/AF e: Procurement of Ammunition, Army/Navy/Air Force
PAAP e: Plastic Area Array
PAAS e: Region 10 External Affairs Labels System
PAAT s: Plan Anual de Asistencia Técnica
PAB e: Personal Address Book
- e: Policy Advisory Body
- e: Public Affairs Bureau
- s: Puesto de Amunicionamiento de Batallón

PABA e: Paraaminobenzoic Acid
PABC e: Physiotherapy Association of British Columbia
PABG e: Pack[ing] a Big Gun [Internet]
Pable f: Payable
PABX e: Private Access Branch Exchange
PAC e: Pacific Area
- e: Pan African Congress
- e: Parachute and Cable
- e: Patient Airlift Center
- e: Pediatric AIDS Canada
- e: Personal Authenticator Card
- e: Photoactive Compound
- e: Planning Advisory Committee
- e: Plasma Arc Cutting
- e: Player Access Control
- e: Plowshare Advisory Committee
- e: Polar Atmospheric Chemistry
- e: Political Action Committee
- e: Pollution Abatement Control
- e: Potentially Adverse Condition
- e: PPTP Access Concentrator
- e: Pre-Admission Certification
- e: Preservation and Conservation
- e: Privilege Access Certificate
- e: Privilege Attribute Certificate
- e: Program Administration and Control
- e: Public Access Catalog
- e: Public Accounts Committee
- f: Politique Agricole Commune
- f: Port Autonome de Cotonou
- f: Programme d'Aménagement Côtier

Pac s: Paracaidista
PAC s: Patrullas de Autodefensa Civil
- s: Puesto Avanzado de Control

PA-C e: Physician Assistant Certified
PAC/A e: Pan-Africanist Congress/Azania
PACAF e: Pacific Air Command Air Forces
PACAIDS e: PACOM Crisis Action Information Distribution System
PACAMS e: Pacific Aircrew Management System
PACBAR e: Pacific Radar Barrier
PACC e: Pacific Airlift Control Center
- e: Pesiticides and Agricultural Chemicals Committee

PACCA e: Airfield Capabilities Application
PACCAT e: Pacific Area Command and Control AUTODIN Terminal
PACCOMMAREA e: Pacific Communications Area
PACCS e: Post Attack Command and Control System
PACD e: Plan of Action to Combat Desertification
PACDIGS e: Pacific Digital Graphics System
PACE e: Pacific Airlift Center
- e: Pacing and Clinical Electrophysiology
- e: Personal Audio Computer Editing
- e: Plant Acquisition and Construction Equipment
- e: Plant And Capital Equipment
- e: Pollution and Abatement Control Expenditures
- e: Priority Access Control Enabled
- e: Processing and Classification of Enlistees
- e: Program Acquisition Cost Estimate
- e: Program for Arrangement of Cables and Equipment

PACEGUA s: Patronato Cultural y Educativo de Guatemala
PACEM f: Propagation en Air Clair Et Meteorologie
PACER e: Program Assisted Console Evaluation and Review
- e: Program for the Acceleration of Commercial Energy Research
- e: Purpose, Agenda, Code of Conduct, Expectations/Introductions and Roles

PACER ACQUIRE e: AFLC Management System Acquisition
PACICOM e: Pacific Coastal Marine Productivity
- f: Productivité des zones côtières du Pacifique
- s: Productividad en las Zonas Costeras del Pacífico

PAC II e: CCSC Project Management System
Pacing Clin Electrophysiol e: Pacing and Clinical Electrophysiology
PACIR e: Propulsion, Aerodynamic, Controls Integrated Research
PACM e: Passive Access Control Module
Pac. Mar. Fish. Comm. Bull./Annu. Rep. e: Pacific Marine Fisheries Commission Bulletin/Annual Report
PACMEDS e: Pacific Meteorological Distribution Systems
PACMS e: Pacific Crisis Management System
PAC[o] s: Puesto de Amunicionamiento de Compañía
Pacode p: Partido do Congresso Democrático
PACOR e: Packet Processor
Pacq f: Pour acquit
PACS e: Pan American Climate Studies
- e: Passenger Automated Check-In System
- e: Pentagon Automated Communications System
- e: Personal Access Communication[s] System
- e: Physics and Astronomy Classification of the AIP
- e: Picture Archive Communication System[s]
- e: Picture Archiving and Communication[s] System
- e: Polar Acquisition and Control Subsystem
- e: Public Access Catalog System[s]

PACSC e: Pesticides and Agricultural Chemicals Standing Committee
Pac. Sci. e: Pacific Science
PACSICOM e: Pan African Conference on Sustainable Integrated Coastal Management
- f: Conférence panafricaine sur la gestion intégrée des zones côtières dans une perspective durable
- s: Conferencia Panafricana sobre la Gestión Integrada y Sostenible de las Zonas Costeras

PACS-L e: Public Access Computer Systems List
- e: Public Access Computers in Libraries-Listservice

PACSU f: Commission du Pacifique Sud pour les géosciences appliquées
PaCT e: PBX and Computer Teaming

pACT e: Personal Air Communications Technology
PACT e: Prefix Access Code Translator
- f: Centre de Propagande et d'Action Contre le Taudis
PACTAIS e: Pacific Theater Air Intelligence System
PACTG e: Pediatric AIDS Clinical Trials Group
PACTIDS e: Pacific TAC Intelligence Data System
PacTOP e: Pacific Tsunami Observation Program
PACWRAC e: PACOM WWMCCS Regional ADP Center
PAD e: Packet Assembler [and] Dissassembler
- e: Packet Assembly/Disassembler (or Disassembly)
- e: Paducah Gaseous Diffusion Plant, Paducah, Kentucky
- e: Physical Acoustics/Dunnegan
- e: Planning and Analysis Division
- e: Plant Apparatus Division [WHC]
- e: Pododermatitis Aseptica Diffusa
- e: Port Air Defense
- e: Pressure Alarm Detector
- e: Prime contractor Address
- e: Program Approval Document
- e: Public Access Device
- e: Pulse Amplifier Discriminator
- f: Parti d'Action pour la Démocratie
- f: Port Autonome de Dakar
- s: Partido Auténtico Democrático
PA&D e: Purchase Authorization and Directive
Päd d: Institut für Philosophie, Pädagogik und Psychologie, Abteilung für Pädagogik
PADAR e: Photoacoustic Detection and Ranging
PADCO e: Pan American Development Corporation
PADDS e: Procurement Automated Data and Document System
Padelimo p: Partido Democrático para a Libertação de Moçambique
PADEM-DNSI f: Projet Amélioration des Enquêtes de Ménage, Direction Nationale de la Statistique et de l'Information
PADEMO p: Partido Democrático de Moçambique
PADER f: Parti pour la Démocratie en Rwanda
PADH p: Partido de Aliança Democrática e Humanitária de Angola
PADI e: Personal Alarm Dose Integrator
- e: Professional Association of Diving Instructors
PaDL e: Pattern Development Language
PADOG f: Plan d'Aménagement et d'Organisation Générale
pADPRP e: poly-ADP-Ribosylpolymerase
Padr f: Pour adresse
PADRE e: Packaging Administration and Inventory Services
- e: Particle Analysis and Data Reduction program
- e: Particulate Data Reduction
PADS e: Pen Application Development System
PADU e: Protected Areas Data Unit
PA[&]E e: Program Analysis and Evaluation
PAE e: Products-Activities-End Products
- e: Project Assurance Engineer
- f: Plan d'Adaptation d'Effectifs
- p: Programa de Ajustamento Estrutural
- s: Proyectil de Alto Explosivo
PAEB e: Pan American EDIFACT Board
Paediatr Anaesth e: Paediatric Anaesthesia
Paediatr Perinat Epidemiol e: Paediatric and Perinatal Epidemiology
PAEK e: Polyaryletherketone
PAET s: Proyectil de Alto Explosivo a Tiempos
PAF d: Pfaffenhofen/Ilm, Bayern
- e: Particle and Field Package
- e: Personnel Action Form
- e: Platelet Aggregation Factor
- e: Postcode Address File
- e: Processing [and] Archiving Facility
- f: Professeur Agrégé à la Faculté de médecine
- f: Programmateur Automatique de Formules
- s: Plan Anual Fiscal
PAFAC e: Plastic And Failure Analysis of Composites
PAFC2N e: PACAF Command and Control Network
PAFCONET e: PACAF Teletypewriter Network
PAFDEF e: PACAF Defense Network
PAFE f: Plan d'Action en Faveur de la Filière Electronique
Pafemo p: Partido Federal de Moçambique
PAFG s: Plan de Acción Forestal para Guatemala
PAF-M s: Plan de Acción Forestal Maya
PAFMECSA e: Pan-African Freedom Movement of East, Central and South[ern] Africa
PAFNA e: Pan-African News Agency
PAFT f: Plan d'Action pour les Forêts Tropicales
- s: Plan de Acción Forestal Tropical
PA-FTIR e: Photoacoustic Fourier Transform Infrared Spectroscopy
PAFVONET e: PACAF Command and Control Voice Network
PAG e: Pesticide Assignment Guidelines
- e: Plant Advisory Group
- e: Prior Austenite Grain
- e: Protective Action Guideline
Pag f: Page
- s: Pagaduría
pag. a: pagina
- a: paginerет
pág. s: página
påg a: pågældende
PAGASA e: Philippine Atmospheric Geophysical and Astronomical Services Administration
PAGD s: Programa de Asesoramiento sobre Gestión de Desechos Radiactivos
PAGE e: Programmable Aerospace Ground Equipment
PAGEL e: Priced AGE List
PAGES e: Past Global Changes [program]
- f: Les changements planétaires survenus dans le passé
- s: Los Cambios Mundiales Pretéritos
PAGES-EXCOMM e: PAGES Executive Committee
PAGES-SSC e: PAGES Scientific Steering Committee
PAGIS e: Pastoral and Agricultural Geographic Information System
PAGM e: Permit Applicants (or Applications) Guidance Manual
på gr. af a: på grund af
PAH d: Panzerabwehrhubschrauber
- d: Polyzyklische Aromatische Kohlenwasserstoffe
- e: [Control] Alarm High
- e: Paraaminohippuric acid
- e: Payload Accommodation Handbook
- e: Polycyclic Aromatic Hydrocarbons
- e: Polynuclear Aromatic Hydrocarbons
pAH e: phenylalanine-hydroxylase
PAHO e: Pan American Health Organization
PAHR d: Photosynthetisch Aktive Strahlung
- e: Postaccident Heat Removal
PAHs e: Polycyclic Aromatic Hydrocarbons
PA&I e: Project Analysis and Integration
PAI e: Performance Audit Inspection
- e: Plasminogen Activator Inhibitor
- e: Polyamide-Imide
- e: Population Action International
- e: Primary Aircraft Inventory
- e: Provisional Acceptance Inspection
- e: Pure Active Ingredient compound
- f: Parti Africain de l'Indépendance
- p: Partido Angolano Independente
- r: Personaladministrativt Informationssystem
- s: Plan de Aseguramiento Ingeniero
PAICV p: Partido Africano da Independência de Cabo Verde
PAID e: Plutonium Air Inhalation Dose
PAIGC p: Partido Africano da Independência da Guiné e Cabo Verde
PAIGE e: Patient Instruction Generator
PAIH e: Public-Access Internet Host
PAI/LIPAD f: Parti Africain de l'Indépendance/Ligue Patriotique pour le Développement
PAIM f: Parti Africain pour l'Indépendance des Masses
PAIR e: Polarization-Agile Instrumentation Radar
- e: Preliminary Assessment Information Rule
PAIRO e: Professional Association of Interns and Residents
PAIRS e: Pushbroom Airborne Infrared Remote Sensor
PAIS e: Public Affairs Information Service
- e: Public-Access Internet Site
PAIVM e: Passive Accessory Inter-Vertebral Movement
PAIVS e: Pulmonary Atresia with Intact Ventricular Septum
PAJOCA p: Partido da Aliança Juventude Operéria Camponesa de Angola

PAJU e: Pan-African Journalists Union
PAK d: Polyzyclische aromatische Kohlenwasserstoffe
- e: Packed
- e: Packet
- e: Polyester Alkyd
- e: Product Authorization Key
Pakete e: Daten-Pakete
Paketlng r: paketladdning
PAKS N: Programmert Apparatur for Konstruerte Svar
PAL e: Palette
- e: Paradox Application[s] Language
- e: Parcel Airlift
- e: Permission Active Link. Device to trigger nuclear bombs from the cockpit
- e: Personal Answer Line
- e: Phase Alternate (or Alternating) Line
- e: Phase Alternating Loop
- e: Phase Alternation Line-rate
- e: Point, Area and Line source air quality mode[l]
- e: Polyanaline
- e: Privileged Architecture Library
- e: Process Asset Library
- e: Process Automation Language
- e: Processing and Analytical Laboratories
- e: Program Array Logic
- e: Programmable Array [of] Logic
- e: Programmed Array Logic
- e: Programming Application [Language]
- e: Programming Assembly Language
- e: Progressive Alliance of Liberia
- e: TV standard used in Britain, Germany and most of the rest of western Europe excluding France
- f: Plan d'Action de Lagos
- p: Partido Angolano Liberal
Pal s: Palomar
PAL+ e: Enhanced-fidelity PAL
PALA e: N-[PhosphoAcetyl]-L-Aspartic acid
PALACE e: Profiling ALACE [Automatic Lagrangian Circulation Explorer]
Palaeogr. e: Palaeography
pal[a]eoHeb e: written in palaeo-Hebrew
Palaeont. e: Palaeontological
PALASM e: C-like language for coding PALs
PALC e: Plasma-Addressed Liquid Crystal
PALCC e: Pacific Airlift Control Center
PALCTS e: Processing Analytical Laboratories Commitment Tracking System
PALDS e: PAL model with Deposition and Settling of pollutants
PALE e: Palaeoclimate from Arctic Lakes and Estuaries
- e: Popular Army for the Liberation of Eritrea
PALIR f: Parti Anarchiste pour les Libertés Individuelles dans la République
PALIS e: Polarized Airborne Laser Imaging Sensor
Palliat Med e: Palliative Medicine
PALM e: Pro Audio, Light & Music
PALMER FLD e: Palmer Field
PALMO p: Parti Libéral et Démocratique de Mozambique, Partido Liberal e Democrático de Moçambique
PALOPS p: Paises Africanos de Lingua Oficial Portuguêsa
Palr f: Palmier
PALS e: Pediatric Advanced Life Support
- e: Portable Airfield Lighting System
- e: Principles of the Alphabet Literacy System
- e: Public Access Library System
Pals f: Palais
PALT e: Procurement Acquisition Lead Times
PALU f: Parti Lumumbiste Unifié
PAM e: Package for Analogue Modeling
- e: Paging Area Memory
- e: Palestine Archaeological Museum
- e: Pan Africanist Movement
- e: Parallel Associative Memory
- e: Partitioned Access Method
- e: Personal Application Manager
- e: Pesticide Analytical Manual
- e: Phono Amplifier
- e: Plant Available Moisture
- e: Pluggable Authentication Module
- e: Portable Alpha Meter
- e: Postaccident Monitoring
- e: Preliminary Aerosol Monitor
- e: Process Application Module
- e: Procurement and Acquisition Management
- e: Program Analysis Memorandum
- e: Programmable Attribut Maps
- e: Puls Amplitude Modulation
- e: Pulse Amplification Modulation
- f: Plan Agricole Minimum
- f: Plan d'Action pour la Méditerranée
- f: Port Autonome de Marseille
- f: Premier Acte Médical
- f: Prestation d'Assurance Maladie
- f: Programme Alimentaire Mondial
- f: Programme des Applications Météorologiques
- s: Piloto de Aviación Militar
- s: Plan de Acción para el Mediterráneo
- s: Plan de Ayuda Militar
- s: Programa de Aplicaciones de la Meteorología
PAM 250 e: Accepted Point Mutation 250
PAMA e: Pan American Medical Association
- e: Preferred Acquisition Method of Analysis
- e: Processing Algorithm for Maintenance Actions
PAMCS e: Panel on Air Space Management and Control System
PAM[-]D e: Payload Assist Module [Delta class]
Pa Med e: Pennsylvania Medicine
PAMELA e: Process Abstraction Method for Embedded Large Applications
PAMIR e: Passive and Active Microwave and Infrared Radiometer [system]
PAMIRASAT e: Passive Microwave Radiometer Satellite
PAMIS e: PSYOP Automated Manpower Information System
PAMM e: Procurement, Accounts payable and Materials Management
- e: Program Against Micronutrient Malnutrition
PAMP s: Programa de Acompañamiento al Movimiento Popular
PAMS e: Pacific AUTODIN Multiplex System
- e: Passive Multifrequency Microwave Scanner Radiometer
- e: Pentax Analytical Measurement System
- e: Photo and Audio/visual Management System
- e: Photochemical Assessment Monitoring Stations
- e: Photogrammetric Analytical Measurement System
- e: PMEL Automated Management System
PAM SCAD e: Program of Action to Mitigate the Social Cost of Adjustment
PAN d: Hochauflösende, panchromatische Zeilen-Kamera
- d: Personennamen der Antike
- d: Rottal-Inn in Pfarrkirchen, Bayern
- e: Peroxiacetylnitrat
- e: Peroxyacetyl Nitride
- e: Personal Area Network
- e: Physical Association Number
- e: Primary Alerting Network
- f: Pacte Atlantique-Nord
- s: Partido de Acción Nacional
- s: Partido de Avanzada Nacional
Pan s: Panadería
- s: Panificación
PANA s: Proyecto de Alimentación y Nutrición del Altiplano
Panade p: Partido Nacional Democrático
Panafe p: Partido Nacionalista Federativo de Moçambique
PANAFTEL e: Pan-African Telecommunications Network
PANAFU e: Pan African Union
Panamo p: Partido Nacional de Moçambique
PANASH e: Pal[a]eoclimates of the Northern and Southern Hemispheres
PANDORA e: Prototyping a Navigation Database of Road Network Attributes
PANFERT e: Pregnancy After Infertility
PAngG d: Gesetz zur Gegelung der Preisangaben
PANGIS e: Pan African Network for a Geological Information System
PAngV d: Preisangabenverordnung
PAni e: Polyaniline
PANIC e: Potential and Needs, Investments and Capabilities
Panminerva Med e: Panminerva Medica
Pann f: Par année
PANs e: Peroxyacyl Nitrides
PANS e: Pretty Advanced New Stuff
- e: Pretty Amazing New Stuff
Pansarinsp r: inspektör för Pansartrupperna
PANSS e: Positive And Negative Syndrome Scale
PANT[EX] e: Pantex Plant, Amarillo, Texas

PAO e: Palo Alto CA
- e: Pediatric Assessment Online
- e: Performance Assessment and Oversight
- e: Pinellas Area Office
- e: Princeton Area Office
- e: Program Action Officer
- e: Property Accountable Officer
- e: Property Accounting Office
- e: Public Affairs Office[r]
- f: Publication Assistée par Ordinateur

PAOC e: Pacific Air Operations Center
PAOWU e: Port Authority and Other Workers Union
PAP e: Packet-level Procedure
- e: Participatory Anthropic Principle
- e: Password Authentication Protocol
- e: Peoples Action Party
- e: Plant Acquisition Plan
- e: Popular Alliance Party
- e: Pre-Approved Procedures
- e: Printer Access Protocol
- e: Priority Action Programme
- e: Process/Application Protocol
- e: Production Assurance Program
- e: Program Advanced Planning
- e: Prostatic Acid Phosphatase
- e: Public Access Profile
- f: Paris-Afrique Presse
- f: Parti d'Action Populaire
- f: Prêt d'Accession à la Propriété
- f: Programme d'Actions Prioritaires
- p: Partido para Aliança Popular
- s: Programa de Acciones Prioritarias

Pap f: Papier
PAPADIGUA s: Patronato de Pacientes Diabéticos de Guatemala
PAPC e: President's Accident Prevention Council
PAP/CAR f: Programme d'Actions Prioritaires/Centre d'activités régionales
Papet f: Papeterie
PAPMS e: Poly[aminopropyl]methyl-siloxane
pa. pple. e: passive participle
- e: past participle

PAPR e: Powered Air Purifying Respirator
PAP/RAC e: Priority Action Programme/ Regional Activity Centre
PAPS e: Portable Ada Programming System
- e: Proceedings of the American Philosophical Society

PAQ e: Pending Action Queue
PAR d: Panzerabwehrregiment
- e: Partners
- e: Peak to Average Ratio
- e: Performance and Availability Report
- e: Personal Animation Recorder
- e: Personnel At Risk
- e: Phased Array Radar
- e: Photosynthetically Active Radiation
- e: Photosynthetically Available Radiation
- e: Planning Activity Report
- e: Planning and Allocation of Resources
- e: Policy and Administrative Reform
- e: Population At Risk
- e: Positive Acknowledgement with Retransmission
- e: Preadmission Review
- e: Preliminary Analysis Review
- e: Procurement Advisory Release
- e: Program Acquisition Request
- e: Program ADP Requirement
- e: Programmed And Remote
- e: Project Activity Report
- e: Project Analysis and Reporting system
- e: Project Appraisal Report
- e: Project Approval Review
- e: Projected Automation Requirement
- e: Purchasing Approval [authorization] Request
- f: Poste Automobile Rurale
- f: Puissance Apparente Rayonnée
- s: Partido de Acción Revolucionario

Par e: Participating provider
- f: Paragraphe
- s: Paracaidismo

par. a: paragraf
PARA e: polyaryl amide
PARABOLA e: Portable Apparatus for Rapid Acquisition of Bidirectional Observations of Land and Atmosphere
PARACDS e: Missing Parameter Codes
PARAGON e: Processing and Archiving of Radar and Gauge Data Off-Line and in Near real time
ParaQuad NSW e: Paraplegic and Quadriplegic Association of NSW
Parasite Immunol e: Parasite Immunology
Parasitol Res e: Parasitology Research
PARAT e: A Study of the Transfer of Particulate and Dissolved Phases from Southern South America to the SW Atlantic Ocean
PARC e: Palo Alto Research Center (or Centre)
- e: Polarimetric Active Radar Calibrator
- e: Prison Activist Resource Center

PARCA e: Patient Access to Responsible Care Act
PARCC e: Precision, Accuracy, Representativeness, Completeness [and] Comparability
PARCOM e: Paris Commission for the Environmental Protection of the North East Atlantic
- e: Paris Commission for the Prevention of Marine Pollution from Land-based Sources
- f: Commission de Paris pour la prévention de la pollution marine d'origine tellurique
- s: Comisión de París para la Prevención de la Contaminación Marina des de Fuentes Terrestres

PARCS e: Perimeter Acquisition Radar Attack Characterization System
PARD e: Personal Alarm Radiation Dose
- e: Phased Array Radar Detection/Track

PARDEM e: Participatory Democracy
PARDPS e: PAR Data Processing System
PARE e: Physiological and Anatomical Rodent Experiment
PARERWA f: Parti Républicain du Rwanda
PARES e: Preprocessing of Airborne Remote Sensing Data
Par ext f: Par extension
PARI f: Parti Africain pour la Renaissance Ivoirienne
PARINFO f: Programme d'Aide à la Recherche en Information
PARIS e: Polarized Angle-Resolved Infrared Spectroscopy
Parlacen e: Central American Parliament
PARLE e: Parallel Architectures and Languages Europe
Parlem f: Parlementaire
PARMEHUTU f: Parti du Mouvement de l'Emancipation Hutu
PARMIS e: Planning and Resource Management Information System
PARN s: Partido de Acción y Reconstrucción Nacional
P450-AROM e: p450-aromatase
PARP e: poly[ADP-Ribose] Polymerase
- e: Pre-Engineered AUTOVON Restoral Plan Pacific
- f: Plan d'Aménagement de la Région Parisienne

PARPRO e: Peacetime Aerial Reconnaissance Program
PARR e: Program Analysis and Resource Review
PARS e: Performance Analysis Reports System
- e: Precision and Accuracy Reporting System
- e: Project Analysis Reporting System

PARSA e: Postgraduate and Research Students Association
Part f: Particulier
- f: Partie

PartGG d: Partnerschaftsgesellschafts-gesetz
PARTS e: Parts Analysis and Review Technique for Spares
- e: Parts Automated Repairable Tracking System

ParÜB d: Pariser Verbandsübereinkunft zum Schutze des gewerblichen Eigentums

PAS e: ParaAminoSalicylic Acid
- e: Paris Air Show
- e: PASCAL source code
- e: Payload Attach Structure
- e: Perceptual Aberration Scale
- e: Performance Analysis and Systems development
- e: Performance Assessment Scientific support
- e: Personnel Accounting Symbol
- e: Photoacoustic Spectroscopy
- e: Physician-Assisted Suicide
- e: Policy Analysis Staff
- e: Pre-Award Survey
- e: Premorbid Adjustment Scale
- e: Pressure Alarm Switch
- e: Product Acceptance test Specification
- e: Professor of Aerospace Studies
- e: Psychiatric Assessment Scale
- e: Public Announcement Service
- e: Publicly Available Specification[s]
- e: PUREX Acidified Sludge
- f: Passagers
- f: Programme d'Ajustement Structurel
- s: Portaaviones Antisubmarino

Pas f: Passage
pas. N: pasient
PAS[/]2 e: Personal Application System/2
PASA e: Paraaminosalicylic Acid
- e: Participating Agency Service Agreement
- e: Polymerase chain reaction Amplification of Specific Alleles

PASC e: Pan-African Student Conference
- e: Planning Advisory Subcommittee
- e: Polar Atmospheric and Snow Chemistry

PASCA e: Positron Annihilation Spectroscopy for Chemical Analysis
PASCAL e: Programming Language named for Blaise Pascal inventor of the 1st computing engine
Pasd s: Pasadera
PASE e: Polar Air-Snow Experiment
PASFIS e: Philippines Aquatic Sciences and Fisheries Information System
- f: Système philippin d'information sur les sciences aquatiques et la pêche
- s: Sistema Filipino de Información sobre las Ciencias Acuáticas y la Pesca

PASH e: Palaeoclimates of the Southern Hemisphere
- e: Polycyclic Aromatic Sulfur Heterocycle[s]

PA/SI e: Preliminary Assessment/Site Inspection
- e: Preliminary Assessment/Site Investigation

PASIPS e: Parallel Signal and Image Processing System
PASM e: Pan-African Student Movement
- e: Partitionable Simd-Mimd Machine-Lars Parallel Processor

PASO e: Pan Africanist Students Organization
PASOA e: Pan-African Students Organization in the Americas
PASOCO f: Parti Socialiste des Comores
PASP e: Program in Aegean Scripts and Prehistory
PASR e: Project Authorization Status Report
PASS e: Performance Assessment Scientific Support
- e: Photogrammetric Archival Storage System
- e: Prenotification Analysis Support System
- e: Primary Avionics Software System
- e: Prince Albert Satellite Station
- e: Pure Acoustic Sounder

Pass f: [bureau des] Passages
pass. e: passive[ly]
Passer f: passager
PaßG d: Paßgesetz
PASTEL f: Passager SPOT de Télécommunications Laser
pa.t. e: past tense
PAT e: Passive Angle-Tracking
- e: Patch
- e: Performance Assessment Team
- e: Permit Assistance Team
- e: Plutonium Air Transportable
- e: Point-After Try
- e: Population, Affluence and Technology
- e: Port and Address Translation
- e: Power Alarm Test
- e: Process Action Team
- e: Profit After Tax
- e: Public Access Terminal
- f: Paiement Avant Terme
- s: Patrulla

Pat f: Patente
PATA e: Pacific Asia Travel Association
PATAC f: Prévision Ameliorée et Techniques d'affinement de la Climatologie
PAT/ARM e: Passive Angle Tracking/ Anti-Radiation Missile system
PATC e: Page Address Translation Cache
PAT&E e: Production Acceptance Test and Evaluation
PATE e: Programmable ATEC Terminal Element
PATH FINDER e: Computer software program based upon the Critical Path Method
Pathol Annu e: Pathology Annual
Pathol Biol f: Pathologie Biologie
Pathol Int e: Pathology International
Pathol Res Pract e: Pathology, Research and Practice
PATHS e: Pacific Transport of Heat and Salt
- f: Transport de chaleur et de sel dans le Pacifique
- s: Transporte del Calor y la Sal en el Pacífico

PATM s: Plan de Aseguramiento Técnico-Material
PATN e: Pattern Analysis Package
- e: Promotional Port Access Telephone Number

PATO s: Proyectil Atómico
patol. a: patologisk
Patol Fiziol Eksp Ter R: Patologicheskaia Fiziologiia i Eksperimentalnaia Terapiia
Patousa e: Police and Traffic Officers Union of South Africa
patr. N: patrulje
PATRAM e: Packaging and transporting of radioactive materials
PATRIOT e: Air Defense SAM System
PATROL e: Program for Administrative Traffic Reports On Line
PATS e: Pacific Animal Therapy Society
- e: Parameterised Abstract Test Suite
- e: Pesticide Action Tracking System
- e: Pesticides Analytical Transport Solution
- e: Prototype Automated Telecommunications System

PATX e: Private Automatic [Telex] Exchange
PAU f: Personnel Autonome de l'Université
Pau f: Poteau
PAULA e: Port, Audio, UART [Universal Asynchronous Receiver Transceiver] and Logic
Pauly d: Paulys Realencyclopädie der classischen Altertumswissenschaft
Pauschaltarif d: Abrechnungsmodus eines Internet-Providers mit seinen Kunden
PAV e: Pressure Aging Vessel
- e: Prototype availability
- e: Stock Pavilion
- f: Point d'Accès Vidéotex
- s: Puesto Avanzado

Pav f: Pavillon
PAv s: Portaaviones
PAVA s: Programa de Ayada para los Vecinos del Altiplano
PAv As s: Portaviones Antisubmarino
pavd N: panseravdeling
PAVE e: Preparation for AIDS Vaccine Evaluation
PAv Esc s: Portaaviones de Escolta
PAv Hel s: Portaaviones con Helicópteros
PAVI f: Point d'Accès Vidéotex
PAv Lco s: Portaaviones Lanzacohetes
PAW e: Performance Automotive Warehouse
- e: Physics Around the World
- e: Plasma Arc Welding
- e: Princeton Alumni Weekly
- e: Protect Appalachian Wilderness, Earth First
- e: PUREX Acid Waste

PAWA e: Power and Water Authority
PAWC e: Pan-African Womens Conference
PAWS e: Portable Acoustic Wave Sensor
- e: Portable AN/UYS-1 Work Station
- e: Portable ASAS/ENSCE Work Station
- e: Protection Against Wrapped Sequence numbers
- e: Prototype Analyst Work Station

PAX e: Parallel Architecture Extended
- e: Passengers
- e: Patuxent River Office Field Support
- e: Pixel Addressing Extension
- e: Portable Archive Exchange
- e: Private Automatic [telephone] Exchange
- e: Voice-powered phone system

PAX DOC e: Passenger Documentation
PAY e: Payson Forest Service special upper-air observing site
- f: Payé
- f: Payement

PAYLOAD e: Sensing equipment carried on a platform
PAYM e: Pan-African Youth Movement
PAYO e: Pan-African Youth Organization

Pay TV d: Kodierte Fernsehprogramme
PB e: Petabyte [10^{15} bytes]
- e: Pipeline Burst
- e: Plasma reactor, Barrel type
- e: Polubutylene
- e: Polybutadiene
- e: Portable
- e: Presidents Budget
- e: Product Baseline
- f: Pacte de Bruxelles
- f: Patente Brute
- f: Pays-Bas
- f: Prêt-Bail
- f: Produit Brut
- f: Propriété Bâtie
- N: Panserbrytende
- r: Planeringsbyrå
- r: Plutonbefälhavare
- R: Pikuriyushi Bombardirovshchikdive
- R: Plovuchaya Baza
Pb d: Blei
- e: lead
- e: Plumbum
- r: Pansarbil
- s: Población
pb N: panserbil
p.b. r: på befallning
- r: på begäran
P&B e: Planning and Budgeting
PB 1 r: Pansarbrigad nr 1
Pb-212 e: lead-212
PBA e: Physical Block Address
- e: Preliminary Benefit Analysis
- e: Pyrenebutyric Acid
PBAN e: Polybutadiene Acrylonitrile
Pbat r: Pansarbataljon
PBB d: Polybromierte Biphenyle
- e: Polybrom[in]ated Biphenyl[s]
- e: Project Budget Base[line]
PBBH N: Panserbrytende med Ball Hette
PBBL e: Project Budget Baseline Log
PBBS e: Physical Baseline Build Standard
PBC a: Pengeinstitutternes Betalingsformidlingscenter
- e: Processor Bus Controller
- e: Produce Buying Company
- s: Pasta Básica [de Coca]
- s: Población Civil
PBC-L s: Pasta Básica [de Coca] Lavada
PBD e: Pacific Bell Directory
- e: Plans and Budget Division
- e: Polybutadine
- e: Production Buy Decision
- e: Program Budget Decision
PBDA p: Partido Bendito Democrático de Angola
PBDBM e: Peripheral Backup Detailed Billing Control Element
PBDP e: Protective Barrier Development Program
PBE e: Piggyback Experiment
- e: Plain Both Ends
- e: Prompt By Example
PBEIC e: Point Beach Energy Information Center
PBET e: Performance-Based Equipment Training
PBFA II e: Particle Beam Fusion Accelerator
PBFD e: Psittacine Beak and Feather Disease
PBGA e: Plastic Ball Grid Array
PBGC e: Pension Bension Guarantee Corporations
PBGM e: Brazilian Marine Geology Programme
- f: Programme Brésilien de Géologie Marine
- s: Programa de Geología Marina del Brasil
PBH N: Panserbrytende med Hette
PBHBH N: Panserbrytende med Hette og Ball Hette
PBI e: Parental Bonding Index
- e: Phone Based Interface
- e: Program Baseline Integration
- e: Push Button Indicator
- f: Programme Biologique International
- s: Peace Brigades International
- s: Programa Biológico Internacional
PBIA e: Pacific Brain Injury Association
Pbil r: Pansarbil
pbil. N: panserbil
PBIMR e: Push Broom Imaging Microwave Radiometer
PBL e: Planetary Boundary Layer
- e: Poly-buffered LOCOS
Pble f: préambule
PBLSH e: Publish
PBLSQ e: The Lead Line Source Model
PBM e: Peribacteroid Membrane
- e: Pipelined Burst Mode
- e: Play By Mail [game]
- e: Port Bypass Module
- e: Portable Bitmap
- e: Poskanzer Portable BitMap
- e: Program Business Management
- e: Project Bridge Model[l]er
PBMA e: Poly Butylmethacrylate
PBMR e: Push Broom Microwave Radiometer
PBN e: Peroxybutryl Nitride
- e: Polybutylene Napthalate
PBNSC e: Peripheral Backup NSC Control Element
PBO e: Placebo
PBP e: Packet Burst Protocol
- e: Performance Based Pay
PBR e: Patrol Boat, River
- e: Performance-Based Regulation
- e: Program Budgetary Review
- e: Program Business Representative
- f: Preneurs des Baux Ruraux
Pbr r: Pansarbrytande
PBRC e: Program Budget Review Committee
Pbrig r: Pansarbrigad
PBS e: Phosphate Buffered Saline
- e: Photon Backscattering
- e: Polybutadiene Styrene
- e: Portable Batch System
- e: Power Blending System
- e: Protective Breathing System
- e: Public Broadcast System
- e: Public Broadcasting Service
PBSC e: Peripheral Blood Stem Cell
PBSCT e: Peripheral Blood Stem Cell Transplantation
PBS/IS e: Public Buildings Service/ Information System
PBT e: Performance-Based Training
- e: Permeable-Base Transistor
PBTF e: Protective Barrier Test Facility
PBTS e: Business-Oriented Technology Promotion Programme
PBU e: Page Buffer Unit
- e: Peribacteroid Unit
PBV s: Peso Bruto Vehicular
Pbv r: Pansarbandvagn
P.B.V. a: På Bestyrelsens Vegne
PBW e: Particle Beam Weapons
PBWS e: Performance [Based] Work Statement
PBWU d: Projektgruppe Bayern zur Erforschung der Wirkung von Umweltschadstoffen
PBX e: Private Branch Exchange[s]
- e: Pull box
PBX-M e: Princeton Beta Experiment Modified
PC d: Persönlicher Computer
- e: Diacyl-glycerophosphocholine
- e: Paramagnetic Center[s]
- e: Patrol Corvette
- e: Peace Corps
- e: Percent[age]
- e: Personal Communication
- e: Personal Computer[s]
- e: Personal Computing
- e: Personnel Coordinator
- e: Phase Conjugate
- e: Phase Conjugation
- e: Photo Cell
- e: Photonic Crystal
- e: Physical Contact
- e: Piece of Crap
- e: Pin Cutting
- e: Planned Commitment
- e: Planning Commitee
- e: Plant Computer - performance monitoring and calculations
- e: Plasmatic Cell
- e: Player Character
- e: Pocket Calculator
- e: Pocket Computer
- e: Position Classification
- e: Post Card
- e: Potential Carcinogen
- e: Power Center
- e: Primary Center
- e: Prime contract
- e: Priority Control
- e: Privileged Character
- e: Process Condensate
- e: Professional Computer
- e: Progenitor Cryptocides
- e: Program Change
- e: Program Contract
- e: Program Control
- e: Program Council
- e: Programmable Channel
- e: Progressive Conservative party
- e: Protective Clothing
- e: Protocol Control
- e: Provincial Commissioner
- e: Pulverized Coal
- f: [de] Premier Choix
- f: [Ecole des] Ponts et Chaussées
- f: Papier Court
- f: Parité des Cours

- f: Parti Communiste
- f: Partie Civile
- f: Parties Contractantes
- f: Pas-de-Calais
- f: Passage Clouté
- f: Pays Chauds
- f: Pêche Côtière
- f: Peine Capitale
- f: Peine Corporelle
- f: Peines Correctionnelles
- f: Pension Civile
- f: Permis de Chasse
- f: Permis de Circulation
- f: Permis de Conduire
- f: Permis de Construire
- f: Personne Civile
- f: Petite Ceinture
- f: Pièces Comptables
- f: Pilote Civil
- f: Plan Cadastral
- f: Pli Chargé
- f: Poinçon de Contrôle
- f: Points Cardinaux
- f: Police Correctionnelle
- f: Police Criminelle
- f: Ponts et Chaussées
- f: Portion Centrale
- f: Pour Cent
- f: Pratique Courante
- f: Premier Clerc
- f: Premier Cours
- f: Président de la Confédération
- f: Président du Conseil des ministres
- f: Prêt à la Construction
- f: Prime à la Construction
- f: Prime de Capture
- f: Prise de Corps
- f: Prise en Charge
- f: Prison Civile
- f: Prix Courant
- f: Prix Coûtant
- f: Procédure Civile
- f: Professeur Certifié
- f: Professions Commerciales
- f: Protection Civile
- f: Puissances Contractantes
- l: Post Cibum
- r: Plutonchef
- s: Panadería de Campaña
- s: Partido Catolico
- s: Preparación Combativa
- s: Puesto Costero
- s: Puesto de Control
- s: Puesto de Mando
- s: Pulgada Cuadra

pc e: after meals
Pc e: Photocopy
- f: Parc
- f: Pas coté

P.C. e: Privy Counselor
p.C. s: post Christum
p/c s: por cuenta
PCA e: Packaging Council of Australia
- e: Panama Canal Authority
- e: Parallel Channel Adapter
- e: Policy Certification Authority
- e: Polycycliche Aromaten
- e: Porsche Club of America
- e: Power Calibration Area
- e: Power Controller Assemblies
- e: Presbyterian Church in America
- e: Pressure Control Assembly
- e: Primary Control Assembly
- e: Prime Candidate Alloy
- e: Principal Component[s] Analysis
- e: Principle Component[s] Analysis [technique]
- e: Procedure Change Authorization
- e: Process Change Authorization
- f: Police des Communications Aériennes
- f: Pool Charbon-Acier
- f: Poste de Contrôle Administratif
- f: Prix Courant Autorisé
- s: Proyectos Comunitarios Abaj

PC&A e: Property Classification and Accounting
PC/ABS e: Polycarbonate/Acrylonitrile Butadiene Styrene Blend
PCAC e: Personal Computer Acquisition Contracts
- e: Primary Control and Analysis Center

PCACIAS e: Personal Computer Automated Calibration Interval Analysis System
PCAD e: Packaging Computer-Aided Design
- e: Programs in Computer Applications Development

PCALT s: Puesto de Mando Alternativo
PCAPS e: Plant Control and Protection System
PCARS e: Partially Coherent Anti-Stokes Raman Scattering
PCAS e: Personnel Cost Accounting System
PCASS e: Parts Control Automated Support System
PCAST e: President's Council of Advisors on Science and Technology
- e: Presidents Committee of Advisors on Science and Technology
- e: Presidents Council of Advisors on Science and Technology

PC-AT e: Personal Computer-Advanced Technology
PC/AT e: Personal Computer/Advanced Technology
PCAV s: Puesto de Mando Avanzado
PCB e: I-GOOS Panel on Capacity Building
- e: Polychlorinated Biphenols
- e: PolyChlorinated Biphenyl[s]
- e: Possibly Carcinogenic Substance
- e: Pressure Core Barrel
- e: Printed Circuit Board
- e: Program Change Board
- e: Proposed Committee Bill
- e: Protocol Control Block
- f: Groupe de l'I-GOOS sur le renforcement des capacités
- f: Parti Communiste Belge
- s: Panel del I-GOOS sobre Aumento de Capacidades

Pcb f: Praticable
PC&B e: Personnel Compensation and Benefits
PCBC e: Plain Cipher Block Chaining
- e: Pretoria Consumer Boycott Committee
- e: Propagating Cipher Block Chaining

PCBS e: Portable Cascade Bottle System
PCC d: Papas Computer-Club
- e: PC Cluster Controller
- e: Peoples Computer Company
- e: Plastic Chip Carrier
- e: Policy Coordination Council
- e: Portable C Compiler
- e: Power Consumption Charge
- e: Primary Command Center
- e: Production Control Center
- e: Professional Computer Corporation
- e: Program Coordinating Centre
- e: Project Change Control
- e: Project Coordination Committee
- f: Pour Copie Conforme
- s: Programa de Capacitación Campesina [San Marcos]
- s: Punto de Concentración de [la] Correspondencia

PCCa e: Pheochromocytoma
PCCAL e: Pharmacy Consortium for Computer Aided Learning
PCCH e: Physical Control Channel
- s: Partido Comunista Chino

PCCIE e: Power Conditioning and Continuation Interfacing Equipment
PCC/OES e: Policy Coordinating Committee for Oceans and International Environmental and Scientific Affairs
PCCS e: PACOM Command and Control System
- e: Personnel Command and Control System
- e: Project Change Control System

PCCTV e: Portable Closed-Circuit Television
PCCW e: Public Citizens Congress Watch

PCD e: Pacific Communications Division
- e: Payload Correction Data
- e: Phase-Change Drive
- e: Photo CD [Compact Disc (or Disk)]
- e: Postcode
- e: Preliminary Conceptual Design
- e: Prime Contractor Department
- e: Principal Critical Document
- f: Parti Communiste du Dahomey
- f: Parti de la Convergence Démocratique
- f: Plate-forme de collecte de données
- f: Poste de Contrôle Douanier
- s: Punto Cero Deseado
- s: Punto de Control Dosimétrico

PCDA e: Program Controlled Data Acquisition
PCDC e: Professionals Committee for Democratic Change
PCDD d: Polychlorierte Dibenzodioxine
- e: Polychlorinated Dibenzodioxin

PCDE e: Photon-induced Chemical Dry Etching
PCDF d: Polychlorierte Dibenzofurane, Dioxin
- e: Polychlorinated Dibenzofuran

PCD-GR p: Partido de Convergência Democrática-Grupo de Reflexão
PCDM f: Parti Centre Démocratique Mauritanien
PC[-]DOS e: Personal Computer DOS [Disk Operating System]
PCDR e: Preconceptual Design Report
- e: Preliminary Conceptual Design Report

PCDS e: Pilot Climate Data System
PCE e: Perchloroethene
- e: Phase Change Erasable
- e: Physical Completion Estimate
- e: Pollution Control Equipment
- e: Preliminary Closure Estimate
- e: Present Company Excluded
- e: Process Control Engineering
- e: Professional Continuing Education
- e: Project Cost Estimate

Pce f: Pièce
PCEB e: PCI to EISA Bridge
PCEC e: Personal Computer Enhanced Connectivity
PCEE e: Presidents Commission on Executive Exchange
PCE/RG f: Programme Coopératif Européen pour la conservation et l'échange des Ressources Génétiques
Pces f: Pièces
PCESP e: Program Communication Electronic Support Program
Pcesse f: Princesse
PCF e: Personal Computing Facility
- e: Plutonium Chemistry Facility
- e: Point Coordination Function
- e: Portable Compiled Font
- e: Power Coupling Factor
- e: Procedure Completion Form
- e: Process Control File
- e: Proposed Change File
- f: Parti Communiste Français

PCFC e: Polytechnics and Colleges Funding Council
PCFIS e: PACAF Combat Fuels Information System
PCFS e: Pacific Coast Fertility Society
PCG e: Preconditioned Gradient
- e: Protein Crystal Growth experiment
- f: Parti Communiste de Grèce
- f: Parti Communiste Guadeloupéen
- s: Partido Comunista de Guatemala

P&CG e: Planning and Coordination Group
PCGG e: PCG Glovebox
PCGM e: Preconditioned Gradient Method
PCGRAFAX e: PC-based High Frequency Facsimile System
PCGRIDS e: Personal Computer Gridded Interactive Display and Diagnostic System
PCGS e: Professional Coin Grading Service
PCH d: Parchim, Mecklenburg-Vorpommern
- e: Packaging, Crating and Handling
- e: Paging Channel
- e: Parallel Channel
- s: Proyectil de Carga Hueca

PCh f: Permis de Chasse
PCHB e: Pollution Control Hearings Board
P[-]Chem e: Physical Chemistry
PCI e: Patent Citation Index
- e: Peripheral Component Interconnect
- e: Peripheral Component Interface
- e: Peripheral Connection Interface
- e: Phase-Conjugate Interferometry
- e: Program Controlled Interruption
- e: Programmable Communication Interface
- e: Protocol Control Information
- f: Parti Communiste Italien
- f: Parti Communiste Ivoirien
- f: Permis de Conduire International
- f: Prestations Collectives Indivises
- s: Proyecto Concern International Guatemala

pCi e: picocurie
PCIAC e: Petro-Canada International Assistance Corporation
PCIC e: PC-Card Interrupt Controller
PCIE e: Presidents Council of/on Integrity and Efficiency in Government
PC-III e: Personnel Concept III
PCI LBS e: Peripheral Component Interconnect Local Bus Specification
PC-I/O e: Program Controlled I/O
PCIO e: PC Information Officer
- e: Program Controlled I/O

PC I/O channel e: Personal Computer Input/Output channel
PCIOS e: Processor Common Input/Output System
PCIP e: Permitting Compliance Implementation Plan
PCIPS e: Personal Computer Image Processing System
Pcirc f: Permis de circulation
PCIS e: Proposals and Contracts Information System
PCI/SWCI e: Product Configuration Item /Software Configuration Item
PC[/]IXF e: Personal Computer Intergrated Exchange Format
PCJ e: Planning Commissioners Journal
- f: Parti Communiste Japonais

PCL e: Partial Core Loading
- e: Peripheral Control Line
- e: Plasma Cell Leukemia
- e: Portable Common Loops
- e: Positive Control Launch
- e: Printer Control Language
- e: Procedure Control Language
- e: Process Control Laboratories (or Laboratory)
- e: Process Control Language
- e: Product Computing module Load
- e: Programmable Command Language
- f: Parti Communiste Luxembourgeois
- s: Puesto de Clasificación
- s: Puesto de Control de Líneas

PCLDI e: Prototype Closed-Loop Development Installation
PCLM s: Patronato Contra la Mendicidad
PCLS e: Prototype Closed-Loop System
PCLT e: Prototype Closed-Loop Test
PCM e: Packet Compress when Matched bit in DCR2
- e: Personal Computer Manufacturer
- e: Personnel Contamination Monitor
- e: Phase Comparison Monopulse
- e: Phase Contrast Microscopy
- e: Phase-Conjugate Mirror
- e: Physical Connection Management
- e: Planar Camping Model
- e: Plug Compatible Manufacturers
- e: Plug-Compatible Mainframe
- e: Power Center Multiplexer
- e: Power Coolant Mismatch
- e: Printer Cartridge Metric
- e: Process Control Manual
- e: Proposed Corrective Measures
- e: Puls-Code-Modulation
- f: Parti Communiste Malgache
- f: Parti Communiste Marocain
- f: Programme Climatologique Mondial
- s: Puesto Central de Mando

Pcm f: Pour cent mille
PCma e: Plasmacytoma
PCMAS e: Portable Computer-Based Maintenance System
PCMC e: PCI, Cache, Memory Controller
PCMCIA e: PC Memory Card International Association
- e: People Can't Memorize Computer Industries Acronyms

PCMD e: Procurement and Contracts Management Division
PCMDI e: Program for Climate Model Diagnosis and Intercomparison
PCMI e: Pellet-Cladding Mechanical Interaction
- e: Presidents Council on Management Improvement

PCMIA e: PC Memory-card International Association
PCMIM e: Personal Computer Media Interface Module
PCML e: Pseudo Current Mode Logic
- f: Parti Communiste Marxiste-Leniniste

PC-MOS e: Personal Computer-Modular Operating System
PCMOV s: Puesto de Mando Móvil
PCMP e: Post Chemical-Mechanical Polishing

PCMS e: Plasma Chemistry Monte-Carlo Simulation
- e: Project and Configuration Management System
PCN e: Personal Communication[s] Network
- e: Policy Criteria Notice
- e: Production Control Number
- e: Program Change Number
- e: Public Communications Network
- f: Programme Climatologique National
- p: Parti de Convention Nationale, Partido da Convenção Nacional
PCNA e: Proliferating Cell Nuclear Antigen
PCNE e: Protocol Converter for Non-SNA Equipment
PCNFS e: Personal Computer Network File System
PCNM e: Packet Compress when Not Matched bit in DCR2
PCO e: Parent Company
- e: Peak Conservation Organization
- e: Peg Count and Overflow
- e: Photocatalytic Oxidation
- e: Point of Control and Observation
- e: Polycystic Ovaries
- e: Polycystic Ovary disease
- e: Principle Contracting Officer
- e: Process Change Order
- e: Procuring Contracting Officer
- e: Professional Conference Organiser
- e: Program Contracting Officer
- s: Proyectil Cohete
pCO_2 e: partial pressure of Carbon dioxide
PCOD e: Polycystic Ovarian (or Ovary) Disease
P-Code e: GPS precision transmission code
PCOM e: Planning Committee
PCOMP e: Packet Compression pin
PCON e: Potential Contractor
- e: Primary Care Optometry News
PCOS e: Polycystic Ovarian Syndrome
PCP d: Pentachlorphenol
- e: Parker Consultant Panel
- e: Pentachlorophenol
- e: Pentachlorophenyl
- e: Phenylcyclidine
- e: Pneumocystis Carinii Pneumonia
- e: Primary Care Physician
- e: Primary Care Provider
- e: Process Control Package
- e: Process Control Plan
- e: Program Change Proposals
- e: Program Control Program
- e: Programmable Data Processors
- s: Puesto de Control de Paso
PCPC e: Personal Computers Peripheral Corporation
- s: Prevención de Catástrofes y Protección Civil
PC-PLC e: Phosphatidylcholine-specific Phospholipase C
PCPM e: Per Contract Per Month
PCPP e: Parts Control Program Plan
- f: Prévention des Catastrophes et Planification Préalable
PC Pral s: Puesto de mando Principal
PCR e: Peak Cell Rate
- e: Phase Change Recording
- e: Physician Contingency Reserve
- e: Polymeran Chain Reaction
- e: Primary Control Rod
- e: Prime Contractors Region code
- e: Principle Component Regression
- e: Problem/Change Report
- e: Procedure Comment Record
- e: Processor Configuration Register
- e: Procurement Center Representative
- e: Product Change Request
- e: Program Clock Reference
- e: Projected Communications Requirement
- e: Protein Catabolic Rate
- f: Parti Communiste Réunionnais
- f: Parti du Centre
- s: Punto Cero Real
PC Ret[r] s: Puesto de mando Retrasado
PCRF e: Parallel Computing Research Facility
PCR-RFLP e: Polymeran Chain Reaction-Restriction Fragment Length Polymorphism
PCR-SSCP e: Polymeran Chain Reaction-Single-Strand Conformation Polymorphism
PCRT s: Puesto de mando Retrasado
PC Rt Div s: Puesto de mando Retrasado Divisonario
PCRV f: Parti Communiste Révolutionnaire Voltaïque
PCS e: Parts Collection Survey
- e: Patchable Control Store
- e: Payload Correction Subsystem
- e: PC Service
- e: PCL State
- e: Permit Compliance System
- e: Personal Communication[s] Service[s]
- e: Personal Communication[s] System
- e: Personal Conferencing Specification
- e: Physical Coding Signalling
- e: Physical Control Space
- e: Pinball Construction Set
- e: Plastic Coding System
- e: Port Concentrator Shelf
- e: Power Center Substation
- e: Pressure Core Sampler
- e: Primary Cooling System
- e: Process Control Systems
- e: Production Cost Savings
- e: Program Coordination Staff
- e: Programmable Drive Table
- e: Programme on Chemical Safety
- e: Project Control System
- e: Promotion of Chemical Safety
- e: Protected Cable System
- f: Parti Chrétien Social
- f: Parti Communiste Sénégalais
- s: Partido Cristiano Socialista
- s: Punto de Control y Selección
Pcs f: Pièces
PCS-1900 e: Personal Communication Service in 1.9 GHz band
PCSA e: Personal Computer (or Computing) System Architecture
PCSC e: Pacific Coast Science Center
- e: PC Site Coordinator
PCSD e: Presidents Council on Sustainable Development
- p: Partido de la Coalición Social Demócrata
PCSMS e: PACAF Combat Supplies Management Subsystem
PCSN e: Private Circuit Switching Networks
PCSO e: Presidential Communications Support Office
PCSP e: Preliminary Cruise Sampling Plan
- e: Program Communications Electronic Support Program
PCS/PCN e: Personal Communications Services/Networks
PCSSD e: Philippines Council for Sustainable Development
PCT e: Parity, Charge conjugation, Time-reversal
- e: Partition Control Table
- e: Patent Co-operation Treaty
- e: PCL Transition
- e: Pest Control Team
- e: Photometric Calibration Test
- e: Polycyclohexyl[ene] Terephthalate
- e: Post Coital Test
- e: Principal Component Transform
- e: Private Communication[s] Technology
- e: Process Change Teams
- e: Product Consistency Test
- e: Pulse Compression Test
- f: Parti Communiste Togolais
- f: Parti Communiste Tunisien
- f: Parti Congolais du Travail
- f: Programme concernant les Cyclones Tropicaux
- s: Programa de Ciclones Tropicales
- s: Puesto Central de Tiro
PCt f: Prix Courant
- f: Prix Coûtant
pct N: prosent
Pct f: Pour cent
pct. a: procent
PCTA e: PC Text-Assist
- e: Personal Computer Terminal Adapter
- e: Plastic-Cased Telescoped Ammunition
PCTC e: Panama Canal Transition Commission
- e: Pentagon Consolidated Telecommunications Center
PC[/]TCP e: Personal Computer/ Transmission Control Protocol
PCTE e: Portable Common Tool[s] Environment
PCTF e: Plant Components Test Facility
- e: Process Condensate Treatment Facility
PCTFE e: Polymonochlorotrifluoro-ethylene
PCT-G e: Glycol modified Polycyclohexyl Terephthallate
PCTMSL e: Permanent Committee on Tides and Mean Sea-Level
- f: Comité permanent sur les marées et le niveau moyen des mers
- s: Comité Permanente sobre Mareas y Nivel Medio del Mar
PCTP e: Plant Completion Task Proposal
PC TR s: Punto de Control de Tránsito

PCTR e: Physical Constants Testing Reactor
- e: Protocol Conformance Test Report
PC/TS e: Performance Criteria/Test Standard
PCTs e: Poly-Chlorinated Triphenyls
PCTS e: POSIX Conformance Test Suite
PCTSCM e: Piezoelectric, Continuously Twisted, Structurally Chiral Medium
PCTV e: Personal Computer Television
- e: Program Controlled TransVerters
PCU e: Platinum Cobalt Unit
- e: Premises Control Unit
- e: Programme Coordination Unit of the Black Sea Environmental Programme
- f: Unité de coordination du Programme de l'environnement de la mer Noire
- p: Posto de Comando Unificado
- s: Unidad de Coordinación del Programa del Medio Ambiente del Mar Negro
PCUCDGIS e: Personal Computer University College Dublin Geographic Information System
PCUS f: Parti Communiste de l'Union Soviétique
- p: Partido Comunista da União Soviética
- s: Partido Comunista de la Unión Soviética
PCV e: Peace Corps Volunteer
- e: Positive Crankase Ventillation [system]
- e: Process Cell Ventilation
- f: Paiement Contre Valeur
- f: Programme de Coopération Volontaire
- s: Programa de Cooperación Voluntaria
PCV[ES] f: Programme d'Équipement et de Services
- s: Programa de Equipo y Servicios
PCV[F] f: Fonds de Coopération Volontaire
- s: Fondo de Cooperación Voluntaria
PCV[FEM] f: Programme de Formation Technique, d'Équipement et de Maintenance
PCVIL s: Protección Civil
PCVS e: Point-to-point Switched Virtual Connections
- e: Professional Credentials Verification Service
PCV[TEM] s: Programa de Formación Técnica, Equipos y Mantenimiento
PCW e: Post Consumer Waste
PCWG e: Personal Conferencing Work Group
PCX d: PCX ist das native Format des Zsoft Corporation PC Paintbrush TM Malprogramms
- e: Picture Exchange
- e: Picture Image
PC-XT e: Personal Computer-Extended Technology
PCY e: Country code for the Prime Contractors county
PCYLIHOONL e: Please Cross Your Legs, I Have Only One Nail Left [Internet]

PD e: Packet Driver
- e: Packetization Delay
- e: Panic Disorder[s]
- e: Payload Developer
- e: Per Diem
- e: Peripheral Decoder
- e: Peritoneal Dialysis
- e: Permits Division
- e: Perphenazine Decanoate
- e: Personality Disorder[s]
- e: Phase Change Dual
- e: Phase Detection
- e: Phase Distortion
- e: Phase-change Dual
- e: Photo-Detection (or Detector)
- e: Photodesorption
- e: Physical Disability
- e: Physically Disabled
- e: Planar-Doped
- e: Plant Deficiency
- e: Position Description
- e: Position Document
- e: Press Division
- e: Priority Directive
- e: Probability of Damage
- e: Probability of Destruction
- e: Procedures Description
- e: Procurement Document
- e: Program Director
- e: Program Document
- e: Project Description
- e: Project Development
- e: Project Directive
- e: Project Director
- e: Propagation Delay
- e: Protocol Driver
- e: Public Domain
- e: PUREX NCRW solids or sludge remaining after decanting
- f: Paiement Différé
- f: Parti Démocratique
- f: Partie Défaillante
- f: Partie Double
- f: Passavant Descriptif
- f: Passe-Debout
- f: Passeport Diplomatique
- f: Peine Disciplinaire
- f: Pension Définitive
- f: Personnel Domestique
- f: Personnes Déplacées
- f: Petite Distance
- f: Point de Départ
- f: Port dû
- f: Poste de Distribution
- f: Pouvoir Discrétionnaire
- f: Premier Degré
- f: Prescription Décennale
- f: Président Désigné
- f: Prime de Déménagement
- f: Prise Directe
- f: Prix de Détail
- f: Puy-de-Dôme
- r: Pansardivision
- s: Partido Democrático
- s: Pieza Directriz
- s: Punto de Distribución
Pd f: Point de distance
pd. a: pund
p.d. r: på dagen
P&D e: Price and Delivery

P/D e: Pitch-to-Diameter
PDA d: Partei der Arbeit
- e: Personal Digital Assistant[s]
- e: Phased-Doppler Anemometer
- e: Polydiacetylene
- e: Private Doctors of Australia
- e: Probabilities Decision Analysis
- e: Procedure Departure Authorization
- e: Procurement Defense Agency
- e: Public Display of Affection
- e: Public Displays of Alcohol[ism]
- e: Push Down Automaton
- p: Partido Democrático Angolano
PDAD e: Proposed Draft Addendum
PDA J Pharm Sci Technol e: PDA Journal of Pharmaceutical Science and Technology
PDAM e: Proposed Draft Amendment
PDAR e: Product Disposition Authorization Report
PDAS e: Paid Associate
- e: Photodiode Array Spectrophotometer
- e: Physiological Data Acquisition System
p.d.a.s. a: på den anden side
PDAU e: Physical Data Access Unit
PDB e: Pee Dee Belemnite isotope standard for carbon
- e: Planar-Doped Barrier
- e: Process Data Base
- e: Product Description Block
- e: Protein Data Bank
PDC e: Center for Parallel Computers
- e: Packaging Design Criteria
- e: Participatory Design Conference
- e: Passive Data Collection
- e: Peoples Defense Committee
- e: Personal Digital Cellular
- e: Polycrystalline Diamond Compact drill bit
- e: Power Demand Charges
- e: Power Disc Cartridge
- e: Primary Development Contract
- e: Primary Domain Controller
- e: Professional Developers Conference
- e: Programming, Design and Construction [system]
- e: Project Design Criteria
- e: Proteus Digital Channel
- e: Pulse Discharge Cleaning
- f: Parti Démocrate Chrétien
- f: Parti Démocrate Chrétien Suisse
- f: Parti Démocratique Congolais
- f: Parti des Démocrates Camerounais
- f: Pas-de-Calais
PD&C e: Plug Drive and Control
PDCC s: Proyecto de Detección del Cambio Climático
PDCH e: Physical Data Channel
PDCI e: Product Data Call-In
- f: Parti Démocratique de la Côte d'Ivoire
PDCI-RDA f: Parti Démocratique de la Côte d'Ivoire-Rassemblement Démocratique Africaine
PDCM e: Positive Displacement Coring Motor
- f: Parti Démocrate Chrétien de Madagascar
PDCN s: Patido Democrático de Cooperación Nacional

PDD e: Patients Dental Dictionary
- e: Pervasive Developmental Disorder
- e: Phorbol-12,13-Didecanoate
- e: Physical Device Driver
- e: Portable Digital Document
- e: Presidential Decision Directive
- e: Process Distillate Discharge
- e: Product Delivery Device
- e: Program Development Department
- e: Proposal Due Date
- e: Protable Digital Document
- f: Plan de Développement Durable
- s: Posición Defensiva Divisionaria

PDDA s: [unidad de] Proyectiles Dirigidos de [para] la Defensa Area
PDDF e: Portable Document Delivery Format
- e: Post-Decision Document File

PDDI e: Product Definition Data Interface
PDDM f: Petit Déjeuner Du Matin
PDDP e: Parallel Data Distribution Preprocessor
PDDU e: Process Development and Demonstration Unit
PDE e: Partial Differential Equation[s]
- e: Particle Dispersion Experiment
- e: Phosphodiesterases
- e: Photon Drag Effect

P-de-C f: Pas-de-Calais
P-de-D f: Puy-de-Dom
PDED e: Program Development and Evaluation Division
PDEF e: Protean Data Exchange Format
PDES e: Product Data Exchange Specification[s]
- e: Product Data Exchange Standard
- e: Product Data Exchange using STEP
- e: Product Description Exchange Standard

p.d.e.s. a: på den ene side
PDESO f: Projet de Développement de l'Elevage au Sénégal-Oriental
PDF e: Package Definition File
- e: Parkinsons Disease Foundation
- e: Patient Data Form
- e: Polar Density Function
- e: Portable Document File
- e: Portable Document Format
- e: Power Distribution Frame
- e: Precision Direction Finder
- e: Pretty Damn (or Darn) Fast [Internet]
- e: Primary Direction Of Fire
- e: Printer Description
- e: Probability Density Function
- e: Processor Defined Function
- e: Program Development Facility
- s: Plano Director de Fuego

PDFID e: Preconcentration Direct Flame Ionization Detection
- e: Preconstruction Direct Flame Ionization Detection

PDFo d: Prodekan für Forschung und Wissenschaftsentwicklung
PDFT e: Programmable Data Formatter
PDG e: Pendant Drop Growth
- f: Parti Démocratique de Guinée
- f: Parti Démocratique Gabonais
- f: Président-Directeur Général
- s: Partido Democrata Guatemalteco

PDGE p: Partido Democrático de Guinea Ecuatorial
PDH e: Plesiochronous Digital Hierarchy
- e: Project Development History
- s: Procuraduría de Derechos Humanos

PDI d: Paul-Drude-Institut für Festkörperelektronik, Berlin
- e: Packet Driver Interface
- e: Personal Data Interchange
- e: Power and Data Interface
- e: Pressure Differential Indicator
- e: Product Data Interchange
- f: Parti Démocratique Islamique

PDIAL e: Public Dialup Internet Access List
PDIC e: Professional Diving Instructors Corp.
PDII in: Pusat Dokumentasi dan Informasi Ilmiah
PDIP e: Plastic Dual In-line Package
- e: Program Development Increment Package

pDISP e: Proposed Draft International Standardized Profile
Pdiv r: Pansardivision
PDJ f: Parti Démocratique de Justice
PDL e: Page Description Language
- e: Page Design Language
- e: Permanent Duty Location
- e: Polarization Diversity Lidar
- e: Primary Defect List
- e: Program Description Language
- e: Sierra Leone Peoples Democratic League
- s: Ponce De Leon

PDLA p: Partido Democrático Liberal de Angola
PDLC e: Polymer Dispersed Liquid Crystal
- e: Polymer-Dispersed Liquid Crystal

PDLD e: Polymer-Dispersed Liquid-crystal Display
PDLE e: Process Development Laboratory East
PDLM f: Direction des Pêches Lagunaires et Maritimes
PDLW e: Process Development Laboratory West
PDM d: Produktdatenmanagement
- d: Projekt- und Datenmanagement
- d: Pulsdauer-Modulation
- e: Parlance Document Manager
- e: Physical Medium Dependent
- e: Planned Depot Maintenance
- e: Positive Displacement [drilling] Motor
- e: Probabilistic Dilution Model
- e: Product and Document Management
- e: Product Data Management
- e: Program Decision Meeting
- e: Program Decision Memorandum
- e: Program Development Manager
- e: Programmable Data Monitor
- e: Programmed Depot Maintenance
- f: Progrès et Démocratie Moderne

PDMP e: Project Data Management Plan
PDMS e: Payload Data Management System
- e: Pesticide Document Management System
- e: Plasma-Desorption Mass Spectrometry
- e: Point-Defense Missile System
- e: Program Data and Management System

PDN e: Physics Data Notebook
- e: Pre-Discharge Notification
- e: Procedure Departure Notice

PDNA p: Partido Democrático Nacional Angolano
PDO e: Packet Data Optimized
- e: Plasma-Deposited Oxide
- e: Portable Distributed Object[s]
- e: Procedure Description Overview
- e: Publications Distribution Office

pdo. s: pasado
PDOS e: Publishing Distribution Office System
- e: Purple Dinosaur Operating System

PDP e: DEC Computer System Designation
- e: Pan-African Democratic Party
- e: Parallel Distributed Processing
- e: Pasture Development Plan
- e: People Data Planning
- e: Peoples Democratic Party
- e: Power-Delay Product
- e: Preconceptual Design Phase
- e: Professional Developers Program
- e: Professional Development Program
- e: Progammable Data Processor
- e: Program Decision Package
- e: Project Data Package
- f: Parti pour la Démocratie et le Progrès
- p: Partido Democrático de Progresso
- p: Partido Democrático para o Progresso

PDPA p: Partido Democrático Pacífico Angolano
PDP-ANA p: Partido Democrático para o Progresso-Aliança Nacional Angolana
PDPC p: Partido Democrático do Povo Cabinda
PDPF e: Project Data Processing Facility
PDPL f: Programme pour le Développement de la Production Laitière
PDPS e: Planning and Data Production System
- e: Project Data Processing System

PDPU s: Posición Defensiva de Pequeña Unidad
PDQ e: Parallel Data Query
- e: Personal Data Query
- e: Personality Diagnostic Questionnaire
- e: Personality Disorders Questionnaire
- e: Physician Data Query
- e: Position Description Questionnaire
- e: Pretty Damn[ed] Quick
- e: Pretty Darn[ed] Quick

PDR e: Packet Drop Rate
- e: Particulate Data Reduction
- e: Peninsula Development Road
- e: Phase-Delay Rectifier
- e: Photon-Dominated Region
- e: Physicians Desk Reference
- e: Pitch-to-Diameter Ratio
- e: Previous Document Reference
- e: Procedure Development Request
- e: Procurement Data Reporting
- e: Property Disposal Request
- f: Parti pour la Démocratie et le Renouveau
- f: prime de Déménagement et de Réinstallation

P/DR e: Performance/Development Review

PDRC e: Positron Diagnostic Research Center
- e: Program Development Review Committee
PDRCA e: Participatory Development Resource Centre for Africa
PDRE e: People's Democratic Republic of Ethiopia
PDRM e: Post-Depositional Remanent Magnetization
PDRP e: Program Data Requirement Plan
- e: Program Document Requirement Plan
PDRR e: Program Definition and Risk Reduction
PDRS e: Payload Deployment [and] Retrieval System
PDRY e: Peoples Democratic Republic of Yemen
PDS d: Partei der Sozialisten
- d: Partei des Demokratischen Sozialismus
- e: Pacific Distribution System
- e: Packaging Data Sheet
- e: Packet Driver Specification
- e: Partitioned Data Set
- e: Passenger Documentation System
- e: Patient Distribution System
- e: Persistent Data Services
- e: Persistent Data Store
- e: Personnel Decontamination Station
- e: Philips Development System
- e: Photo Densitometer System
- e: Physicians Data Query
- e: Planar Diffusion Source
- e: Planar Dopant Source
- e: Planetary Data System
- e: Plant Data System
- e: Portable Display Shell
- e: Portable Document Software
- e: Positional Data System
- e: Premise Distribution System
- e: Processing and Display System
- e: Processor Direct Slot
- e: Program Design Specification
- e: Program Distribution System
- e: Protected Distribution System
- e: Public Domain Software
- f: Parti Démocratique Sénégalais
- f: Parti Démocratique Social
- f: Parti Démocratique Socialiste
- f: Parti pour le Développement Solidaire
PDSA e: Paid Scientific Associates
- e: Peroxydisulfuric Acid
PDSAR e: Public Document Status of Assessment Report
PDSC e: PACOM Data Systems Center
- f: Parti Démocrate et Social Chrétien
PDSF e: Parallel Distributed Simulation Facility
PDSI e: Palmer Drought Severity Index
PDSL e: Pressure Differential Switch Load
PDSP e: PACOM Data System Program
- e: Peripheral Data Storage Processor
PDS-R f: Parti Démocratique Sénégalais-Renouveau
PDSS e: Post Development [and] Software Support
- e: Post Development Support System
- e: Program Development Support System
PDST e: Personnel Data System for Training
PDSt d: Prodekan für Studienangelegenheiten
PDSU e: Plant Data System Upgrade
PDSZ e: Party of Convinced Social Democrats of Zaire
PDT d: Produktionsdatentechnik
- e: Performance Diagnostic Tool
- e: Procurement Data Transmittal
- e: Professional Development and Teaching
- e: Programmable Drive Table
- e: Propagation Delay Time
- f: Parti Démocratique Togolais
- s: Plano Director
Pdt f: Pendant
- f: Président
PDT. e: Printer Definition Table
PDTE e: Packet-mode DTE
pDTR e: Proposed Draft Technical Report
PDU e: Packet Data Unit
- e: Philippines Digital Upgrade
- e: Plug Distribution Unit
- e: Product Development Unit
- e: Protocol Data Unit[s]
- f: Parti Démocratique Unifié
- f: Parti des Démocrates pour l'Unité
PDUS e: Primary Data User[s] Station[s]
PDV e: Path Delay Value
- e: Peculiar Dorsal Vertebra[e]
P.D.V. a: På Doreltpratets Vegne
PDVC e: Phase-Dependent Voltage Contrast
PDWE e: Pulse Detonation Wave Engine
PDWS e: Passive Doppler Wind Sensor
PDX e: MicroSoft PC Paint Brush
- e: Paradox files
- e: Printer Description Extension
PE e: Parity Error
- e: Parity Even
- e: Percent bit Error probability
- e: Performance Evaluation
- e: Perkin-Elmer
- e: Personnel Division
- e: Phase Encoded
- e: Physical Education Building
- e: Plasma Etch[ing]
- e: Plutonium Equivalent
- e: Pneumatic E-something
- e: Polyethylen
- e: Portable Error
- e: Potential Electrode
- e: Power Exchange
- e: Precipitation Efficiency
- e: Preliminary Edition
- e: Pressurized Element
- e: Printers Error
- e: Process Engineering
- e: Processor Element
- e: Procurement Executive
- e: Proposed Endangered
- e: Protect[ion] Enable[d]
- e: Prototype Event
- e: Pulmonary Embolus
- f: paiement en espèces
- f: Palais de l'Elysée
- f: Papiers d'Etat
- f: Par Express
- f: Parties Égales
- f: Permanence Électorale
- f: Permis d'Exportation
- f: Petite Entreprise
- f: Petits-Enfants
- f: Police Économique
- f: Poste Extérieur
- f: Pouvoir Exécutif
- f: Presse Étrangère
- f: Prestation en Espèces
- f: Prime d'Emménagement
- f: Prime d'Exportation
- f: Problèmes Économiques
- f: Promesse Écrite
- f: Propriété d'Etat
- f: Protection de l'Enfance
- f: Provenance Étrangère
- f: Puissance Étrangère
- f: Pupille d'Etat
- s: Patrulla de Exploración
- s: Patrullero-escolta
- s: Puesto de Enganche
- s: Puesto Especial
pe e: Peru [Internet]
Pe f: Par exemple
PEA e: Pocket Ethernet Adapter
- e: Proposal Expansion Award
PEAB r: Philips Elektronikindustrier
PEACAMPOT e: Perturbation by East Asia Continental Air Mass to Pacific Oceanic Troposphere
PEACE e: Plasma Electron And Current Analyzer
- e: Protection of Environment for Assuring Cleaner Earth
- f: Protection de l'environnement pour une Terre plus saine
- s: Protección del Medio Ambiente para lograr una Tierra más Limpia
PEAM e: Personal Electronic Aid for Maintenance
PEARL e: Portable Environmental Assessment and Research Laboratory
- e: Processor and Experiment Automation Realtime Language
PEARS e: Porcine Epidemic Abortion and Respiratory Syndrome
PEAS e: Policy and External Affairs Staff
PEAS/SOAS e: PSYOP Effects and Analysis/SOAS
PE B e: Physical Education Building [New Gymnasium]
PEB d: Projekt Elektronisches Buch
- e: Partial EtchBack
- e: Post Exposure Bake
- f: Prêt Entre Bibliothèques
PEBA e: Polyester (or Polyether) Block Amide
PEBCO e: Port Elizabeth Black Civic Organization South Africa
PEC e: Parents Education Committee
- e: Perfect Electric Conductor
- e: Performance Evaluation Committee
- e: Physical Education Centre
- e: Pre-Existing Condition
- e: Primary Environment Care
- e: Program Execution Control
- e: Program Executive Council
- e: Protocol Engineering Center
- f: Parti pour l'Evolution des Comores
- f: [Société de] Potasse et Engrais Chimique
- p: Plano Estatal Central
- s: Patrulla de Exploración Combativa

PECDAR e: Palestinian Economic Council for Development And Reconstruction
PECH e: Polyepichlorohydrin
PECHIKA f: Pêche Industrielle de Kaloum
PE-CI e: Plutonium Equivalent Curies
PECI e: Productivity Enhanced Capital Investment
PECIG f: Pêcheries Industrielles Gabonaises
PECM e: Passive ECM [Electronic Countermeasures]
PECO e: CEC-COPERNICUS Program
- f: Parti des Ecologistes
- f: Pays de l'Europe Centrale et Orientale
PECOS e: Picture Enhancement Computer Operating System
PECOSY e: Purged Exclusive COSY [Correlation Spectroscopy]
PECP e: Preliminary Engineering Change Proposal
PECTS e: Performance Evaluation Commitment Tracking System
PECVD e: Plasma-Enhanced CVD
PED e: Physical Education
- e: Plastic-Encapsulated Device
- e: Professional Education
- e: Program Evaluation Division
- e: Protective Equipment Decontamination
- e: Purchase Early Development
- f: Pays En Développement
PE&D e: Plant Engineering and Design
PEDF e: Protective Equipment Decontamination Facility
Pediatr Allergy Immunol e: Pediatric Allergy and Immunology
Pediatr Ann e: Pediatric Annals
Pediatr Cardiol e: Pediatric Cardiology
Pediatr Clin North Am e: Pediatric Clinics of North America
Pediatr Dent e: Pediatric Dentistry
Pediatr Dermatol e: Pediatric Dermatology
Pediatr Emerg Care e: Pediatric Emergency Care
Pediatr Hematol Oncol e: Pediatric Hematology and Oncology
Pediatr Infect Dis J e: Pediatric Infectious Disease Journal
Pediatr Med Chir e: Pediatria Medica e Chirurgica
Pediatr Nephrol e: Pediatric Nephrology
Pediatr Neurol e: Pediatric Neurology
Pediatr Neurosurg e: Pediatric Neurosurgery
Pediatr Pathol Lab Med e: Pediatric Pathology and Laboratory Medicine
Pediatr Pol P: Pediatria Polska
Pediatr Pulmonol e: Pediatric Pulmonology
Pediatr Pulmonol Suppl e: Pediatric Pulmonology. Supplement
Pediatr Radiol e: Pediatric Radiology
Pediatr Res e: Pediatric Research
Pediatr Rev e: Pediatrics in Review
PEDIN e: Petroleum Exploration Data Index
PEDITOR e: Portable Editor
PEDL e: Prototype Equipment Development Laboratory
PedNSS e: Pediatric Nutrition Surveillance System
PEDS e: Plasma-Enhanced Deposition System
- e: Program Element Descriptive Summary
- e: PUREX/UO3 Effluent Data System
PED[V] e: Porcine Epidemic Diarrhea [Virus]
PEEK e: Partners Early Experience Kit
- e: Polyetheretherketone
PEELS e: Parallel Electron Energy Loss Spectrometry (or Spectroscopy)
PEEM e: Panel of Experts on Environmental Management for Vector Control
- f: Groupe d'experts de la gestion de l'environnement pour la lutte antivectorielle
- s: Grupo de Expertos sobre Ordenación del Medio Ambiente para la Lucha contra los Vectores
PEER e: Partnerships for Environmental Education and Research
- e: Performance Efficiency Evaluation Report
PEF d: persönliches Entwicklungs- und Förderseminar
PEFOS e: Program Evaluation and Field Operations Staff
PEFP f: Programme d'Enseignement et de Formation Professionnelle
- s: Programa de Enseñanza y Formación Profesional
PEFU e: Panel of Experts on Fish Utilization
- f: Groupe d'experts de l'utilisation du poisson
- s: Panel de Expertos de la Utilización del Pescado
PEG e: Polyethylene Glycol-1000, also Carbowax
- e: Program Evaluation Group
- e: Program Execution Guidance
- e: Project Execution Guidelines
- f: Parti des Ecologistes Guinéens
- s: Proyectil Especial de Gas
PEGAB f: Pêcheries Gabonaises
PEGASUS e: a current velocity profiling instrument (or system)
PEH e: [Programme for the] Promotion of Environmental Health
- s: Proyectil Especial de Humos
PEI e: Plant Engineering Introduction
- e: Polyetherimide
- e: Preparedness Exercise Issue
- e: Prince Edward Island
- s: Patrulla de Exploración Ingeniera
- s: Proyectil Especial Incendiario
PEIN e: Pan European Intelligent Network
PEINFS e: Pan European Intelligent Network Freephone Service
PEIS e: Programmatic Environmental Impact Statement
p. ej. s: por ejemplo
PEK e: Polyetherketone
PEKEKK e: Polyetherketone Etherketone Ketone
PEKK e: Polyetherketoneketone
PEL e: Payroll Expenditure Listing
- e: Permissible Exposure Level
- e: Permissible Exposure Limit[s]
- e: Personal (or Personnel) Exposure Limit
- e: Physics and Engineering Laboratory
- e: Picture Element
- e: Programmable Electronics Load
- f: Plan d'Épargne-Logement
- s: Proyectil Especial de Iluminación
Pel e: Pixel
- s: Pelotón
pel. N: pelotong
PELSS e: Precision Emitter Locator Strike System
PEM e: Partial Equilibrium Multimarket Model
- e: Particle Environment Monitor
- e: Particle Modulator Radiometer
- e: Payload Electronics Module
- e: Personal Exposure Model
- e: Plastic Epoxy Molded
- e: Polioencephalomalacia
- e: Privacy Enhanced Mail
- e: Privacy Enhanced Message
- e: Privacy Enhancement for Electronic Mail
- e: Privacy Enhancement for Internet Mail
- e: Privacy-Enhanced Mail
- e: Program and Environmental Management
- e: Properties/Events/Methods
- e: Proton Exchange Membrane
- f: Programme pour l'Environnement dans la Méditerranée
PEMARZA f: Société des Pêches Maritimes du Zaire
PEM-Central e: Pacific Exploratory Mission - Central
PEMD e: Personalized Educational Materials Development
PEMEX s: Petroleos Mexicanos
PEMG f: Première Expérience Mondiale du GARP
- s: Primer Experimento Mundial del GARP
PEMRC e: Program Executive Management Review Committee
PEMS e: Paris Evangelical Missionary Society
PEM WEST e: Pacific Exploratory Mission - West
PEM-West e: Pacific Exploratory Mission - West
PEN e: Pacific Exchange Network
- e: Peninsula Environmental Network
- e: Plasma-Enhanced silicon Nitride
- e: Poets, Playwrights, Editors, Essayists and Novelists
- e: Polyethylene Napthalene
- s: Proyectil Especial de Nieblas
Pen s: Planeador
Pén f: [code] Pénal
PENMS e: Pan European Network Management System
Penn e: University of Pennsylvania
Penn. e: Pennsylvania
PennSAHIC e: Pennsylvania Substance Abuse and Health Information
Penn. State e: Pennsylvania State University

PENPAL GREETINGS e: A virus warning that proagates itself by scaring well-meaning emailers into warning their friends
PENS e: Pediatric Endocrinology Nursing Society
- e: Pensacola Naval Air Station
Pens f: Pensions
pens. a: pension[at]
- a: pensioneret
- a: pensionist
PEN SDK e: Pen Computing Software Development Kit
Penta e: Pentachlorophenol
PENTAFLUX e: Fifth Flux Experiment
Pentium e: Produces Erroneous Numbers Through Incorrect Understanding of Mathematics
PEO e: Patrol Emergency Officer
- e: Patrol Emergency Operator
- e: Program Executive Office[r][s]
- e: Progressive External Ophthalmoplegiai
PEOC e: Presidential Emergency Operations Room
PEOLE e: Preliminary Eole
PEP d: Projekt Elektronisches Publizieren
- e: Packet Exchange Protocol
- e: Parkinsons Educational Program
- e: Plasma Exhaust Process
- e: Pole-Equator-Pole [transects]
- e: Precipitation Enhancement Program
- e: Preliminary Evaluation Plan
- e: Producibility Engineering and Planning
- e: Productivity Enhancement Program
- e: Productivity Enhancement Project
- e: Project Element Plan
- e: Protocol Extension Protocol
- e: Pupil Evaluation Program
- f: Plan d'Epargne Populaire
PEPAT e: Protocol on Environment Protection of the Antarctic Treaty
PEPCase e: Phosphenol-Pyruvate Carboxylase
PEPCI e: Providing Emergency Presidential Communications Interface
PEPE e: Pepe Est Presque Emacs
- e: Perkin-Elmer Performance Evaluation
- e: Persistant Elevated Pollution Episodes
- e: Prolonged Elevated Pollution Episode
PEPGB e: Plasma Exhaust Pumping Glove Box
PEPICO e: Photoelectron Photoion Coincidence
Pépre f: Pépinière
PEPS e: Preliminary Evaluation Programme for SPOT
- f: Programme d'Évaluation Préliminaire [de] SPOT
- s: Programa de Evaluación Preliminar de SPOT
PEPSY e: Scandinavian Pedagogical Publications Bibliography
Pept Res e: Peptide Research
PEQ e: Palestine Exploration Quarterly
PEQUOD e: Pacific Equatorial Ocean Dynamics [program]
- f: Dynamique de l'océan Pacifique équatorial
- s: Dinámica del Océano Pacífico Ecuatorial
PER d: Perleberg, Brandenburg
- e: Percussion
- e: Program for Ecological Research
- e: Program for Ecosystem Research
- e: Public Environment Report
- f: Plan d'Epargne Retraite
Per s: Periscopio [de submarino]
- s: Persecución
PER7 e: Region 7 Personnel System
PERA s: Proyectos Educativos Regionales de Autoayuda
PERC e: Preliminary Editorial Review Checklist
Percept Mot Skills e: Perceptual and Motor Skills
PERD e: Panel for Energy Research and Development
PERF e: Petroleum Environmental Research Forum
PERFECT e: Performance Evaluation for Cost Effective Transformations
- e: Productivity Erosion Runoff Functions to Evaluate Conservation Techniques
Perfect f: Perfectionnement
Performance d: Leistungsfähigkeit und Verarbeitungsgeschwindigkeit von Hard- und Software
Perfprof f: Perfectionnement professionnel
Perf Rot s: Perforadora Rotatoria
PERGS e: Portable Earth Resources Ground Station
perh. e: perhaps
PERI e: Periodicals
- e: Pharmaceutical Education and Research Institute
Perit Dial Int e: Peritoneal Dialysis International
PERL e: Pathologically Eclectic Rubbish Lister
- e: Practical Extraction and Report[ing] Language
- e: Pre-positioned Equipment Requirements List
PERM d: Programmgesteuerte Elektronische Rechenanlage TH München
perm N: permisjon
Perm s: Permanente
PERMARS e: MACOM Military Personnel Management Reports System
PERMDATA e: PERMDATA Management System
PERMIS e: Preference Management Information System
PERMIT e: Polar Orbiter Effective Rainfall Monitoring Integrative Technique
PERNNS f: Pôle d'Expertise et de Référence National des Nomenclatures de Santé
PEROM e: Programmable [and] Erasable Read-Only Memory
Perpét f: perpétuel
PERRY e: Perry Building
PERS e: ACD Personnel Subsystem
- e: ERL-Ada Personnel System
- e: Proactive Error Reduction System
Pers f: Persan
- f: Personne[l]
pers N: personell
pers. a: personer
- a: personlig[t]
- e: person[al]
PERSAS e: Personnel Accounting System
Pers. avd./Bilkol. 2 N: Person bilavdelingen ved Biltrenkolonne 2
P[ers]BefG d: Personenbeförderungsgesetz
PersBG d: Personalrechtliches Begleitgesetz zum Telekommunikationsgesetz vom 17.12.1997
PERSCO e: Personal Support for Contingency Operations
PERSCOM e: Personnel Command
PERSDEP e: Personnel Deployment Report
PERSGA e: Red Sea and Gulf of Aden Environment Programme
- f: Programme pour l'environnement de la mer Rouge et du golfe d'Aden
- s: Programa para el Medio Ambiente del Mar Rojo y del Golfo de Adén
PERSINSCOM e: Personnel Information System Command
PERSITREP e: Personnel Situation Report
PERSNET e: Personnel Communications Network
Perspect Biol Med e: Perspectives in Biology and Medicine
Perspect Dev Neurobiol e: Perspectives on Developmental Neurobiology
PERSPROP e: Personal Property Data Entry and Report System
Perstpftg r: Personaltransportfartyg
PersZulV d: Verordnung über die Personenzulassung zum Aufbauen, Anschalten, Ändern und Instandhalten von Telekommunikationsendeinrichtungen [Personenzulassungsverordnung]
PERT e: Program Evolution and Review Technique
PERTH e: Western Australian Herbarium International Acronym
PES e: PACOM Exercise Schedule
- e: Planning and Evaluation Staff
- e: Polyethersulfone
- e: Positioning Error Signal
- e: Processor Enhancement Socket
- e: Processor Upgrade Socket
- e: Proposed Encryption Standard
- e: Public Enquiry System
- f: Permanence d'Entraide Sociale
- f: Pôle Européen de Santé
PEs e: Phorbol Esters
Pes s: Pesado
PESA e: Percutaneous Epididymal Sperm Aspiration
PESC e: EMSL-Cinci Performance Evaluation System
PESCANOVA p: Empresa Mista Moçambicana-Espanhola
PESO e: Product Engineering Services Office
PE SS & RM e: Department of Physical Education, Sports Science and Recreation Management

PEST e: People for Environmentally Sustainable Transport
- e: Pesticide Evaluation Summary Tabulation System
- e: Region 7 Nebraska Pesticide
PESTAN e: Pesticides Analytical Transport Solution
PET a: Politiets Efterretningstjeneste
- d: Positronenemmisionstomographie
- e: Payroll Expense Transfer
- e: Peritoneal Equilibration Test
- e: Personal Electronic Transaction computer
- e: Polyethylene Terephthalate
- e: Poly[ethylenterephtalat]ester
- e: Positron Electron Tomography
- e: Positron Emission Tomographie
- e: Post-Etch Treatment
- e: Potential-Effect Transistor
- e: Print Enhancement Technology
- e: Progressive Educational Technology
- e: Prototype Evaluation Testing
- s: Proyectil Especial Trazador
- s: Proyectil Explosivo Trazador
Pet f: Petit
PETA s: Proyectil perforante Explosivo Trazador [con] Autodestruccin
Pet C f: Ponts et Chaussées
PETE e: Partnership for Environmental Technology Education
- e: Polyethylene Terephthalate
PETEOS e: Plasma-Enhanced Tetraethylosilicate
PET[-]G e: Glycol-modified Polyethylene Terephthalate
Pet I f: Principal et Intérêts
PETNAIS e: Pacific Essential Telecommunications Network and Information System
Pet P f: Profits et Pertes
pet po f: pour et par ordre
PETRANGOL p: Companhia de Petróleos de Angola
PETROCI f: Société Nationale d'Opérations Pétrolières de la Côte d'Ivoire
Petrogr. e: Petrography
Petrol. e: Petrology
PETROL RAM e: Petroleum Resource Automated Management
PETROMOC p: Empresa Nacional Petróleos de Moçambique
PETS e: Parameterised Executable Test Suite
PETT f: Professeurs d'Enseignement Technique Théorique
Pet T f: Postes et Télécommunications
PEV e: Pressure Equalization Valves
- f: Programme Elargi de Vaccination
PeV e: Parameter Equals Value
- e: Petaelectron Volt
PEWS e: Portable EDS Workstation
PEWSX e: PEWS Q-D Subsystem
PEX e: PHIGS Extension to X
Pex f: Par exemple
Pexc f: Par exception
PEX-SI e: PEX-Sample Implementation
Pext f: Par extension
PEYCO e: Port Elizabeth Youth Congress
PF a: Pressefotografforbundet
- e: Parity Flag
- e: Path Finder
- e: Patriotic Front
- e: Patrol Frigate
- e: Physical File
- e: Polycenter Framework
- e: Polyurethane Foam
- e: Potency Factor
- e: Prostatitis Foundation
- f: Pacte Fédéral
- f: Patronat Français
- f: Père de Famille
- f: Personnel Féminin
- f: Petit-Fils
- f: Petite-Fille
- f: Petit Format
- f: Peuple Français
- f: Plantes Fourragères
- f: Pompes Funèbres
- f: Port-Franc
- f: Porte Fermée
- f: Porte-Fenêtre
- f: Possessions Françaises
- f: Préfecture
- f: Presse Française
- f: Prestation Familiale
- f: Prix de Fabrique
- f: Prix de Facture
- f: Prix de Fret
- f: Prix Fixe
- f: Prix Forfaitaire
- f: Prix Fort
- f: Production Française
- f: Produits Fabriqués
- f: Propriété Familiale
- f: Propriété Foncière
- N: Polyteknisk Forening
- r: Posttjänstemännens Förening
- s: Proyectil De Fundición [metralla]
- s: Proyectil Fumígeno
- s: Proyecto Fátima
pF e: micro-microfarad
pf e: Polynesia [French] [Internet]
Pf f: Puissance fiscale
- r: Pansarfordon
pf. e: perfect
pF. e: picofarad
PF. e: Program Function
Pf. e: Proof
- r: Postfack
p.f. s: punto de fusión
P&F e: Program and Financial plan
P/F e: Pass/Fail
pf- r: pansarförstörnings-
PFA a: Pensionsforsikringsanstalten
- e: Perfluoroalkoxy polymer
- e: Power FORTRAN Accelerator
- e: Predictive Failure Analysis
- e: Probability of False Alarm
- f: Presse Française Associée
- f: Prestation Familiale Agricole
- f: Protêt Faute d'Acceptation
p.fag a: produktive [skole]fag
PFAW e: People For the American Way
PFB e: Postscript Font Binary
- e: Printer Font Binary
PFBC e: Pressurized Fluidized Bed Combustor
PFC e: Perfluorated Carbon
- e: Perfluoro Carbon
- e: Perfluoro Chemical
- e: Perfluorocompound
- e: Portable Four Channel [radiometer]
- e: Power Factor Controller
- e: Power Factor Correction
- e: Prefrontal Cortex
PFCRA e: Program Fraud Civil Remedies Act
PFCs e: Perfluorinated hydrocarbons
- e: Perfluorocarbon Compounds
PFD d: Pfund
- e: Programming Facility for Display users
PFDL e: Parameterized Format Description Language
PfDL d: Pflegedienstleiter[in]
- d: Pflegedienstleitung
PF/DOS e: Product Improvement
p.f.e. a: produktionsfoderenhed
PFE e: Portable Fire Extinguisher
- e: Programmer's File Editor
PFES e: Portable Field Emission Spectrometer
PFF e: Planning Factors File
- e: Presidential Faculty Fellow
- e: Project Formulation Framework
- s: Marco de Formulación del Proyecto
PFG e: Pulsed Field Gradient
PFI N: Papirindustriens Forskningsinstitutt
PFIS e: Personnel Feedback Information System
- e: Property Fabric Information System
PFK e: Program Function Key[board]
PFL e: Pre-Fader Listening
- f: Propriété Foncière Libre
PFLE e: Popular Front for the Liberation of Eritrea
PflegeVG d: Pflege-Versicherungsgesetz vom 26.5.1994
PflR d: Pflegerecht. Zeitschrift für Rechtsfragen in der stationären und ambulanten Pflege
PFLT e: Paint Filter Liquids Test
Pflugers Arch d: Pflugers Archiv. European Journal of Physiology
PflVersG e: Gesetz über die Pflichtversicherung für Kraftfahrzeuge [Pflichtversicherungsgesetz]
PFM e: Patient Flow Model
- e: Postscript Font Metric[s]
- e: Printer Font Metric[s]
- e: Process Facility Modification
- e: Proto-Flight Model
- e: Pure Fantastic Magic
- e: Pure Friggin' Magic [Internet]
- f: Plateforme Multimission
PFMDS e: PACAF Force Management Display System
PFN f: Père de Famille Nombreuse
PFO N: Program for Oljevernberedskap
pfo s: párrafo
PFP d: Privat Finanzierte Preise
- e: Partnership For Peace
- e: Progressive Federal Party
- f: Protêt Faute de Paiement
- s: Punto de Fuego Permanente
PFPE e: Perfluorinated Polyether
PFP-EIS e: Plutonium Finishing Plant Environmental Impact Statement

PFPPEI f: Parti Fraternel des Planteurs des Parents d'Eleves et Industriels Ivoiriens
PFPU e: Processor Frame Power Unit
PFR e: Plug-Flow Reactor
- e: Portable Font Resource
- e: Power Fail Restart
- e: Program Financial Review
PFRA e: Prairie Farm Rehabilitation Association
PFRC e: Pacific Forest Research Centre
PFRS e: Portable Field Reflectance Spectrometer
PFS e: Progression-Free Survival
Pfs e: Proportionalitätsfaktor für die Schwächung der spektralen CCD-Empfindlichkeit
Pfsmax e: Proportionalitätsfaktor der minimalen zulässigen Schwächung der CCD-Empfindlichkeit bei l max
PFSS e: Pesticide Farm Worker Safety Staff
PfsVmx e: Proportionalitätsfaktor der minimalen zulässigen Schwächung der CCD-Empfindlichkeit bei l Vmx
PFT e: Permanent Full Time
- e: Physical Fitness Test
- e: Plant Functional Type[s]
PFTBA e: Perfluorotributylamine
PFTE e: Permanent Full Time Equivalent
PfTQ e: Partnering for Total Quality
PFTs e: Perfluorocarbon Tracers
PFU e: Plaque-Forming Unit
- e: Programme Funding in millions of european currency Units
- f: Parti Feministe Unifié
P.F.V. a: På Formandens Vegne
PFXP e: Plan Failure explanation Pattern
PFZ e: Polar Front Zone
- s: Zona del Frente Polar
PG e: GTOS Ad Hoc Scientific and Technical Planning Group
- e: Perge Notice
- e: Planning Group
- e: Planning Guidance
- e: Product Group
- e: Prostaglandin[s]
- f: Paierie Générale
- f: Partie Gauche
- f: Plantes Grasses
- f: Poinçon de Garantie
- f: Police Générale
- f: Pourvoi en Grâce
- f: Prêt Gratuit
- f: Prix de Gros
- f: Procuration Générale
- f: Procureur Général
- p: Pesos da Guiné-Bissau
- s: Prisionero de Guerra
pg e: Anzahl der durchzurechnenden Generationen
Pg e: Pentagram
- e: Petagrams
- f: Passage
pg. a: piger
Pg. e: Portuguese
P+G e: Proctor and Gamble
PGA e: Program Global Area
- e: Programme Acronym
- e: Pure Grain Alcohol
pga a: på grund af
Pga r: på grund av
PGC e: Program Grande Carajas
- e: Program Group Control
PGCE e: Postgraduate Certificate of Education
PGCS e: Photogrammetric Graphics Compiling System
PGD e: Policy and Grants Division
PG de I s: Plan General de Instrucción
PGDF e: Product Generation and Dissemination Facility
PGDN e: Page Down
PGDP e: Paducah Gaseous Diffusion Plant
PGE e: Pacific Great Eastern
- e: Persian Gulf Evacuee
- e: Petroleum and Geosystems Engineering
- e: Platinum-Group Elements
- e: Product Generation Executive
Pge f: Page
- f: Passage
PGED e: Point-contact Germanium Diode
PGF e: Plant Growth Facility
- e: Pressure Gradient Force
- f: Paierie Générale de la France
- f: Permission de Grandes Fêtes
PGH e: Patrol Gunboat, Hydrofoil
PGI e: Physicians Guide to the Internet
- e: Purge Gas Inlet
PGI2 e: Prostaglandin I2
PGII s: Plan General de Intercambio de Inteligencia
P-GILD e: Projection Gas Immersion Laser Doping
PGILD e: Projection Gas Immersion Laser Doping
PGK d: Personalgemeinkosten
PGL e: Graphics
- e: Peer Group Leader
pgl. a: pagældende
PGM e: PGM-11 Redstone Missile
- e: Planning Guidance Memorandum
- e: Platinum-Group Metals
- e: Platinum-Group Minerals
- e: Portable Gray Map
- e: Precision Guided Missile
- e: Precision Guided Munitions
- e: Processing Graph Methodology
- s: Primera Guerra Mundial
PGN s: Procuradoria General de la Nación
PGOC e: Payload Ground Operation Contract
- e: Payload Ground Operations Control
PGP e: Pretty Good Piracy (or Privacy)
- e: Program Parameter
- f: Parti Gabonais du Progrès
- f: Parti Guinéen du Progrès
PGPE s: Plan General y Programa de Ejecución
PGR e: Plant Genetic Resources
- e: Programme Relationships
Pgr r: Pansargranat
p. g[r]. a. N: på grunn av
PGRA e: Plant Genetic Resources for Agriculture
PGRC e: Plant Genetic Resource Centre
PGRFA e: Plant Genetic Resources For Agriculture
PGS e: Physicians for Global Survival
- e: Product Generation System
- e: Programme Status
- e: Publications Gravure Server
PGSE e: Processing Graph Support Environment
PGT f: Parti Guinéen du Travail
- s: Partido Guatemalteco de Trabajadores
PGTA e: Pharmaceutical Grade Transfer Adhesive
PGUD f: Parti Guinéen pour l'Union et la Démocratie
PGUP e: Page Up
P.G.V. a: På Gruppens Vegne
PH d: Physik
- d: Pädagogische Hochschule
- e: Parity High bit
- e: Philippines
- e: Physics
- e: Prentice Hall
- e: Process Hold
- f: Paie Hebdomadaire
- f: Péninsule Hispanique
- f: Prêt d'Honneur
- f: Prêt Hypothécaire
- f: Priorité d'Hypothèque
- f: Prix d'Honneur
- f: Prix Homologué
- f: Prud'homme
- f: Publication Hebdomadaire
- f: Purge d'Hypothèque
- s: Portahelicópteros
- s: Proyectil de Hidrógeno
ph e: Phillipines [Internet]
pH e: refers to acidity-alkalinity scale
Ph f: Par heure
P/H s: Por Hora
PHA e: Physics and Astronomy
- e: Process Hazard Analysis
- e: Public Health Association of Australia
- e: Pulse Height Analysis (or Analyzer)
PHAA e: Pritikin Health Association of Australia
PHABSIM e: Physical Habitat Simulation Model
PHAR e: Pharmacology
PHARE e: Poland and Hungary Assistance for Reconstruction of the Economy
- f: Pologne-Hongrie: Aide à la Reconstruction Économique
- s: Plan de Acción para una Ayuda Coordinada a Polonia y Hungría
PHARM e: Pharmacy Building
Pharm. e: in Pharmacology
Pharmacol Biochem Behav e: Pharmacology, Biochemistry and Behavior
Pharmacol Res e: Pharmacological Research
Pharmacol Rev e: Pharmacological Reviews
Pharmacol ther e: Pharmacology and therapeutics
Pharmacol Toxicol e: Pharmacology and Toxicology
Pharm Acta Helv l: Pharmaceutica Acta Helvetiae
Pharm Biotechnol e: Pharmaceutical Biotechnology
Pharm Res e: Pharmaceutical Research

Pharm Unserer Zeit d: Pharmazie in Unserer Zeit
Pharm World Sci e: Pharmacy World and Science
Pharos e: Pharos of Alpha Omega Alpha Honor Medical Society
PHARUS e: Phased Array Universal Sar
PHAs e: Public Housing Agencies
PHASE f: Physique et Applications des Semi-Conductor
PHASE IV e: Base Level Data Automation Program
PHC e: Prairie Home Companion
- e: Primary Health Care
- e: Principal Hazardous Constituent
- e: Public Health Committee
PHCB e: Page Header Control Bit
PHCDS e: Public Health Common Data Set
PHCI f: Plantations et Huileries de Côte d'Ivoire
PHCSG e: Primary Health Care Specialist Group
PHCVD e: Photon-induced CVD
PHD l: Philosophiae Doctor
Ph.D e: Doctor of Philosophy
PHE e: Public Health Evaluation
PHED e: Physical and Health Education
- e: Physical Education
PHEM e: Public Health Evaluation Manual
PHEMT e: Pseudomorphic High Electron Mobility Transistor
PHERMEX e: Pulsed High Energy Radiation Machine Emitting X-Rays
PHERP e: Public Health Education and Research Program
PHF d: Philosophische Fakultät
- e: Paired Helical Filaments
- e: Patriotic Health Front
- e: Public Health Foundation
PHGY e: Physiology
PHI e: Archie Phinny Hall
- e: Pacific Health Information Network
- e: Packard Humanities Institute
- e: Philadelphia, PA
- e: Pre-Harvest Interval
- f: Programme Hydrologique International
- s: Programa Hidrológico Internacional
PHICs e: Poorest and Most Heavily Indebted Countries
PHIGS e: Programmers Hierarchi[c]al Interactive Graphical (or Graphics) Standard[s]
- e: Programmers Hierarchi[c]al Interactive Graphics System
- e: Programmers Hierarchical Interface Graphics Standard
PHIGS PLUS e: Programmers Hierarchical Interactive Graphics System Plus Lumiere Und Surfaces
PHIL d: Philosophie
- e: Philosophy Library
Philol. e: Philological
Philos d: Institut für Philosophie, Pädagogik und Psychologie, Abteilung für Philosophie
Philos. e: Philosophic
- e: Philosophy
Philos Trans R Soc Lond B Biol Sci e: Philosophical Transactions of the Royal Society of London. Series B: Biological Sciences
PHIS e: Population and Housing Information System
PHL e: Philosophy
PHLI e: Public Health Leadership Institutes
PHLIS e: Public Health Laboratory Information System
PHM e: Hydrofoil Patrol Craft, Medium
- e: Missile-armed Patrol Hydrofoil
- e: Pyroheliometer
PHN e: Population, Health and Nutrition
- e: Public Health Network
PHO e: Phoenix, AZ
- e: Phone List
- e: Physician-Hospital Organization
- f: Programme d'Hydrologie Opérationnelle
- s: Programa de Hidrología Operativa
Pho e: Photonic
PHONEBOOK e: Region 3 Telephone Directory
phonet. e: phonetic[ally]
PHOTO-CD e: Compact Disk for storing still photographs
PHOTOCD e: Photographic Compact Disk
Photochem Photobiol e: Photochemistry and Photobiology
Photodermatol Photoimmunol Photomed e: Photodermatology, Photoimmunology and Photomedicine
Photogr d: Institut für Photogrammetrie
Photogr. e: Photographical
PHOXA e: Photochemical Oxidant and Acid deposition model application within the framework of control strategy development
- e: Photochemical Oxidant and Acid deposition model application within the framework of PIC
PHP e: Parents Helping Parents
- e: Peoples Heritage Party
- e: Personal Handy Phone
- e: Personal Home Page
- e: Physicians Home Page
- f: Presse Hebdomadaire et Périodique
PHPA in: Perlindungan Hutan dan Pelestrian Alam
PHPES e: Program for Health Policy in Economies under Stress
PHPPO e: Public Health Practice Program Office
PHQ e: Peacetime Headquarters
PHR e: Pharmacy
phr. e: phrase
PHRE f: Programme d'Hydrologie et de mise en valeur des Ressources en Eau
PHRED e: Public Health Risk Evaluation Data Base
PHRH s: Programa de Hidrología y Recursos Hídricos
PHRIR e: Pointable High Resolution Imaging Radiometer
PhRMA e: Pharmaceutical Research and Manufacturers of America
PH S e: Physical Science Building
PHS e: Personal Handy-Phone System
- e: Personal Health Summary
- e: Photo Hydrography System
- e: Pulmonary Hypertension Syndrome
PHSA e: Public Health Service Act
PHSS e: Population Health Summary System
PHS&T e: Packaging, Handling, Storage and Transportation
PHT e: Packaging, Handling and Transportation
- e: Port Hold Time
PHTD e: PNNL HWVP Technology Development
PHTLS e: Pre Hospital Trauma Life Support
PHTM e: Public Health and Tropical Medicine
PHTN e: Public Health Training Network
PHTS e: Packaging, Handling, Transportation, Storage
PHU e: Power Holdover Unit
pHVA level e: Plasma level of Homovanillic Acid
PHX e: Phoenix, Arizona
PHY e: Physical access
- e: Physical layer [control]
- e: Physical layer medium independent
- e: Physical layer of the OOSI model
- e: Physical layer protocol
PHyp f: Prêt Hypothécaire
PHYS e: Physical Science Building
- e: Physics Library, a member of PSL
Phys d: Physikalisches Institut
Phys. e: Physical
PhysCh e: Institut für Physikalische Chemie
PhysEl e: Institut für Physikalische Elektronik
Physiol. e: Physiological
Phys[iol]. e: Physiology
Physiol Behav e: Physiology and Behavior
Physiol Chem Phys Med NMR e: Physiological Chemistry and Physics and Medical NMR
Physiol. Ecol. Jpn. e: Physiology and Ecology Japan
Physiol Meas e: Physiological Measurement
Physiol Res e: Physiological Research
Physiol Rev e: Physiological Reviews
PHYSLAC e: Physical Sciences Library Advisory Committee
Phys Med Biol e: Physics in Medicine and Biology
PhysRev e: Physical Review Journal

PI d: Psychologisches Institut
- e: Division of Plant Industry
- e: Photographic Intelligence
- e: Plan Information
- e: Polyimide
- e: Polyisoprene
- e: Practice Interception
- e: Predicted Impact
- e: Preliminary Injunction
- e: Primary Infertility
- e: Prime Item
- e: Principal (or Principle) Investigator
- e: Profil Infirmier
- e: Program Information
- e: Program Interrupt[ion]
- e: Public Involvement
- f: Paquet International
- f: Passage Inférieur
- f: Passage Interdit
- f: Pêche Interdite
- f: Peine Infamante
- f: Péninsule Ibérique
- f: Pension d'Invalidité
- f: Perception d'Impôts
- f: Péremption d'Instance
- f: Péréquation de l'Impôt
- f: Période d'Instruction
- f: Permis d'Importation
- f: Personne Interposée
- f: Pièce d'Identité
- f: Plaque d'Immatriculation
- f: Point Initial
- f: Poteau Indicateur
- f: Première Instance
- f: Prêt d'Installation
- f: Prière d'Insérer
- f: Prix Imposé
- f: Production Industrielle
- f: Professions Industrielles
- f: Profits Illicites
- f: Propriété Immobilière
- f: Propriété Indivise
- f: Propriété Industrielle
- f: Responsable de recherche
- r: Personalinstruktionen
- s: Investigador Principal
- s: Pelotón de Infantería
- s: Plan Internacional
- s: Préstamo Interbibliotecario
- s: Punto Inicial

pI e: maximale Individuenanzahl in einer Generation

Pi f: Par intérim
- r: Pionjär

PI. e: Pacing indicator

pi. N: pionér

P. I N: Personellstabens organisasjonsavdeling

P&I e: Physics and Irradiation
- e: Piping and Instrumentation
- e: Planning and Integration

P³I e: Pre-Planned Product Improvement

PI-3 e: Parainfluenza-3 virus

PIA e: Parallel Interface Adapter
- e: Payload Interface Agreement
- e: Personal Intelligent Agent
- e: Piano
- e: Plastics Industry Association
- e: Plug In Administrator
- f: [Société des] Plastiques Industrielles Automobiles
- f: Police Internationale de l'Air

PIAMA s: Programa de Investigación de la Atmósfera y el Medio Ambiente

PIAS e: PLSS Intelligence Augmentation System

PIAT e: Projector, Infantry, Anti-Tank
- e: Public Information Assist Team

PIB e: Particle In a Box
- e: Plans Intelligence Branch
- e: Potential-Induced Breathing
- e: Program Information Block
- f: Produit Intérieur Brut
- s: Producto Interior Bruto

PIBA f: Programme International de Bouées dans l'Antarctique
- s: Programa Internacional de Boyas en el Antártico
- s: Programa Internacional sobre las Polinias en el Artico

PIBM f: Produit Intérieur Brut Marchand

PIC e: Microcontrollers based MicroChip Technology
- e: Pacific Imagery Processing and Interpretation Center
- e: Pacific Island Country
- e: Paired-Ion Chromatography
- e: Parent Indicator Code
- e: Particulate Inorganic Carbon
- e: Personal Information Carrier
- e: Personal Information Communicator
- e: Personal Intelligent Communicator
- e: Pig Improvement Company
- e: Point In Call
- e: Polyolefin Insulated Cable
- e: Position-Independent Code
- e: Pressurized Ion Chamber
- e: Primary Independent Carrier
- e: Prior Informed Consent [procedure]
- e: Product Information Catalog
- e: Products of Incomplete Combustion
- e: Program Interrupt Controller
- e: Programmable Integrated Circuit
- e: Programmable Interrupt Controller
- e: Programme Information Circular
- e: Public Information Center
- f: [Société française de] Préparation Industrielle des Combustibles
- f: Circulaire d'Information sur le Programme
- f: Pêcheries Industrielles du Congo
- p: Polícia de Investigação Criminal
- s: Circular de Información sobre el Programa
- s: Punto de Intercambio de Correspondencia

PICA e: Portable Image Computation Architectures
- e: Posterior Inferior Cerebellar Artery

PICDMS e: Picture Database Management System

PICES e: North Pacific Marine Science Organization[s]
- f: Organisation des Sciences de la Mer pour le Pacifique Nord
- s: Organización del Pacífico Norte para las Ciencias del Mar

PICG f: Programme International de Corrélation Géologique
- s: Programa Internacional de Correlación Geológica

PICK e: An operating system for library packages

PICO e: Pacific Islands Contact Office
- e: Polio Information Center Online

PICS e: Partnership in Computational Science
- e: Payload Integration Control Schedule
- e: PC Image (or Imaging) Communication[s] System
- e: Photographic Integrated Control System
- e: Platform for Internet Content Selection
- e: Plug in Inventory Control System
- e: Protocol Implementation Conformance Statement[s]
- e: Protocol Interoperability Conformance Statement

PICS[/]DCPR e: PICS Detailed Continuing Property Record

PICT e: Format for images used by Apple Macintosh

Pict. e: Pictorial
- e: Picture

PICTS e: Photo-Induced Current Transient Spectroscopy

PID d: Photo-Ionisations-Detektor
- d: Privater Informationsdienst
- e: Particle-Induced Desorption
- e: Persistent Identifier
- e: Personal Identifier
- e: Photo-Induced Desorption
- e: Photoion[ization] Detector
- e: Plan Identification Number
- e: Planning and Integration Division
- e: Poisons Information Database
- e: Position Identifier
- e: Primitive Interface Definition
- e: Process ID [Identification] [number]
- e: Process Identifier
- e: Process-Induced Defect
- e: Product Identifier
- e: Program Identifier
- e: Proportional Integro-Differential
- e: Proportional, Integral, Derivative
- e: Protocol Identifier
- f: Parti Ivoirien pour la Démocratie
- s: Partido Institucional Democrático

P&ID e: Piping and Instrumentation Design
- e: Piping and Instrumentation Drawing
- e: Process and Instrumentation Diagram

PIDAC f: Projet Intégré de Développement Agricole en Casamance

PIDAS e: Perimeter Intrusions Detection Analysis System
- e: Portable Instant Display and Analysis Spectrometer
- e: Portable Intelligent Display and Analysis Spectrometer

PIDE p: Polícia Internacional de Defesa do Estado

Pidemo p: Partido Internacionalista Democrático de Moçambique
Pidico e: Project Industrial Development and Investment Corporation
PIDP II e: Second Generation PIDP Hardware
PIDS e: Prime Item Development Specification
- e: Process Integration, Devices and Structures
PIDX e: Petroleum Industry Data Exchange
PIE e: Personal Interactive Electronics
- e: Polar Ice Extent
- e: Postirradiation Examination
- e: Proto-Indo-European
- s: Patruila Independiente de Exploración
PIEA s: Programa Internacional de Educación Ambiental
PIEE f: Programme International d'Éducation relative à l'Environnement
PIER e: Pierpont Commons
- e: Procedures for Internet/Enterprise Renumbering
PIERC e: Pacific Island Ecosystems Research Center
PIERS-4 e: Pressure-Induced Extra Resonance in Four-wave mixing
PIES e: Pollution prevention Information Exchange System
PIE syndrome e: Pulmonary Infiltration with Eosinophilia syndrome
PIF d: Programminformation
- e: Paid In Full
- e: Periodicals catalogue
- e: Photo Interpretation Facility
- e: Picture Interchange Format file
- e: Postscript Interchange Format
- e: Process Interchange Format
- e: Productivity Investment Fund
PIFRS e: Prototype Increase Frequency Reporting System
PIFS e: Point Coordination Inter Frame Space
- e: Point Interframe Space
Pig e: drill-pipe internal rust scraper
PIG e: Process Ink Gamut
PIGADD e: Permanent Intergovernmental Authority on Drought and Development
PIGB f: Programme International sur la Géosphère et la Biosphère Une étude de l'évolution à l'échelle mondiale
PIGLET e: Personalized Intelligently Generated Explanatory Text
PIGME e: Particle-Induced Gamma ion Emission
Pigment Cell Res e: Pigment Cell Research
PigmLa d: Forschungsinstitut für Pigmente und Lacke e.V.
PIGS e: Pesticides in Groundwater Strategy
PIH e: Population and International Health
PII e: Plasma Ion Implantation
- e: Program Integrated Information
P. II N: Personellstabens utdannelsesavdeling
PIII e: Plasma Immersion Ion Implantation
P. III N: Personellstabens administrasjonsavdeling
PIIX e: PCI IDE/ISA Xcelerator
PIJS f: Parti Ivoirien de Justice et de Solidarité
PIK d: Paritätische Interpretations-kommission
- d: Potsdam-Institut für Klimafolgen-forschung e.V., Potsdam
- e: Programmers Imaging Kernel
PIL e: PackRat Integration Library
- e: Page Interchange Language
- e: Preferred Item List
- e: Product Identifier Label
- e: Public Image Limited
- s: Proyectil de Iluminación
- s: Punto Inicial
pi. Lag N: pionérlag
PILOT d: Pilotierung [Aufstieg für Windmessungen]
- e: Piloted Lowspeed Tests
- e: Programmed Inquiry Learning Or Teaching
PilotACE e: Pilot Automatic Computing Engine
PILP e: Program for Industry/Laboratory Projects
PILPS e: Program for Intercomparison of Land Surface Parameterization Schemes
- e: Project for Intercomparison of Land-surface Parameterization Schemes
PIM e: Patient Interface Module
- e: Payload Integration Manager
- e: Personal Information Management (or Manager)
- e: Position of Intended Movement
- e: Primary Interface Module
- e: Programmatic Interchange Meetings
- e: Protocol Independent Multicast [protocol]
- e: Protocol Independent Multicasting
- f: Prescriptions internationales
PIMB e: Portable Common Tool Environment
PIMO f: Programme des Instruments et des Méthodes d'Observation
- s: Programa de Instrumentos y Métodos de Observación
Pimo p: Partido Independente de Moçambique
PIMR e: Polar Ice Mapping Radiometer
- e: Pushbroom Imaging Microwave Radiometer
PIMRIS e: Pacific Islands Marine Resources Information System
- f: Système d'information sur les ressources marines des îles du Pacifique
- s: Sistema de Información sobre Recursos Marinos de las Islas del Pacífico
PIMS e: Pesticide Incident Monitoring System
- e: Procurement Information Management Services
- e: Profit Impact of Marketing Strategy
PIMT s: Programa de Investigación sobre Meteorología Tropical
Pi Mu Epsilon e: National Mathematics Honor Society
PIN d: Persönliche Identifikations-Nummer
- e: P-doped, Intrinsic, N-doped
- e: Pacific Island Network
- e: Particle Induced Noise
- e: Personnel Increment Number
- e: Pesticide Information Network
- e: Plan Identification Number
- e: Positive Intrinsic Negative Diode
- e: Pre-Invitation Notice
- e: Procedure Interrupt Negative
- e: Process Identification Number
- e: Processor Independent Netware
- e: Procurement Information Notice
- s: Partido de Integración Nacional
- s: Planificación Industrial Naval
PInd f: Production Industrielle
PINE e: Pine Is Not Elm [Internet]
- e: Program for Internet News and E-mail
PINEA f: Produits Industriels Nécessaires aux Exploitations Agricoles
PINELLAS e: Pinellas Plant, Largo, Florida
PINET e: Physics Information Network
PING e: Packet Internet Grouper
PInt f: Par Intérim
PINVE s: Programa de Investigación
PINX e: Private Integrated Services Network Exchange
PIO e: Parallel I/O [Input/Output]
- e: Parallel Interface Output
- e: Programmable I/O [Input/Output][-chip]
- e: Programmed Input/Output
- e: Project Implementation Order
PIO/C e: PIO [Project Implementation Order] for Commodities (or Commodity)
PIOFS e: Parallel I/O File System
PIO/P e: PIO [Project Implementation Order] for Participant Training
PIOS e: Planning Information Overlay System
- e: State Inspectorate for Environmental Protection
PIO/T e: PIO [Project Implementation Order] for Contract Technicians
P3IP e: Pre-Planned Product Improvement Program

PIP e: IGOSS Plan and Implementation Programme
- e: P-Internet-Protocol
- e: Packet Interface Port
- e: Parallel I/O Port
- e: Parks In Peril
- e: Performance Improvement Program
- e: Periodical Informational Posting
- e: Periphal Interchange Program
- e: Personal Information Processor
- e: Phosphatidylinositol-4-phosphate
- e: Picture In Picture
- e: Planning Integration Process
- e: Plot Inconsistencies Project
- e: Problem Isolation Procedure
- e: Process Improvement Proposal
- e: Process-Induced Particles
- e: Product Integration Program
- e: Programmable Interconnect Point
- e: Public Interest Parole[e]
- e: Public Investment Program
- e: Public Involvement Plan
- e: Public Involvement Program
- e: Wetnet Precipitation Intercomparison Project
- f: Plan et programme de mise en oeuvre du SMISO
- f: Programme d'Investissements Publics
- p: Plano Indicativo Prospectivo
- s: Plan y Programa de Ejecución del IGOSS

PIP_2 e: PhosphatidylInositol-4,5-bisPhosphate
PIP2 e: Phosphatidylinositoldiphosphate
PIPD e: Planning Input for Program Development
PIPICO e: Panel for International Program[me]s and International Cooperation in Ocean Affairs
Piplut r: Pionjärpluton
PIPO e: Parallel In[put]/Parallel Out[put]
PIPOR e: Programme for/of International Polar Oceans Research
PIPPQ s: Programa Internacional de Protección frente a los Productos Químicos
PIPPS e: Publication Information Processing and Printing System
PIPQUIC e: Program Integration Project Queries Use in Interactive Command
PIPS e: PCAC Information Processing System
- e: Personal Image Processing System
- e: Plans Integration Partition System
- e: Polymerization-Induced Phase Separation
- e: Portable Interactive Planning system

PIR d: Pirna, Sachsen
- e: Party of Independent Republicans
- e: Photon-Induced x-Ray
- e: Polyisocyanurate
- e: Precision Infrared Radiometer
- e: Primary Intelligence Requirements
- e: Project Internal Result
- e: Protein Identification Resource[s]
- s: Parte de Interpretación Fotográfica

PIRATA e: Pilot Research [moored] Array in the Tropical Atlantic
PIRDES f: Programme Interdisciplinaire de Recherche et de Développement en Energie Solaire
PIRE s: Protección de la Información Reservada
PIRG e: Public-Interest Research Group
PIRLA e: Paleoecological Reconstruction of Recent Lake Acidification
PIRMAT f: Programme Interdisciplinaire de Recherches sur les Matériaux
PIRO f: Programme Interdisciplinaire de Recherche Océanologique
PIRT e: Pretreatment Implementation Review Task Force
- f: Project Inventaire des Ressources Terrestres

PIRU e: Public Information Reference Unit
PIS e: Pointable Imaging Spectrometer
- e: Public Information Specialist

PISA e: Prototype Image Spectrum Analyzer
Pisapa p: Partido Islámico para a Salvação do Povo Angolano
Pisc f: Pisciculture
PISCES e: PoISson and Continuity Equations Solver
PISD e: Pacific Information Systems Division
PISO e: Parallel-In, Serial-Out
PISP e: PCG-insensitive Streptococcus Pneumoniae
- e: Polar Ice Sheet Program

PIST e: Prévention Information Sida Toxicomanie
Pist r: Pistol
- s: Pistola

PISYS e: Photographic Interpretation Data System
PIT e: Permit Improvement Team
- e: Pittsburgh, PA
- e: Powder-In-Tube
- e: Principal, Interest and Taxes
- f: Parti de l'Indépendance et du Travail
- f: Parti Ivoirien des Travailleurs
- s: Proyectil de Instrucción

Pit r: Pionjärtjänst
PITA e: Pain In The Accronym
- e: Pain In The Anatomy
- e: Pain In The Arse [Internet]
- e: Pain In The As[s] [Internet]
- e: Pine In The Ass [Internet]

Pi Tau Sigma e: National Honorary Mechanical Engineering Society
PITI e: Principal, Interest, Taxes [and] Insurance
PITMOS e: Post Mode Optimizing System
pi tropp N: pionértropp
PITS e: Photo-Induced current Transient Spectroscopy
- e: Project Information Tracking System
- e: Projected Intensity Triplet Space

Pitt. e: Pittsburgh
PIU. e: Path Information Unit
PIUS e: Process Inherent Ultimate Safety
PIV e: Particle Image Velocimeter (or Velocimetry)
- s: Puesto Independiente de Vigilancia

PIW e: Photo Imaging Workstation
PIWG e: Permanent Interoperability Working Group
PIXE e: Particle Induced X-ray Emission
- e: Proton Induced X-ray Emission analysis

PIXEL e: Picture Element
PIXIT e: Protocol Implementation Extra Information[s] for Testing [proforma]
PIXSYS e: Pictorial Information Extraction System
PIZ d: Patentinformationszentrum
PIZO f: Société de Distribution de Produits Pétroliers
PJ d: Projekt
- e: Petajoule[s]
- f: Palais de Justice
- f: Parcs et Jardins
- f: Pièce Jointe
- f: Pièce Justificative
- f: Police Judiciaire
- f: Poursuite Judiciaire
- f: Pouvoir Judiciaire

Pj r: Pjäs
PJAS e: Project Associates
Pjavd r: Pjäsavdelning
PJD f: Parti de la Justice Démocratique
- f: Parti des Jeunes Démocrates

Pje s: Pasaje
PJH e: PLRS/JTIDS Hybrid System
PJL e: Printer Job Language
PJNF e: Project Join Normal Form
PJP f: Presse Judiciaire Parisienne
Pjplut r: Pjäspluton
Pjplutch r: Pjäsplutonchef
PJS e: The status of the project
- s: Socialist Justice Party

PJST e: Peace and Justice Studies
PK d: Präsidentenkonferenz
- e: Primary Key
- f: Poteau-Kilomètre

pK d: Anzahl der Kinder nach einer Rekombination
pk. a: pakke
P[k] e: Probability of kill
PKAS r: Pansartruppernas Kadett- och Aspirantskola
Pkb r: Pansarkanonbåt
PKB s: Proyecto Kaibil Balam
PKCS e: Public Key Cryptography Standards [group]
PKD e: Proliferative Kidney Disease
PkDB d: Paläoklima-Datenbank [Universität Stuttgart-Hohenheim]
PKG-POL e: Packaged POL
PKI d: Philips Kommunikationsindustrie AG
pkl. r: poliklinik
PKM e: Panevëys Museum of ethnography
p-kort a: pensionistkort
PKP d: Praktische Krankenpflege
pkr r: pistolkrut
PKR R: Protivolodochniy Kreyser
PKS d: Paritätische Kommission Spitalleistungskatalog
- d: Polizeiliche Kriminalstatistik
- r: Pansartruppernas Kadettskola

PKSDD e: Probability Kriging to Spatially Distributed Data System
PKS DIV e: Parks Division
pkt. a: punkt[um]
- N: punkt

PKU d: Schweizerische Vereinigung privater Kranken- und Unfallversicherer
PKV d: Privatkundenvertrieb
PL e: Parity Low bit
- e: Payload Length
- e: Photolithography
- e: Photoluminescence
- e: Physical Layer
- e: Piecewise-Linear
- e: Pomraning-Livermore
- e: Presentation Layer
- e: Project Leader
- f: Paquet-Lettre
- f: Parti Libéral
- f: Poids Lourd
- f: Pouvoir Législatif
- f: Prescription Libératoire
- f: Prescriptions Légales
- f: Prestations Légales
- f: Principal Locataire
- f: Prix Licite
- f: Prix Limite
- f: Procuration Limitée
- f: Produits de Luxe
- f: Professions Libérales
- f: Projet de Loi
- f: Purge Légale
- N: Patentlov
- N: Poståpnernes Landsforbund
- R: Podvodnaya Lodka
Pl f: Papier long
- f: Place
- f: Plafond
- f: Plan[te]
- r: Plansch
- r: Plats
pl N: plass
Pl. a: Plads
- r: Postlåda
pl. d: Plappart
- N: plansje
P.L. e: Public Law
P&L e: Profit-and-Loss
- e: Projects and Logistics
PL1 e: Programming Language no. 1
P.L. 480 e: Public Law 480
PLA e: Panchromatic Linear Array
- e: Patent License Agreement
- e: Peoples Liberation Army
- e: Plain Language Addressee
- e: Programmable Logic Array
- e: Public Library Assocation
- e: Pulsed Laser Ablation
- f: Prêt Locatif Aidé
- p: Partido Liberal Angolano
- R: Atomnaya Podvodnaya Lodka
- s: Pelea
PLACE e: Post-LANDSAT-D Advanced Concept Evaluation Program for Satellite Modelling
PLAD e: Plain Language Address Directory
PLADES s: Plan de Acción de Desarollo Social
PLADS e: Plain Language Address System
- e: Pulsed Light Airborne Depth Sounder
PLAI e: Potsdam Land-Atmosphere Interaction Model
PLAIL e: Public Libraries and Independent Learners
PlaKo d: Institut für Entwerfen und Konstruieren, Fachgebiet Grundlagen der Planung und Konstruktion im Hochbau
PLAM e: Peoples Liberation Army of Malawi
PL/AMOS e: Phillips Laboratory Air Force Maui Optical Station
PLAN e: Peoples Liberation Army of Namibia
- e: Public Lands Action Network
PLANAN s: Proyecto Latinoamericano Niño a Niño
PlanArch d: Institut für Grundlagen der Planung
PLANOX e: Planar Oxide
Planta Med e: Planta Medica
Plant Cell Physiol e: Plant and Cell Physiology
Plant Foods Hum Nutr e: Plant Foods for Human Nutrition
Plant J e: Plant Journal
Plant Mol Biol e: Plant Molecular Biology
Plant Physiol e: Plant Physiology
PLAP R: Atomnaya Protivolodochnaya Podlodka
PLARB R: Atomnaya Podlodka Raketnaya Ballisticheskaya
PLARK R: Atomnaya Podlodka Raketnaya Krilataya
PLAS e: Professional Library Automation System
PLASA e: Professional Lighting and Sound Association
Plasma d: Institut für Plasmaforschung
Plast Reconstr Surg e: Plastic and Reconstructive Surgery
PLAY e: Telematics project on music services for the blind
PLB d: Pfälzisches Landesbibliothek in Speyer
- e: Payload Bay
- e: Picture Level Benchmark
PLBB e: Patent Licensing Bulletin Board
PLBD e: Payload Bay Doors
PLBSI e: Picture Level Benchmark Sample Implementation
PLC e: Plague Locust Commission
- e: Planar Lightwave Circuit[s]
- e: Polymer Liquid Crystal
- e: Programmable Logic Controller[s]
- e: Public Limited Company
PLCC e: Plastic Leaded (or Leadless) Chip Carrier
PLCE e: Peripheral and Loading Control Element
PLCI f: Parti Libéral de Côte d'Ivoire
PLCO e: National Cancer Institute Prostate, Lung, Colorectal and Ovarian Cancer Screening Trial
PLCP e: Physical Layer Convergence Procedure
- e: Physical Layer Convergence Protocol
PLD e: Presentation Logic Domain
- e: Program Load Disc
- e: Pulsed-Laser Deposition
- f: Parti Libéral Démocratique
- f: Parti pour les Libertés et la Démocratie
- p: Partido Liberal Democrático
- s: Plan de Dispersión
PLDEF s: Plan de Defensa
PLDIS e: Public Libraries Development Incentive Scheme
PLDS e: Pilot Land Data System
- e: Planetary Land Data System
PLDST s: Plan de Destrucciones
PLDT e: Philippine Long Distance Telephone Company
PLE e: Photoluminescence Excitation spectroscopy
- e: Piecewise-Linear Expression
PLED e: Polymer Light Emitting Diode
Pleist. e: Pleistocene
PLEJ f: Parti pour la Liberté, l'Egalité et la Justice
PLES s: Plan de Empleo de Servicios
PLEV s: Plan de Evacuación
PLF s: Plan de Fuego[s]
plf e: pounds per foot
PLFA e: Phospholipid Ester-linked Fatty Acid
- e: Primary Level Field Activity
- s: Plan de Fuegos de Artillería
PLFC s: Plan de Fuego[s] Conjunto[s]
PLFI s: Plan de Fuegos de Infantería
PLFM s: Proyecto Lingüístico Francisco Marroquín
PLG e: Planning Liaison Group
PL/GP e: Phillips Laboratory, Geophysics Directorate
PL G Per I s: Plan General Permanente de Información
PLGR e: Position Location Ground Receiver
PLH e: Pharyngeal lymphoid hyperplasia
- f: Programme Local d'Habitat
- s: Plan de Hopitalización
PLHDWU e: Port Louis Harbor and Dock Workers Union
PLI e: Paternal Leukocyte Immunization
- e: Programming Language 1
PLIC e: Programmable Logic Integrated Circuit
PLICE e: Programmable Low-Impedance Circuit Element
PLIF e: Planar Laser-Induced Fluorescence
PL I J s: Plan de Información del Jefe
Plim f: Page liminaire
PLIN s: Plan de Información
Plio. e: Pliocene
PL[/]IP e: Parallel Line IP [Internet Protocol]
PLIRRA e: Pollution Liability Insurance and Risk Retention Act
PLK R: Podlodka raketnaya Krilataya
PLL e: Permanent Logical Link
- e: Phase Locked Loop
- e: Plasma Lockload
- e: Prelinked Library
PLM e: Polarized Light Microscopy
- f: Paris-Lyon-Méditerranée
Pl M s: Plana Mayor
PL[/]M e: Programming Language for Microcomputers
PLMA s: Plan de Maniobra
PLMM s: Plana Mayor de Mando
PLMN e: Public Land Mobile Network
PLMR s: Plana Mayor de Regimiento
- s: Plana Mayor Reducida
PLN e: Program Line Number

PLNAG e: Public Libraries Networking Advisory Group
PLNASG e: Public Libraries (or Library) Network Awareness Steering Group
Plnd s: Planeador
PLNG e: Planning
PLO a: Pædagogisk Landsforening for Orientering
- a: Praktiserende Lægers Organisation
- e: Palestina Liberty Organisation
- e: Palestine (or Palestinian) Liberation Organization
- e: Please Leave On
- N: Praktikantenes Landsorganisasjon
PLÖ d: Plön/Holstein, Schleswig-Holstein
PLOKTA e: Press Lots Of Keys To Abort [Internet]
Plon f: Plantation
Plonk d: Geräusch, das ein ins Killfile fallender User erzeugt
PlottL r: Plottingledare
PL[-]OU e: Physical Layer Overhead Unit
PLP e: Packet Layer Protocol
- e: Peoples Liberation Party
- e: Physical Layer Protocol
- f: Parti de la Liberté et du Progrès
- s: Partido Liberador Progresista
- s: Plan a [de] Largo Plazo
PLPC e: Physical Layer Convergence Protocol
PLPT e: Pretreatment Local Program Tracking System
PLR e: Packet Lost Ratio
- e: Public Law Review
- f: Paquet-Lettre Recommandé
PLRF e: Program[me] on Long-Range Forecasting Research
- f: Programme de recherche sur la prévision météorologique à longue échéance
PLRG e: Public Libraries Research Group
PLS e: Palletized Load System
- e: Palletized Loading System
- e: Partial Least Squares [method]
- e: Personal Library Software
- e: Physical [Layer] Signaling
- e: Pre-Launch Survivability
- e: Primary Link Station
- e: Principles and Practice of Land Surveying
- e: Program Liaison Staff
- e: Programmable Logic Sequencer
- e: Projection of Latent Structures
- e: Provincial Land Surveyor
- e: Publishers Licensing Society
pls e: Please [Internet]
PLSC e: PACAF Logistics Support Center
- e: Political Science
PLSE e: Port Louis Stock Exchange
pLSI e: programmable Large Scale Integration
PLSO e: Phonetic Letters Spelled Out
PLSPC e: Research Programme on Land-Surface Processes and Climate
PLT e: Panduit Locking Tie
- e: Platelet
- e: Power Line Transient
- e: Program List Table
- e: Project Learning Tree
- f: Plan à Long Terme
PLTB s: Plan de Trabajos
PLTCI f: Parti pour la Libéralisation Totale de la Côte d'Ivoire
PL-TLM e: Payload Telemetry
PLTP s: Plan de Transportes
PLTRS s: Plan de Transmisiones
Plufs r: Planerings- och uppföljningssystem
Plug-In d: Hilfsprogramm von Web-Browser
PLUM e: Planning Land Use Model
PLUME e: Outfall PLUME Model
PLUME2D e: Two-Dimensional Plumes in Uniform Ground Water Flow
PLUME3D e: Three-Dimensional Plumes in Uniform Ground Water Flow
pl[ur]. e: plural
PLUS e: Parent Loans for Undergraduate Students
Plus f: Plusieurs
Plut[ch] r: Pluton[chef]
Plutstab r: Plutonstab
PLUVUE e: Plume Visibility Model
PLV e: Presentation Level Video
- e: Production Level Video
- f: Plan de Vol
- f: Pour la Vie
PLY e: Photolimited Yield
PLYF e: Port Louis Youth Federation
plz e: Please [Internet]
PM a: Politimester
- d: Phasenmodulation
- e: Pagemaker
- e: Patrol Missile Boat
- e: Performance Management
- e: Performance Measure
- e: Peripheral Module
- e: Peripheral Modules
e: PHIGS Monitor
- e: Photosynthetic Measurements
- e: Physical Medium
- e: Plant Management
- e: Plasma Membrane[s]
- e: Police Magistrate
- e: Popular Mechanics
- e: Postmortem
- e: Presenation Manager
- e: Presentation Manager
- e: Pressurized Module
- e: Preventative Maintenance
- e: Preventive Maintainance
- e: Preventive Medicine
- e: Private Mail [Internet]
- e: Procedure Memory
- e: Process Manager
- e: Process Module
- e: Product Management
- e: Product Manager
- e: Program Manager
- e: Program Memorandum
- e: Project Manager
- e: Protected Mode
- e: Provost Marshal
- e: Saint Pierre and Miquelon
- f: [de] Petit Modèle
- f: [de] Première Main
- f: Paie Mensuelle
- f: Papier-Monnaie
- f: Pêche Maritime
- f: Personne Morale
- f: Personnel de Maîtrise
- f: Personnel Masculin
- f: Plantes Médicinales
- f: Police Mobile
- f: Police Municipale
- f: Pour Mémoire
- f: Pouvoir Marital
- f: Première Marque
- f: Première Mutation
- f: Préparation Magistrale
- f: Prescription Médicale
- f: Presse Maritime
- f: Presse Médicale
- f: Presse Mondiale
- f: Prestation-Maladie
- f: Prix Marchand
- f: Prix Marqué
- f: Prix Modéré
- f: Produits de la Mer
- f: Produits Manufacturés
- f: Produits Médicamenteux
- f: Propriété Mobilière
- f: Publication de Mariage
- f: Puissance Médiatrice
- r: Promemoria
- R: Plovuchaya Masterskaya
- s: Plana Mayor
- s: Pleamar
- s: Policía Militar
- s: Policía Municipal
- s: Por Minuto
- s: Proyectil de Metralla
- s: Puesto de Mando
pm e: post meridiem [Internet]
Pm e: Promethium
- f: Par minute
- f: Par mois
p.m. e: post meridian
P.M. s: Patrulla Marítima
PM-2.5 e: Particulate Matter of diameter less than 2.5 micrometers
- e: Particulate Matter smaller than 2.5 micrometers in diameter
PM4 e: Purdue Multi-Mode Multimicroprocessor
PM-10 e: Particulate Matter less than 10 micrometers diameter
- e: Particulate Matter-less than 10 microns
- e: Particulate Matter nominally 10 microns and less
- e: Particulate Matter of aerodynamic diameter equal to or less than 10 micrometers
PM15 e: Particulate Matter nominally 15 microns and less
Pm-147 e: Promethium-147
PMA d: Pauschale Minderausgaben
- d: Personennamen des Mittelalters
- e: Performance Management and Accounting
- e: Physical Medium Attachement
- e: Pressurized Mating Adapter
- e: Program Memory Area
- f: Pays les Moins Avancés
- s: Policía Militar Ambulante
- s: Puesto de Mando Avanzado
PMAC e: Peripheral Module Access Controller
PMAD e: Portable Maintenance Aid Device

PMAe s: Programa de Meteorología Aeronáutica
PMAé f: Programme de Météorologie Aéronautique
PMAEA p: Associação dos Portos da Africa Oriental
PM Aer s: Puesto de Mando Aéreo
PMAG e: Program Management Advisory Group
PMAg f: Programme de Météorologie Agricole
- s: Programa de Meteorología Agrícola
P-MAIL e: Paper Mail
PMAN e: Polymethactylonitrile
P Manp s: Punto de Manipulación
PMAP e: Port Mapper
pMAP e: Professional Map Analysis Package
PMAPS e: Predirected Munitions Automated Planning System
PMAS e: Photochemical Assessment Monitoring Stations
PMASC f: Programme Mondial des Applications et des Services climatologiques
- s: Programa Mundial de Aplicaciones y Servicios Climáticos
PM Aux s: Puesto de Mando Auxiliar
PMB e: Performance Management Baseline
- e: Performance Measurement Baseline
- s: Puesto Médico de Batallón
Pmb r: Pansarmotorbåt
- r: Patrullmotorbåt
PMBI e: Pardon My Butting in [Internet]
PMBR e: Preliminary Multistate Bar Review
PMC e: Pacific Marine Center
- e: Partial Mission Capable
- e: Permanently Manned Capability
- e: Planning Ministers Council
- e: Polar Mesospheric Cloud
- e: Port Management Center
- e: Post Ministerial Conference
- e: Postmodern Culture
- e: Pressure-Modulator Cell
- e: Process Module Controller
- e: Program Management Charter
- e: Program Manager Charter
- s: Programa Mundial sobre el Clima
pMC d: Anzahl der durch spontane Mutation entstandenen Monte-Carlo-Kinder
P&MC e: Procurement and Material Control
PM&C e: Department of Prime Minister and Cabinet
p.m/c s: por mi cuenta
PMC [B,M,S] e: Partially Mission Capable [Both, Maintenance, Supply]
PMCC e: Pensky-Martens closed cup
- e: Platforms Mission Control Center
- e: Port Management Control Center
PMCCN e: Performance and Management of Complex Communication Networks
PM&C[E] e: Plant Maintenance and Construction [Engineering]
PMCL e: Proposed Maximum Contaminant Level
PMCLG e: Proposed Maximum Contaminant Level Goal
PMCM e: Partial Mission Capable, Maintenance
PMCS e: Partial Mission Capable, Supply
- e: Preventive Maintenance Checks and Services
PMD e: Packet Mode Data
- e: Personnel Management Division
- e: Physical Layer, Medium Dependent
- e: Physical Media (or Medium) Dependent
- e: Planning and Management Division
- e: Pre-Metal Dielectric
- e: Product Marketing Division
- e: Program Management Directive
- e: Program Management Document
- e: Project Management Division
- s: Puesto Médico de División
PMDC e: Permanent Magnet Direct Current
PMDF e: Pascal Memo Distribution Facility
- e: Project Master Data File
P Mdo s: Puesto de Mando
PMDP e: Project Manager Development Program
PMDSC f: Programme Mondial des Données climatologiques et de Surveillance du Climat
PMDVC s: Programa Mundial de Datos y Vigilancia Climáticos
PM&E e: Performance Measurement and Evaluation
PME e: Personnel Management Evaluation
- e: Precision Measurement Equipment
- e: Prime Mission Equipment
- f: Petites et Moyennes Entreprises
PMEC s: Programa Mundial de Estudios del Impacto del Clima
PMEG e: Page Map Entry Group
PMEIERC s: Programa Mundial de Evaluación de Impacto y Estrategias de Respuesta sobre el Clima
PMEL e: Pacific Marine Environment[al] Laboratories (or Laboratory)
PME/PMI f: Petites et Moyennes Entreprises/Petites et Moyennes Industries
PMF e: Parameter Management Frame[s]
- f: Police Municipale Féminine
PMFBI e: Pardon Me For Butting In [Internet]
PMFJI e: Pardon Me For Jumping in [Internet]
PMFJIB e: Pardon Me For Jumping In But [Internet]
PMFJOT e: Pardon Me For Jumping Off Topic [Internet]
PM-FTIR e: Polarization Modulation Fourier Transform InfraRed
PMG e: Page Map Group
- e: Primary Medical Group
- e: Program Maintenance Group
- s: Puesto de Mando General
PMGM e: Program Manager Guidance Memorandum
PMGSE e: Provisional Military Government of Socialist Ethiopia
PMH e: Petroleum Movement History
PMHEMT e: Pseudomorphic High Electron Mobility Transistor
PMI e: Passive Microwave Imager
- e: Phase Measuring Interferometer
- e: Presidential Management Intern
- e: Programmable Multispectral Imager
- f: Petite[s] et Moyennes Industries
- f: Protection Maternelle et Infantile
PMIC f: Programme Mondial concernant l'étude des Incidences du Climat
- s: Programa Mundial de Investigaciones Climáticas
PMIC-O s: Componente Oceanográfico del PMIC
PMICSP f: Programme Mondial d'évaluation des Incidences du Climat et de formulation de Stratégies de Parade
PMIGBOM e: Put Mind In Gear, Before Opening Mouth [Internet]
PMIP e: Paleoclimate (or Paleoclimatic[al]) Model[ing] Intercomparison Project
- e: Presidential Management Intern Program
PMIPK e: Poly Methyl Isopropenyl Ketone
PMIR e: Pressure Modulator Infrared Radiometer
PMIRR e: Pressure Modulated (or Modulator) Infrared Radiometer
PMIS e: Passive Microwave Imaging System
- e: Personnel Management Information Systems
- e: Production Management Information System
PMISP e: Personnel Management Information System Plan
PMJEG e: Performance Measurement Joint Executive Group
PMJI e: Pardon My Jumping In [Internet]
PML e: Perfect Matched Layer
- e: Plymouth Marine Laboratory
- e: Production Measurement Log
P M L e: Phoenix Memorial Laboratory
PMLM f: Partie Marxiste Léniniste Mauricien
PMM e: Professional Media Magazine
- f: Professeur de la Marine Marchande
- f: Programme de Météorologie Maritime
- s: Programa de Meteorología Marina
- s: Puesto de Mando Móvil
PMMA e: Poly[methyl methacrylate]
PMMA/MAA e: A copolymer of Methyl Methacrylate and Methacrylic Acid
PMMC e: Precious Metals Marketing Company
PMMO e: Projects and Missions Management Office
PMMRS e: Passive Microwave Mapping and Retrieval Simulator
PMMU e: Page Memory Management Unit
- e: Paged Memory Management Unit
PMMW e: Passive Millimeter Wave
PMN e: Polymorphonuclear leucocyte[s]
- e: Program Model Number
- s: Predicción Meteorológica Numérica
PMNF e: Premanufacture Notification Form

PMO e: Port Meteorological Officer
- e: Prime Ministers Office
- e: Program Management Office
- e: Purple Mountain Observatory, Academia Sinica, Nanjing
- f: Agent météorologique dans les ports
- f: Prix Marqué Obligatoire
- s: Agente Meteorológico de Puerto
- s: Puesto de Mando y Observación

Pmob r: Partiell mobilisering
PMOC e: Prototype Mission Operations Center
PMOD d: Physikalisch-Meteorologisches Observatorium Davos
PMOS e: P-channel MOSFET
- e: P[ositive] Channel MOS [Metal Oxide Semiconductor]
- e: Program Management and Operation Staff

P Mov S s: Puesto Móvil de Servicios
PMP e: Parallel Microprogrammed Processor
- e: Patient of the Month Program
- e: Performance Management Process
- e: Performance Monitoring Program
- e: Plan of Actions and Milestones
- e: Point to Multipoint
- e: Post-Metal Programming
- e: Prevention, Mitigation and Preparedness
- e: Preventive Maintenance Package
- e: Prime Mission Product
- e: Program Management Directive
- e: Project Management Plan
- e: Project Management System
- f: Parti du Mouvement Populaire
- s: Puesto de Mando Principal

PMPM e: Per Member Per Month
PMPR s: Puesto de Mando Principal de Reserva
PMPRB e: Program to Manage Pesticide Residues in Beef
PMPY e: Per Member Per Year
PMR e: Passive Microwave Radiometer
- e: Pollutant Mass Rate
- e: Precision Microwave Radiometer
- e: Pressure Modulated Radiometer
- e: Primary Mission Readiness
- e: Private Mobile Radio
- e: Problem Management Report
- e: Program Management Responsibilities
- e: Program Management Review
- e: Program Modification Request
- e: Proportionate Mortality Ratio
- e: Provisioning Master Record
- f: Parti Mauritanien de Renouveau
- s: Puesto de Mando de Reserva

PMRC e: Parents Music Resource Center
- e: Proctor Maple Research Center
- f: Programme Mondial de Recherche sur le Climat

PMRC-O f: Composante océanographique du PMRC
PMRK e: NEIC Permit Ranking System
PMRS e: Performance Management and Recognition System
- e: Performance Monitoring and Reporting System

PMRT e: Program Management Responsibility Transfer
PMRTD e: Program Management Responsibility Date
- e: Project Management Responsibility Date

PMRTP e: Program Management Responsibility Transfer Plan
PMS e: PACOM Munitions Storage
- e: Paramethylstyrene
- e: Particle Measuring System[s]
- e: Performance Monitoring System
- e: Periodical Management System
- e: Personnel Management Specialist
- e: Pipeline Management System
- e: Planned Maintenance Support
- e: Planning and Management Staff
- e: Position Management System
- e: Post-Modern Slut
- e: Power-Management System
- e: Premenstrual Syndrome
- e: Preventive-Maintenance System
- e: Printer Management Software
- e: Product Monitoring System
- e: Production Management Support
- e: Program Management Staff

PMSD e: Program Management and Support Division
- f: Parti Mauricien Social Démocrate

PMSE e: Polar Mesopause Summer Echo
PMSG e: Pregnant Mare Serum Gonadotropin
PMSI f: Programme de Médicalisation des Systèmes d'Information Hospitaliers
PMSO e: Program Management and Support Office
PMSR s: Puesto de Mando de Servicios de Retaguardia
PMSS e: Policy and Management Support Staff
- e: Program Management Support System

PMSSC f: Programme Mondial de Surveillance du Système Climatique
PMSV e: Pilot to Metro Services
PMSW e: Polymer Microstructure Waveguide
PMT e: Photomechanical Transfer [photostat]
- e: Program Management Team

Pmt f: paiement
PMTC e: Point Mugu Test Center
PMTF e: Product Management Task Force
PMTP e: Program Management Transfer Point
PMTPD e: Program Management Transfer Point Date
PMTPM e: Program Management Transfer Point Milestone
PMTPP e: Program Management Transfer Point Plan
Pmts f: Paiements
PMU e: Police Mobile Unit
- e: Power Meter Unit
- f: Pari Mutuel Urbain

P.M.V. a: Pa Ministerens Vegne
PMW e: Passive Microwave
PM[/]X e: Presentation Manager for the X window[s] [system]
PMxAS d: Primär-Multiplex-Anschluß
PMYMHMMFSWGAD e: Pardon Me, You Must Have Mistaken Me For Some One Who Gives A Damn [Internet]
Pmyr s: Plana mayor
PN e: Path Number
- e: Petri Net
- e: Pitcairn
- e: Primary Node
- e: Processing Node
- f: Paiement en Nature
- f: Paiement en Numéraires
- f: Parc National
- f: Patente Nette
- f: Pavillon National
- f: Pays Natal
- f: Pays Neutre
- f: Pays Nordiques
- f: Permis de Navigation
- f: Personne Naturalisée
- f: Personnes Nécessiteuses
- f: Petit Nom
- f: Peuple Nordique
- f: Peuplement Normal
- f: Plaque de Nationalité
- f: Police de la Navigation
- f: Pont
- f: Premier-Né
- f: Prière de Noter
- f: Prix Net
- f: Prix Nobel
- f: Prix Normal
- f: Produits Naturels
- f: Programme National
- f: Propriété Nationale
- f: Pupille de la Nation
- r: Permanenta Nätet
- s: Policía Nacional
- s: Policía Naval
- s: Puesto Naval

Pn s: Pelotón
PNA e: Pacific North Atlantic
- e: Pacific/North American
- e: Parallel Network Architecture
- e: Pennsylvania Nurses Association
- e: Polynuclear Aromatic Hydrocarbon[s]
- e: Programmable Network Access
- e: Protein equivalent of Nitrogen Appearance
- f: Parti National Africain

PNAD e: Personnel Nuclear Accident Dosimeter
PNAS e: Proceedings of the National Academy of Sciences
PNB e: Pacific Northwest Bell
- f: Produit National Brut
- s: Producto Nacional Bruto

PNBA e: Poly[n-Butyl Acrylic]
PNBH e: Petri-Net-Based Hypertext
PNC e: Japans Power Reactor and Nuclear Fuel Development Corporation
- e: Paging Network Controller
- e: People, Networks and Communication
- e: Peoples National Convention
- e: Power Reactor and Nuclear Fuel Development Corporation
- f: Professions Non Commerciales
- s: Policía Nacional Civil
- s: Programa Nacional sobre el Clima

PNCP e: Peripheral Network Control Point
PNCS e: Pakistani National Conservation Strategy
Pn CSR s: Pelotón de Canoñes Sin Retroceso

PND d: Personennamendatei
- e: Position-Navigation Device
- f: Parti National Démocratique
- f: Parti National Djiboutien
- f: Programme national déclaré
- s: Personal No Docente
PNDC e: Provisional National Defense Council
- p: Partido Nacional Democrático de Cabinda
PNdS d: Prüfung zum Nachweis deutscher Sprachkenntnisse
PNE f: Programme National d'Engrais
PNEA i: Programma Nazionale di Ricerche in Antartide
- p: Partido Nacional Ecologista de Angola
PNEDC f: Programme National d'Étude[s] de la Dynamique du Climat
Pneumonol Alergol Pol P: Pneumonologia i Alergologia Polska
PNF e: Partial Neutralization Feed
PNFE e: Prototype Network Front End
PNFI e: Petawawa National Forestry Institute
PNFPP e: Philippines National Family Planning Program
PNG e: Papua New Guinea
- e: Portable Network Graphic[s]
P N G Med J e: Papua New Guinea Medical Journal
PNGV e: Partnership for a New Generation of Vehicles
PNHP e: Physicians for a National Health Program
PNI e: Picture Network International
PNK d: Personalnebenkosten
PNL e: Pacific Northwest Laboratories (or Laboratory)
PNLF e: Planetary-Nebula Luminosity Function
PNLS f: Programme National de Lutte contre le Sida et les Maladies Sexuellement Transmissibles
PNM e: Periodicals/News/Microfilm unit
- e: Portable Anymap
- e: Pyranometer
- f: Programme National Manioc
PNMP e: Preliminary NMCS Master Plan
PNMR e: Proton Nuclear Magnetic Resonance
PNN s: Profesor No Numerario
P[-]NNI e: Private Network-to-Network Interface
PNNI e: Private Network-to-Network Interface
- e: Private Node-to-Network Interface
- e: Public Network Node Interface
PNNL e: Pacific Northwest National Laboratory
PNO e: Public Network Operator
pN$_2$O e: partial pressure of Nitrous Oxide
PNP d: Passauer Neue Presse
pnp e: BJT with net n-doped base and p-doped collector and emitter
PNP e: Plug and Play
- e: Positive Negative Positive
- f: Parti National pour le Progrès
PNPE f: Parti National pour le Progrès Economique
PNPG f: Pour une Nouvelle Politique à Gauche
PNPN e: Positive Negative Positive Negative [devices]
PNR e: Polish National Railways
- s: Partido Nacional Renovador
p.nr. r: personnummer
PNRI e: Philippines Nuclear Research Institute
PNRO e: DOE Pittsburgh Naval Reactors Office, West Mifflin, Pennsylvania
PNRP e: Pollution and Natural Resources Program
PNS e: Peacenet Sweden
- e: PPTP Network Server
- f: Parti National Socialiste
- f: Professeurs Non Salariés
PNSS e: Pregnancy Nutrition Surveillance System
- e: Pseudonoise Spread Spectrum
PNTC f: Parti National Travailliste Camerounais
PNU e: Protein Nitrogen Units
PNUD f: Programme des Nations Unies pour le Développement
- s: Programa de las Naciones Unidas para el Desarrollo
PNUE f: Programme des Nations Unies pour l'Environnement
PNULAD e: United Nations Program for Drug Abuse Control
PNUMA e: Chilean National Union for the Environment
- s: Programa de las Naciones Unidas para el Medio Ambiente
PNV d: Personennahverkehr
- e: Present Net Value
PNW e: Personal Netware
PO e: Office of Policy, Planning and Program Evaluation
- e: Paper Out
- e: Per Orally
- e: Per Os
- e: Persistent Object
- e: Polar-Optical
- e: Primitive data holding Object
- e: Program Office[r]
- e: Program Operations
- e: Project Office
- e: Purchase Order
- e: Purchasing Office
- f: [de] Premier Ordre
- f: Paiement Ordonnancé
- f: Par Ordre
- f: Paris-Orléans
- f: Pays d'Origine
- f: Permission Officielle
- f: Personnage Officiel
- f: Plantes Officinales
- f: Police Ouverte
- f: Préparation Officinale
- f: Présomption d'Origine
- f: Prix Obligatoire
- f: Produits Oléagineux
- f: Publications Officielles
- f: Pyrénées-Orientales
- s: Proyectil Ordinario
- s: Puesto de Observación
Po r: Personalorganisation
p.o. s: por orden
P/O e: Purchase Order
P&O e: Pipe and Operating
PO4 e: Phosphate
PO4-3 e: Phosphate
Po-212 e: Polonium-212
Po-216 e: Polonium-216
POA e: Plan Of Action [Internet]
- e: Program Office Approvals
- e: Programme Of Action
POAA s: Puesto de Observación Antiaéreo
PO AER s: Puesto de Observación Aérea
POAM e: Polar Ozone Aerosol Measurement
- e: Polar Ozone Aerosol Monitor
POB e: Periphal Order Buffer
- e: Port of Benton
Pob. s: Población
POBA e: Person[s] Overwhelmed By Acronyms
PO Box e: Post Office Box
POBS s: Puesto de Observación
POC e: Panel of Counselors
- e: Paralympic Organizing Committee
- e: Physical Oceanography Committee
- e: Picosecond Optical Calorimetry
- e: Point Of Compliance
- e: Point Of Contact
- e: Policy Oversight Committee
- e: Products of Conception
- e: Program Office Contacts
- e: Program-Operator Communication
- f: Comité d'Océanographie Physique
- p: Plataforma de Oposición Común
- s: Comité de Oceanografía Física
- s: Puesto de Observación Costero
POCC e: Payload Operations [and] Control Center (or Centre)
- e: Platform Operations Control Center
- e: Project Operations Control Center
POCCAR e: Programme on Physical Oceanography and Climate
- f: Programme sur l'océanographie physique et le climat
- s: Programa sobre Oceanografía Física y Clima
POC-ET e: Proof of Concept Experimental Testbed
POCM e: Parallel Ocean-Climate Model
POCS e: Passive Ocean Colour Subsystem
POCSAG e: Post Office Standardisation Advisory Group
POCT e: Periodically Oscillating Crystal Temperature
POCW e: Park Operations and Capital Works
POD e: Plain Old Documentation
- e: Plan Of the Day
- e: Planet Of Death
- e: Point Of Distribution
- e: Precise Orbit Determination
- e: Precision Orbit Determination [experiment]
- e: Printing On Demand
- e: Probability of Detention
- e: Programmable Option Devices
- e: Project Organization Demographic
- e: Proposed Operational Dates
- e: Proton Omnidirectional Detector
- e: Publishing On Demand

PODA e: Priority Oriented Demand Assignment
PoDAG e: Polar DAAC Advisory Group
PODI e: Print On Demand Initiative
PODS e: Pilot Ocean Data System
- e: Precise Orbit Determination System
POE e: Point Of Exposure
- e: Port Of Embarkation
- e: Port Of Entry
- e: Power Open Association
- e: Power Open Environment
- e: Program Office Estimate
- e: Property Owning Expenses
POEM e: Physical Oceanography of the Eastern Mediterranean
- e: Polar Orbiting Earth Monitoring
- e: Polar Orbit[ing] Earth[-observation] Mission
- e: Portable Object-oriented Entity Manager
- f: Océanographie physique de la Méditerranée orientale
- s: Oceanografía Física del Mediterráneo Oriental
POEMS e: Positron Electron Magnet[ic] Spectrometer
POEP e: Primary Operand Execution Pipeline
POES e: Polar-Orbiting [operational] Environmental Satellite[s]
POET e: Partially Overlapped Echo Cancelled Transmission
- e: Persistent Objects and Extended Database Technology
- e: Primer Oscillator Expendable Transponder
poet. e: poetic[al]
Poetry e: Prose typeset in unjustified lines
POEWGP e: Operational Experiments Working Group
POF e: Plastic Optic[al] Fiber
- e: Premature Ovarian Failure
- e: Programmable Operator Facility
POGO e: Polar Orbiting Geophysical Observatories (or Observatory)
e: Privately-Owned/Government-Operated
PoGP e: Pearl of Great Price
POH e: Path Section Overhead
- e: Power-On Hours
POHC e: Principal Organic Hazardous Constituent
POI e: Path Overhead Identifier
- e: Path Overhead Indicator
- e: Period Of Interest
- e: Point Of Information
- e: Point Of Interaction
- e: Point Of Interception
- s: Puesto de Observación Ingeniero
POIC e: Payload Operations Integration Centre
Point d: Programm für den automatischen Datenaustausch
POK e: Power on Okey Signal
PoK e: Pakistan-occupied Kashmir
POL e: Petroleum, Oil[s] and Lubricants
- e: Physicians OnLine
- e: Point Of Learning
- e: Political Science
- e: Problem Oriented Language
- e: Proudman Oceanographic Laboratory
Pol d: Institut für Politikwissenschaft
Pol f: Police
- f: Politique
Pol. e: Polish
- e: Political
- e: Politics
pol. N: politi
- N: politisk
- r: poliklinik
POLA e: Pointable Optical Linear Array
Polar e: Global Geospace Science Polar Spacecraft
POLAR e: Polar Plasma Laboratory
Pol Arch Med Wewn P: Polskie Archiwum Medycyny Wewnetrznej
POLARIS e: Photochemistry of Ozone Loss in Arctic Region In Summer
POLARSCAT e: Polarimetric Scatterometer
Polarstern e: Research vessel
POLCAP e: Petroleum Capability
POLDER e: Polarization and Directionality of [the] Earth Reflectances
PolDi d: Politische Didaktik
POLDIRAD e: Polarimetric Diversity [Doppler] Radar
PolDok d: Politische Dokumentation
POLE e: Polar Ozone Lidar Experiment
Pol. Econ. e: Political Economy
POLES e: Polar Exchange at the Sea Surface
POLEX e: An International Polar Experiment
- f: Nord Expérience polaire dans l'hémisphère Nord
POLEX Norte s: Experimento Polar Hemisferio Norte
POLEX South e: Polar Experiment Southern Hemisphere
POLEX Sud f: Expérience polaire dans l'hémisphère Sud
POLEX Sur s: Experimento Polar Hemisferio Sur
Polim Med P: Polimery w Medycynie
POLISARIO s: Popular para la Liberación de Saguia el Hamra y Rio de Oro
Polit. e: Political
- e: Politics
pol[it]. a: politisk
politiass. a: politiassistent
politibetj a: politibetjent
Pol J Pathol e: Polish Journal of Pathology
Pol J Pharmacol e: Polish Journal of Pharmacology
Pollen d: Daten mit einer Mailbox austauschen
Pol M s: Policía Militar
POL/MIL e: Political/Military
POLRE e: Pollution Report
POLREP e: PACOM Petroleum Report
- e: Pollution Report
POLS e: Politics
Pol Tyg Lek P: Polski Tygodnik Lekarski
POLW d: Politische Wissenschaft
Poly f: Polytechnique
PolyCenter e: Digitalis new name for DECmcc
POLYDAT e: Polymetallic Nodule Deposits Database
- f: Base de données sur les gisements de nodules polymétalliques
- s: Base de Datos sobre los Depósitos de Nódulos Polimetálicos
POLYGON e: Oceanographic Experiment in the North-East Atlantic
- f: Expérience océanographique dans l'Atlantique du Nord-Est
- s: Experimento Oceanográfico en el Atlántico Noreste
Polynya e: Wind or current driven opening in winter sea-ice
POM e: Particulate Organic Matter
- e: Persistent Object Manager
- e: Phase Of Moon
- e: Point of Presence
- e: Princeton Ocean Model
- e: Polycyclic Organic Matter
- e: Production/Operations Management
- e: Program Objective Memorandum
- e: Program Operating Memorandum
- e: Project Operations and Milestones
PoM e: Plan of Management
POMA e: Pointable Optical Mosaic Array
POMC e: Period of Minimum Change
POMCUS e: Prepositioned Overseas Material (or Materiel) Configured to Unit Sets
PO Min s: Puesto de Observación de Minas
POMINS e: Portable Mine Neutralization System
pomo e: post-modern
POMO e: Production Oriented Maintenance Organization
- e: Program Operations and Management Office
POMOA e: Provisional Office for Mass Organizational Affairs
PO Mov s: Puesto de Observación Móvil
POMS e: Pilot Ocean Monitoring Study
- e: Polar Operational Meteorological Satellite
- e: Polar Orbiting Meteorological Satellite
- f: Etude pilote de surveillance continue des océans
- s: Estudio Piloto de Vigilancia del Océano
POMSS e: Prepositioned Operational Material Storage Site
POMWG e: Program Objective Memorandum Working Group
p-on e: phoshorylation
PON e: Particulate Organic Nitrogen
- e: Project Originator Number
- e: Pump Octane Number
- f: Programme Océanographique National
Pon f: Pavillon
PONC e: Price Of Non-Conformance
Pon Red s: Pontón para Redes
PONT s: Pontoneros
POO e: Portland, OR
POOH e: Pull Out Of Hole
POOL e: Parallel Object Orientated Language
POOP e: Proper Object Oriented Programming
POOT s: Puesto de Observación de la red de Observadores Terrestres

POOZ e: Permanently Open Ocean Zone
POP e: International Polar Orbiting Platform
- e: Package for Online Programming
- e: Parallel Ocean Program
- e: Period Of Performance
- e: Permanently Out of Print
- e: Persistent Organic Pollutant
- e: Point Of Presence
- e: Point Of Production
- e: Point-of-Presence
- e: POP from Stack
- e: Processor On Plug in
- e: Profiler Online Program
- f: Parti Ouvrier et Paysan du Congo
- f: Parti Ouvrier Populaire
- f: Polluant Organique Persistant
- p: Polícia da Ordem Pública
- R: PO Polyot
- s: Patrulla de Orden Público
Pop f: Populaire
pop. e: popular[ly]
POP3 e: Post Office Protocol, version 3
POPA e: POP All registers
POPCI f: Parti Ouvrier et Paysan de Côte d'Ivoire
POPCRU e: Police and Prison Service Civil Rights Union
POPE e: Polar Orbiting Platform Element
POPF e: POP Flags
POPIN e: United Nations Population Information Network
POPLINE e: Population Information Online
POPS e: Paperless Order Processing System
- e: Primary Oceanographic Prediction System
- e: Principles of Operations
POPs e: Points Of Presence
POPSAT e: Precise Orbit Positioning Satellite-Geodetic Satellite
PopSci e: Popular Science
POP-Server e: Post Office Protocol-Server
Popul Rep A e: Population Reports. Series A; Oral Contraceptives
Popul Rep B e: Population Reports. Series B; Intrauterine Devices
Popul Rep C e: Population Reports. Series C; Female Sterilization
Popul Rep D e: Population Reports. Series D; Male Sterilization
Popul Rep E e: Population Reports. Series E; Law and Policy
Popul Rep H e: Population Reports. Series H; Barrier Methods
Popul Rep J e: Population Reports. Series J; Family Planning Programs
Popul Rep L e: Population Reports. Series L; Issues in World Health
Popul Rep M e: Population Reports. Series M; Special Topics
Popul Trends e: Population Trends
POR e: Point Of Receive
- e: Point Of Return
- e: Power-On-Reset
- e: Price On Request
- e: Prime contractor Organization
- e: Process-Of-Record
- e: Program Of Requirements
- f: Région de l'Océan Pacifique
- s: Región del Océano Pacífico
PORBASE e: On-line Union Catalogue of Portuguese Libraries
Porc. e: Porcelain
PORP e: Partial Ossicular Replacement Prosthesis
PORS e: Project Officer Record System
PORSEC e: Pacific Ocean Remote Sensing Conference
port r: portion
Port. e: Portlandian
PORTAL e: Polar Orbiter Remapping and Transformation Application Library
PORTS e: Physical Oceanographic Real-Time System
- e: Port Objective for Real-Time Systems
- e: Ports Characteristics File
- e: Portsmouth Gaseous Diffusion Plant, Portsmouth, Ohio
port[ug]. a: portugisisk
POS e: Peacetime Operating Stocks
- e: Permanently Out of Stock
- e: Persistence (or Persistent) Object Service
- e: Point-Of-Sale[s]
- e: Polar Orbiting Satellite
- e: Primary Operating Stocks
- e: Product-Of-Sums
- e: Program Operations Staff
- e: Programmable Object Select
- e: Programmable Option Select
- f: Période d'Observation Spéciale
- f: Plan d'Occupation des Sols
- r: Pansarofficersskola
- s: Por Orden Superior
pos N: posisjon
POS[] e: Position in AA or NA sequence
POSC e: Petrotechnical Open Software Corporation
- e: Primary Operations System Control
POS[C] s: Posición
POSE e: Portable Operating System Environment
- e: Portable Operating System Extension
POSEIDON e: French component of TOPEX POSEIDON
- e: Positioning Ocean Sea Earth and Ice Dynamics and Orbiting Navigator
- e: Satellite Program[me]
- f: Projet d'Observation (ou Observatorie) de Surveillance et d'Études Intégrées de la Dynamique des Oceans
POSF e: Ports of Support File
POSH e: Probability Of Severe Hail
Poshaub r: Positionshaubits
POSI e: Promoting conference for OSI
- e: Promotion of OSI
POSIT e: Position Report
- e: Profiles for Open Systems Interconnection Technologies
- e: Profiles for Open Systems Interworking Technologies
POSIX e: Portable Operating System Interface for Unix
POSNA e: Pediatric Orthopaedic Society of North America
POSNAV e: Position/Navigation
POSS e: Plant Occurrence and Status System
- e: Program Operations Support Staff
- e: Prototype Optical Surveillance Subsystem
Poss f: Possédant
poss. e: possessive
POSSLQ e: Person of Opposite Sex Sharing Living Quarters [Internet]
POST e: Parliamentary Office of Science and Technology
- e: Periodically Oscillating Source Temperature
- e: Point-Of-Sale Transaction
- e: Power-On Self Test
- e: Proposed Optical Stinger
- e: Prototype Ocean Surveillance Terminal
Posten d: eine Nachricht in eine Newsgroup verschicken
Postepy Biochem P: Postepy Biochemii
Postepy Hig Med Dosw P: Postepy Higieny I Medycyny Doswiadczalnej
PostG d: Postgesetz vom 22.12.1997
PostGiroO d: Postgiroordnung
Postgrad Med e: Postgraduate Medicine
Postgrad Med J e: Postgraduate Medical Journal
POSTGRES e: POSTGRES Object-Relational Database Mangement System
POSTNET e: Postal Numeric Encoding Technique
PostO d: Postordnung
PostPersRG d: Postpersonalrechtsgesetz vom 14.9.1994
PostSpO d: Postsparkassenordnung
PostUmwG d: Postumwandlungsgesetz vom 14.9.1994
postv. a: postvæsenet
POSYBL e: Programming System For Distributed Applications
POT e: Periodic Orbital Theory
- e: Potential Temperature
- e: Pre-production Operational Test
- e: Program for Operational Trajectories
- s: Puesto de Observación Técnica
pot. a: potentiel[t]
POTAD e: Program for Operational Transport and Dispersion
POTC e: Peck on the cheek [Internet]
POTDR e: Polarization Optical Time-Domain Reflectometry
POTO e: Potomac River Basin
POTRK e: Region 3 Purchase Order Tracking System
POTS e: Plain Old Telephone Service
- e: Plain Old Telephone System
- e: Plain Ordinary Telephone Service
- e: Plain Ordinary Telephone System
- e: Purchase of Telephones and Systems
Pott. e: Pottery
POTV e: Personnel Orbit Transfer Vehicle
POTW e: Publicly Owned Treatment Works
POTWA e: Post and Telecommunications Workers Association

POU[CG] e: Point-of-Use [Chemical Generation]
Poult Sci e: Poultry Science
POV e: Persistence Of Vision
- e: Point Of View [Internet]
POVMR e: Problem-Oriented Veterinary Medical Record
POV-ray e: Persistence Of Vision Ray Tracer
POWC e: Program Out-of-Window Collision
POWER e: Performance Optimized (or Optimization) With Enhanced RISC
- e: Priority Output Writer, Execution processors and input Readers
- e: PROFS Operational Weather Education and Research
POWER2 e: A line of single-chip IBM POWER products
POWER CTR e: Power Center for the Performing Arts
PowerPC e: Performance Optimization with Enhanced RISC-Performance Computing
POWS e: PROFS Operational Work Station
POY e: Partially Oriented Yarn
POZC f: Programme pour les Océans et les Zones Côtières
- s: Programa para los Océanos y las Zonas Costeras
PP d: Publishing Praxis [product]
- e: Packet Processor
- e: Painting Processor
- e: Parallel Plate
- e: Parallel Port
- e: Participatory Program
- e: Passage Point
e: Path Provisioning
- e: Payload Pointer
- e: Payphone
- e: Peering Point
- e: Perfect Program
- e: Physical Partition
- e: Physical Plane
- e: Plasma Processing
- e: Point-to-Point
- e: Portable Part
- e: Post Pay
- e: Predictive Pipelining
- e: Presentation Position
- e: Pressure Plate
- e: Primary Point
- e: Principles and Practice
- e: Private Pilot
- e: Program Planning
- e: Projection Plane
- e: Proposed Plan
- f: Paiement Partiel
- f: Par Pièce
- f: Par Préciput
- f: Par Prescription
- f: Par Priorité
- f: Par Procuration
- f: Parc Public
- f: Partage Proportionnel
- f: Parti du Peuple
- f: Parti Paysan
- f: Parti Politique
- f: Partie Publique
- f: Parties Prenantes
- f: Pas de Porte
- f: Payable au Porteur
- f: Pays de Provenance
- f: Pays Producteur
- f: Peine Pécuniaire
- f: Pension Provisoire
- f: Permis Périmé
- f: Permis Permanent
- f: Personne Physique
- f: Personne Privée
- f: Pétition de Principe
- f: Pilote Professionnel
- f: Plaque de Police
- f: Pleins Pouvoirs
- f: Police Parisienne
- f: Port Payé
- f: Portion Principale
- f: Poste de Police
- f: Pouvoirs Publics
- f: Préfecture de Police
- f: Premier Président
- f: Presse Périodique
- f: Prise de Possession
- f: Prison Préventive
- f: Prisonnier Politique
- f: Prix de Pension
- f: Prix de Place
- f: Prix pour le Public
- f: Proches Parents
- f: Procuration Partielle
- f: Profits et Pertes
- f: Programme Parisien
- f: Publication Périodique
- f: Puissance Paternelle
- p: Partido del Progreso
- p: Polícia de Protecção
- s: Partido Populista
- s: Proyectil Perforante
Pp f: Passeport
- f: Pères
- f: Publié par
P.P. s: Por Poder[es]
- s: Porte Pagado
P/P e: Pre/Planned
P&P e: Planning and Programming
- e: Policy and Procedure Statements
- e: Preservation and Packaging
P2P e: Peer to Peer [net]
PPA e: PACAF Product Performance Agreement Center
- e: Parent of a Person with Autism
- e: Peoples Progressive Alliance
- e: Per Procura
- e: Planned Program Accomplishment
- e: Pollution Prevention Act
- e: Polymer Processing Additive
- e: Polyphenylacetylene
- e: Polyphthalamide
- e: Post Pack Audit
- e: Program[me] Project [and] Activity
- f: Parti Populaire Africain
- f: Parti Progressiste Africain
- R: Pastanovchik-Pamech Aktivnii
PPAG e: Pediatric Pharmacy Advocacy Group
PPARC e: Particle Physics and Astronomy Research Council
PPAS e: Personal Property Accountability System
PPAs e: Purchase Power Agreements
PPB e: Parts Per Billion
- e: Police Paramilitary Battalion
PPB[a] e: means atoms per billion atoms
PPBES e: Planning, Programming, Budgeting and Execution System
ppb[k] e: paperback
PPBS e: Planning, Programming and Budgeting System[s]
- e: Program Planning and Budget Staff
ppbv e: parts per billion by volume
PPB[w] e: means parts per billion of weight
PPC e: Chlorinated Polypropylene
- e: Parallel Plate Chambers
- e: Personal Protective Clothing
- e: Pin-to-Pin Compatible
- e: Policy and Planning Committee
- e: Polyphthalate Carbonate
- e: Postal code for the Prime Contractor
- e: Power PC
- e: Process Proximity Correction
- e: Production Planning and Control
- e: Program-to-Program Communication
- e: Project Preparation Committee
PPCA e: Preliminary Physical Configuration Audit
PPCH s: Proyectil Perforante de Carga Hueca
PPCP e: People to People Contact for Peace
- e: Power PC Platform
- s: Programa de Investigación sobre Predicción Meteorológica a Corto y Medio Plazo
PPCS e: Person to Person, Collect, Special
PPD e: I-GOOS Panel on Products and Distribution
- e: Peripheral Pulse Distributor
- e: Postscript Printer Description
- e: Project Planning Document
- f: Groupe de l'I-GOOS sur les produits et la diffusion
- f: Programme Populaire de Développement
- s: Panel del I-GOOS sobre Productos y Distribución
PPDD e: Peoples Party for Democracy and Development
PPDM p: Partido do Progresso do Povo Moçambicano
PPDO e: Provincial Planning and Development Office
PPDPC f: Parti pour la Démocratie et le Progrès Comorien
PPDU e: Presentation Protocol Data Unit[s]
PPDVWG e: Population Dynamics and Physical Variability Working Group
- f: Groupe de travail sur la dynamique des populations et la variabilité physique
- s: Grupo de Trabajo sobre Dinámica de Poblaciones y Variabilidad Física
PP&E e: Production Process and Equipment

PPE e: Particle Physics Experiments Division
- e: Party for the Protection of the Environment
- e: PAVE PAWS East
- e: People, Places and Events
- e: Personal Protective Equipment
- e: Personnel Protection Equipment
- e: Peruto Peñasco special upper-air observing site
- e: Polyphenylene Ether
- f: Papeete
- f: Parti pour la Protection de l'Environnement
- r: Pjäspresentationsenhet

PPEI s: Proyectil Perforante Explosivo Incendiario
Ppens f: Prix de pension
PPET s: Proyectil Perforante Explosivo Trazador
PPF e: Columbus Polar Platform
- e: Plain Paper Fax
- e: Print Production Format

PPFD e: Photosynthetic Photon Flux Density
PPFS e: Pre-Printed Forms System
- e: Product Performance Feedback System

PPG e: Performance Partnership Grant
- e: Priority Planning Grid
- e: Pulses Per Group
- f: Parti du Peuple Guinéen

Ppg e: Pounds per gallon
PPGA e: Plastic Pin Grid Array
PPGE f: Parti du Progrès de la Guinée Equatoriale
PPGM e: Planning, Programming and Guidance Memorandum
Ppgn f: Perpignan
PPI d: Pape und Partner Informationssysteme GmbH
- e: Pre-planned Product Improvements
- e: Professional Press Inc.
- e: Publishing Partners International
- f: Parti Progressiste Ivoirien
- s: Proyectil Perforante Incendiario

PPIC e: Pesticide Programs Information Center
PPID e: Parent PID
- e: Parent Process Identification number
- e: Process Program Identification

PPIF e: Photographic Processing and Interpretation Facilities
PPI gauge e: international loading gauge
p-pille a: præventivpille
PPIRG e: PAVE PAWS Interface Requirements Group
PPIS e: Pesticide Product Information System
- e: Pollution Prevention Incentives for States

PPIVM e: Passive Physiological Intervertebral Movement
PPJR f: Parti Progressif pour la Jeunesse du Rwanda
PPL d: Projekt Pro Lehre, offizielles Evaluationsprojekt der FU Berlin
- e: PC-board Programming Language
- e: People
- e: Point-to-Point Link
- e: Power Plate Loading
- e: Private Pilots License
- e: Process-to-Process Link

Ppl e: People [Internet]
p-plads a: parkeringsplads
PPLAN e: Programming Plan
pple. e: participle
pp[le]. a[dj]. e: participial adjective
PPLEK p: Partido Popular Liberal Etu Mudyetu
PPLFCRM p: Partido do Progresso, Liberal e Federal das Comunidades Religiosas de Moçambique
PPLP s: Programa de Investigación sobre Predicción Meteorológica a Largo Plazo
P-PLUS e: Polar Platform Utilisation Study
PPM e: Page[s] Per Minute
- e: Parts per Mille
- e: Peoples Pharmacy of Mali
- e: Perfect Prognosis Method
- e: Portable Pix Map
- e: Positions and Proper Motions
- e: Prediction by Partial Matching
- e: Presidential Review Memorandum
- e: Principal Period of Maintenance
- e: Product Portfolio Management
- e: Project Programmer Number
- f: Méthode de Prévision Parfaite
- f: Pharmacie Populaire du Mali
- p: Polícia Popular de Moçambique
- s: Método de Predicción Perfecta

ppm e: part[s] per million
P/PM e: Program/Project Manager
PPMC e: Petroleum Products Marketing Company
- e: Physician Practice Management Companies

PPME e: Protection and Preservation of the Marine Environment
PPMF e: Pulp and Paper Manufacturers of Australia
PPMIS e: Personal Property Management Information System
PPMMS e: Polaris/Poseidon Material Management System
PPM/PPB e: Parts Per Million/Parts Per Billion
PPMS e: Personal Property Management System
- e: Personal Property Movement and Storage System
- e: Polyphenylmethylsiloxane

ppmv e: parts per million by volume
PPN e: Parameterized Post Newtonian formalism for general relativity
- e: Partial Parenteral Nutrition
- e: Peroxypropionyl Nitride
- e: Programmer Project Number
- e: Project Programmer Number
- e: Public Packet Switching
- f: Parti Progressiste Nigérien

PPND s: Proyectil Perforante de Núcleo Duro
PPNG e: Penicillinase-Producing Neisseria Gonorrheas
PPO e: Preferred Provider Organization
- e: Prepaid Purchase Order
- f: Petit Paquet Ordinaire

PPOA e: Pollution Prevention Opportunity Assessment
Ppol f: Poste de police
Ppon f: Par procuration
PP-ORMNT e: Open Partial Agreement on the Prevention of, Protection Against and Organization of Relief, in Major Natural and Technological Disasters
- f: Accord partiel ouvert en matière de prévention, de protection et d'organisation des secours contre les risques naturels et technologiques majeurs
- s: Acuerdo Parcial Abierto sobre Prevención, Protección y Socorro en caso de Catástrofes Naturales y Tecnológicas Importantes

P Posc s: Punto de Posición
PPOX e: Polypropylene Oxide
PPP d: Punkt-zu-Punkt-Protokoll
- e: Palmoplantar Pustulosis
- e: Peoples Progressive Party
- e: Point-to-Point Protocol
- e: Preposition Procurement Package
- e: Progressive Peoples Party
- e: Public Policy Program
- e: Purchasing Power Parities
- f: Parti Patriotique du Peuple
- f: Parti Progressiste du Peuple
- f: Points Par Pouce
- f: Préfecture de Police de Paris

PPPB in: Pusat Penelitian dan Pengembangan Biologi
PPPL e: Princeton Plasma Physics Laboratory, Princeton, New Jersey
PPPM p: Partido do Progresso do Povo de Moçambique
PPPO e: Personal Property Processing Office
- in: Pusat Penelitian dan Pengembangan Oseanologi

PPPP e: Previous Paragraph was Polemical Position
PPQ e: Plant Protection and Quarantine
- e: Possible Parliamentary Question
- e: Pulses Per Quarternote
- s: Pulgada[s] Quadrada[s]

PPR e: Photopolarimeter Radiometer
- e: Planar Plate Reactor
- e: Prepublication Review
- f: Petit Paquet Recommandé
- s: Pedido de Piezas de Repuesto

PPRC e: Physician Payment Review Commission
- e: Pollution Prevention Resource Center

PPRef e: Process Program Reference
PPRF e: Pontine Paramedian Reticular Formation
Pprj r: Pansarprojektil
PProv f: Permis Provisoire
PPRR e: Program Pre-Release Reviews

PPS d: Printing and Publishing Service GmbH & Co. KG
- e: Packets Per Second
- e: Party for Progress and Socialism
- e: Party for Progress and Solidarity
- e: Policies and Procedures
- e: Post Post Scriptum
- e: Post-Polio Syndrome
- e: Postpostscriptum
- e: Postproduction Support
- e: Power Personal Systems
- e: Precision Processing System
- e: Product Performance Surveys
- e: Program Performance Specification
- e: Proposal Pricing System
- e: Provincial Peace Structure
- e: Public Packet Switching [network]
- e: Pulses Per Second
- f: Parti Populaire Sénégalais
- f: Parti pour le Progrès et la Solidarité
- f: Parti pour le Progrès et le Socialisme
- f: Parti Progressiste Soudanais

PPSI d: poröses polykristallines Silizium
PPSM f: Profils par Pays dans le domaine des Sciences de la Mer
PPSP e: Pollution Prevention and Safety Panel
- e: Postproduction Support Plan

PPSU e: Polyphenylene Sulfone
PPT e: Parts Per Trillion
- e: Permanent Part Time
- e: Polypropylene Terephthalate
- e: Polysilicon Pressure Transducer
- e: Port Processing Time
- e: Printer Pass-Through
- e: Program Processing Table
- e: Pulse Pair Timing
- e: Temple University, Physics Library
- f: Parti pour le Progrès Togolais
- s: Proyectil Perforante-Trazador

Ppt f: Passeport
PptDipl f: Passeport Diplomatique
Ppté f: Propriété
PPTP e: Point-to-Point Tunnel[l]ing Protocol
pptv e: parts per trillion by volume
PPU e: Pollution Prevention Office
- e: Pool Planned Unit

PPUC e: Probably Permanently Under Construction
PPV e: Pay Per View
- e: Poly[p-PhenyleneVinylene]

PPW e: Pave Paws West
PPWSM e: Post Production Weapon System Management
PQ f: Personnes Qualifiées
- f: Plan Quinquennal
- f: Première Qualité
- f: Presse Quotidienne
- f: Prime de Qualification
- s: Puesto Quirúrgico

Pq s: Parque
PQA[v] s: Puesto Quirúrgico Avanzado
PQ/CEI s: Parque Central de Incendios
PQF f: Plan Quinquennal Français
PQFP e: Plastic Quad Flat Pack
PQL e: Practical Quantitation Limit
PQLI e: Physical Quality of Life Index
PQP e: Priorities, Quality, Productivity
PQRP f: Plan Quinquennal pour le Relèvement de la Production
PQT e: Preliminary Qualifications Test
- e: Production Qualification Testing

PQUA e: Preliminary Quantitative Usage Analysis
Pque s: Parque
PQV e: Program Quality Verification
PR e: Deputy Chief of Staff, Requirements and Programs
- e: Directorate of Programs and Resources
- e: Pacing Response
- e: Party of Reunification
- e: Path Replacement
- e: Pesticide Regulation Notice
- e: Pharmaceutical Representative
- e: Photoresist
- e: Phrasal Representation
- e: Polarization Ratio
- e: Precipitation Radar
- e: Product Removal
- e: Puerto Rico
- e: Purchase Request
- e: Purchase Requisition
- f: Parti Radical
- f: Pension de Retraite
- f: Pension Réversible
- f: Place Réservée
- f: Police Routière
- f: Poste Recommandée
- f: Poste Restante
- f: Président de la République
- f: Prévention Routière
- f: Prime de Rapidité
- f: Prime de Remboursement
- f: Prix de Revient
- f: Prix Réduit
- f: Procuration Révocable
- f: Procureur de la République
- f: Propriétaires Riverains
- s: Partido de Reunificación
- s: Proyectil Rompedor
- s: Pulsorreactor
- s: Punto de Referencia

Pr f: Pour
- f: Précédent
- f: Prime
- f: Prochain
- f: Production
- f: Professeur
- f: Propre
- s: Prisionero

P R e: Public Relations
pr. a: per
- e: present
- N: presis
- r: procent
- r: prov
- r: prövning

Pr. e: Provencal
P&R e: Planning and Resources
- e: Plans and Requirements

PR- r: Pansarreglementen
P[R] e: Receive sequence number of Packet
PRA e: Popular Resistance Army
- e: Preliminary Reserve Analysis
- e: Primary Rate Access
- f: Parti de Regroupement Africain
- i: Potencia Radiada Aparente
- p: Partido da República Angolana
- s: Partido Revolucionario Auténtico

PRAC s: Programa para Ciegos Ancianos
- s: Punto de Resistencia Antitanque de [la] Compañía

PRACCI f: Pour la Réflexion et l'Action des Certifiés Cadres Infirmiers
PRACDA e: Polarimetric Radar Control and Data Aquisition
PRACSA e: Public Remote Access Computer Standards Association
Pract. e: Practical
- e: Practice

P Rad s: Puesto de Radio
PrAdj f: Professeur Adjoint
PRAE f: Programme consacré à la Recherche Atmosphérique et à l'Environnement
PRAI e: Project Research Applied in Industry
PRAIEN f: Centre de préparation pratique aux applications industrielles de l'énergie nucléaire
P[-]RAM e: Parameter Random Access Memory
PRAM e: Parallel Random-Access Machine
- e: Parameter Random Access Memory
- e: Permanent Random-Access Memory

PRAMS e: Paperwork Reduction Act Management System
- e: Passenger Reservation and Manifesting System
- e: Pregnancy Risk Assessment Monitoring System
- e: Programmatic Risk Assessment Model System

PRAP e: Pylos Regional Archaeological Project
PRARE e: Precise (or Precision) Range and Range Rate Equipment
PRAREE e: Precise Range and Range Rate Equipment, Extended version
PRAS e: Particle Reactor Analysis Services
- e: Principle Research Areas

Prat f: Pratique
PRATS e: Pesticides Regulatory Action Tracking System
PRAU e: Public Relations Association of Uganda
Prax Kinderpsychol Kinderpsychiatr d: Praxis der Kinderpsychologie und Kinderpsychiatrie
Prax Kinderpsychol Kinderpsychiatr Beih d: Praxis der Kinderpsychologie und Kinderpsychiatrie. Beiheft
PRB e: Pay Review Body
- e: Physical Review B
- e: Pseudo-Random Binary
- f: Parti de la Renaissance du Bénin

PRBS e: Pseudo-Random Binary Sequences
- e: Pseudo-Random Bit Sequence

PRC e: Litton/PRC, Incorporated
- e: Peoples Redemption Council
- e: Planning Research Corporation
- e: Portuguese Civilization
- e: Precipitation
- e: Primary Reference Clock
- e: Prime contractor Region Code
- e: Projects Review Committee
- e: Provisional Ruling Council
- f: Parti de la Révolution Centrafricaine
- f: Parti Républicain Centrafricain

PrC f: Présidence du Conseil
P.R.C. s: Punto de Referencia de Comunicaciones
pr[c]. a: præcis
Prcdt s: Paracaidista
PRCI f: Parti Républicain de Côte d'Ivoire
PRCP e: Programmed Communication Electronic Support Program
PrCt f: Prix Courant
PRD e: Party for Renovation and Development
- e: Party for Reviving Democracy
- e: Presidential Review Directive
- e: Product Requirements Document
- f: Parti du Renouveau Démocratique
- f: Parti pour Revivre la Démocratie
- f: Parti Radical-Démocratique Suisse
- p: Partido para a Renovação e o Desenvolvimento

PR&D e: Process Research and Development
PRDA e: Program Research and Development Agreement
P3RDB e: Prelaunch/Postlanding Processing Requirements Data Base
PRDI f: Parti Réformiste Démocratique Ivoirien
PRDMD e: Private Directory Management Domain
PRDS f: Parti Républicain Démocratique et Social
PRE e: Partial Reflection Experiment
- e: Preformatted
- e: Prime Contractors Region
- f: Fédération européenne des fabricants de produits réfractaires
- p: Programa de Reabilitação Econômica

Pre f: Préfecture
PréAO f: Présentation Assistée par Ordinateur
Préap f: Préapprentissage
Préc f: Précédemment
- f: Précédent
- f: Précision

PRECIS e: Preserved Context Index System
PRECP e: Processing of Emissions by Clouds and Precipitation
pred. e: predicative
Pref. e: Preface
pref. e: prefix
Préf f: Préfecture
- f: Préfet

Préfre f: Préfecture
PREG s: Partido Reformador Guatemalteco
P Reg s: Puesto de Regulación
PREGUA s: Por la Reivindicacion del Ejercito de Guatemala
PRELOG e: Peoples Revolutionary League of Ghana
PRELOGE e: Preliminary Logistics Evolution
PREMIS e: Premises Information System
PREMO e: Presentation Environment for Multi-Media Objects
PREMPT e: Program For Electromagnetic Pulse Testing
PREN e: European draft standard
Prenat Diagn e: Prenatal Diagnosis
PREOS e: Performance and Range of Electro-Optical Systems
PREP e: Pediatrics Review and Education Program
- e: Penaeid Prawns Recruitment Project
- e: Pre-Edit Processor
- e: Programmable Electronics Performance Corp.
- f: Projet sur le Recrutement des Pénéidés
- s: Proyecto de Repoblación de Camarones Peneidos

PReP e: Power PC Reference Platform
Prep s: Preparación
prep. e: preposition
PREPA f: [Société de] Prospection et Exploitation Pétrolière en Alsace
Prep Biochem Biotechnol e: Preparative Biochemistry and Biotechnology
PREPCOM e: Preparatory Commission for the International Seabed Authority and for the International Tribunal for the Law of the Sea
- e: Preparatory Consultations Meeting
- f: Commission préparatoire de l'Autorité internationale des fonds marins et du Tribunal international du droit de la mer
- s: Comisión Preparatoria de la Autoridad Internacional de los Fondos Marinos y del Tribunal Internacional del Derecho del Mar

PREPNET e: Pennsylvania Research and Economic Partnership Network
PREP/RAMP e: Pre-Edit Processor/ Report and Message Processor
PREPRZM e: Pesticide Root Zone Model
PRES e: Promotion Recommendation Forms
pres. e: present
Prés f: Président
Presq f: Presqu'île
PRESS e: Prediction Error Sum of Squares
Press f: Pression
Presse Med i: Presse Medicale
PREST e: Policy Research in Engineering, Science and Technology
Prest f: Prestation
PRESTO e: Precursory Research for Embryonic Science and TechnOlogy
- e: Programmable Etalon Spectrometer for Twilight Observations

PRESTO-E e: PA A Low-Level Radioactive Waste Environmental Transport and Risk
PRE-STORM e: Preliminary Regional Experiment for STORM [Stormscale Operational and Research Meteorology]
PRET e: Program for the Recovery of the Economy in Transition
Pret. e: Pretoria
Preu. s: Preuniversitario
PREVENT e: Pacific Northwest Regional Visibility Experiment using Natural Tracers
Prévis f: Prévision
PREVLINE e: Prevention Online
Prev Med e: Preventive Medicine
PRF e: Parliament Reference
- e: Personnel Resources File
- e: Preferences
- e: Problem Report Form

PRFV s: Plástico Reforzado por Fibra de Vidrio
PRG e: PAL Programming Department
- e: Program file
- e: PROM Programming Department

Prg f: Progression
PRGC s: Proyectil Rompedor de Gran Capacidad
PRGE p: Partido Republicano de Guinea Ecuatorial
P R Health Sci J e: Puerto Rico Health Sciences Journal
PRI e: Periodic Reinvestigation
- e: Primary Rate Interface
- e: Public Radio International
- f: Parti Radical Indépendant
- f: Prévention Routière Internationale
- s: Partido de la Revolución Islámica
- s: Punto de Reunión de Informes

pri N: prioritet
PRICAI e: Pacific Rim International Conference on Artificial Intelligence
PRICE e: Peroxy Radicals Intercomparison Exercise
- e: Productivity Improvement and Cost Effectiveness
- e: Programmed Review of Information for Costing and Estimation

PRICE-S e: Software Costing Model
PRIDE e: People and Resources Identified for Distributed Environments
- e: Performance Readiness Initiatives Decision Evaluation System
- e: Policy based Routing, Implementation and Deployment in Europe

Prikl Biokhim Mikrobiol U: Prikladnaia Biokhimiia i Mikrobiologiia
Prim f: primaire
PRIMA e: Primary Care Information Management across Anglia
PRIMAS e: Precipitation Radar Image Management System
Prim Care e: Primary Care; Clinics in Office Practice
PRIME e: Program for Innovation in Microenterprise
PRIMI s: Proyecto de Rescate del Indioma Maya Itzaj
PRIMIR e: Product Improvement Information Report
PRIMIS e: McGraw Hill digitised book selection and printing initiative
PRIMO e: International Research Programme in the Western Mediterranean
- f: Programme de recherche international en Méditerranée occidentale
- s: Programa de Investigación Internacional en el Mediterráneo Occidental

Prin. e: Princeton

Princ. e: Principle[s]
Princess Takamatsu Symp e: Princess Takamatsu Symposia
Princeton e: Princeton University
PRINTF e: Print with Formatting
PRIO e: International Peace Research Institute
PR/IR e: Public Relations/Investor Relations
PRIR s: Proyectil Rompedor Instantáneo y con Retardo
PRIRODA e: Multisensor module of the space station MIR
PRIS e: Personnel Resource Impact System
Pris s: Prisionero
PRISE e: Primary Care Sharing the Evidence
PRISM e: OCLCs new online service
- e: Parallel Reduced Instruction Set Multiprocessing (or Multiprocessor)
- e: Photo-Refractive Information Storage Material
- e: Prioritized Requirements, Impacts and Schedule Milestones
- e: Projection and Integrated Stand alone Monitor
PRISMS e: Phoenix Real-time Instrumentation for Surface Meteorological Studies network
- e: Plan of Research for Integrated Soil Moisture Studies
Priv f: privilégié
priv. e: privative
PRIZ e: Prizes, Committee on
PRJ e: Project
Prj r: projektil
PRJCI f: Parti du Rassemblement du Peuple pour la Jeunesse de Côte d'Ivoire
PRK e: Photorefractive Keratectomy
PRL e: Physical Research Laboratory, Navrangpura
- e: Physical Review Letters
- e: Polar Research Laboratory
- e: Private Relocatable Library
- e: Prolactin
- f: Parti Républicain de la Liberté
- f: Pouvoir Révolutionnaire Local
P R L e: Physics Research Laboratory
PRL ADMIN e: Parks, Records and Libraries Administration
PRLG e: Prologue
pr. Ltn N: premierløytnant
PRM e: Pressure Modulated Radiometer
- e: Prevention Reference Manuals
- e: Process Resource Manager
- e: Program Review Meetings
- e: Programmers Reference Manual
- e: Project Review Meeting
- e: Pulse Repetition per Minute
- e: Pyrradiometer
- f: Parti Républicain de Madagascar
- f: Prêt Remboursable par Mensualité
- p: Partido Revolucionario de Moçambique
- p: Polícia da República de Moçambique
PRMC e: Pacific Regional Monitoring Center
- f: Parti Révolutionnaire Marxiste Congolais
PRMD e: Private Management Domain
pr min N: per minutt
PRMIS e: Printing Resources Management Information System
PRML e: Partial Response Maximum Likelihood
PRMM s: Programa Regional Meteorológico Marítimo
PRMS e: Plans Review Management System
PRMT f: Programme de Recherche en Météorologie Tropicale
PRN e: Pesticide Registration Notice
- e: Post-Rotatory Nystagmus
- e: Program Revision Number
- f: Parti de Regroupement National
- s: Plan de Reorganización Nacional
- s: Punto de Reunión
Prn f: Prochain
pRNA e: Plasmid RNA
PRND f: Parti pour la Reconstruction Nationale et la Démocratie
PRNG e: Pseudo Random Number Generator
PRO e: Digital Equipment Corporation
- e: Plastic Repair Optimized
- e: Precision RISC Organization
- e: Profile
- e: Public Records Office
- e: utilization and quality control Peer Review Organization
- R: Protivoraketnaya Oborona
Pro f: Procédure
- f: Professionnel
PROAP f: Bureau principal régional pour l'Asie et le Pacifique
- s: Oficina Principal Regional para Asia y el Pacífico
prob. e: probably
PROBAR e: Programmable Band Radiometer
PROBE e: Pilot Radiation Observation[s] Experiment
- e: Portable Remote Observation of the Environment
- e: Prototype Radiation Observation Experiment
Probem p: Associação dos Proprietérios Moçambicanos de Bens Nacionalizados
Probl Khig U: Problemi Na Khigienata
Probl Kosm Biol R: Problemy Kosmicheskoi Biologii
Probl Med Wieku Rozwoj P: Problemy Medycyny Wieku Rozwojowego
Probl Sotsialnoi Gig Istor Med R: Problemy Sotsialnoi Gigieny i Istoriia Meditsiny
Probl Tuberk R: Problemy Tuberkuleza
proc e: Prozedur
PROCAARE e: Program for Collaboration Against AIDS and Related Epidemics
PROCAFOR s: Programa Regional Forestal para Centro América
Proc Annu Meet Med Sect Am Counc Life Insur e: Proceedings, Annual Meeting of the Medical Section of the American Council of Life Insurance
Proc Annu Symp Comput Appl Med Care e: Proceedings of the Annual Symposium on Computer Applications in Medical Care
Proc Assoc Am Physicians e: Proceedings of the Association of American Physicians
Proc Inst Mech Eng [H] e: Proceedings of the Institution of Mechanical Engineers. Part H, Journal of Engineering in Medicine
Proc Natl Acad Sci USA e: Proceedings of the National Academy of Sciences of the United States of America
Proc Natl Sci Counc Repub China B e: Proceedings of the National Science Council, Republic of China. Part B, Life Sciences
Proc Nutr Soc e: Proceedings of the Nutrition Society
ProCom e: IEEE-SA Standards Board Procedures Committee
PROCOM e: Projection Compositor
PROCOPA e: Peruvian-German Cooperative Program for Fisheries Investigations
- f: Programme coopératif germano-péruvien de recherches sur la pêche
- s: Programa Cooperativo Peruano-Alemán de Investigación Pesquera
Proc R Soc Lond [Ser]B Biol Sci e: Proceedings of the Royal Society of London. Series B: Biological Sciences
Proc Soc Exp Biol Med e: Proceedings of the Society for Experimental Biology and Medicine
PROCURE e: Command Procurement System
Proc West Pharmacol Soc e: Proceedings of the Western Pharmacology Society
Proc World Congr Anaesthesiol e: Anaesthesiology. Proceedings of the World Congress of Anaesthesiologists [journ.]
PRODESA s: Proyecto de Desarollo Santiago
ProdHaftG d: Gesetz über die Haftung für fehlerhafte Produkte [Produkthaftungsgesetz]
PRODOS e: Professional Disk Operating System
ProdTAut d: Fraunhofer-Institut für Produktionstechnik und Automatisierung [IPA] und Institut für Industrielle Fertigung und Fabrikbetrieb der Universität Stuttgart
PROE f: Programme Régional Océanien de l'Environnement
PROECODI s: Promoting Ecological and Integral Development
PROF e: Premixed One-Dimensional Flame Code
Prof f: Professeur
prof. a: professionel
- N: profesjonelt
Prof. s: Profesor
Profagr f: Professeur agrégé
PROFFENS e: Prototype Flash Flood Estimation and Nowcasting System
PROFILE e: An on-line search service
PROFIS e: Professional Officer Filler Information System
PROFIT e: Planners Remote Offline Form Image Tape
- e: Propulsion/Flight Control Technology

PROFRUTA s: Proyecto de Desarrollo de la Fruticultura y Agroindustria
PROFS e: Professional Office System
- e: Program for Regional Observing and Forecasting Service
- e: Prototype [for] Regional Observing and Forecasting Service
PROG e: Peer Review Oversight Group
- e: Program[mer]
Prog f: Progrès
Prog AIDS Pathol e: Progress in AIDS Pathology
Prog Behav Modif e: Progress in Behavior Modification
Prog Biochem Pharmacol e: Progress in Biochemical Pharmacology
Prog Biophys Mol Biol e: Progress in Biophysics and Molecular Biology
Prog Brain Res e: Progress in Brain Research
Prog Cardiovasc Dis e: Progress in Cardiovascular Diseases
Prog Cardiovasc Nurs e: Progress in Cardiovascular Nursing
Prog Clin Biol Res e: Progress in Clinical and Biological Research
Prog. Clin. Biol. Res. e: Progress in Clinical and Biological Research
Prog Clin Parasitol e: Progress in Clinical Parasitology
Prog Drug Res e: Progress in Drug Research
PROGERT f: Projet de Recherche pour l'Observation et la Gestion des Ressources Terrestres
Prog Exp Pers Psychopathol Res e: Progress in Experimental Personality and Psychopathology Research
Prog Exp Tumor Res e: Progress in Experimental Tumor Research
Prog. Fish-Cult. e: Progressive Fish-Culturalist
Proggi e: Programm
Prog Histochem Cytochem e: Progress in Histochemistry and Cytochemistry
PROG.INV. s: Programa de Investigación
Prog Lipid Res e: Progress in Lipid Research
Prog Liver Dis e: Progress in Liver Diseases
Prog Med Chem e: Progress in Medicinal Chemistry
Prog Med Virol e: Progress in Medical Virology
Prog Mol Subcell Biol e: Progress in Molecular and Subcellular Biology
Prog Nucleic Acid Res Mol Biol e: Progress in Nucleic Acid Research and Molecular Biology
Prog. Oceanogr. e: Progress in Oceanography
Prog Pediatr Surg e: Progress in Pediatric Surgery
Progress in Neurobiology e: Prog Neuropsychopharmacol Biol Psychiatry
Prog Urol e: Progres en Urologie
PROHUZA f: [Centre d'études et d'informations sur les problèmes] humains en zones arides
Proj f: Projet
- r: Projektil
PROKLA d: Probleme des Klassenkampfes
PROLIFIC e: Programme Ligurean France, Italy, Canada
PROLOG e: Programming Logic
PROM e: Pockels Read-Out Optical Modulator
- e: Programable Read Only Memory
- f: Mémoire morte programmable
- s: Memoria de lectura solamente programable
PROMACX e: Production Management and Control System
PROMAR e: Program[me] on the Promotion of Marine Sciences
- f: Programme de promotion des sciences de la mer
- s: Programa de Promoción de las Ciencias del Mar
PROMARE e: Programme on Marine Arctic Ecology
- N: [det Nasjonale] Program for Marin Arktisk Ekologi
PROMATS e: Programmable Magnetic Tape System
ProMED e: Program for Monitoring Emerging Diseases
PROMESS e: Penmarch Radar Oceanographic Microwave Experiment on Signature Studies
PROMIS e: Procurement Management Information System
- e: Programme for an Operational Meteorological Information System
PROMMIS e: Programmed Restructuring of Maintenance Management Information System
PROMOGABON f: Agence Gabonaise de Promotion Industrielle et Artisanale
Promot Educ f: Promotion et Education
PROMS r: Produktionsfaktorskatt
pron. e: pronoun
PRONAM f: Programme National Maïs
pronunc. e: pronunciation
PROP e: Protected Areas Programme
- f: Programme pour les aires protégées
- s: Programa de Areas Protegidas
prop. e: properly
- r: proportionell
- r: proposition
Prop. s: propietario
PROPAL e: Programmed PAL
Propé f: Propédeutique
PROPEPH s: Programa de Pequeños Productores de Huehuetenango
Pror f: Prorogé
PRO-RADAM s: Projecto Radargrammetrico del Amazonas
PRORADAM p: Projeto Radar Amazonas
PROS e: ROSAT analysis system developed within IRAF
Pros. e: Prosody
PROSA a: Programmørernes og Operatørernes Landssammenslutning
PROSEA e: Plant Resources of Southeast Asia
Prostaglandins Leukot Essent Fatty Acids e: Prostaglandins Leukotrienes and Essential Fatty Acids
Prostate Suppl e: Prostate. Supplement
Prosthet Orthot Int e: Prosthetics and Orthotics International
Prot f: Protocole
- s: Protección
prot. a: protestantisk
- a: protokol
ProTech e: Prospective Technology Communication System
Protein Eng e: Protein Engineering
Protein Expr Purif e: Protein Expression and Purification
Protein Sci e: Protein Science
PROTEL e: Procedure Oriented Type Enforcing Language
PROTEUS e: Profile Telemetry of Upper Ocean Currents
- e: Prototype Real-Time Operational Test, Evaluation and User Simulation
Protocol/Internet d: Protokoll regelt den Daten- und Informationsaustausch [Kommunikation] im Internet
Protokoll d: Verhaltensregeln und Formate bei der Daten-Kommunikation zwischen entfernten Partnern gleichen logischen Niveaus
prov r: proviant
- r: provisorisk
Prov. e: Provencal
prov. N: proviant[ering]
- r: provins
prov. förr. r: proviant förråd
Prov I r: Provisorisk Instruktion
Provider d: kommerzieller AnbieterIn von Datenübertragungsdienstleistungen im Internet
Provis f: Provisoire
PROVORG e: Providing Organization
prov. pl. N: provianteringsplass
PrOWP d: Prüfungsordnung für Wirtschaftsprüfer
Prox s: Próximo
Proxy d: Zwischenstation für das Abrufen von Interntdaten
Proxy-Server d: Zwischenspeicher
Proy s: Proyector
Proy O s: Proyector para Observación en el mar
PRP e: Peer Review Panel
- e: Peoples Redemption Party
- e: Precision Registration Processor
- f: Parti pour la Réconciliation du Peuple
- f: Parti Républicain pour le Progrès
- f: Parti Révolutionnaire du Peuple
- s: Proyectil Rompedor de Percusión
PRPB e: Peoples Revolution Party of Benin
- f: Parti de la Révolution Populaire du Bénin
PRQ f: Parti Républicain de Québec
PRR e: Project Requirements Review
- f: Pension de Retraite Réversible
- f: Poste Restante Recommandée
PrR f: Présidence de la République
PRRD e: Proprietary Rights to Registration Data

PRRS e: PACAF Readiness Reporting System
- e: Porcine Reproductive and Respiratory Syndrome
PRS e: Persian
- e: Personal Response System
- e: Print Management Service
- e: Printer
- e: Procurement Tracking System
- e: Production Reporting System
- f: Parti Républicain Sénégalais
- p: Partido de Renovação Social
- p: Partido para a Renovação Social
Prs f: Paris
- f: Persan
PRSC e: Parametric Response Surface Control
PRSE f: Parti pour la Réhabilitation Ivoirienne du Social et de l'Economie
PRSIG e: Pacific Rim SMDS Interest Group
PRT e: Polysilicon Resonant pressure Transducer
- s: Proyectil Rompedor a Tiempos
PRTA s: Puesto de Recolección de la Técnica
prTBR e: Draft Technical Basis for Regulations
PRTC e: Puerto Rico Telephone Company
PRTI e: Professional Realtime Interface
PRTLS e: Powered RTLS
PRTYPOLS e: Priority Pollutants
PRU e: Percent Reduction of Urea
- e: Personnel Readiness Unit
PRUCOL e: Permanently Residing Under Color Of Law
PRUN s: Partido Republicano de Unidad Nacional
PRV r: Patent- och Registreringsverket
Prvt f: Provisoirement
Prx f: Prix
Przegl Epidemiol P: Przeglad Epidemiologiczny
Przegl Lek P: Przeglad Lekarski
PRZM e: Pesticide Root Zone Model
PS a: Politiken Søndag
- d: Proportionalschrift
- e: Parenteral Society
- e: Particle Simulation
- e: Performance Status
- e: Personal System
- e: Phosphatidylserine
- e: Physics Specification
- e: Planning Staff
- e: Point Source
- e: Polar Stereographic
- e: Polystyrene
- e: Popular Science
- e: Porous Silicon
- e: Portable Station
- e: Postscript[um]
- e: Power Steering
- e: Preparedness Staff
- e: Privilege Service
- e: Probability of Survival
- e: Process Solvent
- e: Process stop Ship
- e: Production System
- e: Programmed Symbols
- e: Proportional Spacing
- e: Proton Synchrotron Division
- e: Public School
- e: Public Service
- f: Partage Successoral
- f: Patente de Santé
- f: Patente Suspecte
- f: Patronage Scolaire
- f: Perte Sèche
- f: Police Sanitaire
- f: Population Salariée
- f: Port Supplémentaire
- f: Préfecture de la Seine
- f: Prestation Supplémentaire
- f: Prix de Souscription
- f: Professeur Stagiaire
- N: Panserskolen
- r: Pansartruppskola
- R: Passazhirskii i Syaz
- s: Puesto de Socorro
Ps e: Proseminar
P S e: Physical Science
PS. e: Programmed symbols
Ps. e: Psalm[s]
P.S. e: Post Script[um]
- l: Postscriptum
P&S e: Personnel and Security
P[S] e: send Sequence number of Packet
PS/2 d: Persönliches System 2
- e: Personal Computer/2
- e: Personal System 2
- e: Programming System 2
PSA e: Pacific Science Association
- e: Pacific Sciences Association
- e: Particularly sensitive Sea Area
- e: Physical Sector Address
- e: Precipitation Series Algorithm
- e: Prices Surveillance Authority
- e: Principal Staff Assistants
- e: Project Support Agreement
- e: Project-Specific Analysis
- e: Prostate-Specific Antigen
- e: Protocol-Specific Annex
- e: Public Servants Association
- f: Association Scientifique du Pacifique
- f: Parti Sans laisser d'Adresse
- f: Parti Socialiste Africain
- f: Peugot SA
- p: Partido Socialista Angolano
- s: Asociación Científica del Pacífico
PSAA e: Pacific Special Activities Area
PSAC e: President's Science Advisory Council
PSAG e: Probabilistic Systems Assessment Group
PSAG-CD e: Public Service Advisory Group/Collection Development
PSAG-PS e: Public Service Advisory Group/Public Services
PSAG-TS e: Public Service Advisory Group/Technical Services
PSAM e: Point Source Ambient Monitoring
PSAP e: Pennsylvania Student Assistance Program
- e: Pharmacotherapy Self-Assessment Program
PSaS e: Personally Safe and Sound
- e: Personally Safe and Sound [IBM product]
PSAT e: Predicted Site Acquisition Table
PSB d: Projektstatusbericht
- e: Path State Block
- e: Phase Shifting Blank
- e: Polished Silica Block
- e: Product Symbology Block
- e: Program Status Book
- e: PS Booster
- f: Parti Socialiste Belge
PS-b-PI e: Polystyrene/Polyisoprene block copolymer
p.s/c s: por su cuenta
PSC e: Periodic Surveillance Committees
- e: Personal Service Contract
- e: Personal Services Contractor
- e: Personal Super Computer
- e: Phase Shift Keying
- e: Pittsburg Supercomputer Center
- e: Pittsburgh Scientific Center
- e: Pittsburgh Supercomputer (or Supercomputing) Center
- e: Plant Simulation Code
- e: Point Stress Criterion
- e: Polar Science Center
- e: Polar Stratospheric Cloud
- e: Polar Stratospheric Clouds
- e: Porous Silicon Capacitor
- e: Prime Service Contractor
- e: Print Server Command
- e: Product Service Center
- e: Program Service Center
- e: Program Site Coordinator
- e: Public Safety Calling system
- f: Parti Social-Chrétien
- f: Parti Socialiste Camerounais
- f: Parti Socialiste des Comores
- f: Produits de Substitution des Céréales
PSCM e: Pilot Scale Ceramic Melter
PSCN e: Program Support Communication Network
PSCNET e: Pittsburgh Supercomputing Center Network
PSCP e: Public School Construction Program
PSCRC e: Portable Small Computer Requirements Contract
PSCT e: Peripheral Stem Cell Transplant
- e: Polymer-Stabilized Cholesteric Texture
PSCU e: Private Sector Casemix Unit
PSD e: Page Setup Dialog
- e: Page Setup Dialogue
- e: Parallel Shift and Dwell
- e: Performance and Security Document
- e: Phase-Sensitive Detection
- e: Photo-Sensitive Detection (or Detector)
- e: Platform Specific Driver
- e: Port Status Display
- e: Position-Sensitive Detection (or Detector)
- e: Program System Description
- e: Program Systems Division
- e: Programmers Supplementary Documents
- f: Parti Social Démocrate
- f: Parti Social Démocratique
- N: Partistridsdommer
- p: Partido Social e Democrático
- s: Pesado
PSDA e: Post-Storm Data Acquisition
- p: Partido Social Democrata Angolano

PSDC e: Pennsylvania State Data Center
- f: Parti Social Démocrate Congolais
PSDD e: Planetary Science Data Dictionary
PSDI f: Parti Social-Démocrate Italien
PSDL e: Region 4 PSD Log System
PSDN e: Packet Switched (or Switching) Data Network
- e: Packet Switched (or Switching) Digital Network
- e: Public Switched (or Switching) Data Network
- e: Public Switched (or Switching) Digital Network
PSDPC e: Polar Satellite Data Processing Centre
PSDS e: Packet-Switched Data Service
PSDT f: Parti Social Démocrate Togolais
PSE d: Personen-Such-Einrichtung
- e: Packet Switch Exchange
- e: Pale Soft Exudative pork
- e: Port Switched Ethernet
- e: Producer Subsidy Equivalent
- e: Program Sub-Element
- e: Project Requirements Review
- s: Proyectil Sin Estallar
P Sec s: Puesto Secreto [de observación]
ps[ec] e: picosecond
PSECU e: Pennsylvania State Employees Credit Union
PSEF e: Plastic Surgery Educational Foundation
PSEMA f: Parti Social d'Education des Masses Africaines
PSERVER e: Print Server
PSES e: Pretreatment Standards for Existing Sources
PSF e: Performance Shaping Factors
- e: Permanent Swap File
- e: Polysulfone
- e: Print Services Facility
- e: Program Support Facility
Psfc e: surface Pressure
PSFG e: Permanent Service for/on the Fluctuation of Glaciers
PSFS e: Performance Shaping Factors
PSG e: Pershing
- e: Production Systems Group
- e: Professional Service Group
- f: Parti Socialiste Gabonais
- f: Parti Socialiste Guinéen
Psg f: Passager
PSG f: Produit Social Global
- s: Partido Socialista de Guatemala
PSGAG e: Polysulfated Glycosaminoglycan
PSGE p: Partido Socialista de Guinea Ecuatorial
Psgr r: Pansargranat
PSH. e: Physical Services Header
PSHTM e: Pilot Scale High Temperature Melter
PSI d: Prozeßsteuerungs- und Informationssysteme AG
- e: Pacific Stratus Investigation
- e: Pacific Sulfur Investigation
- e: Packet Switch Interface
- e: Packetnet System Interface
- e: Parapsychology
- e: Parapsychology, Psi
- e: Policy Studies
- e: Pollutant Standards Index
- e: Post-tensioning Institute
- e: PostScript Interpreter
- e: Pressure per Square Inch
- e: Process Safety Information
- f: Parti Socialiste Italien
- f: Parti Socialiste Ivoirien
- f: Petit Système Individuel
psia e: pounds per square inch, absolute
PSIC e: Planning and Systems Integration Center
PSID e: Postscript Image Data
psid e: pounds per square inch, differential
psig e: pounds per square inch, gauge
PSII e: Plasma Source Ion Implantation
PSIMPLE e: Problem-Solving In Medical Physiology, Logically Explained
PSIS e: Plastic Surgery Information Service
PSIU e: Packet Switch Interface Unit
PSJ e: Policy Studies Journal
PSK e: Phase Shift Key
Psk r: Pansarskepp
- r: Pansarskytt
Pskimp r: Pansarspaningskompani
p-skive a: parkeringsskive
Pskott r: Pansarskott
PSKR R: Pogranichniy storozhevoy korabl
Pskybil r: Pansarskyttebil
Pskyddskomp r: Pansarskyddskompani
Pskyttebv r: Pansarskyttebandvagn
PSL e: Environmental Study Center
- e: Permanent Seat License
- e: Physical Science Libraries
- e: Polystyrene Latex
- e: Problem Statement Analyser
- e: Program Supply Logic Unit
- e: Program Support Letter
- e: Program Support Library
- e: Public Software Library
PSLP e: Private Sector Linkages Program
PSL/PSA e: Problem Statement Language/ Problem Statement Analysis
PSLR e: Peak Side Lobe Ratio
PSLS e: Polystyrene Latex Sphere
PSLU e: Power Supply Logic Unit
PSM e: Packet Service Module
- e: Persistent Stored Modules
- e: Personal Software Marketing
- e: Personnel Systems Manager
- e: Phase Shift[ing] Mask
- e: Point Scattering Model
- e: Point Source Modeling
- e: Point Source Monitoring
- e: Position Switching Module
- e: Print Management System
- e: Printing Systems Manager
- e: Process Safety Management
- e: Professional Service Manager
- e: Professional Staff Member
- e: Program Security Manager
- e: Project Security Manager
- e: Psychosomatische Medizin
- f: Parti Socialiste Malgache
- f: Parti Socialiste Mauricien
PSMO s: Puesto de Servicio de Mantenimiento del Orden
PSMP e: Program[me] on Short and Medium-Range Weather Prediction Research
- e: Public Sector Management Programme
- f: Programme de recherche sur la prévision météorologique à courte et à moyenne échéance
- s: Programa de Servicios Meteorológicos al Público
PSMSL e: Permanent Service for Mean Sea Level
- f: Service permanent du niveau moyen des mers
- s: Servicio Permanente del Nivel Medio del Mar
PSN e: Packet Switch Node
- e: Packet Switching Network
- e: Penicillin, Streptomycin and Neomycin
- e: Private Switching Networks
- i: Piano Spaziale Nazionale
PSNH e: Public Service Company of New Hampshire
PSNP e: Partial Sequence Number Packet
- e: Primary School Nutrition Programme
PSNS e: Pretreatment Standards for New Sources
PSNZ e: Pharmaceutical Society of New Zealand
PSO e: Pending Service Order
- e: Polysulfone
- e: Project Science Office
- N: Politiets Sentralorganisasjon
p.s/o s: por su orden
P/solde f: Pour solde
PSOM e: Persistency framework in SOMobjects
Pson f: Prison
Ps or Pss e: Psalm or Book of Psalms
PSOS e: Personal Services Operating System
PS + P e: Political Science and Politics

PSP e: Paint Shop Pro
- e: Paralytic Shellfish Poisoning
- e: Payload Support Plan
- e: Payroll Savings Plan
- e: Peoples Solidarity Party
- e: Personal Software Process
- e: Personal Software Products
- e: Phased Support Plan
- e: Pierced Steel Planking
- e: Precision Spectral Pyranometer
- e: Priority Strike Plan
- e: Probability-of-Single-Pulse
- e: Product Support Plan
- e: Program Segment Prefix
- e: Program Support Plans
- e: Progressive Socialist Party
- e: Public Security Police
- f: Intoxication paralysante par les coquillages
- f: Parti Socialiste Pan-Africain
- p: Polícia de Segurança Pública
- s: Parálisis tóxica de los moluscos
PSPA e: Physician Services Practice Analysis
PSPC e: Permanent South Pacific Commission
- e: ROSAT Position Sensitive Proportional Counters
PSPD e: Permits and State Programs Division
PSPDC e: Polar Satellite Precipitation Data Centre
PSPL e: Puget Sound Power and Light
PSPS e: Passive Surface Pressure Sounder
PSQ e: Personnel Security Questionnaire
PSR e: Personal Service Radio
- e: Positive Support Review
- e: Post-Strike Reconnaissance
- e: Product Specific Realizations
- e: Product Specific Release
- e: Program Status Review
- f: Parti Socialiste du Rwanda
- f: Parti Socialiste Réunionnais
psr s: por sustitución reglamentaria
psrak[et] r: pansarskyddsraket
P-SRAM e: Pseudo-Static Random Access Memory
PSRAM e: Pseudo SRAM [Static Random Access Memory]
PSRG e: Privacy and Security Research Group
PSRPP e: Public Safety and Resource Protection Program
PSRR e: Power Supply Rejection Ratio
PSRS i: Partito Socialista Rivoluzionario Somalo
PSRT e: Postscript Round Table
PSS A: Private Sektor-Stigting
- e: Packet Switch[ed] Services
- e: Packet Switch[ing] Stream
- e: Personnel Security Staff
- e: Personnel Staffing Specialist
- e: Physical Security Specialist
- e: Physiologic Saline Solution
- e: Private Signalling System
- e: Program Support Staff
- e: Project Summary Sheet
- e: PUREX Sludge Supernatant liquid
PSSC e: Packaging, Shipping and Safety Committee
PSSG e: Protocol Standards Steering Group
PSSI e: Pacific/Sulfur/Stratus Investigation
PSSK e: Probability of Single-Shot Kill
PSSL e: Private Source Statement Library
PSSM e: Photographic Systems Simulation Model
- e: Planar Supercell Stack Method
PST d: Patentstelle für die Deutsche Forschung
- e: Pre-Stretched Tubing
- e: Pre-Structured Technology
- f: Parti Suisse du Travail
- f: Promotion Supérieure du Travail
pst. N: prosent
P-staben N: Personellstaben
PSTART e: Page Start register
PSTC f: Pour Solde de Tout Compte
pstg r: postering
PSTN e: Pesticide Safety Team Network
- e: Public Service Telephone Network
- e: Public Swiched (or Switching) Telephone Network
Psto s: Puesto
PSTOL e: Spacecraft Test Operations Language
PSTOP e: Page Stop register
PSTP e: Protocol Standards Technical Panel
PSTV e: Protocol Specification, Testing and Verification
PSU e: Page Storage Unit
- e: Penn State University
- e: Polysulfone
- e: Public Service Union
- e: Publications Services Unit
PS[U] e: Power Supply [Unit]
PSUA p: Partido Socialista Unificado da Alemanha
PSUN e: Public Service Union of Namibia
PSV e: Platform Supply Vessel
- e: Preliminary Site Visit
PSVT e: Pilot Secure Voice Terminal
PSW e: PUREX Sludge Waste
PSWM e: Postscript Window Manager
PSWU e: Public Services Workers Union
PSY d: Psychologie
- e: Psychology
Psy f: Psychologique
PSYC e: Psychology
Psych d: Institut für Philosophie, Pädagogik und Psychologie, Abteilung für Psychologie
Psychiatr Clin North Am e: Psychiatric Clinics of North America
Psychiatr Enfant f: Psychiatrie de l'Enfant
Psychiatr Genet e: Psychiatric Genetics
Psychiatr Pol P: Psychiatria Polska
Psychiatr Prax d: Psychiatrische Praxis
Psychiatr Q e: Psychiatric Quarterly
Psychiatr Serv e: Psychiatric Services
Psychiatry Clin Neurosci e: Psychiatry and Clinical Neurosciences
Psychiatry Res e: Psychiatry Research
Psychoanal Q e: Psychoanalytic Quarterly
Psychoanal Rev e: Psychoanalytic Review
Psychoanal Study Child e: Psychoanalytic Study of the Child
Psych[ol]. e: Psychology
Psychol. e: Psycholog[ical]
Psychol Aging e: Psychology and Aging
Psychol Bull e: Psychological Bulletin
Psychol Issues e: Psychological Issues
Psychol Med e: Psychological Medicine
Psychol Med Monogr Suppl e: Psychological Medicine. Monograph Supplement
Psychol Rep e: Psychological Reports
Psychol Res e: Psychological Research
Psychol Rev e: Psychological Review
Psychopharmacol Bull e: Psychopharmacology Bulletin
Psychopharmacol Ser e: Psychopharmacology Series
Psychosom Med e: Psychosomatic Medicine
Psychother Psychosom e: Psychotherapy and Psychosomatics
Psychother Psychosom Med Psychol d: Psychotherapie, Psychosomatik, Medizinische Psychologie
PS y Cl s: Puesto de Socorro y Clasificación
PSYD e: Doctor of Psychology
PsyD, LCP e: Doctor of Psychology, Licensed Clinical Psychologist
psyk r: psykologisk
psyk. a: psykologi[sk]
Psyres d: Psychotherapeutische Forschungsstelle in Stuttgart
PSYWAR e: Psychological Wafare
ps.Z e: compressed postscript
PSZ d: Psychosoziale Medizin
- e: Partially Stabilized Zirconia
PT d: Patient
- d: Produktionstechnik
- d: Projekttutorium
- e: Packet Type
- e: Patrol Torpedo
- e: Payload Type
- e: Personnel and Training
- e: Perturbation Theory
- e: Phase Transition
- e: Physical Therapy
- e: Plain Text
- e: Plasma Triode
- e: Posts and Telecommunications
- e: Pro Tempore
- e: Program Timer
- e: Publication Type
- e: Test Plan
- e: Torpedo Boat
- f: Pacte Tacite
- f: Parti du Travail
- f: Parti Travailliste
- f: Pays Transitaire
- f: Pension Temporaire
- f: Permis Temporaire
- f: Perte Totale
- f: Petit Taille
- f: Piastre
- f: Plantes Textiles
- f: Police Temporaire
- f: Postes et Télécommunications
- f: Préavis de Traitement
- f: Presse Technique
- p: Polícia de Trânsito
- s: Patrullero
- s: Proyectil a Tiempos

Pt e: Plantinum
- e: Platelet
- f: Petit
- f: Président
- f: Printemps
- f: Provisoirement
- s: Portátil
pt e: Portugal [Internet]
pt. a: patent
Pt. e: Portuguese
P&T a: Post- og Telegrafvæsenet
- e: Packaging and Transportation
- e: Personnel and Training
- e: Pharmacy and Therapeutics
- e: Privilege and Tenure Committee
P/T e: Precision-to-Tolerance ratio
P[/]T e: Part Time
PTA e: Parent Teacher Association
- e: Preferential Trade Area for/of East[ern] and South[ern] Africa[n] [states]
- e: Procrustes Target Analysis
- f: Parti du Travail Albanais
- f: Prise Totale Autorisée
- s: Captura Total Permisible
PTAB e: Project Technical Advisory Board
PTAC e: Procurement Technical Assistance Centers
PTAL e: Payment Transaction Application Layer
PTAP e: Profiler Triangle Analysis Package
PTAT e: Pesticides, Toxics and Air Team
- e: Private Trans Atlantic Telecommunications
PTB d: Physikalisch-Technische Bundesanstalt, Braunschweig
- e: Powers That Be [Internet]
- f: Parti Travailliste Brésilien
- s: Punto de Travesía en Balsa
PTC e: Peace Through Confrontation
- e: Pentagon Telecommunications Center
- e: Plastics Technology Centre
- e: Post and Telecommunications Corp.
- e: Pre- and post-process Treatment Chambers
- e: Primary Technical Control
- f: Permis de Transports en Commun
- p: Polícia dos Transportes e Comunicações
Pt/CrNe e: Plantinum Cromium Neon-onboard spectral calibration
PTCS e: Pentagon Telecommunication Center System
PtCT e: Patient Care Technologies
PtCt f: Paiement au Comptant
PTD e: Parallel Transfer Disk
- e: Parallel Transfer Disk Drive
- e: Parallel Transfer Drive
Ptd e: Phosphatidic
- e: Phosphatidyl
PtdCho e: Phosphatidylcholine
PtdEtn e: Phosphatidylethanolamine
PtdGro e: Phosphatidylglycerol
PtdIns e: Phosphatidylinositol
PtdIns 4-P e: Phosphatidylinositol-4-Phosphate
PTDIS e: Single Stack Meteorological Model in EPA UNAMAP Series
PtdSer e: Phosphatidylserine
PTE e: Path Terminating Equipment
- e: Potential to Emit
- s: Punto de Tratamiento Especial
Pte f: porte
- s: presente
- s: puente
P-te f: Poste
P-te de Dne f: Poste de Douane
PTemp f: Permis Temporaire
Pte Mtar s: Puente Militar
Pte Pont s: Puente de Pontones
Pte Rte f: Poste Restante
Pte Tab s: Puente de Tableros
PTF e: P-Thread Framework
- e: Part-Time Force
- e: Phase Transfer Function
- e: Pool Transmission Facility
- e: Problem Temporary/Trouble Fix
- e: Program Trouble Fix
- e: Prototype and Test Facility
- s: Policía de Tráfico
P&TF e: Prototype and Test Facility
PTFE e: Polytetrafluo[o]rethylene
PTG e: Torpedo Boat, Guided missile armament
ptg. a: portugisisk
ptgb[il] r: personalterrängbil
PTH d: Philosophisch-Theologische Hochschule
- d: Psychotherapeutische Medizin
- e: Pin-Through-Hole
- e: Plated Through-Hole
- f: Pression, Température, Humidité
- s: Presión, Temperatura, Humedad
PTHrP e: PTH-related Proteins
PTI e: I-GOOS Panel on Technical Implementation
- e: Packet Type Identifier
- e: Payload Type Identifier
- e: Payload Type Indicator
- e: Portable Test Instrument
- e: Press Trust of India
- e: Product Test Interface
- f: Groupe de l'I-GOOS sur la réalisation technique
- s: Panel del I-GOOS sobre Ejecución Técnica
PTIS e: Photothermal Infrared Spectroscopy
- e: PhotoThermal Ionization Spectroscopy
Ptjs f: Pour toujours
PTK r: Privattjänstemannakartellen
PTL e: Process Tolerance Limit[s]
- e: Project Title
PTLLA s: Patrulla
PTM e: Missile-armed Torpedo Boat
- e: Packet Transfer Mode
- e: Packet Transport Mode
- e: Pilot Training Missile
- e: Professional, Technical and Management
- e: Pulse Tie Modulation
PTMAX e: Single Stack Meteorological Model in EPA UNAMAP series
PTMM e: Please, Tell Me More [Internet]
PTMP e: Point To Multi Point
PTMTP e: Multistack Meteorological Model in EPA UNAMAP Series
PTN e: Public Telephone Network
- f: Pression et Température Normales
- s: Pelotón
- s: Proyectil Termonuclear
Ptn Cbl Mltp s: Pelotón de Cable Múltiple
Ptn Cbl Ord s: Pelotón de Cable Ordinario
Ptn Hrtz s: Pelotón de Cable Hertziano
Ptn Rd s: Pelotón de Radio
Ptn Tf s: Pelotón de Telefonía
PTO e: Pacific Theater of Operations
- e: Parent-Teacher Organization
- e: Patent and Trademarks Office
- e: Postal Town of the prime contractor
- e: Public Telecommunication [network] Operator
Pto s: Puerto
- s: Puesto
PTOCA e: Presentation Text Object Content Architecture
PTOM[A] f: Pays et Territorires d'Outre-Mer [Associés]
Pton s: Pelotón
Pton Amet s: Pelotón de Ametralladoras
Pton Amet AA s: Pelotón de Ametralladoras Antiaéreas
Pton Aut s: Pelotón de Automotrices
Pton CD s: Pelotón Contradesembarco
Pton Comones s: Pelotón de Comunicaciones
Pton Const s: Pelotón de Construcción
Pton Expl s: Pelotón de Exploración
Pton Fgo s: Pelotón de Fuego
Pton I s: Pelotón de Infantería
Pton IM s: Pelotón de Infantería de Marina
Pton Ind TA s: Pelotón Independiente de Transporte Automóvil
Pton Mdo s: Pelotón de Mando
Pton Mort s: Pelotón de Morteros
Pton Mot s: Pelotón de Motociclistas
Pton Rd s: Pelotón de Radio
Pton Sum s: Pelotón de Suministro
Pton T s: Pelotón de Tanques
Pton Trans s: Pelotón de Transporte
Pton Z[ap] s: Pelotón de Zapadores
PTP e: Peripheral Target Position
- e: Preoperational Test Procedure
- e: Probability To Penetrate
- e: Programmable Telemetry Processor
- e: Progress of Theoretical Physics
- f: Parti Togolais du Progrès
- f: Perte Totale Présumée
PTPLU e: Point Source Gaussian Diffusion Model
PTPRM e: human Receptor-like Protein Tyrosine Phosphatase
PTPS e: Photothermal Phase-shift Spectroscopy
PTR e: Platinum Resistance
- e: Pointer Record
- e: Post-Tenure Review
Ptr e: Pointer
- r: Patron
- r: Patrull
Ptrb r: Patrullbåt
Ptrbrote r: Patrullbåtrote
Ptrh r: Patronhylsa
Ptrlåda r: Patronlåda

PTRM e: Partial Thermoremanent Magnetization
Ptrp r: Pansartrupper
Ptrrör r: Patronrör
PTRRS e: Pipeline Time Recording and Reporting Systems
PTS d: Papiertechnische Stiftung
- d: Parktronic System
- e: Phillips Theological Seminary
- e: Photothermal Spectroscopy
- e: Plasma Technology Section
- e: Pressure Tuning Spectroscopy
- e: Project Tracking System
- e: Put To Sleep [Internet]
- n: Post- och Telestyrelsen
- s: Punto de Tratamiento Sanitario
Pts f: Points
- f: Printemps
P-ts f: Puits
PTSA e: Parent-Teacher Student Association
PTSD e: Pesticides and Toxic Substances Division
- e: Post Traumatic Stress Disorder
PTSE e: PNNI Topology State Element
PTSED e: Pesticides and Toxic Substances Enforcement Division
PTSR e: Preliminary Technical Survey Report
PTT a: Post, Telegraf, Telefon
- d: Post, Telephon, Telegraph
- e: Petroleum Authority of Thailand
- e: Platform Transmitter Terminal
- e: Post[al], Telegraph and Telephone [administration (or agency)]
- e: Prenegotiated Target Time
- f: Plate-forme de Transmission
- f: Postes, Téléphones et Télégraphes
- s: Plataforma de Transmisión Terminal
PTt e: Pressure-Temperature-time
PTTA s: Punto de Travesía en Transporte Anfibio
PTTC e: Petroleum Technology Transfer Council
PTTEM r: Preliminär Taktisk-Teknisk-Ekonomisk Målsättning
PTU e: Pressure, Temperature, Humidity
PTUF a: Parallellteknologi for Undervisning og Forskning
PTV e: Passenger Transport Vehicle
PTVCR e: Peak-to-Valley Current Ratio
PTW e: Primary Translation Word
PTWC f: Centre d'alerte aux tsunamis dans le Pacifique
- s: Centro de Alerta contra los Tsunamis en el Pacífico
PTX e: Packet transmitted bit in TCR
- e: Pressure-Temperature-composition
P-T-X-T e: pressure-temperature-composition-time
PTY e: Pseudo-Terminal driver
PU e: Participating Unit
- e: Polyurethane
- f: Paix Universelle
- f: Prix Unique
- r: Proviantuppbördsman
- s: Pequeña Unidad
- s: Policía Uniformada
Pu e: Plutonium
- f: Pourvoyeur
- r: Personal- och utrustningstabell
- r: Personalutbildning
P/U e: Pick up
Pu-238 e: Plutonium-238
Pu-239 e: Plutonium-239
Pu-240 e: Plutonium-240
Pu-241 e: Plutonium-241
PU. e: Physical Unit
PUAE e: Agreement on Peaceful Uses of Atomic Energy
PUB e: Public [directory]
- e: Publications
Pub f: Publicité
Pubbl Stn Zool Napoli II i: Pubblicazioni della Stazione Zoologica di Napoli. Section II: History and Philosophy of the Life Sciences
PUBDIS e: Publishing Distribution System
PUBL e: Public Health Library
Publ. e: Publications
PublG d: Gesetz über die Rechnungslegung von bestimmten Unternehmen und Konzernen [Publizitätsgesetz]
PUBLICA e: Public Libraries Concerted Action
Public Health Nurs e: Public Health Nursing
Public Health Rep e: Public Health Reports
Public Health Rev e: Public Health Reviews
PUBMAR e: Marine Science Publications Unit
- f: Unité des publications relatives aux sciences de la mer
- s: Unidad de Publicaciones sobre Ciencias del Mar
PUC d: Projektgruppe Umweltgefährdungspotentiale von Chemikalien
PUCHE e: Potchefstroom University for Christian Higher Education South Africa
PUCHO A: Potchefstroomse Universiteit vir Christlike Hoër Onderwys, South Africa
PUCO s: Partido Unificación de Campesinos y Obreros
PUCs e: Public Utility Commissions
PUD e: Party of Union and Development
- f: Parti de l'Union et du Développement
- s: Partido Unificación Democrática
PUDE f: Programme pour l'utilisation durable des espèces sauvages
PUDECO e: Putu Development Corporation
PU de Inf s: Pequeña Unidad de Infantería
PUDEMO e: Peoples United Democratic Movement
Puéric f: Puériculture
PUF e: Planning Units File
- e: Polyurethane Foam
PUGT e: Puget Sound Basin
PUHCA e: Public Utility Holding Company Act
PUJC e: Priority, Urgency, Justification and Category
PUJO e: Peoples Union of Justice and New Order
PUK a: Politiets Ungdoms Klubber
- f: Pechiney-Ugine-Kuhlmann
PUL e: Packet Underway Limit
- e: Press Union of Liberia
Pulm Pharmacol e: Pulmonary Pharmacology
pulv e: pulverized, powder
PUMA e: PACAF Unit Munitions Allocation System
- e: Parallel Universal Message Passing Architectures
PUMS e: Public Use Microdata Samples
PUMSUD f: Pour un Mouvement Syndical Uni et Démocratique
PUN e: Physical Unit Number
- p: Partido Unico Nacional
Pun s: Puente
PUNA f: Parti d'Unité Nationale
PUNC e: Possibly Uncommitted
Punj. e: Punjab [Lahore, Pakistan]
Punji. e: Punjabi
PUNT p: Partido Unico Nacional de Trabajadores
PUO e: Princeton University Observatory
PUP e: PARC Universal Packet [protocol]
- e: Peoples Unity Party
- e: Popular Unity Party
- e: Princeton University Press
- e: Principal User Processor
- e: Product Update Program
- f: Parti de l'Unité du Peuple
PUPM e: Party for Unity and Progress in Mali
- f: Parti pour l'Unité et le Progrès au Mali
PUR e: Polyurethane
- e: Purpose
PURE e: POMCUS Unit Residual Equipment
PUREX e: Plutonium-Uranium Extraction
PURL e: Permanent Uniform Resource Locator
- e: Permanent Universal Resource Locator
- e: Persistent Uniform Resource Locator
PURPA e: Public Utilities Regulatory Policies Act
PURSIS e: Peking University Real-world Spatial Information System
- e: Peking University Remote Sensing Information System
PUS a: Praktisk U-lands Samarbejde
- e: Processor Upgrade Socket
- e: Public Understanding of Science
- e: Section for Staff Development at Lund University
PU3S e: Power-Up 3-State
PUSD p: Partido Unido Social Democrático
PUSHA e: Push All Registers
PUSHF e: Push Flags
PUSPICS in: Penginderaan Jauh Untuk Sumberdaya Dengan Pendekatan Interpretasi Citra Dan Survai Terpadu
PUSPITEK e: Training Facilities in the Centre for Research, Science and Technology
PUSVS s: Programa de Utilización Sostenible de la Vida Silvestre
PUT f: Parti de l'Unité Togolaise
PUTCO e: Public Utility Transport Corporation

P.U.V. a: På Udvalgets Vegne
PV d: Paketvermittelt
- e: Photo-Voltaic
- e: Physical Volume
- e: Potential Vorticity
- e: Present Value QFs qualifying facilities
- e: Principal Value
- e: Private Venture
- e: Privileged information
- e: Project Verification
- f: Payable à Vue
- f: Plantes Vivaces
- f: Plus-Value
- f: Point Vernal
- f: Prise de Vue
- f: Prix de Vente
- f: Procès-Verbal
- f: Promesse Verbale
- N: Panservogn
- r: Pansarvärns-
- R: Pulemot Vozdushny
- s: Puesto de Vigilancia
- s: Punto Vulnerable
Pv r: Pansarvärn
p.v. a: på ...s vegne
p/v s: pequeña velocidad
P[/]v N: Panservern
PVA e: Polyvinyl Acetate
- e: Population Viability Analysis
PVAc e: Polyvinyl Acetate
PVAS e: Primary Voice Alerting System
Pvavd N: Panservernavdeling
PVB e: Polyvinyl Butyral
- e: Polyvinyl Butyryl
pv. batt. N: panservernbatteri
p.v.c. a: pro vera copia [bekræftelse af afskrift]
PVC d: Polyvinylchlorid
- e: Permanent Virtual Channel
- e: Permanent Virtual Circuits
- e: Permanent Virtual Connection[s]
- e: Poly n-Vinyl Carbazole
- e: Polyvinyl Chloride
- e: Private Virtual Circuit
- e: Pro-Vice-Chancellor
- f: Procès-Verbal de Carence
- f: Procès-Verbal de Contravention
PVC. e: Permanent Virtual Circuit
PVCA e: Polyvinyl Chloride Acetate
PVCC e: Permanent Virtual Channel Connection
PVCR e: Peak-to-Valley Current Ratio
PVCT e: Posterior Vena Caval Thrombosis
PVD e: Plasma Vapor Deposition
- e: Primary Volume Descriptor
- f: Paquet Valeur Déclarée
- f: Pays en Voie de Développement
Pvd s: Proveedor
PVDA e: Polyvinylidene Acetate
PVDF e: Polyvinylidene Fluoride
PVE f: Procès-Verbal d'Essai
Pvg r: Pansarvärngevär
pvgjengen N: Panservernrakettkasterg-jengen
PVI e: Perpendicular Vegetation Index
PVIL s: Protección Civil
PVIS e: Program Verification Information System
PVK d: Paritätische Vertrauens-kommission
- e: Polyvinyl Carbazole
- N: Panservernkanon
pv. kan. N: panservernkanon
Pvkan r: Pansarvärnkanon
pvk[anon] r: pansarvärnkanon
pvkv r: pansarvärnkanonvagn
PVL e: Parameter Value Language
pv. Lag N: panservernlag
PVM d: Projektverfolgungssystem
- e: Parallel Virtual Machine
- e: Pass-through Virtual Machine
- e: Passthru Virtual Machine
- e: Performance Verification Matrix
- e: Protein-Vitamin-Mineral
- s: Patrullera (or Patrulleros) Vigilancia Marítima
PVMe e: Parallel Virtual Machine
PVN e: Private Virtual Network
PVO e: Pioneer Venus Orbiter
- e: Private Voluntary Organization
- f: Prix de Vente Obligatoire
- R: Protivo-Vosdushniya Oborona
PVOH e: Polyvinyl Alcohol
PVO[S] R: Protivo-Vozdushnoy Oborony Strany
PVP e: Packet Video Processor
- e: Packet Video Protocol
- e: Parallel Vector Processing
- e: Performance Verification Plan
- e: Poly[2-Vinylpyridine]
PVPC e: Permanent Virtual Path Connection
pvpjbv r: pansarvärnpjäsbandvagn
PVR e: Permanent Virtual Route
- e: Plant Variety Right[s]
- e: Product Verification Review
- f: Procès-Verbal de Réception
pvrakettkaster N: panservernrakettkaster
pvrb r: pansarvärnrobot
pvrbplut r: pansarvärnrobot pluton
pvrbvagn r: pansarvärnrobotvagn
p. v. r. k. N: panservernrekylfrikanon
pvrobot r: pansarvärnrobot
pvrobotfordon r: pansarvärnrobotfordon
pvrobotkomp r: pansarvärnrobotkompani
PVRQ f: Procès-Verbal de Réception Qualitative
Pvrt f: Provisoirement
PVS d: Politische Vierteljahrsschrift
- e: Parallel Visualization Server
- e: Photo-Voltage Spectroscopy
PVT e: Permanent Virtual Terminal
- e: Photo-Voltaic/Thermal
- e: Physical Vapor Transport
- f: Provisoirement
PVTD e: PNNL Vitrification Technology Department
PVUL s: Punto Vulnerable
PVV f: Procès-Verbal de Vérification
pvvapen r: pansarvärnvapen
pvvåpen N: panservernvåpen
PW e: Palau. USPS abbreviation
- e: Password
- e: Precipitable Water
- f: Parti Wallon
PWA d: Papierwerke Waldhof-Aschaffenburg
- e: People With AIDS coalition
- e: Printed Wiring Assemblies
- e: Process Waste Assessment
P-wave e: Primary compressional wave
- e: primary seismic wave
PWB e: Printed Wire Board
- e: Printing Wiring Board
- e: Programmers Work Bench
PWBA e: Pension and Welfare Benefits Administration
PWBS e: Program Work Breakdown Structure
PWC e: National Organization of Physicians Who Care
- e: Pacific Weather Centre
- e: Personal Watercraft
- e: Public Work Center
pwc e: piecewise constant
PWC/MIS e: Public Work Centers/MIS
PWD e: Participatory Work Design
- e: Present Working Directory
- e: Print Working Directory
- e: Programmed Warhead Detonation

PWDS e: Protected Wire Distribution System
PWDTS e: Peacekeeper Warranty Data Tracking System
PWE e: Present Working Estimate
PWF e: Personnel Working File
PWG e: Printer Working Group
PWGCA e: Permanent Working Group on Cooperation in Antarctica
- f: Groupe de travail permanent sur la coopération dans l'Antarctique
- s: Grupo de Trabajo Permanente sobre Cooperación en la Antártida
PWGS e: Pacific Weather Graphics System
PWH e: Person infected With HIV
PWI e: Public Windows Initiative
- e: Public Windows Interface
PWIC e: Prisoner of War Information Center
PWIS e: Prisoners of War Information System
PWL e: P-Wave Logger
- e: Password List
PWM e: Puls Width Modulation
PW/MSF e: Prairie Warrior / Mobile Strike Force
PWP e: Particles per Wafer Pass
- e: Presidents Working Party
PWR e: Passive Warning Receiver
- e: Pressurised (or Pressurized) Water Reactor
PWRC e: Patuxent Wildlife Research Center
PWRI e: Public Works Research Institute
PWRMR e: Pre-positioned War Reserve Materiel Requirement
PWRMS e: Prepositioned War Reserve Materiel Stocks
PWRS e: Pre-Positioned War Reserve Stocks

PWS e: Pore-Water Sampler
- e: Power Web Suite
- e: Private Web Server
- e: Programmable Work Station
- e: Public Water Supply
- e: Public Water System

PWSA e: Prader-Willi Syndrome Association
PWSA NZ e: Prader-Willi Syndrome Association of New Zealand
PWSCS e: Programmable Workstation Communication Services
PWSP e: Public Weather Services Programme
- f: Programme des services météorologiques destinés au public

PWSS e: Public Water Supply System
PWT f: Parti Wallon des Travailleurs
P W/U e: Public Works and Utilities
PWU e: Plantation Workers Union
PWV e: Precipitable Water Vapor
- e: Pretoria, Witwatersrand, Vereeniging Industrial Area

PWW e: Planar Wing Weapon
PX e: Primary Index
Px e: Physican examination
px e: Pixels
Px64 e: Low-resolution video compression for teleconferencing
PY e: Prior Year
- e: Publication Year

py e: Paraguay [Internet]
Py e: Pyrimidine
- f: Psychologique

P-Y e: Percus-Yevick
PYI e: Presidential Young Investigator
PYLL e: Potential Years of Life Lost
PyrOr f: Pyrénées-Orientales
PZ f: [Ecole] Polytechnique de Zürich
- f: Parc Zoologique
- f: Passage Zébré
- s: Punto Zero

Pza. s: Plaza
Pza[s] s: Pieza[s]
PZJA e: Protected Zone Joint Authority
PZLE f: Petite Zone de Libre Échange
PzM d: Punkt-zu-Mehrpunkt
PZÖ d: Projektzentrum Ökosystemforschung, Universität Kiel
PzP d: Punkt-zu-Punkt
PZS d: Punktzugriffsstruktur
PZT e: Lead Zirconate Titanate

Q

Q a: Kristelig Folkeparti
- d: [Gesamt]- Strahlungsbilanz
- e: Qualified
- e: Transistor
- f: Qualité
- f: Quantité
- f: Quartier
- f: Quotidien
- s: Químico
q d: Symbol einer beliebigen Funktion
- d: Vertikaler Drehwinkel der Positioniervorrichtung
Q.931 e: digital subscriber signaling system No. 1 for ISDN users
Q1 e: First quarter [of the year]
Q2 e: Second quarter [of the year]
Q3 e: Third quarter [of the year]
Q4 e: Fourth quarter [of the year]
Q375 e: High-quality image Q374 with scale
QA e: Q-Adapter
- e: Quality Audit
- e: Quality Assessment
- e: Quality Assurance
- e: Query Analyser
Q&A e: Questions and Answers
QAA e: Quality Assurance Audit
QAA/P e: Quality Assurance Assessment and Plan
QAC e: Quality Assurance Control
- e: Quality Assurance Coordinator
- e: Quality Assurance Council
QA&C e: Quality Assurance and Control
QACISIG e: Quadrilateral Army Communications and Information Systems Interoperability Group
QAE e: Quality Assurance Engineer
QAF e: Q-Adapter Function Block
- e: Quality Air Force
QAFA e: Quality Air Force Assessment
QAFILE e: Ambient Quality Assurance Data Base
QAI e: Quality Assurance Inspection
- e: Quality Assurance Institute
QAL e: Quality Activities List
- e: Quality Assurance Level
QALE e: Quality-Adjusted Life Expectancy
QAM d: Quadraturamplitudenmodulation
- e: Quadratur Amplitude Modulation
QAMP e: Quality Assurance Management Plan
QANGO e: Quasi Autonomous Non Government Organisation
QAO e: Quality Assurance Office[r]
QAP e: Quality Assurance Policy
- s: Químico Artificiero-Polvorista
QAPD e: Quality Assurance Program Description
QAPE e: Quality Assurance Program Engineer
QAPM e: Quality Assurance Program Manager
QAPMP e: Quality Assurance Performance Monitoring Plant
QAPP e: Quality Assurance Project Plan
QAPRM e: Quality Assurance Program Requirements Manual
QA[/]QC c: Quality Assistance/Quality Control
- e: Quality Assurance/Quality Control
QAR e: Quality Assurance Representative
QA/R e: Quality Assurance/Reliability
QARD e: Quality Assurance Requirements and Description
- e: Quality Assurance Requirements Document
QAS e: Quality Affecting Software
- e: Quasi Associated Signaling
QA/SAC e: Quality Assurance/Science Activity Centre
- f: Centre d'activité scientifique chargé de l'assurance de la qualité des données
QASP e: Quality Assurance Support Plan
QASPR e: QUALCOMM Automatic Satellite Position Reporting
QASRCE e: EMSL-RTP Quality Assurance Data System
QAT e: Quality Action Team
QATS e: Quality Assurance Test System
QAU e: Quality Assessment Unit
QAWINAQEL s: Asociación Cultural Poqomam Qawinaqel
QB e: Queens Bishop
QBD e: Charge [Q] to BreakDown
QBF e: Query by Form
QBFP e: Queensland Boating and Fisheries Patrol
QBIC e: Query By Image Content
Q-bit e: Qualified data bit
QBO f: Oscillation Quasi Biennale
- s: Oscilación Cuasibienal
QBP e: Queens Bishops Pawn
QBS e: Qualifications-Based Selection
QBTU e: Quadrillion British Thermal Units
QC e: Quality Control [of Data]
- e: Quebec
- e: Queens Counsel
- e: Qumran Chronicle, published in Poland
- f: Contrôle de la Qualité [des données]
- s: Control de Calidad [de los Datos]
- s: Quinta Columna
Q.C. e: Queens Council
QCA e: Quality Compliance Assurance
- e: Quantum[-dot]
QCAL e: Qualified Contractor Access List
QCBSDB e: Quality Control Blind Standards Database
QCC e: Quadrupole Coupling Constant
- e: Queensland Conservation Council
QCCEC e: Queensland Conservation Council and Environment Centre
QCD d: Quantenchromodynamik
- e: Quantum Chromo Dynamics
QCDP e: Quality Color Dithering Modus
QCE e: Quality Control Experts
QCESR e: Quarterly Cost Estimate and Schedule Review
QCFO e: Queensland Commercial Fishermens Organisation
Qch f: Quelque chose
Q Channel e: Quadrature Channel
QCI e: Quality Cost Index
QCIF e: Quarter Common Intermediate Format
QCM e: Quartz Crystal Microbalances
- f: Questionnaire à Choix Multiple
QCP e: Queens Capital Program
Qcq f: Quelconque
QC/QA e: Quality Control/Quality Assurance
QCRN e: Quality Control Record Number
QCSDS e: Quality Control Sample Distribution
QCSE e: Quantum-Confined Stark Effect

QD e: Quadruple Density
- e: Quantum Dot
- e: Quasi-Distributed
- e: Querty/Dvorak
- e: Queuing Delay
- f: Quantité Disponible
- f: Quotité Disponible

Q1D e: Quasi-One-Dimensional
Q2D e: Quasi-Two-Dimensional
Q-DATA e: Greenland Computer and Statistics Center
QDCS e: Quality, Deliverly, Cost, Safety
- e: Quality, Deliverly, Cost, Service

Qdd f: [à] Qui de droit
QDE e: Qualified Designated Entity
- e: Quantum Detection (or Detector) Efficiency

QDEH e: Queensland Department of Environment and Heritage
QDhG d: Quellen und Darstellungen zur hansischen Geschichte
QDI e: Quicken Dictionary
.qdi e: quicken dictionary [filename extension]
QDL e: Quadri Data Layer
QDOS e: Quick and Dirty Operating System
QDPI e: Queensland Department of Primary Industries
QDR e: Quick Dump Rinse
QDRs e: Quality Deficiency Reports
QDT e: Quicken Data
QE e: Quality Engineer
- e: Quality Equalizing
- e: Quantum Efficiency
- e: Quick Estimation
- f: Quartiers Excentriques
- f: Question Écrite

QECPR e: Quarterly Energy Conservation Performance Report
QED l: Quod Erat Demonstrandum
QELS e: Quantum Electronics and Laser Science conference
QEMM e: Quarterdeck EMM [Expanded (or Extended) Memory Manager]
QES e: Quality Education System
QET e: Quantum Effect Transistor
QF e: Quality factor
- e: Quasifree
- f: Quotient Familial

QFA e: Quick File Access
QFC e: Queens Fund Council
QFE e: Quick-Fix Engineering
QFMA e: Queensland Fish Management Authority
QFP e: Quad Flat Pack
- e: Questions Frequencement Possess

QFT d: Querfurt, Sachsen-Anhalt
- e: Quantum Field Theoretic[al] (or Theory)

Qgco s: Quirúrgico
QGP e: Quark-Gluon Plasma
QHD e: Quantum Hydrodynamic
QHERB e: Queensland Herbarium
QHR e: Quality History Records
QI e: Quality Inspection
- e: Quality Instruction
- f: Qualité Inférieure
- f: Quantité Indéterminée
- f: Quantité Infinitésimale

QIC e: Quality Information Center
- e: Quality Information using Cycle time
- e: Quarter Inch Cartridge

QID d: Quanteninformationsdynamik
q.i.d. e: four times daily [Internet]
QIF e: Quicken Import File
- e: Quicken Interchange Format

.qif e: quicken interchange format [filename extension]
QIP e: Quality improvement plan
QISAM e: Queued Index Sequential Access Method
QIT e: Quality Improvement Team
Q.I.T. e: Queensland Institute of Technology
QI TQM e: Quality Improvement Total Quality Management
Q J Exp Psychol A e: Quarterly Journal of Experimental Psychology. A, Human Experimental Psychology
Q J Exp Psychol B e: Quarterly Journal of Experimental Psychology. B, Comparative and Physiological Psychology
QJM e: QJM: Monthly Journal of the Association of Physicians
Q J Nucl Med e: Quarterly Journal of Nuclear Medicine
QKt e: Queens Knight
QKtP e: Queens Knights' Pawn
ql e: As much as desired [Internet]
QL e: Quant[ific]ation Limit
- e: Queensland Lawyer
- f: Quantité Limitée
- f: Quartier Latin
- f: Quittance de Loyer

Q Law Soc J e: Queensland Law Society Journal
QLB d: Quedlinburg, Sachsen-Anhalt
QLCI e: Quick Look Capabilities Index
QLI e: Query Language Interpreter
QLIC e: Queensland Land Information Council
QLIRS e: Quick-Look Image Recording System
QLIS e: Queensland Land Information System
QLLC. e: Qualified Logical Link Control
QLM e: Quartz-feldspathoid-olivine and ore Minerals plot
- e: Quasi-Lagrangian Model

QLN e: Quantitative Links Networks
Qlq f: Quelques
QLRO e: Quasi-Long-Range Order
QLS e: Quiet Line State
QM e: Qualification Model
- e: Quality Management
- e: Quantum Mechanical
- e: Queensland Museum, Brisbane
- f: Qualité Marchande
- f: Quantité Maximum

QMA e: Quality Management Assessment
QMAS e: Quality Measurement Advisory Service
QMB e: Qualified Medicare Beneficiary
- e: Quartz Microbalance

QMBA e: Quasi-Moving-Boundary Approximation
QMD d: Qualitätsmanagement Druckindustrie
QME e: Quality Measurement Experiment
q-message e: Classified message relating to navigational dangers
QMF e: Query Management System
QML e: Qualified Manufacturer Lists
- e: Qualified Materials Lists

QMP e: Quality Measurement Plan
- f: Quantité Maximum Permise

QMPR e: Quality Management Policies and Requirements
QMR e: Quick Medical Reference
QMRL e: Quartermaster Radiation Laboratory, Natick
QMS e: Quadrupole Mass Spectrometer
QMSM e: Quantum Molecular Similarity Measures
QMT e: Quicken Memorized List
.qmt e: quicken memorized list format [filename extension]
QMW e: Queen Mary and Westfield College
qn e: every night [Internet]
QN e: Quasineutral
- e: Quinoline
- f: Quantité Négative
- f: Quantité Négligeable

Qn f: Quelqu'un
QNC e: Queensland Naturalists Club
QNE e: Altimeter reading at airfield when subscale is set to 1013.2 millibars
QNH e: Altimeter subscale setting to destination airfield altitude
QNM e: Queuing Network Model
Qnnn e: High-quality image No. with scale [selected]
QNPWA e: Queensland National Parks and Wildlife Authority
QNPWS e: Queensland National Parks and Wildlife Service
QNR e: QuasiNeutral Region
QNX e: Quicken Indexes to Data
.qnx e: quicken index to data [filename extension]
QO f: Quai des Orfèvres
QOT&E e: Qualification, Operational Test & Evaluation
QP e: Quick Pascal
- f: Quantité Positive
- f: Question Préalable
- f: Quote-Part
- f: Quotidiens Parisiens

QPC e: Quantum Point Contacts
qPCR e: quantitative Polymerase Chain Reaction
QPG e: Quantum Phase Gate
QPL e: Qualified Product Lists
QPM e: Quasi-Phase Matching
QPNC e: Queensland Place Name Committee
QPR e: Qualitative Performance Requirements
- e: Quarterly Progress Report

QPRI e: Queensland Pharmaceutical Research Institute
QPRP e: Quality Performance Reporting Program
QPS e: Quark Publishing System
QPSK e: Quad[rature] Phase Shift Key[ing]

QPSX e: Queue Packet and Synchronous Circuit Exchange
- e: Queued Packet Switch Exchange
- e: Queued Packet Synchronous Exchange
QQ f: Quantité Quelconque
Qq[ch] f: Quelque [chose]
Qqn f: Quelqu'un
Qqs f: Quelques
QR e: QA/QC Requirements
- e: Quality assurance Review
- f: Quotidiens Régionaux
Q-r f: Quartier
QRA e: Qualitative Risk Assessment
QRAC e: Queensland Radiation Advisory Council
Q ratio e: Koenigsberger ratio
QRB e: Quality Review Bulletin
QRC e: QRC Advisor/Managing Hospital Quality Risk & Cost
QRC CAC e: Quick Reaction Center Combat Analysis Center
Q Rev Biol e: Quarterly Review of Biology
Q. Rev. Biol. e: Quarterly Review of Biology
Q Rev Biophys e: Quarterly Reviews of Biophysics
QRF e: Quick Reaction File
QRI e: Queue Response Indicator
QRP e: Q sign: Reduce Power
- e: Quick Reaction Packages
- e: Quick Response Package
QRS e: Australian National Herbarium, Atherton International Acronym
- e: Quick Release System
QRS complex e: The principal deflection in the electrocardiogram, representing ventricular depolarization
QRSIN e: Queensland Remote Sensing Information Note
QRSS e: Quasi Random Signal Source
QRT e: Quad Routing Table[s]
QS e: Quality Services
- e: Quick Splice
- f: Qualité Supérieure
- f: Quantité Suffisante
qs e: sufficient quantity
QSAR e: Quantitative Structure Activity Relationship[s]
- f: Relations Quantitatives Structure-Activité
- s: Relaciones de actividad y estructura cuantitativas
QSD e: Quality Systems Division
- f: Quotidien Sauf Dimanche
QS&D e: Quality Systems and Development
QSDPCM d: Quadtree-Strukturierte Differenz-Puls-Code-Modulation
QSE e: Quad Switch Element
QSFS e: Queensland Space Frontier Society
QSG d: Quellen zur Schweizer Geschichte
QSH e: Quality, Safety and Health programs division
Q-ship e: Decoy ship
QSIG d: Quersignalisierung
- e: Unified International Corporate Network Signalling Standard
QSMTP e: Queued SMTP
QSO e: Q sign: Send On
- e: Quasi Stellar Object
QSP e: Questionnaire for Sensitive Position
Qsp f: quantité suffisante pour
QSR e: Query and Reporting Process
QSS e: Quality Surveillance System
QSSOP e: Quartersize Small Outline Package
QST e: Quality [and] Safety Trending
- e: Quality Steering Committee
Q Star e: A Schweizer sailplane
QSTS e: Quality, Safety and Trend System
Qsuff f: Quantité suffisante
QT e: QRS-onset-to-end-of-T-wave interval
- e: Quick Turn
- e: Quicktime
- s: Qualquier Terreno
QTA e: Quick-Turn Area
QTC e: Quantitative Technology Corporation
- e: Quicktime Conferencing
QTc e: QT interval corrected for heart rate
Qtd. e: Quarter-sawed
QT&E e: Qualification Test and Evaluation
Qte e: Quartzite
Qté f: Quantité
QTF e: Quantitative Temperature Forecast
QTL e: Quantum Theory of Laundry
QTML e: Quicktime Media Layer
QTO e: Quasi-Triennial Oscillation
QTP e: Qualification Training Package
QTRC e: Quality Training and Resource Center
- e: Quality Training Research Center
QTV e: Queens Television
QTVR e: Quicktime Virtual Reality
QTY e: Quantity
Qu. e: Queens [Canada]
QUA e: Qualitative Use Assessment
QUADS e: Canada, EC, Japan and United States
Quads e: Quadrillion
Qual f: Qualificatif
- f: Qualité
Qual Assur e: Quality Assurance
QUAL2E e: Enhanced Stream Water Quality Model
QUAL2EU e: Enhanced Stream Water Quality Model with Uncertainty Analysis
QUAL-II e: Stream Quality Model
Qual Life Res e: Quality of Life Research
QUANGO e: Quasi Non Governmental Organization
Quant f: quantité
QUANTAS e: Quantas Empire Airways
quasar e: quasi star
qude fin f: quart de finale
QUESSI e: Queens University Engineering Society Services Inc.
QUEST e: Quality, Environment[al] and Safety Tracking [system]
- e: Queens University Evaluation System for Teaching and Courses
QUFA e: Queens University Faculty Association
QUI e: Quadratic Upstream Implicit
QUICK e: Quick Reacting General War Gaming System
QUIF e: Queens University Initiation Fund
QUIPE e: Quarterly Update for Inspector in Pesticide Enforcement
QUISAM e: Queued Index Sequential Access Method
Quoten d: aus einer fremden Nachricht zitieren
quot[s]. e: quotation[s]
QUSA e: Queens University Staff Association
QUTLJ e: Queensland University of Technology Law Journal
Q-V e: Charge-Voltage
qv e: As much as desired [Internet]
QVD e: Quality Verification Department
QWERTY e: first six keys from left on top alphabetic row of standard keyboard QL Quick Look
QWG e: Quality Working Group
QWI e: Quantum-Well Intermixing
QWIP e: Quantum-Well Infra-red Photodetector
- e: Quantum-Well Intersubband Photodetector
QWire e: Quantum Wire
QWITT e: Quantum-Well Injection Transit Time
QWL e: Quality of Work Life
- e: Quantum Well
QWR e: Quantum Wire
QWTMA e: Queensland Wet Tropics Management Authority
QWW e: Quantum Well Wire

R

R a: Radikale [Venstre]
- a: rekommanderet [på forsendelse]
- a: Ridder af Dannebrog
- d: Reflexionsfaktor [des ersten Farbfilters bei einer Vollfilterung]
- d: Reflexionsfaktor des ersten Farbfilters des Lichteinfalls
- d: Reflexstrahlung
- d: Registrierendes Gerät
- e: Organic group
- e: Reference point
- e: Ramsar
- e: Rare earth element
- e: Rawl Building
- e: Receive
- e: Reference
- e: Republican
- e: Resolving power
e: Rolemaster
- e: Rookie
- e: Router
- e: Routine Precedence
- e: Row
- f: Rebut
- f: Récif
- f: Recommandé
- f: Récupérable
- f: Réduit
- f: Réformé
- f: Registre
- f: Règlement[ation]
- f: Rente
- f: Renversé
- f: Réponse
- f: Reprise
- f: Républicain
- f: République
- f: Réserve
- f: Résistance
- f: Round
- f: Route
- f: Rupture
- N: resymé
- N: Rør
- r: Rekrytskola
- r: Revolver
- r: Rör
- s: Regimiento
- s: Regulación
- s: Remolcador
- s: Remolque
- s: Reparación
- s: Reservado
- s: Retirado
- s: Rumbo

r d: Reflexionskoeffizient
- d: Streuvektor

°R e: degrees Rankine

r0 d: Laufradius in der Lichtaustritts-pupille des Kameraobjektivs. Abstand eines Punktes in der H'-Ebene vom festgelegten Koordinatenursprung

R0 d: Maximaler Radius der Lichtaustrittspupille des Kameraobjektivs
- e: Record zero

R1 r: Radartrupp nr 1
- r: Sidriktare

R,1 d: Reflexionsfaktor des ersten Vollfilters der ersten Partialfläche einer Partialfilterung

R2 r: Höjdriktare

R,2 d: Reflexionsfaktor des ersten Vollfilters der zweiten Partialfläche einer Partialfilterung

r_2 e: coefficient of determination

R-4 e: Results Review and Resource Requirement

RA d: Regulärer Ausdruck
- e: Race Announcement
- e: Radiation Alarm
- e: Radiation Area
- e: Radiology Assembly of Medical Group Management Association
- e: Real Audio
- e: Reasonable Alternative
- e: Regional Administrator
- e: Regional Association
- e: Regular Army
- e: Regulatory Alternative[s]
- e: Removal Action
- e: Requirements Analysis
- e: Research Assistant
- e: Research Associate
- e: Resident Advisor
- e: Resource Allocation
- e: Retinoic Acid
- e: Return Authorization
- e: Risk Analysis
- e: Routing Arbiter
- f: Rapport Annuel
- f: Recette Auxiliaire
- f: Recherches Aéronautiques
- f: Recherches Agronomiques
- f: Réforme Agraire
- f: Régime Alimentaire
- f: Réglementation Administrative
- f: Renseignements Administratifs
- f: Rente Amortissable
- f: Répertoire Alphabétique
- f: Réponse Affirmative
- f: Retraite Anticipée
- f: Retraite d'Ancienneté
- f: Revenu Annuel
- f: Révision Annuelle
- f: Route Agricole
- s: Radiogoniómetro
- s: Regimiento de Artillería
- s: Región Aérea
- s: Remolcador de Altura

Ra r: Radar[skola]

Rå r: Riksåklagarämbetet

R.A. e: Royal Academician
- e: Royal Artillery

R&A e: Reliability and Availability

RA-2 e: Advanced Radar Altimeter
- e: POEM altimeter

Ra-226 e: Radium-226

Ra-228 e: Radium-228

RA5-0 e: Register address pins that are set to identify the addresses of the registers in the IBM Adapters for Ethernet

RAA e: Rapid Access Archive
- e: Resource Apportionment Aid
- e: Rice Growers Association of Australia
- e: Rotation Adaptor Assembly

RAAA s: Regimiento de Artillería Antiaérea

RAAC e: Remote Active Archive Center

RAAFAS e: Report of Accounting and Finance Activities

RAAN e: Right Ascention of the Ascending Node

RAAP e: Rapid Application of Air Power

RAAS e: Remedial Action Assessment System

RAAS/TIS e: Remedial Action Assessment System/Technology Information System

RAATS e: RCRA Administrative Action Tracking System

ra. avd. N: radioavdeling

RAAWS e: Radar Altimeter and Altitude Warning System

RAB e: RAID Advisory Board
- e: Readiness Assessment Board
- f: Red Africana de Biociencias
- f: Réseau Africain des Biosciences
RABATS e: Rapid Analytical Block Aerial Triangulation System
Rab[att]G d: Rabattgesetz
Rabef r: Radiobefäl
RABIN n: Raad van Advies voor Bibliotheekwezen en Informatieverzorging
RABn N: Rakettbataljon
RAC d: Reallexikon für Antike und Christentum
- e: Radiation Advisory Committee
- e: Radio Amateurs of Canada
- e: Recirculating Air Cooler
- e: Regional Activity Centre
- e: Regional Adjudication Center
- e: Regional Asbestos Coordinator
- e: Remote Access and Control
- e: Remote Access Computer
- e: Resource Assessment Commission
- e: Response Action Coordinator
- e: Risk Assessment Council
- e: Rotating-Arm Collector
- e: Royale Automobile Club
- f: Radio Amateurs du Canada
- i: Rivista di Archeologia Cristiana
Rac s: Ración
RACA s: Regimiento de Artillería de Campaña
RACC e: Repairable Asset Control Center
RACE e: Random Access Card Equipment
- e: Research [and development program] in[to] Advanced Communication[s] [technologies] in Europe
RACES e: Radio Amateur Civil Emergency Services
- e: Remote Acoustic Electronic Sounding System
RACF e: Resource Access Control Facility
RACH e: Random Access Channel
RACHADE f: Rassemblement pour le Changement et la Démocratie
RACKS e: Receptors for Activated C Kinase
- e: Reusable Application Interface for Communicating Real-Time Kernels
RACM e: Reasonably Available Control Measures
- e: Regulated Asbestos-Containing Materials
RACR e: Royal Australasian College of Radiologists
RACS e: Reiter Acoustic Startle
- e: Rotating Antenna C-Band Scatterometer
RACT e: Reasonably Achievable Control Technology
- e: Reasonably Available Control Technology
RACTA s: Regimiento de Artillería de Costa
RAD e: Radiation Adsorbed Dose
- e: Radioactive Acid Digestion
- e: Radioecology Programme
- e: Rapid Application[s] Development
- e: Request for Authority to Develop
- e: Required Availability Data
- e: Requirements Action Document
- s: Regimiento de Artillería Divisionario
Rad d: Radioaktivität
- e: Radiation absorbed dosis
- e: Radiolaria
Rada s: Radi[o]activo
RADAM e: Radar for the Amazon
- p: Radar Na Amazonia
RADAP e: Radar Data Processor
RADAR e: Radio Association Defending Airwave Rights
- e: Regional Alcohol and Drug Awareness Resource [Network]
- e: Rights, Availabilities, Distribution Analysis and Reporting
RADAREXREP e: Radar Exploitation Report
Radarls r: Luftbevakningsradarstation
Radarpbr r: Radarpansarbandvagn
RADARSAT e: [Canadian Synthetic Aperature] Radar Satellite
Radarspafpl r: Radarspaningsflygplan
RADB e: Routing Arbiter Data Base
RADCAL e: Radar Calibration
- e: Radiation Calibration Laboratory
RADCOMP e: Radar Data Communications and Processing [System]
RAD[/]CON e: Radiological[/]Control
RADDES f: Rassemblement pour la Démocratie et le Développement Economique et Social
RADE e: Request for Analytical Data Evaluation
RADEC e: Radiation Detection Capability
RADES e: Radar Evaluation Squadron
RADES-MR e: Radar Evaluation Squadron Minicomputer Replacement
RADESO f: Rassemblement pour la Démocratie Sociale
Rad Fo s: Radiofaro
RADHARD e: Radiation Hardness
RADHO f: Réseau Africain pour les Droits de l'Homme
RADIAC e: Radiological Detection Identification And Computation
Radiak r: Radioaktiv
Radiat Environ Biophys e: Radiation and Environmental Biophysics
Radiat Med e: Radiation Medicine
Radiat Res e: Radiation Research
Radiats Biol Radioecol R: Radiatsionnaia Biologiia, Radioecologiia
RADID e: Radar Information Display
RADIL e: ROCC/AWACS Digital Information Link
RADIMT s: Regimiento de Artillería de División De Infantería Motorizada
Radiol. e: Radiology
Radiol Clin North Am e: Radiologic Clinics of North America
Radiol Med [Torino] i: Radiologia Medica
Radiol Technol e: Radiologic Technology
Radiother Oncol e: Radiotherapy and Oncology
RADISAR e: Radiometer/Synthetic Aperture Radar
RADIUS e: Remote Authentication Dial-In User Service
- e: Residue Arithmetic Digital Image Understanding System
RADM e: Random walk Advection and Dispersion Model
- e: Regional Acid Deposition Model
RADM [LH] e: Rear Admiral-Lower Half
RADM [UH] e: Rear Admiral-Upper Half
RADPATH e: Biochemical Pathways of Artificial Radionuclides
Rad Rec s: Radiorrecepción
RADRISK e: Radionuclide Dose Rate/Risk
RADS e: Radar and Algorithm Display Model
- e: Real-time Analysis and Display System
RADSL e: Rate Adaptive Digital Subscriber Line
Rad-soc f: Radicaux-socialistes
Rad Sond [Met] s: Radiosonda [Meteorológica]
RADTRAN e: Radioactive material Transportation
Rad Trn s: Radiotransmisión
RADTU e: Radioactive Acid Digestion Test Unit
RADX e: ORD-Cinci Radiation Exposure Tracking System
RAE e: Research Assessment Exercise
- e: Residue After Evaporation
- e: Resistive Anode Encoder
- e: Royal Aerospace Establishment
RAEBNC e: Read And Enjoyed, But No Comment [Internet]
RAEP e: Radioactive Air Emissions Program
- e: Radiological Airborne Emission Program
RAeS e: Royal Aeronautical Society
RAES e: Rural Afforestation and Extension Services
RAF d: Rote Armee Fraktion
- e: Red Army Fraction
- e: Remote Airborne Fluorosensor
- e: Resource Allocation Frame
R.A.F. e: Royal Air Force
RA/FA e: Report of Accounting and Finance
RAFAL f: Radar A Faisceau Lateral
RAFAN e: Royal Air Force Airmove Network
RAFAS e: Report of Accounting and Finance Activities System
RAFC e: Regional Area Forecast Centre
RAFG e: RAF Germany
RAFKADPS e: RAF Kemble ADP Support
RAFOS e: Listening floats
- e: SOFAR spelled backwards
RAFS e: Regional Analysis and Forecast System
- e: Regional Analysis Forecast System
RAF-T e: Reconnaissance Attack Fighter Trainer
RAFTS e: Reconnaissance-Attack-Fighter Training Systems
RAFUS e: Radio Frequency Usage Reporting System
RAFVR e: RAF Volunteer Reserve

RAG d: Ruhrkohle Aktiengesellschaft
- e: Radiological Analysis Group
- e: Rainforest Action Group
- e: Replenishment Air Group
- e: Requirements Assessment Group
- e: Row Address Generator
RAH e: Radiation Alarm High
RA I e: Regional Association I - Africa
RAI e: Rainfall Anomaly Index
- e: Remote Alarm Indication
- e: Research Access, Inc.
- e: Royal Albert Institution or Royal Anthropological Institute
- i: Radio Audizioni Italiane
RAID e: Redundant Array[s] of Independent Discs (or Disks)
- e: Redundant Array[s] of Independent Drives
- e: Redundant Array[s] of Inexpensive Discs (or Disks)
- e: Robotics and Artificial Intelligence Database
- f: Rassemblement d'Associations Indépendantes et Démocratiques
RAIDS e: Real-time AUTODIN Interface Distribution System
- e: Remote Atmospheric and Ionospheric Detection System
RA II e: Regional Association II - Asia
Rainforests LAP e: North Queensland Rainforests Labour Adjustment Package
RAIP e: Request for Approval-In-Principle
- e: Requester's Approval-In-Principle
RAIPR e: Royal Australian Institute of Parks and Recreation
RAIRS e: Reflection-Absorption InfraRed Spectroscopy
RAIS e: Redundant Arrays of Inexpensive Systems
RA IV e: Regional Association IV - North and Central America
RAJ f: Rassemblement-Actions-Jeunesse
RAJEUNIR f: Rassemblement de la Jeunesse de l'UNIR
RAJS f: Réseau Africain des Journalistes Scientifiques
RAK d: Regelwerk für die alphabetische Katalogisierung
- r: Rikets Allmänna Kartverk
Rak N: Rakett
- r: Raket
Rakan r: Riktare vid Automatkanon
RAK-AV d: RAK Audio-visuelle Medien
Ra. kp. N: Radiokompani
RAK-PB d: RAK Parlaments- und Behördenbibliotheken
Rakpj r: Raketpjäs
RAKR R: Atomnaya Raketniy Kreyser
RAK-WB d: RAK Wissenschaftliche Bibliotheken
RAL e: Radiation Level Alarm
- e: Rutherford[-]Appleton Laboratory
R Alj s: Remolque Aljibe
RALS e: Remote Afterloading System
- e: Remote Augmentation Lift System
RALSA e: Random Access Line Scanned Array
RALU e: Register [File and] ALU [Arithmetic Logic[al] Unit]
Ra. lys. kp. N: Radio- og lyskasterkompani
RAM d: Radioaktivitätsmeßnetz
- e: Radar Absorbent Material
- e: Radiation Air Monitors
- e: Radiation Area Monitor
- e: Random Access Machine
- e: Rarely Adequate Memory
- e: Readiness Assessment Module
- e: Rechargeable Alkaline Manganese dioxide battery
- e: Reliability, Availability and Maintainability [analysis]
- e: Remote Area Monitor
- e: Requirements Allocation Matrix
- e: Resource Allocation [and] Management
- e: Responsibility Assessment Matrix
- e: Retailers Association of Massachusetts
- e: Urban Air Quality Model for Point and Area Source[s] in EPA UNAMAP Series
- s: Reserva de Alto Mando
Ram r: Radarmatros
R.A.M. e: Royal Academy of Music
RAMAN e: Regional Atmospheric Measurement and Analysis Network
RAMAS e: Radioactive Maintenance and Storage
- e: Reliability, Availability, Maintainability Analysis System
RAMB e: Redstone Arsenal Military Base
R.A.M.C. e: Royal Army Medical Corps.
RAMC e: Regional Assessment Management Committee
RAMCAD e: Reliability and Maintainability in Computer-Aided Design
RAMDAC e: Random Access Memory Digital[to]Analog Converter
RAMDIS e: RAMM Branch Advanced Meteorological Satellite Demonstration and Interpretation System
RAMEAU f: Répertoire d'Autorité-Matières Encyclopédique et Alphabétique Unifié
RAMI f: Rassemblement des Mobutistes Indépendants
RAMIS e: Reliability and Maintainability Information System
RAMM e: Regional and Mesoscale Meteorology
RAMMS e: Regional Atmospheric Mesoscale Modeling System
RAMOS e: Remote Automated Meteorological Observing System
RAMP e: Radiation Area Monitor Probe
- e: Ratings And Maintenance Phase
- e: Reliability Analysis and Modeling Program
- e: Remote Access Maintenance Protocol
- e: Remote Atmospheric Measurements Program
- e: Remote Atmospheric Monitoring Project
- e: Report and Message Processor
- e: Rural Abandoned Mine Program
RAMPAC e: Radioactive Material Packaging
RAMPAK e: Radioactive Material Packaging
RAMP-C e: Requirements Approach to Measuring Performance in COBOL
RAMPOST e: Radioactive Material Postnotification Reports
RAMPS e: Recipe and Menu Pricing System
RAMQ f: Régie de l'Assurance Maladie du Québec
RAM/RMA e: Reliability, Availability and Maintainability
RAMRT e: Radioactive Material Routing Report
RAMS e: Range Measurement System
- e: Recruiting Automated Management System
- e: Regional Agricultural Management System
- e: Regional Air Monitoring System
- e: Regional Atmospheric Modeling System
- e: Remote Atmospheric Measurements Systems
- e: Representative Association of Medical Schemes
- e: Reproduction Administrative Management System
- e: Reprographics Automated Management System
- e: Research Aircraft Measurement System
- e: Results Analysis and Management System
- f: Système de mesure avec accès sélectif
- s: Sistema de medida de acceso aleatorio
RAMSAR e: Convention on Wetlands of International Importance
RAMSES f: Radar Multifréquences Sol pour l'Étude des Signatures Spectrales
RAN f: Régie des Chemins de Fer Abidjan-Niger
RAND e: Radio Array Detection of Neutrinos
- e: Randall Laboratory
- e: Rural Area Network Design
RANH n: Rijksarchief in Noord-Holland in Haarlem
RAO e: Remedial Action Objective
- e: Revenue Accounting Office
RAOB e: Radio Atmospheric Observation Balloon
- e: Radiosonde Observation
- e: Rawindsonde Observation
RAOC e: Rear Area Operations Center
R.A.O.C. e: Ordnance Corps
RAOU e: Royal Australian Ornithologists Union
RAP e: Radio Local Loop Access Profile
- e: Radon Action Program
- e: Rapid Application Prototyping
- e: Regional Analysis and Prediction
- e: Remedial Accomplishment Plan
- e: Remedial Action Plant
- e: Remedial Action Program
- e: Remedial Action Project
- e: Remote Advisory Panel
- e: Report Analysis Program
- e: Research Applications Program
- e: Resource Allocation Process
- e: Response Action Plan
- e: Restricted Access Processor
- f: Régie autonome des pétroles
- f: Règlement d'Administration Publique
RAPD e: Random[ly] Amplified Polymorphic DNA Technique

RAPDOC e: PICAs individual article supply service
- e: Rapid Document Delivery Project involving 19 major libraries and PICA in the Netherlands

RAPI e: Royal Australian Planning Institute Inc.

RAPID e: Remote Atmospheric Probing Information Display
- e: Riparian Aerial Photographic Inventory of Disturbance

Rapid Commun Mass Spectrom e: Rapid Communications in Mass Spectrometry

RAPIDE e: Relocatable Army Processors for Intelligence Data-Europe

RAPIDS e: Real-time Automated Personnel Identification System
- e: Regional Air Pollutant Inventory Development

RAPID SAR e: Real-Time Acousto-Optic Programmable Imaging and Display for Synthetic Aperture Radar

RAPIDSIM e: Rapid Intertheatre Deployment Simulator

RAPIER e: Rapid Emergency Reconstitution
- e: Rapid Emergency Relocation

RAPPORT III e: ECM System

RAPS e: ASRL-RTP Regional Air Pollution Study
- e: Real Application on Parallel Systems
- e: Remedial Action Priority System

RAPs e: Recommended Areas for Protection

RAQ e: Rarely Asked Questions

RAR f: Recette Auxiliaire Rurale

RARA e: Radiation Area Remedial Action

Rå rd r: Rådmansö radiostation

RARDE e: Royal Armaments Research and Development Establishment

RARE e: European Association of Research Networks
- f: Réseaux Associes de/pour la Recherche en Europe
- f: Réseaux Associes pour la Recherche Européene

RARP e: Radiation Area Reduction Program
- e: Reverse ARP [Address Resolution Protocol]

RARS e: Robot Auto Racing Simulation

RAS e: Radar Absorptive Structures
- e: Radar Altimeter System
- e: Rapid Assault Squads
- e: Reactor Analysis and Safety
- e: Reader Admission System
- e: Refractive Aberration Simulator
- e: Reliability, Availability and Serviceability
- e: Remote Access Server (or Service)
- e: Remote Analysis Station
- e: Reuel A. Stallones Building
- e: Routine Analytical Service[s]
- e: Row Access Strobe
- e: Row Address Select
- e: Rural Adjustment Scheme
- e: Russian Academy of Sciences

RASAC e: Rural Assistance Scheme Advisory Committee

RASAM e: Radiometer-Scatterometer

RASAPI e: Remote Access Service Application Programming Interface

R.A.S.C. e: Service Corps

RASC e: Radio Atmospheric Science Center
- e: Radio Atmospheric Study Centre
- e: Regional AUTODIN Switching Center
- e: Residence Account Service Center

RASCAL e: Remote Airborne Sensing Computer Analysis Link

RASCO e: Regional African Satellite Communications Organisation

RASCOM e: Regional African Satellite Communication System for Development
- f: Système régional africain de communications par satellite au service du développement
- s: Sistema Regional Africano de Comunicación por Satélite para el Desarrollo

RASD f: République Arabe Saharienne Démocratique
- s: República Arabe Saharuí Democrática

RASMED e: Redundant Array of Slightly More Expensive Disks

RASP e: Random Access Stored Program Machine
- e: Regional Assimilation and Prognosis System
- e: Remote Alarm Setpoint Panel

RASS e: Radar Altimeter Simulation System
- e: Radar Altimeter System Simulator
- e: Radio-Acoustic Sounding System

RAS/STADES e: Record Association System/Standard Data Elements System

RAST e: Radio Allergosorbent Test

Rast r: Radarstyrman

rast[n] r: radiostation

RASVY e: Royal Australian Survey Corps

RAT e: Radiation Air Thermometer
- e: Radiation Alarm Transmitter
- e: Radiological Assistance Team
- e: Ram-Air Turbine
- e: Response Action Team
- f: Rente Accident du Travail
- s: Radar de Tiro
- s: Región Antitanque
- s: Reserva Antitanque

RATA s: Reserva Antitanque de Artillería

RATAC s: Radar de Tiro de Artillería de Campaña

RATAN e: Radio And Television Aid to Navigation

RATCF DAIR e: Radar Air Traffic Control Facility Direct Altitude and Identity Readout

RATCHET e: Regional Atmospheric Transport Code for Hanford Emission Tracking

RatjL r: Radiotjänstledare

RATO e: Rocket-Assisted Take Off

RATP e: Reliable Asynchronous Transfer Protocol

RATS e: Radio Amateur Telecommunications Society
- e: Retrieval and Transmission System

RATTAWU e: Radio, Television and Theater Workers Union of Nigeria

RATWU e: Road, Air-Service and Transport Workers Union

RAU A: Randse Afrikaanse Universiteit, South Africa
- e: Radio Access Unit
- e: Rand Afrikaans University
- e: Regional Administrative Unit
- f: Recette Auxiliaire Urbaine
- s: República Arabe Unida

RAuaZulPrG d: Gesetz zur Prüfung von Rechtsanwaltszulassungen, Notarbestellungen und Berufungen ehrenamtlicher Richter vom 24.7.1992

RaumfAntr d: Institut für Raumfahrtsysteme

Raumordn d: Institut für Raumordnung und Entwicklungsplanung

RAV s: Radar de Vigilancia

R.A.V.C. e: Veterinary Corps

RAVE e: Rendering Acceleration Virtual Engine

RAVEN e: Radar Analysis and Visualization Environment

RAW e: Replenishment Agricultural Worker

RAWG/Hydrology e: Regional Associations Working Groups on Hydrology

RAWINSONDE e: Radar Wind Sounding and Radiosonde
- e: Radio and Wind Sounding

RAWINSONES e: Radiosonde and RAWIN

RAWP e: Remedial Action Work Plan

RAWS e: Radar Altitude Warning System
- e: Radar Attack and Warning System
- e: Radar Wind Sounder
- e: Remote Automated Weather Station

RAXON s: Fundación RAXON

RB e: Red border
- e: Research Board
- e: Revue Biblique
- e: Right Button
- e: Rule Based
- f: Race Blanche
- f: Remboursement
- f: Renouvellement de Bail
- f: Revenu Brut

rb. a: rubel
- a: rubler

R/B e: Red/Black
- e: Release/Birth

RBA e: Radiological Buffer Area
- e: Rapid Biodiversity Assessment
- e: Receive Buffer Area
- e: Relative Block Address
- e: Relative Byte Adresse

RBB e: Base Sheet Resistance
- e: Really Big Button
- e: Redland Braas Building Group

Rbbas r: Robotbas

Rbbåt r: Robotbåt

Rbbatt r: Robotbatteri

RBBS e: Residential Broadband Service

RBC e: Red Blood Cell[s]
- e: Relocatable Binary Code
- e: Request for Budget Change
- f: Réserve Biologique Communale

RBCB e: Run-Beyond-Cladding-Breach

RBCCN f: Répertoire des Bibliothèques participant au CCN

RBCS e: Remote Bar Code System

RBD e: Receiver Backward Channel Data
- e: Relative Absorption Band Depth
- f: Réserve Biologique [domaniale] Dirigée
RBDS e: Radio Broadcast Data System
RBE e: Red Brown Earth
RBECS e: Revised Battlefield Electronic CEOI System
RBER e: Residual Bit Error Rate
RBez. D: Regierungsbezirk
RBF e: Random Block Filemanager
- e: Record Blueprint file System
- e: Rockefeller Brothers Fund
Rbfre r: Robotfregatt
RBG d: Rechnerbetriebsgruppe
rBGH e: recombinant Bovine Growth Hormone
RBH e: Remote Bridge Hub
RBHC e: Regional Bell Holding Company
RBI e: Radiation Biology Institutes
- e: Rock Bit International
- f: Réserve Biologique [domaniale] Intégrale
Rbjag r: Robotjagare
RBL e: Reliability Based Logistics
RBM e: Rule Base Memory
RBMA e: Radiology Business Management Association
RBMK e: Light boiling water cooled graphite moderated pressure tube reactor
RBMS e: Resource Base Monitoring System
RBOF e: Receiving Basins for Offsite Fuels Savannah River Plant
RBOR e: Request Basic Output Report
R Bot s: Remolque Botes
RBPE s: Red de Bibliotecas Públicas del Estado
RBR e: Rockwell Budget Report
Rbradar r: Robotradar
Rbrak r: Robotar och raketer
RBRVS e: Resource Based Relative Value Scale
RBS e: Radar Bomb Scoring
- e: Radar Bomb Site
- e: Rare Book School
- e: Reactor Service Building
- e: Refractive Backscattering
- e: Remote Batch Station
- e: Rutherford Backscattering Spectroscopy
- r: robotsystem
RBSN e: Regional Basic Synoptic Network
RBSWG e: Risk Based Standards Working Group
RBT e: Reliable Broadcast Toolkit
- f: Remboursement
RBTL e: Read Between The Lines
RBTS e: Reverse Bias and Temperature Stress
Rbub r: Robotubåt
RBV A: Regse Blanke Volksfront
- e: Retrun Beam Vidicon Camera
RBWC e: Remaining Buffer Word Count register
RBWC0,1 e: Remaining Buffer Word Count 0,1
RC d: Republikanischer Club e.V.
- e: Reactor Compartment
- e: Receive[r] Clock
- e: Receiver Signal Element Timing
- e: Record Copy
- e: Recoverable
- e: Red Cross
- e: Refrigerated Chiller
- e: Regional Council (or Counsel)
- e: Repair Center
- e: Request for Change
- e: Residential College-East Quad
- e: Resistance Committees
- e: Resistor-Capacitor circuit
- e: Resource Center
- e: Response Cell
- e: Routing Control
- e: Run Command
- f: Rachat des Cotisations
- f: Raison Commerciale
- f: Rédacteur en Chef
- f: Reddition de Comptes
- f: Registre Cadastral
- f: Registre Central
- f: Registre du Commerce
- f: Règlement Collectif
- f: Relevé de Compte
- f: Rendre Compte
- f: Renseignements Complémentaires
- f: Représentant de Commerce
- f: Retraite des Cadres
- f: Révision de la Constitution
- f: Rôle des Contributions
- f: Rupture de Contrat
- r: Reservofficerarnas Centrala Organisation
- s: Radiofaro Circular
- s: Reconocimiento
- s: Regimiento de Caballería
Rc r: Rotechef
RC. e: Return Code
R/C e: Residue Crop ratio
s: Radio Circuito
RCA d: Steckernorm
- e: Radio Corporation of America
- e: Radiological Consultants Association
- e: Radiological[ly] Controlled Area
- e: Reactivity Cavity Access
- e: Refugee Cash Assistance
- e: Regina Coeli Academy
- e: Regional Co-operative Agreement for research, development and training related to nuclear science and technology
- e: Responsibility Center A
- e: Riot Control Agent
- e: Ripple Carry Adder
- e: Royal College of Anaesthetists
- f: Régie Commerciale des Alcools
R.C.A. e: Royal College of Art
RCAC s: Regimiento Acorazado de Caballería
RCAF e: Rail Cost Adjustment Factor
- e: Rail Cost-Adjustment Factor
R Caj Cer s: Remolque Caja Cerrada
RC ALJ e: Regional Chief Administrative Law Judge
RCAM e: Refueling Capability Assessment Model
RCAR e: Root Cause Analysis and Corrective Action Report
RCAS e: Reserve Component Automated Systems
- e: USREDCOM Command & Control System
RC AUD e: Residential College Auditorium
RCB e: Responsibility Center B
- e: Rotary Core Barrel
- f: Rationalisation des Choix Budgétaires
- s: Regimiento de Caballería Blindada
RCC e: Radiation Coordinating Council
- e: Radio Communications Coordinator
- e: Reduced Complexity Computer
- e: Regional Climate Center (or Centre)
- e: Regional Coordinating Council
- e: Remote Cluster Controller
- e: Remote Control Center
- e: Remote Control Console
- e: Rescue Coordination Center
- e: Reverse Command Channel
- e: Reverse Control Channel
- e: Revolution Command Council
- e: Routing Control Center
- e: Routing Control Channel
- f: Centre de coordination des opérations de sauvetage
- f: Registre central du commerce
RCCA e: Root Cause and Corrective Action
R.C.Ch. e: Roman Catholic Church
RCD e: Reactor components design
- e: Reference Cladding and Duct [program]
- e: Reference Conceptual Design
- e: Remote Change Directory
- f: Rassemblement pour la Culture et la Démocratie
- f: Receveur des Contributions Directes
RCDATA e: Replaceable Character Data
RCDR e: Reference Conceptual Design Report
RCDS e: Railcar Demurrage System
RCE e: Remote Channel Extenders
RCF e: Radiometer Calibration Facility
- e: RCRA Facility assessment
- e: Readers Comment Form
- e: Remote Concentration Facility
- e: Responsibility Center F
- e: Routing Control Field
RCFA e: Root Cause Failure Analysis
RCFM f: Régie des Chemins de Fer du Mali
RCfTA e: Research Centre for Theoretical Astrophysics
RCG e: Radioactive Concentration Guide
RCGMS e: Reactor Cover Gas Monitoring System
RCGP e: Royal College of General Practitioners
RCH e: Route Characteristics File
RCHP e: Reactor Component Handling Plan
- e: Reactor Core Handling Plan
RCH-WP e: Reactor Component Handling Work Package
- e: Reactor Core Handling-Work Package
RCI e: Relativistic Configuration Interaction
- e: Rodent Cage Interface
- f: Receveur des Contributions Indirectes

RCIRS e: Recoverable Consumption Item Requirements System
R Cis s: Remolque Cisterna
RCIS e: Research Centre for Injury Studies
Rck r: Radiocertifikatkurs
RCL e: Radioactive Corrosion Loops
- e: Rotate Carry Left
- e: Rotator Cuff Ligament
- e: Rubber Corporation of Liberia
- e: Runtime Control Library
RCLAC s: Regimiento de Caballería Ligera
RCLDN e: Retrieval of Calling Line Directory Number
RCLib e: Package from RCI
RCLK e: Rack Clock
RCM e: Radar Coded Message
- e: Radar Countermeasure
- e: Radioactive Convective Model
- e: Range Cell Migration
- e: Recording Current Meter
- e: Regional Circulation Model
- e: Remote Carrier Module
- e: Remote Control Module
- e: Resource-Centered Management
- e: Responsibility Center Monthly
- e: Ring-Closing Metathesis
- f: Rassemblement Chrétien de Madagascar
- s: Regimiento de Caballería Mecanizada
RCMAC e: Recent Change Memory Administration Center
RCMI e: Research Centers in Minority Institutions
RCMs e: Radar Coded Messages
RCN e: Refueling Change Notice
- e: Royal College of Nurses
- e: The Netherlands
- n: Reactor Centrum Nederland
RCO e: Radiological Control Organization
- e: Record Communications Operator
RCOA e: Refugee Council of Australia
R Coc s: Remolque [de] Cocina
RCOC e: Region Communications Operations Centers
R.C.O.&G. e: Royal College of Obstetricians and Gynaecologists
R Comb s: Remolque Combinado
RCP e: Reaction Control Panel
- e: Regional Contingency Plan
- e: Registration Control Points
- e: Remote Copy
- e: Remote Copy Program
- e: Republican Colored Party
- e: Request Change Proposal
- e: Research Centers Program
- e: Restore Cursor Position
- e: Retransmission Control Procedure
- e: Right hand Circular Polarization
- s: Resucitación Cardiopulmonar
RCPA e: Royal College of Pathologists of Australasia
RCPAC e: Reserve Component Personnel and Administration Center
R.C.P.&S. e: Royal College of Physicians and Surgeons
RCR e: Receive Control Register
- e: Remote Control Rack
- e: Review Comment Record
- e: Rotate Carry Right
- e: Royal College of Radiologists
- e: Runway Condition Rating
RCRA e: Resource Conservation and Recovery Act
RCRC e: Regional Cost Recovery Coordinator
RCRE e: Required Collaborative Research Experience
RCRIS e: Resource Conservation and Recovery [act docket] Information System
RCS e: Computer Timesharing Resource Control System
- e: Rainforest Conservation Society
- e: Reaction Control System[s]
- e: Reactor Containment System
- e: Reactor Coolant System
- e: Reactor Core Supply
- e: Record[s] Communication[s] Switching System
- e: Reference Climate (or Climatological) Station
- e: Remote Characterization System
- e: Remote Cloud Sensing
- e: Reports Control System
- e: Resource Construction Set
- e: Revision Control System
- e: Rural Counselling Service
- f: Régiment de Commandement et de Soutien
- f: Station Climatologique de Référence
RCSC e: Remote Spooling Communications Subsystem
RCS[Eng] e: Royal College of Surgeons of England
RCSF f: Réseau Canadien pour la Santé des Femmes
RCSM e: Reactor Cladding and Structural Materials
RCSSMRS e: Regional Centre for Services in Surveying, Mapping and Remote Sensing
- f: Centre régional de services spécialisés dans le domaine des levés, des cartes et de la télédétection
- s: Centro Regional de Servicios en Levantamientos, Cartografía y Teledetección
RCT e: Radioactivity Control Technology
- e: Radiometric Calibration Target
- e: Radomized Control Trial
- e: Randomized Clinical Trial
- e: Record Type
- e: Reference Cluster Tool
- e: Regents Competency Test
- e: Repair Cycle Time
RCTC e: Rewritable Consumer Time Code
RCT-COSY e: Relayed Coherence Transfer Correlation Spectroscopy
RCTM s: Régimen de Control de Tecnología de Misiles
RCTS e: Reactor Component Test Section
- e: Regulatory Compliance Tracking System
- e: Remote Commercial Television Services
RCTs e: Radiological Control Technicians
RCTV e: Remote-Controlled Transport Vehicle
RCU e: Radio Channel Unit
- e: Regional Coordinating Unit
- e: Relative Capacity Unit
- e: Reserve Component Unit
- f: Récépissé de Camionnage Urbain
- f: Unité Régionale de Coordination
- s: Unidad Regional de Coordinación
RCV e: Reaction Control Valves
RCVR e: Receiver [Sequence Number]
RCW S: Rapidly Changing World
- e: Red-Cocaded Woodpecker
- e: Regulatory Code of Washington
- e: Revised Code of Washington
RCWP e: Rapidly Changing World
RCWT f: Réseau des Chemins de Fer et du Wharf du Togo
RD e: Data Requirements Document
- e: Division of Reactor Development
- e: Receive
- e: Recursive Design
- e: Registration Division
- e: Remedial Design
- e: Remote Disconnect
- e: Remove Directory
- e: Replicated Data
- e: Request Disconnect
- e: Resource Discovery
- e: Rocketdyne
- e: Route Descriptor
- e: Routing Domain
- f: Récépissé de Déclaration
- f: Recette-Distribution
- f: Receveur des Douanes
- f: Receveur-Distributeur
- f: Reconnaissance de Dette
- f: Recouvrement à Domicile
- f: Régime Douanier
- f: Règlement Différé
- f: Rente Différée
- f: Rente Différentielle
- f: République Démocratique
- f: République Dominicaine
- f: Rhin-Danube
- f: Rive Droite
- f: Route Départementale
- s: Radar de Distancia
- s: Radiofaro Direccional
rd d: Einheitenkurzzeichen für Rad
Rd N: Radar
- r: Radio[skola]
RD. e: Receive Data
R&D e: Research and Development
R/D e: Research and Development
- e: Rig Down
- s: Radiodirección
R4D e: Response, Resistance, Resilience and Recovery of arctic ecosystems to Disturbance
RDA e: Radar Data Acquisition
- e: Receive Descriptor Area
- e: Regional Development Account
- e: Remote Data[base] Access [protocol]
- e: Research, Development and Acquisition
- e: Right Displacement of Abomasum
- f: République Démocratique d'Allemagne
RDAAC e: Remote Data, Analysis and Archive Center
RDAC e: Research and Development Associated with Construction

R-DAT e: Rotary [head] Digital Audio Tape
RDB A: Reddingdaadbond
- d: Rechtsdatenbank
- e: Rare Disease Database
- e: Receive Data Buffer
- e: Red Data Book
- e: Relational Database
RdB e: Relationale Datenbank
RDB e: Rigid Disk Block
RDBG f: Rassemblement Démocratique des Bâtisseurs de Guinée
RDBMS d: Relationales Datenbank-Management-System
- e: Relational Database Mangement System
RDBMSMIB e: Relational DataBase Management System-Management Information Base
RDBS e: Relational Database System
RDC e: Reconstruction and Development Corporation
- e: Refugee Documentation Center
- e: Regional Data Center
- e: Regional Dispatch Centre
- e: Research and Development Corporation
- f: Rayer Des Contrôles
- s: Centro Regional de Mensajería
R&DC e: Research and Development Contractor
RD&C e: Research, Development and Commercialization
R2DC3 e: Rapid Reaction Deployable Command and Control Center
RDCC f: Rassemblement Démocratique des Cadres Tchadiens
RDCLK e: Received Timing Clock
RDD e: Random Digit Dialed
- e: Replaceable Database Driver[s]
- e: Responsibility Driven Design
- f: Rassemblement pour la Démocratie et le Développement
- s: Radio De Dispersión
RD&D e: Research Development and Demonstration
RDDP e: Radar Data Development Program
RDDT&E e: Research, Development, Demonstration Testing and Evaluation
R&D/EA e: Research and Development and Experimental Activities
R-de-C f: Rez-de-Chaussée
RDEC e: Research, Development and Engineering Center
RDES e: Remote Data Entry System
RDEU e: Reaction Diagnostic Experimental Unit
RDEX e: Readiness Exercise
RDF e: Rapid Deployment Forces
- e: Rate Decrease Factor
- e: Request Denied Frame
- e: Resource Description Framework
- e: Route Designator Field
- f: Radiodiffusion Française
Rdfn f: Radiodiffusion
R 'dg e: Reading
Rd G s: Radiogoniómetro
RDG d: Ribnitz-Damgarten, Mecklenburg-Vorpommern
Rdg f: Risque de guerre
RDGE f: Réforme Démocratique de Guinée Equatoriale
RDI e: Remote Defect Identification
- e: Remote Defect Indicator
- e: Replication, Distribution, Installation and Training
RDI-P e: Remote Defect Indicator-Path Level
RDIT e: Remote Defect Identification
- e: Remote Defect Indication
RDL e: Remote Digital Loopback
- f: Rassemblement des Démocrates Libéraux
RD[-]LAP e: Radio Data Link Access Protocol
RDLGE s: Reagrupamiento Democrático para la Liberación de la Guinea Ecuatorial
RDM d: Ring Deutscher Makler
- e: Reliably Delivered Message
- f: République Démocratique de Madagascar
Rdm r: Radiomästare
RDMF e: Rapid Deployable Medical Facility
RDMS e: Rapid Deployable Mobile System
RDMSS e: Rapidly Deployable Mobile SIGINT System
RdN f: Rien de Nouveau
RDN p: Radiodifusão Nacional
rDNA e: recombinant DNA
- e: ribosomal DNA
RDO e: Reconnaissance Duty Officer
- e: Remote Data Object[s]
- e: Resource Definition On-line
RDP e: Radar Development Project
- e: Rapid Deployment Planning
- e: Reconstruction and Development Plan
- e: Reconstruction and Development Program
- e: Reconstruction and Development Programme
- e: Regional Development Program
- e: Reliable Data[gram] Protocol
- e: Remote Desktop Protocol
- f: Rassemblement Démocratique du Peuple
- f: Rassemblement pour la Démocratie et le Progrès
- f: Révolution Démocratique et Populaire
- s: Radio De Peligro
RDPC e: Regional Data Processing Center
- f: Rassemblement Démocratique des Peuples Comoriens
- f: Rassemblement Démocratique du Peuple Camerounais
RDPI e: Remote Data Processing Installation
RDPS f: Rassemblement pour la Démocratie et le Progrès Social
Rd Pt f: Rond-Point
RDPT f: Rassemblement Démocratique des Populations Tahitiennes
RDR e: Rotating Disk Reactor
- f: Rassemblement des Républicains
- f: Rassemblement pour la Démocratie et le Retour des Réfugiés à Rwanda
Rdr f: Rive droite
RDR n: Rijksdienst voor Radiocommunicatie
RD/RA e: Remedial Design/Remedial Action
RDRAM e: Rambus DRAM [Dynamic Random Access Memory]
RDRC e: Ross Dependency Research Committee, Wellington
RDS d: Radio Daten Service
- d: Register Deutscher Spendenorganisationen
- e: Related Deviation Standard
- e: Remote Data Service[s]
- e: Risk Data Sheet
- s: Radio De Seguridad
RDSG e: Replicated Data Serialization Graph
RDS-HWDMS e: National Hazardous Waste Data Management Systems
Rdsi r: Radiosignalskola
RDSI s: Red Digital de Servicios Integrados
RDSR e: Receiver Data Service Request
RDSS e: Radio Determination Satellite Service
RDST e: Rotating Head Digital Storage Tape
R.D.S.T.P. p: Republica Democrática de São Tomé e Príncipe
RDT s: Rendimiento normal
RD&T e: Research, Development and Testing
RDTD e: Restricted Differential Time Delay
RDT[&]E e: Research, Development, Test[ing] and Evaluation
RDT&E/A/N/AF e: RDT&E, Army/Navy/AF
Rdtg r: Radiotelegrafist
RDTM e: Rated Distribution Training Management
RDTO e: Receive Data Transfer Offset
RDU e: RADEC Data Unit
- e: Regional Decision Unit
- f: Rassemblement pour la Démocratie et l'Unité
RDV e: Raman Doppler Velocimetry
Rdv f: Rendez-vous
RDV s: Ración Diaria de Víveres
RDWP e: Remedial Design Work Plan
RDX e: Directorate of Program Installation
- e: explosive used in pipe-severing system
RDYLD e: Ready to Load Data

RE a: Regningsenhed
- e: Radiochemical Engineering
- e: Reasonable Efforts
- e: Recall
- e: Refugee
- e: Regarding
- e: Regular Exposure is a scientific exposure
- e: Report Engine
- e: Reportable Event
- e: Reporting Subsystem
- e: Reproductive Endocrinologist
- e: Resonance Energy
- e: Reunion [French]
- f: Raison d'Etat
- f: Rapport d'Expert
- f: Receveur de l'Enregistrement
- f: Réforme Économique
- f: Réfugié Étranger
- f: Région Économique
- f: Règle d'Escompte
- f: Relations Extérieures
- f: Réponse Écrite
- f: Retrait d'Emploi
- f: Révision Électorale
- s: Ración de Emergencia
- s: Radioenlace

re e: regards
Re e: Reinsurance company
- e: Reply [Internet]
- f: Roche
- s: Reactivo

R E e: Real Estate
R.E. e: Engineers
R&E e: Rules and Elections Committee
R/E e: Reentry
REA e: Rail Express Agency
- e: Request for Equitable Adjustment
- e: Resource End Address register
- e: Restriction Enzyme Analysis

Reabast s: Reabastecimiento
REACT e: Rain Echo Attenuation Compensation Technique
- e: Revolutionary Action Committee

REACTS e: Radiation Emergency Assistance Center Training Site
- e: Readiness Exercise and Crisis Transportation System

READ e: Regulatory and Economics Analysis Division
REAG e: Reproductive Effects Assessment Group
REAL f: Recherches et Expansion d'Alimentation
RealAudio d: Technik von Progressive Networks, über die sich Audiodaten [Ton] in Echtzeit übertragen lassen
REALP e: Real Property Inventory
REAP e: Regional Enforcement Activities Plan
- e: Regional Environment Assessment Program
- e: Resource Evaluation Aerial Photography [project]
- e: Rural Education Access Program

REAS e: Regional Enforcement and Superfund System
REB e: Resist Etchback
Reb f: Rebelle
REBIUN s: Red de Bibliotecas Universitarias
REBM e: Radio Electronic Battle Management
REBS e: Radiation Energy Balance System
REBUS f: Réseau des Bibliothèques utilisant SIBIL
REC d: Rechtswissenschaften
- e: Radio Electronic Combattranslation
- e: Radiochemical Engineering Cell
- e: Radiochemical Engineering Complex
- e: Reconstitutable Enduring Communication
- e: Recreation[al or Entertainment Activities]
- e: Regional Engineering Center (or Centre)
- f: Réclamé

Rec f: Récupéré
- s: Recepción
- s: Receptor
- s: Recibir
- s: Reconocer
- s: Reconocimiento

REC. e: Recreation
RECA e: Residual Capability Assessment
RECALLDB e: Recall Data Base
RECAN s: Registro Nacional de Cáncer
RECAS e: Residual Capability Assessment System
RECAU a: Regionale EDB-Center ved Aarhus Universitet
Rec Blind s: Reconocimiento Blindado
RECC e: Regional Emergency Communications Coordinator
REC DIV e: Recreation Division
RECEIV-I e: I Receiving Water Model
Recent Dev Alcohol e: Recent Developments in Alcoholism
Recenti Prog Med i: Recenti Progressi in Medicina
Recent Prog Horm Res e: Recent Progress in Hormone Research
Recent Results Cancer Res e: Recent Results in Cancer Research
Recep s: Recepción
- s: Receptor

Receptors Channels e: Receptors and Channels
RECFMS. e: Record Formatted Maintenance Statistics
Rech f: Recherches
Recht d: Institut für Sozialforschung, Abteilung für Privatrecht
REC3I e: Reconstitutable Enduring Communications Command, Control and Intelligence
Récip f: Réciproque
Reco s: Reconocimiento
RECOA s: Reconocimiento Aéreo
Recoc e: Returned Exile Coordinating Committee
RECOM s: Reconocimiento Marítimo
RECON e: Reconnaissance [[Information] System]
Reconstr Surg Traumatol e: Reconstruction Surgery and Traumatology
RECOSCIX-CEA e: Regional Cooperation in Scientific Information Exchange in the Central Eastern Atlantic
- f: Coopération régionale en matière d'échanges d'information scientifique dans l'Atlantique du Centre-Est
- s: Cooperación Regional en el Intercambio de Información Científica en el Atlántico Centrooriental

RECOSCIX-WIO e: Regional Cooperation in Scientific Information Exchange in the Western Indian Ocean
- f: Coopération régionale en matière d'échanges d'information scientifique dans l'océan Indien occidental
- s: Cooperación Regional en el Intercambio de Información Científica en el Océano Indico Occidental

RECOTAC s: Reconocimiento Táctico
RECREP e: Reconnaissance Report
RECS e: Radio Electronic Combat Support
- e: Rear Echelon HF COMINT Systems Radio Electronic Combat Support

RECSHIP e: Receiving Ship
RECSTA e: Receiving Station
RECTAS e: Regional Centre for Training in Aerospace Surveys
Rectif f: Rectificatif
RECTO/T s: Reconocimiento Aéreo Táctico
Recup s: Recuperación
Recup Mat s: Recuperación De Material
RED e: RCRA Enforcement Division
- e: Risk Estimate Distance

red. a: redaktør
- a: redigeret [af]
- N: redaksjon

REDAR e: Radiation and Environmental Data Acquisition and Recording System
- s: Red de Agro-Industría Rural

REDCOM e: U.S. Readiness Command
REDHINTEGR s: Fundación de Recostrucción y Desarollo Humano Integral
RED HORSE e: Rapid Engineer Deployable, Heavy Operational Repair Squadron, Engineer
REDIA f: Recherches et Exploitation Diamantifères
redig. a: redigeret [af]
REDN e: Red River Of North Basin
REDOX e: Reduction-Oxidation
REDP e: Redeployment
- e: Regional Energy Development Program

REDR e: Recombination-Enhanced Defect Reactions
REDSO e: Regional Economic Development Support Office
redupl. e: reduplicating
REE e: Rare Earth Element[s]
- e: Russian, East European and Eurasian Studies
- f: Relations Économiques Extérieures

REECO e: Reynolds Electrical and Engineering Co., Inc.
REEF e: Great Barrier Reef
- f: Répertoire des Ensembles et Éléments Fabriqués du bâtiment

REEP e: Reasonable Extra Efforts Program
- e: Regional Environment Employment Program
- e: Review of Environmental Effects of Pollutants

REF e: Reference
- f: Référence
- r: Riksnämnden för Ekonomisk Försvarsberedskap

Ref s: Referencia
- s: Reforzado

ref. a: refererende
- a: refereret [af]
- N: referanse
- N: refselse

Réf f: Réforme
- f: Réfugié

refash. e: refashioned
- e: refashioning

REFILES e: JCS Standard Reference Files

REFIPS f: Réseau Francophone International pour la Promotion de la Santé

refl. e: reflexive

REFLEX e: Remote Fluvial Experimental [Series]
- e: Remote Fluvial Experiments

REFM e: Resource Ecology and Fisheries Management

Ref-no e: Literature No. in References

REFORM e: Reengineering Effort for Organization and Management

Réfr f: Réfrigérateur

REFRIBEL f: Régie des services Frigorifiques de Belgique

REFSPEC e: Barringer Reflectance Spectrometer

Reftel e: Reference Telegram

Refz s: Refuerzo

REG d: Regen/Bayrischer Wald, Bayern
- e: Regular
- e: Regular Languages
- f: Réglementation

Reg f: Registre
- r: Regemente
- s: Regimiento

reg N: regiment

reg. a: regerende
- a: regering
- a: registreret
- N: registrert

Rég f: Régénération
- f: Régent
- f: Région
- f: Régulier

REGA e: Rettungsflugwacht

REGÆND a: Registerændringssystemet [ved militæret]

REGAL e: Rigid Epoxy Glass Acrylic Laminate

Reg Anesth e: Regional Anesthesia

Reg Art s: Regimiento de Artillería

Regch r: Regementschef

RegCM2 e: Regional Climate Model, Version 2

RegelT d: Institut für Automatisierungs- und Softwaretechnik

REGENCY NET e: USEUCOM HF/SSB communications system

REGEX e: Regular Expression

Regförbpl r: Regementsförbandsplats

REGIDESO f: Régie de Distribution d'Eau et d'Electricité

REGIFERCAM f: Régie des Chemins de Fer du Cameroun

Reg Immunol e: Regional Immunology

Reg Inf s: Regimiento de Infantería

Regint r: Regementsintendent

REGIS e: Regional Geographical Information System
- e: Regional Information System for Microcomputers
- f: Système régional d'information pour micro-ordinateurs
- s: Sistema Regional de Información para Microcomputadoras

Regl r: Reglemente

Règl f: Règlement

Regläk r: Regementsläkare

Reg Lig Ac Cab s: Regimiento Ligero Acorazado de Caballería

regn. a: regning

Reg-no e: SEM negative-film No. registered in NUM

REGO e: Reinventing Government

Regoff r: Regementsofficer

regsjef N: Regimentssjef

RegTP d: Regulierungsbehörde für Telekommunikation

Regul Pept e: Regulatory Peptides

Regul. Rivers Res. & Manage. e: Regulated Rivers: Research and Management

Regul Toxicol Pharmacol e: Regulatory Toxicology and Pharmacology

RegVBG d: Registerverfahrens-beschleunigungsgesetz vom 20.12.1993

REH f: Relevé Épidémiologique Hebdomadaire

REHI e: Hi, too [Internet]

REI d: Richtlinie zur Emissions- und Immissionsüberwachung kerntechnischer Anlagen
- f: Régie des Exploitations Industrielles

REIIF e: Renewable Energy Innovation Investment Fund

REI-L e: Remote Error Indicator-Line Level

REI-P e: Remote Error Indicator-Path Level

REIS e: Reconstitutable and Enduring Intelligence System

REIT e: HERL-RTP Reiter Maze System

rek N: rekognosere
- N: rekognosering
- N: rekommandert

Rek r: Rekognos[c]ering
- r: Rekognosera
- r: Rekommenderat [brev]

REKLIP d: Regio[nales] Klimaprojekt
- e: Regional Climate Project

rekr. N: rekrut[t][ering]

RekrI KA r: Rekrytinstruktion för kustartilleriet

rekv. N: rekvisisjon

REL e: Recommended Exposure Limit
- e: Recovered Energy Logic
- e: Relative Exposure Limit
- e: Religious Studies

rel e: related to

Rel f: Relié
- f: Religieux

rel. a: relativ

Rel. e: Relativ[er]

RELAB e: Latin American Biological Network
- e: Reflectance Experiment Laboratory

REL COPAN e: USAF HQ Advance Network

RELMAP e: Regional Lagrangian Model of Air Pollution

RELOC e: Relocation File

RELP e: Regular-Pulse Excited LPC

RELROK e: Releasable to Republic of Korea

RELS e: Religion

RELSECT e: Relative Sector

REM e: Radioactivity Environmental Monitoring
- e: Random Entry Memory
- e: Raster Electron Microscope (or Microscopy)
- e: Rat Enclosure Module
- e: Rational Economic Man
- e: Remark
- e: Remote [Equipment Module]
- e: Request for Engineering Memorandum
- e: Ring Error Monitor
- e: Roentgen Equivalent in Man
- e: Route Evaluation Module
- f: Recherches et Exploitation Minières
- f: Remise
- f: Reseau d'Enseignement Multimedia

Rem e: Roentgen equivalent man
- f: Remarquable
- f: Remarque
- s: Remolcador
- s: Remolque
- s: Remonta

rem. r: remiss
- r: remittera

REMA s: Remolcado con Motor Auxiliar

REMAB e: Remote Marshaling Base

Rem AM s: Remolcador de Alta Mar

REMAP e: Regional Environmental Mapping and Analysis Process

REMAPP e: Remote Sensing Array Processing Procedures

REMARC e: Resource Entry, Management, Accountability and Reporting for Contracts
- e: Resource Management Accounting Control System

Remb f: Remboursement

Rem Bah s: Remolcador de Bahía

REMBASS e: Remotely Monitored Battlefield Sensor System

REMCO e: Reference Materials Committee
- f: Comité pour les Matériaux de Référence
- s: Comité de Materiales de Referencia

REM-CR e: Remoxipride in Controlled-Release formulation

REM/FIT e: Remedial/Field Investigation Team

Rem G s: Remolque-Grúa

REMHI s: Proyecto Interdiocesano, Recuperación de la Memoria Histórica

REMI e: French Coastal Microbiology Monitoring Network
- f: Réseau de surveillance Microbiologique du littoral français
- s: Red de Vigilancia Microbiológica de las Zonas Costeras Francesas
REMIDAS e: Resource Evaluation and Monitoring Integrated Data Analysis System
Reminisc. e: Reminiscence[s]
REM-IR e: Remoxipride in Immediate-Release formulation
REMIS e: Reliability and Maintainability Information System
- e: Remote Imaging System
REML e: Radiological Environmental Monitoring Laboratory
- e: Rapid Eye Movement Latency
REMO r: Renovering och Modifering
REMOBS e: Remote Observation System
REMPEC e: Regional Marine Pollution Emergency Response Centre for the Mediterranean Sea
- f: Centre régional méditerranéen pour l'intervention d'urgence contre la pollution marine accidentelle
- s: Centro Regional de Operaciones Realizadas en los Casos de Urgencia Ocasionados por Contaminación Marina para el Mar Mediterráneo
REMPT e: Resonance-Enhanced Multi-Photon ionization spectroscopy
REMS e: RCRA Enforcement Management System
Rem T s: Remolque Taller
REMUS e: Resource Management Utilisation System
REN e: Rename
- e: Renfrew Hall
- e: Reno, NV
- e: Ring Equivalence Number
- e: Ringer Equivalence Number
Ren f: Rennes
RENAB p: Reparações Navais de Beira
RENACER s: Hogar el Renacer
RENAMO p: Resistência Nacional de Moçambique
- s: Resistencia Nacional de Mozambique
Renass f: Réseau National de Surveillance Sismique
RENATER f: Réseau National de Télécommunications pour la Technologie, l'Enseigement et la Recherche
RENE e: Rehearsal Experiment ERS 1-Validation, Northern Europe
RENES f: Redéploiement Energétique du Sénégal
RENEW e: Republican Network to Elect Women
R Enf s: Remolque Enfriadero
Ren Fail e: Renal Failure
RENFE f: Rénovation Wallon
Rénov f: Rénovation
RENPAP e: Regional Network on the Production, Marketing and Control of Pesticides for Asia and the Pacific
Ren Physiol Biochem e: Renal Physiology and Biochemistry
Ren-Ref e: Renaissance-Reformation
Rens f: Renseignements
REO e: Removable, Erasable, Optical [Disk]
- e: Reoriented
Réorg f: Réorganisation
REP d: Repertorium
- e: Reasonable Efforts Program
- e: Repeat
- e: Replacement
- e: Reply to Message Number
- e: Report[s]
- e: Reserve Enlistment Program
- e: Resource Estimating Procedure
- e: Robots Exclusion Protocol
Rep d: Repetitorium
- f: Représentant
- f: Représentation
- r: reparera
Rép f: Répertoire
- f: Républicain
- f: République
rep. a: republik
- N: reparasjon
Rep. e: Report[s]
REPE e: Repeat while Equal
- e: Resonance Energy Per Electron
REPGEN e: FLOGEN Report Generator
Rep Group Adv Psychiatry e: Report / Group for the Advancement of Psychiatry
Repgrp r: Reparationsgrupp
Rep Health Soc Subj [Lond] e: Reports on Health and Social Subjects
Répind f: Républicains indépendants
Rep. Int. Whaling Comm. e: Report of the International Whaling Commission
REPNE e: Repeat while Not Equal
REPNZ e: Repeat while Not Zero
repo e: repossession
REPOL e: Reporting Emergency Petroleum, Oils and Lubricants
rep. øv. N: repetisjonsøvelse
Repövning r: Repetitionsövning
Repr f: Représentant
repr. a: Repræsentant
- a: Repræsentere
- e: Representative
REPRISK e: High Level Radioactive Waste-Repository Risk Model
REPRO e: Regional Programs Office
Reprod Fertil Dev e: Reproduction, Fertility and Development
Reprod Nutr Dev e: Reproduction, Nutrition, Development
Reprod Toxicol e: Reproductive Toxicology
ReproLine e: Reproductive Health Online
REPS e: Regional Emissions Projection System
REPSAM e: Repository Sampling computer program
RépSoc f: Républicains Sociaux
reptj N: reparasjonstjeneste
Repvst r: Reparationsverkstad
REPZ e: Repeat while Zero
Req s: Requisa
- s: Requisición
REQCAP e: Requirement Capability
Reql s: Remolque
REQMS e: Record Maintenance Statistics
REQMS. e: Request Maintenance Statistics
REQSPEC e: Requirements Specification
RER e: Ranging Exciter Receiver
- e: Refocusing Environmental Restoration
- e: Reseau Express Regional
- e: Resting Energing Requirement
Rer f: Rocher
REREG e: Re-Register
RERIC e: Regional Energy Resources Information Center
Rers f: Rochers
RERV e: Radar Estimated Rain Volume
RES e: Radiated Electromagnetic Susceptibility
- e: Radio Equipment and Systems
- e: Recruitment Service
- e: Reformed Ecumenical Synod
- e: Regional Evaporation Study
- e: Remote Execution Server
- e: Remote Execution Service
- e: Reset
- e: Reset Resolution
- e: Restricted
- e: Robot Exclusion Standard
- f: Rachat de l'Entreprise par ses Salariés
Res r: Reserv
- s: Reserva
Rés f: Réseau
- f: Réserve
- f: Résidence
- f: Résistance
- f: Résolution
res. a: residerende
R&Es e: Rare and Endangered Plants
RESA e: Scientific Research Society of America
RESACENT e: Remote Sensing Application Centre
res. bilkol. N: reservebiltrenkolonne
RESC e: Reconstitutable Enduring Satellite Communications System
RESCO e: Review of Employment and Social Conditions
Res Commun Mol Pathol Pharmacol e: Research Communications in Molecular Pathology and Pharmacology
RESCU e: Remote Emergency Satellite Cellular Unit
- e: Resource Cost and Utilization System
RESDA e: Remote Sensing Data
- e: Remote Sensing Development Agency
Res Dev Disabil e: Research in Developmental Disabilities
RESEDA e: Remote Sensor Data Analysis
RESENA e: Remote Sensing of Nature
RESERVES e: Responsive Replacement Vehicles
RESET e: Regional Service Team
- e: Reset pin
Res Exp Med [Berl] e: Research in Experimental Medicine
RESHELL e: Remote Sensing Expert System Shell
Res Immunol e: Research in Immunology
RESIN s: Resumen de Información
RESIS e: Remote Sensing Information System
Reskap r: Reservkapacitet
RESL e: Radiological and Environmental Sciences Laboratory

Res Microbiol e: Research in Microbiology
Res Nurs Health e: Research in Nursing and Health
RESO f: Réseau Européen de Stations Océaniques
- N: Folkebevegelsens Reise- og Ferieorganisasjon i Sverige
Resoff r: Reservofficer
RESOLVE e: Center for Environmental Conflict Resolution
RESORS e: Remote Sensing On-Line Retrieval System
R Esp s: Remolque Especial
Resp f: Respectivement
- f: Respiration
- f: Responsable
resp. s: respectivamente
RESPAS e: Remote Sensing Preprocessing and Archiving System
RESPECT e: Rapid Electromagnetic Scattering Predictor for Extremely Complex Targets
Respir Med e: Respiratory Medicine
Respir Physiol e: Respiration Physiology
Res Publ Assoc Res Nerv Ment Dis e: Research Publications-Association for Research in Nervous and Mental Disease
Res Q Exerc Sport e: Research Quarterly for Exercise and Sport
Res Rep Health Eff Inst e: Research Report/Health Effects Institute
RESSAC e: The Consequence of a Major Nuclear Accident on the Surrounding Environment: Soil and Surface Rehabilitation
- f: Radar pour l'étude du Spectre des Surfaces par Analyse Circulaire
RESSFOX e: Recessed Sealed Sidewall Field Oxidation
Resst r: Reservstat
RESSYS e: Regional Ecosystem Simulation System
RESTEC e: Remote Sensing Technology Center
Results Probl Cell Differ e: Results and Problems in Cell Differentiation
RESV e: RSVP Reservation Request
Res Vet Sci e: Research in Veterinary Science
Res Virol e: Research in Virology
RET e: Renewable Energy Technologies
- e: Resolution Enhancement Technology
- e: Returning
- f: Relations Extérieures
- f: Retrait
Ret s: Retaguardia
- s: Retirada
RETA e: Regional Environment[al] Technical Assistance [project]
- f: Projet régional d'assistance technique pour l'environnement
- s: Proyecto Regional de Asistencia Técnica sobre el Medio Ambiente
Ret D f: Rhin et Danube
RETEC f: Registre des Essais Therapeutiques Européens
Retg s: Retaguardia
RETM e: Rare Earth/Transition Metal
RETMA e: Radio Electronics Television Manufacturers Association
RETN e: Request for Engineering Test Notice
retn N: retning
RETRF e: Rural Electrification and Telephone Revolving Fund
Retro e: Retrospective Rate Derivation
RettAssAPrV d: Ausbildungs- und Prüfungsordnung für Rettungsassisteninnen und Rettungsassitenten vom 7.11.1989
RettAssG d: Gesetz über den Beruf der Rettungsassistentin und des Rettungsassistenten [Rettungsassistentengesetz] vom 10.7.1989
REU e: Research Experience[s] for Undergraduates
REUN f: Réseau Européen Unifié des Nivellements
REV e: Details of the revisions to the programme
- e: Reticuloendotheliosis Virus
Rev d: Innenrevision
- f: Revenu
rev r: revolver
Rév f: Révision
rev. a: revideret
- a: revision
- a: revisor
- e: revised
- N: revidert
- r: revisor
Rev Alerg Mex s: Revista Alergia Mexico
Rev Argent Microbiol s: Revista Argentina de Microbiologia
Rev Assoc Med Bras p: Revista Da Associacao Medica Brasileira
Rev Belge Med Dent f: Revue Belge de Medecine Dentaire
Rev Biol Trop s: Revista de Biologia Tropical
Rev Bras Biol p: Revista Brasileira de Biologia
REVC e: Reversed charging
Rev Chil Obstet Ginecol s: Revista Chilena de Obstetricia y Ginecologia
Rev Chir Orthop Reparatrice Appar Mot f: Revue de Chirurgie Orthopedique et Reparatrice de l'Appareil Moteur
Rev Clin Esp s: Revista Clinica Espanola
RevCom e: IEEE-SA Standards Board Standards Review Committee
Rev Cubana Med Trop s: Revista Cubana de Medicina Tropical
Rev Elev Med Vet Pays Trop f: Revue d'Elevage et de Medecine Veterinaire des Pays Tropicaux
Rev Environ Contam Toxicol e: Reviews of Environmental Contamination and Toxicology
Rev Environ Health e: Reviews on Environmental Health
Rev Epidemiol Sante Publique f: Revue d'Epidemiologie et de Sante Publique
Rev Esp Anestesiol Reanim s: Revista Espanola de Anestesiologia y Reanimacion
Rev Esp Cardiol s: Revista Espanola de Cardiologia
Rev Esp Enferm Dig s: Revista Espanola de Enfermedades Digestivas
Rev Esp Fisiol s: Revista Espanola de Fisiologia
Rev Esp Oncol s: Revista Espanola de Oncologia
Rev Fac Cien Med Univ Nac Cordoba s: Revista de la Facultad de Ciencias Medicas/Universidad Nacional de Cordoba
Rev Fr Gynecol Obstet f: Revue Francaise de Gynecologie et d'Obstetrique
Rev Gastroenterol Mex s: Revista de Gastroenterologia de Mexico
Rev Gastroenterol Peru s: Revista de Gastroenterologia del Peru
Rev Hosp Clin Fac Med Sao Paulo p: Revista do Hospital das Clinicas; Faculdade de Medicina da Universidade de Sao Paulo
REVIC e: Revised Version of Intermediate CoCoMo
revid. a: revideret
- r: reviderad
Rev Inst Med Trop Sao Paulo p: Revista do Instituto de Medicina Tropical de Sao Paulo
Rev Int Trach Pathol Ocul Trop Subtrop Sante Publique f: Revue Internationale du Trachome et de Pathologie Oculaire Tropicale et Subtropicale et de Sante Publique
Rev Invest Clin s: Revista de Investigacion Clinica
Rev Laryngol Otol Rhinol [Bord] f: Revue de Laryngologie Otologie Rhinologie
Rev Latinoam Microbiol s: Revista Latinoamericana de Microbiologia
Rev Mal Respir f: Revue des Maladies Respiratoires
Rev Med Brux f: Revue Medicale de Bruxelles
Rev Med Chil s: Revista Medica de Chile
Rev Med Chir Soc Med Nat Iasi ru: Revista Medico-Chirurgicala a Societatii de Medici si Naturalisti din Iasi
Rev Med Interne f: Revue de Medecine Interne
Rev Med Liege f: Revue Medicale de Liege
Rev Med Panama s: Revista Medica de Panama
Rev Med Suisse Romande f: Revue Medicale de la Suisse Romande
Rev Med Univ Navarra s: Revista de Medicina de la Universidad de Navarra
Rev Neurol s: Revista de Neurologia
Rev Neurol [Paris] f: Revue Neurologique
Rev Neurosci e: Reviews in the Neurosciences
Rev Oculomot Res e: Reviews of Oculomotor Research
Rev Paul Med s: Revista Paulista de Medicina
Rev Physiol Biochem Pharmacol e: Reviews of Physiology Biochemistry and Pharmacology
Rev Pneumol Clin f: Revue de Pneumologie Clinique

Rev Port Cardiol p: Revista Portuguesa de Cardiologia
Rev Rhum Engl Ed e: Revue du Rhumatisme. English Edition
Rev Roum Virol f: Revue Roumaine de Virologie
Rev Sanid Hig Publica [Madr] s: Revista de Sanidad e Higiene Publica
Rev Saude Publica s: Revista de Saude Publica
Rev Sci Tech f: Revue Scientifique et Technique
Rev Soc Bras Med Trop p: Revista da Sociedade Brasileira de Medicina Tropical
Rev Stomatol Chir Maxillofac f: Revue de Stomatologie et de Chirurgie Maxillo-Faciale
REWA e: Revolutionary Ethiopia Womens Association
REX d: Deutscher Rentenindex
- e: Radiation Exposure System
- e: Rainfall Emigration Experiment
- e: Readiness Exercise
- e: Relocatable Executable
- e: Remote Execution
- f: Expérience sur les migrations liées aux pluies
- s: Experimento de Emigración por Lluvias
REXEC e: Remote Execution
REXMAN e: Relax, Experiment and read the Manual [Internet]
REXS e: Report on Executive Staffing
REXX e: Restructured Extended Executive Language
- e: Restructured Extended Executor [language]
REYA e: Revolutionary Ethiopia Youth Association
REZATELSAT f: Réseau Zairois de Télécommunications par Satellite
RF e: Bibliographic Reference
- e: Radio Frequency
- e: Registered Forester
- e: Remote File
- e: Representative Fraction
- e: Research Foundation of
- e: Resonance Frequency
- e: Resume Flag
- e: Rocky Flats Office
- e: Russian Federation
- f: Receveur des Finances
- f: Réforme Fiscale
- f: Réfugié Français
- f: Régie Française
- f: Régime Forestier
- f: Renaissance Française
- f: Rente Foncière
- f: Rente Française
- f: Représentant de Fabrique
- f: Résistance Française
- f: Resources Familiales
- f: Retrait de Fonds
- f: Revenu Fixe
- f: Revenu Foncier
- f: Roulement de Fonds
- r: Regional Förråd
Rf f: Récif
r.f. r: registrerad förening
- r: relativ fuktighet
R-F e: Resorcinol Formaldehyde
R/F N: Redningsfartøy
RF-4 e: Phantom II
RFA e: RCRA Facility Assessment
- e: Regional Forest Agreement
- e: Restrictive Fire Area
- f: Règlements Forfaitaires et Anticipés
- f: République Fédérale d'Allemagne
R.F.A. e: Royal Field Artillery
RFAA e: Request for Accounting Assistance
RFB a: Retsforbundet
- e: Request for Bids
- e: Room, Food and Beverage
RFB&D e: Recording for the Blind and Dyslexic
RFBR e: Russian Foundation for Basic Research
RFC e: Remote Function Call
- e: Request for Comment[s] [Internet]
- e: Response For a Class
- e: River Forecast Center[s]
RFD e: Range Focused Doppler Display
- e: Reference Dose values
- e: Request For Discussion [Internet]
- e: Resident Flash Disk
- e: Royal Forestry Department
- f: Rassemblement des Forces Démocratiques
- f: Répression des Fraudes Douanières
RfD e: Reference Dose
RFDB e: RED FLAG Data Base
RfDdt e: Developmental RfD
RFDP e: Rehoboth Free Democratic Party
RFE e: Radio Free Ethernet
- e: Regional Finite Element
- e: Request For Enhancement
- e: Requisition For Expenditure
RFE/RL e: Radio Free Europe/Radio Liberty
RFETS e: Rocky Flats Environmental Technology Site
RFF e: Remote File Facility
- e: Request For Funds
- e: Required Form Feed
RFG a: Råstofforvaltningen For Grønland
- e: Reformulated Gasoline
RFH d: Rheinische Fachhochschule, Köln
- e: Remote Frame Handler
RFHL r: Riksförbundet för Hjälp åt Läkemedelsmissbrukare
RFI e: Radio Frequency Interface
- e: RCRA Facility Investigation
- e: Relative Fluorescence Intensity
- e: Remedial Facility Investigation
- f: Radio-France Internationale
RFIC f: République Fédérale Islamique des Comores
RFID e: Radio Frequency Identification
RFI/RI e: RCRA Facility Investigation/ Remedial Investigation
RFL e: Restrictive Fire Line
RFM e: Radio Frequency Monitoring
- e: Reconfigurable Furnace Module
RFMC f: Revues de Formation Médicale Continue
RFO e: Restricted Flow Orifice
- e: Rocky Flats Office
RFOV e: Resolution Field Of View
RFP a: R. Frimodt Pedersen A/S
- e: Radio Fixed Parts
- e: Radio Frequency Probe
- e: Reasonable Further Progress
- e: Request For Papers
- e: Request For Plan
- e: Request For Proposals
- e: Rocky Flats Plant
RFQ e: Request For Quotation[s] (or Quotes)
- e: Requisition For Quote
RFR e: Radio Frequency Radiation
R Frig s: Remolque Frigorífico
RFRIKANGR N: Rekylfrikanon Gruppe
RFRL e: Remote Fuels Refabrication Laboratory
Rfro s: Refuerzo
Rfrs e: Roofers
RFS e: Remote File Service
- e: Remote File Sharing
- e: Request for Service
- e: Richmond Field Station
RFS. e: Ready For Sending
RFSH e: Rust Federal Services of Hanford Inc.
RFSNW e: Rust Federal Services Northwest
RFSU N: Riksförbundet för Sexuell Upplysning
RFT d: Rundfunk-Fernsehen Telekommunikation Electronic GmbH
- e: Request For Technology
- e: Retrievable Formation Tester
- e: Revisable Form Text
- e: Rich Text Format
- f: Régie Française de Tabacs
RFTAS400 e: OfficeVision/400 revisable form document
RFTDCA e: Revisable Form Text/ Document Content Architecture
RFTF e: Radio Frequency Test Facility
RFU e: Reserved For Future Use
RFÜ d: Reaktorfernüberwachung, Mainz
RFW e: Radiofrequency Weapons
RG e: Radio Grade
- e: Ragsdale Hall
- e: Reconnaissance Group
- e: Renormalization Group
- e: Republican Guard
- e: Research Group
- e: Ringing Generator
- e: Rio Group
- f: Renseignement Gratuit
- f: Renseignements Généraux
- f: Rive Gauche
- f: Rôle Général
- s: Reserva General
Rg r: Raketgevär
- s: Regimiento
RG-59 e: 75-ohm coax cable
RG-62 e: 93-ohm coax cable
RGA e: Reduction Gas Analyzer
- e: Relative Growth Rate
- e: Residual Gas Analysis
- s: Reserva General de Artillería
R.G.A. e: Royal Garrison Artillery
RGB d: Rot, Grün, Blau
- e: Red, Green, Blue [colour space]
- e: Red-orange, Green, Blue-violet signal
RGBI e: Red Green Blue Intensity

RGCFTP f: Régie Générale de Chemins de Fer et de Travaux Publics
RGCT e: Reflective Gross Cloud Test
RGD e: Radiation-Generating Device
- e: Rarefied Gas Dynamics
- f: Rassemblement Guinéen pour le Développement
RGDI e: Renaissance Graphics Device Interface
RG-DW e: Recursive Grid - Dynamic Window
RGE f: Regroupement des Guinéens de l'Extérieur
Rge f: Refuge
RGF e: Regional Growth Forecast
RGFC e: Republican Guard Forces Command
RGI e: Regional-Global Interactions
RGICS e: Region 7 Version of Region 5 Grants Information and Control System
RGIG s: Reserva General de Ingenieros
RGIS e: Regional Geographic Information System
RGISM e: Raster Geographic Information System for Mapping
RGL f: Régional
RGM e: Red, Green, Blue [Monitor]
- e: Ship-launched, guided ground-to-ground Missile
Rgmto s: Regimiento
- s: Reglamento
RGN A: Raad vir Geestewetenskaplike Navorsing, Pretoria
Rgn s: Región
RGP e: Raster Graphics Processor
- f: Régie Générale de Presse
RGR f: Rassemblement des Gauches Républicaines
R-gruppe N: Rekognoserings gruppe
RGS e: Readiness Groups
- e: Reflection Grating Spectrometer
- e: Research Grants Staff
- e: Royal Geographical Society, England
- f: Répartiteur de Groupes Secondaires
Rgt s: Regimiento
Rgt Aert s: Regimiento Aerotransportado
Rgt Art s: Regimiento de Artillería
Rgt Av s: Regimiento de Aviación
Rgt Ind s: Regimiento Independiente
Rgto s: Regimiento
RG/U e: Radio Guide/Utility
RGWG e: Resource Generating Working Group
RH e: Remote[ly] Handled
- e: Resistance Heating
- e: Right-Hand[ed]
- f: Race Humaine
- f: Repos Hebdomadaire
- N: Rikshospitalet
RH-53 e: See Sea Stallion
RH. e: Request Header
- e: Response Header
RHA e: Residence Hall Association
Rha f: Rendement à l'hectare
R.H.A. e: Marines
RHAW e: Radar Homing And Warning system
RHBS e: Rehabilitation Therapy
RHC e: Regional [Bell] Holding Company
- f: Regiment d'Helicopteres de Combat
RH_c e: critical relatice humidity
RHCC e: Rural Health Care Corporation
RHD e: Removable Hard Disk
RHDP e: Right-Hand Decimal Point
RHEED e: Reflection High Energy Electron Diffraction
RHESSys e: Regional Hydro-Ecological Simulation System
Rhet. e: Rhetoric
Rheumatol Int e: Rheumatology International
Rheum Dis Clin North Am e: Rheumatic Diseases Clinics of North America
RHF e: Relativistic Hartree-Fock
R-hFSH e: Recombinant Human Follicle Stimulating Hormone
Rhgr r: Rökhandgranat
RHIC e: Relativistic Heavy Ion Collider
- e: Relativistic Heavy-Ion Collider
RHIESSys e: Regional Hydro-Ecological Simulation System
RHIMS e: Red Horse Information Management System
RHIN e: Regional Health Information Network
Rhinol Suppl e: Rhinology. Supplement
RHIP e: Rank Has Its Privileges [Internet]
RHIS e: REUTERS Health Information Services
RHMF e: Research in High Magnetic Fields
RHNT e: Registration Hierarchical Name Tree
RHO e: Remanufactured High Output motor
- e: Rockwell-Hanford Operations
RHOB e: Rho- and B for bulk density
RHOSA e: Regional Health Organisation of Southern Africa
R-HOUR e: Release Hour
RHP e: Rainforest Health Project
- e: Right-Handed Pitcher
- e: Rödelheim Hartreim Projekt
RHPR f: Revue d'Histoire et de Philosophie Religieuses
RHR f: Revue de l'Histoire des Religions
RHS e: Right Hand Side
RH-TRU e: Remote Handled Transuranic Waste
RHTRU e: Remotely Handled TRU [transuranic] waste
RHUDO e: Regional Housing and Urban Development Office
RHV e: Reid Hillview Intergalactic, San Jose
RhVjBll d: Rheinische Vierteljahrsblätter
RHWR e: Radar Homing and Warning Receiver
RHWS e: Radar Homing and Warning System
RI e: Radar Interferometry
- e: Reconnaissance Inspection
- e: Recordable Injury
- e: Reduction Index
- e: Referential Integrity
- e: Reliability Improvement
- e: Remedial Investigation
- e: Reproductive Immunologist
- e: Request for Information
- e: Ring In
- e: Ring Indicate
- e: Rockwell International Corp.
- f: Radicaux Indépendants
- f: Rappel d'Intérêts
- f: Réduction d'Impôt
- f: Régime Intérieur
- f: Règlement Intérieur
- f: Réglementation Internationale
- f: Règlements Internationaux
- f: Remise d'Impôt
- f: Rente Immédiate
- f: Rente Incessible
- f: Rente Insaisissable
- f: Reproduction Interdite
- f: Républicains Indépendants
- f: Revenu Imposable
- s: Regimiento de Infantería
- s: Resumen de Información
Ri e: Richardson number
- r: Riktning
RI. e: Ring Indicate
RIA e: Red Ink Award
- e: Regulatory Impact Analysis
- e: Regulatory Impact Analysis Facility Profile
- e: Regulatory Impact Assessment
- e: Research Institute on Addictions
- f: Régiment Interarmes d'Appui
RIACS e: Research Institute for Advanced Computer Science
RIAM f: Régiment Interarmes Malgache
RIAS e: Radio In the American Sector, Berlin
- f: Radar A Impulsion et Antenna Synthetiques
RIATT e: Research Institute for Assistive and Training Technologies
RIB e: Recoverable Item Breakdown
- e: Renderman Interface Bytestream
- e: Routing Information Base
- f: Réseau International des Biosciences
- s: Red Internacional de ciencias Biológicas
RIBA e: Recombinant Immunoblot Assay
RIBE e: Reactive-Ion-Beam Etching
RIBS e: Readiness In Base Services
- e: Research Information Bulletins

RIC e: Radio Identify Code
- e: Radon Information Council
- e: Railway Industry Council
- e: Regional Information Center (or Centre)
- e: Remote Intelligent Communications
- e: Repeater Interface Controller
- e: Requirement Identity Code
- e: Resource Identification Code
- e: Resource Implementation Committee
- e: Routing Indicator Code
- e: RTP Information Center
- f: Recherches Industrielles et Commerciales
- f: Régie Industrielle de la Cellulose
- f: Régiment d'Infanterie Commando
- f: Règlement International de la Circulation
- f: Règlements Internationaux de Charge

RICA e: Radar Image Classification Aid
- i: Regole Italiane di Catalogazione per Autori

RICAE e: Riga Institute of Civil Aviation Engineers

RICAI f: Réunions Interdisciplinaires de Chimiothérapie Anti-infectieuse

RICC e: Retirement Information and Counseling Center

RICE e: Regional Interactions of Climate Ecosystems
- e: Rest, Ice, Compression and Elevation
- e: Rice Cultivation and Trace Gas[es] Exchange

RICEP s: Red del Impacto del Clima y Estrategias de Respuesta

RICH e: Richland, Washington
- e: Ring Imaging Cherenkov

RICHS e: Rural Information Center Health Service

RICO e: Racketeer Influenced and Corrupt Organizations [act]

RID e: Record [row] Identifier
- e: Regulatory Integration Division
- e: Remote Isolation Device
- e: Royal Irrigation Department, Thailand
- f: Règlement International concernant le transport des matières Dangereuses par chemin de fer

RIDB e: Readiness Integrated Data Base

RIDDLE e: Rapid Information Display and Dissemination in Library Environments

RIDGE e: Ridge Inter-Disciplinary Global Experiment[s]

RIDS e: Radar Image Display System
- e: Readiness Improvement Data System
- e: Records Inventory and Disposition Schedule and file index

RIE d: Riesa, Sachsen
- e: Reactive-Ion Etch[ing]
- e: Receiver Interrupt
- f: Revente Interdite à l'Exportation

RIF e: Reduction-in-Force
- e: Resource Interchange File
- e: Routing Information Field

RIFF e: Raster Image File Format
- e: Resource Interchange File Format

RiFlEtikettG d: Gesetz zur Durchführung der Rechtsakte der Europäischen Gemeinschaft über die besondere Etikettierung von Rindfleisch und Rindfleischerzeugnissen

RI/FS e: Remedial Investigation/ Feasibility Study
- e: Risk Investigation/Feasibility Study

RIG e: Regional Inspector General
- e: Related Interest Group

RiG DDR d: Richtergesetz vom 5.7.1990

RIGES e: Renewable Intensive Global Energy System

RIGGS e: Ross Ice Shelf Geophysical and Glaciological Survey

RIGIS e: Relational Image-Based Geographic Information System

RIG-PMC s: Reunión Intergubernamental sobre el Programa Mundial sobre el Clima

RIH e: Run In Hole

RIHMI-WDC e: All-Russian Research Institute of Hydrometeorological Information-World Data Center

RII e: Recordable Injury Incidence
- e: Request for Intelligence Information
- e: Routing Information Indicator

RIITTSS e: Reality Impaired Individuals with Tendencies Toward Satiric Surrealism

RIIXS e: Remote Interrogation Information Exchange Subsystem

RIKES e: Raman-Induced Kerr Effect Spectroscopy

RILM e: Repertoire International de Literature Musicale

RILWAS e: Regional Integrated Lake-Watershed Acidification Study

RIM e: a Public-domain relational database
- e: Reaction Injection Mold[ed] (or Molding)
- e: Regional Ice Model
- e: Regulatory Interpretation Memorandum
- e: Remote Installation and Maintenance
- e: Request Initialisation Mode
- e: Ship-launched SAM. Examples
- s: Regimiento de Infantería [de] Marina
- s: Resumen de Información Militar

RIMD e: Regulation and Information Management Division

RIME e: Relaynet International Message Exchange

R I Med e: Rhode Island Medicine

RIMI e: Riso Integrated Environmental Project

RIMS e: Recognition Information Management System
- e: Records Information Management System
- e: Regional Input-Output Modeling System
- e: Rescue Information Management System
- e: Resonant Ionization Mass Spectroscopy
- e: Resource Information Management System

RIMSTOP e: Retail Inventory Management and Stockage Policy

RIN e: Regulatory Identifier Number
- e: Relative Intensity Noise
- e: Remote Intelligent Network
- f: Rassemblement pour l'indépendance Nationale

R.I.N. e: Royal Indian Navy

RINEX e: Receiver Independent Exchange Format

RING s: Regimiento de Ingenieros

Rinsho Byori j: Rinsho Byori. Japanese Journal of Clinical Pathology

Rinsho Ketsueki j: Rinsho Ketsueki. Japanese Journal of Clinical Hematology

Rinsho Shinkeigaku j: Rinsho Shinkeigaku. Clinical Neurology

RINT e: Unintentional radiation intelligence

RIOG e: Rio Grande Valley

RIOS e: River Inputs to Ocean Systems Working Group 46
- f: Apports des fleuves aux océans Groupe de travail 46
- s: Aportes Fluviales a los Sistemas Oceánicos Grupo de Trabajo 46

RIP e: Radar Image Processor
- e: RCRA Implementation Plan
- e: Remote Image (or Imaging) Processing
- e: Remote Image (or Imaging) Protocol
- e: Report on Individual Personnel
- e: Reproductive Immunophynotype
- e: Requiescat In Pace
- e: Research-In-Progress
- e: Rest In Peace
- e: Router (or Routing) Information Protocol
- e: Routing Interface Protocol
- f: Route d'Intérêt Provincial
- l: Requiescat In Pace

RIPAS e: Regional Integrated Protected Areas System

RIPC e: Remote Information Processing Center

RIPE e: Réseaux IP Européen[e]/[s]

RIPEM e: Riordan's Internet Privacy Enhanced Mail

RIPE NCC e: RIPE Network Coordination Centre

RIPL e: Reconnaissance and Interdiction Planning Line
- e: Remote Initial Program Load[ing]

RIPQPT s: Registro Internacional de Productos Químicos Potencialmente Tóxicos

RIPS e: Raster Image Processing System
- e: Raster Image Processing Systems Corp.
- e: Reconnaissance Information System
- e: Regional Institute for Population Studies [University of Ghana]
- e: Remote Image Processing System
- e: Remote Information Processing System

RI/RA e: Remedial Investigation/Remedial Action

RIRDC e: Rural Industry Research and Development Corporation

RIS e: Radiological Information System[s]
- e: Range Information System
- e: Reflector In Space
- e: Regional Information Service
- e: Registered Image Sequence
- e: Regulatory Impact Statement
- e: Regulatory Innovations Staff
- e: Resonant Ionization Spectroscopy
- e: Retroreflector In Space
- e: Rural Information System

RISC e: Radiology Information System Consortium
- e: Reduced Instruction Set Code
- e: Reduced Instruction Set Computers
- e: Regulatory Information Service Center
- e: Relegate Important Stuff to the Compiler
- e: Research Institute for Symbolic Computation

RISCOS e: RISC Operating System
RISCPT f: Registre International des Substances Chimiques Potentiellement Toxiques
R.I.S.D. e: Rhode Island School of Design
RISD e: Research and Acquisition Information Systems Division
RISE e: Readiness Improvement Through Systems Engineering
- e: Remote Sensing, Image and Signal Engineering

Rise f: Remise
RISF e: Resupply and Replacement Intermediate Storage File
RISI e: Resource Information Systems, Inc.
Risk Anal e: Risk Analysis
RISLU e: Remote Integrated Services Line Unit
RISOPS e: Rapid Interval Scan Operations
RISP e: Ross Sea Ice Shelf Project
- f: Projet sur la barrière de glace de la mer de Ross
- s: Proyecto de la Barrera de Hielo del Mar de Ross

RISS e: Rural Injury Surveillance Study
R Ist s: Remolque Isotermo
RIST e: Rule Induction and Statistical Testing
RIT e: Raw Input Thread
- e: Readiness Initiative Team
- e: Red Interface Terminal
- e: Renewals In Time
- e: Resonant Interband Tunneling
- e: Rochester Institue of Technology
- f: Régiment Interarmes Togolais
- f: Relations Internationales du Travail

RITA e: Reduced Interim Type Approval
- e: Relocation Income Tax Allowance

RITL e: Radio In The Loop
RITS e: Radiatively Important Trace Species
RITZ e: Regulator and Treatment Zone Model
RIU e: Remote Interface Unit
Riv Biol i: Rivista di Biologia
Riv Eur Sci Med Farmacol i: Rivista Europea per le Scienze Mediche e Farmacologiche
Riv. Idrobiol. i: Rivista di Idrobiologia
RIVM e: National Institute of Public Health and Environmental Protection
- e: Netherlands National Institute of Public Health and Environmental Protection

RIVP f: Régie immobilière de la ville de Paris
RIW d: Recht der Internationalen Wirtschaft
- f: Règlement International pour l'emploi réciproque des Wagons en trafic international

RIXT e: Remote Information Exchange Terminal
RJ e: Registered Jack
- f: Race Jaune
- f: Rapport Journalier
- f: Ration Journalière
- f: Recette Journalière
- f: Repris de Justice

RjF d: Reichsbund jüdischer Frontsoldaten
RJ III e: Rivet Joint
RJO f: Rectificatif au Journal Officiel
RJOS e: Ruth Jackson Orthopaedic Society
RJPT f: Rassemblement de la Jeunesse Patriotique du Tchad
RK e: Radial Coefficient
- e: River Kilometer
- r: Röda Korset

rk. a: række[r]
R,k d: Reflexionsfaktor des ersten Vollfilters seits des Lichteinfalles, zugeordnet zur k-ten Partialfläche einer Partialfilterung
R/K N: Redningskutter
R[-]K e: Runge Kutta
RKA R: Raketniy Kater
- R: Rossiskoye Kosmicheskoye

RKB e: Rotary Kelly Bushing
- r: Riksnämnden för Kommunal Beredskap

RKKY e: Ruderman-Kittel-Kasuya-Yosida
RKL N: Reklamekonsulentenes Landsforening
RKM e: Rokiðkis Museum of ethnography
RKR R: Raketniy Kreyser
RKRM e: ROM Kernel Reference Manual
RKS a: Røde Kors Samaritter-forening[erne]
RKW d: Rationalisierungs-Kuratorium der Deutschen Wirtschaft e.V.
RL e: Radioluminescent Light
- e: Raman Lidar
- e: Real Life [Internet]
- e: Receive Location
- e: Remote Loopback
- e: Richland Operations Office, U.S. Department of Energy
- f: Reconstruction et Logement
- f: Relève
- f: Réparations Locatives
- f: Reprise Légale
- f: Réserve Légale
- f: Responsabilité Limitée
- f: Retenue Légale
- f: Révision Limitée
- f: Risque Locatif
- r: Relä
- r: Riksdagsledamot

Rl r: Radiolänk
r.l. e: random lengths
RLA e: Random Loading Analysis
- e: Reaction-Limited Aggregation

RLAA s: Red Local de vigilancia Antiaérea
RLAB e: Region 4 Labels System
RLAC s: Regimiento Ligero Acorazado de Caballería
RLAF f: Réseau des Lignes Aériennes Françaises
RLAO f: Revue de Langues de l'Afrique de l'Ouest, at the University of Dakar
R Lav s: Remolque Lavadero
RLB e: Relocatable Binary Code
Rlb N: Rullebane
RLBJ e: Random Light Bulb Joke
RLCM e: Remote Line Concentrating Module
- e: Remote Line Control Module

RLCO e: Real Life Conference
RLD e: Ready-to-load date
- e: Received Line Detect
- e: Reload

RLDS e: Recovery Log Data Set
- e: Reorganized Church of Jesus Christ of Latter Day Saints

RLE e: Run Length Encoded [file]
- e: Run Length Encoding

RLFCM e: Radioactive Liquid Fed Ceramic Melter
RLG e: Ring Laser Gyrometer
RLI e: Resist Lithography
RLL e: Radio in the Local Loop
- e: Rapid and Large Leakage
- e: Row Level Locking
- e: Run-Length Limited

RLM d: Reichsluftfahrtministerium
RLN e: Remote LAN Node
RLOGIN e: Remote Login
RLP e: Radio Link Protocol
- e: Remote Line Printer
- e: Resource Location Protocol

RLPROP e: USAF Real Property Inventory Accounting System
RLRC e: Regional Logistic Readiness Center
RLS e: Raman Lidar System
- e: Received Line Signal
- e: Restless Legs Syndrome
- f: Radiobalise de Localisation des Sinistres
- R: Radiolokatisonnaya Stantsiya
- s: Radiobaliza de Localización de Siniestros

RLSBO R: Radial Okazionnaja Sistema Bokowo Obzora
RLSD e: Received Line Signal Detect[ed] (or Detector)
RLSI e: Ridiculously Large-Scale Integration
RLSTK e: Rolling Stock Summary System
RLU e: Relative Light Unit
RLV d: Ringleitungsverteiler
- e: Recordable Laser Videodisc

RLV TD&D e: RLV Technology Development and Demonstration
RLWS e: Radioactive Liquid Waste System

RM e: Deputy Commander for Resources
- e: Radiation Monitoring
- e: Random Matrix
- e: Record Marking
- e: Recovery Manager
- e: Reference Model
- e: Reflection Mode
- e: Reject Message
- e: Remanent Magnetism
- e: Resource Maintenance
- e: Resource Management
- e: Ressource Mangement
- e: Returned Missionary
- e: Risk Management
- e: River Mile
- f: Références Morales
- f: Régime Matrimonial
- f: Registre Matricule
- f: Registre Mortuaire
- f: Règne Minéral
- f: Relevé Mensuel
- f: Républicains Modérés
- f: Retraite Mutualiste
- p: Radio Moçambique
- s: Reclutamiento y Movilizaci6n
- s: Regi6n Militar

Rm a: Rummeter
- r: Radiomatros

R&M e: Reliability and Maintainability
R/M e: Repair and Maintenance
r/m r: varv per minut
RM2 f: Remarié depuis Moins de 2
RM8 e: Reference Model 8
RMA e: Refugee Medical Assistance
- e: Reliability, Maintainability and Availability
- e: Return Manufacturer Authorization
- e: Return Material[s] Authorization
- e: Return Merchandise Authorization
- e: Return to Manufacturer Authorization
- e: Revolution in Military Affairs
- e: Riparian Management Area
- f: Régularisation des Marchés Agricoles

R.M.A. e: Royal Marine Artillery
RMAO e: Resources Management and Administrative Office
R-MAP e: Regional Environmental Mapping and Assessment Program
RMATS e: Remote Maintenance Administration and Traffic System
RMB e: Requirements Management Board
- e: Right Mouse Button

RMC e: Rating Method Code
- e: Resource Management Center
- e: Returned to Military Control
- e: River Murray Commission

R-MC e: Randolph-Macon College
RMCD e: Radaiation Monitoring Container Device
RMCGP e: River Murray Corridor of Green Program
RMCL e: Recommended Maximum Contaminant Levels
RMDCN e: Regional Meteorological Data Communication Network
RMDHS e: Regional Model Data Handling System
RMDIR e: Remove Directory
RMDS e: Resource Management Directives System
RME e: Radiation Monitoring Experiment
- e: Reasonable Maximum Exposure

RMENR e: Russias Ministry of Environment
RMES e: Relay Mirror Experimental Satellite
RMF e: Read Me First
- e: Remote Management Facility
- e: Rich Music Format

RMG d: Rheinische Missionsgesellschaft, Windhoek
- e: Resource Management Group
- e: Routine Morning Grumpiness

RMI e: Guaranteed Minimum Income
- e: Java Remote Method Invocation
- e: Reactive Metals, Incorporated
- e: Remote Method Invocation
- e: Resource Management (or Manager) Interface
- e: Root Modeling Interface
- f: Revenue Minimum d'Insertion

RMIB e: Royal Meteorological Institute of Belgium
RMID e: Resource Manager ID
- e: Risk Management Implementing Directive

RMIS e: Reprographics Management Information System
- e: Resource[s] Management Information System

RMIS/BAR e: Resource Management Information System/Budget Analysis System
RMISTB e: Resource Management Information System Test Bed
R.M.I.T. e: Royal Melbourne Institute of Technology
RMIT e: Royal Melbourne Institute of Technology
RMK e: Remarks
RML e: Regional Medical Libraries
- e: Remote Microwave Link

RMLA e: Rocky Mountain Llama and Alpaca association
R.M.L.I. e: Royal Light Infantry
RMM e: Ring Management Module
RMMP e: Regional Marine Meteorological Programme
- e: Reliability and Maintainability Management Plan
- e: Risk Management Master Plan
- f: Programme Météorologique Maritime Régional

RMMT e: Rail Movements Management Team
RMN e: Remote Multiplexer Node
RMO e: Records Management Officer
- e: Requirements Management Office
- e: Resource Management Office
- f: Recrutement de la Main-d'Œuvre

RMODE e: Residency Mode
RM-ODP e: Traffic Matrix-The Reference Model for Open Distributed Processing
RMON e: Remote [Network] Monitor[ing]
RMP e: ASHP Resident Matching Program
- e: Radon Progeny Measurement Proficiency program
- e: Remote Maintenance Processor
- e: Requirements Management Plan
- e: Reserve Mobility Plan
- e: Resource Management Plan

RMPP e: Risk Management and Prevention Program
RMR e: Rain Mapping Radiometer
- e: Reliability, Maintainability and Recovery
- e: Remote Message Registers
- e: Resting Metabolic Rate
- e: Return Material Receipt

RMS e: RAID Management Software
- e: Rate Monotonic Scheduling
- e: Reason Maintenance System
- e: Recipe Management Standard
- e: Record Management Services
- e: Region 10 Resource Management System
- e: Remote Manipulating System
- e: Resource Management Staff
- e: Resource Measurement System
- e: Revised Management Scheme
- e: Root Mean Square[d]
- r: Arméns Radar- och Luftvärnsmekanikerskola

Rms f: Reims
rms. e: root-mean-square
RM&S e: Reliability, Maintainability and Supportability
RMSa e: Rhabdomyosarcoma
RMSEP e: Root Mean Square Error of Prediction
R&MSG e: Robotics and Mechanical Systems Group
R/M/T e: Reliability, Maintainability and Testability
RMT e: Random Matrix Theory
- e: Raw Magnetic Tape
- e: Rectangular Midwater Trawl
- e: Required Monthly Test
- e: Ring Management
- f: Chalut pélagique rectangulaire
- f: Registre Morphologique des Tumeurs au Grand-Duché de Luxembourg
- s: Red de Arrastre Pelágico Rectangular

Rmta s: Remonta
RMTC e: Regional Meteorological Training Centre
RMTF e: Recipe Management Task Force
RMTN e: Regional Telecommunications Network
RMTO f: Référence mon télégramme officiel
RMTP e: Regional Marine Turtle Management and Conservation Program
- e: Revised Medium-Term Plan
- f: Plan à Moyen Terme Révisé
- s: Plan a Plazo Medio Revisado

RMV d: Rhein-Main-Verkehrsverbund
RMVL e: Redundant Multi-Valued Logic
RMW e: Radioactive Mixed Waste

RN a: Rustning og Nedrustning
- e: Read News
- e: Reference Noise
- e: Regency Net
- e: Relay Node
- e: Routing Number
- f: Refus Net
- f: Régie Nationale
- f: Registre des Naissances
- f: Réglementation Nationale
- f: Réponse Négative
- f: Représentation Nationale
- f: Ressources Naturelles
- f: Revenu National
- f: Revenu Net
- f: Route Nationale
- r: Riksskattenämnden
- s: Renovación Nacional

Rn d: Randnummer
- e: net Radiometer
- e: Radon
- s: Reorganización

R.N. e: Royal Navy
Rn-220 e: Radon-220
Rn-222 e: Radon-222
RNA d: Rufnummernanzeige
- e: Remote Network Access
- e: Research Natural Areas
- e: Ribo-Nucleic Acid
- e: Ring, No Answer
- p: Radio Nacional de Angola

RNAP e: RNA Polymerase
RNARS e: Royal Navy Amateur Radio Society
RNAS e: Royal Naval Air Service
- e: Royal Navy Air Station

RNAV e: Navigation system that makes use of airway beacons
RNB e: Rassemblement National-Bucherons
- f: Radio Nationale Belge

RN-C e: Registered Nurse Certified
RNC e: Republian National Committee
RNCFM f: Réseau National de Chemins de Fer Malagasy
RNCP e: Rural Nature Conservation Program
RNCS e: Registered Nurse Clinical Specialist
RND f: Rassemblement National Démocratique
- f: Rassemblement National pour la Démocratie
- s: Redención Nacional Democrática

RNDB e: Road Network Data Base
RNDP e: Rural National Democratic Party
- f: Rassemblement National Démocratique et Populaire
- f: Rassemblement National pour la Démocratie et le Progrès

RNE e: Register of the National Estate
Rné f: Ruiné
RNEG e: UCR Conserv
Rnes f: Ruines
RNG e: Range
- e: Ranging
- f: Revenu Net Global

RNGC e: Revised New General Catalog
RNHF e: Royal Navy Historical Flight
RNI f: Rassemblement National des Indépendants
RNIIGMI e: All Russia Research Institute of Hydrometeorological Information
RNIS f: Réseau Numérique à Intégration de Services
RNL e: Required New Line
- e: Risoe National Laboratories

RNM e: Rwandese National Movement
- f: Rassemblement National Malgache
- p: Resistência Nacional Moçambicana
- s: Reserva Naval Movilizada

RNMI e: Royal Netherlands Meteorological Institute
RNN e: Recurrent Neural Network
- e: Regional NOCN Node

RN/NP e: Registered Nurse, Nurse Practitioner
RNO e: Regional Nuclear Options
- e: Reoxidized Nitrided Oxide

RNOC e: Regional Network Operations Center
RNODC e: Responsible National Oceanographic Data Centre[s]
- s: Centro Nacional Responsable de Datos Oceanográficos

RNODC-INDO e: Responsible National Oceanographic Data Centre for the Indian Ocean
- s: Centro Nacional Responsable de Datos Oceanográficos para el Océano Indico

RNP e: Regional Network Provider
- e: Remote Network Printer

RNR e: Receive Not Ready
RNR. e: Request Not Ready
RNS e: Respiratory Nursing Society
- e: Russian Numismatic Society

RNSI e: Remote Node Service for ISDN
RNSP e: Residual Nuclear Strike Plan
- f: Réseau National de Santé Publique

RNT f: Radiodiffusion Nationale Tchadienne
RNUI e: Remote Network User Identification
RNUJ f: Rassemblement National pour l'Unité et la Justice
RNUR f: Régie Nationale des Usines Renault
RNV d: Regional-Nahverkehrsverbund Darmstadt Dieburg
- f: Réserve Naturelle Volontaire

RNZCGP e: Royal New Zealand College of General Practitioners
RNZFB e: Royal New Zealand Foundation for the Blind
RO e: Read-Only
- e: Recruitment Office
- e: Ring Out

Ro N: Radio
Rø a: Rødovre
R-O e: Read Only
ROA e: Record Of Approval
- e: Regional Office for Africa
- e: Return On [total] Assets

ROADCHEM e: Roadway Version that Includes Chemical Reactions of NO, NO_2 and O_3
ROADS e: Resource Organisation and Discovery in Subject-based services
ROADWAY e: A Model to Predict Pollutant Concentrations Near a Roadway
ROAR e: Recovery of Overpayments, Accounting and Reporting
ROB e: Remote Order Buffer
- e: Reorder Buffer
- e: Research Operations Building
- e: Royal Observatory of Belgium

ROBCO e: Readiness Objective Code
ROBIN n: Rotterdam Bibliotheek en Informatienetwerk
ROBO e: Remote Office Branch Office
ROBY e: Results, Output, Benefits or Yield
ROC e: Receiving Operator Characteristics
- e: Reconnaissance Operations Center
- e: Record Of Communication
- e: Regional Operating Company
- e: Regional Operations Center
- e: Remote Object Communications
- e: Representative Organization for a Country
- f: Coordonnateur national
- f: Reconnaissance Optique de Caractères
- s: Coordinador Nacional

R.O.C. e: Republic Of China
ROCC e: Regional Operations Control Centers
ROCLANT e: Regional Operations Center Atlantic
ROCOF e: Rate Of Occurrence Of Failures
ROCOZ e: Rocket Ozone Sonde
ROCRAM e: Regional Network on Cooperation among Maritime Authorities
- f: Réseau Régional de Coopération Entre Autorités Maritimes
- s: Red de Cooperación Regional entre Autoridades Marítimas

Rocz Akad Med Bialymst P: Roczniki Akademii Medycznej w Bialymstoku
Rocz Akad Med Bialymst Supl P: Roczniki Akademii Medycznej w Bialymstoku. Suplement
Rocz Panstw Zakl Hig P: Roczniki Panstwowego Zakladu Higieny
ROD e: Record Of Decision[s]
- e: Required Operational Dates
- e: Rewritable Optical Disk
- e: Rich-Oil Demethanizer

RODC e: Regional Oceanographic Data Centre
- s: Centro Regional de Datos Oceanográficos

RODE s: Robot de recogida y transporte de artefactos explosivos
RODM e: Resource Object Data Manager
RoD/PO e: Reading on Demand/Paper Option
ROE e: Rate Of Exchange
- e: Return on common Equity
- e: Rotating-frame Overhauser Enhancement

Roe f: Rouen
RØF a: Rask-Orsted Fonden
ROFF e: Run OFF
ROFI e: Region of Freshwater Influence
- f: Zone d'influence de l'eau douce
- s: Zona de Influencia del Agua Dulce

ROFL e: Roll[ing] On [the] Floor Laughing [Internet]
ROFLASTC e: Rolling On Floor Laughing And Scaring The Cat [Internet]
ROFLBTC e: Rolling On the Floor Biting The Carpet [Internet]
ROFLGO e: Rolling On Floor Laughing Guts Out [Internet]
ROFLMAO e: Rolling On The Floor Laughing My Ass Off [Internet]
Rofo Fortschr Geb Röntgenstr Neuen Bildgeb Verfahr d: Rofo. Fortschritte auf dem Gebiete der Röntgenstrahlen und der Neuen Bildgebenden Verfahren
ROG e: Reactive Organic Gases
RogFdi N: Rogaland Forsvarsdistrikt
ROH e: Regular Overhaul
- e: Restricted Overhaul
ROHF e: Restricted Open-shell Hartree-Fock
ROHO e: Regional Oral History Office
ROI e: Rate On Investment
- e: Rectangle Of Interest
- e: Region Of Influence
- e: Return Of Investment
ROICC e: Resident Officer in Charge of Construction
ROK e: Republic Of Korea
- r: Kurs i Radartjänst för Officerar
Rok r: Reservofficerskurs
r.o.k. r: rum och kök
rökdet r: rökdetonator
rökgr r: rökgranat
rökhgr r: rökhandgranat
röklng r: rökladdning
ROKMC e: Republic of Korea Marine Corps
rökmtrllåda r: rökmateriallåda
röksgr r: rökspränggranat
ROKUS e: Republic Of Korea/United States
rökvinggr r: rökvinggranat
ROL e: Rotate Left
- e: Rule of Law
ROLAP e: Relational OLAP [On-Line Analytical Processing]
ROLC e: Routing Over Large Clouds
ROLLBACK e: A Proportional Reduction Model
ROLTP e: Remote On-Line Transaction Processing
ROM d: Romanistik
- e: RAID On the Motherboard
- e: Read-Only-Memory
- e: Regional Oxidant Model
- e: Rupture Of Membrane
- f: Mémoire morte
- s: Memoria de lectura solamente
rom. a: romersk
ROM. e: Read-Only Memory
Rom. e: Roman[ce]
- e: Romanic
ROMAD e: Radio Operator, Maintenance and Driver
ROMADS e: Radio Operators and Drivers
ROMBI f: Résultats des recherches de biologie marine
- s: Resultados de las Investigaciones de Biología Marina
ROMCOE e: Rocky Mountain Center on Environment
ROMIS e: Research on the Molecular Immunology of STDs
Rom J Intern Med e: Romanian Journal of Internal Medicine
Rom J Morphol Embryol e: Romanian Journal of Morphology and Embryology
Rom J Neurol Psychiatry e: Romanian Journal of Neurology and Psychiatry
Rom J Physiol e: Romanian Journal of Physiology
rom.kat. a: romersk katolsk
ROMP e: Remote Operations Microprocessor
- e: Ring-Opening Metathesis
ROM-X e: Remote Opacity Monitor
RON e: Remain Over Night
RONA e: Return On Net Assets
RONABIT e: Return On Asset Before Income Tax
RONEWWYTARS e: Research On Nationwide Estimation of Winter Wheat Yield Through Aerospace Remote Sensing
RONSARD f: Recherche sur les Orages et les Nuages par un Système A Deux Radars Doppler
ROOM e: Real-Time Object-Oriented Modeling
ROP e: Raster Operation
- e: Receive Only Printer
- e: Record Of Proceedings
- e: Refined Oil Products
- e: Regional Oversight Policy
- e: Remote Operations Service
- e: RISC[-like] Operation
- f: Retraites Ouvrières et Paysannes
ROPA e: Realising Our Potential Award[s]
- e: Record Of Procurement Action
ROPES e: Route and Penetration Evaluation
ROPME f: Organisation Régionale pour la Protection du Milieu Marin
- s: Organización Regional para la Protección del Medio Marino
ROPOS e: Remotely Operated Platform for Ocean Science
ROPS e: Readiness, Operations and Planning Systems Division
- e: Remote Operations Service
- e: Roll Over Protection Standards
ROQ e: Return on Qualification
ROR e: Rotate Right
rör r: rörande
RO2, RO3 e: RISC Operating System 2, RISC OS 3
RORO e: Roll On, Roll Off
RORSAT e: Radar Ocean Reconnaissance Satellite[s]
ROS d: Rostock/Landkreis, Mecklenburg-Vorpommern
- e: Radarsat Optical Sensor
- e: Reactive Oxygen Species
- e: Record On Silicon
- e: Remote Operations Service
- e: Research Optical Sensor
- f: Radar à Ouverture Synthetique
ROS. e: Read-Only Storage
ROSA e: RACE Open Service Architecture
- e: Regional Office for Southern Africa
- f: Bureau régional pour l'Afrique australe
ROSAR e: Rotating SAR [Synthetic Aperture Radar]
- e: Rotor-SAR [Synthetic Aperture Radar]
ROSAT e: Roentgen Satellite
ROSCA e: Rotation Savings and Credit Association
ROSCOP e: Report of Observations/Samples Collected by Oceanographic Programmes
- f: Compte rendu des observations/échantillons recueillis en exécution de programmes océanographiques
- s: Informe sobre las Observaciones Realizadas y las Muestras Recogidas en los Programas Oceanográficos
ROSE e: RATS Open System Environment
- e: Remote Operation[s] Service Element
- e: Remote Operations Service Entity
- e: Remote Optical Sensing of Emission
- e: Reporting On the State of the Environment
- e: Rural Oxidants in the Southern Environment
ROSELT e: Network of Observatories for Long-term Ecological Monitoring
- f: Réseau d'observatoires de surveillance écologiques à long terme
- s: Red de Observatorios de Vigilancia Ecológica a Largo Plazo
ROSIS e: Reflective Optics System Imaging Spectrometer
ROSKOMNEDRH e: The Committee on Geology and the Use of Underground Resources
ROSS e: RAND Object Simulation System
ROSSWG e: Radarsat Optical Sensor Science Working Group
ROSTA f: Bureau régional de science et de technologie pour l'Afrique
- s: Oficina Regional de Ciencia y Tecnología para Africa
ROSTAS f: Bureau régional de science et de technologie pour les pays arabes
- s: Oficina Regional de Ciencia y Tecnología para los Países Arabes
ROSTLAC f: Bureau régional de science et de technologie pour l'Amérique latine et les Caraïbes
- s: Oficina Regional de Ciencia y Tecnología para América Latina y el Caribe
ROSTSCA f: Bureau régional de science et de technologie pour l'Asie du Sud et l'Asie centrale
- s: Oficina Regional de Ciencia y Tecnología para Asia Meridional y Central
ROSTSEA e: Regional Office for Science and Technology for South-East Asia
- f: Bureau régional de science et de technologie pour l'Asie du Sud-Est
- s: Oficina Regional de Ciencia y Tecnología para Asia Sudoriental
ROT e: Running Object Table
- r: Oktantal [recherchenmetoden]
- s: Red de Observadores Terrestres

ROTAP e: Rare or Threatened Australian Plants Data Base
ROTBA e: Reality On The Blink Again [Internet]
ROTC e: Reserve Officer[s] Training Corps
ROTD e: Release Of The Day
ROTF e: Rolling On The Floor [Internet]
ROTFFNAR e: Roll[ing] On The Floor For No Apparent Reason [Internet]
ROTFL e: Rolling On The Floor Laughing [Internet]
ROTFLACGU e: Rolling On The Floor Laughing And Can't Get up [Internet]
ROTFLAHMS e: Rolling On The Floor Laughing And Holding My Side[s] [Internet]
ROTFLASTC e: Rolling On The Floor Laughing And Scaring The Cat [Internet]
ROTFLBTC e: Rolling On The Floor Laughing Biting The Carpet [Internet]
ROTFLBTCACGU e: Rolling On The Floor Laughing Biting The Carpet And Can't Get up [Internet]
ROTFLBTCASCTC e: Rolling On The Floor Laughing Biting The Carpet And Scaring The Cat [Internet]
ROTFLBTCUTS e: Roll[ing] Unable To Stop
ROTFLGO e: Rolling On The Floor Laughing Guts Out [Internet]
ROTFLMAO e: Rolling On The Floor Laughing My Ass Off [Internet]
ROTFLMFHO e: Rolling On The Floor Laughing My Fucking Head Off [Internet]
ROTFLOL e: Rolling On The Floor Laughing Out Loud [Internet]
ROTFLOLVH e: Roll[ing] Out Loud Very Hard [Internet]
ROTFLSTC e: Roll[ing] Scaring The Cat [Internet]
ROTFLSTCIIHO e: Rolling On The Floor Laughing Scaring The Cat If I Had One [Internet]
ROTFLTC e: Rolling On The Floor, Licking The Glue [Internet]
ROTFTPOF e: Roll[ing] Trying To Put Out Flames [Internet]
ROTL e: Remove Office Test Line
ROTS e: Rotary Out Trunks Selectors
ROTs e: Rare Or Threatened plants
ROUE e: Revolutionary Organization of Uganda
Roum Arch Microbiol Immunol e: Roumanian Archives of Microbiology and Immunology
ROUS e: Rodents Of Unusual Size
Router-Port d: Anschlußbuchse an der Hardware des Providers für eine Standleitung des Kunden
ROV e: Remotely Operated Vehicle
- f: Véhicule télécommandé
- s: Vehículo teledirigido
ROVE e: Radar Observation of Vegetation
ROW d: Rotenburg/Wümme, Niedersachsen
ROWS e: Radar Ocean Wave Spectrometer
- e: Remote Ocean Wave Spectrometer
- e: Remote Orbital Wave Spectrometer
ROX e: Run Of Experiments
Roy f: Royaume
ROYGBIV e: Red, Orange, Yellow, Green, Blue, Indigo, Violet
ROZ e: Restricted Operating Zone
Rozhl Chir U: Rozhledy v Chirurgii
RP e: Rapid Prototyping
- e: Recreation Park
- e: Relay Party
- e: Reporting Post
- e: Republican Party
- e: Repurchase agreement
- e: Respirable Particulates
- e: Response Planning
- e: Responsible Party
- e: Rocket Propelled
- e: Rolling Plan
- e: Route Package
- f: Rassemblement des Patriotes
- f: Réadaptation Professionnelle
- f: Recette Principale
- f: Recette-Perception
- f: Receveur des Postes
- f: Receveur-Percepteur
- f: Reclassement Professionnel
- f: Réduction de Peines
- f: Réduction de Prix
- f: Réduction Progressive
- f: Régime Postal
- f: Région Parisienne
- f: Régions Polaires
- f: Règlement de Police
- f: Remboursement Partiel
- f: Rente Perpétuelle
- f: Réponse Payée
- f: Représentation Proportionnelle
- f: Républicains Populaires
- f: Républicains Progressistes
- f: République Populaire
- f: Réservation de Places
- f: Retrait de Permis
- f: Retraite Proportionnelle
- f: Revenu Professionnel
- f: Revenus Publics
- f: Révocation d'une Procuration
- f: Ristourne Partielle
- f: Rond-Point
- r: Riksprovplats
- s: Relaciones Públicas
- s: Reserva Permanente
Rp r: Riktpunkt
R.P. s: Respuesta Pagada
R&P e: Reception and Placement
RP2 f: Remarié depuis Plus de 2 ans
RPA e: Relative Percent Accuracy
- e: Remote Passphrase Authentication
- e: Remote Passphrase Authorization
- e: Remote Preamp[lifier]
- e: Renal Physicians Association
- e: River Purification Authority
- e: Route Planning Aid
- e: Rwandan Patriotic Army
- f: Reclassement professionnel des adultes
- p: República Popular de Angola
- s: República Popular de Alemania
RPAL e: Rubber Planters Association of Liberia
RPAM e: Recruiting Policy Analysis Model
RPB d: Radio-Praktiker-Bücherei
- e: Reverse Path Broadcasting
- f: Rassemblement du Peuple du Burundi
RPC e: Radiation Protection Committee
- e: Real Procedure Call
- e: Remote Procedure Call[ing]/[s]
- e: Resistive Plate Chambers
- e: Root Path Cost
- f: Rassemblement du Peuple Centrafricain
- f: Régiment Parachutiste Commando
- f: République Populaire Chinoise
- p: Regras Portuguesas de Catalogação
RPCA e: Remotely Programmable Conference Arranger
RPCL e: Remote Procedure Call Language
RPD e: Rapid Prototyping Development
- e: Redox Potential Discontinuity
- e: Regulatory Policy Division
- e: Relative Percent Difference
- e: Revolutions Per Day
RPDC f: République Populaire Démocratique de la Corée
RPE e: Office of OECD, European Community and Atlantic
RPE-LPC e: Regular Pulse Excited Linear Predictive Coder
RPE-LTP e: Regular Pulse Excitation with Long-Term Prediction Loop
RPF e: Remote Processor Function
- e: Report Format
- e: Report Program Facility
- e: Rwandan Patriotic Front
- f: Recette-Perception des Finances
RpflAnpG d: 2. Gesetz zur Anpassung des Rechtspflegeanpassungsgesetzes und anderer Gesetze vom 20.12.1996
RPFO e: Resupply Planning Factors Office
RPG e: Project Reference within the programme
- e: Radar Product General
- e: Radar Product Generator
- e: Receiver Processing Group
- e: Record Program Generator
- e: Refugee Policy Group
- e: Remote Product Generator
- e: Report Generator
- e: Research Planning Guide
- e: Research Project Grant
- e: Role Playing Game[s] [Internet]
RPI e: Restoration Priority Indication
- e: Rolling Plan Indices
- f: Républicains Populaires Indépendants
RPIE e: Real Property Installed Equipment
- e: Real Property Inventory Account System
RPIF e: Regional Planetary Image Facility
RPIFS e: Real Property Industrial Fund System
RPIO e: Responsible Planning and Implementation Officer
RPIS e: Real Property Inventory System
RPITB e: Royal Pain in the Butt
RPJ e: Project reference
RPL d: Rheinland-Pfalz Landesregierung und Landtag
- e: Recognition of Prior Learning
- e: Recurrent Pregnancy Loss
- e: Requested Privilege Level
- e: Resident Programming Language
- f: Remploi

RPLOT e: Resource Plot
RPM e: Office of NATO and Atlantic Political-Military Affairs
- e: Radio Packet Modem
- e: Rapid Prototyping Module
- e: Reactive Plume Model
- e: Red Hat Package Manager
- e: Regional Particulate Model
- e: Register of Preservation Microfilms
- e: Remedial Project Manager
- e: Remote Process Management
- e: Removable Peripheral Module
- e: Resales Price Maintenance
- e: Retro Propulsion Module
- e: Revolution[s] Per Minute
- e: Revolutionary Peoples Movement
- e: Risk-Based Prioritization Method
- f: Revenu des Personnes Morales
- p: República Popular de Moçambique
- s: Revoluciones Por Minuto
RPMA/MIS e: Real Property Maintenance Activities/MIS
RPMT e: Radiant Power Measuring Instrument
RPN e: Research programme
- e: Reversed Polish Notation
- e: Risk Priority Number
- f: Division de la Recherche en Prevision Numerique
RPO e: Regional Planning Officer
- e: Regional Program Officer
- e: Reproduction Rights Organisation
- e: Responsible Property Officer
RPOA e: Recognized Private Operating Agencies (or Agency)
RPP e: Red Patch Panel
- e: Rural Policy Plan
- f: Rassemblement Populaire pour le Progrès
- f: Revenu des Personnes Physiques
RPPC e: Research Programme Planning Committee
RPPROM e: Reprogrammable PROM
RPQ. e: Request for price quotation
RPR d: Rheinland-Pfalz Radio
- e: Replacement Production Reactor
RPRINTER e: Remote Printer
RPS e: Parcel Service
- e: Redundant Power Supply
- e: Remedial Planning Staff
- e: Remote Procedure Services
- e: Remote Processing Station
- e: Renal Pathology Society
- e: Renewables Portfolio Standard
- e: Revolution[s] Per Second
- e: Ring Parameter Server
- e: Rotational Positional Sensing
- e: Route Partner Set
- e: Routine Product Set
- f: Rassemblement pour le Progrès Social
- f: Riz-Pain-Sel
- s: Revoluciones Por Segundo
RPSD f: Rassemblement du Parti Social-Démocrate
RPSGB e: Royal Pharmaceutical Society of Great Britain
RPSO f: Région Parisienne, Seine-et-Oise
RPSS e: Relative Partial Sum of Squares
RPT e: Radiation Protection Technician
- e: Radiation Protection Technologist
- e: Regular Public Transport
- e: Renormalized Perturbation Theory
- f: Rassemblement du Peuple Tchadien
- f: Répétez
- f: Réseau Principal de Télécommunications
- s: Red Principal de Telecomunicaciones
RPTF e: Red Patch and Test Facility
Rptns f: Répétons
RPTOR e: Reporting Organization
RPTS e: Regional Priority Tracking System
RPU s: Refugio Público
RPV e: Remotely Piloted Vehicles
- e: Replacement Plant Value
- s: Refugio Privado
RPx f: Réponse Payée x francs
RQ e: Relational Query
- e: Reportable Quantity
- e: Runequest
- f: Rapport Quotidien
- f: Recensement Quinquennal
- f: Revue de Qumran
RQBE e: Relational Query By Example
RQE e: Responsive Quantum Efficiency
RQF e: Rescue Flight
RQG e: Rescue Group
RQS e: Rate/Quote System
- e: Request To Send
- e: Rescue Squadron
RQSM e: Regional Quality Service Management
RQUERY e: RISOP Query
RQW e: Rescue Wing
RR e: Radio Resource Management
- e: Readiness Reports
- e: Real Reality
- e: Receive[r] Ready
- e: Relative Risk
- e: Removal Rate
- e: Repeatable Read
- e: Requirements Review
- e: Resource Record
- e: Resource Reservation protocol
- e: Reverse Recovery
- e: Risk Reduction
- e: Rolls Royce
- e: Round Robin
- e: Route Relay
- f: Radioreportage
- f: Régie Renault
- f: Règlements Routiers
- f: Retraite Réversible
- f: Route Rurale
- r: Rådhusrätten
- r: Regeringsrätten
- s: Remolcador de Rada
Rr r: Radar
RR. e: Receive Ready
- e: Request Ready
R&R e: Removal and Reassembly
- e: Rest and Relaxation
- e: Resupply and Return
R2R e: Run-to-Run
RRA e: Rahawein Resistance Army
- e: Receive Resource Area
- e: Regular Reports of Audit
RRAM e: Ruhengeri Resource Analysis and Management
RRAMS e: Renewable Resource and Management Statistics
RRAS e: Routing and Remote Access Service
RRB/ATM e: ReRouting Banyam ATM
Rrbled r: Radarrobotledare
Rrbv r: Radarbandvagn
RRC e: National Relief and Rehabilitation Commission
- e: Regional Research Centre
- e: Regional Response Center
- e: Residency Review Committee
- e: Richland Research Complex
- e: Richmond Research Center
R.R.C. e: Red Cross
RRCDM s: Red Regional de Comunicacón de Datos Meteorológicos
RRCM e: Reservation Request Control Mechanism
RRCT e: Reflectance Ratio Cloud Test
RRD e: Reactor Research and Development
RRDA e: Range-Resolved Differential Absorption
RRDC e: Regional Research and Development Center
RRDP e: Regional Rural Development Project
RRDS e: Relative Record Data Sets
RREA e: Renewable Resource Extension Act
- e: Renewable Resources Extension Act
R Rec s: Rumbo Recomendado
RRF e: Ready Reserve Fleet
- e: Regional Radio Facility
- e: Regional Relay Facility
- e: Regional Research Funds
- e: Research Reference
- e: Residual Renal Function
- e: Risk Reduction Factor
RRG d: Rentenreformgesetz
RRHSI f: Registre des Ressources Humaines en Santé Internationale
RRIP e: Rock Ridge Interchange Protocol
RRITA e: Rapid Response Intra-Theater Airlifter
RrjaL r: Radarjaktledare
Rrjal r: Radarjaktledning
RRLS e: Resonance Rayleigh Light Scattering
Rrls r: Radarluftbevakningsstation
Rrlvled r: Radarluftvärnsledare
RRM e: Red Resource Monitoring
RRMSEP e: Relative Root Mean Square Error of Prediction
RRN e: Regional Research Network[s]
RRN ANT e: RRN Antarctic
RRN ART e: RRN Arctic
RRN CAA e: RRN Central Arid Asia
RRN CAR e: RRN Caribbean
RRN ESA e: RRN Equatorial South America
RRN MED e: RRN Mediterranean
RRN NAF e: RRN Northern Africa
RRN OCE e: RRN Oceania
RRN SAF e: RRN Southern and Eastern Africa

RRN TAM e: RRN Tropical Asian Monsoon Region
RRN TEA e: RRN Temperate East Asia
RRN TNH e: RRN Temperate Northern Hemisphere
RRN TSA e: RRN Temperate South America
RRO e: Rate and Route Operator
- e: Reports Receiving Office
- e: Reproduction Rights Organisation
RRP e: Reconfigurable Ring of Processors
- e: Reconnaissance Reference Point
- e: Resource Read Pointer register
RRQ e: Return Request
R/R/R e: Reassignment/Retargeting/Reconstitution
RRR e: Remove, Repair and Replace
- e: Repatriation, Resettlement, Reconstruction
- e: Review of Local Government Residential Regulations
RRRS e: Royal Rife Research Society
RRS e: Radio Relay Site
- e: Readiness Reporting System
- e: Ready Reserve Status
- e: Regional Research Site[s]
- e: Regulatory Reform Staff
- e: Resonant Raman Scattering
- e: Resupply Requirement System
- r: Rikspolisstyrelsen
- s: Buque Real de Investigación
R&RSA e: Restoration and Remediation Safety Analysis
RRS A/S a: Randers Rustfri Stål-Industri A/S
RRSF e: Regional Remote Sensing Facility
Rrsiss r: Radarsignalspaningsstation
RRSP e: Regional Remote Sensing Programme
RRSS e: Resources and Remote Sensing Systems Association
RRSSC e: Regional Remote Sensing Service Centres
Rrstn r: Radarstation
RRT e: Regional Response Team
- e: Requisite Remedial Technology
- e: RRT: Canadian Journal of Respiratory Therapy
RRTDM f: Réseau Régional de Transmission des Données Météorologiques
RRTE e: Request for Reimbursement of Travel Expenses
RRTM e: Rapid Radiative Transfer Model
Rrto r: Radartrupp
Rrttn r: Rådhusrätten
RRV f: Revalorisation de rente viagère
- r: Riksrevisionsverket
RRWDS e: Radar Remote Weather Display System
RRZ e: Restricted Release Zone
RRZE d: Regionales Rechenzentrum Erlangen
RRZN d: Regionales Rechenzentrum für Niedersachsen, Hannover

RS a: Rederiaktieselskab
- e: Radio Shack
- e: Read Stability
- e: Record Status
- e: Red Switch
- e: Reed Solomon [code]
- e: Registery Service
- e: Registry Service
- e: Relief Society
- e: Remoting Subsystem
- e: Requirements Steering
- e: Research Scientist
- e: Ring Station
- e: Russian Segment
- f: Radicaux Socialistes
- f: Radio Sorbonne
- f: Raison Sociale
- f: Règle de Société
- f: Relevé de Salaires
- f: Rente de Survie
- f: Rente Saisissable
- f: Républicains Sociaux
- f: Responsabilité Solidaire
- f: Rien à Signaler
- r: Arméns Rid- och Körskola
- s: Resolución Suprema
Rs a: Råpoints
- f: Ruines
- r: Ringsikte
R S e: Religious Studies
r,s e: indizes
R/S e: Remote Sensing
r/s N: redningsskip
RS/6000 e: IBM RISC System/6000
RSA d: Rivest, Shamir und Adleman
- e: Random Scheduling Algorithm
- e: Rapid Sediment Analyzer
- e: Recurrent Spontaneous Abortion
- e: Resource Start Address register
- e: Rivest-Shamir-Adleman [cryptosystem]
- e: Russian Space Agency
R.S.A. e: Royal Scottish Academy
RSA129 e: Digit cryptographic security number named after its inventors: Drs. Ronald L. Rivest, Adi Shamir and Leonard M. Adelman
RSAA e: Remote Sensing Association of Australia
RSAC e: Recreational Software Advisory Council
- e: Remote Sensing Applications Center [China]
- e: Remote Sensing Applications Committtee
RSAL e: Remote Sensing Applications Laboratory
RSAS e: RAND Strategic Assessment System
RSAT e: RADARSAT Project
- e: Rapid Slide Agglutination Test
RSAW e: Resistive Shear Alfven Wave
RSB d: Regionalschnellbahn
- e: Regional Science Branch
- e: Repair Service Bureau
- e: Reservation State Block
- s: Dirección Regional de Ciencias
RSBMS e: Rotating Shadow Band Multispectral Radiometer

RSBS e: Research School of Biological Sciences
- e: Rotating Shadow Band Spectrometer
RS232C d: Norm für serielle Schnittstelle
RSC e: Receive Sequence Counter register
- e: Regional Service Center (or Council)
- e: Remote Sensing Center (or Committee)
- e: Remote Switching Center
- e: Request for Software Change
- e: Residence Service Center
- e: Restaurants Supervisory Committee
- e: Royal Society of Chemistry
RSCAT e: Radarsat Scatterometer
RSCC e: Regional Sample Control Center
RSCD e: Reference Services and Collection Development
RSCE e: Reverse Short-Channel Effect
RSCS e: Remote Source Control System
- e: Remote Spooling and Communication[s] [Sub]system
RSCT e: Regents Society
RSCV e: Route Selection Control Vector
RSD e: Reflex Sympathetic Dystrophy
- e: Residual Standard Deviation
- e: Route Server Daemon
- f: Rassemblement des Sociaux-Démocrates
RSDB e: Remote Sensing Data Base
RSDG f: Rassemblement National Démocratique Social
RSDI e: Retirement and Survivors Disability Insurance
Rsd orlstn N: Rasmund orlogsstasjon
RSE e: Reactive Sputter Etch
- e: Reflection Secondary-electron Emission
- e: Remote Streams Environment
- e: Research Scientists and Engineers
- e: Resolution
R Sec s: Remolque Secador (or Secadero)
RSED e: Research and Scientific Exchanges Division
RSEP e: Resource Supply Expansion Program
RSER e: Residual Soft-Error Rate
RSET e: Reset
RSET. e: Received Signal Element Timing
RSF a: Revolutionære Socialisters Forbund
- e: LANL computer code
- e: Recreational Sports Facility
- e: Remote Support Function
- f: Reporters Sans Frontières
- r: Radarspaningsflygplan
RSFP e: Rivet Switch Follow-on Program
RSFQ logic e: Rapid Single-Flux-Quantum Logic
RSFTA f: Réseau du Service Fixe des Télécommunications Aéronautiques
RSG e: Ready, Set, Go
- e: Remote Sensing Group
- e: Retail Solutions Group
- s: Reglamento de Servicio en Guarnición
RSGDAPS e: Remotely-Sensed and Ground Data Automated Processing System
RSH e: Remote Shell
- e: Restricted Shell
- e: The Royal Society for the Promotion of Health

RSHI e: Russian State Hydrometeorological Institute
- f: Institut Hydrométéorologique d'Etat de Russie
- s: Instituto Hidrometeorológico Estatal de Rusia

RSHMI e: Russian Hydrometeorological Institute

RSHsb N: Rekruttskole for Hærens samband

RSI e: RADARSAT International Inc.
- e: Rationalization, Standardization and Interoperability
- e: Remote Sensing Institute
- e: Repetitive Strain Injuries (or Injury)
- e: Repetitive Stress Injury
- e: Retirement and Survivors Insurance
- e: Review of Scientific Instruments
- e: Ripple Symmetry Index
- e: Roll Systems, Inc.
- e: Rough Sleepers Initiative
- f: Recherche en Soins Infirmiers

RSIAL e: Remote Sensing and Image Analysis Laboratory

RSIC e: Radiation Shielding Information Database

RS-IGIS e: Remote Sensing Integrated Geographic Information System

RSIMS e: Remote Sensing Information Management System

RSIP e: Remote Sensing and Image Processing

RSIS e: Relocateable Screen Interface Specification

RSITA f: Règlement de la Société Internationale des Télécommunications Aéronautiques

RSK f: République Serbe de Krajina

RSkC r: Reservskyddscentral

RSKERL e: Robert S. Kerr Environmental Research Laboratory

RSL d: Rosslau/Elbe, Sachsen-Anhalt
- e: Remote Learning System
- e: Request and Status Link

RSLE e: Remote Subscriber Line Equipment

RSLM e: Remote Subscriber Line Module

RSLMF e: Republic of Sierra Leone Military Force

RSM e: Reciprocal-Space Map
- e: Research Scale Melter
- e: Response Surface Matrix
- e: Response Surface Methodology
- f: Renseignements sur la Sécurité Maritime
- f: Révolution Socialiste Malgache

RSMAS e: Rosenstiel School of Marine and Atmospheric Science[s]

RSMC e: Regional Specialized Meteorological Center[s] (or Centre)

RSMD e: Resource Systems Management Division

RSM/EA e: Regional Support Mission for East Asia

RSMIP e: Remote Sensing Management Information Program

RSMIS e: a Real property or Computerized Data Base
- e: Real Estate and Space Management Information System

RSN e: Radio Supernova
- e: Real Soon Now [Internet]
- f: Rassemblement du Salut National

R S NC REC e: Rec Sports N. C. Recreation Building

RSO e: Radiation Safety Officer
- e: Reconnaissance System[s] Operator
- f: Radar à Synthèse d'Ouverture

RSOC e: Radiation Safety Officers Committee

RSOF e: Required Special Optional Features

RSOH e: Regenerator Section Overhead

RSP e: Radial Structure Plot
- e: Red Switch Program
- e: Regional Seas Programme
- e: Required Space Character
- e: Research and Surveys Program
- e: Response
- e: Responsible Person
- f: Programme pour les mers régionales
- s: Programa de los Mares Regionales

RSPA e: Research and Special Program Administration

RS&PAA e: Remote Sensing and Photogrammetry Association of Australia Ltd.

RS/PAC e: Regional Seas Programme Activity Centre
- f: Centre d'activité du Programme pour les mers régionales
- s: Centro de Actividad del Programa de los Mares Regionales

RSPAC e: Remote Sensing Public Access Center

RSPAS e: Research School of Pacific and Asian Studies

RSPB e: Royal Society for the Protection of Birds

RSPC e: Reed Solomon Product Code

RSPCA e: Royal Society for the Protection of Animals

Rspl r: Rapportsamlingsplats

RSPP e: Remote Sensing Preparatory Programme

RSPS e: Research School of Pacific Studies

RSPX e: Remote Sequenced Packet Exchange

RSR e: Rain Scattering Radiometer
- e: Rain Scatterometer and Radiometer
- e: Republican Subordination Raison
- e: Rotating Shadowband Radiometer

RSRB e: Remote Source Route Bridging

RSRC e: Resource System Review Committee

RSRP e: Remote Sensing Research Program

RSRS f: Régime Supplémentaire de Retraites des Salariés

RSS e: Radiation Safety Section, Department of Health and Community Services
- e: Regional Services Staff
- e: Regional Stratigraphic Scale
- e: Relaxed Static Stability
- e: Reliability, Supportability and Survivability
- e: Remote Switch Subsystem
- e: Return Signal Simulator
- e: Rockland Simpson Scale
- e: Rotating Shadowband Spectrometer
- e: Runway Supervisory Sets
- N: Regimentets Stabssersjant

rss e: root-sum-square

Rß d: Rest-Beta-Aktivität

RSSA e: Resources Support Services Agreement

R&SSD e: Radiation and Safety Surveys Department

RSSD e: Responses of Savannas to Stress and Disturbance

RSSI e: Received Signal Strength Indicator
- e: Russian Space Science Internet project

RSSP e: Radarsat Sar Simulation Package

RSST e: Reactive System Scanning Tool

RST e: Readability, Signal Strength and Tone
- e: Receive Start Threshold
- e: Remote Switching Terminal
- e: Reset [and Restart]
- e: Resolution Syntax Tree
- e: Rotary Shifting Tool
- f: Réponse à un avis de Service Taxé

RSTA e: Reconnaissance, Surveillance and Target Acquisition

RSTC e: Recreational Scuba Training Council

RSTE e: Regenerator Section Terminating Equipment

RSTN e: Radio Solar Telescopic Network

Rstngrp r: Radiostationsgrupp

RSTS e: Resource Sharing and Time Sharing
- e: Resource System Time Sharing

RSTSE e: Resource System Time Sharing/Enhanced

RSTS/E e: Resource Sharing Time-Sharing System/Extended

R-STW e: Robotic-Storage Tank West

RSU e: Remote Sensing Units
- e: Remote Switching Units
- e: Runway Supervisory Unit

R&SU e: Repair and Salvage Unit

RSUM e: Research Supplements for Underrepresented Minorities

RSV e: Revised Standard Edition
- r: Riksskatteverket
- r: Riktad Sprängverkan

RSVO d: VO zur Durchführung des $ 22 des Bundessozialhilfegesetzes [Regelsatzverordnung]

RSVP e: Resource Reservation Protocol
- f: Répondez, S'il Vous Plaît

RSWG e: Red Sea Working Group
- e: Response Strategies Working Group
- f: Groupe de travail des stratégies d'adaptation
- s: Grupo de Trabajo de Estrategias de Respuesta

RSWGSC e: RSWG Steering Committee

R-SWIMS e: Richland Solid Waste Information Management System
RSWK d: Regeln für den Schlagwortkatalog
RSX e: Real Time
- e: Resource Sharing Executive
RSX11D e: Resource Sharing Executive-11D
RSX11M e: Resource Sharing Executive-11M
RT e: Radiation Transport
- e: Radiative Transfer
- e: Rapid-Thermal
- e: Realtime
- e: Red Team
- e: Regional Total
- e: Register-Transfer
- e: Relocatable Targets
- e: Report Test
- e: Right Tackle
- e: RISC Technology
- e: Room-Temperature
- e: Rotation Time
- e: Routing Type
- e: Run Time
- e: The HP/RT-743 Board
- f: Réduction de Tarif
- f: Région Territoriale
- f: Remboursement Total
- f: Reprise du Travail
- f: Retrait de fonds par Télégraphe
- f: Révocation d'un Testament
- f: Ristourne Totale
- N: Registertonn
- R: Reydovoy Tralshchik
- s: Radiotelégrafo
- s: Radiotelémetro
- s: Recuperador de Torpedos
- s: Regimiento de Transmisiones
- s: Regulación de Tránsito
- s: Retaguardia
- s: Retrasado
- s: Riego [directo desde] Tierra
Rt. r: Rikstelefon
R.T. N: Radiotelefon
R/T e: Receive/Transmit
RTA d: Runder Tisch Amateurfunk
- e: Rapid Thermal Anneal
- e: Rapid Thermal Anneal[ing]
- e: Request for Travel Authorization
- e: Residual Threat Assessment
- e: Right torsion of the abomasum
- f: Régie des Tabacs et Allumettes
RTAC e: Regional Technical Assistance Center
RTACS e: Real-Time Adaptive Control System
RTAS e: Real-Time Advisory System
RTASS e: Remote Tactical Airborne SIGINT System
RTB d: Regionale Testbeds im Rahmen des DFN
- e: Real-Time Backplane
- e: Regional Testbed
- f: Radiodiffusion et Télévision Belge
RTBISC f: Radiodiffusion-Télévision Belge-Institut des Services Communs
RTBT e: Resonant Tunneling Bipolar junction Transistor
RTC e: Radiative Transfer Code
- e: Reactor Technology Center
- e: Real Time Compression
- e: Regional Trading Corporations
- e: Religious Technology Centre
- e: Remote Terminal Concentrator
- e: Right To Copy
- e: Run Time Clock
- f: Radiodiffusion Télévision Congolaise
- f: Radiotechnique Compelec
- f: Réseau Téléphonique Commuté
RTCM e: Reasonable Transportation Control Measures
RTCON e: Real-Time Connect
RTCS e: Radarsat Topographic Correction System
RTD e: Radio Television Djibouti
- e: Real-Time Data
- e: Research and Technological Development
- e: Resistance Temperature Device
- e: Resonant Tunneling Diode
- e: Retraction date
- e: Returns To Duty
- f: Radio-Télédiffusion
- f: Rassemblement du Travail pour la Démocratie
Rtda s: Retirada
RTDC e: Real Time Data Compressor
RTD[&]D e: Research, Technological Development and Demonstration
RTDM e: Real-Time Data Migration
- e: Rough Terrain Diffusion Model
RTDS e: Rapid Thermal Decomposition of precursors in Solution
RTE e: Radiative Transfer Equation
- e: Remote Terminal Equipment
- e: Repairs-To-Extend
- e: Run-Time Environment
- s: Región de Tratamiento Especial
rte. s: Remite[nte]
RT&E e: Research, Test and Evaluation
R/Técnico s: Radiotécnico
Rte Déple f: Route Départementale
RTEID e: Real Time Executive Interface Definition
RTEL e: OA-RTP FTS Telephone System
- e: Reverse Telnet
Rte Natle f: Route Nationale
RTF e: Radio-Television-Film
- e: Read This First
- e: Reconnaissance Technical Flight
- e: Remote Task Force
- f: Radio[diffusion]-Télévision Française
Rtf s: Radioteléfono
RTFAQ e: Read The FAQ [Frequently Asked Questions] list [Internet]
RTFB e: Read The Funny Binary
RTFF e: Read The Fantastic FAQ [Frequently Asked Questions]
RTFI e: Read The Fucking Instructions [Internet]
RTFL s: Radioteléfono Ligero
RTFM e: Read The Fabulous (or Fantastic, or Farging , or Fershlugginer, or Ficticious, or Fine, or Flaming, or Freaking, or Friggin', or Fucking, or Funky, or Funny) Manual [Internet]
- s: Radioteléfono Medio
RTFMA e: Read The Fine (or Friggin', or Fucking) Manual Again [Internet]
RTFMS e: Read The Fucking Manual, Sir [Internet]
RTFOH e: Read The Fine (or Fucking) Online-Help [Internet]
RTFP s: Radioteléfono Pesado [Internet]
RTFS e: Read The Funny Source [Internet]
RTG e: Radioisothope Thermoelectric Generator
- e: Radioisotope Thermal Generator
- e: Reconnaissance Technical Group
- e: Research Training Grant
- f: Radio Télévision Guinéenne
- f: Rassemblement du Travail Guinéen
Rtg r: Röntgen[ogram]
- s: Radiotelegrafista
Rtgdia s: Retaguardia
RTI e: Routing and Targeting Instructions
- e: Routing Indicator
- f: Radio-Télévision Ivoirienne
RTIC e: Real-Time Intelligence in the Cockpit
RTIF e: RAMP Test and Integration Facility
r/tim r: röntgen per timme
RTIP e: Real Time Interactive Processor
- e: Regional Transportation Improvement Program
RTK e: Real Time Kernel
RTK Net e: Right-To-Know Network
RTL d: Radio Tele Lustmolch
- e: Radio Television Luxemburg
- e: Ray Trace Language
- e: Register-Transfer Level
- e: Research Title
- e: Resistor Transfer Logic
- e: Right-To-Left
- e: Run Time Library
RT-LPMOCVD e: Rapid Thermal Low-Pressure Metal-Organic Chemical Vapor Deposition
RTM e: Radiative Transfer Model
- e: Rapid Thermal Multiprocessing
- e: Read The Manual [Internet]
- e: Real Time Modeling
- e: Realtime Modeling
- e: Regional Telecommunications Management
- e: Regional Transport Model
- e: Remote Test Module
- e: Requirements Traceability Matrix
- e: Response-Time Monitor
- e: Runtime Manager
- f: Radio-Télévision Malgache
- f: Régie des Transports du Mali
- s: Radioteléfono Medio móvil de campaña
Rtm s: Radiotelémetro
R&TM e: Road and Track Maintenance
RTMP e: Routing Table Maintenance Protocol
RTMS e: Real Time Monitoring System
RTN e: Rapid Thermal Nitridization
RTNEPH e: Real Time Nephanalysis
RTO e: Regenerative Thermal Oxidation (or Oxidizer)
- e: Responsible Test Organization
- e: Retransmission Time Out
- f: Référence au Télégramme Officiel
Rto s: Retiro

Rto. s: Regimiento
RTOC e: Rear Area Tactical Operations Center
RTOH e: Read The Online-Help [Internet]
RTOP e: Research [and] Technology Objectives and Plans
- e: Research Topic Operating Plan
RTOSUH e: Real Time Operating System-Universitaet Hannover
RTP e: Rapid Thermal Processor
- e: Rapid Transport Protocol
- e: Real Time Photogrammetry
- e: Real Time Processor
- e: Real Time [Transport] Protocol
- e: Regional Transportation Plan
- e: Regulatory Toxicology and Pharmacology
- e: Rendezvous and Termination Protocol
- e: Repository Technology Program
- e: Routing Update Protocol
RTPA e: Regional Transportation Planning Agency
RT PC e: First IBM workstation
RTPC e: Real-Time and Parallel Computing
RT-PCR e: Reverse Transcriptase Polymerase Chain Reaction
RTPDC e: Real-Time Performance Data Collection
RTPN e: Remote Transaction Program Name
RTPS e: Remote Tape Processing System
RTPSC e: Real-Time Power System Control
RTR e: Real-Time Reporting
- e: Reel-To-Reel
- e: Return and Restore Condition Codes
R Tr Amas s: Remolque Tren Amasador
RTRS s: Regimiento de Transmisiones
RTRSOC e: Real Time Reporting System in Oceanic Conditions
RTS e: Real Time Systems
- e: Reconnaissance Technical Squadron
- e: Regulation Tracking System
- e: Relative Time Sequence
- e: Remote Takeover System
- e: Residual Time Stamp
- e: Return from Subroutine
- e: Return To Stock
- f: Radio-Télévision Seychelles
- f: Régie des Transports du Sénégal
- f: Régiment des Transmissions et Services
RTS. e: Request To Send
RTSE e: Remote Transfer Service Element
RTSL e: Read The Source Luke
RTSP e: Real Time Sar Processor
- e: Real-Time Streaming Protocol
RTSPC e: Real-Time Statistical Process Control
RTSS e: Reserve Training Support System
RTT e: Round Trip Time
- f: Régie des Télégraphes et Téléphones
- N: Rådet for Teknisk Terminologi
- s: Radioteletipo
RTTF e: Range in Times To Failure
RTTI e: Run Time Type Identification
- e: Run Time Type Information
RTTS e: Real-Time Transmission Subsystem
RTTY e: Radio Teletype[writer]
RTU e: Real Time Unix
- e: Remote Telemetry Unit
- e: Remote Trunking Unit
- e: Replacement Training Unit
RTV e: Real-Time Video
- e: Room-Temperature Vulcanized
- N: Rikstrygdeverket
RT-VBR e: Real-Time Variable Bit Rate
RTVF e: Room-Temperature Void-Formation
RTWFQ e: Read The Whole Friggin' (or Fucking) Question [Internet]
RTY e: Result Type
- e: Retry Limit
.ru e: Russian Federation [Internet]
RU a: Radikal Ungdom
- A: Rhodes-Universiteit Grahamstown, South Africa
- e: Receive Unit
- e: Reporting Unit
- e: Request Unit
- e: Response Unit
- e: Rhodes University Grahamstown, South Africa
- e: Rig Up
- f: Reconstruction et Urbanisme
- f: Règlement d'Usine
- f: Réseau Urbain
- f: Royaume-Uni
Ru r: Repetitionsutbildning
Ru-103 e: Ruthenium-103
Ru-106 e: Ruthenium-106
RUA e: Remote User Agent
RUB e: Ruhr-University Bochum, Germany
RUBADA e: Rufiji Basin Development Authority
RuBipy e: Rutheniumbipyridyl-[based photovoltaic system]
Rubr r: Rubricerad
RUC a: Roskilde Universitets Center
- e: Rapid Update Cycle
- e: Reservation Upon Collision
- e: Roskilde University Center
- f: République Uni du Cameroun
RUCA e: State of University Antwerp
RÜD d: Rheingau-Taunus-Kreis in Rüdesheim, Hessen
RUDI e: Resource for Urban Design Information
RUDT f: Rassemblement pour l'Unité et de Démocratie au Tchad
RUEI e: Ranger Uranium Environmental Inquiry
RUF e: Revolutionary United Front
RUG e: RADARSAT Users Guide
RÜG d: Rügen in Bergen, Mecklenburg-Vorpommern
RUGU a: Det Rådgivende Udvalg for de Gymnasiale Uddannelser
RUH a: Radikal Ungdom i Hovedstaden
RUIC e: Reporting Unit Identification Code
Ru Kyst a: Rungsted Kyst
RUL e: Right Unilateral
RUM d: Rechenzentrum Universität Mannheim
- e: Ranger Uranium Mines Pty Ltd.
- e: Remote User Multiplex
Rum. e: Rumanian
RUMS e: Resource Use [and] Management Subgroup
- f: Sous-groupe de l'exploitation et de gestion des ressources
- s: Subgrupo de Explotación y Gestión de los Recursos
RUMSEY e: Rumsey House West Quad
RUN f: Rassemblement pour l'Unité Nationale
RUO r: Reservunderofficerskola
RUOK e: Are You OK? [Internet]
- r: Kurs i Radartjänst för Underofficerare
RUOW e: Remote Unit of Work
RUP e: Remote Users Package
- e: Restricted Use Pesticide
- f: Reconnu d'Utilité Publique
RUR e: Resource Utilization Report
- e: Ship-launched, anti-submarine rocketthe torpedo
RUR-5A e: antisubmarine rocket
RURPIS e: Resource Utilization and Regional Planning Information System
RURPOP e: Rural Population File
RUS e: Rechenzentrum der Universität Stuttgart
RUSN e: Russian
russ. a: russisk
Russ. e: Russian
RUT e: Reflectance Uniformity Test
RUV s: Residencia Universitaria Verapaz
RUW e: Remote Unit Of Work
RV e: Rabies Vaccine
- e: Real Virtuality
- e: Research Volume
- e: Right Valve
- f: Recherches Vétérinaires
- f: Relais de Voie
- f: Rendezvous
- f: Rente Viagère
- f: Résolution de Vente
- f: Retraite-Vieillesse
- N: Rød Valgallianse
- s: Buque de investigación
- s: Rumbo-proa
r.v. e: Random Variable
Rv a: Retvisende
RVA f: Régie des Voies Aériennes
- s: Red de Vigiliancia del Aire
R-value e: Resistivity and Resistance
RVAM f: Référence à Votre Avis de Mutation
RVAU e: Royal Veterinary and Agricultural University
RVB e: Return-Beam Vidicon
RVC e: Research Volume Coat
RVCAT e: Research Vessel Catch
RVD e: Remote Virtual Disk Protocol
- e: Rendezvous and Docking
- f: Rente Viagère Diférée
Rvda. s: Reverenda
Rvdo. s: Reverendo
RVF e: Right Ventricular Failure
- f: Régie des Voies Fluviales
RVI e: Ratio Vegetation Index
- f: Renault Véhicules Industriels
RVK a: Radikale-Venstre-Konservative
RVL d: Ringverteilungsleiter
- f: Radar à Visée Latérale

RVM e: Rotation-Vibration Model
- f: Régie des Voies Maritimes
RVN e: Regional Voice Network
RVO N: Rådet for Videregående Opplæring
RVOC e: Research Vessel Operators Council
RVPS e: Remnant Vegetation Protection Scheme
RVRE e: Radio Voice of Revolutionary Ethiopia
RVS e: Ramped-Voltage Stress
- f: Radio Vallée de Seine
- r: Riktad Sprängverkan
RVSN R: Rakentye Voiska Strategityes-ko[vo] Naznacheniya
RVSVP f: Répondez Vite, S'il Vous Plaît
RVT f: Référence à Votre Télégramme
- f: Retraite des Vieux Travailleurs
RVTO f: Référence à Votre Télégramme Officiel
RVV d: Regensburger Verkehrsverbund
RVVT e: Russells Viper Venom Time
RVWS e: Reviews
RW d: Rechnungswährung
- d: Rechtswert
- e: Office of Civilian Radioactive Waste Management
- e: Radioactive Waste
- e: Receiver Window
- e: Reconnaissance Wing
- e: Report Writer
- e: Requried Warming
- e: Rivet Workforce
r.w. e: random widths
R[/]W e: Read-Write
RWA e: radiation work authorization
RWAGP e: Rural Womens Access Grant Program
R-wave e: Rayleigh wave
RWB e: Responsive Workbench
RWC e: Regional Warning Centre
- e: Relative Water Content
- e: Remote Work Center
- e: Residential Wood Combustion
- e: Rural Water Commission
RWCC e: Recycled Water Coordination Committee
RWCEP e: Real World Crisis Evaluation Plan
RWCT e: Radioactive Waste Collection Tank
RWE d: Rheinisch-Westfälische Elektrizitätswerke, Dortmund
- e: Remote Wall Enclosure
RWI d: Rheinisch-Westfälisches Institut für Wirtschaftsforschung, Essen
- e: Radio-Wire Integration
- e: Reset Window Indicator
RWL d: Rheinland-Westfalen-Lippe
RWM e: Read Write Memory
RWMC e: Radioactive Waste Management Center
RWO e: Ramstein Warning Office
RWP d: Rechts- und Wirtschaftspraxis
- e: Radar Wind Profiler
- e: Radiation Work Procedure
- e: Radiological Work Permit
- e: Radiological Work Procedure
- e: Recurring Work Program
- e: Resource Write Pointer register
- e: RIP While Printing
RWP50 e: 50-MHz Radar Wind Profiler
RWP915 e: 915-MHz Radar Wind Profiler
RWPP e: Rural Water Pricing Policy
RWRW e: Rescue and Weather Reconnaissance Wing
RWS n: Rijkswaterstaat
RWS. e: Remote Workstation
RWT e: Required Weekly Test
RWVL e: Raman Water Vapor Lidar
RWW e: Railway Workers Union
RX e: Receive Data
RX75 e: Region 10 Laboratory Management System
RXD e: Receive Data
RXDIS e: Receiver Disable
RXEN e: Receiver Enable
RXTE e: Rossi X-ray Timing Explorer
R/Y e: Message Traffic Type Designation
RY a: Ryvang
RYFM e: Read Your Fine (or Friggin', or Fucking) Manual [Internet]
RYMOM e: Rural Youth Movement of Mauritius
RYO e: Roll Your Own
RYR e: Ryanodine Receptor
RYS e: Read Your Screen [Internet]
RZ d: Rechenzentren
- d: Regionales Rechenzentrum der Universität Stuttgart
- e: Range Zone
- e: Recovery Zone
- s: Regimiento de Zapadores
Rzdo s: Rezagado
RZS e: Range Zero Set

S

S a: Sekund
- a: Socialdemokratiet
- d: Seminar
- d: Staatsexamen
- d: Studienleistung
- e: Quality mark of SPIB
- e: S reference point
- e: Senate
- e: Sleeve
- e: Sulfur
- e: Supersaturation
- e: Supplement[al]
- f: Salariés
- f: Salle
- f: Samedi
- f: Sanitaire
- f: Seconde
- f: Seine
- f: Semestriel
- f: Sépulture
- f: Siècle
- f: Société
- f: Soir
- f: Sortie
- f: Sous
- f: Spécial
- f: Stère
- f: Supplément
- f: Sûreté
- f: Suspect
- N: Samvirkelag
- N: Spreng
- r: Sekond
- r: Sida
- r: Signalregemente
- r: Söder
- r: Södra
- r: Spaningsflygplan
- r: Syd
- r: Sydlig
- s: Sección
- s: Secreto
- s: Segundo
- s: Seguro
- s: Soldado
- s: Su
- s: Submarino
- s: Sur

s d: Streumatrix
- d: Streuzahl
- e: smiling [Internet]
- r: spets-

s. d: siehe
- e: Shilling
- e: succeeded
- N: sørlig
- s: siglo

S. e: Sequenced
- e: Side
- N: Sjøforsvaret
- N: Syd
- N: Sør
- N: Sørlandet

s/ f: sur

S+ e: S-Plus statistical software package

s- r: Spanings-

S- r: Spräng-

S1 d: 1. Heu- bzw. Silageschnitt
- e: Personnel
- r: 1. Signalregemente
- r: Upplands Signalregemente

S-1 e: Secretary of Energy

S2 d: 2. Heu- bzw. Silageschnitt
- e: Intelligence
- r: Göta Signalregemente

S3 e: Sealift Support System
- e: Semiconductor Serial Storage
- e: Starboard Truss Section 3
- e: Training/Operation
- r: Norrlands Signalbataljon

S-3 e: Viking

S4 e: Supply/Logistics

S5 e: Civil Affairs

S-5 s: Armys Civilian Affairs Unit

S1100 e: Sperry 1100 series mainframe computer

SA d: Kurzschreibweise für die spektrale Strahlungsfunktion der Normlichtart A
- d: Staatsarchiv
- e: Anonymous Society
- e: Republic of South Africa
- e: Sabreliner
- e: Safety Analysis
- e: Scientific Area
- e: Scientific Atlanta
- e: Security Architecture
- e: Seismic Analysis report
- e: Self-Activating
- e: Semen Analysis
- e: Sensitivity Analysis
- e: Service Assessment
- e: Service Assistant
- e: Service Availability
- e: Simple-Adjoint
- e: Single Access
- e: Sino-Atrial
- e: Situation Assessment
- e: Situation[al] Awareness
- e: Small Animal
- e: Source Adress
- e: Space Available
- e: Spaced Antenna
- e: Speech Article
- e: Sport Assembly
- e: Sputter-Anneal
- e: Storage Analyzer
- e: Strike Assessment
- e: Structured Analysis
- e: Student Assistant
- e: Student Association
- e: Subarea
- e: Subresolution Assist
- e: Sunshine Act
- e: Supportability Analysis
- e: Surface Attack
- e: Surface to Air
- e: System Administration
- e: Systems Automation
- f: Saisie-Arrêt
- f: Salarié Agricole
- f: Sans Adresse
- f: Sarre
- f: Scrutin d'Arrondissement
- f: Secrétaire d'Administration
- f: Secrétaire d'Ambassade
- f: Section Administrative
- f: Service Administratif
- f: Service Agricole
- f: Service Ambulant
- f: Service des Arbitrages
- f: Servitude d'Alignement
- f: Solde d'Absence
- f: Somme Avancée
- f: Sous-peuplement Accentué
- f: Statistique Agricole
- f: Surface Agraire
- f: Surpeuplement temporairement Admissible
- f: Syndicat Agricole
- f: Syndicat Autonome
- N: Sjøforsvarets Artilleri
- N: Statiske Analyser
- r: Svårt Artilleri

- s: Sector Aéreo
Sa e: Sarcoma
- f: Sans année indiquée
SÅ a: Socialstyrelsen-Åndssvageforsorgen
sa. a: samme
SA. e: Selection Addressing
Sa. s: sábado
.sa e: Saudi Arabia [Internet]
s.a. s: sin año
s.å. N: samme år
- r: samma år
S&A e: Sampling and Analysis
S/A e: Sensor/Actuator
- e: Services and Agencies
- f: Sur Aube
- s: Semiautomático
- s: Sin Anexos
s:a r: södra
SA3 e: Security Assistance Automation Army
SA90 e: IPCC Scenario A, 1990
SAA d: System-Anwendungsarchitektur
- e: Department of Agriculture, Adelaide
- e: Satellite Accumulation Area
- e: Satellite Active Archive
- e: Single Access Antenna
- e: Single Agency Action
- e: Single Assignment Approach
- e: Solution-Annealed and Aged
- e: South Atlantic Anomaly
- e: Standard Application Architecture
- e: Standards Association of Australia
- e: Standards Australia
- e: Static Automated Analysis
- e: Sub-Aqua Association
- e: System[s] Application Architecture
- f: Service des Abonnés Absents
- f: Service des Activités Aériennes
- s: Sección de Apoyo Aéreo
SAAA e: Simulated Air-to-Air Attack
SAAAPA e: Student Academy of the American Academy of Physician Assistants
SAAB a: Sammenslutningen af Aften- og Ungdomskolelærere i Arbejdsteknik, Afspænding og Bevægelselære
- f: Syndicat des Agents de l'Administration du Burundi
- r: Svensk[a] Aeroplan AktieBolaget
SAABTS e: South African Army Bantu Training School
SA ADC e: Successive-Approximation ADC
SAADTA e: SA-ALC Damage Tolerance Analysis
SAAEB e: South African Atomic Energy Board
SAAFR e: Standard use Army Aircraft Flight Route
SAAHS e: Stability Augmentation & Altitude Hold System
SAAIF e: Sequential Access Auxiliary Interface File
SAAL e: Signal[l]ing AAL [ATM Adaptation Layer]
SAALA e: Southern African Applied Linguistics Association South Africa
SAALC e: San Antonio Air Logistics Center
SAAM e: Special Assignment Airlift Mission
SAAMEX e: Surface and Atmosphere Airborne Microwave Experiment
SAAMI e: Sporting Arms and Ammunition Manufacturers Institute
SAAMP e: Separate Absorption and AMplification Photodiode
SAAMS e: Sample Analyses and Management System, Region 4
SAANETS e: South African Army Non-Effective Troops Section
SAAQS e: State Ambient Air Quality Standards
SAAR e: Seasonally Adjusted at Annual Rates
SAARC e: South Asian Association for Regional Co-operation
SA'ARI e: Southern African Atmosphere Research Initiative
SAASS e: South African Airborne Scanner Survey
SAATN e: Shuttle/Atlas Advanced TIROS-N Satellite
SAAWU e: South African Allied Workers Union
SAB e: Sensor/Actuator Bus
- e: Software Abort
- e: Standards Activities Board
- e: Synoptic Analysis Branch
- p: Sector Autônomo de Bissau
Sab N: Sabotasje
SABC e: South Africa Broadcasting Corporation
SABEL e: State Administrative Board of Election Laws
SABEN f: Société Anonyme Belge d'Études Nucléaires
SABENA f: Société Anonyme Belge d'Exploitation de la Navigation Aérienne
SAbfG d: Saarländisches Abfallgesetz
SABINI e: Commercial library system
SABIT e: Special American Business Internship Training
SABLE e: Simulator Atmospheric Boundary Layer Environment
SABM. e: Set Asynchronous Balance Mode
SABMAWU e: South African Black Municipal and Allied Workers Union
SABME e: Set Asynchronous Balanced Mode Extended
SABRA A: Suid-Afrikaanse Buro vir Rasse Aangeleenthede
SABRE e: Semi Automated Business Research Environment
- e: South Atlantic Bight Recruitment Experiment
- e: Support Airlift By Remote Entry
- e: Sweden and Britain Radar Experiment
SABREX e: Satellite-Borne Radiometer Experiment
SABS s: Servicio Auxiliar de Bienestar Social
sabs d: Absolute spektrale Strahlungsempfindlichkeit
SABSWA e: South African Black Social Workers Association
SABTA e: South African Black Taxi Association
SAC e: Sacramento Area Office
- e: Satellite Application Centre
- e: Scientific Advisory Committee
- e: Secretarial Advisory Board
- e: Secretarial Advisory Committee
- e: Security Applications Center
- e: Security Approval Certificate
- e: Service Access
- e: Service Access Code
- e: Service Area Code
- e: Short-run Average Cost
- e: Single Attached Concentrator
- e: Single Attachment Concentrator
- e: Society of African Culture
- e: Space Advisory Committee
- e: Space Application Centre
- e: Special Access Clearance
- e: Special Agent [in] Charge
- e: Special Analysis Center
- e: Special Analysis Centre
- e: Strategic Air Command
- e: Support Agency Coordinator
- f: Centre Spécial d'Analyse
- f: Conseil consultatif scientifique
- f: Secrétariat général à l'Aviation Civile
- f: Service d'Action Civique
- f: Service d'Approvisionnement des Coopératives
- s: Centro Especial de Análisis
- s: Consejo Consultivo Científico
SAC ADA e: SAC [Strategic Air Command] Airborne Data Automation
SAC ADVON e: SAC [Strategic Air Command] Advanced Echelon
SAC APS e: SAC [Strategic Air Command] Adaptive Planning System
SACARMS e: SAC [Strategic Air Command] Aircrew Resource Management System
SACBD e: Scientific Advisory Committee on Biological Diversity
SACC e: Shore ASW Command Centers
SACCA e: Strategic Air Command Communications Area
SACCAIM e: South African Climate Change, Analysis, Interpretation and Modelling
SACCAWU e: South African Commercial Catering and Allied Workers Union
SACCH e: Slow Associated Control Channel
SACCOLA e: South African Consultative Committee on Labor Affairs
SACCOP e: SAC [Strategic Air Command] Continuity of Operations Plan
SACCS DTS e: SAC [Strategic Air Command] Automated Command and Control System Data Transmission Subsystem
SACEADM e: System Auxiliary Control Element Administration
SAC ECMP e: SAC [Strategic Air Command] Engine Condition Monitoring Program
SACECP e: System Auxiliary Control Element Common Pair

SACEP e: South Asia Cooperative Environment Program
- f: Programme Coopératif sur l'Environnement pour l'Asie du Sud
- s: Programa Cooperativo de Asia Meridional para el Medio Ambiente

SachBerG d: Sachbereinigungsgesetz
SachBezV d: Verordnung über den Wert der Sachbezüge in der Sozialversicherung [Sachbezugsverordnung]
SachenRAndG d: Sachenrechtsänderungsgesetz
SächsWG d: Sächsisches Wassergesetz
SACI e: Space Allocation and Capital Improvements [Committee]
SACINTNET e: SAC [Strategic Air Command] Intelligence Network
SACK e: Scientific Advisory Committee on Kangaroos
- e: Selective Acknowledgement

SACL e: South African Confederation of Labor
SACLA e: South African Christian Leadership Assembly
- e: South African Christian Leadership Association

SAC/LGL e: SAC/LGL Contingency Planning and Management System
SACLOS e: Semi-Active Command Line Of Sight guidance
SACM e: SAC [Strategic Air Command] Manual
- f: Société pour les Assurances des Coopératives et de leurs Membres

SACMA f: Société d'Applications et de Constructions pour le Matériel Automobile
SACNET e: Secure Automatic Communication Network
SACOB e: South African Chamber of Business
SACOL e: South African Confederation of Labor
SACOM e: Surveillance for Aircraft and Cruise Missile
SACOP e: SAC [Strategic Air Command] Continuity of Operations Plan
SACOS e: Safeguards Accountability Computer System
- e: South African Council on Sports
- e: Strategic Air Combat Operations Staff

SACP e: SAC [Strategic Air Command] Pamphlet
- e: South African Communist Party

SACPAS e: SAC [Strategic Air Command] Primary Alerting System
SACR e: SAC [Strategic Air Command] Regulation
- e: Sacramento Basin

SACRD e: Safety Analysis Core Related Data
SACRIN e: Strategic Air Command Readiness Information Network
Sacro e: Swaziland Association for Crime Prevention and Rehabilitation of Offenders
SACRS e: South Australia[n] Centre for Remote Sensing
SACS e: Small Area Census Studies
- e: Structuring and Composition System
- e: Surveillance Analysis Computer System

SACSA e: South African Conservative Students Association
- e: Special Assistant for Counter-Insurgency and Special Activities

SACSO e: SAC [Strategic Air Command] Security Office
SACSS e: South African Central Statistics Office
SACTELNET e: SAC [Strategic Air Command] Telephone Network
SACTTYNET e: SAC [Strategic Air Command] Teletypewriter Network
SACTW e: South African Council of Transport Workers
SACTWU e: South African Clothing and Textiles Workers Union
SACU e: South African Customs Union
SACV f: Société Anonyme à Capital Variable
SACVD e: Sub-Atmospheric Chemical Vapor Deposition
SACWARNS e: Strategic Warning System
SACWU e: South African Chemical Workers Union
SACZ e: South Atlantic Convergence Zone
SAD d: Schwandorf, Bayern
- d: Sonderabfalldeponie
- e: EMSL-RTP Acid Rain System
- e: Safety Analysis Diagram
- e: Safety Assessment Document
- e: Spaced Antenna Drift
- e: Special Activities Division
- e: Strategic Air Defense
- f: Sans Achat de Devises
- s: Sistema Automatizado de Dirección

SAd f: Service Administratif
s'ad f: s'adresser à
SADAA e: Science Assistant to the Deputy Assistant Administrator
SADC e: Southern Africa[n] Development Community
SADCC e: Southern African Development Coordination Conference
- e: Southern African Development Coordination Council

SADCO e: South African Data Centre for Oceanography
- f: Centre Sud-Africain de Données pour l'Océanographie
- s: Centro Sudafricano de Datos Oceanográficos

SADD e: Students Against Drunk Driving
SADE e: Signal Analysis and Dissemination Equipment
- f: Société Africaine pour le Droit de l'Environnement
- f: Société d'Adduction et de Distribution d'Eau

S.A. de C.V. s: Sociedad Anónima de Capital Variable
SADELM e: South Australian Department of Environment and Land Management
SADEP e: South Australian Department of Environment and Planning
SADF e: South African Defense Force
- e: South Australian Department of Fisheries

SADHA e: Students American Dental Hygienists Association
SADHS e: South African Demographic and Health Survey
SADIFS e: SAC [Strategic Air Command] Automated Deployable Intelligence Fusion System
SADL e: Situation Awareness Digital Link
- e: Synchronous Data Link Control

SADO e: Senior Air Defense Officer
- f: Systèmes, Aides et dispositifs pour l'acquisition de Données Océaniques

SADRS e: Simplified Autonomous Data Receiving Station
SADS e: Safety Assessment Documents
- N: Sjøforsvarets Administrasjonsskole

SADT e: Structured Analysis and Design Techniques
SADTU e: South African Democratic Teachers Union
SAE e: Standard American English
- e: Standard Average European [languages]
- f: Service des Affaires Économiques
- s: Sección Avanzada de Etapas

SÄE d: Schweizerischer Ärztlicher Ehrenrat
SA&E e: Safety Analysis and Engineering
SAEB e: Schweizerische Arbeitsgemeinschaft zur Eingliederung Behinderter
SAEC e: Sport Assembly Executive Committee
SAED e: Selective Area Electron Diffraction
- f: Société d'Aménagement et d'Exploitation du Delta du Fleuve Sénégal

sædv. a: sædvanlig[vis]
SAEI f: Service des Affaires Economiques et Internationales
SAEM e: Society for Academic Emergency Medicine
SAEMA f: Société Alsacienne d'Électronique et de Mécanique Appliquée
SAEN f: Syndicat Autonome de l'Éducation Nationale
SAEOC e: Second Alternate Emergency Operations Center
særsk a: særskilt
SAES f: Syndicat Autonome de l'Enseignement Supérieur
SAEWG e: Standing Air Emission Work Group
SAF e: Safety and Environmental Protection Division
- e: Safety Appraisal/audit Finding/observation
- e: Sample Analysis Form
- e: Sample Authorization Form
- e: Satellite Application Facility
- e: Schizophrenia Australia Foundation
- e: Second Amendment Foundation
- e: Security Authorization Facility
- e: Service Access Facilities
- e: Southern, Central and Eastern Africa
- e: Spec Airfield File
- e: Special Applications File
- e: Strategic Army Force
- f: Service Administratif et Financier
- f: Sommation Avec Frais
- N: Senter for Anvendt Forskning

- N: Skipsfartens Arbeidsgiverforening
- r: Svenska Arbetsgivarföreningen
- s: Servicio Aerofotogrammetrico de la Fuerza aerea

SAFAA f: Société Anonyme Française des Appareils Automatiques
SAF/AL e: Secretary AF/Acquisition and Logistics
SAFARI e: Southern African Fire Atmosphere Research Initiative
- e: Swedish and French Auroral Radar Investigation
- f: Système Automatisé pour les Fichiers Administratifs et le Répertoire des Individus

SAFARI-92 e: South African Fire-Atmosphere Research Initiative
SAFCOC e: Southern African Federated Chambers of Commerce
SAFCOM e: Southern, Central and Eastern Africa Regional Committee for START
SAFE e: Secure Analyst File Environment
- e: Selected Area For Evasion
- e: Smokefree Air For Everyone
- e: Support to Analysts File Environment
- e: Supportability Analysis Forecasting and Evaluation
- e: System Analyst File Environment

SAFENET e: Survivable Adaptable Fiber Optic Embedded Network
SAFER e: Spectral Application of Finite Element Representation
- e: Streamlined Approach For Environmental Restoration
- f: Société d'Aménagement Foncier et d'Établissement Rural

SAFETY e: Sun Awareness For Educating Todays Youth
SAFF e: Sodium Aerosol Filter Fixture
SAF/FM e: Secretary AF/Financial Management
SAF/FMD e: Department Asstistent Secretary AF
SAFIC e: South Australian Fishing Industry Council
SAFIRE e: Spectroscopy of the Atmosphere Using Far Infrared Emission
SAFISY e: Space Agency Forum for /on [the] International Space Year
SAFNETU e: South African Federation of Non-European Trade Unions
SAFPA e: Society of Air Force Physician Assistants
SAFR r: Svenska Arméns och Flygvapnets Reservofficersförbund
S.Afr. e: South Africa[n]
SAFREP f: Société Anonyme Française de Recherches et d'Exploitation du Pétrole
S Afr J Commun Disord e: South African Journal of Communication Disorders. Die Suid-Afrikaanse Tydskrif Vir Kommunikasieafwykings
S Afr J Surg e: South African Journal of Surgery
S. Afr. J. Wildl. Res. e: South African Journal of Wildlife Research
S Afr Med J e: South African Medical Journal
SAFRS e: Saudi Arabian Facility for Remote Sensing
SAFSTOR e: Safe Storage
SAFT e: Simple Asynchronous File Transfer Protocol
SAG d: Software AG
- e: Scientific Advisory Group
- e: SQL Access Group
- e: Street Address Guide
- e: Sub-Activity Group
- f: Salaire Annuel Garanti
- f: Service de l'Administration Générale
- f: Service des affaires générales

SAg f: Société Agréée
SAGA e: Solar Array Gain Augmentation
- e: Soviet-American Gas[es] and Aerosol[s] [experiment]

SAGA-3 e: Third Soviet-American Gas and Aerosol
SAGACE f: Système Automatique pour la Gestion et l'Echange des Comptes Economiques
SAGAES e: Safeguards Automated Graphics Analysis and Evaluation System
SAGASCO e: South Australian Gas Company
SAGE e: [UNIX] System Administrators Guild
- e: Stratospheric Aerosol [and] Gas Experiment
- e: System Administrators Guild

SAGE-II e: Stratospheric Aerosol and Gas Experiment II
SAGE III e: Stratospheric Aerosol and Gas Experiment III
sagf. a: sagfører
SAGK d: Schweizerische Arbeitsgemeinschaft der Gewerkschaften des Krankenpflegepersonals
SAGRI e: Substance Abuse, Growth and Recovery Institute
SAGSTRAM e: Standing Advisory Group on Safe Transportation of Radioactive Material
SAH d: Schweizer Archiv für Heraldik
SAHB d: Schweizerische Arbeitsgemeinschaft Hilfsmittelberatung für Behinderte und Betagte
SAHC e: South Asia Hydrological Cycle
SAHFOS e: Sir Alister Hardy Foundation for Ocean Science
- s: Fundación Sir Alister Hardy de Ciencias Oceánicas

SAHRS e: Standard Attitude and Heading Reference System
SAHS e: School of Allied Health Sciences
SAHSSO e: South African Health and Social Services Organisation
SAHSU e: Small Area Health Statistics Unit
SAHWCO e: South African Health Workers Congress
SAI e: Scanning Auroral Imager
- e: Serving Area Interface
- e: Software Architects Inc.
- e: Solution Assay Instrument
- e: Special Assistance Initiative
- e: Supreme Audit Institution
- e: Systems Application Institute
- f: Service des Affaires Internationales
- f: Société Astronautique Internationale
- f: Syndicat des Agences d'Information
- N: Sjøforsvarets Artilleriinspeksjon

2 SAI e: Second South African Battalion Group
SAIAC f: Société Anonyme Internationale de l'Asbeste-Ciment
SAIC e: Science (or Scientific) Applications International Corporation
- e: South African Indian Council
- e: Special-Agents-in-Charge

SAID e: Safety Action Item Database
SAIDAS s: Sistema Integral de Adquisicion de Datos de Satelite
SAIDCOR e: South Africa Inventions Development Corporation
SAIDI e: Spaced Antenna Imaging Doppler Interferometer
SAIE e: Special Acceptance and Inspection Equipment
SAIF e: Spatial Archive Interchange Format
- e: Spec Automated Installation File

SAIGN A: Suid-Afrikaanse Instituut vir Geskiedenisnavorsing, Pretoria
SAIH N: Studentenes og Akademikernes Internasjonale Hjelpefond
SAIIA e: South African Institute of International Affairs
SAIKN A: Suid-Afrikaanse Instituut vir Kommunikasienavorsing, Pretoria
SAIL e: Scattering from Arbitrarily Inclined Leaves
- e: Serial ASCII Instrumentation Loop
- e: Stanford AI Laboratory
- e: Steel Authority of India Limited

SAILS e: Standard Army Intermediate Level Supply System
SAINO A: Suid-Afrikaanse Instituut vir Navorsingsontwikkeling
SAION A: Suid-Afrikaanse Instituut vir Opvoedkundige Navorsing
SAIR A: Suid-Afrikaanse Indienraad
- e: Synthetic Aperture Interferometric Radiometer

SAIRE e: Scaleable Agent-based Information Retrieval Engine
SAIRR e: South African Institute of Race Relations
SAIRS f: Système Automatisé d'Information sur la Recherche en Santé
SAIS e: Science and Applications Information System
- e: Site Alarm and Information System
- e: South African Information Service
- e: Standard Automated Information Systems
- e: Staring Array Imaging Spectrometer

SAISAIU e: South African Iron, Steel and Allied Industries Union
SAIT f: Société Alsacienne d'Installations Techniques

SAITALEK A: Suid-Afrikaanse Instituut vir Taal, Lettere en Kuns, Pretoria
Saitex e: South African International Trade Exhibition
såk. a: såkaldt
SAKAB r: Svenska Avfallskonvertering AB
SAKDC e: Swiss Army Knife Data Compression
sakf. N: sakfører
SAKFØS a: Sammenslutningen af Kontrol-foreninger for Økonomisk Svinefodring
SäkI r: Säkerhetsinstruktionen för vapen och ammunition
SAKK A: Suid Afrika Kaapse Korpsopleidingseenheid
Säktj r: Säkerhetstjänst
Säkv r: Säkerhetsvinkel
SAKVA A: Suid-Afrikaanse Konfederasie van Arbeid
SAL A: Suid-Afrikaanse Lugdiens
- d: Saarland Landesregierung und Landtag
- e: Self Addressed Envelope
- e: Semware Applications Language
- e: Shielded Analytical Laboratory
- e: Shift Arithmetic Left
- e: Strategic ALCM launcher
- e: Synthetic Aperture Ladar
Sal f: Social
Sal. s: Salida
SALA a: Sammenslutningen af Landbrugets Arbejdsgivere
- e: Sound and Light Asia
SalArgos e: Argos-located buoy equipped with temperature and conductivity sensors
SALCC e: Satellite Airlift Control Center
SALDRU e: South African Labour and Development Research Unit
SALDS e: State-Approved Land Disposal Structure
SALF e: Somali Abbo Liberation Front
- f: Société Anonyme anversoise de Liaisons Fluviales
- r: Sveriges Arbetsledarförbund
Salf. e: Salford
SALH e: Semi-Active Laser Homing
Sal Hebd f: Salaire Hebdomadaire
SALI e: Standalone Automatic Location Identification
- e: Subnetwork Adjacent Layer Interface
SALIC e: South Australian Land Information Council
SALID e: Solar and Lunar Illumination Data
SALIN e: Self-Aligned Laser Inertial Navigator
SALIS e: Security Assistance Logistics Information System
- e: Set-Aside Land Information System
säll. r: sällan
SALM A: Suid-Afrikaanse Lugmag
SalMens f: Salaire Mensuel
SALMON e: Satellite remote-sensing for Lake Monitoring
- e: Semianalytic Monte Carlo Radiative Transfer Simulation Model
SALog e: Sent Alarm Log
SALS e: Special Analysis and Laboratory Systems
sal/t f: salaire à la tonne
SALT e: Savannas in the Long Term
- e: SPINSAT Altimeter
- e: Strategic Arms Limitation Treaty
- e: Strategic Arms Limitting Talks
SALU d: Stadtarchiv Luzern
Salud Publica Mex s: Salud Publica de Mexico
SALUTE e: Size, Activity, Location, Uniform (or Unit), Time, Equipment
Salv s: Salvamento
SALVA s: Ejército de Salvación de Guatemala
Salvpj r: Salvpjäs
SAM a: Selskabet for Arbejdsmiljø
- e: Sample Analysis Method
- e: Scanning Auger Microprobes
- e: Scientific American Medicine
- e: SCSI Architecture Model
- e: SCSI-3 Architecture Model
- e: Security Account[s] Manager
- e: Sensing with Active Microwaves
- e: Sequential Access Mode/Method
- e: Shipboard sampling computer program
- e: Shuttle Activation Monitor
- e: Single Application Mode
- e: Smart ATM Modul
- e: South African Museum, Capetown
- e: South Australian Museum
- e: Staging Area Manager
- e: Stationing Analysis Model
- e: Stereo Amplifier
- e: Stratospheric Aerosol Measurement[s] [experiment]
- e: Substitute Alloy Materials
- e: Surface Area Model
- e: Sustainability Assessment Module
- e: Symantec Antivirus for Macintosh
- e: System Administration Manager
- f: Service des Affaires Musulmanes
- f: Société Africaine de Mines
- f: Soins par Auxiliaires Médicaux
- r: Självgående Akustiskt-magnetiskt Minsvep
SAMA e: Scientific Apparatus Maker's Association
- e: South African Musicians Alliance
- e: South Australia Museum, Adelaide
- e: Step-by-step Automatic Message Accounting
- f: Syndicat national des Appareils de Mesure et d'Automatisme
SAMAC e: Standing AIRS Management Advisory Committee
SAMANGOKY f: Société Malgache d'Aménagement et la Mise en Valeur de la Vallée du Bas Mangoky
SAMAS e: Semi-Automated Mesoscale Analysis System
- f: Société Anonyme des Machines à Statistiques
samb N: samband
SAMBA a: Samfundsvidenskabelige Basisstuderende
- e: Society of Ambulatory Anesthesia
samb c r: sambandschef
SAMBO e: Simultaneous Auroral Multi-Balloons Observations
sambtjC r: sambandstjänstchef
SAMDS e: Serbian American Medical and Dental Society
SAMEER e: Society for Applied Microwave Electronics Engineering and Research
SAMER f: Société d'Armement et de Manutention de la Mer Rouge
- f: Système Automatique de Mesure de l'Évaporation Regionale
SAMEX e: Shuttle Active Microwave Experiment
SAMF s: Sindicato de Acción y Mejoramiento Ferrocarrilero
samf. a: samfund
SAMFARS e: Stand-Alone Automated Message Formatting
SAMH e: Standard Automated Message Handling
SAMHS e: Standard Automated Message Handling System
SAMHSA e: Substance Abuse and Mental Health Services Administration
SAMI e: System Acquisition Management Inspections
SAM II e: Stratospheric Aerosol Measurement
SAM-II e: Stratospheric Aerosol Measurement II
SAMIR e: Satellite Microwave Radiometer
SAMKAT e: Catalogue cooperation project
saml. a: samling
samlpl. N: samlingsplass
SAMM p: Serviços de Assistência Médica Militar
SAMMA e: Spectral-Agronomic Multisite-Multicrop Analyses
SAMO e: Sensitivity Analysis of Model Output
SAMOB e: Surface-to-Air Missile Order of Battle
SAMOFOR p: Sociedade Argelino-Moçambicana de Florestas
SAMOS e: Stacked-Gate Avalanche-injection Metal-Oxide-Semiconductor
SAMP e: Single Acquisition Management Plan
SAMPEX e: Solar, Anomalous and Magnetospheric Particle Explorer
SAMPI A: Suid-Afrikaanse Mielieprodusente-Instituut
- e: South Africa Maize-Producing Institute
SAMS e: South African Medical Service
- e: Sow Abortion and Mortality Syndrome
- e: Space Acceleration Measurement System
- e: Spatial Analysis and Modeling System
- e: Standard Army Maintenance System
- e: Standard Automated Maintenance System
- e: Stratospheric and Mesopheric Sounder
- e: Surface Automated Meteorological System[s]
- e: System Application Management System
sams r: sammansatt
SAMSA e: Substance Abuse and Mental Health Services Administration

SAMSARS e: Satellite Based Maritime Search and Rescue System
SAMSIM e: SAM Simulation
SAMSO A: Suid-Afrikaanse Mieliespesialiteite Organisasie
- e: South Africa Maize Specialty Organization
samsp r: sammansättningsparti
SAMSS e: Space Assembly Maintenance and Servicing Study
samt. a: samtidig
SAMTRAC e: Sample Analysis Management Tracking system
SAMU e: South African Mining Union
- f: Service d'Assistance Médicale d'Urgence
samv r: samverkan
samv. a: samvirkende
SAMW d: Schweizerische Akademie der Medizinischen Wissenschaften
- e: Subantarctic Mode Water
SAMWU e: South African Mineworkers Union
SAN d: Schulen ans Netz e.V.
- d: Schweizerische Arbeitsgemeinschaft Nichtraucher
- e: Satellite Access Network
- e: Senior Advocate of Nigeria
- e: Standard Address Number[ing] [system]
- e: Styrene/Acrylonitrile
- e: System Advisory Notices
- e: System Area Network
- f: Serveur d'Archivage Numérique
- f: Société Aéronautique Normande
- s: Sanidad
san N: sanitet
SANA d: Sanierung der Atmosphäre
- e: Santa Ana Basin
SANAC e: South African Native Affairs Commission
SANAE d: Südafrikanische Nationale Antarktische Expedition
- e: South African National Antarctic Expedition
SANAGA f: Système d'Acquisition Numérique pour l'Analyse des Grains Africains
SANAS e: South African National Accreditation System
SANC e: South African Native Congress
- e: South African Nursing Council
SANCO e: South African National Civics Organisation (or Organization)
SAND e: Spectro-Angular Normalized Difference
SANDF e: South African National Defence (or Defense) Force
SANDIA e: Sandia National Laboratory
SANDRAMINE e: Haut Sassandra Mining Company
- f: Compagnie Minière Haut Sassandra
SANE e: Standard Apple Numeric Environment
- e: Sulfur and Nitrogen Emissions
SANEL e: South African National Epilepsy League
Sangrp HFT N: Sanitetsgruppen av Hærens Forsyningstropper
SANIL f: Système d'Analyse des Images LANDSAT
SANJ e: San Joaquin-Tulare Basins
San. kp. N: sanitetskompani
SANLAM e: South African National Life Assurance Company
SANNC e: South African Native National Congress
SANOS e: Stabilization Airborne Night Observation System
SANPU e: Southern African Newspaper Press Union
SANROC e: South African Non-Racial Open Committee for Olympic Sports
SANS e: Scale for the Assessment of Negative Symptoms
- e: Small Angle Neutron Scattering
SANSCO e: South African National Students Congress
SANT e: Santee Basin and Coastal Drainages
Sant. e: Santonian
SANU e: Somali African National Union
SANZ d: Schweizerische Arzneimittel-nebenwirkungszentrale
SAO e: Safety Assessment Office
- e: Security Assistance Office
- e: Southern Association for Oncology
- e: Surface Aviation Observation
- f: Syndicat Autonome Ouvrier
SAOCC e: Sector Air Operations Coordination Centers
SA-OCRE e: Stand-Alone Optical Character Recognition Equipment
SAOERN f: Société Anonyme d'Organisation d'Études et de Réalisations Nuclaires
SAOEU e: Strike/Attack Operational Evaluation Unit
SAP a: Socialistisk Arbejderparti
- d: Sammelstelle arbeitsloser Physiker
- d: Schrecken, Angst, Panik
- d: Software Aus Persien
- d: Systeme, Anwendungen, Produkte [AG], Walldorf
- d: Systems, Applications and Products in Data Processing
- e: Sample (or Sampling) Analysis Plan
- e: Send And Pray
- e: Sensor AVE Package
- e: Service Access Point
- e: Service Access Protocol
- e: Service Advertising Protocol
- e: Service Assessment Pool
- e: Seychelles Press Agency
- e: Simple Asynchronous Protocol
- e: Simplified Acquisition Procedures
- e: South African Party
- e: State Assessment Panels
- e: Strong Anthropic Principle
- e: Structural Adjustment Program
- e: Subassembly Profilometer
- e: Sustainable Agricultural Practices
- e: Symbolic Assembler Program
- e: Symbolic Assembler Programm
- e: Syndrome of Arthrogryposis and Palatoschisis
- e: System Access Protocol
- e: Systems, Applications and Products [in Data Processing]
- f: Sans Avis Préalable
- r: Socialdemokratiska Arbetarpartiet
SAP. e: Service Access Point
S&AP e: Sampling and Analysis Plan
SAPA A: Suid-Afrikaanse Pers-Assosiasie
- e: Society of Army Physician Assistants
SAPC e: Southern Association for Primary Care
SAPCD e: Sacramento Air Pollution Control District
SAPE e: Survivable Adaptive Planning Experiment
SAPEHDAK f: Salon de l'Agriculture, de la Pêche, de l'Elevage et de l'Hydraulique de Dakar
SAPF e: South African Permanent Force
SAPFU e: Surpassing All Previous 'Foul' Ups
SAPH f: Société Africane de Plantations d'Heveas
SAPHYDATA e: System for Acquisition, Transmission and Processing of Hydrological Data
- e: Système d'Acquisition, de Transmission et de Traitement des Données Hydrologiques
- s: Sistema de Adquisición, Transmisión y Tratamiento de Datos Hidrológicos
SAPI e: Service Access Point Identifier
- e: Service Access Protocol Interpreter
- e: Speech API [Application Program Interface]
SAPIG f: Syndicat des Agences de Presse d'Informations Générales
SAPL e: South African Public Libraries
SAPMER f: Société Anonyme des Pêches Maritimes et de Ravitaillement
SAPO e: Subordinate Area Petroleum Officer
SÄPO r: Säkerhetspolisen
SAPOCO e: Safety Policy Committee
SAPPI e: South African Paper and Pulp Industries
Sapp. kp. N: Sappørkompani
SAPS e: Scale for the Assessment of Positive Symptoms
- e: Shippingport Atomic Power Station
- e: South African Police Service
SAPSE e: South African Post-Secondary Education
SAPU e: South African Police Union
SAPWU e: Swaziland Agriculture and Plantation Workers Union
SAQ e: Self Assessment Question
- e: South Atlantic Quarterly
- e: SQL Access Group
SAQA e: South African Qualifications Authority

SAR e: International Convention on Maritime Search and Rescue
- e: Safeguards Analysis Report
- e: Safety Analysis Report
- e: Safety Analysis Review
- e: Sample Analysis Request
- e: Schedule Add Requests
- e: Search and Rescue [Instruments]
- e: Second Assessment Report
- e: Segmentation And Reassembling
- e: Segmentation And Reassembly [sub-layer]
- e: Selected Acquisition Report
- e: Selected Acquisition Review
- e: Shift Arithmetic Right
- e: Significant-Activity Report
- e: Spares Adjustment Request
- e: Start Action Request
- e: Structural Activity Relationship
- e: Subcommittee on Atmospheric Research
- e: Supplied Air Respirator
- e: Synthetic Aper[a]ture Radar
- f: Convention Convention internationale sur la recherche et le sauvetage maritimes
- f: Deuxième rapport d'évaluation
- f: Recherche et sauvetage
- s: Búsqueda y salvamento
- s: Convenio Internacional sobre Búsqueda y Salvamento Marítimos
- s: Radar de Apertura Sintética
- s: Segundo Informe de Evaluación

Sar r: Spetsanslagsrör

SA&R e: Safety Analysis and Regulation

SAR-580 e: Erim X-L-Band SAR Flown on CCRS Convair 580

SARA e: Satellite pour Astronomie Radio Amateur
- e: South African Republican Army
- e: State Archives and Records Administration
- e: Superfund Amendment[s] [and] Reauthorization Act
- e: Synthetic Aperture Radar Antenna

SARA-H e: System Analysis and Resource Accounting for Honeywell

SARAH e: Standard Remote to AUTODIN Host
- f: Satellite Radar Hydrologie

SARAH-Lite e: Miniaturized Administrative Version of SARAH

SARAM f: Société d'Applications Radio-électriques à l'Aéronautique et la Marine

SAR and H e: South African Railways and Harbors

SARAS e: Synthetic Aperture Radar Advanced Simulator

SARB e: South African Reserve Bank
- e: Surface and Atmospheric Radiation Budget

SARC e: Space and Atmospheric Research Center
- e: Systems Acquisition Review Council

SARCCUS e: Southern African Regional Commission for the Conservation and Utilization of the Soil

SARCS e: Southeast Asia Regional Committee for START

SARCS-SAP e: SARCS Scientific Advisory Panel

SARD e: Sustainable Agriculture and Rural Development
- e: Swedish Agency Research Cooperation with Developing Countries

SARDA e: Assistant Secretary of the Army for Research, Development and Acquisition
- e: Sodium Removal Development Apparatus

SARDAB e: Search and Aerospace Rescue Data Base

SARDAPF e: Synthetic Aperture Radar Data Acquisition and Processing Facilities

SARDO e: Search and Rescue Deputy Office

SARE e: Segmentation And Reassembly Element
- f: Système Automatique de Radiosondage Embarque en Francais

SAREC e: Science of Anticipation Recognition Evaluation Control of Health Risks
- e: Swedish Agency for Research Cooperation with Developing Countries
- f: Agence suédoise de coopération scientifique avec les pays en développement
- s: Agencia Gubernamental Sueca de Cooperación Científica con los Países en Desarrollo

SAREMCI f: Société Anonyme de Recherches et d'Exploitation Minières en Côte d'Ivoire

SAREX e: Search and Rescue Exercise
- e: Shuttle Amateur Radio Experiment
- e: South American Radar Experiment

SARF e: Seal application and removal form
- e: Spectral Analysis of Resistance Fluctuations
- e: Spectrally Analyzed Resistance Fluctuations

SARG e: Solvent Abuse Resource Group

SARH e: Secretariat of Agriculture and Water Resources of the United Mexican States
- e: Semi-Active Radar Homingmissile
- f: Service d'Armement et de Ravitaillement Hôteliers

SARHP e: South African Railway and Harbor Police

SARHWU e: South African Railway and Harbor Workers Union

SARI e: Synthetic Aperture Radar Industrial
- f: Système Automatique de Réservation Intégrée
- N: Sjøforsvarets Asdic-Radar Inspeksjon

SARIPS e: Search and Rescue Information Processing System

SARIS e: Safeguards Active Response Inventory System
- e: Statutory And Regulatory Information Service

SARJ e: Solar Array Rotary Joint

SARL f: Société à Responsabilité Limitée

SARLAB e: Synthetic Aperture Radar Laboratory

SARM f: Société Aérienne de Recherches Minières

SARME e: Set Asynchronous Response Mode Extended

SAROAD e: Storage and Retrieval of Aerometric Data/AIRS

SARP e: Safety Analysis Report for Packaging
- e: Sardine-Anchovy Recruitment Project
- e: Satellite Acquisition Request Programming
- e: Search and Rescue Processor
- e: Ship Alteration and Repair Package
- f: Projet sur le Recrutement des Sardines et des Anchois
- s: Proyecto de Repoblación de Sardinas y Anchovetas

SARPDU e: Segmentation And Reassembly Protocol Data Unit

SARPMA e: San Antonio Real Property Management Agency

SAR QSAR Environ Res e: SAR [Search And Rescue] and QSAR in Environmental Research

SARS e: Safety Analysis Recall System
- e: Safety Analysis Reports
- e: System Architecture Requirements Specification

SARSAT e: SAR [Search and Rescue] Satellite[-Aided Tracking] [system]
- e: Synthetic Aperture Radar Satellite
- f: Aide à la Recherche et au Sauvetage par Satellite
- s: Localización con ayuda de satélite a efectos de búsqueda y salvamento

SARSDM e: State Agency Remote Sensing Data Management

SARSEX e: SAR [Search And Rescue] Internal Wave Signature Experiment
- e: Synthetic Aperture Radar Signature Experiment

SART e: Safeguards and Accountability Remote Terminal
- e: Search And Rescue Radar Transponder
- e: Society for/of Assisted Reproductive Technology

Sart f: Sarthe

SARTF e: Search and Rescue Task Force

SARTG e: Scandinavian Amateur Radio Teleprinter Group

SARTOC e: Southern African Regional Tourism Council

SAS e: Safeguards And Security
- e: Sales Accounting System
- e: Scientific American Surgery
- e: Single Attached Station
- e: Single Audio System
- e: Solid-State Array Spectroradiometer
- e: Source Assessment System
- e: Southern All Stars
- e: Southern Asian Region
- e: Space Activity Suit
- e: Special Analytical Services
- e: Special Analytical Studies
- e: Statistical Analysis Software
- e: Sub-Arctic Summer
- e: Supply Analysis System
- e: Support Analysis Section
- s: Servicio de Aprovisionamiento Sanguíneo

sas r: så att säga

S As s: Submarino Antisubmarino

s.a.s. r: sist anförda stället
SASANEE e: South and Southeast Asia Network for Environmental Education
SASAWU e: South African Security and Allied Workers Union
SASC e: Semi-Automatic Switch Center
SASCON e: Southern Africa Solidarity Conference
SASCO, Sasco e: South African Students Congress
SASE e: Self Addressed Stamped Envelope
- e: Space Shuttle (or Spacelab) Atmospheric Science Experiment
- e: Specific Application[s] Service Element[s]
SAS en H A: Suid-Afrikaanse Spoorwee en Hawens
SASFA e: South Australian Shark Fishermen's Association
SASI e: Shugart Associates System Interface
- e: Statistical Analysis Systems Institute
SASIAM e: School for Advanced Studies in Industrial and Applied Mathematics
SASJ e: Southern African Society of Journalists
SASM A: Suid-Afrikaanse Staande Mag
- e: South[ern] African Students Movement
- e: Special Assistant for Strategic Mobility
- e: Synthetic Aperture Strip Map
SASO e: South African Students Organisation
SASOL A: Suid-Afrikaanse Steenkool-Olie-en Gaskorporasie
SASP e: SJCS Administrative Support Programs
SASPRSC e: South African Society for Photogrammetry, Remote Sensing and Cartography
SASQUA e: Southern African Association for Quaternary Research
SAS&R e: Satellite Aided Search and Rescue
SASR e: Spare Available Space Required System
SASS e: Seasat A Scatterometer System
- e: SEASAT-A Satellite Scatterometer System
- e: Self-Actuated Shutdown System
- e: Semi-Automatic Switching System
- e: SIOP Additive Spare Support
- e: Source Assessment Sampling System
- e: South Area Safety Support
- e: Subsonic Assessment
SASSSA e: South African Senior Schools Sports Association
SAST e: Safeguards And Security Transition
- e: Scientific Assessment and Strategy Team
SASWC e: SACLANT Anti-Submarine Warfare Center
SAT e: Safety Assessment for Transportation
- e: Satisfiability Problem
- e: Saturday
- e: Scholastic Achievement Test
- e: School of Advanced
- e: Security Alert Team
- e: Simplified Acquisition Threshold
- e: Situational Awareness Terminal
- e: Special Access Termination
- e: Spray Acid Tool
- e: Surface Air Temperature
- e: Synthetic Aperture Telescope
- e: Systems Acceptance Test
- f: Secrétariat Administratif et Technique
- f: Service Autonome de Télédétection
- r: Signaturanpassningstechnik
Sat f: Secrétariat
S At s: Submarino Atómico
SATA e: Society for Advanced Telecommunications in Anesthesia
- e: South Africa Teachers Association
- e: Southern Africa Telecommunications Administration
SATAN e: Security Administrators Tool for Analyzing Networks
- e: System Administrator Tool for Analyzing Networks
SATARCH e: Satellite Archive
SATBVC e: South Africa, Transkei, Bophuthatswana, Venda, Ciskei
SATC e: South Africa Truth Commission
SATCC e: Southern African Transportation and Communications Commission
SATCOM e: Satellite Communications System
SATCOMA e: Satellite Communications Agency
SATCOM System e: Satellite Communications System
SATDAS e: Satellite Data Acquisition System
SATEC p: Sociedade Angolana de Tecidos Estampados
SATEL f: Salon des Télécommunications
SATELIT f: Société des Télécommunications Internationales du Togo
SATEM e: Satellite remote upper-air soundings of pressure, temperature and humidity
- e: Satellite Temperature Sounding
- f: Société d'Applications Thermiques, Électriques et Mécaniques
- f: Sondages Aérologiques de la pression, de la Température et de l'humidité effectués par satellites
- s: Sondeos de altitud mediante teledetección por satélite de presión, temperatura y humedad
SATET f: Société Africaine de Travaux et d'Études Topographiques
SATF e: Shared Access Transport Facility
SATIN e: Survivability Augmentation for Transport Aircraft Now
SAT IV e: Standard AUTODIN Terminal IV System
SATLID e: Single Frequency Atmospheric LIDAR
SATMA f: Société Anonyme de Traitement des Métaux et Alliages
SATMOS e: Service for Archiving and Processing of Meteorological Observations From Space
SATNEB e: Satellite Nebulosity
SATNUC f: Société pour les Applications Techniques dans le domaine de l'énergie Nucléaire
SATO e: Scheduled Airline Traffic Office
SATOB e: Satellite Observations of wind, surface temperature, cloud, humidity and radiation
- f: Observations du vent, de la température en surface, des nuages, de l'humidité et du rayonnement effectuées par satellites
- s: Observaciones mediante los satélites del viento, la temperatura de superficie, nubosidad, humedad y radiación
SATOBS e: Satellite Observations
SATOR e: Stratospheric and Tropospheric Ozone Research
SATRAN e: Satellite Reconnaissance Advance Notice
- e: Satellite Transmission
SATREC e: Satellite Technology Research Center
SATS e: Software Accountability Tracking System
- e: South African Transport Services
- e: Southwest Asia Telecommunications System
- e: Standard Army Training System
SATSCAM e: Satellite Snow Covered Area Measurement
SATSS e: SEASAT Antenna Temperature Software System
SATUC e: South African Trade Union Council
SATUCC e: Southern African Trade Union Coordination Council
SATV e: South African Television
SAU e: Search and Attack Unit
- e: Station Antenna Utilization
- e: Support Analysis Unit
- f: Section Administrative Urbaine
- f: Service d'Architecture et d'Urbanisme
- f: Superficie Agricole Utilisée
- f: Surface Agricole Utilisée
SAV A: Stigting Afrikanervryheid
- d: Schweizer Apothekerverein
- d: Sonderabfallverbrennungsanlage
- e: Saved
- e: Scan And Verify
- e: Service Appointed Validator
- e: State-of-the-Atmosphere Variables
- r: Statens Arbetsgivarverk
- r: Statens Avtalsverk
Sav f: Savoie
- r: Stormartillerivagn
- r: Stridsartillerivagn
SA/V e: Surface Area-to-Volume ratio
SAVANNA e: A process-oriented model of pastoral ecosystems
SAVAR e: Standard Army Validation and Reconciliation System
SAVDM e: Single Application VDM
SAVE e: Satellite Validation Experiment
- e: South Atlantic Ventilation Experiment
- e: Systematic Verification for Entitlements

Savéco f: Savoir économiser
SAVI e: Soil-Adjusted Vegetation Index
SAVIEM f: Société de Véhicules Industriels et d'Equipements Mécaniques
SAVPP f: Société d'Achat et de Vente de Produits Pétroliers
SAVS e: Southern Association of Vascular Surgery
SAW A: Suid-Afrikaanse Weermag
- d: Salzwedel, Sachsen-Anhalt
- e: Sawlog Harvest
- e: Spark Angle Word
- e: Special Agricultural Worker
- e: Sub-Arctic Winter
- e: Surface Acoustic Wave [electronics]
- e: Surface Acoustic Wave[-filter]
SAWB e: South African Weather Bureau
- f: Bureau météorologique sud-africain
- s: Oficina Meteorológica de Sudáfrica
SAWCAA e: Soil and Water Conservation Association of Australia
SAW-CTD e: Surface Acoustic Wave Charge Transfer Device
SAWEGS e: System for Automatic Wind Extraction from Geostationary Satellite Data
SAW&F e: Woods and Forests Department of South Australia
SAWNET e: South Asian Womens NETwork
SAWP e: Society of American Wood Preservers Inc.
SAWPS e: Stand Alone Ware Plan System
SAW/V e: Surveillance and Warning/ Verification
SAWVS e: Satellite Attack Warning and Verification System
SAX e: Selective-Area X-ray Photoelectron Spectroscopy
SaxB r: Saxfjärdsbas
SAXON e: SAR X-Band and Ocean Nonlinearities
SAXON-FPN e: Synthetic Aperture Radar X-Band Ocean Nonlinearity-Forschungs Platform Nordsee
SAXS e: Small-Angle X-ray Scattering
SAYC e: Southern Association of Youth Clubs
SAYRCO e: South African Youth Revolutionary Council
SÄZ d: Schweizerische Ärztezeitung
SÄzK d: Schweizerische Ärztekommission für Notfallhilfe und Rettungswesen
SB a: Statsbibliotek Århus
- e: Salomon Islands
- e: Schottky Barrier
- e: Semi-Balance
- e: Signal Battery
- e: Sound Blaster
- e: Sound Board
- e: South Bend
- e: Specific Behaviour
- e: Standby Base
- e: Standby indicator
- e: State University of New York at Stony Brook
- e: Stop Band
- e: Strong Base ion exchange
- e: Strongback
- e: Support Battalion
- f: Saisie-Brandon
- f: Salaire de Base
- f: Scrutin de Ballotage
- f: Séparation de Biens
- f: Station Balnéaire
- f: Supplément de Bagages
- r: Samband
- r: Sjukvårdsbyrå
- r: vakt Styrbord
Sb e: antimony
sb e: smiling back [Internet]
sb. e: substantive
s.b. N: styrbord
- N: svar betalt
- N: sydlig bredde
S/B e: Set Back
- e: Soil/Bentonite
SB3 e: System Base 3
Sb-125 e: antimony-125
SBA e: Scene Balance Algorithms
- e: Shipboard Analysis
- e: Small Business Act
- e: Small Business Administration
- e: Spina Bifida and Anencephaly
- e: Synchronous Bandwidth Allocation
- f: Société Belge de l'Azote
- n: Stadsbibliotheek Antwerpen
SBag f: Sans Bagages
s-bälte r: säkerhetsbälte
SBAM e: Single Base Aircraft Maintenance Model
SBAP e: Signal Bias and Adjustment Panel
- e: Small Business Assistance Programs
SBAS s: Sección de Buques Antisubmarinos
SBASE e: Database of protein domain sequences maintained by ICGEB
SBASS e: Space-Based Atmospheric Surveillance System
SBAVD/HOK N: Sambandsavdelingen Hærens Overkommando
SBB e: Subtract With Borrow
SBBE f: Société Biblique Britannique et Étrangère, London
SBC e: Brazilian Society for Cartography, Geodesy, Photogrammetry and Remote Sensing
- e: S Bus Circuit
- e: SCSI Block Commands
- e: Space Based Communication
- e: Sum of Base Cations
- r: Sifferbehandlingscentral
Sbc r: Sambandscentral
SBCCI e: Southern Building Code Congress International, Inc.
SBC Communications e: SBC [Southwestern Bell Corporation]
SBCCS e: Single Byte Command Code Set
Sbckomp r: Sambandscentralkompani
SBCN e: Sitting Behind the Computer Naked [Internet]
SBCS e: Single Byte Character Set
- e: Single Byte Command Set
SBCX e: S Bus Circuit Extended
SBD e: Schottky-Barrier Detector
- e: Smart Battery Data
- f: Société de Bauxite de Dabola
SBDC e: Seal Bore Drill Collar
- e: Small Business Development Centers
SBE e: Society of Broadcast Engineers, Inc.
SBEE f: Société Belge d'Études et d'Expansion
- f: Société Béninoise d'Electricité et d'Eau
S-bend distortion e: S-curve distortion
SBF a: Skandinavisk Bogforlag A/S
- a: Sola Brettseilerforening
- d: Schiffbauforschung
- e: School Book Fairs
- e: Science Books and Films
- e: Scientific Balloon Flight
- e: Sea Breeze Front
- e: Search By Form
- e: Sequential Block Filemanager
- e: Simulated Body Fluid
- e: Single Barrier Failure
- e: Single Black Female
- e: Small Business Finance
- e: Small Business Friendly
- e: Small But Fast
- e: Spark, Burn Firefighter
- e: Spherical Bessel Function
- e: Spirobifluorene
- e: Standard Broadband Format
- e: Steel Bridge Forum
- e: Structure Behavior Function
- e: Students Broadcasting Federation
- e: Subic Bay Freeport
- e: Surface Brightness Fluctuation
- f: Secretariat des Bailleurs de Fonds
- f: Société Batrachologique de France
- f: Société Belge de Filtration
- f: Société des Bourses Francaises
- l: Studium Biblicum Franciscanum
- N: Sjøforsvarets Befals-forbund
- p: Sociedade Brasileira de Fruticultura
- r: Stiftelsen Bioteknisk Forskning
- r: Svenska Betong Foreningen
- r: Svenska Botaniska Förening
- r: Svenska Brand Forsvarsforeningen
- r: Sveriges Begravningsbyraers Forbund
- z: Siyasal Bilgiler Fakultesi
- z: Siyasal Bilimler Fakultesi
Sbf f: Sauf bonne fin
SBFF e: Scanned Blood Flow Field[s]
Sbfkomp r: Sambandsförstärkningskompani
SBG d: Strasburg, Mecklenburg-Vorpommern
- e: Strategic Business Group
SBGRL d: Schweizerischer Berufsverband der Geriatrie-, Rehabilitation- und Langzeitpflege

SBH e: Severe Behavior Handicap
SBI e: Sound Blaster Instrument
- e: Space Based Interceptor
- e: Subsidiary Body for Implementation
- e: Synchronous Backplane Interconnect
- e: Synchronous Backplane Interface
- f: Organe subsidiaire de mise en oeuvre
- s: órgano Subsidiario de Ejecución
SbI r: Sambandsinstruktion
SBIA e: Saskatchewan Brain Injury Association
SbI A r: Sambandsinstruktion för Armén
SBIGFET e: Schottky-Barrier Insulate-Gate Field-Effect Transistor
SBII f: Société Belge des Ingénieurs et des Industriels
SBIR e: Small Business Innovation Research [program]
SBIS e: Sustaining Base Information Services
- e: Swaziland Broadcasting and Information Service
SBIT f: Société Belge des Ingénieurs et Techniciens
SBJCT e: Subject
SBK a: Skandinavisk Bygningskemi A/S
- d: Schönebeck/Elbe, Sachsen-Anhalt
- e: Fast Breeder Nuclear Power Federal Station Co.
SBKB d: Staatsbibliothek Preußischer Kulturbesitz
SBKKV e: Space-Based Kinetic-Kill Vehicle
Sbkompani r: Sambandskompani
Sbkp N: Sambandskompani
sbkp/KR N: sambandskompani av Kombinert Regiment
SBL e: Serials in the British Library
- e: Similarity-Based Learning
- e: Space-Based Lasers
- i: Sistema Beni Librari
- N: Skipsbyggerienes Landsforening
- r: Statens Bakteriologiska Laboratorium
SBLC e: Small Business Legislative Council
- e: Standard Base Level Computer
SBLC/S e: Standard Base Level Computer/Systems
Sb Lek c: Sbornik Lekarsky
SBM e: Scientific Basis of Medicine
SBML e: Special Bauxite Mining Lease
SBMS e: Southwestern Bell Mobile Service
SBN a: Standard Bog Nummer
- e: Satellite Broadcast Network
- e: Semiconductor Business News
- e: Standard Book Numbering agency
- i: Servizio Bibliotecario Nazionale
- r: Medicinalstyrelsens Sjukvårdsberedskapsnämnd
SBO e: Senior Budget Officer
Sboff r: Sambandsofficer
S Bol s: Submarino de Bolsillo
SBP e: SCSI-3 serial Bus Protocol
- e: Serial Bus Protocol
- e: Space Bandwidth Product
- e: Strategic Business Plan
- e: Structured Buffer Pool
SBPK d: Staatsbibliothek Preußischer Kulturbesitz
sb. pl. N: sambandsplass
Sbplut r: Sambandspluton
SBPTC f: Société Belge de Photogrammetrie, de Télédétection et de Cartographie
S. br. N: Sørlig bredde
s.br. a: sydlig bredde
- r: sydlig bredd
SBR e: Source Browser
- e: Space Based Radar
- e: Special Boat Squadron
- e: Styrene Butadiene Rubber
- N: Sambandsregiment
SBRB e: Site Baseline Review Board
SBRC e: Senate Budget Review Committee
- e: Sheep Red Blood Cells
SBRS e: Sonic Boom Reporting System
SBS e: Senior Battle Staff
- e: Sick Building Syndrome
- e: Siemens Business Services
- e: Skyline Business Systems
- e: Small Business Server
- e: Special Broadcasting Service Corp.
- e: Styrene Butadiene Styrene block copolymer
- e: Submarine Broadcast System
- e: Submerged Bed Scrubber
- e: Support Battle Staff
- N: Sambandskolen
SBS rubber e: Styrene-Butadiene-Styrene Rubber
SBSS e: Smart Battery Surveillance System
SBSTA e: Subsidiary Body for Scientific and Technological Advice
- f: Organe subsidiaire de conseil scientifique et technologique
- s: órgano Subsidiario de Asesoramiento Científico y Tecnológico
SBSTTA e: Subsidiary Body on Scientific, Technical and Technological Advice
- s: Organe subsidiaire chargé de fournir des avis scientifiques, techniques et technologiques
SBSVC e: Selective Broadcast Signaling Virtual Channel
SBT a: Statens Bibliotekstjenste
- e: Sothern Bluefin Tuna
sbto r: sambandstrupp
sb tpfpl r: sambands- och transportflygplan
sb. tr. N: sambandstropp
Sbt[t]e s: Subteniente
SBTTL e: Schottky-Barrier Transistor-Transistor Logic
SBU e: Special Boat Unit
SBUF e: The Development Fund of the Swedish Construction Industry
SBUP e: System Backup Partition
SBUV e: Solar Backscatter Ultraviolet [experiment]
- e: Solar Backscatter Ultraviolet [instrument]
- e: Solar Backscatter Ultraviolet [radiometer]
- e: Solar Backscattered Ultraviolet [system]
SBUV-2 e: Solar Backscatter Ultraviolet 2-2
SBUV-TOMS e: Solar Backscatter Ultraviolet Total Ozone Mapper System
- e: Solar Backscatter Ultraviolet-Total Ozone Mapping Spectrometer
Sb Ved Pr Lek Fak Karlovy Univerzity Hradci Kralove c: Sbornik Vedeckych Praci Lekarske Fakulty Karlovy Univerzity v Hradci Kralove
Sb Ved Pr Lek Fak Karlovy Univerzity Hradci Kralove Suppl c: Sbornik Vedeckych Praci Lekarske Fakulty Karlovy Univerzity v Hradci Kralove. Supplementum
SBWAS e: Space Based Wide Area Surveillance
SBWP e: Space BandWidth Product
SBX e: Borate Oxide
- e: Slow Burning Explosive
SC a: Socialdemokratisk Centrum
- e: Scanner Controller
- e: Science Council
- e: Science Councils
- e: SCSI Stream Commands
- e: Secondary Channel
- e: Self-Consistent
- e: Session Connector
- e: Single Crystal[line]
- e: Small submarine reactor prototype
- e: Southern California
- e: Southwest College
- e: Space Council
- e: Special Condition
- e: Specification Compliance
- e: Square Cut
- e: State and Country Code
- e: Structural Change
- e: Sub Committee
- e: Switching Center
- e: System Compliance
- e: Systems Code (or Control)
- f: Saisie Conservatoire
- f: Sans Changement
- f: Santé Canada
- f: Seconde Classe
- f: Sens Commun
- f: Séparation de Corps
- f: Service Commercial
- f: Services Centraux
- f: Services Communs
- f: Slogan Commercial
- f: Société Chimique
- f: Société Civile
- f: Société Commerciale
- f: Société Coopérative
- f: Société en Commandite
- f: Son Compte
- f: Sous Caution
- f: Sous-Commission
- f: Statistique Commerciale
- f: Suivant le Cours
- f: Surface Corrigée
- f: Surpeuplement Critique
- f: Syndicat Corporatif
- r: Stabschef
Sc e: Stratocumulus clouds
- f: Scientifique
- s: Sección
S C e: Serbian/Croatian
SC. e: Status code

s.c. e: self-contained
- e: single-column
- e: small caps
S&C e: Safety and Control
- e: Sand, slag and Crucible
- e: Skull and Crossbones
SC1 e: Standard Clean 1
SC2 e: Standard Clean 2
SC3 e: Surveillance, Command, Control and Communications
SC4 e: Systems for Command, Control, Communications and Computers
SCA e: Scandinavian
- e: Security Control and Audit
- e: Service Cryptologic Agency
- e: Sexual Compulsives Anonymous
- e: Shared-Cost Action
- e: Single Connector Attachment
- e: Single County Authority
- e: Snow-Covered Areas
- e: Spatial Correction Analysis
- e: Standing Committee on Agriculture
- e: Sunlink Channel Adapter
- e: Surface Charge Analysis (or Analyzer)
- e: Surface Contamination Area
- e: Switching Center AUTOVON
- e: System Communication Architecture
- e: System Configuration Audit
- f: Service Central des Approvisionnements
S.C.A. e: Sydney College of the Arts
SC & A e: Security Certification and Accreditation
SCAA e: School Curriculum and Assessment Authority
SCAAP e: Supercomputer Automotive Applications Partnership
SCAB e: South Coast Air Basin
- r: Sparbankernas Centralaktiebank
SCAC e: Soil Conservation Advisory Committee
- e: Standard Carrier Accounting Code
- e: Support Careers Advisory Committee
SCAD e: Senate Committee on Academic Development
SCADA e: Supervisory Control and Data Aquisition
S.C.A.E. e: Sydney College of Advanced Education
SCAF e: Service Control Agent Function
- f: Société Centrale des Agriculteurs de France
SCAG e: Special CLEM Assembly Grapple
SCAI e: Institute for Algorithms and Scientific Computing
- e: Switch-to-Computer Application[s] Interface
SCALE e: SEMATECH Cell Application Learning Environment
- e: Shuttle Coherent Atmospheric Lidar Experiment
SCALOP e: Standing Committee on Antarctic Logistics and Operations
- f: Comité permanent sur la logistique et les opérations dans l'antarctique
- s: Comité Permanente sobre la Logística y las Operaciones en el Antártico
SCALP e: Scenario-Determined, Computer Assisted Logistics
- e: Solar Calibration Package
SCAM e: SCSI Configuration (or Configured) Automatically
- e: Strip Chart Analysis Manual
- e: Support Capability Assessment Model
SCAME e: Screen oriented Anti Misery Editor
SCAMP e: Single Chip A series Mainframe Processor
SCAMPI e: Network of Leased Wideband Carriers
SCAMS e: Security Combat Automated Management Subsystem
SCAN e: Shipboard Communications Area Network
Scand. e: Scandinavia[n]
Scand Audiol [Suppl] e: Scandinavian Audiology. [Supplementum]
Scand J Clin Lab Invest [Suppl] e: Scandinavian Journal of Clinical and Laboratory Investigation. [Supplement]
Scand J Gastroenterol [Suppl] e: Scandinavian Journal of Gastroenterology. [Supplement]
Scand J Immunol [Suppl] e: Scandinavian Journal of Immunology. [Supplement]
Scand J Infect Dis [Suppl] e: Scandinavian Journal of Infectious Diseases. [Supplementum]
Scand J Med Sci Sports e: Scandinavian Journal of Medicine and Science in Sports
Scand J Plast Reconstr Surg Hand Surg [Suppl] e: Scandinavian Journal of Plastic and Reconstructive Surgery and Hand Surgery. [Supplementum]
Scand J Prim Health Care [Suppl] e: Scandinavian Journal of Primary Health Care. [Supplement]
Scand J Psychol e: Scandinavian Journal of Psychology
Scand J Rehabil Med [Suppl] e: Scandinavian Journal of Rehabilitation Medicine. [Supplement]
Scand J Rheumatol [Suppl] e: Scandinavian Journal of Rheumatology. [Supplement]
Scand J Soc Med [Suppl] e: Scandinavian Journal of Social Medicine. [Supplementum]
Scand J Thorac Cardiovasc Surg [Suppl] e: Scandinavian Journal of Thoracic and Cardiovascular Surgery. [Supplementum]
Scand J Urol Nephrol [Suppl] e: Scandinavian Journal of Urology and Nephrology. [Supplementum]
Scand J Work Environ Health e: Scandinavian Journal of Work, Environment and Health
SCANFAX e: A disk-to-fax service from HMSO
SCANIQ f: Système Conjoint d'Analyse Numérique d'Images de Québec
Scanner d: Gerät zum Einlesen von Texten, Grafiken und Bildern
Scanning Microsc e: Scanning Microscopy
Scanning Microsc Suppl e: Scanning Microscopy. Supplement
SCANOR f: Société Anonyme de Constructions Aéronautiques du Nord
SCANS e: Status Configuration and Alarm Notification System
SCANSAR e: Scan Beam Synthetic Aperture System
SCANSCAT e: Scanning Scatterometer
SCAP e: Senate Committee on Academic Procedures
- e: Special Climate Applications Programme
- e: States Cooperative Assistance Program
- e: Superfund Comprehensive (or Consolidated) Accomplishments Plan
- e: Surface Combat Air Patrol
- f: Programme Spécial d'Applications Climatologiques
- s: Programa Especial de Aplicaciones Climáticas
SCAPE e: Slantwise Convective Available Potential Energy
ScAppl f: Sciences Appliquées
SCAPTL e: Senate Committee on Appointment, Promotion, Tenure and Leave
SCAR e: Scientific Committee for/on Antarctic Research
- e: Smoke, Clouds, [Aerosols] and Radiation [experiment]
- e: Society for Computer Applications in Radiology
- f: Comité Scientifique pour les Recherches Antarctiques
- f: Société des Chantiers et Ateliers du Rhin
- s: Comité Científico de Investigaciones Antárticas
SCARA e: Scanner for Radiation Budget
SCARAB e: Scanner for Radiation Balance
- e: Scanner for [the] Radiation Budget
- e: Scanning Radiometer for Radiation Balance
- R: Scanner Radiatsionnogo Balansa
SCARABE e: Scanner for Radiative Budget Experiment
- e: Scanning Radiation Budget Experiment
SCAR-B e: Smoke, Clouds and Radiation over Brazil
SCARD e: Society of Chairmen of Academic Radiology Departments
SCARF e: Standing Committee of the Australian Forestry Council
SCARM e: Standing Committee on Agriculture and Resource Management
SCARP e: Skylab Concentrated Atmospheric Radiation Project
SCARS e: Status, Control, Alerting, Reporting System
SCARS II e: SCARS Upgrade
SCART f: Syndicat des Constructeurs d'Appareils Radio Recepteurs et Techniciens
- f: Syndicat des Constructeurs d'Appareils Radio Récepteurs et Téléviseurs
SCAS e: Scan String
SCA[SS] e: Scientific Council for Africa South of the Sahara London

SCAT e: radar scatterometer
- e: Scout-Attack version
- e: Secure CATCOMS
- e: Sodium Containment And Transfer
- e: Standing Corrective Action Team
- e: Stromberg Carlson Assistance Team
- e: System Control of AUTODIN Traffic
- e: Systematic Comparison of Alternate Task

SCATA e: Society for Computing and Technology in Anaesthesia
SCATRA[D] e: Scatterometer/Radiometer
SCATS e: Safeguard Control Accountability Transaction System
SCAV e: Soil Conservation Association of Victoria
SCAWG e: Systems Control Architecture Working Group
SCAWU e: Swaziland Commercial and Allied Workers Union
SCAZ e: South Atlantic Convergence Zone
SCB e: Santa Cruz Basin
- e: Society for the Conservation of Biology
- e: SPO Control Board
- e: Subsystem Control Block
- e: System Control Block
- f: Séparation de Corps et de Biens
- f: Société de Chimie Biologique
- f: Sous-Chef de Bureau
- f: Syndicat Chrétien du Burundi

SC[B] r: Statistiska Centralbyrån
SCBA e: Self Contained Breathing Apparatus
SCBT/MR e: Society of Computed Body Tomography and Magnetic Resonance
SCC e: SCSI Controller Commands
- e: Secure Communications Controller
- e: Serial Communication[s] Control[ler][s]
- e: Serial Controller Chip
- e: Serial Controller Chip Synchronous Channel Check
- e: Set According to Condition
- e: Source Classification Code[s]
- e: Soweto Community Council
- e: SPADOC Computation Center
- e: Special Computation Center
- e: Special Coordination Committee
- e: Specialized Common Carriers
- e: Specialized Communication Carriers
- e: Species Survival Commision
- e: Standards Council of Canada
- e: Standing Concertation Committee
- e: State Construction Corporation
- e: State Consultative Councils
- e: Steering Committee Chair
- e: Strategic Cell Controller
- e: Student Coordinating Committee
- e: Summary Class Code
- e: Switching Control Center
- e: Synchronous Channel Check
- f: Service Central du Cadastre
- f: Société Canadienne du Cancer
- f: Société Centrale des Coopératives
- f: Société Chimique des Charbonnages
- f: Sur Compte Courant
- f: Système du Courant des Canaries
- s: Sistema de la Corriente de las Canarias

s/c.c. s: su cuenta corriente
SCCA e: Surface Contamination Control Area
- s: Servicios para el Control de la Calidad de los Análisis

SCCCPMA e: Society for Computing in Critical Care, Pulmonary Medicine and Anesthesia
SCCD e: Short Circuit Current Delay
- f: Service du Contrôle des Contributions Directes

sCCD d: Relative spektrale Empfindlichkeit der CCD-Kamera
SCCE f: Service du Contrôle et des Constructions d'État
SCCM3 e: Single-column version of the Community Climate Model
SCCN e: Secure Command and Control Net
SCCO e: Special Circuit Control Officer
SCCP e: Signaling Connection [and] Control Part
- e: Signaling Connection Control Point
- e: Signalling Connection Control Port

SCCS e: SAC Command Conference System
- e: Safeguards Classified Computer System
- e: Safeguards Control and Communications System
- e: Source Code Control System
- e: Southern California Computer Society
- e: Specialized Common Carrier Service
- e: Standard Communications-Computer Summary
- e: Subcommission on Carboniferous Stratigraphy
- e: Swarthmore College Computer Society

SCCSB e: Software Computer Configuration Sub-Board
SC&CU e: Signal Conditioning and Control Unit
SCD e: Santa Cruz Operation
- e: Scientific Computing Division
- e: Segmented-array, Charge-coupled device Detector
- e: Silicon Capacitor Diodes
- e: Software Change Description
- e: Software Configuration Description
- e: Standard Color Display
- e: Strategic Communications Division
- e: Strathclyde
- e: Structered Complex Object
- e: Structured Charts Diagram
- e: System Correction Data

SC&D e: Stock Control and Distribution System
SCD1 e: Stearyl-CoA-desaturase
SCDB e: Sub-Committee on Digital Bathymetry
- f: Sous-Comité de la Bathymétrie numérique
- s: Subcomité sobre Batimetría Numérica

SCDDB e: Satellite Climate Diagnostics Data Base
SCDHEC e: South Carolina Department of Health and Environmental Control
SCDI p: Satelite de Coleta de Dados
S-CDMA e: Synchronous Code Division Multiple Access
SCDP f: Société Camerounaise des Dépôts Pétroliers
SCDR e: Subsystem Critical Design Review
SCDS f: Service Central des Documents Secrets
SCE e: Saturated-Calomel Electrode
- e: School of Civil Engineering
- e: Service Creation Environment
- e: Service in Conservation Education
- e: Software Capability Evaluation
- e: Structure Chart Editor
- e: System Cost Estimate
- f: Sous-Comptoir des Entrepreneurs

Sce f: Service
- f: Source

SCEC e: Small Computer Engineering Center
- e: Southern California Earthquake Center
- e: Sports Council Executive Committee
- e: Sunshine Coast Environment Council

SCED e: SCSI/Ethernet/Diagnostics
- f: Société Centrafricaine d'Exploitation Diamantifère

SCEEMAS e: Small Companies EC Eco-Management and Audit Scheme
SCEES f: Service Central des Enquêtes et Études Statistiques
SCEF e: Service Creation Environment Function
SCEM e: Single-Channel Electron Multiplier
SCENARIO e: Exercise Scenario [Schedule] System
SCEO e: Strategic Connectivity Engineering Office
- f: Syndicats de Cadres, d'Employés et d'Ouvriers

SCEP e: Service Creation Environment Point
- e: Special Committee on Antarctic Research

SCEPA f: Société de Construction d'Équipement pour l'Aéronautique
SCEPP e: Southern California Earthquake Preparedness Project
SCERA e: Senate Committee on Environment, Recreation and the Arts
SCERI f: Service Centrale d'Études et de Réalisation de l'Intendance
SCERT e: Systems and Computer Evaluation and Review Technique
SCETA f: Société du Contrôle et de l'Exploitation des Transports Auxiliaires
SCF e: Science Computing Facility
- e: Selective Call Forwarding
- e: Sequential Character Filemanager
- e: Service Control Function
- e: Standing Committee on Fishing
- e: Standing Committee on Forestry
- e: Strategic Concertation Forum
- e: Super Critical Fluid
- f: Syndicat des Corporations Françaises

SCFA e: South Carolina Forestry Association
scf/d e: standard cubic feet per day
SCFEL e: Standard Cryptologic Facility Equipment List
SCFL e: Schottky Diode FET Logic

SCGN e: Sub-Committee on Geographical Names and Nomenclature of Ocean Bottom Features
- f: Sous-Comité des noms géographiques et de la nomenclature des formes du relief océanique
- s: Subcomité para los Nombres Geográficos y la Nomenclatura de las Formas del Relieve Oceánico

SCGPC f: Service des Cartes Grises et Permis de Conduire
SCH d: Beginn des Schießens bzw. Längenwachstums
- e: Schizophrenia
- e: Separate-Confinement Heterostructure
- e: Student Credit
- e: Subcortical Hyperintensity
- f: Service Central Hydrographique

Sch. e: School
S-Ch f: Sous-Chef
SCHA f: Station Centrale d'Hydrobiologie Appliquée
Sch-a e: Schwannoma
SCHBT e: Schottky-Collector Heterojunction Bipolar Transistor
SC-HDP e: Steering Committee of the HDP
SCHEMB HALL e: Glenn E. Schembechler Hall
SchiedsVfG d: Gesetz zur Neuregelung des Schiedsverfahrensrechts [Schiedsverfahrens-Neuregelungsgesetz] vom 22.12.1997
SchiffbMstrV d: Schiffbaumeisterverordnung vom 18.9.1996
SchiffsBG d: Gesetz über die Schiffspfandbriefbanken [Schiffsbankgesetz]
Sch Inq Nurs Pract e: Scholarly Inquiry for Nursing Practice
Schizophr Bull e: Schizophrenia Bulletin
Schizophr Res e: Schizophrenia Research
SCHLEICH e: Scattering of Heavy, Low-Energy Ions with Channeling
SCHM e: State Committee for Hydrometeorology
SchRAnpG d: Schuldrechtsanpassungsgesetz
Schriftenr Neurol d: Schriftenreihe Neurologie. Neurology Series
Schriftenr Ver Wasser Boden Lufthyg d: Schriftenreihe des Vereins für Wasser-, Boden-, und Lufthygiene
SCHS e: School Squadron
SchuldRÄndG d: Schuldrechtsänderungsgesetz
SchuR d: Schulrecht. Informationsdienst für Schulleitung und Schulaufsicht
SchutzbG d: Schutzbaugesetz
SchWAG d: Bundesschienenwegeausbaugesetz vom 15.11.1993
SchwArbG d: Gesetz zur Bekämpfung der Schwarzarbeit
SchwAwV d: Ausweisverordnung Schwerbehindertengesetz
SchwbAV d: Verordnung zur Durchführung des Schwerbehindertengesetzes [Ausgleichsabgabeverordnung - Schwerbehindertengesetz]
SchwBG d: Gesetz zur Sicherung der Eingliederung Schwerbehinderter in Arbeit, Beruf und Gesellschaft [Schwerbehindertengesetz]
Schweiz Arch Neurol Psychiatr d: Schweizer Archiv für Neurologie und Psychiatrie
Schweiz Arch Tierheilkd d: Schweizer Archiv für Tierheilkunde
Schweiz Med Wochenschr d: Schweizerische Medizinische Wochenschrift. Journal Suisse de Medecine
Schweiz Med Wochenschr Suppl d: Schweizerische Medizinische Wochenschrift. Supplementum
Schweiz Monatsschr Zahnmed d: Schweizer Monatsschrift für Zahnmedizin
Schweiz Rundsch Med Prax d: Schweizerische Rundschau für Medizin Praxis
SCI e: Scalable Coherent Interface
- e: Scaleable Coherent Interface
- e: Science
- e: Secret Compartmented Information
- e: Sensitive Compartmented Information
- e: Sequential Circuits Inc.
- e: Spacecraft Inc.
- e: Special Compartmented Information
- e: Special Compartmented Intelligence
- e: Surface Charge Imaging
- e: Systems Control Inc.
- e: World Multiconference on Systemics, Cybernetics and Informatics
- f: Service Central d'Identification

SCI. e: Science
SCIA e: Society for Computers in Anesthesia
- e: Supreme Council for Islamic Affairs

Sci Am e: Scientific American
SCIAMACHY e: Scanning Imaging Absorption Spectrometer for Atmospheric Cartography
- e: Scanning Imaging Absorption Spectrometer for Atmospheric Chemistry

SCIB e: Ships Characteristics and Improvement Board
SCIBP e: Scientific Committee for the International Biological Programme
- e: Special Committee on the International Biological Program

SCIC e: Satellite Cloud Information Chart
- e: Suppressed-Conductivity Ion Chromatography

Sci China B e: Science in China. Series B, Chemistry, Life Sciences and Earth Sciences
SCID e: Severe Combined Immunodeficiency Disease
- e: Structured Clinical Interview for DSM-III-R

SCIDAR e: Scintillation Detection and Ranging
SCIDS e: Severe Combined Immuno-Deficiency Syndrome
SCIE e: Science Hall
- e: Scientific Computing Information Exchange Council

SCIEH e: Scottish Centre for Infection and Environmental Health
SCIF e: Sensitive Compartmented Information Facility
SC [-]IGBP e: Scientific Committee for IGBP [International Geosphere-Biosphere Programme]
SC-IGBP e: Special Committee for/of the IGBP [International Geosphere Biosphere Programme]
- f: Comité Scientifique pour le Programme International sur la Géosphère et la Biosphère
- s: Comité Científico para el Programa Internacional sobre la Geosfera y la Biosfera

Sci Justice e: Science and Justice
SCILAC e: Science Library Advisory Committee
SCIM e: Self-Consistent Interstitial Method[s]
SCIMITAR e: System for Countering Interdiction Missiles and Target Acquisition Radars
SCIMPEX f: Syndicat des Commerçants Importateurs et Exportateurs de l'Ouest Africain
SCIMPEXTO f: Syndicat des Commerçants Importateurs et Exportateurs du Togo
SCINET e: Sensitive Compartmented Information Network
SCIP e: Site Configuration and Installation Plan
- e: Society of Competitive Intelligence Professionals
- e: Systems Control Improvement Program

SCIPMIS e: Standard Civilian Personnel Management Information System
Sci Prog e: Science Progress
SC2ISA e: Standard Command and Control Information Systems Architecture
SCISR f: Société Canadienne des Infirmières en Santé Respiratoire
SCIT e: Storm Cell Identification and Tracking
- e: System Concept and Integrated Technology

SCITNET e: SOUTHCOM Intel Teletype Network
Sci Total Environ e: Science of the Total Environment
SCITS e: Single Channel Injection Transponder System
SCITT e: School-Centred Initial Teacher Training
SCL e: Separately Controlled Link
- e: Solar Cycle Length
- e: Space-Charge Limited
- e: Subject Class

SCLM e: Software Configuration and Library Management
SCLSC e: Ship Configuration and Logistics Support Control

SCM e: Satellite Communications Module
- e: Scanning Capacitance Microscopy
- e: Screencam Format
- e: Single-Chip Module
- e: Single-Column Model
- e: Society for Cardiovascular Management
- e: Sonic Core Monitor
- e: Space Cargo Modified
- e: Station Class Mark
- e: Stochastic and Computational Mechanics
- e: Sub-Carrier Modulation
- e: Subscriber Carrier Module
- f: Service Central Météorologique
- f: Société de Chirurgie de Montréal
- f: Société de Constructions Mécaniques

S&CM e: Surfaced and Center Matched
SCMCC e: State Catchment Management Coordinating Committee
SCMG e: Sub-Committee on Meteorology and Geophysics
- f: Sous-Comité pour la Météorologie et la Géophysique
- s: Subcomité sobre Meteorología y Geofísica

SCMI f: Société continentale de matériel industriel
SCMO f: Société Canadienne de Météorologie et d'Océanographie
- s: Asociación Canadiense de Meteorología y Oceanografía

SCMP e: System Console Monitor Printer
SCMPL f: Société de Constructions Mécaniques Panhard et Levassor
SC-MRI e: Marine Science Related Issues
- f: Questions liées aux sciences de la mer
- s: Cuestiones relacionadas con las Ciencias del Mar

SCMS e: Serial Copy Management System
- f: Société de Construction du Matériel de Servitude

SCMU e: Species Conservation Monitoring Unit
SCN e: Shipbuilding and Conversion Navy
- e: Software Change Notice
- e: System Change Number
- e: System Commit Number

SCNA e: South Carolina Nurses Association
ScNat f: Sciences Naturelles
Sc. Nat. Dict. e: Scottish National Dictionary
SCNC e: Southern Cameroon National Council
SCNEA e: Sealing Commission for the North-East Atlantic
- f: Commission du phoque de l'Atlantique Nord-Est
- s: Comisión de la Foca para el Atlántico Nororiental

SCNR e: Sorry Could Not Resist [Internet]
SCNT1 e: Single Chip Network Termination 1
SCNWA e: Sealing Commission for the North-West Atlantic
- f: Commission du phoque de l'Atlantique Nord-Ouest
- s: Comisión de la Foca para el Atlántico Noroccidental

SCO e: Santa Cruz Operation[s], Inc.
- e: Seasonal Climate Outlook
- e: Serving Central Office
- e: State Coordinating Officer
- e: Structured Complex Object
- e: Surface Contaminated Objekt
- e: System Control Officer

SCOAS e: Small Computer/Office Automation Services
SCOASB e: Small Computer/Office Automation Service Branch
SCOASO e: Small Computer/Office Automaton Service Organization
SCOC e: System Control Operations Center
- f: Service Central d'Organisation et de Gestions

SC/OCE e: Division of Marine Sciences
- f: Division des Sciences de la Mer
- s: División de Ciencias del Mar

SCOCE e: Special Committee on Compromising Emanations
SCOD e: Subcontract and Other Direct Costs
S codes e: Army staff at BDE and BN levels
SCOE e: SEMATECH Center of Excellence
SCOI e: Southern California Orthopedic Institute
- f: Service de Consultance en Observation de la Terre

S.Con.Res e: Senate Concurrent Resolution
SCoop f: Société Coopérative
SCOOPS e: Scheme Object Oriented Programming System
SCOP e: Structural Classification of Proteins
SCOPE d: Wissenschaftliches Komitee für Umweltprobleme
- e: San Clemente Ocean Probing Experiment
- e: Scientific Commitee on Problems of the Environment
- e: Simple Communications Programming Environment
- f: Comité Scientifique pour/sur les Problèmes del'Environnement
- s: Comité Científico sobre los Problemas del Medio Ambiente

SCOPE DIAL e: Base Telecommunications Modernization
SCOPE SIGNAL e: USAF Program For Worldwide HF Ground-to-Air Broadcast
SCOPIL e: Southern California Occasional Papers in Linguistics, University of Southern California, Los Angeles
SCOR e: Scientific Committee for Oceanographic Research
- e: Special Committee on Ocean Research
- f: Comité Scientifique de la Recherche Océanique
- s: Comité Científico de Investigaciones Oceánicas

SCORE e: Scenario-Oriented Recurring Evaluation System
- e: Spacecraft Operational Readiness Exhibit System

SCORIF e: Stimulated Cycle Oocyte Retrieval In Fertilization
SCORPIO e: Oceanographic Cruise in the South Pacific
- f: Croisière Océanographique dans le Pacifique Sud
- s: Crucero Oceanográfico en el Pacífico Sur

SCORRAD e: Standing Committee on Resources Research and Development
- f: Comité permanent sur la recherche et le développement des ressources
- s: Comité Permanente de Investigaciones y Desarrollo de Recursos

SCOS e: Southern COST-43 Operational [Drifting Buoy] System
- f: Système opérationnel [de bouées dérivantes] de COST-43 pour le Sud
- s: Sistema Operacional [Boyas a la Deriva] COST-43 en el Sur

SCOSA e: Senate Committee on Student Affairs
SCOSTEP e: Scientific Committee on Solar-Terrestrial Physics
- f: Comité Scientifique de Physique Solaire et Terrestre
- s: Comité Científico sobre Física Helioterrestre

Sc[ot]. e: Scottish
SCOT e: Services Consultance en Observation de la Terre
- e: Soweto Committee of Ten
- e: Special Command, Control and Communication Operations and Testing
- e: Stepper Central Office Tester

Scotl. e: Scotland
SCOTS e: Surveilance and Control Of Transmission Systems
SCOTT e: Single Channel Objective Tactical Terminals
Scott Med J e: Scottish Medical Journal
SCOTVEC e: Scottish Vocational Education Certificate
SCOWAR e: Scientific Committee on Water Research
- f: Comité Scientifique pour les Recherches sur l'Eau
- s: Comité Científico de Investigaciones Hídricas

SCP e: Computer Security Program Office
- e: Save Cursor Position
- e: Secondary Communications Processor
- e: Secure Conferencing Program
- e: Secure Conferencing Project
- e: Serial Clock Pulse
- e: Service Control Point
- e: Session Control Protocol
- e: Signal Control Point
- e: Signal Conversion Point
- e: Single-Chip Package (or Packaging)
- e: Software Control Procedure
- e: Soil Conservation Program
- e: Special Committee on Parking
- e: Species Conservation Program
- e: Stacked-Chip Packaging
- e: Standard Change Package
- e: Subnet Communications Processor
- e: Subsystem Control Port
- e: Sunlink Communications Processor
- e: Support Center Pacific

- e: Survivable Command Post
- e: System Change Package
- f: Société Canadienne de Pédiatrie
SCPC e: Signal Channel Per Carrier
- f: Société Centrale de Plastique et de Caoutchouc
SCPC-Technologie e: Single Carrier Per Channel
SCPD e: Supplementary Central Pulse Distributor
SCPDM e: Suppressed Clock Pulse Duration Modulation
SCPH f: Société Canadienne des Pharmaciens d'Hôpitaux
SCPI e: Session Control Programmable Interface
- e: Standard Commands for Programmable Instruments
- f: Société Civile de Placement Immobilier
SCPIM f: Service Civique, Patriotique, Idéologique et Militaire
SCPP e: Seasonal-to-Interannual Climate Prediction Program
- e: Sierra Cooperative Pilot Project
SCPS e: Sun Yat-sen Center for Policy Studies
SCPS-M e: Survivable Collective Protection System-Medical
SCQS f: Services de Contrôle de la Qualité des analyses
SCR e: Scanning Radiometer
- e: Schedule Change Request
- e: Selective Catalytic Reduction
- e: Senate Concurrent Resolution
- e: Silicon Control Rectifier
- e: Space-Charge Region
- e: Standard Class Rate
- e: Standard Context Routing
- e: Surface Contour Radar
- e: Sustainable (or Sustained) Cell Rate
- f: Société de la Croix-Rouge
SC&RA e: Specialized Carriers and Rigging Association
SCRAM e: emergency reactor shutdown
- e: State Consolidated RCRA Authorization Manual
- e: Static Column RAM [Random Access Memory]
SCRAPE f: Service Central de Recherche et d'Action Pour l'Enfance
SCRB e: SPO Control Review Board
SCRC e: Small Computer Requirements Contract
- e: Superfund Community Relations Coordinator
SCRGR f: Service de Contrôle des Radiations et de Génie Radioactif
SCRI e: Supercomputer Computations Research Institute
SCRIBE e: Stratosphere Cryogenic Balloon Experiment
SCRIPPS e: Scripps Institution of Oceanography
SCRP e: Superfund Community Relations Program
SCRS e: Scalable Cluster of RISC Systems
SCRT f: Service Central des Recherches Techniques
SCRTY e: Security
SCS e: Science Commanding System
- e: Security Control Station
- e: Silicon Controlled Switch
- e: Single-Channel Seismic
- e: SNA Character String
- e: SNA Communication Service
- e: Software Consulting Service
- e: Standard Communications-Computer System
- e: State Conservation Strategy
- e: Supplementary Control Strategy
- e: Supreme Council of State
- e: Swiss Calibration Service
- e: System Communication Services
SCSA e: Signal Computing System Architecture
- e: Sun Common SCSI Architecture
SC-SAR e: Signum Coded Synthetic Aperture Radar
SCSB e: Small Computer Support Bulletin
SCSC e: Standing Committee on Soil Conservation
SCSD e: Systems Control and Support Directorate
SCSDI f: Service Central de Statistique et de Documentation Industrielles
SCSI e: Small Computer Systems Interface
- f: Société Canadienne de Santé Internationale
SCSMEX e: South China Sea Monsoon Experiment
- f: Expérience sur la mousson dans le sud de la mer de Chine
- s: Experimento sobre los Monzones del Sur del Mar de la China
SCSMO e: Systems Control Software Management Office
SCSN f: Société Centrale de Sauvetage des Naufragés
SCSP e: South China Sea Fisheries Development and Coordinating Program[me]
- e: Storm and Combined Sewer Program
- f: Programme de mise en valeur et de coordination des pêcheries de la mer de Chine méridionale
- f: Société Canadienne des Sciences Pharmaceutiques
- s: Programa de Desarrollo y Coordinación de la Pesca en el Mar del Sur de China
SCSRSS e: Standard Command Supply Review System-SAILS
SCSS e: Shoe Cove Satellite Station
- e: Specialist Computer Systems and Software
SCST e: State Commission for Science and Technology
SCSW e: Systems Communications Software
SCT e: Satellite Communications Terminal
- e: Schmidt-Cassegrain Telescope
- e: Short Contact Time
- e: Society for Clinical Trials
- e: System Compatibility Test
- f: Sociéte Canadienne de Télédétection
Sct f: Section
Sct. a: Sankt
SCTA f: Service Central des Travaux de l'Aéronautique
SC4TAP e: System for Command, Control, Communications and Computer Technology Application Program
SCTC e: Serial Channel-To-Channel
- e: Small Computer Technical Centers
- e: System Console Transfer Channel
SCTE e: Serial Clock Transmit External
SCTIS e: Single Channel Transponder Injector System
SCTL e: Storage Control
SCTM s: Sindicato Central de Trabajadores Municipales
ScTP e: Screened Twisted Pair
SCTS e: Secondary Clear To Send
SCTSS e: Shoe Cove Satellite Station
SCTU e: Ukrainian Science and Technology Center
SCTX e: South Central Texas
SCU e: Ocean Sciences Unit
- e: Scud missile
- e: Selector Control Unit
- f: Unité d'océanologie
- s: Unidad de Ciencias Oceánicas
SCUBA e: Self Contained Underwater Breathing Apparatus
SCUD e: Septicemic Cutaneous Ulcerative Disease
SCUFN e: Sub-Committee on Undersea Feature Names
- f: Sous-Comité des noms du relief sous-marin
- s: Subcomité sobre Nombres de los Accidentes Topográficos Submarinos
sculps. a: sculpsit [hugget af; stukket af]
SCUMMRA f: Société Centrale de l'Uranium, des Minérais et Métaux Radio-Actifs
SC-US e: Save the Children USA
SCV e: Software Compatible Vendor
- f: Service de la Carte de la Végétation
SCVIR e: Society of Cardiovascular and Interventional Radiology
SCVM f: Service Central des Ventes du Mobilier de l'état
SCVP e: Society for Cardiovascular Pathology
SCW S: Slowly Changing World
- e: Special Case Waste
- e: Supercritical Water [oxidation]
SCWG e: Security Certification Working Group
SCWP P: Slowly Changing World
SCX e: Specialized Communications Exchange
SCZ d: Schleiz, Thüringen

SD a: Socialdemokratiet
- a: Socialdemokratisk Debat
- e: Directorate of Software Development
- e: Doctor of Science
- e: Scan Duration
- e: Science Data
- e: SCSI Disk
- e: Shine-Dalgarno
- e: Small Dual in-line package
- e: Speed Density
- e: Staff Development
- e: Standard Dispatcher
- e: Standing Document
- e: Starting Delimiter
- e: Statistical Division
- e: Structured Design
- e: Sudan
- e: Sunshine Duration
- e: Super Density
- e: Syntactic Description
- e: System Description
- f: Salaire Départemental
- f: Sauf Dimanche
- f: Scrutin Départemental
- f: Second Degré
- f: Secrétaire de Direction
- f: Semi-Direct
- f: Service Départemental
- f: Service des Domaines
- f: Service des Douanes
- f: Sous-Directeur
- f: Sous-Direction
- f: Statistique Douanière
- N: Stridsdommer

sd d: absolute spektrale Empfindlichkeit der Eichdiode
- d: relative spektrale Empfindlichkeit der Vergleichsphotodiode

Sd f: Sans date

s.d. a: samme dag
- r: samma dag
- s: sin datos

S&D e: Space and Defense

S/D e: Source/Drain

S/d f: Sur demande

SDA e: Secure Domain Authority
- e: Send Data with immediated Acknowledge
- e: Service Delivery Arrangement
- e: Seventh-Day Adventist
- e: Sialodacryoadenitis
- e: Software Disk Array
- e: Somali Democratic Alliance
- e: Source Data Automation
- e: Start Date of the programme
- e: Start of Domain Authority
- e: Swappable Data Area
- e: System Design Automation
- e: System Display Architecture
- f: Sous-Direction Administrative
- f: Système de Detection Aeroportee

S[-]DAB e: Satellite-Digital Audio Broadcasting

SDAC e: Satellite Data Analysis Center

SDA/CAP e: Spaceborne Distributed Aperture/Coherent Array Processing

SDAI e: Standard Data Access Interface
- f: Société de Développement Agricole et Industriel

SDAP e: SAR Data Aquisition Plan

SDAS e: System Design and Analysis Support

SDAT e: Department of Assessments and Taxation

S-Day e: times

S&DB e: Small and Disadvantaged Businesses

SDB e: Safe-Deposit Box
- e: Safety Deposit Box
- e: Small and Disadvantaged Business
- e: Standby Dispersal Base
- e: Symbolic Debugger
- e: Systems Development Branch

SDBE e: Small and Disadvantaged Business Enterprise

SDBP e: Society for Developmental and Behavioral Pediatrics

SDBUP e: Small Disadvantaged Business Utilization Program

SDBUS e: Small and Disadvantaged Business Utilization Specialist

SDC a: Sparekassernes Datacentraler
- a: Sønderjyllands Data-Center
- e: Secondary Domain Controller
- e: Slave Display Console
- e: Society for Democratic Culture
- e: Software Development Center
- e: Source Data Collection
- e: Strategic Defense Command
- e: Strategic Direction Center
- e: Swiss Development Corporation
- e: Systems Decision Plan
- f: Sous-Direction du Contentieux et des pensions
- f: Systèmes Défense et Contrôle

SDCCC e: Standard Desktop Computer Companion Contact

SDCCH e: Stand Alone Dedicated Control Channel

SD-CD e: Super Density Compact Disc

SDCD e: Secondary Data Carrier Detect

SDCIPS e: SDC [System Development Corporation] Image Processing System

SDCM e: Standard Dry Cubic Meter

SDCS e: Survivable DSCS Control Segment

SDCSS f: Service Départemental de Coordination des Services Sociaux

SDD e: Scitech Display Doctor
- e: Software Definition Document
- e: Software Description Database
- e: Super Density Disk
- e: System Design Document

SDDBMS e: Secure Distributed Database Management System

SDDI e: Shielded [twisted pair] Distributed Data Interface
- e: Shielded twisted-pair Digital Data Interface
- e: SONY Digital Data Interface

SDDM e: Secretary of Defense Decision Memorandum

SDDN e: Secure Data Defense Network

SDDS e: Satellite Data Distribution System

SDE e: Standard Data Element
- e: Standard Data Entry
- e: Support details
- e: System Development Environment
- s: Sudeste

S de A y M s: Servicio de Armamento y Material

Sde b f: Salle de bain

SDECE f: Service de Documentation Extérieure et de Contre-Espionage

SDEI f: Société de Distribution des Eaux Intercommunales

S de M s: Servicio de Municionamiento

SDEP e: Staff Diversity Enhancement Program

SDEPT e: Selective Distortionless Enhancement by Polarization Transfer

S Desc s: Submarino Deconocido

SDET s: Asociación Saulo de Tarso

SDF e: Association of Danish Research Libraries
- e: Secondary Distribution Frame
- e: Service Data Function
- e: Signal Delay Format
- e: Social Democratic Front
- e: Software Development Facility
- e: Space Delimited File
- e: Space Delimited Format
- e: Standard Distance File
- e: Structured Display File
- e: Sub Distribution Frame
- e: SubDistribution Frame
- e: Synchronous Diaphragmatic Flutter
- e: Synthetic Discriminant Function
- e: System Data Format
- f: Sauf Dimanches et Fêtes

SDFL e: Schottky Diode Field-Effect Transistor

s*D,F,rel d: vollständige Schreibweise der speziellen relativen spektralen Empfindlichkeit bei Partialfilterung

SDG e: San Diego
- e: Simulated Data Generator
- e: Software Development Group
- e: Spin-Dependent Generation
- e: System Development Group

Sdg. e: Siding

SDH d: Sondershausen, Thüringen

SdH e: Shubnikov-de Haas

SDH e: Synchronous Digital Hieararchy
- f: Service des Distinctions Honorifiques

SDHPT e: State Department of Highways and Public Transportation

SDHS e: Satellite Data Handling System

SDHS-COMNET e: SDHS Communications Network

SDHT e: SDH Transceiver

SDI e: Serial Data Input
- e: Serial Device Interface
- e: Single Document Interface
- e: Software Development Interface
- e: Space Defense Initiative
- e: Standard Disk Interconnect
- e: Steel Door Institute
- e: Storage Device Interconnect
- e: Storage Device Interface
- e: Strategic Defence Institute
- e: Subchronic Daily Intake
- f: Service de Documentation et d'Information

SDIA f: Syndicat de Défense des Intérêts Agricoles

SDigN d: Vorschriftensammlung für digitale Netze

SDIN e: Special Defense Intelligence Notice
SDIPC e: Software Digital Image Processing Calculator
SDIS e: Satellite Data Input Subsystem
- e: Surveillance and Display Interface System
- e: Switched Digital Integrated Service
SDIT f: Service de Documentation et d'Information Technique
SDITA f: Service de Documentation et d'Information Technique de l'Air
S D J Med e: South Dakota Journal of Medicine
SDK d: Schweizerische Sanitätsdirektoren-Konferenz
- e: Software Developers Kit (or Development) Kit
- e: System Developers Kit
- R: Sredniy Desantniy Korabl
SDL d: Stendal, Sachsen-Anhalt
- e: Satellite Data Laboratory
- e: Semiconductor Diode Laser
- e: Shared Dataspace Language
- e: Software Development Laboratory
- e: Software Development Library
- e: Solidarity
- e: Specification and Design Language
- e: Symbolic Description Language
- f: Service Départemental du Logement
SDL 92 e: Object-Oriented version of Specification and Design Language
SDLC e: Service Data Link Control
- e: Software Development Life Cycle
SDLC. e: Synchronous Data Link Control
SDL[/]GR e: Specification and Description Language/Graphic[al] Representation
SDLP e: Standard Device Level Protocol
SDL/PR e: Specification and Design Language Phrase Representation
S[-]DLTS e: Scanning DLTS
SDM e: Schematic Data Model
- e: Secure Digital Modem
- e: Short Data Message
- e: Site Defense of Minuteman
- e: Somali Democratic Movement
- e: Space Division Multiplexing
- e: Specific Device Model for SAB
- e: Sub-rate Data Multiplexer
- e: System Development Multitasking
SDMBIM e: Sensor-Driven, Model-Based Integrated Manufacturing
SDME e: Software Development and Maintenance Environment Software
- f: Sous-Direction des Moyens d'Essais
SDMM f: Service des Données sur le Milieu Marin
SDMMC e: Software Development, Modification and Maintenance Contract
SDMO f: Service Départemental de la Main-d'Œuvre
S/DMS e: SONET/Digital Multiplex System
SDMS e: SCSI Device Management System
- e: Society of Diagnostic Medical Sonographers
SDMX e: Space Division Matrix
SDN e: Schedule Delete Notification
- e: Software Defined Network
- e: Software Delivery Notice
- e: Software Development Notebook
- e: Software Digital Network
- e: Software Distribution Net
- e: Switched Digital Network
- e: System Development Notification
Sdnl f: Sans date, ni lieu
SDNP e: Sustainable Development Networking Programme
SDNS e: Secure Data Network Service
SDO e: Staff Duty Officer
- e: Standards-Developing Organization
- f: Station de Données Océaniques
Sdo s: Soldado
SDOC e: Selective Dynamic Overload Controls
SDP e: Sample Distribution Policy
- e: Science Data Processing
- e: Scitex Digital Printing
- e: Service Data Point
- e: Service Delivery Point
- e: Seychelles Democratic Party
- e: Social Democratic Party
- e: Sozialdemokratische Partei
- e: Specialized Data Point
- e: Standard Directory Package
- e: State Data Program
- e: Swaziland Democratic Party
- e: System Decision Package
- e: System Development Plan
- f: Sous-Direction des Pensions
- N: Stillingsdisponeringsplan
SDPA e: Society of Dermatology Physician Assistants
SDPC f: Service Départemental de la Protection Civile
SDPES f: Sous-Direction Prospective d'Études de Systèmes
SDPF e: Satellite Data Processing Facility
- e: Science Data Processing Facility
SDPK e: Social Democratic Party of Kenya
SDPL e: Sensor Data Processing Laboratory
SDPS e: SAR Data Processing System
- e: Science Data Processing System
SDR d: Süddeutscher Rundfunk
- e: Segment Design Review
- e: Sensor Data Record
- e: Session Director
- e: Software Design Review
- e: Somali Democratic Republic
- e: Special Drawing Right[s]
- e: Spin Dipole Resonance
- e: Spin-Dependent Recombination
- e: Staff Development Review
- e: Store Data Register
- e: Streaming Data Request
- e: Surrogate Data Radio
- e: System Design Review
- f: Service De Recherches
Sdr. a: Sønder
- a: Søndre
SDRA e: Software Development Resources Architecture
- f: Société pour le Développement de la Riviera Africaine
SDRAM e: Synchronous DRAM [Dynamic Random Access Memory]
SDRB e: Software Design Review Board
s*D,rel d: vollständige Schreibweise der speziellen relativen spektralen Empfindlichkeit bei Vollfilterung
SDRI e: Sealed Double-Ring Infiltrometer
SDRM f: Société pour l'administration du Droit de Reproduction Mécanique des auteurs, compositeurs et éditeurs
SD-ROM e: Super Density ROM
SDRPC e: Satellite Data Receiving and Processing Center
SDRPS e: Satellite Data Receiving and Processing System
SDRS e: Seaward-Dipping Reflector Sequence
SDRSI e: South Dakota Remote Sensing Institute
SDRT f: Sous-Direction des Recherches Techniques
SDS a: Sammensluttede Danske Sparekasser
- a: Socialdemokratisk Samfund
- e: Safety Disconnect Switch
- e: Satellite Data System
- e: Schema Definition Set
- e: Scientific Data Set
- e: Site Data System
- e: Smart Distributed System
- e: Social Development Service
- e: Societatis Divini Salvatoris
- e: Software Development Standards
- e: Software Distribution Service
- e: Solstice Disk Suite
- e: Staff Development Summary
- e: Standard Depot System
- e: Strategic Defense System
- e: Striped Domain Structure
- e: Students for a Democratic Society
- e: Submerged Demineralizer System
- e: Surface Discharge Spectroscopy
- e: Surface Distribution System
- e: Switched Data Service
- e: Switched Digital Service
- e: Synchronous Data Set
- e: Sysops Distribution System
- N: Stridsdommersjef
SDSC e: Standards and Data Services Committee
- e: Synchronous Data Set Controller
SDSCNET e: San Diego Supercomputer Center Network
SDSD e: Satellite Data Services Division
SDSF e: Somali Democratic Salvation Front
SDSI e: Synchronous Data-Link control
SDSL e: Single Line Digital Subscriber Line
- e: Symmetric Digital Subscriber Line
S[-]DSMA e: Slotted Digital Sense Multiple Access
SDSMA e: Slotted Digital Sense Multiple Access
SDSM&T e: South Dakota School of Mines and Technology
SDSP e: Small Data Set Packing
- e: Software Development Support Plan
- e: Space Defense System Program

SDS-PAGE e: Sodium Dodecyl Sulfate Polyacrylamide Gel Electrophoresis
SDSS e: Sequential Data Set Striping
- e: Spatial Decision Support System
SD_STB e: Streaming Data Strobe
SDSU e: SMDS Data Service Unit
- e: Switched multi-megabit Data Service Unit
SDT d: Schwedt/Oder, Brandenburg
- e: Sigma Delta Theta
- e: Solar Diameter derived Temperature variations
- e: Soldier Data Tag
- e: Sonic Digital logging Tool
- e: Structured Data Transfer
- e: Synchronous Data Transport
- e: Synchronous Digital Transmission
- f: Social-Démocratie Togolaise
SDTR f: Syndicat Départemental des Transports Routiers
SDTS e: Spatial Data Transfer Specification
- e: Spatial Data Transfer Standard
SDTT e: Similarly Dependence Textural Transformation
SDTV e: Standard Definition Television
SDU e: Satellite Delay Units
- e: Secure Data Unit
- e: Signalling Data Unit
- e: Slave Display Unit
- e: Software Distribution Utilities
- e: Subsurface Disposal Unit
SDUO e: Satellite Data Utilization Office
SDUS e: Satellite Data Users Station
- e: Small-Scale Data Utilization Station
SDV e: Shuttle Derived Vehicle
- e: Switched Digital Video
SDVGA e: Super-Duper VGA
SDWA e: Safe Drinking Water Act
SDX e: Storage Data Acceleration
SDXC e: Synchronous Digital Cross Connect
SE e: Shared Everything
- e: Sheet Extrusion
- e: Special Edition
- e: Standard Edition
- e: Staphylococcal Entero toxin
- e: Status Enquiry
- e: Switching Element
- e: Systems Engineer[ing]
- f: Saisie-Exécution
- f: Sauf Erreur
- f: Sceau de l'État
- f: Sciences Exactes
- f: Sciences Expérimentales
- f: Secrétaire d'État
- f: Service Économique
- f: Service Extérieur
- f: Services Exceptionnels
- f: Session Extraordinaire
- f: Société d'Encouragement
- f: Société française des Électriciens
- f: Sous-Entendu
- f: Suppression d'Emploi
- f: Sûreté de l'État
- f: Syndicat d'Éditeurs
- f: Syndicat d'Employés
- f: Système d'Exploitation
- f: Système Électoral
- f: Système Expert
- r: Skandinaviska Enskilda bank
Se r: skeppseldare
SE s: sudeste
.se e: Sweden [Internet]
S.E. s: Su Excelencia
S&E e: Salaries and Expenses
- e: Science and Engineering
- e: Surfaced one side and one Edge
S/E s: Servicio Especial
.sea e: Computer filename extension indicating a self-extracting archive
SEA a: Salgs- og Eksportorganisationen for Avlssvin
- e: Seattle, WA
- e: Self Extracting Archive
- e: Senior Enlisted Advisor
- e: Site Evaluation Accomplished
- e: Society for Electronic Access
- e: Standard Extended Attribute
- e: State Earthquake Administration
- e: State Enforcement Agreement
- e: State/EPA Agreement
- e: System for Environmental Analysis
- e: Systems Effectiveness Analysis
- f: Administration sismique d'État
- f: Service de l'Environnement Atmosphérique
- f: Société d'Électronique et d'Automatisme
- p: Secretéria de Estado do Algodão
- s: Administración Antisísmica Estatal
- s: Sociedad El Adelante
SEAADSA e: Sea Automated Data Systems Activity
SEABEAM e: Shipboard multi-transducer swath echo sounding system
Seabee e: CB, Construction Battalion
SEABEE e: Sea Barge System
SEAC e: Safety and Environmental Advisory Council
- e: Standard Eastern Automating Computing
- e: Standards Eastern Automatic Calculator
- e: State Energy Advisory Council
- e: Student Environmental Action Coalition
- p: Secretéria de Estado dos Antigos Combatentes
SEACAMP e: South-East Asian Centre for Atmospheric and Marine Prediction
- f: Centre de l'Asie du Sud-Est pour la prévision atmosphérique et maritime
- s: Centro de Asia Sudoriental para la Predicción Atmosférica y Marítima
SEACAT e: Seabird Conductivity and Temperature Recorder
SEACOP e: Strategic Sealift Contingency Planning System
SEAD e: Suppression of Enemy Air Defense[s]
SeaDAS e: SeaWiFS Data Analysis System
SEADCUGNAV e: SEA Data Communications User Group
SEADEX e: Shoreline Environment Atmospheric Dispersion Experiment
SEAFARER e: Surface ELF Antenna for Addressing Remotely EMP Receivers
SEAFDEC e: South-East Asian Fisheries Development Center
- f: Centre de développement des pêches de l'Asie du Sud-Est
- s: Centro de Desarrollo Pesquero de Asia Sudoriental
SEAFIRE e: South East Asia Fire Experiment
- e: South-East Asia Fire Research Experiment
SEAFIS e: South-East Asian Fisheries Information System
- f: Système d'information sur la pêche en Asie du Sud-Est
- s: Sistema de Información sobre la Pesca de Asia Sudoriental
SEAFLOE e: Southeast Florida Outfalls Experiment
SEAGS e: Southeast Asian Geotechnical Society
SEAHP e: Southeast Asian Health Project
SEAJ e: Semiconductor Equipment Association of Japan
SEAL e: Screening External Access Link
- e: Sea, Air, Land
- e: Segmentation and Reassembly Layer
- e: Simple and Efficient Adaptation Layer
- e: Software Engineering and Analysis Laboratory
SEALION e: Sea Level Instrumentation and Observation Network
SEAL team e: Sea-Air-Land team
SEAM e: Sedimentary Evolution in Active Margins
- e: Superfund Exposure Assessment Manual
- e: Surface, Environment and Mining
- f: Evolution Sédimentaire sur les Marges Actives
- s: Evolución Sedimentaria en Márgenes Activas
SEAMAP e: Southeast Area Monitoring and Assessment Program
SEAMEO e: South East Asian Ministers Education Organization
- e: Southeast Asian Ministers Education Organisation
SEAMO e: Southeastern Ontario Academic Medical Organization
SEAPOL e: South-East Asia Programme on Ocean Law, Policy and Management
- f: Projet relatif au droit, aux politiques et à la gestion maritimes en Asie du Sud-Est
- s: Proyecto de Asia Sudoriental sobre Derecho, Política y Ordenación Oceánicos
SEAREX e: Sea-Air Exchange [program]
- e: Study on Sea-Air Exchanges
- f: Étude sur les échanges mer-air
- s: Estudio sobre el Intercambio Mar-Aire
SEAR GOOS e: South-East Asian Regional GOOS
- f: Composante régionale du GOOS pour l'Asie du Sud-Est
- s: GOOS Regional para el Sureste de Asia

SEAS a: Sydøstsjællands Electrictets Aktieselskab
- e: Shipboard Environmental [data] Acquisition System
- e: Signaling Engineering and Administration System
- e: Strategic Environmental Assessment System
- e: Survey of Environment Assisted by Satellite
- e: Systems, Engineering and Analysis Support
- f: Surveillance de l'Environnement Assistée par Satellite
- f: Système Embarqué pour l'Acquisition de données sur l'environnement
- p: Secretéria de Estado dos Assuntos Sociais
- s: Sistema de Adquisición de Datos Ambientales desde Buques
- s: Vigilancia del Medio Ambiente mediante Satélite

SEASAR e: Sea Synthetic Aperture Radar
SEASAT e: amerikanischer Satellit zur Meeresbeobachtung
- e: Earth Satellite Dedicated to Oceanographic Applications
- f: Satellite de la terre destiné à des applications océanographiques
- s: Satélite terrestre para aplicaciones oceanográficas

Seasat e: Sea Satellite
SEASPAN e: SEAREX South Pacific Aerosol Network
SEASTAG e: SEATO Standardization Agreement
SeaStar e: Spacecraft flying SeaWiFS
SEASTRAT e: Sealift Strategic Contingency Planning System
SEATAR e: Joint CCOP-IOC Working Group on Post-IDOE Studies of East Asian Tectonics and Resources
- f: Groupe de travail mixte CCPM-COI pour les études postérieures à la DIEO sur la tectonique et les ressources de l'Asie orientale
- s: Grupo Mixto de Trabajo CCOP-COI sobre Estudios Posteriores al DIEO relativos a los Recursos y la Tectónica de Asia Oriental

SEATO e: Southeast Asian Treaty Organization
SEA[-]WIFS e: Sea [viewing] Wide Field [of view] Sensor[s second]
- f: Instrument à grand champ d'observation des mers
- s: Detector océanico de amplio campo de visión

SEB d: Sebnitz, Sachsen
- e: Small and Emerging Business
- e: State Electricity Board

S E B e: School of Education Building
SEBACAM f: Société des Bauxites du Cameroun
SEBEX e: Sahelian Energy Balance Experiment
SEBI e: Securities and Exchange Board of India
SEBINA e: Library system
SEbS e: Southeast by South
SEBS e: Styrene Ethylene Butylene Styrene block copolymer
SEBSCC e: Southeast Bering Sea Carrying Capacity
SEC e: Office of the Secretary
- e: Science Executive Committee of the IWG
- e: Securities (or Security) [and] Exchange Commission
- e: Security Clearance
- e: Senior Enforcement Counsel
- e: Size Exclusion Chromatography
- e: South Equatorial Current
- e: Space Environment[al] Center
- e: Space Event Conference
- e: Standard Ethernet Controller
- e: State Enterprises Commission
- f: Section d'Enquête et de Contrôle
- f: Service de l'État Civil
- f: Service des Échanges Culturels
- f: Société d'Études Camerounaises
- p: Secretéria de Estado do Caju
- s: Sección

sec e: secant
Sec e: Section
- f: seconde
- f: secrétariat
- f: secteur

SECA f: Service d'Études et Constructions Aéronautiques
- f: Société Européenne des Carburants

Sec A Ap s: Sección de Armas de Apoyo
Sec Abast s: Sección de Abastecimiento
Sec Acopto s: Sección de Acompañamiento
SECAM e: Sequential Color And Memory
- f: Séquentiel [en] Couleur Avec Mémoire
- f. Système Electronique Couleur Avec Memoire
- f: Système Séquentiel Couleurs à Mémoire

SECAN e: Communications Security and Evaluation Agency
- e: military committee communications Security and evaluation Agency
- e: Science and Engineering Committee on Advisory to NOAA

Sec Ap A s: Sección de Apoyo Aéreo
SECAS e: Ship Equipment Configuration Accounting System
SECAT f: Société d'Études et de Construction d'Appareils de Télémécanique
SECB e: Severely Errored Cell Block
- e: Systems Engineering Control Board

SECC e: Survivable Enduring Command Center
Secc s: Sección
Sec Cbl Mltp s: Sección de Cable Múltiple
Sec Cbl Ord s: Sección de Cable Ordinario
Sec CCL s: Sección de Carros [de Combate] Ligeros
Sec CCM s: Sección de Carros [de Combate] Medios
Sec CCP s: Sección de Carros [de Combate] Pesados
SecE f: Secrétaire d'État
Secede III e: Release of a barium cloud in near space by a rocket
SECES f: Syndicat des Enseignants et Chercheurs dans l'Education Suprieure
Sec Fus s: Sección de Fusiles
SECGRU e: Security Group
SECHO e: Supervised Extraction and Classification of Homogeneous Objects
Sec Hrtz s: Sección de Cable Hertziano
SECIG f: Service d'Études et de Coordination de l'Information Gouvernementale
Sec LT s: Sección de Lanchas Torpederas
SECM e: Scanning Electrochemical Microscopy
- f: Société pour l'Étude du Chargement Mécanique dans les mines de fer de Lorraine

SECMAI f: Société d'Études et de Construction de Machines et Appareils Industriels
Sec Mec s: Sección Mecanizada
Sec Mot s: Sección Motorizada
SECO r: Sveriges Elevers Centralorganisation
SECOFI e: The Ministry of Commerce and Industrial Development
SECOM e: Special Elections Commission
SECOMP e: Secure Enroute Communications Package
SECON s: Servicios Comunitarios de Norte
SECORD e: Secure Cord Switch Board
SECOTER f: Société d'Exploitation et de Construction d'Engins de Réparation navale
SECPIA f: Société d'Études Chimiques Pour l'Industrie et l'Agriculture
SECPLN e: Security Plan
SECR e: Standard Embedded Computer Resource
Secr f: Secrétaire
- f: Secrétariat

Sec Rd s: Sección de Radio
SECRE f: Société d'Études et de Recherches de Constructions Électroniques
SECREN f: Société d'Exploitation pour la Construction et la Réparation Navale
Secrgen f: Secrétaire Général
SECS e: Semiconductor Equipment Communication Standards
SECSTATE e: Secretary of State
SECT f: Syndicat des Enseignants Catholiques du Togo
Sect f: Secteur
SECTA f: Syndicat Belge des Employés Techniques et Cadres de l'Aéronautique
Sec Tf s: Sección de Telefonía
Sec Tg s: Sección de Telegrafía
SECTIONLOG e: curatorial database for individual core-section histories
SECTIONS e: Energetically Active Zones of the Ocean [EAZO] and Climate Variability
- e: Energetically Active Zones of the Ocean and Climate Variability
- e: Oceanographic Program[me] of the USSR
- f: Zones énergétiquement actives de l'océan [EAZO] et variabilité climatique
- s: Zonas Energéticamente Activas del Océano [EAZO] y Variabilidad Climática

SECTO f: Syndicat des Employés du Commerce du Togo
Secu f: Sécurité sociale
SECYT e: Secretariat for Science and Technology
Sec Zap s: Sección de Zapadores
Sec Zap Ac s: Sección de Zapadores Acorazada
SED e: Safety Evaluation Document
- e: School of Education
- e: Sensor Evolutionary Development
- e: Simulative Electronic Deception
- e: Smoke Emitting Diode
- e: Special Education
- e: Spherical Equivalent Diameter
- e: Stream[-Oriented] Editor
- f: Service d'Études des Débouchés
- f: Service des Études et de la Documentation
- f: Système d'Exploitation de Disquettes
SEDA e: Staff and Educational Development Association
- f: Société d'Études pour le Développement de l'Afrique
SEDAC e: Socio-Economic Data and Applications Center
SEDAS d: Standardregelungen einheitlicher Datenaustausch
- e: Satellite Environmental Data Acquisition System
Sedco e: Swaziland Enterprise Development Corporation
SEDCO/BP 471 e: registered name of drilling vessel used by ODP
s-EDDA e: symmetrical EDDA
SEDEFITA f: Société d'Étude et de Développement des Périmètres du Fiherenana et de la Taheza
SEDES f: Société d'Études pour le Développement Economique et Social
SEDESOL e: Mexican environmental agency
SEDF f: Service d'Entraide et de Documentation Familiale
SEDI f: Société d'Études de Développement Industriel
- f: Société Européenne de Développement Industriel
SEDIAC f: Société pour l'Étude et de Développement de l'Industrie, l'Agriculture et le Commerce
Sedichar f: Service de Distribution de Charbon
SEDIME f: Service d'Échange et de Diffusion d'Images Médicales Européen
SEDIMENT e: Soil Erosion Direct Measurement
SEDIS e: Site Exercise Data Injection System
SEDOC e: Service de Documentation
SEDOM e: Smallholder Enterprise Organization of Malawi
SEDS e: Students for the Exploration and Development of Space
SEDT e: Synthetic Environment Dynamic Terrain
SEE d: Seelow, Brandenburg
- d: Seewesen
- e: Sample Exchange Evaluation
- e: Senior Environmental Employment
- e: Signing Exact English
- e: Single-Event Effects
- e: Surface Energy Exchange
- f: Service de l'Expansion Économique
- f: Société des Electriciens, des Electroniciens et des Radiolectriciens
S&EE e: Safety and Environmental Engineering
SEED e: Seed House
- e: Self Electrooptical Effect Device
- e: Support for East European Democracy
- e: U.S. Support for Eastern European Democracy Act
SEEE f: Société d'Études et d'Entreprises Electriques
SEEG f: Société d'Energie et d'Eau du Gabon
SEEK BUS e: High Data Rate, Anti-Jam Communications System
SeeLotUntV 1998 d: Seelotsenuntersuchungsverordnung vom 12.3.1998
SEEMG s: Sindicato de Educadores de Enseñanza Media
SEEP e: Shelf Edge Exchange Processes
- e: Shelf-Edge Exchange Processes
SEER e: Surveillance, Epidemiology and End Results
- e: Surveillance, Epidemiology and End Results Program
SEES e: Support Equipment and Engineering Services
SEET f: Société Européenne d'Équipement de Turboforage
seet o f: Sauf Erreur Et Omission
SEF e: Satellite Evaluation Facilities
- e: Source-Explicit Forwarding
- e: State Employees Federation
- e: Student Evaluation of Faculty
- f: Société Électrique Française
- p: Saneamento Econômico e Financeiro
SEFA e: South East Forest Alliance
- e: State Employees Federated Appeal
SEFC e: South-East Fisheries Centre
- e: Southeast Fisheries Center
- f: Société Européenne de Fibres de Carbone
SEFCAR e: Southeast Florida and Caribbean Recruitment
SEFD e: System Equivalent Flux Density
SEFI e: Sequential Electronic Fuel Injection
SEFT f: Section d'Études et de Fabrication des Télécommunications
SEG e: Scientific Ecology Group, Inc.
- e: Scientific Evaluation Group
Seg s: Seguridad
SEGA f: Société d'État de Gestion des Abattoirs
Seg Comb s: Seguridad Combativa
SEGEPLAN s: Secretaría General de Planificacón Ecnómica
SEH e: Structured Exception Handling
SEHS e: Student and Employee Health Services
SEI e: Software Engineering Institute
- e: Space Exploration Initiative
- e: Special Education-Emotionally Impaired
- e: Stockholm Environment Institute
- e: Strategic Environmental Initiative
- f: Institut de Stockholm pour l'environnement
- f: Service Européen d'Information
- f: Système Électrique International
- s: Instituto de Estocolmo para el Medio Ambiente
SEIA e: Socioeconomic Impact Analysis
SEIB f: Société d'Études d'Invention et de Brevets
Seicom e: ECRC-POP in Pfullingen
SEIDAM e: System of Experts for Intelligent Data Management
SEIE f: Société d'Études et d'Informations Économiques
Seikagaku j: Seikagaku. Journal of Japanese Biochemical Society
SEIM e: Software Engineering Improvement Method
SEIP e: Standard Engineering Installation Package
SEIS e: Senior Executive Information System
- e: Supplemental Environmental Impact Statement
Seishin Shinkeigaku Zasshi j: Seishin Shinkeigaku Zasshi. Psychiatria et Neurologia Japonica
SEISMED e: Earthquake Hazard Disaster Management in the Mediterranean Region
- f: Réduction des risques sismiques dans la région méditerranéenne
- s: Lucha contra los Desastres provocados por Terremotos en la Región del Mediterráneo
SEJF e: Sri Lankan Environmental Journalists Forum
SEK e: Swedish Electrical Commission
- r: Svenska Kronor
Sek r: Sekund
sek. N: sekund[er]
sek. Ltn N: sekondløytnant
sekr. N: sekretær
Sekt r: Sektion
Sektch r: Sektionschef
SEL a: Statens Erhverspædagogiske Læreruddannelse
- d: Standard Elektrik Lorenz
- e: Science and Engineering Library
- e: Selector
- e: Software Engineering Laboratory
Sel. e: Select
- e: Selection[s]
SELA e: Sistema Economico Latinoamericana
- e: System Error Log Analyzer
SELAF f: Société des Études Linguistiques et Anthropologiques de France
SELC e: Southern Environmental Law Center
SELCAL e: Selective Call
SELDADS e: SEL Data Acquisition and Display System
- e: Space Environmental Laboratory Data Acquisition and Display System

SELF e: Submarine Extremely Low Frequency
SELINCOR e: Selective Inverse detection of carbon-hydrogen Correlation
SELL e: Spectral Ellipsometry
SELMA e: Southeastern Lumber Manufacturers Association
SELOR e: Shipboard Emitter Location Group
SELPER e: Society of Latin American Specialists in Remote Sensing
- f: Société des Spécialistes Latino-Américains de Télédétection
- s: Sociedad de Especialistas Latinoamericanos en Percepcion Remota
- s: Sociedad de Especialistas Latinoamericanos en Percepción Remota
SELRAS e: SEL Research and Analysis System
SELREL e: Selective Release
SELRES e: Selective Reserves
SELRIP e: Selective Release Improvement Program
SELSCAN e: Automatic Frequency Scanning/Channel Selecting
selsk. a: selskab
SELT f: Syndicat d'Enseignants Laïcs du Togo
SELV e: Small Expendable Launch Vehicle
SEM d: Semester
- e: Scanning Electron Microscope
- e: Server Enhancement Module
- e: Significant Events Message
- e: Society for Experimental Mechanics
- e: Software Engineering Manual
- e: Solar Environmental Monitor
- e: Specific Equipment Model
- e: Standard Equipment Modules
- e: System Environment Monitor
- f: Appareil d'Étude du Milieu Spatial
- f: Société d'Électricité et de Mécanique
- f: Son Excellence Monsieur
- s: Observador del Medio Espacial
- s: Servicio de Estado Mayor
Sem e: Seminar
- f: Semaine
SEMA f: Syndicat des Équipements et Matériels Aéronautiques
SEMAM e: Secretariat for the Environment of the Presidency of the Federative Republic of Brazil
SEMATECH e: A US government-industry-academic consortium for Semiconductor Technology
SEMA-X e: Special Electronic Mission Aircraft
SEMC e: Scaled Ensemble Monte Carlo
- e: Schrödinger Equation-based Monte Carlo
SEM/EDA e: Scanning Electron Microscope/Energy Dispersive Analysis
SEM/EDXA e: Scanning Electron Microscope/Energy Dispersive X-Ray Analysis
SEMG e: Systems Engineering Management Group
Semin Arthritis Rheum e: Seminars in Arthritis and Rheumatism
Semin Cancer Biol e: Seminars in Cancer Biology
Semin Cell Biol e: Seminars in Cell Biology
Semin Dermatol e: Seminars in Dermatology
Semin Diagn Pathol e: Seminars in Diagnostic Pathology
Semin Gastrointest Dis e: Seminars in Gastrointestinal Disease
Semin Hematol e: Seminars in Hematology
Semin Immunol e: Seminars in Immunology
Semin Liver Dis e: Seminars in Liver Disease
Semin Nephrol e: Seminars in Nephrology
Semin Neurol e: Seminars in Neurology
Semin Nucl Med e: Seminars in Nuclear Medicine
Semin Oncol e: Seminars in Oncology
Semin Oncol Nurs e: Seminars in Oncology Nursing
Semin Pediatr Surg e: Seminars in Pediatric Surgery
Semin Perinatol e: Seminars in Perinatology
Semin Respir Infect e: Seminars in Respiratory Infections
Semin Roentgenol e: Seminars in Roentgenology
Semin Speech Lang e: Seminars in Speech and Language
Semin Surg Oncol e: Seminars in Surgical Oncology
Semin Thorac Cardiovasc Surg e: Seminars in Thoracic and Cardiovascular Surgery
Semin Thromb Hemost e: Seminars in Thrombosis and Hemostasis
Semin Ultrasound CT MR e: Seminars in Ultrasound, CT and MR
Semin Urol Oncol e: Seminars in Urologic Oncology
Semin Vasc Surg e: Seminars in Vascular Surgery
Semin Vet Med Surg [Small Anim] e: Seminars in Veterinary Medicine and Surgery [Small Animal]
SEMIP e: Secretariat of Energy, Mines and Parastatal Industry
SemiR s: Semirremolque
SemiSPIN e: Semiconductor Software Process Improvement Network
SEMM e: Systems Engineering Management Meeting
SEMO f: Saint-Etienne Machine-Outil
SEMP e: Standard Embarkation Management System
SEMPA e: Society of Emergency Medicine Physician Assistants
SEMPAO f: Syndicat des Entreprises de Manutention de Ports d'Afrique Occidentale
SEMPEL f: Société d'Économie Mixte Pour l'Équipement du Loiret
SEMPER e: Secure Electronic Marketplace for Europe
SEMS e: Seafloor Earthquake Measurement System
- e: Space Environment Monitor System
- e: Systems Engineering and Management Support
- f: Société Européenne de Mini-informatique et de Systèmes
SEMT f: Société d'Études de Machines Thermiques
SEN d: Sammelentsorgungsnachweis
- e: System Evaluation Network
sen. a: senere
Señ s: Señal
Sén f: sénateur
SENA f: Société d'Énergie Nucléaire des Ardennes
S. en C. s: Sociedad en Comandita
SEN CTR e: Senior Center
SENELEC f: Société Nationale d'Electricité
Señ Enf Manp s: Señal de Enfilación de Manipulación
SENFAS e: Support and Education Network for Fetal Alcohol Syndrome Parents and Caregivers
SENIR e: Brazils Irrigation Ministry
SENN f: Société Électronucléaire Nationale
SENSAT e: Sensor-Atmosphere-Target
SENTAR e: Senior Engineering Technology Assessment Review
s.e.&o. a: salvo errore et omissione
SEO e: Sequences Number
- e: Socio-Economic Objective
- f: Sauf Erreur[s] ou Omission[s]
SEOCS e: Sun Earth Observatory and Climatology Satellite
SEOI p: Secretéria de Estado da Ordem Interna
SEON e: Solar Electro-Optical [Observing] Network
SEOS e: Synchronous Earth Observation Satellite
- f: Satellite synchrone d'observation de la Terre
- s: Satélite sincrónico de observación de la Tierra
SEP e: [Color] Separation
- e: Scorched Earth Party
- e: Self-Employed Pension
- e: September
- e: Software Engineering Process
- e: Somebody Elses Problem
- e: Someone Elses Problem
- e: Special Emphasis Panel
- e: Special Emphasis Program
- e: System Engineering Process
- e: System Enhancement Program
- f: Société Européenne de Propulsion
- s: Septiembre
SEPA e: Science Education Program for Africa
- e: Scottish Environment Protection Agency
- e: State Environmental Policy Act
- e: System Evaluation and Performance Analysis
SEPAC e: South-Eastern Pacific
- f: Pacifique Sud-Est
- s: Pacífico Sudoriental
SEP BDE e: Separate Brigade

SEPDU e: Southern Ethiopian Peoples Democratic Union
SEPEC e: Seychelles Petroleum Company
SEPEMA f: Société d'Études Pour l'Électronique et les Mécanismes Automatiques
SEPESCA e: Secretariat of Fisheries
SEPFA e: South-East Professional Fishermen's Association
SEPG e: Software Engineering Process Group
SEPHOS f: Société d'Études et de Participations Phosphatières
SEPI e: School Educational Provision Improvement
SEPM e: Stochastic Entraining Parcel Model
SEPP e: Secure Electronic Payment Protocol
- e: Secure Encryption Payment Protocol
- e: Society for the Education of Physicians and Patients
- f: Société d'Études de Procédés Pharmaceutiques
SEPPIC f: Société d'Exploitation de Produits Pour les Industries Chimiques
SEPR e: Senior Enlisted Performance Report
- f: Société d'Étude pour la Propulsion par Réaction
SEPSU e: Science Engineering Policy Studies Unit
Sept f: septembre
SEPT f: Service d'Études communes de la Poste et de France Télécom
- f: Syndicat des Enseignants Protestants Togolais
SEPU e: Southern Ethiopian Peoples Union
SEPWC e: Senate Environment and Public Works Committee
Seq# e: Sequence Number
SEQDB e: Semiconductor Equipment Database
SEQUAL e: Seasonal Equatorial Atlantic Experiment
- e: Seasonal Response of the Equatorial Atlantic program
- f: Expérience saisonnière dans l'Atlantique équatorial
- s: Experimento Estacional en el Atlántico Ecuatorial
SER e: Satellite Equipment Room
- e: Scanning Emission Radiometer
- e: Society for Epidemiologic Research
- e: Soft Error
- e: Solar Extinction Radiometer
- e: State of the Environment Report
- e: System Engineering Report
- f: Service d'Études et [de] Recherches
- f: Service Européen de Recherche [des disparus]
- f: Société Equatoriale de Raffinage
- f: Syndicat des Électroradiologistes
Sér f: Série
SERA e: Software Engineering Review and Audits
SERAE f: Société d'Études et Réalisations Aérothermiques
Serapp r: Sektorrapportering
SERAT f: Société d'Études, de Réalisations et d'Applications Techniques
SERATOM f: Société d'Études et de Recherches Atomiques
SERB f: Société d'Études et de Recherches Balistiques
SERBEP e: Southeastern Regional Biomass Energy Program
SERBS e: Surface Energy and Radiation Balance System
SERC e: Science and Engineering Research Council
- e: Science Executive Research Council
- e: Smithsonian Environmental Research Center
- e: Software Engineering Research Center
- e: State Emergency Response Commission
- e: Swedish Energy Research Commission
SERCAA e: Support of Environmental Requirements Cloud Analysis and Archive
SERCATE s: Servicios de Capacitación Tecnica y Educativa
SERCE f: Syndicat des Entrepreneurs de Réseaux de Centrales et d'equipement industriel Électrique
SER/CED e: Socio-Economic Council, Committee of Economic Experts
SERD e: School of Environment, Resources and Development
- e: Support Equipment Recommendation Data
- e: Support Equipment Requirements Data
- e: Support Equipment Requirements Document
SERDP e: Strategic Environmental Research and Development Program
SERE e: Spatial Effective Resolution Element
SEREHI s: Secretaria de Recursos Hidraulicos
SEREM f: Société d'Études, de Recherches et d'Exploitations Minières
SEREPCA f: Société de Recherches et d'Exploitation des Pétroles du Cameroun
SERES-1 e: Sedimentation in Reservoirs
SERF e: Solar Electromagnetic Radiation Flux
- e: Special Environmental Research Facility
- e: Subsurface Environmental Research Facility
- e: Systems Enhancement Request Form
SERFIN a: Skade- och Erfarenhetsuppföljning på Internet
SERGG e: Southeastern Regional Genetics Group
SERHOLD e: Serials Holdings
SERI e: System for Earth Resource Information
- e: Systems Engineering Research Institute
SERIES e: Satellite Emission Radio Interferometric Earth Surveying
SERIES-X e: Satellite Emissions Range Infrared Earth Surveying Experiment
SERIS e: State of the Environment Reporting Information System
SERJUS s: Servicios Jurídicos y Sociales
SERL e: System Engineering Reference Library
SERM e: Structured Entity Relationship Model
- f: Service des Essais et Recherches sur la sécurité dans les Mines
- f: Société d'Études et de Recherches Minières de Madagascar
SERMIS e: Support Equipment Rework MIS
SEROCC e: Southeast Region Operations Center
SERODS e: Surface-Enhanced Raman Optical Data Storage
SERON e: Southeastern Regional Oxidant Network
SERP e: Sedimentary Rock Program
- f: Service d'Édition et de Routage de la Presse
- f: Syndicat d'enseignement de la Région parisienne
SERPAJ s: Servicio Paz y Justicia
SERPI f: Société d'études et de recherches de production industrielle
SERS e: Science and Engineering Research Semester
- e: Selective Equipment Removal System
- e: SES and Executive Resources Staff
Ser Salv s: Servicio de Salvamento
SERSID f: Société d'Études et de Réalisations de la Sidérurgie
sersj N: sersjant
SERT e: Sertraline
- e: Ship Electronic Readiness Team
SERUH f: Société d'Études et de Recherches pour l'Utilisation d'Hélicoptères
Serv s: servicio
Server d: Rechner, der Datenbestände, Ressourcen und Peripherie [Drucker usw.] verwaltet und sie auf Anforderung den angeschlossenen Rechnern [Clients] zur Verfügung stellt
SERVFOR e: Service Force
SES e: Seagrass Ecosystem Study
- e: Secondary Emissions Standard
- e: Secondary-Electron Spectroscopy
- e: Senior Executive Service
- e: Service Evaluation System
- e: Sidewall-Entry Sub
- e: Smart Energy System
- e: Spin Echo Spectroscopy
- e: Surface Environmental Surveillance
- e: System Engineering Specification
- f: Étude des écosystèmes de zostères
- f: Section Experimental et de Servitude
- f: Société des Écrivains Suisses
- f: Société des Étudiants Suisses
- f: Société Européenne des Satellites
- f: Syndicat de l'Enseignement Supérieur
- f: Syndicat des Enseignants du Sénégal
- s: Estudio de los Ecosistemas de Zosteras
- s: Servicio de Estrategias de Sustenabilidad
SESA e: State Employment Security Agency
SESAM e: System for Electronic Support of Academic Material

SESAME e: Second European Stratospheric Arctic Middle Latitude Experiment
- e: Solid Earth Science and Application Missions for Europe
- e: Structure and Energetics of the Stratosphere and Mesophere

SESAP e: Surgical Education and Self-Assessment Program

SESC e: SGN Eurisys Services Corp.
- e: Space Environment Service Center

SESCAN e: Senior Staff Consultation Association of Nigeria

SE-SCSI e: Single Ended SCSI

SESD e: Science and Engineering Systems Division

SESI e: System Engineering and System Integration

SESIN f: Service Central de Sûreté des Installations Nucléaires

SESP e: Space Environmental Satellite Program
- e: Surface Environmental Surveillance Project

SESRA e: Seattle Employee Services Recreation Association

SESS e: Space Environment Sensor Suite

SESSIA f: Société pour l'Étude des Souffleries Supersoniques de l'Industrie Aéronautique

SEST e: Swedish ESO [European Space Organisation] Submillimeter Telescope

SESTS e: Security Evaluation Space Transportation System

SESYM r: Strids-, Eldlednings- och Sensorsystem för Ytattack i Marinen

SET e: Driver Set
- e: Image Settings
- e: Science, Engineering and Technology
- e: Secure Electronic Transaction
- e: Secure/Standard Electronic Transaction
- e: Security Evaluation Team
- e: Sediments, Environment and Tectonics
- e: Setkey
- e: Simultaneous Engineering Teams
- e: Single Electron Tunneling
- e: Software Engineering Technology
- e: Student Evaluation of Teaching
- e: Supreme Electoral Tribunal
- e: Systems Engineering Team
- f: Sédiments, Environnement et Tectonique
- f: Société Européenne de Téléguidage
- f: Standard d'Echange et de Transfert
- f: Systeme d'Exchange et de Transfer
- s: Sedimentos, Medio Ambiente y Tectónica

Set. e: Signal element timing

SETA e: System[s] Engineering and Technical Assistance

SETAC e: Society of Environmental Toxicology and Chemistry

SETAG e: Software Engineering Technology Advancement Group

SETCA f: Syndicat des Employés Techniciens et Cadres

SETCB f: Syndicat des Employés, Techniciens et Cadres de Belgique

SETEC e: Semiconductor Equipment Technology Center

SETEL f: Société Européenne de Téléguidage

SETEXT e: Structure Enhanced Text

SETF e: South East Trawl Fishery

SETFIA e: South East Trawl Fishery Industry Association

SETI e: Science, Education and Technical Information
- e: Science, Engineering and Technology Institute (or Institutions)

Set L f: Saône-et-Loire

SETL e: Set Theory Language

Set M f: Seine-et-Marne

SETMAC e: South East Trawl Ministerial Advisory Committee

Set O f: Seine-et-Oise

SETR e: Sedimentary Environments, Eustatic Sea-Level Changes, Tectonics and Resources
- f: Environnements Sédimentaires, variations eustatiques, Tectonique et Ressources
- s: Medios Sedimentarios, Modificaciones Eustáticas, Tectónica y Recursos

Set S f: Sain et Sauf

SETS e: Site Enforcement Tracking System
- e: Statistical Export and Tabulation System
- e: Superfund Enforcement Tracking System

SETT e: Space Environmental Technology Transition

Set-Top-Box d: Gerät zur Umwandlung der digitalen Programme

SEU e: Single Event Upset[s]
- e: Smallest Executable Unit

s.e.u.o. s: salvo error u omisión

SEV e: Stockpile Emergency Verification

SEVIRI e: Spinning Enhanced Visible and Infra-Red Imager

SEVOC e: Secure Voice

SEVOCOM e: Secure Voice Communications

SEVUPS e: South Eastern Virginia Urban Plume Study

SEW d: Sozialistische Einheitspartei Westberlin
- e: Space and Electronic Warfare

SEWC e: Space And Electronic Warfare Commander

S/EWCC e: Signals Intelligence/Electronic Warfare Coordination Center

SEWCC e: Signals Intelligence Electronic Warfare Coordination Center

SEW-HG e: SEW-Hochschulgruppe

SEWP e: Scientific and Engineering Workstation Procurement

SEWP BOWL e: SEWP Business Operations and Workstation Laboratory

SEWP IPOC e: SEWP Installation Point of Contact

SEWS e: Satellite Early Warning System
- e: Scientific Engineering Work Station

SEX e: Software Exchange

SEXAFS e: Surface Extended X-ray Absorption Fine-Structure

Sex Transm Dis e: Sexually Transmitted Diseases

SEZID f: Société d'Equipement des Zones d'Industrialisation Décentralisée

SF a: Samfundsvidenskabeligt Forskningsråd
- a: Socialistisk Folkeparti
- a: Sømændenes Forbund i Danmark
- e: San Francisco [operations office]
- e: Satellite Fix
- e: Science Fiction
- e: Screening Factor
- e: Serum-Free
- e: Service Feature
- e: Slope Factor
- e: Spayed Female
- e: Special Forces
- e: Speculative Fantasy
- e: Speculative Fiction
- e: State Forest
- e: Stock Fund
- e: Subfield
- e: Superframe Format
- e: Superfund
- e: Support Facility
- e: Surface Freight
- e: Switching Fabric
- f: Sage-Femme
- f: Saisie Foraine
- f: Salaire Forfaitaire
- f: Sans Feuille d'avis
- f: Service Financier
- f: Sexe Féminin
- f: Sinistrés de France
- f: Société Fermière
- f: Soutien de Famille
- f: Sursalaire Familial
- f: Syndic de Faillites
- f: Syndicat de Fonctionnaires
- N: Sidefart
- N: Sivilforsvar
- N: Sjømanns Forbundet
- N: Sosialistisk Folkeparti
- N: Studentenes Fellesutvalg
- r: Statsanställdas Förbund

Sf f: Sans faute
- f: Sans formalités

s.f. N: syd for

s/f s: su favor

S/F s: Subfusil

S+F d: VJS für Sicherheit und Frieden

SF6 e: Sulphur Hexaflouride

SF-52 e: SF-52 Personnel Management Tracking System

SFA d: Soltau-Fallingbostel, Niedersachsen
- e: Silver Flag Alpha
- e: Space Flight Awareness
- e: Spectral Flame Analyzers
- f: Service des Fabrications de l'Aviation
- f: Service Fixe Aéronautique
- f: Société Française d'Astronautique
- s: Servicio Fijo Aeronáutico

SFAAB r: Svenska Flygmotor AB

SFAAT f: Société Financière pour l'Aviation d'Affaires et de Tourisme

SFAC f: Société de Forges et Ateliers du Creusot

SFAé f: Service des Fabrications de l'Aéronautique

SFATAT f: Service de la Formation Aéronautique, du Travail Aérien et des Transports

SFAusl d: Vorschriftensammlung Fernmeldeverkehr Ausland
Sfavc f: Sauf avis contraire
SFB d: Senftenberg, Brandenburg
- d: Sonderforschungsbereich[e]
Sfb e: Sonderforschungsbereich
SFBT N: Sivilforsvarets Befals- og Tjenestemannsforening
SFC a: Statens Filmcentral
- e: Space Forecast Center
- e: State Fishing Corporation
- e: Sub Frame Connectors
- e: Surface [Forcing]
- f: Société Française du Cancer
SFCA e: Space Transportation System
- e: System Flow Control Assembly
SFCF e: Standby Functional Check Flight Crew
SFCG e: Space Frequency Coordination Group
- f: Groupe de Coordination des Fréquences Spatiales
- s: Grupo de Coordinación de Frecuencias Espaciales
SFCI e: Singapore Federation of the Computer Industry
SFCP e: Strategy Force and Capabilities Plan
SFCS e: Shop Floor Control System
SFCS I/F e: Shop Floor Control System Interface
SFD e: Staffordshire
- e: Start [of] Frame Delimiter
- e: Sun Federal Division
SFDF e: Satellite Field Distribution Facility
SFDK e: Stepped-Frequency Delta-K
SFDM e: Statistical Frequency Division Multiplexing
SFDR e: Spurious Free Dynamics Range
SFDS a: A/S Sydfyenske Dampskibsselskab
SFDU e: Special Formatted Data Unit
- e: Standard Format[ted] Data Unit
SFE e: Special Facilities Equipment
SFEC f: Société de Fabrication des Éléments Catalytiques
SFECMAS f: Société Française d'Étude et de Construction de Matériel Aronautique Spécial
SFEL e: Standard Facility Equipment Lists
SFEM e: Second-Tier Foreign Exchange Market
SfEou O f: Sauf Erreurs ou Omissions
SFEP e: Society of Freelance Editors and Technical Information
SFERMA f: Société Française d'Entretien et de Réparation du Matériel Aéronautique
SFERT f: Système Fondamental Européen de Référence pour la transmission Téléphonique
SFES e: Small Explorer Front-end System
SFF a: Sammenslutningen af Firma-Funktionærer
- e: Small Form Factor
- e: Strategic Fuel Fund
- f: Secrétariat Faune-Flore
SFFAS e: Superfund Financial Assessment System
SFFS e: Satellite Freeze (or Frost) Forecast System
SFGL f: Société Française de Génie Logiciel
SFH e: Slow Frequency Hop
SFI e: Strategic Facilities Initiative
- f: Service Français d'Information
- N: Skipsteknisk Forskningsinstitutt
- N: Statlig Forvaltningsselskap for Industri
SFIA e: Surplus Facilities Inventory and Assessment Plan
SFIB f: Syndicat National des Fabricants d'Ensembles d'Informatique et de Machines de Bureau
SFIM f: Société de Fabrication d'Instruments de Mesure
SFIP f: Société Française d'Investissements Pétroliers
- f: Société Francophone d'Imagerie Pédiatrique
SFIR e: Sequential Filter Imaging Radiometer
SFIREG e: State FIFRA Issues Research and Evaluation Group
SFITV f: Société Française des Ingénieurs et Techniciens du Vide
SFK a: Slagteriernes Fællesindkøbsforening
- e: Störfallkommission
- e: Suffolk
- N: Sjøforsvarets Forsyningskommando
SFL e: Spherically Focused Log
SFLA e: Stupid Four Letter Acronym
SFLM f: Société Française des Lasers Médicaux
SFLU e: Spherically Focused Resistivity Measurement
Sflygning r: Spaningsflygning
SFM e: Sidama Freedom Movement
- e: Solar Flux Monitor
- e: Sum Frequency Mixing
- f: Société Française des Mécaniciens
Sfm f: Sauf mieux
SfM N: Sjefen for Marinen
SFMC e: Satellite Facility Management Center
SFMJI e: Sorry For My Jumping In
SFMP e: Surplus Facilities Management Program
- f: Société Française de Médecine Périnatale
SFMR e: Stepped Frequency Microwave Radiometer
- e: Swept Frequency Microwave Radiometer
SFN e: Single Frequency Network
- e: Society for Neuroscience
SFNAé f: Société Française de Navigation Aérienne
SFO e: San Francisco CA
- e: Satellite Field Office
- e: Servicing Finance Office
SFOAE e: Stimulus-Frequency Oto-Acoustic Emmission[s]
SFOB e: Special Forces Operations Base
SFOF a: Socialistisk Folkepartis Oplysnings-Forbund
SFOG f: Société Française d'Oncologie Gynécologique
SFO-S f: Société Français d'Organo-Synthèse
SFOSRC e: South Florida Oil Spill Research Center
SFOV e: Sounding Field of View
SFP d: Schriftenreihe des Senatsinstitut für Politikwissenschaft
- e: Scanning Fabry-Pérot Interferometer
- e: Single Function Processor
- e: Super Fine Positioning
- e: Supplementary Feeding Program
- f: Service de Formation des Pilotes
- f: Service de la Formation Professionnelle
Sfpl r: Spaningsflygplan
SFPS e: Secure Fast Packet Switching
SFPSM f: Société Française de Prophylaxie Sanitaire et Morale
SFQL e: Structured Full-text Query Language
SFR e: San Francisco, CA
- e: Spherically Focused Resistivity
- e: System Functional Review
- f: Société Française de Radiologie
- f: Société Française de Rhumatologie
- f: Société Française Radiotechnique
SFRD e: Secret, Formerly Restricted Data
- e: System Functional Requirements Description
SFRO r: Svenska Flottans Reservofficersförbund
SFRP e: SAC Inflight Refueling Program
SFRPG e: Science Fiction Role Playing Game
SFS e: Finnish Standards Association
- e: Saskatchewan Forest Service
- e: Secure File System
- e: Solar Flux Spectrometer
- e: Start of Frame Sequence
- e: Suomem Standardisoimisliitto
- e: System File Server
- f: Supplément Familial de Solde
SFSA f: Société Française de Secours aux Animaux
SFSC e: Southeast Fisheries Science Center
SFSS e: Satellite Field Service Site, Station
SFT d: Stassfurt, Sachsen-Anhalt
- e: System File Table
- f: Société des Fonctionnaires Techniques des chemins de fer suisses
- N: Statens Forurensningstilsyn
SFTA f: Société Française de Toxicologie Analytique
SFTAH e: Society For The Autistically Handicapped
SFTM e: Switched Frame Transfer Mode
SFTP e: Simple File Transfer Protocol
- f: Société Française de Transports Pétroliers
SFTU e: Swaziland Federation of Trade Unions
SFU a: Socialistisk Folkepartis Ungdom
- N: Sosialistisk Folkepartis Ungdom
- N: Studentenes Friundervisning
SfU r: Socialförsäkringsutskottet
SFUL e: Summary Fault Unit
SFV d: Standard-Festverbindung
SFW e: Sensor Fused Weapon
SFWA e: Science Fiction and Fantasy Writers of America

SFWRR e: Stock Fund Ware Reserve Requirements
SFX e: Sound Effect[s]
- e: Suffix
SG d: Sondergenehmigung
- e: Directorate of Medical Systems
- e: Earth system science division
- e: Scharfetter-Gummel
- e: Semiconductor-Grade
- e: Silicate Glass
- e: Star Guard
- e: State Graph
- e: Step-Graded
- f: Saisie-Gagerie
- f: Sans Garantie
- f: Secrétaire Général
- f: Service de la carte Géologique de France
- f: Service Général
- f: Société de Géographie
- f: Soins Gratuits
- f: Subvention du Gouvernement
- f: Sûreté Générale
- f: Syndicat Général
.sg e: Singapore [Internet]
SGA e: Shared Global Array
- e: Society of Geology Applied to mineral deposits
- e: Subjective Global Assessment
- e: System Global Area
SGACC f: Secrétariat Général de l'Aviation Civile et Commerciale
SGAM f: Service de Groupement des Achats de Matériels
SGAP e: Society for Growing Australian Plants
SGATC e: Second Generation Advisory Technical Committee
SGAUA e: Scitex Graphic Art User Association
SGB f: Société Générale du Bénin
SGBC f: Société Générale de Banques au Cameroun
SGBCI f: Société Générale de Banques en Côte d'Ivoire
SGBD f: Système de Gestion de Base[s] de Données [relationnelles]
SGBI f: Source Généralement Bien Informée
SGBS f: Société Générale de Banque du Sénégal
SGBT d: Schweizerische Gesellschaft für Biomedizinische Technik
SGB VI-ÄndG d: Gesetz zur Änderung des Sechsten Buches Sozialgesetzbuch
SGC e: SCSI Graphic Commands
- e: Symbolic Gray Code
- f: Service Général de Contrôle des semences et des plantes agricoles et horticoles
- f: Société Générale de Construction
- s: servicio de guardacosta
S-GCOS e: Space-based Global Change Observation System
SGCR e: Subcommittee on Global Change Research
SGCS e: Standard Global Stratigraphic Scale
SGCTMA f: Syndicat Général des Constructeurs de Tracteurs et de Machines Agricoles
Sgd s: Seguridad
SGDE e: Sustainable Gross Domestic Expenditure
SGDF f: Société du Gaz De France
SGDP e: Sustainable Gross Domestic Product
SGDRAM e: Synchronous Graphics DRAM
SGDRC e: Subgroup on Data Representation and Codes
- f: Sous-Groupe de la Répresentation des Données et des Codes
- s: Subgrupo sobre Representación de Datos y Claves
SGDT e: Store Global Descriptor Table
SGdWK d: Schulen des Gremiums der Wiener Kaufmannschaft
SGE f: Société Générale d'Entreprises
- f: Société Générale des Eaux
- f: Société Guinéenne d'Equipement
SGEMP e: Systems Generated Electromagnetic Pulse
SGEN e: Signal Generator
- e: System Generator
SGEOS e: Second Generation Earth Observation Satellite
sges d: [nicht]normierte spektrale Strahlungempfindlichkeit des gefilterten Empfängers
sgesD,F,Vmx d: Wert der gesamten spektralen Empfindlichkeit bei Vmx
SGF e: Screen Generation Facility
- e: Secretary to the Government of the Federation
- f: Société Générale de Fonderie
- f: Statistique Générale de la France
SGFET e: Suspended-Gate Field-Effect Transistor
SGFS e: Special Group on Functional Standardization
SGG d: Schweizerische Gesellschaft für Gerontologie
- e: Atmospheric experimental branch
- f: Secrétariat Général du Gouvernement
SGH d: Sangerhausen, Sachsen-Anhalt
SGI d: Systematisch-Geobotanisches Institut, Universität Göttingen
- e: Silicon Graphics, Inc.
- e: Sun-Glint Imager
SGIC f: Syndicat Général des Ingénieurs-Conseils
SGIGB f: Syndicat Général des Industries des Goudrons et Benzols
SGIM e: Society of General Internal Medicine
- f: Syndicat Général des Industries Mécaniques
SGK d: Sachgemeinkosten
SGL e: Standard General Ledger
- e: Static Groundwater Level
SGl f: Secrétariat Général
Sgl r: Segelfartygsutbildning
SGLDM e: Spatial Grey Level Dependence Matrices
SGLT e: Space to Ground Link Terminal
SGM e: Second Generation Meteosat
- e: Second-Generation Model
- e: Segmentation Message
- e: Shaded Graphics Modeling
- f: Société Générale de Minéraux
- f: Syndicat des Gens de Mer
SGML e: Standard General[ized] Mark-up Language
- e: Standard Graphic Markup Language
SGMM e: Semiconductor Generic Manufacturing Model
- f: Secrétariat Général de la Marine Marchande
SGMP e: Simple Gateway Management Protocol
SGMRS e: Semiconductor Generic Manufacturing Requirements Specification
SGMTP e: Service de Gestion du Matériel des Travaux Publics
SGN f: Société Générale pour les Techniques Nouvelles
SGNR f: Sans Garantie Ni Responsabilité
SGNTS e: Signatures
SGOE/TS e: Shuttle Ground Operations Efficiencies/Technologies Study
SGOM f: Service Géographique d'Outre-Mer
SGP e: Single Gauss Point
- e: Southern Great Plains
SGPF e: Spencer Gulf Prawn Fishery
SGPP e: Sedimentary and Geochemical Processes Panel
SGPV e: Sheep and Goat Poxvirus
SGR e: Society of Gastrointestinal Radiologists
- e: Specific Growth Rate
- f: Service Général du Renseignement
Sgr r: Spränggranat
SGRAM e: Synchronous Graphic[al]/[s] RAM [Random Access Memory]
SG REQ e: Surgeon General Requirements
Sgrp r: Spaningsgrupp
- r: Stridsgruppering
SGRS e: Spaceborne Geodynamics Ranging System
S-gruppe N: Stridsgruppe
SGS e: Site Ground Station
- e: SPSS Graphic Scheduling Interface
- f: Secrétariat général de la santé
SGT e: Sigma Gamma Tau
- e: Sonic Gate Tone
- e: Surrounding Gate Transistor
Sgt f: Segment
- s: Sargento
SGT s: Subagrupación Táctica
SGTD f: Société Générale des Transports Départementaux
SGTM f: Société Générale des Transports Maritimes
- s: Sindicato General de Trabajadores de Malaria
Sgto s: Sargento
Sgto Ayte s: Sargento Ayudante
SGUHM d: Schweizerische Gesellschaft für Unterwasser- und Hyperbarmedizin
S-gun e: Sputter gun
SGV d: Schweizerischer Fachverband für Gemeinschaftsverpflegung

SGVOG d: Schweizerische Gesellschaft für ein volkswirtschaftlich orientiertes Gesundheitswesen
SGW e: Simulated Groundwater
SGWCPFA e: Spencer Gulf and West Coast Prawn Fishermens Association
SGX e: Spacecraft Glow Experiment
SGZBB d: Schweizerische Gesellschaft für die zahnmedizinische Betreuung Behinderter und Betagter
SH a: Statens Husholdningsråd
- a: Statens Hørecentral
- a: Søderjyllands Højspændingsværk andelsselskab
- e: Second-Harmonic
- e: Shit Happens
- e: Spherical Harmonic[s]
- f: Salaire Hebdomadaire
- f: Salaire Horaire
- f: Service de l'Heure
- f: Surface Habitable
- f: Surtaxe d'Habitation
s.h. N: sjukehus
- N: sykehus
S&H e: Safety and Health
- e: Shipping and Handling
s/h a: svart-hvid
S/H e: Sample-and-Hold
SHA d: Schwäbisch Hall, Baden-Württemberg
- d: Stätische Handelsakademie
- e: Secure Hash Algorithm
- N: Skyteskolen for Hærensartilleri
SHAC e: Simple Hands-free Add-on Circuit
SHAEF e: Supreme Headquarters Allied Expeditionary Forces
SHAFT e: Society to Help Abolish FORTRAN Teaching
SHAFTS e: Semi-Hardened Facilities Tracking System
SHA-LCD e: Super High Aperture
S Hål Fdi N: Sør-Hålogaland Forsvarsdistrikt
SHAPE e: Supreme Headquarters Allied Powers, Europe
SHAPIRO e: Shapiro Undergraduate Library [Formerly UGLI]
SHAR e: Shell Archive[r]
SHARC e: Super Harvard ARchitecture Computer
SHARE e: Software Help for/in Applications, Research and Education
- e: Space Station Heat Pipe Advanced Radiator Element
SHARES e: Sharing of Experience in Space
SHARK e: Station HRPT Archiving and Reprocessing Kernel
SHARP e: Self-Healing Alternate Route Protection
- e: Short-Term Automated Weather Prediction
- e: Standard-Family HRPT Archiving and Request Product
- e: Surface Hugging Aid for Route Planning
- e: Synthetic Aperture Radar Processing System
SHARS e: Subtle Heavy Rainfall Signatures
SHASE e: Shareware Search Engine
SHAZAM e: Semi-Heuristic Algorithm by Zemel And Miller
SHB d: Sehen, hören, bilden. Beiträge zur Medienpädagogik
- d: Sozialistischer Hochschulbund
- e: Spectral Hole Burning
SHBG e: Sex Hormone-Binding Globulin
SHBT e: Single-Heterostructure Bipolar Transistor
SHC e: Shipping Coordination Committee
- e: State Hotels Corporation
- e: Student Health Center
- e: Supreme High Command
- f: Service Hydrographique du Canada
- s: Servicio Hidrográfico del Canadá
SHCA e: Society for Healthcare Consumer Advocacy
SHCD e: Sherlock Holmes Consulting Detective
SHCO e: Sexual Harassment Contact Officer
SHCS e: Sand Hill Crane Survey
SHDOM e: Spherical Harmonic Discrete Ordinate Method
S-HDSL e: Single-pair High bit-rate Digital Subscriber Line
SHE e: Safety, Health and the Environment
SHEA e: Society for Healthcare Epidemiology of America
SHEBA e: Surface Heat Budget of the Arctic [ocean]
SHED e: Segmented Hypergraphic Editor
- f: Service Hydrographique d'Études et de Documentation
SHEF e: Sheffield, Illinois
- e: Standard Hydrologic Exchange Format
- e: Standard Hydrometeorological Exchange Format
SHEFC e: Scottish Higher Education Funding Council
Sheff. e: Sheffield
SHEiiB e: Safety, Health and Environment Intra Industry Benchmarking Association
SHEL e: Software, Hardware, Environment, Liveware
SHELF e: Super Hard Extremely Low Frequency
SHEM e: Safety, Health and Environmental Management
SHEMT e: Single-channel High Electron Mobility Transistor
Sheng Li Hsueh Pao ch: Sheng Li Hsueh Pao [Acta Physiologica Sinica]
Sheng Li Ko Hsueh Chin Chan ch: Sheng Li Ko Hsueh Chin Chan [Progress in Physiological Sciences]
SHEPS e: Society of Health Education and Health Promotion Specialists
SHER p: Sociedade Hidroelétrica do Revue
SH-2F e: Seasprite
SHF e: Super High Frequency [radar and microwave radiometers]
SHG d: Schaumburg in Stadthagen, Niedersachsen
- d: Sozialhilfegesetz
- e: Second-Harmonic Generation
- e: Segmented Hypergraphics
Shgr r: Språnghandgranat
SHHS d: Städtische Höhere Handelsschule
SHI e: Social Health Insurance
- e: State Hydrological Institute
- f: Institut Hydrologique d'État
- s: Instituto Estatal de Hidrología
SHIA e: Strongly Hedonistic Internet Area
S Hid s: Sector Higrográfico
SHIFOR e: A climatology and persistence model
SHIFT e: Scalable, Heterogeneous, Integrated Facility
SHIHATA e: Tanzania News Agency
Shih Yen Sheng Wu Hsueh Pao ch: Shih Yen Sheng Wu Hsueh Pao [Acta Biologiae Experimentalis Sinica]
Shinrigaku Kenkyu j: Shinrigaku Kenkyu. Japanese Journal of Psychology
SHIP e: Report of Surface Observation from a Sea Station
- f: Message d'observation en surface provenant d'une station en mer
- s: Informe de Observación de Superficie proveniente de una Estación en el Mar
SHIPCAD/CAM e: Ship Computer Aided Design/Computer Aided Manufacturing
SHIPS e: Statistical Hurricane Intensity Prediction Scheme
SHIRA e: Ships Radar
SHIRL e: Serial Holdings in Irish Libraries
Shiva e: Laser System at Lawrence Livermore National Laboratory
ShKAS R: Shpitalny-Komaritsky Aviatsionny Skorostrelny
SHL d: Suhl, Thüringen
- e: Shift Logical Left
SHLF N: Sør-Hålogaland Land Forsvar
SHLW e: Solidified High-Level Waste
SHLWS e: Simulated High-Level Waste Slurry
SHLWS T/S e: Simulated High-Level Waste Slurry Treatment/Storage
SHM e: Semantical Hierarchy Model
- e: Shit Happens, Mate
- e: Short Hold Mode
- e: Sunspot derived solar constant variations
Shm e: Shared Memory
SHN e: Hydrographic Naval Service of the Argentine Army
- e: Simple Hysteresis Network
- f: Service d'Hydrographie Navale de l'armée argentine
- s: Servicio de Hidrografía Naval de la Armada Argentina
Shn f: Saharien
SHNS e: Self-Healing Network Service
SHO e: Super High Output
SHOA e: Hydrographic and Oceanographic Service of the Navy
- f: Service Hydrographique et Océanographique de la Marine
- s: Servicio Hidrográfico y Oceanográfico de la Armada
SHOALS e: Scanning Hydrographic Operational Airborne LIDAR Survey
SHOFM f: Système Hydrologique Opérationnel à Fins Multiples
SHOH e: Small Office/Home Office

SHORAD e: Short Range Air Defense
- e: Short Range Radar
SHORADEZ e: Short-Range Air Defense Engagement Zone
SHOROC e: Shore Required Operational Capability
SHORT ORD e: SAC Short Order Network
SHORTZ e: Short Term Terrain Model
SHOS f: Segmentation Hierarchique d'Image basé sur l'Optimisation Sequentielle
SHP e: Standard Header Packet
- e: Subhost Processor
- f: Supplément d'Honoraires aux Pharmaciens
SHPL e: Southern High Plains
SHPO e: State Historic Preservation Officer
- e: State Historical Preservation Office
SHR e: Shift Logical Right
- e: Standardized Hospitalization Rates
- f: Station Halte-Repas
SHRI e: Stereo High Resolution Imager
SHRIKE e: Anti-Radiation Missile
SHRIMP e: Sensitive High-mass Resolution Ion Microprobe
SHRIP e: Stereo High Resolution Imager Panchromatic
SHRPG e: Super Hero Role Playing Game
SHRS d: Sehen, Hören, Riechen, Schmecken [Multimedia]
- e: School Health Resource Services
SHS e: Secure Hash Standard
- e: Society for Human Sexuality
- e: Swollen Head Syndrome
SHSB e: Safety and Health Standards Board
SHSG e: Spherical Harmonic Spatial Grid
SHSRR e: Society for Health Services Research in Radiology
SHTML e: Server side include HTML
S[-]HTTP e: Secure HTTP [Hypertext Transfer Protocol]
SHV e: Safe High Voltage
- e: Standard High Volume
- N: Sjøheimevern
SHV-15 e: Safe High Voltage-15kV
SHV-5 e: Safe High Voltage-5kV
ShVAK R: Shpitalny-Vladimirov Aviatsionnaya Krupno-Kalibernaya
SHV-B e: Safe High Voltage-Bakeable
SHV-kretssjef N: Sjøheimevernetskrets-sjef
SHW e: Show
SHWL e: Seasonal High Water Level

SI e: Information Systems
- e: International System[s of Units]
- e: Safety Index
- e: Science Instrument
- e: Secondary Infertility
- e: Secondary Investigator
- e: Semantic Interpretation
- e: Sensitive Intelligence
- e: Service Interaction
- e: Signalling [rate Indicator]
- e: Site Inspection
- e: Socialist International
- e: Society of Indexes
- e: Solar Influences
- e: Solidification Index
- e: Source Index
- e: Source Index
- e: Space Images
- e: Staten Island
- e: Status Indicator
- e: Still Image
- e: Studia Iranica
- e: Sustainable Indicators
- e: System Information
- e: System Integrater
- e: System Interface
- e: System International
- f: Saisie Immobilière
- f: Sauf Imprévu
- f: Scrutin Individuel
- f: Secours Internationaux
- f: Sécurité Intérieure
- f: Service d'Information
- f: Service des Inventions
- f: Service Inclus
- f: Service Industriel
- f: Sinistrés Immobiliers
- f: Socialistes Indépendants
- f: Société Industrielle
- f: Somme Indue
- f: Souscription Irréductible
- f: Stationnement Interdit
- f: Statistique Industrielle
- f: Syndicat d'Instituteurs
- f: Système International[e] [d' Unités]
- N: Sentralinstituttet for Industriell Forskning
- N: Sjøforsvarets Intendantur
- N: Statens Informasjonstjeneste
- r: Siktinställare
- r: Stabsingenjör
- r: Sveriges Industriförbund
- s: Sección de Infantería
- s: Servicio de Información
- s: Sistema Internacional de Unidades
Si f: Sidi-
- r: Signalskola
.si e: Slovenia [Internet]
SI. e: Suppress Index
si- r: signal-
S&I e: Stabilization and Isolation
S/I a: Selvejende Institution
- e: Speaker/Interrupt
s'/i d: gewichtete spektrale Empfindlichkeit der Kamera

SIA d: Schweizer Ingenieur- und Architektenverein
- e: Safety Institute of Australia
- e: Satellite Industry Association
- e: Semiconductor Industries Association
- e: Social Impact Assessment
- e: Special Interface Adapter
- e: Systems Interconnect Architecture
- i: Società Interbancaria per l'Automazione
- s: Servicio de Información de Artillería
SIAA f: Société des Ingénieurs de l'Automobile et de l'Aviation
- s: Servicio de Información Antiaérea
SI Ac s: Sección de Infantería Acorazada
SIACONGO f: Société Congolaise Agro-Industrielle
SIADH e: Syndrome of Inappropriate Antidiuretic Hormone Secretion
SIADIS s: Sistema de Interpretación Automatica De Imagenes de Satelite
SIAK d: Schweizerisches Institut für angewandte Krebsforschung
SIAL e: Silica- and Alumina-rich portion of Earths crust
SIAM d: Supraleitendes Intelligentes Antennen-Modul
- e: Shipboard Ice Alert and Monitoring System
SIAN f: Société Industrielle et Agricole du Niari
SIAP e: Strategic Intelligence Architecture Program
SIAS e: The Scandinavian Institute of African Studies Uppsala, Sweden
SIASL e: Stranger In A Strange Land
SIASRNR e: Scientific and Industrial Association for Space Research of Natural Resources
SIAT e: Scene Image Annotation Tape
SIAWS e: SAC Intelligence Advanced Workstation Software
SIB e: Scaled Index Based
- e: Service Independent Building Block
- e: Special Investigations Board
- f: Société Internationale de Biométrie
- i: Salone Internazionale delle Attrezzature e Tecnologie per Discoteche e Locali da Ballo
SiB e: Simple Biosphere [model]
SiB2 e: Simple Biosphere model
SIBAC e: Simplified Intragovernmental Billing and Collection System
SIBAN i: Sistema Informativo Biblioteche Ateneo Napoletano
SIBC f: Société Internationale de Biologie Chimique
SIBEX e: Second International Biological Experiment
- e: Second International BIOMASS Experiment
- e: Submillimeter Infrared Balloon Experiment
- f: Deuxième expérience internationale BIOMASS
- s: Segundo Experimento Internacional de BIOMASS
SIBI e: Sistemas de Información de las Bibliotecas Cientificas Interconectadas y Abiertas

SIBIL f: Système Intégré pour les Bibliothèques universitaires de Lausanne
SIBILO e: An integrated library system
SIBO e: Small Intestinal Bacterial Overgrowth
SIBS e: Special Broadcasting Service
SIC d: Sozialarbeit im Cyberspace
- e: Small Intelligence Computer
- e: Spacecraft Identification Code
- e: Standard Industrial Code
- e: State Industrial Codes
- e: Subject Index Classification Codes
- f: Service de l'Information Commerciale
- f: Service Industriel et Commercial
- f: Société Industrielle et Commerciale
- s: Sistema Integrado de Combate
SICAB e: Sistema de Información para Catálogos Automatizados de Bibliotecas
SICAM e: Sortie Integration Capability Assessment Model
SICAR e: Sea Ice Chart Archival System
SICAS e: Semiconductor International Capacity Statistics
SICB e: SPO Interface Control Board
SICBM e: Small ICBM
SICCIM e: System Input from Command and Control to Information Management
SICCNA f: Société Industrielle et Commerciale de Constructions Navales
SICD e: System Interface Control Document
- s: Sistema Internacional de Acopio de Datos
SICE f: Service International du Commerce Extérieur
SICEA e: Steel Industry Compliance Extension Act
SICF f: Société des Ingénieurs Civils de France
SICI e: Serial[s] Item and Contribution Identifier
SICK e: Single Income, Coupla Kids
SICLOPS e: Scanner Imagery Correction and Line-By-Line Orientation Programs
SICO f: Société Internationale de Chirurgie Orthopédique
SICOB f: Salon International d'informatique, télématique, Communication, Organisation de bureau et Bureautique
SICOPP f: Société d'Importation et de Commercialisation de Produits Pétroliers
SICORP e: SPOT Image Corporation
SICOS e: Sidewall base-Contact Structure
SICOSOI e: Sidewall base-Contact Structure based on Silicon-on-insulator substrate
SICP e: Society of Invasive Cardiovascular Professionals
SICR e: Sorry I Couldn't Resist [Internet]
SICS e: Sitting In Chair Snickering [Internet]
- e: Swedish Institute for Computer Science
- s: Sociedad Internacional de Ciencias del Suelo
SICTA f: Société Ivoirienne de Contrôle Technique de l'Automobile
SID e: Sample Investigations computer Database
- e: Sealift Information Data Base
- e: Secondary Image Dissemination
- e: Security Identifier
- e: Selective Imagery Dissemination System
- e: Signaling Identifier
- e: Silence Descriptor
- e: Society for Investigative Dermatology
- e: Standard Information Display
- e: State Inpatient Database
- e: Station Identification
- e: Symbolic Interactive Debugger
- e: System ID number
- e: System Identification
- f: Service d'Information et de Documentation
sid. r: sida
- r: sidor
SIDA e: Single Integrated Data Base
- e: Swedish International Development Authority
- f: Syndrome Immuno-Déficitaire Acquis
- p: Sindroma de Inmunidad Deficiente Adquerido
- s: Síndrome de Inmunodeficiencia Adquirida
SIDAAC e: Snow and Ice Distributed Active Archive Center
SIDAC e: Single Integrated Damage Assessment Capability
SIDAM s: Sistema de Información sobre Derecho Ambiental
SIDEC e: Seychelles Industrial Development Corporation
SIDELOR f: Union Sidérurgique lorraine
SIDEMA f: Société Industrielle pour le Développement du Machinisme Agricole
SIDEN f: Syndicat Intercommunal de Distribution d'Eau du Nord
sidepatr. N: sidepatrulje
SIDERGUA s: Servicios Integrales para el Desarollo Rural de Guatemala
SIDF e: System Independent Data Format
SIDH e: System Identification for Home Systems
SIDL e: Standard Intelligence Document List
SIDONI f: Serveur d'Images et de Données en Oncologie Nucléaire sur Internet
SIDP e: Sputter Ion Depth Profiling
- e: Structured Interview for DSM Personality Disorders
SIDR f: Société Immobilière de la Réunion
SIDRO f: Société Internationale d'Énergie Hydro-Électrique
SIDS e: Satellite Image Display System
- e: Small Island Developing States
- e: Suddern Infant Death Syndrome
- e: Sustainable Development of Small Island Developing States
- f: Petits États Insulaires en Développement
- f: Société Internationale de Défense Sociale
- f: Société Internationale de Droit Social
- s: Pequeños Estados Insulares en Desarrollo
SIDS Network e: Sudden Infant Death Syndrome Network
SIDT e: Special Interest Group
- e: Store Interrupt Descriptor Table
SIDU e: Sorry I Don't Understand [Internet]
SIE e: International Secretariat for Water
- e: Selective Ion Electrode
- e: Sounding Interactive Evaluation
- f: Secrétariat International de l'Eau
- f: Service des Importations et des Exportations
- f: Société Internationale des Électriciens
- s: Sección Intermedia de Etapas
- s: Secretariado Internacional del Agua
SIEC e: Stress-Induced Excess Current
SIECA e: Permanent Secretariat of General Treaty of Central America Economic Integration
SIECUS e: Sexuality Information and Education Council of the U.S.
SiedlWB d: Institut für Siedlungswasserbau, Wassergüte- und Abfallwirtschaft
SIEFA e: Source Inventory Emission Factor Analysis
SIEL f: Salon International de l'Equipement des Lieux de Loisirs et de Spectacles
SIEP f: Syndicat Indépendant de l'Enseignement Public
SIES e: Sugar Industry Efficiency Bill
SIF e: SONET Interoperability Forum
- e: Source Input Format
- e: Standard Interchange Format
- e: Status Information Frame
- f: Syndicat Indépendant de Fonctionnaires
- N: Sjøforsvarets Intendanturoffiserers Forening
- r: Svenska Industritjänstemannaförbundet
SIFA e: Subsequent Injury Fund Agency
SIFCI f: Société Industrielle et Forestière de Côte d'Ivoire
SIFE e: Second ISLSCP Field Experiment
SIFES e: SAR Image Feature Extraction System
SIFF N: Statens Institutt for Folkehelse
- r: Statens Institut för Folkhälsan
SIFS e: Short Inter Frame Space
- e: Supply Interface System
SIFT e: Stanford Information Filtering Tool
SIG d: Sigmaringen, Baden-Württemberg
- e: Selective Integral Ground
- e: Signal Ground
- e: Signature
- e: SMDS Interest Group
- e: Software Interest Group
- e: Special Interest Group
- f: Service d'Information Géologique
- f: Service Industriel de Geneve
- f: Service International du système GPS pour la Géodynamique
- f: Système d'Information Géographique
- p: Sistema de Informações Geograficas
- s: Servicio Internacional del Sistema Mundial de Localización
- s: Sistema de Información Geográfica
SIG Ada e: Special Interest Group on Ada
SIGAL f: Système Intégré de Gestion Automatisée du Livre
SIGBIT e: Special Interest Group on Business Information Technology

SIGCAT e: Special Interest Group for/on CD-ROM Applications and Technology
SIGE f: Société Internationale de Gastroentérologie
SIG-EVENTS e: Significant Events Message System
SigFA r: Signalförbindelseanvisningar
SigG d: Signaturgesetz
SIGHT e: Slide Image and Graphics Hardcopy Transmission
SIGINT e: Signal Intelligence
SIGMA e: Force Level and Maneuver Control System
- f: Société Industrielle Générale de Mécanique Appliquée
- f: Système d'Information Géographique pour la Mise en Valeur Agricole
SIGMALOG e: Simulation and Gaming Method for Analyses and Logistics
SIGN e: Special Interest Group Network
Sign r: Signal[ist]
- r: Signaltrupper
sign. a: signeret
Signalavd N: Signalavdeling
Signalptrlåda r: Signalpatronlåda
Signbat r: Signalbataljon
Signinsp r: inspektör för Signaltrupperna
SignKAS r: Signaltruppernas Kadett- och Aspirantskola
SignKS r: Signaltruppernas Kadettskola
Signplut r: Signalpluton
Signreg r: Signalregemente
SignS r: Signalskola
SIGPLAN e: Special Interest Group for Programming LANguages
SIGRID e: Format for the archival and exchange of sea-ice data in digital form
- e: Sea Ice Grid System
- f: Forme de présentation numérique pour l'archivage et l'échange des données relatives aux glaces de mer
- s: Forma de presentación para el archivo e intercambio de datos de hielos marinos en forma numérica
SIGU d: Staatliches Institut für Gesundheit und Umwelt
SigV d: Verordnung zur digitalen Signatur vom 22.10.1997
SIG-WAIS e: Special Interest Group-Wide Area Information Server
SIH e: Super Information Highway
SIHRS f: Système d'Information Hydrogéologique a Référence Spatiale
SII e: Seiko Instruments Inc.
- e: Seriously Ill or Injured
- e: Special Interest Item
- e: Static Invocation Interface
- e: System Integrators, Inc.
- f: Support Individuel d'Informations
SIIB p: Sistema Integrado de Informação Bibliográfica da Região Centro
SIIEQ f: Syndicat des Infirmières et Infirmiers de l'Est du Québec
SIIG e: Site Installation/Implementation Guide
SIIP e: Site Installation and Integration Plan
SIIS e: Sea Ice Information System
SIJ e: Stand In Jammer
SIK a: Sulinermik Inutigssarsiutigdlit Katuvfiat
- a: Sulinermil Inutissarsliutegartut
Sikr N: Sikring
Sikr. pt. N: Sikringspatrulje
SILASI e: Superintense Laser pulse-Solid Interaction
SILK e: Speech, Image, Language, Knowledge
Silk Purse e: Boeing EC-135H, ABNCP aircraft
SILMU e: Finnish Research Programme on Climate Change
- f: Programme finlandais de recherche sur le changement climatique
- s: Programa de Investigación Finlandés sobre el Cambio Climático
SILO e: Sealed-Interface Local Oxidation
- e: Signal Location
- e: Strain-Induced Layer Ordering
SILS e: School of Information and Library
SILU e: Sugar Industry Laborers Union
SILW e: Solidified Intermediate-Level Waste
SILWF e: Sugar Industry Labor Welfare Fund
SIM d: Rhein-Hunsrück-Kreis in Simmern, Rheinland-Pfalz
- e: Satellite Interpretation Message
- e: Set Initialisation Mode
- e: Society for the Internet in Medicine
- e: Solar Irradiance Monitor
- e: Student Integration Movement
- e: Subscriber Identification Module
- f: Service des Instruments de Mesure
- f: Société de l'Industrie Minérale
- s: Servicio de Información Militar
SIMA e: Silica- and Magnesium-rich portion of Earths crust
- e: Stanford Institute of Manufacturing and Automation
- e: Support Initiative for Multimedia Applications
SIMAS e: Ship Intermediate Maintenance Activities
SIMBA e: Selective Inverse Multiple-Bond Analysis
- e: Simula Interactive Multispektral Billedanalyse
- e: Simultaneous Block Adjustment of Models Program
- e: Solar Irradiance Monitoring by Balloon
SIMC f: Société Internationale de Médecine Cybernétique
SIMCA e: Sitting In My Chair Amused [Internet]
- f: Société Industrielle et Mécanique des Constructions Automobiles
SIMCE e: Simulation, Communication-Electronics
SIMD e: Single Instruction [stream] Multiple Data [stream] [machine]
SIMDB e: Systems Integration and Management Data Base
SIMDE e: Simplified Method of Damage Estimation
SIMDUT f: Système d'Information sur les Matières Dangereuses Utilisées au Travail
SIMEX e: Simulation Exercise
SIMHA f: Société Internationale de Mycologie Humaine et Animale
SIMI e: Sea Ice Mechanics Initiative
- p: Serviço de Informações Militar
SIMIS d: Sicheres Mikroprozessor System
SIMM e: Single Inline Memory Module
- e: System Integrated Memory Module
SIMMS e: Shipboard Integrated Maintenance Management System
SIMNET e: Large Scale Interactive Simulator Networking
- e: Simulations Network
SIMO e: Single Input, Multi Output
- f: Société industrielle des minérais de l'Ouest
SIMOX e: Separation by Implantation of Oxygen
SIMP f: Service Itinérant de Médecine Préventive
SIMPLE e: Sheer Idiots Monopurpose Programming Language Environment
- e: Spatial and Inductive Modelling Package for Land Evaluation
SIMQ e: SIM Quarterly
SIMRISA e: Simulation of Radar Images using an Image Space Algorithm
SIMS e: A PC package
- e: Secondary Ion-Mass Spectrometry
- e: Support Information Management System
- e: Surface Impoundment Model System
SIMS-X e: Selected Item Management System-Expanded
SIMTEL e: Simulation and Teleprocessing
SIMULA e: Simulation
SIMURA f: Société Industrielle et Minière de l'Uranium
SIMUVIMA s: Sistema Mundial de Vigilancia del Medio Ambiente
SIMWG e: Site Installation Management Working Group
SIN d: Schweizer Institut für Nuklearforschung
- e: Service Indicator
- f: Sinistre
SiN e: Silicon-Nitride
SINA d: SISY Integriertes FTP-Server Archiv, Karlsruhe
SINAD e: Signal, Noise And Distortion
SINAES f: Société Industrielle des Applications de l'Energie Solaire
SINAPS e: System of Integrated Acquisition Procedures for Satellite data
- f: Système Intégré d'Acquisition par Satellite
SINAXE e: Secondary Initiated Non-Activation Xid Exchange
SINBIB e: An integrated library system
SINDICATE e: Study on the Indirect and Direct Climate influences of Anthropogenic Trace gas Emissions
SINDINUSAC s: Sindicato de Trabajadores Docentes e Investigadores de la Universidad de San Carlos
SINE e: SINE is not EINE
SINELAC f: Société Internationale d'Electricité des Pays des Grands Lacs
Sing r: Stabsingenjör
Singapore Med J e: Singapore Medical Journal

SINIX e: Siemens UNIX
SINODE e: Surface Indian Ocean Dynamic Experiment
- f: Expérience sur la dynamique de l'océan Indien en surface
- s: Experimento sobre la Dinámica Superficial del Océano Indico
SINR e: Signal-to-[Interference-plus-Noise] Ratio
Sint r: Stabsintendent
SINTEF e: Norwegian Institute of Technology
- N: Selskapet for Industriell og Teknisk Forskning
- N: Stiftelsen for Industriell og Teknisk Forskning ved NTH
SINTP f: Société d'Intérêt National des Travaux Publics
SINTRACUDE s: Sindicato de Trabajadores del Ministerio de Cultura y Deportes
SINTRAFIP s: Sindicato de Trabajadores del Ministerio de Finanzas Públicas
SINTRAINFOM s: Sindicato de Trabajadores del Instituto de Fomentación Municipal
SINTRAMA s: Sindicato de Trabajadores de Mantenimiento y demas Dependencias del Ministerio de Finanzas Publicas
SIO e: Scripps Institute of Oceanography, University of California at San Diego
- e: Scripps Institution of Oceanography, La Jolla
- e: Serial I/O [Input/Output]
- e: Service Information Octet
- e: Signal Intelligence Officer
- e: Site Intelligence Officer
- e: Skidaway Institute of Oceanography
- e: System Integration Office
- f: Institut Scripps d'océanographie
- s: Instituto Oceanográfico de Scripps
- s: Servicio de Información Operativa de la VMM
- s: Servicio de Operación
SiO_3 e: Silicate
SiON e: Silicon-Oxide-Nitride
SIOP e: Society for Industrial and Organizational Psychology
- f: Société Ivoirienne d'Opérations Pétrolières
SIOP-ESI e: SIOP-Extremely Sensitive Information
SIOP-MATE e: SIOP-Matrix Automation Through the Emergency Message Automatic Transmission System
SIP e: Other Subject Indexes Provided
- e: Safe Internet Programming Team
- e: SCSI-3 Interlocked Protocol
- e: Segment Interface Protocol
- e: SEMATECH Internal Program
- e: Simple Internet Protocol
- e: SINCGARS Improvement Program
- e: Single Inline Package
- e: Single Inline Pin
- e: SMDS [Switched Multimegabit Data Service] Interface Protocol
- e: Software Instrumentation Package
- e: Solar Influences Program
- e: Solar Irradiance Photometer
- e: Standard Implementation Plan
- e: Stewardship Incentives Program
- e: Strategic Implementation Panel
- e: System Implementation Plan
- e: System Improvement Plan
- e: Systems Implementation Plan
- e: Systems Improvement Program
- e: Systems Information Package
- f: Groupe Stratégique de mise en oeuvre
- f: Société Indigène de Prévoyance
- r: Stategisk Instituttprogram
- s: Panel Estratégico de Ejecución
SIPB e: Scientific Information Policy Board
SIPC e: Securities Investment Protection Corporation
- e: Simple Integrated Personal Computer
- e: Simple Interface Personal Computer
- e: Simply Interactive Personal Computer
- e: Simulated Information Processing Center
- s: Proyecto de Servicios de Información y Predicción del Clima
SIPE f: Service de l'Information et de la Propagande Économique
- f: Société Ivoirienne pour la Poste et l'Epargne
SIPh f: Société Internationale de Photogrammétrie
SIPHES s: Sistema de Prediccion de Heladas por Satelites
SIPI f: Société Ivoirienne de Promotion Industrielle
SIPO e: Serial-In, Parallel-Out
- e: System Integration Project Office[s]
SIPOS e: Semi-Insulating Polycrystalline Oxygen-doped
SIPP e: Simple Internet Protocol Plus
- e: Simple Polygon[e] Processor
- e: Single Inline Package Pin
- e: Single Inline Pin Package
- e: Strategic Industry Partners Program
SIPREP e: Special Intelligence Report
SIPRI e: Stockholm International Peace Research Institute
SIPRNET e: Secret Internet Protocol Network
SIPROCI s: Sistema de Protección Ciudadana
SIPS e: Satellite Image Processing System
- e: Satellite Imagery Processing System
- e: Satellite Input Processing System
- e: Solvent-Induced Phase Separation
- e: Spatial Information Processing System
- e: Spectral Image Processing System
SIPT e: Security Integrated Product Team
- e: Slurry Integrated Performance Test
- f: Société Internationale des Programmes pour Télédiffusion
SIPWR e: System Integrated Pressurised Water Reactor
SIR e: Save Instruction Recognition
- e: Scientific Information and Retrieval
- e: Selective Information Reporting
- e: Serial Infrared [communication]
- e: Shuttle Imaging Radar
- e: Software Installation Request
- e: Student Instructional Ratings
- e: Substained Information Rate
- e: Subsurface Interface Radar
- e: Surface Imaging Resist
- e: Sustained Information Rate
- e: System Industrial Relations
- f: Société Internationale Rorschach
- f: Syndicat Indépendant Renault
SIR-A e: Shuttle Imaging Radar-A
SIRB f: Société Internationale de Radiobiologie
SIR-B e: Shuttle Imaging Radar-B
SIR-C e: Shuttle Imaging Radar-C
SIRDS e: Single-Image Random Dot Stereogram
SIR-E e: Spaceborne Imaging Radar-EOS
SI-REFLECTAR e: Sea Ice Reflectance Database for the Arctic Basins
s/irel d: relative spektrale Strahlungsempfindlichkeit nach DIN 5032
Siremo p: Sistema de Regadio Eduardo Mondlane
SIRENE e: Supplementary Information REquest at the National Entry
SIRIAC e: CERN Commercial Purchasing System
SIRIJ e: Semiconductor Industry Research Institute of Japan
SIRIS e: Single FOV [Field Of View] Infrared Intelligent Spectrometer
SIRMO e: Senior Information Resources Management Officer
SIROS e: Solar and Infrared Observation (or Observing) System
- e: Solar and Infrared Radiation Observation Stations
SIRP f: Service de l'Information et des Relations Publiques
SIRS e: Satellite Infra-Red Spectrometer
- e: Satellite Infrared Sounder
- e: Single Integrated Reporting System
- e: Soils Information Retrieval System
- e: Solar Infrated Radiation Stations
- e: Status Inventory Reporting System
- e: Systemic inflammatory response syndrome
- f: Satellite équipé d'un Spectromètre à Infrarouge
- f: Systeme d'Information A Reference Spatiale
- s: Satélite dotado de Espectrómetro Infrarrojo
SIRVES e: SIGINT Requirements Validation And Evaluation Subcommittee

SIS d: Stelleninformationssystem
- e: National Library of Medicine Specialized Information Services
- e: Safety Interlock System
- e: Scanning Image Spectroradiometer
- e: Science Integration Staff
- e: Scientific Instrument Services
- e: Secretarial Information System
- e: Selective Inquiry System
- e: Shuttle Imaging Spectrometer
- e: Software Information Services
- e: Solar Irradiance Spectrometer
- e: Spatial Information System
- e: Special Interest Section
- e: Standard Information System
- e: Stay in School
- e: Styrene Isoprene Styrene block copolymer
- e: Submarine, Fleet Ballistic Missile
- e: System Interface Specification
- f: Société des Ingénieurs Soudeurs
- r: Standardiseringskommissionen I Sverige
- s: Sistema Inercial Solidario

SiS r: Signalverkstaden i Sundbyberg
SISAC e: Serials Industry System Advisory Committee
SISAL e: Streams and Iteration in a Single Assignment Language
SI/SAO e: Special Intelligence/Special Actions Office
SISC e: Standard Information Systems Center
SISCOM e: Satellite Information System Company
SISCR e: SISC Regulation
SISD e: Single Instruction [on] Single Data [stream]
- e: Standards and Interface Specification Document
- e: Strategic Information Systems Division

SISDIM e: Software for Integrating Spatial Data By Images
SISDU e: Sorry I still don't understand [Internet]
SISE p: Serviço de Informação e Segurança
SISEX e: Satellite Imaging Spectrometer Experiment
SISIS e: Package for Siemens computers
Siskol r: Signalskola
SISM e: Standard System Systems Manager
SISMER e: Marine Scientific Information Systems
- f: Systèmes d'Informations Scientifiques pour la Mer
- s: Sistemas de Información Científica sobre el Mar

SISO e: Serial Imput/Serial Output
- e: Single-Input, Single-Output
- n: Schema voor de indeling van de systematische catalogus in openbare bibliotheeken

SI[/]SO e: Serial In/Serial Out
- e: Shift In/Shift Out

SISP e: Standard Information Systems Panel
- e: Support Information System Panel
- e: Surface Imaging and Sounding Package
- e: System Interface Support Plan

SISR e: Standard Information Systems Review
SISSA i: Scuola Internazionale Superiore di Studi Avanzati
SISTERRA e: Spatial Information System for Terrain Evaluation
Sisu r: Skräp in, skräp ut
SISY d: Software Informations-System, Karlsruhe
SIT d: System und Informationstechnik GmbH
- e: Scanner Interface Trace
- e: Special Information Tones
- e: Special Interest Target
- e: Strategy Implementation Team
- e: Stuff-It [file]
- e: System Integration Testing
- f: Service de l'Inspection du Travail
- f: service des Installations Télégraphiques
- f: Spectro-Imageur Transportable
- f: Syndicat des Industries des Télécommunications
- i: Servizio di Taratura in Italia

Sit N: Situasjon
- s: Situación

SITA e: Society of International Aeronautical Telecommunications
- f: Societe Internationale Telecommunications Aeronautiques

SITAC e: Spectral Imagery Technology Applications Center
SITAP e: Simulation for Transportation Analysis and Planning
SITCARD e: Card-to-Tape Activity
SITCEN e: NATO Situation Center
SITD e: Still In The Dark [Internet]
SITE e: Sensor and Instrument Test and Evaluation
- e: Space Influences on the Terrestrial Environment
- e: Superfund Innovative Technology Evaluation

SITEL f: Société des Ingénieurs de Télécommunications
SITEX f: Société Industrielle des Textiles
SITP f: Société Internationale de Thérapie Psychomotrice
SITRABANVI s: Sindicato de Trabajadores del Banco Nacional de la Vivienda
SITRACERCA s: Sindicato de Trabajadores de la Cerveceria Centroamericana
SITRADIGEPON s: Sindicato de Trabajadores de la Dirección General de la Policía Nacional
SITRADIGESA s: Sindicato de Trabajadores de la Dirección General de Servicios Agricolas
SITRAEMSA s: Sindicato de Trabajadores Embotelladora la Mariposa S.A.
SITRAINTEXSA s: Sindicato de Trabajadores de Internacional Export S.A.
SITRALU s: Sindicato de Trabajadores de Lunafil SA.
SITRAM f: Société Ivoirienne de Transport Maritime
Sitrap N: Situasjonsrapport
SITRASEG s: Sindicato de Trabajadores Administrativos y de Servicio de la Educación Guatemalteca

SITRASS s: Sindicato de Trabajadores de Servicios de Salud
SITRB e: System Integration Test Review Board
SITREP e: Situation Report
SITRI e: Integral System for the Treatment of Industrial Residues
SITS e: Single Integrated Telecommunications System
- e: Site Investigation Tracking System
- f: Syndicat des Ingénieurs et Techniciens du Sénégal

SITSA f: Société Industrielle de Thérapeutique Scientifique Appliquée
SITURN e: Galileo carries some rather sensitive instruments aboard
SIU A: Senter for Internasjonalt Universitetssamarbeid
SIUP e: Seychelles Islands United Party
SIUPM f: Système d'Intervention d'Urgence en cas de Pollution de la Mer
SIV e: Sensors In Vacuum
- e: Special Interest Vehicle
- e: Swine Influenza Virus

SIVA N: Selskapet for Industrivekstanlegg
SIVORG N: Sivilorganisasjonen
SIVVP e: Software Independent Verification and Validation Plan
SIW e: Smithsonian Institute, Washington
SIWG e: Special Interest Working Group
SIX e: Share with Intent Exclusive
s'/ix d: Wert der s'/i bei einer beliebigen Wellenlänge l
SIXA e: Silicon X-ray Array
- e: Spectrum-X-Gamma solid state Si[Li] detectors

SiXi f: Soins Infirmiers et Informatique
SIZ e: Seasonal Ice Zone
SIZEX e: Seasonal Ice Zone Experiment
SJ e: Sample Job
- e: Society of Jesus
- e: Somali Jet
- f: Séquestre Judiciaire
- f: Service de formation des Jeunes
- f: Soeurs Jumelles
- f: Solde Journalière
- r: Statens Järnvägar

sj- r: sjuk-
sj. a: sjælden[t]
- N: sjef

S/J e: Signal to Jammer (or Jamming) Ratio
självst. r: självständig
Sjb r: Sjukbärare
- r: Sjunkbomb

SJC e: San Jose Ca.
- e: Supreme Judicial Court of Massachusetts

sjd. r: sjukdom
SJF e: Scheduler JCL Facility
SjFV N: Sjefen for Flyvåpnet
SjFVst N: Sjefen for Flyvåpnets stab
SjH N: Sjefen for Hæren
sjh r: sjukhus
sjkd. r: sjukdom
SjL N: Sjefen for Luftforsvaret
Sjlegen Fv N: Sjefslegen for Flyvåpnet
Sjlegen H N: Sjefslegen for Hæren
Sjlegen L N: Sjefslegen for Luftforsvaret
Sjlegen S N: Sjefslegen for Sjøforsvaret

SJLH N: Sjefslegen for Hæren
SJLL N: Sjefslegen for Luftforsvaret
SJLM N: Sjefslegen for Marinen
sjöbevplut r: sjöbevakningspluton
Sjø-Hv N: Sjøheimevern
SJØKOM NORD N: Sjøkommandør Nord-Norge
SJØKOM SØR N: Sjøkommandør Sør-Norge
SjöS r: Sjöfartsstyrelsen
Sjötp r: sjötransport
Sjøtrsp N: sjøtransport
sjöv. r: sjöväsendet
S.J.Res. e: Senate Joint Resolution
SJS e: Social Justice Strategy
SjS N: Sjefen for Sjøforsvaret
sjst r: sjukstall
SjSy e: Sjögrens syndrome
Sjt N: Sersjant
sjtpb r: sjuktransportbåt
sjtpbil r: sjukvårdstransportbil
sjtpkomp r: sjuktransportkompani
sjuktr r: sjuktransport
sjv r: sjukvård
- r: sjukvårdare
sjvbef r: sjukvårdsbefäl
SJVF e: Veterinary Agricultural Research Council
sjvgrp r: sjukvårdsgrupp
sjvkomp r: sjukvårdskompani
sjvmtrl r: sjukvårdsmateriel
Sjvspeckomp r: Sjukvårdsspecialkompani
Sjvs[tyr] r: Sjukvårdsstyrelse
sjvt[j] r: sjukvårdstjänst
sjvto r: sjukvårdstrupp
sjvtrp r: sjukvårdstrupper
SK a: Storkors af Dannebrog
- d: Sozialistische Korrespondenz
- d: Synchronknoten
- e: Automated Telecommunications Systems Directorate
- N: Sjøforsvaretskommando
- N: Stabskaptein
- N: Statens Kantiner
- r: Sjökurs
- r: Stockholms Kustartilleriförsvar
Sk r: Skarp
- r: Skyddskupol
- r: Skyddsvakt
sk r: specialistkurs
sk- r: skytte-
Sk. a: skæppe
- e: Sketch
sk. N: skole
- N: skudd
.sk e: Slovakia [Internet]
s.k. N: såkalt[e]
- r: så kallad
S-K e: Saltonstall-Kennedy Promote and Develop American Fisheries
Skar. r: Skaraborg
SKAS N: Sjøforsvarets Kystartilleriskole
SKB a: Ingeniørfirmaet Storm & Kristensen
- N: Sjømilitære Korps befal
Skb N: Skipsbygger
Skbat r: Skyttebataljon
Skbd r: Skinnband
SKBS r: Svensk Kärnbränsleförsörjning
SkC r: Skyddscentral
SKCT e: Skeleton Cursor Table
SkD a: Skånska Dagbladet
SKD a: Skattedepartementet
Skdsb a: Skodsborg
ske 1 r: värnpliktiga med skepparexamen av l:a klass
Skeletal Radiol e: Skeletal Radiology
SKER r: Skjutelementräknare
SKF r: AB Svenska Kullagerfabriken
SKG d: Snellkampfgeschwader
Skgrp r: Skyttegrupp
Skhv r: Skeppshantverkare
SKI e: Nuclear Power Inspectorate
SKi e: Swedish Nuclear Power Inspectorate
SkI r: Skyddsingenjör
SKI r: Statens Kärnkraftinspektion
Skidl r: Skidlöpare
Skil N: Skiløper
Skin Pharmacol e: Skin Pharmacology
SKIP e: Association of Library and Information Professionals
- e: Simple Key-Management for Internet Protocols
SKIPS e: COMP Kernel Interface Package
Skj r: Skjutning
Skjuti r: Skjutinstruktion
Skkomp r: Skyttekompani
SKM e: Finnish National Museum
SkMC e: Skeletal Muscle Cell
Skmd a: Storkommandør af Dannebrog
sk/min N: skudd i ett minutt
SK/min r: Skott/minut
SKN N: Sjøforsvarets Kommando Nord-Norge
SKO N: Sjøforsvarets Kommando Østlandet
SkO r: Skyddsofficer
skogsv. r: [i] skogsväsendet
skolfpl r: skolflygplan
skomp r: spaningskompani
skpsb. r: i skeppsbyggeriet
SKPT e: Skeleton Package Table
sk ptr r: skarp patron
skr r: skyddsrum
SKR R: Storozhevoy korabl
skr. a: skriftlig[t]
- a: svenske kroner
SKS a: Sparekassen København Sjælland
- N: Sjøforsvarets Kommando Sørlandet
- r: Sjökrigsskola
SKS[K] N: Sjøkrigsskolen
sk. st. N: skytestilling
SKT e: Skill Knowledge Test
- N: Sjøforsvars Kommando, Trøndelag
Skt. a: Sankt
sktl. N: skattelov[en]
SKU r: Skatteutskottet
SKU# e: Stock Keeping Unit Number
SkUO r: Skyddsunderofficer
Skutk r: Skyddsutkik
SKV N: Sjøforsvarets Kommando Vestlandet
Skv r: Skvadron
SKWG e: Spar-, Konsolidierungs- und Wachstumsprogramm-Gesetz
SKY e: Sky Network
Sky r: Skyddstjänst
SkydS r: Försvarets Skyddsskola
Skyddso r: Skyddsområde
Skyddsptr r: skyddspatrull
SKYHI e: A general circulation model
SKYNET e: United Kingdom Communications Satellite
Skyskol r: Skyddstjänstskolan
SL a: Skandinavisk Linietrafick A/S
- a: Skolepsykologernes Landsforening
- a: Statens Ligningsdirektorat
- e: Session Layer
- e: Shoemaker Levy [comet]
- e: Single Layer
- e: Space Launch
- e: Specification Limit
- e: Spin Lock
- e: SPS and LEP Division
- e: Stable precipitation over Land
- e: Subsidiary Ledger
- e: Superlattice
- f: Sans Logis
- f: Saône-et-Loire
- f: Scrutin de Liste
- f: Service du Logement
- f: Sous-Locataire
- f: Superluxe
- f: Syndicat Libre
- N: Stor-Oslos Lokaltrafikk
- r: Stabsläkare
- r: Strafflag för Krigsmakten
Sl r: Stridsledning
S.L. N: Startlinje
- s: Sociedad Limitada
S&L e: Savings and Loan companies
S/L N: Salgslag
sl- r: spårljus-
SL-1 e: Stationary Low Power Reactor-1
Sl ,A e: Spektrale Strahlungsfunktion der Normlichtart A nach DIN 5032
SLA e: Software Library Administrator
- e: Somali Liberation Army
- e: Statistical Local Area
- e: Stereo Lithographic Apperatus
- f: Société Canadienne de la Sclérose Latérale Amyotrophique
- N: Sammenslutningen av Landbrukets Arbeidsgiverforeninger
SLAB e: Subsonic Low Altitude Bomber
- r: Svenska Lastbil Ab
SLAC e: Stanford Linear Accelerator Center, Palo Alto
- e: Support List Allowance Cards
SLAD e: Survivability/Lethality Analysis Directorate
SLALOM e: Scalable Language independent Ames Laboratory One minute Measurement
SLAM e: Sierra Leone Alliance Movement
- e: Single Layer Alumina Metallization
- e: Single-Layer Alumina, Metalized
- e: Stand-off Land Attack Missile
- e: Strategic Lift Analysis Model
- s: Sistema de Lanzacohetes de Artillería Múltiple
SLAMMR e: Side-Looking Airborne [Modulated] Multi-Mission Radar
SLAMS e: State/Local Air Monitoring Station
SLAN l: Sine Loco, Anno, Nomine
SLANG e: Selected Letter and Abbreviated Name Guide

SLAO f: Société Linguistique de l'Afrique de l'Ouest
SLAP e: Space Link Access Protocol
SL-API e: Salutation Application Programming Interface
SLAPP e: Strategic Litigation Against Protesting People
SLAPS e: St. Louis Air Pollution Study
SLAR e: Side-Looking Airborne Radar
- e: Side-Looking Array Radar
- f: Radar Aéroporté à visée Latérale
- s: Radar de exploración Lateral Aerotransportado
SLAV e: Slavistik
Slav. e: Slavonic
SLAW e: California State Library [CSL] Law
SLB e: Shallow Land Burial
Slbef r: Stridsledningsbefäl
SLBM e: Submarine Launched Ballistic Missiles
SLBMAN e: SLBM Alerting Network
SLC e: Salt Lake City Weather Forecast Office
- e: Salt Lake City, Utah
- e: Side Lobe Canceling
- e: Simple Line Code
- e: SLAC Linear Collider
- e: Station Logistics Carrier
- e: Strategic Laboratory Council
- e: Super Low Cost
- e: Surface Laminar Circuit
- e: Synchronous Link Control
Slc r: Stridsledningscentral
SLCM e: Sea Launched Cruise Missile
S Lco s: Submarino con Lanzacohetes
SLCS e: Shallow-Level Centers in Semiconductors
SLCSE e: Software Life Cycle Support Environment
SLD e: Satellite Launch Dispenser
- e: Second Level Domain
- e: Shetland
- e: Source Level Debugger
- e: Stanford Large Detector
Sld s: Soldado
SLDB e: Sierra Leone Development Bank
SLDC e: Synchronous Data Link Control
SLDCOM e: Satellite Launch Dispenser Communications
SLDF e: Sierra Leone Defense Force
SLDMS e: SAR/LANDSAT Data Merging (or Mapping) System
SLDP e: Sierra Leone Democratic Party
SLDS e: Swedish Land Information System
- e: System Log Data Set
SLDT e: Store Local Descriptor Table
SLE e: Screen [-based] Line Editor
- e: Screening Line Editor
- e: Student Library Employee
SLED e: Ship Launched Electronic Decoy
- e: Single Large Expensive Disk
- e: Single Large Expensive Drive
SLEF e: Sierra Leone Employers Federation
SLEJF e: Sri Lanka Environmental Journalists Forum
SLENA e: Sierra Leone News Agency
SLEUTH e: System for Locating Eruptive Underwater Turbidity and Hydrography
SLF a: Socialistisk Lytter- og Fjernseerforbund
- d: Saalfeld, Thüringen
- e: Somali Liberation Front
Slf f: Selon la Formule
SLFMS e: SAC Logistics Force Management System
SLFP e: Serial Line Framing Protocol
- e: Shared Frame Buffer Interconnect
SLG d: Schiessleistungsgruppe
- e: Soda-Lime Glass
slgn. a: samenligne
SLHC e: Standards for Long Haul Communications
Sli r: Sjukvårdsmateriel och läkemedelsinstruktionen
SLIAG e: State Legalization Impact Assistance Grants
SLIC e: Science Learning Instruction Center
- e: Serial Link and Interrupt Controller
- e: State Land Information Council
- e: System Level Integration Circuit
- e: System Link and Interrupt Controller
SLICP e: State Land Information Capture Programs
SLIDES e: Database program for smear-slide descriptions
SLIES e: Stratospheric Limb Infrared Emission Spectrometer
SLIM e: Software Life Cycle Management
- e: Subscriber Line Interface Module
SLIMOS e: Shift Leader in Matters of Safety
SLIN e: Subline Item Number
SLIP e: Serial Line Interface Protocol
- e: Serial Line IP [Internet Protokoll]
SL-J e: Spacelab Japan
Slj r: Spårljus
Sljp r: Spårljusprojektil
SLK d: Schnell, Leicht, Kurz
- d: Spitalleistungskatalog
- e: Severely Labeled Kid
- N: Selvladekarabin
- r: Sveriges Lottakår
SLM e: Salutation Manager
- e: Scribe-Line Monitor
- e: See Last Mail
- e: Service Logic Management
- e: Sidama Liberation Movement
- e: Standard Laboratory Module
- e: Steel-Line Measurement
SLMM e: Submarine Launched Mobile Mine
SLMR e: Silly Little Mail Reader
- e: Silly Little Modem Reader
SLN d: Schmölln, Thüringen
- f: Société Le Nickel
s.l.n.a. s: sin lugar ni año
Slnd f: Sans lieu, ni date
Slng r: Sprängladdning
SLO e: Service Level Objective
- e: Squadron Liaison Officer
- r: Stridsledningsofficer
SLOA e: Sloan Hall
SLOC e: Sea Lanes of Communications
SLOPCC e: Sierra Leone Opposition Parties Consultative Committee
SLOSH e: Sea, Lake and Overland Surge[s] from Hurricane[s]
Slövngr r: Spårljusövningsgranat
Slövnprj r: Spårljusövningsprojectil
SLP e: Sea Level Pressure
- e: Seaward Launch Point
- e: Service Logic Processing Program
- e: Service Logic Program
- e: Single-Link Procedures
- e: Speech [and] Language Pathologist
- e: System Load Partition
- f: Société Linguistique de Paris
Slpgr r: Spårljuspansargranat
SLPMP e: Sierra Leone Produce Marketing Board
SLPP e: Sierra Leone People's Party
Slpprj r: Spårljuspansarprojektil
SLPP-SO e: Sea Level Pilot Project-Southern Ocean
Slprj r: Spårljusprojektil
Slpsgr r: Spårljuspansarspränggranat
SLR e: Sea Level Rise
- e: Side Looking [Reconnaissance] Radar
- e: Sidelooking radar
- e: Small Lattice Relaxation
- e: Straight Leg Raise
- f: Elévation du niveau de la mer
- s: aumento del nivel del mar
- s: Elevación del nivel del mar
- s: Radar de exploración lateral
SLRG d: Schweizerische Lebensrettungs-Gesellschaft
SLS a: Sammenslutningen af Smågrisesalg
- d: Saarluis, Saarland
- e: Single Level Storage
- e: Society of Laparoendoscopic Surgeons
- e: Softlanding Linux System
- e: Solution-Liquid-Solid
- e: Space[lab] Life Sciences
- e: Stereo Line Scanner
- e: Streptococcal Lymphadenitis
- e: Superfund Label System
- e: SWALCAP Library Services Ltd
Slsgr r: Spårljusgranat
SLSI e: Super Large Scale Integration
Slsk r: Slagskepp
SLSP e: Surface Layer Scientific Panel
SLTB f: Syndicat Libre des Travailleurs du Burundi
sltj C r: stridsledningstjänstchef
sltj L r: stridsledningstjänstledare
slto r: stridsledningstrupp
SLTU e: Sierra Leone Teachers Union
SLU e: Spoken Language Understanding
SLUFAE e: Surface Launched Unit, Fuel Air Explosive
SLUGGER e: Small Light-weight GPS Receivers
SLURP e: Simple Linear Unit Reservoir Program
slutn. a: slutning
SLV r: Statens Livsmedelsverk
sl. Vægt a: slagtet Vægt
SLVSV f: Section de Liaison et de Vol Sans Visibilite
SLW e: Supercooled Liquid Water
SLWG e: Sea Level Working Group
- e: Student Life Working Group
SLZ d: Bad Salzungen, Thüringen

SM d: Schwermetalle Pb, Cd, Cu, Fe, Mn
- e: Deputy Chief of Staff, Maintenance and Modification
- e: Deputy Commander for Maintenance and Modification
- e: Ground-to-ground missile. Examples
- e: Russian Service Module of ISS
- e: Security Management
- e: Senate Memorial
- e: Service Mark
- e: Situation Monitoring
- e: Sports Medicine
- e: Standard Model
- e: Stationary Media
- e: Stress Migration
- e: Switching Module
- f: Saisie Mobilière
- f: Salaire Mensuel
- f: Salaire Minimum
- f: Sciences Mathématiques
- f: Scrutin Majoritaire
- f: Seconde Main
- f: Secrétaire de Mairie
- f: Seine-et-Marne
- f: Seine-Maritime
- f: Service Médical
- f: Service Municipal
- f: Sexe Masculin
- f: Sinistrés Mobiliers
- f: Société Mutualiste
- f: Soins Médicaux
- f: Solde Mensuelle
- f: Sourd-Muet
- f: Sous Mandat
- f: Sous-peuplement Modéré
- f: Spécialités Médicales
- f: Station Minérale
- f: Substances Médicamenteuses
- f: Succursales Multiples
- f: Sûreté Militaire
- f: Syndicat de Mandataires
- f: Syndicat Maritime
- f: Syndicat Médical
- N: [Institutt for] Skipsmaskineri
- N: Sjøforsvarets Minevesen
- s: Sanidad Militar
- s: Servicio Militar
- s: Submarino [de acción] Mediana
- s: Submarino Minador

sm r: signalmatroser
- r: specialmedicinska sektionen

.sm e: San Marino [Internet]

S M e: School of Music

s.m. a: samme måned
- a: sammen med
- r: samma månad

S&M e: Scheduling and Movement
- e: Superlattices and Microstructures
- e: Surfaced and Matched
- e: Surveillance and Maintenance

S+M e: Siemens-Matsushita

S2M d: 2MBit/s ISDN Wählverbindung mit 20 Kanälen a 64 KBit/s

Sm³ e: Standard meter³

SM3 e: SONET 3:1 Multiplexer

SMA e: S-band Multiple Access
- e: Shape Memory Alloy
- e: Sharing with Minimum Allocation
- e: Southern Medical Association
- e: Specially Managed Area
- e: Spectrum Manufacturers Association
- e: State Meteorology Administration
- e: Stone Matrix Asphalt
- e: Styrene Maleic Anhydride copolymer
- e: System Message Alert
- f: Administration météorologique d'État
- s: Administración Meteorológica Estatal

S&MA e: Safety and Maintainability Assurance

S/Ma f: Sur Marne

Sm-A e: Smectic-A phase

SMAC e: State Minerals Advisory Council

SMADTTA e: SM-ALC Damage Tolerance Analysis

SMAE e: System Management Application Entity

SMAF e: Service Management Agent Function

SMAG f: Salaire Minimum Agricole Garanti

SM-ALC e: Sacramento Air Logistics Center

SMAN f: Système Mondial d'Avertissements de Navigation

SMAP e: System Management Application Part

SMAPP e: System Managed Access Path Protection

SMARP e: SIOP Mating and Ranging Program

SMART e: Self-Monitoring Analysis and Reporting Technology
- e: Simulation Model for Allocation of Resources for Training
- e: Software Metering and Resource Tracking
- e: Spares Management Analysis Review Technique
- e: Student-Managed Academic Resource Time

Smart BIT e: Smart Built-In Test

SMAS e: Standard Materiel Accounting System
- e: Supplementary Main Store

SMASCH e: Specimen Management System for California Herbaria

SMase e: Shingomyelinase

SMASE e: System Management Application Service Element

SMASF e: SMAS Frame

SMASPU e: SMAS Power Unit

SMAWU e: Swaziland Manufacturing and Allied Workers Union

S&MB e: Surveys and Mapping Branch

Sm-B e: Smectic-B phase

SMB e: Service Message Block
- e: Seychelles Marketing Board
- e: Single-Mask Bumping
- e: System Message Block
- e: Systems Management Branch
- f: Service Matériel et Bâtiment
- f: Société Minière des Bandamas

smb. r: samband
- r: sammanbunden

SMBA e: Scottish Marine Biological Association
- e: Shared Memory Buffer Architecture

S.M.B.A. a: Selskab Med Begrænset Ansvar

SMBL e: Stratocumulus-topped Boundary Layer

smbmtrl r: sambandsmateriel

SMBP e: Sever Message Block-Protocol
- f: Société Maritime de Brute-Pétrole

SMBS e: School of Medicine and Biomedical Sciences

Sm-C e: Smectic-C phase

SMC a: Skandinavisk Motor Compagni A/S
- e: Saint Marys College
- e: Savings and Mortgage Corp.
- e: SCSI-3 Medium changer Commands
- e: Ship-of-opportunity programme Management Committee
- e: Sleep Mode Connection
- e: Space and Missile systems Center
- e: Spin Muon Collaboration
- e: Standard Microsystems Corp.
- e: System Management Center
- e: System Monitoring Center
- e: Systems, Man and Cybernetics
- f: Comité de gestion du Programme de navires occasionnels
- s: Comité de Gestión del Programa de Buques que Colaboran Ocasionalmente
- s: Stratégie Mondiale de la Conservation
- s: Servicio Militar de Construcción[es]

SMCA e: Monochloroacetate

SMCC e: Subsurface Microbial Culture Collection
- e: Sun Microsystems Computer Corp.
- e: System Monitor Coordination Center
- e: System Monitoring and Coordinating Center

SMCO e: S-methyl-L-cysteine-sulfoxide

SMD e: Ship Manpower Document
- e: Silicone Microstrip Detector
- e: Standard Military Drawing[s]
- e: Standardized Military Drawing Program
- e: Storage Modul Device
- e: Surface Mounted Design
- e: Surface-Mount[ed] Device
- f: Salaire Moyen Départemental
- f: Société de Mèdecine Dentaire
- f: Société Maritime de Dégazage

SMDB f: Stratégie Mondiale pour la Diversité Biologique

SMDC e: Short Monel Drill Collar

SMDD e: Storage Module Disk Drive

SMDF e: Subscriber Main Distributing Frame

SMDI e: Subscriber Message Desk Interface

SMDM e: Society for Medical Decision Making

SMDO e: Special Mission Duty Officer

SMDPS e: Strategic Mission Data Preparation System

SMDR e: Station Message Detailed Recording
- e: Storage Management Data Requester
- e: Subscriber Message Detail Recording

SMDS e: Switched Multi-megabit Data Service[s]
- e: Switched Multimedia Data Service
SMDSCNM e: SMDS Customer Network Management
SMDSM f: Système Mondial de Détresse et de Sécurité en Mer
SMDSU e: Switched Multi-Megabit Data Service Unit
SME e: Science-Mathematics Education
- e: Small and/or/to Medium[-sized] Enterprise[s]
- e: Society of Mining Engineers
- e: Software Maintenance Engineer
- f: Système Européen Monétaire
- s: Sistema Monetario Europeo
SMEGMA e: Sophisticated Modern Editor with Gloriously Magnificent Abilities
Smek r: Signalmekanikerutbildning
Smepavu e: Swaziland Media, Publishers and Allied Workers Union
SMEs e: Small and Medium-Sized Enterprises
SMES e: Strategic Missile Evaluation Squadron
- N: Sjøforsvarets Maskin og Elektrikerskole
SMES-ETM e: Superconducting Magnetic Energy Storage-Engineering Test Model
SMET e: ARCS Meteorological Instruments
- e: Surface Meteorology
SMEX e: Small Explorer
SMF e: Service Management Function
- e: Shielded Materials Facility
- e: Simple Message Format
- e: Single-Mode Fiber
- e: Software Maintenance Facility
- e: Special Mobile Force
- e: Standard Message Format
- e: Standard MIDI File[s]
- e: Status Message Facility
- e: Surface-Mount Fuse
- e: System Management (or Manager) Facility
- f: Syndicat Médical Français
- r: Souschef För Marinförvaltningen
SMFA e: Special Management Functional Element
- e: System Management Functional Area
SMFE s: Sistema Modular de Fijación por Enganche
SMG e: Screen Management Guide
- e: Screens Using Machine Graphics
- e: Special Mobile Group
- e: Super-Master Group
- f: Salaire Minimum Garanti
- f: Soins Médicaux Gratuits
SM G e: Smith Gymnasium
SMGL f: Scieries Modernes de Grand-Lahou
SMH f: Salaire Minimum Horaire
- f: Service Municipal d'Hygiène
SM/HD e: Survey Meter and Historical Dosimetry Data Base
SMHI e: Swedish Meteorological and Hydrological Institute
- f: Institut Suédois d'Hydrologie et de Météorologie
- r: Sveriges Meteorologiska och Hydrologiska Institut
- s: Instituto Sueco de Hidrología y Meteorología
SMI e: Satellite Monsoon Index
- e: Smithsonian Institution
- e: Structure and identification of Management Information
- e: Sun Microsystems Inc.
- e: Supplementary Medical Insurance [program]
- e: System Management Interrupt[s]
- f: Service Médical Interentreprises
- f: Société des Mines d'Ity
- N: Sjøforsvarets Mineinspeksjon
- r: Sveriges Militära Idrottsförbund
SMIA s: Sistema Mundial de Información y Alerta sobre la Alimentación y la Agricultura
SMIAR f: Système Mondial d'Information et d'Alerte Rapide sur l'alimentation et l'agriculture
S-MIC e: STP Medium Interface Connector
SMIC e: Surveying and Mapping Industry Council
- f: Salaire Minimum Interprofessionnel de Croissance
SMIFK d: Schweizerische Medizinische Interfakultätskommission
SMIG f: Salaire Minimum Interprofessionnel Garanti
S Min s: Submarino Minador
S-MIME e: Secure MIME
SMIP e: Spares Management Improvement Program
SMIS e: Safety Management Information System
- e: Security Management Information System
- e: Ships Management Information System
SMISO f: Système Mondial Intégré de Services Océaniques
SMIT e: System Management Interface Tool
SMJ e: Southern Medical Journal
SMK e: Shared Management Knowledge
- N: Sjømilitære Korps
- r: Speciell Minkurs
SML e: Service Management Layer
- e: Shared Memory Link
- e: Siemens Modular Link
- e: Simulation Model Library
- e: Standard Machine Language
- e: Standard Material List
sml. a: sammenlagt
- a: sammenligne
S&ML e: Sampling and Mobile Laboratories (or Laboratory)
SMLA e: Soviet Maritime Law Association
- f: Association Soviétique de Droit Maritime
- s: Asociación Soviética de Derecho Marítimo
SMLNJ e: Standard Meta Language/New Jersey
sml. pl. N: samlingsplass
SMLS e: Shelterized Microwave Landing System
SMLV e: Standard Memory Load Verifier
SMM e: Solar Maximum Mission [satellite]
- e: STDN Mission Manager
- e: System Managers Manual
- e: System Mangement Mode
- f: Service des Métaux et Minérais
S&MM e: Safeguards and Materials Management
SMMA e: Styrene Methyl Methacrylate
SMMD f: Salaire Mensuel Moyen Départemental
SMME e: Small, Medium and/or Micro Enterprise
SM-MIMD e: Shared Memory, Multiple Instruction and Multiple Data
SMMR e: Scanning Multichannel Microwave Radiometer
- e: Scanning Multifrequency Microwave Radiometer
- e: Scanning Multispectral Microwave Radiometer
- f: Radiomètre à hyperfréquences à balayage multivoies
- s: Radiómetro de microondas multicanal de exploración
SMMS e: Support Maintenance Management System
- e: Support Ships Material Management System
SMN f: Service de la Météorologie Nationale
f: Service Médical de Nuit
- f: Société des Mines du Niger
- N: Samarbeidsgruppen for Natur og Miljøvern
SMNF f: Service des Métaux Non-Ferreux
SMNIG f: Salaire Minimum National Interprofessionnel Garanti
SMO e: Samoa Observatory
- e: Sample Management Office
- e: Santa Monica Ca.
- e: Science Management Office
- e: Senior Meteorological and Oceanographic officer
- e: Senior METOC Officer-specifically
- e: Serious Mode On
- e: Shelter Management Office
- e: Support To Military Operations
- e: Systems Management Office[r]
- f: Système Mondial d'Observation
- p: Serviço Militar Obrigatório
- s: Servicio [Militar] Obligatorio
- s: Servicio de Mantenimiento de Orden
- s: Sistema Mundial de Observación
- s: Sociedad Mexicana de Ortopedia
Smo r: Smörolja
SMOC e: Small Matter Of Commitment
- f: Système Mondial d'Observation du Climat
- s: Sistema Mundial de Observación del Clima
SMOff e: Serious Mode Off
SMON e: Switched Monitoring Network

SMOO f: Système Mondial d'Observation de l'Ozone
SMOOS e: Shipboard Meteorological and Oceanographic Observing Sensor
SMOOTH e: Shipboard navigation-plotting program
SMOP e: Simple Matter Of Programming
- e: Small Matter Of Programming
SMOS e: Surface Meteorological Observation System
- e: Surface Meteorological Observing Station
SMOV e: Service Mission Orbital Verification
SMOW e: Standard Mean Ocean Water [isotope standard]
SMP d: Symmetrischer Multiprozessor
- e: SBSS Master Plan
- e: Service Management Point
- e: Shipboard Measurements Panel
- e: Simple Management Protocol
- e: Small Mechanical Parts
- e: Software Modernization Plan
- e: Software Motion Picture
- e: Soil Management Program
- e: SONET [Synchronous Optical Network] Physical Layer Mapping
- e: Standard Message Package
- e: Sub Motor Pool
- e: Switching Module Processor
- e: Symbolic Manipulation Program
- e: Symmetric[al] Multi-Processing
- e: Symmetric[al] Multi-Processor
- e: System Management Plan
- e: Systems Management Policy
- f: Service Maritime Postal
smp. a: smeltepunkt
SMPA f: Service des Marchés et de la Production Aéronautique
SMPC e: Shared Memory Parallel Computer
SMPD s: Sistema Mundial de Proceso de Datos
SMP/E e: System Modification Program/Extended
SMPF f: Service Municipal des Pompes Funèbres
SMPM e: SECS Message Protocol Machine
SMPS e: Switched-Mode Power Supply
SMPT e: 4-Succinimidyloxycarbonyl-a-methyl-a-[e-pyridylidithio]toluene
Smpt r: Smältpunkt
SMPTE d: professionelles System für die Synchronisierung von Video- und Soundquellen
- e: Society of Motion Picture[s] und Television Engineers
SMPZ f: Système Mondial de Prévisions de Zone
SMR e: Selective Message Router
- e: Semiconductor Mask Representation
- e: Service Marketing Representative
- e: Society of Magnetic Resonance
- e: Software Maintenance Responsibility
- e: Source, Maintenance and Recoverability
- e: Special Management Review
- f: Société Malgache de Raffinerie
SMRAM e: System Management Random Access Memory
SMRC e: Source Maintenance Recoverability Code
SMRI e: Service Message Routing Indicator
SMRN s: Servicio Mundial de Radioavisos Náuticos
SMRSS e: Surveys, Mapping and Remote Sensing Sector
SMRT e: Software Maintenance Responsibility Transfer
SMS e: School of Mathematical Sciences
- e: School of Mathematics and Statistics
- e: Science Mission Specification
- e: SECS Message Service
- e: Security Management Section
- e: Security Management Station
- e: Security Management System
- e: Service Management Station
- e: Service Management System
- e: Short Message Service
- e: Short Messaging System
- e: Single Molecule Spectroscopy
- e: Site Management System
- e: Small Messaging System
- e: Standard Management System
- e: Storage Management Services
- e: Storage Management Subsystem
- e: Storage Management Subsystem/System
- e: Storage Management System
- e: storage-management system
- e: Styrene/a-methyl Styrene
- e: Subsystem Management Service
- e: Synchronous Meteorological Satellite[s]
- e: System[s] Management Server
- f: Satellite Météorologique Synchrone
- f: Société Minière de Sabodala
- r: Svenska Mekanförbunds Standardcentral
- s: Satélite Meteorológico Síncrónico
sms. a: sammensætning
SMSC e: Short Message Service Center
- e: Standard Multi-user Small Computer
SMS-CB e: SMS Cell Broadcast
SMSCRC e: Standard Multi-user Small Computer Requirements Contract
SMSD e: Superconducting Microstrip Detectors
SMS/GOES e: SMS [Synchronous Meteorological Satellite]/Geostationary Operational Environmental Satellite
SMS-MO e: SMS [Synchronous Meteorological Satellite] Mobile Station Originated
SMS-MT e: SMS [Synchronous Meteorological Satellite] Mobile Station Terminated
SMSS e: Sealift Management Support System
- e: Shuttle Modular Scanning Spectroradiometer
SMSSM s: Sistema Mundial de Socorro y Seguridad Marítimos
smst. a: samme steds
SM/SW e: Situation Monitoring/Strategic Warning
SMSW e: Store Machine Status Word
SMT e: Special Maintenance Team
- e: Station Management
- e: Strategic Management Team
- e: Surface Mount Technology
- f: Service Médical du Travail
- f: Système Mondial de Télécommunications
- r: Smålands Tygstation
- s: Sistema Mundial de Telecomunicación
SMTD e: STOL/Maneuvering Technology Demonstrator
- f: Système Mondial de Traitement des Données
SMTF f: Société Minière de Tenke-Fungurume
SMTI e: Systems Management Tools Initiative
SMTM f: Société Malgache de Transports Maritimes
SMTP e: Simple Mail Transfer Protocol
- e: Simple Mail Transport Protocol
SMTP-Server e: Simple Mail Transfer Protocol-Server
SMTS e: Space and Missile Training Squadron
- e: Strategic Material Transport System
SMTWGWLAF e: Slapping My Thighs While Giggling Wildly Like A Fool [Internet]
SMTWTFS e: Slapping My Thighs While Terribly Foolishly Snickering [Internet]
SMU e: Source/Measure Unit
- e: Southern Methodist University
- e: Special Mission Unit
- e: System Management Utility
SMUD e: Standoff Munitions Dispenser
SMUR e: Spectrum Management Update and Review System
SMV e: Science Museum of Victoria
- e: Surveying and Mapping Victoria
- e: Symbolic Model Verifier
- f: Salaire Minimum Vital
SMVU e: Survey of Motor Vehicle Use
SMWG e: Synthesis and Modeling Working Group
SMXQ e: Sharing with Maximum Queues
SMZ e: Streamside Management Zone
- e: Sulfamethazine
SN a: Studienævn
- e: Satellite Navigation
- e: Service Node
- e: Space Network
- e: Subnetwork Number
- e: Subscriber Number
- e: Super Nova
- e: Supertwisted Nematic
- f: Sans Numéro
- f: Sciences Naturelles
- f: Secours en Nature
- f: Sécurité Nationale
- f: Sécurité Nucléaire
- f: Service de Normalisation
- f: Service Normal
- f: Soie Naturelle
- f: Souscription Nationale
- f: Syndic des Notaires
- l: Supernova[e]
- N: Sjøforsvarets Navigasjonsvesen
- s: Submarino Nodriza

Sn a: Sogn
- e: Tin
- f: Sans nom
- f: Section
- r: Sn-linje
S N e: School of Nursing
s.n. a: se nedenfor
- r: särskild nämnd
S&N a: Simonsen & Nielsen A/S
S/N e: Signal/Noise
- s: Sector Naval
- s: Sin Neumáticos
- s: Sin Novedad
- s: Sin Número
SN95 e: Refers to the 94+ Mustang Platform
SNA e: San Antonio, Tx.
- e: Santa Ana
- e: Somali National Alliance
- e: Somali National Army
- e: Systems Networked Architecture
- f: Service de la Navigation Aérienne
- f: Société Nationale d'Allumettes
- f: Syndicat National Autonome
- f: Syndicat National des Automobilistes
SNA. e: Systems Network Architecture
SNAAQS e: Secondary National Ambient Air Quality Standards
SNACC e: Society for Neuroanesthesia and Critical Care
SNACR e: SNA [System[s] Network Architecture] Condition Report
SNACS e: Swaziland National Association of Civil Servants
SNA[/]DS e: SNA [System[s] Network Architecture][/]Distribution Services
SNAé f: Service de la Navigation Aérienne
SNA/FS e: SNA [System[s] Network Architecture]/File Services
SNAFU e: Situation Normal All Fed (or Fouled, or Fucked) Up [Internet]
SNAG e: SDSC NSSN Advisory Group
- e: Student Nurses Association of Guam
Snail Mail d: Bezeichnung für die herkömmlichen Postdienste
SNAMI i: Sindacato Nazionale Autonomo Medici Italiani
SNA/MS e: SNA [System[s] Network Architecture]/Management Services
SNAP e: Shipboard Non-tactical Automated data Processing [program]
- e: Significant New Alternatives Policy
- e: Soluble NSF [N-ethylmaleimide-sensitive fusion] Attachment Protein
- e: Streamlined Network Accelerated Protocol
- e: Student Nurses Association of Pennsylvania
- e: Sub-Net[work] Access Protocol
- e: Sub-Network Attachment Point
- e: System and Network Administration Program
- f: Syndicat National des Agents de Presse
SNAPDS e: SNA [System[s] Network Architecture] Distribution Services
SNAPI e: Soft Switch Network Application Programming Interface
SNAPII e: Shipboard Nontactical Automated Data Processing II
SNAPS e: Strategic Nuclear Attack Planning System
SNAR e: Space Network Anomaly Report
SNARE e: SNAP Receptor
SNARL e: Suggested No Adverse Response Level
SNARS e: Spill Notification and Response System
- e: Switched Network Automatic Routing System
SNAS e: Strategic National Alternatives Study
SNASC e: Student Nurses Association of South Carolina
SNAT e: Swaziland National Association of Teachers
SNBF f: Société Nationale Botanique de France
SNBU e: Switched Network Backup
SNC e: Significant Noncompliers
- e: Site Nodal Controller
- e: Sub-Network Connection
- e: Swazi National Council
- f: Société Nigérienne des Ciments
- f: Stratégie Nationale de la Conservation
SNCA f: Société Nationale des Constructions Aéronautiques
SNCAC f: Société Nationale des Constructions Aéronautiques du Centre
SNCAN f: Société Nationale des Constructions Aéronautiques du Nord
SNCAO f: Société Nationale des Constructions Aéronautiques de l'Ouest
SNCASE f: Société Nationale des Constructions Aéronautiques du Sud-Est
SNCASO f: Société Nationale des Constructions Aéronautiques du Sud-Ouest
SNCC e: Student Nonviolent Coordinating Committee
- f: Syndicat National de Ciments et Chaux hydrauliques
SNCDR f: Société Nationale Congolaise de Développement Rural
SNCF f: Societe National des Chemins de Fer
SNCFV f: Société Nationale des Chemins de Fer Vicinaux
SNCM f: Société Nationale des Constructions de Moteurs
SNCN f: Service de Normalisation de la Construction Navale
SNCP e: Single Node Control Point
- e: System Services Control Point
SNCS e: Satellite Network Control Station
SNCTP f: Société Nationale de Construction et de Travaux Publics
SNCZ f: Société Nationale des Chemins de Fer du Zaire
SND e: San Diego, Ca.
- e: STDN Network Directive
s.n.d. e: sap no defect
S.N.D. e: Scottish National Dictionary
Snd s: Sanidad
SNDP e: Sustainable National Domestic Product
- e: Sustainable Net Domestic Product
sndrift r: snorkeldrift
SNE f: Société Nationale d'Electricité
- f: Syndicat National de l'Enseignement
- p: Sistema Nacional de Educação
SNEA f: Société Nationale Elf-Aquitaine
SNEAHV f: Syndicat National des Enseignants Africains de Haute Volta
SNEC f: Syndicat National de l'Education et de la Culture
SNECMA f: Société Nationale d'Études et de Constructions de Moteurs d'Avions
SNED f: Service National des Enquêtes Douanières
SNEEL f: Syndicat National de l'Enseignement Elémentaire
SNEG e: UCSB Conserv
- f: Syndicat National des Enseignants de Guinée
SNEL f: Société Nationale d'Electricité
SNEM s: Servicio Nacional de Erradicación de la Malaria
SNEMAG f: Station nationale d'essais des matériels de génie civil
SNEN f: Syndicat National des Enseignants du Niger
SNEPPCI f: Syndicat National de l'Enseignement Primaire Public de Côte d'Ivoire
SNES f: Station Nationale d'Essais de Semences
SNESS f: Syndicat National des Enseignants du Secondaire et du Supérieur
SNET e: Southern New England Telephone
SNEWS e: Secure News Server
SNF e: Server Natural Format
- e: Server Normal Format
- e: Session sequence Number Field
- e: Somali National Front
SNFE e: Standard Network Front End
SNFGE f: Société Nationale Française de Gastro-Entérologie
SNG R: Sodruzhestva Nezabisimikh Gosudarst
SNGE f: Stratégie Nationale de Gestion de l'Environnement
SNGF f: Service du Nivellement Général de la France
SNH f: Service de Normalisation des Houillères
- f: Service National Hospitalier
- f: Société Nationale de l'Habitat
SNHB e: Service of Navigation, Hydrology and Buoyage
- p: Serviço de Navegação Hidrografia, e Balizagem
SNHFA f: Service National Hors Forces Armées
SNHU f: Société Nouvelle des Huileries
SNI d: Siemens Nixdorf Informationssysteme AG, Paderborn
- e: SNA [System[s] Network Architecture] Network Interconnect[ion]
- e: Subscriber Network Identifier
- e: Systems network architecture Network Interconnection
- f: Service National d'Intelligence
- N: Studieselskapet for Norsk Industri
SNIA e: Syndicat National des Infirmiers Anesthésistes
SNIBODE f: Syndicat National des Infirmiers de Bloc Opératoire Diplomés d'État

SNIC e: System Network Interface Controller
SNICAF f: Syndicat National des Informaticiens Conseils
SNID e: Standard Network Interface Device
SNIE f: Syndicat National des Industries d'Équipement
S N II e: School of Nursing Annex II
SNIIL f: Syndicat National des Infirmières et Infirmiers Libéraux
SNIMAB f: Syndicat National des Importateurs de Matériels de Bureau et d'Informatique
SNIP f: Service National d'Information et de Protection
- f: Société Nationale d'Investissements de Pétrole
SNIR f: Syndicat National des Industries Radio-électriques
SNIT e: Streams NIT
SNJ e: San Jose, Ca.
SNL d: Sandia National Laboratorium
- e: Sandia National Laboratories/New Mexico
- e: Space Network Loading
SNLA e: Sandia National Laboratories, Albuquerque, New Mexico
SNL/CA e: Sandia National Laboratories, Livermore, California
SNLI f: Service National de Lutte contre l'Incendie
SNLL e: Sandia National Laboratories, Livermore, California
SNL/NM e: Sandia National Laboratories, Albuquerque, New Mexico
SNL/PPG e: SNL/Parallel Processing Group
SNM e: Seychelles National Movement
- e: Somali National Movement
- e: Sun[Net] Manager
- e: System Notification Message
- N: Samarbeidsgruppen for Naturlig Miljøvern
SNMP e: Simple Network Management Protocol
- e: Small Network[s] Management Protocol
SNMPV2 e: Simple Network Management Protocol Version 2
SNNE e: Sustainable Net National Expenditure
SNNS e: Stuttgart Neuronal Network Simulator
SNOBOL e: String Oriented Symbolic Language
SNOE e: Student Nitric Oxide Experiment
SNOM e: Scanning Near-field Optical Microscopy
SNOMED e: Systematized Nomenclature of Human and Veterinary Medicine
SNORE e: Sleep apnea Online Resource for Education
SNOS e: Scanning Near-Field Optical Spectroscopy
SNOSS f: Secrétariat National des Œuvres Sanitaires et Sociales
SNOTEL e: Snow [survey and remote] Telemetry
SNOWCAT e: Support of Nuclear Ops with Conventional Air Tactics
SNP e: Sequence Number Packet
- e: Sequence Number Protection
- e: Serial Number/Password
- e: Somali National Police
- e: Sub-Net[work] Protocol
- f: Sans Nom Patronymique
SNPA e: Sub-Network Point of Attachment
SNPC f: Service National de la Protection Civile
SNPL f: Syndicat National des Pilotes de Ligne
SNPN f: Société Nationale de Protection de la Nature
SNPNAC f: Syndicat National du Personnel Navigant de l'Aviation Civile
SNPQ f: Syndicat National de la Presse Quotidienne
SNR e: Short Notice Requirement
- e: Signal to Noise Ratio
- f: Société Nationale de Roulements
- f: Société Nationale Renault
SNRM. e: Set Normal Response Mode
SNRME e: Set Normal Response Mode Extended
SNRO e: Schenectady Naval Reactors Office, Schenectady, New York
SNRS e: Southern Nursing Research Society
SNRT f: Syndicat National de la Radio[diffusion] et Télévision
SNS e: Secondary Network Server
- e: Secure Network Service
- e: Session Network Services
- e: Shomron News Service
- e: Software Network Solutions
- f: Service National des Statistiques
- r: Signalskyddsskolan
- r: Studieförbundet för Näringsliv och Samhälle
SNSB e: Senior Navy Steering Board
SNSF N: Sur Nedbørs virkning på Skog og Fisk
SNSM f: Société Nationale de Sauvetage en Mer
SNT a: Store Nordiske Telegrafselskab
- e: Super Twisted Nematic
SNTEMP e: Stream Network/Stream Segment Temperature Model
SNTFP p: Sindicato Nacional dos Trabalhadores Ferro-Portuários
SNTP e: Simple Network Time Protocol
SNTPT f: Syndicat National des Travailleurs des Postes et Télécommunications
SNTS e: Studiorum Novi Testamenti Societas which publishes the quarterly international journal, New Testament Studies
SNTSC e: Super NTSC
SNU a: Sikkerhed og Nedrustning Udvalget
- e: Solar Neutrino Units
SNUG e: SN Users Guide
SNUMB e: System Number
SNUR e: Significant New Use Rule
SNV e: Netherlands Development Organization
- f: Syndicat National des Vétérinaires
SNVC f: Société Nationale des Chemins de Fer Vicinaux
SNWSC e: Space and Naval Warfare System Command
SO d: Lehramt an Sonderschulen
- e: Sellers Option
- e: Significant Other [Internet]
- e: Southern Oscillation [index]
- e: Space Opera
- e: Special Options
- e: Spin-Orbit
- e: Split-Off
- e: Staff Office
- e: Steward Observatory, University of Arizona
- e: Strategic Objective
- e: Strike Out
- e: Sulfate radical
- e: Sulfur dioxide
- f: Sans Objet
- f: Seine-et-Oise
- f: Sens Obligatoire
- f: Service d'Orientation [professionnelle]
- f: Statistiques Officielles
- f: Sud-Ouest
- f: Syndicats Ouvriers
- N: Stående Ordre
So N: Sydøst
- r: Sydost
SØ a: Sydøst
- a: Søborg
SÖ r: Skolöverstyrelsen
- r: Sveriges Överenskommelser
Sö r: Särskild övning
sø. a: søndag
s.o. a: se ovenfor
- e: someone
S.O. s: Sistema Operativo
S.Ø. N: Sørøst
S/O e: Split Off
S[/]O s: Suboficial
SO_2 d: Schwefeldioxid
- e: Sulfur dioxide
SO_4 e: Sulfate radical
So_x e: oxides of Sulfur
SOA e: Separate Operating Agency
- e: Slow Outdiffuser Approximation
- e: Southern Orthopaedic Association
- e: Special Operating Agency
- e: Start Of Authentication
- e: Start Of Authority [record]
- e: State Oceanic Administration
- e: Status of Actions System
- f: Administration océanique d'État
- s: Administración Oceánica Estatal
SOAC e: Service Order Analysis Control
SOAE e: Spontaneous Oto-Acoustic Emmission[s]
SOAI f: Service des Organisations Aéronautiques Internationales
SOAP e: Symbolic Optimizing Assembler Program
SOAR e: Satellite Ocean Analysis for Recruitment
- f: Analyse Satellitaire de l'Océan pour l'étude du Recrutement
- s: Análisis de los Océanos mediante Satélites para la Repoblación
SOARS e: Satellite Onboard Attack Reporting System

SOAS e: School of Oriental and African Studies University of London
- e: Special Operations ADP System
SOB e: Strap-On Booster
SöB r: Särskild övning Befäl
SOBA e: System of Operational Buoys in the [North] Atlantic
- f: Système Opérationnel de Bouées dans l'Atlantique [Nord]
- s: Sistema Operacional de Boyas en el Atlántico [Septentrional]
SOBCOT f: Société Belge de Chirurgie Orthopédique et de Traumatologie
SOBELAIR f: Société Belge de transport par Air
SOBEM f: Société de Brevets et d'Études de Moteurs
SOBEMAC f: Société Béninoise des Matériaux de Construction
SOBI f: Société de Banque et d'Investissements
SOBP e: Society Of Biological Psychiatry
SOBT e: Submarine On-Board Training
SOC e: Section Operations Center
- e: Secure Operator Console
- e: Security Operations Center
- e: Service Oversight Center
- e: Soil Organic Carbon
- e: Solar Constant variations
- e: Span Of Control
- e: Special Operations Command
- e: Specialised (or Specialized) Oceanographic Center (or Centre)
- e: Sphere Of Control
- e: State-Owned Corporations
- e: Statement Of Compatibility
- e: Synthetic Organic Chemicals
- e: System On a Chip
- e: Systems Operations Center
- f: Centre Océanographique Spécialisé
- f: Service Officiel de Contrôle
- f: Société
- f: Société d'Oncologie du Canada
- s: Centro Oceanográfico Especializado
Soc f: Socialiste
S Oc s: Submarino Oceánico
soc. a: socialdemokratisk
- a: socialistisk
SOC. e: Social Issues
SocAn f: Société Anonyme
Soc Appl Bacteriol Symp Ser e: Society for Applied Bacteriology Symposium Series
SOCARAHVE f: Société Camerounaise de Raffinage des Huiles Végétales
Soc Biol e: Social Biology
SOCC e: Satellite Operation Command and Control
- e: Satellite Operations Control Center
- e: Sector Operations Control Center
- e: Submarine Operations Command Center
- e: Subordinate Operations/Command Center
SOCCT e: Special Operations Combat Control Team
soc.dem. a: socialdemokratisk
SOCEF f: Société de Construction et d'Exploitation des Installations Frigorifiques en Côte d'Ivoire
SOCEUR e: Special Operations Command Europe
SOCEX e: Southern Ocean Cloud Experiment
Soc Gen Physiol Ser e: Society of General Physiologists Series
SOCIMA f: Société de Cimenteries du Mali
Sociol. e: Sociological
- e: Sociology
SOCIT e: Satellite Orbital Debris Characterization Impact Test
SOCKER e: SR Origin Communication Kernel
SOCKS e: Socket Secure
SOCL e: Sociology
SOCMI e: Synthetic Organic Chemical Manufacturing Industry
SOCOCIM f: Société Ouest-Africaine des Ciments
SOCOGIB f: Société de Construction et de Gestion Immobilière du Burkina
SOCOMAB f: Société Congolaise de Manutention des Bois
SOCOTRAM p: Sociedad de Comercialização e Transformação de Madeira
Soc Psychiatry Psychiatr Epidemiol e: Social Psychiatry and Psychiatric Epidemiology
SOCR e: Stand-Alone Optical Character Recognition
SOCRATES e: Special Operations Command, Research, Analysis And Threat System
- e: Strategy for Organisations Concerned in Rural Advanced Telecommunications Experiments
SOCS e: SAC Operational Communications System
- e: SAC Operational Conference System
- e: Social and Contextual Studies
SocSav f: Société Savante
SocSci f: Société Scientifique
SOCSCILAC e: Social Sciences Library Advisory Committee
Soc Sci Med e: Social Science and Medicine
SOCSE e: Special Operations Communications Supply Element
Soc Secur Bull e: Social Security Bulletin
Soc Secur Bull Annu Stat Suppl e: Social Security Bulletin. Annual Statistical Supplement
SOCW e: Social Welfare Library
Soc Work e: Social Work
Soc Work Health Care e: Social Work in Health Care
SOCY e: Sociology
SOD e: Spin-On-Dielectric
- f: Société d'études pour l'Obtention du Deutérium
SODAF e: Somali Democratic Action Front
SODAR e: Sonic Detection and Ranging
SODART e: X-ray telescope on Spectrum-X-Gamma
SODAS e: SEMATECH Organized Damage Analysis Software
SODECAO f: Société de Développement du Cacao
SODEFEL f: Société pour le Développement de la Production des Fruits et Lgumes
SODELAC f: Société de Développement du Lac Tchad
SODEMO f: Société pour le Développement de la Région de Morondava
SODESUCRE f: Société pour le Développement des Plantations de Canne à Sucre, l'Industrialisation et Commercialisation du Sucre
SODETEG f: Société d'Études Techniques et d'Entreprises Générales
SODIC f: Société de Développement Industriel et Commercial
SODIFAG s: Sociedad para el Desarollo Integral de la Familia Guatemalteca
SODIMIZA f: Société de Développement Industriel et Minière du Zaire
SO[-]DIMM e: Small Outline DIMM
SODIPRESSE f: Société de Distribution de la Presse et du Livre
SODO e: Senior Operations Duty Officer
SODSSWG e: Satellite Ocean Data Systems Scientific Working Group
SODT e: Senior Operations Duty Technician
S&OE e: Systems and Operations Evaluation
SOE e: Special Operating Expense
- e: Standard Operating Environment
- e: State-Owned Enterprises
- s: Suroeste
SoE e: State of the Environment
SOEKOR A: Suidelike Olie-Eksplorasie-Korporasie
SOEP e: Secondary Operand Execution Pipeline
SOER e: State of the Environment Report
SOEST e: School of Ocean and Earth Science and Technology
SOF e: Sales Order Form
- e: Service Order Form
- e: Special Operating (or Operations) Force[s]
- e: Supervisor of Flying
- e: Sustainability of Forces
- f: Système Officiel Français
- r: Svenska Officersförbundet
S/Of s: Suboficial
SOFA e: State Office For the Aging
- e: Status of Forces Agreements
SOFADS/SOAS e: Special Operating Foreign Analysis Data System/SOAS
SOFAPIL f: Société de Fabrication des Piles
SOFAR e: Sound[ing] Fixing and Ranging [floats]
- f: Dispositif acoustique de localisation et de détermination des distances
- p: Sociedade de Farinhas
- s: Exploración y localización acústicas
SOFCAB f: Société anonyme Française des Câbles d'Acier
SOFCEP f: Société Française des Chirurgiens Esthétiques Plasticiens
SOFCPRE f: Société Française de Chirurgie Plastique Reconstructrice et Esthétique
SOFD e: Social Foundations

SOFDI s: Sociedad para el Fomento del Desarollo Integral
S of E e: Secretary of Energy
SOFIBEL f: Société Forestière Industrielle de Belabo
Soficar f: Société Fibre Carbone
SOFIDAK f: Société de la Foire Internationale de Dakar
SOFIDE f: Société Financière de Développement
SOFIMEC f: Société Française pour l'Industrie Minière au Cameroun
SOFISEDIT f: Société Financière Sénégalaise pour le Développement de l'Industrie et du Tourisme
SOFITEX f: Société des Fibres Textiles
SOFL e: Southern Florida
SOFR e: State of the Forests Report
SOFRADEL f: Société Française de Documentation Électronique
SOFRA-TP f: Société Française de Travaux Publics
Sofrecco e: Southern Free State Regional Civics Congress
SOFRED f: Société Française d'Études et de Développement
SOFRERAIL f: Société Française d'Études et de Réalisations Ferroviaires
SOFTCON e: WWMCCS Site Software Configuration System
SOGA e: SNA Open Gateway Architecture
SOGACCO f: Société Gabono-Coréenne de Commerce
SOGACEL f: Société Gabonaise de Cellulose
SOGAF f: Société des grands abattoirs de France
SOGAFERRO f: Société Gabonaise de Ferro-Alliages
SOGAFINEX f: Société Gabonaise de Financement et d'Expansion
SOGAPECHE f: Société Gabonaise de Pêche
SOGAREM f: Société Gabonaise de Recherche et d'Exploitation Minières
SOGC e: Society of Obstetricians and Gynaecologists of Canada
SOGERCA f: Société Générale d'Entreprise de Centrales Atomiques
SOGESTA f: Société civile de Gestion et de Soutien du personnel navigant de l'Aviation marchande
SOGETHA f: Société Générale des Techniques Hydroagricoles
SOGETRAG f: Société Générale des Transports de Guinée
SOGITS e: Senior Officials Group-Information Technologies Standardization
SOGREAH f: Société Grenobloise d'Études et d'Applications Hydrauliques
SOGS e: Science Operations Ground System
SOGT e: Senior Officials Group on Telecommunications
SOH e: Section Overhead
- e: Sense Of Humor
- e: Service Order History
- e: Start of Header
SOHO e: A satellite
- e: Single Owner Home Operated
- e: Small Office, Home Office
- e: Solar and Heliospheric Observatory
- e: Solar Heliosphere Observatory
SOHP e: Sediments and Ocean History Panel
SOI e: Semiconductor On Insulator
- e: Signal Operating Instruction
- e: Southern Oscillation Index
- e: Stunk On Ice
- e: Survey Of India
- f: Système Opérationnel Initial
- R: Stratigicheskaya Oboronitelnaya Initsiativa
- s: Sistema Operativo Inicial
SOIC e: Space Operations Intelligence Center
SOI-CMOS e: Silicone-On-Insulator, Complementary Metal-Oxide-Semiconductor
SOIF e: Summary Object Interchange Format
SOJ e: Small Outline package with J-leads
- e: Stand-Off Jammer
SOJT e: Structured On-the-Job-Training
SOJUGMA s: Sociedad para la Juventud Guatemalteca del Mañana
SOK a: Søværnets Operative Kommando
- N: Sjøforsvarets Overkommando
Sok r: Signal officerskurs
SOL e: Office of the Solicitor
- e: Science On-Line education project
- e: Sequence Order List
- e: Short On Landing
- e: Shut Outta Luck [Internet]
- e: Siemens Optical Link
- e: Site Operations Log
- e: Special Object Library
- e: Suda On-Line
- f: Solde à nouveau
Sol p: Partido Social, Liberal e Democrático
SOLA e: Scholars Online Academy
SOLAGRAL f: Solidarites Agricoles et Alimentaires
- f: Solidarités Agro-Alimentaires
SOLAR e: Solar/Lunar Calculator
SOLARS e: SAC On-Line Analysis Retrieval System
SOLAS e: [International Convention for the] Safety Of Life At Sea
- f: Convention internationale pour la sauvegarde de la vie humaine en mer
- s: Convenio Internacional sobre la Seguridad de la Vida Humana en el Mar
Sold N: Soldat
- s: Soldado
SoldI r: Soldatinstruktion
SOLERS22 e: Solar Electromagnetic Radiation Study for Solar Cycle 22
Solid State Nucl Magn Reson e: Solid State Nuclear Magnetic Resonance
SOLIS e: SIGINT On-Line Information System
Sol J e: Solicitors Journal
SOLLAC f: Société Lorraine de Laminage Continu
SOLP e: Standards of Laboratory Practice
SOL e: Shut Outta Luck [Internet]
SOLSTICE e: Satellite OnLine Searching Training Interactive Conferencing Experiment
- e: Solar Stellar Irradiance Comparison Experiment
SOLTIP e: Solar Connections to Transient Interplanetary Processes
SOLVL e: Suda On-Line Volunteer List
SolZG d: Solidaritätszuschlaggesetz
SOM e: Scanning Optical Microscopy
- e: School Of Management
- e: Start Of Message
- e: Structered Object Method
- e: Sulfuric acid-Ozone Mixture
- e: System Object Model
Som f: Somme
SÖM d: Sömmerda, Thüringen
SOMADEC f: Société Malgache d'Electricité
SOMAFCO e: Solomon Mahlangu Freedom College
SOMALAC f: Société Malgache d'Aménagement du Lac Alaotra
SOMAPALM f: Société Malagasy pour le Palmier à Huile
SOMASAKA f: Société Malagasy d'Aménagement de la Sakay
Somat Cell Mol Genet e: Somatic Cell and Molecular Genetics
Somatosens Mot Res e: Somatosensory and Motor Research
SOMER e: State of the Marine Environment Reporting
SoMi d: Sonstige[r] Mitarbeiter[in]
SOMI e: Start of Message In
- f: Stratégie d'Observation Mondiale Intégrée
SOMIBA f: Société Minière de Moba
SOMIKA d: Société Minière de Karongo
SOMINKI f: Société Minière et Industrielle du Kivu
SOMIVAC f: Société pour la Mise en Valeur du Fleuve Casamance
SOMM e: Stop On Micro Match
Somm f: Sommaire
SOMMA e: Single-Operator Multiparameter Metabolic Analyzer
SOMPF e: Special Operations Mission Planning Folder
SOMSA e: Space Object Measurement and Signal Analysis
SON d: Sonneberg, Thüringen
- e: School Of Nursing
- e: Standards Organization of Nigeria
- e: Statement of [Operational] Need[s]
Son f: Section
- f: Station
SONABANQUE f: Société Nationale de Banque
SONACEB f: Société Nationale de Commercialisation et d'Exportation du Bénin
SONACI f: Société Nationale Commerciale et Industrielle
SONACO f: Société Nationale de Commerce
SONACOS f: Société Nationale de Commercialisation des Oléagineux du Sénégal

SONACOTRAP f: Société Nationale de Construction et de Travaux Publics
SONAD e: Subcommittee on Non-Academic Discipline
SONADIG f: Société Nationale d'Investissement du Gabon
SONADIS f: Société Nationale d'Investissement du Sénégal
SONAE f: Société Nationale d'Equipement
SONAFEL f: Société Nationale pour le Développement des Fruits et Légumes
SONAGECI f: Société Nationale de Génie Civil
SONAGRI f: Société Nationale pour la Production Agricole
SONAM f: Société Navale Malienne
SONAMEL f: Société Nationale de Matériel Electrique et Electroménager
SONAPA f: Société Nationale de la (ou des) Production Animale[s]
SONAPAL f: Société Nationale de Papeterie et de Librairie
SONAPECHE f: Société Nationale d'Armement et de Pêche
SONAPH f: Société Nationale pour le Développement des Palmeraies et des Huileries
SONAR e: Service Order Negotiation And Retrieval
- e: Sound Detection And Ranging
- f: Société Nationale d'Approvisionnement du Monde Rural
- f: Société Nationale d'Assurance et de Réassurance
SONARA f: Société Nationale de Raffinerie
- f: Société Nigérienne de Commercialisation de l'Arachide
Sonar Cost s: Sonar de Costa
SONAREM f: Société Nationale de Recherches et d'Exploitation des Ressources Minières du Mali
SONAREX f: Société Nationale de Recherches et d'Exploitations Minières
SONATEL f: Société Nationale de Télécommunications
SONATRA f: Société Nationale des Transports Aériens
SONATRAC f: Société Nationale de Transit et de Consignation
SONATRAD f: Société Nationale de Trading
SONATRAM f: Société Nationale des Transports Maritimes
Sonderb Z Strahlenther Onkol d: Sonderband. Zeitschrift Strahlentherapie und Onkologie
SONDS e: Small Office Network Data System
SONEES f: Société Nationale des Eaux et Electricité du Sénégal
SONEFE p: Sociedade Nacional de Estudos e Financiamentos de Empreendimentos Ultramarinos-SARL
SONEL f: Société Nationale de l'Electricité
SONEPI f: Société Nationale d'Études et de Promotion Industrielle
SONET/SDH e: Synchronous Optical Networks/Synchronous Digital Hierarchy
SONETT d: Soziales Netz Telecommunikation
SONIA f: Société Nigérienne d'Industries Alimentaires
SONIAH f: Société Nationale d'Irrigation et d'Aménagement Hydro-Agricole
SONIB f: Société Nationale d'Importation du Bénin
SONIC e: SAC On-Line Interactive Controller
SONICHAR f: Société Nigérienne de Charbon
SONIC-T e: Systems-Oriented Network Interface Controller with Twisted Pair
SONIDEP f: Société Nigérienne de Produits Pétroliers
SONIMACI f: Société Nippo-Malgache Commerciale et Industrielle
SONITEXTIL f: Société Nigérienne de Textile
SONMET e: Special Operations Naval Mobile Environment Team
SOO e: Senior Operation Officer
- e: Statement Of Objective[s]
SOONSPOT e: SOONs Solar Patrol on Tape
SOOP e: Ship Of Opportunity Programme
SOOPIP e: IGOSS Ship-of-Opportunity Programme Implementation Panel
- f: Groupe de mise en oeuvre du Programme de navires occasionnels du SMISO
- s: Equipo de Ejecución del Programa de Buques que Colaboran Ocasionalmente del IGOSS
SOP e: Senior Officer Present
- e: Service Object Pair
- e: Ship of Opportunity Program[me]
- e: Southern Oceans Panel
- e: Space-based Observation Panel
- e: Special Observing Period
- e: Standard (or Standing) Operating Procedure[s]
- f: Programme de navires occasionnels
- s: Periodo de Observación Especial
- s: Programa de Buques que Colaboran Ocasionalmente
SOPAC e: South Pacific Applied Geoscience[s] Commission
- e: South Pacific Countries
- s: Comisión de Geociencias Aplicadas del Pacífico Sur
SOPC e: Shuttle Operations and Planning Complex
SOPECAM f: Societe de Presse et d'Éditions du Cameroun
SOPEMEA f: Société pour le Perfectionnement des Matériels d'Équipement Aéronautique
SOPICOMA f: Société pour la Promotion Industrielle et Commerciale de Madagascar
SOPO d: Sozialistische Politik
SOPP e: State Oil Palm Plantation
SOPROCA f: Société Provinciale de Commercialisation des Produits Agricoles
SOPROTHEL f: Société de Promotion Touristique et Hôtelière
SOPS e: Select Committee on Ocean Policy
SOQ e: Statement Of Qualifications
SOR e: Sales Order Request
- e: Squadron Operations Room
- e: Struck Off Records
- e: Systems Operations Review
SörB r: Södertörnsbas
SORCE e: Software Resource Center
SORD e: Special Operations and Research Division
- e: Submerged Ordnance Recovery Device
- e: System Operational Requirements Document
SOREM f: Société pour la Réalisation et l'Étude de Monocristaux
SOREMA f: Société Rhénane d'Exploitation et de Manutention
SOREMIB f: Société de Recherches et d'Exploitation des Mines du Burkina
SORIFEMA f: Société Rizicole et Feculière Malagasy
SORIN f: Société de Recherches et Installations Nucléaires
SORS e: SIGINT Overhead Reconnaissance Subcommittee
SORSA e: Spatially Orientated Referencing Systems Association
SORSI e: Sacro Occipital Research Society International
sort. a: sortering
SortenschutzG d: Sortenschutzgesetz
SORTI e: Solar Radiance Transmittance (or Transmission) Interferometer
SORTS e: Status of Resources and Training System
SORTST e: Region I Source Test File
SORU a: Landsforeningen til Syges og Rekonvalescenters Underholdning
SOS e: Satellite Observing System
- e: Schedule Optimization Study
- e: Sealift Operating System
- e: Software Obsolescence Study
- e: Southern Oxidant[s] Study
- e: Squadron Officers School
- e: Stabilized Sight
- e: Standards and Open Systems
- f: Système d'Observation du SMISO
- f: Système d'Observation par Satellite
- N: Sonarskolen
- r: Sveriges Officiella Statistik
- s: Sistema de Observación por Satélite
SOSCO e: Soweto Students Congress
SOSG e: Senior Officer Steering Group
SOSIDER f: Société d'Exploitation Sidérurgique
SOSIG e: Social Science Information Gateway
SOSLC e: Southern Ocean Sea Level Centre
SOSMED e: Study of the South Mediterranean Sea
- f: Étude de la Méditerranée méridionale
- s: Estudio del Mediterráneo Meridional
SO4=SO42 e: Sulfat-Ion
SOSS e: Service on Strategies for Sustainability
SOS/SONIA e: SOS/Southeast Oxidant and Nitrogen Intensive Analysis
SOSTF e: SPADCO Off-Site Test Facility
SOSUHO f: Société Sucrière du Haut-Ogooue
SOSUNIARI f: Société Sucrière du Niari

SOSWG e: Sampling and Observational Systems Working Group
- f: Groupe de travail sur les systèmes d'échantillonnage et d'observation
- s: Grupo de Trabajo sobre Sistemas de Muestreo y Observación
SOT e: STS-1 Overhead Terminator
- f: service des objets trouvés
SOTA e: SIGINT Operational Tasking Authority
SOTACA e: State-Of-the-Art Contingency Analysis
SOTDAT e: Source Test Data
SOTER e: Soi[s]l and Terrain [digital] Data[base]
- e: Soil and Terrain Survey
SOTERA f: Société Tchadienne d'Exploitation des Ressources Animales
SOTEXIM f: Société Togolaise d'Importation et d'Exportation
SOTFA e: Special Operations Task Force Atlantic
SOTFE e: Support Operations Task Force Europe
SOTOA e: Sacro Occipital Technique Organisation Australasia
SOTRA f: Société de Transports d'Abidjan
SOTRACOB f: Société de Transit et de Consignation du Bénin
SOTRAD f: Société de Transport Dakarois
SOTWG e: SACDIN Offutt Transition Working Group
SOU f: Souscription
Soud Lek c: Soudni Lekarstvi
SOUNDING ROCKETS e: Rockets launched up to the upper atmosphere for soundings of the earth
SOUP e: Simple Offline Usenet Packet
SOUTHAG e: Southern Army Group
SOUTHCOM e: U.S. Southern Command
Southeast Asian J Trop Med Public Health e: Southeast Asian Journal of Tropical Medicine and Public Health
South Med J e: Southern Medical Journal
SOV e: Single Occupancy (or Occupant) Vehicle
- e: Start Of Visibility
SOVETCO f: Société pour la Vente de Thons Congélés
sovj. a: sovjetisk
Sov. J. Mar. Biol. [Engl. Transl. Biol. Morya] e: Soviet Journal of Marine Biology [English Translation of Biologiya Morya]
SOW d: Sozialwesen
- e: Speaking Of Which
- e: Special Operations Wing
SOWER e: Southern Ocean Whale and Ecosystem Research Programme
- f: Programme de recherches sur les baleines et les écosystèmes dans l'océan Austral
- s: Programa de Investigación sobre las Ballenas y el Ecosistema del Océano Austral
SOWEX e: Southern Ocean Waves Experiment
SOWI d: Sozialwissenschaftliche Informationen
SOWK e: Social Work
SOWT e: Special Operations Weather Team
SOWT/TE e: Special Operations Weather Team/Tactical Element
SOX e: Sound Exchange
SOx e: Sulfur (or Sulphur) Oxides
SOZ d: Sozialwissenschaften
- d: Soziologie
SOZACOM f: Société Zairoise de Commercialisation des Minéraux
SOZIR f: Société Zairo-Italienne de Raffinage
SozÖk d: Wirtschaftswissenschaftliche Bibliothek
SozPl d: Institut für Sozialforschung, Abteilung für Soziologie
Soz Präventivmed d: Sozial- und Präventivmedizin
SP a: Skolepatrulje
- e: Intelligent Network Service Plane
- e: Sagittal Polarization
- e: Segment Priority
- e: Service Pack
- e: Short Play
- e: Shotpoint
- e: Signaling Point
- e: Small Polaron
- e: Southern Pine
- e: Speight Building
- e: Sponsored Projects
- e: Stack Pointer
- e: Stack Pointer System Product
- e: Standard and Performance
- e: Star Patrol
- e: Startpoint
- e: Strictly Private
- e: Strongpoint
- e: Substance PSP
- e: Summating Potential
- e: Switch Processor
- e: System Product
- f: Saint-Père
- f: Salaire-Pourboire
- f: Salubrité Publique
- f: Salut Public
- f: Sanction Pénale
- f: Sans Profession
- f: Sans Protêt
- f: Santé Publique
- f: Sciences Politiques
- f: Séance Publique
- f: Secours Public
- f: Secret Professionnel
- f: Secrétaire Particulier
- f: Secteur Postal
- f: Section du Personnel
- f: Sécurité Publique
- f: Semaine Prochaine
- f: Séparation
- f: Service de Pensions
- f: Service de Placement
- f: Service de Presse
- f: Service des Prix
- f: Service Payeur
- f: Service Permanent
- f: Service Postal
- f: Services Publics
- f: Simple Police
- f: Société française de Physique
- f: Sous-Préfecture
- f: Souscription Publique
- f: Souverain Pontife
- f: Spécialités Pharmaceutiques
- f: Surtaxe Progressive
- f: Suspension de Paiement
- f: Syndicat Patronal
- f: Syndicat Pharmaceutique
- f: Syndicat Professionnel
- f: Système Protecteur
- N: Sperreområde
- r: Stabspastor
- r: Statens Provningsanstalt
- s: Semirremolque Plataforma
- s: Submarino de tipo Pequeño
Sp e: Sialoprotein
- f: Sans prix
- f: Spécial
- N: Sperre
sp. a: spansk
Sp. e: Speech[es]
sp. e: spelling
- r: spalt
S/P e: Subpoint
s/p f: sur préavis
s/P s: su Pagaré
sp? e: spelling in question [Internet]
SPA e: Service Point Application
- e: Small Program Assistance
- e: Society of Pediatric Anesthesia
- e: Software Publisher's Association
- e: Special Program of Assistance for Africa
- e: Specialist in Arts
- e: Specially Protected Area
- e: Specific Poll Address
- e: Sperm Penetration Assay
- e: Strategy and Policy Assessment
- e: Subprogramme
- e: System Performance Assessment
- e: Swedish association of Practicing Architects
- f: Service des Produits Agricoles
spa- r: spanings-
SPAAG e: Self-Propelled Anti-Aircraft Gun
SPAC e: Salinity Program Advisory Council
SPACC e: Small Pelagic Fish and Climate Change Group
- e: Space Command Center
- f: Groupe sur les petits poissons pélagiques et les changements climatiques
- s: Grupo sobre Pequeños Peces Pelágicos y Cambio Climático
SPACECMD e: Air Force Space Command
SPACETRACK e: USAF Space Tracking System
SPACHEE e: South Pacific Action Committee for Human Ecology and Environment
- s: Comité de Acción del Pacífico Sur para la Ecología Humana y el Medio Ambiente
SPADA d: Sprache und Daten
SPADE e: Stratospheric Photochemistry, Aerosols and Dynamics Expedition
SPADOC 4 e: Space Defense Operations Center, block 4

SPADS e: Staff Planning and Decision Support System
SPAF e: Seychelles Peoples Air Force
SPAFE f: Société des Pétroles d'Afrique Equatoriale
SPAG e: Standards and Practices Action Group
- e: Standards Promotion and Applications Group [address]
SPAIS e: South Pacific Air Intelligence System
SPA-LEED e: Spot Profile Analysis of Low-Energy Electron Diffraction
SPALS e: Subsequent Pregnancy After a Loss Support
SPAM e: Send Phenomenal Amounts of Mail
- e: Stupid person's advertisement
SPAN e: Small Publishers Association of North America
- e: Space Physics [and] Analysis Network
- e: System Performance Analyzer
- s: Red de Análisis de Física Espacial
SPANS e: Simple Protocol for ATM Network Signaling
S/Pap f: Sans Papiers
SPAP e: Shiva Password Authentication Protocol
spa ptr r: spaningspatrull
SPAR e: Software Programmable Advanced Receiver
- e: Space Available/Required system
- e: SPAR Aerospace Limited
- e: Status of Permit Application Report
- e: Stock Points Adp Replacement
SPARC e: Scalable Performance Architecture
- e: Standards, Planning And Requirements Committee
- e: Stratospheric Processes and their Role in Climate
- e: Systems Planning and Requirements Committee
- f: Les processus stratosphériques et leur rôle dans le climat
- s: Los Procesos Estratosféricos y su Función en el Clima
SPaRCE e: Schools of the Pacific Rainfall Climate Experiment
SPARCLE e: Spaceborne Aerosols and Cloud Lidar Earthprobe
SPARK e: Solid Propellant, Advanced Ramjet, Kinetic energy
SparPG d: Spar-Prämiengesetz
spärrbat r: spärrbataljon
spärrkomp r: spärrkompani
SPAS e: Security Police Automation System
- e: Space Pallet Satellite
- e: Space VLBI Assistance Software
- f: Syndicat des Professeurs Africains du Sénégal
SPAST f: service de préparation aux activités saisonnières et temporaires
SPA SUR e: Space Surveillance Network
SPATS e: Small Purchase Automated Tracking System
Spat Vis e: Spatial Vision
SPAW e: Specially Protected Areas and Wildlife
- f: Protocole concernant les zones spécialement protégées et la faune et la flore sauvages
- s: Protocolo sobre Zonas y Fauna y Flora Silvestres Especialmente Protegidas
SPAWAR e: Space and [naval] Warfare system command
SPAWAR/30 e: Space and naval Warfare systems command-naval warfare systems architecture and engineering
SPAWARCOM e: Space and naval Warfare systems Command
SPAWARS e: Space and naval Warfare Systems command
SPB d: Spremberg, Brandenburg
- e: Science Planning Branch
- e: Synchronous Pipelined Bus
SPBB f: Système de Planification, Programmation, Budgétisation
SPBOPS e: SPB GSSS Operations Environment
SPBS e: Standard Property Book Systems
SPC e: Satellite Processing Center
- e: Scientific Policy Committee
- e: SCSI[-3] Primary Commands
- e: Semi-Permanent Circuit
- e: Shuttle Processing Contract
- e: Shuttle Processing Contractor
- e: Software Productivity Center
- e: Space Planning Council
- e: Spanish Civilization
- e: Special Political Committee
- e: Special Premiers Conference
- e: Specific fuel Consumption
- e: Statistical Process Control
- e: Storm Prediction Center
- e: Sub-Process Control
- e: Sulfur Polymer Cement
- e: System Planning Corporation
- f: Service de la Politique Commerciale
- f: Service de Poste et Colis
- f: Service des Permis de Conduire
- f: Société des Potasses du Congo
SPCC e: Ship Parts Control Center
- e: Soweto Parents Crisis Committee
- e: Spill Prevention, Containment and Countermeasures
- e: State Pollution Control Commission
SPCDRO f: Secrétariat Permanent du Comité Directeur de la Recherche Opérationnelle
SPCE s: Superficie Cubierta de Edificación
SPCF e: Service Point Command Facility
SPCK e: Society for the Promotion of Christian Knowledge London
SPCL e: Spectrum Cellular [Corp.]
SPCP e: Society of Permanent Cosmetics Professionals
SPCS e: Short Ported Coaxial System
- e: Stored Program Control Systems
- e: Stored Program Controlled Switch
SPD d: Sozialdemokratische Partei Deutschlands
- e: Serial Presence Detect
- e: Software Products Division
spd. N: speider
SPDF e: Seychelles Peoples Defense Force
SPDIF e: Sony Philips Digital Interface
SPDL e: Standard Page Description Language
SPDM e: Shekash Peoples Democratic Movement
- e: Special Purpose Dextrous Manipulator
SPDO e: Sidama People's Democratic Organization
SPDP e: N-succinimidyl-3-[2-pyridydithio]propionate
SPDR e: Short-Pulse Doppler Radar
- e: Subsystem Preliminary Design Review
- e: System Program Directors Review
SPDT e: Single Pole, Double Throw
spd. Ving N: speiderving
SPE e: Sampler Performance Evaluation
- e: Senior Procurement Executive
- e: Shropshire
- e: Solid Phase Extraction
- e: Space-Tracking and Pointing Experiment
- e: Special Programming Enhancement
- e: Speech
- e: Symbolic Programming Environment
- e: Synchronous Payload Envelope
SPEAK Test e: Speaking Proficiency English Assessment Kit Test
SPEAR e: Signal Processing Evaluation and Reporting
- e: Strike Projection Evaluation and Anti-Air Research
SPEC e: System[s] Performance Evaluation Corporation
spec. a: specielt[t]
- e: specifically
Spec. e: Specification
Spéc f: Spécialiste
specfpl r: specialflygplan
SPECI e: Special METAR
speciallng r: specialladdning
SPECIATE e: VOC/PM Chemical Speciation Database Management System
SPECMAP e: Mapping Spectral Variability in Global Climate Project
SPECNet e: Semiconductor Process and Equipment Network
SPECT e: Single Photon Emission CT
SPECTA f: Société de Promotion et de Commercialisation des Technologies Avancées
SPECTRE e: Spectral Radiance (or Radiation) Experiment
Spectrum-X-Gamma e: The Spectrum X-Gamma mission
Spectrum-X-Gamma-CF e: Spectrum X-Gamma Coordinating Facility
spec. v. a: specifik vægt
sped. a: spedition, speditør
SPEED e: Snoopy Protocol Enhanced and Extended with Directory
SPEEDEX e: System-Wide Project for Electronic Equipment at Depots
SPEI e: Simulation Parameters Electronic Interface
SPELDA e: Structure Porteuse External pour Lancements Doubles Ariane
SPEM e: Semispectral Primitive Equation Model
- e: Smooth Pursuit Eye Movement

SPEN f: Syndicat des Professeurs d'Écoles Normales
SPER f: Syndicat Professionnel des industries de matériel Électronique et Radio-électrique
spes N: spesial
S Pet s: Submarino Petrolero
SPF e: Shortest Path First
- e: SIDPERS Personnel File
- e: Structured Programming Facility
- e: Sun Protection Factor
- e: System Programming Facility
- e: Systems Productivity Facility
- f: Forum du Pacifique Sud
- r: Styrelsen för Psykologiskt Försvar
- s: Foro del Pacífico Meridional
SPG e: Service Protocol Gateway
- e: Space Group
- e: Subdivisions of the programme
- f: Système des Préférences Généralisé
SPG[FP] e: Society for the Propagation of the Gospel in Foreign Parts London
SPGN e: Signal Processor Graph Notation
- e: Special Education-General Courses
SPH e: School of Public Health
SPHEM e: Superfund Public Health Evaluation Manual
SPHI e: OSW Sludge Program-Health Impacts
- e: Special Education-Hearing Impaired
SPI e: Scientific Placement Inc.
- e: SCSI[-3] Parallel Interface
- e: Security Parameter[s] Index
- e: Service Provider Interface
- e: Single Process Initiative
- e: Slagging Pyrolysis Incinerator
- e: Society of the Plastics Industry, Inc.
- e: Software Patent Institute
- e: Software Process Improvement
- e: Special Education-Physically and Otherwise Impaired
- e: Standard Process Inc.
- e: Swiss Performance Index
- e: Syncronization Programming Interface
- f: Comité de liaison des secrétariats professionnels internationaux
- f: Service de la Propagande et de l'Information
- f: Service de la Propriété Industrielle
- f: Service Presse-Information
SPIAF r: Svenska Papperindustriarbetarförbundet
SPICE e: Spacecraft, Planetary and Probe Ephemeris, Instrument, C-Matrix, Event File
SPIC MACAY e: Society for Promotion of Indian Classical Music and Culture Amongst Youth
SPID e: Service Profile Identifier
- e: Service Protocol Identifier
- e: Service Provider Identifier
SPIDER e: SEMATECH Process Induced Damage Effect Revealer
SPIDER-MEM e: SPIDER-Manufacturing Equipment Monitor
SPIDS f: Syndicat Patronal des Industries de Dakar et du Sénégal
SPIE e: International Society for Optical Engineering
- e: Society of Photo-optic Instrumentation Engineers
- e: Society of Photometric and Instrumentation Engineers
- f: Système de Paiement Intereuropéen
SPIFIN e: SCI PACOM Intelligence File
SPII e: Support Programme for Industrial Innovation
SPIKE e: An AI scheduling tool used on HST, ASCA and XTE
- e: Science Planning Intelligent Knowledge-base[d] Environment
SPIL e: Spillman Hall
- f: Syndicat Professionnel des Infirmières Libérales
spild N: sperreild
SPILS e: Spin Prevention and Incidence Limiting System
- e: Stall Protection and Incidence Limiting System
SPIM e: Sun Product Information Meeting
SPIMOX e: Separation by Plasma-Implanted Oxygen
SPIN e: Software Process Improvement Network
- e: Special Instruction Referenced in the FRAG
- e: Standard Prescriber Identification Number
- f: Section des Prévisions Ionosphériques du Centre National des télécommunications
SPINS e: Special Instructions
- e: Sponsored Projects Information System
SPINTCOM e: Special Intelligence Communications Handling System
SPINTCOMM e: Special Intelligence Communications Network
SPIR e: Strategic Partnership for Industrial Resurgence
- f: Syndicat Professionnel des Industries Radio-électriques
SPIRAT e: Strategic Program for Innovative Research on AIDS Treatment
SPIREP e: Special Intelligence Report
- e: Special Report
SPIRES e: Stanford Public Information Retrieval System
SPIRS e: Super-JANET Project on Information Resources
SPISD e: Space Information Systems Division
SPIT e: Site Performance and Integration Test
- f: Société de Prospection et d'Inventions Techniques
S6PK e: ribosomal S6 Protein Kinase
SPK e: Software Porting Kit
- r: Statens Pris- och Kartellnämnd
SPKM e: Simple Public-Key Mechanism
SPL e: Set Priority Level
- e: Spell Checker
- e: Spooler [Programming Language]
- e: Spooling
- e: Supplies, Procurement And Logistics Division
Spl N: Samlingsplass
Spl r: Sammanställningsplats
SPL/1 e: Signal Processing Language
SPLA e: Seychelles Peoples Liberation Army
- e: Sudanese Peoples Liberation Army
Splats r: Stridsledningsplats
SPLC e: Standard Payload Computer
SPLD e: Simple PLD
Sple f: Simple
SPLI e: Special Education-Learning Impaired
SPLICE e: Stock Point Logistics Integrated Communications Environment
SPLM e: Sudanese Peoples Liberation Movement
SPLMD e: Soil-Pore Liquid Monitoring Device
SPLT e: South Platte Basin
SPLTS e: Special Private Line Telephone System
S-Plus e: Statistical software package
SPM e: Scanning-Probe Microscope (or Microscopy)
- e: Secure Program[s] Manager
- e: Security Protection Module
- e: Seychelles Peoples Militia
- e: Site Program Manager
- e: Skill Projection Model
- e: Social and Preventive Medicine
- e: Software Programmers Manual
- e: Solar Proton Monitor
- e: Somali Patriotic Movement
- e: SONET Physical layer Mapping
- e: Space Switch Module
- e: Spatial Phase Modulator
- e: Special Purpose Monitoring
- e: Statistical Parametric Map[s]
- e: Statistical Parametric Modeling
- e: System Performance Monitor
- e: System Program Manager
spm. a: spørgsmål
SPM/2 e: System Performance Monitor /2
SPMB e: Security and Property Management Branch
SPMC e: Southern Pine Marketing Council
SPMD e: Single Program Multiple Data
SPMI e: Special Education-Mentally Impaired
SPMS e: Special Purpose Monitoring Stations
SPMW e: Subpolar Mode Water
SPN d: Spree-Neisse, Brandenburg
- e: Shared Processing Network
- e: Shipment Performance Notice
- e: Society of Pediatric Nurses
SPNC f: Syndicat du Personnel Navigant Commercial
SPNFZ e: South Pacific Nuclear Free Zone
SPNFZT e: South Pacific Nuclear Free Zone Treaty
SPNH e: Spanish
SPNSR e: Sponsor
SPNT e: Society for the Promotion of Nutritional Therapy

SPO d: Sport[medizin]
- d: Studien- und Prüfungsordnung
- e: Servicing Personnel Officers
- e: South Pole Observatory
- e: Special Position Operator
- e: Sports
- e: Station Priority Order
- e: System Project Organization
- f: Section des Publications Officielles
- N: Studieselskapet for Personalopplæring

SPOC e: Science Payload Operations Center
- e: Single Point Of Control
- e: Single Point[s] Of Contact

SPOCC e: South Pacific Organizations Coordinating Committee

SPOD e: Seaport Of Debarkation
- e: Surface Port Of Debarkation

SPOE e: Seaport Of Embarkation
- e: Surface Port Of Embarkation

SPOF e: Single Point Of Failure
- e: Single Point Of Faulure

SPOOL e: Simultaneous Peripheral[s] Operation On-Line
- e: Standby Power System

SPORTS e: Intramural Sports Building

Sports Med e: Sports Medicine

Sportverletz Sportschäden d: Sportverletzung Sportschäden

SPOS e: Solid Phase Organic Synthesis

SPOSC e: Shared Processing Operations Steering Committee

SPOT e: Earth Observation Satellite
- e: Satellite Positioning and Tracking
- e: Shared Product Object Tree
- f: Satellite Probatoire d'Observation de la Terre
- f: Système Pour l'Observation de la Terre
- f: Système Probatoire d'Observation de la Terre
- s: Satélite de Observación de la Tierra

SPOUT e: Spark Output

S/PP e: Subsystem Project Plans

SPP e: Scalable Parallel Processing
- e: Science Power Platform
- e: Score Perfect Professional
- e: Security Policy and Procedures
- e: Sequencial Packet Protocol
- e: Single-Phase Printing
- e: Society for Pediatric Pathology
- e: Species Plantarum Project
- e: Standard Parallel Port
- e: Standard Practices and Procedures
- e: Subproject Program Plan
- e: Swaziland Progressive Party
- e: System Package Program
- e: System Project Plan
- f: Service du Plan de Paris
- f: Service Presse et Publicité
- f: Syndicat de la Presse Parisienne

SPPAC e: Salinity Pilot Program Advisory Council

SPPC e: Self-Pumped Phase-Conjugate

SPPCo e: Sierra Pacific Power Company

SPPF f: Front Progressiste du Peuple Seychellois

sppl e: social support pregnancy loss newsgroup

SPPS e: Scalable Power Parallel System

S-pr f: Sous-préfecture

SPR e: Secretary of Air Force Program Review
- e: Semiconductor Process Representation
- e: Single Point Resistance
- e: Software Problem Reports
- e: Special Purpose Register
- e: Statistical Pattern Recognition
- e: Strategic Petroleum Reserve [project office]
- e: System Problem Reports
- f: Syndicat Patronal Régional

Spr N: Spreng-

spr r: spärr

spr- r: späng-

spr. a: sproglig[t]

SPRAA e: Strategic Plans and Resource Analysis Agency

SPRAM e: Special Purpose Recoverables Authorized to Maintenance

sprbat r: spärrbataljon

SPRD e: Surplus Production Reactor Decommissioning

SPREAD e: Systems Programming, Research, Engineering and Development

S-préf f: sous-préfecture

SPREP e:South Pacific Reginal Environment[al] [Protection] Program[me]
- s: Programa Regional del Medio Ambiente para el Pacífico Sur

SPREX e: Spring Runoff Experiment

SPRI e: Scott Polar Research Institute, Cambridge

Springer Semin Immunopathol e: Springer Seminars in Immunopathology

SPRINTEL e: Speedy Retrieval of Information on the Telephone

SPRL f: Société de Personnes à Responsabilité Limitée

SPRO e: Strategic Petroleum Reserve Office

SPROCAS e: Study Project on Christianity in Apartheid Society South Africa

SPROGSAM a: Sammenslutningen af de Regionale Sproglærerforeninger i Danmark

SPRP e: Sector Policy Reform Program
- e: Spin Phase Reference Pulse

sprv. a: Sprogvidenskab[elig][t]

SPS d: Speicherprogrammierbare Steuerung[stechnik]
- d: Spezialseminar
- e: Safety Plan System
- e: Scandinavian Political Studies
- e: Security Policy Squadron
- e: Sense Point Scanner
- e: Serial Port Select
- e: Service and Power Support
- e: Session Presentation Services
- e: Short-Period Superlattice[s]
- e: Standby Power System
- e: Sulfonated PolyStyrene
- e: Surface Preparation System

SPSG e: Space Propulsion Synergy Group

SPSH e: Six Phase Soil Heating

SPSI e: Special Education-Speech and Language Impaired

spsk. a: spiseskefuld

SPSLC e: SPS and Lear Committee

SPSO e: Science Processing Support Office

SPSP e: Site Preparation Support Plan

SPSR e: Service Planning Segment Replacement

SPSRM f: Service de Physique du Solide et de la Résonance Magnétique

SPSS e: Science Planning and Scheduling Subsystem
- e: Statistical Product and Service Solutions

SPSTU e: Secondary and Preparatory School Teachers Union

SPSW e: Support Software

SPT e: Shortest Processing Time
- e: Substrate-Plate Trench

SPTG e: Support Group

SPTN e: Single Protocol Transport Network

SPTR e: South Pole TDRS Relay

sptr r: sprängpatron

SPTS e: Single Program Transport Stream
- e: Small Purchase Tracking System

SPTW e: Support Wing

SPU e: Scientific Programs Unit
- e: Society for Pediatric Urology
- e: Special Projects Unit
- e: Strategic Planning Unit
- e: Summa Plus Ultra

SPÜ d: Schulpraktische Übung[en]

SPUCDL e: Serial Peripheral Unit Controller/Data Link

SPUD e: Storage Pedestal Upgrade Disk
- e: Storage Pedestal Upgrade Drive

SPUP e: Seychelles Peoples United Party

SPUR e: Software Package for Unique Reports
- e: Supercomputing Program for Undergraduate Research
- e: Systech Pluraxial Unplug Repeater
- e: System for Predetermining Unit Requirement

sp.v. N: spesifikk vekt

SPV d: Semipermanente Verbindung
- e: Sampler Performance Validation
- e: Sheep Poxvirus
- e: Surface Photo Voltage

SPVC e: Senior Pro-Vice-Chancellor

SPVP e: Somali Peoples Vanguard Party

SPVS e: Society of Practising Veterinary Surgeons

SPW e: Space Wing

SPWT e: SSM/I Land Products Working Team

SPX e: Secured Packet Exchange
- e: Sequenced (or Sequential) Packet Exchange

SQ d: Signal-Quantisierungsgeräusch
- e: Signal Quality [detector]
- e: Significant Quantity
- e: State Question
- f: Service Quotidien
- f: Source Qualifiée

S Q e: South Quad

SQA e: Semiquantitative
- e: Statistical Quality Assurance

S&QA e: Safety and Quality Assurance

S&QAA e: Safety and Quality Assurance Administration

SQAE e: Senior Quality Assurance Evaluator

SQAO e: Reg 5 Toxics Monitoring System
SQBE e: Small Quantity Burner Exemption
SQC e: Squamous Cell Carcinoma
SQE e: Signal Quality Error
Sq. E&S e: Square Edge and Sound
SQFP e: Shrink Quad Flat Pack
SQL e: Software Quality Language
- e: Structured Query Language [for DBMS]
SQL3 e: Proposed definition of 'structured query language with object support
SQLCA e: SQL Communication Area
SQLCLI e: SQL Call Level Interface
SQL[/]DS e: Structure Query Language/Data System
SQMD e: Squadron Manpower Document
SQ OC e: Squadron Operations Center
SQPMM e: Software Quality and Process Maturity Model
SQR e: Special Qualifications/ Requirements
SQUADS e: Stanford QUAntum Device Simulator
SQUEL e: Spatial Query Langauage
SQuID e: Source Quench Introduced Delay
SR a: Sikkerhedsrådet
- a: Studenterrådet
- d: Semainarraum
- e: Safety Requirements
- e: Sales Representative
- e: Savannah River
- e: Schedule Request
- e: Scientific Results of the proceedings of ODP
- e: Search and Retrieve
- e: Selector
- e: Short-Range
- e: Signalling Rate [selector]
- e: Silicon Revision register
- e: Size-Resolving
- e: Source Routing
- e: Space Required
- e: Special Reconnaissance
- e: Special Relativity
- e: State Road
- e: State Route
- e: Stockholders Report
- e: Street Regal
- e: Sum Rule
- e: Suriname
- e: Sustained Response
- e: System Readiness
- f: Radiomètre à balayage, Scanneur
- f: Saisie-Revendication
- f: Salaire de Référence
- f: Sans Ressources
- f: Sans Résultat
- f: Sans Retard
- f: Second Rang
- f: Secours Routier
- f: Service des Redevances
- f: Service Routier
- f: Services Rapides
- f: Sucre Raffiné
- f: Suffrage Restrient
- f: Suisse Romande
- N: Senkningsreaksjon
- r: Sidriktare
- r: Statstjänstemännens Riksförbund
- r: Strategisk
- s: Radiómetro de exploración
Sr e: Seminar
Sr. s: Señor
S/R e: Storage Retrieval
- e: Support Requirements
- s: Sin Retroceso
- s: Su Referencia
SR-71 e: Blackbird
Sr-89 e: Strontium-89
Sr-90 e: Strontium-90
- e: Strontiumisotop 90Sr
SRA e: Scanning Radar Altimeter
- e: Scientific Review Administrator
- e: Selected Restricted Availability
- e: Selective Routing Arrangement
- e: Specialized Repair Activity
- e: Specially Reserved Area
- e: Standard Requirements Code
- e: State Recreation Area
- e: Supply Readiness Assessment
- e: Systems Research and Application Corporation
- f: Service de Réglementations Administratives
- f: Service des Recherches de l'Aéronautique
- r: Statens Reproduktionsanstalt
- r: Svenska Reproduktionsaktiebolag
Sra. s: Señora
SRAC e: Supplier Relations Action Council
- f: Service des Relations avec les Autorités Civiles
SRAD e: Spectral Radiometer
Sradar r: Spaningsradar
SRAEN f: Système de Référence pour la détermination de l'Affaiblissement Équivalent par la Netteté
SRAG e: Sydney Rainforest Action Group
SRAIS e: Statewide Resource Information and Accounting System
SRALCC e: Southern Region Airlift Control Center
SRALD e: Southern Region Airlift Division
SRAM e: Scratchpad RAM [Random Access Memory]
- e: Shadow Random Access Memory
- e: Static Random Access Memory
SRAP e: Superfund Remedial Accomplishment Plan
SRAPI e: Speech Recognition API
SRARM e: Short-Range Anti-Radiation Missile
SRAS e: State Rural Assistance Scheme
S-rating e: Sustainability rating
SRB d: Strausberg, Brandenburg
- e: Solid [fuel] Rocket Booster
- e: Source Route Bridging (or Bridge)
- e: Surface Radiation Budget
- e: System[s] Review Board
- f: Balance de Radiación de Superficie
- f: Bilan Radiatif en Surface
S R B e: Space Research Building
SRBCP e: Satellite Radiation Budget Climatology Project
SRBG f: Société Royale Belge de Géographi[qu]e Brussells
SRBL e: Solar Radio Burst Locator
SRBOC e: Super Rapid Blooming Offboard Chaff
SRC e: Science Research Center program
- e: Scientific Review Committee
- e: Semiconductor Research Corporation
- e: Source [Resource Controller]
- e: Standard Context Routing
- e: Standard Requirement Code
- e: Strategic Reconnaissance Center
- e: Student Relations Committee
- e: Students Representative Council
- e: Suluh Reconciliation Committee
- e: Supply Readiness Center
- e: Supreme Revolutionary Council
- e: Survival and Recovery Cell
- e: Survival Recovery Center
- e: System Reference Code
- e: System Resource Controller
- e: System Review Committee
- f: Service des Renseignements Commerciaux
- f: Station Radio Côtière
SRC. e: System Reference Code
SRCC f: Société Nationale pour la Rénovation et le Développement de la Cacaoyere et de la Caféière Togolaises
SRCF e: Subregional Control Facility
SRCL s: Servicio de Reclutamiento
SRCT f: Service des Recherches et du Contrôle Technique
SRD e: Screen Reader System
- e: Secondary Receive[d] Data
- e: Short Range Device
- e: Super-Radiant Diode
- e: System Requirements Document
S/RD e: Secret/Restricted Data
SRDC e: Southern Rural Development Center
- e: State Rural Development Councils
- e: Subrate Data Cross-connect
- e: Sugar Research and Development Corporation
- e: Surface Reference Data Center (or Centre)
SR-DPG e: Sedimented Ridges Detailed Planning Group
SRDRAM e: Self-Refreshed DRAM
SRE e: Self Routing switch Element
- e: Serum-Responsive Element
- e: Society for Radiological Engineering
- f: Service des Recherches et Expérimentation
- s: Sección de Retaguardia de Etapas
S/Ref[a] s: Su Referencia
SREH e: Storm-Relative Environmental Helicity
s*rel d: Kurzschreibweise der speziellen relativen spektralen Empfindlichkeit nach DIN 5032
srel d: relative spektrale Empfindlichkeit der angepaßten Kamera
- d: relative spektrale Empfindlichkeit des gefilterten Empfängers
srelD, d: vollständige Schreibweise der relativen spektralen Empfindlichkeit bei Vollfilterung
srelD,F, d: vollständige Schreibweise der relativen spektralen Empfindlichkeit bei Partialfilterung

SREM e: School of Resource and Environmental Management
SREML e: Shortened Rapid Eye Movement Latency
SREMP e: Source Region Electromagnetic Pulse
SREP e: State Rivers and Estuaries Policy
- f: Société Royale d'Économie Politique
S.Res. e: Senate Resolution
SRES e: Signed Result
SRET f: Service Régional d'Exploitation des Transmissions
SREX e: Strontium Extraction
SRF e: Secure Reserve Force
- e: Service Resource Function
- e: Short-Range Force[s]
- e: Solar Radiation Facility
- e: Specialised Resource Function
- e: Specifically Routed Frames
- e: Spectral Response Function
- e: Status Report Frame
- e: Strategic Reserve Force
- e: Summary Reference File
- f: Service de Répression des Fraudes
- f: Station de Recherches Forestières
- r: Sveriges Redarförening
SRF. e: Specifically Routed Frames
SRFD f: Service de Répression des Fraudes Douanières
SRF FOR r: Skolreglemente för Förberedande Officerskurs
SRFS e: Session Reservable File System
SRFWG e: Stand Replacement Fire Working Group
SRG d: Schweizer Radio- und Fernsehgesellschaft
- e: Scientific Review Group
- f: Système de Référence Géocentrique
SRGB e: Sustained RGB
SRI e: Short-Range International
- e: SIP Relay Interface
- e: Special Request for Information
- e: Stanford Research Institute
- e: Surveillance, Reconnaissance and Intelligence
- f: Service de Recherches et d'Investigations
- f: Service des Relations Internationales
SRIC e: Short-Rotation Intensive Culture
S/RID e: Standards/Requirements Identification Document
SRIS e: Science and Reference Information Service
SRK e: Sark
- e: Strain Release Kit
- r: Svenska Röda Korset
SRL d: Systematische Rechtssammlung des Kantons Luzern
- e: Scanning Raman Lidar
- e: Shuttle Research Laboratory
- f: Section Reacteur Leger
- f: Société à Responsabilité Limitée
SRLF e: Southern Regional Library Facility
Srlpbv r: Stridsledningspansarbandvagn
SRLZ e: Southern Rock Lobster Zone
SRM e: Security Reference Monitor
- e: Send Routing Message
- e: Social Religion Mormon
- e: Solid Rocket Motor
- e: Standard Reference Material[s]
- e: Standard Reference Method
- f: Material Patrón de Referencia
- f: Matériel de Référence Normalisé
s.r.m. e: social religion mormon
SRMA e: Split-Channel Reservation Multiple Access
SRMU e: Subrate Multiplexer Unit
SRN e: Source/Recipient Node
sRNA e: Soluble RNA
SR-NETT e: Nordic Search and Retrieve Project
SRNP e: Secure Remote Network Processor
SRO d: Stadtroda, Thüringen
SROSS e: Stretched Rohini Satellite Series
SRP e: Script
- e: Sealift Readiness Program
- e: Sealift Reserve Program
- e: Seaward Recovery Point
- e: Solicitation Review Panel
- e: Soluble Reactive Phosphorus
- e: Source Routing Protocol
- e: Spreading-Resistance Probe
- e: Spreading-Resistance Profining
- e: Strategic Reconnaissance Plan
- e: Suggested Retail Price
- e: Superimposed Renewal Process
- e: Supplemental Retirement Program
SRP< e: Signal Recognition Particle
Srp Arh Celok Lek se: Srpski Arhiv za Celokupno Lekarstvo
SRPC e: Society of Rural Physicians of Canada
SRPI e: Server Requester Programming Interface
SRPJ f: Service de Recherche de la Police Judiciaire
SRPT e: Shortest Remaining Processing Time
SRP-UDP e: Standard Review Plan Update and Development Program
SRQ e: Seconds
- e: Serially Reusable Resource
- e: Single Sided
- e: Solid State Disk
- e: Stack Segment
SRR e: Secondary Receiver Ready
- e: Serially Reusable Resource
- e: Survivability, Recovery and Reconstitution
- e: System[s] Requirement[s] Review
- f: Service Régional des Redevances
SrR e: Strontium Recovery
Srr r: Spaningsradar
SR&R e: Staff Rules and Regulations
SRRB e: Solar Radiation Research Branch
- e: Surface Radiation Research Branch of NOAA
Srrtro r: Spaningsradartrupp
SRS e: Savannah River Site
- e: Savannah River Site, Aiken, South Carolina
- e: Secondary Request to Send
- e: Sequence Retrieval
- e: Service Repair System
- e: Situation Reporting System
- e: Software Requirement Specification
- e: Solar and Infrared Radiation Station
- e: Spatial Reference System
SRS1 e: South Research Station 1
S/R-SATELLITE e: Search and Rescue Satellite
SRSP e: Somali Revolutionary Socialist Party
SRT e: Scheduled Return Time
- e: Secure Request Technology
- e: Source Routing Transparent
- e: Standard Remote Terminals
Srt d: Schrittzahl der Iteration
Srta. s: Señorita
SRT/AMME e: Standard Remote Terminal/Automated Multi-Media Message Exchange
SR-TB e: Source Routing-Transparent Bridging
SRTC e: Savannah River Technology Center
Srtmax d: Maximale Schrittzahl in einer Iteration
SRTP e: Sequenced Routing Table Protocol
SRTS e: Secondary Ready To Send
- e: Synchronous Residual Time Stamp
SRU d: Der Rat von Sachverständigen für Umweltfragen, Wiesbaden
- e: Ocean Services Unit
- e: System Resource Unit
- f: Société de Raffinage d'Uranium
- f: Unité des Services Océaniques
- s/: Unidad de Servicios Oceánicos
SRUC e: Split Reservation Upon Collision
Srum r: Sammanställningsrum
SRV f: Sur Rendez-Vous
SRVIFS e: Server Installable File System
SRWBR e: Short Range Wideband Radio
SRWCP e: Short Rotation Woody Crop Program
SRY e: Surrey
SRZ d: Satz-Rechen-Zentrum, Berlin

SS d: Semesterstunden
- e: Deputy chief of Staff, Systems Support
- e: Deputy Commander for Systems Support
- e: Sales Support
- e: Sample Stabilizer
- e: Seconds
- e: Self-Sustaining
- e: Service Segment
- e: Session Services
- e: Settleable Solids
- e: Ship Shore
- e: Simulator System
- e: Single Seconds
- e: Sparc Station
- e: Special Services
- e: Standard Sequential delivery
- e: Steel Shelves
- e: Stop Ship
- e: Subscriber Station
- e: Substation
- e: Supply Support
- e: Switching System
- e: System Support CSCI
- f: Sans Surtaxe
- f: Sciences Sociales
- f: Scrutin Secret
- f: Secteur Social
- f: Section Scientifique
- f: Sécurité Sociale
- f: Service de Sécurité
- f: Service Sanitaire
- f: Service Social
- f: Services Secrets
- f: Signature Sociale
- f: Sous Scellé
- f: Sous Séquestre
- f: Sous-Seing
- f: Sous-Sol
- f: Statut d'une Société
- f: Supplément de Solde
- N: Seilskip
- r: Sidställning
- r: Signalstation
- s: Secciones

Ss f: Sans suite
- f: Sous
- r: Stockholms stabskompani

S S e: Social Science

SS. e: Surge Suppressor

ss. r: såsom

S&S e: Safeguards and Security

S/S e: Ship-to-Shore
- e: Shuttle to Station
- s: Subsector

S1S e: Surfaced one Side

S2S e: Surfaced two Sides

SS6 e: Signaling System 6

SS7 e: Signaling System 7

SSA e: Selective Service Administration
- e: Semiconductor Industry Association
- e: Serial Storage Architecture
- e: Service Support Adapter
- e: Singular Spectrum Analysis
- e: Small-Signal Analysis
- e: Software Support Agency
- e: Sole Source Aquifer
- e: Somali Salvation Alliance
- e: Specific Surface Area
- e: Staff Support Agency
- e: Station Select Address
- p: Serviços de Saúde e Assistência
- r: Sveriges Sändareamatörer

s/Sa f: sur Saône

SSAB e: Source Selection Advisory Board
- r: Svenska Stål Aktiebolag

SSAC e: Senior Staff Advancement Committee
- e: Soil Site Assimilated Capacity
- e: Source Selection Advisory Committee

SSADM e: Structured System Analysis and Design Method

SSAE f: Service Social d'Aide aux Émigrants
- f: Société Shell de l'Afrique Equatoriale

SSA-F e: S-band Single Access-Forward

SSALT e: Solid-State [radar] Altimeter

SSAM e: Session Service Access Memory

Ssan N: Sjøforsvarets sanitet

SSAP e: Session Service Access Point
- e: Source link Service Access Point

SSA-R e: S-band Single Access-Return

SSB d: Schwäbischer Standard-Benchmark
- e: Ballistic Missile Submarine
- e: Select Standby
- e: Single Side Band
- e: Space Science Board
- e: Space Studies Board
- e: Spinning Sidebands
- e: Subsurface Barrier

SSBA f: Suite Synthetique des Benchmarks de l'AFFU

SSBB e: Service Segment Building Block

Ssbl N: Sjøforsvarets sambandsinspeksjon

SSBR e: Social Statistics Briefing Room

SSBU e: Shuttle Solar Backscatter Ultraviolet

SSBUV e: Shuttle SBUV [Solar Backscatter Ultraviolet] [experiment]
- e: Shuttle Solar Backscatter Ultraviolet radiometer

SSBW e: Surface-Skimming Bulk Wave

SSC e: I-GOOS Strategy Sub-Committee
- e: Science (or Scientific) Steering Committee
- e: Scientific Support Coordinator
- e: SCSI Stream Commands
- e: Space Surveillance Center
- e: Special Scholarships Committee
- e: Special Sensor C
- e: Special Services Center
- e: Specialized Systems Consultants
- e: Species Survival Commission
- e: Standard Speech Circuit
- e: Standard System Center
- e: State Security Council
- e: State Shipping Corporation
- e: State Superfund Contracts
- e: Strategy Sub-Committee
- e: Superconducting Super Collider
- e: Supply Systems Command
- e: Support Service Contract
- e: System Support Contractor
- f: Commission de la sauvegarde des espèces
- f: Sous-Comité de la stratégie de l'I-GOOS
- f: Surveillance du Système Climatique
- s: Subcomité sobre Estrategia del I-GOOS

SSCC e: Serial Shipping Container Code
- e: Space Station Control Center
- e: Special Security Communications Center
- e: State Stewardship Coordinating Committee
- s: Servicio Social Casa Central

SSCE e: Sodium chloride-Saturated Calomel Electrode
- e: Solid Surface Combustion Experiment

SSCF e: Service Specific Coordination Function

SSC-GLOBEC e: Scientific Steering Committee of the Global Ocean Ecosystems Dynamics Core Project of IGBP
- f: Comité directeur scientifique du projet de base du PIGB Dynamique des écosystèmes océaniques à l'échelle mondiale
- s: Comité Científico de Dirección del Proyecto Central del IGBP sobre Dinámica de los Ecosistemas Oceánicos Mundiales

SSCL e: Superconducting Super Collider Laboratory

SSCOP e: Service Specific Connection Orien[ta]ted Protocol

SSCP e: Service Switching and Control Point
- e: Society for a Science of Clinical Psychology
- e: Subsystem Services Control Point
- e: System Services Control Program
- e: System[s] Service Control Point

SSCP-PU e: System Service Control Point-to-Physical Unit

SSCR e: Standard Systems Center Regulation

S4SCS e: Surfaced four sides with Chalking Seam on each edge

SSCS e: Service Specific Convergence Sublayer

SSCSAP e: Standard Small Computer Software Acquisition

SSD d: Stahlschiebdach
- e: Sample Server Disk
- e: Scientific Support Division
- e: Secondary Send Data
- e: Shared Service Domain
- e: Simultaneous Signal Detection
- e: Source Summary Database
- e: Special Security Detachment
- e: Special Sensor D
- e: Statistical Subdivision
- e: Synthesis Solid Diffusion
- e: Systems Support Database
- f: Section de Soutien de Dugny
- f: Service des Stratégies de Durabilité

SSDB e: Site Survey Databank
SSDC e: Space and Strategic Defense Command
- e: Stack Segment Descriptor Cache

SSDD e: Same Stuff, Different Day
SSDI e: Social Security Death Index
SSDS a: System af Social og Demografisk Statistik
- e: System of Social and Demographic Statistics

S1S1E e: Surfaced one Side and one Edge
SSE e: Software Support Engineer
- e: South by South East
- e: Specific Schedule Event
- e: Sum Squared Error
- e: System Security Engineering
- f: Sécurité Sociale Étudiante
- f: Sud-Sud-Est

Sse f: Suisse
SSEA f: Société Suisse des Études Africaines, Bern
SSEALS e: South/Southeast Asia Library Service
SSEC e: Space Science and Engineering Center
- s: Centro de Ciencias e Ingeniería del Espacio

SSED e: Software Systems Engineering Directorate
SSEG e: Source Selection Evaluation Group
S Seg A s: Servicio de Seguridad en Alto
SSEIS e: Standard Support and Environmental Impact Statement
s-seksjon N: sikkerhetsseksjon
SSEL e: Session Service Selector
S-sele[r] a: Sikkerhedsseler
SSEM e: Stepper Specific Equipment Model
SSEP e: Source Selection Evaluation Plan
SSER e: System Soft Error Rate
s/sersj N: stabssersjant
SSES e: Ship Signal Exploitation Space
- e: Suite of Survivability Enhancement Systems

SSESCO e: Supercomputer Systems Engineering and Services Company
SSET e: Source Selection Evaluation Team
SSF a: Sydslesvigsk Forening
- e: Sampled Servo Format
- e: Semi-lagrangian and Semi-geostrophic Finite element
- e: Service Switching Function
- e: Southern Shark Fishery
- e: Switch Supervisory Function

SSFA N: Skyteskolen for Feltartilleriet
SSFDC e: Solid State Floppy Disc Card
SSFDR e: Solid-State Flight Data Recorder
SSFF e: Showcase Software Factory of the Future
- N: Skogbrukets og Skogindustriens Forskningsforening

SSFL e: Santa Susana Field Laboratory, Canoga Park, California
SSFMAC e: Southern Shark Fishery Management Advisory Committee
SSFR e: Solar Spectral Flux Radiometer
SSG e: Scientific Steering Group for WOCE
Ssg r: Sammansättning
SSGA e: System Support Gate Array
SSG-TOGA e: TOGA Scientific Steering Group
- s: Grupo Científico de Dirección sobre TOGA

SSG-WOCE e: WOCE Scientific Steering Group
- s: Grupo Científico de Dirección sobre WOCE

SSH e: Special Sensor H
- f: Service Social à l'Hôpital

SS&H e: Social Science and Humanities
SSHA e: System Safety Hazards Analysis
- N: Skyteskolen for Hærensartilleri

SSHD d: Stahl-Schiebe-Hebe-Dach
SSHLAC e: Social Sciences and Humanities Library Advisory Committee
SSHRCC e: Social Sciences and Humanities Research Council of Canada
SSHT e: Sea Surface Height
SSHTM e: Small Scale, High Temperature Melter
SSI d: Schwedisches Strahlenschutzinstitut
- e: Scuba Schools International
- e: Server-Side Include[s]
- e: Silicon Systems, Inc.
- e: Small Scale Integration
- e: Social Security Insurance
- e: Software Support Instruction
- e: Solid State Imager (or Imaging)
- e: Space Services Inc.
- e: Spectral Statistical Interpolation
- e: Supplemental Security Income [program]
- e: Supply System Item
- r: Sektionen för Särskild Inhämtning

SSIA e: Soviet Scientists Immigration Act
SSID e: Secondary Station Identifier
SSIES e: Icon Scintillation Monitor on board the DMSP satellites
SSIF f: Syndicat des Sociétés Immobilières Françaises
SSIGE f: Société Suisse de l'Industrie du Gaz et des Eaux
SSIGR f: Société Suisse des Ingénieurs du Génie Rural
SSII f: Société de Services d'Ingénierie Informatique
SSILS e: Solid-State Instrument Landing System
S[-]SIMM e: Single RAS SIMM [Single Inline Memory Module]
SSIMS e: Scanning Secondary-Ion Mass Spectroscopy
SSIMT e: Suppressed Sidewall-Injection MagnetoTransistor
SSIPS/LDS e: Ship Support Improvement Projects/Log Data System
SSIS e: Store Stock Inventory/Accounting System
SSIT e: The IEEE Society on Social Implication of Technology
SSIXS e: Submarine Satellite Information Exchange Subsystem
SSJ e: Sole Source Justification
SSK e: Hunter-Killer Submarine
- e: Sociology of Scientific Knowledge

ssk. r: särskilt
SSKA N: Kystartilleriets Skyteskole
Sskai e: Skalaaufteilung
SSKN e: Hunter-Killer Submarine, nuclear powered
sskv r: spaningsskvadron
SSL a: Statsskatteloven
- e: Secure Socket[s] Layer
- e: Soweto Students League
- e: Spin SuperLattice

SSLA e: Shuttle Scanning Laser Altimeter
SSL fault e: Single Stuck Line fault
SSLSM e: Single Service Logistics Support Manager
SSL-Verschlüsselung d: Secure Socket Layer-Verschlüsselung
SSM e: Magnetometer on board the DMSP satellites
- e: Small Scale Melter
- e: Strategic Sourcing Methodology
- f: Société de Secours Mutuels

SSMA e: Small-Scale Miners Association
SSMB f: Syndicat Suisse des Marchands de Bétail
SSMC e: Silver Spring Metropolitan Complex
- e: Sports Science and Medical Council

SSMCEC e: Sports Science and Medical Council Executive Committee
SSME e: Space Shuttle Main Engine
SSMF e: Space Station Millimeter Facility
SSMI e: Special Sensor Microwave Imager
- e: Special Sensor Microwave Instruments
- f: Imageur à hyperfréquences à capteur spécial
- s: Sensor especial de imágenes por microondas

SSM/I e: Special Sensor Microwave/ Imager [on board the DMSP satellites]
SSMIA e: Senior Managers Information Analysis
SSMIS e: Special Sensor Microwave Imager/Sounder
SSMJ/4 e: Precipitating Electron and Icon Spectrometer on board the DMSP satellites
SSMR-SSM/I e: Scanning Multichannel Microwave Radiometer-SSM/I
SSMS e: Spark-Source Mass Spectrometer
- e: Spark-Source Mass Spectrometry
- e: Spark-Source Mass Spectroscopy
- e: Standard Survivable Message Set

SSMT e: Special Sensor Microwave Temperature Sounder
SSM/T e: SSM/Temperature

SSM/T1 e: Special Sensor Microwave Temperature sounder on board the DMSP satellites
SSM/T2 e: Special Sensor Microwave water vapor sounder
SSMT/2 e: Special Sensor Microwave water vapor profiler on board the DMSP satellites
SSN e: Satellite Sports Network
- e: Service Subscriber Number
- e: Sortie Sequence Number
- e: Sunspot Number
- i: Servizio Sanitario Nazionale
- N: Sentrale Sjefsnemnd
SSNC e: Stargroup Server for Netware Clients
SSNIT e: Social Security and National Insurance Trust
SSNM e: Southern Somali National Movement
s1 s nn d: Elemente der Streumatrix
SSNNP e: Sustainable Social Net National Product
SS No.7 e: Signaling System No. 7
SSNR e: Society for the Study of Neuronal Regulation
SSO d: Südsüdosten
- e: Oiler Submarine
- e: Satellite Switching Office
- e: Source Selecting Official
- e: Source Selection Office
- e: Special Security Office[r]
- e: Spurious Shut-off
- e: Standard System Support Office
- e: State Services Organization
- e: Subsistence Support Officer
- e: System Security Officer
- f: Sud-Sud-Ouest
SSØ a: Sydsydøst
SSÖ r: Sydsydöst
SSOC e: Solid State Optoelectronics Consortium of Canada
SSOEG e: Satellite Systems Operations Evaluation Group
SSOF e: Standby Supervisor of Flying
SSOP e: Shrink Small-Outline Package
- f: Service de Sélection et d'Orientation Professionnelle
SSP a: Samarbejde mellem skole, Social-forvaltning og Politi
- e: Scanning Spectral Polarimeter
- e: Security Support Plan
- e: Serial Storage Protocol
- e: Service-Specific Part
- e: Service Subscriber Profile
- e: Service Switching Point
- e: Siemens Switching Processor
- e: Signal[ing] Switching Point
- e: Silicon Switch Processor
- e: Site Survey Panel
- e: Software Standard and Procedure
- e: Source Selection Plan
- e: Spectrally Scanning Polarimeter
- e: Sponsor Selective Pricing
- e: Standard Printer Port
- e: Strategic Strike Plan
- e: Subsurface Science Program
- e: Summer Scholarship Programme
- e: Switch-to-Switch Protocol
- e: System Service Processor
- e: System Stack Pointer
- e: System Support Processor
- e: Vehicle Transport Submarine
- f: Service Social des Prisons
- f: Société de Statistique de Paris
- f: Sous-Seing Privé
SSp f: Solde Spéciale
SSPA e: Russian equivalent to the RSPCA
SSPC e: SSP Controller
SSPF e: Space Station Processing Facility
SSPH e: Single-Shot Probability of Hit
SSPI e: Security Service Provider Interface
s.s.p.l e: social support pregnancy loss newsgroup
SSPM e: Software Standards and Procedures Manual
SSPO e: Strategic Systems Program Office
SSPP e: Solar Stellar Pointing Platform
SSPRU e: SSP Relay Unit
S&SPS e: Sensor and Signal Processing Systems
SSPS e: Sky Survey Prototype System
SSPT f: Société Sénégalaise de Publicité et de Tourisme
SSQA e: Standardized Supplier Quality Assessment
SSR d: Sozialistische Sowjetrepubliken
- e: Radar Picket Submarine
- e: Solid State Recrystalization
- e: Staff/Student Ratio
- e: Subsea Release cementing system
- e: Supplemental Security Income Record
- e: Supply Support Request
- e: Support System Requirement
- e: System Software Review
- e: System Specification Review
- e: System Support Request
- f: Service Social Rural
SSr f: Sous-Secteur
SSRC e: Standard Software Requirements Contract
SSRG s: Secretario Social de Refugiados de Guatemala
SSRI e: Selective Serotonin Re-uptake Inhibitor
SSRL e: SEMATECH Software Reuse Library
SSRMS e: Space Station Remote Manipulating System
SSRN e: Radar Picket Submarine, nuclear powered
SSRP e: SEMATECH Software Reuse Program
- e: Simple Server Redundance Protocol
- f: Services Sociaux de la Région Parisienne
SSRR e: Subsystem Requirements Review
SSRS e: Standard Service Request Specification
- N: Selskapet for Sosialt Ansvar i Videnskapen
SSRT e: Single Stage Rocket Technology
SSRV e: Solar-Sail Race Vehicle
SSS e: Sea Surface Salinity
- e: Security Support Staff
- e: Ship Maintenance Support System
- e: Solid State Spectrometer
- e: Source Selection Sensitive
- e: Special Study Section
- e: Standard Stratigraphic Scale
- e: State Security Service
- e: Station Satellite System
- e: Storage Serviceability Standards
- e: Strategic Satellite System
- e: Subsystem Specification
- e: Switching Subsystem
- e: System Support Specialist
- f: Service de Sondages et Statistiques
SSSD e: State Soil Survey Data Base
SSSI e: Site of Special Scientific
SSSNA e: Server to Server Systems Network Architecture
SSSO e: Standard Systems Support Office
SSSP e: Source Selection Support Plan
SS/SSS e: System Subsystem Specification/Signal Simulator
SSST e: Single Subscriber Terminal
- e: Southeastern Symposium on Systems Theory
SST e: Sea Surface Temperature
- e: Self Service Terminal
- e: Ship Systems Team
- e: Simple SIPP Transition
- e: Single-Shell Tank
- e: Spread Spectrum Technology
- e: Spread Spectrum Transmission
- e: Synchronous Service Transport
- e: Systems Services and Technology
- f: Service de la Surveillance du Territoire
- f: Service Social du Travail
- f: Température de Surface de la mer
- N: Sjøforsvarets Skipstilsyn
- N: Sjøforsvarsstaben
Sst r: Signalstyrman
SST. e: System Service Tools
s.st. a: samme sted
S.St. a: Sønderjydsk Studenterforening
SSTA e: Sea-Surface Temperature Anomaly
SS/TDMA e: Satellite-Switched TDMA
SSTF f: Service de la Surveillance des Travaux et des Fabrications
SSTI e: Small Spacecraft Technology Initiative Program
SSTIX e: Small Ship Teletypewriter Information Exchange System
SSTM e: Single Service Tanker Manager
SSTOL e: Super[sonic]-Short Take-Off and Landing
SSTP e: Screened Shielded Twisted Pair
SSTR e: Subsystem Test Review
SSTS e: Space-Based Surveillance and Tracking System
- N: Sjøforsvarets Stabsskole
SSTTSS e: Space Space Time Time Space Space network
SSTV e: Slow Scan Television

SSU e: Session Support Utility
- e: Sources of Support
- e: Special Support Unit
- e: Species Services Unit
- e: Star Sensor Unit
- e: Stratospheric Sounder Unit
- e: Swaziland Students Union
- e: System Support Unit

SSUPS e: Solid State Uninterruptable Power Source
- e: Station Static UPS

SSURGO e: Soil Survey Geographic Data Base

SSURO e: Stop Sale, Use and Removal Order

SSUV d: Sammelstelle für die Statistik der Unfallversicherung

SSV a: Sydslesvigsk Vælgerforening
- a: Sydsydvest

SSW d: Schnittstellenwandler
- d: Südsüdwesten
- e: South by South West
- e: Sudden Stratospheric Warming
- e: System Software

SSWAHC e: Society for Social Work Administrators in Health Care

S S W B e: School of Social Work Building

SSWEAT e: Simplified SWEAT [Standard Wafer-level Electromigration Acceleration Test]

SSWG e: Supplementary Strategies Working Group

SSWQC e: Site-Specific Water Quality Criteria

SSWS e: Standard Sea Water Service

SSY e: Small-Scale Yielding

ST a: Sjællands Tidende
- d: Schwebstaub
- e: Seagate Technologies
- e: Segment Type
- e: Self-Trap[ping]
- e: Send Timing
- e: Signaling Terminal
- e: Sixteen-Thirty-Two [bits]
- e: Southern Tablelands
- e: Start Timer bit
- e: Straight Terminus
- e: Stratosphere-Troposphere
- e: Stream Protocol
- e: Subscriber Terminus
- e: Technical Support division
- e: Telescope Science institute
- f: Sans Travail
- f: São Tome et Principe
- f: Service du Trésor public
- f: Service Temporaire
- f: Siège du Tribunal
- f: Somme Totale
- f: Station Thermale
- f: Subrogé Tuteur
- f: Surveillance du Territoire
- l: Studia Theologica
- r: Skeppteknisk
- r: Statstjänstemannaförbundet
- r: Stockholms Tygstation
- s: Sección Topográfica
- s: Sistema de Transmisión
- s: Submarino Torpedero

St f: Strasbourg
- N: Stab
- r: Stål

st. a: i stedet for
- a: stuen
- a: stødt
- a: størrelse
- N: stasjon
- r: styck[e]

St. a: Smørsyretal
- a: Store
- e: Saint

s.t. a: samme tid
- e: something
- e: such that
- r: samma tid

S/T e: Sales Tax

S3T e: Sequentially Sampling Sediment Trap

STA d: Starnberg, Bayern
- e: Science and Technology Advisor
- e: Science and Technology Agency
- e: SCSI Trade Association
- e: Service Terminal Arrangement
- e: Society for Technology in Anesthesia
- e: Strategic Transportation Analysis
- e: Student Affairs Committee
- e: System Threat Assessment
- f: Agence de Science et de Technologie
- s: Organismo de Ciencia y Tecnología

StA d: Staatsarchiv

Sta. s: Santa

STAAIM f: Syndicat des Travailleurs Agricoles et Artisans de l'Ile Maurice

STAARS e: Standard Army After Action Review System

STAB f: Service des Transports Aériens du Burundi

staben/B 3 r: 3. basbataljons stab

STABEX e: Export Stabilization Fund
- e: Stabilisation of Export earnings

stabsbatt r: stabsbatteri

Stabsbs r: [Arméns] Stabs- och Sambandsskola

stabsch r: stabschef

stabskomp r: stabskompani

stabsplut r: stabspluton

stabs/sjt N: stabssersjant

stabsuoff r: stabsunderofficer

STAC e: Satellite Tracking Angle Calculations
- e: Supersonic Transport Aircraft Committee

STACCS e: Standard Theater Army C2 System

STACK e: Start Acknowledgement

STACS e: Sub-Tropical Atlantic Climate Studies (or Study)
- f: Études sur le Climat de l'Atlantique Subtropical
- s: Estudios sobre el Clima del Atlántico Subtropical

ST/ÄD d: Stellvertreter des Ärztlichen Direktors

STADD e: Ship Towed Acoustic Deception Device

STADIR e: Station Director

StadtA d: Stadtarchiv

StadtB d: Stadtbibliothek

StaDynL d: Institut für Statik und Dynamik der Luft- und Raumfahrtkonstruktionen

STAF e: SDH Transceiver And Framer
- e: Systems Technical Applications Facility

STAFAST e: Standard Frequency and Standard Time

STAFF e: Spatio-Temporal Analysis of Field Fluctuations

STAFISKO N: Statens Ferskfiskkontor

Stafo N: Statstjenestemannsforbundet

STAG e: Soils, Trees and Grass Program

Stag f: Stagiaire

STAGE e: Structured Text And Graphics Editor

StahlB d: Institut für Konstruktion und Entwurf

STAJ e: Science and Technology Agency of Japan

STALAPCO e: State and Local Air Pollution Control Officials

STALOG e: System to Automate Logistics

StALU d: Staatsarchiv des Kantons Luzern

S.T.A.M.P.E.D. e: Size, Temperature, Application, Medium, Pressure, Ends

STAN e: Space Telescope Analysis Newsletter

STANA e: Fisheries Statistics on the North Atlantic
- f: Statistiques des pêches de l'Atlantique Nord
- s: Estadísticas de Pesca del Atlántico Norte

STANAG e: Standard NATO Agreement

STANAVFORLANT e: Standing Naval Force, North Atlantic

STANAVFORMED e: Standing Naval Force Mediterranean

STANAV-SFORLANT e: Standing Naval Force Atlantic

St.And. e: St. Andrews

STANEVAL e: Standardization and Evaluation

Stanf. e: Stanford Dictionary of Anglicised Words and Phrases

STANFINS e: Standard Finance System

Stan L Rev e: Stanford Law Review

STANO e: Surveillance, Target Acquisition and Night Operations

STANTEK r: Standardiseringstekniska Föreningen

StAnz d: Staatsanzeiger des Landes Hessen

STA&P e: System Threat Assessment and Projections

STAP e: Scientific and Technical Advisory Panel
- e: Strategic Target Aim Point
- f: Groupe consultatif Scientifique et Technique
- s: Panel de Asesoramiento Científico y Técnico

STAR e: ISRS with library modules
- e: Joint CCOP/SOPAC-IOC Working Group on South Pacific Tectonics and Resources
- e: Self defining Text Archival
- e: Shareware Trade Association and Resources
- e: Simultaneous Transmitted and Reflected
- e: Society for the Technological Advancement of Reporting
- e: Solenoidal Tracker at RHIC
- e: Something like System for Tuning at Random
- e: Space Three-Axis Accelerometer for Research
- e: Special Telecommunication Action Regional Development
- e: Stability wind Rose
- e: Storage and Retrieval System
- e: Store-And-report
- e: System for Training of Aviation Regulations
- s: Groupe de travail mixte CCPM/PACSU-COI sur la tectonique et les ressources du Pacifique Sud
- s: Grupo Mixto de Trabajo CCOP/ SOPAC-COI sobre Tectónica y Recursos del Pacífico Sur

STAR 1100 e: Tape library system
STARA e: Studies on Toxicity Applicable to Risk Assessment
STARAP a: Statistik- og Rapportsystemet
STARC e: State Area Command
- e: State Army Reserve Center

STARCAL e: Star Calibration
STARCIPS e: Standard Army Civilian Payroll System
STARE e: Southern Tropical Atlantic Region[al] Experiment
STARFIARS e: Standard Army Financial Inventory and Reporting System
STARNET e: Standard Army Network
STARS e: Software Technology for Adaptable Reliable Systems
- e: Superfund Transactions Automated Retrieval System

STAR-T e: Super High Frequency Tri-Band Advanced Range Extension Terminal
START e: [Global Change] System for Analysis, Research and Training
- e: Strategic Arms Reduction Talks
- f: Système d'Analyse, de Recherche et de formation concernant le changement au niveau mondial
- s: Sistema de Análisis, Investigación y Capacitación para hacer frente al Cambio Mundial

STARTEX e: Start of Exercise
START I e: Strategic Arms Reduction Talks I Treaty
START II e: Strategic Arms Reduction Talks II Protocol
Startrrjal r: Start- och jaktledningsradar
START-SC e: START Standing Committee
STAS e: Space Transportation Architecture Study
STAT e: Slotted-Tube Atom Trap
- e: Stratospheric Tracers of Atmospheric Transport

Stat. e: United States Statutes at Large
StatBerG d: Statistikbereinigungsgesetz
Stat Bull Metrop Insur Co e: Statistical Bulletin/Metropolitan Insurance Companies
STATE e: Department of State
Stat/Fst N: Statistikkoffiseren ved Forsvarsstaben
Stat Med e: Statistics in Medicine
Stat Methods Med Res e: Statistical Methods in Medical Research
STATPRO e: USAREUR Stationing Program
STATSGO e: State Soil Geographic
STATUS e: Region 4 Status of GSA Orders
STAYSEG s: Sindicato de Trabajadores Administrativos y de Servicio Educación de Guatemala
STB d: Sternberg, Mecklenburg-Vorpommern
- e: Save the Bush Project
- e: Set-Top Box
- e: Software Technical Bulletin
- e: Strobe
- e: Strut Tower Brace
- f: Société des Transports de Brazzaville
- r: AB Svenska Telegrambyrån

Stb r: Studsbrisad
Stbg f: Strasbourg
StBGebV d: Steuerberatergebühren-verordnung
STBK e: Staubkonzentration
STBL e: Stratocumulus-Topped Boundary Layer
STC e: Science and Technology Commercialization
- e: Scientific and Technical Computing
- e: Security Transition Component
- e: Self-Test Connector
- e: Set Carry Flag
- e: Set Top Computer
- e: Short-Term Concentration
- e: Society of Theatre Consultants
- e: Software Technology Center
- e: Space Threat Conference
- e: Space Tracking Center
- e: State Transport Corporation
- e: Stennis Test Center
- e: Subtechnical Committee
- e: Switching Technical Center
- f: Solde de Tout Compte
- f: Station Terrienne Côtière
- f: Station Terrestre Côtière

S.T.C. e: Sydney Technical College
STCCS e: Strategic Theater Command and Control System
STCN e: Short Title Catalogue Netherlands
STCRP f: Société des Transports en Commun de la Région Parisienne
STCU e: Science and Technology Center in Ukraine
STCW e: International Convention on Standards of Training, Certification and Watchkeeping for Seafarers
- f: Convention internationale sur les normes de formation des gens de mer, de délivrance des brevets et de veille
- s: Convenio Internacional sobre Normas de Formación, Titulación y Guardia para la Gente del Mar

STCW-F e: International Convention on Standards, of Training, Certification and Watchkeeping for Fishing Vessel Personnel
- f: Convention internationale sur les normes de formation de personnel des navires de pêche, de délivrance des brevets et de veille
- s: Convenio Internacional sobre Normas de Formación, Titulación y Guardia para el Personal de los Buques de Pesca

STCZ e: Subtropical Convergence Zone
STD d: Stade, Niedersachsen
- e: Salinity, Temperature, Depth [probe]
- e: Salinity, Temperature, Depth profiler
- e: Secondary Transmit[ted] Data
- e: Set Direction Flag
- e: Sexually Transmitted Disease
- e: Software Test Description
- e: Standard Deviation
- e: Standard Test Description
- e: Subscriber Trunc Dialing

Std. e: Standard
STDA e: StreetTalk Directory Assistance
STDAS f: Système de Traitement des Données et d'Assistance du SMISO
STDAUX e: Standard Auxillary
STDB e: Science and Technology Development Board
- e: Space Telescope data base
- e: Strategic Target Data Base

STDCE e: Surface Tension Driven Convection Experment
STDCRC e: Sexually Transmitted Diseases Cooperative Research Centers
STDERR e: Standard Error
STDF e: STGT Tracking Data Formatter
STDIN e: Standard Input
STDIOH e: Standard Input/Output Header
ST DIV e: Street Division
STDJ 9 e: F. García Martínez, Qumran and Apocalyptic. Studies on the Aramaic Texts of Qumran [Leiden, 1992]
STDM e: Statistical Time Division Multiplexer
- e: Synchronous Time Division Multiplexer

STDMS e: Sample Tracking and Data Management System
STDN e: Space-flight Tracking and Data Network
STDOUT e: Standard Output
STDPRN e: Standard Printer
ST&E e: Security Test and Evaluation
STE e: Satellite Tracking Element
- e: Section Terminating Equipment
- e: Signaling Terminal [Equipment]
- e: Spanning-Tree Explorer
- f: Société
- N: Sjøforsvarets Tekniske État

Ste f: Sainte
- r: Ställföreträdare

Sté arl f: Société à responsabilité limitée
Sté Ayme f: Société Anonyme
STEC e: Science Team Executive Committee
STECAMOSA s: Sindicato de Trabajadores de la Empresa CAMOSA
STECSA s: Sindicato de Trabajadores de Embotelladora Central S.A.
STEE f: Société Tchadienne d'Eau et l'Electricité
STEG s: Sindicato de Trabajadores de la Educación de Guatemala
STEGAC s: Sindicato de Coca Cola
STEGT s: Sindicato de Trabajadores de la Empresa Guatemalteca de Telecomunicaciones
STEGUATEL s: Sindicato de Trabajadores de la Empresa de Telecomunicaciones
STEL e: Short-Term-Exposure Limit
Stell. D: Stellenbosch
STELLA e: Software for the Teaching of English Language and Literature and its Assessment
STEM e: Scanning Transmission-Electron Microscopy
- e: Scanning Tunneling Electron Microscope (or Microscopy)
- e: Solar Terrestrial Environment Model
- e: Southern Technical and Economic Movement
- e: Systems Telecommunications Engineering Manager
STEP e: Science and Technology for Environmental Protection
- e: Solar Terrestrial Energy Program[me]
- e: Standard for [the] Exchange of Product[s] [model] data
- e: Standardized Tactical Entry Point
- e: Stratosphere Troposphere Exchange Project
- e: Stripes for Exceptional Performers
- e: Supercomputer Teacher Enhancement Program
- e: Surveillance and Tracking Experiment Program
Stereotact Funct Neurosurg e: Stereotactic and Functional Neurosurgery
ST&ETR e: Security Test and Evaluation Test Report
SteuerEntlG d: Gesetz zur Steuerentlastung und Familienförderung [Steuerentlastungsgesetz]
STEX e: Stimulation Test and Exercise
STF e: Structured File
Stf d: Studienfach
Stf C r: Ställföreträdare för Chef
STFC f: Service du Transport Fluvial Centrafricain
STFO f: Société Technique de la Forêt d'Okoume
STFR N: Sjøforsvarets Tekniske Fellesråd
STG e: State Transition Graph
Stg f: Strasbourg
St.g. d: bestätigtes Lehrangebot für das Studium generale
STGB f: Service Technique des Grands Barrages
Stgr r: Stålgranat
StGRat d: Städte- und Gemeinderat
STGT e: Second TDRSS Ground Terminal
STGWG e: State and Tribal Government Working Group
Sth r: Största hastighet
Sth. a: Stockholm
- e: Something
SThM e: Scanning Thermal Microscope
- e: Scanning Thermal Microscopy
STI d: Schulung, Training und Information
- e: Set Interrupt Flag
- e: Sexually Transmitted Infection
- e: Shallow Trench Isolation
- e: Space Telescope Science Institute
- e: Sparc Technology Business
- e: Standard Tape Interface
- e: Standby Indicator
- e: Strategic Transition Initiatives Division
- e: Super Twisted Nematic
- r: Stockholms Tekniska Institut
STIA f: Service Technique et Industriel de l'Aéronautique
STIARP f: Service des Travaux et Installations Aéronautiques de la Région Parisienne
STIB e: Stratosphere-Troposphere Interaction[s] and the Biosphere
STIC e: Science, Technology and Industry Collection
- e: Scientific and Technical Intelligence Committee
- e: Scintillation Tile Calorimeter
STICAVSA s: Sindicato de Trabajadores de Industria Centro Americana de Vidrio
STICS e: Scalable Transportable Intelligence Communications System
STID e: Standard Identification Number
Stidar r: Spetstidsanslagsrör
Stidr r: Spetstidrör
STIET f: Service Technique de l'Installation et Exploitation des Transmissions
STIF e: Short-Term Investment Fund
STIGGS s: Sindicato de Trabajadores del Instituto Guatemalteco de Seguridad Social
STIII/ST-SFS e: Star Trek, third movie/'Search for Spock'
STII/ST-WOK e: Star Trek, second movie/'Wrath of Khan'
STIKSCAT e: Stick Scatterometer
STILO e: Scientific and Technical Intelligence Liaison Office
STIMAD f: Société des Télécommunications Internationales de la République Démocratique Malgache
STIMS e: Shared Technical Information Management System
STINDE s: Sindicato de Trabajadores del Instituto Nacional de Electricidad
STINFO e: Scientific and Technical Information Program
STING e: Software Technology Interest Group
STINTA s: Sindicato de Trabajadores del Instituto Nacional de Transformación Agraria
STIP e: State Transportation Improvement Program
STIRAP e: Stimulated Raman Adiabatic Passage
STIS e: Space Telescope Imaging Spectrometer
- e: Space Telescope Infrared Spectrograph
STI/ST-TMP e: Star Trek, first movie/The Motion Picture
STITSE s: Sindicato de Trabajadores del Tribunal Supremo Electoral
STIV/ST-TVH e: Star Trek, fourth movie/'The Voyage Home'
STIWG e: Satellite Telemetry Interagency Working Group
STIX e: SmallTalk Interface to X
Stj. d: Studienjahr
STJD 15 e: New Qumran Texts and Studies. Proceedings of the First Meeting of the International Organization, Paris 1992, G. J. Brooke [ed.] [Leiden 1994]
STJD 11 or MQC e: The Madrid Qumran Congress: Proceedings of the International Congress on the Dead Sea Scrolls, Madrid 18-21 March, 1991
S-tjeneste N: Sikkerhetstjeneste
ST. JO HOSP e: St. Joseph Mercy Hospital
STK a: Statstjenstemændens Kartel
- e: Satellite Tool Kit
- f: Société des Transports de Kinshasa
- r: Sveriges Tekniska Kontrollinstitut
stk. a: Styk[ke]
S/TK e: Sectors Per Track
STKFND e: Stock Fund and Equipment Requirements System
Stkp N: Stabskompani
STL d: Stollberg, Sachsen
- e: St. Louis [airport]
- e: Standard Template Library
- f: Service Territorial Local
STLD e: Society of Television Lighting Directors
STLDD e: Software Top-Level Design Document
StlMindÜbk d: Übereinkommen über die Minderung der Staatenlosigkeit
STLS e: Surface Tension Lowering Substance
StLÜbk d: Staatenlosen-Übereinkommen
STM e: International Association of Scientific, Technical and Medical Publishers
- e: Scanning Tunneling Microscope
- e: Scientific, Technical [and] Medical
- e: Service Test Model
- e: SGS-Thomson Microelectronics
- e: Station Management
- e: Storage Management
- e: Streams Manager
- e: Structural and Thermal Model
- e: Synchrone Transfer Mode
- e: Synchronous Transfer Modul
- e: Synchronous Transport Mode
- e: Synchronous Transport Modul
- f: Service Technique du Métropolitain
- f: Service Technique Municipal
- f: Services de Trafic Maritime
- s: Servicio de Tráfico Marítimo
- s: Sindicato de Trabajadores de Migración
STM-1 e: Synchronous Transport Module-level 1
StMBG d: Steuerbereinigungs- und Mißbrauchbekämpfungsgesetz

St. Meld N: Stortingsmelding
StMELF d: Staatsministerium für Ernährung, Landwirtschaft und Forsten
STMEQ s: Sindicato de Trabajadores de la Municipalidad de Quetzaltenango
St-mina r: Stötmina
STMU s: Sindicato de Trabajadores Municipales Unidos
STMW e: Sub-Tropical Mode Water
STN e: Scientific and Technical Information Network
- e: Super Twist[ed] Nematic [field effective material]
- f: Société des Terres Neuves
- f: Station Terrienne de Navire
- N: [Avlslaget for] sidet Trønderfe og Nordlandsfe
- s: Station Terrestre de Navire
Stn N: Stasjon
- r: Station
STNF N: Superdatamaskin Til Norsk Forskning
STNG e: Star Trek Next Generation
STNLCD e: Super Twisted Nematic Lyquid Crystal Display
Stnplut r: Stationspluton
STO e: Maryland State Treasurer Office
- e: Source Translation and Optimization
- e: Subcontractor Task Order
Sto. s: Santo
Stø N: Støtte
STÖ d: Störung
STOA e: Shock Time-Of-Arrival
Støavder N: Støtteavdelinger
STOCKWELL e: Stockwell Hall
STOF e: Standard Tactical Operations Facility
STOH e: Since Top Overhaul
STOJ s: Sindicato de Trabajadores del Organismo Judicial
stø. kp. N: støttekompani
STOL e: Short Takeoff and Landing [aircraft]
STOMP e: Subsurface Transportation Over Multiple Phases
Ston f: Station
STONE e: Structured and Open Environment
STON[s] e: Short Ton[s]
Stop d: Strafprozeßordnung
STOP e: SCOMP Trusted Operating Program
STORAGE e: CSSD-Cinci [EMSAC] Disk Storage Annual Report
STORE e: Businesses Offering Goods
STORET e: Storage and Retrieval [of water quality information]
storl. r: storlek
STORM e: Storm-scale Operational and Research Meteorology
STORM-FEST e: STORM [Storm-scale Observations Regional Measurement Program]-Fronts Experiment Systems Test
- e: Stormscale Operational Research and Meteorology Fronts Experiment Systems Test
stormkankomp r: stormkanonkompani
STOS e: Store String
støttetr N: støttetropp
STOVL e: Short Take Off, Vertical Landing
STOW e: Synthetic Theater of War
STP e: Salinity, Temperature and Pressure
- e: Science and Technology Policy
- e: Secure Transfer Protocol
- e: Self Test Pass
- e: Service Traffic Position
- e: Service[s] Transaction Program
- e: Sewage Treatment Plant
- e: Short-Term Plan[ning]
- e: Short-Term Prediction
- e: Signal[ing] Transfer Point
- e: Software Test Plan
- e: Software Through Pictures
- e: Space Test Program
- e: Spanning Tree Protocol
- e: Stop timer bit
- e: Stream Protocol
- e: Subsurface Transport Program
- e: Synchronized Transaction Processing
- e: System Transition Plan
- f: S'il Te Plaît
- f: Service du Trésor Public
- f: Sonde de mesure de la salinité, de la Température et de la Profondeur
- s: Sonda de medida de la salinidad, la Temperatura y la Profundidad
STPC e: Southern Timber Purchasers Council
STPF f: Société Technique des Parfumeurs de France
stpftg r: sjuktransportfartyg
STPG e: Scientific and Technical Planning Group
STPI e: Services Transaction Program Interface
STPP f: Service Technique du Port de Paris
STPR e: Software Test Procedures
St.pr.nr. N: Stortingsproposisjon nr
St.prp. N: Stortingsproposisjon
STPS e: S-band Tracking Processor System
STR e: Colorado Basin River Forecast Office
- e: Service Transaction Record
- e: Short Tandem Repeats
- e: Society of Thoracic Radiology
- e: Store Task Register
- e: Systems Test Review
- f: Service Territorial Régional
- f: Sous Toutes Réserves
STr f: Sans Travail
Str f: Secteur
str r: streck
str. a: stræde
- a: størrelse
- e: strong
- N: strid[ende]
- r: strids-
Str. A N: Stridsdyktig A
STRA s: Sindicato de los Trabajadores de la Aduana
STRAC s: Sindicato de los Trabajadores de Caminos
STRAD e: Strategic Air Division
StraFo d: Strafverteidiger-Forum
Strahlenther Onkol d: Strahlentherapie und Onkologie
StrahlPh d: Institut für Strahlenphysik
strålk. r: strålkastare
StrÄndG d: Strafänderungsgesetz
STRAPP e: Standard Tank, Rack, Adapter and Pylon Package
STRASCO e: Southern Transvaal Students Congress
Straßenw d: Institut für Straßen- und Verkehrswesen
STRATCOM e: Strategic Command
STRAT FAX e: Strategic Facsimile Network
Strath. e: Strathclyde
STRATMOB e: Strategic Mobility Support System
STRATOPS e: Strategic Operations
STRATOZ e: Stratospheric Ozone
STRATPRO e: Strategic Programs
STRATUS e: Strategic Mobility Support System
StraWerkz d: Institut für Strahlwerkzeuge
Str. B N: Stridsdyktig B
StrbEG d: Gesetz über die strafbefreiende Erklärung von Einkünften aus Kapitalvermögen und von Kapitalvermögen
STRC e: Scientific, Technical and Research Commission
- e: Strategic Target Range Complex
- f: Commission de Recherches Scientifiques et Techniques
- s: Comisión Científica, Técnica y de Investigación
STREAMLINER e: Special Intelligence Message Switching and Terminal System
STRESS e: Sediment Transport Events on Shelves and Slopes
STRG d: Steuerung
StRG d: Strafrechtsreformgesetz
strgrp r: stridsgruppering
STRICOM e: Simulations, Training and Instrumentation Command
STRIDE e: Science and Technology Reporting Information Distribution Enhancement program
STRIKEFOR-SOUTH e: Naval Striking and Support Forces, Southern Europe
stril r: stridsledning
strilch r: stridsledningschef
strilförb r: stridsledningsförband
STRIL-systemet r: Stridslednings- och Luftbevakningssystemet
Stripbv r: Stridsledningspansarbandvagn
STRIS e: System Trouble Report Information System
- r: Flygvapnets Stridsledningsskola
STRIVE e: Science Teachers Research Involvement for Vital Education
STRIX e: ISRS system with dLIB components
Strkr r: Stridskrafter
strl. N: straffeloven
STrl Fdi N: Sør-Trøndelag Forsvarsdistrikt
Strlk r: Strålkastare
Strlkavd r: Strålkastaravdelning
Strlktro r: Strålkastartrupp
STRR e: [Sub]system Test Readiness Review
StrRG [1.] d: 1. Strafrechtsreformgesetz
STRT e: Start Initialization

Struct. e: Structural
- e: Structure
STRV e: Space Technology Research Vehicle
Strv N: Stridsvogn
- r: Stridvagn
Strvavd N: Stridsvognavdeling
Strvbat r: Stridsvagnsbataljon
Strvkan r: Stridsvagnskanon
Strvkanon N: Stridsvognskanon
Strvksp r: Stridsvagnskulspruta
Strvmina r: Stridsvagnsmina
Strvreg r: Stridsvagnsregemente
Strvtorn r: Stridsvagnstorn
Strvtp r: Stridsvagnstransport
STS a: Snedkermestrenes Træ- og Finerskæreri A/S
- e: Sales Technical Support
- e: Scanning Tunneling Spectroscopy
- e: Science and Technology Studies
- e: Science, Technology and Society
- e: Shared Tenant Service
- e: Shuttle Transport System
- e: Space Time Space network
- e: Specialty Training Standard
- e: Synchronous Transfer Signal
- e: Synchronous Transport Signal [level]
- s: Sector de Tropas Serranas
STSA e: Secure Telephone System Architecture
STSB e: Scientific and Technical Services Branch
STSC e: COPUOS Scientific and Technical Subcommittee
- e: Software Technology Support Center
STSCC e: Standing Transportation Support Coordinating Committee
STSDAS e: Space Telescope Science Data Analysis System
STSK N: Skandinaviske Telesatelittkomité
STSP e: Secure Telephone Systems Program
- e: Solar Terrestrial Science Program
St Sy-mina r: Stötsyftmina
STT e: Secure Transaction Technology
- e: Secure Tunnel Transistor
- e: Small Tactical Terminal
- f: Service de Téléinformatique Touristique
Stt f: Statut
Stte s: Subteniente
STTM f: Société Tchadienne de Transformation des Métaux
ST-TNG e: Star Trek, The Next Generation
STTO e: Start-Taxi-Takeoff
ST-TOS e: Star Trek, The Original Series
STTR e: Small Business Technology Transfer
STU d: Schere-Tippex-Uhu[-Verfahren]
- f: Service de Topographie et d'Urbanisme
- r: Styrelsen för Teknisk Utveckling
STUB d: Stadt- und Universitätsbibliothek, Frankfurt
- f: Service de Transport Urbain Brazzavillois
Stub r: Skeppteknisk utbildning
StückAG d: Stückaktiengesetz vom 25.3.1998
STUCSC s: Sindicato de Trabajadores de la Universidad de San Carlos
stud. a: studerende
Stud Fam Plann e: Studies in Family Planning
Stud Hum Ecol e: Studies in Human Ecology
Stue f: Statue
STUFIA e: Students Forum of African and International Affairs
StUG d: Stasi-Unterlagen-Gesetz vom 20.12.1991
STUIII e: Secure Telephone Unit Third Generation
STUK e: Centre for Radiation and Nuclear Safety
STUN e: Serial Tunnel[ing]
StuPa d: StudentInnenparlament
StuRa d: StudentInnenrat
STUW e: Subtropical Underwater
STV d: Stellvertretend
- e: Single Transferable Vote
- e: Space Transfer Vehicle
- e: Star Trek Voyager
- e: Steerable Low-Light TV
- r: Sluten Television
Stvgr F N: Stavanger Festning
Stvgr SA N: Sjøforsvarsavsnitt Stavanger
STVI/ST-TUC e: Star Trek, sixth movie/'The Undiscovered Country'
STVNS e: Spherical Tokamak Volume Neutron Source
StVO d: Straßenverkehrsordnung
STVP f: Service Technique de la Voirie Publique
STV/ST-TFF e: Star Trek, fifth movie/'The Final Frontier'
StVZO d: Straßenverkehrs-Zulassungsordnung
STW-2 e: Synchronous Telecommunications and Broadcasting Satellite
STWA e: Short Trailing Wire Antenna
STY e: Style
Sty N: Styrke
Styckj r: Styckjunkare
SU a: Samarbejdsudvalg
- a: Statens Uddannelsesstøtte
- a: Sundby
- e: Scheduling Unit
- e: Set UID [User Identification]
- e: Stockholm University
- e: Subresolution attenuated
- e: Supporting Unit
- e: Switching Unit
- f: Salaire Unique
- f: Scrutin Uninominal
- f: Secours d'Urgence
- f: Sens Unique
- f: Suffrage Universel
- f: Surintendante d'Usine
- f: Système d'Unités [de mesure]
- N: Socialistisk Ungdom
- s: Servicios Universitarios y de Extensión Educativa
.su e: Soviet Union [Internet]
s.u. a: svar udbedes
S.u. N: Svar utbes
SUA e: Stored Upstream Address
- f: Syndicat Unifié des Administratifs
SUADPS-RT e: Shipboard Uniform ADP System-Real Time
SUAVE e: Submersible system Used to Assess Vented Emissions
SUAWACS e: Soviet Union Airborne Warning And Control System
SUB d: Systemunabhängige Betriebsführungstechnik
- e: Subaddressing
- e: Subtract [Binary]
- r: Ubåt med Strategiska kärnvapenrobotar
sub e: subpopulation of
Sub f: Subordonné
Sub. s: Submarino
Sub Amet s: Sub-Ametralladora
Subcell Biochem e: Sub-Cellular Biochemistry
SUBIM e: Subsidiary Body on Implementation
SUBLANT e: Submarine Force Atlantic
Suboff r: Subalternofficer
subord. cl. e: subordinate clause
SUBPAC e: Submarine Force, Pacific
SUBR e: Subroutine
Subs s: Subsistencias
subseq. e: Subsequent[ly]
SUBSIST e: Subsistence
subst. e: substantively
SUBSTA e: Subsidiary Body for Scientific and Technological Advice
Subst Use Misuse e: Substance Use and Misuse
Subt s: Subteniente
- s: Subterráneo
Sub Tte s: Sub-Teniente
SubvAbG d: Subventionsabbaugesetz
SUC f: Société des Usines Chimiques
suc. s: sucesor
SUCAP e: Surface Combat Air Patrol
Succ f: Successeur
SUCCESS e: Subsonic [aircraft] Contrail and Cloud Effects Special Study
- e: Sun Corporate Catalyst Electronic Support Service
- e: Synthesized UHF C2 and Exploitation Subsystem
Suchmaschine d: Programm im Internet, erleichtert dem Internet-Anwender die Orientierung
Sucie f: Sucrerie
SUCOMA e: Sugar Corporation of Malawi
SUD e: Subject Descriptors
SUDECO e: Sugar Development Corporation
SUDES f: Syndicat Unique et Démocratique des Enseignants du Sénégal
Sudhoffs Arch Z Wissenschaftsgesch d: Sudhoffs Archiv; Zeitschrift für Wissenschaftsgeschichte
Sudhoffs Arch Z Wissenschaftsgesch Beih d: Sudhoffs Archiv; Zeitschrift für Wissenschaftsgeschichte. Beihefte
Sud Med Ekspert R: Sudebno-Meditsinskaia Ekspertiza
SUDS e: Software Update and Distribution System
SUE e: Systems Utilization Enhancement
Sue f: Suède
SUEL f: Syndicat Unifié des Enseignants Laïcs

SUF a: Socialistisk Ungdoms Forum
- e: Somali United Front
- N: Sosialistisk Ungdomsforbund
Suff f: Suffrage
suff. e: suffix
SUFIAW e: Swaziland Union of Finance Institutions and Allied Workers
SUFID e: Screwing Up Face In Disgust
SUF [m-l] N: Sosialistisk Ungdomsforbund [marxismen-leninismen]
SUG e: Sun User[s] Group
- e: System Users Guide
SÜG d: Sicherheitsüberprüfungsgesetz vom 20.4.1994
SUGIA d: Sprache und Gesichte in Afrika
Suicide Life Threat Behav e: Suicide and Life-Threatening Behavior
SUID e: Set UID
SUK e: Central Educational Committee at NTH
SuK d: Sozial- und Kulturwissenschaften
SUKP a: Sovjetunionens Kommunistiske Parti
SULA e: Gulf of Finland Environment Society
- f: Société pour l'environnement du golfe de Finlande
- s: Asociación para el Medio Ambiente del Golfo de Finlandia
SULM d: Schweizerische Union für Laboratoriums Medizin
SUM e: Simple User Model
- e: Single User Mode
- e: Software Users Manual
- e: Symantec Utilities for Macintosh
- f: Syndicat Unique des Mineurs
- f: Système d'Unités de Mesure
Sum s: Suministro
SUMA e: Satellite Universal Modulation Adapter
- f: Supermarché
SUMER e: Solar UV Measurements of Emitted Radiation
SUMR e: Satellite User Mapping Register
Sum Rep s: Suministros y Reparaciones
SUM-SPARE e: Spare Engine Report
SUMT e: Sequential Unconstrained Maximization Technique
- e: Sequential Unconstrained Minimization Technique
- e: Susceptibility Magnetic Tool
SUN e: Serum Urea Nitrogen concentration
- e: Stanford University Network
- e: Sun Microsystems, Inc.
- e: Symbols, Units and Nomenclature
- f: Symboles, Unités et Nomenclature
- s: Símbolos, Unidades y Nomenclatura
SUNA e: Society of Urologic Nurses and Associates
SUNBURN e: Way beyond a Tangent
SUNDIAG e: SUN system Diagnostics software
SUNET e: Swedish University Network
SUNIST f: Serveur Universitaire National de/pour l'Information Scientifique et Technique
SUNT e: Specialized Undergraduate Navigator Training
SUNVIEW e: SUNs Visual Integrated Environment for Workstations
SUNY-AMP e: State University of New York Alliance for Minority Participation
SUOF r: Svenska Underofficersförbundet
SUP e: Society for Ultrastructural Pathology
- e: Soluble Unreactive Phosphorus
- e: Student Unification Party
- e: Supplemental Dictionary
Sup f: Supérieur
SUPC e: Southern Universities Purchasing Consortia
Super JANET e: [High-Speed] Joint Academic Network
SuperJANET e: Super Joint Academic Network
superl. e: superlative
SUPERMAX e: Unix-based library system from DDE
SUPIDEN e: NCC Support Identification Code
SUPIR e: Supplemental Photo Interpretation Report
Sup[p] f: Supplément
Suppl f: Suppléant
suppl. a: suppleant
- a: supplerende
Suppl Eur J Neurosci e: Supplement to the European Journal Neuroscience
Support Care Cancer e: Supportive Care in Cancer
SUPP SERV e: Support Services
Supra e: Supra Corporation
SUPRAS f: Services Universitaires pour la Recherche et les Applications en Supraconductivité
SUPS e: Static Uninterruptible Power Source
SUPSHIPS e: Superintendent of Shipbuilding
SUPTRK e: Superfund Site Tracking Information System
SUPU a: Sammenslutningen af Undervisere ved den Prøveforberedende Undervisning
SUR e: Shell Update Release
- e: Small Unit Radio
- e: Society of Uroradiology
SURAnet e: Southeastern Universities Research Association Network
SURE e: Sulfate Regional Experiment [program]
SURF e: Sequent Users Resource Forum
- e: Surface Storage Facility
SURFnet e: Dutch academic network
- e: Dutch ISDN university network
SURFRAD e: Surface Radiation [budget network]
Surg. e: Surgery
- e: Surgical
Surg Annu e: Surgery Annual
Surg Clin North Am e: Surgical Clinics of North America
SURGE e: Supply User Reports Generator
Surg Endosc e: Surgical Endoscopy
Surg Laparosc Endosc e: Surgical Laparoscopy and Endoscopy
Surg Neurol e: Surgical Neurology
Surg Oncol e: Surgical Oncology
Surg Oncol Clin N Am e: Surgical Oncology Clinics of North America
Surg Radiol Anat e: Surgical and Radiologic Anatomy
Surg Today e: Surgery Today
SûrNa f: Sûreté Nationale
SURP e: School of Urban and Regional Planning
SURSATCOM e: Survivable Satellite Communications
SURTAC e: Surveillance and Tactical Warning Network
SURTAC TTY e: Surveillance and Tactical Action Teletype
SURTASS e: Surface Towed Array Surveillance System
SURTROPAC e: Survey of the Tropical Pacific Experiment
Surv Ophthalmol e: Survey of Ophthalmology
SUS e: Shelf and Upper Slope
- e: State University System
- f: Plate-forme et pente supérieure
- s: Plataforma y Talud Superior
Sus. e: Sussex
SUSD e: Shelf and Upper Slope Dynamics
- f: Dynamique de la plate-forme et de la pente supérieure
- s: Dinámica de la Plataforma y el Talud Superior
SUSDP e: Standard for the Uniform Scheduling of Drugs and Poisons
SUSE d: Software Und System-Entwicklung
SUSP e: System Use Sharing Protocol
Susp f: Suspendu
SUSY e: Super Symetry
SUT f: Syndicat Unifié des Techniciens
SUTELEC f: Syndicat Unique des Travailleurs de l'Electricité
SUTT e: Single User Test Tools
SUU e: Suspended Underwing Unit
SUUG e: Soviet UNIX User Group
SUV e: Sport Utility Vehicle[s]
- e: Standard Uptake Value
SuVit f: Sud Viet-nam
Su VPL r: Särskildutbildningsanvisning för Värnpliktiga
SUW e: Surface to Underwater Missile
SÜW d: südliche Weinstrasse in Landau, Rheinland-Pfalz
SUWP e: Sustainable Use of Wildelife Programme
SUZI e: Sub-Zonal Insertion

SV a: Sydvest
- d: Standortverteiler
- e: Sampling Visit
- e: Satellite Vehicle
- e: Small Volume water samples
- e: Subvector
- e: Survivability
- f: Sans Valeur
- f: Secours aux Vieillards
- f: Secours Viager
- f: Sergent de Ville
- f: Service de la Voirie
- f: Succession Vacante
- N: Sosialistisk Valgforbund
- r: Statens Vattenfallsverk
- r: Sydväst
- s: Servicio en Vuelo
- s: Submarino de Vigilancia

Sv e: Sverdrup
- r: Sidvinkel
- s: Servicio

sv r: svensk

sv. N: svensk
- r: svår

s.v. a: svart/hvit
- l: sub verbo
- l: sub voce

S.V. N: Sørvest

SVA e: Styrene Vinyl Acrylonitrile

SVAT e: Soil-Vegetation-Atmosphere Transfer [model]

SVATS e: Surface Vegetation Atmosphere Transfer Scheme

SVBA d: Sinnvolle Vierbuchstaben-abkürzung

SVBG d: Schweizerischer Verband der Berufsorganisationen im Gesundheitswesen

SVC e: ATM Switched Virtual Circuit
- e: Shadow Village Council
- e: Signaling Virtual Channel
- e: Superior Vena Cava
- e: Switched Virtual Channel
- e: Switched Virtual Circuit[s]
- e: Switched Virtual Connection[s]

SVCC e: Switched Virtual Channel Connection

SVCCN e: Secure Voice Command Control Network

SVCS e: Services

SVD e: Singular-Value Decomposition
- e: Software Version Description
- s: Societas Verbi Divini

SVE e: Special Vehicle Engineering
- e: Special Vehicle Equipment

Sveptj L r: Sveptjänstledare

SVEQUS d: Schweizerischer Verband qualifizierter Spitexdienste

SVF e: Simple Vector Format
- r: Sjövärnsflottilj
- r: Ståendevågförhållande

SVG d: Schweizerische Vereinigung Gesundheitstechnik
- d: Sozialversicherungsgesetz

SVGA e: Super VGA [Video Graphic[s] Array]

SV/GC e: Secure Voice and Graphics Conferencing

SVGO d: Schweizerische Vereinigung der Gemeindekranken- und Gesundheitspflege

SVGR d: Schweizerische Vereinigung für Gesundheitsrecht

SVH f: Sécurité de la Vie Humaine

SVHBL d: Schweizerischer Verein Hauswirtschaftlicher Betriebsleiterinnen

SVHM d: Schweizerische Vereinigung für hirnverletzte Menschen

SVHO d: Schweizerische Vereinigung der Hauspflegeorganisationen

S-VHS e: Super VHS [Video Home Standard]
- e: Super Video Home System

SVI e: Smithonian Volcanic activity Index
- e: Subvector Identifier
- N: Skyte- og Vinterskolen for Infanteriet

SVID e: System V Interface Definition
- e: System V Interface Device
- e: UNIX System V Interface Definintion[s]

SVID Unix e: System V Interface Definition

SVIMS e: Short VIMS

Svinggr r: Sprängvinggranat

SVIP e: Secure Voice Improvement Plan

S/VISSR e: Stretched Visible and Infrared Spin-Scan Radiometer

SVK d: Schweizerischer Verband für Gemeinschaftsaufgaben der Krankenkassen
- N: Stridsvogn[s]kanon
- r: Sjövärnskåren

Svkr r: Svartkrut

SVL a: Skandinavisk Væg- og Loftsmontage A/S
- N: Studentvenstrelaget

SVLBI e: Space VLBI

SVLK d: Schweizerische Vereinigung der Leiterinnen und Leiter von Kranken-pflegeschulen

SVLPPK d: Schweizerische Vereinigung des leitenden Personals der psychiatrischen Krankenpflege

SVM e: Separation of Variable Method
- e: System Virtual Machine

SVMA d: Schweizerische Vereinigung für medizinische Ausbildung

SVMB e: Society of Vascular Medicine and Biology

SVN e: Secure Voice Network
- e: Society for Vascular Nursing
- e: Switched Virtual Network
- f: Sud Viet-nam

SVOC e: Semi-Volatile Organic Compounds
- e: Semivolatile Organic Chemical

Svoffsekt r: Samverkansofficerssektion

SVP d: Schweizerische Volkspartei
- e: Scan-line Video Processor
- e: Schematic Verification Program
- e: Secure Voice Program
- e: Surface Velocity Program[me]
- f: Programme sur la Vitesse Superficielle
- f: S'il Vous Plaît
- s: Programa sobre la Velocidad Superficial

Svp r: Svepskola

SVPIS e: Space VLBI Principal Investigator Software

SVPK d: Schweizerische Vereinigung der Privatkliniken

SVPL d: Schweizerische Vereinigung der Pflegedienstleiterinnen und -leiter

Svpm r: Svepmatroser

SVPW-Jahrbuch d: Schweizerisches Jahrbuch für politische Wissenschaft

SVR e: Server
- e: Session-based Virtual Route
- e: Shop Visit Rate
- e: Slow Vertex Response
- e: Swedish Society of Civil and Structural Engineers
- e: System V Unix Release
- e: System Validation Review
- e: System Verification Review

SVR4 e: System V Release 4

SVR# e: System V Release Number

SVRA e: State-owned Vehicular Recreation Areas

S V Radar s: Submarino de Vigilancia [con radar] para descubrir objetivos submarinos

SVRN e: System V Release Number

SVS d: Schweizerische Vereinigung der Spitaldirektoren
- e: Switched Voice Service
- f: Section de Vol Sportif

SVSI e: Secure Voice Systems Interface
- N: Skyte- og Vinterskolen for Infanteriet

SVSLVA N: Luftvernartilleriets Skyte- og Våpenskole

SVT e: Special Vehicle Team

SVT 4 f: Volume du Congrés Strasbourg 1956 [Leiden: Brill, 1957]

SVTS e: Secure Video Teleconferencing System

sv-v r: svartvit

SVV e: Subvector Value

SVVP f: Services Vétérinaires de la Ville de Paris

SVVS e:[UNIX] System V Validation Suite
- e: [UNIX] System V Verification Suite

SVVS [UNIX] e: System V Validation Suite

SVW d: Stifterverband für die Deutsche Wissenschaft e.V.

SVWG e: Space Surveillance Working Group

SW d: Sämtliche Werke
- d: Schwarz-Weiß
- e: Shareware
- e: Short Wave
- e: Short Wheelbase
- e: Social Work[er[s]
- e: South West
- e: Strategic Warning
- e: SubWoofer
- e: Super World

S W e: Social Work

S.W. e: Sound Wormy

s.w. e: south-western [dialect]

S&W e: Stone and Webster Engineering Corporation
- e: Surveillance and Warning

SWA e: Shortwave Albedo
- e: Slit Wheel Assembly

SWAAF e: Wetterlagenklasse Nr. 09 nach der Objektiven Wetterlagenklassifikation

SWAAT e: SDIO Working and Action Team

SWAAU e: Southwest Africa Agricultural Union

SWABC e: South-West African Broadcasting Corp.
SWABOU e: Southwest African Building Society
SWAC e: Mobile version of SEAC, built for deployment to White Sands and named Standards Western Automatic Calculator
- e: Standards Western Automatic Calculator
SWACOM e: Standard Water Column Model
SWADE e: Surface Wave Dynamic Experiment
SWAG e: Scientific Wild-Ass Guess
- e: Simulated Waste Access To Ground Water
- e: Super wild Ass Guess [Internet]
SWAGM d: SWA-Gebietsmacht
SWAIS e: Simple Wide Area Information Server
SWAL e: Survey of West African Languages Survey University of Dakar
SWALCAP e: South West Academic Libraries Co-operative Automated Project
SWALIMO e: Swaziland Liberation Movement
SWAMA e: South-West Africa Municipal Association
SWAMP e: Southwest Area Monsoon Project
SWAMSA e: South-West African Municipal Staff Association
SWAN e: Service Wide Area Network
- e: Solar Wind Anisotropy [experiment]
- e: Sun Wide Area Network
SWANAFRO e: Swaziland National Front
SWANLA e: South West African Native Labo[u]r Association Namibia
SWANOV e: South-West African Black Teachers Union
SWANP e: South-West Africa National Party
SWANTITER e: South-West Africa Anti-Terrorism Fund
SWANU e: South West African National Union Namibia
SWANUF e: South-West Africa National United Front
SWANU-O e: South-West African National Union-Opposition
SWANU-P e: South-West African National Union-Progressives
SWAOU e: South-West African Teachers Union
SWAP e: Standard WWMCCS Accounting Program
SWAPDUF e: SWA Peoples Democratic United Front
SWAPO e: South-West African Peoples Organisation
SWAPO-D e: SWAPO Democrats
SWAPOL e: South West African Police
SWAPOU e: South-West African Professional Teachers Union
SWAS e: Short-Wave Astronomical Satellite
SWASS e: South West African Scientific Society Windhoek
SWAT e: Software Action Team
- e: Software Analysis and Test
- e: Soil and Water Assessment Tool
SWATF e: SWA Territorial Forces
SWATL e: South-West Atlantic Region
- f: Région de l'Atlantique du Sud-Ouest
- s: Región del Atlántico Sudoccidental
SWATRE d: Wasserhaushaltsmodell
SWATS e: Soil Water And Temperature Sensors
SWAUNIO e: South-West African United National Independence Organization
SWAWEK e: South-West Africa Water and Electricity Supply Commission
SWAWG d: Südwestafrikanische Wissenschaftliche Gesellschaft
SWAWV A: Suidwes-Afrikaanse Wetenskaplike Vereniging
Swayoco e: Swaziland Youth Congress
SWB d: Südwestdeutscher Bibliotheksverbund in Konstanz
- e: Software Bus
- e: South-Western Bell
SWBT e: Southwestern Bell Telephone Corporation
SWBTU e: Scheduler Work Block Text Unit
S&WC e: Surveillance and Warning Center
SWC e: Switch [Controller]
SWCA e: Soil and Water Conservation Act
SWCAA e: Soil and Water Conservation Association of Australia
S W C B e: Social Work Center Building
SWCC e: SAC Warning Control Console
- e: Second World Climate Conference
- f: Deuxième Conférence mondiale sur le climat
- s: Segunda Conferencia Mundial sobre el Clima
SWCCB e: Software CCB
SWCF e: Shortwave Cloud Radiative Forcing
SWCI e: Software Configuration Item
SWCS e: Soil and Water Conservation Society
SWD d: Schlagwortnormdatei
- e: Software Development
SWDO e: Somali Womens Democratic Organization
SWDS e: Software Distribution System
SWE e: Shallow-Water Equations
- e: Snow Water Equivalent
- e: Society of Women Engineers
- e: Software Engineer[ing]
SWEAT e: Standard Wafer-level Electromigration Accelerated (or Acceleration) Test
SWEDAC e: Swedish Board for Technical Accreditation
Swed Dent J e: Swedish Dental Journal
Swed Dent J Suppl e: Swedish Dental Journal. Supplement
SWEG e: Scientific Wild Eyed Guess
SWEM e: Status of Women and Ethnic Minorities
SWERD e: Solid Waste and Emergency Response Division
SWERLS e: South West Regional Library System
SWETS e: Safe Water Enforcement Tracking
SWF e: Short Wave Fading
- e: Social Welfare Function
- e: Strategic Weapons Facilities
SWG d: Saarländisches Wassergesetz
- e: Science Working Group
- e: Security Working Group
- e: Standards Working Group
SWI e: Software Interrupts
- e: Static Walk through Inspection
SWIE e: Southern Waste Information Exchange
S.W.I.F.T. e: Society for Worldwide Interbank Financial Telecommunication
SWIFT e: Society for Worldwide Interbank Financial Transfers
SWIM e: Semiconductor Workbench for Integrated Modeling
- e: Simulation Workbench for Integrated Modeling
- e: Solar Wind Interplanetary Measurements
SWIMS e: Solid Waste Information Management System
SWIN e: Swine Center
SWING e: Subscriber Wireless Network Gateway
SWIP e: Single-Wafer Integrated Process
- e: Super Weight Improvement Program
- e: Systems and Weapon Improvement Program
Swipnet e: Swedish IP Network
SWIR e: Short Wavelength Infrared
SWIRLS e: Stratospheric Wind Infrared Limb Sounder
SWISH e: Simple Web Indexing System for Humans
Swiss Surg [Suppl] e: Swiss Surgery. [Supplement]
SWITCH d: Wissenschaftsnetz der Schweiz
- e: Swiss Academic and Research Network
SWITS e: Solid Waste Information [and] Tracking System
SWJ e: Student-Wide Judiciary
SWL e: Short Wave Laser
SWLC e: Salt Water Leak Check
SWM e: Solbourne Window Manager
SWMBO e: She Who Must Be Obeyed [Internet]
SWMM e: Software Maintenance and Modification
- e: Storm Water Management Model
SWMP e: Storm Water Monitoring Program
SWMU e: Solid Waste Management Unit
SWnT e: Single-Walled NanoTub[ul]e
SWO e: Sales Work Order
SWOG e: Southwest Oncology Group
SWOT e: Strengths, Weakness, Opportunities, Threats analysis
SWP e: Short-Wavelength Pass
- e: Simple Web Printing
- e: Single Wafer Processing
- e: South-west Pacific
- e: Special Work Permit
- e: Swap

SWPA e: South West Pacific Area
SWPPP e: Storm Water Pollution Prevention Plan
SWQI e: South West Queensland Initiative
SWQO e: Statutory Water Quality Objective
SWR d: Südwestrundfunk
- e: Semiconductor Wafer Representation
- e: Shortwave Radiation
- e: State Wildlife Reserve
SWRCB e: State Water Resources Control Board
SWRI e: Site Wide Remedial Investigation
SWRK e: Social Work
SWRRB e: Simulator for Water Resources in Rural Basins
SWRSC e: SAC WIS Requirements Steering Committee
SWS e: Scientific Workstation Support
- e: Silly Window Syndrome
SWSA e: Solid Waste Storage Area
SW,S/W e: Software
SWT e: Science Working Team
- s: Grupo de Trabajo Científico
SWTS e: Solid Waste Technology Support
S&WUO e: Steam and Water Utility Operations
SWV e: Square Wave Voltammetry
SWW e: Soil, Water, Waste
SX e: Simplex signaling
SXE e: East Sussex
SXI e: Solar X-ray Imager
SXRD e: Surface X-Ray Diffraction
SXRP e: Stellar X-Ray polarimeter
SXS e: Step by Step switching
SXW e: West Sussex
SXXM e: Unextended Numbering Set Mode
SY e: Scientist-year
- e: Seasonal Year
- e: System Release
Syb N: Sykebærer
SYBETRA f: Syndicat Belge d'Entreprise a l'Etranger
SYC e: System Control
SYCA f: Syndicat d'études des Centrales Atomiques
SYCOMIMPEX f: Syndicat des Commerçants Importateurs-Exportateurs
Sy-cy e: Synoviocyte
SYD e: John Ray Herbarium, University of Sydney International Acronym
Syd LR e: Sydney Law Review
SYDMIS e: Shipyard MIS
SYDP e: Six Year Defense Plan
Syd. Soc. Lex. e: Sydenham Society, Lexicon of Medicine and Allied Sciences
Syd.Teach.Coll. e: Sydney Teachers College
SYERS e: Senior Year Electro-optical Relay System
Syh N: Sykehus
Sykeavd./Bilkl. 2 N: Sykebilavdelingen ved Biltrenkolonne 2
sykeb. N: sykebærer
sykeh. N: sykehus
SYL e: See You Later [Internet]
- e: SWAPO Youth League
SYLK e: Symbolic Link
syll. e: syllable
SYM e: Symbols
SYMETAIN f: Syndicat Minière de l'Etain
SYMIGETO f: Syndicat des Mines et de la Géologie du Togo
Sy-mina r: syftmina
SYMPL e: Systems Programming Language
Symp Soc Exp Biol e: Symposia of the Society for Experimental Biology
SYN e: Synchronizing [segment]
- e: Synchronous
SYN. e: Synchronous idle
SYNACAAB f: Syndicat des Agents de la Caisse Autonome d'Amortissement du Bénin
SYNACIB f: Syndicat National des Commerçants et Industriels Africains du Bénin
SYNAD f: Syndicat des Administrateurs Civils
SYNAEM f: Syndicat National de l'Enseignement Moyen Général Publique
SYNAESS f: Syndicat National des Professeurs des Enseignements Secondaire et Supérieur
SYNAMAB f: Syndicat National des Magistrats du Bénin
SYNAPAG f: Syndicat National des Personnels de l'Administration Générale
SYNARES f: Syndicat National de la Recherche et de l'Enseignement Supérieur
SYNASEG f: Syndicat National des Agents du Secteur de l'Electricité et du Gaz
SYNAT-CIE f: Syndicat National des Travailleurs de la Compagnie Ivoirienne de l'Electricité
SYNATRACOM f: Syndicat National des Travailleurs du Commerce et Assimilés
SYNATRATHO f: Syndicat National des Travailleurs du Tourisme et de Hôtellerie
SYNBANK f: Syndicat des Employés et Cadres des Banques, des Etablissements Financiers et des Assurances du Togo
SYNC e: Synchronous [mission name with continuous visibilities]
Sync e: Syncronization
SYNCOBENI f: Syndicat des Cheminots de l'OCBN
SYNDARC e: Synoptic Data Archiving Format
- f: Format d'archivage de données synoptiques
- s: Formato de Archivo de Datos Sinópticos
SYNDONI f: Système de Documentation National Informatisé
SYNDUSTREF f: Syndicat des Industries
SYNEBACI f: Syndicat National Unique des Entrepreneurs en Bâtiment de Côte d'Ivoire
SYNEBANC f: Syndicat National des Employés de Banques et Etablissements de Crédits
SYNECECI f: Syndicat National des Educateurs et Conseillers d'Education de Côte d'Ivoire
SYNECTO f: Syndicat des Employés de Commerce du Togo
SYNELS f: Syndicat National de l'Enseignement Laïc du Sénégal
SYNEPPCI f: Syndicat National de l'Enseignement Primaire Public de Côte d'Ivoire
SYNERCAU f: Syndicat national d'études et de recherches pour les coopératives agricoles et leurs unions
SYNERVA f: Syndicat national d'études, de révision et de vulgarisation des coopératives agricoles
SYNESCI f: Syndicat National des Enseignants du Second Degré
SYNINFO f: Syndicat National des Agents de l'Information
SYNOP d: Code für die weltweite, gleichzeitige Beobachtung der meteorologischen Elemente
- e: Surface Weather Observations
- f: Observations météorologiques en surface
- s: Observaciones Meteorológicas de Superficie
SYNPAF f: Syndicat du Personnel de l'Administration Financière
SYNPAG f: Syndicat du Personnel de l'Administration Générale
SYNPICS f: Syndicat National des Professionnels de l'Information et de la Communication du Sénégal
SYNTER f: Syndicat National des Travailleurs de l'Education et de la Recherche
SYNTES f: Syndicat National des Travailleurs de l'Electricité de Sénégal
SYNTRAMHYG f: Syndicat National des Travailleurs des Mines, des Hydrocarbures et de la Géologie
SYR e: Screw Your Roommate
Syr. e: Syracuse
- e: Syrian
SYRUP e: System Resource Utilization Package
SYS e: Scientist-years
- e: System Configuration
- e: System Device Driver
Sys f: Système
SYSAD e: System Administrator
SYSADMIN e: System Administrator
SYSADS e: Systems Administration Branch
SYSCON e: Systems Control
- e: Systems Control Center
SYSDATA e: AFSC Data Tie Line Network
SysEx e: System Exclusive
SYSGEN e: System Generation
SYSMOD e: System Modification
SYSOP e: System Operator
SYSPLEX e: System Complex
Syspro f: Systèmes-Productique-Robotique
SYSREQ e: System Request
Syst. e: System[atic]
SystDyn d: Institut für Systemdynamik und Regelungstechnik
SYSTEM e: Students and Youth into Science, Technology, Engineering and Mathematics
SYSTEM-80 e: Automatic Program Generator
Syst. Zool. e: Systematic Zoology

SYSVOICE e: AFSC Voice Tie Line Network
Syte r: System- och teleteknisk
SZ d: Softwarezentrum Böblingen
- d: Sprachenzentrum
- e: Send Z-Modem
SZABIST e: Shaheed Zulfikar Ali Bhutto Institute of Science and Technology
SZB d: Schwarzenberg, Sachsen
SZF f: Société Zoologique de France
SZG d: Schweizerische Zeitschrift für Geschichte
SZR d: Sonderziehungsrecht
SZW e: Ministry of Social Affairs and Employment

T

T d: symbol transponiert
- d: Tagesprobe
- d: Teilnahmetestat
- d: Temperatur
- d: Testat
- e: T reference point
- e: Tasmania
- e: Tera [10^{12}]
- e: Tesla
- e: Traveler
- f: Tarif
- f: Téléphon[i}e
- f: Teneur
- f: Terme
- f: Tête
- f: Tout
- f: Traite
- f: Tranche
- f: Travail
- f: Tronçon
- N: Tenn
- N: Trøndelag
- r: Tecken
- r: Tid
- r: Torpedflygplan
- r: Torpedskola
- r: Trotyl
- r: Tung
- R: Tyazhely
- s: Tanque
- s: Tarde
- s: Teclado
- s: Terrorista
- s: Torpedero
- s: Transmisiones
- s: Transporte
- s: Trazador
- s: Tren

t d: Bezugstransmissionskoeffiziente der 1-ten bzw. der n-ten Schicht
- d: relativer spektraler Transmissionsgrad [des ersten Farbfilters mit der Dicke D1]
- d: spektraler Transmissionskoeffizient
- e: metric ton

T. e: Taler

t. N: til
- N: time
- N: tonn
- r: tidigare
- r: till
- r: timmar
- r: timme

T 1 r: 1. Trängregemente
- r: Svea Trängregemente

T 2 r: Göta Trängregemente

t 2 d: relativer spektraler Transmissionsgrad des zweiten Farbfilters mit der Dicke D2

T_2 e: Technology transfer

T3 e: Triiodothyronine

T 3 r: Norrlands Trängregemente

T 4 r: Skånska Trängregementet

T-33A e: Shooting Star

T-38A e: Talon

T-41 e: Mescalero

TA d: Terminaladapter
- e: Messick Theatre Arts Center
- e: Target Analysis
- e: Targeted Assistance
- e: Task Authorization
- e: Technical Applications
- e: Technical Architecture
- e: Technical Assessment
- e: Temperature Alarm
- e: Terminal Adapter
- e: Test Assembly
- e: Total Alpha
- e: Trace Analytical
- e: Traffic Analysis
- e: Training Agency
- e: Transaminase
- e: Transfer Allowed
- e: Travel Authorization
- e: Turnover Agreement
- f: Taux d'Allocation
- f: Taxe d'Apprentissage
- f: Taxes Additionnelles
- f: Témoin Auriculaire
- f: Terre Arable
- f: Test d'Aptitude
- f: Testament Authentique
- f: Textiles Artificiels
- f: Tiers Arbitre
- f: Tonne d'Affrètement
- f: Total de l'Actif
- f: Travailleur Agricole
- f: Tribunal Arbitral
- r: Taktiska Anvisningar
- s: Transportador Anfibio
- s: Transporte Aéreo
- s: Transporte Automóvil
- s: Transporte de Ataque

Ta s: Trampa

T.A. e: Territorial Army

T-A s: Tierra-Aire

T/A e: Test Assembly

TA1 e: Tumor Antibody 1

TA2 e: Tumor Antibody 2

TAA e: Technical Area Advisor
- e: Total Active Aircraft Authorization
- e: Trans-Atlantic Agreement
- e: Transitional Administrative Authority

TAAA e: Total Active Aircraft Authorization

TAAC of S e: Theater Army Assistant Chief of Staff

TAACOM e: Tank Automotive and Armament Command
- e: Theater Army Area Command

TAACS e: Tactical Airlift Accomplishments and Commitments System

TAADS e: The Army Authorizations Documentation System

TAAF e: Test, Analyze and Fix
- f: Terres Australes et Antarctiques Françaises
- f: Territoires Australes et Antarctiques Français

TAAG e: Technical Assessment and Advisory Group

TAAN e: Tactical Aircrew Alert Net

TAANOS e: The American Academy of Neurological and Orthopedic Surgeons

TAAO e: Time Available at Action Office

TAAS e: Triple Azimuth Angle Scan

TAB d: Büro für Technikfolgen-Abschätzung
- e: Tabular Alphanumerics Block
- e: Tabulator
- e: Tape Array Bonding
- e: Tape Assembly Bonding
- e: Technical Adisory Board
- e: Technology Assessment Bureau
- e: Total Allocated Budget
- e: Training Aid Booklet

Tab f: table[au]

tab. a: tabel
- N: tabell

Ta Bbs s: Trampa para Bobos

TABD e: Trans-Atlantic Business Dialogue

t a, bl d: spektrale Transmissionskoeffizienten bei den Indizes a bzw. b

TABLE e: Tsukuba Atmospheric Boundary Layer Experiment
TABR e: Targeted Area Background Report
TAB-VEE e: Hardened Aircraft Shelter Half-Round Steel/Concrete
TABWS e: Tactical Air Base Weather System
TAC e: Teachers Action Committee
- e: Technical Advisory Committee
- e: Technical Advisory Council
- e: Technical Advisory Group
- e: Terminal Access Center
- e: Terminal Access Circuit
- e: Total Area Coverage
- f: Tables Annuelles internationales de Constantes et données numriques
- f: Taux des Allocations de Chômage
- s: Total Admisible de Capturas
Tac s: Táctica
TAC. e: Transmit Activity Check
TAC2 e: Theater Army Command and Control
TAC-A e: Tactical Air Coordinator-Airborne
TACACPS e: Theater Army Construction Automated Planning System
TACACS e: TAC [Terminal Access Controller] Access Control System
TACAIRCG e: Tactical Air Control Group
TACALS e: TAC [Terminal Access Controller] Command and Control Alerting System
TACAM e: Technical Advisory Committee on Antarctic Mapping
TACAMO e: Take Charge And Move Outaircraft
TACAS e: Terminal Access Controller Access System
TACB e: Texas Air Control Board
TACBEAMS e: TAC [Terminal Access Controller] Base Engineering Automated Management System
TACC e: Tracking And Control Center
TACCIMS e: Theater Automated C2 Information Management System
- e: Theater Automated Command and Control System
TACCO e: Tactical Coordinator
TAC CP e: Tactical Command Post
TACCS-K e: Theater Automated Command and Control System Korea
TACD e: Tactical Deception
TA/CE e: Technical Analysis/Cost Estimate
TACE e: Technical Analysis and Cost Estimate
TACELINT e: Tactical ELINT
TACF e: Tactical Air Control Flight
TACG e: Tactical Air Command Guard
TACIDN e: Tactical Information Distribution Network
TACIES e: Tactical Imagery Exploitation System
- e: Tactical Intelligence Information Exchange Subsystem
TAC INTEL e: Tactical Intelligence
TAC2IS e: Theater Army Command and Control Information System
TACIS e: Technical Assistance to the Commonwealth of Independent States and Georgia
TACMET e: TAC [Terminal Access Controller] Forces Control Management System
TACMIS e: Tactical Management Information System
TACNET e: TAC [Terminal Access Controller] Force Control Management Network
TACOM e: Tank and Automotive Command
- e: Tank-Automotive and Armaments Command
- e: Theater Army Command
TACOPS e: Tactical Air Combat Operations
TACP e: Theater Ammunition Control Point
TAC PACERS e: Peacetime Analysis of the Combat Effectiveness of Repairable Spares
TACPLAN e: Tactical Plan Generation
TACR e: TAC [Terminal Access Controller] Regulation
TACREP e: Tactical Report
TACRON e: Tactical Air Control Squadron
TACS e: Target Acquisition and Control System
- e: Test Assembly Conditioning Station
- e: Theater Air Control System
- e: Total Access Communications
TACSIM e: Tactical Simulation
TACS I&P e: Test Assembly Conditioning Station Inert and Preheat
TACSIT e: Tactical Situation
TACSM e: Tactical Air Combat Simulation Model
TACS NS e: Test Assembly Conditioning Station New Storage
TACS RC e: Test Assembly Conditioning Station recycled CLIRA
TACS RS e: Test Assembly Conditioning Station Radioactive Storage
TACS/TADS e: Tactical Air Control System/Tactical Data System
TACSY e: Tailored Correlation Spectroscopy
TACT e: The Automated Controller Test
TAC/TADS e: Tactical Air Control System/Tactical Air Defense System
TACTICS e: Tactical Information Control System
TACW e: Tactical Air Control Wing
TACWX e: Tactical Weather
TAD e: Technology Area Descriptions
- e: Temporary Duty
TADAT e: Tauern-Dachstein Testsite
TADB e: Terrain Analysis Data Base
TADIXS e: Tactical Data Information Exchange [Sub]system
TADNET e: Test and Development Network
TADS e: Target Acquisition And Designation System
- e: Target Acquisition and Detection System
- e: Test Assembly Decay Storage
- e: Text Adventure Development System
TADSS e: Training Aids, Devices, Simulators and Simulations
Tadv e: Temperature Advection
TAE d: Telekommunikationsanschluß-einheit
- e: Toroidal Alfven Eigenmodes
- e: Transportable Applications Environment
- e: Transportable Application[s] Executive
TAEDP e: Total Army Equipment Distribution Plan
TAF a: Træindustriarbejderforbundet i Danmark
- e: Tanzanian Action Front
- e: The Asia Foundation
- e: Tumor Angiogensis Factor
- f: Téléspectateurs et Auditeurs de France
TAFB e: Tinker Air Force Base
TAFE e: Technical and Further Education
TAFEMIS e: TAFE Management Information System
TAFF e: Thermally Assisted Flux-Flow
TAFHUG e: Tactical Air Forces Tactical Host Upgrade
TAFIC e: Tasmanian Fishing Industry Council
TAFIIS e: Tactical Air Forces Integrated Information System
TAFIM e: Technical Architecture [Framework] for Information Management
TAFN e: That's All For Now [Internet]
TAFS e: Terminal Aerodrome Forecast
TAFT e: Test, Analyze, Fix and Test
- e: That's a Frightening Thought [Internet]
TAFTRAMS e: TAC Automated Flying Training Management System
TAFWG e: Tactical Air Force Working Group
Tag d: Steueranweisung in HTML-Dokumenten
TAG e: Technical Assessment Group
- e: Thalassemia Action Group
- e: TM Archive Generation
- e: Topic Area Guide
- e: Treatment Action Group
TAGA e: Trace Atmospheric Gas Analyzer
TAGB p: Transportes Aéreos de Guiné-Bissau
TAGO e: The Adjutant General Staff
TAHC e: Texas Animal Health Commission
TAHL e: Temperature Alarm High-Low
TAHRP e: Trinational Animal Health Research Project
TAI e: International Atomic Time
- e: Thermal Alteration Index
- e: Trip After Instability
- f: Transports Aériennes Intercontinentaux
TAII s: Tanque de Apoyo Inmediato de la Infantería
TAIS e: Test Assembly Inert Storage
TAJA e: Tanzanian African Journalists Association
TAKR R: Takticheskoye Avianosniy Kreyser
takt. N: taktisk

TAL e: Target Analyte List
- e: Thanks A Lot [Internet]
- e: Trans-Atlantic Abort Landing
tal. r: talare[n]
TALAE e: Taligent Application Program
TALC e: Textile Apparel Linkage Council
TALCE e: Tanker/Airlift Control Element
- e: Transportable Airlift Control Element
TALCM e: Tomahawk Air-Launched Cruise Missile
TALD e: Tactical Air-Launched Decoy
- e: Tactical Airborne Laser Designator
TALDE e: Taligent Development Environment
TALES e: The Association of Library Equipment Suppliers
TALFF e: Total Allowance Level of Foreign Fishing
TALI e: Tactical Analysis Logistics Information
- e: Transfer Adjacent Layer Interface
TALIS e: BLS Ltd client server library management system
TALI-WRM e: TALI-War Readiness Materials
TAlk e: Total Alkalinity
Tal[l] s: Taller
Tall Ap Nav s: Taller de Aparatos e instrumentos para la Navegación
Tall Mat Hid s: Taller de Material Hidrográfico
TALMS e: Tunable Atomic Line Molecular Spectroscopy
TALO e: Tactical Airlift Liaison Officer
- e: Theater Airlift Liaison Officer
TALON e: Theater Application: Launch On Notice
TALOS e: Taligent Object Services
- e: Tropical Atmospheric Lidar Observations from Space
- e: Working Group on Technical Aspects of the Law of the Sea
- f: Groupe de travail sur les aspects techniques du droit de la mer
- s: Grupo de Trabajo sobre Aspectos Técnicos del Derecho del Mar
TALSO e: Theater Army Logistics System Office
T-ALT e: Topex Altimeter
TA [Luft] d: Technische Anleitung [zur Reinhaltung der Luft]
TA&M e: Texas A&M University
TAM d: Telefon, Anrufbeantworter, Modem
- e: International manufacturer of packers used in ODP operations
- e: Tactical Airlift Model
- e: Tampa, Fl.
- e: Temporal Associative Memory
- e: Theater Airlift Manager
- e: Timer Active Monitor
- e: Tropical Asian Monsoon Region
- f: Transports Automobiles Municipaux
TAMCA e: Theater Army Movement Control Agency
TAMCOM e: Transvaal Association of Management Committees
TAME e: Tactical Air-to-Air Mission Evaluation Model
- e: TEMPEST Advanced Multiplexer Equipment
TAMEX e: Taiwan Area Mesoscale Experiment
TAMMC e: Theater Air Material Management Center
TAMMS e: The Army Maintenance Management System
- e: Transportation Movements Management System
Tamofa e: Tanzania-Mozambique Friendship Association
TAMP e: Tactical Air Missile Program
- e: Theater Army Maintenance Program
TAMPER e: Antitampering and Fuel-Switching Information System
TAMRF e: Texas A&M Research Foundation
TAMS e: Test and Monitoring System
- e: Theater Airlift Management System
TAMSAT e: Tropical Agricultural Meteorology using Satellite
TAMSS e: Target Area Meteorological Sensor System
TAMTAC e: Toxic Air Monitoring System Advisory Committee
TAMU e: Texas A & M University
TAN e: Table Area Network
- e: Tananarive, Madagascar
- e: Tangential[ly]
- e: Tasman Air Services
- e: Test Advisory Notice
- e: Test Area North, Idaho National Engineering Laboratory
- e: Total Area Network[ing]
- e: Transaction Number
- f: Traité de l'Atlantique Nord
tan e: tangent
Tan s: Tanque
TANAPA e: Tanzania National Parks
T Anf s: Tanque Anfibio
TANGO e: Command Post for UNC/CFC/CGCC
Tanie f: Tannerie
TANJ e: There Ain't No Justice [Internet]
TANK d: Tanker
Tanpakushitsu Kakusan Koso j: Tanpakushitsu Kakusan Koso. Protein, Nucleic Acid, Enzyme
TANSEED e: Tanzania Seed Company
TANSTAAFL e: There Ain't No Such Thing As A Free Lunch [Internet]
TANU e: Tanganyika Peoples Liberation Front
TAO e: Technical Area Owner
- e: Telephony Application Objects
- e: Thermal Array in the Ocean
- e: TOGA Automated Observations
- e: Track At Once
- e: Tradeable Absorption Obligation
- e: Tropical Atmosphere-Ocean [array]
- f: Réseau pour l'observation océan-atmosphère dans les mers tropicales
- s: Red para la Observación Océano-Atmósfera en los Mares Tropicales
TAOCP e: The Art Of Computer Programming
TAOM e: Tactical Air Operations Module
TAP a: Tekniske og Administrative Personale
- d: Theater am Platanenhain
- e: Tank Advisory Panel
- e: Tanks Advisory Panel
- e: Tappan Hall
- e: Target Analysis and Planning System
- e: Target Approach Point
- e: Technical Acceptance Plan
- e: Technology Apprenticeship Program
- e: Telelocator Alphanumeric Protocol
- e: Telephone Assistance Plan
- e: Test Access Port
- e: The American Psychoanalyst
- e: Tool Application Program
- e: Toxic Air Pollutant
- e: Trace Analysis Program
- e: Transboundary Air Pollution
- e: Tuition Assistance Plan
- f: Transports Aériens Privés
- p: Transportes Aéreos Portuguêses
TA&P e: Theater Assessments and Planning
TAP/2 e: Time and Place/2
TAPA e: Tanganyika African Parents Association
TAPC/2 e: Time and Place Connectivity/2
TAPCIS e: The Access Program for the CompuServe Information Service
TAPDB e: The Army Personnel Data Base
TAPDS e: Toxic Air Pollutant Data System
TAPER e: The Army Personnel Roll-Up System
TAPES e: Tactical Automated Planning and Engineering System
TAP[h]A e: Transactions of the American Philological Association
TAPI e: Telephony API[Application[s] Program[ming] Interface]
TAPIRO e: Source Reactor Used for Shielding
TAPO e: Tanzania Association of Parastatal Organizations
TAPP e: Time and Attendance, Personnel [and] Payroll
TApp f: Taxe d'Apprentissage
TAPPI e: Technical Association of Pulp and Paper Industries
TAPR e: Tucson Amateur Packet Radio Group
TAPS e: Task Authorization and Planning System
- e: Terrain Analysis Production System
- e: Trans-Alaskan Pipeline
- e: Trans-Atlantic Pollution Survey
- e: Tropical Assimilation and Prognosis System
- f: Etude de la pollution à travers l'Atlantique
- s: Estudio de la Contaminación Trans-Atlántica
TAR e: Tape Archive Format
- e: Tape Archiver
- e: Technical Aspects Report
- e: Temporary Alternative Route
- e: Threat Assessment Report
- f: Transports Automobiles Routiers
tar e: tape archive
TARA e: Tanzanian Resistance Army

TARDIS e: Titles Automated Register and Document Information System
TAReC e: Tasmanian Accreditation and Recognition Committee
TARECU e: Tabora Region Cooperative Union
TA reg r: Tung Attack regmente
TARES e: Tactical Reconnaissance Evaluation System
TARFOX e: Tropical Aerosol Radiative Forcing Observational Experiment
- e: Tropospheric Aerosol Radiative Forcing Observation Experiment
TARFU e: Things Are Really Fouled Up [Internet]
TARG e: Telescoped Ammunition Revolver Gun
Targa e: a bitmap format developed by Truevision
TARGET e: Targeted Audience Electronic Communication System
Targeted Diagn Ther e: Targeted Diagnosis and Therapy
TARH e: Terminal Active Radar Guidance
TARIC f: Tarif Intégré Communautaire
TARM e: TIROS Atmospheric Radiance Module
TARMAG d: Vereinigung medizinischer und betriebswirtschaftlicher Tarifexperten
Tarn-et-Gar f: Tarn-et-Garonne
TARP e: Theater Army Rebuild Program
- e: TID Address Resolution Protocol
TARRS e: Tactical Air Reconnaissance Results Reporting System
TARS e: Tactical Aerial Reconnaissance and Surveillance Needs
- e: Tactical Air Reconnaissance System
- e: Tactical Air Reconnaissance/ Surveillance
- e: Tethered Aerostat Radar System
- e: Transportation Automated Routing System
- s: Tanque de Almacenamiento Recuperable Submarino
TARUS e: Training assessment reporting and utilization system
tar.Z e: Compressed Archived files
TAS e: Technical Area Specialist
- e: Telephone Access Server
- e: Tenure Administration System
- e: Test And Set an operand
- e: Test Assembly Strongback
- e: Thallium Arsenic Selenide
- e: Time and Attendance System
- e: Trace Analysis System
- e: Tritium Analysis System
- e: True Air Speed
- f: Textiles Artificiels et Synthétiques
- N: Taktikkskolen
TASA e: Total Aircraft Status Accounting
TASC e: Centre for Technology and Social Change
- e: Technical Assistance Service Center
- e: Telecommunications Alarm Surveillance and Control system
- e: Test Assembly Shipping Container
- e: The American Surrogacy Center
- e: The Analytic[al] Sciences Corp.
- e: Theater ADP Service Center
TASCAM e: Tactical Aerial System Model
- e: Tactical Airlift System Comparative Analysis Model
TASCFORM e: Technique for Assessing Comparative Force Modernization
TASCS e: Training Analysis Support Computer System
TASD e: Technical and Administrative Support Division
TASE f: Traité de l'Asie du Sud-Est
TASES e: Tactical Airborne Signal Exploitation Space
TASI e: Time Assigned (or Assignment) Speech Interpolation [system]
TASKA e: Conceptual fusion power plant
Task Force e: Consumer Education Advisory Task Force
TASM e: Tomahawk Anti-Ship[ping] Missile
TASO e: Terminal Area Security Officers
TASP e: Tactical Advanced Signal Processor
TASPAWS e: Tasmanian Parks and Wildlife Service
TASQUE e: University of Tasmania consultative unit
TASS e: Tactical Air Support Squadron
- e: Tactical Air-to-Surface System
- e: Tactical Analysis Support System
- e: Total Army School System
TA/SSB e: Test Assembly/Shipping Strongback
TASSEL e: TASS Element
TA SSR e: Tadzhik Soviet Socialist Republic
TASTA e: Tasmanian State Training Authority
TAT e: Theoretical Arrival Time
- e: To Accompany Troops
- e: Trans Atlantic Trunk
- e: Transatlantic [cable]
- e: Transportation and Ammunition Tracking
- e: Turn Arround Time
TATA e: Tanzania Agricultural Trainers Association
TATC e: Test Article Transfer Container
TA/TF e: Technical Assessment/Technical Forecasting
TATP e: Texas Assistive Technology Partnership
TATRC e: Telemedicine and Advanced Technology Research Center
TATS e: South Australias computerised titles database
- e: Tactical Assessment Teams
- e: Task Analytic Training System
TAU e: Thousand Astronomical Unit
Tau Beta Pi e: National Engineering Honor Society
TAUVEX e: a UV images on Spectrum-X-Gamma
TAV e: Total Asset Visibility
- e: Trans-Atmospheric Vehicle
T Av s: Transporte de Aviones
TAVP f: Travaux d'Architecture de la Ville de Paris
TAW e: Tactical Airlift Wing
TAWDS e: Transportable Automated Weather Distribution System
TAWU e: Transport and Allied Workers Union
TAXI e: Transparent Asynchronous Transceiver[/Receiver] Interface
- e: Transparent Asynchronous Transmitter/Receiver Interface
Taxon. e: Taxonomical
- e: Taxonomy
TAZ e: Taphronomically Active Zone
TAZARA e: Tanzania Zambia Railway Authority
TB a: Torpedobåd
- d: Tarifbestimmungen in der Kraftfahrtversicherung
- e: Brightness Temperatures
- e: Technical Bases
- e: Terabyte
- e: Total Beta
- e: Translation Buffer
- e: Transparent Bridging
- e: Trolley Black
- f: Tarif des Bagages
- f: Taxe de Balayage
- f: Traité de Bruxelles
- f: Troupeau de Bétail
Tb r: Torpedbåt[sskola]
- s: Trabajo
T/B e: Top and Bottom
tb- r: tungbomb-
TBA e: To Be Arranged
T[-]BACT e: Toxic[s-]Best Available Control Technology
TBAMS e: Test-Bed Aircraft Multispectral Scanner
T-bana r: Tunnelbana
TBAP R: Tyazhyolobombardirovotschnii Aviatsonii Polk
T-bas r: Tillfällig bas
Tbatt r: Tungt batteri
TBB d: Main-Tauber-Kreis in Tauberbischofsheim, Baden-Württemberg
- e: Tsunami Bulletin Board
- f: Tableau d'affichage électronique sur les tsunamis
- s: Tablero Electrónico de Anuncios sobre los Tsunamis
TBBS e: The Bread Board System
TBC e: Texas Bicycle Coalition
- e: Time-Based Competitiveness
- e: To Be Considered
- e: Track Descriptor Block
Tbc r: Tuberkulos
TBCF e: The Breast Cancer Fund
TBD e: To Be Decided
- e: To Be Defined
- e: To Be Delivered
- e: To Be Designed
- e: To Be Determined
- e: Transmitter Backward Channel Data
TBE e: Threaded Both Ends
- f: [Fédération européenne des fabricants de] Tuiles et de Briques
TBEP e: Training Base Expansion Plan
TBF d: Technisch betriebliche Funktionsbedingungen
- e: Tick-Borne Fever
Tbfpl r: Tungt Bombflygplan

TBGA e: Tape Ball Grid Array
- e: Tape Ball Grid Array illustration at right and link are from IBM
TBH e: To Be Honest
TBI e: Throttle Body Injection [systems]
- e: To Be Issued
TBII e: Tau Beta Pi National Engineering Honor Society
TBIM e: Trigger Based Item Management
TBINET e: The Brain Injury Information Network
Tbioch d: Institut für Technische Biochemie
TBIRD e: Traumatic Brain Injury Resource Directory
TBK e: Toolbook
TBL d: Technik für Betriebslenkung
TBM e: Tactical Ballistic Missiles
- e: Terrestrial Biogeochemical Model
- e: Theater Ballistic Missile
- e: Theater Battle Management
- e: Training Baseline Materials
TBMS e: Thirteen Band Multispectral Scanner
TBO d: Technische Betriebsordnung
- e: Technical work Order
TBOAA e: Tuna Boat Owners Association of Australia
TBP e: TATA-Binding Protein
TBq e: Terabecquerel
TBR e: Technical Basis for Regulation[s]
T/BR N: Transportberedskapsrådet
TBRTS e: Triple-[quantum] Barrier Resonant Tunneling Structure
TBS e: Technitium Bone Scan
- e: To Be Supplied
TBT e: Tributyl Tin
TBU e: Tape Backup Unit
TBVC A: Transkei, Bophuthatswana, Venda, Ciskei
TBX e: Terminal box
TBYB e: Try Before You Buy
Tbyte e: Terabyte [10^{12}]
TC e: Target Concentration
- e: Teachers College
- e: Teaching Company
- e: Technical Corrigendum
- e: Technological Committee
- e: Telephone Conversation report
- e: Test Cell
- e: Test procedures/instructions
- e: Thermocompression
- e: Think C
- e: Timing Counter
- e: Total Color
- e: Toxic Concentration
- e: Training Coordination
- e: Training Council
- e: Transaction Capabilities
- e: Transit Corridor
- e: Transition Components
- e: Transmission Convergence [sub-layer]
- e: Transmitted signal element timing
- e: Transportation Coordinator
- e: Tumour Complement
- e: Turks & Caicos Islands
- f: Tables Chronologiques
- f: Taux de Cotisation
- f: Taux du Change
- f: Taxe de Compensation
- f: Taxes Comprises
- f: Témoin à Charge
- f: Temps Compensé
- f: Tout Compris
- f: Traité de Commerce
- f: Transports Commerciaux
- f: Transports en Commun
- f: Travail Collectif
- f: Tribunal Civil
- f: Tribunal Consulaire
- f: Tribunal Correctionnel
- f: Tribunal de Commerce
- f: Troisième Classe
- f: Trottoir Cyclable
- s: Taller de Comunicaciones
- s: Telecomunicaciones
- s: Teniente Coronel
- s: Tren de Combate
Tc s: Táctica
- s: Tractor
Tc e: committed rate measurement interval
T C e: Tutorial Course
TC. e: Transmission Control
T/C e: Temperature/Conductivity sensor chains on Drifters
- e: Thermocompression
- e: Thermocouple
- e: Thrust Chamber
t/c e: thickness/chord ratio
T/c f: Transfert de contrôle
T2C3 e: Theater/Tactical Command, Control and Communications
TC2-21 e: 21st Century Tactical Command and Control Study
TC4 e: Transportable Command and Control Computer Center
Tc-99 e: Technetium-99
TCA e: 1,1,1-Trichloroethane
- e: 1,1,1-Trichloroethate
- e: Target Class Assessment
- e: Telecommunication[s] Adapter
- e: Telecommunications Association
- e: Terminal Controlled Airspace
- e: Test Calibration Assembly
- e: Tightly Coupled Architecture
- e: Trichloracetic Acid
- e: Trichloroethane
TCAA s: Tropas Coheteriles de Artillería Antiaérea
TCAC e: Tactical Collection and Analysis Center
- e: Technical Committee on Agricultural Chemicals
TC[-]ACCIS e: Transportation Coordinator[s] Automated Command and Control Information System
TCAC[D] e: Technical Control and Analysis Center [Division]
TCA cycle e: Tricarboxylic Acid cycle
TCAD e: Technology CAD
- e: Technology Computer-Aided Design
- e: Tradeable Carbon Absorption Obligation
TCAF e: Target Cost Award Fee
TCAIMS e: Transportation Coordinators Automated Information for Movements System
T-CAN e: Transportation-Capability Assessment Net
TCAP e: Tank Characterization Advisory Panel
- e: Transaction Capabilities (or Capability) Application[s] Part
TCAPA e: Tri-County Air-Pollution Authority
TCAS e: T Carrier Administration System
- e: Technical Control and Analysis Section
- e: Telephone Call Analysis System
- e: Temporary Criticality Alarm System
- e: Traffic alert and Collision Avoidance System
- e: Transportable Criticality Alarm System
TCB e: Test Cell Basement
- e: Thermocompression Bonding
- e: Trouble Came Back
- e: Trusted Computer Base
TCB. e: Task Control Block
TCC e: Tactical Cell Controller
- e: Tactical Command Centers
- e: Technological Change Committee
- e: Telecommunications [Control] Center
- e: Tocix Chemicals Committee
- e: TRANSCOM Control Center
- e: Transmission Control Codes
- e: Transportation Component Command
- e: Transportation Coordination Center
- e: Trunk Class Code
- s: Tabla de Conversación de Comunicaciones
TCC[A] e: Theater Communications Command
TCCB e: digital telemetry tool
TCCC e: Technical Committee on Computer Communications
TCCC IEEE e: Technical Committee on Computer Communications of IEEE
TCCD e: Test Cell Closure Device
TCCP e: Telecommunications Command Control Projects
- e: Tri Cities Commercialization Partnership
TCC Upgrade e: Telecommunications Center AUTODIN Terminal Equipment Replacement and Upgrade Program
TCD e: Tactical Control Director
- e: Tank Characterization Database
- e: Technical Control Documents
- e: Thermal Conductivity Detector
TCDA e: Tank Calibration Data Acquisition
TCDBCP e: Technical Coordinator of the Data Buoy Cooperation Panel
- f: Coordonnateur technique du Groupe de coopération pour les programmes de bouées de mesure
- s: Coordinador Técnico del Panel de Cooperación sobre Boyas de Acopio de Datos
TCDD e: Tetrachlordibenzo-p-dioxin
TCDIS e: Transportation Coordinator Deployment Information System
TCDM e: Tool Development Cost Model
TCE e: Targeted Client Environment
- e: Temperature Coefficient of Expansion
TCEC e: Test Cell Environment and Cooling [FMIT]
TCEx f: Tarif Commun international pour le transport des colis Express

TCF e: Tactical Combat Force
- e: Technical Control Facility
- e: Technical Cooperation Fund
- e: Total Chlorine Free
- e: Transient Correction Factor
- f: Transport par Chemin de Fer

TCF5 e: IL-6 depdt. DNA-binding protein
TCFC e: Transec Communications Front-End Controller
Tcf/d e: Trillion cubic feet per day
TCFEP e: Telemetry and Command Front-end Processor
TCFS e: Transparent Cryptographic File System
TCG e: Geocentric Coordinate Time
- e: Tactical Control Group
- e: Target Constant Generation
- e: Test Call Generation
- e: Total Combustible Gas
- e: Transportation Coordination Group
- f: Temps Civil de Greenwich
- f: Temps Coordonné Géocentrique

Tch[éc] f: Tchécoslovaqu[i]e
TCHEP e: Technical Committee on High Energy Physics
Tchra s: Trinchera
TCHTG e: Technical Training Group
TCI e: Tactical Computer Interface
- e: Target Class Identification
- e: Tele-Communications Inc
- e: Test Cell Input
- e: Tungsten Carbide Inserts

TCID e: Truckee-Carson Irrigation District
TCIE e: The Center for Industrial Effectiveness at UB
TCIF e: Telecommunications Industry Forum
TCIO e: Task Control and Information Office
TCIP e: Technical Control Improvement Program
- e: Technical Control system Improvement/ control technician improvement Plan

TCiv f: Temps Civil
TCJ e: The Chiropractic Journal
TCL e: Target Compound List
- e: TM Control Point Library
- e: Tool Command Language
- e: Tool Control Language
- e: Tsumeb Corporation Limited

TCLP e: Toxicity Characteristic Leach Procedure
Tcl/Tk e: Tool command language/Tool kit
TCM e: Temperature Control Module
- e: Termal Conduction Module
- e: Terminal Control Mechanism
- e: Terminal Controller Module
- e: Tetrachlormercurat
- e: Time Compression Multiplexer
- e: Time Compression Multiplexing
- e: Total Catchment Management
- e: Traditional Chinese Medicine
- e: Trajectory Control Maneuver
- e: Transportation Control Measure
- e: Trellis Coded Modulation
- e: Trellis Coding Modulation
- e: Tropical Cyclone Motion
- e: Tunneling Current Microscopy

TCMD e: Transportation Control Movement Document
TCMP e: Transportation Control and Movement Plan
TCMS e: Test Control and Monitoring System
TCMWU e: Textile and Clothes Manufacturing Workers Union
TCN e: Topology Change Notification
- e: Transparent Content Negotiation

TCNIC e: Tri-Cities Nuclear Industrial Center
- e: Tri-City Nuclear Industrial Council

Tcnl s: Teniente coronel
TCNRDI e: Transmission Corrected Normalized Reflectance Difference Index
TCNS e: Thomas Conrad Network[ing] System
TCNZ e: Telecom New Zealand
TCO d: Strahlenschutz-Norm für Bildschirme
- e: Tactical Combat Operation System
- e: Tactical Command and Operations
- e: Technical Control Operator
- e: Technical Cooperation Activities
- e: Telecommunications Certification Office
- e: Telecommunications Control Office
- e: Terminal Clean-Out [Operation]
- e: Test Cell Output
- e: Total Cost of Ownership

TCO_2 e: Total CO_2
TCOL s: Teniente Coronel
TC/OPC e: Technical Committee on Ocean Processes and Climate
T COR s: Teniente Coronel
Tcor e: cortical TL
TCOS e: Technical Committee on Operating Systems
TCOS IEEE e: Technical Committee on Operating Systems of IEEE
TCOSS e: Technical Control Operations Support System
TCOT e: Tactical Control Operations Teams
TCP e: Tank Characterization Plan
- e: Tape Carrier Package
- e: Telecommunications Protocol
- e: Test Coordination Procedure
- e: TeX Code Page
- e: Transaction Control Protocol
- e: Transformer-Coupled Plasma
- e: Transmission Control Protocol
- e: Transport Control Protocol
- e: Trichloroethylene
- e: Tropical Canine Pancytopenia
- e: Tropical Cyclone Program[me]

TCPA e: Tantawangalo Catchment Protection Association
- e: Time of Closest Point of Approach

TCPACO e: TCP Alternate Checksum Option
TCPBEUI e: Transmission Control Protocol BIOS Extended User Interface
TCPC e: Technical Committee on Privatization and Commercialization
- e: The Clean Personal Computer group
- e: Transportation Claims Prevention Council
- e: Tri-Cities Professional Center

TCPI e: To-Complete Performance Index
TCPIP e: Transmission Control Protocol/ Internet Protocol
TCP/IP e: Transfer Control Protocol over Internet Protocol
- e: Transmission Control Program/Internet Protocol
- e: Transmission[s] Control Protocol/ Internet Protocol
- e: Transport Control Protocol/Interface Program
- e: Transport Control Protocol/Internet Protocol

TCPLDP e: TCP extensions for Long Delay Paths
TCP/UDP e: Transmission Control Protocol/User Datagram Protocol
TCR e: Tank Characterization Reporting
- e: Technical Control Rebuild
- e: Technical Control Room
- e: Technical Report
- e: Testchart Reader
- e: Thyristor-Controlled Reactor
- e: Tonnes of Coal Replacement
- e: Transient Call Record
- e: Transmit Configuration register

TCr f: Tarif Criminel
TCR f: Taxe de Coopération Régionale
- f: Témoignages Chrétiens de la Réunion

TCRI e: Toxic Chemical Release Inventory
TCRP e: Tactical Command Readiness Program
- f: Transports en Commun de la Région Parisienne

TCRS e: Tanganyika Christian Refugee Service
- e: Tennessee Career Reporting System

TCS e: Tactical Command System
- e: Tactical Computer System
- e: Telemetry Command Station
- e: Temporary Criticality Specifications
- e: Terminating Call Screening
- e: Theater Communications System
- e: Thermal Control System[s]
- e: Time Card System
- e: Toroidal Confinement Systems
- e: Torus Cleanup System
- e: Transmission Convergence Sublayer

TC2S e: Transportation Command and Control System
TCSB e: Traffic Control State Block
TCSH e: Trusted C SHell
TCSP e: Technical Control Systems Program
- e: Tourism Council of the South Pacific

TCSS e: Technical Control Software Support
- e: Telecommunications Service System

TCSTP e: Tri Cities Science & Technology Park
TCSTS e: Tactical Communications Systems Technical Standards
TCSUH e: The Texas Center for Superconductivity at the University of Houston
TCT e: Tandem Connecting Trunk
- e: Tasmanian Conservation Trust
- e: To Challenge Tomorrow

Tct s: Táctica

TCTE e: Trace Constituents in the Troposphere over Europe
TC/TF e: Training Calendars/Training Files
TCTS e: EMSL-Vegas Time Card Tracking System
TCU e: Transit Control Unit
- e: Transportation Communications Union
- e: Transportation Control Unit
- e: Tri-City University [Washington State University]
- e: Trunk Coupling Unit
TCUC e: Tri-Cities University Center
TCUL p: Transportes Coletivos Urbanos de Luanda
TCV e: Tokamak with Variable Configuration
- e: Total Cell Volume
- f: Tarif Commun international pour le transport des Voyageurs et des bagages
TCW e: Tactical Control Wing
- e: Tagged Control Word
TCWF e: Toxic Custard Workshop Files
TCWG e: Tactical Communications Working Group
TCY e: Target partner Country
TD e: Technical Data
- e: Technical Delegate
- e: Technology Development
- e: Threading Dislocation[s]
- e: Time and Date
- e: Time-Dependent
- e: Total Depth [of a hole]
- e: Training Developer
- e: Turbo Dispatcher
- f: Tarif Différentiel
- f: Tarif Douanier
- f: Télédiffusion
- f: Témoin à Décharge
- f: Timbre de Dimension
- f: Tous Départements
- f: Train Direct
- f: Traitement à Domicile
- f: Trajet Direct
- f: Travail à Domicile
- f: Travailleur du Dimanche
- f: Travaux Dangereux
- f: Très Difficile
- f: Tuteur Datif
Td f: Titre définitif
T D e: Theatre and Dance
T.D. e: Territorial Decoration
t.d. N: til døme[s]
td. a: tonde
TD. e: Transmit Data
T&D e: Training and Development/Test and Diagnostic
T/D e: Top Drive
TDA e: Table of Distribution and Allowances
- e: Table of Distribution and Authorization
- e: Tactical Decision Aids
- e: Tea Development Authority
- e: Tornado Detection Algorithm
- e: Trade and Development Agency
- e: Transmit Descriptor Area
- e: Transportation Development Act
- e: US Trade and Development Agency
TDA. e: Test Design Automation
TD&A e: Time Distribution and Attendance
T-DAB e: Terrestrial Digital Audio Broadcasting
T[-]DAB e: Terrestrial Digital Audio Broadcasting
TDAC e: Tracking, Data, Acquisition and Communication
TDAE e: Tetrakis [Diethylamino] Ethylene
TDAS e: Tracking and Data Acquisition System
T-day e: Times
TDB e: Barycentric Dynamical Time
- e: Technical Data Base
- e: Technology Development Branch
- e: Transportation database
TDB. e: Test Data Base
TDC e: Tabular Data Control
- e: Tape Data Controller
- e: Technical Data Checklist
- e: Technology Development Contractor
- e: Temperature Display Controller
- e: Terrestrial Data Circuit
- e: Time Distance Calculator
- e: Total Direct Costs
- e: Town Development Committee
- e: Transient Damage Correlation
- e: Transportation Data Channel
- e: Tropical Data Center (or Centre)
TDCA e: Tactical Deployment and Control Aircraft
- e: Tactical Deployment Coordination Aircraft
TDCC e: Transportation Data Coordinating (or Coordination) Committee
TDD e: Target DGZ Designation
- e: Technical Directive Documents
- e: Telecommunication[s] Device for [the] Deaf
- e: Telephone (or Telephonic) Device for the Deaf
- e: Time Division Duplex[ing]
TDDB e: Time-Dependent Dielectric Breakdown
TDDO e: Time Delay on Drop Out
TDDR e: Total Direct Diffuse Radiometer
TDDSG d: Teledienstdatenschutzgesetz
TDD/SP e: Top Down Design and Structured Programming
TDE e: Telescript Development Environment
- e: Terminal Display Editor
- e: Top Disengager
- e: Transition Diagram Editor
- f: Taxe de Déversement à l'Égout
TDEAT e: Tetrakis [Diethylamino] Titanium
TDEC e: Tennessee Department of the Environment and Conservation
TDEG e: Two-Dimensional Electron Gas
T&DER e: Tracking and Data Relay Experiment
Tde rd r: Tingstede radiostation
T Des s: Transporte de Desembarco
TDEV e: Transport-level Deviation
TDF e: Trace Definition File
- e: Tracking Data Formatter
- e: Transkei Defense Force
- e: Typeface Definition File
TDG d: Telekommunikationsdienstgesetz
TDH d: bodennahe Lufttemperatur, Flächemittel für Deutschland aus wahrscheinlich homogenen Stationen
- e: Texas Department of Health
- e: Total Discharge Head
TDI e: Telecommunications for the Deaf, Inc.
- e: Test/Debug Interface
- e: Thomson Digital Image
- e: Time-Delay[ed] Integration
- e: Transport Device Interface
- e: Transport Driver Independence
- e: Transport Driver Interface
TDICMS e: Redesign of Technical Data/Configuration Management System
TDIF e: Topex/Poseidon Data Information Facility
TDIS e: TEM Disk Identification System
TDJ e: Transfer Delay Jitter
TDK e: Tension Device
- e: Transmitter Data Relay
TDL e: Techniques Development Laboratory
- e: Test Deficiency List
- e: Transition Documentation Library
tdl. a: tønde land
TDLAS e: Tunable Diode Laser [Absorption Spectrometer]
TDLC. e: Twinaxial Data Link Control
TDLDA e: Time-Dependent Local Density Approximation
TDLHS e: Tunable Diode Laser Heterodyne Spectrometer
TDLS e: Tunable Diode Laser Spectroscopy
TDLWG e: Tactical Data Link Working Group
TDM e: Technical Direction Memorandum
- e: Technical Document Management
- e: Telekom Designed Networks
- e: Test Definition Memorandum
- e: Testbed Dismantling Machine
- e: Time Division Multiplexer
- e: Too Damn Many [Internet]
- e: Tracking Data Message
- e: Transportation Demand Management
TDMA e: Tandem Differential Mobility Analyzer
- e: Time Division Multiple[x] Access
TDMAT e: Tetrakis [Dimethylamino] Titanium:
TDMS e: Terminal Display Management System
- e: Thermal-Desorption Mass Spectroscopy
TDN e: Total Digestable Nutrients
TDN-EBB e: Tactical Data Network-Echelons Below Battalion
TDO e: Table of Denial Orders
- e: Technical Directors Organization
- e: Technical Work Order
TDOP e: Total DCS Operations Plan
TDOs e: Tactical Deception Officers

TDP e: Tag Distribution Protocol
- e: TAM Drilling Packer
- e: Technical Data Package
- e: Technology Development Program
- e: Telelocator Data Protocol
- e: Time Domain Reflectometry
- e: Training Development Plan
- e: Trigger Detection Point
- e: Triton Data Path

TDPI e: Tasmanian Department of Primary Industry
TDPO e: Technology Development Program Office
TDPU e: Time-delay on pickup
TDR e: Global Tropical Disease Research Program
- e: Tape Dump Run
- e: Technical Design Report
- e: Time Distribution Record
- e: Time Distribution Report
- e: Time Domain Reflectometer
- e: Time-Delay Relay
- e: Toxicology Desk Reference
- e: Traffic Distribution Record
- e: Transportation Discrepancy Report
- e: UNDP/World Bank/WHO Special Programme for Research and Training in Tropical Diseases
- f: Tous Droits Réservés

tdr. a: tonder
- a: tønder

TDRC e: Temperature Differential Recorder Controller
TDRI e: Thailand Development Research Institute
TDRS e: Test Data collection and Reduction System
- e: Tracking and Data Relay Satellite
- e: Tracking and Data Relay Systems
- f: Satellite-relais de poursuite et de retransmission de données
- s: Satélite de Seguimiento y Retransmisión de Datos

TDRSS e: Tracking [and] Data Relay Satellite System
TDRU e: Tropical Disease Research Unit
TDS e: Tactical Data Station
- e: Tape Duplication System
- e: Technical Data Server
- e: Technical Documentation Standard
- e: Temperature Display System

TDSC e: Technical Data Service Center
- e: Tinker Data Services Center

TDSE e: Time-Dependent Schrödinger Equation
TDSF e: Tasmanian Department of Sea Fisheries
TDSM e: Time Division Switching Matrix
TDSQ e: Training Development Squadron
TDSR e: Transmitter Data Service Request
TDSV d: Telekomm[munikations]-Datenschutzverordnung
TDT e: Terrestrial Dynamical Time
TDTC e: Test, Development and Training Center
TDtoDP e: Tablet Coordinates to Display Coordinates
TDTOX e: Tetradichloroxylene
TDU e: Topology Database Update
TDW e: Triply Distilled Water
TDWG e: Taxonomic Databases Working Group for Plant Sciences
TDX e: Time Delay to X
- e: Time Domain Crosstalk

TDY e: Temporary Duty
- e: Time Delay to Y

TE e: Echo Time
- e: ISDN Terminal Equipment
- e: Threatened and Endangered
- e: Total Exposure
- e: Transferred Electron
- e: Transmitted Electron
- e: Transpiration Efficiency
- f: Taux d'Escompte
- f: Terme Échu
- f: Titre Étranger
- f: Tour Eiffel
- f: Train Express
- f: Travailleur de l'Etat
- f: Travailleur Étranger
- f: Tutelle de l'Etat
- s: Tratamiento Especial

Te r: Tekniker
- r: Teknisk

te r: till exempel
T&E e: Test and Evaluation
- e: Threatened and Endangered
- e: Training and Education

TE1 e: Terminal Equipment type 1
TE2 e: Terminal Equipment type 2
TE/2 e: Terminal Emulator/2
TEA e: Targeted Export Assistance
- e: Temperate East Asia
- e: Torque Equilibrium Attitude
- e: Total Exposure Assessment
- e: Totally Egregious Acronym
- e: Transverse Excited Atmosphere
- e: Transverse-Excited Atmospheric pressure
- e: Triethanolamine
- f: Tribunal pour Enfants et Adolescents

TEACC e: Technical Engineers and Administrative Careers Committee
TEACCH e: Treatment and Education of Autistic and Related Communication Disorders
TEA-COM e: Temperate East Asia-Planning Committee
TEACOM e: Temperate East Asia Regional Committee for START
TEA-ER e: Traffic Executives Association, Eastern Railroads
TEAL e: The Electronic Anesthesiology Library
TEAM e: Technology Base Enhancement for Autonomous Machines
- e: Tele-Education And Medicine
- e: Total Environment Analysis and Management
- e: Transportation Emergency Assi
- f: Télécommunications Électroniques, Aéronautiques et Maritimes

TEAMA e: Taiwanese Electric Appliance Manufacturers Association
- e: Terrestrial Ecosystems in Monsoon Asia

TEAMS e: Tactical EA6B Mission Planning System
- e: Terrestrial Ecosystem Analysis and Modeling System

TEARS e: Thermal model for electromigration
- e: Traffic Engineering for Automated Route Selection

TEB e: Tactical Exploitation Battalion
- e: Teterboro NJ
- e: Thread Environment Block

TEBA e: The Employment Bureau of Africa
TEBALDI e: Télécommande Ballon Longue Distance
TeBG e: Testosterone-estradiol-Binding Globulin
TEC e: Monitor Terminal Equipment Configuration
- e: Tasmanian Environment Centre
- e: Technology Evaluation and Coverage Program
- e: Technology Exchange Company
- e: Tennessee Eastman Corporation
- e: Test and Electrical Characterization
- e: The Energy Consortium
- e: Thermal Elongation Coefficient
- e: Thermoelectric Coolers
- e: Tokyo Electronics Corporation
- e: Topographic Engineering Center
- e: Total Construction line item
- e: Total Electron Content
- e: Total Environment Centre
- e: Training Executive Committee
- f: Tarif Extérieur Commun

Tec f: Technique
- s: Técnica

TECAP e: Transistor Electrical Characterization And Analysis Program
TECDA e: Thai Environmental and Community Development Association
TECDOC e: IMPACT study on CD-ROMs for technical documentation
- e: Technical Documentation

TECH e: Technical Students
TECHLIB e: A package for VAX, MVS or Unix
TECHMOD e: Technology Modernization
techn. e: technical[ly]
TechnCh d: Institut für Technische Chemie
Technicatome f: Société Technique pour l'Energie Atomique
Technion e: Israel Institute of Technology, Israel
Technol Health Care e: Technology and Health Care
TechnOpt d: Institut für Technische Optik
Techsec e: Technical Secretariat
- f: Secrétariat Technique
- s: Secretaría Técnica

TECO e: Tape Editor and Corrector
- e: Terrestrial Ecological Research Initiative
- e: Terrestrial Ecosystems

TECOMAC e: Technical Conf. on Observation and Measurements of Atmospheric Contaminants
TECP e: Tectonics Panel
- f: Groupe Chargé de la Tectonique
- s: Panel sobre Tectónica

TECRO e: Taiwans Economic and Cultural Representative Office

TECUP e: Test-bed Implementation of the ECUP Framework
TED d: Telefonischer Ermittlungsdienst
- e: Technological and Economic Dynamics
- e: Testability and Embedded Diagnostic
- e: Thermal Expansion Difference
- e: Tiny Editor
- e: Total Electron Detector
- e: Trailing-Edge Detector
- e: Transient Enhanced Diffusion
- e: Transmission Electron Diffraction
- e: Transmission Electron Diffractometry
- e: Transmitted Electron Detection
- e: Trawl Efficiency Device
TEDB e: Timber Export Development Board
TEDCOM e: Technology and Engineering Development Committee
TEDDY e: TERI Energy Data Directory Yearbook
TEDF e: Treated Effluent Disposal Facility
TEDIS e: Trade Electronic Data Interchange Systems
TEDREP e: Type Unit Equipment Detail Report
TEDSS e: Tactical Environmental Data Support System
TEDU e: Technology Education
TEE f: Trans-Europe-Express
TEEA e: Thai Environmental Engineers Association
TEED e: Tactical End-to-End Encryption Device
TEEM e: Techno-Economic-Environmental Model
TEES e: Thermochemical Environmental Energy System
- e: Thermoelectric Effect Spectroscopy
TEESE e: TERI Energy Economy Simulation Evaluation
TEFC e: Totally Enclosed, Fan-Cooled
TEFLON e: Polytetrafluorethylene
TEFSA e: Territorial Forces Effectiveness File
TEG d: Fraunhofer-Technologie-Entwicklungsgruppe
- e: Technical Exchange Group
- e: Test and Evaluation Group
- N: Trafikketterforskningsgruppen
TEGFET e: Two-dimensional Electron Gas FET
TEHIP e: Toxicology and Environmental Health Information Program
TEHM e: Terrestrial Ecosystem Hydrology Model
TEHO e: Tail End Hop Off
TEI e: Text Encoding and Interchange
- e: Text Encoding Initiative
Teint f: Teinturerie
TEJ e: Tropical Easterly Jet
Tekn N: Teknikk[er]
tekn. a: teknisk
teknr r: tecknare
t.eks. a: til eksempel
TEL e: Telephone [number]
- e: Transporter Erector Launcher
Tel s: Telefón[ic]o
TELA e: The Electronically Linked Academy
TEL[AR] e: Transporter-Erector Launcher [And Radar]
TELCOM e: WWMCCS Site Telecommunications Configuration System
Telearbeit d: Informationsverarbeitung mit entsprechenden Arbeitsplätzen zu Hause
Telebanking d: Erledigung von Bankgeschäften mit Hilfe von Personal Computern, EDV oder Bildschirmtext
TELEFAX e: Telecommunications Facsimile
teleförr r: telematerielförråd
Telefs. s: Teléfonos
Telegr. e: Telegraphy
telegrp r: telegrupp
telehv r: telehantverkare
Tele I r: Teleingenjör
Telekommunikationsdienst d: Gesamtheit der Funktionen, Protokolle und Eigenschaften, die für die Abwicklung einer bestimmten Kommunikation
Telekonsultation d: Sprechstunde am Bildschirm
Telekooperation d: Zusammenarbeit einer Gruppe, die sich ggf. über den ganzen Globus verteilen kann. Alle Teilnehmer der Arbeitsgruppe haben dabei auf dieselben Ressource Zugriff
Telelearning d: Online-Nutzung von interaktiven, multimedialen Lernprogrammen am PC oder am Fernseher mit Set-Top-Box
Telem s: Telémetro
TELEMAIL e: Electronic Mail
- f: Messagerie électronique; courrier électronique
- s: Correo electrónico
TELEMALI f: Télécommunications Internationales du Mali
TELEMAN e: Telephone Management System
Telemotm r: Telemotmedel
TELENET e: Telecommunications Network Protocol
- e: Teletypewriter Network
Teleph. e: Telephony
TELER e: Telecommunication Requirements
TeleSCAN e: Telematics Services in Cancer
TELESENEGAL f: Société de Télécommunications Internationales du Sénégal
Teleshopping d: Auswahl und Einkauf von Warenangeboten mit dem PC oder einem TV-Gerät mit Set-Top-Box
TELESPO e: Computer Operator Console Training
TELETECH NRW d: Landesinitiative zur Förderung der Telekommunikation in NRW
Teletek r: Teletekniker
TELETHESES e: French catalogue of theses
Teletj r: Teletjänst
TELEX e: Teletypewriter Exchange [Service]
Telf N: Telefon
- s: Telefón[ic]o
telf. a: telefon
Telfs. s: Teléfonos
Telg s: Telegrafico
TELIA e: Swedish Telecom
TELIAS e: Telidon Image Analysis System
TelNet d: Software, die den Dateientransport und den Anschluss an Hostsysteme über das FTP [File Transfer Protocol]-Protokoll ermöglicht
Telnet d: Dienst im Internet
- d: Terminalemulationsprotokoll
TELNET e: GTE Commercial Data Network
- e: Telecommunication[s] Network [access]
- e: Telephone Network
- e: Terminal emulation protocol, a TCP/IP application protocol for remote connection service
- e: Terminal to Host Communication Service Protocol
TELPAK e: Telecommunications Package
TELRIC e: Total Element Long-Run Incremental Cost
TELSAR e: Tracking and Evolution of Solar Active Regions
TELUB d: Telemetrisches Luftgütemeßnetz Brandenburg
TEM e: Terrestrial Ecosystem Model
- e: Trajectory Ensemble Model
- e: Transverse Electro-Magnetic
TeMa d: Technomathematik
TEMA e: Terrestrial Ecosystems in Monsoon Asia
- e: Training, Education and Mutual Assistance
- e: Tubilar Exchanger Manufacturers Association
TEMA-CB e: TEMA Group of Experts for Capacity Building
- f: Groupe d'experts TEMA pour le renforcement des capacités
- s: Grupo de Expertos de TEMA sobre Aumento de Capacidades
TEME e: Thromboembolic Meningoencephalitis
TEMES d: Telemetrisches Immissionsmeßnetz Nordrhein-Westfalen
Temex e: Telemetry Exchange
TEML e: Turbo Editor Macro Language
TEMO e: Training Exercise and Military Operations
TEMP e: Upper-level temperature, humidity and wind report from a land station
- f: Message d'observation en altitude de pression, température, humidité et vent provenant d'une station terrestre
- s: Informe de observación en altitud de la presión, la temperatura, la humedad y el viento, proveniente de una estación terrestre
TEMPEST e: Transient Electro Magnetic Pulse Emanation STandard
- e: Transient Energy Momentum and Pressure Equation Solutions
TEMPO e: Technical Military Programming Organization
- e: Tyrrhenian Eddy Multi-Platform Observation

TEMS e: Technical Engineering and Management Support
- e: Terrestrial Ecosystem Monitoring Sites
TEMSE e: Technical and Managerial Support Environment
TEN e: Total Environment Network
- e: Trans European Network[s]
TENAMIT s: Ac' Tanamit / Pueblo Nuevo
TENCAP e: Tactical Exploitation of National Capabilities Program
TENIS e: Terminal Environment Network Information System
TEntgV d: Telekommunikations-Entgeltregulierungsverordnung vom 1.10.1996
TENV e: Totally Enclosed, NonVentilated
TEO e: Technology Enhancement Office
- e: Thermoplastic Elastic Olefin
TEOA e: Test and Evaluation Objective Annex
TEOAE e: Transient-Evoked Otoacoustic Emmission[s]
TEOC e: Technical Objective Camera
- e: Tromso Earth Observation Center
TEOM f: Taxe d'Enlèvement des Ordures Ménagères
TEOS e: Tetraethelorthosilicate
TEP e: Team Execution Plan
- e: Technical Evaluation Panel
- e: Technical Evaluation Program
- e: Technology Enhancement Program
- e: Terminal Emulation Processor
- e: Thermal Exchange Pump
- e: Traffic Engineering Program
- e: Transequatorial Propagation
- f: Tonnes Équivalent Pétrole
- s: Tonelada[s] Equivalente[s] de Petróleo
TEPA e: Taiwan Environmental Protection Agency
TER e: Australian Territory
- e: Tardeable Emmission Rights
- e: Technical Education Resources
- e: Terminal Interface
- e: Thermal Eclipse Reading
- e: Training Event Report
- e: Travel Expense Report
Ter f: Terrain
TERA e: Toxicology Excellence for Risk Assessment
Ter Arkh R: Terapevticheskii Arkhiv
Teratog Carcinog Mutagen e: Teratogenesis, Carcinogenesis and Mutagenesis
TERB e: Time Enroute Bulletin
TEREC e: Tactical Electronic Reconnaissance
Terec e: Tactical electronic reconnaissance sensor with real time-datalink capability
TEREC GP e: TEREC Ground Processor
TERENA e: Trans-European Research and Education Networking Association
TERF e: Terrain Flight
TERI e: Tata Energy Research Institute
TERIS e: Teratogen Information System
TERM e: Temperature and Emissivity measurements by Reflection Method
TERMCAP e: Terminal Capability
TERMINFO e: Terminal Information
TERMPWR e: Terminator Power
TERMS e: Terminal Management System
- e: Test Equipment Reporting Management System
TERN e: The Terrestrial Ecosystem Research Network of Germany
TERPES e: TAC Electronic Reconnaissance Processing Evaluation System
Ter[r] f: territorial
TERRA e: Terrestrial Ecosystem Regional Research and Analysis [laboratory]
- i: Tecniche di Elaborazione e Rilevamento delle Risorse Ambientali
TERRAIN e: Terrain Analysis File
terr lv r: territoriellt luftvärn
terrlvdiv r: territorialluftvärnsdivision
TERS e: Tromso ERS-1 Station
- e: Tropical Earth Resources Satellite
- e: Tropical Environments Remote Sensing
TERSE e: Tunable Etalon Remote Sounder for/of [the] Earth
TERSS e: Tasmanian Earth Resources Satellite Station
TERS-WG e: Tropical Environments Remote Sensing Working Group
TERVIHT e: Trace-Element Reference Values in Human Tissues
TES e: Thermal Emission Spectrometer
- e: Tilting Etalon Spectrometer
- e: Tropospheric Emission Spectrometer
TESA e: Testicular Sperm Aspiration
- e: The Events Suppliers Association
TESAC e: Temperature, Salinity and Currents Report from a Sea Station
- f: Message d'observation de température, salinité et courants provenant d'une station en mer
- s: Informe de Observación de Temperatura, Salinidad y Corrientes, proveniente de una Estación en el Mar
TESE e: Testicular Sperm Extraction
TES/EMRS e: Terrain Edit System/ Elevation Matrix Processing System
TESL-EJ e: Teaching English as a Second or foreign language
TESOL e: Teaching English to Speakers of Other Languages
TESS e: Tactical Environmental Support System [Third Generation]
- e: The Extended Simulation System
- e: TRW Environmental Safety Systems, Inc.
TEST e: Telemedicine and Education Support Team
TEst f: Territoire Est
TESTLAB e: Testing Systems using Telematics for Library Access for Blind and Visually Handicapped Readers
TESWG e: TRAMCON Engineering Subworking Group
TET d: Teterow, Mecklenburg-Vorpommern
- e: Transient Energy Transfer
- e: Tubal Embryo Transfer
- e: Turbine-Entry Temperature
TETAP e: Test and Evaluation Tasking Plan
Tet G f: Tarn-et-Garonne
TeTherm d: Institut für Technische Thermodynamik und Thermische Verfahrenstechnik
TETRA e: Trans European Trunked Radio [System]
TEW e: Transport of Equatorial Waters
TEWES e: Tactical Electronic Warfare Environment Simulator
TEWG e: Test and Evaluation Working Group
- e: Transitional Environmental Working Group
TEWU e: Teachers and Educational Workers Union
t. ex. r: till exempel
TEX e: Text Executive Processor
TeX e: Typesetting language developed by Donald Knuth
Texcep f: à Titre exceptionnel
Tex Heart Inst J e: Texas Heart Institute Journal
TEXIN e: Texas Intersection Air Quality Model
Tex Med e: Texas Medicine
TextilCh d: Deutsche Institute für Textil- und Faserforschung Stuttgart, Institut für Chemiefasern und Institut für Textilchemie
TextilFCh d: Institut für Textil- und Faserchemie
TextilT e: Deutsche Institute für Textil- und Faserforschung Stuttgart, Institut für Textil- und Verfahrenstechnik
Texx e: A Macintosh implementation of the language REXX
TEY e: Total Electron Yield
TEZ e: Total Exclusion Zone
TF d: Theologische Fakultät
- d: Toleranzfaktor
- e: French Southern Teritories
- e: Technical Feasability
- e: Thomas-Fermi
- e: Throughfall
- f: Tarif Fédéral
- f: Tarif Ferroviaire
- f: Taxation des Frais
- f: Taxe des Frais
- f: Taxe Fiscale
- f: Territoire Français
- f: Timbre Fixe
- f: Travailleuse Familiale
- f: Travaux Forcés
- f: Tribunal Fédéral
- N: Teknologisk Forlag
- r: Taktiska Föreskrifter
- r: Truppföring
- s: Teléfono
tf a: teknisk forberedelseseksamen
t.f. r: tillförordnande, tjänstförrättande
T/F e: Thermal-to-fast
TFA e: Tanks Focus Area
- e: Tennessee Forestry Association
- e: Thyroid Foundation of America
- e: Transparent File Access
- s: Tipo de Fuerza Armada
TFAP e: Tropical Forest Action Plans
- e: Tropical Forestry Action Programme
TFB e: Thai Farmers Bank
TFC a: Teknisk Film Compagni Produktionsaktieselskab
- e: Tasmanian Forestry Commission
- e: Total Fault Coverage
- e: Total Full Color

Tfc f: Trafic
TFCA e: Total Force Capability Assessment
TFCC e: Tactical Flag Command Center
- e: Tactical Fleet Control
TFCC/FDDS e: Tactical Flag Command Center/Flag Data Display System
TFCS e: Treasury Financial Communications System
TFD a: Tjenerforbundet i Danmark
- e: Thin Film Diffractometer
- e: Thin Film Diffractometry
- e: Thin Film Diode
- e: Transcription Factor Database
- r: Transportforskningsdelegationen
TFDD e: Text File Device Driver
T1FE e: T1 carrier Front End
TFE d: Türfreisprecheinrichtung
- e: Tetrafluorethylene
- e: Thermionic Field Emission
TFEL e: Thin-Film Electroluminescent
Tfer s: Teleférico
TFG e: TM Final Product Generation
- e: Transportation Facility Guide
TFH d: Technische Fachhochschule Berlin
TFI e: Terminal Facility Identifier
- e: Tropical Forest Initiative
TFIIV e: Thick Film Integrated IV Ignition
TFIS e: Thai Aquatic Sciences and Fisheries Information System
- f: Système thaîlandais d'information sur les sciences aquatiques et la pêche
- s: Sistema Tailandés de Información sobre las Ciencias Acuáticas y la Pesca
TFIV e: Task Force IV
TFL e: Timing and Frequency Laboratory
TFLAP e: T carrier Fault Locating Applications Program
Tflops e: One teraflop equals 1 trillion floating point operations per second
TFM e: Tagged Font Metric
- e: TeX Font Metric
- e: Total Forest Management
TFMRDBS e: Total Force Manpower Requirements Data Base
TFMS e: Treasury File Management System
Tfn r: Telefon[era]
Tfo s: tráfico
TFODM e: Task Force on Observations and Data Management
TFOM f: Territoires Français Outre-Mer
TFOV e: Total Field Of View
TFP e: Therapeutic Feeding Program
- e: Time and Frequency Processing
- e: Tops Filing Protocol
- e: Trees on Farms Program
- f: Travaux Forcés à Perpétuité
Tfpl r: Transportflygplan
TFR r: Tekniska Forskningsråd
TFS e: Thomas-Fermi-Scott
- e: Three Finger Salute
- e: Translucent File System
- e: Trick Flow Specialties
- f: Timbres Fixes Spéciaux
TFSH N: Tungt Feltsykehus
TFSOI e: Thin-Film SOI
TFT e: Task File Table
- e: Temporary Full Time
- e: The Fantasy Trip
- e: Thin Film Transistor[s]
- e: Trifluorotoluene
TFTE e: Temporary Full Time Equivalent
TFTG e: Tactical Fighter Training Group
TFTHAOT e: Thanx For The Help Ahead Of Time [Internet]
TFTR e: Thermonuclear Fusion Tokamak Reactor
- e: Tokomak Fusion Test Reactor
TFTS e: Tactical Fighter Training Squadron
TFTT e: Thanks For The Thought
TFTW e: Tactical Fighter Training Wing
TFX e: Toxic Effects
TG d: Transportgenehmigung
- e: Thieves Guild
- e: Trace Gas
- e: Transmission Group
- e: Triacylglycerol
- f: Tarif Général
- f: Tarn-et-Garonne
- f: Titre Gratuit
- f: Trésorerie Générale
- s: Telégrafo
- s: Teniente General
- s: Tiempo Guerra
Tg e: Teragrams
- f: Télégramme
- f: Télégraphe
- r: Tempering
- r: Terräng
tg. a: temmelig godt
T&G e: Tongued and Grooved
TGA e: Therapeutic Goods Administration
- e: Thermal Gas Analysis
Tga r: Telegramadress
TGAC e: Technical Grade Active Constituent
TGAL e: Think Globally, Act Locally [Internet]
tg. am. avd. N: tung artilleriammunisjonsavdeling
Tg. art. batj. N: Tung artilleribataljon
TGB e: Temporary Guide Base
- e: Twisted Grain Boundary
Tgbil r: Terrängbil
TGBM e: Tide-gauge Bench Mark
- f: Niveau de référence de marégraphes
- s: Marca de referencia para los mareómetros
TGBS e: Third Generation Beam Former System
tg. btt. N: tung batteri
TGC e: Tender Gentle Care
- e: Terminal Group Controller
- e: Traffic Guidance Computer
TGCR e: Tactical Generic Cable Replacement
- e: Thermal Gross Cloud Restoral
TGCT e: Thermal Gross Cloud Test
TGD e: Transmission and General Dispatch
TGDDIS e: Trace Gas Dynamics Data Information System
TGE e: Thats Good Enough
Tgen f: Tarif général
t ges d: absoluter spektraler Gesamttransmissionsgrad
TGF N: Tekstilgrossistenes Forbund
- p: Tropas Guarda Fronteira
- s: Tropas Guardafronteras
Tgf r: Telegraf[era]
TGFA p: Tropas Guarda Fronteira de Angola
TGI e: Temperature Growth Index
TGIF e: Tactical Ground Intercept Facility
- e: Thank God Its Friday [Internet]
TGIF II e: Tactical Ground Intercept Facility
TGL d: bodennahe Lufttemperatur, global
TGM e: Timetable Generation Module
Tgmaskin r: Temperingsmaskin
TG/MS e: Thermal Gravimetric/Mass Spectrometry
TGN e: Transmission Group Number
Tgnyckel r: Temperingsnyckel
TGO e: Total Gross Output
TgOff f: Télégramme Officiel
TGOJ r: Trafikaktiebolaget Grägenberg-Oxelösunds Järnvägar
Tgr f: Télégramme
Tgrk r: Tung granatkastare
Tgrkkomp r: Tungt granatkastarkompani
T-gruppe N: Trengruppe
TGS e: Ticket Granting Server (or Service)
- e: TOPEX Ground Data System
- e: Triglyceride Sulfate
Tgsprint r: Temperingsprint
Tgstr r: Temperingsstreck
TGT e: TDRSS Ground Terminal
- e: Thin Gap Turbine
Tgt r: Tungt
TgV d: Transportgenehmigungsverordnung vom 10.9.1996
TGV e: Two Guys and a Vax
- f: Train[s] à Grande Vitesse
TGV-A f: Train[s] à Grande Vitesse Atlantique
TGV-PSE f: Train[s] à grande vitesse Paris-Sud-Est
TGV-R f: Train[s] à grande Réseau
TH a: Tekstet for Hørehæmmede
- d: Technische Hochschule oder Theologische Hochschule
- e: Theoretical Physics Division
- e: Transportation and Handling
- e: Trouble History
- f: Table d'Hôte
- f: Taxe d'Habitation
- f: Titre Honorifique
- N: Tekstet for Hørselssvekkede
- s: Tren Hospital
Th e: helper TL
- r: Tändhatt
- r: Termit huv
th. a: til højre
Th. e: Thornton's American Glossary
TH. e: Transmission header
.th e: Thailand [Internet]
t.h. N: til høyre
- r: till höger
T&H e: Thermal and Hydraulic
T/H e: Thermohydraulic
T[/H e: Track-and-Hold
Th-232 e: thorium-232
THA e: Tanzania Harbors Authority

THAAD e: Theater High Altitude Air Defense
THAALES e: Trajectory prediction Handling Airspace and Aircraft models Linked to Evaluation Software
THAIS e: Type Commanders, Headquarters Automated Information System
THANX e: Thanks [Internet]
THAP e: Tactical High Altitude Penetrator
THC e: Tetra-Hydro Cannabinol
- e: Total Hydrocarbon[s]
THCV e: Tetra-Hydro Cannabivarin
THD d: Technische Hochschule Darmstadt
THE e: Theologie und Religionspädagogik
Theatr. e: Theatre
- e: Theatrical
THENET e: Texas Higher Education Network
The NINE e: Senior Staff Advisory Committee
THEO e: System for predicting solar flare probabilities
Theol. e: Theological
- e: Theology
Theor. e: Sir Frank Packer Theoretical Physics Department, U. Sydney
TheorAngPh d: Institut für Theoretische und Angewandte Physik
Theor. Appl. Genet. e: Theoretical and Applied Genetics
TheorCh d: Institut für Theoretische Chemie
TheorElT d: Institut für Theorie der Elektrotechnik
Theoret. e: Theoretical
Theor Med e: Theoretical Medicine
TheorPh d: Institut für Theoretische Physik
Theor Popul Biol e: Theoretical Population Biology
THEP e: TOGA Heat Exchange Program[me]
- e: TOGA Heat Exchange Project
- f: Programme TOGA sur l'échange de chaleur
- s: Programa TOGA sobre Intercambio de Calor
Ther Drug Monit e: Therapeutic Drug Monitoring
Ther Hung l: Therapia Hungarica
Ther Immunol e: Therapeutic Immunology
THERMINIC e: Thermal Investigations of ICs and microstructures
ThermL d: Institut für Thermodynamik der Luft- und Raumfahrt
ThermStrö d: Gemeinsame Bibliothek des Instituts für Thermische Strömungsmaschinen und des Instituts für Luftfahrtantriebe
Ther Umsch d: Therapeutische Umschau
THETA e: Tunneling Hot Electron Transfer Amplifier
THF d: Theologische Fakultät
- e: Tetrahydrofuran
THGA e: Transverse-Heated Graphite Atomizer
THIC e: The Heart Institute for Children
THIEF e: This Here Isn't Even Fine [Internet]
THIR e: Temperature/Humidity Infrared Radiometer
THIRSPEC e: Thermal Infrared Spectrometer
THIS e: Transportable Hemispherical Illumination System
THITT f: Thalweg de la Haute Troposphère Tropicale
THK z: Turk Hava Kuvvetleri
THkp r: Tungt Helikopter
THL d: Thüringen Landesregierung und Landtag
THM e: Tuhmbnail
THMT e: Tactical High Mobility Terminal
THO N: Tuberkulosens Hjelpeorganisasjon
THOM e: Thompson Hall
THOMAS e: The [U.S.] House [of Representatives] Open Multimedia Access System
THOR e: Tandy High-intensity Optical Recording
- e: Tandy High-performance Optical Recording
- e: Thermal Optical Recording
Thorac Cardiovasc Surg e: Thoracic and Cardiovascular Surgeon
THOREX e: Thorium Extraction
THORP e: The Thermal Oxide Reprocessing Plant
T-HOST e: WWMCCS Transportable Host
THP e: Terminal to Host Protocol
ThR d: Theologische Rundschau
THR e: Transmit Holding Register
- r: Trumhinneruptur
THRIP e: Technological Human Resources for Industry Programme
Thromb Haemost e: Thrombosis and Haemostasis
Thromb Res e: Thrombosis Research
THRUSH e: Technological Hierarchy for the Removal of Undesireables and Subjugation of Humanity
THRUST e: Tsunami Hazard Reduction Using Systems Technology
- e: Tsunami Hazard Reduction Utilizing Systems Technology
- f: Réduction des risques dus aux tsunamis à l'aide de la technologie des systèmes
- s: Reducción de los Riesgos de Tsunamis mediante la Tecnología de Sistemas
THS e: Test Host Substitute
- e: The Hydrographic Society
- e: Thesaurus
THSP e: Theatre and Speech Communication
THT e: Token Holding Time[r]
THURSIS e: The Human Roles in Space
ThürWG d: Thüringisches Wassergesetz
Thv r: Torpedhantverkare
THWAPS e: Temperature, Humidity, Winds and Pressure System
THWLAIAS e: The Hour Was Late, And I Am Senile [Internet]
THX e: Thanks [Internet]
- e: Tom Holman Xover
- e: Tomlinson Holman[s] Experiment
ThZ d: Theologische Zeitschrift
THZ e: Tera Hertz
TI e: Temporary Intermittent
- e: Texas Instruments, Inc.
- e: Thermal Imager[y]
- e: Title
- e: Transaction Identifier
- e: Transport Independent
- e: Trials Installation
- f: Taux d'Impôts
- f: Taux d'Intérêt
- f: Taux d'Invalidité
- f: Taxe d'Importation
- f: Témoin Instrumentaire
- f: Trafic International
- f: Travail Insalubre
- f: Travailleur Intellectuel
- s: Territorio del Interior
ti. a: tirsdag
T/I e: TPFDD Interface
ti0 d: relativer spektraler innerer Transmissionsgrad des ersten Farbfilters für seine im Schott-Mainz-Katalog angegebenen Bezugsdicke D0
t 2i0 d: relativer spektraler innerer Transmissionsgrad des zweiten Farbfilters für seine im Schott-Mainz-Katalog angegebenen Bezugsdicke D0
Ti-208 e: Thallium-208
TIA e: Taxation Institute of Australia
- e: Telecommunications Industries (or Industry) Association[s]
- e: Telephone Information Access
- e: Temperature Inversion Analysis
- e: Territorial and International Affairs
- e: Thanks In Advance [Internet]
- e: The Internet Adapter (or Adaptor)
TIAA/CREF e: Teachers Insurance and Annuity Association/College Retirement Equity Fund
TIAC e: The Internet Access Company, Inc.
- e: Tourism Industry Association of Canada
TIADD e: Texas Instruments Automatic Data Display System
TIALD e: Thermal Imaging and Laser Designation
TIAP e: Theater Intelligence Architecture Program
TIAS e: Tokai Image Analysis System
TIATLG e: Truly, I Am The Living God [Internet]
TIB d: Technische Informationsbibliothek, Hannover
- e: Tag Information Base
- e: Target Intelligence Branch
- e: Technical Information Library
TIBA e: Triisobutlaluminum
TIBS e: Tactical Information Broadcast System
- e: Through Ice Bathymetry System
- e: Trends in Biochemical Sciences

TIC a: Teknologisk Informationscenter
- e: TDRSS Interface Channel
- e: Technical Information Coordinator
- e: Technical Interface Committee
- e: Tentatively Identified Compound[s]
- e: Term-Info Compiler
- e: Theater Intelligence Center
- e: Token Ring Interface Coupler
- e: Tongue In Cheek [Internet]
- e: Total Installed Cost
- e: Troops In Contact
- f: Tarif International à Coupons
- f: Taxe Intérieure de Consommation
- s: Trabajo de Investigación Científica

TICARRS e: Tactical Interim CMS and REMIS Reporting System
TICE e: Training Information and Communication Enhancement
TICH e: Total Identifiable Chlorinated Hydrocarbons
TICS e: Thermally Induced Copper Separation
t.i.d. e: three times a day
TID e: Target ID [Identifier]
- e: Technical Implementation Directive
- e: Total Ionizing Dose
- f: Transaction Inter-Divisionnaire

tid. r: tidigare
tida r: tid eller anslag
TIDCS e: Transitional Integrated DCS
TIDL e: Tool Integration Description Language
tidl. N: tidligere
TIDM s: Tribunal Internacional del Derecho del Mar
TIDPTE e: Technical Interface Design PlanTest Edition
tidr r: tidrör
TIDS e: Telemetry Interface Data System
Tidsskr Nor Laegeforen N: Tidsskrift for den Norske Laegeforening
TIDSV d: Telekommunikations- und Informationsdienstunternehmen-Datenschutzverordnung
TIDY e: Teletype Integrated Display System
TIE e: Telemedicine Information Exchange
- e: Total Ion Electropherogram
- e: Transmitter Interrupt
- e: Transponder Interface Equipment
- e: Treatment Improvement Exchange

Tie f: Tuilerie
Tierärztl Prax [Suppl] d: Tierärztliche Praxis [Supplement]
TIES e: Telecom Information Exchange Services, Genf
- e: Time Independent Escape Sequence

TIF e: Tactical Interface Facility
- e: Tag[ged] Image File [format]
- e: Technical Integration Forum
- e: Tumor Induction Factor
- f: Transports Internationaux Ferroviaires

TIFF e: Tag[ged] Image File Format
TIFF/IT e: Tag Image File Format/ Information Technology
TIG e: Tactical Intelligence Group
- e: Time In Grade
- e: Time until Ignition
- e: Tm Initial Product Generation
- e: Track Initiation and Grouping
- e: Transmission Identification Generator
- e: Tungsten Inert-Gas
- f: Société de Télécommunications Internationales Gabonaises

TIGA e: Texas Instruments Graphics Adapter
- e: Transport Issues Group Australia

T[-]IGBT e: Trench-IGBT [Insulating-Base Junction Transistor]
TIGER e: Terrestrial Initiative for/in[to] Global Environmental Research
- e: Thermal Infrared Ground Emission Radiometer
- e: Topologically Integrated Geographic Encoding [and] Reference (or Referencing)

TIGER/GICS e: TIGER/Geographic Identification Code Scheme
TIGER/GRF-N e: TIGER/Geographic Reference File-Names
TIGGER e: Earth sciences dimension to TIGER
TIGR e: The Institute for Genomic Research
- e: TOVS initial guess retrieval

TIGRE f: Traitement d'Image Génie Rurale Ecublens
TIGRIS e: Topologically Integrated Geographically Referenced Information System
TIH e: Trip Into Hole
TIIAP e: Telecommunications and Information Infrastructure Assistance Program
TIIPS e: TASCS Interactive Image Processing System
TIIS e: Thermal Infrared Imaging Spectrometer
t i0j d: innerer Transmissionsgrad des j-ten Farbfilterglases
Tijdschr Diergeneeskd n: Tijdschrift voor Diergeneeskunde
Tijdschr Gerontol Geriatr n: Tijdschrift voor Gerontologie en Geriatrie
TIKA r: Tidskrift för Kustartilleriet
TIL e: Technical Information Library
- e: Transportation Information Library

TiLE e: Trends and Technology in Leisure and Entertainment
tilf. a: tilfælde
- a: tilføjelse
- a: tilføjet

tilh. a: tilhørende
TILL e: The Institute for Liberian Languages
till. a: tillæg
- N: tillegg

TILLMA r: Tillsatsmaskinerisystem
tills. r: tillsammans
tillsv r: tillsvidare
tilltp r: tilltransport
tillv. r: tillverkad
- r: tillverkning

TILMC M: Timber Industry Labor Management Committee
TILO e: Technical and Industrial Liaison Office
TILS e: TMDE Inventory List System
tilsv. N: tilsvarende
TILT e: Teaching with Independent Learning Technologies
TIM e: Target Identification Markers
- e: TCP/IP Inverse Multiplexing Protocol
- e: Technical Information Management
- e: Technical Information Memo
- e: Technical Interchange Meeting
- e: Technical Interface Meeting
- e: Technical, Interactional and Managerial
- e: Two Intensity Measurements

tim r: timme
- r: timmerman

TIMACH s: Tijolb'al Mayab' Adrián Inés Chávez
TIME e: Tsunami Inundation Model[l]ing Exchange Project
- f: Projet d'échange sur la modélisation des inondations dues aux tsunamis
- s: Proyecto de Intercambio de Modelos de Inundaciones provocadas por Tsunamis

TIMED e: Thermosphere, Ionosphere, Mesosphere Energy and Dynamics
TIMEZONES AST e: Atlantic Standard Time
TIMI e: Technology Independent Machine Interface
TIMIT e: Time Interval
TIMO e: Tissue Inhibitor of Metalloproteinases
TIMS e: Text Information Management Systems
- e: Theater Information Management System
- e: Thermal Infrared Mapping Spectrometer
- e: Thermal Infrared Multispectral Scanner
- e: Thermal Ionization Mass Spectrometry
- e: Tuberculosis Information Management System

TIMSS e: Thermal Infrared Spectral Scanner
TiN e: Titanium Nitride
TIN e: Transportation Information Network
- e: Triangular Irregular Network

TINA e: Telecommunication Information Network Agent
- e: Telecommunication Information Network[ing] Architecture
- e: Telecommunications Intelligent Network Architecture
- e: Truth in Negotiation[s] Act

TINAC e: TINA Consortium
TINALO e: This Is Not A Legal Opinion [Internet]
TINAR e: This Is Not A Recommendation [Internet]
- e: This Is Not A Review [Internet]

TINLEND e: A ILL system related to TINLIB
TINLIB e: An integrated library system of British origin
TINS e: Thermal Imaging Navigation Set
TINWIM e: That Is Not What I Meant [Internet]
TINWIS e: That Is Not What I Said [Internet]

TIOP e: Tactical Interface Operational Procedures
tip e: pitch
TIP e: TAO Implementation Panel
- e: Target Impact Prediction
- e: Team Improvement Program
- e: Technology Integration Program
- e: Telecommunications Implementation Plan
- e: Template-Controlled Image Processor
- e: Terminal IMP
- e: The Industrial/Organizational Psychologist
- e: TIROS Information Processor
- e: Tool Interface Plate
- e: Total Incidence Pyranometer
- e: Transition and Integration Plan
- e: Transportation Improvement Program
- e: Transputer Image Processing
- f: Groupe de mise en oeuvre du TAO
- s: Equipo de Ejecución del TAO
TIPE e: TIAS Image Processing Executive
TIPER e: Tanganyika-Italian Petroleum Company
TIPI e: Tactical Image Processing Interpretation
TIPI-IIS e: Tactical Information Processing and Interpretation-Imagery Interpretation Segment
TIPIL e: Turn In Point Item List
TIPMEX e: Tibetan Plateau Meteorology Experiment
Tipp f: Tax intérieure sur les produits pétroliers
TIPPSA e: Technical Industrial Pharmacists and Pharmaceutical Scientists Association
TIPRO e: Texas Independent Producers and Royalty Association
TIPS e: Thematic Mapper Image Processing System
- e: Transjugular Intrahepatic Portosystemic Shunt
- e: Transportation Information Processing System
- e: Tutorial Image Processing System
tips e: external fuel tanks
TIPTOP e: The Internet Pilot To Physics
TIR d: Tirschenreuth, Bayern
- e: Thermal Imaging Radar
- e: Thermal Infrared [Radiometer]
- e: Total Indicated Runouts
- e: Total Indicator Runout
- e: Tracking and Illumination Radar
- e: Transportation Item Reporters
- f: Transports Internationaux Routiers
Tir s: Tirador
T-I-R e: Title I Report
TIRAS e: Task Identification and Resource Allocation System
TI-RCP e: Transport-Independent Remote Procedure Call
TIREC e: TIROS Ice Reconnaissance
TIREX e: TIROS Ice Experiment
TIRKS e: Trunk Integrated Record Keeping System
TIRL e: Telematique International Research Laboratories
TIRLS e: Thermal Infrared Linescan Imagery
TIRONET e: Trans-Ireland Optical Network
TIROS e: Television [and] Infrared Observation[al] Satellite[s]
- e: Television and Infrared Observation Satellite Sounder
- e: Television Infrared Operational System
- e: Topographic Infrared Operations Satellite
- f: Satellite d'Observation par Télévision Sensible à l'Infrarouge
- s: Satélite de Observación por Televisión Sensible al Infrarrojo
TIROS-N e: Advanced Television Infrared Observation Satellite
TIR/VIS e: Thermal Infrared and Visible Region Opto-Electronic Scanner
TIS e: TAC Information System
- e: TACOPS Information System
- e: Tactical Intelligence Squadron
- e: Tactical Intelligence Support
- e: Tate Integrated Systems
- e: Technical Inspection and Safety Commission
- e: Thematic Information Systems
- e: Timber Industry Strategy
- e: Tool Induced Shift
- e: Tools Interface Standard
- e: Total Induced Shift
- R: Tyazhely Istrebitel Soprovozdeniya
- r: Typinflygningsskola
TISA e: TAC Information Systems Architecture
TISCA e: Technical Information Storage and Control Application
TISD e: Tactical Information Systems Division
TISE e: Take It Somewhere Else [Internet]
TISGLOW e: TERI Information Service on Global Warming
TISI e: Thermal Infrared Spectral Indices
TISL e: Target Indicator System Laser
TISLL e: The Institute for Sierra Leonean Languages
TISS e: Tanzanian Intelligence and Security Service
- e: TEWS Intermediate Support System
- e: Transportation Information Subsystem
- e: Troop Issue Subsistence System
Tiss f: Tissage
TISSS e: Tester Independent Support Software System
Tissue Cell e: Tissue and Cell
TISURS e: Trainer Inventory Status Utilization Reporting System
TIT e: Theater Indoctrination Trainingfamiliarization
- e: Treponema-pallidum Immobilisation Test
- f: Télécommunications Internationales du Tchad
- f: Titre
Tit f: Titulaire
TITAN f: Terminal Interactif par Téléphone et Appel par Numérotation
TITC e: TSCA Interagency Testing Committee
Title II-A e: U.S. Department of Education Access to Research Library Collections Program
Title II-C e: U.S. Department of Education Research and Demonstration Project
TITOS e: Television Infrared Orbital Satellite
TIU e: Transponder Interface Unit
- e: Trusted Interface Unit
TIV e: Torus Interface Valve
TIVPC e: Transitional Immigrant Visa Processing Center
TIWADS e: TIP Interface with Automated Data System
TIWE e: Tropical Instability Wave Experiment
TIWG e: Test Integration Working Group
TIWU e: Tea Industry Workers Union
TJ e: Tadjikistan
- f: Tarif Journalier
- r: Tungt Jaktflygplan
t j d: spektraler innerer Transmissionsgrad
tj d: Bezugstransmissionskoeffizient der i-ten Schicht bei der Dicke D
- N: tjeneste
Tj r: Tjänst
TJC e: Tandem Junction Cell
TJF r: Tjänsteföreskrift
Tjf r: Tjänsteförrättande
Tjg r: Tjänstgöra[nde]
tjg. a: tjenstgørende
tjgj N: tjenestegjørende
Tjgr r: Tjänstegrensskola
t ji0 d: innerer spektraler Transmissionsgrad des j-ten Farbfilters für die entsprechende Bezugsdicke Dj0
TJI e: This Just In
Tjkl r: Tjänsteklass
tjm. r: tjänsteman
TjRk r: Tjänstereglemente för Krigsmakten
Tjs f: Toujours
TK d: Titelkommission
- e: Tokelau
- e: Tool Kit
- r: Trafikkommissionen
- r: Tystnadspliktskomittén
Tk r: Tändkolv
- r: Tjänstemeddelanden för krigsmakten
TK2 e: Toolkit 2 delivery
TK3 e: Toolkit 3 delivery
TK4 e: Toolkit 4 delivery
TK5 e: Toolkit 5 delivery
TKA R: Torpedniy Kater
TKG d: Telekommunikationsgesetz vom 25.7.1996
- e: Terminal Key Generators
TKL d: Tierklinik
TKLGebV d: Telekommunikations-Lizenzgebührenverordnung vom 28.7.1997
TKO e: Technical Knock Out
TKR e: Tons/Kilometers
TKSC e: Tsukuba Space Center
TK/TK e: Track to Track
TKV d: Telekommunikations-Kundenschutzverordnung
TKZulV d: Telekommunikations-zulassungsverordnung

TL e: Industry and Technology Annex
- e: T-lymphocyte
- e: Technical Letter
- e: Technical Liaison
- e: Thermal Lensing
- e: Thermoluminescence
- e: Thermoluminescent
- e: Total Loss
- e: Transit Level
- e: Transport Layer
- e: Tree Locking
- f: Taxe de Luxe
- f: Taxe Locale
- f: Télégramme-Lettre
- f: Temps Légal
- f: Temps Local
- f: Titre Légal
- f: Traduction Libre
- f: Tribunal des Loyers
- r: Trafikledare
Tl r: Tjänstemeddelanden rörande landförsvaret
TL1 e: Transaction Language 1
TL2 e: Taxonomic Literature, edition 2
Tl-1223 e: Thalium Barium Calcium Copper Oxide
TLA e: Three Letter Abreviation
- e: Three Letter Acronym
- f: Taxe Locale Additionnelle
TLAN e: Transportable Payload Operations Control Center LAN
TLAP e: TokenTalk Link Access Protocol
TLB e: Table
- e: Translation Look Aside Buffer
TLC e: Technology, Literacy and Culture
- e: Telecommunications Line Center
- e: Telecommunications Live Controller
- e: The Laser Center
- e: The Learning Channel
- e: Thin Layer Chromatography
- s: Tratado de Libre Comercio
TLCC e: Theater Logistics Coordination Center
TLCF e: Team Leader Computing Facility
- e: Teleconferencing
TLCs e: Transitional Local Councils
TLCSC e: Top-Level Computer Software Component
TLD e: Thermoluminescent Dosimetry Data Base
- e: Top Level Domain
- e: Toxics Litigation Division
Tldir s: Teledirigido
TLE e: Template Extension Language
- e: Tool Loading Elevator
- e: Treaty-Limited Equipment
TLeLv F: Tiedustelulentolaivue
Tles H f: Toutes les Heures
Tles J f: Tous les Jours
TLEV e: Transitional Low-Emissions Vehicle
TLF e: Temporary Lodging Facilities
- e: Two-Level Fluctuations
- N: Tobakkhandlernes Landsforbund
- N: Troms Landforsvar
- s: Tren de Laminación en Frío
tlf. a: telefon
TLG e: The Living God [Internet]
Tlg f: Télégramme
- N: Telegraf
- s: Telegrama
tlg.adr. a: telegramadresse
Tlg. kp. L N: det første Telegrafkompani 1
tlgr. a: telegram
TLH e: Transport Layer Header
- f: Tréfileries et Laminoirs du Havre
TLI e: Thin-Layer Imaging
- e: Transfer Layer Interface
- e: Transfer Library Interface
- e: Transport Layer Interface
- e: Transport Level Interface
- e: Transport Library Interface
- e: Trypsin-Like Immunoreactivity
TLIB e: The Teaching Library
TLJ e: Torts Law Journal
TLKM f: Télégramme-Lettre Mandat
TLL l: Thesaurus Linguae Latinae
T/LLM e: Temperature/Liquid Level Monitor
TLM e: Transition Line Model
- e: Transmission Length Model
- e: Transmission Line Method
- e: Transmission Line Model
- e: Two-Layer Method
TLMP e: Tongass Land Management Plan
TLNKG d: Teilnehmerkennung
Tloc f: Taxe locale
TLP e: Test Level Point
TLQ e: Top Letter Quality
TLR e: Telecommunications Leasing Request
TLRC e: The Lunar Resources Company
TLRN e: Technical Logistics Reference Network
TLRS e: Transportable Laser Ranging System
TLS e: Thailand Landsat Station
- e: Transparent LAN Service
- e: Transport Layer Security
- e: Two-Level Store
TLSC e: The Llama Steering Committee
Tlse f: Toulouse
T/LSEs e: Transects and Large-Scale Land Surface Experiments
TLS-tjänst r: Tätnings-, Länsnings- och Stöttningstjänst
TLT e: Temperature-Logging Tool
Tlt s: tubo lanzatorpedos
TLTN e: Two Layer Thin Tape
TLTP e: Teaching and Learning Technology Programme
- e: Trunk Line and Test Panel
TLTSU e: Teaching and Learning Technology Support Unit
TLU e: Table Look-Up
TLV e: Type, Length, Value
TLV-TWA e: Threshold Limit Value-Time Weighted Average
TLX e: Telex [number]
TM e: Task Memory
- e: Tasking Memorandum
- e: Test, Measurement and Diagnostic Equipment
- e: Thread Mangler
- e: Trace Metal
- e: Transaction Monitor
- e: Transmission Mode
- e: Transport Module
- e: Trimethyl
- f: Front des Travailleurs
- f: Instrument de cartographie Thématique
- f: Table des Matières
- f: Taxe de Mainmorte
- f: Taxe Municipale
- f: Télégramme Multiple
- f: Territoire Métropolitain
- f: Transport de Marchandises
- f: Transports Maritimes
- f: Travail Malsain
- f: Travail Manuel
- r: Tygmästare
- s: Instrumento de cartografía Temática
- s: Transporte Militar
- s: Tropas de Montaña
Tm e: Thul[l]ium
- r: Torpedmästare
T&M e: Time and Materials
T-M e: Time-to-Market
TM2 e: Tracer Model 2
TMA e: Technology Modeling Associates
- e: Terminal Area
- e: Thermal Mechanical Analyzer
- e: Training Mission Area
- e: Tycho Magnetic Anomaly
Tma s: Termonuclear
TMAA e: TriMethylAmine Alane
TMAC e: Time Accounting System
- e: Treasury departments Multiuser Acquisition Contract
TMACS e: International Mathematics and Computer Simulations
TMAl e: Trimethyl Aluminum
TMAP e: Thermal Modeling and Analysis Project
TMARP e: Tanker Mating and Ranging Program
TMAS e: Transaction Monitoring and Analysis System
TMAX e: Maximum TTRT
Tmax f: Tarif maximum
TMB e: Trimethyborate
TMBPS e: Thematic Mapper Bulk Processing System
TMC e: Technical Management Committee
- e: Thinking Machine[s] Corporation
- e: Tourism Ministers Council
- e: Traffic Message Channel
- e: Transfer Module Controller
- e: Transport Module Controller
- f: Tribunal Maritime Commercial
TMCA e: Transportation Management Control Agency
TMCP e: Theater Munitions Control Point
TMCS e: Tactical Maintenance Control System
- e: Telecommunications Management and Control Subsystems

TMCs e: Transitional Metropolitan Councils
TMD e: Theater Missile Defense
- e: Thematic Mapper Derivative
- e: Time Division Multiplexing
TMDB e: Tivoli Management Data Base
TMDE e: Test Measurement and Diagnostic Equipment
TMDF e: Trunk Main Distributing Frame
TMDS e: Theater Missile Defense System
TME e: Tivoli Management Environment
- e: Tropical Mountain Ecosystems
TMF e: Time Management Facility
- e: Tivoli Management Framework
TMG e: Tactical Multinet Gateway
- e: Trimethyl Gallium
- s: Tiempo Medio de Greenwich
TMGR s: Tabla de Material de Guerra Reglamentaria
TMH e: Tasmania Museum and Art Gallery, Tasmania
- e: Tiermedizinische Hochschule
TMI e: Thai Medical Informatics Society
- e: Three Mile Island [reactor site] Middletown, Pennsylvania
- e: TRMM Microwave Imager
TMICS e: Transportation Management Information Control System
T-MIL e: Transportation-Mini Information Library
TM-IMAGERY e: Thematic Mapper imagery
TMIn e: Trimethyl Indium
Tmin f: Tarif minimum
TMI RAD e: Three Mile Island Environmental Radiation
TMIS e: Technical Management Information System
- e: Telecommunications Management Information System
TML e: Template
TMLF N: Telegrafmennenes Landsforbund
TMM e: Museum of Theater and Music
- e: Thermoset Microwave Material
- e: Total Materials Management
- e: Transfer Matrix Method
TMMS e: Telephone Message Management System
TMN e: Telecom Managed Network
- e: Telecommunication[s] Management Network
- e: The Movie Network
- e: True Mach Number
TMO e: Telecommunications Managed Object
- e: Telephone Money Order
- e: Theater of Military Operations
- e: Thermal Magnetic Optical
- e: Transportation Movement Office
Tmob e: 2,4,6-Trimethoxybenzyl
TMOD e: Treatment and Management of Ocular Disease
T-MOP e: Transitional Meteosat Operational Programme Satellite
TMOS e: Transportation Management/ Optempo System
TMP e: TAC Mission Processor
- e: Technical Management Plan
- e: Test Management Protocol
- e: Texts and Monographs in Physics
- e: The Morrow Project
- e: Thermo Mechanical Pulp
- e: Transmembrane Pressure
- e: Trimethylphosphate
TMPC e: Theater Mission Planning Center
TMR a: Tillidsmandsrådet
- e: The Medical Reporter
- e: Thrust Measuring Rig
- e: TOPEX/Poseidon Microwave Radiometer
- e: Transient Memory Record
- e: Triple Modular Redundancy
- p: Tribunal Militar Revolucionário
- r: Transportabel Markradiostation
TMRA s: Taller Móvil de Reparación de Automóviles
TMRC e: Tech Model Railroad Club
- e: Transportation Modeling Research Center
- e: Tropical Medicine Research Center
- e: Tropical Medicine Research Cooperation
TMRI e: Topics in Magnetic Resonance Imaging
- s: Taller Móvil de Reparación Ingeniera
TMRP e: Tropical Meteorology Research Programme
TMRS e: Traffic Measurement and Recording System
- e: Traffic Metering Remote System
TMRT s: Taller Móvil de Reparación de Tanques
TMR/TL e: Table of Manpower Requirements/Troops List System
TMS e: FMD Travel Management System
- e: Time-Multiplexed Switch[ing]
- e: Trace Mineral Salt
- e: Transport Management System
- e: Truth Maintenance System
- e: Typical Shape Functions, Maximum A Posteriori Simultaneous
- N: Torpedo/Mineskolen
TMSC e: Tape Mass Storage Control
TMSI e: Temporary Mobile Subscriber Identity
TMSL e: Test and Measurement Systems Language
TMT e: TRAMCON Master Terminal
TMTS e: Thematic Mapper Transcription System
T-MUX e: Terminal Multiplexer
TmV N: Trondhjems Mekaniske Verksted
TMVC e: Travellers Medical and Vaccination Centre
TMx f: Télégramme Multiple
TMXP e: Trace Meta-explanation Pattern
TN e: Technical Needs
- e: TelNet
- e: Transaction Number
- e: Transit Node
- e: Trigeminal Neuralgia
- e: Twisted Nematic
- f: Tarif Normal
- f: Titre Nominatif
- f: Travailleur de Nuit
Tn s: Tren
t n d: spektraler Transmissionsgrad des n-ten Farbfilters
.tn e: Tunisia [Internet]
t.n. r: till natten
T/N r: Teknisk/Naturvetenskaplig
TNA d: Temex-Netzabschluß
- e: Technology Needs Assessment
- e: Tennessee Nurses Association
- e: Texas Nurses Association
- e: Transitional National Assembly
TNAS e: TELCON Network Administration System
TNBS e: Trinitrobenzene Sulfonic acid
TNC e: Threaded Navy Connector
- e: Threaded Neill Concelman
- e: Trade Negotiating (or Negotiations) Committee
- e: Transnational Corporations
- e: Trunk Node Control
- r: Tekniska Nomemklaturcentralen
TNE e: Transient Nematic Effect
TNEF e: Transportation-Neutral Encapsulation Format
TNF e: Tumor Necrosis F-r
TNF-ƒ¿ e: Tumor Necrosis Factor-alpha
TNFC3 e: Tactical Nuclear Force Command, Control and Communications
TNFEM e: Twisted Nematic Field Effective Material
TNFIP e: Theater Nuclear Forces Improvement Plan
TNG e: The Next Generation
TNH d: bodennahe Lufttemperatur, nordhemisphärisch
- e: Temperate Northern Hemisphere
TNHD e: The New Hackerss Dictionary
t ni0 d: relativer spektraler innerer Transmissionsgrad des n-ten Farbfilters für seine im Schott-Mainz-Katalog angegebenen Bezugsdicke
TNIP e: Transkei National Independence Party
TNL e: Target Nomination List
TNMCS e: Total Not Mission Capable-Supply
TNMOC e: Total Nonmethane Organic Carbon Content
TNN e: Transport Network Node
- e: Trunk Network Number
TNO e: Netherlands Organisation for Applied Scientific Research
- e: The Netherlands Organization
- e: The Network Observer
- n: Nederlandse Organisatie voor Toegepast-Natuurwetenschappelijk Onderzoek
TNO-FEL e: Netherlands Organization for Applied Scientific Research-Physics and Electronics Laboratory
TNOP e: Total Network Operation Plan
TNOTVS e: Theres Nothing on TV So
TNP f: Traité de Non-Prolifération
- f: Traité de Non-Prolifération [nucléaire]
TNPAL e: W-trinitrophenylaminolauric acid
TNPC e: Taiwan[ese] New PC Consortium
- e: Traffic Network Planning Center
TNPP e: Transkei National Progressive Party

TNRSP e: Thailand National Remote Sensing Programme
TNRSWG e: Turkish National Remote Sensing Working Group
TNS e: Transcultural Nursing Society
Tns f: Tunisien
TNSS e: Region 7 TOSCA Neutral Selection
TNSTAAFL e: There's No Such Thing As A Free Lunch [Internet]
TNT d: Trinitrotoluene
- e: 2,4,6-Trinitrotoluol
- e: Telephone Nursing Telezine
- e: The News Toolkit
- e: Trinitro Toluene
TNV d: Technische Nachverbrennung
- e: Total Network Visibility
TNW e: Theater Nuclear War
- e: Theater Nuclear Weapons
TNWS e: Tactical Nuclear Weapons
TNX e: Thanks [Internet]
TNX 1.0E6 e: Thanks a million [Internet]
TO A: Transvaalse Onderwysvereniging
- e: Task Order
- e: Technical Officer
- e: Telecommunication Organisation
- e: Test Object
- e: Thin-Outline
- e: Timeout
- e: Timestamp Ordering
- e: Transistor Outline [package]
- e: Transverse Optic[al]
- f: Table d'Orientation
- f: Taille Ordinaire
- f: Télégramme Officiel
- f: Télégramme Ouvert
- f: Témoin Oculaire
- f: Test d'Orientation
- f: Testament holographe
- f: Tiers Opposant
- f: Titre Onéreux
- f: Tous Objets
- f: Train Omnibus
- f: Travail Obligatoire
- r: Torpedofficer
- s: Teatro de Operaciones
To r: Torpedteknisk
- r: Tropp
to. a: torsdag
TÖ r: Taktisk Övning
T&O e: Testing and Operations
T/O e: Take Off
TOA e: Table of Allowance
- e: Take-Off Angle
- e: Task Order
- e: Terre Ocean Atmosphere
- e: Total Obligation Authority
- e: Trace Organic Analysis
- e: Transfer Of Authority
- e: Transfer of Operational Authority
- e: Transportation Operating Agency
- e: Tropical Ocean-Atmosphere
- e: Type Of Address
- s: Transporte Oruga Acorazado
TO A d: Tarifordnung für Angestellte
TOAA e: Trawler Owner's Association of Australia
Toa Amb s: Transporte oruga acorazado Ambulancia
TOADS e: Transportable Optical Atmospheric Data System
TOAN e: Tropical Ocean-Atmosphere Newsletter
Toa Pers s: Transporte oruga acorazado de Personal
TOB e: Thrown Overboard
- s: Transporte Oruga Blindado
TO B d: Tarifordnung für Lohnempfänger des öffentlichen Dienstes
TOBAL e: There Oughta Be A Law [Internet]
TOBG e: This Oughta Be Good [Internet]
TOBIAS e: Integrated on-line library system
TOC e: Table Of Content
- e: Tactical Operations Center
- e: TAF OPS Center
- e: Total Organic Carbon
- e: Total Organic Compound
- e: Total Oxidizable Carbon
ToC e: Table of Contents
TOCA e: Tactical Operations Center Automated System
TOCE e: Tonne Of Carbon Equivalent
Toch r: Troppchef
TOCl e: Total Organic Halogens
TOCS e: Tropical Ocean Climate Study
TOCSE e: Tactical Operations Center Support Element
TOCSEEG e: Tradeoffs Among Competing Social, Environmental and Economic Goals
TOCS-IN e: Tables Of Contents of journals of Interest to classicists
TOD e: Time Of Day
- e: TIROS Oceanographic Drifter
TODR e: Take-Off Distance Required
TO&E e: Table of Organization and Equipment
TOE e: Tons of Oil Equivalent
- e: Truncated driven nuclear Overhauser Enhancement
- e: Type Of Effort
TOEM e: Technical Original Equipment Manufacturer
TOF e: Task Oriented Forms
- e: Theoretical Orthogonal Function
- e: Time Of File
- e: Time Of Flight
- e: Time-On Factor
TOFAC e: Technical Office Automation and Communication System
TOFI e: Time-Of-Flight Isochronous
TOF[/]MS e: Time Of Flight Mass Spectrometry
TOF SIMS e: Time Of Flight Secondary Ion Mass Spectrometry
TOG e: Total Organic Gases
- e: Transactions On Graphics
TOGA e: Tropical Ocean[s] [and] Global Atmosphere [program[me]]
- e: Tropical Ocean[s] Global Atmosphere experiment
- f: Les océans tropicaux et l'atmosphère globale
- s: Los Océanos Tropicales y la Atmósfera Mundial
TOGA[-]COARE e: TOGA [Tropical Ocean Global Atmosphere]-Coupled Ocean Atmosphere Response Experiment
TOGA NEG e: TOGA Numerical Experiment Group
TOGA[-]NEG e: TOGA [Tropical Ocean Global Atmosphere]-Numerical Experimentation Group
- f: Groupe de TOGA chargé de l'expérimentation numérique
- s: Grupo de TOGA sobre Experimentación Numérica
TOH e: Transmission Overhead
- e: Transport Overhead
- e: Trip Out Of Hole
Tohoku J Exp Med e: Tohoku Journal of Experimental Medicine
TOI e: Pacific Oceanological Institute, Far East Department of the USSR Academy of Sciences [Vladivostok]
- e: Track Of Interest
TØI N: Transportøkonomisk Institutt
TOJ e: Track On Jam
TOK e: Technological Development Credit
Tokai J Exp Clin Med e: Tokai Journal of Experimental and Clinical Medicine
Tokh. e: Tokharian
TOKTEN e: Transfer of Knowledge through Expatriate Nationals
- f: Transfer des connaissances par l'intermédiaire des nationaux expatriés
- s: Transferencia de Conocimientos por intermedio de Profesionales Expatriados
Tokushima J Exp Med e: Tokushima Journal of Experimental Medicine
TOL e: Tactical Operating Location
- e: Temporary Occupation Licence
- e: Type Organization List
Tol f: Toulouse
TÖL d: Bad Tölz-Wolfratshausen, Bayern
TOLD e: Take Off and Landing Data
TOLIMAC e: Total Library Management Concept
TOLIMO e: Togoland Liberation Movement
TOLS e: TERMS-On-Line System
TOM e: Temperature Oscillating Method
- e: Theory Of Mind
- e: Triple Output Module
- f: Territoire[s] d'Outre-Mer
- s: Teatro de Operaciones Militares
Tom f: Tome
t.o.m. N: til og med
- r: till och med
TOMARR e: Total Manpower Requirements and Resources
Tomb f: Tombeau
TOMCAT e: Toulouse Off-Line Model of Chemistry and Transport
TOME e: Task, Operator, Machine, Environment
TOMIS e: Task Oriented Multipurpose Information System

TOMS e: Total Ozone Mapping [experiment] Spectro[photo]meter
- e: Total Ozone Measuring Satellite
- e: Transactions on Mathematical Software
- f: Spectromètre pour la cartographie de l'Ozone Total
- s: Espectrógrafo para la cartografía del Ozono Total

TOMS-EP e: Total Ozone Mapping Spectrometer-Earth Probe
TOMUIS e: 3-D Ozone Mapping with Ultraviolet Imaging Spectrometer
TON e: Total Organic Nitrogen
- e: Transactions on Networking

T-Online e: Telekom Online-Dienst
TOOIS e: Transactions on Office Information Systems
TOP a: Træbranchens Oplysningsråd
- e: GCOS/GTOS Terrestrial Observation Panel
- e: Tactical Operations Planner
- e: Technical [and] Office Protocol
- e: Technical Operations Procedure
- e: Thunderstorm Overflight Program
- e: Time of Possession
- e: Time of Preparation
- e: Time-sensitive Operation Planning Procedures
- e: Topeka Weather Forecast Office
- f: Technicien de l'Organisation Professionnelle
- f: Test d'Orientation Professionnelle

Top s: Topografía
top. N: topografisk
TO&P e: Technical Objectives and Plan
TOPAS e: Terra Observation Pattern Analysis System[-Next Generation CCRS Analysis System]
TOPC e: Terrestrial Observation Panel for Climate
TOPEX e: Ocean Topography Experiment POSEIDON
- e: The Ocean Topography Experiment
- e: Topography Experiment [satellite]
- e: Topography Ocean Experiment
- e: Typhoon Operation Experiment
- f: Expérience Opérationnelle sur les Typhons
- f: Expérience relative à la Topographie de l'Océan
- s: Experimento de Topografía Oceánica
- s: Experimento Operativo sobre Tifones

TOPEX[/]POSEIDON e: Topography Experiment/Positioning Ocean Sea Earth and Ice Dynamics and Orbiting Navigator
TOPFET e: Temperature- and Overload-Protected FET
TOPIAS e: Topographic Information and Archiving Software
TOPICS e: Total On Line Program and Information Control System
TOPINT e: Technical Operational Intelligence
TOPIS e: South Australia's digital topographic database
TOPLAS e: Transactions on Programming Languages and Systems
Top Magn Reson Imaging e: Topics in Magnetic Resonance Imaging
TOPMODEL e: Topographic Model
TOPMS e: Telemarketing Operation Performance Monitoring System
TOPO e: Tri-n-Octylphosphine Oxide
TOPOG e: CSIRO land-wetness model
topogr r: topografisk
TOPOGULF e: Effects of the Mid-Atlantic Ridge upon the General Circulation and its Transients at Mid-Latitudes
- f: Effets de la dorsale médio-Atlantique sur la circulation générale et ses phénomènes transitoires aux latitudes moyennes
- s: Efectos de la Dorsal Mesoatlántica en la Circulación General y sus Efectos Transitorios a Latitudes Medias

TOPO-250K e: 1:250,000 Topographic Map Series
Topo Svy e: Topographical Survey of Canada
TOPREP e: Time-sensitive Operation Planning and Reporting
TOPS e: Tera-Operations Per Second
- e: Timesharing Operating System
- e: Total Ocean Profiling System
- e: Toward Other Planetary Systems
- e: Transcendental Operating System

TOPSAR e: Topographic Synthetic Aperture Radar
TOPSEC e: Top Secret
TOPSY e: Toxics and Pesticides Management System
TOQ e: Technical Operations Questionnaire
TOR e: Tactical Operation Room
- e: Target Partner Organization
- e: Terminal Owning Region
- e: Top Of Range

Tor. e: Toronto
t.o.r. r: tur och retur
TORCH e: Toxoplasmosis, Other, Rubella, Cytomegalovirus and Herpes test
TOR DB e: TOR-Datenbank
TORES e: Text Oriented Editing System
TORMS e: TAC Operations Resources Management System
torp r: torped
Torpbef r: Torpedbefäl
Torp I r: Torpedingenjör
torr e: pressure in millimeters of mercury with respect to vacuum
TORR e: Take-Off Run Required
Tort LR e: Tort Law Review
TOS e: Tactical Operations Squadron
- e: Technology Option Study
- e: The Operating System
- e: TIROS Observational System
- e: Tramiel[s] Operating System
- e: Transfer Orbit Space
- e: Transfer Orbit Stage
- e: Type Of Service
- r: Trängofficersskolan

TOSCANE f: Travaux d'Oceanographie Spatiale Capteurs Actifs dans l'Atlantique Nord-Est
TOSM e: Tenant Operations Systems Manager
TOSP e: Tailored Ocean Surveillance Product
TOSPICS e: Toshiba Pattern Information Cognitive System
TOT e: Time of Transmission
- e: Time Of Travel
- e: Turbo Object Toolkit

TOT BLOCK e: Time Over Target Block Time
TOTE e: Tropical Ozone Transport Experiment
TOTIDO e: Turn on, Tune In, Drop out
TOTS e: Transportable Orbital Tracking Station
TOU d: Technologieorientierte Unternehmensgründungen
- e: Transmission Operating Unit

Touch-Screen d: Berührungs-empflindlicher Bildschirm
TØUT N: Taktiske Øvelser Uten Tropper
TOV A: Transvaalse Onderwysersvereniging
- e: Teleoperated Vehicle

TOVS e: TIROS N Operational Vertical Sounder
- f: Sondeur de profils verticaux de TIROS opérationnel
- s: Sonda vertical operacional para TIROS

TOW e: Tube launched, Optically tracked [and] Wire guided [[anti-tank] missile]
TOWARD e: Tower Ocean Wave and Radar Dependence
TOWG e: Technical Order Working Group
TOW missile e: Tube-launched, Optically tracked, Wire-guided missile
TOWSLEY e: Towsley Center
TOX e: Molecular and Cellular Toxicology
- e: Tetradichloroxylene
- e: Total Organic Halides
- e: Total Organic Halogens
- e: Toxaphene

TOXFLO e: Urban Wastewater Toxics Flow Model
Toxicol Appl Pharmacol e: Toxicology and Applied Pharmacology
Toxicol Ind Health e: Toxicology and Industrial Health
Toxicol Lett e: Toxicology Letters
Toxicol Pathol e: Toxicologic Pathology
TOXLIT e: Toxicology Literature
toxo e: Toxoplasmosis
TOY e: Time Of Year
TOYA e: Ten Outstanding Young Airmen
TOYM e: Turn On Your Mind [Internet]
Tøym N: Tøymester
TP d: Teilprüfungsleistung
- d: Tiefpass
- e: Telecommunications Processing
- e: Teleprocessing [monitor]
- e: Test Purpose
- e: Timber Products inspection
- e: Transaction Process[ing] [system]
- e: Transaction Processor
- e: Transition Plan
- e: Transport Protocol
- e: Tropospheric Profiler
- e: Turbo PASCAL
- e: Twisted Pair
- f: Tarif Préférentiel
- f: Taux des Prêts
- f: Taxe à la Production
- f: Taxe des Prestations

- f: Taxe Piscicole
- f: Taxe Postale
- f: Taxe Proportionnelle
- f: Temps Probable
- f: Temps Prohibé
- f: Terres Polaires
- f: Testament Public
- f: Tierce Personne
- f: Tiers Payant
- f: Tiers Provisionnel
- f: Timbre Proportionnel
- f: Timbre-Poste
- f: Titre au Porteur
- f: Titre de Propriété
- f: Titre Posthume
- f: Titre Provisoire
- f: Total du Passif
- f: Tout Paris
- f: Traité de Paix
- f: Transport Public
- f: Transports Parisiens
- f: Travail par Pièces
- f: Travaux [forcés] à Perpétuité
- f: Travaux Pratiques
- f: Travaux Publics
- f: Travaux Publiques
- f: Trésor Public
- f: Tribunal des Pensions
- f: Trop Perçu
- N: Trafikkpost
- s: Telegrama Postal
- s: Tiempo Paz

Tp f: Tout payé
- N: Torpedo
- r: Transport[era]
- s: Transporte

TP0 e: [OSI] Transport Protocol Class 0
TP4 e: [OSI] Transport Protocol Class 4
TPA e: 12-O-Tetradecanoyl-phorbol-13-acetate
- e: Target Planning Aide
- e: Target Prioritization Aid
- e: Terminal Processor Assembly
- e: Trade Practices Act
- e: Trading Partner Agreement
- e: Tri-Party Agreement
- e: Two-Phonon Absorption
- f: Taxe Postale d'Affranchissement

tPA e: tissue Plasminogen Activator
Tpa s: Tropa
TPAD e: Terminal PAD [Packet Assembler/Disassembler]
TPAR e: Transnational Project Authorization Request
TPAS e: Term Paper Advisory Service
TPAU e: Twisted Pair Attachment Unit
Tpb r: Transportbåt
TPBAR e: Tritium-Producing Burnable Absorber Rod
tpb[m] r: transportbåt [mindre]
tpb[s] r: transportbåt [större]
TPC e: Testing Priorities Committee
- e: The Phone Company
- e: Trade Practices Commission
- e: Transaction [Processing] [Performance] Council
- e: Tropical Prediction Center
- f: Tribunal de Police Correctionnelle

TPCE e: Tank Pressure Control Experiment
TPD e: Technical Product Documentation
- e: Temperature Program Desorption
- e: Terminal Protection Device
- e: Threat Planning Document
- e: Toxics and Pesticide Division
- f: Tribunal Populaire de District

TPDC e: Tanzanian Petroleum Development Corporation
TPDDI e: Twisted Pair Distributed Data Interface
TPDF e: Tanzania Peoples Defense Forces
TPDFS e: Teampack Direction Finding System
TPDMP e: Training Program Development Master Plan
TPDP e: Tanzanian Peoples Democratic Party
TPE e: Tracking and Pointing Experiments
- e: Twisted Pair Ethernet
- f: Travaux Publics de l'Etat

TPEB e: Technical Proposal Evaluation Board
TPEG e: Technical Planning and Evaluation Group
TPE-O e: Thermoplastic Elastomer-Olefinic
TPE-S e: Thermoplastic Elastomer-Styrenic
TPEW e: Technology Panel For Electronic Warfare
TPF e: Tanzanian Peoples Front
- e: Terminal Presentation Facility
- e: Transaction[s] Processing Facility
- e: Transmission Priority Field

TPFCC e: Trans-Pacific Fisheries Consultative Committee
TPFDD e: Time-Phased Force and Deployment Data
TPFDD maintenance e: Time-Phased Force and Deployment Data Maintenance
TPFDD refinement e: Time-Phased Force and Deployment Data Refinement
TPFDL e: Time-Phased Force and Deployment List
TPFE e: Tibetan Plateau Field Experiment
TPFO e: TOPEX/Poseidon Follow-On
tpfpl r: transportflygplan
TPG e: TEREC Ground Processor
- e: Test-Pattern Generation

TPGF p: Tropas Populares de Guarda-Fronteiras
TPH e: Transport Packet Header
- e: Tropical Public Health

tphkp r: transport helikopter
TPI e: Technical Proposal Instructions
- e: Tritium Pellet Injector
- e: Tuned Port Induction
- e: Twisted-Pair interface
- f: Tribunal de Première Instance

TPICK e: Telecommunication Plan for Improvement of Command in Korea
TPIN e: Trading Partner Identification Number
TPIP e: Time Phased Implementation Plan
tpklass r: transportklass
TPL f: Tribunal Populaire Local
tpl r: timplan
- r: transportledare

TPLG d: Gesetz über die Spende, Entnahme und Übertragung von Organen [Transplantationsgesetz] vom 5.11.1997
TPLJ e: Trade Practices Law Journal
TPLOT e: Terra-Plot
tplut r: tung pluton
TPM e: Technical Program Manager
- e: Thermo-Power under Magnetic field
- e: Third Party Maintenance
- e: Total Productive Maintenance
- e: Total Productive Manufacturing
- e: Transactions Per Minute
- e: Transparent Prolog Machine
- f: Taxe de Première Mutation
- s: Tonelada[s] de Peso Muerto

TPMCM e: Total Partial Mission Capable-Maintenance
TPMCS e: Total Partial Mission Capable-Supply
tpmedel r: transportmedel
TPMP e: Total network data system Performance Measurement Plan
TPN e: Tactical Packet Network
- e: The Perspectives Network
- e: Transaction Program Name

TPO e: Technical Program Office
- e: Technical Project Officers
- e: Telecommunication Program Objective
- e: Test Plan Outline
- e: Thermoplastic elastomer-Olefinic
- e: Traveling Post Office
- e: Tree Preservation Order
- e: TRIMIS Program Office
- e: Turn Power Off
- e: Turn Power On
- e: Turn Power On/Off

Tpo s: Teletipo
TPOC e: Technical Point Of Contact
TPOCC e: Transportable Payload Operations Control Center
TPON e: Telephony Passive Optical Networks
T-POP e: TOGA Program on Prediction
TPOP e: Telco Point of Presence
TPORT e: Twisted Pair Port Transceiver
TPP e: Taxpayers and Producers Party
- e: Technical Program Plan
- e: Theater Planning Package
- f: Tribunal Populaire de Province
- p: Tribunal Popular Provincial

TPP2 e: Tripeptidyl-Peptidase II
tpplut r: transportpluton
TP-PMD e: Twisted Pair Physical Medium Dependent
TPPMD e: Twisted Pair Physical layer Medium Dependent
TPPN e: Trans-Pacific Profiler Network
TPQ e: Threshold Planning Quantity
TPR f: Transport Public Routier
- f: Tribunaux Populaires de la Révolution
- p: Tribunal Popular Revolucionário

tpreg r: transportflygregemente
Tprop f: Timbre proportionnel
TPRS e: Temperature Programmed Reaction Spectroscopy

TPS e: Tactical Product Set
- e: Technical Performance Specifications
- e: Technical Publication Software
- e: Teleprocessing Services
- e: Temporary Protected Status
- e: Test Package Set
- e: Test Pilots School
- e: Test Program Set
- e: Trans-Pacific Sections
- e: Tritium Purification System
- f: Taxe sur les Prestations de
- s: Tecnologia para la Salud [Chimaltenango]

TPSE e: Transaction Processing Service Element
TPSN e: Troop Program Sequence Number
TPSO e: Teleprocessing Services Center
TPSR e: Transmit Page Start Address register
TPS[S] e: This Program Sucks [Severely]
T/P SWT e: TOPEX/Poseidon Science Working Team
TPT e: Training Planning Team
- e: Twisted Pair Transceiver
- f: Travail Pour Tous
- s: Tropas Radiotécnicas

TPTB e: The Powers That Bee [Internet]
Tpte s: Transporte
TPTH e: Triphenyltinhydroxide
TPTMS e: Tropical Pacific Thermal Monitoring System
tpto r: transporttrupp
tptr r: tändpatron
TPTT f: Travaux Publics, Transports et Tourisme
TPU e: Terminal Processor Unit
- e: Text Processing Utility
- e: Thermal Processing Unit
- e: Threatened Plants Unit
- e: True Proof Unit

TPUMF e: Total Package Unit Materiel Fielding
TPUS e: Troop Program Unit
TPVP f: Travaux Publics de la Ville de Paris
TPW e: Turbo Pascal for Windows
TPX e: Tokamak Physics Experiment
TQ f: Tel Quel
Tq s: Tanque
T&Q e: Training and Qualification
TQA e: Teaching Quality Assessment
TQC e: Total Quality Control
TQFC e: Total Quality Flea Collar
TQFP e: Thin Quad Flat Pack
TQL e: Total Quality Leadership
TR a: Tjæreborg Rejser A/S
- d: Beginn der Teigreife
- e: Repetition Time
- e: Technical Response
- e: Technology Readiness
- e: Technology Refreshment
- e: Technology Review
- e: Test Register
- e: Text Retrieval
- e: Theater Reserve
- e: Thomas Register of American Manufacturers
- e: Todd Rundgren
- e: Token Ring
- e: Transaction Register
- e: Transit Satellite
- e: Transmit/Receive
- e: Travan Cartridge
- f: Tacite Reconduction
- f: Tarif Réduit
- f: Taux du Report
- f: Taxe Radiophonique
- f: Titres de Rente
- f: Transports Routiers
- f: Turquie
- r: Tekniska Rapporter
- r: Torpedrum
- r: Truppradio
- s: Tren Regimental

tr r: tändrör
- r: tross
- r: trupp

Tr f: Tour
- f: Travail
- f: Très
- f: Tribunal
- N: Tropp
- r: Tidrör
- r: Träng-
- s: Turborreactor

tr. a: træffes
- r: trappa

T-R e: Transformer-Rectifier
t/r a: tur-retur
T/R e: Transmit/Receive
TR-1 e: Tactical Reconnaissance Aircraft
TRA e: Traffic Routing Administration
- e: Tumor Rejection Antigen
- f: Thermomètre à Rayonnement Aéroporté
- s: Taller de Reparación de Automóviles

TRAC e: Teacher Research Associate
- e: Technical Report Awareness Circular
- e: Total Risk Analysis Calculation
- e: Tracks RadioActive Components
- e: TRADOC Analysis Center
- e: Training and doctrine command Analysis Center
- e: Transvaal Regional Action Committee

TRACE e: Temperature and Resistance Analysis to Characterize Electromigration
- e: Total Risk Assessing Cost Estimate
- e: Total Risk Assessing Cost Estimating
- e: Transition Region and Coronal Explorer
- e: Transport and Atmospheric Chemistry near the Equator
- e: Tripartite Advisory Committee on Conditions of Employment

TRACE-A e: Transport and Atmospheric Chemistry near the Equator-Atlantic
TRACER e: Tropospheric Radiometer for Atmospheric Chemistry and Environment[al] Research
TRACIR e: Tracking Air with Circular[ly] Polarized Radar
TRACKOB e: Report of Marine Surface Observations along a Ship's Track
- f: Message d'observation de la mer en surface le long de la route d'un navire
- s: Informe sobre Observaciones Marinas de Superficie a lo Largo de una Ruta Marítima

TRACY e: Data base on toxic metals in human tissues and fluids
Trad f: Traducteur
TRAD s: Taller de Reparación de Artillería de División
TRADACOMS e: Trading Data Communications
TRADE e: Training Services
TRAE s: Taller de Reparación de Artillería del Ejército
TRAF1 e: TNFR-Associated f-r 1
TRAF2 e: TNFR-Associated f-r 2
TRAFDIST e: Traffic Distribution System
TRAFFIC e: Trade Records Analysis of Fauna and Flora in Commerce
- f: Analyse des registres du commerce de la faune et de la flore
- s: Análisis de los Registros del Intercambio de Fauna y Flora en el Comercio

TRAGB e: Tritium Receiving and Analytical Glove Box
TRAGEX e: Trace Gas Exchange between Midlatitude Terrestrial Ecosystems and Atmosphere
TragKonstr d: Institut für Tragkonstruktionen und Konstruktives Entwerfen
TRAGNET e: Trace Gas Network
TRAIN e: Australian Training Information Network
TRAIS e: Transportation Reporting and Inquiry System
Trak r: Traktor
TRAM e: Translocon-Associated Membrane
TRAMAR e: Tropical Rain Mapping Radar
TRAMAS e: Transportable Microwave Active Spectrometer
TRAMCOM e: Transmission Monitoring and Control System
TRAMED e: Traditional Medicines Programme
TRAMICO f: Transformations et Applications Modernes à l'Industrie et au Commerce
TRAMP e: Translocon-Associated Membrane Protein
TRAMS e: Transportation Management System
- e: UCCIS Transportation and Movement Subsystem

TrängKAS r: Trängtruppernas Kadett- och Aspirantskola
Trans. e: Transactions
trans. e: transitive

Trans Am Clin Climatol Assoc e: Transactions of the American Clinical and Climatological Association
Trans. Am. Fish. Soc. e: Transactions of the American Fisheries Society
Trans Am Ophthalmol Soc e: Transactions of the American Ophthalmological Society
Transat f: [Compagnie générale] Transatlantique
TRANSCAM e: Trans-Cameroonian Railway
- f: Chemin de Fer Transcamerounais
TRANSCOM e: Transportation Command [management information system]
- e: Transportation Communications
transf. e: transferred sense
Transfus Clin Biol f: Transfusion Clinique et Biologique
Transfus Med e: Transfusion Medicine
Transfus Med Rev e: Transfusion Medicine Reviews
Transgenic Res e: Transgenic Research
TRANSIT e: Satellite Navigation [position Location] System
tr[ansl]. e: translated
- e: translation
Trans[m] s: Transmisión
Trans Med Soc Lond e: Transactions of the Medical Society of London
TRANSMOD e: Transportation Mobility Development
- e: Transportation Module
Transp s: Transporte
TRANSPAC e: North Pacific Ocean Monitoring [XBT] for Climate Research
- e: Trans-Pacific Experiment
- f: Surveillance de l'océan Pacifique Nord [XBT] aux fins de recherches sur le climat
- s: Vigilancia del Océano Pacífico Norte [XBT] para Investigación Climática
Transplant Proc e: Transplantation Proceedings
Transplant Sci e: Transplantation Science
Transpl Immunol e: Transplant Immunology
Transpl Int e: Transplant International
Transp Mot s: Transporte Motorizado
Trans R Soc Trop Med Hyg e: Transactions of the Royal Society of Tropical Medicine and Hygiene
Trans Stud Coll Physicians Phila e: Transactions and Studies of the College of Physicians of Philadelphia
TRANSVANS e: Transportable Radar and Communications Simulator Vans
TRAP e: Tactical Related Applications
- e: Tanks, Racks, Adapters and Pylons
TRAP1 e: TNFR-Associates Protein 1
TRAPIL f: [Société pour] Transports Pétroliers par Pipe-Line
TRASCO e: Transvaal Students Congress
TRAU e: Transcoder and Rate Adaption Unit
TRAV e: Travel System
Trav f: Travail
Trav. e: Travel[s]
Travpub f: Travaux publics
Travx f: Travaux
trb. a: trinbræt
Trbatt r: Trossbatteri
TRB[C] s: Toneladas de Registro [Compensado]
TRbF e: Technische Regeln brennbare Flüssigkeiten
tr./Btt. 2 N: tropp av 2. batteri
TRC e: Tactical Reconnaissance Center
- e: Tanzania Railway Corporation
- e: Technical Review Committee
- e: Technology Repair Center
- e: TEREC Remote Terminal
- e: Transmission Release Code
Trc f: Turc
TRCEI s: Taller de Reparación Central de Equipos Ingenieros
TRCR e: Technical Review Committee on Draft Resolutions
- f: Comité d'examen technique des projets de résolution
- s: Comité de Examen Técnico de Proyectos de Resolución
TRCVR e: Transceiver
- e: Transmitter-Receiver
TR&D e: Tropical Research and Development
TRD e: Technical Resources Document
- e: Transition Radiation Detector
TRDMC e: Tears Running Down My Cheeks [Internet]
TRE e: Tactical Receive Equipment
- e: Target partner Region
- e: Tracer Release Experiment
Tre f: Territoire
TREA e: Trap Enrichment Eddy Accumulation
Treas. e: Treasury
TREASURE e: Timber, Recreation, Environment, Aesthetics for a Sustained, Usable, Resource
TREAT e: TRE Advanced Terminal
- e: TREe Associative Temporally
- e: Trouble Report Evaluation Analysis Tool
Treat. e: Treatise
Treatm. e: Treatment
TREC e: Target Recognition for Electronic Combat
- e: Tracking Radar Echoes by Correlation
TRECAMS e: Tritium Remote Control and Monitoring System
TREDAT e: CSIRO forestry database
TREDMAR e: Marine Science Training and Education Programme
- f: Programme de formation et d'enseignement en sciences de la mer
- s: Programa de Capacitación y Enseñanza sobre Ciencias del Mar
TREDS e: Tactical Reconnaissance Exploitation and Dissemination System
- e: TR-1 Exploitation Development System
TREE e: Transient Radiation Effects on Electronics
- e: Tropical Rainforest Ecology Experiment
TREES e: CSIRO Tree Seed Centre, Canberra, database
- e: Tropical Ecosystem Environment[al] Observation[s] by Satellite
TREF e: Tripartite Employment Conditions Forum
TREIS e: Tropical Radar Environmental Information System
TREM e: Tropical Rainfall Experimental Mission
TREMI e: Tracking Radar Echoes by Moment Invariants
TREMORS e: Tsunami Risk Evaluation through Seismic Moment from Real-Time System
- f: Système d'évaluation en temps réel des risques de tsunamis par le moment sismique
- s: Sistema de Evaluación en Tiempo Real del Peligro de los Tsunamis por el Momento Sísmico
Trends Biochem Sci e: Trends in Biochemical Sciences [Cambridge [Eng]]
Trends Ecol. & Evol. e: Trends in Ecology and Evolution
Trends Genet e: Trends in Genetics
Trends Microbiol e: Trends in Microbiology
Trends Neurosci e: Trends in Neurosciences
Trends Pharmacol Sci e: Trends in Pharmacological Sciences
Trenkol N: Hestetrenkolonne
TREQ e: Requested TTRT
TRES e: Tomsk Radiation Experiment in Siberia
Trés f: Trésorier
TRET e: TIROS Retrieval Module
TRETS e: Transportation, Energy and Troop Support
TreuhLÜÄndV d: Treuhandliegenschaftsübertragungsänderungs-Verordnung vom 24.6.1996
TRF e: Teacher Rating Forms
- e: Tropical Forests Program
- f: Transfert
Trf r: Trafik
- r: Trafikera
Trfreglering r: Trafikreglering
TRFS e: Transparent Remote File System
TRG d: Technische Regeln Druckgase
- e: Technical Review Group
Trgrp r: Trossgrupp
TRH e: Thyroid-Releasing Hormone
TRI e: Toxic Release Inventory
- e: Trinidad
T/RIA e: Telemetry/Range Instrumented Aircraft
TRIAC e: Triode AC
TRIAD e: TeleGen2 Real-Time Integrated Ada Development
- e: Timesharing Research Image Analysis and Display
TRIANA e: Triana Medical Claims Information System
TriASP e: Three ASP units installed in a single enclosure used in BQQ-5 and BSY-1
TribCiv f: Tribunal Civil
TribCom f: Tribunal de Commerce
TRIBLE e: Transaction Image Builder/Editor
TRIC e: Tank Riser Interface Containment
- e: Time-Resolved Ion Correlation
TRICOMS e: TRIAD Computer System
TRICS e: Transaction Identification Codes

TRIDEC e: Tri-Cities Industrial Development Council
TRIDENT e: program named for Neptunes/Poseidons three-pronged spear
Trienn f: Triennal
TRIFAR e: Three Frequency Airborne Radar
- e: Three-Frequency Frequency-Agile Radar
TRIFOOD e: Tri-Service Food Service System
TRIG e: Tank Riser Interface Glovebox
Trig. e: Trigonometry
TRIGA e: Training Reactor, Isotopes, General Atomic
TRIGAT e: Third Generation Anti-Tank Missile
Tri H f: Trihebdomadaire
TRILOG e: Tri-Service Logistics System
TRIM e: Terrain Resource Information Management
- e: Time-Related Instructional Management System
- e: Trails, Roads, Interdiction Multisensor
Trim f: Trimestre
TRIN e: Trinity River Basin
Trin s: Trinchera
Trinsp r: inspektör för Trängtrupperna
Tr Inst Im Pastera R: Trudy Instituta Imeni Pastera
TRIO e: Telecommunications Research Institute of Ontario
- e: Transplant Recipients International Organization
tr-ion e: translocation
TRIP e: Text Retrieval, Input and Presentation
- e: Toxic Release Inventory Program
- f: Transports Interrégionaux de Personnes
TRIPAD e: DoD Standard Patient Administration System
Tri-Party Agreement e: Hanford Federal Facility Agreement and Consent Order
TRIPP e: Teaching Resource for Instructors in Prehospital Pediatrics
TRIPS e: Trade Related [aspects of] Intellectual Property Rights
TRIS e: Toxic [chemical] Release Inventory System
- e: Transportation Reporting and Inquiry System
- e: Tropical Region Imaging Spectrometer
TRITAC e: Tri-Service Tactical
TRITON e: Triangle Trans-Ocean Buoy Network
TRK d: Technische Richtkonzentration
tr./Kp. 7 N: tropp av 7. Kompani
TRL e: Target Recommendation List
- e: Transistor-Resistor Logic
- e: Turbulence Research Laboratory
TRLF e: Time Resolved Lidar Fluorosensor
Trl ingbn N: Trøndelag ingeniørbataljon
TRLN e: Triangle Research Library Network
Trl sbbn N: Trøndelag sambandsbataljon
TRM e: Technical Reference Model
- e: Thermoremanent Magnetization
- f: Titulaire d'une Retraite Mutualiste
Trm r: Truppregistreringsmyndighet
TrMan f: Travail Manuel
TRMM e: Tropical Rainfall Measurement Mission
- e: Tropical Rainfall Measuring Mission
- e: Tropical Rainfall Monitoring Mission
- f: Mission de Mesure des pluies Tropicales
- s: Misión de Medición de las Lluvias Tropicales
TRMPWR e: Terminator Power
TRMTR e: Transmitter
TRN e: Terrain-Reference Navigation
- e: Threaded Read News
- e: Token-Ring Network
- e: Turn-around Actions
Trn e: Token Ring Network
- f: Tarn
- s: Transmisión
Trng s: Triangulación
TRNOE e: Transferred Nuclear Overhauser Enhancement
TRNSM e: Transmission
- e: Transmittal
TRO e: Trombone
- f: Thérapie par Réhydration Orale
TROA e: Truckee River Operating Agreement
TROC e: Tactical Relay Operations Center
TRODERP e: Tropical Demersal Recruitment Project
- f: Projet sur le recrutement dans les communautés démersales tropicales
- s: Proyecto de Repoblación de las Comunidades Demersales Tropicales
T-ROESY e: Transverse Rotating-frame Overhauser Enhancement SpectroscopY
TROFF e: Text Run OFF
- e: Typsetter new ROFF
trol. r: troligen
TROLI e: Token Ring Optimized Link Interface
Troms Fdi N: Troms Forsvarsdistrikt
TRON e: The Real time Operating Nucleus
TROOP e: Transportation Operational and Organizational Plan
Trop. e: Tropical
Trop Anim Health Prod e: Tropical Animal Health and Production
Trop Doct e: Tropical Doctor
TROPEX f: Expérience Tropicale
- s: Experimento Tropical
Trop Gastroenterol e: Tropical Gastroenterology
Trop Geogr Med e: Tropical and Geographical Medicine
TROPIC HEAT e: Tropical Pacific Upper Ocean Heat and Mass Budgets
- f: Bilans thermiques et de masse dans les couches supérieures de l'océan Pacifique tropical
- s: Balances Térmicos y de Masa de las Capas Superiores del Océano Pacífico Tropical
Trop Med Int Health e: Tropical Medicine and International Health
TROPO e: Tropospheric Scatter
TROPOZ e: Tropospheric Ozone [campaign]
TRP e: Technology Reinvestment Program
- e: Technology Research Partnerships
- e: Theater Response Package
- e: Tiepoint Registration Processor
- e: Token Ring Processor
- e: Token Ring serial Port
- e: Toxicology Review Panel
- e: Trapping, Aldus/Liminous
- e: Trapwise
- f: Timbre pour Réponse
Trp f: Transport
- r: Träffpunkt
trp r: trupp[er]
trp. a: transport
TRPA e: Tahoe Regional Planning Agency
TRPL e: Time-Resolved Photoluminescence
Trplut r: Trosspluton
Trpminering r: Truppminering
TrPr f: Travaux Pratiques
TRR e: Technical Readiness review
- e: Technical Report Request
- e: Technical Response Request
- e: Tip Ring Reverse
- e: Token Ring Repeater
Trr s: Terrestre
TRRJ e: Thermal Radiator Rotary Joint
TRRMS e: Training Requirements and Resource Management System
TRRP e: Trends Report
TRS d: Technische Richtlinien zur GGVS oder TR-GGVS
- e: Tactical Reconnaissance Squadron
- e: Teacher Retirement System
- e: Telephone Relay Service
- e: Terminal Reservation System
- e: Topology and Route Selection
- e: Tritiation and Recovery System
- e: Tritium Receiving System
- e: Trunked Radio System
TRSA e: Timesharing Retrieval System
TRSC e: Thailand Remote Sensing Centre
- e: Total Reduced Sulfur Compounds
TRSIURS e: Trainer Status, Inventory and Utilization Reporting System
Trsm s: Transmisión
Trsp N: Transport[ere]
Trspgrp HFT N: Transportgruppen av Hærens Forsyningstropper
Trspkp N: Transportkompani
TRSS e: Token Ring Subsystem
Trstab N: Troppsstab
TRT d: Technische Richtlinien Tanks
- e: Technical Review Team
- e: TEREC Remote Terminal
- e: Token Ring Terminal
- e: Token Rotation Time[r]
- f: Temps Réel de Travail
- s: Taller de Reparación de Tanques
- s: Tropas Radiotécnicas
TRTC d: Technische Richtlinien Tankcontainer
TRU e: Toys-R-Us
- e: Transuranic Waste
- e: Trumpet
- r: Tungregneutvalget

TRUCE e: Tropical Urban Climate Experiment
- f: Expérience sur le climat urbain dans les régions tropicales
- s: Experimento sobre el Clima Urbano en las Regiones Tropicales

TRUSAF e: Transuranic Waste Storage and Assay Facility
TRUSIX e: Trusted UNIX
TRUW e: Transuranic Waste
TRV e: Tower Restoral Vehicle
- s: Taller de Reparación de Vestuario

Trvx f: Travaux
TRW e: Thompson, Ramo and Woolridge
TRX e: Transceiver
- e: Transmitter-Receiver

Trz s: Trazadora
TS A: Transvaalse Separatiste
- d: Teilstudienleistung
- e: Table Space
- e: Technical Support
- e: Telekom Service
- e: Telemetry Simulator
- e: Test Suite
- e: Top Secret
- e: Top Secret Time Stamp
- e: Top Sub
- e: Tourettes Syndrome
- e: Toxic Substances
- e: Transvaal Separatists
- e: Type of Signaling
- f: Talon de Souche
- f: Tarif Syndical
- f: Taxe de Séjour
- f: Taxe de Stationnement
- f: Taxe Successorale
- f: Taxe Supplémentaire
- f: Technicien de Secrétariat
- f: Techniciens Sociaux
- f: Textiles Synthétiques
- f: Travail de Sortie
- f: Travailleur Syndiqué
- f: Travailleurs Sociaux
- N: Telesetting
- N: Tørrstoff
- r: Telesystem

Ts f: Tous
- r: Skjuttid
- r: Tjänstemeddelanden rörande Sjöförsvaret

TS. e: Transmission Services
T/S e: Temperature/Salinity
- e: Thermosonic
- e: Troubleshooting

T/s N: Tankskip
TS3 e: Time Sensitive Support System
- e: Top Secret Support System

TSA e: Target Service Agent
- e: Technical Support Alliance
- e: Technical Systems Audit
- e: Temperate South America
- e: Theater Storage Area
- e: Troops Support Agency
- r: Totalförsvarets Signalskyddsavdelning

TSAC e: Time Slot Assigner Circuit
TSAE e: Troop Support Agency Europe
TSAF e: Transparent Services Access Facility
TsAGI R: Tsentralny Aerogidrodinamichesky Institut
TSAP GOOS e: Technical and Scientific Advisory Panel for GOOS
TSAPI e: Telephony Server (or Service[s]) Application Program[mer] (or Programming) Interface
TSAR e: Theater Simulation of Airbase Resources
- e: Time-series Storage And Retrieval

TSAT e: T-1 rate Small Aperture Terminal
TSAVI e: Transformed Soil Adjusted Vegetation Index
TSB e: Termination Status Block
- e: Thin-Section Billet
- e: Totaler Sauerstoffbedarf

Ts-befal N: Tekniske sersjantbefal
TSBF e: Tropical Soil Biology and Fertility Programme
TSBR e: Total Solar Broadband Radiometer
Ts/c e: supressor/cytotoxic TL
TSC e: Tactical Support Center
- e: Technical Students Selection Committee
- e: Teleservice Centers
- e: Thermally Stimulated Current
- e: Thyristor-Switched Capacitor
- e: Time Stamp Counter
- e: Transportation Systems Center
- e: Triton System Controller

TSCA e: Toxic Substance[s] Control Act
- e: Toxic Substances Control Agreement

TSCAI e: TSCA Incinerator
TSCAP e: Thermally Stimulated Capacitance
TSCATS e: TSCA [Toxic Substances Control Act] Test Submissions Database
TSCC e: Toxic Substances Coordinating Committee
TSCIXS e: Tactical Support Center Information Exchange System
TSCO e: Target Selection Confusion of Operator
TSCPF e: Time Switch and Call Processor Frame
T-SCRC e: TEMPEST Small Computer Requirements Contract
TSCT e: AF Small Computer
- e: Tactical Support Center
- e: Tandem Switch Center
- e: Technical Support Center
- e: Technical Support Contractor
- e: Transportation Systems Center

TSD e: Team Start Date
- e: Technical Support Division
- e: Total System Design
- e: Treatment, Storage and/or Disposal

T/S/D e: Treatment, Storage and Disposal
TSDC e: TOGA Sub[-]surface Data Center (or Centre)
- f: Centre TOGA de Données de Subsurface
- s: Centro TOGA de Datos Subsuperficiales

TSDCS e: Tritium Storage and Delivery Cleanup System
TSDF e: Treatment Storage and Disposal Facilities (or Facility)
TSDGB e: Tritium Storage and Delivery System Glove Box
TSDS e: Tritium Storage and Delivery System
TSE e: TDRS H,I,J Spacecraft Simulator
- e: Technical Support Engineer
- e: Test of Spoken English
- e: The Semware Editor
- e: Transmission Secondary electron Emission
- s: Tribunal Supremo Electoral

TSEL e: Transport Service Selector
TSET. e: Terminal set
- e: Transmitted Signal Element Timing

TSF f: Télévision Sans Frontières
- N: Tyskstudentenes Forening

Tsfc e: surface Temperature
TSFC e: Thrust Specific Fuel Consumption
TSG e: Terrestrial Solar Grade
- e: TMDE Support Group

TSGCE e: Technology Sub-Group: Communications and Electronics
TSGCEE e: Tri-Service Group on Communications and Electronic Equipment
TSH d: bodennahe Lufttemperatur, südhemisphärisch
- e: Thyroid Stimulating Hormone
- e: Thyrotropin
- s: Telegrafía Sin Hilos

TSI e: Terminal Equipment System Initialization
- e: Thermo Systems, Inc.
- e: Time Slice Intervals
- e: Time Slot Interchange[r]
- e: Top-Surface Imaging
- e: Transport Service Interface

TsIAM R: Tsentralny Institut Aviatsionnove Motorostroeniya
TSIC e: Thermally Simulated Ionic Current
TSIE e: Transformed Spectral Index of the External standard
TSIM e: Total Solar Irradiance Mission
TSINFO e: ODP program to track condition and location of ODP thin sections
TSIP e: Task Force XXI Systems Integration Plan
- e: Trimble Standard Interface Protocol

TSIR e: Total System Integration Responsibility
Tsitol Genet R: Tsitologiia i Genetika
Tsjv r: Tjänstemeddelanden från Försvarets Sjukvårdsstyrelse
tsk. a: teskefuld
TsKB R: Tsentralnoye Konstruktorskoye byoro
Tskr r: Tändskruv
TSL e: Teaching English to Speakers of Other Languages
- e: Technical Services Laboratory
- e: Template Style Language
- e: Two-Stage Liquifaction

TSl f: Tchécoslovaque
TSLC e: TOGA Sea Level Center (or Centre)
- f: Centre TOGA de données sur le niveau de la mer
- s: Centro TOGA de Datos sobre el Nivel del Mar

TSLRIC e: Total Service Long-Run Incremental Cost

TSM e: Tech Support Manager
- e: Test Schedule Message
- e: Time Switch Module
- e: Timer Standby Monitor
- e: TIROS Stratospheric Mapper
- e: Transportation Simulation Model
- e: Transportation System Management
- e: Traveling Solvent Method
- f: Tarif Syndical Minimum
- f: Témoignage de Satisfaction du Ministre
- s: Temperatura de la Superficie del Mar

TSMA e: Theater of Strategic Military Action
TSMK N: Telegrafstyrets Militærkontor
TSMP e: Threat Simulator Master Plan
- e: True Symmetric Multiprocessor

TSMRI s: Taller Semimóvil de Reparación Ingeniera
Tsn r: Telestyrelsen
TSNS e: Technical Staff Negotiating Sub-Committee
TSO e: Tailored Systems Offering
- e: Technical Standing Order
- e: Technical Supply Officer
- e: Technical Support Online
- e: Technical Support Order
- e: Terminating Screening Office
- e: Territorial Safety Officer
- e: Time-Share (or Sharing) Option
- f: Température de la Surface de l'Océan
- r: Teknisk Stabsofficer

TSOCMS e: Time Sharing Option/ Conversational Monitor System
TSO/E e: Time-Sharing Option/Extension
TSOHF e: Total Sense Of Humor Failure
TSOM e: Time Series of Ocean Measurements
- f: Séries chronologiques de mesures océaniques
- s: Series Cronológicas de Medidas Oceánicas

TSOP e: The Society for Organic Petrology
- e: Thin Small-Outline Package

TSOW e: Technical Scope of Work
TSP e: Tactical Strike Program
- e: TAM Straddle Packer
- e: Technical Steering Panel
- e: Technical Support Plan
- e: Telecommunications Services Priority
- e: Temperature-Sensitive Parameter
- e: Test Suite Parameter
- e: Test Supervisor
- e: Texture and Shading Processor
- e: Thrift Savings Plan
- e: Time Synchronization Protocol
- e: Timely Spares Provision
- e: Total Soluble Phosphorus
- e: Total Suspended Particles
- e: Training Support Package
- e: Training Support Plan
- e: Twisted Shielded Pair
- f: Tribunal de Simple Police

TSp[ec] f: Tarif Spécial
TSPI e: TAPI Service Provider Interface
TSPIRS e: Timber Sale Program Information Reporting System
TsPK R: Tsentr Podgotovka Kosmonavti
TSPR e: Total System Performance Responsibility
T&S programme e: Technology and Society Programme
TSPZ e: Torres Strait Protected Zone
TSPZA e: Torres Strait Protected Zone Authority
TSQAP e: Tasmanian Shellfish Quality Assurance Program
TSR e: Technical Safety Requirements
- e: Technical Support Request
- e: Technical Support Requirement
- e: Telecommunications Service Requests
- e: Telephone Service Request
- e: Terminate [and] Stay Resident [program]
- e: Time Sensitive Reporting
- e: Totally Stupid Rules
- e: Training Safety Requirements
- e: Transmit Status Register
- f: Témoignage de Satisfaction à l'ordre de la Région

TS/RD e: Top Secret/Restricted Data
TSRS e: Technical Staff Review Sub-Committee
TSRT e: Transportable Standard Remote Terminal
TSS e: Tactical Shelter System Program
- e: Targeted Search System
- e: Task State Segment
- e: Technical Services Staff
- e: Technical Support Staff
- e: Telecommunication Standardization Sector
- e: Telecommunications Security System
- e: Telecommunications Standards Sector
- e: Test Suite Structure
- e: Tethered Satellite System
- e: TFTR Shift Supervisor
- e: Threatened Species Strategy
- e: Time Sharing System
- e: Times Square Stores
- e: Top Side Sounder
- e: Total Surveying Systems
- e: Traditional Stratigraphic Scale
- e: Traffic Service Section
- e: Transmission Subsystem
- e: TRIDENT Support System
- e: Tritium Shift Supervisor
- e: Tromso Satellite Station
- e: Trunk and Signaling Subsystem

TSSDC e: Task State Segment Descriptor Cache
TS/SI e: Top Secret/Sensitive Information
TSSI e: Time Slot Sequence Integrity
TSSMS e: Time Sharing Services Management System
TSSOP e: Thin Shrink Small-Outline Package
TSSR e: TDAC Services Status Review
- e: Tropo-Satellite Support Radio

TSSS e: Time-Sensitive Support System
TSST e: Time Space Space Time network
TSST[-1] e: Toxic Shock Syndrome Toxin
TSSW e: Technical Document Support Software
- e: Technical Support Software

TST e: Stewart [Canada] [Airport]
- e: Technical Standards
- e: Technical Support Team
- e: Test an Operand
- e: Time Space Time [network]
- e: Torres Strait Treaty
- e: Transition State Theory
- e: Transmit Start Threshold
- e: Traveling Wave Tube
- e: Triple-Sorbent Trap

Tst r: Förutvarande Tygstaben
TSTFA e: Tasmania Sashimi Tuna Fishermens Association
TSTN e: Triple Supertwisted Nematic
TSTO e: Two Stage To Orbit
Tstrvbat r: Tunga stridvagnsbataljon
TSTS e: Time Space Time Space [network]
TSU e: Terminal Service Unit
- e: Trouble Shooting Utility

TsUP R: Tsentr Upravileniya Polyoti
TSV e: Tab Separated Values
- e: Total Support Vessel

TSVMS e: Tactical Secure Voice Management System
TSVP e: Tactical Secure Voice Program
- f: Tournez, S'il Vous Plaît

TSW e: Tele-Software
TT d: Taupunkt im SYNOP
- d: Temperatur im SYNOP
- e: Microtesla
- e: Task Team
- e: Technology Trasnsfer
- e: Text Telephone
- e: Thirtytwo/Thirtytwo
- e: Trinidad & Tobago
- e: Trunk Type
- e: Turn Time
- e: Typewriter Text
- f: Taux d'une Taxe
- f: Taxe de Télévision
- f: Taxe perçue [pour la Trésor]
- f: Taxe sur les Transactions
- f: Télégramme Téléphoné
- f: Titre de Transport
- f: Titre Temporaire
- f: Transfert des Titres
- f: Transfert Télégraphique
- N: Trygg Trafikk
- r: Tidningarnas Telegrambyrå AB
- r: Torpedtub
- s: Teletipógrafo
- s: Todo Terreno

Tt f: Tout
.tt e: Trinidad and Tobago [Internet]
T&T e: Travel and Tourism
- e: Trip and Throttle
- e: Tunnels and Trolls

t.t. e: technical term[s]

TTA e: Teacher Training Agency
- e: Technical Task Agreement
- e: Telecommunications Technology Association
- e: Teletex Attachment
- e: Transport-Triggered Architecture
- e: Transvaal Teachers Association
- e: Treasury Telecommunications Architecture
- f: Transmission Temporelle Asynchrone
- p: Empresa Nacional de Transporte e Trabalho Aéreo

TTAC e: Tasmanian Training Agreements Committee
TTAD e: Telemedicine Technology Area Directorate
TTAE e: Total Time, Airframe and Engine
TTAP e: Training Technology Applications Program
TTB e: Tanker-Transport-Bomber
TTBOMK e: To The Best Of My Knowledge [Internet]
TT&C e: Telemetry, Tracking and Command[ing] [facility]
TTC e: Teachers Training College
- e: Technical Training Center
- e: Telecommunications Technology Committee (or Council)
- e: Telemetry, Telecommand and Control
- e: Terminating Toll Center
- e: The Triplet Connection
- e: Timing, Trigger and Control
- f: Toutes Taxes Comprises

TTCE e: Transportation Terminal Center, Europe
Tt Cft f: Tout Confort
TTCN e: Tree and Tabular Combined Notation
TTCP e: The Technology Cooperative Program
TTCS e: Telemetry, Tracking and Command Stations within the MCS
TTD e: Touch Tone Director
- e: Training and Technical Data

TTE f: Taxe de Transmission à l'Exploitation
Tte f: Toute
- f: Traite

Tte Cor s: Teniente Coronel
Tte Corb s: Teniente de Corbeta
TTEM r: Taktisk-Teknisk-Ekonomiska Målsättningar
Ttes f: toutes
TTF e: Threat Training Facility
- e: Time To Live
- e: Time-To-Fail[ure]
- e: Timetable File[s]
- e: Training Test and Ferry
- e: TrueType Font

TTFN e: Ta Ta For Now [Internet]
TTF&T e: Technology Transfer, Fabrication and Test
TTG e: Time To Go
TTH r: Teknisk-Taktisk Hotbild
TTHS e: Trainees, Transients, Holdees and Students
T THTR e: Trueblood Theater, Frieze Building
TTI e: Teaching, Training and Information
TtjL r: teletjänst ledare
TTL e: Test Tone Level
- e: Time-To-Live
- e: Transistor[-to-]Transistor Logic

TT-linjen a: Trelleborg-Travemunde-Linjen
TTL4N e: That's The Lot For Now [Internet]
TTMA e: Tennessee Tech Microcomputer Association
TTMM e: Tamil Telugu Marathi Movement
TTMP e: Transit Time Magnetic Pumping
TTMS e: Talk/Type to Me Soon [Internet]
TTO e: Total Toxic Organic[s]
- e: Transient Tracers in the Ocean[s]
- f: Traceurs Transitoires dans l'Océan
- s: Trazadores Transitorios en el Océano

T Tor s: Transporte de Torpedos
TTP e: Tactics, Techniques and Procedures
- e: Technical Task Plan
- e: Telephone Twisted Pair
- e: Thermal-Transfer Printing
- e: Timed Token Protocol
- e: Trail Termination Point
- e: Tropospheric Transport of Pollutants
- e: Trunk Test Panel
- e: Trusted Third Party
- f: Transport Troposphérique des Polluants
- r: Tungt Transportflygplan
- s: Transporte Troposférico de Contaminantes

TT-QCAS e: Task Team on Quality Control Procedures for Automated Systems
- f: Equipe spéciale sur les procédures de contrôle de qualité pour les systèmes automatiques
- s: Equipo de Trabajo sobre Procedimientos de Control de Calidad de Sistemas Automatizados

TTR e: Target Tracking Radar
- e: Track Record
- e: Training-Through-Research
- f: Formation par la recherche
- s: Capacitar mediante la investigación

TTRP e: Time Token Rotation Protocol
TTRT e: Target Token Rotation Time
TTS e: TAFE Training Services
- e: Text-To-Speech
- e: TRI Tracking System
- e: Trouble Ticketing System
- e: Trunk Time Switch

TTSP r: Taktiskt-Tekniskt Systemprov
TTT e: Technojocks Turbo Toolkit
- e: Teleteaching and Training

TTTE e: Tornado Tri-national Training Establishment
TTTF e: Thermal Treatment Technology Test Facility
TTTN e: Tandem Tie Trunk Network
TTU e: Tennessee Technological University
TTUL e: Talk To You Later [Internet]
TTW e: Technical Training Wing
- e: Transtracheal Washes (or washing)

T&TW e: Test and Training Wing
TTWG e: Turnover and Transfer Working Group
TTX d: Videotext
- e: Tetanus Toxin

TTY d: Terminal eines grossen IBM-Systems
- e: Tele-Type
- e: Typewriter Telephone
- f: Télétype

TTYAWFN e: Talk To You A While From Now [Internet]
TTYL e: Talk To You Later [Internet]
- e: Type To You Later [Internet]

TTYNET e: Teletype Network
TTYRS e: Talk To You Real Soon [Internet]
TTYS e: Talk To You Soon [Internet]
TU e: Delft University of Technology Delft
- e: Temporary Unit
- e: Thammasat University, Thailand
- e: Transceiver Unit
- e: Tributary Unit
- e: Trunk Unit
- f: Taxe Unitaire
- f: Téléphone Urbain
- f: Transport Urbain
- r: Trupputbildning

Tu f: Tunisie
TU- R: Tupolev. Prefix for some Soviet aircraft
TU[B] d: Technische Universität Berlin
TÜB d: Temex-Übertragungsbaugruppe
TUBA e: TCP and UDP with Big Addresses
TUBE e: TDRSS User Base Expansion
Tuber Lung Dis e: Tubercle and Lung Disease
TUC d: Technische Universität Clausthal
- e: Total User Cell [count]
- e: Trades Union Congress
- f: Temps Universel Coordonné
- s: Transporte Urbano Colectivo

TUCC e: Triangle University Computer Center
TUCD e: Total User Cell Difference
TUCHA e: Type Unit Characteristics Data/File
TUCOWS e: The Ultimate Collection of Winsock Software
TUCSA e: Trade Union Congress of South Africa
TUCSIN e: The University Center for Studies in Namibia
TUD d: Technische Universität Darmstadt
- e: The date the record was last verified

TUDE e: Teletype Users Data Entry
TUDET e: Type Unit [Equipment] Detail
TUDRAD e: Technical University of Denmark Radiometer
TUE e: Tuesday
TUF e: Terminal User Function
- e: Turnhalle United Front

TUFD e: The User File Died
TUFI e: Tactical Unit Financial Interface
TUFIMS e: The University Financial Information Management System
TUFMIS e: Tactical Unit Financial Management Information System
TUFT e: Trans-Uterine Fallopian Transfer
TUFTEAUF e: The Universal Format to End All Universal Formats

TUG e: Tactical UTC Generator
- e: Technical Users Group
- e: TeX User[s] Group
- e: Tributary Unit Group
TUGP f: Taxe Unique Globale à la Production
TUI e: Tactical User-Oriented Intelligence System
- e: Text[-based] User Interface
TUIS d: Transport Unfall Informations- und Hilfeleistungssystem
TUK N: Trafikkulykkeskommando
TUL e: Tulsa Weather Forecast Office
TULIP e: TCP and UDP over Lightweight IP
TULIP[+] e: [Elseviers] The University Licensing Program
TULIPS e: Terrain Understanding and Land Information Processing System
Tumour Biol e: Tumour Biology
TUMS e: Table Update and Management System
Tun f: Tunisie
TUN r: Totalförsvarets Upplysningsnämnd
TUNA e: Trans-Urethral Needle Ablation
TUNES e: Thomas Washingtons Unexpected Nautical Extravaganza South-north
TUNIS e: Toronto University System
Tunis Med f: Tunisie Medicale
TUOC e: Tactical Unit Operations Center
TUP e: Telephone User Part
- e: Telephone User Path
tuples e: rows, or records of a relational database
TUR e: Thermal Uniformity Restoral
- e: Traffic Usage Recording
- e: Trunk Utilization Report
Tur. e: Turonian
TURFTS e: TDRSS User RF Test Set
Turk. e: Turkish
Turk J Pediatr e: Turkish Journal of Pediatrics
TURN e: Turnaround
TURTLE e: The Universal Reflection and Transmission Model for Layered Crop Experiments
TUS e: NWS WSO in Tucson, Arizona
- e: Trailing Umbilical System
- e: Transmission Unattended Site
TUs e: Technical Universities
tus. r: tusen
TUSA e: Transnet Union of South Africa
TUSTEP d: Tübinger System von Textverarbeitungsprogrammen
TUT d: Tuttlingen, Baden-Württemberg
- e: Tampere University of Technology, Tampere, Finland
- e: TDRSS Unscheduled Time
- e: Thermal Uniformity Test
- f: Taxe Unique sur les Transactions
TUTT e: Tropical Upper-tropospheric Trough
TÜV d: Technischer Überwachungs-Verein
TUVP f: Topographie et Urbanisme de la Ville de Paris
TV e: Television [camera (or display)]
- e: Turbo-Vision
- f: Taxe de Visa
- f: Taxe Vicinale
- f: Télévision
- f: Textiles Végétaux
- f: Travailleur Volontaire
- r: Tullverk
- s: Televisión
tv r: terrängvinkel
- r: tillsvidare
- r: timverk
tv. a: til venstre
t.v. N: til venstre
- r: till vänster
TVA e: Target Value Analysis
- e: Tennessee Valley Authority, Knoxville, Tennessee
- e: Time Variant Amplifier
- f: Taxe à la Valeur Ajoutée
- N: Teglverkenes Arbeidsgiverforening
TVA/ADF e: Target Value Analysis and Allocation and Distribution of Fire
Tvapen r: Tungt vapen
TVB f: Tout Va Bien
- s: Tren de Víveres y Bagajes
TVBC e: Transkei, Venda, Bophuthatswana [and] Ciskei
TVC e: The Vice Chancellor
TVC&P e: The Vice Chancellor and Provost
TVCS e: Tritium Vault Cleanup System
TVD e: Tank-Vapor Database
- e: Theater of Military Operations
TVE e: Television Trust for the Environment
- e: Town and Village Enterprise
- p: Televisão Experimental
- s: Consorcio Televisivo para el Medio Ambiente
TVF e: Table of contents Verbosely from File
- e: Time Variant Filter
TVFS e: Thermal Video Frame Scanner
- e: Toronto Virtual File System
TVI e: Transformed Vegetation Index
TVM e: Tactical Video Mapping
- e: Test Verification Matrix
- e: Thanks Very Much
- e: Track Via Missile
TVO e: Thermoplastic Vulcanites
- e: TIROS Operational Vertical Sounder
- e: TV Ontario
TVOC e: Total Volatile Organic Compounds
TVOL e: Television On-Line
- e: TV Online Cable Service
TVP r: Teknisk-Vetenskaplig Prognos
TVP-NCR e: Thermoviscoplastic Nonlinear Constitutive Relations
TVPS e: Torus Vacuum Pumping System
TVR d: Temex-Vermittlungsrechner
TVRO e: TV Receive Only
TVS e: Tornado Vortex Signature
- e: Triangular Voltage Sweep
TVSt d: Teilnehmer-Vermittlungsstelle
TVVL r: Tverrvitenskapelig Visualiseringslaboratorium
Tvx f: Travaux
TW e: Teleware
- e: Thieves World
- e: Training Wing or Test Wing
- e: Transmitter Window
- e: Twisted Twedge
.tw e: Taiwan [Internet]
T&W e: Targeting and Weaponeering
T/W e: Thrust/Weight ratio
TWA d: Technisch-wissenschaftliche Anwendungen
- e: Time-Weighted Analysis
- e: Time-Weighted Authority
- e: Time-Weigthed Average
- e: Trailing Wire Antenna
- e: Trans World Airlines
- e: Two Way Alternate
TW/AA e: Tactical Warning/Attack Assessment
- e: Threat-Warning and Attack Assessment
TW/AC e: Tactical Warning/Attack Characterization
TWAIN e: De facto industry standard for scanner interfaces
- e: Technology Without An Important Name
- e: Technology Without Any Interesting Name
TWAP e: Target-Weapon Association Program
TWARS e: Third World Association for Remote Sensing
TWAS e: Third World Academy of Science
TWAT e: Traveling Wave Amplifier Tube
TWC e: Texas Water Commission
- e: Tsunami Warning Centre
- f: Centre d'alerte aux tsunamis
- s: Centro de Alerta contra los Tsunamis
TWCP e: Tank Waste Characterization Plan
TWCS e: Tomahawk Weapons Control System
TWD e: Technical Work Document
TWEAT e: Ternary Waste Envelope Assessment Tool
TWERLE e: Tropical Wind Energy [conversion] and Reference Level Experiment
TWF d: Technisch-Wissenschaftliches Forum
TWG e: Technology Working Group
- e: The Wollongong Group
- e: Tidal Working Group
- e: Turnover Working Group
- f: Groupe de travail sur les marées
- s: Grupo de Trabajo sobre Mareas
TWHA e: Western Tasmanian Wilderness National Parks World Heritage Area
TWI e: Threat Warning Indicator
- e: Trade Weighted Index
TWICO e: Tanzania Wood Industry Corporation
TWINS e: Tank-Waste Information Network System
TWINSPAN e: Two-Way Indicator Species Analysis
TWIRP e: Two Way Infrared Protocol
TWIU e: Textile Workers Industrial Union
TWN e: Third World Network
TWNU e: Tigray Werji Nationality Union

TWO ATAF e: Second Allied Tactical Air Force
TWOS e: Tropical Wind Observing Ship
- f: Navire d'observation des vents tropicaux
- s: Buque de Observación de los Vientos Tropicales
TWOV e: Transit Without Visa
TWP e: Tropical Western Pacific
TWPO e: Tropical Western Pacific Ocean
TWR e: aerodrome control Tower
- e: Tyne and Wear
TWRA e: Tennessee Wildlife Resources Agency
TWRAPS e: Traffic Management Workload Reporting and Productivity System
TWRS e: Tank Waste Remediation System
TWRS-EIS e: Tank Waste Remediation System Environmental Impact Statement
TWS e: Tactical Weather System
- e: Tasmanian Wilderness Society
- e: The Wilderness Society
- e: Two Way Simultaneous
- e: Two-Way Simultaneous
- f: Système d'alerte aux Tsunamis
- s: Sistema de Alerta contra los Tsunamis
TWSC e: Team World Service Center
TW/SD e: Tactical Warning and Space Defense
TWSO e: Transuranic Waste Systems Office
TWT e: Traveling Wave Tube
- e: Two-Way Traveltime
TWTA e: Travel[l]ing Wave Tube Amplifier
TWTF e: Tank Waste Task Force
TWTTY e: Tactical Warning Teletypewriter
TWX e: Joint Message Form
- e: TeletypeWriter Exchange [service]
- e: Twinax
TWXXPPC e: TOGA/WOCE XBT XCTD Programme Planning Committee
- f: Comité de planification du Programme TOGA/WOCE de XBT et de XCTD
- s: Comité de Planificación del Programa TOGA/WOCE sobre XBT y XCTD
TWzNT d: Theologisches Wörterbuch zum Neuen Testament
TX e: Transuction
TXA e: Textiles and Apparel
TXC e: Telephone Exchange, Crossbar
- e: Transmit Clock
- e: Triton Xcellerated Controller
TXD e: Telephone Exchange, Digital
TXE e: Telephone Exchange, Electronic
- e: Transmit Error bit
TXF e: Tax Exchange Format
TXID e: Transuction Identifier
TXL e: Thanxalot! [Internet]
TXP e: Target Partner Expertise
- e: Transmission Packet
Tx Pcs f: Travaux Publics
TXRF e: Total X-Ray [Reflection] Fluorescence
TXT e: Text Document
Txt f: texte
TXTHDR e: Text Header
TXT2STF e: Text To Structured File
TxU e: University of Texas at Austin
TY e: Ten Year
ty. a: tysk
TYCLO e: Turn Your Caps Lock Off [Internet]
TYCOON e: Typed Communicating Objects in Open Environments
Tygavd r: Tygavdelning
Tygförv r: Tygförvaltning
TygK r: Tygtekniska Kåren
Tygmtrl r: Tygmateriel
Tygoff r: Tygofficer
TygS r: Tygförvaltningsskolan
Tygte r: Tygtekniker
Tyguoff r: Tygunderofficer
Tygvst r: Tygverkstad
Tygvstkomp r: Tygverkstadskompani
TYKKI e: North Pacific Ocean Observation and Research Initiative
- f: Initiative d'observation et de recherche pour le Pacifique Nord
- s: Iniciativa de Observación e Investigación para el Pacífico Septentrional
TYP e: Ten Year Plan
TYPE MSN e: Type of Mission
Typog[r]. e: Typography
TYPREP e: Type Unit Report
TYR e: County Tyrone
TYS e: Tayside
TZ e: Treatment Zone
TZS d: Fern- und Teilzeitstudiengänge

U

U e: U reference point
- e: Uranium [total]
- f: Unité
- f: Université [de Paris]
- R: Uluchshennyi
- r: Uppsättning
- r: Utan
- s: Unidad
- s: Urgente

u d: Schlupfvaribalen-Vektor
- e: micro [10^6]
- e: unified atomic mass unit
- r: undervattens-
- r: upp[gång]

Ü d: Übung

u. a: uden
- a: under

U. e: Unprotected

U/ N: uten

u- r: understöds-

U-2 e: Strategic Reconnaissance Aircraft

U-234 e: Uranium-234

U-235 e: Uranium-235

U-236 e: Uranium-236

U-238 e: Uranium-238

UA d: Archiv der Universität Göttingen
- d: Universitätsarchiv
- e: Unbalance[d] Asynchronous
- e: United Artits
- e: Unnumbered Acknowledgement
- e: Urban Affairs
- e: User Agent
- e: User Area
- f: Union Agricole
- f: Unité Administrative

UA. e: Unnumbered Acknowledgement

.ua e: Ukraine [Internet]

u.a. d: und andere
- r: utan anmärkning

u.ä. d: und ähnlich
- r: utom äktenskapet

u.å. a: uden år

UAA e: Universally Administered Address
- e: University of Alaska, Anchorage
- s: Unidad Antiaérea

UA&A e: User Authentication and Authorization

UAAI f: Union Africaine Agricole et Industrielle

UAAP e: USAREUR Automation Architecture Plan

UA-ARE e: University of Arizonas Aerosol and Radiation Experiment

UAB e: UNIX Appletalk Bridge

UAC e: Uganda Action Convention
- e: Unbalanced Asynchronous Class
- e: Unit Advisory Council
- e: Universal Access Control
- e: Upper Atmosphere Composition

UACB e: Undergraduate Admissions Coordination Board

UACOM f: Union Africaine de la Communication

UADE f: Union Africaine des Distributeurs d'Eau

U. Adelaide e: University of Adelaide

UADPS-ICP e: Unified ADP System Inventory Control Points

UADPS-SP e: Unified ADP System Stock Points

UAE d: Universelle Anschlußeinrichtung
- e: United Arab Emirates
- e: Unrecoverable Application Error

UAF e: University of Alaska-Fairbanks
- e: User Authorization Facility

uafh. a: uafhængig

UAFT f: Union Africaine des Télécommunications

UAG d: Umweltauditgesetz vom 7.12.1995
- e: Ugandan Action Group
- e: User Advisory Group

UAH e: University of Alabama at Huntsville

UAI f: Union Académique Internationale
- s: Unión Astronómica Internacional

UAIC f: Unité d'Afforestation Industrielle du Congo

UAK d: Augenklinik und Poliklinik

UA[L] e: United Airlines

U. Alberta e: University of Alberta, Canada

UALS f: Union Agricole pour les Libertés Sociales

UAM e: Urban Airshed Model
- e: User Authentication Method
- e: User Authentication Module

UAMCE f: Union Africaine et Malgache de Coopération Economique

UAMD f: Union Africaine et Malgache de Défense

UAMHS e: User-Oriented Automated Message Handling System

UAMPT f: Union Africaine et Malgache des Postes et Télécommunications

UAMS e: USEUCOM Alert Management System

UAMU e: User Assistance and Marketing Unit

UAN e: User Access Node

UÄndG d: Gesetz zur Änderung unterhaltsrechtlicher, verfahrensrechtlicher und anderer Vorschriften [Unterhaltsänderungsgesetz]

U and non-U e: Upper-class and other class

UAO e: Undergraduate Admissions Office

UAOB e: Upper Air Observation

UAP e: Universal Availability of Publication
- e: Ununited Anconeal Process
- e: Uranyl Ammonium Phosphate
- e: User Acceptance Program

UAPA e: Union of African Press Agencies

UAPDU e: User Agent Protocol Data Unit

UAPSP e: Utility acid precipitation study program

UAPT f: Union Africaine des Postes et des Télécommunications

UAQI e: Uniform Air Quality Index

UAR e: User Action Routine

UARC e: University Archives, division of the Bancroft Library

UARK e: Upper Arkansas River Basin

UARP e: Upper Atmosphere (or Atmospheric) Research Program

UARS e: Upper Atmosphere (or Atmospheric) Research Satellite
- f: Satellite de recherche dans la haute atmosphère
- s: Satélite de Investigaciones en la Alta Atmósfera

UART e: Universal Asynchronous Receive [and] Transmit[ter]
- e: Universal Asynchronous Receiver Transceiver

UARV e: Unmanned Air Reconnaissance Vehicles

UAS d: Unter Aller Sau [Internet]
- e: Upper Address Strobe
- e: Users Accountability System
- s: Academia de Ciencias de Ucrania

UASB e: Upflow Anaerobic Sludge Blanket
UASI e: Union of Artisans of the Sugar Industry
UASP s: Unión de Acción Sindical y Popular
UASV e: User Access Service Validation
UAT f: Union Aéromaritime des Transports
UATI e: User Applications and Technology Insertion
- f: Union des Associations Techniques Internationales
UATRAP s: Unidad y Acción Sindical y Popular de Trabajadores del Petén
UAV e: Unmanned Aerial Vehicle[s]
- e: Unmanned Aerospace Vehicle
- e: Unmanned Airborne Vehicle[s]
- e: Unpiloted Aerial Vehicle
- e: Upper Atmosphere Vehicle
- e: User Antenna View
UAV-AERI e: Miniaturized version of the ground-based atmospheric emitted radiance interferometer
u-avdeling N: underavdeling
UAVM e: UAV Maneuver
UAVO d: VO [EWG] Nr. 1836/93 des Rates vom 29.6.1993 über die freiwillige Beteiligung gewerblicher Unternehmen an einem Gemeinschaftssystem für das Umweltmanagement
UAV-SR e: Unmanned Aerial Vehicle-Short Range
UAW e: United Aerospace Workers
- e: Upper Atmosphere Winds
UB d: Universität Bremen
- e: Ungermann Bass
- e: United Breweries
- e: University at Buffalo
- e: University of Benin
- f: Université du Bénin
- N: Undervannsbåt
- N: Universitetsbiblioteket
- N: Utskrevet Befal
ub r: undervattensbåt
Ub s: ubicación
ÜB d: Übertragungseinheit mit Basisbandverfahren
UB2 e: Lund University Library Building 2. Lund Institute of Technology, Mathematics and Natural Sciences and Faculty of Medicine
UB-92 e: Uniform Billing Code of 1992
UBA d: Umweltbundesamt Berlin
- d: Universitätsbauamt
- e: Unibus Adapter
- s: Universidad de Buenos Aires
UBAC f: Union Bancaire en Afrique Centrale
U[-]båd a: Undervandsbåd
UBAF N: utskrevet befal administrasjon og forsyning
UBAMeN d: UBA Meßnetz
UBAT N: Utskrevet Befal Administrasjon og Transport
ubat r: understödsbataljon
u-båt N: undervannsbåt
UBB f: Unité Baleine Bleue
UBCE e: Uniform Building Code-Earthquake
UBCIM e: Universal Bibliographic Control and International MARC
UBD e: User Brain Damage
ubdiv r: ubåtsdivision
UBDP f: Union belge pour la défense de la paix
UBE e: Unsolicited Bulk Electronic Mail
ubef r: underbefäl
u-ben r: underben
ÜbersV d: Verordnung über die Übersetzung europäischer Patentschriften vom 2.6.1992
Ubesk a: Ubeskåret
UBEsk r: Ubåtseskader
ubesl. a: ubeslægtet
ubest. a: ubestemt
UBF e: UB Foundation
Ubf N: Ubefestet
UBFlj r: ubåtsflottilj
UBGG d: Gesetz über Unternehmensbeteiligungsgesellschaften
Ubi e: Ubiquitin
UBIM f: Unité de Biostatistique et d'Informatique Médicale
UBIW e: Union of Bus Industry Workers
Ubj r: Ubåtsjakt
Ubjesk r: Ubåtsjakteskader
Ubjflyg r: Ubåtsjaktflyg
Ubjftg r: Ubåtsjaktfartyg
UbjO r: Ubåtsjaktofficer
UbJ [skol] r: Ubåts- och Ubåtsjaktskola
Ubjskol r: Ubåts- och ubåtsjaktskolan
Ubjub r: Ubåtsjaktubåt
UBKV N: Utskrevet Befal Kontrol og Varsling
UBLK N: Utskrevet Befal Lett Kanon
UBLS e: University of Botswana, Lesotho and Swaziland
UBM e: Under-Bump Metallurgy
- e: Unpressurized Berthing Mechanism
Ubma r: Ubåtsmaskinist
UBN e: Unisource Business Networks
UBNI e: Ungermann-Bass Network Interface
UBO f: Université de Bretagne Occidentale
U. Bordeaux e: University of Bordeaux, France
UBOS f: Union de la Bijouterie et de l'Orfèvrerie Suisse
UBOT f: Union Belge des Ouvriers du Transport
UBR e: Unassigned Bit Rate
- e: Undefined Bit Rate
- e: Unfocused Boss Rambling
- e: Unspecified Bit Rate
U/BR N: Utan Brannrør
UBRBDA e: Upper Benue River Basin Development Authority
UBS e: United States Biological Survey
- f: Union des Banques Suisses
UBT e: Universal Bus Transceiver
UBTFPL r: Ubåtsjaktflygplan
UBVA a: Udvalget til Beskyttelse af Videnskabeligt Arbejde
UBW e: Ultra-Wide-Band
UC e: The University of California
- e: Unclassified [category]
- e: Union Catalogue
- e: University Council
- e: University of California
- f: Union Camerounaise
- f: Union Commerciale
- f: Unité Centrale
- f: Unité de Compte
Uc r: Statens Upplysningscentral
UC. e: Unit Check
U.C s: Unidad Central
u.c. e: upper case
U&C e: Usual and Customary
UC2 e: Compressed File
UCA e: Under Cover Addition
- e: Universal Cheerleaders Association
- e: Utility Communications Architecture
- f: Union Chimique Africaine
UCAA e: University Committee on Affirmative Action
UCABEL e: Union Catalogue of Art Books in Edinburgh Libraries
UCAD f: Université Cheikh Anta Diop
UCAF e: University Committee on Academic Freedom
UCAFORMEC f: Union de la Région Centre des Associations de Formation Médicale Continue
UCAMIS e: Uniform Chart of Accounts Management Information System
UCANS f: Union des Caisses Nationales de Sécurité sociale
U. Canterbury e: University of Canterbury, New Zealand
UCAO f: Unité de Compte de l'Afrique de l'Ouest
UCAP e: University Committee on Academic Personnel
U. Capetown e: University of Cape Town, South Africa
UCAR e: University Corporation for Atmospheric Research
UCAS e: Universities and Colleges Admissions Service
UCASA e: Urban Councils Association of South Africa
UCB e: University College of Belize
- e: University of California at Berkeley
- e: University of Colorado at Boulder
- f: Union de Crédit pour le Bâtiment
- f: Union des Coopérateurs Borins
- s: Unidad de Cazabombardero
UCC e: Ultra Clean Coal
- e: Uniform Code Council, Inc.
- e: Unit Control Center
- e: University Classroom Center
UCCA e: Union Cotonnière Centrafricaine
- f: Union des Chambres de Commerce Aéronautiques
UCCAO e: Union Centrale des Coopératives Agricoles de l'Ouest
UCCEGA f: Union des Chambres de Commerce et d'Etablissements Gestionnaires d'Aéroports
UCCFS e: United Council of Cultural Fraternities and Sororities
UCCIS e: USAREUR Command and Control Information System

UCCLO e: United Committee of Central Labor Organizations
UCCM f: Union des Chambres de Commerce Maritimes
UCCND e: Union Carbine Corporation Nuclear Division
UCCQ e: University College of Central Queensland
UCCS e: USEUCOM Command Center System
UCD e: University College, Dublin
- e: University of California [at] Davis
- e: Use Case Diagram
- f: Union des Croyants pour la Démocratie
UCDP e: United Christian Democratic Party
UCE e: Unité de Compte Européenne
- e: Unsolicited Commercial Electronic Mail
- f: Union Canadienne des Étudiants
UCEA e: Universities and Colleges Employers Association
UCEAP e: University Committee on Education Abroad Program
UCEE s: Unidad de Construcción de Edificios Educativos
UCEP e: University Committee on Educational Policy
UCEPHA f: Union Chimique et Pharmaceutique
UCF e: UNICODE 16-bit character set
- e: Uniform Contract Format
- e: Union Catalogue File
- e: Union Circulation File
- e: Universal Conductance Fluctuations
- f: Unité de Compte FAD
- s: Unión Cívica Femenina
UCFS e: Unix Connectivity File Service
UCFW e: University Committee on Faculty Welfare
UCG s: Unión Católica Guatemalteca
UCHG e: Universities and Colleges Hardware Group
UCHI e: University of Chicago
UCI e: University of California at Irvine
- e: University of California, Irvine
UCID p: União Caboverdiana Independente e Democrática
UCIN f: Union pour le Commerce et l'Industrie Nationale
UCISA e: Universities and Colleges Information System Association
- e: Universities and Colleges Information Systems Administrators
UCJF f: Union Chrétienne de Jeunes Filles
UCJG f: Union Chrétienne de Jeunes Gens
UCL e: Underwater Sound Laboratory
- e: Universal Communications Language
- e: University College London
- e: University Committee on Library
- f: Union Centrale des Locataires
- f: Université Catholique de Louvain
UCLA e: University of California at Los Angeles
UCLA L Rev e: UCLA Law Review
UCL[GLOR] f: Université Catholique de Louvain: Département d'études Grecques, Latines et Orientales
UCLRL e: University of California, Lawrence Radiation Laboratory
UCM f: Union Culturelle Musulmane
UCMISG e: Universities and Colleges Management Information Services Group
UCMR s: Unidad de Construcciones Militares de Reserva
UCN s: Unión del Centro Nacional
UCNG e: Universities and Colleges Network Group
UCNI e: Unclassified [Controlled] Nuclear Information
UCNS e: University Committee for Non-Teaching Staffs
UCNW e: University College of North Wales [Bangor]
UCO e: Unstructured Complex Object
UCOL e: Upper Colorado River Basin
UCOP e: University of California Office of the President
UCOPE e: University Committee on Preparatory Education
UCORP e: University Committee on Research Policy
UCoSDA e: Universities and Colleges Staff Development Agency
UCOSIDER f: Union Commerciale de Sidérurgie
UCP e: Uniform Customs and Practices
- e: United Christian Party
- f: Union Comores pour le Progrès
- f: Union Comorienne pour le Progrès
- s: Unidad Campesina del Petén
UCPA e: United Cerebral Palsy Association
UCPB e: University Committee on Planning and Budget
UCPBRP f: Union des Chambres de la Propriété Bâtie de la Région Parisienne
UCPS e: United Concrete Products Seychelles
UCR e: Under Color Removal
- e: Union of Cameroonian Republicans
- e: University of California at Riverside
- e: Usual, Customary and Reasonable
UCRG e: Uniform Contractor Reporting Guideline
UCRL e: University of California Lawrence Radiation Laboratory
UCRSG e: Uniform Contractor Reporting System Guide
UCRSU e: United Copperbelt Regional Students Union
UCS e: Secretariat to the U.S. National Commission
- e: Unicode Conversion Support
- e: Universal Card Slot
- e: Universal Communications Standards
- e: Universal Component System
- e: Universal Computer Science
- e: University Counselling Service
- f: Union des Centrales Suisses d'électricité
UCSB e: University of California at Santa Barbara
UCSC e: University Classified Staff Council
- e: University of California at Santa Cruz
UCSD e: University of California at San Diego
UCSF e: University of California, San Francisco
- f: Union des Chambres Syndicales de France
UCSG e: Universities and Colleges Software Group
UCSIP f: Union des Chambres Syndicales de l'Industrie du Pétrole
UCSRP f: Union Confédérale des Syndicats de la Région Parisienne
UCT e: University Center Tower
UCTG e: Universities and Colleges Training Group
UCTLIG e: Universities and Colleges Teaching, Learning and Information Group
UCU s: Unión Cultural Universitaria
UD e: Ulcerative Dermatosis
- f: Union Départementale
- f: Union Douanière
- N: Utenriksdepartementet
- p: União Democrática
- s: Unión Democrática
ud N: udyktig
UDA e: Urban Dwellers Association
- p: Unificação Democrata Angolana
UDAAV s: Unidad de Defensa Atómica Activa
UDAC e: Unified Dose Assessment Center
Udad Mar s: Unidad de Marcha
UDAE f: Union Douanière d'Afrique Equatoriale
- s: Unidad de Defensa Aérea
UDAETO s: Unidad de Defensa Aérea del Teatro de Operaciones
UDA-RDA f: Union Démocratique Africaine-Rassemblement Démocratique Africain
UDB e: Unified Data Base [for Acquisition Logistics]
Udb[et]. a: Udbetaling
UDBPR p: Unidade de Direcção de Barragens e Pequenos Regadios
UDC e: Uganda Development Company
- e: Union Démocratique Centraficaine
- e: Universal Digital Channel
- e: User Defined Command[s]
- f: Union Démocratique Comorienne
- f: Union pour la Démocratie Congolaise
UDD e: Union for Democracy and Development
- f: Union Démocratique Djiboutienne
- f: Union pour la Démocratie et le Développement
udd. a: uddannelse
- a: uddannet
UDDS s: Unión para la Democracia y el Desarrollo Social
UDE e: Universal Data Exchange
UDEA f: Union Démocratique des Etats Africains
UDEAC e: Central African Customs and Economic Union
- f: Union Douanière et Economique de l'Afrique Centrale
UDEAO f: Union Douanière des Etats de l'Afrique de l'Ouest
UDEC e: Universal Digital Electronic Computer
- f: Union d'Entreprises de Constuctions
UDECMA f: Union Démocratique Chrétienne Malgache
UDECO f: Union Démocratique Comorienne

UDEF f: Union des Exploitants Familiaux
Udemo p: União Democrática de Moçambique
UDEMU p: União Democrática das Mulheres da Guiné
UDEN f: Union Démocratique des Enseignants
UDENA s: Unión Democrática Nacional
UDENAMO s: União Democratico Nacional de Mozambique
udenl. a: udenlandsk
UDES f: Union Démocratique des Etudiants Sénégalais
- f: Union pour le Développement Economique et Social
UDF e: UAV Demonstration Flights
- e: Un-Ducted Fan
- e: United Democratic Front
- e: Universal Disc (or Disk) Format
- e: Unshielded Developers Forum
- e: User Defined Function
- e: User Defined Functions
- f: Union Démocratique des Femmes
- f: Union démocratique Française
- f: Union pour la Défense de la République
UDG e: User Defined Gateway
- f: Union Démocratique de la Guinée
udg. a: udgår
- a: udgave
- a: udgiver
- a: udgivet [af]
UDGE s: Unión Democrática de Guinea Ecuatorial
UDGRP f: Union des Démocrates Gaulistes et Républicains de Progrès
UDI e: Underseas Drilling, Inc.
- e: Universal Document Identifier
- e: User Debug Interface
- f: Union Démocratique des Indépendants
UDINOV s: Asociación Unidad de Desarollo Integral 'La Novena'
UDIP p: Unidade de Direcção de Indústria Pesqueira
UDISTP p: União Democrática e Independente de São Tomé e Príncipe
Udiv r: Understödsdivision
UDL e: Unique Data Log
- e: Unit Designation List
UDL/I e: Unified Design Language for Integrated Circuits
UDM d: Unternehmensdatenmodell
- e: Union Démocratique Mauricienne
Udm N: Utskrevet dekksmann
UDMA e: Ultra DMA
UDMPS e: Ultra-fine Differential Mobility Particle Sizer
udn. a: udnævnt
UDOA f: Union Douanière Ouest-Africaine
UDOPSS f: Union Départementale des Œuvres Privés Sanitaires et Sociales
UDP e: Unitary Development Plan
- e: United Democratic Party
- e: Universal Datagram Protocol
- e: User Data[gram] Protocol
- e: User Defined Protocol
- f: Union des Progressistes
- f: Union pour la Démocratie et la Paix
- f: Union pour la Démocratie Populaire
UDP[/]IP e: User Datagram Protocol/ Internet Protocol
UDPS f: Union pour la Démocratie et le Progrès Social
UDPT f: Union Démocratique pour le Progrès du Tchad
UDR e: Union for Democracy and the Republic
- f: Union Démocratique des Réfractaires
- f: Union des Démocrates pour la République
- f: Union pour la Démocratie et la République
UDRF e: U.S. Data Reception Facility
UDROC e: U.S. Data Reception Operations Centre
UDRS f: Union Démocratique pour le Renouveau Social
UDRU e: Ugandan Democratic Redemption Union
UDS d: Unfalldatenspeicher
- e: Union for Democracy and Solidarity
- e: User Display System
- f: Union pour la Démocratie et la Solidarité
- s: Unión Democrática Social
uds a: udsalg
uds. a: udsendelse
- a: udsendt
- a: udsolgt
Uds. s: Ustedes
UDSG f: Union Démocratique et Sociale Gabonaise
UDSM f: Union Démocratique et Socialiste de Madagascar
UDS[-]R f: Union Démocratique Sénégalaise-Renouveau
UDSU e: Uganda Democratic Salvation Union
UDSV d: Teledienst-Unternehmen-Datenschutz-Verordnung
UDT e: Uniform Data Transfer
- e: Universal Dataflow and Telecommunication
- e: Unstructurede Data Transfer
- e: User Defined [Data]type
- f: Union Démocratique du Travail
- f: Union Démocratique Tchadienne
udt. a: udtale[s]
udtr. a: udtryk
UDTS f: Union Démocratique des Travailleurs du Sénégal
UDU e: Uganda Democratic Union
U dub e: Yoo Double-yoo
UdUnits e: Unidata Units
U. Durham e: University of Durham, UK
UDUSA e: Union of Democratic University Staff Associations
udv. a: udvidet
- a: udvikling
UD/VR f: Union Démocratique pour la Ve République
UDZIMA f: Union Comorienne pour le Progrès
UE d: Unterhaltungs-Elektronik
- e: Undergraduate Education
- e: Unit Equipage
- e: Unusual Event
- e: Uranium Enrichment
- e: Urban and Environment credit program
- e: User Elements
- e: User Equipment
- f: Union Économique
- f: Union Européenne
- s: Unidad de Enlace con aire
- s: Unidad de Explotación
- s: Unión Europea
Ue r: Ubåtseldare
UEA f: Union des Etats africains
- s: Unidad de Enlace con Aire
UEAC e: Union of Central African States
- f: Union des Etats de l'Afrique Centrale
ÜEB d: Übertragungseinheit mit Basisbandverfahren
UEC e: User Environment Component
UECB f: Union Européenne des Commerces du Bétail
UECG f: Union Européenne des Commerces de Grains
UECL f: Union Européenne des Constructeurs de Logement
UECRI s: Unidad de Estudios de las Cuencas de los Ríos Internacionales
UED f: Union des Etudiants de Dakar
u-EDDA e: unsymmetrical EDDA
UEEB f: Union des Exploitations Électriques en Belgique
UEF f: Union des Étudiants de France.703
ÜEG d: Übertragungseinheit nach ITU G.703
UEI f: Union Européenne d'Ichtyologie
- s: Unión de Estudiantes Independientes
- s: Unión Europea de Ictiología
ÜEI d: Übertragungseinheit nach ITU I.430
UEJF f: Union des Étudiants Juifs de France
UEM d: Uckermünde, Mecklenburg-Vorpommern
- e: United Evangelical Mission Namibia
- f: Union Économique et Monétaire
- p: Universidade Eduardo Mondlane
- s: Unión Económica y Monetaria
ÜEM d: Übertragungseinheit mit Modemverfahren
UEMP f: Union Évangélique Médicale et Paramédicale
UEMS f: Union Européenne de Médecins Spécialistes
UEN e: United European Nations
UENDC f: Union Européenne des Négociants Détaillants en Combustibles
UEO f: Union de l'Europe Occidentale
- s: Unión Europea Occidental
UEOA f: Union des Etudiants Ouest-Africains
UEPO e: Unit Emergency Power Off
UEPR e: Unsatisfactory Equipment Performance Report
UER e: University-Emeriti Relations Committee
- f: Unité d'Enseignement et de Recherche
UES e: UCCIS Engineer Subsystem

UESM f: Union des Etudiants Socialistes Malgaches
U Esp s: Unidad Especial
UESTS f: Union des Etudiants et Stagiaires Togolais au Sénégal
UETB f: Union des Etudiants Tchadiens dans le Bénélux
UEV e: User End of Volume
U. Exeter e: University of Exeter, UK
UEXT e: University Extension Committee
UEZa f: Union des Ecrivains Zairois
UF e: Utilization Flight
- f: Union des Fabricants
- f: Union Fédérale
- r: Underställd Förband
U/F s: Unidad de Fuego
UFA d: Universum Film-AG
- e: Uganda Freedom Army
- e: Unsaturated Flow Apparatus
Ufa r: Utgångsläge för anfall
UFAC e: Uninstrumented Fast reactor Accelerated Component
- e: Uninstrumented Fueled Accelerated
UFACNC e: Ultrafine Aerosol Condensation Nucleus Counter
UFAP f: Union Fédérale Autonome Pénitentaire
- f: Union Française d'Action Patriotique
UFAS e: Unified File Access System
UFB f: Union des Femmes Burkinabe
- f: Union Française de Banques
UFBR e: Universal r Fast Breeder Eactor
UFC e: Uniform Fire Code
- e: Upflow Cooling
- f: Union des Forces pour le Changement
- f: Union Fédérale de la Consommation
UFCA f: Union Féminine Centrafricaine
UFCE f: Union Française pour le Commerce Extérieur
UFD e: User Functional Description
- f: Union des Forces Démocratiques
UFDM e: Ugandan Federal Democratic Movement
UFDS f: Union des Forces Démocratiques du Sénégal
UFDT e: User File Descriptor Table
UFE e: Undergraduate Faculty Enhancement
- f: Union des Femmes Européenne
UFEA e: Underground Facility Expansion Area
- f: Union Financière pour l'Europe et l'Afrique
UFEMAT f: Union des Fédérations nationales de négociants en Matériaux de construction
UFERI f: L'Union des Fédéralistes et des Républicains Indépendants
UFF e: Uganda Freedom Fighters
- f: Union des Femmes Françaises
- f: Union Française des Fédéralistes
UFI e: User Friendly Interface
- f: Union des Foires Internationales
- f: Union Fédéraliste Interuniversitaire
UFIC f: Union des Forces Indépendants pour le Changement
- f: Union pour Financement de l'Industrie Cinématographique
UFIDA f: Union Financière Internationale pour le Développement de l'Afrique
UFIE f: Union Française des Industries Exportatrices
UFIMAR f: Union Financière et Maritime
UFIS d: Umweltforschungs-Informationssystem
UFK e: Frauenklinik und Poliklinik
UFL e: Unbuffered FET Logic
- e: University of Florida
- f: Union des Femmes Luxembourgeoises
UFM e: Ugandan Freedom Movement
- e: USAFE Forms Mode
- f: Union Fédérale des Magistrats
- f: Union Fédéraliste Mondiale
- f: Union Franco-Musulmane
UFMAT f: Union des Fédérations nationales des négociants en Matériaux de construction
UFN f: Union des Femmes Nigériennes
UFO e: UHF Follow-On
- e: United Functions and Objects
UFONAL f: Union des Forces Nationalistes et Lumumbistes
u.forf. a: uden forfatter[angivelse]
UFP e: Unit Focal Point
UFPDG f: Union des Femmes du Parti Démocratique Gabonais
UFR e: Ultrafiltration Rate
- e: Unique Factorization Ring
- f: Union des Femmes de la Réunion
- f: Union des transports Ferroviaires et Routiers
UFS d: Unbemanntes Feuerschiff
- e: Ultra Fast Switching
- e: United Farmers and Stockowners
- e: Universal File System
- e: UNIX File System
U.F.S.B f: Union Franèaise pour la Santé Bucco-Dentaire
UFSE f: Union Française pour le Sauvetage de l'Enfance
UFST e: Universal Font Scaling Technology
UFT e: Unified File Transfer
- f: Union des Fédérations des Transports
UFU f: Union Fédéraliste Universitaire
UFZ d: Umweltforschungszentrum Leipzig-Halle
UG d: Universitätsgesetz
- e: Micro Grip
- e: User[s] Group
- e: Users Guide
- s: Unidad de Guerrilleros
- s: Universidad de Guanajuato, Mexico
.ug e: Uganda [Internet]
ug a: udmærket godt [karakter]
UGAMP e: Universities Global Atmospheric Modelling Programme
UGB f: Unité de Gros Bétail
UGBE f: Union Générale Belge d'Électricité
UGC e: Uppsala General Catalog
UGDO f: Union Générale pour le Développement d'Ouidah
UGE f: Union des Grandes Écoles
UGEAN f: Union Générale d'Etudiants de l'Afrique Noire
UGEAO f: Union Générale des Etudiants de l'Afrique Occidentale
UGEB f: Union Générale des Etudiants Burkinabes
UGECI f: Union Générale des Etudiants de Côte d'Ivoire
UGECR f: Union Générale des Etudiants Créoles de la Réunion
UGEEC f: Union Générale des Elèves et Etudiants Congolais
UGES f: Union Générale des Etudiants Sénégalais
Ugeskr Laeger r: Ugeskrift for Laeger
UGEV f: Union Générale des Etudiants Voltaïques
UGFF f: Union Générale des Fédérations de Fonctionnaires
UGGI f: Union Géodesique et Géophysique Internationale
UGH d: Universität-Gesamthochschule
- e: Upper Guide Horn
UGI s: Unión Geográfica Internacional
UGIB f: Union Générale des Infirmières de Belgique
UGIC f: Union Générale des Ingénieurs et Cadres
UGICT f: Union Générale des Ingénieurs, des Cadres, et des Techniciens
UGL e: Undergraduate Library
- r: Utveckling av Grupp och Ledare
UGM e: Underwater-launched, guided Ground-to-ground Missile
- e: University of Gadjah Mada, Indonesia
Ug, Og d: UntereR bzw. ObereR Grenz-vektor [Schranke] der Variablenwerte
UGOV e: U.S. Government
UGP f: Union Générale des Pétroles
UGR f: Union des Gauches Républicains
Ugr r: Understödsgrupp
UGRS i: Unione della Gioventù Rivoluzionaria Somala
UGTA s: Unión General de Trabajadores Argelinos
UGTAN f: Union Générale des Travailleurs de l'Afrique Noire
UGTC f: Union Générale des Travailleurs Camerounais
- f: Union Générale des Travailleurs Centrafricains
- f: Union Générale des Travailleurs Congolais
UGTCI f: Union Générale des Travailleurs de la Côte d'Ivoire
UGTD f: Union Générale des Travailleurs Djiboutiens
UGTG f: Union Générale des Travailleurs de Guinée
ugtl. a: ugentlig
UGTRF f: Union Générale des Travailleurs Réunionnais en France
UGTS f: Union Générale des Travailleurs du Sénégal
UH e: University of Hawaii, Honolulu
- s: Unidad de Helicópteros
uh r: underhåll
u.h. a: under havet[s overflade]
- r: under havsytan
UH-1 e: Iroquois helicopter
UH-60 e: Black Hawk Helicopter
- e: Blackhawk helicopter
UHA a: Underholdningsafdelingen
- r: Allmänn Underhåll
UHAP e: University Handbook for Appointed Personnel

UHB e: Ultra-High Bypass
UHD e: University of Houston, Downtown
- s: Unidad de Helicópteros Divisionaria
U H E C e: University Hospital Education Center
UHF e: Ultra High Frequency
UHF Radios e: Ultra High Frequency Radios
Uhk r: Underhållskurs för arméingenjörer
Uhkomp r: Underhållskompani
Uhkp r: Underhållshelikopter
UHL e: User Head[er] Label
UHMW e: Ultra High Molecular Weight
UHMW-PE e: Ultra High Molecular Weight Polyethylene
U HOSP e: University Hospital
UHP e: Upper Half-Plane
Uhplut r: Underhållspluton
UhrenT d: Institut für Zeitmeßtechnik, Fein- und Mikrotechnik
UHS e: University Health Services
u.h.t. a: uden hensyn til
UHT d: Umwelt- und Hygienetechnik
- e: Undergraduate Helicopter Training
Uhtjänst r: Underhållstjänst
UHTRE e: Ultra-High Temperature Reactor Experiment
UHU d: Unterstützungshubschrauber
UHV e: Ultra High Vacuum
UHVCF e: Ultra-High Vacuum Cleaning Facility
UHWM e: Uniform Hazardous Waste Manifest
UI e: University of Indiana
- e: University of Iowa
- e: UNIX International
- e: Unnumbered Information
- f: Union Interalliée
- f: Union Internationale
- f: Usage Interne
U/I e: Urban/Industrial
UIA f: Union Internationale contre l'Alcoolisme
- f: Union Internationale des Avocats
UIAM f: Union Internationale des Assureurs Maritimes
UIAP e: UNIX International Asia/Pacific
- f: Union Internationale des Associations de Presse
UIAT f: Union Internationale d'Assurances Transport
UIB f: Union internationale des Maîtres Boulangers
UiB N: Universitet i Bergen
UIC e: Ukpeaguik Inupiat Corp.
- e: Ukpeagvik Inupiat Corporation
- e: University of Illinois at Chicago
- e: Urban Industrial Complex
- e: User ID Code
- f: Union Internationale de Chimie
- f: Union Internationale des Chemins de Fer
UICA e: Underground Injection Control Act
UICM f: Union Internationale des Classes Moyennes
UICN f: Union Internationale pour la Conservation de la Nature et de ses ressources
- s: Unión Mundial para la Naturaleza
UI/CSRD e: University of Illinois/Center for Supercomputing Research and Development
UICTS e: UIC Tracking System
UID e: Unit IDentifier
- e: User Identifier
- e: User ID [Identity]
- f: Union Internationale de Dendrologie
- f: Unique Identifier
UiD e: Union in Deutschland
uid e: uniformly independently distributed
UIE f: Union Internationale d'Électrothermie
- f: Union Internationale d'Etudiants
- f: Union Internationale de l'Enfance
UIES f: Union Internationale d'Études Sociales
- f: Union Internationale pour l'Éducation Sanitaire
UIF e: UAV Interim Flight
- f: Union des Ingénieurs Français
- f: Union Infirmières France
UIFB e: Uninsured Employers Fund Board
UIFN e: Universal International Freephone Numbers
UIFRAME e: Unnumbered Information Frame
UIG d: Umweltinformationsgesetz vom 8.7.1994
UIGGebV d: Umweltinformationsgebührenverordnung vom 7.12. 1994
UII f: Union Internationale des Interprètes
UIL e: User Interface Language
UILE f: Union Internationale pour la Liberté de l'Enseignement
U. Illinois e: University of Illinois
UIM e: User Interface Management
UIMF f: Union Indépendante des Maires de France
UIMM f: Union des Industries Métallurgique et Minière
UIMP f: Union Internationale pour la protection de la Moralité Publique
UIMS e: User Interface Management System
UIMT f: Union Internationale de Ministres de Transports
UIMTCT f: Union Internationale de Médecine Thermale et de Climatothalassothérapie
UIORF f: Union Internationale des Organisations de Recherches Forestières
U. Iowa e: University of Iowa
UIP e: Undegraded Intake Protein
- e: United International Pictures
- e: User Interface Program
- f: Union Internationale de Physique pure et appliquée
UIPDEE f: Union Internationale des Producteurs et Distributeurs d'Énergie Électrique
UIPE f: Union Internationale de Protection de l'Enfance
UIPFB f: Union Internationale de la Propriété Foncière Bâtie
UIPV f: Union Internationale contre le Péril Vénérien et tréponématoses
UIR e: User Interface Requirements
UIRB e: University of Illinois Research Board
- e: Upper Illinois River Basin
UIS d: Umweltinformationssystem
- e: Upper Internal Structure
- e: Urban Information System
- e: User Interface Specification
UISAE f: Union Internationale des Sciences Anthropologiques et Ethnologiques
UISG e: Utility Industry Standards Group
UISM f: Union Internationale des Sociétés de Microbiologie
UISP f: Union Internationale de Sciences Physiologiques
UIT f: Union Internationale des Télécommunications
- f: Union Internationale des Traducteurs
- s: Unión Internacional de Telecomunicaciones
UITA e: Union of International Technical Associations
- s: Unión Internacional de Trabajadores de Alimentación
UITAM f: Union Internationale de Mécanique Théorique et Appliquée
UITP e: Universal Information Transport Plan
- f: Union Internationale des Transports Publics
UITR f: Union Internationale des Transports Routiers
UITS e: Unacknowledged Information Transfer Service
UIUC e: University of Illinois at Urbana [and] Champaign
UIUC-RB e: University of Illinois Research Board
UJA f: Union des Jeunes Avocats
- p: União de Jornalistas Angolanos
UJD f: Union des Juristes Démocrates
UJDSR e: Nuclear Regulatory Authority of the Slovak Republic
UJNR e: United States-Japan Cooperative Program in/on National Resources
UJP f: Union des Jeunes pour le Progrès
UJPM f: Union des Jeunes pour le Progrès de Mayotte
UJRB f: Union de Jeunesse Révolutionnaire Burundaise
UJRD f: Union des Jeunesses pour le Renouveau Démocratique
UJT e: UniJunction Transistor
UK A: Uitvoerende Komitee
- A: Universiteit van Kaapstad
- r: Ultrakortvågsanläggning
Uk r: Underkäke
- r: Underkant
.uk e: United Kingdom [Internet]
U.K. e: United Kingdom
UKADR e: United Kingdom Air Defence Region
UKAEA e: United Kingdom Atomic Energy Agency
- e: United Kingdom Atomic Energy Authorities
UKAIR e: United Kingdom Air Command and Control
- e: United Kingdom Air Forces
Ukam r: Undertryckskammare

U[-]Kat d: Ungeregelter Katalysator
UKAWRE e: United Kingdom Weapons Research Alder Establishment, Maston, Foulness
UKB a: Ultrakorte Bølger
UKE d: Universitätskrankenhaus Eppendorf, Hamburg
UKERNA e: United Kingdom Education[al] and Research Networking Association
UK HE e: United Kingdom Higher Education
UKJ d: Kinder- und Jugendklinik
UKM e: Urea Kinetic Modeling
UK MARC e: Tape format of BL Bibliographic Services
UKMARC e: United Kingdom MARC [Machine Readable Cataloguing] format
UKMO e: United Kingdom Meteorological Office
UKMRC e: United Kingdom Medical Research Council
UK-NII e: United Kingdom Nuclear Installation Inspectorate
UKOLN e: United Kingdom Office for Library and Information Networking
UKOLUG e: United Kingdom Online User Group
UKOOA e: United Kingdom Offshore Operators Association
UKOP e: Catalogue of UK Official Publications
U-kort r: Utskrivningskort
UK-PA e: Urokinase-like Plasminogen Activator
UKR e: Ukraina
Ukr f: Ukrainien
Ukr Biokhim Zh U: Ukrainskii Biokhimicheskii Zhurnal
UKRV d: Universitätsklinikum Rudolf-Virchow
UKS d: Universitätsklinikum Steglitz
- e: University of Kansas
UKT d: Universitätsklinik Tübingen
UKTS e: United Kingdom Treaty Series
UK[V] r: Ultrakorta Vågor
UKV R: Udarnye Kosmicheskie Vooruzheniya
UL e: Underwriters Laboratories, Inc.
- e: Underwriters Laboratory
- e: Unequal Length
- e: University Librarian
- e: University of Liberia
- e: Unordered List
- e: Urban Legend
- f: Union Locale
- f: Usufruit Légal
U/L e: Universal/Local
U[/]L e: Upload
ULA e: Uncommitted Logic Array
- e: Unit Level Automation Program
ULAF f: Union Locale des Associations de Familles
Uläge r: Undervattensläge
ULANA e: Unified LAN [Local Area Network] Architecture
ULC e: Unit Level Code
- e: Unit Level Computer
- e: Unpressurized Logistics Carrier
ULCM f: Union Libanaise de Commerçants
ULD f: Union pour la Liberté et le Développement
ULDPE e: Ultra-Low-Density Polyethylene
ULEU s: Fundación Agropecuaria de Guatemala
ULEV e: Ultra-Low-Emission Vehicle
ULF e: European Ultraviolet Laser Facility
- e: Ultra Low Frequency
- e: United Leukodystrophy Foundation
- r: Universitetslärarförbundet
ULFWS e: United Liberation Front for Western Somalia
ULIMO e: United Liberation Movement of Democracy for Liberia
ULIMO-J e: United Liberation Movement for Democracy in Liberia
ULIMO-K e: United Liberation Movement for Democracy in Liberia-Koromah
ULLC e: Ultra-Long-Life Core
ULM e: Ultra Light Motorized
ULN e: Unit Line Number
- e: Universal Link Negotiation
ULP e: Unfair Labor Practice[s]
- e: Unit in the Last Place
- e: Upper Layer Protocol[s]
- e: Upper Level Protocol
ULPGC e: University of Las Palmas de Gran Canaria
- f: Université de Las Palmas de Grande Canarie
- s: Universidad de Las Palmas de Gran Canaria
ULR e: Union of Liberals for the Republic
- e: USAREUR Logistics Review
- f: Union des Libéraux pour la République
ULS e: Union List of Serials
- e: Union List Subsystem from OCLC
- e: Unit Level Switch
- e: Upward-Looking Sonar[s]
ULSA e: Ultra Low Sidelobe Antenna
ULSDS e: Ultra-Large Screen Display System
ULSEL e: Ultra-Long-Spaced Electric Log
ULSI e: Ultra Large Scale Integration
ULSIC e: Ultra-Large-Scale Integrated Circuit
ULSP e: USAREUR Logistics Support Plan
ULST e: Uniform Low Stratus Test
Ulster Med J e: Ulster Medical Journal
ult. a: ultimo
- e: ultimately
Ultraschall Med d: Ultraschall in der Medizin
Ultrason Imaging e: Ultrasonic Imaging
Ultrasound Med Biol e: Ultrasound in Medicine and Biology
Ultrasound Obstet Gynecol e: Ultrasound in Obstetrics and Gynecology
Ultrastruct Pathol e: Ultrastructural Pathology
ULTRIX e: DECs version of UNIX
ULTT e: Upper Limb Tension Test
UM a: Udenrigsministeriet
- e: Object Manager
- e: Unified Model
- e: University of Miami, Florida
- e: University of Montana
- e: US Minor outlying Islands
- e: Users Manual
- e: Utilization Management
- f: Union Minière
- f: Unité Monétaire
Um N: Utskrevet Mannskap
- r: Ubåtsmatros
ÜM d: Übertragungseinheit mit Modemverfahren
U-M e: University of Michigan
U[/]M s: Unidad Militar
UMA e: Unified Memory Access
- e: Unified Memory Architecture
- e: Universal Measurement Architecture
- e: Upper Memory Area
- e: Urgent Message Application
- f: Union du Maghreb
UMADD f: Union Malienne pour la Démocratie et le Développement
UMADS e: Universal Measurement Architecture Data Storage
UMAF f: Union Médicale Arménienne de France
UMAP e: University Mobility in Asia and the Pacific
Umaskm N: Utskrevet maskinmann
UMB e: Ultrasound in Medicine and Biology
- e: Universal Naming Convention
- e: University of North Carolina
- e: Upper Memory Block[s]
- f: Union Margarinière Belge
UMBC e: University of Maryland, Baltimore Count
UMBRELLA e: ODP database containing descriptions of all ODP databases [science]
UMBSM e: University Marine Biological Station, Millport
UMC e: Under Ministry Control
- e: Unit Movement Coordinator
- e: United Microelectronics Corporation
- e: University Medical Center
- e: University of Missouri, Columbia
UMCM s: Unidad de Medidas Contraminas
UMCP e: Ununited Medial Coronoid Process
UMD e: Union for Multiparty Democracy of Tanzania
- e: Unit Manning Document
- e: University of Michigan-Dearborn
UME e: UNI Management Entity
UMEG d: Gesellschaft für Umweltmessungen und Umwelterhebungen mbH, Karlsruhe
UMEGREM s: Unión de Maestros Guatemaltecos Refugiados en México
U. Melbourne e: University of Melbourne
UMERS e: Uniform Medical Expense
UMFA e: United Mineworkers Federation of Australia
UmformT d: Institut für Umformtechnik

UMG e: Upgraded Metallurgical Grade
- s: Universidad Mariano Gálvez de Guatemala
UMGAP e: Upper Midwest Gap Analysis Program
UMHE e: Ununited Medial Humeral Epicondyle
UMI e: University Microfilm[s] [of theses and dissertations] International
- e: University of Miami
- e: User MAN [Metropolitan Area Network] Interface
- f: Union Mathématique Internationale
UMi f: Université de Paris, minéralogie
U-MIC e: UTP Medium Interface Connector
UMIF f: Union des Maîtres-Imprimeurs de France
UMIMA f: Union Maritime Industrielle Malienne
UMIS e: Upper Mississippi River Basin
U.M.I.S.T. e: University of Manchester Institute of Science and Technology
UMIST e: University of Manchester, Institute of Science and Technology
UML e: Unified Modeling Language
UMLA e: Universal Multi Line Adapter
UMLS e: Unified Medical Language System
UMM e: Unidirectional Multipoint-to-Multipoint
- f: Université Maritime Mondiale
- s: Universidad Marítima Mundial
UMMIPS e: Uniform Material (or Materiel) Movement and Issue Priority System
UMMIS e: UCCIS Medical Management Information Subsystem
UMN e: University of Minnesota
- e: University of New Mexico
- e: Upper Motor Neurons
UMNE e: University of Maryland teaching reactor
UMO e: University of Missouri
UMOA f: Union Monétaire Ouest-Africaine
Umopar s: Unidad móvil de patrullaje rural
- s: Unidad móvil policial para el área rural
UMOS e: U-shaped groove
UMP e: Uridine 5'-monophosphate
UMPE s: Unión Misionera para Enfermos [Jalapa]
UMPLIS d: Umweltdokumentations- und Informationssystem
- d: Umweltplanungs-Informationssystem
UMPS e: Unit Mission Planning System
U5MR e: Under 5 Mortality Rate
UMR e: University of Missouri in Rolla
UMS d: Umweltschutzarzt
- e: Universal Military Simulator
UMSA e: United Municipalities of South Africa
UMSC e: Uganda Muslim Supreme Council
- e: Upper Mississippi Science Center
UMSLG e: University Medical School Librarians Group
UMSWU e: Uganda Mines and Smelter Workers Union
UMT e: Union Musulmane Togolaise
- e: Universal Military Training
- e: Universal Moynihan Theory
- e: Upper Management Team
- e: Uranium Mill Tailings
- e: User Maintained Data Table
UMTRA e: Uranium Mill Tailings Remedial Action[s]
UMTRCA e: Uranium Mill Tailings Radiation Control Act
UMTS e: Universal Mobile Telecommunication[s] System
UMW e: United MineWorkers [union]
UmweltHG d: Umwelthaftungsgesetz
UmwG d: Gesetz über die Umwandlung von Kapitalgesellschaften und bergrechtlichen Gewerkschaften [Umwandlungsgesetz]
UmwStG d: Gesetz über steuerliche Maßnahmen bei Änderung der Unternehmensform [Umwandlungs-Steuergesetz]
UMWU e: Uganda Medical Workers Union
UN e: Unbalanced Normal
- e: United Nations
- e: Unplanned release not to an existing disposal facility
- f: Union Nationale
- f: Université Nouvelle
U.N. e: United Nations
UNA e: Upstream Neighbor Address
- e: Utah Nurses Association
- f: Union Nationale de l'Alimentation
UNAB f: Union Nationale des Agriculteurs Belges
- f: Union Nationale des Anciens du Burkina
UNAC f: Union Nationale des Artistes Congolais
UNACA p: União Nacional dos Camponeses Angolanos
UNACE e: Universal Aircraft Command/ Navy Evaluation
UNADEF f: Union Nationale des Associations de Défense des Économiquement Faibles
UNAFORMEC f: Union Nationale des Associations de Formation Médicale Continue
UNAG e: National Union of Farmers and Cattle Owners
UNAGRO s: Unión Nacional de Agricultores
UNAGUA s: Unión Nacional de Asociaciones Ganaderas
UNAIBODE f: Union Nationale des Associations d'Infirmiers de Bloc Operatoire Diplomés d'État
UNAIDS e: Joint United Nations Programme on HIV/AIDS
- e: United Nations Acquired Immunodeficiency Syndrome
UNALM s: Asociación Union Nacional de Lisiados
UNAM e: National Autonomous University of Mexico
- f: Union Nationale pour l'Avenir de la Médecine
- f: Université Nationale Autonome du Mexique
- s: Universidad Nacional Autónoma de México
UNAMAP e: Users Network for Applied Modeling of Air Pollution
UNAM-G s: Unión Nacional de Mujeres Guatemaltecas
UNAMIR e: United Nations Assistance Mission for/to Rwanda
UNAMO p: União Nacional Moçambicano
UNANGOLA p: União dos Angolanos no Exterior
UNAPE f: Union Nationale des Associations de Parents d'Élèves
UNAPEI f: Union Nationale des Associations des Parents d'Enfants Inadaptés
UNAPES f: Union Nationale Patriotique des Etudiants Sénégalais
UNAR f: Union Nationale des Associations Régionales pour le sauvetage de l'enfance
- f: Union Nationale Rwandaise
Unar p: União Nacional da Rombezia
UNARS e: Upper Atmosphere Research Satellite
UNASGUA s: Unidad Nacional de Asentamientos de Guatemala
UNASIIF f: Union Nationale des Associations et Syndicats Infirmiers et des Infirmiers[ères] Français
UNATRAT f: Union Nationale des Travailleurs du Tchad
UNAVCO e: University NAVSTAR Consortium
UNAVEM e: United Nations Angola Verification Mission
UNAVEM II e: United Nations Angola Verification Mission II
UNAVEM III e: United Nations Angola Verification Mission III
UNAZA f: Université Nationale du Zaire
UNB f: Université Nationale du Bénin
UNC e: Unbalanced Normal Class
- e: UNC Nuclear Industries
- e: Unified National Course thread
- e: Unique Naming Convention
- e: United Nuclear Corporation
- e: Universal Naming Convention
- e: Universal Network Convention
- e: Uuencoded Netnews Collator
- f: Union Nationale des Coopératives
- s: Universidad Nacional de Córdoba, Argentina
UNCA f: Union Nationale Centrafricaine
UNCAA e: UNC Nuclear Industries Activities Association
UNCAC f: Union Nationale des Coopératives Agricoles de Céréales
UNCAIRCOMP e: United Nations Command Air Component
UNCAM f: Union Nationale des Cadres et Agents de Maîtrise
UNCC f: Union Nigérienne de Crédit et de Coopération

UNCDF e: United Nations Capital Development Fund
UNCED e: United Nations Conference on [the] Environment and Development, held at Rio de Janeiro, June 1992
UNCHE e: United Nations Conference on Human Environment
- f: Conférence des Nations Unies sur l'environnement humain
- s: Conferencia de las Naciones Unidas sobre el Medio Humano
UNCHS e: United Nations Centre for Human Settlements [Habitat]
- e: United Nations Commission for Human Settlement[s]
UNCI f: Union Nationale du Commerce et de l'Industrie
UNCLE e: Unified Nuclear Coalition of Latvian Emigrants
- e: United Network Command for Law Enforcement
UNCLOS e: United Nations Conference on the Law of the Sea
- e: United Nations Convention on the Law of the Sea
- f: Convention des Nations Unies sur le droit de la mer
- s: Convención de las Naciones Unidas sobre el Derecho del Mar
UNCM f: Union Nationale des Cadres et de la Maîtrise
UNCN e: United Nations Council for Namibia
UNCOL e: Universal Compiler Oriented Language
- e: Universal Computed Oriented Language
UNCRD e: United Nations Centre for Regional Development
UNCRO e: United Nations Confidence Restoration Operation in Croatia
UNCSTD e: United Nations Center for Science and Technology for Development
- e: United Nations Conference for/on Science and Technology for Development
- f: Conférence des Nations Unies sur la science et la technique au service du développement
- s: Conferencia de las Naciones Unidas sobre Ciencia y Tecnología para el Desarrollo
UNCTAD e: United Nations Commission on Trade and Development
- e: United Nations Conference on Trade and Development
- s: Conferencia de las Naciones Unidas sobre Comercio y Desarrollo
UNCTC e: United Nations Centre for Transnational Corporations
UNCTT f: Union Nationale des Chefs Traditionnels du Togo
UND f: Union Nationale Démocratique
- f: Union Nigérienne Démocratique-Sawaba
- p: União Nacional para a Democracia
Und r: undantag
- r: underrättelse
UNDC f: Union Nationale Démocratique du Cameroun
- f: Union Nationale pour la Démocratie dans les Comores
UNDCP e: United Nations International Drug Control Programme
UNDD f: Union Nationale pour la Défense de la Démocratie
- f: Union Nationale pour le Développement et la Démocratie
UNDE e: Moffitt Undergraduate Library
unders. r: undersåte
Undersea Hyperb Med e: Undersea and Hyperbaric Medicine
underv. r: undervisning
UNDHA e: United Nations Department of Humanitarian Affairs
UNDICHO f: Union pour la Défense des Intérêts des Chômeurs
undmtrl r: underrättelse materiel
UNDP e: United Nation[s] Development[al] Program[me]
- f: Union Nationale pour la Démocratie et le Progrès
UNDR f: Union Nationale pour le Développement et le Renouveau
UNDRC e: Office of the United Nations Disaster Relief Coordinator
UNDS f: Union Nationale pour la Démocratie et la Solidarité
undt. a: undtagelse
- a: undtagen
UNDTCD e: United Nations Department for/of Technical Cooperation for Development
UNEBA f: Union Nationale des Etudiants Burundi
UNEC e: United Nations Economic Committee
UN/ECA e: United Nations Economic Commission for Africa
UNECA f: Union Nationale des Etudiants Centrafricains
UN-ECE e: United Nations Economic Commission for Europe
UN/ECE e: United Nations Economic Commission for Europe
UNECE e: UN Economic Commission for Europe
UNECI f: Union Nationale des Etudiants de la Côte d'Ivoire
UN/ECLAC e: United Nations Economic Commission for Latin America and the Caribbean
Unedic f: Union nationale interprofessionnelle pour l'emploi dans l'industrie et la commerce
UN/EDIFACT e: United Nations Electronic Data Interchange For Administration, Commerce and Transport
UNEDIFACT e: United Nations EDIFACT
UNEEM f: Union Nationale des Elèves et Etudiants du Mali
UNEF/Renouveau f: Union Nationale des Etudiants de France de Renouveau
UNEK f: Union Nationale des Etudiants du Kamerun
UNEL f: Union Nationale des Étudiants de Luxembourg
UNEMF f: Union Nationale des Étudiants en Médecine de France
UNEP e: United Nations Environment[al] Program[me]
UNEPCOM e: United Nations Commission of the USSR
UNEP-HEM e: United Nations Environment Programme - Harmonization of Environmental Measurement
UNEPNET-LAC e: UNEP Network for Latin America and the Carribean
UNES f: Union Nationale des Étudiants de Suisse
- f: Union Nationale des Etudiants Sénégalais
UNESCO e: United Nations Education[al] Science (or Scientific) and Cultural (or Culture) Organization, Paris
- f: Organisation des Nations Unies pour l'éducation, la science et la culture
- s: Organización de las Naciones Unidas para la Educación, la Ciencia y la Cultura
UNETO f: Union Nationale des Etudiants Togolais
UNETPSA e: United Nations Educational and Training Program for Southern Africa
UNF e: Unified National Fine thread
- f: Union Nationale des Femmes
UNFC f: Union Nationale des Femmes Congolaises
UN-FCCC e: United Nations Framework Convention on Climate Change
- f: Convention-cadre des Nations Unies sur les changements climatiques
UNFCCC e: United Nations Framework Convention on Climate Change
UNFD f: Union Nationale des Femmes de Djibouti
UNFM f: Union Nationale des Femmes du Mali
UNFN e: United Nations Fund for Namibia
UNFPA e: United Nations Fund for Population[s] Activities
UNFR f: Front Ougandais du Salut National
UNFSTD e: United Nations Fund for Science and Technology Development
UNFT f: Union Nationale des Femmes Togolaises
UNG e: Uracil N-Glycosylase
ung r: ungefär
UNGA f: Assemblée générale des Nations Unies
- s: Asamblea General de las Naciones Unidas
UNGCRP e: United Nations Global Change Research Program
UNGEGN f: Groupe d'experts des Nations Unies pour les noms géographiques
- s: Grupo de Expertos de las Naciones Unidas para los Nombres Geográficos
UN[/]GTDI e: United Nations Guidelines for Trade Data Interchange
UNH e: Uranyl Nitrate Hexahydrate
UNHCR e: United Nations High Commission[er] for Refugees
UNHHSF f: Fondation des Nations Unies pour l'habitat et les établissements humains
- s: Fundación de las Naciones Unidas para el Hábitat y los Asentamientos Humanos

UNI d: Universität
- e: UNC Nuclear Industries, Inc.
- e: Union of Independent Democrats
- e: United Nuclear Industries
- e: United Nurses International
- e: Universal Network Interface
- e: Uranium Trioxide
- e: User[-to-]Network Interface
- f: Union Nationale pour l'Indépendance
- i: Ente Nazionale Italiano di Unificazione
- s: Unión Nacional Independiente

UNIA f: Union Nationale des Ingénieurs Agricoles
UNIBIBLIO e: An integrated library system
UNIBO e: University of Bophuthatswana
UNICA e: Network of Universities from the Capitals of Europe
UNICAF f: Union d'Importations Industrielles et Commerciales Africaines
UNICAN s: Unidad Indígena y Campesina del Norte
UNICC e: United Nations International Computing Center
UNICE f: Union des Industries de la Communauté Européenne
UNICEF e: United Nations International Children[s] Emergency Fund
UNICHAL f: Union Internationale des Distributeurs de Chaleur
UNICHAR f: Union Charbonnière Rhénane
UNICOM e: Universal Integrated Communication
UNICOP f: Union de Coopératives Nationales du Mali
UNICOR e: Federal Prison Industries, Inc.
UNICORN e: Library management software from SIRSI Ltd
UNICOS e: Universal Compiler FORTRAN compatible
UNICS e: Uniplex Information Computer Services
UNIDATA e: University Data System
Unidata e: University Data Systems
UNIDO e: United Nations Industrial Development Organization
UNIEF e: USEUCOM Nuclear Interface Element
UNIENET e: United Nations International Emergency Network
- f: Réseau international d'information des Nations Unies sur les situations d'urgence
- s: Red Internacional de las Naciones Unidas para casos de Emergencia

UNIGES f: Union des Groupements Economiques du Sénégal
UNI HD d: Universität Heidelberg
UNII e: Unlicensed National Information Infrastructure
UNIK N: Universitetsstudiene ved Kjeller
UNIKOM f: Union des Komoriens
UNILOR f: Union textile Lorraine
UNIMA f: Union Internationale des grands Magasins
UNIMARC e: Universal MARC [Machine Readable Cataloguing] format
UNIN e: United Nations Institute for Namibia
UNION e: Michigan Union
UNIP e: United National Independence Party
UNIPA e: United Nations Peace Action
UNIPAC e: UNICEF Packing and Assembly Center
UNIPEDE e: International Union of Producers and Distributors of Electrical Energy
Uni.Platz d: Universitäts Platz
UNIR f: Union Nationale des Indépendants Républicains
- f: Union Nationale pour l'Indépendance et la Révolution

UNIS a: Universitetsstudiene på Svalbard
UNISA e: Universiteit van Suid-Afrika/ University of South Africa Pretoria
UNISIST e: ICSU-UNESCO Joint Project to Study the Feasibility of a World Information System
- e: Intergovernmental Program for Cooperation in the Field of Scientific and Technological Information
- e: United Nations System for Scientific and Technological Information

UNISTAR e: Universal Single call Telecommunications Answering and Repair
UNISWA e: University of Swaziland
UNIT e: University of Trondheim
- f: Union Nationale des Ingénieurs et des Techniciens

UNITAR e: United Nations Institute for/of Training and Research
- f: Institut des Nations Unies pour la formation et la recherche
- s: Instituto de las Naciones Unidas para la Formación Profesional y la Investigación

UNITCHADIENNE f: Union Tchadienne de Transports
UNITEC f: Union des Ingénieurs et des Techniciens français
UNITEGUA s: Unión de Transportistas Terrestres de Guatemala
UNITRA e: University of Transkei
- f: Union pour l'Industrie et les Travaux Publics

UNITRACK e: Unit Tracking Function
UNITREP e: Unit Status and Identity Report
UNITREPCSRC e: UNITREP System Chief of Staffs Readiness Chart
Univ f: Universel
- f: Université

univ. a: universitet
Univ. e: University
UNIVAC e: University Automatic Computer
UNIVERBEL f: Union des Verreries Mécaniques Belges
Univté f: Université
UNIX e: AT&T Bell Laboratories Operating System
UNJM f: Union Nationale des Jeunes du Mali
UNJS f: Union Nationale de la Jeunesse du Sénégal
UNJSPF e: United Nations Joint Staff Pension Fund
unkn. e: unknown
UNL e: University of Nebraska, Lincoln
- f: Union des Navigants de Ligne

UNLA f: Union Nationale de la Lutte contre l'Analphabétisme
UNLC f: União Nacional para a Libertação de Cabinda
UNLDDA p: União Nacional da Luz da Democracia e Desenvolvimento de Angola
UNLF e: Ugandan National Liberation Front
UNLP s: Universidad Nacional de La Plata, Argentina
UNLV e: University of Nevada, Las Vegas
UNM e: Ugandan National Movement
- f: Union Nationale Malgache

UNMA e: Unified Network Management Architecture
- e: Universal Product Code

UNMIH e: United Nations Mission in Haiti
UNMOT e: United Nations Military Observers in Tajikistan
- e: United Nations Mission of Observers in Tajikistan

UN/NA e: United Nations/North American Hazardous Materials Code
UN-Nr. e: Nummer eines Stoffes
UNO e: Ugandan Nationalist Organization
- e: United Nations/Organization
- e: Universal Networked Objects
- s: Unidad Nacionalista Organizada

UNOASD e: United Nations Outer Space Affairs Division
UN-OLA e: United Nations-Office of Legal Affairs
- f: Nations Unies-Bureau des affaires juridiques
- s: Naciones Unidas-Oficina de Asuntos Jurídicos

UNOLS e: University National Oceanographic Laboratory System
- e: Upgrade Nuclear Ordnance Logistics System

UNOMIG e: United Nations Observer Mission in Georgia
UNOMIL e: United Nations Observer Mission in Liberia
UNOMOZ e: United Nations Operation in Mozambique
UNOMUR e: United Nations Observer Mission Uganda-Rwanda
UNOSOM e: United Nation[s] Operations in Somalia
UNOSOM II e: United Nations Operation in Somalia II
UNP e: Uluru National Park
- e: United Nationalist Party
- e: United Nations Political Affairs

UNPA e: United Nations Postal Administration
UNPAAERD e: United Nations Program of Action of African Economic Recovery and Development
UNPAZA f: Union Nationale des Producteurs Agricoles et Artisans du Zaire
UNPDAC e: United Nations Program for Drug Abuse Control
UNPEEL f: Union Nationale des Parents d'Élèves d'Enseignement Libre
UNPP e: United Namibia People Party

UNPREDEP e: United Nations Preventive Deployment Force
UNPZA f: Union Nationale de Presse du Zaire
UNR e: Union for the New Republic
- f: Union pour la Nouvelle République
- f: Université Nationale du Rwanda
UNRD f: Union Nationale Républicaine Démocratique
UNRF e: Ugandan National Rescue Front
UNRFNRE e: United Nations Revolving Fund for Natural Resources Exploration
- f: Fonds autorenouvelable des Nations Unies pour l'exploration des ressources naturelles
- s: Fondo Rotatorio de las Naciones Unidas para la Exploración de los Recursos Naturales
UNRWA e: United Nations Relief [and] Works Agency [for Palestine Refugees in the Near East]
UNS e: Under Secretary of the Navy
- f: Union Nationale du Spectacle
- i: Unione Nazionale Somalo
UNSAS f: Union Nationale des Syndicats Autonomes du Sénégal
UNSB f: Union Nationale des Sinistrés de Belgique
UNSCEAR d: Wissenschaftliches Komitee der Vereinten Nationen über die Wirkung von atomarer Strahlung
- e: United Nations Scientific Committee on [the] Effects of Atomic Radiation
- f: Comité scientifique des Nations Unies pour l'étude des effets des rayonnements ionisants
- s: Comité Científico de las Naciones Unidas para el Estudio de los Efectos de las Radiaciones Atómicas
UNSITRAGUA s: Unión Sindical de Trabajadores de Guatemala
UNSO e: Office to Combat Desertification and Drought
- e: United Nations Sahelian Office
- e: United Nations Statistical Office
- f: Bureau de la lutte contre la désertification et la sécheresse
UNSP f: Union Nationale pour la Solidarité et le Progrès
UNS-R e: Unclassified National Security-Related
UNSTB f: Union Nationale des Syndicats des Travailleurs du Bénin
UNSW e: University of New South Wales, Australia
UNSWLJ e: University of NSW Law Journal
UNSWP e: United National South-West Party
UNT e: Undergraduate Navigation Training
UNTA p: União Nacional dos Trabalhadores de Angola
UNTC e: Union Nationale des Travailleurs Camerounais
UNTCI f: Union Nationale des Travailleurs de Côte d'Ivoire
UNTG p: União Nacional de Trabalhadores da Guiné
UNTM f: Union Nationale des Travailleurs Maliens
UNTN f: Union Nationale des Travailleurs Nigériens
UNTRAP e: Uranium Mill Tailings Remedial Action Program
UNTS f: Union Nationale des Travailleurs du Sénégal
UNTSFA e: United Nations Trust Fund for Southern Africa
UNTSP s: Unidad Nacional de Trabajadores de Salud Pública
UNTZa f: Union Nationale des Travailleurs du Zaire
UNU e: Uganda National Union
- f: Université des Nations Unies
- s: Universidad de las Naciones Unidas
UNWRTF e: Utah Nuclear Waste Repository Task Force
UNYOUG e: Upstate New York Oracle Users Group
UO e: Unusual Occurrence
- e: Users Office
- e: Utilization Officer
- f: Union Occidentale
- f: Union Ouvrière
Uo f: Unité Oxford
UO r: Underofficer
UOAC e: United Ostomy Association of Canada
UOB f: Université Omar Bongo
UOBB d: Unabhängiger Orden Bne Briss
UOC e: Unit Operations Center
UOD f: Union Ouvrière et Démocratique
Uoff r: Underofficer
uoff 2 [gr] r: underofficer av andra graden
U of OK e: University of Oklahoma
UOFS e: University of the Orange Free State Bloemfontein, South Africa
U of SC e: University of South Carolina
UOIF f: Union des Organisations Islamiques de France
UOLF e: United Oromo Liberation Front
UOME f: Union des Opposants Malgaches Extérieurs
UON e: University Of Nairobi
UOP e: UCAR [University Corporation for Atmospheric Research] Office of Programs
- e: Upper Ocean Panel
- f: Groupe sur les couches supérieures de l'océan
- p: Unidade Operativa de Proteção
- s: Panel sobre la Capa Superior del Océano
UOPLO e: United Oromo Peoples Liberation Organization
uopsk. a: uopskåret
UOR e: Undulating Oceanographic Recorder
- e: Urgent Operational Requirements
- f: Enregistreur remorqué à immersion variable
- s: Registrador Oceanográfico Ondulante
UO&S e: Utility Operations and Services
UO [school] r: Underofficersskola
UOT e: Upper Ocean Thermal [project]
- f: Projet sur les conditions thermiques dans les couches supérieures de l'océan
- s: Proyecto Térmico de la Capa Superior del Océano
UOT-CAL e: UOT/TREAT Calibration experiments
UOTDAC e: Upper Ocean Thermal Data Assembly Centre
- f: Centre de collecte de données sur les conditions thermiques dans les couches supérieures de l'océan
- s: Centro de Acopio do Datos Térmicos de la Capa Superior del Océano
UOVS A: Universiteit van die Oranje-Vrystaat, Bloemfontein, South Africa
UOW e: Unit Of Work
U. Oxford e: University of Oxford, UK
UP A: Universiteit van Pretoria South Africa
- e: Uninterruptible Power
- e: Union Pacific Railroad
- e: Unit Process
- e: United Party
- e: Unity Party
- e: University of Pretoria South Africa
- e: Unnumbered Poll
- e: Unpressurized
- e: Update Processor
- e: User Plane
- f: reconnu d'Utilité Publique
- f: Union Patriotique
- f: Uniprix
- f: Unité Pratique
- f: Université de Paris
- f: Université Populaire
- N: Utrykningspolitiet
- s: Unidad de Pontoneros
- s: Unión Popular
up e: up to, increases
- r: utgångspunkt
Up N: Utgangspunkt
ÜP d: Übergabepunkt
u.p. a: uden for particrnc
- a: uden portefølje
U3P f: Union pour la Promotion de la Propulsion Photonique
UPA e: Uganda Peoples Alliance
- e: Ultra Port Architecture
- f: Union Populaire Africaine
- p: União do Povo Angolano
- s: Unidad de Policía
upa r: utan personligt ansvar
UPAA e: Unit Process Accountability Area
UPADS f: Union Pan-Africaine pour la Démocratie Sociale
UPAMS e: Unpriced Action Management System
UPANG p: União Patriótica Anti-Neocolonialista de Guiné-Bissau
UPANIC e: Nicaraguan Union of Agricultural Producers
UPAP f: Union Pan-Africaine de Postes
UPAS e: Unpaid Associate
UPAUD e: Update Audit file
UPB f: Union de la Propriété Bâtie
upb r: utgångspunkt på flygbild
UPBF f: Union de la Propriété Bâtie de France

UPC e: Uganda Peoples Congress
- e: Unit Product Code
- e: Universal Printer Controller
- e: Usage Parameter Control
- e: User Parameter Control
- e: User Permit Code
UPCDA e: Uganda Peoples Christian Democratic Army
UPCF f: Union des Pilotes Civiles de France
UPCL f: Union Professionnelle des Créateurs de Lotissements et de logements
UPCONE e: Upconing of a Salt-Water/ Fresh-Water Interface Below a Pumping Well
UPCS e: Unit Process Control System
- e: United Presbyterian Church of Scotland
UPD d: User-Parameterdaten
- e: date this record was updated
- s: Unidad Popular por la Paz, Democracia, Desarrollo y Derechos Humanos
UPDA e: Uganda Peoples Democratic Army
UPDB f: Union Patriotique des Démocrates Béninois
UPDC p: União Popular para a Democracia Científica
UPDEA f: Union des Producteurs, Transporteurs, et Distributeurs d'Energie Electrique des Pays Africains, Malgache et Mauricien
UPDES f: Union du Peuple pour le Développement Economique et Social
UPDF e: Uganda Peoples Defense Forces
UPDM e: Uma Peoples Democratic Movement
UPDR f: Union Démocratique du Peuple Rwandais
UPDS e: Uninterruptible Power Distribution System
- f: Union pour la Démocratie et le Progrès Social
UPE A: Universiteit van Port Elizabeth South Africa
- e: University of Port Elizabeth South Africa
- e: Unsaturated Polyester
- e: User Portability Extension
upers. a: upersonlig[t]
UPEU e: Uganda Public Employees Union
UPF e: Uganda Popular Front
- e: United Peoples Front [of Lebowa]
- e: User Productivity Facility
UPG s: Unión Patriótica Guatemalteca
UPI e: UCL Primary rate Interface
- e: United Press International
- e: Universal Programmable peripheral Interface
- e: University Physicians, Inc.
- e: Upper Prediction Interval
- f: Union Postale Internationale
UPIN e: Universal Physician Identification Number
UPJU e: Uganda Vernacular, Primary and Junior Secondary Teachers Union
upk r: utgångspunkt på karta
UPL e: User Program Language
upl r: uppehållsplats
uplats r: uppehållsplats
uplomg r: uppehållsplatsområdegrupp
UPM e: Undefined Protocol Machine
- e: Unidirectional Point-to Multipoint
- e: Universal Permissive Modules
- e: Unix Programmers Manual
- e: User Profile Management (or Manager)
- s: Universidad Politécnica de Madrid
UPMAG s: Unión de Pueblo Maya de Guatemala
UPMC e: Urban Planning Ministers Conference
UPML f: Unions Professionnelles des Médecins Libéraux
UPMLRA f: Union Professionnelle des Médecins exerçant à titre Libéral de la région Rhône-Alpes
UPN d: Umgekehrte Polnische Notation
- e: Unity Party of Nigeria
- e: Universal Program Number
UPNA s: União de População do Norte de Angola
UPNM f: Union Professionnelle des Négociants en Meubles
UPOSA e: United Peoples Union of South Africa
UPOSCI f: Union des Paysans, Ouvriers, et des Salariés de Côte d'Ivoire
UPOV e: International Union for the Protection of New Varieties of Plants
- f: Union internationale pour la Protection des Obtentionsvégétales
UPOWU e: Uganda Petroleum Oil and Chemical Workers Union
UPP e: Unidirectional Point-to-Point
- e: United Peoples Party
- f: Union de Pêche et de Pisciculture
UPPD f: Union des Patriotes Progressistes et Démocrates
uppf. r: uppfödd
- r: uppförd
uppl. r: upplaga
- r: upplöst
UPPN e: United Peoples Party of Nigeria
UPPS e: Universal Portable Protocol Stack
upps. r: uppsats
UPR d: Umwelt- und Planungsrecht
UPRA e: University Postgraduate Research
U-process e: Umklapp process
UPRONA f: Unité et Progrès National
UPRP f: Union des Propriétaires de la Région Parisienne
UPS e: Unattended Power Source
- e: Under Secretary of the Army
- e: Uninterruptible Power Source
- e: Uninterruptible Power Supply
- e: Uninterruptible Power System
- e: Union of Progressive Students
- e: Univeral Press Syndicate
- e: User Planning System
- e: User Profile System
- f: Union des Producteurs Suisses
- f: Union pour le Progrès Social
- f: Union Progressiste Sénégalaise
UPSA e: Unpaid Scientific Associate
UPSD f: Union Populaire Socialiste et Démocratique
- f: Union pour le Progrès Social et la Démocratie
UPSIT e: University of Pennsylvania Smell Identification Test
Ups J Med Sci [Suppl] e: Upsala Journal of Medical Sciences. [Supplement]
UPST e: Union of Primary School Teachers
UPSTAIRS e: Universal Processing System for Treasuring Up and Analyzing Images in Remote Sensing
UPT e: Universal Personal Telecommunication[s] [service]
- e: Universal Personal Telephony
upt r: utgångspunkt i terrängen
UPTC f: Union Panafricaine et Malgache des Travailleurs Croyants
UPTCAGP e: University of the Philippines Training Center for Applied Geodesy and Photogrammetry
UPTEU e: Uganda Posts and Telecommunications Employees Union
UPTRI f: Union Professionnelle belge des Transporteurs Routiers Internationaux
UPTRN f: Union du Peuple Tchadien pour la Reconstruction Nationale
UPUSA e: United Peoples Unions of South Africa
UPW e: Ultra-Pure Water
UPZA f: Union de Presse du Zaire
UQLJ e: University of Queensland Law Journal
UR e: Unit of Recovery
- e: Universal Reference
- e: Uranium Recovery
- e: Uranium Registry
- e: User Response
- f: Union Régionale
- f: Union Républicaine
- f: Union Routière
- f: Usage Restreint
- f: Usagers de la Route
- N: Undervisningsrådet
- s: Unidad de Recuperación
ur r: urkopplad
URA e: User Requirements Analysis
URAC e: Utilization Review Accreditation Commission (or Committee)
URANIT e: Uranium enrichment company
uraniuma e: uranium total
URANU e: Unar Rombezia African National Union
URAS e: [a commercial CO_2 analyzer]
Urb f: urbanisme
URBA f: Compagnie des Mines d'Uranium de Bakouma
URC e: Coordinated Universal Time
- e: Uniform Resource Characteristic[s]
- e: Uniform Resource Citation
- e: Universal Resource Characteristic
- f: Union des Républicains Camerounais
URCERE e: USSR Research Center for Earth Resources Exploration
URD a: Underretsdom
- f: Union pour le Renouveau et la Démocratie
- f: Union Républicaine Démocratique
- s: Unión Revolucionaria Democrática
URDA e: Upper Receive Descriptor Address register
URDB e: User Requirements Data Base
URDU e: Urban Regional Development Unit
- e: Urban Resource Development Unit

URE e: Uranium Resources and Enrichment
UREAL p: União Revolucionaria dos Estudantes de Angola Livre
UREP e: UNIX RSCS Emulation Program
- e: Unix RSCS Emulation Protocol
URER f: Union Régionale d'Expansion Rurale
URF e: Unit Risk Factor
- e: United Revolutionary Front
- f: Union pour le Renouveau Français
- f: Union Routière de France
- r: Ubåtsräddningsfarkost
Urf r: Undervattensrättningsfarkost
URG e: Universal Radio Group
- e: University Research Grant
Urg f: Uruguayen
URG s: Unión de Radioperiódicos de Guatemala
URGE p: Unión Revolucionaria de la Guinea Ecuatorial
UrhG d: Gesetz über Urheberrecht und verwandte Schutzrechte [Urheberrechtsgesetz]
URI e: Underground Research Inc.
- e: Unfavorable Rotation Index
- e: Uniform Resource Identifier
- e: Universal Resource Identifier
- e: University of Rhode Island, Narragansett
- e: University Research Institute
- f: Union Radio-électrique Internationale
URICA e: Universal Real-time Information Control Administration
- e: Commercial library system from McDonnell Douglas
URIFT e: University of Regina Image File Translator
URIP e: University Research Instrumentation Program
URISA e: American Urban and Regional Information Systems Association
URK N: Ungdommens Røde Kors
URL e: Underground research laboratory
- e: Uniform Resource Locator
- e: Universal Record Locator
- e: Universal Resource Locator
- e: Universal Resource Protocol
- e: Upper Range Limit
- s: Universidad Rafael Landivar
URM d: Unbegrenzte Registermaschine
- e: Underground Radiation Material
- e: Unit Release Method
URN e: Uniform Resource Name
- e: Uniform Resource Number
- e: Unit Requirement Number
- e: Universal Resource Name
- e: Universal Resource Number
- f: Union pour la République Nouvelle
URND f: Union Républicaine Nationale Démocratique
Urol Clin North Am e: Urologic Clinics of North America
Urol Int l: Urologia Internationalis
Urol Nefrol [Mosk] R: Urologiia i Nefrologiia
Urologe A d: Urologe. Ausgabe A
Urol Res e: Urological Research
URP e: Unit Record Processors
- f: Union des Républicains Progressistes
URPE f: Union des Résistants Pour une Europe unie
URPIS e: Urban and Regional Planning Information Systems
URPOP e: Urban Population File
UR/QA e: Utilization Review/Quality Assurance
URR e: Ultra Reliable Radar
- e: Urea Reduction Ratio
- f: Union pour le Rénovation de la République
URRA e: Upper Receive Resource Address register
URSI e: International Union of Radio Science
- f: Union Radio Scientifique Internationale
- s: Unión Radiocientífica Internacional
URSIL f: Union des Responsables des Soins Infirmiers du Luxembourg
URTNA f: Union des Radio-Télévisions Nationales Africaines
URV e: Upper Range Value
Urv r: Urverk
URW e: Union of Rubber Workers
- e: You are welcome [Internet]
URWU e: Uganda Railways Workers Union
US A: Universiteit van Stellenbosch, South Africa
- e: On-orbit Segment
- e: United States
- e: University of Stellenbosch, South Africa
- f: Union Sénégalaise
- f: Urbanisme Souterrain
- r: Underhållsskola
- r: Uppställning
- s: Usted
Us f: Usuel
- r: Utgångsställning
US. e: Unit Separator
us. r: undersökning
.us e: United States [Internet]
U.S e: United States Code
U&S e: Unified and Specified
u/s e: ultrasound
USA e: Union of South Africa
- e: United States Army
- e: United States Attorney
- f: Union Syndicale Agricole
U.S.A. e: United States of America
USAAA e: United States Army Audit Agency
USAAG e: United States Army Artillery Group
USAAS e: United States Army Air Service
USAC e: United States Answer Center
- f: Union des Syndicats Autonomes Camerounais
- f: Union Syndicale de l'Aviation Civile
- s: Universidad de San Carlos
USACC e: United States Army Communications Command
USACDC e: United States Army Combat Development Command
USACE e: United States Army Corps of Engineers
USACEEIA-CONUS e: United States Army Communications Engineering Electronics Installation Agency - Continental United States
USACE INET e: United States Army Corps of Engineers Information Network
USACGSC e: United States Army Command and General Staff College
USACSA e: United States Army Communications System Agency
USACSG e: United States Army Command Support Group
USACSLA e: United States Army Communications Security Logistics Activity
US-AEP e: United States-Asia Environmental Partnership
USAEP e: United States-Asia Environmental Partnership Initiative
USAESEIA e: United States Army Electronic Systems Engineering and Installation Activity
USAETL e: United States Army Engineering Topographic Laboratories
USAEUR e: United States Army Europe
USAF e: United States Air Force
USAFAD e: United States Army Field Artillery Detachment
USAFADWC e: United States Air Force Air Defense Weapons Center
USAFA MID e: United States Air Force Academy Micros in the Dormitory
USAFBMD e: United States Air Force Ballistic Missile Division
USAFE e: United States Air Forces in Europe
USAFETAL e: United States Air Force Environmental Technical Applications Center
USAF/MP EAB e: United States Air Force Manpower Personnel Emergency Actions Book
USAFPP e: United States Air Force Personnel Plan
USAFR e: United States Air Force Regulation
USAFSAM e: United States Air Force School of Aerospace Medicine
USAFSBSS e: United States Air Force Standard Base Supply System
USAFSS e: United States Air Force Security Service
USAIC e: United States Army Intelligence Center
USAIC&FH e: United States Army Intelligence Center and Fort Huachuca
USAICS e: United States Army Intelligence Center And School
US-AID e: United States Agency for International Development
USAID e: United States Agency for International Development
- f: Agence des Etats-Unis pour le développement international
- s: Organismo de los Estados Unidos para el Desarrollo Internacional
USAISEC e: United States Army Information Systems Engineering Center
USAISMA e: United States Army Information Systems Management Activity
USAISSSC e: United States Army Information Systems Software Support Command

USAMMA e: United States Army Medical Materiel Agency
USAMP e: United States Automotive Materials Partnership
USAMSSA e: United States Army Management Systems Support Agency
USAN e: University Satellite Network
USANCA e: United States Army Nuclear Chemical Agency
USANP e: United South African National Party
USAO e: United States Attorneys Office
USAPC e: United States Argos Processing Center
- f: Centre de traitement Argos, Etats-Unis
- s: Centro de Tratamiento Argos, Estados Unidos
Usapl r: Uppsamlingsplats
USA/RC e: Universities Space Automation/Robotics Consortium
USAREUR e: United States Army in Europe
USARIEM e: United States Army Research Institute of Environmental Medicine
USARP e: United States Antarctic Research Program
USARPAC e: United States Army, Pacific
USARSO e: United States Army, South
USART e: Universal Synchronous Asynchronous Receiver Transmitter
USAS e: Unpaid Associate with Subsistence Allowance
- i: Unione Sindicale Africana della Somalia
USASAC e: United States Army Security Assistance Center
USASATCOMA e: United States Army Satellite Communications Agency
USASDC e: United States Army Strategic Defense Command
USASOC e: United States Army Special Operations Command
USASSC e: United States Army Space and Strategic Command
USASSDC e: United States Army Space and Strategic Defense Command
USAT e: Ultra Small Aperture Terminal
USATESAE e: United States Army Test Equipment Support Agency
USATHAMA e: United States Army Toxic and Hazardous Material Agency
USAV f: Union Syndicale des Agences de Voyages
USB e: United Soybean Board
- e: Universal Savings Bank
- e: Universal Serial Bus
- e: University at Stony Brook
- e: University of California, Berkeley
- e: Upper Side Band
- e: Useless Serial Bus
- f: Union Sénégalaise de Banque
- f: Union Syndicale du Bâtiment
USBC e: United States Bureau of Census
USBCF e: United States Bureau of Commercial Fisheries
USBE e: United States Book Exchange
USBGN e: United States Board on Geographic[al] Names
USBM e: United States Bureau of Mines, Department of Interior
USBOC e: United States Bureau of the Census
USBR e: United States Bureau of Reclamation
USBRO e: United States Bases Requirements Overseas
USC e: Unified Soil Classification
- e: Unit Source Listing
- e: United Somali Congress
- e: United States Citizen
- e: United States Code [of laws]
- e: Universal Service Carrier
USCA e: United States Court of Appeals
- e: United States Code, Annotated
USCAP e: United States and Canadian Academy of Pathology
USCAR e: United States Council for Automobile Manufacturers
- e: United States Council for Automotive Research
USCCEMC e: United States Coordinating Committee on Electromagnetic Compatibility
USCE e: United States [Army] Corps of Engineers
USCEEIA e: United States Army Communications-Electronics Engineering Installation Agency
USCEL e: United States Council of EMC Laboratories
USCENTAF e: United States Central Command Tactical Air Force
USCG e: United States Coast Guard
USCGNET e: United States Communications Grid Network
USC&GS e: United States Coast and Geodetic Survey United States Department of Commerce
USCINCAFRED e: United States Commander-in-Chief, Air Force Readiness Command
USCINCARRED e: United States Commander in Chief, Army Forces Readiness Command
USCINCCENT e: United States Commander In Chief, Central Command
USCINCEUR e: United States Commander-in-Chief, European Command
USCINCLANT e: United States Commander In Chief, Atlantic Command
USCINCPAC e: United States Commander-in-Chief, Pacific Command
USCINCRED e: United States Commander-in-Chief, Readiness Command
USCINCSO e: United States Commander-in-Chief, Southern Command
USC/ISI e: University of Southern California/Information Sciences Institute
USCJ e: United Synagogue of Conservative Judaism
USCL e: United Society for Christian Literature London
USCOM e: United States Communications Board
USC-PM e: United Somali Congress-Peace Movement
USCS e: United States Customs Service
- e: Unified Soil Classification System
USC-SNA e: United Somali Congress-Somali National Alliance
USCV f: Union Scientifique Continentale du Verre
USD e: Underwriters Safety Device
- e: United States Dollars
- e: Users Supplementary Documents
- f: Union des Sociaux Démocrates
- f: Unités Speciales de Désarmement
- p: União Social Democrata
USDA e: United States Department of Agriculture
- e: United States Disarmament Administration
- e: Under Secretary of Defense for Acquisition
USDA/APHIS e: United States Department of Agricultures Animal and Plant Health Inspection Service
USDA-CRGP e: United States Department of Agriculture, Competitive Research Grants Program
USDA-CSRS e: United States Department of Agriculture, Cooperative State Research Education and Extension Service
USDA-LISA e: United States Department of Agriculture Low-Input/Sustainable Agriculture
USDA-NRI e: United States Department of Agriculture National Research Initiatives
USDA-NRIGC e: United States Department of Agriculture
USDA-SCS e: United States Department of Agriculture-Soil Conservation Service
USD[A&T] e: Under Secretary of Defense [Acquisition and Technology]
USD[C] e: Under Secretary of Defense [Comptroller]
USDC e: United States Department of Commerce
USDCFO e: United States Defense Communications Field Office
USDEL e: United States Delegation
USDI e: United States Department of [the] Interior
USDL e: United States Department of Labor
USDLA e: United States Distance Learning Association
USDOC e: United States Department Of Commerce
USDOD e: United States Department of Defense
USDOE e: United States Department of Energy
USDOI e: United States Department of the Interior
U.S. DOT e: United States Department of Transportation
USD[P] e: Under Secretary of Defense [Policy]
USDP e: Under Secretary of Defense for Policy
USD[P&R] e: Under Secretary of Defense [Personnel and Readiness]
USDU e: Universities' Staff Development Unit

USDW e: Underground Source[s] of Drinking Water
- e: Usable Supply of Drinking Water
USE e: Ultrasonic Scanning Equipment
- e: UNIVAC Systems Exchange
USEADS e: Ultrasonic Ranging and Data Acquisition System
USEC e: United States Enrichment Corp.
USEIT e: User System Evaluation and Integration Tool
USEMAP e: Urban Survey Experimental Method for Analysing Photo-Data
USEMARCON e: User-controlled generic MARC converter
USENET e: User[s] Network
Usenet d: Users Network [öffentliches weltweites Netzwerk]
USENIX e: Advanced Computing Systems Professional and Technical Association
US[/]EPA e: United States Environmental Protection Agency
USEPL f: Union Syndicale de l'Enseignement Privé Laïque
USERID e: User Identification
USERP e: United Scientists for Environmental Responsibility and Protection
USERREG e: User Registration Interface
USES e: United States Employment Service[s]
USETI e: United States Environmental Training Institute
USEUCOM e: United States European Command
Usext f: Usage externe
USF e: Upper Support Fixture
USF[&]DA e: United States Food and Drug Administration
U.S. Fish Wildl. Serv. Biol. Rep. e: United States Fish and Wildlife Service Biological Report
U.S. Fish Wildl. Serv. Spec. Sci. Rep. e: United States Fish and Wildlife Service Special Scientific Report
USFJ e: United States Forces, Japan
USFK e: United States Forces, Korea
USFMG e: United States Foreign Medical Graduates
USFS e: United States Forest[ry] Service
USFWS e: United States Fish and Wildlife Service
USG d: Umweltgutachter- und Standortregistrierungsgesetz
- e: United States Government
- e: Undoped Silicate Glass
- e: UNIX Support Group
- e: UNIX System Group
- f: Union Socialiste Gabonaise
- s: Unión Solidarista Guatemalteca
USGAO e: United States General Accounting Office
USGC e: United States Global Change
US/GCRP e: United States Global Change Research Program
USGCRP e: United States Global Change Research Program
USGCRPO e: United States Global Change Research Program Office
USGS e: United States Geodetic Survey
- e: United States Geological Survey, Washington
- e: United States Ground Segment
USH e: Universal Sleeve Housing
USHC e: United Seniors Health Cooperative
USHCN e: United States Historical Climatology Network
USHMM e: United States Holocaust Memorial Museum
USI e: User-System Interface
USIA e: United States Information Agency
USIABP e: United States Interagency Arctic Buoy Program
USIAé f: Union Syndicale des Industries Aéronautiques
USIAF f: Union Syndicale des Industries Aéronautiques de France
USIAS f: Union Syndicale des Industries Aéronautiques et Spatiales
USIB e: United States Intelligence Board
USIC e: United States International Carrier
USICA e: United States International Communication Agency
USIE f: Union Suisse des Installateurs Électriciens
USIEF e: United States-Israel Educational Foundation
USIEM f: Union des Syndicats d'Intérêt Economique de Madagascar
USIF e: United States-India Fund for Cultural, Educational and Scientific Cooperation
US-IGBP e: United States Geosphere/ Biosphere Program
USIMA f: Union Sénégalaise d'Industries Maritimes
Usint f: Usage interne
USIPE s: Unidad Sectorial de Investigación y Planificación
USIRF f: Union des Syndicats de l'Industrie Routière Française
USISTC e: United States-Israel S&T Commission
USJCMSP e: United States-Japan Cooperative Medical Science Program
USJF e: Ugandan Struggling Joint Front
USJGOFS e: United States Joint Global Ocean Flux Study
Usk r: Uppbördsskeppare
- r: Utredning om Hälso- och Sjukvården i Krig
USKA d: Union Schweizer Kurzwellenamateure
USKAC e: United States Key Access Code
USL e: United States Laboratory
- e: Universal Sign Language
- e: UNIX System Laboratories
USLE e: Universal Soil Loss Equation
- e: UNIX System Laboratories Europe
USM e: Ultrasonic Scanning Mechanism
- e: Unscheduled Maintenance
- e: Unsharp Masking
- e: User Schedule Message
- e: User Security Matrix
- e: User Support Management
- f: Union Socialiste Malgache
- ma: Universiti Sains Malaysia
USMA e: United States Military Academy
USMAB e: United States Man and the Biosphere
USMAPS e: United States Map Analysis Package System
USMARC e: United States MARC [Machine Readable Cataloging] format
USMB f: Union Suisse des Marchands de meubles de Bureau
USMC e: United States Marine Corps
USMCC e: United States Mission Control Center
USMCEB e: United States Military Communications-Electronics Board
USMLE e: United States Medical Licensing Examination
USMP e: United States Microgravity Payload
USMTF e: United States Message Test Formatting
- e: United States Message Text Format
USN e: United States Navy
- e: Update Sequence Number
- f: Union Scolaire Nigérienne
USNA e: United States Naval Academy
U.S. Natl. Mar. Fish. Serv. Mar. Fish. Rev. e: United States National Marine Fisheries Service Marine Fisheries Review
USNATO e: United States Mission to NATO
USNAVCENT e: United States Naval Forces Central Command
USNAVEUR e: United States Naval Forces Europe
USNC/IHP e: United States National Committee for IHP
USNC/SH e: United States National Committee on Scientific Hydrology
USNEPEC e: United States National Earthquake Predictions Evaluation Council
USNK e: Upper Snake River Basin
USNMAS e: United States National Map Accuracy Standard
USNMR e: United States National Military Representative
USNO e: United States Naval Observatory [library]
USNODC e: United States National Oceanographic Data Center (or Centre)
- f: Centre National de Données Océanographiques des Etats-Unis
- s: Centro Nacional de Datos Oceanográficos de los Estados Unidos
USNPS e: United States National Park[s] Service
USNR e: United States Navy Reserve[s]
USNRC e: United States Nuclear Regulatory Commission
USNSSD e: United States National Seed Storage Department
USNSTN e: United States Naval School, Transportation Management
USN&WR e: United States News and World Report
USO e: Ultra-Stable Oscillator
- e: Universal Service Obligation
- e: Universal Service Order
- e: UNIX Software Operation

USOC e: Universal Service Order[ing] Code
- e: User Support and Operations Centre
USOE e: United States Office of Education
US-OPM e: United States Office of Personnel Management
USOR f: Union Sacrée de l'Opposition Radicale
USORP f: Union des Syndicats Ouvriers de la Région Parisienne
USOS e: United States On-orbit Segment
- e: User Support and Operations Centre
USP e: United States Pharmacopeia
- e: Unified Simulation Program
- e: Universal Sampling Plan
- f: Union des Syndicats Patronaux
- f: Union Syndicale de Province
- f: Université du Pacifique Sud
- s: Universidad del Pacífico Sur
USPA f: Union Syndicale Panafricane
USPACOM e: United States Pacific Command
USPADA s: Unidad Sectorial de Planificación Agricola, MAGA
USP DI e: United States Pharmacopeial Dispensing Information
USPEA e: United States Poultry and Egg Association
USPF f: Union des Syndicats Professionnels Féminins
Usp Fiziol Nauk R: Uspekhi Fiziologicheskikh Nauk
USPG e: United Society for the Propagation of the Gospel London
USPHS e: United States Public Health Service
USPS e: United States Park Service
- e: United States Postal Service
USPTO e: United States Patent and Trademark Office
USQ e: University of Southern Queensland
- e: Unreviewed Safety Question
- e: Unsqueezed
USQD e: Unreviewed Safety Question Determination
USR e: Universities Statistical Record
- e: Unsatisfactory Service Report
- e: US Robotics Inc
- e: User [Service Routines]
- f: Union Sacrée Radicale
USRA f: Union des Services de Renseignements et d'Action
USRADS e: Ultrasonic Ranging And Data System
US-RDA f: Union Soudanaise-Rassemblement Démocratique Africain
USRDS e: United States Renal Data System
USREQ e: United States Request for information
USRP f: Union des Syndicats de la Région Parisienne
USRRO e: United States Refugee Resettlement Office
USRS e: United States Reclamation Service
USRT e: Universal Synchronous Receiver Transmitter
USS e: United States Senate
- e: United States Ship
- e: User Support Subsystem
USSA e: United States Security Authority for NATO Affairs
- e: United States Standard Atmosphere
- e: Universal Storage System Architecture
- e: Unpaid Scientific Associate with Subsistence Allowance
- e: User Supported Software Association
USSAC e: United States Science Advisory Committee
USSALEP e: United States-South Africa Africa Leadership Exchange Program
USSB e: United States Satellite Broadcasting
- e: Upper Single SideBand
US-SDTS e: United States Spatial Data Transfer Specification
USSG e: United States Standard Gauge
USSIE e: Ultraviolet Solar Spectral Irradiance Experiment
USSR e: Union of Soviet Socialist Republics
USSR/EE e: Union of Soviet Socialist Republics/Eastern Europe
USSS e: United States SIGINT System
- e: User Services and Systems Support
USSSP e: United States Science Support Program
USSTRATCOM e: United States Strategic Command
u.st. a: uden stedsangivelse
UST e: United States Testing Company
- e: UHF SATCOM Terminals
- e: Uncontrolled Stationary Tank
- e: Undergraduate Space Training
- e: USAFE Sortie Tables
- f: Union des Syndicats du Tchad
- f: Union Sénégalaise du Travail
Ust f: Ustensile
- r: Ubåtsstyrman
USTA e: United States Telephone Association
UstA d: Unabhängiger StudentInnen-ausschuß
UStÄR d: Umsatzsteuer-Änderungs-richtlinien
USTC e: University of Science and Technology of China
- f: Union Syndicale des Travailleurs Centrafricains
USTF e: Uniformed Services Treatment Facilities
USTG s: Unión Sindical de Trabajadores de Guatemala
USTICA f: Union des Syndicats des Techniciens de l'Industrie, du Commerce et de l'Agriculture
UST[-]ID e: Underground Storage Tank Integrated Demonstration
USTM f: Union Syndicale des Travailleurs de la Métallurgie
USTPO e: United States TOGA Project Office
USTR e: United States Trade Representative
- e: United States Transuranium Registry
USTS e: Ultra High-Frequency Satellite Terminal System
UST[y]M s: Unidad de los Servicios de Talleres y Municionamiento
usu. e: usually
USUAA e: University Students Union of Addis Ababa
USUC e: United States Unilateral Control
USUD e: University Students Union of Dar es Salaam
Usuf f: Usufruit
USUTC e: United States Unified Transportation Command
USV d: Unterbrechungsfreie Stromversorgung
- e: Undersodium Viewing
- e: Unmanned Surface Vehicle
USW e: University of Sweden
usw. d: und so weiter
USWB e: United States Weather Bureau
USWOCE e: United States component of WOCE
USWRP e: United States Weather Research Program
USWU e: Uganda Sugar Workers Union
U. Sydney e: University of Sydney
UT a: Udenrigsministeriets Tidsskrift
- e: Ultrasonic Test
- e: Universal Terminal
- e: Universal Time
- e: University of Tasmania
- e: University of Texas [at Austin]
- e: Upper Tester
- e: User Terminal
- e: User Terminal Universal Thread
- f: [Conférence internationale pour] l'Unité Technique [des chemins de fer]
- f: Unité de Travail
- f: Unité Togolaise
- s: Unidad de Transmisiones
U T e: University Towers
U&T e: Utilities and Telecommunications
- e: Utilities and Transportation
UTA f: Union de Transports Aériens
UTAA s: Unidad de Transmisiones para el Apoyo Aéreo
UTACCS e: USAREUR Tactical Command and Control System
UT-AEC e: University of Tennessee - Atomic Energy Commission
UTAG e: University Teachers Association of Ghana
UTAIN e: USAFE Tactical Air Intelligence Network
UTAIS e: USAFE Tactical Air Intelligence System
UTAM e: Unlicensed PCS Ad Hoc Committee for 2 GHz Microwave
UTASA e: Union of Teachers Associations of South Africa
U Tas LR e: University of Tasmania Law Review
U. Tasmania e: University of Tasmania
UTB f: Union des Travailleurs du Burundi
- f: Union Togolaise de Banque
Utbavd r: Utbildningsavdelning
Utboff r: Utbildningsofficer
UTC e: Under The Counter
- e: Unified Transportation Command
- e: Universal Time Coordinate[d]/[s]
- s: Tiempo Universal Coordinado

UTD d: Untertägige Sonderabfalldeponie im Salzgestein
- f: Union Togolaise pour la Démocratie
UTDA e: Upper Transmit Descriptor Address register
UTE e: Aircraft Utilization Rate
- e: Flying Hour Utilization
- e: User Terminal Element
- s: Unidad Técnica de Explotación
- s: Unión de Trabajadores de Escuintla
UTEN e: Upper Tennessee River Basin
UTEP e: University of Texas, El Paso
UTEPYMI s: Unidad Tecnológica para el Desarrollo Pequeña y Mediana Industria
UTE RATE e: Utilization Rate
UTESP s: Unidad de Trabajadores Estatales y Sectores Populares
UTEXBEL f: Usines Textiles réunies de Belgique
UTEXI f: Union Textile Industrielle de Côte d'Ivoire
UTF a: Udvidet Teknisk Forberedelseeksamen
- e: Unicode Translation Format
utg. N: utgave
- r: utgift
- r: utgiven
UTH e: Upper Troposphere (or Tropospheric) Humidity
- e: USAREUR Transportable Host
UT-H e: University of Texas-Houston Health Science Center
U ThP d: Institut für theoretische Physik der Universität
UTHS e: University of Texas-Houston Health Services
UTI e: Universal Text Interchange
- e: Universal Text Interface
- f: Union des Télécommunications Internationales
- f: Union des Travailleurs Indépendants
UTIAS e: University of Toronto Institute for Aerospace Studies
UTIC e: Under Temperature-Indicating Controller
- e: USAREUR Theater Intelligence Center
UTIG e: University of Texas Institute for Geophysics
UTJ f: Union pour le Travail et la Justice
- s: Unión de Trabajadores de Jalapa
UTK e: University of Tennessee, Knoxville
UTL e: User Trailer Label
- e: Utility Billing
utl. r: utlåtande
UTLB e: Unified TLB
UTL OPS e: Utility Operations
UTM e: Underground Tank Management
- e: Unified Transport Model
- e: Universal Transverse Mercator [map projection]
- e: University Transverse Mercator
- ma: Universiti Teknologi Malaysia
UTMB e: University of Texas Medical Branch at Galveston
UTMDACC e: University of Texas M.D. Anderson Cancer Center
UTM-grid e: Universal Transverse Mercator grid
UTMMI f: Union des Transporteurs et Manutentionnaires de Masses Indivisibles
UTM projection e: Universal Transverse Mercator projection
UTMS e: Universal Mobile Telephone System
UTN e: Universal Tracking Number
U. Tokyo e: University of Tokyo, Japan
UTOPIA e: Universal Test and Operations PHY Interface for ATM
UTor e: Astronomy Library, University of Toronto
UTP e: Universal Personal Telephone Service
- e: Unshielded Twisted-Pair [cable]
- e: USAREUR Theater Plan
utp. N: utkikspost
UTPC e: Universal Television Program Code
utp[ost] r: utkikspost
UTPT e: Unshielded Twisted Pair Transceiver
UTPUR f: Union des Transports Publics Urbains et Régionaux
UTQ s: Unión de Trabajadores de Quetzaltenango
UTR e: Uptime Ratio
- f: Union pour le Triomphe du Renouveau Démocratique
- f: Union Togolaise pour la Réconciliation
utr r: utrustning
UTRA p: Unidade Técnica para a Reestruturação das Alfandegas
UTRAP s: Unidad de Trabajadores del Petén
UTRE p: Unidade Técnica para a Reestruturação de Empresas, Unité Technique pour la Réorganisation des Entreprises
utrobj r: utrustningsobjekt
UTS f: Union des Travailleurs du Sénégal
- f: Union des Travailleurs Scientifiques
U.T.S. e: University of Technology, Sydney
UTS-20 e: Universal Terminal System-20
UTS-40 e: Universal Terminal System-40
UT-SAB p: União dos Trabalhadores do Sector Autônomo de Bissau
UTSE f: Union Technique des Syndicats de l'Électricité
utsk N: utenskjærs
utskr N: utskrevet
- N: utskrivning
UTSL e: Use The Source, Luke [Internet]
U.T.S. [N.Y.] e: Union Theological Seminary, New York
UTT a: Tillægget til Udenrigsministeriets Tidsskrift
- e: Under The Table
UT&T e: Utilities, Telecommunications and Transportation
uttr. r: uttryck
UTUC e: Uganda Trade Union Congress
UTWU e: Uganda Textile Workers Union
UU a: Uafhængig Ungdom
- a: Ungdommens Uddannelsesfond
- e: Urinary Uranium Program
- e: Uuencode/Uudecode
uu r: utrikesutskottet
u.U. d: unter Umständen
UUAB e: University [Student] Union Activities Board at UB
UUCICO e: Unix to Unix Copy Incoming Copy Outgoing
UUCP e: UNIX-to-UNIX Copy [Program]
- e: UNIX-UNIX Copy Protocol
UUD e: UUDecoding
UUE e: UUEncoding
UUG e: UNIX User Group
UUI e: User to User Information
- e: User-To-User Information
UUID e: Univerersally unique IDs
- e: Universal Unique Identifier
UUK d: Urologische Klinik und Poliklinik
UULP e: Unified User Level Protocol
UUM e: Submarine-launched, anti-submarine missile
UUM-44A e: Submarine rocket
UUP e: Ulster Unionist Party
- e: United University Professions
UUS e: User to User Signalling
UUT e: Unit Under Test
U. Utrecht e: University of Utrecht, The Netherlands
UUU e: Unknown UNIX User
UUV e: Unmanned Undersea Vehicle
UUX e: UNIX-to-UNIX-Execute
UV a: Undervands-
- d: Ultraviolett
- e: Ultraviolet [radiation]
u.v.a. d: und vieles andere
- r: utan väsentlig anmärkning
UV A d: Und Viele Andere
- e: UV light in the wavelength range 400 nm to 320 nm
- f: Union Viticole d'Alsace
uv-A e: ultraviolet type A
UV-A d: Ultraviolette Strahlung mit einer Wellenlänge zwischen 400 und 313 Nanometer
UV B e: UV light in the wavelength range 320 nm to 290 nm
- N: Undervannsbåt
uv-B e: ultraviolet type B
UV-B d: Ultraviolett, biologisch wirksame Strahlung mit einer Wellenlänge zwischen 313 und 289 Nanometer
UV-befalet N: Undervisnings- og Velferdsbefalet
UVBI N: Undervannsbåtinspeksjon
UVBV N: Undervannsbåtvåpenet
UV-C d: Ultraviolette Strahlung mit einer Wellenlänge zwischen 289 und 180 Nanometer
UVCF f: Union des Villes et Communes de France
UVEG d: Unfallversicherungs-Einordnungsgesetz vom 7.8.1996
UVF d: Umlandverbund Frankfurt
- f: Union des Vieux de France
- f: Union des Volontaires Français
UVG d: Unfallversicherungsgesetz
- e: Ultraviolet Germicidal Irradiation
- s: Universidad del Valle de Guatemala
uvi r: urinväginfektion
UVI s: Unidad de Vigilancia Intensiva
UVICO-CI f: Union des Villes et Communes de Côte d'Ivoire
U VIENNA e: University of Vienna-Austria

UV/IR e: Ultraviolet/Infrared Line Scanner Imaging Radiometer
U. Virginia e: University of Virginia
UVL e: User Volume Label
UVLS e: Ultraviolet Limb Spectrometer
UVM d: Und Vieles Mehr
Uvm a: Undervisningsmidler
UVMCS e: Ultraviolet Multiple Channel Spectrometer
UVNO e: Ultraviolet Nitric Oxide Spectrometer
UVP d: Unverbindliche Preisempfehlung
- d: Unverträglichkeitsprüfung
- f: Union pour la Vente des Produits
UVPG d: Gesetz über die Umweltverträglichkeitsprüfung
UVPI e: Ultraviolet Plume Instrument
UV-Strahlung d: Ultraviolette Strahlung mit einer Wellenlänge zwischen 180 und 400 Nanometer
UVTP f: Union des Véhicules de Transport Privé
UVV e: Usenet Volunteer Votetakers
UV/VIS e: Ultra Violet/Visible
UV-VS R: Upravlenie Voenno-Vozdushnikh Sil
UVVS e: Ultraviolet Visible Spectrometer
UW A: Universiteit van die Witwatersrand Johannesburg
- e: Ultra Wide
- e: Unconventional Warfare
- e: Under Way
- e: Underwriter
- e: Unit of Work
- e: University of the Witwatersrand Johannesburg
- e: University of Wisconsin
- e: University of Wyoming
UWALR e: University of Western Australia Law Review
UWC e: Underwater Communications
- e: United Workers Congress
UWCSA e: University of Washington Computer Science Affiliates
UWEL e: University Welfare Committee
UWG d: Gesetz gegen den unlauteren Wettbewerb
- e: Underway-Geophysics laboratory
- e: User Working Group
UWis e: Washburn Observatory Library, University of Wisconsin
UWK e: University of West Cape Province
UWM e: Universal Window Manager
- e: University of Wisconsin-Milwaukee
UWOA e: Unconventional Warfare Operational Area
UWRA e: Urban Water Research Association
UWS e: Ultrix Workstation Software
UWSLF e: United Western Somali Liberation Front
UW/UD e: University of Wisconsin/ University of Denver
UWUSA e: United Workers Union of South Africa
UXO e: Unexploded Ordinance
U. Yamagata e: University of Yamagata, Japan
UYVDRA e: Upper Yarra Valley and Dandenong Regional Authority
UZB f: Union Zairoise de Banques
UZM e: Unsaturated Zone Monitoring
UZOPASE f: Union Zonale pour la Promotion des Activités Socio-Educatives
UZS d: Uhrzeigersinn

V

V a: [Radikale] Venstre
- a: Validitet
- a: Vest
- d: Anpassung eines bildauflösenden Lichtempfängers
- d: Seminarvortrag
- d: Vorlesung
- d: Vertikale Höhe der CCD-Matrix
- e: Victoria
- e: Volt[s]
- e: Volume of distribution. When referring to urea, this is total body water
- f: [chèque à] Vue
- f: Vatican
- f: Vendredi
- f: Vice-
- f: Vieillards
- f: Ville
- f: Virement
- f: Visa
- f: Visite
- f: Votre
- f: Voyageurs
- f: Voyez
- N: Vakt
- N: Vestlandet
- r: Vänster
- r: Varv
- r: Väster
- r: Väte
- r: Växelpunkt

v d: Schlupfvariablen-Vektor
- d: Verkleinerungsfaktor
- f: véhicule
- N: vers
- N: våpenverker
- N: vestlig

V- r: Vice
- r: Vikt

v. a: ved
- e: village
- N: vekt
- N: vest
- r: vänd!
- r: volym
- s: véase

V. a: Vester
- e: Viscount
- e: Volt

V1 e: takeoff decision speed
V2 e: minimum takeoff safety speed at 35'
V6 e: A six-cylinder engine with the cylinders in a vee pattern
V8 e: An eight-cylinder engine with the cylinders in a vee pattern
V12,5 r: hastighet på 12,5 m
V-22A e: Osprey
V.24 d: CCITT-Norm für eine serielle Schnittstelle
V.32 e: CCITT standard for 9600 bps modems

VA a: Valby
- d: Virtuelle Akustik
- d: Volladdierer
- e: Department of Veterans Affairs (or Administration)
- e: USN attack squadron
- e: Vatican City State
- e: Veterbi Algorithm
- e: Virginia
- e: Virtual Acoustic
- e: Visual Age
- e: Volcanic Ash
- e: Volt-Amp[ere][s]
- e: Vulnerability Assessment
- f: [par] Voie Aérienne
- r: Vetenskapsakademien
- s: Vicealmirante

Va e: Vanadium
- r: Anslagshastighet
- r: Vakuumtorkad

Va. e: Virginia
V.A. e: Lady of the Royal Order of Victoria & Albert
VAAFB e: Vandenberg Air Force Base
VAB a: Varebestemmelsesreglementet
- a: Vejle Amts Boligselskab
- e: Value Added Business
- f: véhicule de l'avant blindé

VAbd e: Van Allen belt dosimeter
VABIS r: Vagn Aktie Bolaget i Soedertaelje
Va Bk e: Virginia Book Company
VABM e: Vertical-Angle Bench Mark
VAC e: Vacuum and component
- e: Visual Age C++
- e: Volts [of] Alternating Current
- f: Voie Aérienne Civile

V ac e: Volts alternating current
VACAB e: Veterans Administration Contract Appeals Board
Vaccine e: Vaccine
VAC DATA e: Vacuum and Component Data
VACM f: Courantomètre fournissant des Valeurs Moyennes
- s: Corrientímetro de promedio vectorial

VACR s: Vehículo Aéreo de Control Remoto
VACRP e: Victorian Association for the Care and Resettlement of Prisoners
VACS e: Visual Access Control System
Vac. Sci. Technol. e: Vacuum Science and Technology
VAD e: Value Added Dealer
- e: Value Added Desaster (or Disaster)
- e: Value Added Distributor
- e: Value Added Driver
- e: Variable Angle Drive
- e: Vascular Access Device
- e: Velocity Azimuth Display
- e: Vertical Assembly/Disassembly
- e: Voice Activity Detection
- e: Voluntary Aid Detachment
- e: Volunteer Aid Detachment

VAd e: Veterans Administration
VADB e: Visitor Access Database
vädbtr r: väderleksbiträde
VADC e: Voltmeter Analog-to-Digital Converter
VADD e: Value Added Disk Driver
VADS e: Value Added Data Service
- e: Vulcan Air Defense System

VÄDS r: Väderskola
VADSL e: Very-High-Rate Asymmetric Digital Subscriber Line
Vädte r: Väderlekstekniker
VAE e: Visible Airglow Experiment[-Atmospheric Explorer-C]
vær. N: værelse
VAFB e: Vandenberg Air Force Base
VAFC e: VESA Advanced Feature Connection (or Connector)
VAFI e: Victorian Association of Forest Industries
VAFSTF e: Vaccine Associated Feline Sarcoma Task Force
VAFTAD e: Volcanic Ash Forecast Transport and Dispersion

VAG d: Gesetz über die Beaufsichtigung der Versicherungsunternehmen [Versicherungsaufsichtsgesetz]
- d: Volkswagen-Audi Gruppe
- e: Value-Added Grids
- e: VRML Architecture Group
- e: VRML Author Group
- f: Veille de l'Atmosphère Globale
- s: Vigilancia de la Atmósfera Global
VAGA e: Visual Artists and Galleries Association
vag charge e: vagrancy charge
Vage f: Village
VAGI e: VESA Advanced Graphic Interface
VAH d: Verbundprojekt Agrarökosysteme Halle, Universität Halle
- e: Veterans Administration Hospital
VA HOSP e: Veterans Administration Hospital
VAI e: Video Artists International
VAID e: Variable Air Inlet Duct
VAIO e: Video Audio Integrated Operation
VAIS e: Virginia Association of Independent Schools
VAKA d: Vereinigung Aargauischer Krankenhäuser
VAKSA A: Vakbondraad van Suid-Afrika
vaktt r: vakttrupp
Val a: Valby i København
VAL e: USN light attack squadron
- e: Validity Checks
- e: Virtual Access Method
- e: Voice Application Language
val e: valyl
- f: valeur
Val. e: Valanginian
VALA e: Viewers and Listeners Association
VALB e: Veterans of the Abraham Lincoln brigade
VALCO e: Volta Aluminum Company
VALFAC e: Validation Facility
VALIC e: Variable Annuity Life Insurance Company
VALLEY e: Meteorological Model to Calculate Concentrations on Elevated Terrain
VAM d: Verband der Anbieter von Mobiifunkdiensten, Bonn
- e: Variational Analysis Method
- e: Vault Air Monitor
- e: Vesicular-Arbuscular Mycorrhizae
- e: Virtual Access Method
VAMC e: Veterans Administration Medical Center
Va Med Q e: Virginia Medical Quarterly
VAMOSC e: Visibility and Management of Operating and Support Costs
vamp e: A nefarious, manipulative seductress
VAMP e: Validation of Model Project
- e: Vesicle-Associated Membrane Protein
vampire e: The mosquitos of mammalia
Van a: Vanløse
VAN e: EYCK Visual Arts Network for the Exchange of Cultural Knowledge
- e: HERL-RTP Van System
- e: Vaccine Awareness Network
- e: Value Added Network
- f: Réseau à Valeur Ajoutée
- s: Red de Valor Añadido
VANEA e: Voucher Accounting for Net Expenditures Accrued
VAN EYCK e: Visual Arts Network for the Exchange of Cultural Knowledge
VANG s: Vanguardia
vanl. r: vanligen
van. pub. e: vanity publisher
van. pub. e: vanity publishing
VANS e: Value Added Network Services
Van Sym e: Vancouver Symphony Orchestra
VAO e: Valence Atomic Orbital
VAP e: Vaccine Action Program
- e: Value Added Procedures
- e: Value Added Process[es]
- e: Value Added Product
- e: Vas Applications Data Processor System
- e: Vendor Assistancc Phase
- e: Victims Assistance Prograrn
- e: Victims Assistance Project
- R: Vilivnoj Aviatsionnii Pribor
VAPI e: Visual Approach Path Indicator
VAPI e: Virtual [ISDN] Application Programming Interface
- e: Voyetra Applications Program Interface
VAPI e: Visual Approach Path Indicator
VAPOX e: Vapor-deposited Oxide
Vaptj C r: Vapentjänst Chef
VAQ e: USN Tactical Electronic Warfare Navy
- e: Visiting Airman[s] Quarters
VAR e: Value Added Remarketer
- e: Value Added Reseller
- e: Value Added Retailer
- e: Voltage Amp Reactive
- e: Vulnerability Assessment Review
var. a: variabel
- a: variant
- a: varierende
- a: varietet
- e: varietas
- e: variety
Var. f: Variété botanique
VarAnz d: Anzahl der Optimierungsvariablen
VARCS e: Vehicle Application and Replacement Categorical System
VARIS e: Virginia Resource Information System
varistor e: variable resistor
värnkomp r: värnkompani
VARP e: Veterans Administration Procurement Regulations
var[r]. e: variant[s] of
vars e: varieties
vars. e: variant[s] of
varv/min r: varv i minuten
VAS e: Value Added Service[s]
- e: Vehicle Allocation and Summary
- e: Victorian Agricultural Strategy
- e: VISSR [Visible Infrared Spin Scan Radiometer] Atmospheric Sounder
- e: Visual Art Studies
- e: Voluntary and Alternative Service
- f: Sondage de l'atmosphère par radiomètre à balayage par rotation dans le visible et l'infrarouge
- s: Sonda Atmosférica de Radiómetro de Barrido en el Visible y el Infrarrojo
Vasa e: Vasa
VASCAR e: Visual Average Speed Computer And Recorder
VASCOM e: Value Added Services Network Communication
VASP e: Vas Processor
VASSP e: Virginia Assodation of Secondary School Principals
VAST e: Variable Array Storage Technology
- e: Vector Alignment Search Tool
- e: Vector and Array Syntax Translator
- e: Virtual Archive Storage Technology
VAT e: Value Added Tax
- e: Vertical Axis Turbine
- e: Video Audio Tool
- e: Virtual Allocation Table
- f: Valeur Agronomique et Technologique
Vat. a: Vatikanet
V-ATE e: Vertical AnisoTropic Etch
VATESI e: Lithuanias Nuclear Power Safety Inspectorate
VATS e: Value Added Telecommunications Service
- e: Video-Assisted Thoracoscopic Surgery
- e: Virginia Assistive Technology System
VATTR e: Value-Added-Tax Tribunal[s]
Vauc f: Vaucluse
vaud f: vaudeville
VAUS e: Value Added Utilisation System
VAV e: Value Application Venders
- e: Variable Air Vent
- e: Variable Air Volume[s]
VAVI e: Advanced Video Interface
VAW e: USN airborne early warning squadron
VAW/ETHZ d: Versuchsanstalt für Wasserbau, Hydrologie und Glaziologie, Eidgenössische Technische Hochschule Zürich
VAwS d: Verordnung über Anlagen zum Lagern, Abfüllen und Umschlagen wassergefährdender Stoffe
VAX e: Virtual Access Extension
- e: Virtual Address Extension
- e: Virtual Address[ing] Extended
VAXBI e: VAX Bus Interconnect
VAXen e: Two or more VAXes
VAX Prof. e: VAX Professional
VAX/VMS e: Virtual Address Extension/ Virtual Memory System

VB e: Vacuum Bottoms
- e: Valid Behaviour
- e: Vehicle Barrier
- e: Venture Business
- e: Visual Basic
- e: Void Box
- f: Valeurs Bancaires
- f: Virement Bancaire
vb r: vedettbåt
vb. e: verb
v.b. r: vid behov
VBA e: Visual Basic for Application[s]
VBAP e: Voice-Band Audio Processor
VBB d: Verbund der Bibliotheken der obersten Bundesbehörden
VBC e: Vancouver British Columbia
VBD d: Verband der Bayerischen Druckindustrie e.V.
VBE e: Valence Band Edge
- e: VGA standard BIOS Extentions
- e: VESA [Video Electronics Standards Association] BIOS [Basic Input-Output System] Extension
VBE/AI e: VESA BIOS Extension/Audio Interface
VBF e: USN fighter-bomber squadron
VBFS e: Very Big File System
VBI e: Vertebro-Basilar Insufficiency
- e: Vertical Blank Interval
- e: Vertical Blanking Interrupt
V.32bis e: CCITT standard for 14400 bps modems
VBITS e: Video Based Information Transmission System
VBK d: Verband Bernischer Krankenhäuser
VBL e: Vertical Blanking
- e: Vertical Bloch Line
vbl. sb. e: verbal substantive
VBN d: Vemitteltndes Breitbandnetz
- d: Vermitteltes Breitbandnetz
- d: Vorläufer Breitband Netz
VBNS e: Very high[speed] Backbone Network Service
VBOK e: EMSL-Vegas Library Book System
V-bomb r: Vätebomb
VBP e: Vietnam Boat People
VBR e: Variable Bit Rate
- e: Ventricular Brain Ratio
VBR delay sensitive e: Variable Bit Rate delay sensitive
VBR delay tolerant e: Variable Bit Rate delay tolerant
VBR non-interactive e: Variable Bit Rate non-interactive
VBR[-]NRT e: Variable Bit Rate-Non-Real-Time
VBR[-]RT e: Variable Bit Rate-Real-Time
VBRUN e: Visual Basic Runtime
VBS e: Vacation Bible School
VBS d: Verteiltes Betriebssystem
- e: Valence Band Splitting
VBSG d: Verband Bernischer Schulen im Gesundheitswesen
VBSZ d: Verband der Blutspendezentren
VBU d: Verband der Betriebsbeauftragten für Umweltschutz e.V.
- d: Vereinigung deutscher Biotechnologie-Unternehmen
VBWR e: Vallecitos Boiling-Water Reactor
VBX e: Visual Basic Extension[s]
VC e: acuity of Color Vision
- e: St Vincent & Grenadines
- e: USN composite squadron
- e: Vacant Code
- e: Vacuum Cleave
- e: Vapor Compression
- e: Vice-Chairman
- e: Vice-Chamberlain
- e: Vice-Chancellor
- e: Vice-Consul
- e: Video Camera
- e: Video Conferencing
- e: Viet Cong
- e: Virtual Channel
- e: Virtual Circuit
- e: Virtual Connection
- e: Virtual Console
- e: Viscount
- f: Vice-Consul
- f: Victimes Civiles
- f: Visa Consulaire
- f: Voix Consultative
- f: Vote de Confiance
- f: Voyage Circulaire
- f: Voyageur de Commerce
- s: Vehículo de Combate
- s: Vicario Castrense
Vc e: Design cruise speed
V C e: Violoncello
V.C. e: Victoria Cross
VC++ e: Visual C++
VCA e: Vacant Code Announcement
- e: Verification Control Authority
- e: Vermont Council on the Arts
- e: Virginia Correctional Association
- e: Voltage Controlled Amplifier
- e: Volunteer Civic Association
- s: Vehículo de Colchón de Aire
- s: Vehículo de Combate Acorazado
VCAC e: Vice Chancellors Advisory Council
VCAS e: Vice-Chief of Air Staff
VCAT e: Veterinary College Admission Test
VCC e: Virtual Channel Connection[s]
- e: Virtual Circuit Connection
VCC e: Venture Capital Corporation
VCCI e: Voluntary Control Council for Intensive
VCCL e: Victoria Council for Civil Liberties
VCCñ s: Vehículo de Combate Cañón [autopropulsado 155 mm]
VCCR e: Vienna Convention on Consular Relations
VCCS e: Voltage-Controlled Current Source
VCD d: Verkehrsclub Deutschland
- e: Variable Center Distance
- e: Vibrational Circular Dichroism
- e: Virtual Communications Driver
- e: Visual Core-Description form
VCE e: Variational Cumulant Expansion
VCE e: Venice, Italy [airport]
VCEIL e: Vaisala Ceilometer
VCEM e: EMSL-Vegas Chemical Inventory
VCF e: Voltage Controlled Filter[s]
VCF&L e: Department of Conservation, Forests and Lands
VCFR e: Vice Chancellor for Research
VCG e: Vice-Consul General
VCI e: Video Cursor Interface
- e: Virtual Channel Identification
- e: Virtual Channel Identifier
- e: Virtual Channel Indicator
- e: Virtual Circuit Identifier
- e: Virtual Connection Identifier
VCIC e: Verrnont Crime Information Center
VCIF e: VESA Display Information Format
VCIGS e: Vice-Chief of the Imperial General Staff
VCILP e: Villanova Center for Information Law and Policy
VCIP e: Veterans Cost-of-Instruction Program
VCIP e: Visual Communication and Image Processing
VCJCS e: Vice Chairman Joint Chiefs of Staff
VCL e: Vegetation Canopy Lidar
- e: Virtual Channel Link
- e: Visual Component Library
- e: Visual Computing Library
Vcl f: Vaucluse
VCLF e: Vertical Cask Lifting Fixture
VCM e: Version Compatibility Matrix
- e: Vibrating Coil Magnetometer
- e: Victorian Chamber of Mines
- e: Vinyl Chloride Monomer
- e: Voice Channel Module
- e: Voice Coil Motor
- e: Volatile Condensed Material
VC/MTM e: Variable Constellation/Multi-Tone Modulation
VCN e: Vendor Contact Notice
VCNA e: VTAM [Virtual Telecommunication Access Method] Communication Network Application
VCNS e: Vice-Chief of Naval Staff
VCO e: Voltage Controlled Oscillator
- e: Voluntary Conservation Organisation
- r: Värnpliktsvägrarnas Centrala Organisation
VCOAD e: Voluntary Committee on Overseas Aid and Development
VCOC e: Venture-Capital Operating Company
VCOI e: Veterans Cost Of Instruction
VCOS e: Visible Caching Operating System
- e: Visual Caching Operating System
VCP e: Vacuum Cup Pencil
VCP e: Vector Control Processor
- e: Voluntary Contribution Plan
- e: Voluntary Cooperation Programme
VCPC s: Vehículo de Combate Puesto Comando
VCP [ES] e: Voluntary Cooperation Program[me] [Equipment and Services]
VCP [F] e: Voluntary Cooperation Program [Fund]
VCP[F] e: Voluntary Cooperation Fund
VCPI e: Virtual Control Program Interface

VCP[TEM] e: Technical Training, Equipment and Maintenance Programme
VCPU e: Video Compression Processor Unit
VCR e: Variable Call Routing
- e: Video Cassette Recorder
- e: Voluntary Conten Rating
VCR e: Voluntary Compliance Resolution
VCRM e: Verification Cross Reference Matrix
VCRT s: Vehículo de Combate Recuperador de Tanques
VCS e: Vice Chief of Staff
- e: Vidicon Camera System
- e: Virtual Circuit Switch
- e: Virtual Circuit System
VCSA e: Vice Chief of Staff, Army
VCSAF e: Vice Chief of Staff
VCSEL e: Vertical-Cavity Surface-Emitting Laser
VCT e: Vent Catch Tank
V&CT e: Verification and Control Technology
VCTM s: Vehículo de Combate de Transporte de Mortero
VCTP s: Vehículo de Combate de Transporte de Personal
VCU e: Voice Communication Unit
VCVS e: Voltage-Controlled Voltage Source
VCXO e: Voltage Controlled Crystal Oscillator
VD e: Vandyke
- e: Valentines Day
- e: Vapor Deposition
- e: Venereal Disease
- e: Volume Descriptor
- e: Voluntary Departure
- f: Valeur Déclarée
- f: Vente du Détail
- f: Vérification Douanière
- f: Versement Différé
- f: Victime du Devoir
- f: Visa de Douane
- R: Vertikalnye Dvigateli
Vd e: Maximum dive speed
- f: Vendée
Vd. s: Usted
V+D e: Voice + Data
V.D. e: Volunteers Officers Decoration
- r: Verkställande Direktör
v.d. r: vardera
VDA e: Variable Digital Amplifier
- e: Variational Data Assimilation
- e: Velocity Dealiasing Algorithm
Vda. s: viuda
VDAC e: Video Digital Analog Converter
v-dagen N: victory dagen
Vdb a: Vedbæk
VDB a: Vestjydsk Data Bureau
- e: Validation Data Base
- e: VISSR [Visible Infrared Spin Scan Radiometer] Data Base
VDBB e: Validation Data Base Board
VDBV f: Vous Demande de Bien Vouloir
VDC e: Civil Engineering Computimng Center at Lund Institute of Technology
- e: Virtual Device Coordinates
- e: Visual Development Center
- e: Volts [of] Direct Current
Vdc e: Volts [of] direct current
VDCE e: Department of Conservation and Environment, Victoria
VDD d: Verein Deutscher Druckingenieure e.V.
- e: Virtual Device Driver
- e: Virtual Display Driver
VDDM e: Virtual Device Driver Manager
VDE d: Verband Deutscher Electrotechniker
- e: Video Display Editor
- e: Video-Decompression Engine
- e: Visual Development Environment
VDEU d: Vertiefung Deutsch
VDF d: Variables Digitalfilter
- e: Vacuum Drum Filter
- e: Variable Data Field
- e: Variable Data File
- e: Variable Digital Filter
VdF i: Vigili del Fuoco
VDFM e: Virtual Disk File Manager
- e: Virtual Disk File Manager
vdH d: vor der Höhe
VDHA e: Virginia Dental Hygienists Association
VDI d: Verein Deutscher Ingenieure e.V., Düsseldorf
- e: Video Device Interface
- e: Video Display Interface [system]
- e: Voice-Data-Image
VDI Ber. d: VDI Berichte [R]
VDIF e: VESA Display Information File
V di R i: Virtuosi di Roma
VDK d: Verband der Kriegsbeschädigten, Kriegshinterbliebenen und Sozialrentner
VDL e: Van Diemans Land [Tasmania]
- e: Vendor Data List
- e: Vienna Definition Language
VDM e: VAS Digital Multiplexer
- e: Virtual Device Metafile
- e: Virtual DOS Machine
- i: Verbi Divini Ministro
VDMAD e: Virtual Direct Memory Access Device
VDMK N: Veidirektoratets Militærkontor
VDMOS e: Vertical Double-diffused MOS power transistor
vdm['s] e: video disc machine[s]
VDM-SL e: Vienna Development Method-Semantic Language
VDN d: Verband der Druckindustrie Nordrhein e.V.
- e: Virtual Data Network
- r: Varudeklarationsnämnden
vdo e: video
Vdo. s: Viudo
vdP A: van der Pauw
VDP e: Vehicle Down for Parts
- e: Video Datagram Protocol
- e: Voice Data Protection
VDR e: VHF Data Radio
- e: Video Data Recorder
VDRL e: Venereal Disease Research Laboratory
VDS d: Verband Deutscher Studentenschaften
- e: Vapor Distribution System
- e: Variable Depth Sonar
- e: Vehicular Data System
- e: Virtual DMA Service[s]
- e: Voice Distribution System
- N: Vesterålens Dampskibsselskab
Vds. s: Ustedes
VDSI e: Video Distribution Subsystem Interface
VDSL e: Very high data/bit rate Digital Subscriber Line
- e: Very High-speed Digital Subscriber Line
VDT e: Video Dial Tone
- e: Video Display Terminal[s]
- e: Visual Display Terminal
vdt's e: video display terminals
VDTT e: Velocity-Density Temperature Tree
VDU e: Vertical Diffusion
- e: Visual Display Unit
VDUC e: VAS Data Utilization Center
- e: VAS Data Utilization Console
VDV d: Vorbestellte Dauerwählverbindung
vdW A: van der Waals
VDZ d: Verband Deutscher Zeitungs- und Zeitschriftenverleger
VE a: Varmeenhed[er]
- a: Vester
- d: Verantwortliche Erklärung
- d: Vermittlungseinrichtung
- d: Volksempfänger
- e: Vertical Exaggeration
- e: Visual Emissions
- e: Vitro Engineering Division Automation Industries
- e: Volumetric Efficiency
- f: Vérification d'Écritures
- f: Ville Éternelle
- f: Visite d'Embauchage
VEA d: Verein für Europäisch-Arabische Zusammenarbeit e.V.
VEB e: Bolivar, Venezuela
VEB Verlag Tech. Mon. Tech. Rev. d: VEB Verlag Technik, Monthly Technical Review
VEC d: Vechta, Niedersachsen
- e: Vascular Endothel Cell
- e: Virtual Embedded Circuitry
VEDA e: Victorian Eastern Development Association
VEDÄG d: Vereinigung deutsch-schweizerischer Ärztegesellschaften
V-E Day e: Victory in Europe Day-May 8, 1945
VEDH d: Vereinigung eidgenössischer diplomierter Haushaltleiterinnen
ved hj. af a: ved hjælp af
vedk. a: vedkommende
vedl. N: vedlagt
vedr. a: vedrørende
- N: vedrorende
vedv. a: vedvarende
Vée f: Vallée
VEEC e: Victorian Environmental Education Council
vee dee e: venereal disease
vee dee e: visiting dignitary

VEEGA e: Venus Earth Earth Gravity Assist
VEENAF e: Vietnamese Air Force
Veep e: Vice President
VEG e: Very Evil Grin
veg e: very evil grin [Internet]
VEGA e: Video-7 Enhanced Graphics Adapter
Vegetatio e: Vegetatio [R]
VEGETATION e: A program to monitor, on a daily basis, terrestrial vegetation over through remote sensing at regional and global levels
Veg Soc e: Vegetarian Society
Ve[h] s: Vehículo
Véh f: Véhicule
vehic manslgtr e: vehicular manslaughter
VEI e: Volcanic (or Volcano) Explosivity Index
VEIC e: Value Engineering Incentive Clause
VEINS e: Variability of Exchanges in Northern Seas
vejl a: vejledning
vejl. a: Vejledende
Vel s: Velocidad
VELKSWA d: Vereinigung Evangelisch-Lutherischer Kirchen Südwestafrika
Velours d: Veloursausstattung
VEM A: Vereinigste Evangelische Mission
VEMAP e: Vegetation/Ecosystem Modeling and Analysis Project
VEMASS e: Vocational Consultant/ Medical Advisor Services Staff
VEMM e: Virtual Expanded Memory Manager
VEMMI e: Versatile Multimedia Interface
VEN d: Vereinfachter Entsorgungs-nachweis
- e: Viral Erythrocytic Necrosis
ven e: venetian [blind[s]
- f: vendredi
- i: venerdi
- d: venezianisch
Vend f: vendredi
vend. mach e: vending machine
VENTEX e: Vent[ing] Experiment
VENTS e: A NOAA PMEL program established in 1984 to focus research on determining the oceanic impacts and consequences of submarine hydrothermal venting
vents. e: ventilators
VENUS e: Low-flux pressure vessel mockup
VEO e: Visible Emission Observation
- e: Visual Emissions Observation
- e: Voluntary Environmental Organisation
VEP e: Visual Evoked Potential
- f: Vente aux Enchères Publiques
VEPCO e: Virginia Electric Power Company
VEPWMF e: Voivodship Environmental Funds
VER d: Verden/Aller, Niedersachsen
- e: VERONICA
- e: Very Easy Rodent-oriented net-wide index to computer archives
- e: Voluntary Export Restraint
ver e: yerse[s]
Ver f: Versement
- f: Vivier
VerbrennM d: Institut für Technische Verbrennung
VerbrKrG d: Verbraucherkreditgesetz
VERDIN e: VLF/LF Multi-Channel Broadcast System
VerfahrT d: Institut für Verfahrenstechnik und Dampfkesselwesen
VerfGGBbg d: Gesetz über das Verfassungsgericht des Landes Brandenburg
Verh Dtsch Ges Pathol d: Verhandlungen der Deutschen Gesellschaft für Pathologie
Verh. Dtsch. Zool. Ges. d: Verhandlungen der Deutschen Zoologischen Gesellschaft
Verh. Ges. Ökol. d: Verhandlungen der Gesellschaft für Ökologie
Verh. Int. Ver. Theoret. Angew. Limnol. d: Verhandlungen der Internationalen Vereinigung für Theoretische und Angewandte Limnologie
Verh K Acad Geneeskd Belg n: Verhandelingen - Koninklijke Academie voor Geneeskunde van Belgie
Verh. Nat.forsch. Ges. Basel d: Verhandlungen der Naturforschenden Gesellschaft in Basel
Verh. Schweiz. nat.forsch. Ges. d: Verhandlungen der Schweizerischen naturforschenden Gesellschaft
VERIP e: Voluntary Early Retirement Incentive Program
VERITY e: Virtual and Electronic Resources for Information Skills Training for Young People
verk N: verksted
Verk. avd./Bilkol. 2 N: Verkstedsavdelingen ved Biltrenkolonne 2
VERK/BFS d: Sektion Verkehr
Verkgrp HFT N: Verkstedgruppen av Hærens Forsyningstropper
Verkh R: Verkhniy
VerkPBG d: Verkehrswegeplanungs-beschleunigungsgesetz
Verkskola r: Verkstadsskola
verk v n: verkorting van
verkst. r: verkställande
VerlG d: Gesetz über das Verlagsrecht
VermBDV e: Verordnung zur Durchführung des Dritten Vermögensbildungsgesetzes
VermBG e: Gesetz zur Förderung der Vermögensbildung der Arbeitnehmer [Vermögensbildungsgesetz]
Vermess. Photogramm. Kult.tech. d: Vermessung, Photogrammetrie, Kulturtechnik
VermG d: Vermögensgesetz
VermRÄndG d: Gesetz zur Änderung des Vermögensgesetzes und anderer Vorschriften
Vern e: Vernons Law Reports
VERNET e: Virginia Educational Research Network
Veröff. Geobot. Inst. Eidgenöss. Tech. Hochsch., Stift. Rübel Zür. d: Veröffentlichungen des Geobotanischen Institutes der Eidgenössischen Technischen Hochschule, Stiftung Rübel in Zürich
Veröff. Landesstelle Nat.schutz Landsch.pfl. Baden-Württ. d: Veröffentlichungen der Landesstelle für Naturschutz und Landschaftspflege in Baden-Württemberg
Veröff. Nat.-Mus. Luzern d: Veröffentlichungen Natur-Museum Luzern
VERONICA e: Very Easy Rodent-Orientated Netwide Index of/to Computer[ized] Archives
VerpackV d: Verpackungsverordnung
VERR e: Vehicle Emissions Reduction Responsibility
- e: Verify Read Access
Vers.anst. Wasserbau Hydrol. Glaziol. d: Versuchsanstalt für Wasserbau, Hydrologie und Glaziologie
Versicherungsmedizin d: Versicherungs-medizin
VERSN e: Version
Ver St d: Vereinigte Staaten
Vertebr. e: Vertebrate[s]
VERTEX e: Vertical Transport and Exchange of material in the upper ocean performed in the California Current
- e: Vertical Transport Experiment
VERW e: Verify Write Access
Verwaltungsblatt PLV d: Verwaltungs-blatt des Preußischen Landesverbandes jüdischer Gemeinden Vol. 1 [1923] - Vol. 12, Nr. 2 [1934], dann: Gemeindeblatt für die jüdischen Gemeinde
VerwZG d: Verwaltungszustellungsgesetz
VES e: Vacuum Exhaust System
- e: Value Engineering Study
- e: Veterans Employment Service
- e: Video Encoding Standard
- s: Vestuario, Equipo y Subsistencias
ves e: very evil smile [Internet]
VESA e: Video Electronics Standards Association
VES/ENS s: [almacén de] Vestuario y Enseres
VESID e: [Office of] Vocational and Educational Services for Individuals with Disease
VESP e: Video Enhanced Service Providers
VESS e: Voluntary Euthanasia Society of Scotland
VEST e: VAX Environment Software Translator
- e: Vehicle Electrical System Test
Vest s: Vestuario
VESTA s: [almacén de] Vestuario
vestl. N: vesentlig
Vestn Akad. Navuk BSSK Biyal. Navuk R: Vestsi Akademii Navuk BSSR, Seriya Biyalogichnykh Navuk
Vestn Akad. Navuk BSSR, Fiz. Energ. Navuk R: Vestsi Akademii Navuk BSSR, Seriya Fizika Energetychnykh Navuk
Vestn Akad. Navuk BSSR, Fiz.-Mat R: Vestsi Akademii Navuk BSSR, Seriya Fizika-Matematichnykh Navuk
Vestn Akad. Navuk BSSR, Fiz.-Tekh. R: Vestsi Akademii Navuk BSSR, Seriya Fizika-Technichnykh Navuk

Vestn Akad. Navuk BSSR, Khim. Navuk R: Vestsi Akademii Navuk BSSR, Seriya Khimichnykh Navuk
Vestn Akad. Navuk BSSR Ser. Biyal. Navuk R: Vestsi Akademii Navuk BSSR, Seriya Biyalogichnykh Navuk
Vestn Akad. Navuk BSSR, Ser. Fiz. Energ, Navuk R: Vestsi Akademii Navuk BSSR, Seriya Fizika Energetychnykh Navuk
Vestn Akad. Navuk BSSR, Ser. Fiz.-Mat. R: Vestsi Akademii Navuk BSSR, Seriya Fizika-Matematichnykh Navuk
Vestn Akad. Navük BSSR, Ser. Fiz.-Tekh. R: Vestsi Akademii Navuk BSSR, Seriya Fizika-Technichnykh Navuk
Vestn Akad. Navuk BSSR, Ser. Kkim. Navuk R: Vestsi Akademii Navuk BSSR, Seriya Khimichnykh Navuk
Vestn Khir Im I I Grek R: Vestnik Khirurgii Imeni I. I. Grekova
Vestn oftalmol R: Vestnik Oftalmologii
Vestn Ross Akad Med Nauk R: Vestnik Rossiisknoj Akademii Meditsinskikh Nauk
VESTO s: [almacén de] Vestuario
VESV e: Voluntary Euthanasia Society of Victoria
VES[V] e: Vesicular Exanthema of Swine [Virus]
VET e: Vocational Education and Training
Vet f: Vétérinaire
vet N: veterinær
vet. a: veterinær
Vet. e: [as label] in Veterinary Science
- e: [in titles] Veterinary
Vet Clin North Am Small Anim Pract e: Veterinary Clinics of North America. Small Animal Practice
veter. N: veterinær
VETLA e: Very Enhanced TLA
Vet Med [Praha] c: Veterinarni Medicina
vetr r: veterinärvård
Vetr. avd. N: Veterinæravdeling
Vet Res e: Veterinary Research
Vet Res Commun e: Veterinary Research Communications
VETS e: Veterans Employment and Training Service
Vet. Sci. e: in Veterinary Science
Vets Info e: Veterans Information Service
VEU e: Vaccine Evaluation Unit
- e: Volume End User
VEV e: Vietnam Era Veterans
- e: Voice Excited Voicecoder
VEW d: Vereinigte Edelstahlwerke
- d: Vereinigte Edelstahlwerke, Austria
VEX e: Video Extension[s] to X[WS]
VF a: Værkstedsfunktionær
- a: Vestre Fængsel
- e: Fixed-wing fighter airplane
- e: USN Fighter Squadron
- e: Vacuum Fluorescent [display]
- e: Vienna Fortran
- e: Virtual Floppy
- e: Voice Frequency
- e: Volunteer Fireman
- f: Vente Forcée
- f: Version Française
- f: Vieux de France
- f: Ville Franche
- f: Virement de Fonds
- f: Voie de Fait
- f: Volontaire Français
- i: Vigili del Fuoco
- s: Valor Fuerza
vf r: vertikalfart
v.f. a: vest for
- N: vest for
VFA e: Fighter-Attack squadron of the US Navy
- e: Virginia Forestry Association
- e: Volatile Fatty Acids
- r: Volvo Flygmotor AB
VfA e: Voice for America
VFAT e: Virtual File Allocation Table
VFBP e: Vacuum Furnace Brazing Process
V.FC e: Version. First Class
VFC e: USN Fighter Composite Squadron
- e: V. Fast Class
- f: Vous Fais Connaître
VFCS e: Vienna Fortran Compilation System
VFCT e: Voice Frequency Carrier Telegraph
VFD e: Vacuum Fluorescent Display
- e: Vanishing Fiber Diameter
- e: Variable Frequency Drive
- e: Veterinary Feed Directive
VFDMIS e: Vertical Force Deployment Management Information System
VFE e: Maximum Flap Extended speed
VFEA e: VMEbus Futurebus and Extended Architecture
VFEI e: Virtual Factory Equipment Interface
VFF e: Victorian Farmers Federation
VFinm avsn N: Vest-Finnmark avsnitt
VFIP e: Voice File Interchange Protocol
VFIS e: Vegetation Fire Information System
VFK e: Variable Function Key
vfl e: variable focal length
- e: variable field length
VFM e: Vehicle Fleet Management
VFMP e: Vehicle Fleet Management Program
VFMX e: Navy Fighter attack
VFN e: Vendor Feature Node
- e: Vendor Feature Note
VFO e: Variable Frequency Oscillator
VFOX e: Visual FoxPro
VFP e: USN photo-reconnaissance squadron
- e: Visual FoxPro
VFP e: Volunteers For Peace
VFPI d: Verein zur Förderung der Pädagogik der Informationstechnologie
VFPYP e: Visual FoxPro Yellow Pages
VFR e: Virulent Foot Rot
- e: Visual Flight Rules
- f: Vidicon à Faisceau de Retour
VfR d: Verein für Raumschiffahrt
vfr e: very far radiation
vfr's e: visiting friends and relatives
VFS e: Virtual File System
VFSS e: Voice Frequency Special Services
VFT e: Visual Function Tester
VFTG e: Voice Frequency Telegraph Group
Vfw d: Vereinigung für Wertstoffrecycling
VFW d: Video für Windows
- e: Veterans of Foreign Wars
- e: Video For Windows
- e: Volunteer Fire Department
VfW e: Video for [Microsoft] Windows
vfZ d: vorlesungsfreie Zeit
VG e: Vacuum Generators
- e: Vertical-Gradient
- e: Very Good
- e: Vicar General
- e: Virgin Islands [British]
- e: Voice Grade
- e: Voltage Gain
- e: Volume Group
- f: Vaccination Gratuite
- f: Veuve de Guerre
- f: Vieilles Gens
- N: Verdens Gang
- s: Vanguardia
Vg f: Vosges
vg s: verbigracia
V.G. e: Vertical Grain
VGA e: Variable-Gain Amplifier
- e: Video Gate Array
- e: Video Graphic[s] Accelerator
- e: Video Graphics Adapter
- e: Video Graphic[s] Array
VGA e: Victor Gruen Associates
VG/BFS d: Sektion Volkswirtschaftliche Gesamtrechnung
VGB Kraftwerkstech. d: VGB-Kraftwerkstechnik. Vereinigung der Großkraftwerksbetreiber
VGB Kraftwerkstech. (Ger. Ed.] d: VGB-Kraftwerkstechnik (German Edition]. Vereinigung der Großkraftwerksbetreiber
VGC s: Vicario General Castrense
VGDA e: Volume Group Descriptor Area
VGDC e: Victorian Geographic Data Committee
VG/DG e: Vertical Gyro/Directional Gyro
Vgdia s: Vanguardia
Vge f: Village
VGF e: Voice Grade Facility
VGH e: Vancouver General Hospital
VGI e: Virtual Graphics Interface
VGIK R: Vsesoyuznyi Gosudarstvenyi Institut Kinematografi
VGK e: Supreme High Command
VGL d: vergleiche
- e: Very Good-Looking
Vgl d: Vergleich
VGLI e: Veterans Group Life Insurance
VGM e: Video Graphics Monitor
VGML e: Vienna Group for Multiple-valued Logics

VGNA e: Video Graphics Network Adapter
VGNV d: Verkehrsgebührennach-verarbeitung
VGO d: Geographische Gesellschaft, Moskau
Vgo e: Vigo
VGP d: Van Gelder Papier
vgr. s: verbigracia
VGS d: Verordnung über die Genehmigungspflicht für das Einleiten von Abwasser mit gefährlichen Stoffen in öffentliche Abwasseranlagen
- e: Virtual Group Switching Subsystem
- e: Virtual Groupselector Service
VGSA e: Viola da Gamba Society of America
VGT e: Variable Geometry Truss
- e: Vegetation
- e: Vegetation sensor planned for SPOT-4
v.g.v. r: var god vänd
Vh f: Véhicule
V&H e: Vertical and Horizontal
V/H e: ratio Velocity/Height ratio
VHA e: basalt Very High Alumina basalt
- e: Volunteer Hospital Association
VHA e: Vermont Headmasters Association
vha. a: ved hjælp af
VHADSL e: Very High-bit-rate Asymmetric Digital Subscriber Line
VHCMA e: Vietnam Helicopter Crew-Members Association
VHD e: Video High Density
- e: Viral Hemorrhagic Disease
VHDCI e: Very High Density Cable Interconnect
VHDL e: Verilog Hardware Description Language
- e: Very High Density Logic
- e: Vhsic Hardware Description Language
VHDSL e: Very High-bit rate Digital Subscriber Line
- e: Very High-bit rate Digital Subscriber Loop
VHDU e: Visual Handcopy Display Unit
VHE e: Virtual Home Environment
VHEMT e: The Voluntary Human Extinction Movement
VHES e: Vitro Hanford Engineering Service
VHEU e: Very High Enriched Uranium
VHF e: Very High Frequency
VHFCN e: Vietnam Helicopter Flight Crew Network
VHF Commun. e: VHF [Very High Frequency] Communications
VHF Radios e: Very High Frequency Radios
v.hj.a. a: ved hjælp af
v.hj.af a: ved hjælp af
VHL-FA e: VHL [Von Hippel-Lindau] Family Alliance
VHLW e: Vitrified High-Level Waste
VHM e: Vehicle Health Management
VHO e: Volunteering Hydrographic Office
- f: Service Hydrographique Volontaire
- s: Servicio Hidrográfico Voluntario
VHORG e: Vertically and Horizontally Organized Grid
VHP e: Very High Productivity
VHPA e: Vietnam Helicopter Pilots Association
VHPCC e: Very High Performance Computing and Communication
VHR e: Very High [Resolution] Radiometer
VHRA e: Very High Radiation Area
VHRCD e: Very High Resolution Color Display
VHRFC e: Very High Resolution Framing Camera
VHRR e: Very-High Resolution Radiometer Experiment
- f: Radiomètre à très Haute Résolution
- s: Radiómetro de muy alta resolución
VHS e: Van Hove Singularity
- e: Variable Hard Sphere
- e: Vertical and Horizontal Spread model
- e: Video Home Standard
- e: Video Home System
- e: Virtual Health Store
- e: Virtual Host Storage
- r: Veterinärhögskolan
Vhs f: Véhicules
VHSA e: Variable Hydrologic Source Areas
VHSIC e: Very High Speed IC [Integrated Circuit[s]]
VHT e: Rabbit Virus
- e: Vehicle-Hours of Travel
Vhus F N: Vardøhus Festning
VHW e: Village Health Worker
VI e: Vector Identifier
- e: Vendor Information
- e: Vibration Isolation
- e: Virgin Islands [US]
- e: Visual Interactive
- f: Valeur Immobilière
f: Valeur Intrinsèque
- f: Vente Immobilière
- R: Vysotny Istrebitel
Vi. s: Viernes
V/i d: spektraler Hellempfindlichkeitsgrad des menschlichen Auges für Tagessehen bei einem 2°-Gesichtsfeld nach DIN 5031
V24I e: V24 Interface
VIA e: Vaccine Information and Awareness
- e: VAX Information Architecture
- e: Vendors ISDN Association
- e: Versatile Integrated Avionics
- e: Viola
VIABLE e: Vertical Installation Automated (or Automation) Baseline [system]
VIAD e: VESA Image Area Definition
VIAL d: Verein der Ingenieur AgronomInnen und der LebensmittelingenieurInnen
vi antigen e: virulence antigen
VIAs e: Vocational Infornnation Agencies
VIB e: Video Interface Bus
VIBTS e: VTAM Integrated Bulk Data Transfer System
VIC e: V.35 Interface Cable
- e: Vector In Command Model
- e: Video Interface Card
- e: Video Interface Controller
- e: Visitor Information Centre
- e: Voice Interface Card
- e: Voice Interface Controller
- s: Vehículo de Ingeniería de Combate
VICA s: Vehículo de Ingeniería de Combate Aligerato
VICAR/IBIS e: Video Image Communication and Retrieval/Image Based Information System
VICBAR e: Barotropic hurricane track prediction model
Vic.B.C. e: Victoria [British Columbia]
Vic.I.C. e: Victorian Institute of Colleges
VICORP e: Virgin Islands Corporation
Vic Pk e: Victoria Park
VICS e: Voluntary Interindustry Communications Standards for EDI
vic[s] e: convict[s]
Vic Sta e: Victoria Station
vid. s: viuda
VID e: Volunteers for International Development
VIDAP e: Video Data Processor
VIDC e: Virgin Islands Department of Commerce
VIDD e: Virgin Islands Development Department
Vide Couches Minces f: Vide, les Couches Minces. Société Française du Vide
Videokonferenz d: über Kamera und Bildschirm werden Konferenzteilnehmer für alle jederzeit sicht- und hörbar zusammengeschaltet
Video-on-Demand e: Film auf Abruf
videot[s] e: video [television] idiot[s]
Videotex Viewp. e: Videotex Viewpoint
VIDL e: Virus Description Language
vidst. r: vidstående
VIE d: Viersen, Nordrhein-Westfalen
- e: Virtual Information Environment
VIEC e: Valve Instrument Equipment and Connector [list]
Vierteljahrsschr. Nat.forsch. Ges. Zür. d: Vierteljahrsschrift der Naturforschenden Gesellschaft in Zürich
Vietsyn e: Vietnam syndrome
VIEW e: Visibility Investigative Experiment in the West
VIF e: Virtual Interface
- e: Virtual Interrupt Flag
VIFA e: Victoria Free-Net Association
VIFIS e: Variable Interference Filter Imaging Spectrometer
VIFRED e: Visual Forms Editor
VIG e: Video Image Generator
- e: Virgin Islands Government
Vig s: Vigilancia
Vig Com e: Vigilance Committee
VIGL e: Visualization Graphic Library
VIGOPRI e: Virgin Islands Government Office of Public Relations and Infonnation
vigs e: vigilantes
VIGS e: Virtual Gap States

VIH f: Virus d'Immunodéficience Humaine
- s: Virus de Immunodeficiencia Humana
VIHIS e: Vancouver Island Head Injury Society
VII e: Vaccine Independence Initiative
VIJANA e: Tanzania Youth Organization
Viking Pr e: Viking Press
VIKOLUM d: Automatisches Vielkomponenten Luftmeßnetz Baden-Württemberg
VIL e: Vertical In-Line
- e: Vertically Integrated Liquid
- e: Volume Imaging Lidar
VILE e: VI Like Emacs
Vile f: Vieille
VILIB e: Virtual Library Libraries
VIM e: Vendor Independent Mail
- e: Vendor Independent Messaging
- e: Video Interface Module
VIMF d: Vierer [kabel] in Metallfolie
VIMHEX e: Venezuelan International Meteorological and Hydrological Experiment
VIMS e: Victorian Institute of Marine Science
VIMSIS e: Victorian Institute of Marine Sciences Information System
VIM/VAR e: Vacuum Induction Melting/Vacuum Arc furnace
VINB e: Virgin Islands National Bank
VINCE e: Vendor Independent Network Control Entity
Vince e: St Vincent, West Indies,
VINE e: Vine Is Not Emacs
VINES e: Virtual Network[ing] [operating] System
- e: Virtual Network[ing] Software
vinggr r: vinggranat
Vinle f: Vinicole
Vinson e: Vinson Massif
vint. r: vintertiden
VIO e: Video Input Output
- e: Violin
- e: Virtual Input Output
VIOLD e: Vision Impaired On-line Documentation
Violence Vict e: Violence and Victims
VIP a: [medlem af] det Videnskabelige Personale
- e: Variable Information Printing
- e: Venda Independence Party
- e: Very Important Person
- e: Video Information Processor
- e: Video Information Provider
- e: Video Input Processor
- e: VINES Internet Protocol
- e: Virgin Islands Police
- e: Virtual Interrupt Pending
- e: Virtual IP Network
- e: Visual Impact Performance
- e: Volunteer in Parks
- e: Volunteers in Probation
ViP e: Visual Programming
VIP e: VLB-ISA-PCI [board]
V.I.P. e: Vitiligo Information Pages
VIPA e: Virtual IP Addressing
VIPAC e: Virgin Islands Public Affairs Council
VIPC e: Vegetation Image Processing Centre
VIPER e: Verifiable Integrated Processor for Enhanced Reliability
VIPI e: Volunteers in Probation, Inc.
VIPNET e: Voice Interactive Paging Network
vipre e: visual precision
VIPS d: Vereinigung der Importeure pharmazeutischer Spezialitäten
- e: Video Image Processor System
- e: Video-Image Processing System
- e: Vizir Image Processing System
VIQR e: Vietnamese Quoted-Readable
VIR e: Visible and Infrared Radiometer
- e: Visual Information Retrieval
- f: Virement International
VIRAF e: Visible Infrared Reflectance Absorbance Fluorescence spectrometer
Viral Immunol e: Viral Immunology
VIRB e: Virginia Insurance Rating Bureau
Virchows Arch d: Virchows Archiv
viremia e: virus in the blood
VIRGS e: VISSR Interactive Registration and Gridding System
VIRI e: Visible/Infrared Imager
VIRIM e: Virtual Reality in Medicine
VIRIS e: Visible Infrared Intelligent Spectrometer
VIRLAB e: Visible and Infrared Spectroscopy Lab
Virology e: Virology
VIROS e: Virtual memory Operating System
VIRR e: Visible and IR Radiometer
VIRS e: Vegetation Image Receiving Station for SPOT-4
- e: Visible [and] Infrared Scanner
VIRSR e: Visible [and] Infrared Scanning Radiometer
VIRT e: Visible to Infrared Transducer
Virus Genes e: Virus Genes
Virus Res e: Virus Research
VIS d: Verläßliche Informationssysteme
- e: Virtual Instruction Set
- e: Visible [Band]
- e: Visual Instruction Set
- e: Visual Interactive Simulation
- e: Voice Information System
- r: Varvsindustrins Standardcentral
vis f: visibilité
VISA e: Virtual Instrument Software Architecture
- e: Visualization Accelerator
- e: VXI System Alliance
VISC e: Video Instruction Set Computing
VISCII e: Vietnamese Standard Code for Information Interchange
Vis. Comput e: Visual Computer
VISCOUNT e: British interlibrary loan project
Visible Lang. e: Visible Language
VISIDEP e: Visual Image Depth Enhancement by Parallax Induction
Vis Neurosci e: Visual Neuroscience
Vis. Res. e: Vision Research
VISSR e: Visible [and] Infrared Spin Scan Radiometer
- e: Visible Infrared Spin-Scale Radiometer
- f: Radiomètre à balayage par rotation dans le visible et l'infrarouge
- s: Radiómetro de Barrido en el Visible y el Infrarrojo
VISTA e: Employee I.D. No. Data Entry at Parklawn Computer Center
- e: Vertical Integration of Science, Technology and Applications
VISTA e: Volunteers in Service to America
VISTAnet e: Vermont Internet Services and Technology Access Network
VISTTA e: Visibility Impairment from Sulfur Transformation and Transport in the Atmosphere
VIS-UV e: Visible/Ultraviolet Spectrometer
VISYON d: Variables Intelligentes Synchrones Optisches Netz
VIT e: Vibration-Isolated Television
Vit f: Vitesse
- f: Viticulture
VITA e: VME[bus] International Trade Association
VITA e: Volunteers In Tax Assistance
Vital e: VHDL Initiative Toward ASIC Libraries
VITAL e: Virtually Integrated Technical Architecture Lifecycle
Vital Health Stat 1 e: Vital and Health Statistics. Series 1: Programs and Collection Procedures
Vital Health Stat 2 e: Vital and Health Statistics. Series 2: Data Evaluation and Methods Research
Vital Health Stat 3 e: Vital and Health Statistics. Series 3: Analytical and Epidemiological Studies
Vital Health Stat 4 e: Vital and Health Statistics. Series 4: Documents and Committee Reports
Vital Health Stat 5 e: Vital and Health Statistics. Series 5: Comparative International Vital and Health Statistics Reports
Vital Health Stat 10 e: Vital and Health Statistics. Series 10: Data From the National Health Survey
Vital Health Stat 11 e: Vital and Health Statistics. Series 11: Data From the National Health Survey
Vital Health Stat 13 e: Vital and Health Statistics. Series 13: Data From the National Health Survey
Vital Health Stat 14 e: Vital and Health Statistics. Series 14: Data From the National Health Survey
Vital Health Stat 21 e: Vital and Health Statistics. Series 21: Data on Natality, Marriage and Divorce
Vital Health Stat 23 e: Vital and Health Statistics. Series 23: Data From the National Survey of Family Growth
VITAP e: Voluntary Income Tax Assistance Program
vit. ass. N: vitenskapelig assistent

VITC e: Vertical Integrated Time Code
- e: Vertical Interval Time Code
Vitr. Enameller e: Vitreous Enameller
VITRO e: Vitro Engineering Corporation
VIU e: Voiceband Interface Unit
VIUF e: VHDL International User Forum
VIUS e: Virgin Islands of the United States
VIV e: Virgin Islands View
Viv s: Víveres
viv i: vivace
VIVA e: Virgin Islands Visitors Association
VIVA e: Voices In Vital America
VIVB e: Virgin Islands Visitors Bureau
vix. l: vixit
VIZ e: humidity sensor used on National Weather Service radiosonde
viz. l: videlicet
VJ e: Van Jacobson
- e: Video Jockey
- f: Vente Judiciaire
v J d: vorigen Jahres
VJHC e: Van Jacobson Header Compression
VJ HEADER e: Van Jacobsen Header
VJhZG d: Vierteljahreshefte für Zeitgeschichte
VJO e: Virtual Journal of Orthodontics
vj's e: video jockeys
VK d: Verbundkatalog maschinenlesbarer Katalogdaten
vk r: vicekorpral
V&K a: Vej og Kloak
VK 2 r: andra Världskriget
VKA d: Verband Kommunaler Aktionäre
VKF d: Vereinigung Kantonaler Feuerversicherungen, Bern
- d: Verkaufsförderung
- F: Viikinkiaapeli Oy
VKI a: Videnskaben om Kreativ Intelligens
VKK r: Vapenkompanikurs
VKL d: Verkehrseinschränkungsklasse
VKM e: Van Kampen Merritt
VKN r: Norra Värnpliktskontor
VKO R: Vnukovo, Moscow [Airport]
vkomp r: värnkompani
VKP R: Vozduzhnyi Komandnii Punkt
VKR a: Venstre-Konservative-Radikale
- e: Congress of Representatives for the Colored
VKR R: Voderinaya Kontr Rozvedka
vkr r: världskrig
VKS d: Verband Kommunaler Stätereiniger
- d: Vereinigung der Kantonsärzte der Schweiz
VKU e: Verband kommunaler Unternehmen
VL e: Vaasa Line
- e: Vector Length
- e: Veronica Lodge
- e: VESA Local Bus
- e: Virtual Library
- e: Virtual Life
- f: Valeur Locative
- f: Véhicule
- f: Véhicule Léger
- f: Vente Libre
- f: Voie Légale
- f: Voiture Légère
- F: Vaasan Laiva
- r: Värnpliktslag
v. l. N: vestlig lengde
V/L f: Votre Lettre
V/l e: Vapor-liquid ratio
VLA e: Very Large Array [radio telescope in Arizona or New Mexico]
- e: Very Low Activity
- e: Veterans Land Administration
- e: Volume Licence Agreement
- e: Volunteer Lawyers for the Arts
Vla f: Villacoublay
VLACC n: Vlaamse Geautomatiseerde Centrale Catalogus
V. lag N: Vaktlag
VLAGES e: Very Low Altitude Gravity Extraction System
VLAM e: Vertical Lambertson
VLAN e: Virtual LAN [Local Area Network]
VLASROC e: Vertical Launch Anti-Submarine Rocket
VLB d: Verzeichnis liefbarer Bücher
- e: VESA Local Bus
VLBI e: Very Long Baseline Inferometry
- f: Interférométrie à très Longue Base
- s: Interferometría de muy larga Línea de Base
VLBR e: Very Low Bit Rate
VL[-]BUS e: VESA Local[-]Bus
VLC e: Variable Length Code
- e: Voice Line Card
VLC/VLD e: Variable Length Coding, Variable Length Decoding
VLD a: Vestre Landsrets Dom
- e: Variable Length Decoder
VLDB e: Very Large Data Base
VLDL e: Very-Low-Density
VLDW e: Very Long Data Word
VLE e: Vapor Levitation Epitaxy
- e: Vapor-Liquid Equilibrium
- e: Vertical Laminar Flow
- e: Very Large Format
- e: Very Low Frequency [band]
- e: Virtual Lookaside Facility
Vle f: Véhicule
v. lgd r: västlig längd
v.lgd. a: vestlig længde
VLH e: Volatile Liquid Hydrocarbons
VLHPC e: Variable-Length Hydraulic Piston Corer
VLI e: Variable Length Integer
VLIB e: EMSL-Vegas Library Inventory
VLINK e: Visited Link
VLIR e: Flemish Inter-University Council
- f: Conseil Flamand Interuniversités
- s: Consejo Interuniversitario Flamenco
VLIS d: Verkehrs-Leit- und Informations-System
VLIW e: Very Large Instruction Word
- e: Very Long Instruction Word[s]
VLL e: Video Lead Locator
VLM e: Virtual Loadable Module
- e: Volume Logical Module
VLO e: Very Low Observable
VLP e: Very Low Power
V&LPM e: Vibration and Loose Parts Monitoring
VLR a: Vestre Landsret
- e: Visitor Location Register
VLS e: Vapor-Liquid-Solid
- e: Virtual Library System
- N: Vestlandske Luftfartsselskap
VLSI e: Very Large Scale Integration [of circuits on a single chip]
- e: Very Low Scale Integration
VLSIPS e: Very Large Scale Immobilized Polymer Synthesis
VLSISC e: Very Large Scale Integration Socketing Capability
VLSI Technology e: A US micro-electronic fabrication company
VLSI Syst Des. e: VLSI Systems Design
VLSS d: Verein Leitender Spitalärzte der Schweiz
VLSTRACK e: Vapor-Liquid-Solid Tracking
VLT e: Variable List Table
VLTG e: Voltage
V-LTH e: Civil Engineering School at Lund Institute of Technology, Lund University
VLTT s: Vehículo Ligero Todo Terreno
VLWIR e: Very Long Wavelength Infrared
VM e: Victorian Museum
- e: Virtual Machine
- e: Virtual Memory
- e: Voice Messaging
- f: [par] Voie Maritime
- f: Valeur Marchande
- f: Valeurs Mobilières
- f: Vente Mobilière
- f: Visite Médicale
- s: Vivamos Mejor
v/m r: varv per minut
VMA d: Vertiefung Mathematik
- e: Vanilmandelic Acid
- e: Virtual Memory Access
- e: Virtual Memory Address
- e: Virtual Memory Area
- e: Voids in the Mineral Aggregate
VMAG e: Vanderpoel Memorial Art Gallery
vMAC e: virtual MAC Address
VMAL e: EMSL-Vegas Mail List
V-MAN e: Vehicle Management
VMAQ e: Marine Tactical Electronic Warfare Squadron
VMAT e: USMC attack training squadron
vmax d: maximaler Verkleinerungsfaktor
VMB e: USMC Bombardment Squadron
- e: Video Memory Buffer
- e: Virtual Machine Boot
- e: Voice Message Bank

VMC e: Media Channel
- e: Variability Model Coverage
- e: Vermont Monitoring Cooperative
- e: VESA Media Channel
- e: Visual Meteorological Conditions
VMCCA e: Veteran Motor Car Club of America
VMCF e: Virtual Machine Communications Facility
VMH e: Victoria Medal of Honour
VMCJ e: USMC electronic warfare squadron
VMCM e: Vector Measuring Current Meter
VMCMS e: Virtual Machine/ Conversational Monitoring System
VMCP e: Virtual Machine Control Program
VMD e: Veterinary Medical Doctor
VME e: Versa Micromodule Extension
- e: Versa Modular Europe
- e: Versa Module Eurocard
- e: Versa Module Europe
- e: Versa[bus] Module Europe[an]
- e: Versacard Modified for Eurocard
- e: Virtual Machine Environment
- e: Virtual Manufacturing Enterprise
- e: Virtual Memory Environment
- e: Virtual Memory Executive
- f: Visite Médicale d'Embauchage
VMEbus e: Versatile Modular E-bus
Vmek r: utbildning till Vapenmechaniker vid flygvapnet
VM/ESA e: Virtual Machine Enterprise Systems Architecture
VMF e: USMC Fighter Squadron. Examples
- e: Variable Message Format
- R: Voenno-Morskoy Flot
VMFA e: Fighter-Attack squadron of the USMC
VMFAT e: Fighter-Attack Training squadron of the USMC
VMG s: Visión Mundial de Guatemala
VMH e: Ventromedial Hypothalamus
VMI e: Video Module Interface
- e: Virginia Military Academy
- s: Vocabulario Meteorológico Internacional
v/mil e: volts per mil
v/min r: varv per minut
VMIS e: VAS Model Impact Study
VMIU e: Virtual Memory Interface Unit
VML e: Association Vaincre les Maladies Lysosomales
VMLI e: Veterans Mortgage Life Insurance
VMLSCS e: Virtual Machine Loosely Synchronous Communication System
VMM e: Virtual Machine/Memory Manager
- s: Vigilancia Meteorológica Mundial
v & mm e: vandalism and malicious mischief
VMMAP e: VM Monitor Analysis Program
VMO e: USMC observation squadron
- f: Visite Médicale Obligatoire
Vmo e: Maximum permitted operating speed
V[-]MOS e: V-groove MOS [Metal-Oxide Semiconductor]
VMP e: Vegetation Management Program
- e: Virtual Modem Protocol
- e: Virtual Multiporting
- f: Visite Médicale Périodique
VMPPF e: VM Performance Planning Facility
VMPRF e: VM Performance Reporting Facility
VMPT e: Virtual Machine Pass-Through
VMR e: Vertical Market Reseller
- e: Volt Meter Reverse
VMRP e: VLAN Membership Resolution Protocol
VMRS e: Voice Message Relay System
VMS e: Vertical Motion Simulator
- e: Virtual Memory [operating] System
- e: Visual Management System
- e: Voice Mail System
- e: Voice Management System
- e: Volcanogenic Massive Sulfides
VMSC e: Visited Mobile Switching Centre
VM[/]SP e: Virtual Machine/System Product
VMT e: Vehicle Miles Traveled
- e: Vehicular Miles Traveled
- e: Velocity Modulation Transistor
- e: Vertical Magnetotransistor
- e: Visual Modeling Technique
VMTSS e: Virtual Machine Time-Sharing System
VMVM e: Vector Measuring Current Meter
VMW e: Vertically Modulated Well
VMX e: Voice Mail
VM/XA e: Virtual Machine/Extended Architecture
VMZ d: Verkehrsmanagement-Zentrale
- d: Verkehrsmanagement-Zentrum
VN d: Vereinfachter Nachweis
- e: Viet Nam
- f: Valeur Nominale
- f: Viet-nam
- f: Visite de Nuit [à domicile]
- s: Vida Nueva-Asociación Hogar para Niños y Centro de Adopciones
Vn f: Vienne
VNA e: Vector Network Analyzer
- e: Vietnam News Agency
- e: Virginia Nurses Association
- e: Virtual Network Architecture
VNAF e: Vietnamese Air Force
VNB e: Valley National Bank
VNC e: Victorian Naturalists Club
- e: Virtual Nursing College
VNCA e: VTAM Node Control Application
VND e: Video and Networking Division
- e: Viet Nam Dông
Vne e: Never to exceed speed
V[-]Net e: Virtual Network
VNF e: Virtual Network Feature
VNIR e: Visible and Near Infrared [Radiometer]
VNIRO d: Wissenschaftliches Forschungsinstitut für Seefischerei, Moskau
- e: All-Russia Research Institute of Marine Fisheries and Oceanography
- f: Institut russe de recherche scientifique sur la pêche maritime et l'océanographie
- s: Instituto Ruso de Investigaciones sobre Pesca Marina y Oceanografía
VNIRP e: Visible and Near Infrared Polarimeter
VNL e: Via Net Loss [plan]
VNM e: Victoria National Museum [Ottawa]
VNMC e: Vietnam Marine Corps
VNO e: Maximum structural cruising speed
- e: Vomero Nasal Organ
Vno i: Violino
VNODC e: Vietnamese National Oceanographic Data Center (or Centre)
- f: Centre national de données océanographiques du Viet Nam
- s: Centro Nacional de Datos Oceanográficos de Viet Nam
VNP e: Vietnamese Piastre
VNR e: Visible and Near Infrared Radiometer
- e: Vitronectin-R
VNS e: Virtual Network Services
- e: Virtual Notebook System
VNS n: Vereenigde Nederlands Scheepvaartmaatschappij
VNSM e: VINES Network and Systems Management
VNTR e: Variable Number of Tandem Repeats
VNU f: Volontaires des Nations Unies
- s: Voluntarios de las Naciones Unidas
VNV a: Vestnordvest
- r: Västnordväst
VNY e: Van Nuys Ca.
VO e:Victorian Order
VO d: Volksoper
VO d: Verordnung
- e: Validation Office
- e: Virtual Object
- e: Visual Objects
- f: Vaccination Obligatoire
- f: Valeur or
- f: Version Originale
- f: Ville Ouverte
- f: Visa Officiel
- f: Voyages Organisés
- s: Vías y Obras
Vo r: Utgångshastighet
vo p: verso
- r: vakthavande officer
v.O. d: vereinbarter Ort
V/O f: [à] Votre Ordre
v/o p: vossa ordem
VOA e: Voice Of America
- e: Volatile Organic Analysis
- e: Volatile Organic Analyte
- e: Volunteers Of American
VOA e: Virginia Opera Association
VOAS d: Verordnung über die Gewährleistung von Atomsicherheit und Strahlenschutz
VOB d: Verdingungsordnung Bund
- d: Verdingungsordnung für Bauleistungen

VÖB d: Vereinigung österreichischer Bibliothekarinnen und Bibliothekare
VOC e: Creative Voice
- e: Voice Of The Customer
- e: Volatile Organic Chemicals
- e: Volatile Organic Compound[s]
- e: Volatile Organic Contaminant
- e: Volatile Organic Content
- e: Volts, Open-Circuit
VOC n: Vereenigde Oostindische Compagnie
VOCA e: Visiting Orchestras Consultative Association [London]
VOCAR e: Variation Of Coastal Atmospheric Refractivity
VOC-Arid ID e: Volatile Organic Compounds-Arid Integrated Demonstration
VOCEM e: Volatile Organic Compounds Emission Measurements
VOCEPP s: Voluntariado Centro Experimental Psiquiátrico Pedagógico
Voc Foun e: Vocational Foundation
vocg e: verbal orders-commanding general
VOCl e: Volatile Organic Chlorinated compound
voco e: verbal order-commanding officer
VOCODER e: Voice Operated Code
VOCs e: Volatile Organic Compounds
- e: V-Model
VOCOSS e: Voluntary Organisations Cooperating in Overseas Social Service
VOD e: Veno-Occlusive Disease
- e: Video On Demand
- e: Voice Of Democracy
VoD e: Video-on-Demand
vod e: vision of right eye
VODAS e: Voice Over Data Access Station
Voen Med Zh R: Voenno-Meditsinskii Zhurnal
VOF d: Verband der Fabrikanten von Technischen Ölen, Fetten und Verwandten Produkten
- d: Vollzugsordnung für den Funkdienst
- e: Volunteer Observing Fleets
- N: De Vernepliktige Offisers Forening
V Ofj Fdi N: Vestre Oslofjord Forsvarsdistrikt
V of K e: Voice of Kenya
VOFR e: Voice Over Frame Relay
VOG e: Vessel Off-Gas
VoG e: Voice of Germany
Vögel Heim. d: Vögel der Heimat
VOI d: Verband der Optischen Informationssysteme
- e: Voice
Voi f: Voiture
VOIC A: Vereenigde Oost-Indische Compagnie
VOICE e: Vertical Observations Involving Convective Exchange Experiment
Voice Annotation e: Sprachanmerkung, z.B. bei elektronischer Post oder als Zusatz zu Dokumenten
VoIP e: Voice over IP
VOIS e: Voice Operated Information System
Vojnosanit Pregl P: Vojnosanitetski Pregled
VOL d: Verdingungsordnung für Leistungen, ausgenommen Bauleistungen
- d: Verdingungsordnung Land
- e: Voice-Over-IP
- e: Volume
vol. e: volume
- e: volumen
Vol. e: Volume
- e: Volunteer
vol% e: volume percent
VOLAG e: Voluntary Agency
VOLAGS e: Voluntary Agencies
VOLAR e: Volunteer Army
Vol Isl e: Volcano Islands
VOLSTORAGE e: Volatile Organic Liquid Storage Facilities
Volvo e: Swedish car company
VOM e: Volt Ohm Meter
- e: Volt-Ohm-Milliammeter
- f: Veille Océanographique Mondiale
VON Video Over [the] Net
- e: Voice On/Over [the] Net
Von f: Vallon
V.O.N.O. e: Vendor of Oysters in New Orleans
VOODOO e: Versions Of Outdated Documents Organized Orthogonally
VOP e: Valued as in Original Policy
- e: Very Oldest Prectirable
Vo-Po d: Volks-Polizei
VOPM f: vente obligatoire au prix marqué
V Oppl Fdi N: Vest-Oppland Forsvarsdistrikt
Vopr. At. Naukt Tekh. Ser., Fiz. Radiats. Povredhdenii Radiats. Mater. R: Voprosy Atomnoj Nauki i Tekhniki, Seriya Fizika Radiatsionnyk Povredhdenii i Radiatsionnoe Materialovedenie
Vopr Kurortol Fizioter Lech Fiz Kult R: Voprosy Kurortologii, Fizioterapii i Lechebnoj Fizichesknoj Kultury
Vopr Pitan R: Voprosy Pitaniia
VOQ e: Visiting Officer[s] Quarters
VOR e: Vestibulo-Ocular Reflex
- e: VHF Omnidirectional Radio Range
- e: Visions Of Reality
- e: Visual Operating Restrictions
- e: Voice-Operated Recording
VORADEP f: Programme de Développement Agricole de la Région de la Volta
VORG e: Victorian Ornithological Research Group
VORGAN p: Voz de Resistência do Galo Negro
VOR/LOC e: VHF Omnidirectional Radio Range Localizer
VORP e: Victim Offender Reconciliation Program
VORT e: Very Ordinary Rendering Toolkit
vort l: vortex
vort e: vortices
VORTAC e: Combined VOR and Tacan system
- e: VHF Omni Range and Tactical Air Navigation
- e: Visual Omnirange and Tacan
VORTEX e: Verification of the Origin[s] of Rotation in Tornadoes Experiment
VORTEX-95 e: Verification of the Origins of Rotation in Tornadoes Experiment 95
VOS a: Buque que colabora ocasionalmente
- e: Vehicle Operating Survey
- e: Verbal Operating System
- e: Vessel Of Opportunity
- e: Voice Operating System
- e: Voluntary Observing Ship
- e: Volunteer Observing Ship[s]
- e: Vulcan Operating System
- f: Navire d'observation bénévole
- f: Navire occasionnel
VoS e: Voice of the Shuttle
vos e: vision of left eye
vos R: vostok
VOS e: Victims Of Superstition
VOST e: VESA Open Set Top [group]
VOT e: Foreign Operational Center of Soviet Intelligence forces
vot e: voluntary overtime
Vot f: votants
VOTE e: Vortex Ozone Transport Experiment
Vote e: Vote here if you're a student
Vo-Tech e: Vocational-Technical
VOW d: Verkehrsverbund Ostwestfalen
VOW e: Voice of Women
VOWB n: Vlaams Overlegorgaan voor Wetenschappelijk Bibliotheekwerk
VOX e: Voice Extension
- e: Voice Operated Relay Circuit
- e: Voice-Operated Threshold [communications]
VOXEL e: Volume Element
vox pop. l: vox populi
Voy f: voyez
Voy. e: [in titles] Voyage[s]
Voyag f: Voyageur
Voyageurs e: Voyageurs National Park on the Canadian border of Minnesota
VOYCO e: Vosloorus Youth Congress
Voy cop f: Voyez copie
VP e: USN patrol squadron
- e: Valencia Park
- e: Valve Position
- e: Vector Processor
- e: Verb Phrase
- e: Verification Program
- e: Vice President
- e: View Processor
- e: Virtual Path
- e: Virtual President
- e: Virtual Prototyping
- e: Vitrification Plant
- f: Vacances Payées
- f: Vente Publique
- f: Versement Provisionnel
- f: Vice-Président
- f: Visite Périodique
- f: Voirie Parisienne
- f: Voiture Publique
- f: Voix Publique
- f: Vote Public
- N: Venstres Pressekontor
- N: Vognplass
- r: Värmepejlanläggning
- R: Visotny Perekhvatchik
vp r: verkstadspluton

VP-40 e: Vector supercomputer introduced by Fujitsu in 1985
VPA e: VAX Performance Advisor
- e: Virtual Proxy Agent
VPAA e: Vice President for Academic Affairs
v pag e: various paging
VPAK e: Vertical Package
VP/BFS d: Abteilung Volkswirtschaft und Preise
VPC e: Vertebrate Pests Committee
- e: Vibra-Percussive Corer
- e: Virtual Path Connection
- f: Voitures Particulières et Commerciales
VPCD e: Videotex Presentation Control Element
VP-Centralen a: Værdipapircentralen
VPCI e: Virtual Path Connection Identifier
VPCP e: Volunteer Probation Counseling Program
VPD e: Vapor Phase Desorption
- e: Vapor Phase Decomposition
- e: Vapor Phase Dissolution
- e: Vice President for Development
- e: Virtual Printer Device
- e: Vital Product Data
V-PDB e: International scale for expressing C and O isotopic composition relative to PDB carbonate
VPDE e: Videotex Presentation Data Element
VPD-ICPMS e: Vapor Phase Decomposition-Inductively-Coupled Plasma Mass Spectroscopy
VPDS e: Virtual Private Data Service
VPE d: Verpackungseinheit
- e: Vapor Phase Epitaxy
- e: Video Port Extensions
- e: Visual Programming Environment
Vpe a: Varmeproducerende enheder
v.p.e. a: varmeproducerende enheder
VPEC e: Virginia Power Electronics Center
VPF e: Vector Product Format
vpg e: very pregnant guppy
VPGA e: Verrnont Personnel and Guidance Association
VPGA e: Virginia Personnel and Guidance Association
VPI e: Virtual Path Identifier
- e: Virtual Path Indicator
- e: Virtual Private Internet
VPIP e: Vinson/Parkhill Implementation Plan
VPIRG e: Verrnont Public Interest Research Group
VPI&SU e: Virginia Polytechnic Institute and State University
VPK A: Veelpartykonferensie
- r: Vänsterpartiet Kommunisterna
- r: Vapenplutonskurs
VPKM e: Vehicles Per Kilometer
VPL e: Virtual Path Link
- e: Virtual Programming Language
- e: Visible Party Line
- e: Visual Programming Languages
vpl N: verneplikt[ig]
- r: värnpliktig
vpl a r: värnpliktiga i allmänhet
vpl K-sh r: sjömanshusinskrivna värnpliktiga hänförda till grupp K
vpl K-spec r: värnpliktiga tygtekniker
VPLMN e: Visited PLMN
VPLT d: Verband für Professionelle Licht und Tontechnik
VPM e: Vertical Package Module
- e: Video Port Manager
- e: Virtual Parallel Machine
- e: Virtual Protocol Machine
- s: Vueltas Por Minuto
VPMI e: Virtual Protected Mode Interface
VPMO f: Vente au Prix Marqué Obligatoire
VPN d: Virtuelles Privates Netz
- e: Virtual Packet Network
- e: Virtual Private Network
VP [NSC] e: Verification Panel [National Security Council]
VPO e: Vegetation Protection Ordinance
- e: Virtual Projection Origin
VPO e: Vienna Philharinonic Orchestra
VPOD d: Schweizerischer Verband des Personals öffentlicher Dienste
VPP e: OSHAs Voluntary Protection Program
- e: Value Purchase Plan
- e: Vertebrate Pest Program
- e: Volume Purchase Plan
- e: VXI Plug-and-Play
- f: Vote Par Procuration
vppm e: volume parts per million
VPR e: Vice President for Research
- p: Vanguarda Popular Revolucionaria
Vpr f: Ville principale
VPRI e: Plant Research Institute, Burnley
VPS d: Verbindungsprogrammierte Steuerung
- d: Video Programm[ier] System
- e: Vapor Phase Soldering
- e: Variable Print Specification
- e: Video Processing System
- e: Video Program System
- e: Vinyl Polysiloxane
- e: Voice Parameter Study
VPSB d: Verband der Privatspitäler des Kantons Bern
VPSC e: Vault, Process, Structure, Configuration
VP SEM e: Varaible Pressure Scanning Electron Microscopy
VPSK d: Vereinigung der Personalchefs schweizerischer Krankenhäuser
VPT e: Virtual Path Terminator
- e: Virtual Print[ing] Technology
VPU e: Video Processing Unit
VPub f: Vente Publique
VPUG e: Ventura Publishing User Group
VPV r: Värnpliktsverket
VPWPU e: SAFE Four Echelon Wartime Casualty System
VPX e: Video Pixel
VQ e: Shore Based Recon Aircraft
- e: USN electronic reconnaissance squadron
- e: Vector Quantization
VQA e: Vendor Quality Assurance
VQFP e: Very fine pitch Quad Flat Pack[age]
VQMG e: Vice Quartermaster General
V2R2 e: Version 2 Release 2 of OS/400 or Office Vision/400
V2R3 e: Version 2 Release 3 of OS/400 or Office Vision/400
VR a: Verdensrekord
- d: Beginn der Vollreife
- d: Virtuelle Realität
- e: Fixed-wing Transport airplane
- e: Virtual Reality
- e: Virtual Route
- e: Voice Repeaters
- e: Voltage Reference
- e: Voltage Regulated
- e: Voltage Regulating transformer
- e: Voltage Regulator
- f: [par] Voie Routière
- f: Vente à Réméré
- F: Valtionrautatiet
- s: Vehículo de Reconocimiento
Vr e: Rotation Speed
- e: Virus
VRA d: Vollständiger Regulärer Ausdruck
- e: Veterans Readjustment Appointment
- e: Voting Rights Act
VRAI e: Virtual Reality Annual International
VRAM e: Volatile Random Access Memory
V[-]RAM e: Video RAM [Random Access Memory]
VRB e: Verification Review Board
VRC e: Verbatim Reporters Center
- e: Vertical Redundancy Character
- e: Vertical Redundancy Check
- e: Viewing Reference Coordinate
- e: Virtual Route Control
- e: Voice Response Controller
- f: Vous Rends Compte
VRCB e: Virtual Route Control Block
VRCI e: Variable Resistive Components Institute
VRCS e: Viewing Reference Coordinate System
VRD e: Victoria River District
- e: Virtual Reality Display
- e: Virtual Retinal Display
v-r'd e: voluntarily returned [deported]
VRDCA e: Victoria River District Conservation Association
VRDS e: Variable-length Relative Record Data Set
VRE e: Voltage Regulated Extended
VREF e: approach Reference speed
V/REF f: Votre Référence
VRF e: USN transport wing
vrg e: veering
Vrgr batj N: Varanger bataljon
VRH e: Variable-Range Hopping
- s: Vidicón de retorno de haz
Vrie f: verrerie
VRIS e: Variable Reluctance
VRM e: Visual Resource Management
- e: Voltage Reference Module
- e: Voltage Regulation Module
- f: Valeur Reçue en Marchandises
VRML e: Virtual Reality Markup Language
- e: Virtual Reality Modeling Language
VRN e: Virtual Route Number

VROOM e: Standard childrens onomatopoeia for powerful internal combustion engine accelerating
- e: Virtual Realtime Object-Oriented Memory Manager

VROOMM e: Virtual Realtime Object-Oriented Memory Management
- e: Virtual Runtime Object-Oriented Memory Manager

VROT e: Victorian Rare or Threatened Plants

VRP e: Vehicle Recycling Partnership
- e: Viewing Reference Point
- f: Voyageurs [de commerce], Représentants, Placiers

VRPM e: Visible Radiation Polarization Monitor
- e: Volume Rate Planning Model

VRPRQ e: Virtual Route Pacing Request

VRPRS e: Virtual Route Pacing Response

VRR d: Verkehrsverbund Rhein/Ruhr
- e: Veterans Reemployment Rights

VRS d: Verkehrsverbund Rhein/Siegen
- e: Vacuum Resource System
- e: Van Riebeeck Society
- e: Virtual Reality Society
- e: Voice Response System
- e: WordPerfect Graphics Driver

VRSS e: Virginia Remote Sensing Study

VRT e: Vacuum Receiver Tank
- e: Vacuum Receiving Tank
- e: Voltage Regulation Technology

VRTCAL e: Vertical Calibration

VRU e: Voice Response Unit

VS a: venerisk sygdom
- d: Vereinfachter Sammelnachweis
- d: Vermessungsschiff
- e: USN submarine warfare squadron
- e: Vacuum System
- e: Variable Star
- e: VAX Station
- e: Video Switch
- e: VINES Security
- e: Virtual Scheduling
- f: Vêtements Spéciaux
- f: Visite Sanitaire
- f: Vote Secret
- r: Väderleksstation
- r: Växelström
- r: Venerisk Sjukdom

Vs e: Stalling Speed
- f: Vous

vs. a: versus

V-S d: Villingen-Schwenningen

V1S e: V one Side

VS-1 e: Valkyrie Spatula One

VSA d: Verband Schweizerischer Angestelltenverbände
- d: Vereinigung der Schweizerischen Archivare
- e: Validity and Statistical Analysis subsystem
- e: Vessel-Sharing Arrangement
- e: Virtual Server Architecture
- e: Virtual Storage Access

vsa a: ved siden af

VSA e: Victorian Society of America

VSÄ d: Vereinigung Schweizerischer Ärztinnen

VSAM e: Virtual Storage Access Method

VSANP A: Verenigde Suid-Afrikaanse Nasionale Party

VSAO e: Verband Schweizerischer Assistenz- und Oberärzte

VSAR e: Versatile Synthetic Aperture Radar
- e: Visual Arts

VSAT e: Very Small Aperture Terminal
- f: Station Terrienne à très petite Antenne
- s: Terminal de Muy Pequeña Apertura

VSB d: Verein für Studentische Belange
- d: Vereinigten Staatsbiblioteken
- e: Variable Spring Base
- e: Vestigial SideBand [modulation]
- e: VME Subsystem Bus
- f: Voyageurs Sans Bagages

VSBPakt r: Vänskaps-, Samarbets- och Biståndspakten

VSC e: Variable Speed Coupling
- f: Voitures Sans Chauffeurs
- s: Vigilancia del Sistema Climático

V SC e: Veterinary Science

VSCCS e: Visual SCCS

VSCE e: Videotex Service Control Element

VSCU e: Vatican Secretariat for Christian Unity

VSD e: Variable Speed Drive
- e: Ventricular Septal Defect

VSD e: Veteran Services Division

VSD e: Veterans Affairs Department

VSDA e: Video Software Dealers Association

VSDL e: Vienna Sausage Definition Language

VSE e: An ESA operating system from IBM, for the IBM/390
- e: Verification Support Environment
- e: Virtual System Extended

VSE[/]ESA e: Virtual Storage Extended/ Enterprise Systems Architecture

VSELP e: Vector-Sum Excited Linear Prediction (or Predictive)

VSEPR e: Valence State Electron-Pair Repulsion

VSESP e: Virtual Storage Extended /System Product

VSF d: Verband der Softwareindustrie Deutschlands e.V.
- d: Verbund Selbstverwalteter Fahrradbetriebe
- e: Vertical Scanning Frequency
- s: Veterinarios Sin Fronteras

vsff p: volle, se faz favor

VSG e: Viscous Semi-Geostrophic
- f: Vente Sans Garantie

VSH e: Variable Spring Hanger

VSHSB N: Våpenskolen for Hærens Samband

VSHVK N: Våpenskolen for Hærens Våpentekniske Korps

VSI d: Fachgruppe Verläßliche IT-Systeme der GI
- d: Verband der Software-Industrie Deutschland e.V.
- e: Virtual Socket Interface
- e: Visible Security Image
- e: Visual Site Inspection

VSIG e: Vitiligo Support and Information Group

VSII e: Very Seriously Ill or Injured

VSIM e: Virtual Surface Image Memory

VSIO e: Virtual Serial Input Output

VSIS e: Vault Safety and Inventory System

VSI/TRA e: Vertical Speed Indicator/ Traffic, Resolution Advisory

VSIV N: Våpenskolen for Ingeniørvåpenet

VSK d: Video-Sender-Kennung
- e: Verband Schweizer Kurhäuser
- N: Våpenskolen for Kavaleriet

vsk r: verkskydd

Vskc r: Verksskyddscentral

VSL e: Variable Speed Limit
- e: Virtual Socket Library
- e: Virtual Software Library
- e: Vitreous State Laboratory

VSL e: Venture Scout Leader

VSLEP e: Vector Sum, Linear Excited Predictive Coding

VSLI e: Very Large Scale Integration

VSM d: Vereinigung Schweizerischer Medizinalingenieure
- e: Virtual Shared Memory
- e: Visual System Management
- e: Volcano System Monitor

vsm r: vapensmed

VSMSC e: Virtual Short Message Center

VSN e: Virtual Switched Network
- e: Volume Serial Number
- N: Visesjefsnemnd

VSNA f: Voluntaires du Service National Actif

VSO e: Vancouver Symphony Orchestra
- e: Very Small-Outline
- e: Victoria Symphony Orchestra
- e: Victor Symphony Orchestra
- e: Vienna State Orchestra
- e: Vienna Symphony Orchestra
- e: Voluntary Service Organization
- r: Vakthavande Sambandsofficer

VSOG e: VSOP Science Operations Group

VSOP e: Very Small-Outline Package
- e: Virtual Symphony Orchestra Performance
- e: VLBI [Very Long Baseline Interference] Space Observatory Program[me]

VSOP-NA e: VOS Special Observing Project North Atlantic
- f: Projet spécial d'observation pour les navires d'observation bénévoles Atlantique Nord
- s: Proyecto Especial de Observación para Buques de Observación Voluntaria Atlántico Norte

VSOS e: Virtual Storage Operating System

VSP e: Variable Soft Sphere
- e: Vector/Scalar Processor
- e: Vertical Surface Package
- e: Vessel Sanitation Program
- e: Virtual Single Processor

VSPC e: EMSL-Vegas Species System

VSR e: Vertical Safety Rod
- e: Voice Storage and Retrieval
- e: Volume Surveillance Radar

VSRF e: Very Short-Range Forecasting

VSS e: Vancouver Symphony Society
- e: Vapor Sampling System
- e: Variable Soft Sphere
- e: Vehicle Sampling System
- e: Vermont State Symphony
- e: VHDL System Simulator
- e: Visual Source Save
- e: Voice Storage System
- e: Voice Switching System
- e: Volunteer Service Scheme
- e: Voucher and Scheduling System

vss e: versions
VSS e: Voluntary Social Services
VS SNG R: Vooruzhennikh Sil SNG
VSSP e: Voice Switch Signaling Point
VSSSN e: Verification Status Social Security Number
VST e: Vinson Subscriber Terminal
- e: Virtual Studio Technology

vst r: verkstad
v.st/ha a: virksomt stof pr. hektar
VSTOL e: Very Short Take-Off [and Landing]
V/STOL e: Vehicle Short Take-Off Landing
- e: Vertical/Short Take-Off and Landing aircraft

v.str. e: strong verb
VSU e: Video Service Unit
VSV a: Vestsydvest
- s: Vuelo Sin Visibilidad

VSW d: Verein für Sicherheit in der Wirtschaft e.V.
VSWG d: Vierteljahresschrift für Sozial- und Wirtschaftsgeschichte
VSWR e: Voltage Standing Wave Ratio
VSX e: Heavier-than-air antisubmarine warfare carrier-bascd aircraft
VSX e: Verification Suite for X/Open
- e: X/open Verification Suite

VSYNC e: Vertical Sync
VT d: Verfahrenstechnik/Thermischer Maschinenbau und Vertiefungsrichtung Verfahrenstechnik
- d: Vertiefungsrichtung
- e: USN training squadron
- e: Vacuum Trap
- e: Variable Time fuse
- e: VAX Terminal
- e: Vermont
- e: Vertical Tab[ulation]
- e: Vertical Tabulator
- e: Veterinary Technician
- e: Video-Terminal
- e: Virginia Tech
- e: Virtual Terminal
- e: Virtual Tourist
- e: Virtual Tributary
- e: Visiting Teacher
- e: Visiting Teaching
- e: Visual Test
- e: Visualization Tool
- f: [par] Voie Terrestre
- f: Vente à Tempérament
- f: Vente à Terme
- f: Victimes du Travail
- f: Vieux Testament
- f: Vieux Travailleurs
- f: Visa de Transit
- f: Visa de Travail
- r: Vapenteknisk
- r: Västtyskland
- r: Vattentätt
- R: Voyenniy Transport
- s: Vehículo de Transporte
- s: Veterinaria

VT- e: Presumably Video Terminal-
Vt. e: Vermont
V.T. s: Volado Trasero
VT100 e: Series of DEC terminals
- e: Virtual Terminal 100

VTA e: Soviet Military Transport Aviation
- e: Vertical Test Assembly
- e: Virtual Storage Translator
- R: Voenno-Transportnaya Aviatsiya

VTAM e: Virtual Telecom[munication] Access Method
- e: Virtual Terminal Access Method

VTAME e: Virtual Telecommunicaitons Access Method Entry
VTAS e: Visual Target Acquisition System
VTB a: Vandteknisk Brancheforening
- d: Vereinigung des technischen Betriebspersonals der schweizerischen Krankenanstalten

VTBN N: Våpenteknisk Bataljon
VTC e: Video Tape Camera
- e: Video Teleconferencing [system]
- e: Voltage Transfer Characteristic

Vtch f: Versement de chèque
VTCO e: Virtual Terminal Control Object
- e: Voltage-Temperature Cut-Off

VTD e: Virtual Tape Device
VTE e: Virtual Terminal Environment
vt[e] r: veckotimme
VTEP e: Vaccine and Treatment Evaluation Program
VTEU e: Vaccine and Treatment Evaluation Unit
VTHL e: Vertical Takeoff Horizontal Landing
VTI e: Virtual Terminal Interface
VT-II e: Virtual Tourist II
VTIR e: Visible and Thermal Infrared Radiometer
VTKK e: Finnish State Computing Center
vtkp N: våpenteknisk kompani
Vtkp N: verkstedskompani
VTLC e: Video Time-Lapsed Camera
VTLS e: Library automation package
VTLs e: Vehicular Traffic Laws
VTM d: Verband der Telekommunikationsnetz- und Mehrwertdiensteanbieter, Köln
- d: Verband der Telekommunikationsnetz- und Mobilfunkanbieter
- e: Valve test manifold

VTM e: Victorian Tourist Ministry [Australia]
vtmatr N: våpenteknisk materiell
VTNS e: Virtual Telecommunications Network Service
VTO d: Video Theresa Orlowski
VTOA e: Voice and Telephony Over ATM
VTOC e: Visual Table of Contents
- e: Volume Table Of Contents

VtoF e: Voltage-to-Frequency
VTOL e: Vertical Take Off and Landing
- e: Vertical Take-Off [and Landing]

VTOS e: Virtual Time Operating System
VTP e: Vertical Thermal Processor
- e: Virtual Terminal Program
- e: Virtual Terminal Protocol

VTPR e: Vertical Temperature Profiler Radiometer
- f: Radiomètre de mesure du profil vertical de température
- s: Radiómetro de medida del perfil vertical de la temperatura

VTQFP e: Very Thin Quad Flat Pack[age]
VTR e: Vent Thru Roof
- e: Videotape Recorder
- e: Virtual Reality
- e: Voice Tape Recorded

VTS e: Variable Tracking Strategy
- e: Vault Tracking System
- e: Vessel Traffic System
- e: Vibration Test Stand
- e: Video Teleconferencing System
- e: Virtual Terminal Service
- e: Virtual Terminal Support
- e: Virtual Touch Screen
- f: Vieux Travailleurs Salariés

VtSEA e: Vermont Society of Enrolled Agents
VTs SO AN SSSR e: Computing Center of Siberian Department of Academy of Sciences, Novosibirsk, USSR
VTST e: Variational Transition-State Theory
- s: Vaguada en la Troposfera Superior Tropical

VTSup e: Vetus Testamentum, Supplements
VTT e: Technical Research Center
VTTA e: Veterans Trial Association
V/Tte f: Votre Traite
VTU e: Volunteer Training Unit
VTVL e: Vertical Takeoff Vertical Landing
VTVM e: Vacuum-Tube Volt Meter
VTX e: Next generation naval trainer
VU e: Air Ivoire
VU e: Fixed-wing utility airplane
VU a: Venstres Ungdom
- e: Virtual User
- e: Volume Unit [meter]
- n: Vreije Universiteit
- r: Verkställande Utskott

Vu r: Vinterutrustning
VUB e: Free University of Brussels
- f: Université libre de Bruxelles
- n: Vrije Universiteit Brussel[s]
- s: Universidad Libre de Bruselas

VUBIS n: Vrije Univeriteit Brussel Information Systeem
VUD e: Verband der Unterhaltungssoftwareindustrie Deutschlands
VUE e: Visible Ultraviolet Experiment
VUEN e: Virtual University Education Network
VUI e: Video User Interface
- e: Virtual User Interface

VUIT e: Visual User Interface Tool
Vul s: vulnerabilidad
vulg. e: vulgar
VUMA e: VESA Unified Memory Architecture
VUMASBE e: VESA Unified Memory Architecture-System BIOS Extension

VU-meter e: A special kind of meter intended for tracking rapidly varying audio signals
VUP d: Verursacherprinzip
- e: VAX Unit[s] of Performance
- e: Very Unimportant Person
VUU e: Vlasov-Uheling-Uhlenbeck
VUV e: Vacuum Ultra-Violet
- e: Vatu, Vanautu
VUW e: Victoria University of Wellington
VUZ e: Higher educational institution
VV e: Vertical transmit-Vertical receive [polarization]
- e: Vessel Vent
vv d: Windgeschwindigkeit im SYNOP
Vv N: Våpenverker
V V e: Victor Vaughan Building
v.v. l: vice versa
V.v. N: Vice versa
V&V e: Validation and Verification
- e: Verification and Validation
- e: Verifying and Validating
- e: Villains and Vigilantes
V2, V3 e: HST spacecraft coordinates
VVA d: Vereinigte Verlagsanstalten GmbH
- e: Vietnam Veterans of America, Inc.
VV&A e: Validation, Verification and Accreditation
VVB d: Vereinigte Volksbank
VVBAD n: Vlaamse Vereniging voor Bibliotheek-, Archief- en Documentatiewezen
VV[&]C e: Validation, Verification [and] Certification
VVD d: Volkswagen Versicherungs Dienst
VVDS e: VSAM Volume Data Set
VVF N: Verdens Villmarksfond
VVG d: Gesetz über den Versicherungsvertrag
- d: Viersener Verkehrs-GmbH
VVIDD e: VESA Video Interface for Digital
VVK d: Vorverkauf
- r: Väg- och Vattenbyggnadskåren
VVND e: Viscerotropic Velogenic Newcastle Disease
VVO A: Verenigde Volke-Organisasie
- A: Vereniging van Orangewerkers
VVP e: Voluntary Protection Program
VV-rör r: vandringsvågsrör
VVS a: Varme-, Ventilations- og Sanitetsteknik
- e: Vessel Vent System
- N: Varme-, Ventilasjons- og Sanitetsteknikk
- r: Väg- och Vattenbyggnadsstyrelsen
- R: Voenno-Vozdushniye Sily
VVSR e: Vegetation Visible Spectral Reflectance
VVSRF R: Voenno-Vozdushniye Sily Rossisknoj Federatsii
VVSS e: Vertical Volute Spring Suspension
VV:styr r: Väg- och Vattenbyggnadsstyrelsen
VVS-VMF R: Voenno-Vozdushniye SilyVoenno Morskovo Flota
VVW d: Verlag Versicherungswirtschaft e.V.
VW d: Volkswagen
- d: Volkswirtschaftslehre
- e: Vaporware
vw e: very wicked grin [Internet]
v.w. e: weak verb
V&W e: Ministry of Transport, Public Works and Water Management
VWA d: Verwaltungsakademie in Stuttgart
- d: Verwaltungs- und Wirtschaftsakademie Duisburg
VWB e: Visual Workbench
VWD d: Vereinigte Wirtschaftsdienste GmbH
- e: Vomiting and Wasting Disease
VWF e: Von Willebrands Factor
VWFS d: Vermessungs-, Wracksuch- und Forschungsschiff
VWIL e: Virginia Womens Institute for Leadership
VWM e: Virtual Warehouse Manager
VWPP e: Visa Waiver Pilot Program
VWQMN e: Victorian Water Quality Monitoring Network
VW&R e: Van Waters & Rogers, Inc.
VwRehaG d: Verwaltungsrechtliches Rehabilitierungsgesetz
VW&S e: Ministry of Health, Welfare and Sport
VWS d: Vermessungs- und Wracksuchschiff
- e: VAX Workstation Software
vws e: very wicked [Internet]
VWS e: Vibrating-Wire Stressmeter
VwV d: Verwaltungsvorschrift
VX d: Nervengas
- e: USN test unit
Vx f: Vieux
vx r: växel
- r: växelvis
VXCO e: Voltage-Controlled Crystal Oscillator
VXD e: Virtual Device [Driver]
VxD e: Virtual Extended Driver
VXE e: USN development squadron
- e: VESA XGA Extensions
VXI e: Virtual Extension for Instrumentation
- e: Virtual Extension Interface
- e: VME[bus] Extension for Instrumentation
VXIbus e: VMEbus extension for instrumentation
Vx rd r: Växholms radiostation
vxstngrp r: växelstationsgrupp
VZA e: Viewing Zenith Angles
VzFdpbDK d: Verein zur Förderung der privat betriebenen Daten-Kommunikation e.V.
VZK d: Verein Zürcher Krankenhäuser
VZMP e: Vadose Zone Monitoring Plan
VZOG d: Vermögenzuordnungsgesetz
VZPU R: Vozdushnyi Zapastnoj Punkt Upravlenya
VZV e: Varicella-Zoster Virus

W

W d: wahlweise-obligatorische Lehrveranstaltung
- d: West[en]
- d: Wolfram
- e: Tungsten
- e: Waste
- e: West
- e: Wilderness
- e: Window[ing]
- f: Wallonie [Libre]

w d: Anzahl der diskreten Stützwellenlängen
- e: Giving a wink [Internet]

W/ e: With
W3 e: World Wide Web
W4 e: What-Works-With-What
W3A e: World Wide Web Applets
WA e: Waferboard Association
- e: Wide Band Analog
- e: Wilderness Act
- e: Word Address
- e: Work Assignment
- e: Work Authority
- e: Work Authorization
- e: Wright Annex

WA e: Womenwealth Ambika
WAA e: Waste Accumulation Area
- e: Work Authorization and Agreement

WAACS e: Wartime Aircraft Activity Capability System
WAAF e: WARM Access Authorization Form
- e: Womens Auxiliary Air Force

WAAM e: Wide-Area Anti-Armor Munitions
- e: WWMCCS Allocation and Assessment Model

WAA RET e: Wartime Aircraft Activity Reporting System Retrievals
WAARS e: Wartime Aircraft Activity Reporting System
WAAS e: Wide Area Augmentation System
WAATS e: Western ARNG Aviation Training Site
WAB e: Work Authorization Board
WAB? e: What Another Bill?
WABI e: Windows Application Binary Interface
WABNRES e: Worldwide Airborne Resources
WaBoLu d: Institut für Wasser-, Boden- und Lufthygiene des Bundesgesundheitsamtes
WABS e: Wide-Angle Broadband Spectrometer
WAC e: Washington Administrative Code
- e: Waste Acceptance Committee
- e: Western Athletic Conference
- e: Wide Area Connector
- e: WINCO Administrative Council
- e: Womens Affairs Committee
- e: Womens Air Corps

WAC e: Womens Advisory Council
WACAF e: Project on Monitoring of Pollution in the Marine Environment of the West and Central African Region
- f: Projet sur la surveillance continue de la pollution dans le milieu marin de l'Afrique de l'Ouest et du Centre
- s: Proyecto sobre la Vigilancia de la Contaminación en el Medio Marino de la Región de Africa Occidental y Central

WACC e: Washington Area Contracting Center
Waccc E: Waste Acceptance Criteria Certification Committee
WACK e: Wait Acknowledgement
WACLIM e: Climate Data Service for West Africa
- f: Service de données climatologiques pour l'Afrique occidentale
- s: Servicio de Datos Climáticos para Africa Occidental

WACMA e: World Arnold-Chiari Malformation Association
WACMS e: WWMCCS/WIS Automated Configuration Management System
WACO e: Region 7 NPDES Water Compliance System
- e: WWMCCS ADP Contracting Officer

WACOS e: WWMCCS ADP Continuity of Service
WACP e: Western Acceptance Corp.
WACS e: Wireless Access Communication[s] System
WAD e: WWMCCS ADP Directorate
WADABA d: Wattenmeerdatenbank
WADCO e: Westinghouse Advanced Development Company
WADCO e: Westinghouse Hanford Company
WADDS e: WWMCCS ADP Directory of Data Bases and Systems
WADEM e: World Association for Disaster and Emergency Medicine
WADIC e: Wave Direction Measurement Calibration Project
WADIS d: Warndienstinformationssystem
WADR e: Waste Acid Detoxification and Reclamation
WADS e: Wide Area Detection System
WAEF e: When All Else Fails
WAENA e: Wide Area Educational Network
WAEPA e: Western Australian Environment Protection Authority
WAES e: Workplace Adult Education Services
W-AESD e: Westinghouse Electric Corporation Advanced Energy Systems Division
WAF d: Warendorf, Nordrhein-Westfalen
- e: Work Approval Form

WAF/400 e: IBM SAA Workfolder Application Facility/400
WAFC e: World Area Forecast Centre
- s: Centro Mundial de Pronóstico de área

WAFDS e: WWMCCS ADP Fiscal Data Systems
WAFF e: Western Australian Farmers Federation
WAFIC e: Western Australian Fishing Industry Council
WAFR e: World Appraisal of Fishery Resources
- f: Evaluation mondiale des ressources halieutiques
- s: Evaluación Mundial de Recursos Pesqueros

WAFS s: Sistema Mundial de Pronóstico de área
WAG e: Waste Area Group
- e: Wild Ass[ed] Guess [Internet]
- e: Work Area Group

WAGE e: Weapons Aerospace Ground Equipment
WAGICS e: Women and Genetics in Contemporary Society
WAGS e: World Atmosphere Gravity Wave study
WAHEC e: World Association of Health, Environment and Culture

WAHERB e: Western Australian Herbarium Plant Specimen Database
WAHLI e: Indonesian Wildlife Forum
WahlO d: Wahlordnung [nach dem Betriebsverfassungsgesetz]
WährG d: Erstes Gesetz zur Neuordnung des Geldwesens
WAI e: Web Accessability Initiative
- e: Web Application Interface
WAIS e: Wechsler Adult Intelligence Scale
- e: West Antarctic Ice Sheet [initiative (or project)]
- e: Wide Area Information Server[s]
- e: Wide Area Information System
WAIS-R e: Wechsler Adult Intelligence Scale, Revised
WAIT e: Western Australian Institute of Technology
WAITR e: West African Institute of Trypanosomiasis Research
WAITS e: Westcoast Alternative to ITS
WAK d: Wiederaufarbeitungsanlage Karlsruhe Betriebsgesellschaft mbH
WAL e: Wallops Tracking Station
Wald Holz d: Wald und Holz
Wald-Holzwirtsch. d: Wald-und Holzwirtschaft
WALIP e: Western Australian Land Information Program
WALIS e: West Australia Land Information System
WALJOK s: Fundación Rural de Desarollo Integrado Waljok
WALRUS e: Wide Angle Large Reflective Unobscured System
WALS e: West African Languages Survey University of Dakar
Wall St. Compat. Rev. e: Wall Street Computer Review
WALSTIB e: What A Long Strange Trip It's Been
WALT e: Warning Assessment Logic Terminal
WAM e: Warren Abstract Machine
- e: Wave Model
- e: Westinghouse Access Menu
- e: Work Assignment Manager
- e: Workload Analysis Activity Report System
- e: Workstation Application Menu
WAMAP e: Waste Management Advisory Programme
- f: Programme des services consultatifs pour la gestion des déchets
WAMDI e: Wave Model Development and Implementation group
WAMDII e: Wide Angle Michaelson Doppler Imaging Interferometer
WAMEX e: West African Monex
- e: West African Monsoon Experiment
- f: Expérience de la mousson d'Afrique occidentale
- s: Experimento sobre los Monzones de Africa Occidental
WAMIS e: Wireless, Adaptive and Mobile Information Systems
WAMKSAM e: Why Are My Kids Staring At Me [Internet]
WAMM e: Waste Management Model
- e: Waste Management Module
WAMPUM e: Wards Automated Menu Package Using Microcomputers
WamS d: Welt am Sonntag
WAMS e: Wetlands Analytical Mapping System
WAN e: Wide Area Network[s]
- f: Grand réseau
- s: Red de área Extensa
WANAD e: West African News Agency Development
WANC e: Western Australian Naturalists Club
WAND e: Wide Area Network Distribution
WANL e: Westinghouse Astronuclear Laboratory
WANO d: Weltverband der Betreiber von Atomanlagen
- e: World Association of Nuclear Operators
WANS e: West African News Service
WAO d: Warteschlange der ablaufbereiten Objekte
- e: Weapons Assignment Officer
- e: Wet Air Oxidation
WAP d: Wissenschaftliche[r] Arbeitsplatz[rechner]
- d: Wissenschaftler-Arbeitsplatz-Rechner-Programm
- e: Waste Acceptance Plan
- e: Waste Acceptance Process
- e: Weatherization Assistance Program
- e: Wide Area Paging
- e: Wireless Application Protocol
- e: Work Assignment Procedure
WAPA e: Waste Acceptance Processing Activities
WAPD e: World Association of Persons with Disabilities
WAPI e: Windows Application Binary Interface
- e: World Air Photography Index
WAPIS e: Western Australian Pastoral Information System
WAPS e: Waste Acceptance Preliminary Specifications
WAR e: Women Against Repression
- e: Work Acceptance Record
- e: World Affairs Report [Database]
War e: Flavius Josephus, The Jewish War
WARAMS e: Wartime Alignment of Reserve and Active Medical Systems
WARC e: World Administrative Radio Conference
- e: World Aeronautical Radio Conference
WARC-MOB e: World Administrative Radio Conference for the Mobile Services
WARCONFAC e: War Consumables Consumption Factor
WARDA e: West Africa Rice Development Association, Monrovia, Liberia
WARD[ARD] e: Westinghouse Advanced Reactors Division
Warden e: British part of Provide Comfort
WARE e: Worksite Accident Reduction Expert
WARFS e: Water Resources Forecasting System
WARG e: West Africa Regional Group
WARI e: Western Australian Agricultural Research Institute
WARIS e: Water Resources Information System
- e: Western Arid Resource Information System
WARM e: Wartime Reserve Modes
WARMAN e: Warsaw Metropolitan Area Network
WARMAPS e: Wartime Manpower Planning System
WARMARS e: Wartime Manpower Requirements System
WARMCOM e: West Africa Regional Remote Sensing Management Committee
WARMER e: World Action for Recycled Material and Energy from Rubbish
WärmeT d: Institut für Thermodynamik und Wärmetechnik
Wärme- u. Stoffübertrag. d: Wärme- und Stoffübertragung
WARNA e: Worker Adjustment and Retraining Notification Act
WARNORD e: Warning Order
War Op e: Warwicks Opinions
WARP e: Weather and Radar Processor
- e: West African Replenishment Plan
- e: Workstation Asset Reduction Program
- e: Worldwide AUTODIN Restoral Plan
WARPAC e: Warsaw Pact
WARR e: Wide-Angle Reflection And Refraction
WARRC e: Western Aerospace Rescue and Recovery Center
WARS e: Wartime Assessment and Requirements System
- e: Wedge Absorption Remote Sensor
WARSAMS e: WARSAW SAM Missile Simulation System
WARSIDEC e: Western Australian Remote Sensing Industry Development And Education Centre
WARSIM e: Warfighters Simulation 2000
WARSL e: War Reserve Stockage List
WARSMART e: Simulation Model for Allocation of Resources for Training
WART e: Weighted Average Remaining Term
- e: Wenceslaus Anxiely Representation Taxonomy
Warta Geol [Kuala Lumpur] ma: Warta Geologi [Kuala Lumpur]
Warth Code e: West Virginia Code
Warv EI RP e: Warvelles Elements of Real Property
WarvV&P e: Warvelles Vendors and Purchasers of Real Property
Warw. e: Warwick
WAS d: Wissenschaft, Ausbildung und Schule
- e: Ward Atmosphere Scale
- e: War-At-Sea
- e: Washington, DC
- e: Waste Acceptance Specification
- e: Weapon Aiming System
- e: Weapons Alert System
- e: Web Application Server
- e: Weld Assembly Strongback
- e: Wide Analysis Sheet
- e: Wiskott-Aldrich-Syndrom

WASA e: Writers Association of South Africa
WASAR e: Wide-Angle Synthetic Aperture Radar
WASAW e: Western Australian Sewerage and Waste Quality Infrastructure Program
WASC e: Seattle Central Community College
- e: West Africa Supply Centre
- e: West African Students Confederation
- e: Western Administrative Support Center
- e: White Anglo-Saxon Catholic
- e: Williams Awareness Sentence Completion

WASC-N e: North Seattle Community College
WASCO e: Childrens Orthopedic Hospital and Medical Center, Seattle
WASC-S e: South Seattle Community College
WaSC-Sh e: Shoreline Community College, Seattle
WASDA e: Washington State Department of Agriculture
WASH e: Water and Sanitation for Health Project
Wash. e: Washington [Seattle]
WASHCAP e: Washington Operations Capabilities System
Wash. Dep. Fish. Fish. Res. Pap. e: WA Department of Fisheries Fisheries Research Papers
Wash. Dep. Fish. Res. Bull. e: Washington Department of Fisheries Research Bulletin
Wash. Dep. Fish. Tech. Rep. e: Washington Department of Fisheries Technical Report
WASHFAX e: Washington Area High-Speed Facsimile System
WASH PIRG e: Washington Public Interest Research Group
Wash State Hortic Assoc Proc e: Proceedings. Washington State Horticultural Association
Wash State Inst Technol Circ e: Washington State Institute of Technology. Circular
Wash State Inst Technology Bull e: Washington State Institute of Technology. Bulletin
Wash State Inst Technol Tech Rep e: Washington State Institute of Technology. Technical Report
Wash State J Nurs e: Washington State Journal of Nursing
Wash State Univ Agric Exp Stn Tech Bull e: Washington State University. Agricultural Experiment Station. Institute of Agricultural Sciences. Technical Bulletin
Wash State Univ Agric Res Cent Res Bull e: Washington State University. Agricultural Research Center. Research Bulletin
Wash State Univ Coll Agric Res Cent Bull e: Washingten State University. College ofagriculture. Research Center. Bulletin
Wash State Univ Coll Agric Res Cent Tech Bull e: Washington State University. College of Agriculture. Research Center. Technical Bulletin
Wash State Univ Coll Eng Bull e: Washington State University. College of Engineering. Bulletin
Wash State Univ Coll Eng Circ e: Washington State University. College of Engineering. Circular
Wash State Univ Coop Ext Serv Ext Bull e: Washington State University. Cooperative Extension Service. Extension Bulletin
Wash State Univ Ext Ser Ext Bull e: Washington State University. Extension Service. Extension Bulletin
Wash State Univ Ext Serv EM e: Washington State University. Extension Service. EM
Wash State Univ Int Symp Particleboard Proc. e: Washington State University. International Symposium on Particleboard. Proceedings
Wash State Univ Publ Geol Sci e: Washington State University. Publications in Geological Sciences
Wash State Univ Symp Particieboard Proc e: Washington State University. Symposium on Particleboard. Proceedings
Wash St G An Rp e: Washington State Geologist. Annual Report
WASHT e: Washtenaw Ave
Wash Ter NS e: Allens Washington Territory Reports, New Series
Wash T[err] e: Washington Territory Opinions
Wash T[err] e: Washington Territory Reports
WASHTO e: Western Association of State Highway and Transportation Officials
Wash Ty e: Washington Territory Opinions
Wash Ty e: Washington Territory Reports
Wash Univ Bull e: Washington University. Bulletin
Wash Univ Dent J e: Washington University. Dental Journal
Wash Univ Dep Geol Sci Abstr Res e: Washington University. Department of Geological Sciences. Abstracts of Research
Wash Univ Pub G e: Washington University. Publications in Geology
Wash Univ St Hum Ser e: Washington University. Studies. Humanistic Series
Wash Univ St Sci Ser e: Washington University. Studies. Scientific Series
Wash Univ Stud Lang & Lit e: Washington University. Studies. Language and Literature
Wash Univ Stud Sci & Tech e: Washington University. Studies. Science and Technology
Wash Univ Stud Sci & Tech NS e: Washington University. Studies. Science and Technology. New Series
Wash Univ Stud Social & Philos Sci e: Washington University. Studies. Social and Philosophical Sciences
Wash Univ Stud Social & Philos Sci NS e: Washington University. Studies. Social and Philosophical Sciences. New Series
WASI e: Whimbey Analytical Skills Inventory
WaSIF e: International Fisheries Commission, Seattle
WASIH e: Western American Society for Italic Handwriting
WASK e: King County Medical Society, Seattle
WASKC e: King County Library System, Seattle
WASM e: Mountaineers, Inc., Seattle
- e: White Anglo-Saxon Male

WAsM e: Memorial Medical Center, Health Sciences Library, Ashland
WASMAC e: Western Australian Survey and Mapping Advisory Council
Wasmann J Biol e: Wasmann Journal of Biology
WASN e: Northland College, Ashland
WASNH e: Northwest Hospital, Effie M. Storey Learning Center, Seattle
WaSnqE e: Echo Glen Childrens Center, Staff Library, Snoqualmie
WaSnqE-R e: Echo Glen Childrens Center, Resident Library, Snoqualmie
WASO e: WWMCCS ADP Security Officer
WASP e: War Air Service Program
- e: Water Quality Analysis Simulation Program, Version 3.1
- e: Weber Advanced Spatial Perception Test
- e: White Anglo-Saxon Protestant
- e: Wind Atlas Analysis and Application Programme
- f: Programme d'analyse et d'application de l'atlas des vents
- s: Programa de Análisis y Aplicación de un Atlas de Vientos

WASPA e: White Anglo-Saxon Protestant Ambulatory
WASPB e: Water Spectrum
WaSpBM e: United States Bureau of Mines, Mining Research Center, Spokane
WaSpBMW e: United States Bureau of Mines, Western Field Operations Center, Spokane
WASPC e: Seattle Pacific College, Seattle
WaSpCN e: Center for Nursing Education, Spokane
WaSpCo e: Spokane County Library, Spokane
WaSpD e: Deaconess Hospital, School of Nursing, Spokane
WASS e: Wide Area Surveillance System
- e: WWMCCS ADP System Security

WasserB d: Institut für Wasserbau
Wasser Boden d: Wasser und Boden
Wasser Energ. Luft d: Wasser, Energie, Luft
Wasser-Energ.wirtsch. d: Wasser- und Energiewirtschaft
WasserkrM d: Institut für Strömungsmechanik und Hydraulische Strömungsmaschinen
WASSM e: WWMCCS ADP System Security Manager

WASSO e: WWMCCS ADP System Security Officer
WASTAC e: Western Australian Satellite Technology and Applications Centre
- e: Western Australian Satellite Technology [and] Applications Consortium
Waste Manage. e: Waste Management
WAT A: Woordeboek van die Afrikaanse Taal
- e: Wafer Acceptance Test
- e: Waste Adjustment Tank
- e: Weight, Altitude, Temperature
- e: West Africa Time
- e: Work Authorization Tracking
Wat. e: Waterloo
WATADS e: WSR-88D Algorithm and Display System
WATASO e: WWMCCS ADP Terminal Area Security Officer
WAT-BEE e: WOCE/Atlantic/Tropical-Boundary Eastern Equatorial
Wat Bull e: Water Bulletin
WATCH e: Mary Bridge Childrens Health Center, Tacoma
- e: Watchers Against Television Commercial Harrassment
- e: Working Group on the Assessment of Toxic Chemicals
- e: World Against Toys Causing Harm
Watch e: Writers and their Copyright Holders
WATCIM e: Waterloo Centre for Integrated Manufacturing
WATCJ e: Cascadia Juvenile Diagnostic Center, Tacoma
Wat Con e: Watkins on Conveyancing
Wat Cop e: Watkins on Copyholds
Wat Cr Dig e: Watermans Criminal Digest
Wat Cr Proc e: Watermans Criminal Procedure
WATD e: Doctors Hospital, Tacoma
WA Teach[ers] J e: Western Australian Teachers Journal
WATER e: Economic Analysis of Water Supply
- e: West African Tropical Experiment In Radiation
WATER e: Womens Alliance for Theology, Ethics and Ritual
Water Air [and] Soil Pollut e: Water, Air and Soil Pollution
Water Air Soil Pollut. e: Water, Air and Soil Pollution
Water Am Inst Chem Eng e: Water. American Institute of Chemical Engineers
Water and San e: Water and Sanitation
Water A S P e: Water, Air and Soil Pollution
Water Biol Syst **e:** Water in Biological Systems
Water Electrolyte Metab Proc Symp e: Water and Electrolyte Metabolism. Proceedings of the Symposium
Water E & M e: Water Engineering and Management
Water Eng e: Water and Wastes Engineering
Water Eng. Mgmt. e: Water Engineering and Management
Water Int. e: Water International. International Water Resources Association
Water Invest Mich Geol Surv Div e: Water Investigation. Michigan Geological Survey Division
Water Law Newsl e: Water Law Newsletter
Water Manage News e: Water Management News
Water Manage Techn Rep Colorado State Univ e: Colorado State University. Water Management Technical Report
Watermeyer e: Watermeyers Cape Of Good Hope Reports
Water Poll Abstr e: Water Pollution Abstracts
Water Poll. Control e: Water Pollution and Control
Water Poll[ut] Cont[r][ol] Fed J e: Water Pollution Control Federation. Journal
Water Pollut Control e: Water Pollution Control
Water Pollut Control [Don Mills Can] e: Water and Pollution Control [Don Mills, Canada]
Water Pollut Control [London] e: Water Pollution Control [London]
Water Pollut Control Res Ser e: Water Pollution Control Research Series
Water Pollut Res Can e: Water Pollution Research in Canada
Water Pollut Res [Stevenage] e: Water Pollution Research [Stevenage]
Water Purif Liquid Wastes Treat e: Water Purification and Liquid Wastes Treatment [Japan]
Water Qual Instrum e: Water Quality Instrumentation
Water Qual. Int. e: Water Quality International. International Association on Water Pollution Research and Control
Water Res. e: Water Research. International Association on Water Pollution Research and Control
Water Res Cent Notes Water Res e: Water Research Centre. Notes on Water Research
Water Res Found Aust Annu Rep Balance Sheet e: Water Research Foundation ofaustralia Limited. Annual Report and Balance Sheet
Water Res Found Aust Bull e: Water Research Foundation of Australia Limited. Bulletin
Water Res Found Aust Lted Ann Rep Balance Sheet e: Water Research Foundation of Australia Limited. Annual Report and Balance Sheet
Water Res Found Aust Lted Ann Rep Balance Sheet e: Water Research Foundation of Australia Limited. Annual Report and Balance Sheet
Water Res Found of Aust Newsl e: Water Research Foundation of Australia Limited. Newsletter
Water Res Found Aust Rep e: Water Research Foundation of Australia Limited. Report
Water Res Inst W Va Univ Inf Rep e: Water Research Institute. West Virginia University. Information Report
Water Res News e: Water Research New
Water Res R e: Water Resources Research
Water Res. Technol. e: Water Research and Technology. International Association on Water Pollution Control
Water Resour e: Water Resources
Water Resour Bull e: Water Resources Bulletin. American Water Resources Association
Water Resour Bull Nev Div Water Resour e: Water Resources Bulletin. Nevada Division of Water Resources
Water Resour Bull [PR] e: Water Resources Bulletin [Puerto Rico]
Water Resour Circ Arkansas Geol Comm e: Water Resources Circular. Arkansas Geological Commission
Water Resour. Congr. Rep. e: Water Resources Congress Report
Water Resour [Engl Transl Vodnye Resursy] e: Water Resources [English Translation of Vodnye Resursy]
Water Resour Invest e: Water Resources Investigations
Water Resour Invest US Geol Surv e: Water Resources Investigations. United States Geological Survey
Water Resour J Econ Soc Comm Asia Pac e: Water Resources Journal. Econornic and Social Commission for Asia and the Pacific
Water Resour Manag Ser e: Water Resources Monograph Series. American Geophysical Union
Water Resour Newsl e: Water Resources Newsletter
Water Resour Reconnaissance Ser Nev Div Water Resour e: Water Resources. Reconnaissance Series. Nevada Division of Water Resources
Water Resour Rep Ariz State Land Dep e: Water Resources Report. Arizona State Land Department
Water Resour Rep Ont Minist Environ Water Resour Branch e: Water Resources Report. Ontario Ministry of the Environment. Water Resources Branch
Water Resour Res e: Water Resources Research. American Geophysical Union
Water Resour Res Cent VA Polytech Inst State Univ Bull e: Water Resources Research Center. Virginia Polytechnic Institute and State University. Bulletin
Water Resour Rev Streamflow Ground-Water Cond e: Water Resources Review for Streamflow and Ground-Water Conditions [United States-Canada]
Water Resour Ser Tenn Div Water Resour e: Water Resources Series. Tennessee Division of Water Resources
Water Resour Symp e: Water Resources Symposium
Water S.A. e: Water S.A. Water Research Commission of South Africa
Water [S Afr] e: Water [South Africa]
Water Sanit e: Water and Sanitation

Water Sanit Eng e: Water and Sanitary Engineer
Water Sci [&] Technol e: Water Science and Technology
Water Serv e: Water Services
Water Sewage Effl e: Water, Sewage and Effluent
Water & Sewage Works e: Water and Sewage Works
Water Supply Manage e: Water Supply and Management [England]
Water Supply Pap Geol Surv GB Hydrogeol Rep e: Water Supply Papers. Geological Survey of Great Britain. Hydrogeological Report
Water Supply Pap US Geol Surv e: Water Supply Paper. United States Geological Survey
Water Treat Exam e: Water Treatment and Examination
Water Waste e: Water and Wastes Engineering
Water & Waste Engng e: Water and Waste Engineering
Water Wastes Dig e: Water and Wastes Digest
Water Wastes Dig. e: Water and Wastes Digest
Water Wastes Eng e: Water and Wastes Engineering
Water Wastes Eng Ind Water and Wastes Engineering Industrial
Water Waste Treat e: Water and Waste Treatment
Water Waste Treat J Water Waste Treatment Journal
Water Wastewater Treat Plants Oper Newsl e: Water and Wastewater Treatment Plants Operators Newsletter
Water Water Eng e: Water and Water Engineering
Water [WC and IC Staff Journal] e: Water [Water Conservation and Irrigation Commission Staff Journal]
Water Well J[our] e: Water Well Journal
Water Works Eng e: Water Works Engineering
Water Works Wastes Eng e: Water Works and Wastes Engineering
WATF e: Waterford Wedgwood PLC
WATFLOOD e: University of Waterloo Flood Forecasting System
WATFOR e: WATerloo FORtran
WATiS d: Wattenmeerinformationssystem
WATM e: Wireless ATM
WATOX e: Western Atlantic Ocean Experiment
WATQ e: Region 2 Water Quality Analysis Graphics System
WATS e: Wide Area Telecommunications Service
- e: Wide Area Telephone Service
- e: Wireless Access Telecommunication[s] Service
- e: Work Authorization Tracking System
Watson House Bull e: Watson House Bulletin
WATSTORE e: Water Data Storage and Retrieval System
WATSUP e: Watfor Student Utility Program
WATTS e: Western Atlantic Thermohaline Transport Study
W.Aust. e: Western Australia
WAUXCP e: West Auxiliary Airborne Command Post
WAV e: Waveform Audio
WAVE e: Weather Altimeter Voice Equipment
Wave Electron e: Wave Electronics
WAVEOB e: Report of Spectral Wave Information from a Sea Station or from a Remote Platform
- f: Message d'observation du spectre des vagues provenant d'une station en mer ou d'une plate-forme éloignée
- s: Informe del Espectro de las Olas proveniente de una Estación en el Mar o de una Plataforma Alejada
Wave Part Dualism e: Wave. Particle Dualism
WAVES e: Weight and Value Engineering System
- e: Western Australian Voluntary Euthanasia Society
- e: Women Accepted for Volunteer Emergency Service
- e: Women Appointed Volunteer Emergency Services
- e: Worker and Visitor Entrance System
WAVESPEC e: Wave Spectrometer
WAVFH e: World Association of Veterinary Food-Hygienists
WAVHS e: United States Park Service, Fort Vancouver National Historical Site
WAVLD e: World Association of Veterinary Laboratory Diagnosticians
WAVMH e: Vancouver Memorial Hospital
WAVMI e: World Association of Veterinary Microbiologists, Immunologists and Specialists in Infectious Diseases
WAVP e: Western Australian. Votes and Proceedings
WAWA e: Work Anywhere With Anybody
WAWAS e: Washington Area Warning System
WAWRC e: Western Australian Water Resources Council
WAWS e: Washington Area Wideband System
WAWV e: United States Veterans Administration Hospital, Walla Walla
WAWW e: Whitman College, Walla Walla
WAWWC e: Walla Walla Community College
WAX e: Waxman Industries, Inc.
- e: Weak Anion Exchanger
- e: Weapon Assignment and Targel Extermination
WAXD e: Wide-Angle X-Ray Diffraction
WAY e: Wayne State College
- e: Wigner-Araki-Yanase
- e: World Assembly of Youth
WaY e: Yakirna Valley Regional Library
WAZ d: Westdeutsche Allgemeine Zeitung
WB e: WayBill
- e: Weak Base
- e: Welcome Back [Internet]
- e: Wet Bulb
- e: Whichard Building
- e: Whole-Body
- e: Wide Band
- e: Wire Bonded
- e: Workbench
- e: World Bank
W/B e: Wide Band
- e: Wire Bonding
WBA e: World Boxing Association
WBAPL e: Westinghouse Bettis Atomic Power Laboratory
WB/BB e: Wideband/Broadband
WBBCC e: Wide Bay Burnett Conservation Council
WBC e: Water Basin Concept
- e: Western Boundary Current
- e: White Blood cell Count
- e: White Blood Cell[s]
- e: Whole-Body Count and Bioassay
- e: Whole Body Count[er]
- e: Wide Band Channel
- e: Wide-Body Cargo
- e: Wide-Body CRAF
- e: WideBand Channel
- e: World Boxing Council
WB-CDMA e: Wide-Band Code-Division Multiple Access
WBCSD e: World Business Council for Sustainable Development
WBD e: Wide Band Data
WBDCS e: Wide-Band Data Collection System
WBE e: Womens Business Enterprise
WBEAA e: Westinghouse-Boeing Employee Activity Association
WBEL e: Western Bell Communications, Inc.
WBEIH e: Holy Family Convent, Benet Lake
WBeISB e: Saint Benedicts Abbey, Benet Library, Benet Lake
WBEM e: Web Based Enterprise Management
- e: Web-Based Enterprise Management
WBEP d: Wiener Beiträge zur Englischen Philologie
WBer e: Berlin Public Library
WBF d: Wissenschaftliches Beifach
WBFM e: WideBand Frequency Modulation
WBG d: Gesetz zur Bereinigung des Wertpapierwesens [Wertpapierbereinigungsgesetz]
- e: West Rand White Community Guard
WBGT e: Wet Bulb Globe Temperature
WBGU d: Wissenschaftlicher Beirat Globale Umweltveränderungen
WBI e: Web Browser Intelligence
WBIS e: Walker Branch Information System
W-bit e: Wait-bit
WBK d: Weiterbildungskonferenz
WBL A: Witbevrydingsleer
- d: Wissengemeinschaft Blaue Liste
- e: Work-Based Learning
WBM e: Web Based Management

WBMS e: Work Breakdown Management Structure
WBNC e: Washington Bancorp, Inc.
WbN, WbS e: West by North, West by South
WBP e: Water Binding Potential
- e: Weather-and Boil-Proof
- e: Woodwind World - Brass and Percussion
WBR e: Warner Brothers Records
WBRC d: Württembergisch-Badischer Radioclub
WBRC e: Walter Bagehot Research Council on National
WBRE e: Elmbrook Memorial Hospital, Brookfield
WBro e: Brodhead Memorial Public Library
WBRO e: Weather Bureau Regional Office
WBRT e: Weather Bureau Radiotheolite
WBRT e: Whole-Blood Recalcification Time
WBS d: Wissensbasierte Systeme
- d: Worbis, Thüringen
- e: Waterbody System
- e: Wednesday Business School
- e: Work Breakdown Schedule
- e: Work Breakdown Structure
WBSC e: Wide Band Signal Channel-Radio
WBSEM e: Wire Bonder Specific Equipment Model
WBSN e: Work Breakdown Structure Number
WBSO e: Research and Development Act to promote
WBST e: Wideband Secure Terminal
- e: Wideband Subscriber Terminal
WBSVN e: Wideband Secure Voice Network
WBSVVS e: Wideband Secure Voice Via Satellite
WBT e: Wait Before Transmission
- e: Windows Based Terminal
- e: Wycliffe Bible Translators
WBU e: World Boxing Union
WBVT e: Wide Band Video Tape
WBW e: Walker Branch Watershed
- e: World Birthday Web
WB/WAAS e: Wartime Beddown/ Wartime Aircraft Activity System
WC A: Witcommando
- e: Wall Cabinets
- e: Water Closet
- e: Water Column
- e: Waveguide, Circular
- e: Weapons Controller
- e: Wire Center (or Centre)
- e: Wiring Concentrator
- e: Word Count
- e: Workers Compensation
- e: Workgroup Computing
- e: World Coordinate
- e: Written Component
W-C e: Wilkinson-Cohn
W/C e: Water Cooling
W3C e: World Wide Web Committee
- e: World Wide Web Consortium
WC³ e: Warning, Command, Control and Communications
WCA e: Waste Collection Authority
- e: World Chiropractic Alliance
- e: Worst-Case Analysis
WCA e: Warrant Claims Action
- e: Weapon Control Area
- e: Weimaraner Club of America
- e: West Coast of Africa
- e: West Coast Airlines, Inc.
- e: Westair Commuter Airlines
- e: Who Cares, Anyway?
- e: Whole Core Accident
- e: Windmill Class Association
- e: Winston S. Churchill Association
WCAA e: Washington Clean Air Act
WCAB e: Working Committee of the Aeronautical Board
WCAC e: WWMCCS Crisis Alerting Capability
WCAN e: WWMCCS Crisis Alerting Network
WCAP e: World Climate Applications Program
WCAR e: Western Carolina Savings and Loan Association, Inc.
WCAS e: Workload Collection and Analysis System
WCASP e: World Climate Applications and Services Program[me]
WCB e: War Communications Board
- e: Warramunga Array
- e: Water Control Board
- e: Weekly Criminal Bulletin
- e: Wellington County Board of Education
- e: Will Call Back
- e: William C. Brown Publishers
- e: Workmens Compensation Board
- e: Write Combining Buffer
WCBC e: West Coast Bancorp
WCBD [Vic] e: Workers Compensation Board Decisions [Victoria]
WCBD [WA] e: Workers Compensation Board Decisions [Western Australia]
WCB-ISEE e: World Council for the Biosphere-International Society for Environmental Education
WCBK e: Workingmens Co-Operative Bank
WCBL e: World Council of Blind Lions
WCBS e: Wireline Core-Barrel Sampling
WCBSU e: West Coat Base Service Unit
WCC e: Washington Computer Center
- e: Westminster Choir College
- e: White [blood] Cell Count
- e: Wideband Communications Capsule
- e: Womens Classical Caucus
- e: Work Control Coordinator
- e: Workers Compensation Commission
- e: World-Class Competitiveness
- e: World Council of Churches
WCCDC e: Wayne County Child Development Center
WCCE e: World Conference on Computers in Education
WCCI e: Windhoek Chamber of Commerce and Industries
WCCRP e: Wyrtki Center for Climate Research and Prediction
WCCS e: Washington Crisis Communications System
- e: Wing Command and Control System
WCDMP e: World Climate [and] Data Monitoring Program[me]
WCDO e: War Consumables Distribution Objective
WCDOS e: War Consumable Distribution Objectives System
WCDP e: World Climate Data Program[me]
WCDR e: War Consumables Distribution Requirements
WCE e: Western Corporate Enterprises, Inc.
- e: Write Cache Enable
WCF e: Waste Calciner Facility
WCGTC e: World Council for Gifted and Talented Children
WCGZ e: World Confederation of General Zionists
WCH e: Weekly Contact Hours
- e: Working Class Hero
- e: Workshop Conferences Hoechst
WCh e: Chippewa Falls Public Library
WCHEN e: Western Council on Higher Education for Nursing
WCHF e: Wei Crude Handling Facilities
WCHI e: Westworld Community Healthcare, Inc.
WCHI e: Womens Council for the Histadrut in Israel
Wchnbl K K Gesellsch Ärzte Wien d: Wochenblatt. K- K. Gesellschaft der Ärzte in Wien
Wchnschr Ges Heilk d: Wochenschrift für die Gesamte Heilkunde
Wchnschr Tierh u Viehzucht d: Wochenschrift für Tierheilkunde und Viehzucht
WCHODW e: Workshop Conferences Hoechst
WCHO-FM e: Washington Court House
WCI e: Wound Care Institute
WCIA e: Welsh Centre for International Affairs
WCID e: Wire Centre Identification
WC [&] Ins R[ep] e: Workmens Compensation and Insurance Reports
WCIP e: Weapon Control Indicator Panel
- e: World Climate Impact [studies] Program[me]
- e: World Council of Indigenous Peoples
WCIQ e: Mount Cheaha State Park
WC&IR[ep] e: Workmens Compensation and Insurance Reports
WCIRP e: World Climate Impact Assesment and Response Strategies Program[me]
WCIS e: Wisconsin Career Information System
WC-ISA e: Womens Commission of the Iranian Students Association
WCISP e: World Climate Impact Studies Program
WCISW e: Wider Caribbean Initiative for Ship Generated Waste
- s: Iniciativa del Gran Caribe relativa a los Desechos Generados en los Buqes
WCIU e: Workshop Computer Interface Unit

WCJ e: White Cloud Journal of American Indian/Alaska Native Mental Health
WCJA e: Western Canadian Journal of Anthropology
WCJA e: World Council of Jewish Archives
WCJC e: Webster City Junior College
- e: Wharion County Junior College
WCJCC e: World Confederation of Jewish Community Centers
WCJCS e: World Conference of Jewish Communal Service
WCK e: Whiskey Creek Resources
WCL e: Washington College of Law
- e: Weapons Custody List
- e: Widget Creation Library
- e: Work control log
WCLA e: Washington Contract Loggers Association
WCLA e: Workers Compensation Legislation in Australia
WCM e: Weapons Countermeasures and Mines
WCMC e: World Conservation Monitoring Center
- e: World Conservation Monitoring Centre
- f: Centre mondial de surveillance de la conservation
- s: Centro Mundial de Vigilancia de la Conservación
WCN e: Westinghouse Change Notice
- e: Worldwide Communications Network
WCNDR e: World Conference on Natural Disaster Reduction
- f: Conférence mondiale sur la prévention des catastrophes naturelles
- s: Conferencia Mundial sobre la Reducción de los Desastres Naturales
WCO e: Wiring Control Office
WCOAT e: Wolfs Computer Operator Aptitude Test
WCOBL e: Wolfe Programming Language Test: COBOL
WCOMMRGN e: Western Communications Region
WCP e: Warner Computer Systems
- e: Waste Characterization Plan
- e: Waste [form] Compliance Plan
- e: World Climate Program[me]
WCPC e: Wire Center Planning Center
WCPN e: Wireless Customer Premises Network
WCP-WATER e: World Climate Programme-Water
WCR e: West Coast Repository at SIO
WCRP e: World Climate Research Program[me]
WCRP-JSC e: Joint Scientific Committee for the WCRP
WCRP-O e: WCRP, Oceanographic Component
WCRS e: War Consumable Reporting System
WCR[WA] e: Workers Compensation Reports
WCS e: Warehouse Control System
- e: Well Completion Summary
- e: Wireless Communications Service
- e: Workload Control System
- e: Workstation Control and Synchronization
WCSG e: WWMCCS Council Support Group
WCSMP e: World Climate System Monitoring Program[me]
- s: Programa Mundial de Vigilancia del Sistema Climático
WCSS e: Waste Characterization and Sampling System
WCST e: Wisconsin Card Sorting Test
WCT e: Westinghouse Commitment Tracking
- e: Westinghouse Correspondence Tracking
- e: WHC Commitment Tracking system
- e: Workstation Control Table
WCTF e: World Childrens Transplant Fund
WCTU e: Womens Christian Temperance Union
WCTUC e: West Cameroon Trade Union Congress
WCVD e: Tungsten Chemical Vapor Deposition
WCW e: World Championship Wrestling
WCWCD e: Washoe County Water Conservation District
WD e: Walt Disney
- e: Water Depth
- e: Water Division
- e: Weapons Director
- e: Western Digital
- e: White Dwarf
- e: Wide Band Digital
- e: Working Draft
wd. e: word
WDA d: Werdau, Sachsen
- e: Westinghouse Document Accountability
- e: Wisconsin Dental Association
WDAMS e: WIS Data Base Management System
WDBII e: World Data Base II
WDC e: Washington, District of Columbia
- e: Western Digital Corporation
- e: Westinghouse Drawing Control
- e: Work Distribution Chart
- e: Workers Defense Committee
- e: World Data Center
- e: World Data Centre [on Micro-organisms]
- e: World Data Centres
- s: Centro Mundial de Datos
WD&C e: Waste Dislodging and Conveyance
WDCA e: World Data Center A, Washington DC, USA
WDC-A e: World Data Centre A, Meteorology
- e: World Data Centre A, Oceanography
- e: World Data Centre A, Tsunami
- f: Centre mondial de données A, Tsunamis
- s: Centro Mundial de Datos A, Meteorología
- s: Centro Mundial de Datos A, Oceanografía
- s: Centro Mundial de Datos A, Tsunamis
WDC-A-MGG e: World Data Centre A, Marine Geology and Geophysics
- s: Centro Mundial de Datos A, Geología y Geofísica Marinas
WDC-A-R&S e: World Data Center A for Rockets and Satellites
WDCB e: World Data Center B, Obninsk, USSR
WDC-B e: World Data Centre B, Meteorology
- e: World Data Centre B, Oceanography
- s: Centro Mundial de Datos B, Meteorología
- s: Centro Mundial de Datos B, Oceanografía
WDCGG e: World Data Center (or Centre) for Greenhouse Gases
- f: Centre mondial de données relatives aux gaz à effet de serre
- s: Centro Mundial de Datos sobre los Gases de efecto Invernadero
WDCM e: World Data Centre on Microorganisms
WDCP e: WSPRS Data Collection Program
WDDES e: World Digital Database for Environmental Sciences
- f: Base mondiale de données numériques destinées aux sciences de l'environnement
- s: Base Mundial de Datos Numéricos sobre Ciencias Ambientales
WDDM e: Wavelength-Division DeMultiplexer
WDEOAT e: Wolfe Data Entry Operator Aptitude Test
WDF e: Western Direct Federal [Division]
WDI e: Wardair, Inc.
- e: Wood-Destroying Insect
WdI d: Waldbelastungen durch Immissionen
WDK e: Word Developers Kit
WDL e: Windows Driver Library
WDM e: Wavelength Division Multiplexer
- e: Wavelength Division Multiplex[ing], [for optical fibre, ultra gigabit]
- e: Windows Driver Model
WDMC e: Woodmoor Corp.
WDMWG e: WOCE Data Management Working Group
WDNR e: Washington State Department of Natural Resources
WDNS e: Weapons Delivery and Navigation System
WDO e: Wood-Destroying Organism
WDOE e: Washington State Department of Ecology
WDOH e: Washington State Department Of Health
WDOP e: Weighted Dilution of Precision
WDP e: Watchdog Process
WDRAM e: Windows Dynamic Random Access Memory
WDROP e: Distribution register of organic pollutants in water
WDS e: Waste and Decontamination Services
- e: Watchdog Server
- e: Wavelength Dispersive Spectrometry [of X-rays]
- e: Wavelength-Dispersive Spectroscopy

WDSD e: Water Data Sources Directory
WDSIL e: Data Automation Resource Management System
WDSS e: Warning Decision Support System
WDT e: Warning Display Terminal
- e: Western Daylight Time
WDW e: Walt Disney World
- e: Warm Deep Water
WDX e: Wavelength Dispersive X-ray
WDXA e: Wavelength Dispersive X-ray Analysis
WDYM e: What Do You Mean [by that] [Internet]
WDYMBT e: What Did You Mean By That [Internet]
WDYW e: What Do You Want [Internet]
WE d: Wochenende
- e: Worker Equivalent
- e: Wrist Extend
- e: Write Enable
WE. e: Western Electric
W&E e: CSIRO Division of Wildlife and Ecology
W/E e: Week Ending
- e: Write/Erase
WEA e: Weapons Expenditures Authorization
WEAC e: Winchester Engineering and Analytical Center
- e: Womens Employment Action Centre
WEAI e: Western Economic Association International
WEAL e: Womens Equity Action League
WEAP e: World Environment Action Plan
WEARFSFAC e: Weather Research Facility
WEARSCHFAC e: Weather Research Facility
WEASERVCOMM e: Weather Service Command
Weather and Clim e: Weather and Climate
Weather C&M e: Weather, Crops and Markets
Weather Dev Res Bull e: Weather Development and Research Bulletin
Weather Research Bull e: Weather Research Bulletin
WEAVE e: Water and Energy: Atmospheric, Vegetative and Earth Interactions
WEB e: Entities Emphasizing the WWW
- e: Wealthbuilding
WEBB e: Water, Energy and Biogeochemical Budget
Webb Pl & Pr e: Webbs Kansas Pleading and Practice
Web-Browser d: Programm zum Abrufen von Web-Seiten
Webb RR e: Webbs Railroad Laws of Maine
Webb Supr Ct Pr e: Webbs English Supreme Court Practice
WebDAV e: Web Distributed Authoring and Versioning
WEBE e: Western European Basic Encyclopedia
WebNFS e: Web Network File System
Web Pat e: Websters New Patent Law
Web P[at] C[as] e: Websters Patent Cases
Web R e: Webster Review
WEBROCK e: Weather Buoy Rocket
WEBS e: Weapons Effectiveness Buoy System
- e: Webster Clothes, Inc.
Webs e: Websters Patent Cases
WEBSEC e: Western Beaufort Sea Ecological Cruise
Web-Server d: Server, der Dokumente und andere Intranet-Ressourcen speichert und über HTTP versendet oder entgegennimmt
Website d: Angebot einer Organisation, einer Firma oder einer Privatperson im WWW
Webs Pat Cas e: Websters Patent Cases
Webst Dict e: Websters Dictionary
Webst Dict Unab e: Websters Unabridged Dictionary
Webster e: Websters [New International] Dictionary
Webster Pat Cas [Eng] e: Websters Patent Cases [England]
Webster in Sen Doc e: Webster in Senate Documents
Webst Int Dict e: Websters International Dictionary
Webst New Int D e: Websters New International Dictionary
Webta e: Western Cape Black Taxi Association
WEB toilet e: Waterless Electric Biological Toilet
Web Tr e: Trial of Professor Webster for Murder
WEC e: Wafer Environment Control
- e: Westinghouse Electric Corporation
- e: World Energy Conference
- e: World Energy Council
- e: World Environment Centre
WECC e: Western European Calibration Cooperation
WECO e: Weather Current Operations
WED e: Water Enforcement Division
- e: Western Europe Daylight
- e: Wire Equilibrium Device
- e: Womens Equality Day
- e: World Environment Day
- f: Journée mondiale de l'environnement
- s: Día Mundial del Medio Ambiente
WEDC e: Water Engineering and Development Centre within Civil and Building Engineering
WEDCO e: Constructor of Indian Point Nuclear Plants No. 2 and 3
WEDI e: Workgroup for Electronic Data Interchange
WEDOS e: World Environment Disaster Observation System
WEDSS e: Whole Earth Decision Support System
WEE e: Western Equine Encephalomyelitis
- e: Wind Erosion Equation
WEEB e: Western Europe[an] EDIFACT Board
WEETAG e: Womens Employment, Education and Training Advisory Group
WEF e: Water Environment Federation
WEFAX e: Weather Facimile
- f: Fac-similé météorologique
- s: Facsímil Meteorológico
WEFSE e: Washington State Energy Facility Site Evaluation Co.
WEFT e: Web Embedding Font Tool
WEG d: Gesetz über das Wohnungseigentum und das Dauerwohnrecht [Wohnungseigentumsgesetz]
- e: Weapons Evaluation Group
WEGS e: Western European Geological Surveys
WEIA e: Washington Environmental Industries Association
Weight Eng. J. e: Weight Engineering Journal. Society of Allied Weight Engineers
Wei Sheng Wu Hsueh Pao ch: Wei Sheng Wu Hsueh Pao
WELD e: Weldon Sprint Site Office
Weld. Des. Fabr. e: Welding Design and Fabrication
Weld Fabr Des e: Welding Fabrication and Design
Weld. Inf. Newsl. e: Welding Information Newsletter. American Welding Institute
Weld. Innov. Q. e: Welding Innovation Quarterly
Weld. Int. e: Welding International. Welding Institute
Weld. J. e: Welding Journal. American Welding Society
Weld Met[al] Fab[r] e: Welding and Metal Fabrication
Weld. Prod. [USSR] e: Welding Production [USSR] [Published in the UK]
Weld. Res. Abroad e: Welding Research Abroad. Welding Research Council
Weld. Res. Counc. Bull. e: Weiding Research Council Bulletin
Weld. Res. Counc. Prog. Rep. e: Welding Research Council Progress Report
Weld. Res. Int. e: Welding Research International
Weld Res [London] e: Welding Research [London]
Weld. Res. News e: Welding Research News. Welding Research Council
Weld. Rev. e: Welding Review
Weld. Source e: Welding Source. Welding Institute of Canada
Weld. Technol. Can. e: Welding Technology for Canada. Welding Institute of Canada
Weld. World e: Welding in the World. International Institute of Welding
Weleda Korrespondenzbl Ärzte d: Weleda Korrespondenzblätter für Ärzte
WELL e: Whole Earth 'Lectronic Link
- e: Whole Earth 'Lectronic Net
Well. e: Victoria University of Wellington
WELP e: Wisconsin Electric Power Co.
Welsh M e: Welsh Music
WEMC e: Weighted Ensemble Monte Carlo
WEMCO e: Westinghouse Environmental Management Company of Ohio
WEMOS e: International Womens Network on Pharmaceuticals
WEN d: Weiden/Oberpfalz, Bayern

WENDB e: Water Enforcement National Data Base
Wenela e: Witwatersrand Native Labor Association
WENPEX e: Western North Pacific Cloud-Radiation Experiment
WEO e: Western European Organization
- e: WWMCCS Evaluation Office
WEO e: World Economic Outlook
WEOG e: Western European and Others Group
WEP d: Welternährungsprogramm
- e: Water-Extended Polymer
- e: West European Politics
- e: Windows Exit Procedure
- e: Wisconsin Electric Power
- e: WWMCCS Evaluation Program
WEPCo e: Wheeled Electric Power Company
WEPH e: Weapon Phenomenology
WEPOCS e: Western Equatorial Pacific Ocean Circulation Study
- e: Western Equatorial Pacific Ocean Climate Studies
- f: Etudes sur le climat de l'océan Pacifique équatorial occidental
- s: Estudios del Clima en el Océano Pacífico Ecuatorial Occidental
WEPOLEX e: Weddell Polynya Expedition
- e: Weddell Polynya Experiment
WEPR e: Walker Energy Partners
WEPS e: Weapons System
- e: Work Element Planning Sheet
WER e: Weekly Epidemiological Record
WERC e: Warehousing Education and Research Council
- e: Waste-Management, Education and Research Consortium
- f: Conseil mondial de l'environnement et des ressources
- s: Consejo Mundial del Medio Ambiente y los Recursos Naturales
WERF e: Waste Experimental Reduction Facility
WERIC e: Western Region Investigations Committee
Werkstatt Betr. d: Werkstatt und Betrieb
WerkstB d: Institut für Werkstoffe im Bauwesen
Werkst. Konstr. d: Werkstoffe und Konstruktion
Werk Wir d: Das Werk und Wir
WerkzM d: Institut für Werkzeugmaschinen
WerkzMB d: Institut für Steuerungstechnik der Werkzeugmaschinen und Fertigungseinrichtungen
WERL e: Water Engineering Research Laboratory
WERUCA e: West Rand Urban Councils Association
WES d: Wesel in Mörs, Nordrhein-Westfalen
- e: Waterways Experiment Station, Vicksburg
- e: Western Europe Standard
- e: Work Environment Scale
- e: WWMCCS Entry System
WESCOM e: Weapons Effects on Satellite Communications
WESF e: Waste Encapsulation and Storage Facility
WESPDOSE2 e: High Level Radioactive Waste Risk Model Environmental Pathways
WESS e: Weapon System Status
- e: Western European Specialists Section
- e: WWMCCS Environmental Support System
West Afr J Med e: West African Journal of Medicine
WESTHEM e: Western Hemisphere
West Indian Med J e: West Indian Medical Journal
WESTIP e: Westinghouse Technology to Improve Processes
West J Med e: Western Journal of Medicine
West J Nurs Res e: Western Journal of Nursing Research
Westm. e: [in titles] Westminster
Westminster Stud Educ e: Westminster Studies in Education
WESTNET e: Western regional Network
WESTPAC e: IOC-Programme Group on the Western Pacific
- e: IOC Sub-Commission for the Western Pacific
- f: Sous-Commission de la COI pour le Pacifique occidental
- s: Subcomisión de la COI para el Pacífico Occidental
WESTRAX e: Western Tropical Atlantic Experiment
WET d: Wirtschaftsingenieur Elektrotechnik
- e: Waste Extraction Test
- e: Wetland Evaluation Technique
- e: Whole Earth Telescope
- e: Whole Effluent Toxicity
WETD e: Westinghouse Electronic tube division
Wetlands e: Convention on Wetlands of International Importance Especially As Waterfowl Habitat
WETM e: Weather Team
WETNET e: Network for Distribution and Joint Analysis of SSM/I Data
WetNet<A> e: Network for Distribution and Joint Analysis of SSM/I Data
Wetter Leben d: Wetter und Leben
WEU e: Western Europe[an Union]
WEV e: Water Energy-Vegetation
WEVR e: Weaver Arms Corp.
WEVS e: Wetland Ecosystem Vulnerability Study
WEXC e: Wolverine Exploration Co.
WF e: Wallis and Futuna Islands
- e: Weight Factor
- e: Weld Fixture
- e: Wigner Function
- e: Work Factor method
- e: Work Function
- e: Workflow
WFA e: Waiting For Answer
- e: Weight Factor Alarm
- e: Work Flow Automation
WFAS e: Weight Factor Alarm Switch
WFBO Association e: White Fish Boat Owners Association
WFC e: Wait For Caller
- e: Wallops Flight Center
- e: Wide-Field Camera
- e: World Federation of Chiropractic
WFD e: World Federation of the Deaf
WFE e: Wiped-Film Evaporator
WFF e: Wallops Flight Facility
WFG d: Wachstums- und Beschäftigungsförderungsgesetz
- e: Wait For Graph
WFI e: Wait For Input
- e: Weight Factor Indicator
- e: Wide Field Imager
WFII e: Wide Field Planetary Camera 2
WFIR e: Wide Field-of-view Imaging Radiometer
WFIS e: Work Force Information System
WFLD e: Womens Field
WFM e: Wired For Management
- e: Works For Me [Internet]
WFMC e: Welding Filler Material Control
WfMC e: Workflow Management Coalition
WFO e: Waiting For Order
- e: Weather Forecast Office
- e: Work For Others [system]
WFOA e: WFO-Advanced
W[-]FOV e: Wide angle Field Of View
WFP e: Waste Form Producer
- e: World Food Program[me][s]
WF[/]PC e: Wide Field/Planetary Camera
WFPC[-]II e: Replacement for the WP/PC
WFPIS e: Whole Farm Plan Incentives Scheme
WFPM e: Western Fine Particulate Monitoring
WFPP e: Whole Farm Planning Program
WFPPC e: WOCE Float Program Planning Committee
WFQ e: Waste Form Qualifications
WFS e: Women For Sobriety
- e: Woodstock File Server
WFSB e: Washington Federal Savings Bank
WFSC e: Wildlife and Fisheries Science
WFT e: Wafer Fabrication Template
- e: Weight Factor Display System
- e: Weight Factor Transmitter
WFTE e: Wigner Function Transport Equation
WFU e: Withdrawn From Use
WFW[G] e: Windows For Workgroups
WF XXI e: Warfighter XXI
WFZ d: Weiter- und Fortbildungszentrum für medizinisches Personal
- e: Weapons Free Zone
WG d: Windgeschwindigkeit
- e: Wage Grade
- e: Weapons Grade
- e: Weight Gauge
- e: Work[ing] Group
Wg e: Wicked grin [Internet]
W/G e: Wave Guide
- e: Within Guidance
W + G d: Wirtschaft und Gesellschaft
WGA e: Western Governors Association
- e: Working Group on Agriculture
WGAE e: Writers Guild of America, East

WGAF e: West Germany Air Force
WGAM e: Working Group on Antarctic Meteorology
- f: Groupe de travail de la météorologie antarctique
WGAS e: Who Gives A Shit (or Shut) [Internet]
WGASF e: Working Group on Air/Sea Fluxes
WGB n: Weekblad voor Gemeeniebelangen. Orgaan van de Vereniging van Nederlandse Gemeenten
WGBEC e: Working Group on Biological Effects of Contaminants
- f: Groupe de travail sur les effets biologiques des contaminants
- s: Grupo de Trabajo sobre los Efectos Biológicos de los Contaminantes
WGC e: Wisconsin Gas Company
- e: World Games Council
- e: World Gospel Crusades
- e: Worthy Grand Chaplain
- e: Worthy Grand Conductor
WGCA e: Wisconsin Gift Cheese Association
WGCC e: World Games Coordination Committee
WGCCD e: Working Group on Climate Change Detection
WG/CDR e: Wing Commander
WGCDR e: Working Group for Community Development Reform
WGCL e: Window Glass Cutters League of America
WGCOM e: Working Group on the Commercialization of Meteorological and Hydrological Services
- f: Groupe de travail de la commercialisation des services météorologiques et hydrologiques
Wg Cr e: Wing Commander
WGCV e: Working Group on Calibration and Validation
- f: Groupe de travail sur l'étalonnage et l'homologation
WGD e: Waste Generation Database
- e: Working Group on Data
- e: World Games for the Deaf
- f: Groupe de travail sur les données
- s: Grupo de Trabajo sobre Datos
WGDM e: Working Group on Data Management
WGDTB e: Working Group on Digital Television Broadcasting
WGEP e: Working Group on Energy Policy
WGES e: Worlds Greatest Environment Statement
WGGC e: Working Group on Global Change
WG-GGI e: Working Group on Geodesy and Geographic Information
WGI e: Within Grade Increase
- e: World Glacier Inventory
WGISS e: Working Group on Information Systems and Services
- f: Groupe de travail sur les systèmes et services internationaux
- s: Grupo de Trabajo sobre Sistemas y Servicios Internacionales
WGK e: Wassergefährdungsklassen
WGLSA d: Wassergesetz für das Land Sachsen-Anhalt
WGM e: Whispering-Gallery Mode
WGmc. e: West Germanic
WGMS e: Working Group on Marine Sediments
- f: Groupe de travail sur les sédiments marins
- f: Service de surveillance mondiale des glaciers
- s: Grupo de Trabajo sobre Sedimentos Marinos
- s: Servicio Mundial de Vigilancia de Glaciares
WGNE e: Working Group [on] Numerical Experiment[ation]
- f: Groupe de travail de l'expérimentation numérique
- s: Grupo de Trabajo sobre Experimentación Numérica
W'gong e: Wollongong
WGPTDR e: Working Group for the Promotion of Tourism in the Danube Region
WGR e: Waste Generation Rate
WGRF e: Working Group on Radiation (or Radiative) Fluxes
- f: Groupe de travail des flux de rayonnement
- s: Grupo de Trabajo sobre Flujos de Radiación
WGS e: Weather Graphics System
- e: Work[ing] Group Systems
- e: World Geodetic Survey
- e: World Geodetic System
WGS-84 e: WGS-84 Ellipsoid
WGS84 e: World Geodetic Spheroid 1984
WGSAT e: Working Group on Satellites
- f: Groupe de travail des satellites
- s: Grupo de Trabajo sobre Satélites
WGSI e: Working Group on Sea Ice
- f: Groupe de travail des glaces de mer
- s: Grupo de Trabajo sobre Hielos Marinos
WGSIC e: Working Group on Sea-ice Research and Climate
- f: Groupe de travail de la recherche sur les glaces de mer et le climat
- s: Grupo de Trabajo sobre Investigaciones relativas a Hielos Marinos y el Clima
WGSSO e: Working Group on Shelf Seas Oceanography
- f: Groupe de travail sur l'océanographie des mers épicontinentales
- s: Grupo de Trabajo sobre Oceanografía de los Mares Epicontinentales
WGSVG d: Gesetz zur Regelung der Wiedergutmachung
WGWA e: Working Group on Waste Acceptance
WH e: War Hammer
- e: White House
- e: World Heritage
W H e: West Hall
W. H. e: Wystan Hugh
WHA e: Waste Handling Area
- e: World Heritage Area
WHAD e: Worm Holes A Defect
WHA/FH e: Waste Handling Area Fume Hood
Whaling e: International Convention for the Regulation of Whaling
WHAM e: Waveform Hold and Modify
- e: West Hemisphere AUTODIN Management
- e: Womens Health Action and Mobilization
WHANCO e: Westinghouse Hanford Company
WHARC e: Womens Health and Action Research Centre
Wharton Rep. e: Wharton Report
WHAT e: Wind Height and Temperature
- e: Wireless Hybrid Asynchronous Time-bounded
WHATIF e: SWAGMAN program on Water Tables and Irrigation
What's new for. res. e: What's new in forest research
WHBL e: World Home Bible League
WHC e: Westinghouse Hanford Company
- e: Wireline Heave Compensator
- e: Workstation Host Connection
WHD e: Watthour Demand Meter
- e: World Habitat Day
- f: Journée mondiale de l'habitat
- s: Día Mundial del Hábitat /Naciones Unidas/
WHDL e: Watthour Demand Limiter
WHEL e: Roadmaster Industries, Inc.
WHERE e: Women for Healthcare Education, Reform & Equity
WHETS e: Washington Higher Education Telecommunication System
WHFS e: WFO Hydrometeorological Forecasting System
WHG d: Wasserhaushaltsgesetz
WHIF e: Waste Handling and Isolation Facility
WHIRL e: Wide-Angle High Resolution Line-Imager
WHIT e: White Hall
- e: White River Basin
White House e: Official Residence of the US President and site of various high level executive-branch activities
WHMO e: White House Military Office
WHND e: Worm Holes No Defect
WHNF e: Weak Head Normal Form
WHO e: World Health Organisation (or Organization)
WHO/ECEH e: World Health Organization European Centre for Environment and Health
WHO/EURO e: World Health Organization Regional Office for Europe
WHOI s: Instituto Oceanográfico de Woods Hole
WHOLIS e: World Health Organization Library Information System
WHOO e: Waterhouse Investor Services, Inc.
WHO Reg Publ Eur Ser e: WHO Regional Publications. European Series
WHOSIS e: WHO [World Health Organization] Statistical Information System
WHO-WMO e: International Air Data Base

WHP e: WOCE Hydrographic Program[me] [office]
- e: WOCE Hydrographic Survey
- f: Programme hydrographique de WOCE
- s: Programa Hidrográfico de WOCE

WHPA e: Washington [State] Hydraulic Projects Act

WHPF e: Waste Handling and Packaging Facility

WHPIP e: WOCE Hydrographic Program Implementation Panel

WHPO e: WOCE Hydrographic Program[me] Office
- f: Bureau du Programme hydrographique de WOCE
- s: Oficina del Programa Hidrográfico de WOCE

WHPP e: Waste Handling and Packaging Plant, Oak Ridge National Laboratory

WHPPC e: WOCE Hydrographic Program Planning Committee

WHQL e: Windows Hardware Quality Laboratory

WHR e: Westinghouse Human Resources
- e: World Health Report

WHS e: Waste Handling System

WHSE e: Warehouse

WHSR e: White House Situation Room

WHT e: Waste Hold Tank

WHTI e: Wheelabrator Technologies, Inc.

WHU d: Wissenschaftliche Hochschule für Unternehmensführung, Koblenz

WHV d: Wilhelmshaven, Niedersachsen

WHWT e: Water and Hazardous Waste Team

WHYCOS e: World Hydrological Cycle Observing System
- f: Système mondial d'observation du cycle hydrologique
- s: Sistema Mundial de Observación del Ciclo Hidrológico

WI d: Wirtschaftsinformatik
- d: Wirtschaftsingenieur für Produktionstechnik
- e: Weizmann Institute, Israel
- e: Wisconsin

WI e: Wexas International

W.I. e: West Indies

w/i e: within

WI 93 d: Studiengang WMB

WIA e: Weather-Impacted Airspace

Wiad Lek P: Wiadomosci Lekarskie

Wiad Parazytol P: Wiadomosci Parazytologiczne

WIAS d: Weierstraß-Institut für Angewandte Analysis und Stochastik, Berlin

WIB e: Within-Batch

WiB d: Woche im Bundestag. Parlamentskorrespondenz

WIBAMU e: Well, I'll Be A Monkey's Uncle [Internet]

WIBNI e: Wouldn't It Be Nice If [Internet]

WIC e: Washington Information Center
- e: West India Committee
- e: Wireless Indoor Communication
- e: Women, Infants and Children

WICEM e: World Industry Conference on Environmental Management

WICHE e: Washington Interstate Commission for Higher Education

WICHE Publ e: Western Interstate Commission for Higher Education. Publications

WICL e: Work Inspection Characteristics List

WICS e: Warning Information and Correlation System
- e: Worldwide Intelligence Communication System

WICU e: Weather Intercept Control Unit

WID e: Women In Development

WIDA e: Weather Impact Decision Aids

WIDD d: Web Informationsdienst Deutschland

WIDS e: Waste Identification Data System

WIEB e: Western Interstate Energy Board

WIEBS e: Winter Ice Experiment Beaufort Sea

Wien Jahrb Kunstgesch d: Wiener Jahrbuch für Kunstgeschichte

Wien Klin W d: Wiener Klinische Wochenschrift

Wien Klin Wochenschr [Suppl] d: Wiener Klinische Wochenschrift. Supplementum

Wien Klin Ws d: Wiener Klinische Wochenschrift

Wien Landwirtsch Ztg d: Wiener Landwirtschaftliche Zeitung

Wien Med Presse d: Wiener Medizinische Presse

Wien Med Wochenschr [Beih] d: Wiener Medizinische Wochenschrift [Beihefte]

Wien Med Wochenschr Suppl d: Wiener Medizinische Wochenschrift. Supplement

Wien Med Ws d: Wiener Medizinische Wochenschrift

Wien Med Wschr d: Wiener Medizinische Wochenschrift

Wien Mitt Photogr Inhalts d: Wiener Mitteilungen Photographischen Inhalts

Wien Mitt Wasser Abwässer Gewässer d: Wiener Mitteilungen. Wasser, Abwässer, Gewässer

Wien Naturh Mus Annalen d: Wien Naturhistoriches Museum. Annalen

Wien Pharm Wochenschr d: Wiener Pharmazeutische Wochenschrift

Wien Prähist Z d: Wiener Prähistorische Zeitschrift

Wien Stud d: Wiener Studien

Wien Stud Z Klass Philol d: Wiener Studien. Zeitschrift für Klassische Philologie

Wien Tierärztl M[onat]sschr d: Wiener Tierärztliche Monatsschrift

Wien Völkerk Mitt d: Wiener Völkerkundliche Mitteilungen

Wien Z Inn Med [Ihre Grenzgeb] d: Wiener Zeitschrift für Innere Medizin und Ihre Grenzgebiete

Wien Z Kunde Morgenlandes d: Wiener Zeitschrift für die Kunde des Morgenlandes

Wien Z Kunde Süd Ostasiens d: Wiener Zeitschrift für die Kunde Süd-, und Ostasiens und Archiv für Indische Philosophie

Wien Z Nervenheilk Grenzgeb d: Wiener Zeitschrift für Nervenheilkunde und deren Grenzgebiete

Wien Zt d: Wiener Zeitung

WIERD e: Wind Energy Reporl

WIEU e: Weekly Intelligence Estimate Update

WIF d: Wirtschaftsinformatik
- e: Mid-Wisconsin Federated Library System, Fond Du Lac
- e: Weapons Integration Facility
- e: Web Interface Facility
- e: West Indies Federation
- e: Wildfire Resources Ltd.
- e: Wilt-Inducing Factor
- e: Women in Film
- e: Worlds of If
- e: Worldview International Foundation

WiF e: William Faulkner. Materials, Studies, and Criticism

WIFA e: Washington Institute of Foreign Affairs

WIFE e: Women Involved in Farm Economics
- e: Womens Independent Film Exchange

WIFF e: Wang Image File Format

WIFMA e: Advance telegraphic notification relating to the operation of the Marine Meteorological Services
- f: Notification télégraphique préalable de modifications concernant l'Assistance météorologique aux activités maritimes
- s: Notificación telegráfica previa sobre el funcionamiento de los Servicios Meteorológicos Marinos

WIFO d: Monatsberichte/Österreichisches Institut für Wirtschaftsforschung

WIFS e: Wide Field Sensor

WiFS d: Multispektraler Weitwinkelsensor

WIG e: Wing-In-Ground effect

WIGE e: Wax-Impregnated Graphite Electrode

Wig Ev e: Wigram on Extrinsic Evidence

Wight e: Wightwicks English Exchequer Reports

Wight El Cas e: Wights Scottish Election Cases

Wightw e: Wightwicks English Exchequer Reports

Wightw [Eng] e: Wightwicks English Exchequer Reports

Wigm Ev e: Wigmore on Evidence

Wig Wills e: Wigmore on Wills

WIH e: State Historical Society of Wisconsin, Madison
- e: Work in Hand

WIHG e: Waclia Institute of Himalayan Geology

WIHP e: Journal. Wisconsin Association for Health, Physical Education and Recreation

WIHS e: Western Institute for Health Studies
- e: Womens Interagency Health Study

WII e: Beloit College Library
- e: Weatherford International, Inc.
- e: Worldwide Information Infrastructure

WIIAD e: Winrock International Institute for Agricultural Development

WIIP e: Waters Intelligent Information Processor
WIIS e: Wang Integrated Image System
WIIU e: Workers International Industrial Union
WIIW d: Wiener Institut für Internationale Wirtschaftsvergleiche
WIJ e: Warburg Institute. Journal
Wijsgerig Perspect Maatsch Wet n: Wijsgerig Perspectief op Maatschappij en Wetenschap
Wijsig Perspect n: Wijsgerig Perspectief op Maatschappij en Welenschap
WIK d: Deutsches Institut für Wirtschaftsforschung. Wochenbericht
WIK e: Kenosha Public Library
WiKG d: Gesetz zur Bekämpfung der Wirtschaftskriminalität [numeriert]
WIL d: Bernkastel-Wittlich/Mosel, Rheinland-Pfalz
WILASS e: Water Infiltration in Layered Subfreezing Snow
Wildbach-Lawinenverbau d: Wildbach- und Lawinenverbau
Wildlife A e: Wildlife in Australia
Wildl. Monogr. e: Wildlife Monographs
Wild Mango Mating Calls e: Unrequited fruit
WILL e: Western Isles
- e: Willamette Basin
William and Mary Bus R e: William and Mary Business Review
Will LR e: Willamette Law Review
WILPF e: Womens International League for Peace and Freedom
WILS e: Wilson Hall
Wilson Bull. e: Wilson Bulletin
Wilson Lib Bull e: Wilson Library Bulletin
WIM e: Weigh-In-Motion
- e: Wideband Interconnection Module
WiMa d: Studiengang Wirtschaftsmathematik
WIMC e: Women in Managed Care
WiMi d: Wissenschaftliche[r] Mitarbeiter[in] Mittelbau
WIMM e: Weapons Integrated Material Manager
WIMP e: Weakly-Interacting Massive Particle
- e: Window/Icon/Mouse/Pointer
- e: Window, Icon[s], Menu[s], Pointing [device]
- e: Windows, Icons, Mice and Pointers
- e: Windows, Icons, Mouse and Pull-down environment
- e: Windows, Icons, Mouse and Pull-down menus
- e: Windward Island Passages Monitoring Program
WIMPS e: Weakly Interacting Massive Particles
- e: Windows, Icons, Menus, Pointers, Scroll bars
WIMS e: Warehouse Inventory Management System
- e: Work Information Management System
- e: Worldwide Intra-theater Mobility Study
WIMWSS e: Worldwide Integrated Management of Wholesale Subsistence Stock
WIN d: Wissenschaftsnetzwerk des DFN
- e: Forum Wireless Information Networks Forum
- e: German nation-wide x.25 research network
- e: Warfighter Information Network
- e: Waste Information Network
- e: Weight-control Information Network
- e: Westinghouse Internal Network
- e: Whip Inflation Now
- e: Windoof
- e: Windows
- e: Wireless In-building Network
- e: Women in Nuclear
- e: WWMCCS Information Network
- e: WWMCCS Intercomputer Network
WIN95 e: Windows 95
WINA e: WWMCCS Information Needs Analysis
WINCE e: Winter Cloud Experiment
WINCO e: Westinghouse Idaho Nuclear Company
WINCS e: WWMCCS Intercomputer Network Communications Subsystem
WIND e: Global Geospace Science Wind Spacecraft
- e: Interplanetary Physics Laboratory
- e: Satellite
- e: Weather Information Display
- e: Windsor Site, Knolls Atomic Power Laboratory
- e: WWMCCS Intercomputer Network Director
Wind En Rpt e: Wind Energy Report
WINDMATIC e: Wind Measuring Along-Track Image-Correlator
WINDOW e: Western Indian Ocean Waters
Window Inds e: Window Industries
WINDOWS d: Grafische Benutzeroberfläche von Microsoft, MS-DOS basierend
WINDOWS NT e: Grafische Benutzeroberfläche und Betriebssystem von Microsoft
WINDS e: Weather Information Network Display System
WINDSAT e: Wind Satellite
WINDSCAT e: Wind Scatterometer
WINE e: Windows Emulator
- e: Winter in Northern Europe
WINForum e: Wireless In-building Network Forum
- e: Wireless Information Networks Forum
WINGS e: Web Interactive Network of Government Services
Wings Afr e: Wings over Africa
WINHEC e: Windows Hardware Engineering Conference
WINLAB e: Wireless Information Network Laboratory
WINN e: Wedding Information Network, Inc.
WinNT e: Windows NT [New Technology]
WINS e: Warehouse Industry National Standards Guidelines
- e: Weather Information Network Service
- e: Windows Internet Name (or Naming) Service
- e: Wollongong Integrated Networking Solutions
WINS e: Women in Naval Service
WINSOCK e: Windows Open Systems Architecture
- e: Windows Sockets
WINSTATS e: WIN Statistics Systems
WIN-TCP e: Wollongong Integrated Networking/TCP
Winterber. Eidgenöss. Inst. Schnee-Lawinenforsch. d: Winterberichte des Eidgenössischen Institutes für Schnee- und Lawinenforschung
WINTERS e: Wind/Temperature Measurement in Thermosphere and Stratosphere
WINTEX/CIMEX e: Winter Exercise/ Civilian-Military Exercise
WinVN e: A Windows News Client
WINWORD e: Word For Windows
WIODIR e: Western Indian Ocean Directory of Marine Science Institutions and Scientists
- f: Répertoire de scientifiques et institutions spécialisés dans les sciences de la mer de l'océan Indien occidental
- s: Repertorio de Científicos e Instituciones Marinas del Océano Indico Occidental
WIOMAP e: Western Indian Ocean Marine Applications Project
- f: Projet relatif aux applications des sciences de la mer dans l'océan Indien occidental
- s: Proyecto de Aplicaciones Marinas en el Océano Indico Occidental
WIOMSA e: Western Indian Ocean Marine Science Association
- f: Association des sciences de la mer de l'océan Indien occidental
- s: Asociación de Ciencias Marinas del Océano Indico Occidental
WIP d: Wissenschaftler Investitionsprogramm
- d: Wissenschaftlerintegrationsprogramm
- e: WOCE Implementation Plan
- e: Women In Politics
- e: Work In Process
- e: Work In Progress[ion]
WIPO e: World Intellectual Property Organization
- e: World International Patents Organization
WIPP e: Waste Isolation Pilot Plant, Carlsbad, New Mexico
WIPP e: Work Isolation Pilot Project
WIPPTREX e: Waste Isolation Pilot Plant Transportation Emergency Exercise
WIPP-WAC e: Waste Isolation Pilot Plant-Waste Acceptance Criteria
WIPT e: Working-Level Integrated Product Team
WIR d: Wirtschaftswissenschaften
- e: Wideband Recorder Interface
- e: World Resources Institute
WIRE e: Wire Scanner

Wire Technol e: Wire Technology
WIRP e: Watershed Integrated Research Program
Wirtpol d: Institut für Sozialforschung, Abteilung für Wirtschaftspolitik und Öffentliches Recht
Wirt und Wettbewerb d: Wirtschaft und Wettbewerb
WIS d: Wismar/Landkreis, Mecklenburg-Vorpommern
- e: Waste Inventory Sheet
- e: Water Induced Shift
- e: Water-Industry Specifications
- e: Wedge Imaging Spectrometer
- e: Weizmann Institute of Science
- e: Womens Information Switchboard
- e: Worldwide Information System
- e: WWMCCS Information System
Wis. e: Wisconsin
WISAP e: Waste Isolation Safety System Analysis Program
WISC e: Wechsler Intelligence Scale for Children
WIS CMS e: WIS Configuration Management System
WISDIM e: WWMCCS Dictionary for Information Management
WISE e: WordPerfect Information System Environment
- e: Work Improvements in Small Enterprises Programme
- e: World-wide Information System for r&d Efforts
WISHA e: Washington [State] Industrial Safety and Health Act
- e: Washington [State] Industrial Safety and Health Administration
WISI e: World Index of Space Imagery
WISJMPO e: Worldwide Military Command and Control Information Systems, Joint Program Office
WIS JPM e: WIS Joint Program Manager
Wis Law R e: Wisconsin Law Review
Wis Med J e: Wisconsin Medical Journal
WISNAS e: WIS Network Authentication Service
WISO d: Wirtschaftspolitische und sozialpolitische Zeitschrift
WISP e: Wartime Information Security Program
- e: Wide Range Image Spectrophotometer
- e: Winter Icing and Storms Programs
- e: Winter Icing and Storms Project
WISPIT e: WISP Instrument Test
WISR e: Westinghouse Independent Safety Review
WISRA e: WIS Requirements Analysis
WISSA e: Wholesale Interservice Supply Support Agreement
Wiss Beitr Martin Luther Univ [Halle Wittenberg] d: Wissenschaftliche Beiträge. Martin-Luther-Universität [Halle-Wittenberg]
Wissensch Inform d: Wissenschaftliche Information
Wissensch Sitzungen Stochastik 81 d: Wissenschaftliche Sitzungen zur Stochastik 81
Wiss Fortchritt d: Wissenschaft und Fortschritt
WIS SPO e: WIS Systems Program Office
Wiss Z [Berl] d: Wissenschaftliche Zeitschrift. Humboldt-Universität [Berlin]. Gesellschafts- und Sprachwissenschaftliche Reihe
Wiss Z [Jena] d: Wissenschaftliche Zeitschrift. Friedrich-Schiller-Universität [Jena]
Wiss Z Rostock d: Wissenschaftliche Zeitschrift. Universität Rostock
Wiss Z Tech Hochsch [Ilrnenau] d: Wissenschaftliche Zeitschrift. Technische Hochschule [Ilmenau]
Wiss. Z. Tech. Univ. Dresd., Sep.r. 5 d: Wissenschaftliche Zeitschrift der Technischen Universität Dresden, Separatreihe 5
Wistar Inst Symp Monogr e: Wistar Institute Symposium Monograph
WiStG d: Gesetz zur weiteren Vereinfachung des Wirtschaftsstrafrechts [Wirtschaftsstrafgesetz]
WIT e: Web Interactive Talk
- e: Women In Transition
WITI e: Women In Technology International
WITS e: Washington Integrated Telephone System
- e: Washington Interagency Telecommunications System
- e: Women and Infants Transmission Study
- e: World Ionosphere Thermosphere Study
- e: WSGT Intelligent Terminal System
WITSA e: World Information and Technology Services Alliance
WITT e: Workstation Interactive Test Tool
Witw. e: Witwatersrand, Johannesburg
WIU e: Weapons Interface Unit
WIW e: Winter Intermediate Water
- e: Within Wafer
WIW d: Wer Informiert Worüber
WIWNU e: Within Wafer Nonuniformity
WIYN e: Wisconsin/Indiana/Yale/NOAO telescope
WizOp e: Wizard Sysop
WJPEB e: Woodcock-Johnson Psychoeducational Battery
WJU e: Wheeling Jesuit University
WK1 e: Worksheet
WKA d: Warburger Klischee-Anstalt GmbH
WKB e: Wentzel-Kramers-Brillouin
- e: Wickenburg special upper-air observing site
- e: Workbook
WKB-J e: Wentzel-Kramers-Brillouin-Jeffreys
WKCONSUPVR e: Work Control Supervisor
WKE e: Worksheet
WKI d: Fraunhofer-Institut für Holzforschung, Wilhelm-Klauditz-Institut
Wkly Epidemiol Rec e: Weekly Epidemiological Record
WKQ e: Spreadsheet
WKRP e: World-Wide Keypunch Replacement Program
WKS e: Warwickshire
- e: Well Known Service[s]
- e: Well Known Sockets
- e: Worksheet
Wks. e: [in titles] Works
WkSchG d: Gesetz über den Kündigungsschutz für Mietverhältnisse über Wohnraum [Wohnraumkündigungsschutzgesetz]
WKZ e: Compressed Spreadsheet
WL e: Delft Hydraulics Laboratory
- e: Geophysical Well Log Surveys
- e: Warning letter
- e: Wireline
- e: Word Line
- e: Working Level[s]
W&L e: Washington and Lee University
W/L e: Water Load
- e: Wireline
WLac e: La Crosse Public Library
WLacFW e: United States Fish and Wildlife Service, Fish Control Laboratory, La Crosse
WLacL e: La Crosse Lutheran Hospital
WLacSF e: Saint Francis Hospital, La Crosse
WLacU e: University of Wisconsin-La Crosse
WLacVC e: Viterbo College, La Crosse
WLadM e: Mount Senario College, Udysmith
WLag e: Lake Geneva Public Library
WLagB e: Badger Union High School District, Lake Geneva
WLagF e: Franciscan Education Center, Lake Geneva
WlagSD e: Joint School District Number One, Lake Geneva
WLAN e: Wireless LAN [Local Area Network]
WLA/TMDL e: Waste Load Allocation/Total Maximum Daily Load
WLB d: Württembergische Landesbibliothek in Stuttgart
- e: Seagoing Buoy Tender
WLBI e: Wafer-Level Burn-In
WLBS e: Wright Laboratories
WL Bull [Ohio] e: Weekly Law Bulletin [Ohio]
WLC e: Weapons Laboratory Civil Engineering Division
- e: Well Logging Cable
- e: Weilco Enterprises, Inc.
- e: White Light Coronagraph [KSC]
- e: Wildcat
- e: Wine Label Circle
- e: World Literacy of Canada
WLCA e: Godfrey, IL [FM radio station call letters]
WLCAC e: Watts Labor Community Action Committee
WLCC e: Walker-Lybarger Construction Company
WLCE e: White Light Coronagraph Experiment
WIL[CL] e: War Legislation, Civil Liabilities
WLCSS e: Weapon Launch Console Switching Section
WLD e: Water and Land Division

Wld Cem Tech e: World Cement Technology
WLDLF e: Wildlife
WLDND e: Wildland
WLEL e: Wire Line Entrance Link
WLET e: Work Location Enhancement Team
WLF e: White Liberation Front
WLF e: Worklife
WLG d: Wolgast/Usedom, Mecklenburg-Vorpommern
WLI e: World Languages Institute
WLL e: Wireless [Local] Loop
WLM e: Coastal Buoy Tender
- e: Working-Level Minute
- e: Working-Level Month
WLO e: Windows Library Objects
WLP e: Welding Procedure
- e: Wireline Packer
WLP e: Western Legal Publications [Databasel
WLPB e: Woodcock Language Proficiency Battery
WLR e: River Buoy Tender
- e: Wafer-Level Reliability
- e: Wireline Reentry
WLS e: Weighted Least Squares
- e: White Line Skip
- e: Wireline Logging Services [logging]
WLSA e: Wartime Logistics Support Assessment
WLT e: Wafer-Level Test
- e: Weapons Load Training
- e: Wiltshire
WLTM e: Would Like To Meet
WM e: Waste Management [group]
- e: Wastite-Magnetite Buffer
- e: Window Manager
- e: Wireless Manager
- e: Work Manager
W&M d: Weiterbildung und Medien
WMA d: Wirtschaftsmathematik
- e: Waste Management Authority
- e: Waste Minimization Audit
- e: Wildlife Management Area
WMAC e: Wireless Media Access Control
W&M aktuell d: Weiterbildung und Medien aktuell
WMB d: Wirtschaftsingenieur für Maschinenbau
WMB[AM] d: Wirtschaftsingenieur für Maschinenbau [Studienrichtung Allgemeiner Maschinenbau]
WMB[VET] d: Wirtschaftsingenieur für Maschinenbau [Studienrichtung Verfahrens- und Energietechnik]
WMC e: Watershed Management Council
- e: Weighted Monte Carlo
- e: Western Mining Corporation
- e: Workflow Management Coalition
- e: World Meteorological Center[s]
WMCE e: Western Mediterranean Circulation Experiment
- f: Expérience sur la circulation en Méditerranée occidentale
- s: Experimento sobre la Circulación en el Mediterráneo Occidental
WMCO e: Westinghouse Materials Company of Ohio
WMD e: Waste Management Division
- e: Water Management Division
- e: Weapons of Mass Destruction
- e: West Midlands
- e: Westinghouse Marine Division
- e: World Meteorological Day
- f: Journée météorologique mondiale
WMD e: Weapon Mounted Display
WMDW e: Western Mediterranean Deep Water
WMEC e: Medium Endurance Cutter
WM&EC e: Waste Management and Environmental Compliance
WMECICG e: Western Massachusetts Electric Company Industrial Consumer Group
WMECo e: Western Massachusetts Electric Company
WMED e: Waste Management and Economics Division
WMF e: Windows Meta File [Format]
WMI e: Woodland Mountain Institute
WMI e: Work Motivation Inventory
WMIC e: Western Lake Michigan Drainage
w.midl. e: west midland [dialect]
WMIN e: Waste Minimization
WMIS e: Waste Management Information System
- e: Westinghouse Management Information System
WMLA e: Washington Medical Librarians Association
WMM e: Wall-Mounted Manipulator
WMMB e: Wisconsin Milk Marketing Board
WMMOWS e: Wash My Mouth Out With Soap [Internet]
WMMS e: Wesleyan Methodist Missionary Society
WMNS e: Womens Studies
WMNW e: Waste Management Federal Services, Inc., Northwest Operations
WMO e: World Meteorological Organisation
WMONEX e: Winter Monsoon Experiment
WMOP e: Weather Observation Master Plan
WMOSA e: WMO Satellite Activities
WMP e: War and Mobilization Plan
- e: Waste Management Plan
- e: Waste Management Program
- e: Waste Minimization Plan
WMPO e: Waste Management Project Office
WMPRT e: Wartime Manpower and Personnel Readiness Team
WM&R e: Waste Minimization and Reduction
WMRK e: Westmark International, Inc.
WMRLS e: West Midlands Regional Library System
WMS d: Wolmirstedt, Sachsen
- e: Windows Messaging Subsystem
- e: Work Force Management Staff
- e: Workflow Management Systeme
- e: Workload Management System
WMSA e: Saint Anthony Hospital, Milwaukee
WMS/AB e: William Morris Society, American Branch
Wms Ann Reg e: Williams Annual Register
Wms Bank e: Williams on Bankruptcy
Wms&Bruce e: Williams and Bruces Admiralty Practice
WMSC e: White Mountain Scenic Railroad
WMSC e: Weather Message Switching Center
WMSCA e: Waste Management Safety Committee A
WMSCR e: Weather Message Switching Center Replacement
WMSE e: Waste Management Systems Engineering
WMSE e: Milwaukee School of Engineering, Walter Schroeder Library
Wms Ex[ors] e: Williams on Executors
WMSF e: Saint Francis Seminary, Milwaukee
WMSFH e: Saint Francis Hospital, Milwaukee
WMSI e: Williams Industries Inc.
WMSJ e: Saint Josephs Hospital
- e: William Morris Society. Journal
WMSKF e: William Morris Society and Kelmscott Fellowship
WMSL e: Saint Lukes Hospital, Milwaukee
WMSM e: Saint Marys Hospital, Milwaukee
Wms Peere e: Peere-Williams English Chancery Reports
Wms PP e: Williams on Personal Property
W/[M²SR] e: Watts per Square Meter Steradian
WMSRG e: Weather-Modification Statistical Research Groups
Wms RP e: Williams on Real Property
WMSS e: Waste management special studies
Wms Saund [Eng] e: Saunders Reports, Edited by Williams [English Reprint]
WMST e: Womens Studies
Wms VT e: Williams Vermont Reports
WMSWH e: Southeastern Wisconsin Health Systems Agency, Health Science Library, Milwaukee
WMT d: Auslandmärkte/Marches Etrangers
- e: Wet Metric Ton
- e: Waste Management and Transportation
- e: Western Motor Tariff Bureau, Los Angeles
- e: Wisconsin Mastitis Test
WMTADS e: Waste Management Technology Analysis and Decision Support
WMTB e: Western Motor Tariff Bureau
WMTC e: Milwaukee Technical College
- e: Waste Management Technology Center
WMTC-N e: Milwaukee Area Technical College, North Campus Center Library, Mequon
WMTC-S e: Milwaukee Area Technical College, South Campus Center Library, Oak Creek

WMTC-W e: Milwaukee Area Technical College, West Campus Center Library, West Allis
WMTDD e: Waste Management and Transportation Development Division
WMY e: World Mathematical Year
- f: Année mondiale des mathématiques
- s: Año Mundial de Matemáticas
WNA e: Wisconsin Nurses Association
WNAI e: Word and Number Assessment Inventory
WNBF e: West Nile Bank Front
WN-CAELA e: Women's Network of the Council for Adult Education In Latin America
WND d: St. Wendel, Saarland
WNE e: Windows Network Environment
WNFE e: WWMCCS Network Front-End
WNIC e: WAN [Wide Area Network] Interface Coprocessor
WNICD e: Westinghouse Nuclear Instrument and Control Division
WNIM e: WAN [Wide Area Network] Interface Module
WNINTEL e: Warning Notice-Intelligence
WNL e: Within Normal Limits
WNLA e: Witwatersrand Native Labour Association South Africa
WNNR A: Wetenskaplike-Nywerheids-navorsingsraad
WNOHGB e: Where No One Has Gone Before [Internet]
WNP e: Washington Nuclear Power
- e: Waste Neutralization Plant
WNP-2 e: Washington Nuclear Plant Number 2
WNPCW e: Western North Pacific Central Water
WNRC e: Washington National Records Center
WNRE e: Whiteshell Nuclear Research Establishment, Pinawa, Manitoba, Canada
WNRM e: Women in Natural Resources Management Program
WNS e: Worldwide Navigation System
WNW d: Westnordwesten
- e: West [by] North West
WNWAC e: Washington Nuclear Waste Advisory Council
WNWB e: Washington Nuclear Waste Board
WNYNSC e: Western New York Nuclear Service Center
WO e: Waiting on iceberg
- e: Waste Operations
- e: Waste Oxide
- e: Weight on bit
- e: Write Only
wo d: wahlobligatorisch
W/O e: Without
- e: Work Order
- e: Written Off
WOA e: Wartime Operating Airfield
- e: Web Object Application
- e: Work Of Art
- e: World Ocean Atlas
- f: Atlas des océans du monde
- s: Atlas de los Océanos del Mundo
WOB d: Wolfsburg, Niedersachsen
- e: Woman (or Women)-Owned Business[es]
WoBindG d: Wohnungsbindungsgeselz
WOC e: Waiting On Cement
- e: World Oceanographic Centre
- f: Centre océanographique mondial
- s: Centro Oceanográfico Mundial
WOCA e: World Outside Communist Areas
WOCE e: World Ocean Circulation Experiment [program]
- f: Expérience mondiale concernant la circulation océanique [Version de l'OMM]
- f: Expérience sur la circulation océanique mondiale [Version de l'UNESCO]
- s: Experimento Mundial sobre la Circulación Oceánica
WOCE/HP e: WOCE/Hydrographic Program
WOCE-IPO e: WOCE International Project Office
- f: Bureau international du Projet WOCE
- s: Oficina Internacional del Proyecto WOCE
WOCE-NEG e: WOCE Numerical Experimentation Group
- f: Groupe de WOCE chargé de l'expérimentation numérique
- s: Grupo de WOCE sobre Experimentación Numérica
WOCE-SSG e: WOCE Scientific Steering Group
WOCN e: Wound, Ostomy and Continence Nurses Society
WOD e: Web Object Definition
- e: Wind Over the Deckof a carrier, of course
- e: Word Of Day
WO3DC e: World Ozone Data Center
WODC e: World Ozone Data Center (or Centre)
WODITEM e: Watershed Oriented Digital Terrain Model
WOE e: Weight Of Effort
WOES e: Warrant Officer Education System
WOFAT e: Waste Of Film and Time
WOFTAM e: Waste Of Flaming Time And Money
WOG e: Work Order Generator
WohneigFG d: Wohneigentumsförde-rungsgesetz
WOL e: Wedge Open Loading
WOLV RM BA e: Wolverine Room, School of Bus Ad
WOM e: Work Order Maintenance
- e: Write-Only Memory
WOMBAT e: Waste Of Money, Brains And Time
Women Health e: Women and Health
Womens Health Data Book e: Womens Health Data Book
Womens Health Issues e: Womens Health Issues
WOM HOSP e: Womens Hospital
WOMIS e: Wing Operations Management Information System
WoMS e: World of Multiple Sclerosis
WONCA e: World Organisation of Family Doctors
WONDER e: Wide-ranging ON-line Data for Epidemiologic Research
WONG e: Weight On Nose Gear
W.Ont. e: Western Ontario
Wood Fiber Sci. e: Wood and Fiber Science
Wood Prod e: Wood and Wood Products
Wood Sci. Technol. e: Wood Science and Technology
Woof e: Well-Off, Older Folks
Wool Rec e: Wool Record and Textile World
WOP d: Werkstattorientierte Programmierung
- e: Wing Operation Plan
WOPG d: Wohnungsbauprämiengesetz
WOPR e: War Operations Planned Response
WOR e: Water/Oil Ratio
- e: Work Order Record
- e: Work Order Request
WORCS e: Work Order Reporting and Communication System
WORD e: Work Order Record and Details
WORDIA e: WORD Internet Assistent
Word Proc e: Word Processing and Information Systems
Working Group e: Hanford Future Site Uses Working Group
Works Inst Higher Nerv Act Patho-physiol Ser e: Works. Institute of Higher Nervous Activity. Pathophysiological Series
Work Study and Manage Serv e: Work Study and Management Services
World A e: World Archaeology
World Affairs J e: World Affairs Journal
World Bibl Social Security e: World Bibliography of Social Security
World Commod Rep Metal Ed e: World Commodity Report. Metals Edition
World Health Forum e: World Health Forum
World Health Organ Tech Rep Ser e: World Health Organization Technical Report Series
World Health Stat Q e: World Health Statistics Quarterly
World J Surg e: World Journal of Surgery
World J Urol e: World Journal of Urology
World Metal Statis e: World Metal Statistics
World Min Equip e: World Mining Equipment
World Patent Inf e: World Patent Information
World Resour. Rev. e: World Resource Review
World Rev Nutr Diet e: World Review of Nutrition and Dietetics
WorldWIDE e: World Women in the Environment
- f: Les femmes du monde entier et l'environnement
- s: Las Mujeres del Mundo y el Medio Ambiente

WORM e: Earthworm, Inc.
- Write On[c]e Read Many [format]
- e: Write Once, Read Many time[s]
- e: Write Once, Read Multiple
Worm e: White, Older Rich Man
WORM CD e: Write Once, Read Many Compact Disk
WORMS e: World-Wide-Web for Operations Research and Management Science
WORN e: Write Once, Read Never
Wor Pol e: World Politics
WOS e: Web Object Script
- e: Wireless Office System
- e: Workstation Operating System
WOSA e: Windows Open Services Architecture
- e: Windows Open System[s] Architecture
- e: Workers Organisation for Socialist Action
WOSAXCEM e: WOSA Extensions for Control, Engineering and Manufacturing
WOSAXFS e: WOSA Extensions for Financial Services
WOSAXRT e: WOSA Extensions for Real-Time market data
WOSC e: Washington Oregon State College
- e: World Organization of Systems and Cybernetics
WOT e: Wide Open Throttle
- e: Wire and Office Tested
WOTAN e: Weather Observation Through Ambient Noise
- e: Wind Observations Through Ambient Noise
WOTK e: World-Wide Technology, Inc.
WOU e: World ORT Union
WoVermittG d: Gesetz zur Regelung der Wohnungsvermittlung [Wohnvermittlungsgesetz]
WOVO e: World Organization of Volcano Observatories
WOW e: We-Cycle Office Wastepaper
- e: Weight On Wheels
- e: Windows On Win32
- e: Windows On Windows
- e: World Ocean Watch
- e: Worlds of Wonder
- s: Vigilancia Oceánica Mundial
WoW e: Window on the World
wö/Wo d: wöchentlich/Woche
WP d: Wirtschaftspädagogik
- e: Warm Pool
- e: Washington Post
- e: Water Pollution laboratory performance evaluation studies
- e: Wheel Pulser
- e: White Pages
- e: White Paper
- e: Wide Pore
- e: Wind Profilers
- e: Word Perfect
- e: Word Processing
- e: Word Processor
- e: Work Package
- e: Work Plan
- e: Work Procedure
- e: Wrist Pivot
- e: Write Protected
WPA e: Wagner-Peyser Act
- e: Water Jet Propulsion Assembly
- e: Water Pump Assembly
- e: Waterless Printing Association
- e: Webb-Pomerene Act
- e: Western Pistachio Association
- e: Western Provident Association
- e: Wet-Process Phosphoric Acid
- e: Whale Protection Act
- e: Wheelchair Pilots Association
- e: Whiskey Painters of America
- e: William Penn Association
- e: Womens Prison Association
- e: Works Progress Administration
- e: Workshop of the Players Art
- e: World Presbyterian Alliance
- e: World Press Archives
- e: Worst Possible Accident
WPAB e: Word Processing Aptitude Battery
- e: Word Processor Assessment Battery
- e: Work Package Proposal/ Agreement Submittal
WPABX e: Wireless Private Branch Exchange
WPAC e: Walden Pond Advisory Committee
WPAC e: Working Program Advisory Committee
WPAC e: West[ern] Pacific [panel]
WPAFB e: Wright Patterson Air Force Base
- e: Wright-Patterson Air Force Base
W PA Hist Mag e: Western Pennsylvania Historical Magazine
W Pakistan J Agr Res e: West Pakistan Journal of Agricultural Research
WPAQ e: Westra Preschool Assessment Questionnaire
WPARR e: War Plans Additive Requirements Report
WPAS e: Work Package Proposal and Authorization System
- e: Work Proposal Agreement System
- e: Workplace Air Sampling
WPASI e: Western Pacific Air-Sea Interaction Study
- f: Etude sur l'interaction air-mer dans le Pacifique occidental
- s: Estudio sobre la Interacción Aire-Mar en el Pacífico Occidental
WPAT e: Wolfe Programming Aptitude Test
WPATC **e:** Western Pennsylvania Advanced Technotogy Center
WPA-USA e: World Pheasant Association of the USA
WPB e: Large Patrol Boat
- e: Waste Processing Building
- e: West Palm Beach
- e: Wide Pulse Blanking
- e: Wiener Enterprises, Inc.
- e: World Peace Brigade
WPBCWS e: Waste Processing Building Chilled Water System
WPBD e: Westinghouse Proposal and Business Development
WPC e: Wafer Process Chamber
- e: Warrior Preparation Center
- e: Washington Processing Center
- e: Water Plant Communications
- e: Word Processing Center
WPD e: Water Planning Division
- e: Windows Printer Description
- e: WordPerfect Document
WPDA e: Wind Profiler Demonstration Array
WPDDG e: Waste Package Design and Development Group
WPDF e: Welayita Peoples Democratic Front
WPDN e: Wind Profile[r] Demonstration Network
WPDO e: Welayita Peoples Democratic Organization
WPDP e: Waste Processing and Disposal Program
WP-DPG e: Western Pacific Detailed Planning Group
WPE e: Westinghouse Pittsburgh Environmental
- e: Word Processing Equipment
- e: Work and People
- e: Workers Party of Ethiopia
WPF d: Wahlpflichtfach
WPFCC e: Western Pacific Fisheries Consultative Committee
- f: Comité consultatif des pêches pour le Pacifique occidental
- s: Comité Consultivo de Pesca para el Pacífico Occidental
WPG e: WordPerfect Graphics
WPH e: Wafers Per Hour
WPHD e: Write Protect[ed] Hard Disk
WPI e: Watershed Protection Initiative
- e: Worcester Polytechnic Institute
WPID e: WIPP Project Integration Office
WPIE e: Warning Plan Interface Equipment
WPIO e: Waste Isolation Pilot Plant Project Integration Office, Albuquerque, New Mexico
WP/IS e: Word Processing and Information Systems
WPK e: Keyboard
WPL e: Wave Propagation Laboratory
WPL+ e: Word-oriented Language internal to PRODOS Applewriter 2.1
WPLID e: Wind Profile Lidar
WPM e: Macro
- e: Words Per Minute
WPMS e: Work Package Milestone Status
WP/N e: Weapons Procurement, Navy
WPNFCR e: Working Party on Nuclear Fuel Cycle Requirements
WPO d: Wirtschaftsprüferordnung
- d: Wirtschaftsprüfungsordnung
- e: Water Policy Office
WPOS e: Workplace Operating System
WPP e: Web Presence Provider
- e: Work Package Preparer
WPPAS e: Work Package Proposal and Authorization System
WPPC e: Water Products Promotion Council
WPP/IS e: Writing Proficiency Program/Intermediate System

WPPISDC e: Wisconsin Power Plant Impact Study Data Center
WPPL e: Welding Performance Proficiency Log
WPPSI e: Wechsler Preschool and Primary Scale of Intelligence
WPPSS* e: Washington Public Power Supply System
WPR e: Waste Management, Production and Reprocessing Division
- e: Water Resource Planning
WP[REI] e: Wildlife Protection [Regulations and Exports and Imports] Act
WPREQ e: Request for Nuclear Reserve Weapons
WPRS e: War Powers Reporting System
WPS e: Wind Power System
WPS e: Windows Printing System
- e: Word Processing Software
- e: Word Processing System
- e: Work Package Summary
- e: Working Prototype System
- e: Workplace Shell
WPSA e: Wildlife Preservation Society of Australia
- e: Workers Party of South Africa
WPSI e: Work Package Status Input
WPSO e: WIPP Project Site Office
WPSQ e: Wildlife Preservation Society of Queensland
WPT d: Wirtschaftsingenieur für Maschinenbau [Studienrichtung Produktionstechnik]
WPT e: Western Personnel Tests
WPT e: Wolfe Screening Test for Programming Aptitude
WPT e: Word Processing Test
WPT e: Working Point
WPVM e: Windows Parallel Virtual Machine
WPWP e: Western Pacific Warm Pool
WQ e: Wilson Quarterly
W Q e: West Quadrangle
WQ1 e: Spreadsheet
WQ! e: Compressed Spreadsheet
WQA e: Water Quality Act
WQC e: Water Quality Criteria
WQL e: Water Quality Laboratory
WQM e: Water Quality Modification
WQM4 e: Region 4 Water Quality Modeling
WQMSGL e: Water Quality Modeling System for the Great Lakes
WQR e: Waste[-form] Qualification Report
WQWM e: Water Quality and Waste Management
WR d: Werkraum
- d: Windrichtung
- e: Reserve Weapon
- e: Washout Rate
- e: Waste Management, Production and Reprocessing Division
- e: Waste Retrieval
- e: Water Resource[s]
- e: Waveguide, Rectangular
- e: Wear Resistant
- e: Weather Reconnaissance
- e: Western Range
- e: Wide Receiver
- e: Wildlife Reserve
- e: Work Release
- e: Work Request
- e: Wrist Rotate
WRA e: Weapons Replaceable Assemblies
- e: Work Release Authorization
WRAANC e: Womens Royal Australien Army Nursing Corps
WRAB e: West Rand Administration Board
WRAC e: Waste and Recycling Advisory Committee
- e: Water Resources Advisory Committee
WRADTA e: Upgrade of WR-ALC GO89 Damage Tolerance System
WRAF e: Womens Royal Air Force
WR-ALC e: Warner Robins Air Logistics Center
WRAM e: Westinghouse Radiation Area Management
- e: WHC Radiation Area Management
- e: Windows RAM [Random Access Memory]
WRAMC e: Walter Reed Army Medical Center
WRAMSOS e: WR-ALC Multiple System OFP Support System
WRAO e: Western Regional Audit Office
WRAP e: War Reserve Automated Process
- e: Warfighting Rapid Acquisition Program
- e: Waste Receiving and Packaging
- e: Waste Receiving and Processing
- e: Water Resources Assessment Program
- e: Western Region Applications Program
- e: Winter Rainfall Acceptance Potential
WRAPD e: World Rehabilitation Association for the Psycho-Socially Disabled
WRAT e: Wide Range Achievement Test
WRB e: Web Request Broker
WRC e: Wartime Consumption Rate
- e: Water Research Centre
- e: Water Research Commission
- e: Water Resource Center at San Carlos University
- e: World Radiation Center
- e: World Radiocommunications Conference
WRC e: Weekly Readiness Check
WRCA e: Water Resources Center Archive
WRCS e: Weapon Release Computer Set
WRD e: Water Reactors Division
- e: Water Resources Division
- e: Western Resource Development
WRDA e: Water Resources Development Act
WRDB e: Worldwide Water Resources Database
WRDC e: Wool Research and Development Corporation
- e: World Radiation Data Centre
WRDD e: Water Resources Development Department
WRD&T e: Weapon Research, Development and Testing
WRECS e: Weapon Radiation Effects on Communications Systems
WRED e: Wrong Reading Emulsion Down
WREU e: Wrong Reading Emulsion Up
WRF e: Waste Retrieval Facility
- e: Wind Reduction Facility
WRI e: Western Research Institute
- e: World Resources Institute
- e: Write
- f: Institut des ressources mondiales
- s: Instituto de los Recursos Mundiales
WRIF e: Water Resources Information File
WRIMS e: WHC Requirements Information Management System
WRIPS e: Wave Rider Information Processing System
WRIPT e: Wide Range Intelligence-Personality Test
WRIST e: Weather Radar Identification of Severe Thunderstorms
WRIT e: Creative Writing
WRITE e: WONCA Rural Information Technology Exchange
WRK e: Windows Resource Kit
WRL e: War Resisters League
WRM e: War Reserve Materiel
- e: Water Recovery and Management
- e: Westinghouse Radioactive Material
- e: White Resistance Movement
WRMC e: World Radiation Monitoring Center
WRMCMR e: WRM Control and Management Reporting
WRME e: Wood Raw Material Equivalent
WRMLRSSL e: WRM Lists Requirements and Spares Supply List
WRMRS e: WRM Rating System
WRN d: Waren/Müritz, Mecklenburg-Vorpommern
WRO e: Water Resource Operations
WROS e: With Right Of Survivorship
WRP e: Wetland Reserve Program
WRR e: Wide Band Regenerative Repeater
- e: World Radiometer (or Radiometric) Reference
WRRC e: Water Resources Research Center
WRRS e: Wire Wrap Removal Station
WRS e: Reg 5 ADP Workload Reporting System
- e: Water Resources
- e: WIMS Reconciliation System
- e: Worldwide Reference System
WRSK e: War Readiness Spares Kit
- e: War Reserve Support Kit
WRSK/BLSS e: WRSK/Base Level Self-Sufficiency Spares
WRSR e: Water Reactor Safety Research

WRSS e: Waste Retrieval Sluicing System
- e: Water Removal and Storage System
WRT e: Whitewater Ressource Toolkit
- e: With Regard To [Internet]
- e: With Respect To [Internet]
- e: Writing
w.r.t. e: with respect to
WRTS e: Waste Remediation Tracking System
WRU e: Wide Receiver Unit
- e: Work Reporting Unit
WRWC e: Wartime Requirements for War Consumables
WRWP e: Western Regional Waste Processing
WS d: Wegfahrsperre
- d: Wirtschaftsschule
- e: Samoa
- e: Wall Shelves
- e: Wannier-Stark
- e: Wash Solvent
- e: Water Supply Evaluation Studies Laboratory Performance
- e: Weak Safety
- e: Weasel Support
- e: Weather Squadron
- e: Wigner-Seitz
- e: Wilderness Society
- e: Window Size
- e: Windshear
- e: Wire Scanner
- e: Woodcock Survey
- e: WordStar
- e: Work Sampling
- e: Work Station
- e: Work Status
W S e: Womens Studies
WS. e: West Saxon
W.S. e: Writer to the Signet
W/S e: Workstation
WS3 e: Weapon Storage and Security System
WSA e: Weapons Storage Area
- e: Williams Syndrome Association
- e: WWMCCS System Architecture
WSAB e: Weapon System Assessment Briefing
WSA&E e: Warfare Systems Architecture and Engineering
WSAFHWC e: Weather Support for AFHWC
WSAG e: Washington Special Action Group
WSAI e: Wire Society of America, Inc.
WSAM e: Weapon System Assessment Model
WSAP e: Weapon System Acquisition Process
WSAPI e: Web Site Application Program Interface
WSAS e: WWMCCS Standard Application Software
WSB e: Weapons Services Branch
- e: Weather Support Branch
- e: Wheat- Soya-Blend
WSB e: Weekly Statistical Bulletin [Database]
WSBCRCS e: Worldwide Stock Balance and Consumption Report Consolidation System
WSBM e: Wheat Soilborne Mosaic
WSBMV e: Wheat Soilborne Mosaic Virus
WSBS e: White Sea Biological Station
- e: Wide-Sweep Backscatter Sounder
- f: Station biologique de la mer Blanche
- s: Estación Biológica del Mar Blanco
WSBW e: Weddell Sea Bottom Water
WSC e: Wall Street Computer review
- e: War Support Center
- e: Warfighting Support Center
- e: Water Survey of Canada
- e: Weapons System Controller
- e: White Sands Complex
- e: WIN Site Coordinator
- e: Wind Sounding Capability
- e: Wireless Switching Centre
WSC-5 e: Washington Science Center, Building 5
WSCF e: Waste Sampling and Characterization Facility
WSCL e: Waste Stream Components List
WSCM e: Weapon System Compatible Munition
WSCNR e: Wire Scanner
WSD e: Work Station Division
WSDA e: Washington State Department of Agriculture
WSDE e: Washington State Department of Ecology
WSDL e: Weapons System Data Link
- e: Wet Scrap Development Laboratory
WSDOH e: Washington State Department of Health
WSDR e: Waste Storage/Disposal Request
WSDST e: Waste Storage and Disposal Safety Team
WSDW e: Weddell Sea Deep Water
WSE e: WWMCCS System Engineer
WSEO e: WWMCCS System Engineering Organization
WSEP e: Waste Solidification Engineering Prototypes
- e: Weapons System Evaluation Program
WSES e: Surface Effect Ship
WSET e: Women in Science, Engineering and Technology
WSF d: Weissenfels, Sachsen-Anhalt
- d: Wirtschafts- und Sozialwissenschaftliche Fakultät
- e: Water Soluble Fraction
- e: Weapons Systems Fill
- e: Weather Support Force
- e: Work Station Feature
- e: Work Station Function [block]
WSFI e: Wood and Synthetic Flooring Institute
WSFN e: Which Stands For Nothing
WSFO e: Weather Service Forecast [office[s]]
WSG e: International Wool Study Group
WSGS e: Woodcock Singing Ground Survey
WSGT e: White Sands Ground Terminal
- e: WWMCCS Standard Graphics Terminal
WSHI e: Washington Homes, Inc.
WSI e: Wackenhut Security, Inc.
- e: Wafer Scale Integration
- e: Waste Stream Identification
- e: Whole Sky Imager
WSIA Journal e: Water Supply Improvement Association. Journal
WSIG e: Wanted type of Signalling
- e: Weapon Support Improvement Group
WSI-MSTE e: Washington Systemic Initiative in Mathematics, Science and Technology Education
WSIP e: Weapon Systems Improvement Program
WSIR e: Wide Swath Imaging Radar
WSIS e: Whole Sky Imaging System
WSITS e: Weapon System Interface Trade Study
WSL d: Eidgenössische Forschungsanstalt für Wald, Schnee und Landschaft, Birmensdorf
- e: Equipment calibration system
- e: Waterloo Systems Language
WSLC e: WOCE Fast Delivery Sea-Level Data Assembly Center (or Centre)
- f: Centre WOCE de rassemblement et de transmission rapide de données relatives au niveau de la mer
- s: Centro WOCE de Acopio y Distribución Rápida de Datos sobre el Nivel del Mar
WS M e: Watershed Management
WSM e: Wafer Starts per Month
- e: Waste form qualification Simulation Model
- e: Waterside Mall
- e: Weapon System Model
- e: Weapon Systems Manager
- e: Wheat-Soya-Milk
- e: Work Space Management report
- e: Workstation Maintenance
WSMC e: Western Space and Missile Center
WSMIS e: Weapon System Management Information System
WSMO e: Weather Service Meteorological Office
WSMP e: Weapon System Master Plan
WSMR e: Facility wet scrap recycle
- e: White Sands Missile Range
- e: White Sands Missile Range nuclear effects
WSN e: Wirth Syntax Notation
WSO e: Warner and Swasey Observatory
- e: Warning Systems Operator
- e: Weapon System Operator
- e: Weapon[s] System[s] Officer
- e: Weather Service Office
- e: Work Scope Breakdown
WSP e: Washington State Plane
- e: Weapons System Pouch
- e: Wireless Service Provider
- e: Work Simplification Program
- e: Workstation Security Package
WSPA e: Western States Petroleum Association
WSPAR e: Weapon System Program Assessment Review
WSPAT e: Wolfe-Spence Programming Aptitude Test
WSPC e: Wideband Semipermanent Connection
WSPCW e: Western South Pacific Central Water
WSPRO e: Water Surface Profile

WSPRS e: WIN SCF Performance Reporting System
WSPS e: Worldwide Standard Port System
WSQ e: Wavelet/Scalar Quantization
W/sr e: Watt per steradian
WSR e: 88D-Weather Surveillance Radar 1988 Doppler
- e: Warehouse Storage Request
- e: Weapons Status Report
- e: Weather Service Radar
- e: Weather Surveillance Radar
- e: Wet Scrap Recovery
- e: Wet Scrap Recycle
WSRA e: Wild and Scenic Rivers Act
WSRAP e: Weldon Spring Site Remedial Action Project
WSRC e: Westinghouse Savannah River Company
WSR-88D e: Weather Surveillance Radar, 1988 Doppler-1988-Doppler
WSRD e: Waste Site Recovery Demonstration
- e: Waste Specification Record
W/[sr-m2] e: Watt per steradian square meter
WSRO e: Weapon Systems Replacement Operations
WSS e: Warfare Support System
- e: Waste Sampling System
- e: Weapons System Security
- e: Weldon Spring Site
- e: Wide Screen Signalling [line]
- e: WWMCCS Standard System
WSS e: Workpack Scheduling System
WSSAB e: Weller-Strawser Scales of Adaptive Behavior
WSSC e: Weapon System Service Center
WSSG e: WOCE Scientific Steering Group
WSSI e: World Seismic Safety Initiative
- f: Initiative mondiale pour la sûreté sismique
- s: Iniciativa Mundial para la Seguridad Sísmca
WSSIB e: WWMCCS Standard System Information Base
WSSM e: Water Supply Simulation Model
- e: Weapon System Support Model
WSSRAP e: Weldon Spring Site Remedial Action Project, Weldon Spring, Missouri
WSSS e: WWMCCS Standard System Software
WST d: Ammerland in Westerstede, Niedersachsen
- e: Tala, Samoa
- e: Weapons System Trainer
- e: Well Seismic Tool
- e: Whitecap Simulation Tank
- e: Wholesale Sales Tax
Wst d: Wetterstation
WS&T e: Waste Systems and Transportation Department
WSTA e: Weapons System Technical Architecture
- e: World Societies for Technology in Anaesthesia
WSTB e: Water Science[s] and Technology Board
WSTC e: Westinghouse Science and Technology Center
WSTP e: downhole Water Sampler, Temperature and Pressure probe
- e: Wastewater Sewage Treatment Plant
WSTS e: World Semiconductor Trade Statistics
WSTT e: Weather Scenario Test Tape
WSU e: Weather Support Unit
WSUCSC e: Washington State Univ. Computing Service Center
WSUMS e: WWDMS Scheduling and Management System
WSU[-]TC e: Washington State University-Tri Cities
WSV e: Wideband Secure Voice
WSVN e: Worldwide Secure Voice Network
WSVS e: Region 2 Water Supply Violation Assessment Graphics System
WSW d: Weisswasser, Sachsen
- d: Westsüdwesten
- e: West [by] SouthWest
WSWA e: Weed Society of Australia
W.Syd. e: Western Sydney
WSZ e: Wood Supply Zone
WT e: Water Tight
- e: Water Treatment
- e: WaveTable
- e: Wireless Transceiver
- e: Without Thinking [Internet]
- e: Write Through
WT0 e: Watchdog Timer 0 register
WT1 e: Watchdog Timer 1 register
wt% e: weight percent
WTA A: Wit Tuisland Aksie
- e: West Turnaround
- e: Western Timber Association
- e: Willingness To Avoid
- e: Winner-Take-All
WTAC e: Waste Isolation Pilot Plant Technical Assistance Contractor
WTAM e: West Texas A&M University
WTB e: Want[ed] To Buy
- e: Wireless Telecommunications Bureau
WTC e: Wafer Transfer Chamber
- e: Waste Technology Center
- e: Water Transmission and Communications
- e: Wilson [truck] Trailer Company
- e: World Trade Center
WTCC e: Wet Tropics Consultative Committee
WTCV e: Weapons and Tracked Combat Vehicles
WTD e: Weapons Training Detachment
- e: Westinghouse Tampa Division
WTDOS e: Weapon Training Detachments Operating Spares
WTEC e: World Technology Evaluation Center
WTF e: Water Treatment Facility
- e: What The Freep (or Fuck) [Internet]
WTF? e: What The [Flying-Frog] [Internet]
WTFDIK e: What the Freep (or Fuck) Do I Know [Internet]
WTG e: Way To Go [Internet]
- e: Wind Turbine Generator
WTGL e: Wave-pipelined Transmission-Gate Logic
WTH e: What The Heck (or Hell) [Internet]
- e: Whole-Tree Harvest
- e: Why The Hell [Internet]
WTHDTM e: What The Hell Does That Mean [Internet]
WTK e: World Toolkit
WTLC e: Western Trunk Line Committee
WTM d: Wittmund, Niedersachsen
- e: WWMCCS Technical Manual
WTMU e: Wildlife Trade Monitoring Unit
WTNP e: Works Technical New Policy
WTO e: World Tourism Organization (or Organisation)
WToO e: World Tourism Organization
WTOY e: Wisconsin Toy Company, Inc.
WTP e: Willingness To Pay
WTPD e: WESCOM Telephone Products Division
WTPR e: Wetterau Properties, Inc.
WTPS e: Water, Toxics and Pesticides Staff
WTR d: Wissenschaftlich-Technischer Rat
- e: Western Test Range
WTrO e: World Trade Organization
WTR OPS e: Water Operations and Maintenance
WTR PLT e: Water Plant
WTS e: HPOB Automated Workload
WTS e: Weapons Training Site-Management System
- e: Washington Tactical Switch
- e: Waste Tank Safety
- e: Web Transaction Security
- e: Word Technology Services
WTSAP e: Wet Tropics Structural Adjustment Package
WTSHRD e: Water and Toxic Substances Health Research Division
WTSS e: Waste Tank Safety Support
WTTM e: Without Thinking To[o] Much [Internet]
WTU e: Wastewater Treatment Units
- e: Wave-Table Upgradeable
WTUSM e: Wuhan Technical University of Surveying and Mapping
WTW e: Wafer-To Wafer
- e: Wall To Wall
WTWBIR e: Waste Isolation Pilot Plant Transuranic Waste Baseline Inventory Report
WTWNU e: Wafer-To Wafer Non-Uniformity
WU d: Wirtschaftsuniversität
WUA e: Weighted Usable Area
WUCAPS e: Wuhan Combined Adjustment Program System
WUDO e: Western Union Defense Organization
WUFR e: World Union of Free Romanians
WUG d: Weissenburg-Gunzenhausen, Bayern
- e: WWMCCS/WIS Users Group
WUIS e: Water Use Information System
WULF e: Wake Up on Line Feed
WUN d: Wunsiedel, Bayern
WUPPE e: Wisconsin Ultraviolet Photo-Polarimeter Experiment
WUR d: Wurzen, Sachsen

WURST e: Wideband, Uniform Rate and Smooth Truncation
Württemb Ärztebl d: Württembergisches Ärzteblatt
WUS e: Word Underscore
WUSTL e: Washington University in St. Louis
WUT e: Wafer Under Test
wuzzy e: was he?
WV e: West Virginia
W&V d: Werben und Verkaufen in Österreich
WVA e: Wetlands Value Assessment Decision Support Model
WVAS e: Wake Vortex Avoidance System
WVAST e: Washer Visual Acuity Screening Technique
WVDP e: West Valley Demonstration Project, West Valley, New York
WVHLW e: West Valley High-Level Waste
WVI e: World Veterinary Index
WVIOP e: Water Vapor Intensive Observation Period
W3VL e: World Wide Web Virtual Library
W V Med J e: West Virginia Medical Journal
WVN e: Water Vapor Nitrogen
WVNET e: West Virginia Network for Educational Telecomputing
WVNS e: West Valley Nuclear Services Company, Inc., West Valley, New York
WVNSC e: West Valley Nuclear Services Company
WVNY e: West Valley, New York
WVO d: Warschauer Vertragsorganisation
WVP e: Waste Volume Projections
WVR e: Water Vapo[u]r Radiometer
- e: Water Vapo[u]r Regained
WVRAAM e: Within Visual Range Air-to-Air Missile
WVRD e: World Vision Relief and Development Inc.
WVRG e: West Valley Reference Glass
WVSA A: Weidingsvereniging van Suidelike Afrika
WVSG-2 e: West Valley Sludge Glass-2
WVSO d: Wiederverwendbare Softwareobjekte
WVSP e: West Valley Support Program
WVTR e: Water-Vapor Transmission Rate
WW A: Wit Wolwe
- e: MCCS ADP Control Center
- e: Weather Wing
- e: Wide Well
- e: Winter Water
- e: Worldwide Equities Ltd.
W + W d: Wissenschaft und Weltbild
W/W e: World-Wide
WW2 e: World War II
WWA e: Wild Weasel Augmentation
WWA e: Who's Who in Australia
WWABNCP e: World-wide Airborne Command Post
WWB A: Wit Weerstandsbeweging
- e: Womens World Banking
wwb e: Writers WorkBench
WWC e: Wing Work Center
- e: World Wide Computing
WWCB e: Weeky Weather and Crop Bulletin
WWCICS e: Wolfe-Winrow CICS/VS Command Level Proficiency Test
WWCN e: World Wide Computer Network
WWDC e: World Wide Developer Conference
WWDSA e: Worldwide Digital Systems Architecture
WWEMA e: Waste and Wastewater Equipment Manufacturers Association
WWER e: Water cooled and moderated energy reactor
WWF e: World Wide Fund for Nature
- e: World Wildlife Fund
- f: Fonds mondial pour la nature
- s: Fondo Mundial para la Naturaleza
WWFA e: World Wide Fund for Nature
WWFO e: World Wide Field Operations
W4WG e: Windows for Workgroups
WWGD e: World Winter Games for the Deaf
WWI e: World War I [One]
WWII e: World War II [Two]
WWIII e: World War III [Three]
WWIMS e: Worldwide Warning Indicator Monitoring System
WWIMWSS e: Worldwide Management of Wholesale Subsistence Stock
W/WIPP e: Westinghouse/WIPP Project, Carlsbad, New Mexico
WWIS e: World Wide Information System
- e: World Wide Information System
WWMCCS e: World Wide Military Command and Control System
WWMCCSDPCE e: WWMCCS Data Processing Center, Europe
WWMMRD e: Water and Waste Management Monitoring Research Division
WWMS e: Water and Waste Management Staff
WWNWS e: World-Wide Navigational Warning Service
WWO e: Waste Water Outflow
- e: Wing Weather Officer
WWOLS e: World Wide On-Line System
WWOPS e: World Wide Operations
WWPA e: Western Wood Products Association
WWPL e: Worldwide Publishing Consortium
WWR A: Witwatersrand
WWREM e: Weekly Web Review in Emergency Medicine
WWS e: Wife Wants Sex
- e: WIS Workstation
WWSP e: Winter Weddell Sea Project
- f: Projet mené en hiver dans la mer de Weddell
- s: Proyecto Invernal en el Mar de Weddell
WWSP86 e: Winter Weddell Sea Project 1986
WWSV e: Worldwide Secure Voice
WWSVA e: Worldwide Secure Voice Architecture
WWSVCS e: Worldwide Secure Voice Conferencing System
WWTCIP e: Worldwide Technical Control Improvement Plan
WWTP e: Wastewater Treatment Plant
WWT PLT e: Wastewater Plant
WWTR OPS e: Wastewater Operations and Maintenance
WWU A: Blanke Werknemersunie
- d: Wirtschafts- und Währungsunion
- e: Western Washington University
WWW d: Informationssystem als Teil des Internet auf Basis der Seitenbeschreibungssprache HTML [Hypertext Markup Language]
- d: Weltweites Warten
- e: Wide Word Workstation
- e: World Weather Watch of the WMO
- e: World Wide Waiting
- e: World Wide Web of the Internet
- e: Wretched Writers Welcome
WWW<A> e: World Weather Watch
WWWDM e: WWW Data Management
WWWIC e: WWW Implementation Coordination
WWWVL e: World-Wide Web Virtual Library
WWWW e: Who's Who in the World of Women
WWWW e: World Wide Web Wizard
- e: World Wide Web Worm
WX e: Weather
WX Detc. e: Weather Detector
WXRCNSq e: Weather Reconnaissance Squadron
WY e: Wyoming
- e: Wyoming
WYGIWYPF e: What You Get Is What You Pay For [Internet]
WYLABOCTGWTR e: Would You Like A Bowl Of Cream To Go With That Remark [Internet]
WYLASOMWTC e: Would You Like A Saucer Of Milk With That Comment [Internet]
WYNB e: Wyoming National Bancorporation
WYS e: Wyse Technology
WYSBYGI e: What You See Before You Get It [Internet]
WYSIAYG e: What You See Is All You Get [Internet]
WYSIWIS e: What You See Is What I See [Internet]
WYSIWYF e: What You See Is What You Fax [Internet]
WYSIWYG e: What You See Is What You Get [Internet]
WYSIWYHYG e: What You See Is What You Hope You Get [Internet]
WYSIWYNG e: What You See Is What You Never Get [Internet]
WYTYSYDG e: What You Thought You Saw, You Didn't Get [Internet]
WYWO e: While You Were Out [Internet]
WZB d: Wissenschaftszentrum Berlin für Sozialforschung gGmbH
WZL d: Wanzleben, Sachsen-Anhalt
WZN d: Weltzentrum für Niederschlagsklimatologie
WZO e: World Zionist Organization

X

X e: Cross-
- e: Crystal
- d: Eingangsgröße nach DIN 5032
- e: Exciton
- e: Experimental
- e: Extension
- e: Trans-
- e: Transistor
- e: X-Interface

x d: Allgemeiner Variablenvektor
- d: Anzahl der an den Kombinationen beteiligten Gläser
- d: eine beliebige reale Variable
- e: Vektor der Zielfunktionsvariablen
- e: Vektor eines Zufallspunktes

x0 d: Vektor [Gene] des ersten Individuums

x1 d: Vektor des Punktes mit dem besten Zielfunktionswert

X-10 e: original site name for the Oak Ridge National Laboratory

X11 e: X window System version 11

X.12 e: ASC X12

X.25 e: ISO Standard for Interface to Packet Switched Communications Services
- e: Network Access Protocol

X.29 d: Virtuelles Terminal der OSI-Welt

X.200 e: CCITT recommendation X.200 is the basic

X.400 e: Mailsystem der OSI-Welt

X*PRESS e: Around-the-clock keyword-based electronic news service

XA e: Exchange Access
- e: Extended Architecture
- e: Extended Attribute
- e: Failure alarm
- e: Office of External Affairs

XAA e: Call Transfer all Calls Attendant

XACT e: Xantes Accurate Calibration Technology

XAES e: X-ray-induced Auger Electron Spectroscopy

XAF f: Franc de la Communauté financière africaine

XAFS e: X-ray Absorption Fine Structure

XAI e: Call Transfer all Calls Internal

XAN e: Call Transfer [All Calls, No restrictions]

XANADU e: X-ray Analysis system maintained at the HEASARC

XANES e: X-ray Absorption Near-Edge Structure [Spectroscopy]

XAO e: Call Transfer all Calls, Outside

XAOI e: Expanded Analog Output Interface

XAPIA e: X.400 API [Application Program Interface] Association

XAS e: X-ray Absorption Spectroscopy

XATEF e: Crosswalk Air Toxics Emission Factor Database

XAU e: Gold

XB e: X Bar

X2B e: Hexadecimal to Binary

XBA e: European Composite Unit

X-band e: Microwave band around 3 cm wavelength

Xbar e: Crossbar

XBB e: European Monetary Unit

XBBM e: X Length Bunch Monitor

XBC e: European Unit of Account 9

XBD e: European Unit of Account 17

XBM e: Extended Bitmap
- e: X Bit Map
- e: X-Windows Bitmap[s]

XBP e: XBase Parts

XBS e: Xerox Business Center

XBT e: Expendable Bathythermograph[s]
- e: X Bar Tandem
- f: Bathythermographe largable, bathythermographe non Récupérable
- s: Batitermógrafo no recuperable, batitermógrafo desechable

XBT/XCTD e: Expendable Bathythermograph/Expendable Conductivity-Temperature-Depth probe

Xbus e: A packet-switched

XC e: Cross Connect Node
- e: Cross-Country
- e: Extended Configuration

xc d: Vektor des durch Kontraktion erzeugten Punktes

X2C e: Hexadecimal to Character

XCA e: External Communications Adapter

XCB e: Extended Core Barrel

XCB/FC e: Extended Core Barrel Flow-Control System

XCCI e: X Common Client Interface

XCEM e: Extensions for Control, Engineering and Manufacturing

XCF e: Cross-System Coupling Facility
- e: Experimental Computing Facility

XCHG e: Exchange

XCHS e: Extended Cylinder[s], Head[s], Sectors

XCMD e: Extended Mode Command
- e: External Command

XCMOS e: Extended CMOS

XCMS e: X Color Management System

XCOFF e: Extended Common Object File Format

XCOPY e: Extended Copy

XCOR e: X Corrector Magnet

XCP e: Expendable Current Profiler
- f: Courantomètre largable, courantomètre non récupérable
- s: Trazador de perfiles de corrientes no recuperable XCTD

XCTD e: Expendable Conductivity Temperature [and] Depth
- e: Expendable Current Temperature Density profiler
- f: Sonde largable de mesure de la conductivité, de la température et de la profondeur
- s: Sonda no recuperable de medida de la conductividad, la temperatura y la profundidad

XCUTS e: X Commands and Utilities Test Suite

XD e: Cancel Disconnect
- e: X [without] Dividend

X2D e: Hexadecimal to Decimal

XDC e: ARM External Data Center
- e: Cross-Connect Data Conflict
- e: Exchange Default Carrier
- e: Xhosa Development Corporation

XDCS e: X/Open Distributed Computing Services

XDF e: Exchange Data Format
- e: Extended Density Format
- e: Extended Distance Feature

XDIN e: Communications Subsystem of IMAPS

XDLE e: Transparent Data Link Escape

XDM e:X Display [Windows] Display Manager

XDMCP e: X Display [Windows] Manager Control Protocol

XDR e: Extended Data Representation
- e: External Data Representation
- e: Internation Monetary Fund Special Drawing Rights

XDS e: X/Open Directory Service
xDSL e: x-Type Digital Subscriber Line
Xdvi e: X-windows-based dvi file viewer
xe d: Vektor des durch Expansion erzeugten Punktes
XEDS e: X-ray Energy Dispersive Spectroscopy
X el l d: Absolute spektrale Eingangsstrahlungsgröße nach DIN 5032
XEROX PARC e: XEROX Palo Alto Research Center
XES d: Xerox Engineering Systems
XETB e: Transparent End of Transmission Block
XETX e: Transparent End of Text
XEU e: ECU
xe, ye d: Vektoren zweier Individuen mit den zwei besten Zielfunktionswerten
XF e: Cancel First Choice
XFCB e: Extended File Control Block
XFCN e: External Function
XFE e: X Front End
XFER e: Transfer
X-Files e: Files of unknown content
XFL e: Exit Flight Level
XFN e: X-Open Federated Naming [service]
XFR e: [BG Attendant] Call Transfer
XFS e: Extended File System
- e: X11 File System
XG e: Extended General MIDI
XGA e: Extended Graphics Adapter
- e: Extended Graphics Array
X-GDS e: Exempt from General Declassification Schedule
XHA e: Ascom Hasler AG
XHD e: Extended Hard Disk
xi d: i-te Individuum
- d: i-te Unbekannte
- d: Vektor des i-ten Zufallspunkt
XID e: Exchange Identification
- e: Exchange Identifier
- e: Exchange Information
XID. e: Exchange Station Identification
XID3 [NA] e: Exchange Identification Type 3 [Non-Activation]
XID3 [NP] e: Exchange Identification Type 3 [Negotiation Proceeding]
XID3 [PN] e: Exchange Identification Type 3 [Pre-Negotiation]
XIE e: X Image (or Imaging) Extension
- e: X-Ray Imaging Experiment
XIE-DIS e: XIE Document Imaging Subset
XII e: Call Transfer Incoming only, Internal
XIM e: X Input Method
XIMAGE e: Image analysis program in XANADU
XIN e: Call Transfer Incoming only, No Restrictions
XINU e: Xinu Is Not UNIX
XIO e: Call Transfer Incoming only, Outside
- e: Extended Input/Output
XIOS e: Extended Input/Output System
XIP e: Execute In Place
XIPS e: Xenon Ion Propulsion System for nudging satellites around
XIWT e: Cross-Industry Working Team
XKB e: X Keyboard Extension
XL e: Cancel Long Lead Connect
- e: Excel
- e: Exclusive Lock
xl d: Vektor des Punktes mit dem schlechtesten Zielfunktionswert
XLA e: X-Linked Agammaglobulinemia
XLAT[E] e: Translate
XLAU e: Extended Limited AUTODIN Upgrade
XLFD e: X Logical Font Description
XLINK d: Deutscher Serviceprovider
- d: Extended Lokales Informatik Netz Karlsruhe
XLM e: Excel Macro Language
xln e: crystalline
XLPE e: Cross-Linked Polyethylene
XLR d: Professionelle Steckernorm für symmetrische Kabelverbindungen
XLS e: Excimer Laser System
- e: Extended Light Scatterer
XM e: Exchange Manager
XMAPI e: Extended Messaging Application Program Interface
XMAS e: Christmas
xmax, xmin d: Maximum, bzw. Minimum der x-Koordinaten der Meßpunkte Mp 1, Mp 2, Mp 3, Mp 4
xMC d: Vektor eines spontanen Mutanden [Monte-Carlo-Kind]
XMI e: Extended Memory Interconnect
- e: XTendbus Multibus Interface
xmin d: Vektor mit dem momentan minimalen Zielfunktionswert
XMIT e: General Purpose Message Transmission System
- e: Transmit
XML e: Extensible Markup Language
XMM e: Extended Memory Manager
- e: X-ray Multi Mirror [Mission]
X-MOS e: High-speed Metal-oxide Semiconductor
XMP e: X/Open Management Protocol
XMS e: Extended Memory Specification
- e: Extended Multiprocessor operating System
XMTR e: Transmitter
XN e: Cancel Connect
XNOS e: Experimental Network Operating System
XNS d: Xerox-Netzwerk
- e: Xerox Network Service[s]
- e: Xerox Network Standard
- e: Xerox Network System[s]
XNSCP e: [PPP] XNS IDP Control Protocol
XNS[-]IDP e: Xerox Network Services[-] ID [Internet Datagram] Protocol
XNSITP e: Xerox Network Services/Internet Protocol
XO e: Crossover
- e: Extended Office
XOF e: West African Franc
XOFF e: Exchange OFF
- e: Transmitter Off
XOM e: X/Open Management
XON e: Exchange ON
- e: Transmitter On
XOR e: Exclusive
XOS e: Experimental Operating System
XOT e: X.25 Over TCP
XP e: Crosspoint
- e: Explanation Pattern
- e: Xeroderma Pigmentosa
XPC e: X Performance Characterization
- e: X Performance Characterization Group
XPD e: X-ray Photoelectron Diffraction
XPEL e: X Protocol Engine Library
XPF f: Franc des Comptoirs français du Pacifique
XPG e: X/Open Portability Guide
XPG2 e: X/Open Portability Guide Release 2
XPGR3 e: X/open Portability Guide Release 3
XPLA e: Extended Programmable Logic Array
XPM e: X-Windows Pixelmaps
XPORT e: Transport
XPR e: Quark XPress Document
XPRF e: Explosion Proof
XPRM e: Xerox Print Resources Manager
XPS e: Expandable Polystyrene
- e: Extended Parallel Server
- e: X-ray Photoelectron Spectroscopy
- e: Xeroderma Pigmentosum Society
XPSB e: X-Protocol Stream Benchmark
X11R4 e: Version 11 release 4 of the X protocol
X11R5 e: Version 11 release 5 of the X protocol
XR e: Crossreaction
- e: Index Register
xr d: Repräsentative x-Koordinate von Mp, Mp 2, Mp 3, Mp 4
- d: Vektor des durch Reflexion erzeugten Punktes
- d: Vektor eines durch Rekombination erzeugten Individuums
XRC e: X-Ray Rocking Curve
XRD e: X-Ray Diffraction
XREF e: Cross Reference [list]
XREG e: Cross-Registration
XRF e: X-Ray Fluorescence [spectrometry]
- e: X-Ray Fluorescence spectroscopy
- e: X-Ray Fluorometry
XRFA e: X-Ray Fluorescence Analysis
XRFS e: X-Ray Fluorescence Spectroscopy
XRIP e: Extended Routing Information Protocol
XRL e: X-Ray Lithography
XRM e: X Resource Manager
- e: X-Ray Microscopy
XRONOS e: Temporal analysis program in XANADU
XRS e: X-Ray Sensor
- e: X-Ray Spectrometer
XRT e: Extensions for Real-Time [market data]
- e: X-Ray Telescope
- e: X-Ray Topography
XS e: Excess
xs d: Schwerpunktvektor aller Polytop-Ecken bis auf die schlechteste
X[-]SAR e: Xband-Synthetic Aperture Radar
XSC e: Extended Scientific Calculation
XSECT e: Cross Section

x-section e: cross-section
xsek i d: i-te Koordinate des Sekantenvektors
XSELECT e: A high-level tool to manage the FTOOLS
xs/i d: absolute spektrale Strahlungsempfindlichkeit nach DIN 5032
XSI e: X/Open System Interface Specification
XSMD e: Extended Storage Module Drive[r interface]
XSMP e: X Session Manager Protocol
XSPEC e: X-ray/gamma-ray spectral analysis package in XANADU
XSS e: Existing Source System
XSSI e: Extended Server Side Includes
XST e: Experimental Stealth Technology
XSTM e: Cross-sectional STM
XSTX e: Transparent Start of Text
XSW e: X-Ray Standing Waves
XSYN e: Transparent Synchronous Idle
XT e: Extended Technology
- e: Quark XPress Xtension
- e: X-Terminal
Xt e: X Toolkit
XTAL e: Crystal
Xtalk e: Cross-talk
XTC e: Extacy
- e: External Transmit Clock
XTCLK e: External Transmit Clock
XTEM e: Cross-sectional Transmission Electron Microscopy
XTI e: X/Open Transport layer Interface
XTLA e: Extended Three-Letter Acronym
XTP e: Express Transfer Protocol
- e: Express Transport Protocol
- e: XPress Transfer Protocol
XUI e: X[-windows] User Interface
XUIDL e: X-Unique-ID-Listing
XUMA d: Expertensystem Umweltgefährlichkeit von Altlasten
XUTIL e: ISP Utility Program
XUV e: Extreme Ultra-Violet
XVE e: X Visual Extensions
XVGA e: Extended VGA
XVI e: XTendbus VMEbus Interface
XVIEW e: X window system based Visual/Integrated Environment for Workstations
XVS e: X/Open System V Specification
XVT e: Executive Vector Table
- e: Extensible Virtual Toolkit
XWD e: X Window Dump
- e: X-Window[s] Dump
X/WIND e: Crosswind
XWS e: X-Window-System
XWSDS e: X Window System Display Station
x1, x2 d: Zwei beliebige nichtnegative reelle Variablen
XXL e: Extra, Extra Large
x1 ... xn d: Zielfunktionsvariablen
XXS e: Extra, Extra Small
XYY e: CIE color space
x, y, z d: Kartesische Koordinaten
x0, y0, z0 d: Kartesische Koordinaten der dA-Fläche
x1, y1, z1 d: Kartesische Koordinaten von dA
XYZ e: CIE color space
- e: Experimental Geometry Zürich

Y

Y a: Venstresocialisterne
- d: Ausgangsgröße nach DIN 5032
- d: Diagonalmatrix
- e: Emergency Command Precedence Symbol
- e: Pyridine
- e: Yield
- f: Yonne

y d: Zufallszahl
y1 d: y-Koordinate einer kleinen Fläche dA 1 in einem karthesischen [xyz] Koordinatensystem
Y-12 e: Oak Ridge Manufacturing and Development Engineering Plant
- e: Oak Ridge Y-12 Plant, Oak Ridge, Tennessee Contents
- e: Y-12 Plant, Oak Ridge Reservation

Y123 e: Yttrium Barium Copper Oxide
Y211 e: Yttrium Barium Copper Oxide
YA e: Yet Another
Ya r: Ytattack-
YA- e: Yet Another
Y.A.A. e: Youth Achievement of America, Inc.
YAA e: Yet another Acronym [Internet]
- e: Yet Another Addition

YAAS e: Yet Another Acronym Server
YABA e: Yet Another Bloody Acronym [Internet]
YACC e: Yet Another Compiler Compiler
YADE e: Yet Another DSSSL Engine
YADS e: Yet Another DOS Shell
YAF N: Yngre Arkitekters Forening
YAFIYGI e: You Asked For It, You Got It [Internet]
YAG e: Yttrium Aluminum Garnet
YAHOO e: Yet Another Hierarchical[ly] Officious Oracle
Yak- R: Yakovlev
YAK-28 e: Firebar
YAK-38 e: Forger
YAKI e: Yakima River Basin
Yakugaku Zasshi j: Yakugaku Zasshi. Journal of the Pharmaceutical Society of Japan
Yale J Biol Med e: Yale Journal of Biology and Medicine
Yale LJ e: Yale Law Journal
YAM e: Yet Another Modem
YANKEE e: Yankee Atomic Electric Co. Massachusetts
YAODL e: Yet Another Object Description Language
Yao Hsueh Hsueh Pao ch: Yao Hsueh Hsueh Pao
YAOTM e: Yet Another Off-Topic Message
YAP e: Youth Action for Peace
YARC e: Yet Another RPN Calculator
YARD e: Youth Associated with the Restoration of Democracy
YARU e: the Yale Arbovirus Research Unit
YASOS e: Yet Another Scheme Object System
YAST e: Yet Another Setup Tool
YATLA e: Yet Another Three Letter Acronym [Internet]
YAUN e: Yet Another Unix Nerd
YAxxx e: Yet Another xxx
YAY e: Yet Another YACC
YB e: Yearbook
YBCO e: Yttrium Barium Copper Oxide
YBD e: Yankee Book Peddler
YBF N: Yrkesskolens Befalsforening
YCA e: Young Concert Artists
YCAGWYW e: You Cannot Always Get What You Want [Internet]
YCC e: Color Space
- e: Lightness, Colour, Chromatics
- e: Youth Conservation
- e: Youth Crisis Center of Jacksonville

YCOR e: Y Corrector Magnet
YCPS e: Yale Center for Parallel Supercomputing
YD e: Youth Dynamics
yd. e: yard
Yd² e: Square yard
yd³ e: cubic yard
YDD e: South Yemeni Dinar
YEA e: Yield Enhancement Analysis
- e: Youth Education in the Arts

Yearb. Eston. Nat. Soc. e: Yearbook of the Estonian Naturalist Society
Yeast e: Yeast
YEI e: Youth for Energy Independence
YELL e: Yellowstone Basin
Yen Ko Hsueh Pao ch: Yen Ko Hsueh Pao
YEP e: Young Eucalypt Program
YER e: Yemeni Riyal
Yerk e: Yerkes Observatory
YES e: Young European Save
YFC e: Yakima Firing Center
- e: Youth For Christ

YGLT e: You're Gonna Love This [Internet]
YH N: Yrkesorganisasjonenes Hovedsammenslutning
YHGASP e: You Have Got A Serious Problem [Internet]
YHR e: Yakima Herald Republic
YHTBT e: You Had To Be There [Internet]
YHTWFWYWTS e: You Have To Wait For What You Want To See [Internet]
Y I A e: Yost Ice Arena
YIG e: Yttrium Iron Garnet
YIN e: Yakama Indian Nation
YIQ e: color components in NTSC color space
- e: US color space

YIS e: Yamaichi Information Systems Co., Ltd.
YIVO e: YIVO Institute for Jewish Research Archives [New York]
YK r: Yrkeskurs
yk r: yrkeskurser
Y2K e: Year 2000 p. Chr. n
YKYARW e: You Know You're A Redneck When [Internet]
YKYBHTLW e: You Know You've Been Hacking To Long When [Internet]
YLF e: Yttrium Lithium Fluoride
YLI N: Yrkesskoleelevenes og Lærlingenes Interesseorganisasjon
Yl/i de: absolute spektrale Ausgangsgröße nach DIN 5032
YM e: Young Men
Ym e: Young & modern
YMA e: Young Muslim Assciation
ymax d: Maximum der y-Koordinaten von Mp 1, Mp 2, Mp 3, Mp 4
YMCA e: Young Mens Christian Association
YMHA e: Young Mens Hebrew Association
ymin d: Minimum der y-Koordinaten von Mp 1, Mp 2, Mp 3, Mp 4
YMMV e: Your Mileage May Vary [Internet]
YMPO e: Yucca Mountain Project Office

YMSCPO e: Yucca Mountain Site Characterization Project Office
YnO e: Yunnan Observatory, P. R. China
YO e: Year[s] Old
Yonsei Med J e: Yonsei Medical Journal
YOP e: Youth Opportunity Program
YOPDA e: Youth Organization for the Promotion of the Doe Administration
YOS e: Yttrium Orthosilicate
YOU e: Youth Organization of Uganda
You f: Yougoslavie
YP e: Yellow Pages
YPBCO e: Yttrium Praseodysium Barium Copper Oxide
YPDF e: Yem Peoples Democratic Front
YPVS e: Yamaha Power Valve System
YQFP e: Yamamoto Quantum Fluctuation Project at Stanford University
YR e: Yield Ramp
Y&R e: Young and Rubicam
yr e: Repräsentative y-Koordinate von Mp, Mp 2, Mp 3, Mp 4
YRA e: Yacht Racing Association
YRBS e: Youth Risk Behavior Survey
YRS e: Ysgarth Rules System
Yrs. e: [in titles] Years
YS N: Yrkesorganisasjones Sentralforbund
YSA e: Youth Service America
YSAG e: Yttrium Scandium Aluminium Garnet
YSC e: Yokohama Science Center
YSHsb N: Yrkesskole for Hærens samband
YSKOR A: Suid-Afrikaanse Yster en Staal-Industriele Korporasie
YSM e: Yourdon Structured Method
YSN e: North Yorkshire
YSO e: Young Stellar Object
YSP e: Young Shakespeare
YSR e: Seaplane Wrecking Derrick
YSS e: South Yorkshire
- e: You Suck Severely
YSW e: West Yorkshire
YSZ e: Yttria-Stabilized Zirconia
YT e: Mayotte
YTB e: Large Harbor Tug
YTC e: Yakama Tribal Council
YTD e: Year-To-Date
YTG e: Yukon Territorial Goverment
YU f: Yougoslavie
YUD e: Yugoslavian New Dinar
YUKON-RRAMS e: Renewable Resource and Management Statistics
YUM e: U.S. Army Proving Ground at Yuma, Arizona, special upper-air observing site
YUMPIE e: Young Upwardly Mobile Professional
Yuppie e: Young Urban Professional
YUV d: Farbmodell: Luminanz Y, Chrominanzwerte U, V
- e: Color components in SECAM and PAL color spaces
- e: European color space
- e: Luminance/chrominance
YV h: Yad Vashem
YVA /YV Ar. H: Yad Vashem Archives [Jerusalem]
YVI N: Yrkesvalghemmedes Industri
YVO e: Yttrium OrthoVanadate
YW e: Young Women
YWCA e: Young Womens Christian Association
YWHA e: Young Womens Hebrew Association
YWIN e: Your Welcome In Advance [Internet]
YWLS e: Young Womens Leadership School
YWSYLS e: You Win Some, You Lose Some [Internet]
YY e: Year
y11 ynn d: Elemente der Diagonalmatrix
x...z, u...w d: Variablen einer beliebigen Funktion

Z

Z a: Fremskridtspartiet
- d: Zeichen
- d: Zeitschrift für Romanische Philologie
- d: Zürcher Bibel
- d: Zusatzstudium
- e: Administrative Aircraft
- e: Atomic number
- e: Collision number
- e: Compressed file upper case Z
- e: Compression factor
- e: Contraction
- e: Figure of Merit
- e: Flash Precedence
- e: Glutamic Acid
- e: Glutamine
- e: Jet Terminal Area
- e: Partition Function
- e: Partition System
- e: Planning
- e: Stadia
- e: Switzerland
- e: Symbol for Magnetic Reluctance
- e: Tower Control
- e: Unit of sleep
- e: Vertical component of the earth's magnetic field
- e: Z-Axis
- e: Zagreb
- e: Zero Rate
- e: Zerubbabel
- e: Zillion
- e: Zircon
- e: Zivot
- e: Zone marker
- e: Zone meridian
- e: Zora
- e: Zulu [time]
- e: Zwingliana
- e: Zycie
- r: Zon
- s: Zona
- R: Zven'ya

z d: Vektor der Zielfunktionskoeffiziente
- e: Aza
- e: charge number of a cell reaction
- e: packed file lower case z
- e: particle
- e: partition function

Z. e: Zone
Z39.47 e: ANSI extended Latin alphabet coded character set for bibliographic use
Z39.50 e: Z39.50 interchange protocol
ZA d: Zeitschrift für Assyriologie und Verwandte Gebiete
- d: Zeitschrift für Assyriologie und Vorderasiatische Archäologie
- d: Zeitschrift für Astrophysik
- d: Zeitschrift für Deutsches Altertum und Deutsche Literatur
- e: Zentralamt
- d: Zentralarchiv für Empirische Sozialforschung
- d: Zentralarchiv zur Erforschung der Geschichte der Juden in Deutschland [Heidelberg]
- d: Zoologischer Anzeiger
- e: Approach Control Office
- e: Zionist Archives
- e: Ziva Antika
- e: Zone of Action
- e: Zunz Archive. Jewish National and University Library
- s: Zona de Acción
- s: Zona de Alarma

Za e: Zabriskies Reports
.za e: Zambia [Internet]
Z-A e: Zaire-Afrique
Z/A s: Zona de Albergue
ZAA e: Alice Arm/Kitsault [Canada] [Airport]
- e: Zartman Association of America
- e: Zeeman-Effect Atomic Absorption

Z Aach[ener] Gesch[ichts]ver d: Zeitschrift. Aachener Geschichtsverein
ZAAK d: Zeitschrift für Ästhetik und Allgemeine Kunstwissenschaft
ZAAT s: Zona defendida por tiro Antiaéreo Terrestre
ZAB f: Zone Autonome de Brazzaville
Zab e: Zabim
ZABIA e: Zastita Bilja
Zab Land Laws e: Zabriskie on the Public Land Laws of the United States
Zab [NJ] e: Zabriskies Reports
ZABO e: Zimbabwe Association of Business Organizations
ZABS e: Zabs Backyard Hots, Inc.
Zaby Przyr Nieozyw P: Zabytki Przyrody Nieozywienej Ziem Rzeczpospolitej Polskiej
ZAC e: Zero Administration Client
- f: Zone d'Aménagement Concerté

Z Acc s: Zona de Acción
ZACFA d: Fresenius Zeitschrift für Analytische Chemie
ZA Ch d: Zeitschrift für Angewandte Chemie
- d: Zeitschrift für Anorganische Chemie
- R: Zurnal Analiticeskoj Chimii

Z Acker [u] Pfl[anzen]b[au] d: Zeitschrift für Acker- und Pflanzenbau
ZACMA e: Zeitschrift für Anorganische Chemie
ZAD d: Zentralstelle für die Ausbildung im Detektivgewerbe
- e: Zadar [Yugoslavia] [Airport]

Zadeco e: Zambia Democratic Congress
ZADI d: Zentralstelle für Agrardokumentation und -Information
Zad Rev se: Zadarska Revija
ZADS d: Zeitschrift. Allgemeiner Deutsche Sprachverein
ZAE s: Zona Agroecológica
ZAED d: Zentralstelle für Atomkernenergie-Dokumentation beim Gmelin-Institut
ZAED Phys Daten d: ZAED [Zentralstelle für Atomkernenergie- Dokumentation] Physik Daten
ZAEKD P: Zeszyty Naukowe. Akademia Ekonomiczna w Krakowie
Za Ekon Mater R: Za Ekonomiyu Materialov
Za Ekon Topi e: Za Ekonomiyu Topliva
Z Aerosol Forsch Ther d: Zeitschrift für Aerosol Forschung und Therapie
ZAFBA d: Zeitschrift für Ärztliche Fortbildung
ZAFICO e: Zanzibar Fisheries Corporation
ZAG e: Zagreb [Yugoslavia] [Airport]
ZÄg d: Zeitschrift für Ägyptische Sprache und Altertumskunde
Zagad[n] Ekon Roln P: Zagadnienia Ekonomiki Rolnej
Zagadn Eksploatacji Masz P: Zagadnienia Eksploalacji Maszyn
Zagadnienia Drgan Nieliniowych P: Zagadnienia Drgan Nieliniowych
Zagadnienie Dyn Rozwoju Czlowieka Zesz Probl Kosmosu P: Zagadnienie Dynamiki Rozwoju Czlowieka Zeszyty Problemowe Kosmosu

Zagad Tech Fal Ultradziek P: Zagadnienia Techniki Fal Ultradzwiekowych
ZAGGD P: Zeszyty Naukowe Akademii Gorniczo-Hutniczej Imienia Stanislawa Staszica. Gornictwo
Z Agrargesch Agrarsoziol d: Zeitschrift für Agrargeschichte und Agrarsoziologie
Z Agrargesch u -Soziol d: Zeitschrift für Agrargeschichte und Agrarsoziologie
Zagreber Studien d: Zagreber Germanistische Studien
ZÄgSpr d: Zeitschrift für Ägyptische Sprache und Altertumskunde
ZAGV d: Zeitschrift. Aachener Geschichtsverein
Z Ägypt Sprache [Altertumskd] d: Zeitschrift für Ägyptische Sprache und Altertumskunde
ZAH e: Zahedan [Iran] [Airport]
ZAHAL h: Z'va Hagana Le'Israel
Zahnärztebl [Baden-Württemb] d: Zahnärzteblatt [Baden-Württemberg]
Zahnärztl Gesundheitsdienst d: Zahnärztlicher Gesundheitsdienst
Zahnärztl Mitt d: Zahnärztliche Mitteilungen
Zahnärztl Prax d: Zahnärztliche Praxis
Zahnärztl Praxisführ d: Zahnärztliche Praxisführung
Zahnärztl Rundsch d: Zahnärztliche Rundschau
Zahnärztl Welt d: Zahnärztliche Welt
Zahnärztl Welt Zahnärztl Reform d: Zahnärztliche Welt und Zahnärztliche Reform
Zahnärztl Welt Zahnärztl Rundsch d: Zahnärztliche Welt, Zahnärztliche Rundschau
Zahnerh Kd d: Zahnerhaltungskunde
Zahn Inf Die d: Zahnärztlicher Informationsdienst
Zahn-Mund-Kieferheilkd d: Zahn-, Mund-, und Kieferheilkunde
Zahn- Mund- Kieferheilkd Zentralbl d: Zahn-, Mund-, und Kieferheilkunde mit Zentralblatt
Zahn Prax d: Zahnärztliche Praxis
Zahn Rd d: Zahnärztliche Rundschau
Zahntechn d: Zahntechnik[er]
Zahntechn Nachr e: Zahntechnische Nachrichten
ZAI e: Zero Administration Initiative
- h: Zeirei Agudath Israel
ZAICEL d: Zeitschrift für Angewandte Ichthyologie
Zaire-Afr e: Zaire-Afrique
ZAIRETAIN e: Zairian Tin Company
ZAIROM f: Société Zootechnique Zairo-Roumaine
ZAK d: Zeitschrift für Ästhetik und Kunstwissenschaft
- d: Zeitschrift für Schweizerische Archäologie und Kunstgeschichte
- e: Zero Administration Kit
Zakhist Rosl U: Zakhist Roslin
Za Khlopk Nezavisimost R: Za Khlopkovuyu Nezavisimost
Zakhyst Rosl Resp Mizhvid Temat Nauk Zb U: Zakhyst Roslyn Respublikans 'Kyi Mizhvidomchyi Tematychnyi Naukovyi Zbirnyk
Zakonomern Raspred Promesnykh Tsentrov Ionnykh Krist R: Zakonomernosti Raspredeleniya Promesnykh Tsentrov v Ionnykh Kristallakh
Zakonomern Razmeshcheniya Polezn Iskop R: Zakonomernosti Razmeshcheniya Poleznykh Iskopaemykh
Zakupki Sel'skokhoz Prod R: Zakupki Sel'skokhozyaistvennykh Produktov
ZAL d: Zeitschrift für Arabische Linguistik
- d: Zeitschrift für Ausländische Landwirtschaft
- e: Rand
- e: State University of New York, Albany Library School, Albany
- e: Zionist Archives and Library
ZALF d: Zentrum für Agrarlandschafts- und Landnutzungsforschung e.V., Müncheberg
Z Allgemeine Wissenschaftstheorie d: Zeitschrift für Allgemeine Wissenschaftstheorie
Z Allgemeinmed d: Zeitschrift für Allgemeinmedizin
Z Allg Med d: Zeitschrift für Allgemeinmedizin der Landärzt
Z Allg Mikr d: Zeitschrift für Allgemeine Mikrobiologie
Z Allg Mikrobiol d: Zeitschrift für Allgemeine Mikrobiologie. Morphologie, Physiologie, Genetik, und Ökologie der Mikrorganismen
Z Allg Österr Apoth Ver d: Zeitschrift. Allgemeiner Österreichische Apotheker-Verein
Z Allg Physiol d: Zeitschrift für Allgemeine Physiologie
Z Allg Wiss[enschaftstheor] d: Zeitschrift für Allgemeine Wissenschaftstheorie
Z Alternsforsch d: Zeitschrift für Alternsforschung
ZaltorientbiblRg d: Zeitschrift für altorientalische und biblische Rechtsgeschichte
Z Alt W[iss] d: Zeitschrift für die Alttestamentliche Wissenschaft
ZAM d: Zeitschrift für Askese und Mystik
- e: State University of New York, Agricultural and Technical College at Alfred
- e: Zamboanga [Philippines] [Airport]
Zambia Dep Game Fish Fish Res Bull e: Zambia. Department of Game and Fisheries. Fisheries Research Bulletin
Zambia Dep Wildl Fish Natl Parks Annu Rep e: Zambia. Department of Wildlife, Fisheries and National Parks. Annual Report
Zambia Div For Res Annu Rep e: Zambia. Division of Forest Research. Annual Report
Zambia Div For Res Res Pam e: Zambia. Division of Forest Research. Research Pamphlet
Zambia For Res Bull e: Zambia Forest Research Bulletin
Zambia Geogr Assoc Mag e: Zambia Geographical Association. Magazine
Zambia Geol Surv Annu Rep e: Zambia. Geological Survey. Annual Report
Zambia Geol Surv Dep Annu Rep e: Zambia. Geological Survey. Department Annual Report
Zambia Geol Surv Dep Econ Rep e: Zambia. Ministry of Lands and Mines. Geological Survey Department. Economic Report
Zambia Geol Surv Econ Rep e: Zambia. Geological Survey. Economic Report
Zambia Geol Surv Rec e: Zambia. Geological Survey. Records
Zambia Geol Surv Tech Rep e: Zambia. Geological Survey. Technical Report
Zambia J Sci Technol e: Zambia Journal of Science and Technology
Zambia LJ e: Zambia Law Journal
Zambia Minist Lands Nat Resour For Res Bull e: Zambia. Ministry of Lands and Natural Resources. Forest Research Bulletin
Zambia Minist Rural Dev For Res Bull e: Zambia. Ministry of Rural Development. Forest Research Bulletin
Zambia Nurse J e: Zambia Nurse Journal
Zambia Rep Geol Surv e: Zambia. Ministry of Lands and Mines. Report of the Geological Survey
ZA Mech d: Zeitschrift für Angewandte Mechanik
Zam LJ e: Zambia Law Journal
ZAMM d: Zeitschrift für Angewandte Mathematik und Mechanik
- e: Zen and the Art of Motorcycle Maintenance
ZAMNA d: ZFA [Zeitschrift für Allgemeinmedizin]
ZAMP d: Zeitschrift für Angewandte Mathematik und Physik
ZAN e: Anchorage, AK
- e: Zantop International Airlines, Inc.
Z Anal Chem d: Fresenius Zeitschrift für Analytische Chemie
Z Anal Entwicklungsgesch d: Zeitschrift für Anatomie und Entwicklungsgeschichte
Z Analit Chim R: Zurnal Analiticeskoj Chimii
Z Analyt Chem d: Zeitschrift für Analytische Chemie
Zan AT s: Zanja Antitanque
ZANC e: Zambia National Congress - Southern Rhodesia
ZANCA d: Zeitschrift für Analytische Chemie
ZAnCh d: Zeitschrift für Anorganische Chemie
Z An Chim R: Zurnal Analiticeskoj Chimii
ZANCO Ser A e: ZANCO. Scientific Journal of Sulaimaniyah University. Series A. Pure and Applied Sciences
ZANCO Ser A Pure Appl Sei e: ZANCO. Series A. Pure and Applied Sciences
Zane e: Zanes Reports
ZANEWS e: Zanzibar News Service

ZANF d: Zeitschrift für Assyriologie und Vorderasiatische Altertumskunde. Neue Folge
Z Ang & Amerik d: Zeitschrift für Anglistik und Amerikanistik
Z Angew Bäder Klimaheilkd d: Zeitschrift für Angewandte Bäder und Klimaheilkunde
Z Angew Chem d: Zeitschrift für Angewandte Chemie und Zentralblatt für Technische Chemie
Z Angew Entomol d: Zeitschrift für Angewandte Entomologie
Z. Angew. Entomol. d: Zeitschrift für Angewandte Entomologie
Z Angew Geol d: Zeitschrift für Angewandte Geologie
Z Angew Ichthyol d: Zeitschrift für Angewandte Ichthyologie
Z Angew Math Mech d: Zeitschrift für Angewandte Mathematik und Mechanik
Z Angew Math Phys d: Zeitschrift für Angewandte Mathematik und Physik
Z Angew Math und Mech d: Zeitschrift für Angewandte Mathematik und Mechanik
Z Angew Met d: Zeitschrift für Angewandte Meteorologie
Z Angew Mikrosk Klin Chem d: Zeitschrift für Angewandte Mikroskopie und Klinische Chemie
Z Angew Photogr Wiss Tech d: Zeitschrift für Angewandte Photographie in Wissenschaft und Technik
Z Angew Phys d: Zeitschrift für Angewandte Physik
Z Angew Psychol d: Zeitschrift für Angewandte Psychologie und Psychologische Forschung
Z Angew Zool d: Zeitschrift für Angewandte Zoologie
Z Ang Geol d: Zeitschrift für Angewandte Geologie
Z Anglis Am d: Zeitschrift für Anglistik und Amerikanistik
Z Ang Ma Me d: Zeitschrift für Angewandte Mathematik und Mechanik
Z Ang Math d: Zeitschrift für Angewandte Mathematik und Physik
Z Ang Math Mech d: Zeitschrift für Angewandte Mathematik und Mechanik
Z Ang Math Phys d: Zeitschrift für Angewandte Mathematik und Physik
Z Ang Ph d: Zeitschrift für Angewandte Physik
Z Anorg A C d: Zeitschrift für Anorganische und Allgemeine Chemie
Z Anorg Allg Chem d: Zeitschrift für Anorganische und Allgemeine Chemie
Z Anorg Chem d: Zeitschrift für Anorganische und Allgemeine Chemie
ZANPA d: Zeitschrift für Experimentelle und Angewandte Psychologie
ZAnt e: Ziva Antika
Z Antimikrob Antineoplast Chemother d: Zeitschrift für Antimikrobielle und Antineoplastiche Chemotherapie
ZANU e: Zimbabwe African National Union Patriotic Front
ZANZ e: Zanzibar
Zanzibar Protect Ann Rep Med Dept e: Zanzibar Protectorate. Annual Report on the Medical Department
Zanzib Prot LR e: Zanzibar Protectorate Law Reports
Z Ao Ch d: Zeitschrift für Anorganische Chemie
ZA Ö R[V] d: Zeitschrift für Ausländisches Öffentliches Recht und Völkerrecht
Za Ovladenie Tekh Kamenougol'n Promsti R: Za Ovladenie Tekhniknoj v Kamenougol'noj Promyshlennosti
ZAP e: Helionetics, Inc.
- e: Zentralapotheke
- e: Zero Ability to Pay
- e: Znamenity Amerikansky Pisatel
Zap s: Zapador
Zapadne Karpaty Ser Geol P: Zapadne Karpaty. Seria Geologia
Zapadn Karpaty Ser Paleontol P: Zapadne Karpaty. Seria Paleontologia
Zap Arm Otd Vses Mineral Ova R: Zapiski Armyanskogo Otdeleniya Vsesoyuznogo Mineralogicheskogo Obshchesiva
Zap Beloruss Gos Inst Sel'sk Lesn Khoz R: Zapiski Belorusskogo Gosudarstvennogo Instituta Sel'skogo i Lesnogo Khozyaistva
Zap Cukotsk Kraeved Muz R: Zapiski Cukotskogo Kraevedceskogo Muzeja
ZAPE e: Zonal Available Potential Energy
ZaPF d: Zusammenkunft aller Physik-Fachschaften
Zap GO R: Zapiski Geograficeskogo Obscestva
ZAPI e: Integrated Priority Action Zones
- f: Zones d'Action Prioritaire Intégrée
Zap Inst Jaz Lit Ist R: Zapiski Gosudarstvennogo Instituta Jazyka, Literatury, i Istorii
Zap Inst Khim Akad Nauk Ukr RSR R: Zapiski Institutu Khimii Akademiya Nauk Ukrains'knoj RSR
Zapisnici Srp Geol Drus se: Zapisnici Srpskog Geoloskog Drustva
Zap Kalm Nauc Issi Inst Jaz Lit Ist R: Zapiski. Kalmyckij Naucno- Issledovatel'skij Institut Jazyka, Literatury, i Istorii
Zap Khar'k S-Kh Inst R: Zapiski Khar'kovskogo Sel'skokhozyaistvennego Instituta
Zap Kiiv Tov Prirodozn R: Zapiski Kiivs'koge Tovaristva Prirodoznavtsiv
Zap Kirg Otd Vses Mineral Ova R: Zapiski Kirgizskogo Otdeleniya Vsesoyuznogo Mineralogicheskogo Obshchestva
Zap KORGO R: Zapiski Kavkazskogo Otdela Russkogo Geograficeskogo Obscesiva
Zap Leningrad Sel'skokhoz Inst R: Zapiski Leningradskogo Sel'skokhozyaistvennogo Instituta
Zap Leningr Gorn Inst R: Zapiski Leningradskogo Gornogo Instituta
Zap Leningr Sel'-Khoz Inst R: Zapiski Leningradskogo Sel'skokhozyaistvennogo Instituta
Zap Leningr S-Kh Inst R: Zapiski Leningradskogo Sel'skokhozyaistvennogo Instituta
Zap Nauchn Semin R: Zapiski Nauchnykh Seminarov
Zap Nauchn Semin Leningr Otd Mat Inst Akad Nauk SSSR R: Zapiski Nauchnykh Seminarov Leningradskoe Otdelenie Matematicheskii Institut Akademia Nauk SSSR
Zap Nauchn Sem Leningrad Otdel Mat Inst Steklov [LOMI] R: Zapiski Nauchnykh Seminarov Leningradskogo Otdeleniya Matematicheskogo Instituta Imeni V. A. Steklova Akademii Nauk SSSR
Zap Naucn Sem Leningrad Otdel Mat Inst Steklov R: Zapiski Naucnyh Seminarov Leningradskogo Otdelenija Matematiceskogo Instituta Imeni V. A. Steklova Akademii Nauk SSSR
ZAPO e: Zimbabwe African Peoples Organization
Zap Odess Ark Obshch R: Zapiski Odesskoe Arkheologicheskoe Obshchestvo
ZAPPA d: Zentralblatt für Allgemeine Pathologie und Pathologische Anatomie
Za Prog Proizvod R: Za Progress Proizvodstva
Zap Ross Mineral Ova R: Zapiski Rossiiskogo Mineralogicheskogo Obshchestva
ZAPS e: Cooper Life Sciences, Inc.
Zap SKK Gor NII R: Zapiski Severo-Kavkazskogo Kraevogo Gorskogo Naucno-Issledovatel'skogo Instituta
Zap Sverdl Otd Vses Bot Ova R: Zapiski Sverdlovskogo Otdeleniya Vsesoyuznogo Botanicheskogo Obshchestva
Zap Sverdlov Otd Vsesoyuz Bot Obshch R: Zapiski Sverdlovskogo Otdeleniya Vsesoyuznogo Botanicheskogo Obshchestva
Zap Tadzh Otd Vses Mineral Ova R: Zapiski Tadzhikskogo Otdeleniya Vsesoyuznogo Mineralogicheskogo Obshchestva
Zap Tsentr Kavk Otd Vses Bot Ova R: Zapiski Tsentral'no-Kavkazskogo Otdeleniya Vsesoyuznogo Botanicheskogo Obshchestva
Zap Ukr Otd Vses Mineral Ova R: Zapiski Ukrainskogo Otdeleniya Vsesoyuznogo Mineralogicheskogo Obshchestva
ZAPU-PF e: Zimbabwe African People's Union-Patriotic Front
Zap Uzb Otd Vses Mineral Ova R: Zapiski Uzbekistanskogo Otdeleniya Vsesoyuznogo Mineralogicheskogo Obshchestva
Zap Voronezh Sei'-Khoz Inst R: Zapiski Voronezhskogo Sel'skokhozyaistvennogo Instituta
Zap Voronezh S-Kh Inst R: Zapiski Voronezhskogo Sel'skokhozyaistvennogo Instituta
Zap Vost Sib Otd Vses Mineral Ova R: Zapiski Vostochno-Sibirskogo Otdeleniya Vsesoyuznogo Mineralogicheskogo Obshchestva

Zap Vses Mineral Obshchest R: Zapiski Vsesoyuznogo Mineralogicheskogo Obshchestva
Zap Vses Mineral O-Va R: Zapiski Vsesoyuznogo Mineralogicheskogo Obshchestva
Zap Zabaik Fil Geogr Ova SSSR R: Zapiski Zabaikal'skogo Filiala Geograficheskogo Obshchestva SSSR
Zap Zabaik Otd Vses Geogr O-Va R: Zapiski Zabaikal'skogo Otdela Vsesoyuznogo Geograficheskogo Obshchestva
ZAR A: Zuid-Afrikaanse Republiek
- e: Rand
- e: Zeus Acquisition Radar
Z Arbeitsgem Österr Entomol d: Zeitschrift. Arbeitsgemeinschaft Österreichischer Entomologen
Z Arbeitswiss N F d: Zeitschrift für Arbeitswissenschaft. Neue Folge
Z Archäol d: Zeitschrift für Archäologie
Za Rekonstr Tekst Promsti R: Za Rekonstruktsiyu Tekstil'noj Promyshlennosti
ZARSA P: Zeszyty Naukowe. Akademia Rolnicza w Szczecinie
ZarSl e: Zaranie Slaskie
Z Ärztl Fortbild d: Zeitschrift für Ärztliche Fortbildung
Z Ärztl Fortbild [Jena] d: Zeitschrift für Ärztliche Fortbildung [Jena]
ZAS d: Zeitung am Sonntag, Freiburg
- d: Zeitschrift für Ägyptische Sprache und Altertumskunde
- e: Zymosan-Activated Serum
ZÄS[A] d: Zeitschrift für Ägyptische Sprache und Altertumskunde
Z Äs Allg Kunst d: Zeitschrift für Ästhetik und Allgemeine Kunstwissenschaft
Zashch Korroz Khim Promsti R: Zashchita ot Korrozii v Khimichesknoj Promyshlennosti
Zashch Met R: Zashchita Metallov
Zashch Pokrytiya Met R: Zashchitnye Pokrytiya na Metallakh
Zashch Rast [Kiev] R: Zashchita Rastenii [Kiev]
Zashch Rast [Leningrad] R: Zashchita Rastenii [Leningrad]
Zashch Rast [Mosc[ow]] R: Zashchita Rastenii [Moscow]
Zashch Rast Vred Bolez R: Zashchita Rastenii ot Vreditelei i Boleznei
Zashch Rast Vred Bolezn R: Zashchita Rastenii ot Vreditelei i Boleznei
Zashch Rast Vredit Bolez R: Zashchita Rastenii ot Vreditelei i Boleznei
Zashch Truboprovodov Korroz R: Zashchita Truboprovodov ot Korrozii
Z Asiat Studien d: Zentralasiatische Studien
Za Soc Zemed R: Za Socialisticke Zemedelsivi
Za Sots Sel'-Khoz Nauku R: Za Sotsialisticheskuyu Sel'skokhozyaistvennuyu Nauku
Za Sots Sel'skokhoz Nauku Ser A e: Za Sotsialisticheskuyu Sel'skokhozyaistvennuyu Nauku. Seriya A
Z Assyr d: Zeitschrift für Assyriologie
ZAST d: Zeitschrift für Afrikastudien
Zast Bilja b: Zastita Bilia
Z Asthet Al d: Zeitschrift für Ästhetik und Allgemeine Kunstwissenschaft
ZastMat U: Zastosowania Matematyki
Zast Mater b: Zastita Materijala
Zastos Mat P: Polska Akademia Nauk. Instytut Matematyczny. Zastosowania Matematyki
Zastosow Mat U: Zastosowania Matematyki
Z Astrophys d: Zeitschrift für Astrophysik
ZAT ch: Zhaotong [China] [Airport]
- e: Zantop Airways, Inc.
- P: Zydowska Agencja Telegraficzna
Z Ata s: Zona de Ataque
ZATB d: Zeitschrift für die Alttestamentliche Wissenschaft. Beihefte
Za Tekh Prog [Baku] R: Za Tekhnicheskii Progress [Baku]
Za Tekh Prog [Gorky] R: Za Tekhnicheskii Progress [Gorky]
ZATPA R: Za Tekhnicheskii Progress
Za Turf Ind R: Za Turfyanuyu Industriyu
Z Augenheilkd d: Zeitschrift für Augenheilkunde
Z Ausländ Landwirt[sch] d: Zeitschrift für Ausländische Landwirtschaft
Z Ausländ Öff Völkerrecht d: Zeitschrift für Ausländisches Öffentliches Recht und Völkerrecht
Z Ausl Öff R d: Zeitschrift für Ausländisches Öffentliches Recht und Völkerrecht
Z Ausl Öff Recht Völkerrecht d: Zeitschrift für Ausländisches Öffentliches Recht und Völkerrecht
Z Ausl Öff RVR d: Zeitschrift für Ausländisches Öffentliches Recht und Völkerrecht
Zav e: Zavim
ZAVA d: Zeitschrift für Assyriologie und Vorderasiatische Archäologie
Zav[od] Lab R: Zavodskaya Laboratoriya
ZAW d: Zeitschrift für die Alttestamentliche Wissenschaft
- e: Zero Administration [initiative] for Windows
- e: Zirflex Acid Waste
ZAWEA e: Zahnärztliche Welt
ZAWU e: Zimbabwe African Womens Union
ZAZ s: Zaragoza [Spain] [Airport]
ZB e: Repetitive Flight Plan Office
- d: Zeitschrift für Betriebswirtschaft
- d: Zeitschrift für Botanik
- d: Zentralblatt für Bibliothekswesen
- e: Zinkblende
- d: Zürcher Bibel
- e: Zero Beat [Radio]
- e: Zimbabwe Banking Corporation Ltd.
- e: Zinc Borate
- e: Zone Boundary
zB d: zum Beispiel
ZBA d: Zeitschrift für Arbeitswissenschaft
- e: Zero Bracket Amount
- e: Zero-Based Analysis
- e: Zoning Board of Approval
Z Bayer Kg d: Zeitschrift für Bayerische Kirchengeschichte
Z Bayer Ldg d: Zeitschrift für Bayerische Landesgeschichte
Z Bayer Revisions Ver d: Zeitschrift. Bayerischer Revisions Verein
Z Bay Land Gesch d: Zeitschrift für Bayerische Landesgeschichte
ZBB e: Zero Base[d] Budget[ing]
Zb Bioteh Fak Univ Edvarda Kardelja Ljublj Kmetijstvo S: Zbornik Biotehniske Fakultete Univerze Edvarda Kardelja v Ljubljani. Kmetijstvo
Zb Bioteh Fak Univ Ljubljani S: Zbornik Biotehniske Fakultete Univerze v Ljubljani
Zb Bioteh Fak Univ Ljublj Kmetijstvo S: Zbornik Biotehniske Fakultete Univerze v Ljubljani. Kmetijstvo
Zb Bioteh Fak Univ Ljublj Vet S: Zbornik Biotehniske Fakultete Univerze v Ljubljani. Veterinarstvo
Zb Bioteh Fak Univ Ljublj Vet Supl S: Zbornik Biotehniske Fakultete Univerze v Ljubljani. Veterinarstvo. Suplement
Zb Biotehn Fak Univ Ljublj S: Zbornik Biotehniske Fakultete Univerze v Ljubljani
ZBBW d: Zentralblatt für Bibliothekswesen
ZBC e: Zebec Resources
- e: Zimbabwe Broadcasting Corporation
ZBCSA d: Zentralblatt für Chirurgie. Supplement
ZBDC e: Zinc Dibutyldithiocarbamate
ZBDLG d: Zürcher Beiträge zur Deutschen Literatur und Geistesgeschichte
ZBDS S: Zveza Bibliotekarskih Drutev Slovenije
ZBDSS d: Zürcher Beiträge zur Deutschen Sprach- und Stilgeschichte
Z Beih d: Zeitschrift für Romanische Philologie. Beihefte
Z Beleuchtungswes Heizungs- Lüftungstech d: Zeitschrift für Beleuchtungswesen Heizungs- und Lüftungstechnik
Z Berg Gesch V d: Zeitschrift. Bergischer Geschichtsverein
Z Berg Hütten Salinenwes Dtsch Reich d: Zeitschrift für das Berg-, Hütten- , und Salinenwesen im Deutschen Reich
Z Bergrecht d: Zeitschrift für Bergrecht
Z Betriebsw d: Zeitschrift für Betriebswirtschaft
Z Betriebswirtsch d: Zeitschrift für Betriebswirtschaft
Z Bevölkerungswiss d: Zeitschrift für die Bevölkerungswissenschaft
Z Bewasserungswirtsch d: Zeitschrift für Bewasserungswirtschaft
ZBF d: Zeitschrift für Betriebswirtschaftliche Forschung
- d: Zeitschrift für Bücherfreunde
Zb f Bibl d: Zentralblatt für Bibliothekswesen

ZBFL P: Zbornik za Filologiju i Lingvistiku
ZBG d: Zeitschrift. Bergischer Geschichtsverein
Zb Geol Vied Zapadne Karpaty P: Zbornik Geologichych Vied Zapadne Karpaty
ZBGR e: Schweizerische Zeitschrift für Beurkundungs- und Grundbuchrecht
ZBGV d: Zeitschrift, Bergischer Geschichtsverein
Z Bibliothekswes[en und] Bibl[iogr] d: Zeitschrift für Bibliothekswesen und Bibliographie
Z Bibl[iot] u[nd] Bibliog d: Zeitschrift für Bibliothekswesen und Bibliographie
Z Bienenforsch d: Zeitschrift für Bienenforschung
Z Binnenfisch DDR d: Zeitschrift für die Binnenfischerei der DDR
Zb Inst Khim Tekhnol Akad Nauk Ukr RSR U: Zbornik Institutu Khimichnoj Tekhnologii Akademiya Nauk Ukrains'knoj RSR
Z Biochem d: Zeitschrift für Biochemie
Z Biol d: Zeitschrift für Biologie
Z Biol Tech Method d: Zeitschrift für Biologische Technik und Methodik
Zbirka Izbran Poglav Fiz c: Zbirka Izbranih Poglavij iz Fizike
Zbirka Izbran Poglav Mat c: Zbirka Izbranih Poglavij iz Matematike
ZbirP U: Zbirnyk Prac' Naukovoji Sevcenkivs'koji Konferenciji
ZBJV d: Zeitschrift. Bernischer Juristen-Verein
ZBKG d: Zeitschrift für Bayerische Kirchengeschichte
ZBL d: Deutsche Zentralbibliothek für Landbauwissenschaften, Bonn
- d: Zeitschrift für Bayerische Landesgeschichte
- e: Brooklyn Law School
Zbl d: Zentralblatt für Mathematik und Ihre Grenzgebiete
Zbl Allg Path d: Zentralblatt für Allgemeine Pathologie und Pathologische Anatomie
Zbl A Med d: Zentralblatt für Arbeitsmedizin und Arbeitsschutz
ZBLAN e: Zirconium, Barium, Lanthanum, Aluminum, Sodium Fluoride
Zbl Bakt A d: Zentralblatt für Bakteriologie. Reihe A
Zbl Bakt B d: Zentralblatt für Bakteriologie. Reihe B
Zbl DDR d: Zentralblatt der Deutschen Demokratischen Republik
Zb Lek Fak [Kosice] c: Zbornik Lekarskej Fakulty [Kosice]
Zbl f Bibl d: Zentralblatt für Bibliothekswesen
ZBLG d: Zeitschrift für Bayerische Landesgeschichte
Zbl Math d: Zentralblatt für Mathematik und Ihre Grenzgebiete
Zbl Soz Vers d: Zentralblatt für Sozialversicherung und Versorgung
Zbl Vet A d: Zentralblatt für Veterinärmedizin. Reihe A
Zbl Vet B d: Zentralblatt für Veterinärmedizin. Reihe B
ZBM d: Zentralbibliothek der Medizin
- e: State University of New York, College at Oneonta
Zb Meteorol Hidrol Rad U: Zbornik Meteoroloskih i Hidroloskih Radova
ZBMM e: Zenana Bible and Medical Mission
ZBMP e: Zero-Base Media Planning
ZBMS c: Zbornik Matice Srpske
ZBN e: Brookhaven National Laboratory, Upton
Zb Nauk Pr Aspir Kiiv Inzh Budiv Inst U: Zbirnik Naukovikh Prats Aspirantiv Kiivs'kii Inzhenerno-Budivel'nii Institut
Zb Nauk Pr Aspir Kyyiv Univ Pryr Nauky U: Zbirnyk Naukovykh Prats' Aspirantiv Kyyivski Universytet Pryrodni Nauky
Zb Nauk Pr Bilotserk Dos Sel Statsiya U: Zbirnyk Naukovykh Prats' Bilotserkiv'sta Doslidno-Selektvionna Statsiya
Zb Nauk Pr Khim Sil'sk Hospod Ukr Sil'skohosped Akad U: Zbirnyk Naukovykh Prats' Khimicheskoho Sil'skoho Hospodarstva Ukrayinskoyi Sil'skohospodarskoyi Akademiyi
Zb Nauk Pr Kiiv Budiv Inst U: Zbirnik Naukovikh Prats Kiivs'kii Budivel'nii Institut
Zb Nauk Pr L'viv Med Inst U: Zbirnyk Naukovykh Prats'L'viv'kyi Medychyni Instytut
Zb Nauk Pr Umans'kyi Sil'skohospod Inst U: Zbirnyk Naukovykh Prats' Umans'kyi Sil'skohospodarskyi Instytut
Zb Nauk Rob Khark Derzh Med Inst U: Zbirnik Naukovikh Robit Kharkivs'-kogo Derzhavnogo Medichnogo Institutu
ZbNPAF U: Zbirnyk Naukovych Prac' Aspirantiv z Filolohiji
ZBO e: Bowen [Australia] [Airport]
Zbor Arheol Muz se: Zbornik na Arheoloskiot Muzej
Zbor Muz Primenjene Umet se: Zbornik Muzej Primenjene Umetnosti
Zbor Narod Muz Beogradu se: Zbornik Narodnog Muzeja u Beogradu
Zborn Biotehn Fak Univ Ljublj Kmet S: Zbornik Biotehniske Fakultete Univerze v Ljubljani. Kmetijstvo
Zbornik Rad Mat Inst [Beograd] se: Zbornik Radova. Matematicki Institut [Beograd]
Zborn Rad se: Zbornik Radova
Zborn Rad Poljopriv Fak Univ Beogr se: Zbornik Radova. Poljoprivrednog Fakulteta. Universitet u Beogradu
Zborn Slov Nar Muz Prir Vedy S: Zbornik Slovenskeho Narodneho Muzea Prirodne Vedy
Zbor Slov Narod Muz S: Zbornik Slovenskeho Narodneho Muzea
Z Bot d: Zeitschrift für Botanik
ZBPHA d: Zentralblatt für Bakteriologie, Parasitenkunde, Infektionskrankheiten, und Hygiene. Abteilung 1. Medizinisch-Hygienische Bakteriologie, Virusforschung
Zb Prac Chem Fak SVST [Bratislava] c: Zbornik Prac Chemickotechnologickej Fakulty SVST [Bratislava]
Zb Prav Fak Zagrebu S: Zbornik Pravnog Fakulteta u Zagrebu
Zb Pr Belarus Dzyarzh Med Inst R: Zbornik Prats. Belaruski Dzyarzhauny Medychny Instytut
Zb Pr Chem-Technol Fak SVST c: Zbornik Prac Chemickotechnologickej Fakulty SVST
Zb Pr Inst Teploenerg Akad Nauk Ukr RSR U: Zbirnik Prats' Institut Teploenergetiki Akademiya Nauk Ukrains'knoj RSR
Zb Prir Nauke Matica Srp c: Zbornik za Prirodne Nauke Matica Srpska
Zb Pr Nauk Inst Fiziol Kyyiv Univ U: Zbirnyk Prats' Naukovodoslidnyts'koho Instytuta Fiziolohiyi Kyyivs'koho Universytetu
Zb Pr Naukovodosl Inst Fiziol Kyyiv Univ U: Zbirnyk Prats' Naukovodoslidnyts'koho Instytuta Fiziolohiyi Kyyivs'koho Universytetu
Zb Pr Ukr Derzh Inst Nauk Prakt Vet U: Zbirnik Prats'Ukrainskii Derzhavnii Institut Naukovoi ta Praktichnoj Veterinarii
Zb Pr Ukr Inst Eksp Vet U: Zbirnik Prats' Ukrains'kogo Institutu Eksperimental'noj Veterinarii
Zb Pr Zool Muz Akad Nauk Ukr RSR U: Zbirnyk Prats' Zoolohichnoho Muzeyu Akademiyi Nauk Ukrayinskoyi RSR
ZBR e: Chah-Bahar [Iran] [Airport]
ZbR U: Zbirnyk Robit Aspirantiv Romano-Germans'koji i Klazycnoji Filolohiji
ZBR e: Zero Bend Radius
- e: Zero Bug Release
- e: Zone Bit Recording
ZBRA e: Zebra Technologies
Zb Rab Belarus Sel'Ska-Gaspad Inst R: Zbornik Rabot Belaruskaga Sel'ska-Gaspadarchaga Instytuta
Zb Rad Biol Inst [Beograd] se: Zbornik Radova. Bioloski Institut [Beograd]
Zb Rad Biol Inst NR Srbye Beogr se: Zbornik Radova. Bioloski Institut NR Srbye Beograd
Zb Rad Math Inst Beograd NS se: Beograd Matematicki Institut. Zbornik Radova. Nouvelle Serie
Zb Rad Mat Inst [Beograd] se: Zbornik Radova. Nova Serija. Matematicki Institut [Beograd]. Zbornik Radova. Nova Serija
Zb Rad Poljopr Fak Univ Beogradu se: Zbornik Radova. Poljoprivrednog Fakulteta. Universitet u Beogradu
Zb Rad Poljopr Inst S: Zbornik Radova. Poljoprivredni Institut
Zb Rad Poljopr Inst [Osijek] S: Zbornik Radova. Poljoprivredni Institut [Osijek]

Zb Rad Poljopriv Fak Univ Beogradu se: Zbornik Radova. Poljoprivrednog Fakulteta. Universitet u Beogradu
Zb Rad Prir Mat Fak se: Zbornik Radova. Prirodno-Matematickog Fakulteta
Zb Rad Prir-Mat Fak Ser Fiz se: Zbornik Radova. Prirodno-Matematickog Fakulteta. Serija za Fiziku
Zb Rad Prir-Mat Fak Univ Novom Sadu se: Zbornik Radova. Prirodno-Matematickog Fakulteta Univerzitet u Novom Sadu
Zb Rad Rud Geol Fak se: Zbornik Radova. Rudarsko-Geoloskog Fakulteta
Zb Rad Srp Akad Nauka Geol Inst se: Zbornik Radova. Srpska Akademija Nauka Geoloski Institut
Zb Rad Zavod Ratarstvo [Sarajevo] se: Zbornik Radova. Zavod za Ratarstvo [Sarajevo]
Zbraslav Res Inst Land Reclam Improv Sei Monogr e: Zbraslav Research Institute for Land Reclamation and Improvement. Scientific Monograph
ZBRFFZ S: Zbornik Radova. Filozofskog Fakulteta. Svencilista u Zagrebu
ZBRL U: Zbirnyk Robit Aspirantiv L'Vivskij Derzavnyj Universitet
Zb Robit Aspir L'Viv Univ Pryr Nauk U: Zbirnyk Robit Aspirantiv L'Vivs'kyi Universytet Pryrodnykh Nauk
ZBRS e: Zs Briefs. CPSU [Cooperativc Park Studies Unit, University of Alaska] Newsletter
ZBS d: Zentrum für Bild- und Signalverarbeitung e.V., Ilmenau
- e: Zambia Broadcasting Service
ZbS se: Zbornik za Slavistiku
ZBS d: Zeitschrift. Deutscher Verein für Buchwesen und Schrifttum
- e: Zivena Beneficial Society
ZBSAN se: Zbornik Radova. Srpske Akademije Nauke
ZBSB d: Zeitschriftenkatalog der Bayerischen Staatsbibliothek, München
Zb Slov Nar Muz[ea] Prir Vedy c: Zbornik Slovenskeho Narodneho Muzea Prirodne Vedy
ZBST e: ZZZZ Best Co., Inc.
ZBT e: Zeta Beta Tau
- h: Zion Bemishpat Tipadeh
ZBTSI e: Zero Byte Time Slot Interchange
ZBV d: zur besonderen Verwendung
Zb Ved Prac Lesn Fak Vys Sk Lesn Drev Zvolene c: Zbornik Vedeckych Prac Lesnickej Fakulty Vysokej Skoly Lesnickej a Drevarskej vo Zvolene
Zb Ved Pr Vys Sk Tech Kosiciach c: Zbor-nik Vedeckych Prac. Vysokej Skoly Technickej v Kosiciach
Zb Vojnomed Akad se: Zbornik Vojnomedicinske Akademije
Zb Vychodoslovenskeho Muzea Ser AB Prir Vedy c: Zbornik Vychodoslovenskeho Muzea. Seria AB. Prirodne Vedy
Zb Vysk Pr Vysk Ustav Zvaracskeho Bratislave c: Zbornik Vyskumnych Prac Vyskumneho Ustavu Zvaracskeho v Bratislave
Zb Vyzk Pr Odboru Pap Celulozy c: Zbornik Vyzkumnych Prac z Odboru Papiera a Celulozy
ZBW e: Deutsche Zentralbibliothek für Wirtschaftswissenschaften, Kiel
- d: Zeitschrift für Betriebswirtschaft
- d: Zentralblatt für Bibliothekswesen
Zb Zgodovino Naravoslovja Tek c: Zbornik za Zgodovino Naravoslovja in Teknike
ZC e: Zangger Committee
- e: Zone Coordinator
- f: Zone Centrale
- s: Zona de Combate
- s: Zona de Concentración
Zc e: Zechariah
Z-C e: Zapalote-Chico
ZCAD e: Zycad Corp.
Z/CAL e: Zero Calibration
Z-CAV e: Zoned Constant Angular Velocity
ZCAV e: Zone Constant Angular Velocity
ZCB e: Chase Manhattan Bank, New York
ZCC e: Zeppelin Collectors Club
ZCCI e: Zippy Collectors Club
ZCCM e: Zambian Consolidaled Copper Mines
ZCFH e: Zero Core For the Hell of it
ZCG e: Impedance Cardiogram
- e: Zeolite Crystal Growth
Z Chem d: Zeitschrift für Chemie
Z Chem Apparatenkd d: Zeitschrift für Chemische Apparatenkunde
Z Chemie [Lpz] d: Zeitschrift für Chemie [Leipzig]
Z Chem Ind Kolloide d: Zeitschrift für Chemie und Industrie der Kolloide
Z Chemother Verw Geh Teil 1 d: Zeitschrift für Chemotherapie und Verwandte Gebiete. Teil 1. Originale
Z Chemother Verw Geh Teil 2 d: Zeitschrift für Chemotherapie und Verwandte Gebiete. Teil 2. Referate
ZCh[r]K d: Zeitschrift für Christliche Kunst
ZCIT f: Zone de convergence intertropicale
- s: Zona de Convergencia Intertropical
ZCL s: Zacatecas [Mexico] [Airport]
ZCM e: CM Preference Corp.
- e: State University of New York, Agricultural and Technical College at Canton, Canton, NY
ZCP d: Zeitschrift für Celtische Philologie
ZCPS f: Zone de Convergence du Pacifique Sud
- s: Zona de Convergencia del Pacífico Sur
ZCR e: Zero-Temperature Coefficient Resistor
ZCS e: Zero Code Suppression
- e: Zero-Current Switching
ZCV e: Zortech Class Viewer
ZCZ e: Cazenovia College, Witherill Learning Center
ZCzest P: Ziemia Czestochowska
Z/D s: Zona de Dispersión
ZD d: Zielsprache Deutsch
- e: Air Traffic Flow Control Unit
- e: Zero Defects
- e: ZIP Code Distribution
- s: Zona de Desembarco
- s: Zona de Dispersión
ZDA d: Zeitschrift für Deutsches Altertum und Deutsche Literatur
- d: Zu den Akten
- s: Zona de Despliegue de Artillería
ZDA/LDA/CA e: Zinc Development Association/Lead Development Association/Cadmium Association
Z Dampfkessel Maschinenbetr d: Zeitschrift für Dampfkessel und Maschinenbetrieb
Z Dampfkesselunters Versicher Ges d: Zeitschrift. Dampfkesseluntersuchungs- und Versicherungs Gesellschaft
ZDAN s: Zona de Defensas Accesorias Navales
Z [Day] e: Zero Day
ZDB d: Zeitschrift für Deutsche Bildung
- d: Zeitschriften-Datenbank
- e: Zambia Development Bank
ZDC e: Philip Crosby Association
- e: Zeus Defense Center
- e: Zinc Dibenzyldithiocarbamate
ZDCM e: Zeus Defense Center Tape and Buffer System
ZDDB e: Zip Code Demographic Data Base
ZDDL e: Zero Deletion Data Link
Z Def s: zona de defensa
ZDemogr d: Zeitschrift für Demographie und Statistik der Juden
Z Desinfekt Gesundheitswes d: Zeitschrift für Desinfektions- und Gesundheitswesen
Z Deut Alt d: Zeitschrift für Deutsches Altertum und Deutsche Literatur
Z Deut Geol Ges d: Zeitschrift. Deutsche Geologische Gesellschaft
Z Deut Phil d: Zeitschrift für Deutsche Philologie
Z Deuts Morgen G d: Zeitschrift. Deutsche Morgenländische Gesellschaft
Z Deut Ver d: Zeitschrift. Deutschen Verein für Kunstwissenschaft
ZDF e: Zimbabwe Defense Force
- e: Zucker Diabetic Fatty
ZDFALP P: Z Dziejow Form Artystycznych Literaturze Polskiej
ZDG d: Gesetz über den Zivildienst der Kriegsdienstverweigerer [Zivildienstgesetz]
- d: Zeitschrift für Deutsche Geistesgeschichte
- d: Zeitschrift für Deutsche Geisteswissenschaft
- e: Corning Community College
ZDK d: Zentralkommittee der Deutschen Katholiken
ZDKAA R: Zdravookhranenie Kazakhstana
ZDKP d: Zeitschrift für Deutsche Kulturphilosophie
ZDL d: Zeitschrift für Dialektologie und Linguistik
- e: Zero Delay Lockout

ZDM d: Zeitschrift für Deutsche Mundarten
- d: Zentralblatt für Didaktik der Mathematik
ZDMG d: Zeitschrift. Deutsche Morgenländische Gesellschaft
ZD Musik d: Zeitschriftendienst Musik
ZDN s: Zona de Defensa Naval
Z DNA e: Deoxyribonucleic Acid, Zigzag
ZD Net e: Ziff-Davis Network
Z Dnipr INO U: Zapiski Dnipropetrovs'-kogo Institutu Narodnoj Osviti
Zdorov'e Nauch Pop Gig Zhurnal se: Zdorov'e Nauchno Populiarnyi Gigienicheskii Zhurnal
ZDP e: Zimbabwe Democratic Party
ZDPh d: Zeitschrift für Deutsche Philologie
ZDPO e: Zero Defects Program Objective
ZDPV d: Zeitschrift des Deutschen Palästina-Vereins
ZDR e: Zone Density Recording
Zdrav Aktual e: Zdravotnicke Aktuality
Zdrav Delo se: Zdravno Delo
Zdravookhr Beloruss[ii] R: Zdravo-okhranenie Belorussii
Zdravookhr Kaz R: Zdravookhranenie Kazakhstana
Zdravookhr Kirg R: Zdravookhranenie Kirgizii
Zdravookhr Ross Fed R: Zdravo-okhranenie Rossiisknoj Federatsii
Zdravookhr Sov Est Sb R: Zdravo-okhranenie Sovetsknoj Estonii Sbornik
Zdravookhr Tadzh R: Zdravookhranenie Tadzhikistana
Zdravookhr Turkm R: Zdravookhranenie Turkmenistana
Zdrav Prac c: Zdravotnicka Pracovnice
Zdravst Vest[n] se: Zdravstveni Vestnik
Zdray Tech[n] Vzduchotech se: Zdra-votni Technika a Vzduchotechnika
Zdrow Publiczne P: Zdrowie Publiczne
ZDS e: Zenith Data Systems
ZDSF d: Zentrum des Datenverbundnetzes der Schweizer Flora, Chambésy
ZDSJ d: Zeitschrift für Demographie und Statistik der Juden
ZDSTJ d: Zeitschrift für Demographie und Statistik der Juden
Zdt d: Die Zoologie des Talmuds
Z Dt Geol Ges d: Zeitschrift. Deutsche Geologische Gesellschaft
Z Dt Phil d: Zeitschrift für Deutsche Philologie
Z Dtschen Morgenländ Ges d: Zeit-schrift. Deutsche Morgenländische Gesellschaft
Z Dtsch Gemmol Ges d: Zeitschrift der Deutschen Gemmologischen Gesellschaft
Z Dtsch Geol Ges d: Zeitschrift. Deutsche Geologische Gesellschaft
Z Dtschl Druckgewerbe d: Zeitschrift für Deutschlands Druckgewerbe
Z Dtsch Morgenl Ges d: Zeitschrift. Deutsche Morgenländische Gesellschaft
Z Dtsch Öl Fett Ind d: Zeitschrift der Deutschen Öl- und Fett-Industrie
Z Dtsch Philol d: Zeitschrift für Deutsche Philologie
Z Dt Spr d: Zeitschrift für Deutsche Sprache
ZDV e: Zero Dead Volume
- e: Zidovudine
ZDV f Kw d: Zeitschrift. Deutscher Verein für Kunstwissenschaft
ZDVGMS d: Zeitschrift. Deutscher Verein für die Geschichte Mährens und Schlesiens
ZDV Kw d: Zeitschrift. Deutscher Verein für Kunstwissenschaft
ZDWDSU d: Zeitschrift für Deutsch-wissenschaft und Deutschunterricht
ZDWF d: Zeitschrift für Deutsche Wortforschung
Z d Z d: Zeichen der Zeit
ZE d: Poliklinik für Zahnerhaltung
- d: Zentraleinrichtung
- d: Zentrale Einrichtungen in der Fakultät
- e: Flight Information Database
- e: Zenith Electronics Corp.
- e: Zollinger-Ellison
- e: Zone Electrophoresis
- s: Zona de Espera
- s: Zona de Etapas
Z/E s: Zona de Evacuación
ZEA e: Zero Energy Assembly
ZEAS d: Zentrales AbsatzanalyseSystem
ZEASA d: Zeitschrift für Astrophysik
ZEB d: Zeiterfassungsbogen
- e: Zebra
Zeb e: Zebahim
ZEBED d: Zeitschrift für Bergrecht
ZEBES d: Zentrales Teilebestandssteue-rungssystem
ZEBFA R: Zhurnal Evolyutsionnoj Biokhimii i Fiziologii
ZEBLA d: Zeitschrift für Biologie
ZEBRA e: Zebra Energy Breeder Assembly
- e: Zero Balance, Reimbursable Account
ZEBRAS e: Zero-coupon Eurosterling Bearer or Registered Accruing Securities
ZEBS d: Zentrale Erfassungs- und Bewertungsstelle für Umweltchemikalien
ZEC e: Zinc-Electrochemical Cell
- e: Zurich Energy Corporation
ZECC e: Zinc-Electrochemical Cell
- e: Zonal Electric Comfort Council
ZECODK d: Zeitschrift für Experimen-telle Chirurgie. Transplantation und Künstliche Organe
ZED d: Zur Erkenntnis der Dichtung
- e: Zero Express Dialing
- e: Zimbabwe Environment and Design
ZEDAT d: Zentraleinrichtung für Datenverarbeitung
ZEDIS d: Zentrales Teiledispositions-system
ZEE d: Zeitschrift für Evangelische Ethik
- f: Zone Économique Exclusive
- s: Zona de Exclusividad Económica, Zona Exclusiva Económica
- s: Zona Económica Exclusiva
Z EEG-EMG d: Zeitschrift für EEG-EMG
ZEEI e: Exclusive Economic Zone of Indonesia
ZEELA d: Zeitschrift für Elektrochemie
Zeews Fruittelersbl n: Zeeuws Fruittelersblad
ZEF d: Zeitschrift für Erziehungs-wissenschaftliche Forschung
- e: Elkin, NC
ZEG e: Zere Energy Growth
- in: Senggo [Indonesia] [Airport]
ZEGL e: Ziegler Co., Inc.
ZEH d: Zeit. Wochenzeitung
ZEI e: Zero Environment Impact
Zei d: Das Zeichen
Zeichn d: Institut für Darstellen und Gestalten, Lehrstuhl
Zeich Zeit d: Zeichen der Zeit
ZEINA d: Zeitschrift für Instrumentenkunde
Z Eisenbahnwes und Verkehrstech Glasers d: Zeitschrift für Eisenbahn-wesen und Verkehrstechnik. Glasers. Annalen
Z Eisenbahnwes Verkehrstech Glasers Ann d: Zeitschrift für Eisenbahnwesen und Verkehrstechnik. Glasers. Annalen
Z Eis Kälte Ind d: Zeitschrift für Eis-und Kälte-Industrie
Zeiss[-]Inf d: Zeiss Information
Zeiss Mitt e: Zeiss Mitteilungen
Zeiss-Mitt Fortschr Tech Opt d: Zeiss-Mitteilungen über Fortschritte der Technischen Optik
Zeit Angewandte Phot d: Zeitschrift Angewandte Photographie
Zeit f Deutk d: Zeitschrift für Deutschkunde
Zeit f Deut Phil d: Zeitschrift für Deutsche Philologie
Zeit f Rom Phil d: Zeitschrift für Romanische Philologie
Zeit f Volk d: Zeitschrift für Volkskunde
Zeitgeschic d: Zeitgeschichte
Zeit Physik d: Zeitschrift für Physik
Zeitschr Angew Geologie d: Zeitschrift für Angewandte Geologie
Zeitschr Anorg u Allg Chemie d: Zeit-schrift für Anorganische und Allgemeine Chemie
Zeitschr Geomorphologie d: Zeitschrift für Geomorphologie
Zeitschr Geomorphologie Neue Folge d: Zeitschrift für Geomorphologie. Neue Folge
Zeitschr Geophysik d: Zeitschrift für Geophysik
Zeitschr Gletscherkunde u Glazial-geologie d: Zeitschrift für Gletscherkunde und Glazialgeologie
Zeitschr Kristallographie d: Zeitschrift für Kristallographie
Zeitschr Physikal Chemie d: Zeitschrift für Physikalische Chemie
Zeitw d: Zeitwende
Zeitwahr d: Zeitschrift für Wahrscheinlichkeitstheorie
ZEIZA d: Zeitschrift für Elektrische Informations- und Energietechnik
ZEK d: Zeitschrift für Evangelisches Kirchenrecht
ZEKE e: Zero Kinetic Energy
ZEKIA d: Zeitschrift für Kinderheilkunde
ZEKID8 d: Zeitschrift für Kinderchirurgie
Z Eks Klin Med R: Zurnal Eksperimen-tal'noj i Kliniceskoj Mediciny

Z Eksper Teoret Fiz R: Zurnal Eksperimental'noj i Teoreticeskoj Fiziki
ZEL e: Bella Bella [Canada] [Airport]
- e: Equitable Life Assurance Society of the United States, Medical Library, NewYork
- e: Zelovo
- e: Zero-Length Launch
ZELAD d: Zeitschrift für Lärmbekämpfung
Z El Ch d: Zeitschrift für Elektrochemie und Angewandte Physikalische Chemie
Z Elek Informations- und Energietech d: Zeitschrift für Elektrische Informations- und Energietechnik
Z Elektr Inf [&] Energietech d: Zeitschrift für Elektrische Informations- und Energietechnik
Z Elektr Inform[ationstech] Energietech d: Zeitschrift für Elektrische Informations- und Energietechnik
Z Elektrochem Angew Phy Chem d: Zeitschrift für Elektrochemie und Angewandte Physikalische Chemie
Z Elektrotech d: Zeitschrift für Elektrotechnik
Zelezarski Zb c: Zelezarski Zbornik
ZELL e: Zero-Length Launch
Zellbiol d: Institut für Zellbiologie und Immunologie
Zell Papier d: Zellstoff und Papier
Zellstoffchem Abh d: Zellstoffchemische Abhandlungen
ZellstPap d: Zellstoff und Papier
Zellst Pap [Berlin] d: Zellstoff und Papier [Berlin]
Zellst Pap [Leipzig] d: Zellstoff und Papier [Leipzig]
Zellwolle Dtsch Kunstseiden Ztg d: Zellwolle und Deutsche Kunstseiden-Zeitung
ZELMAL e: Zero-Length Launch and Mat Landing
ZELTB d: Zeitschrift für Elektrotechnik
Z El Techn d: Zeitschrift für Elektrotechnik
ZEM e: East Main [Canada] [Airport]
- e: Hobart and William Smith Colleges, Geneva
- e: Zero Electrophoretic Mobility
Zem Beton d: Zement und Beton
Zemed Arch c: Zemedelsky Archiv
Zemed Ekon c: Zemedelska Ekonomika
Zemed Tech c: Zemedelska Technika
Zemed Tech Cesk Akad Zemed Ustav Vedeckotech Inf Zemed c: Zemedelska Technika. Ceskoslovenska Akademie Zemedelska Ustav Vedeckotechnickych Informaci Pro Zemedelstvi
Zemed Zahr c: Zemedeistvi v Zahranici
Zemep Sb c: Zemepisny Sbornik
ZEMGA P: Zeszyty Naukowe Akademii Gorniczo-Hutniczej. Elektryfikacja i Mechanizacja Gornictwa i Hutnictwa
ZEMHA d: Zeitschrift für Erzbergbau und Metallhüttenwesen
ZEMIDI d: Zentralblatt für Mikrobiologie
Zem-Kalk-Gips d: Zement-Kalk-Gips
Zemled e: Zemledelie
Zemled Mekh R: Zemledel'cheskaya Mekhanika
Zemled Zhivotnovod Mold R: Zemledelie i Zhivotnovodstvo Moldavii
Zemlerob Resp Mizhvid Temat Nauk Zb U: Zemlerobstvo Respublikans'kyi Mizhvidomchyi Tematychnyi Naukovyi Zbirnyk
Zemleustroistvo Plan Sel'sk Naselennykh Punktov Geod U: Zemleusiroistvo. Planirovka Sel'skikh Naselennykh Punklov i Geodeziya
Zemlj Biljka se: Zemljiste i Biljka
Zemlya Sib Dal'nevost R: Zemlya Sibirskaya Dal'nevostochnaya
Zeml Zhivot Moldav R: Zemledelie i Zhivotnovodstvo Moldavii
Zem ve Sk c: Zemepis ve Skole
ZEN d: Zeitgeist, Enhancement and Nonglare
zen e: Zenaga
ZEN e: Zenith Laboratories, Inc.
- e: Zero Effort Network
Z Energiewirtsch d: Zeitschrift für Energiewirtschaft
ZENO e: Zenox, Inc.
ZENT e: Zentec Corp.
Zentbl Bakt ParasitKde d: Zentralblatt für Bakteriologie, Parasitenkunde, Infektionskrankheiten, und Hygiene
Zentbl Bakt ParasitKde Abt 1 or 2 d: Zentralblatt für Bakteriologie, Parasitenkunde, Infektionskrankheiten, und Hygiene. Abteilung 1 oder 2
Zentbl Biblioth d: Zentralblatt für Bibliothekswesen
Zentbl Vel Med B d: Zentralblatt für Veterinärmedizin. Reihe B
Zent Eur Giesserei Ztg d: Zentral-Europäische Giesserei-Zeitung
Zent Math d: Zentralblatt für Mathematik und Ihre Grenzgebiete
Zentralb Bakteriol Parasitenkd Infektionskr d: Zentralblatt für Bakteriologie, Parasitenkunde, und Infektionskrankheiten
Zentralbl Allg Pathol d: Zentralblatt für Allgemeine Pathologie und Pathologische Anatomie
Zentralbl Allg Pathol Pathol Anat d: Zentralblatt für Allgemeine Pathologie und Pathologische Anatomie
Zentralbl Arbeitsmed d: Zentralblatt für Arbeitsmedizin und Arbeitsschutz
Zentralbl Arbeitsmed Arbeitsschutz d: Zentralblatt für Arbeitsmedizin und Arbeitsschutz
Zentralbl Arbeitsmed Arbeitsschutz Prophyl[axe] d: Zentralblatt für Arbeitsmedizin, Arbeitsschutz, und Prophylaxe
Zentralbl Arbeitsmed Arbeitsschutz Prophyl Ergon d: Zentralblatt für Arbeitsmedizin, Arbeitsschutz, Prophylaxe, und Ergonemie
Zentralbl Bakteriol [B] d: Zentralblatt für Bakteriologie, Parasitenkunde, Infektionskrankheiten, und Hygiene. Erste Abteilung. Originale Reihe B. Hygiene, Betriebshygiene, Prävention
Zentralbl Bakteriol Mikrobiol Hyg Abt 1 Orig A d: Zentralblatt für Bakteriologie, Mikrobiologie, und Hygiene. Abteilung 1. Originale A. Medizinische Mikrobiologie, Infektionskrankheiten, und Parasitol
Zentralbl Bakteriol Mikrobiol Hyg 1 Abt Orig C d: Zentralblatt für Bakteriologie, Mikrobiologie, und Hygiene. 1 Abteilung. Originale C
Zentralbl Bakteriol Mikrobiol Hyg Ser A d: Zentralblatt für Bakteriologie, Mikrobiologie, und Hygiene. Series A. Medical Microbiology, Infectious Diseases, Virology, Parasitology
Zentralbl Bakteriol Mikrobiol Hyg Ser B d: Zentralblatt für Bakteriologie, Mikrobiologie, und Hygiene. Serie B. Umwelthygiene, Krankenhaushygiene, Arbeitshygiene, Präventive Medizin
Zentralbl Bakteriol Naturwiss d: Zentralblatt für Bakteriologie, Parasitenkunde, Infektionskrankheiten, und Hygiene. Zweite Naturwissenschaftliche Abteilung. Mikrobiologie der Landwirtschaft
Zentralbl Bakteriol Orig A d: Zentralblatt für Bakteriologie, Parasitenkunde, Infektionskrankheiten, und Hygiene. Erste Abteilung. Originale Reihe A. Medizinische Mikrobiologie
Zentralbl Bakteriol [Orig B] d: Zentralblatt für Bakteriologie, Parasitenkunde, Infektionskrankheiten, und Hygiene. Erste Abteilung. Originale Reihe B. Hygiene, Betriebshygiene, Prävention
Zentralbl Bakteriol Parasitenkd Infektionskr Abt 1 d: Zentralblatt für Bakteriologie, Parasitenkunde, und Infektionskrankheiten. Abteilung 1. Medizinische-Hygienische Bakteriologie Virusforschung und Tier
Zentralbl Bakteriol Parasitenkd Infektionskrankheiten Hyg II d: Zentralblatt für Bakteriologie, Parasitenkunde, Infektionskrankheiten, und Hygiene. Naturwissenschaftliche Abteilung
Zentralbl Bakteriol Parasitenkd Infektionskr Hyg Abt 2 d: Zentralblatt für Bakteriologie, Parasitenkunde, Infektionskrankheiten, und Hygiene. Abteilung 2. Allgemeine Landwirtschaftliche und Technische Mikrobiologie
Zentralbl Bakteriol Parasitenkd Infektionskr Hyg Abt 1 Orig d: Zentralblatt für Bakteriologie, Parasitenkunde, Infektionskrankheiten, und Hygiene. Abteilung 1 Originale
Zentralbl Bakteriol Parasitenkd Infektionskr Hyg Abt 1 Ref d: Zentralblatt für Bakteriologie, Parasitenkunde, Infektionskrankheiten, und Hygiene. Abteilung 1. Medizinisch-Hygienische Rakteriologie, Virusforschung
Zentralbl Bakteriol Parasitenkd Infektionskr Hyg Abt 1 Suppl d: Zentralblatt für Bakteriologie, Parasitenkunde, Infektionskrankheiten, und Hygiene. Abteilung 1. Supplementheft

Zentralbl Bakteriol Parasitenk Infektionskr Hyg d: Zentralblatt für Bakteriologie, Parasitenkunde, Infektionskrankheiten, und Hygiene
Zentralbl Bauverwaltung d: Zentralblatt der Bauverwaltung
Zentralbl Bibliothekswesen d: Zentralblatt für Bibliothekswesen
Zentralbl Biochem Biophys d: Zentralblatt für Biochemie und Biophysik
Zentralbl Biol Aerosol-Forsch d: Zentralblatt für Biologische Aerosol-Forschung
Zentralbl Chir d: Zentralblatt für Chirurgie
Zentralbl Chir Suppl d: Zentralblatt für Chirurgie. Supplement
Zentralbl Exp Med d: Zentralblatt der Experimentellen Medizin
Zentralbl Geol Paläontol Teil 1 d: Zentralblatt für Geologie und Paläontologie. Teil 1. Allgemeine, Angewandte, Regionale, und Historische Geologie
Zentralbl Geol Paläontol Teil 2 d: Zentralblatt für Geologie und Paläontologie. Teil 2. Paläontologie
Zentralbl Gesamte Forstwes d: Zentralblatt für das Gesamte Forstwesen
Zentralbl Gesamte Hyg Einschluss Bakteriol Immunitätsl d: Zentralblatt für die Gesamte Hygiene mit Einschluss der Bakteriologie und Immunitätslehre
Zentralbl Gesamte Hyg Ihre Grenzgeb d: Zentralblatt für die Gesamte Hygiene und ihre Grenzgebiete
Zentralbl Gesamte Physiol Pathol Stoffwechsels d: Zentralblatt für die Gesamte Physiologie und Pathologie des Stoffwechsels
Zentralbl Gesamte Radiol d: Zentralblatt für die Gesamte Radiologie
Zentralbl Gesamte Rechtsmed d: Zentralblatt für die Gesamte Rechtsmedizin und ihre Grenzgebiete
Zentralbl Ges Hyg d: Zentralblatt für die Gesamte Hygiene und ihre Grenzgebiete
Zentralbl Gewerbehyg Unfallverhüt d: Zentralblatt für Gewerbehygiene und Unfallverhütung
Zentralbl Gynäkol d: Zentralblatt für Gynäkologie
Zentralbl Hütten Walzwerke d: Zentralblatt der Hütten- und Walzwerke
Zentralbl Hyg Umweltmed d: Zentralblatt für Hygiene und Umweltmedizin
Zentralbl Industriebau d: Zentralblatt für Industriebau
Zentralbl Inn Med d: Zentralblatt für Innere Medizin
Zentralbl Mineral Geol Paläontol d: Zentralblatt für Mineralogie, Geologie, und Paläontologie
Zentralbl Mineral Geol Paläontol Teil 1 d: Zentralblatt für Mineralogie, Geologie, und Paläontologie. Teil 1. Kristallographie und Mineralogie
Zentralbl Mineral Geol Paläontol Teil 2 d: Zentralblatt für Mineralogie, Geologie, und Paläontologie. Teil 2. Gesteinskunde, Lagerstättenkunde, Allgemeine, und Angewandte Geologie
Zentralbl Mineral Geol Paläontol Teil 3 d: Zentralblatt für Mineralogie, Geologie, und Paläontologie. Teil 3. Historische und Regionale Geologie, Paläontologie
Zentralbl Mineral Teil 1 d: Zentralblatt für Mineralogie. Teil 1. Kristallographie und Mineralogie
Zentralbl Mineral Teil 2 d: Zentralblatt für Mineralogie. Teil 2. Petrographie, Technische Mineralogie, Geochemie, und Lagerstättenkunde
Zentralbl Neurochir d: Zentralblatt für Neurochirurgie
Zentralbl Papierind d: Zentralblatt für die Papierindustrie
Zentralbl Pharm d: Zentralblatt für Pharmazie
Zentralbl Pharm Pharmakother Laboratoriumsdiagn d: Zentralblatt für Pharmazie, Pharmakotherapie, und Laboratoriumsdiagnostik
Zentralbl Phlebol d: Zentralblatt für Phlebologie
Zentralbl Physiol d: Zentralblatt für Physiologie
Zentralbl Verkehrs-Med Verkehrs-Psychol Luft- Raumfahrt-Med d: Zentralblatt für Verkehrs-Medizin, Verkehrs-Psychologie Luft-, und Raumfahrt-Medizin
Zentralbl Veterinärmed d: Zentralblatt für Veterinärmedizin
Zentralbl Veterinärmed [A] d: Zentralblatt für Veterinärmedizin. Reihe A
Zentralbl Veterinärmed [B] d: Zentralblatt für Veterinärmedizin. Reihe B
Zentralbl Veterinärmed Beih d: Zentralblatt für Veterinärmedizin. Beiheft
Zentralbl Veterinärmed Reihe A d: Zentralblatt für Veterinärmedizin. Reihe A
Zentralbl Veterinärmed Reihe B d: Zentralblatt für Veterinärmedizin. Reihe B
Zentralbl Veterinärmed Reihe C d: Zentralblatt für Veterinärmedizin. Reihe C
Zentralbt Bakteriol d: Zentralblatt für Bakteriologie
Zentralinst Kernforsch Rossendorf Dresden [Ber] e: Zentralinstitut für Kernforschung Rossendorf bei Dresden [Bericht]
Zentralinst Versuchstierzucht Annu Rep e: Zentralinstitut für Versuchstierzucht. Annual Report
ZentrBibl d: Zentralblatt für Bibliothekswesen
Zentr Org Ges Chir e: Zentralorgan für die Gesamte Chirurgie und Ihre Grenzgebiete
Z Entwick P d: Zeitschrift für Entwicklungspsychologie und Pädagogische Psychologie
ZentZtgOptMech d: Zentral-Zeitung für Optik und Mechanik
ZENXC e: Zenex Synthetic Lubricants, Inc.
ZEO e: Zeolite
ZEOS e: Zeos International Ltd.
ZEP e: Zeppelin
- f: Zone d'Echanges Préférentielle
- f: Zone d'Éducation Prioritaire
ZEPAD d: Zeitschrift für Parlamentsfragen
ZEPHYR e: Zero Energy Plutonium-Fueled Fast Reactor
ZEPL e: Zero Excess Propellanis Line
ZEPRAS d: Zentrales Preisabwicklungs- und Analysesystem
ZER e: Pollsville, PA
Zer h: Zera'im
ZERA e: Zero Energy Critical Assemblies Reactor
Z Erdkundeunterricht d: Zeitschrift für den Erdkundeunterricht
Z Erdk Unt d: Zeitschrift für den Erdkundeunterricht
ZERED d: Zeitschrift für Rechtspolitik
Z Erkr Atmungsorgane d: Zeitschrift für Erkrankungen der Atmungsorgane
ZERLINA e: Zero Energy Reactor for Latlice Investigation and New Assemblies
Z Ernährung[sw][iss] d: Zeitschrift für Ernährungswissenschaft
Z Ernährungswiss Suppl d: Zeitschrift für Ernährungswissenschaft. Supplementa
Z Ernähr Wiss d: Zeitschrift für Ernährungwissenschaft
Zernovoe Khoz R: Zernovee Khozyaistvo
Zernovye Maslichn Kul't R: Zemovye i Maslichnye Kul'tury
Zero Popul Growth Natl Rep e: Zero Population Growth. National Reporter
Zero Un e: Zero Un Hebdo
Z Erzbergbau Metallhüttenwes d: Zeitschrift für Erzbergbau und Metallhüttenwesen
ZES d: Zeitschrift für Eingeborenen-Sprachen
- e: Zil Elwannyen Sesei
- e: Zollinger-Ellison Syndrome
- s: Zona marina Especialmente Sensible
ZESA e: Zimbabwe Electricity Supply Authority
ZESIA d: Zeitschrift für Sinnesphysiologie
ZESOD d: Zentrale Teilesortimentdatei
ZE Spr d: Zeitschrift für Eingeborenen-Sprachen
ZEST e: Zinc, E-Vitamin, Sisan Ginseng, Tumera
ZESTA d: Zeitschrift für Schweißtechnik
Zesz Muz Etnogr Wrocl P: Zeszyty Muzeum Etnofflficznego Wroclawie
Zesz Nank Akad Gorn-Hutn Im Staszica Zesz Spec P: Zeszyty Naukowe Akademii Gorniczo-Hutniczej Imienia Stanislawa Staszica. Zeszyt Specjalny
Zesz Nank Akad Roln Wroclawiu Zootech P: Zeszyty Naukowe Akademii Rolniczej we Wroclawiu. Zootechnika
Zesz Nank Mat Fiz Chem P: Zeszyty Naukowe. Matematyka, Fizyka, Chemia
Zesz Nauk Akad Ekon P: Zeszyty Naukowe Akademii Ekonomicznej w Kalowicach
Zesz Nauk Akad Ekon Krakow P: Zeszyty Naukowe Akademii Ekonomicznej w Krakowie
Zesz Nauk Akad Ekon Poznan P: Zeszyty Naukowe Akademii Ekonomicznej w Poznaniu

Zesz Nauk Akad Ekon Poznaniu Ser 2 P: Zeszyty Naukowe. Akademia Ekonomiczna w Poznaniu. Seria 2. Prace Habilitacyjne i Doktorskie
Zesz Nauk Akad Ekon Wroclaw P: Zeszyty Naukowe Akademii Ekonomicznej w Wroclawiu
Zesz Nauk Akad Gorn Hutn [Cracow] Eletryf Mech Gorn Hutn P: Zeszyty Naukowe Akademii Gorniczo-Hutniczej [Cracow]. Elektryfikacja i Mechanizacja Gornictwa i Hutnictwa
Zesz Nauk Akad Gorn Hutn [Cracow] Geol P: Zeszyty Naukowe Akademii Gorniczo- Hutniczej [Cracow]. Geologia
Zesz Nauk Akad Gorn-Hutn [Cracow] Mal Fiz Chem P: Zeszyty Naukowe Akademii Gorniczo-Hutniczej [Cracow]. Matematyka, Fizyka, Chemia
Zesz Nauk Akad Gorn-Hutn [Cracow] Metal Odiew P: Zeszyty Naukowe Akademii Gorniczo-Hutniczej [Cracow]. Metalurgia i Odlewnictwo
Zesz Nauk Akad Gorn Hutn [Cracow] Rozpr P: Zeszyty Naukowe Akademii Gorniczo-Hutniczej [Cracow]. Rozprawy
Zesz Nauk Akad Gorn-Hutn Im Stanisiawa Staszica Ceram P: Zeszyty Naukowe Akademii Gorniczo-Hutniczej Imienia Stanislawa Staszica. Ceramica
Zesz Nauk Akad Gorn-Hutn Im Stanisiawa Staszica Metal Odiew P: Zeszyty Naukowe Akademii Gorniczo-Hutniczej Imienia Stanislawa Staszica. Metalurgia i Odlewnictwo
Zesz Nauk Akad Gorn-Hutn Im Stanisiawa Staszica Zesz Spec P: Zeszyty Naukowe Akademii Gorniczo-Hutniczej Irnienia Stanislawa Staszica. Zeszyt Specjalny
Zesz Nauk Akad Gorn-Hutn Im Stanislawa Staszica Geol P: Zeszyty Naukowe Akademii Gorniczo-Hutniczej Imienia Stanislawa Staszica. Geologia
Zesz Nauk Akad Gorn-Hutn Im Stanislawa Staszica Gorn P: Zeszyty Naukowe Akademii Gorniczo-Hutniczej Imienia Stanislawa Staszica. Gornictwo
Zesz Nauk Akad Gorn-Hutn Im Stanislawa Staszica Mat Fiz Chem P: Zeszyty Naukowe Akademii Gorniczo-Hutniczei Irnienia Stanislawa Staszica. Matematyka, Fizyka, Chemia
Zesz Nauk Akad Gorn-Hutn Im Stanislawa Staszica Ser Autom P: Zeszyty Naukowe Akademii Gorniczo-Hutniczej Imienia Stanislawa Staszica. Seria Automatyka
Zesz Nauk Akad Gorn-Hutn Im Staszica Gorn P: Zeszyty Naukowe Akademii Gorniczo-Hutniczej Imienia Stanislawa Staszica. Gornictwo
Zesz Nauk Akad Gorn-Hutn Im Staszica Mat Fiz Chem P: Zeszyty Naukowe Akademii Gomiczo-Hutniczej Imienia Stanislawa Staszica. Matematyka, Fizyka, Chemia
Zesz Nauk Akad Gorn-Hutn [Krakow] Ceram P: Zeszyty Naukowe Akademii Gorniczo-Hutniczej [Krakow]. Ceramika
Zesz Nauk Akad Gorn-Hutn [Krakow] Elektryf Mech Gorn Hutn P: Zeszyty Naukowe Akademii Gorniczo-Hutniczej [Krakow]. Elektryfikacja i Mechanizacja Gornictwa i Hutnictwa
Zesz Nauk Akad Gorn-Hutn [Krakow] Geol P: Zeszyty Naukowe Akademii Gorniczo-Hutniczej [Krakow]. Geologia
Zesz Nauk Akad Gorn-Hutn [Krakow] Gorn P: Zeszyty Naukowe Akademii Gorniczo-Hutniczej [Krakow]. Gornictwo
Zesz Nauk Akad Gorn-Hutn Krakowie Rozpr P: Zeszyty Naukowe Akademii Gorniczo-Hutniczej w Krakowie. Rozprawy
Zesz Nauk Akad Gorn-Hutn [Krakow] Mat Fiz Chem P: Zeszyty Naukowe Akademii Gorniczo-Hutniczej [Krakow]. Matematyka, Fizyka, Chemia
Zesz Nauk Akad Gorn-Hutn [Krakow] Metal Odlew P: Zeszyty Naukowe Akademii Gorniczo-Hutniczej [Krakow]. Metalurgia i Odlewnictwo
Zesz Nauk Akad Gorn-Hutn [Krakow] Ses Nauk P: Zeszyty Naukowe Akademii Gorniczo-Hutniczej [Krakow]. Sesja Naukowa
Zesz Nauk Akad Gorn-Hutn [Krakow] Sozologia Sozotechnika P: Zeszyty Naukowe Akademii Gorniczo-Hutniczej [Krakow]. Sozologia i Sozotechnika
Zesz Nauk Akad Gorn-Hutn [Krakow] Zesz Spec P: Zeszyty Naukowe Akademii Gorniczo-Hutniczej [Krakow]. Zeszyty Specjalny
Zesz Nauk Akad Gorn-Hutn Stanisi Staszica Autom P: Zeszyty Naukowe Akademii Gorniczo-Hutniczej Imienia Stanislawa Staszica. Automatyka
Zesz Nauk Akad Gorn-Hutn Stanislawa Staszica Geot e: Zeszyly Naukowe Akademii Gorniczo-Hutniczej Imienia Stanislawa Staszica. Geologia
Zesz Nauk Akad Gorn-Hutn Stanisl Staszica P: Zeszyty Naukowe Akademii Gorniczo-Hutniczej Imienia Stanislawa Staszica. Metalurgia i Odlewnietwo
Zesz Nauk Akad Gorn-Hutn Stanisl Staszica Geol P: Zeszyty Naukowe Akademii Gorniczo-Hutniczej Imienia Stanislawa Staszica. Geologia
Zesz Nauk Akad Gorn-Hutn Stanisl Staszica Mat Fiz Chem P: Zeszyty Naukowe Akademii Gorniczo-Hutniczej Imienia Stanislawa Staszica. Matematyka, Fizyka, Chemia
Zesz Nauk Akad Roln Szczecinie P: Zeszyty Naukowe. Akademia Rolnicza w Szczecinie
Zesz Nauk Akad Roln Szczecinie Ser Rybactwo Morsk P: Zeszyty Naukowe Akademii Rolniczej w Szczecinie. Seria Rybactwo Morskie
Zesz Nauk Akad Roln Tech Olsztynie P: Zeszyty Naukowe Akademii Rolniczej Technicznej w Olsztynie
Zesz Nauk Akad Roln Tech Olsztynie Roln P: Zeszyty Naukowe Akademii Rolniczo- Technicznej w Olsztynie. Rolnictwo
Zesz Nauk Akad Roln Tech Olsztynie Zootech P: Zeszyty Naukowe Akademii Rolniczo-Technicznej w Olsztynie. Zootechnika
Zesz Nauk Akad Roln Tech Otsztynie Technol Zywn P: Zeszyty Naukowe Akademii Rolniczo-Technicznej w Olsztynie. Technologia Zywnosci
Zesz Nauk Akad Roln Warszawie Melior Rolne P: Zeszyty Naukowe Akademii Rolniczej w Warszawie. Melieracje Rolne
Zesz Nauk Akad Roln Warszawie Ogrod P: Zeszyty Naukowe Akademii Rolniczej w Warszawie. Ogrodnictwo
Zesz Nauk Akad Roln Warszawie Technol Drewna P: Zeszyty Naukowe. Akademii Rolniczej w Warszawie. Technologia Drewna
Zesz Nauk Akad Roln Warszawie Technol Rolno Spozyw P: Zeszyty Naukowe Akademii Rolniczej w Warszawie. Technologia Rolno Spozyweza
Zesz Nauk Akad Roln Warszawie Zootech P: Zeszyty Naukowe Akademii Rolniczej w Warszawie. Zootechnika
Zesz Nauk Akad Roln Wroclawiu Melior P: Zeszyty Naukowe Akademii Rolniczej we Wroclawiu. Melioracja
Zesz Nauk Akad Roln Wroclawiu Roln P: Zeszyty Naukowe Akademii Rolniczej we Wroclawiu. Rolnictwo
Zesz Nauk Akad Roln Wroclawiu Weter P: Zeszyty Naukowe Akademii Rolniczej we Wroclawiu. Weterynaria
Zesz Nauk Akad Roln Wrocl Wet P: Zeszyty Naukowe Akademii Rolniczej we Wroclawiu. Weterynaria
Zesz Nauk Gorn P: Zeszyty Naukowe Gornictwo
Zesz Nauk Inst Ciezkiei Synt Org Blachowni Slask P: Zeszyty Naukowe Instytut Ciezkiej Syntezy Organicznej w Blachowni Slaskiej
Zesz Nauk Lesn Akad Roln Warsz P: Zeszyty Naukowe. Lesnictwo-Akademia Rolniczaw Warszawie
Zesz Nauk Mechan Budownictwo Akad Rolo-Tech Olsztyn P: Zeszyty Naukowe. Mechanika i Budownictwo-Akademia Rolniczo Techniczna w Olsztynie
Zesz Nauk Melior Rolne Akad Roln Warsz P: Zeszyty Naukowe. Melioracje Rolne-Akademia Rolnicza w Warszawie
Zesz Nauk Ochr Wod Rybactwo Srodiadowe P: Zeszyty Naukowe. Ochrona Wod i Rybactwo Srodladowe
Zesz Nauk Politech Czestochow P: Zeszyty Naukowe Politechniki Czestochowskiej
Zesz Nauk Politech Czestochow Metal P: Zeszyty Naukowe Politechniki Czestochowskiej. Metalurgia

Zesz Nauk Politech Czestochow Nauki Tech Hutn P: Zeszyty Naukowe Politechniki Czestochowskiej. Nauki Techniczne. Hutnictwo

Zesz Nauk Politech Gdansk Chem P: Zeszyty Naukowe Politechniki Gdanskiej. Chemia

Zesz Nauk Politech Gdansk Elektr P: Zeszyty Naukowe Politechniki Gdanskiej. Elektryka

Zesz Nauk Politech Gdansk Fiz P: Zeszyty Naukowe Politechniki Gdanskiej. Fizyka

Zesz Nauk Politech Gdansk Mat P: Zeszyty Naukowe Politechniki Gdanskiej. Matematyka

Zesz Nauk Politech Gdansk Mech P: Zeszyty Naukowe Politechniki Gdanskiej. Mechanika

Zesz Nauk Politech Krakow Chem P: Zeszyty Naukowe Politechniki Krakowskiej. Chemia

Zesz Nauk Politech Krakow Mech P: Zeszyty Naukowe Politechniki Krakowskiej. Mechanika

Zesz Nauk Politech Lod Budow P: Zeszyty Naukowe Politechniki Lodzkiej. Budownictwo

Zesz Nauk Politech Lodz Chem P: Zeszyty Naukowe Politechniki Lodzkiej. Chemia

Zesz Nauk Politech Lodz Chem Spozyw P: Zeszyty Naukowe Politechniki Lodzkiej. Chemia Spozywcza

Zesz Nauk Politech Lodz Elek P: Zeszyty Naukowe Politechniki Lodzkiej. Elektryka

Zesz Nauk Politech Lodz Elektr P: Zeszyty Naukowe Politechniki Lodzkiej. Elektryka

Zesz Nauk Politech Lodz Fiz P: Zeszyty Naukowe Politechnika Lodzka. Fizyka

Zesz Nauk Politech Lodz Inz Chem P: Zeszyty Naukowe Politechniki Lodzkiej. Inzynieria Chemiczna

Zesz Nauk Politech Lodz Mech P: Zeszyty Naukowe Politechniki Lodzkiej. Mechanika

Zesz Nauk Politech Lodz Wlok P: Zeszyty Naukowe Politechniki Lodzkiej. Wlekiennictwo

Zesz Nauk Politech Poznan Chem Inz Chem P: Zeszyty Naukowe Politechniki Poznanskiej. Chemia i Inzynieria Chemiczna

Zesz Nauk Politech Poznan Elektr P: Zeszyty Naukowe Politechniki Poznanskiej. Elektryka

Zesz Nauk Politech Rzeszowskiej P: Zeszyty Naukowe Politechniki Rzeszowskiej

Zesz Nauk Politech Slask P: Zeszyty Naukowe Politechniki Slaskiej

Zesz Nauk Politech Slaska Energ P: Zeszyty Naukowe Politechnika Slaska. Energetyka

Zesz Nauk Politech Slask Chem P: Zeszyty Naukowe Politechniki Slaskiej. Chemia

Zesz Nauk Politech Slask Energ P: Zeszyty Naukowe Politechniki Slaskiej. Energetyka

Zesz Nauk Politech Slask Gorn P: Zeszyty Naukowe Politechniki Slaskiej. Gornictwo

Zesz Nauk Politech Slask Hutn P: Zeszyty Naukowe Politechniki Slaskiej. Hutnictwo

Zesz Nauk Politech Slask Inz Sanit P: Zeszyty Naukowe Politechniki Slaskiej. Inzynieria Sanitarna

Zesz Nauk Politech Slask Ser Elektr P: Zeszyty Naukowe Politechniki Slaskiej. Seria Elektryka

Zesz Nauk Politech Slask Ser Mat-Fiz P: Zeszyty Naukowe Politechniki Slaskiej. Seria Matematyka-Fizyka

Zesz Nauk Politech Swietokrz Probl Nauk Podst P: Zeszyty Naukowe Politechniki Swietokrzyska. Problemy Nauk Podstawowych

Zesz Nauk Politech Szezecin Chem P: Zeszyty Naukowe Politechniki Szczecinskiej. Chemia

Zesz Nauk Politech Szezecin Pr Monogr P: Zeszyty Naukowe Politechniki Szezecinskiej. Prace Monografiezne

Zesz Nauk Politech Warsz Chem P: Zeszyty Naukowe Politechniki Warszawskiej. Chemia

Zesz Nauk Politech Wroclaw Chem P: Zeszyty Naukowe Politechniki Wroclawskiej. Chemia

Zesz Nauk Roln Akad Roln Warsz P: Zeszyty Naukowe. Rolnictwo Akademia Rolnicza w Warszawie

Zesz Nauk Szk Gl Gospod Wiejsk Akad Roln Warszawie Ogrod P: Zeszyty Naukowe Szkoly Glownej Gospodarstwa Wiejskiego. Akademii Rolniczej w Warszawie. Ogrodnictwo

Zesz Nauk Szk Gl Gospod Wiejsk Akad Roln Warszawie Wter P: Zeszyty Naukowe Szkoly Glownej Gospodarstwa Wiejskiego. Akademii Rolniczej w Warszawie. Weterynaria

Zesz Nauk Szk Gl Gospod Wiejsk Akad Roln Warszawie Zootech P: Zeszyty Naukowe Szkoly Glownei Gospodarstwa Wiejskiego. Akademii Rolniczej w Warszawie. Zootechnika

Zesz Nauk Szk Gl Gospod Wiejsk Akad Roln Warsz Ogrodn P: Zeszyty Naukowe Szkoly Glownej Gospodarstwa Wiejskiego Akademii Rolniczej w Warszawie Ogrodnictwo

Zesz Nauk Szk Gl Gospod Wiejsk Warszawie Melior Rolne P: Zeszyty Naukowe Szkoly Glownej Gospodarstwa Wiejskiego w Warszawie. Melioracje Rolne

Zesz Nauk Szk Gl Gospod Wiejsk Warszawie Ogrod P: Zeszyty Naukowe Szkoly Glownej Gospodarstwa Wiejskiego w Warszawie. Ogrodnictwo

Zesz Nauk Szk Gl Gospod Wiejsk Warszawie Roln P: Zeszyty Naukowe Szkoly Glownej Gospodarstwa Wiejskiego w Warszawie. Rolnictwo

Zesz Nauk Szk Gl Gospod Wiejsk Warszawie Technol Drewna P: Zeszyty Naukowe Szkoly Glownej Gospodarstwa Wiejskiego w Warszawie. Technologia Drewna

Zesz Nauk Szk Gl Gospod Wiejsk Warszawie Zootech P: Zeszyty Naukowe Szkoly Glownej Gospodarstwa Wiejskiego w Warszawie. Zootechnika

Zesz Nauk Szk Glo Gospod Wiejsk Warszawie Lesn P: Zeszyty Naukowe Szkoly Glownej Gospodarstwa Wiejskiego w Warszawie. Lesnictwo

Zesz Nauk Szk Glow Gospod Wiejsk Warszawie P: Zeszyty Naukowe Szkoly Glownej Gospodarstwa Wiejskiego w Warszawie

Zesz Nauk Szk Glown Plan Statystyki P: Zeszyty Naukowe Szkoly Glownei Planowania i Statystyki

Zesz Nauk Szkol Gospod Wiejsk Warsz [Lesn] P: Zeszyty Naukowe Szkola Glowna Gospodarstwa Wiejskiego w Warszawie [Lesnictwo]

Zesz Nauk Szkol Gospod Wiejsk Warsz Technol Drewna P: Zeszyty Naukowe Szkola Glowna Gospodarstwa Wiejskiego w Warszawie. Technologia Drewna

Zesz Nauk Technol Drewna Akad Roln Warsz P: Zeszyty Naukowe. Technologia Drewna-Akademia Rolnicza w Warszawie

Zesz Nauk Technol Zywn Akad Roln-Tech Olsztynie P: Zeszyty Naukowe. Technologia Zywnosci-Akademia Rolniczo-Techniczna w Olsztynie

Zesz Nauk Tech Wyzsza Szk Inz Lublinie P: Zeszyty Naukowo-Techniczny Wyzsza Szkola Inzynierska w Lublinie

ZeszNauKUL P: Zeszyty Naukowe Katolickiego Uniwersytetu Lubelskiego

Zesz Nauk Uniw Jagiellon Acta Cosmol P: Zeszyty Naukowe Uniwersytetu Jagiellonskiego. Acta Cosmologica

Zesz Nauk Uniw Jagiellon Pr Biol Mol P: Zeszyty Naukowe Uniwersytetu Jagiellonskiego. Prace Biologii Molekularnej

Zesz Nauk Uniw Jagiellon Pr Chem P: Zeszyty Naukowe Uniwersytetu Jagiellonskiego. Prace Chemiczne

Zesz Nauk Uniw Jagiellon Pr Etnogr P: Zeszyty Naukowe Uniwersytetu Jagiellonskiego. Prace Etnograficzne

Zesz Nauk Uniw Jagiellon Pr Fiz P: Zeszyty Naukowe Uniwersytetu Jagiellonskiego. Prace Fizyczne

Zesz Nauk Uniw Jagiellon Pr Zool P: Zeszyty Naukowe Uniwersytetu Jagiellonskiego. Prace Zoologiczne

Zesz Nauk Uniw Jagiellon Ser Nauk Mat Przy P: Zeszyty Naukowe Uniwersytetu Jagiellonskiego. Seria Nauk Matematyezno-Przyrodniczych. Matematyka, Fizyka, Chemia

Zesz Nauk Uniw Jagiellonsk Pr Bot P: Zeszyty Naukowe Uniwersytetu Jagiellonskiego. Prace Botaniczne

Zesz Nauk Uniw Jagiellonsk Zool P: Zeszyty Naukowe Uniwersytetu Jagiellonskiego. Prace Zoologiczne
Zesz Nauk Uniw Lodz P: Zeszyty Naukowe Uniwersytetu Lodzkiego
Zesz Nauk Uniw Lodz Fiz P: Zeszyty Naukowe Uniwersytetu Lodzkiego. Fizyka
Zesz Nauk Uniw Lodz Nauki Humanist-Spolecz P: Zeszyty Naukowe Uniwersytetu Lodzkiego. Nauki Humanistyczno-Spoleczne
Zesz Nauk Uniw Lodz Nauki Mat Przyr P: Zeszyty Naukowe Uniwersytetu Lodzkiego. Nauki Matematyczno-Przyrodnicze
Zesz Nauk Uniw Lodz Ser II P: Zeszyty Naukowe Uniwersytetu Lodzkiego. Seria II. Nauki Matematyczno-Przyrodnicze
Zesz Nauk Uniw Lodz Ser III P: Zeszyty Naukowe Uniwersytetu Lodzkiego. Seria III
Zesz Nauk Uniw Mikolja Kopernika Torun P: Zeszyty Naukowe Uniwersytetu Imienia Mikolaja Kopernika w Toruniu
Zesz Nauk Uniw Poznaniu Mat Fiz Chem P: Zeszyty Naukowe Uniwersytetu Imienia Adama Mickiewicza w Poznaniu. Matematyka, Fizyka, Chemia
Zesz Nauk Uniw Slaski Katowicach Seke Chem P: Zeszyty Naukowe Uniwersytet Slaski w Katowicach Sekeja Chemii
Zesz Nauk Weter Akad Roln Warsz P: Zeszyty Naukowe. Weterynaria-Akademia Rolnicza w Warszawie
Zesz Nauk Wydz Mat Fiz Chem Uniw Gdanski Chem P: Zeszyty Naukowe Wydzialu Matematyki, Fizyki, Chemii. Uniwersytet Gdanski. Seria Chemia
Zesz Nauk Wyzs Szk Ekon Poznaniu P: Zeszyty Naukowe Wyzszej Szkoly Ekonomicznej w Poznaniu
Zesz Nauk Wyzs Szk Roln Krakowie P: Zeszyty Naukowe Wyzszej Szkoly Rolniczej w Krakowie
Zesz Nauk Wyzs Szk Roln Krakowie Zootech P: Zeszyty Naukowe Wyzszej Szkoly Rolniczej w Krakowie. Zootechnika
Zesz Nauk Wyzs Szk Roln Olsztynie P: Zeszyty Naukowe Wyzszej Szkoly Rolniczej w Olsztynie
Zesz Nauk Wyzs Szk Roln Szczecinie P: Zeszyty Naukowe Wyzsza Szkola Rolnicza w Szczecinie
Zesz Nauk Wyzs Szk Roln Wreclawiu Wet P: Zeszyty Naukowe Wyzszej Szkoly Rolniczej we Wroclawiu. Weterynaria
Zesz Nauk Wyzs Szk Roln Wroclawiu P: Zeszyty Naukowe Wyzszej Szkoly Rolniczej we Wroclawiu
Zesz Nauk Wyzsza Szk Ekon Poznaniu Ser 2 P: Zeszyty Naukowe Wyzsza Szkola Ekonomiczna w Poznaniu. Seria 2. Prace Habilitacyjne i Doktorskie
Zesz Nauk Wyz Szkol Ekon P: Zeszyty Naukowe Wyzszej Szkoly Ekonomicznej w Katowicach
Zesz Nauk Wyzsz Szk Inz Bialymstoku Mat Fiz Chem P: Zeszyty Naukowe Wyzszej Szkoly Inzynierskiej w Bialymstoku Matematyka, Fizyka, Chemia
Zesz Nauk Wyzsz Szkoly Ekon Katowic P: Zeszyty Naukowe Wyzszej Szkoly Ekonomicznej w Katowicach
Zesz Nauk Wyzsz Szkoly Ekon Poznan P: Zeszyty Naukowe Wyzszej Szkoly Ekonomicznej w Poznaniu
Zesz Nauk Wyzsz Szk Pedagog Gdansku Mat Fiz Chem P: Zeszyty Naukowe Wyzszej Szkoly Pedagogicznej w Gdansku. Malematyka, Fizyka, Chemia
Zesz Nauk Wyzsz Szk Pedagog Katowicach Sekc Fiz P: Zeszyty Naukowe Wyzszej Szkoly Pedagogicznej w Katowicach. Sekcja Fizyki
Zesz Nauk Wyzsz Szk Roln Olsztynie P: Zeszyty Naukowe Wyzszej Szkoly Rolniczej w Olsztynie
Zesz Nauk Wyzsz Szk Roln Szczecinie P: Zeszyty Naukowe Wyzszej Szkoly Rolniczej w Szczecinie
Zesz Nauk Wyzsz Szk Roln Wroclawiu Melior P: Zeszyty Naukowe Wyzszej Szkoly Rolniczej we Wroclawiu. Melioracja
Zesz Nauk Wyzsz Szk Roln Wroclawiu Roln P: Zeszyty Naukowe Wyzszej Szkoly Rolniczej we Wroclawiu. Rolnictwo
Zesz Nauk Wyzsz Szk Roln Wroclawiu Weter P: Zeszyty Naukowe Wyzszej Szkoly Rolniczej we Wroclawiu. Weterynaria
Zesz Nauk Wyzsz Szk Roln Wroclawiu Zootech P: Zeszyty Naukowe Wyzszej Szkoly Rolniczej we Wroclawiu. Zootechnika
Zesz Nauk Wyzsz Szk Roln Wrocl Melior P: Zeszyty Naukowe Wyzszej Szkoly Rolniczej we Wroclawiu. Melioracja
Zesz Nauk Wyzsz Szk Roln Wrocl Roln P: Zeszyty Naukowe Wyzszej Szkoly Rolniczej we Wroclawiu. Rolnictwo
Zesz Nauk Zootech Akad Roln-Tech Olsztyn P: Zeszyty Naukowe. Zootechnika-Akademia Rolniczo-Techniczna w Olsztynie
Zesz Nauk Zootech Akad Roln Warsz P: Zeszyty Naukowe. Zootechnika-Akademia Rolnicza w Warszawie
Zesz NWSP P: Zeszyty Naukowe Wyzsza Szkola Pedagogiczna w Kalowicach
Zesz Probl Gorn P: Zeszyty Problemowe Gornictwa
Zesz Probl Kosmosu P: Zeszyty Problemowe Kosmosu
Zesz Probl Nauki Pol P: Zeszyty Problemowe Nauki Polskiej
Zesz Probl Postepow Nauk Roln P: Zeszyty Problemowe Postepow Nauk Rolniczych
Zesz Prob Postepow Nauk Roln P: Zeszyty Problemowe Postepow Nauk Rolniczych
Zeszty Nauk Uniw Jagiellon Prace Mat P: Zeszyty Naukowe Uniwersytetu Jagiellonskiego. Prace Matematyczne
Zesz Wrocl P: Zeszyty Wroclawskie
Zeszyty Nauk Akad Gorn-Hutniczej Mat Fiz Chem P: Zeszyty Naukowe Akademii Gorniczo-Hutniczej Imienia Stanislawa Staszica. Matematyka, Fizyka, Chemia
Zeszyty Nauk Politech Lodz Mat P: Zeszyty Naukowe Politechniki Lodzkiej. Matematyka
Zeszyty Nauk Politech Slask Automat P: Zeszyty Naukowe Politechniki Slaskiej. Automatyka
Zeszyty Nauk Politech Slask Mat-Fiz P: Zeszyty Naukowe Politechniki Slaskiej. Seria Matematyka-Fizyka
Zeszyty Nauk Politech Szczecin P: Zeszyty Naukowe Politechniki Szezecinskiej
Zeszyty Nauk Szkoly Glown Planowania i Statyst P: Zeszyty Naukowe Szkoly Glownej Planowania i Statystyki
Zeszyty Nauk Uniw Jagiellon Prace Fiz P: Zeszyty Naukowe Uniwersytetu Jagiellonskiego. Prace Fizyczne
Zeszyty Nauk Wyz Szkoly Ped w Opolu Fiz P: Zeszyty Naukowe Wyzszej Szkoly Pedagogicznej w Opolu. Fizyka
Zeszyty Nauk Wyz Szkoly Ped w Opolu Mat P: Zeszyty Naukowe Wyzszej Szkoly Pedagogicznej w Opolu. Matematyka
ZETA e: Zero Energy Thermonuclear Apparatus
- e: Zero Energy Thermonuclear Assembly
ZETF R: Zurnal Eksperimental'noj i Teoreticesknoj Fiziki
ZETFA R: Zhurnal Eksperimentalnoj i Teoretichesknoj Fiziki
ZEthn d: Zeitschrift für Ethnologie
Z Ethnolog d: Zeitschrift für Ethnologie
ZETK e: Zetek, Inc.
ZETUA d: Zeitschrift für Tuberkulose und Erkrankungen der Thoraxorgane
ZEUMD d: Zeitschrift für Umweltpolitik
Zeumer's Q St d: Zeumer's Quellen und Studien zur Verfassungsgeschichte des Deutschen Reichs in Mittelalter und Neuzeit
ZEURA d: Zeitschrift für Urologie
ZEUS e: Zero Energy Uranium System
- e: Zeus Components, Inc.
ZEV e: Zero Emission[s] Vehicle
Zev e: Zevahim
Z Evan Eth d: Zeitschrift für Evangelische Ethik
ZEVBA d: Zeitschrift für Vererbungslehre
ZevE d: Zeitschrift für Evangelische Ethik. Gütersloh
Z Ev Ethik d: Zeitschrift für Evangelische Ethik
ZEV Glasers Ann d: ZEV [Zeitschrift für Eisenbahnwesen und Verkehrstechnik] Glasers. Annalen

Z Ev K[r] d: Zeitschrift für Evangelisches Kirchenrecht
ZEVR d: Zeitschrift für den Evangelischen Religionsunterricht
ZEW d: Zentrum für Europäische Wirtschaftsforschung in Mannheim
ZEWV e: Zambia Electricity Workers Union
Z Exp A[ngew] Psy[chol] d: Zeitschrift für Experimentelle und Angewandte Psychologie
Z Exp Chir d: Zeitschrift für Experimentelle Chirurgie
Z Exp Chir Chir Forsch d: Zeitschrift für Experimentelle Chirurgie und Chirurgische Forschung
Z Exper & Angew Psychol d: Zeitschrift für Experimentelle und Angewandte Psychologie
Z Exp Pathol Ther d: Zeitschrift für Experimentelle Pathologie und Therapie
Z Exp Psychol d: Zeitschrift für Experimentelle Psychologie
ZF
- d: Zahnradfabrik Friedrichshafen AG
- d: Zentralblatt für das Gesamte Forstwesen
- e: Center in charge of a Flight Information Region or an Upper Flight Information Region when the message is relevant to a VFR [Visual Flight Rules] Flight
- e: Zermelo-Fraenkel
- e: Ziegfeld Follies
- e: Zweig Fund
- f: Zone Franche
- f: Zone Frontière
- s: Zona Fasciculata
- s: Zona Franca

ZF0 d: Zielfunktionswert bei Start der Sekantenorientierten Stochastik
ZFA e: Zone First Application
Z f A d: Zeitschrift für Agrargeschichte und Agrarsoziologie
- d: Zeitschrift für Anatomie und Entwicklungsgeschichte

ZfA d: Zeitschrift für Astrophysik
ZFA [Dresden] d: Zeitschrift für Alternsforschung [Dresden]
ZFAL e: Zacherley Fans at Large
Z f Allg Med d: Zeitschrift für Allgemeinmedizin
Z Farben Ind d: Zeitschrift für Farben Industrie
Z Farben Text Chem d: Zeitschrift für Farben- und Textil-Chemie
ZFA [Stuttgart] d: Zeitschrift für Allgemeinmedizin [Stuttgart]
ZFAX e: ZFAX Image Corp.
ZFB d: Zentrales Fachschaftenbüro
ZfB d: Zeitschrift für Bücherfreunde
ZFC e: Zero Failure Criteria [IEEE]
- e: Zero-Field Cooled
- e: Zipp-Forming Cells

ZFD d: Zielfunktionswert, wenn eine Koordinate durch einen kleinen Schritt D verändert wurde
Z f D Altert d: Zeitschrift für Deutsches Altertum und Deutsche Literatur
ZFDF d: Zeitschrift für Freie Deutsche Forschung
ZFDNMR e: Zero-Field Deuterium Nuclear Magnetic Resonance
Z f D Phil d: Zeitschrift für Deutsche Philologie
ZfDSdJ d: Zeitschrift für Demographie und Statistik der Juden
ZfE d: Zeitschrift für Ethnologie
ZFET e: Zionist Federation Educational Trust
ZFEU d: Zeitschrift für Französischen und Englischen Unterricht
ZFFB se: Zbornik Filozofskog Fakulteta
ZfG d: Zeitschrift für Geschichtswissenschaft
ZFGV d: Zeitschrift. Freiburger Geschichtsvereine
ZfHb d: Zeitschrift für Hebräische Bibliographie
ZFIC e: Zinc-Phosphate Coating
ZFI-Mitt d: ZFI [Zentralinstitut für Isotopen- und Strahlenforschung]-Mitteilungen
ZfIR d: Zeitschrift für Immobilienrecht
ZFIS d: Zeitschrift für Innere Sicherheit in Deutschland und Europa
Z Fisch Hilfswiss d: Zeitschrift für Fischerei und deren Hilfswissenschaften
ZF/ix d: Zielfunktion
ZfK d: Zeitschrift für Kunstgeschichte
ZFKPhil c: Zbornik Filozofickej Fakulty Univerzity Komenskeho-Philologica
ZFL d: Zeitschrift für Luftrecht- und Weltraumrechtsfragen
- se: Zbornik za Filologiju i Lingvistiku

Z Fleisch Milchhyg d: Zeitschrift für Fleisch- und Milchhygiene
Z Flugwiss d: Zeitschrift für Flugwissenschaften
Z Flugwiss [und] Weltraumforsch d: Zeitschrift für Flugwissenschaften und Weltraumforschung
ZFM d: Zeitschrift für Musik
- e: Community College of the Finger Lakes, Canandaigua
- e: Fort McPherson [Canada] [Airport]

Zfmin d: minimaler Zielfunktionswert
ZFMK d: Zoologisches Forschungsinstitut und Museum Alexander König, Bonn
ZFMR e: Transformer
Zf Mus Theorie d: Zeitschrift für Musiktheorie
ZfN d: Zentrum für Netze
- d: Zeitschrift für Numismatik

ZFNMR e: Zero-Field Nuclear Magnetic Resonance
ZFNU d: Zeitschrift für Neusprachlichen Unterricht
ZFO d: ZFO. Zeitschrift für Führung und Organisation
Z. Forstgenet. Forstpflanzenzücht. d: Zeitschrift für Forstgenetik und Forstpflanzenzüchtung
Z Forst Jagdwes d: Zeitschrift für Forst- und Jagdwesen
ZFP e: Zyglo-Fluorescent Penetrant
- f: Zone du Front Polaire

ZfP d: Dokumentation Zerstörungsfreie Prüfung
ZFPT e: Zyglo-Fluorescent Penetrant Testing
ZFR e: ZEV Fleet Requirement
Z Französische Spr Lit d: Zeitschrift für Französische Sprache und Literatur
ZfRG d: Zeitschrift für Religions- und Geistesgeschichte
ZFRP[H] d: Zeitschrift für Romanische Philologie
ZFRSL d: Zeitschrift für Französische Sprache und Literatur
ZfRuGg d: Zeitschrift für Religions- und Geistesgeschichte
ZFS e: Zone Field Selection
ZfS d: Zeitschrift für Semitistik und Verwandte Gebiete
- d: Zeitschrift für Soziologie

ZFSC e: Zero-Field Splitting Constant
ZfSchKg d: Zeitschrift für Schweizerische Kirchengeschichte
Z f Schweiz Recht d: Zeitschrift für Schweizerisches Recht/Revue de Droit Suisse/Revista di Diritto Svizzero
ZFSH e: Z&Z Fashions Ltd.
ZFSL d: Zeitschrift für Französische Sprache und Literatur
Zft f Celt Phil d: Zeitschrift für Celtische Philologie
Zft f D Alt d: Zeitschrift für Deutsches Altertum und Deutsche Literatur
Zft f Fr Sp u Lit d: Zeitschrift für Französische Sprache und Literatur
Zft f Rom Phil d: Zeitschrift für Romanische Philologie
ZfU d: Z für Umweltpolitik und Umweltrecht. Beiträge zur recht-, wirtschafts- u. sozialwissenschaftliche Umweltforschung
Z für die Ost Gym d: Zeitschrift für die Österreichischen Gymnasien
ZFV d: Zeitschrift für Volkskunde
- e: Fort Severn [Canada] [Airport]

ZfW d: Zeitschrift für Wasserrecht
ZFW e: Zero Fuel Weight
ZFYZD e: Zhonghua Fangshe Yixue Yu Fanghu Zazhi
ZfZ d: Zeitschrift für Assyriologie
ZF Ziel d: angestrebter Zielfunktionswert
ZG d: Zeitschrift für Geomorphologie. Gebrüder Bornträger
- e: Air Traffic Control
- e: Zinc Gluconate
- e: Zymbal Gland
- s: Zona Gaseada
- s: Zona Glomerulosa

Z-G e: Zapalote-Grande
ZGA e: Zero Grade Air
ZGAKE d: Zeitschrift für Geschichte und Altertumskunde der Ermlands
ZGärungsphysiol d: Zeitschrift für Gärungsphysiologie
Z Gastroent d: Zeitschrift für Gastroenterologie
Z Gastroenterol d: Zeitschrift für Gastroenterologie
Z Gastroenterol Verh d: Zeitschrift für Gastroenterologie. Verhandlungsband
ZGB d: Zivilgesetzbuch [Land]
ZGDJ d: Zeitschrift zur Geschichte des Deutschen Judentums
ZGDV d: Zentrum für Graphische Datenverarbeitung
ZGE e: Zero-Gravity Expulsion

Z Geburtshilfe Gynäkol d: Zeitschrift für Geburtshilfe und Gynäkologie
Z Geburtshilfe Neonatol d: Zeitschrift für Geburtshilfe und Neonatologie
Z Geburtshilfe Perinatol d: Zeitschrift für Geburtshilfe und Perinatologie
ZGEIA R: Zhurnal Gigieny, Epidemiologii, Mikrobiologii, i Immunologii
ZGEMA d: Zeitschrift für die Gesamte Experimentelle Medizin
ZGEN e: ZG Energy Corp.
Z Geol Wiss d: Zeitschrift für Geologische Wissenschaften
Z Geomorph[ol] d: Zeitschrift für Geomorphologie
Z Geomorphol Suppl d: Zeitschrift für Geomorphologie. Supplementband
Z Geophys d: Zeitschrift für Geophysik
Z Ger Ling d: Zeitschrift für Germanistische Linguistik
Z Gerontol d: Zeitschrift für Gerontologie
Z Gerontol Geriatr d: Zeitschrift für Gerontologie und Geriatrie
Z Gesamte Brauwes d: Zeitschrift für das Gesamte Brauwesen
Z Gesamte Exp Med e: Zeitschrift für die Gesamte Experimentelle Medizin
Z Gesamte Exp Med Einschl Exp Chir d: Zeitschrift für die Gesamte Experimentelle Medizin. Einschliesslich Experimenteller Chirurgie
Z Gesamte Forstwes e: Zeitschrift für das Gesamte Forstwesen
Z Gesamte Genossenschaftswes[en] d: Zeitschrift für das Gesamte Genossenschaftswesen
Z Gesamte Getreide Mühlen Bäckereiwes d: Zeitschrift für das Gesamte Getreide Mühlen- und Bäckereiwesen
Z Gesamte Getreidewes d: Zeitschrift für das Gesamte Getreidewesen
Z Gesamte Giessereiprax d: Zeitschrift für die Gesamte Giessereipraxis
Z Gesamte Hyg [Ihre] [Grenzgeb] d: Zeitschrift für die Gesamte Hygiene und Ihre Grenzgebiete
Z Gesamte Inn Med [Grenzgeb] d: Zeitschrift für die Gesamte Innere Medizin und Ihre Grenzgebiete
Z Gesamte Inn Med Grenzgeb Klin Pathol Exp d: Zeitschrift für die Gesamte Innere Medizin und Ihre Grenzgebiete. Klinik, Pathologie, Experiment
Z Gesamte Inn Med Ihre Grenzgeb [Suppl] d: Zeitschrift für die Gesamte Innere Medizin und Ihre Grenzgebiete. [Supplementum]
Z Gesamte Kälte Ind d: Zeitschrift für die Gesamte Kälte-Industrie
Z Gesamte Kälte Ind Beih Ser 1 d: Zeitschrift für die Gesamte Kälte-Industrie. Beihefte. Serie 1
Z Gesamte Kälte Ind Beih Ser 2 d: Zeitschrift für die Gesamte Kälte-Industrie. Beihefte. Serie 2
Z Gesamte Kälte-Ind Beih Ser 3 d: Zeitschrift für die Gesamte Kälte-Industrie. Beihefte. Serie 3
Z Gesamte Kreditwesen d: Zeitschrift für das Gesamte Kreditwesen
Z Gesamte Mühlenwes d: Zeitschrift für das Gesamte Mühlenwesen
Z Gesamte Naturwiss d: Zeitschrift für die Gesamte Naturwissenschaft
Z Gesamte Nervenheilkd Psychother d: Zeitschrift für die Gesamte Nervenheilkunde und Psychotherapie
Z Gesamte Neurol Psychiatr d: Zeitschrift für die Gesamte Neurologie und Psychiatrie
Z Gesamte Phys Ther d: Zeitschrift für die Gesamte Physikalische Therapie
Z Gesamte Schiess Sprengstoffw[es] d: Zeitschrift für das Gesamte Schiess-und Sprengstoffwesen mit der Sonderabteilung Gasschutz
Z Gesamte Staatswiss d: Zeitschrift für die Gesamte Staatswissenschaft
Z Gesamte Textilind d: Zeitschrift für die Gesamte Textilindustrie
Z Gesamte Text-Ind d: Zeitschrift für die Gesamte Textil-Industrie
Z Gesamte Versicherungswiss d: Zeitschrift für die Gesamte Versicherungswissenschaft
Z Gesch Erzieh u Unterr d: Zeitschrift für Geschichte der Erziehung und des Unterrichts
Z Geschichtsw d: Zeitschrift für Geschichtswissenschaft
ZGeschJud d: Zeitschrift für die Geschichte der Juden
Z Gesch Oberrhein d: Zeitschrift für die Geschichte des Oberrheins
Z Gesch Saar d: Zeitschrift für die Geschichte der Saargegend
Z Geschv [Mülheim] d: Zeitschrift. Geschichtsverein Mülheim an der Ruhr [Mülheim, West Germany]
Z Gesch-Wiss d: Zeitschrift für Geschichtswissenschaft
Z Ges Exp Med d: Zeitschrift für die Gesamte Experimentelle Medizin
Z Ges Inn Med d: Zeitschrift für die Gesamte Innere Medizin und Ihre Grenzgebiete
Z Ges Staatswiss d: Zeitschrift für die Gesamte Staatswissenschaft
Z Gesundheitstech Städtehyg d: Zeitschrift für Gesundheitstechnik und Städtehygiene
ZGF e: Grand Forks [Canada] [Airport]
ZGG d: Zeitschrift für das Gesamte Genossenschaftswesen
- e: Zero-Gravity Generator
ZGGJT d: Zeitschrift. Gesellschaft für die Geschichte der Juden in der Tschechoslowakei
ZGGYA d: Zeitschrift für Geburtshilfe und Gynäkologie
ZGH e: Zonal Gravity Harmonie
ZGI e: Gods River [Canada] [Airport]
ZGJ d: Zeitschrift für die Geschichte der Juden
ZGJD d: Zeitschrift für die Geschichte der Juden in Deutschland
ZGJT d: Zeitschrift. Gesellschaft für die Geschichte der Juden in der Tschechoslowakei
ZGK d: Zusatzgemeinkosten
ZGL d: Zeitschrift für Germanistische Linguistik
Z Gletsch[er]k[d] Glazialgeol d: Zeitschrift für Gletscherkunde und Glazialgeologie
ZGM e: City University of New York, Graduate School, New York, NY
- e: Zinc Glycinate Marker
ZGMFA P: Zeszyty Naukowe Akademii Gorniczo-Hutniczej [Cracow]. Matematyka, Fizyka, Chemia
ZGMPA d: Zeitschrift für Geomorphologie
ZGO d: Zeitschrift für die Geschichte des Oberrheins
Z Godschmiede Juwelerie Graveure d: Zeitschrift für Goldschmiede Juwelerie und Graveure
ZGOR d: Zeitschrift für die Geschichte des Oberrheins
ZGOrh d: Zeitschrift für die Geschichte des Oberrheins
ZGP e: Zinc Germanium Phosphide
ZGR e: Little Grand Rapids [Canada] [Airport]
ZGS d: Zeitschrift für die Gesamte Staatswissenschaft
- e: Gethsemani [Canada] [Airport]
- e: Zero-Gradient Synchrotron, Argonne
- e: Zero-Gravity Shower
- e: Zone Gradient Synchrotron
ZGSHG d: Zeitschrift. Gesellschaft für Schleswig-Holsteinische Geschichte
ZGSSA d: Zeitschrift für das Gesamte Schieß- und Sprengstoffwesen mit der Sonderabteilung Gasschutz
ZGT-index e: Zircon-Garnet-Tourmaline index
ZGW d: Zeitschrift für Geschichtswissenschaft
ZGWBS e: Zero-Gravity Whole Body Shower
ZGWS e: Zane Greys West Society
ZGZAE6 e: Chinese Journal of Orthopedics
ZH e: Helicopter Air Traffic Control
- e: Zonal Harmonie
- e: Zone Healer
ZHA ch: Zhangjiang [China] [Airport]
Z Hals Nasen Ohrenheilkd d: Zeitschrift für Hals Nasen- und Ohrenheilkunde
Zh Analit Khim R: Zhurnal Analiticheskoj Khimii
Zh Anal Khim R: Zhurnal Analiticheskoj Khimii
Z Haut-Geschlechtskr d: Zeitschrift für Haut- und Geschlechtskrankheiten
Z Hautkr d: Zeitschrift für Hautkrankheiten
ZHB d: Zeitschrift für Hebräische Bibliographie
zHd d: Zu Händen
ZHE e: Zero Headspace Extractor

Zh Eksp [i] Teor Fiz Pis'ma [v] Red R: Zhurnal Eksperimental'noj i Teoreticheskoj Fiziki. Pis'ma v Redaktsiyu
Zh Eksp Biol Med R: Zhurnal Eksperimental'noj Biologii i Meditsiny
Zh Eksp i Teor Fiz R: Zhurnal Eksperimentalnoj i Teoreticheskno j Fiziki
Zh Eksp Klin Med R: Zhurnal Eksperimentalnoi i Klinicheskno j Meditsiny
Zh Eksp Teo[r Fiz] R: Zhurnal Eksperimentalnoj i Teoreticheskoj Fiziki
Zh Eksp Teor Fiz Pis R: Zhurnal Eksperimentalnoj i Teoreticheskno j Fiziki. Pis'ma
Zh Eksp Teor Fiz Pis'ma R: Zhurnal Eksperimental'noi i Teoreticheskno j Fiziki. Pis'ma
Zheleznodorozhn Transp R: Zheleznodorozhnyi Transport
Zhelezn Splavy R: Zheleznye Splavy
Zhelezobeton Konstr Chelyabinsk R: Zhelezobetonnye Konstruktsii Chelyabinsk
Zh Evol Biokhim Fiziol R: Zhurnal Evolyulsionnoi Biokhimii i Fiziologii
- R: Zhurnal Evoliutsionnoi Biokhimii i Fiziologii
ZHF d: Schmalenbachs Zeitschrift für Betriebswirtschaftliche Forschung
- d: Zeitschrift für historische Forschung
Zh Fiz Khem Tsiklu Vseukr Akad Nauk R: Zhurnal Fizichno-Khemichnogo Tsiklu Vseukrains'ka Akademiya Nauk
Zh Fiz Khirn R: Zhurnal Fiziche sknoj Khimii
Zh Geofiz R: Zhurnal Geofiziki
Zh Gig Epidemiol Mikrobiol Immunol R: Zhurnal Gigieny, Epidemiologii, Mikrobiologii, i Immunologii
Zhidkofazn Okislenie Nepredel'nykh Org Soedin e: Zhidkofaznoe Okislenie Nepredel'nykh Organicheskikh Soedinenii
Zhilishchnoe Kommunal'n Khoz R: Zhilishchnoe i Kommunal'noe Khozyaistvo
Z Hist Fors d: Zeitschrift für Historische Forschung
Zhivot Nauki se: Zhivotnovudni Nauki
Zhivotnov'd Nauki se: Zhivotnov'dni Nauki
Zhivotnovod se: Zhivotnovodstvo
Zhivotnovod Vet se: Zhivotnovodstvo i Veterinariya
Zhivotnovud se: Zhivotnovudstvo
Zhivotnovud Nauki se: Zhivotnovudni Nauki
ZHJID ch: Zhongguo Jiguang
Zh Khim Promsti R: Zhurnal Khimicheskno j Promyshlennosti
ZHKPA R: Zhurnal Khimicheskno j Promyshlennosti
ZHL e: Hofstra University, Law School, Library, Hempstead
ZHM e: Hunter College of the City University of New York
Zh Mikrob E R: Zhurnal Mikrobiologii, Epidemiologii, i Immunobiologii
Zh Mikrobiol Epidemiol Immunobiol R: Zhurnal Mikrobiologii, Epidemiologii, i Immunobiologii
Zh Mikrobiol Immunobiol R: Zhurnal Mikrobiologii i Immunobiologii
Zh Nauchnoi i Prikl Fotogr i Kinematogr R: Zhurnal Nauchnoi i Prikladnoi Fotografii i Kinematografii
Zh Nauchn Prikl Fotogr Kinematogr R: Zhurnal Nauchnoi i Prikladnoi Fotografii i Kinematografii
Zh Nauch Prik Foto Kinematog R: Zhurnal Nauchnoi i Prikladnoi Fotografii i Kinematografii
Zh Neorg Kh R: Zhurnal Neorganicheskno j Khimii
Zh Neorg Khim R: Zhurnal Neorganicheskno j Khimii
Zh Nevropatol Psikhiatr R: Zhurnal Nevropatologii i Psikhiatrii Imeni S. S. Korsakova
Zh Nevropatol Psikhiatr Im S S Korsakova R: Zhurnal Nevropatologii i Psikhiatrii Imeni S. S. Korsakova
ZHNID j: Zhongguo Niangzao
Zh NP Fotog R: Zhurnal Nauchnoi i Prikladnoi Fotografji i Kinematografii
Zh Obs Biol R: Zhurnal Obshchei Biologii
Zh Obshch Biol R: Zhurnal Obshchei Biologii
Zh Obshchei Khim R: Zhurnal Obshchei Khimii
Zh Obshch Khim R: Zhurnal Obshchei Khimii
Zh Obs Kh R: Zhurnal Obshchei Khimii
Zh Opytn Agron R: Zhurnal Opytnoi Agronomii
Zh Org Kh R: Zhurnal Organicheskno j Khimii
Zh Org Kkim R: Zhurnal Organicheskno j Khimii
ZHPMA[T] d: Zentralblatt für Bakteriologie, Parasitenkunde, Infektionskrankheiten, und Hygiene. Erste Abteilung. Originale Reihe B. Hygiene, Betriebshygiene, Prävention
Zh Priki Mekhan Tekh Fiz R: Zhurnal Prikladnoi Mekhaniki i Tekhnicheskei Fiziki
Zh Prikl Fiz R: Zhurnal Prikladnoi Fiziki
Zh Prikl Khim R: Zhurnal Prikladnoi Khimii
Zh Prikl Mekh Tekh Fiz R: Zhurnal Prikladnoi Mekhaniki i Tekhnicheskno j Fiziki
Zh Prikl Spektrosk R: Zhurnal Prikladnoi Spektroskopii
ZHR e: Zero Hop Routing
Zh Rezin Promsti R: Zhurnal Rezinovoi Promyshlennosti
Zh Russ Fiz-Khim Ova R: Zhurnal Russkago Fiziko-Khimicheskago Obshchestva
Zh Russ Fiz-Khim Ova Chast Fiz R: Zhurnal Russkogo Fiziko-Khimicheskogo Obshchestva Chast Fizicheskaya
Zh Russ Fiz Khim Ova Chast Khim R: Zhurnal Russkogo Fiziko-Khimicheskogo Ovshchestva Chast Khimicheskaya
Zh Russ Khim Ova R: Zhurnal Russkago Khimicheskago Obshchestva
Zh Russ Khim Ova Fiz Ova R: Zhurnal Russkago Khimicheskago Obshchestva i Fizicheskago Obshchestva
Zh Russ Metall Ova R: Zhurnal Russkogo Metallurgicheskogo Obshchestva
ZHSA e: Zeiss Historica Society of America
Zh Sakh Promsti R: Zhurnal Sakharnoi Promyshlennosti
Zh Strukt Khim R: Zhurnal Strukturnoi Khimii
ZHT d: Zeitschrift für Historische Theologie
ZHUCA c: Zpravy Hornickeho Ustavu CSAV [Ceskoslovenska Akademie Ved]
Zhurnal Mikrobiol R: Zhurnal Mikrobiologii
Zh Ushn Nos Gorl Bolezn R: Zhurnal Ushnykh Nosovykh i Gorlovykh Boleznei
Zh Ushn Nosov Gorloy Bolez R: Zhurnal Ushnykh Nosovykh i Gorlovykh Boleznei
Zhu Us Nos i Gorl Bol R: Zhurnal Ushnykh Nosovykh i Gorlovykh Boleznei
ZHVNAS e: Animal Science
ZHVNS d: Zeitschrift. Historischer Verein für Niedersachsen
Zh Vopr Neirokhir R: Zhurnal Voprosy Neirokhirurgii Imeni N. N. Burdenko
Zh Vopr Neirokhir Im N N Burdenko R: Zhurnal Voprosy Neirokhirurgii Imeni N. N. Burdenko
ZHVS d: Zeitschrift. Historischer Verein für Steiermark
Zh Vses Khi R: Zhurnal Vsesoyuznogo Khimicheskogo Obshchestva Imeni D. I. Mendeleeva
Zh Vses Khim Obshch R: Zhurnal Vsesoyuznogo Khimicheskogo Obshchestva
Zh Vses Khim Ova R: Zhurnal Vsesoyuznogo Khimicheskogo Obshchestva Imeni D. I. Mendeleeva
Zh Vses Khim O-Va Im D I Mendeleeva R: Zhurnal Vsesoyuznogo Khimicheskogo Obshchesiva Imeni D. I. Mendeleeva
Zh Vychisl Mat i Mat Fiz R: Zhurnal Vychislitel'noi Matematiki i Matematicheske Fiziki
Zh Vychisl Mat Mat Fiz R: Zhurnal Vychislitel'noi Matematiki i Matematicheske Fiziki
Zh Vyssh Nerv Deiat Im I P Pavlova R: Zhurnal Vysshei Nervnoi Deiatelnosti Imeni I. P. Pavlova
Zh Vyssh Nerv Deyat R: Zhurnal Vysshei Nervnoi Deyatel'nosti
Zh Vyssh Nervn Deyat Im I P Pavlova R: Zhurnal Vysshei Nervnoi Deyatel'nosti Imeni I. P. Pavlova
Zh Vyss Ner R: Zhurnal Vysshei Nervnoi Deyatel'nosti Imeni I. P. Pavlova
ZHW n: Ziekenhuis
Z Hyg d: Zeitschrift für Hygiene
ZHYGA d: Zeitschrift für die Gesamte Hygiene und Ihre Grenzgebiete
Z Hyg Infektionskr d: Zeitschrift für Hygiene und Infektionskrankheiten

Z Hyg Infekt Kr d: Zeitschrift für Hygiene und Infektionskrankheiten
Z Hyg InfektKrankh d: Zeitschrift für Hygiene und Infektionskrankheiten
Z Hyg Zool Schädlingsbekämpf d: Zeitschrift für Hygienische Zoologie und Schädlingsbekämpfung
ZI d: Zentrale Instandhaltung
- e: Zentralinstitut für Seelische Gesundheit in Mannheim
- d: Zentralinstitute
- e: Flight Information Center
- s: Zona del Interior
Z I d: Zeitschrift für Instrumentenbau
Z/I e: Zoom In
ZIA e: ZIA Airlines
- e: Zone of Interior Armies
ZIAD e: Ziyad, Inc.
ZI-AEDAS e: Zhengzhou Institute of Surveying and Mapping Automatic Elevation Data Acquisition System
ZIALA d: Zeitschrift für Immunitäts- und Allergieforschung
ZIAM d: Zentrum für Industrielle Anwendungen Massiver parallelitaet GmbH
ZIAVA d: Zeitschrift für Induktive Abstammungs- und Vererbungslehre
ZIAX e: Zantop International Airlines, Inc.
ZIB d: Konrad-Zuse-Zentrum für Informationstechnik, Berlin
- d: Zentralinstitut für Bibliothekwesen
- d: Zentrum für Informationstechnik Berlin
ZIC e: Zimbabwe Investment Centre
ZICON e: Zone of the Interior Consumers Network
ZID e: Indianapolis, IN
- e: Zone of Initial Dilution
ZIE e: Zone Immunoelectrophoresis
ZIEGLER e: Ziegler Co., Inc.
ZIF e: Zenix Income Fund
- e: Zero Inser[ta]tion Force
ZIFT e: Zygote Intra-Fallopian Transfer
ZIG e: Z39.50 Implementors Group
- e: Ziguinchor [Senegal] [Airport]
ZIGO e: Zygo Corp.
ZIH e: Hofstra University, Hempstead
- s: Zihuatanejo [Mexico] [Airport]
ZII d: Zeitschrift für Indologie und Iranistik
ZIID d: Zentralinstitut für Information und Dokumentation
ZI Int d: ZI [Ziegelindustrie] International
ZIK e: Zbornik Istorije Knijizevnosti
ZIL e: Zork Implementation Language
ZILA e: Zila, Inc.
Zilla CD e: Zilla Court Decisions, Bengal, Madras, Northwest Provinces
ZIM d: Zentrum für Interaktive Medien e.V., Köln
ZiM P: Ziemia i Morze
ZIM e: Zimmer Corp.
Zimbabwe Agric J e: Zimbabwe Agricultural Journal
Zimbabwe Div Livest Pastures Annu Rep e: Zimbabwe. Division of Livestock and Pastures. Annual Report
Zimbabwe Eng e: Zimbabwe Engineer
Zimbabwe J Agric Res e: Zimbabwe Journal of Agricultural Research
Zimbabwe J Econ e: Zimbabwe Journal of Economics
Zimbabwe Rhod Nurse e: Zimbabwe Rhodesia Nurse
Zimbabwe Rhod Sei News e: Zimbabwe-Rhodesia Science News
Zimbabwe Sci News e: Zimbabwe Science News
Zimbabwe Vet J e: Zimbabwe Veterinary Journal
Zimb Agric J e: Zimbabwe Agricultural Journal
ZIMBANK e: Zimbabwe Banking Corporation Ltd.
Zimb Eng e: Zimbabwe Engineer
Zimb J Agric Res e: Zimbabwe Journal of Agricultural Research
Zimb Law J e: Zimbabwe Law Journal
Zimb Libr e: Zimbabwe Librarian
Zimb Sci News e: Zimbabwe Science News
ZIMEN d: Zentrales Immissionsmeßnetz für Rheinland-Pfalz
ZIMG d: Zeitschrift. Internationale Musik Gesellschaft
Z Immun-Allergie-Forsch d: Zeitschrift für Immunitäts- und Allergieforschung
Z Immun Exp d: Zeitschrift für Immunitätsforschung. Experimentelle und Klinische Immunologie
Z Immunforsch Exp Ther d: Zeitschrift für Immunitätsforschung und Experimentelle Therapie
Z Immunitäts-Allergieforsch d: Zeitschrift für Immunitäts- und Allergieforschung
Z Immunitätsforsch d: Zeitschrift für Immunitätsforschung
Z Immunitätsforsch Allerg Klin Immunol d: Zeitschrift für Immunitätsforschung. Allergie und Klinische Immunologie
Z Immunitätsforsch Exp Klin Immunol d: Zeitschrift für Immunitätsforschung. Experimentelle und Klinische Immunologie
Z Immunitätsforsch Exp Klin Immunol Suppl d: Zeitschrift für Immunitätsforschung. Experimentelle und Klinische Immunologie. Supplemente
Z Immunitätsforsch Exp Ther d: Zeitschrift für Immunitätsforschung und Experimentelle Therapie
Z Immunitätsforsch Exp Ther 1 d: Zeitschrift für Immunitätsforschung und Experimentelle Therapie. 1. Originale
Z Immunitätsforsch Exp Ther 2 d: Zeitschrift für Immunitätsforschung und Experimentelle Therapie. 2. Referate
Z Immunitätsforsch Exp Ther 1 Abt Orig d: Zeitschrift für Immunitätsforschung und Experimentelle Therapie. 1. Abteilung Originale
Z Immunitätsforsch Immunobiol [Suppl] d: Zeitschrift für Immunitätsforschung. Immunobiology. [Supplemente]
Z Immunitätsforsch Suppl d: Zeitschrift für Immunitätsforschung. Supplemente
Zimofa e: Zimbabwe-Mozambique Friendship Association
ZIMordASSR R: Zapiski Naucno-Issledovatel'nogo Instituta pri Sovete Ministrov Mordovskoj ASSR
ZIMR e: ZIM Energy Corp.
Zinat Raksti Rigas Politeh Inst R: Zinatniskie Raksti. Rigas Politehniskais Instituts [lettisch]
Zinc Abstr e: Zinc Abstracts
Zinc/Cadmium Res Dig e: Zinc/Cadmium Research Digest
Zinc Res Dig e: Zinc Research Digest
Z Indukt Abstammungs-Vererbungsl d: Zeitschrift für Induktive Abstammungs- und Vererbungslehre
Z Indukt Abstamm u Vererblehre d: Zeitschrift für Induktive Abstammungs- und Vererbungslehre
Zink e: Zero Income, No Kids
Zinn Ca Tr e: Zinns Select Cases in the Law of Trusts
Zinn Verwend d: Zinn und Seine Verwendung
ZinsAnpG d: Zinsanpassungsgesetz
Z Instrum[entenk][d] d: Zeitschrift für Instrumentenkunde
Z Int Inst Zuckerrübenforsch d: Zeitschrift. Internationales Institut für Zuckerrübenforschung
Z Int Ver Bohring Bohrtech d: Zeitschrift. Internationaler Verein der Bohringenieure und Bohrtechniker
ZIO e: Zinc Iodide-Osmium
ZION e: Zions Bancorporation
Zion e: Zionism
ZIP e: Compressed File
ZIP e: Zig-zag Inline Package
- e: Zig-zag Inline Pin
- e: ZIP Target Marketing
- e: Zone Information Protocol
- e: Zoster Immune Plasma
- p: Zonas de Influência Pedagógica
ZiP P: Za i Przeciw
ZIP code e: Zone Improvement Plan code
ZIPE d: Zentralinstitut Physik der Erde
ZIPP e: National Reference Publishing, Inc.
Zipp e: Zone of Inhibited Phage Plaques
zippers e: Target dawn and dusk combat air patrol
ZIPRA e: Zimbabwe Peoples Revolutionary Army
Ziraat Derg z: Ziraat Dergisi
Ziraat Fak Derg Ege Univ z: Ziraat Fakultesi Dergisi Ege Universitesi
ZIRST f: Zone pour l'Innovation et les Réalisations Scientifiques et Techniques
ZIS d: Zeitschrifteninformationsservice
- e: Zone Information Socket
ZISC e: Zero Instruction Set Computer
ZISCO e: Zimbabwe Steel Co.
Zisin Seismol Soc Jap [J] j: Zisin/ Seismological Society of Japan. Journal
ZISM e: Zeiss Light Section Microscope
- e: Zhengzhou Institute of Surveying and Mapping
ZIS Mitt d: ZIS [Zentralinstitut für Schweißtechnik] Mitteilungen
ZISS e: Zambian Intelligence and Security Service
- e: Zebulun Israel Seafaring Society

ZIS [Zentralinst Schweisstech DDR] Mitt d: ZIS [Zentralinstitut für Schweißtechnik der Deutschen Demokratischen Republik] Mitteilungen
ZIT d: Interdisziplinäres Forschungszentrum für Informatik und Technik
- d: Zentrum für interdisziplinäre Technikforschung
- d: Zeitschrift Interne Revision
- e: Zone Information Table
ZITI e: Sidari Corp.
ZITL e: Zitel Corp.
Ziv A d: Archiv für die Zivilistische Praxis
Ziva c: Ziva. Casopis pro Biologickou Praci
ZIVAN R: Zapiski Instituta Vostokoveden'ia Akademii Nauk SSSR
Zivocisna Vyroba Cesk Akad Zemed Ustav Vedeckotech Inf Zemed c: Zivocisna Vyroba-Ceskoslovenska Akademie Zemedelska. Ustav Vedeckotechnickych Informaci pro Zemedelstvi
Zivoc Vyroba c: Zivocisna Vyroba
Zivotn Prostr c: Zivotne Prostredie
ZIVP c: Zivotne Prostredie
Ziv Pr d: Archiv für die Zivilistische Praxis
Ziz e: Zizit
ZJ e: Zipper Jacket
- h: Zivi Jezici
- P: Zeszyty Jezykoznawcze
Zj s: zanja
Z Jagdwiss d: Zeitschrift für Jagdwissenschaft
ZJARDK e: Zimbabwe Journal of Agricultural Research
ZJC e: State University of New York, Central Administration, Albany, NY
ZJKF c: Zpravy Jednoty Klasickych Filologu
Z Journ d: Zeitschrift für Journalistik
ZJSTD e: Zambia Journal of Science and Technology
ZK d: Schering AG
- d: Zentralkommitee
- d: Zeitschrift für Keilschriftforschung und Verwandte Gebiete
- d: Zeitschrift für Kirchengeschichte
- d: Zeitschrift für Kunstgeschichte
- d: Zeitschrift für Kunstwissenschaft
- e: Barrage balloon
- e: Zachary Kurintner Books Ltd.
- h: Zera' Kodesh
ZKA d: Zeitschrift für Kulturaustausch
- d: Zentraler Kreditausschuß
- d: Zentralkatalog der Ausländischen Literatur
- d: Zollkriminalamt
Z Kardiol [Suppl] d: Zeitschrift für Kardiologie. [Supplementum]
ZKB d: Zimmer, Küche, Bad
- e: Bomber
- e: Kasaba Bay [Zambia] [Airport]
ZKC e: Keuka College, Lightner Library, Keuka Park
ZK Ch d: Zeitschrift für Kinderchirurgie und Grenzgebiete
ZKD d: Zentrum für Kommunikationstechnik und Datenverarbeitung Stuttgart
ZKE e: Kaschechewan [Canada] [Airport]
- e: Zonal Kinetic Energy
ZKG d: Zeitschrift für Kunstgeschichte
- e: Kegaska [Canada] [Airport]
Z Kindch G d: Zeitschrift für Kinderchirurgie und Grenzgebiete
Z Kinde-Jugendpsychiatr d: Zeitschrift für Kinder- und Jugendpsychiatrie
Z Kinderchir Grenzgeb d: Zeitschrift für Kinderchirurgie und Grenzgebiete
Z Kinderheilkd d: Zeitschrift für Kinderheilkunde
Z Kinder Jugenpsychiatr d: Zeitschrift für Kinder- und Jugendpsychiatrie
Z Kinderpsychiatr d: Zeitschrift für Kinderpsychiatrie
Z Kind Jug d: Zeitschrift für Kinder- und Jugendpsychiatrie
Z Kirch G d: Zeitschrift für Kirchengeschichte
ZKJ se: Zbornik za Knjizevnost i Jezik
ZKKOBW e: Cancer Research and Clinical Oncology
ZKKODY e: Gerontology Extension Lectures
ZKLCA d: Zeitschrift für Klinische Chemie
Z Klin Chem d: Zeitschrift für Klinische Chemie und Klinische Biochemie
Z Klin Chemie d: Zeitschrift für Klinische Chemie
Z Klin Chem Klin Biochem d: Zeitschrift für Klinische Chemie und Klinische Biochemie
Z Klin Med d: Zeitschrift für Klinische Medizin
Z Klin Psychol Psychopathol Psychother d: Zeitschrift für Klinische Psychologie, Psychopathologie und Psychotherapie
Z Klin Psychol Psychother d: Zeitschrift für Klinische Psychologie und Psychotherapie
Z Kl M d: Zeitschrift für Klinische Medizin
ZKM d: Zentrum für Kunst und Medientechnologie Karlsruhe
- d: Zeitschrift für die Kunde des Morgenlandes
ZKN d: Zur Kenntnisnahme
Z Köln Zoo d: Zeitschrift des Kölner Zoo
Z Kompr Flüss Gase Pressluft-Ind d: Zeitschrift für Komprimierte und Flüssige Gase sowie für die Pressluft-Industrie
ZK Ph d: Zeitschrift für Keltische Philologie
ZKPVF d: Zeitschrift für Keltische Philologie und Volksforschung
Z Kr d: Zeitschrift für Kristallographie, Kristallgeometrie, Kristallphysik, Kristallchemie
ZKR d: Zeitschrift für das Gesamte Kreditwesen
Z Krankenpfl d: Zeitschrift für Krankenpflege
Z Krebsf Kl d: Zeitschrift für Krebsforschung und Klinische Onkologie
Z Krebsforsch d: Zeitschrift für Krebsforschung
Z Krebsforsch Klin Onkol d: Zeitschrift für Krebsforschung und Klinische Onkologie
Z Kreislaufforsch d: Zeitschrift für Kreislaufforschung
Z Kreisl Forsch d: Zeitschrift für Kreislaufforschung
Z Kr F d: Zeitschrift für Krebsforschung
Z Krist[all] d: Zeitschrift für Kristallographie, Kristallgeometrie, Kristallphysik, Kristallchemie
Z Kristallogr d: Zeitschrift für Kristallographie
Z Kristallogr Kristallgeom Kristallphys Kristallchem d: Zeitschrift für Kristallographie, Kristallgeometrie, Kristallphysik, Kristallchemie
Z Kristallogr Mineral d: Zeitschrift für Kristallographie und Mineralogie
ZKRU d: Zeitschrift für Katholischen Religionsunterricht
ZKRVD se: Zmiesana Komisia o Rybolove vo Vodach Dunaja
ZKTh d: Zeitschrift für Katholische Theologie
ZKUFAK e: Journal of Rural Engineering and Development
ZKuG d: Zeitschrift für Kunstgeschichte
Z Kult G d: Zeitschrift für Kulturgeschichte
Z Kult-Tech Flurberein d: Zeitschrift für Kulturtechnik und Flurbereinigung
Z Kulturaustausch d: Zeitschrift für Kulturaustausch
Z Kulturtech d: Zeitschrift für Kulturtechnik
Z Kulturtech Flurbereinig d: Zeitschrift für Kulturtechnik und Flurbereinigung
Z Kulturtech Landentwicklung d: Zeitschrift für Kulturtechnik und Landentwicklung
Z Kunstg[es][c][h] d: Zeitschrift für Kunstgeschichte
Z Kunstwis d: Zeitschrift für Kunstwissenschaft
ZKZ d: Zentralkatalog der Zeitschriften und Serien
ZL e: Z-Axis of Spacelab
- P: Zycie Literackie
- R: Zavodskaja Laboratorija
- s: Zona de Lanzamiento
Z Laboratoriumsdiagn d: Zeitschrift für Laboratoriumsdiagnostik
ZLAMP f: Zone de passage des Lignes Aériennes Mondiales Principales
Z Landeskult d: Zeitschrift für Landeskultur
Z Landwirsch Versuchswes Dtsch Österr d: Zeitschrift für das Landwirtschaftliche Versuchswesen in Deutsch-Österreich
Z Land[wirt][sch] Vers [u] Unters[uchungsw][es] d: Zeitschrift für Landwirtschaftliches Versuchs- und Untersuchungswesen
ZLARN f: Zone des Lignes Aériennes Régionales et Nationales
Z Lar[yngol] Rhinol Otol [Ihre] [Grenzgeb] d: Zeitschrift für Laryngologie, Rhinologie, Otologie, und Ihre Grenzgebiete

ZLC e: Zinc, Lead and Cadmium Abstracis
Z-LCF-L e: Creep strength and low-cycle fatigue test
ZLDI d: Zentralstelle für Luft-Raumfahrtdokumentation und Information
ZLE e: Zaba Lee Enterprises
- f: Zone de Libre Échange
ZLEA f: Zone de Libre-Échange des Amériques
Z Lebensmit[[t]elunters] [u] [Forsch] d: Zeitschrift für Lebensmittel-Untersuchung und Forschung
Z Lebensm-Technol-Verfahrenstech d: Zeitschrift für Lebensmittel-Technologie und Verfahrenstechnik
Z Lebensm Unters Forsch d: Zeitschrift für Lebensmittel-untersuchung und -forschung
Z Leder Gerberei Chem d: Zeitschrift für Leder- und Gerberei-Chemie
ZLGIA R: Zapiski Leningradskogo Gomogo Instituta
ZLH e: [F.] Zorell. Lexicon Hebraicum
- e: Lincoln Hospital, Bronx, NY
ZLISP e: Zilog List Processor
ZLit P: Zycie Literackie
Z Lit Wiss d: Zeitschrift für Literaturwissenschaft
ZLJ e: Zambia Law Journal
ZLL e: Zero-Length Launch
ZLM e: State University of New York, College at New Paltz, New Paltz, NY
ZLO e: Manzanillo [Mexico] [Airport]
ZLOG e: Zilog Inc.
ZLR e: Zanzibar Law Reports [1919-50]
- e: Zanzibar Protectorate Law Reports [1868-1950]
ZLROA d: Zeitschrift für Laryngologie, Rhinologie, Otologie
ZLS e: Zero Lavel Sparing
ZLSIA R: Zapiski Leningradskogo Sel'skokhozyaistvennogo Instituta
ZLT f: La Tabatiere [Canada] [Airport]
ZLThK d: Zeitschrift für die Gesamte Lutherische Theologie und Kirche
ZLTO e: Zero-Length Takeoff
Z Luft-Weltraumrecht d: Zeitschrift für Luft- und Weltraumrecht
Z-LV e: Z-Axis along Local Vertical
ZLV e: Zero-Length Vector
ZLW d: Zeitschrift für Luftrecht und Weltraum-Rechtsfragen
Z Lymphol d: Zeitschrift für Lymphologie
ZM d: Zu Mir [Internet]
- e: Zahnmedizin
- d: Zeitschrift für Militärgeschichte
- d: Zeitschrift für Missionskunde
- d: Zeitschrift für Mundartforschung
- e: Zone Marker
- e: Zoom/MODEM [ZOOM Telephonics, Inc.]
- P: Zycie i Mysl
- R: Zagraditel' minniy
- s: Zapadores Minadores
- s: Zona de Marcha
- s: Zona Militar
ZM-AF d: Zeitschrift für Mikroskopische-Anatomische Forschung
ZMAF d: Zeitschrift für Mundartforschung
ZMag e: Z Magazine [Zambia]
Z Markt Meinungs und Zukunftsforsch d: Zeitschrift für Marktforschung, Meinungsforschung, und Zukunftsforschung
Z Math Log[ik Grundlag[en] Math] d: Zeitschrift für Mathematische Logik und Grundlagen der Mathematik
Z Mat Phys d: Zeitschrift für Mathematik und Physik
ZMB e: Zinc Mercaptobenzimidazole
ZMBT e: Zinc Mercaptobenzothiazole
ZMBTA e: Zement und Beton
ZMC e: Manhattan College, Library, Bronx
ZMDC e: Zimbabwe Mining Development Corp.
- e: Zinc Dimethyldithiocarbamate
Z Med Chem d: Zeitschrift für Medizinische Chemie
Z Med Isotopenforsch Deren Grenzgeb d: Zeitschrift für Medizinische Isotopenforschung und deren Grenzgebiete
Z Med Lab Diagn d: Zeitschrift für Medizinische Laberatoriumsdiagnostik
Z Med Laboratoriumsdiagn d: Zeitschrift für Medizinische Laboratoriums-diagnostik
Z Med Labortech d: Zeitschrift für Medizinische Labortechnik
Z Med Mikrobiol Immunol d: Zeitschrift für Medizinische Mikrobiologie und Immunologie
Z Med Mikrosk d: Zeitschrift für Medizinische Mikroskopie
ZMEIA R: Zhurnal Mikrobiologii, Epidemiologii, i Immunobiologii
Z Menschl Vererb-Konstitutionsl d: Zeitschrift für Menschliche Vererbungs-und Konstitutionslehre
Z Metallk d: Zeitschrift für Metallkunde
Z Metallkd d: Zeitschrift für Metallkunde
Z Metallkun d: Zeitschrift für Metallkunde
Z Meteorol d: Zeitschrift für Meteorologie
Z Met Schmuckwaren Fabr Verchrom d: Zeitschrift für Metall- und Schmuckwaren. Fabrikation sowie Verchromung
ZMF d: Zeitschrift für Mundartforschung
ZMFWA P: Zeszyty Naukowe. Matematyka, Fizyka, Chemia
ZMG d: Zentralblatt für Mathematik und Ihre Grenzgebiete
ZMH d: Zeitschrift. Museum Hildesheim
- e: Zone Mail Hour
Z Mikrosk Anat Forsch d: Zeitschrift für Mikroskopische-Anatomische Forschung
Z Mikrosk-Anat Forsch [Leipz] d: Zeitschrift für Mikroskopische- Anatomische Forschung [Leipzig]
Z Militärmed d: Zeitschrift für Militärmedizin
Z Miss[- u Relig] [W[iss]] d: Zeitschrift für Missionswissenschaft und Religionswissenschaft
ZMK c: Zpravodaj Mistopisne Komise CSAV [Ceskoslovenske Akademie Ved]
- e: Klinik und Polikliniken für Zahn-, Mund- und Kieferheilkunde
- d: Zeitschrift für Missionskunde und Religionswissenschaft
ZMKBA6 P: Zeszyty Naukowe Uniwersytetu Mikolaja Kopernika w Toruniu. Nauki Matematyczno-Przyrodniczne Biologia
ZML e: Medical Library Center of New York, Standardized Cataloging Service
ZMM d: Zeitschrift für Marktforschung, Meinungsforschung und Zukunftsforschung
- e: State University of New York, Maritime College, Bronx
ZMMPAO e: Zentralblatt für Bakteriologie, Parasitenkunde, Infektionskrankheiten und Hygiene. Erste Abteilung. Originale Reihe A. Medizinische Mikrobiologie und Parasitol
ZMNP R: Zurnal Ministerstva Narodnogo Prosvescenija
Z Morphol Anthropol d: Zeitschrift für Morphologie und Anthropologie
Z Morph[ol] Ökol Tiere d: Zeitschrift für Morphologie und Ökologie der Tiere
Z Morph[ol] T[iere] d: Zeitschrift für Morphologie der Tiere
ZMÖTA d: Zeitschrift für Morphologie und Ökologie der Tiere
ZMP d: Zeitschrift für Musikpädagogik
- R: Zurnal Moskovsknoj Patriarkhii
ZMPHA d: Zeitschrift für Mathematik und Physik
ZMR d: Zeitschrift für Missionswissenschaft und Religionswissenschaft
- e: Zone-Melt[ing] Recrystallization
ZMS c: Zbornik Matice Srpske
ZMT d: Zentralstelle für Medizinaltarife
- d: Zentrum für Multimediale Telekommunikation
- e: Centre for Tropical Marine Ecology
- e: Masset [Canada] [Airport]
- e: Zoom Telephonics
- f: Centre d'écologie tropicale marine
- s: Centro de Ecología Tropical Marina
Z Mth d: Zeitschrift für Musiktheorie
ZMtheorie d: Zeitschrift für Musiktheorie
ZMUC a: Zoologisk Museum, University of Copenhagen
ZMVKA d: Zeitschrift für Menschliche Vererbungs- und Konstitutionslehre
ZMW d: Zeitschrift für Missionswissenschaft
ZMX e: Zemex Corp.
Z Mykol d: Zeitschrift für Mykologie
ZN d: Zeitschrift für Namenforschung
- d: Zeitschrift für Nationalökonomie
- e: Zentrum für Nervenheilkunde
- e: Ziehl-Neelsen
- f: Zone Nord
- P: Zeszyty Naukowe
- P: Zycie Nauki
Zn e: Zinc
- R: Znamya
- s: Zanjadora
- s: Zona
Zn-65 e: Zinc-65

ZNA e: Nanaimo [Canada] Harbour Airport
ZNa P: Zycie Nauki
ZNACD P: Zeszyty Naukowe Akademii Gorniczo-Hutniczej Imienia Stanislawa Staszica. Matematyka, Fizyka, Chemia
ZNAGA P: Zeszyty Naukowe Akademii Gorniczo-Hutniczej w Krakowie. Rozprawy
ZNAGB P: Zeszyty Naukowe Akademii Gorniczo-Hutniczej [Krakow]. Gornictwo
ZNAGD P: Zeszyty Naukowe Akademii Gorniczo-Hutniczej Imienia Stanislawa Staszica. Geologia
ZNAGDF e: Akademia Gorniczo-Hutnicza Imienia Stanislawa Staszica w Krakowie Zeszyty Naukowe Geologia
ZNAHD P: Zeszyty Naukowe Akademii Gorniczo-Hutniczej Imienia Stanislawa Staszica. Elektryfikacja i Mechanizacja Gornictwa i Hutnictwa
Z Nahrungsm Unters Hyg Warenkd d: Zeitschrift für Nahrungsmittel-Untersuchung Hygiene und Warenkunde
ZNAND i: Zootecnica e Nutrizione Animale
Z Nat F d: Zeitschrift für Naturforschung
Z Natforsch d: Zeitschrift für Naturforschung
Z Nationalö[kon][om] d: Zeitschrift für Nationalökonomie
Z Nat-Ökon d: Zeitschrift für Nationalökonomie
Z Naturf[o][rsch] A d: Zeitschrift für Naturforschung. Teil A. Astrophysik, Physik, und Physikalische Chemie
Z Naturf[o][rsch] B [Chem Org Chem] d: Zeitschrift für Naturforschung. Teil B. Anorganische Chemie, Organische Chemie
Z Naturforsch B Anorg Chem Org Chem Biochem Biophys Biol d: Zeitschrift für Naturforschung. Teil B. Anorganische Chemie, Organische Chemie, Biochemie, Biophysik, Biologie
Z Naturforsch [C] d: Zeitschrift für Naturforschung. Section C
Z Naturf[o][rsch] C [Biochem Biophys Biol Virol] d: Zeitschrift für Naturforschung. Teil C. Biochemie, Biophysik, Biologie, Virologie
Z Naturforsch C Biosci d: Zeitschrift für Naturforschung. Teil C. Biosciences
Z Naturforsch Sect B d: Zeitschrift für Naturforschung. Section B. Inorganic Chemistry, Organic Chemistry
Z Naturforsch Sect C Biosci d: Zeitschrift für Naturforschung. Section C. Biosciences
Z Naturforsch Teil A d: Zeitschrift für Naturforschung. Teil A
Z Naturforsch Teil C d: Zeitschrift für Naturforschung. Teil C. Biosciences
Z Naturforsch Teil C Biochem Biophys Biol Virol d: Zeitschrift für Naturforschung. Teil C. Biochemie, Biophysik, Biologie, Virologie
Z Naturheilk d: Zeitschrift für Naturheilkunde
Z Naturwiss-Med Grundlagenforsch d: Zeitschrift für Naturwissenschaftlich-Medizinische Grundlagenforschung
ZNB d: Zusätzliche Nebenbedingungen
ZNBC e: Zambia National Broadcasting Corporation
ZNC e: New York City Technical College, Library, Brooklyn
- e: Zone of Nonproliferating Cells
ZNCAV c: Zdenku Nejedlemu Ceskoslovenska Akademie Ved
ZN Ch R: Zurnal Neorganiceskoj Chimii
ZND d: Zeitnachrichtendienst
- e: Zinder [Niger] [Airport]
ZNE e: Newman [Australia] [Airport]
- f: Zone Nord-Est
ZNEC e: Zambian National Energy Corporation, Ltd.
Z Neorg Chim R: Zurnal Neorganiceskoj Chimii
Z Neurol d: Zeitschrift für Neurologie
Z Neutral d: Zeitschrift Neutralität
Z Neut W d: Zeitschrift für die Neutestamentliche Wissenschaft
Z Neut Wiss d: Zeitschrift für die Neutestamentliche Wissenschaft und die Kunde der Älteren Kirche
ZNF d: Zeitschrift für Namenforschung
ZNFU e: Zimbabwe National Farmers Union
ZNG e: Negginan [Canada] [Airport]
- P: Zeszyty Naukowe Wydzialu Humanistycznego, Wyzsza Szkola Pedagogiczna w Gdansku
ZNGGA P: Zeszyty Naukowe Akademii Gomiczo-Hutniczej [Krakow]. Geologia
ZNGI e: Zero Net Growth Isocline
ZNIC e: Zonic Corp.
ZNIEFF f: Zone Naturelle d'Intérêt Écologique, Faunistique et Floristique
ZNIO P: Zaklad Narodowy Imeni Ossolinskich
ZNK P: Zeszyty Naukowe, Sekcja Jezykoznawcza, Wyzsza Szkola Pedagogiczna w Katowicach
ZNKUL P: Zeszyty Naukowe Katolickiego Uniwersytetu Lubelskiego
ZNLSA P: Zeszyty Naukowe Politechniki Lodzkiej. Chemia Spozywcza
ZNM e: Nioga Library System, Niagara Falls
ZNO d: Zeitschrift für Nationalökonomie
- e: North Country Community College, Saranac Lake
- e: Zenco Resources, Inc.
- f: Zone Nord-Ouest
ZnO e: Zinc Oxide
ZNOE e: Zinc Oxide-Eugenol
ZNotP d: Zeitschrift für die Notarpraxis
ZNP e: Zanzibar Nationalist Party
- e: Zinc Pyrithione
- e: Zion Nuclear Plant
ZnP e: Zinc Protoporphyrin
ZNPEA P: Zeszyty Naukowe Politechniki Lodzkiej. Elektryka
ZNPED P: Zeszyty Naukowe Politechniki Poznanskiej. Elektryka
ZNPIA R: Zhurnal Nevropatologii i Psikhiatrii Imeni S. S. Korsakova
ZNPPD P: Zeszyty Naukowe Politechniki Swietokrzyskiej. Seria P. Problemy Nauk Podstawowych
ZNR d: Zeitschrift für Neuere Rechtsgeschichte
ZNS d: Zeitschrift für Neuere Sprachen
ZnS e: Zinc Sulfide
ZNSCA P: Zeszyty Naukowe Politechniki Slaskiej. Chemia
ZnSe e: Zinc Selenide
ZNSGA P: Zeszyty Naukowe Politechniki Slaskiej. Gornictwo
ZNSL d: Zeitschrift für Neufranzösische Sprache und Literatur
ZNSPK P: Zeszyty Naukowe Wyzszej Szkoly Pedagogicznej [Katowice]
ZNSPO P: Zeszyty Naukowe Wyzszej Szkoly Pedagogicznej [Opole]
ZNSSA P: Zeszyty Naukowe Akademii Gorniczo-Hutniczej [Krakow]. Sozologia i Sozotechnika
ZNT e: Zenith National Insurance Corp.
- e: Ziffnet
ZnTe e: Zinc Telluride
ZNTFA d: Zeitschrift für Naturforschung
ZNTHA P: Zeszyty Naukowe Politechniki Czestochowskiej. Nauki Techniczne. Hutnictwo
ZNTL e: Zehntel, Inc.
Z Ntl W d: Zeitschrift für die Neutestamentliche Wissenschaft
ZNTSL U: Zapysky Naukovoho Tovarystva Imeny Svecenka [Literature Series]
ZNTW d: Zeitschrift für die Neutestamentliche Wissenschaft
ZNTZA P: Zeszyty Naukowe Akademii Rolniczo-Technicznej w Olsztynie. Technologia Zywnosci
ZNU e: Namu [Canada] [Airport]
- d: Zeitschrift für Neusprachlichen Unterricht
ZNUFA P: Zeszyty Naukowe Uniwersytetu Jagiellonskiego. Prace Fizyczne
ZNUG P: Zeszyty Naukowe Uniwersytetu Gdanskiego
ZNUIA P: Zeszyty Naukowe Uniwersytetu Imienia Adama Mickiewicza w Poznaniu. Matematyka, Fizyka, Chemia
ZNUJ P: Zeszyty Naukowe Uniwersytetu Jagiellenskiego
ZNUL P: Zeszyty Naukowe Uniwersytetu Lodzkiego
ZNULHist P: Zeszyty Naukowe Uniwersytetu Lodzkiego. Nauki Humanistyczno-Spoleczne. Historia
ZNUMD P: Zeszyty Naukowe Uniwersytetu Jagiellonskiego. Prace Biologii Molekularnej
ZNUMK P: Zeszyty Naukowe Uniwersytetu Mikolaja Kopernika
ZNUnWr P: Zeszyty Naukowe Uniwersytetu Wroclawskiego
ZNUP P: Zeszyty Naukowe Uniwersytetu Imienia Adama Mickiewicza w Poznaniu
ZNUPHSzt P: Zeszyty Naukowe Uniwersytetu Imienia Adama Mickiewicza w Poznaniu. Historia Sztuki

ZNUT e: Zambia National Union of Teachers
- P: Zeszyty Naukowe Uniwersytetu Mikolaja Kopernika w Toruniu. Nauki Humanistyczne-Spoleczne
ZNUW P: Zeszyty Naukowe Uniwersytetu Wroclawskiego Imienia B. Bieruta
ZNUZA P: Zeszyty Naukowe Uniwersytetu Jagiellonskiego. Prace Zoologiczne
Z.n.V. D: Zeit nach Vereinbarung
ZNW d: Zeitschrift für die Neutestamentliche Wissenschaft
ZNWFA P: Zeszyty Naukowe Wyzszej Szkoly Pedagogicznej w Katowicach. Sekcja Fizyki
ZNWKÄK d: Zeitschrift für die Neutestamentliche Wissenschaft und die Kunde der Älteren Kirche
ZNWKU d: Zeitschrift für die Neutestamentliche Wissenschaft und die Kunde des Urchristentums
ZNWSPK P: Zeszyty Naukowe Wyzszej Szkoly Pedagogicznej [Katowice]
ZNWSPO P: Zeszyty Naukowe Wyzszej Szkoly Pedagogicznej [Opole]
ZNWSPOp P: Zeszyty Naukowe, Jezykoznawstwo, Wyzsza Szkola Pedagogiczna w Opolu
ZNWVA e: Zimbabwe National War Veterans Association
ZNXPO e: Zeus-Nike X Program Office
ZNZ e: Zanzibar [Tanzania] [Airport]
ZNZSA P: Zeszyty Naukowe Akademii Gorniczo-Hutniczej [Cracow]. Zeszyt Specjalny
ZO d: Zeitschrift für Organisation
- d: Zeitschrift für Ortsnamenforschung
- e: Oceanic Air Traffic Control
- e: Z-Axis of Orbiter
- e: Zanzibar Organization
Z/O e: Zoom Out
Z Oberschles Berg-Hüttenmänn Ver Kat d: Zeitschrift des Oberschlesischen Berg- und Hüttenmännischen Vereins zu Katowice
ZOBIDX e: Zoo Biology
Zobozdrav Vestn R: Zobozdravstveni Vestnik
Z Obsc Biol R: Zurnal Obscej Biologii
Z Obsc Chim R: Zurnal Obscej Chimii
ZOC e: Zap-O-Comm
- e: Zone of Convenience
ZOD e: Zero Order Detector
ZOE e: Zinc Oxide-Eugenol
- e: Zone of Exclusion
ZOEG e: Zeitschrift für Osteuropäische Geschichte
ZOEP e: Zoe Products, Inc.
ZOest G d: Zeitschrift für die Österreichischen Gymnasien
ZOF e: Ocean Falls [Canada] [Airport]
ZOFF d: Zeichenorientierte Fenster-Funktionen
Z Öff Chem d: Zeitschrift für Öffentliche Chemie
Z Öff Gem Wirtsch Unterneh d: Zeitschrift für öffentliche und gemeinwirtschaftliche Unternehmen
ZOFI f: Zone Franche d'Inga
ZOfO d: Zeitschrift für Ostforschung
ZOG e: Paramaribo [Suriname] [Airport]
ZOG e: Zionist Occupational Government
ZoG e: Zinc Gluconate
ZOGAAV d: Zoologische Gärten
ZOI e: Zero Of Incorporation
Z Ö IAV d: Zeitschrift des Österreichischen Ingenieur und Architekten Vereins
Z. Ökol. Nat.schutz d: Zeitschrift für Ökologie und Naturschutz
ZOLD e: Zeroth Order Loprithmic Distribution
ZOLGA e: Zoologica
Zolnierz Pol P: Zolnierz Polski
Zolotaya Promst R: Zolotaya Promyshlennost
ZOM f: Zone d'Outre-Mer
ZOMO P: Zmotoryzowane Oddzialy Milicji Obywatelskiej
ZÖMS d: Zeitschrift für die Österreichischen Mittelschulen
ZON d: Zeitschrift für Ortsnamenforschung
- e: Zone Petroleum Corp.
Zonar e: Zonaras
ZON/COM s: Zona de Comunicaciones
ZOND e: [The] Zondervan Corp.
- e: Russian space platform
ZONEF e: Zone Petroleum Corp.
Zoning and Plan L Rep e: Zoning and Planning Law Report
ZOO e: Compressed File
- e: Minnesota Zoological Garden, Apple Valley
ZOOACT e: Zoological Action Committee
Zoo] Biel Mar e: Zoologia e Biologia Marinha
Zoo Biol e: Zoo Biology
ZOOCHEM e: Zoochemistry
Zoogeogr. e: in Zoogeography
Zooiatr Rev Med Vet Prod Pecu s: Zooiatria Revista de Medicina Veterinaria y Produccion Pecuaria
ZOOL e: Zoological
- e: Zoology
zool. a: Zoologisk
Zool Abh [Dres] d: Zoologische Abhandlungen [Dresden]
Zool Afr e: Zoologica Africana
Zool Ann d: Zoologische Annalen
Zool Anzeiger d: Zoologischer Anzeiger
Zool Anz [Leipzig] d: Zoologischer Anzeiger [Leipzig]
Zool Anz Suppl d: Zoologische Anzeiger. Supplement
Zool B e: Zoological Bulletin
Zool Beitr d: Zoologische Beiträge
Zool Ber d: Zoologischer Bericht
Zool Bidr Upps[ala] r: Zoologiska Bidrag fran Uppsala
Zool Bijdr n: Zoologische Bijdragen
Zool Biol Mar [Sao Paulo] [Nova Ser] p: Zoologia e Biologia Marinha [Sao Paulo] [Nova Serie]
Zool Entomol Listy c: Zoologicke a Entomologicke Listy
Zool Gärt [Lpz] d: Zoologische Gärten [Leipzig]
Zool. Gart. N.F. d: Zoologischer Garten N.F. [Neue Folge]
Zool Inst Fac Sci Univ Tokyo Annu Rep e: Zoological Institute. Faculty of Science. University of Tokyo. Annual Report
Zool Jahrb [Abt Allg Zool Physiol Tiere] d: Zoologische Jahrbücher. Abteilung für Allgemeine Zoologie und Physiologie der Tiere
Zool Jahrb Abt Anat Ontog Tiere d: Zoologische Jahrbücher. Abteilung für Anatomie und Ontogenie der Tiere
Zool. Jahrb. Abt. Syst. d: Zoologisches Jahrbuch, Abteilung für Systematik
Zool Jahrb Abt Syst [Jena] d: Zoologische Jahrbücher. Abteilung für Systematik Ökologie und Geographie der Tiere [Jena]
Zool Jahrb Abt Syst Ökol Geogr Tiere d: Zoologische Jahrbücher. Abteilung für Systematik Ökologie und Geographie der Tiere
Zool Jb d: Zoologische Jahrbücher
Zool Jb Abt Allg Zool Physiol Tiere d: Zoologische Jahrbücher. Abteilung für Allgemeine Zoologie und Physiologie der Tiere
Zool Jb Abt Syst Okol Geog Tiere d: Zoologische Jahrbücher. Abteilung für Systematik Ökologie und Geographie der Tiere
Zool Jhrb Abt Allg Zool Physiol Tiere d: Zoologische Jahrbücher. Abteilung für Allgemeine Zoologie und Physiologie der Tiere
Zool J Linn [Soc] e: Zoological Journal. Linnean Society
Zool Listy c: Zoologicke Listy
Zool Mag e: Zoological Magazine
Zool Mag [Tokyo] e: Zoological Magazine [Tokyo]
Zool Meded [Leiden] n: Zoologische Mededelingen [Leiden]
Zool Meded Rijks Mus Nat Hist Leiden n: Zoologische Mededelingen. Rijks Museum van Natuurlijke Historie te Leiden
Zool Muz Raksti Invertebrata se: Zoologijas Muzeja Raksti. Invertebrata
Zool[ogica] Pol l: Zoologica Poloniae
Zoologica Scr e: Zoologica Scripta
Zool Orient l: Zoologica Orientalis
Zool Publ Victoria Univ Weilington e: Zoology Publications. Victoria University of Wellington
Zool Rec e: Zoological Record
Zool Res e: Zoological Research
Zool Revy r: Zoologisk Revy
Zool Sci e: Zoological Science
Zool Sci [Tokyo] e: Zoological Science [Tokyo]
Zool Scr l: Zoologica Scripta
Zool Soc Egypt Bull e: Zoological Society of Egypt. Bulletin
Zool Soc London Pr[oc] e: Zoological Society of London. Proceedings
Zool Verh [Leiden] n: Zoologische Verhandelingen [Leiden]

Zool Verh Rijksmus Nat Hist [Leiden] n: Zoologische Verhandelingen. Rijksmuseum van Natuurlijke Historie [Leiden]
Zool Zentralbl d: Zoologisches Zentralblatt
Zool Z[h] R: Zoologicheskii Zhurnal
Zoonoses Res e: Zoonoses Research
Zoophysiol Ecol e: Zoophysiology and Ecology
Zoo Rec e: Zoological Record
ZOOREO l: Zoologica Orientalis
Zoo Rev Parque Zool Barc s: Zoo Revista del Parque Zoologico de Barcelona
Zootech Experiment Stn Res Bull e: Zootechnical Experiment Station. Research Bulletin
Zootec Nutr Anim p: Zootecnica e Nutrizione Animale
Zootec Vet p: Zootecnica e Veterinaria
Zootec Vet Agric p: Zootecnica. Veterinaria e Agricoltura
Zootec Vita p: Zootecnia e Vita
ZOPA e: Zinc Oxide Producers Association
Z Oper[ations] Res Ser A-B d: Zeitschrift für Operations Research. Serie A. Serie B
Z Oper Res Ser A d: Zeitschrift für Operations Research. Serie A. Theorie
Z Oper Res [Ser] B d: Zeitschrift für Operations Research. Serie B. Praxis
ZOPFAN e: Zone of Peace, Freedom and Neutrality Declaration
ZOR d: Zeitschrift für Operations Research
- e: Zorah Media Corp.
Z Organ d: Zeitschrift für Organisation
ZORPB d: Zeitschrift für Operations Research. Serie B. Praxis
Z Orthop [Ihre] [Grenzgeb] d: Zeitschrift für Orthopadie und Ihre Grenzgebiete
ZOS e: Zapata Corp.
ZOSC l: Zoologica Scripta
Z Österr Entomol Ver d: Zeitschrift. Österreichischer Entomologe-Verein
Z Österr Ing Archit Ver d: Zeitschrift des Österreichischen Ingenieur und Architekten Vereins
Z Österr Ver Gas Wasserfachmännern d: Zeitschrift. Österreichischer Verein von Gas- und Wasserfachmännern
Z Ostforsch d: Zeitschrift für Ostforschung
ZOTS e: Zoom Optical Target Simulator
Zouch Adm e: Zouches Admirally Jurisdiction
ZOV e: Zinc-Oxide Varistor
ZOVBW d: Zeitschrift. Österreichischer Verein für Bibliothekswesen
ZOW e: State University of New York, College at Old Westbury
ZP d: Zeitschrift für Phonetik
- d: Zeitschrift für Politik
- d: Zwischenprüfung
- e: Air Traffic Services Reporting Office
- e: Z-Axis of Payload
- e: Zadok Perspectives
- e: Zep Energy
- f: Zone de Protection
- i: Zona Pellucida
- s: Zona de Peligro
Zp e: Zephaniah
Z/P s: Zona de Peligro
ZPA e: Zero-Period Acceleration
- e: Zero-Point acceleration
ZPAAD d: Zeitschrift für Physik. Sektion A. Atoms and Nuclei
ZPAlV d: Zeitschrift. Deutscher Palästinaverein
ZPapEpigr d: Zeitschrift für Papyrologie und Epigraphik
Z Pap Pappe Zellul Holzst d: Zeitschrift für Papier. Pappe, Zellulose, und Holzstoff
Z Papyrologie Epigraphik d: Zeitschrift für Papyrologie und Epigraphik
Z Parapsych d: Zeitschrift für Parapsychologie und Grenzgebiete der Psychologie
Z Parasit[en][Kd][e] d: Zeitschrift für Parasitenkunde
Z Parlamentsfr[agen] d: Zeitschrift für Parlamentsfragen
ZPBBD d: Zeitschrift für Physik. Sektion B. Condensed Matter and Quanta
ZPC e: Zero Point of Change
ZPCA e: Zugzwang! Postal Chess Association
ZPCAA d: Zeitschrift für Physikalische Chemie. Abteilung A. Chemische Thermodynamik, Kinetik, Elektrochemie, Eigenschaftslehre
ZPCBA d: Zeitschrift für Physikalische Chemie. Abteilung B. Chemie der Elementarprozesse, Aufbau der Materie
ZPCHA d: Zeitschrift für Physiologische Chemie
ZPCLA d: Zeitschrift für Physikalische Chemie [Leipzig]
ZPD e: Zero Path Difference
Zpd s: Zapador
ZPDBA d: Zeitschrift für Pflanzenernährung, Düngung, und Bodenkunde
ZPDP e: Zimbabwe Peoples Democratic Party
Zpds s: Zapadores
ZPE d: Zeitschrift für Papyrologie und Epigraphik
- e: Zere Point Energy
- e: Zeta Phi Eia
Z PflKrankh d: Zeitschrift für Pflanzenkrankheiten, Pflanzenpathologie, und Pflanzenschutz
ZPFL e: Zanzibar and Pemba Federation of Labour
Z Pflanzenernähr Bodenkd d: Zeitschrift für Pflanzenernährung und Bodenkunde
Z Pflanzenernähr Düng d: Zeitschrift für Pflanzenernährung und Düngung
Z Pflanzenernähr Düng Bodenkd d: Zeitschrift für Pflanzenernährung, Düngung, und Bodenkunde
Z Pflanzenkr d: Zeitschrift für Pflanzenkrankheiten
Z. Pflanzenkrankh. Pflanzenschutz d: Zeitschrift für Pflanzenkrankheiten und Pflanzenschutz
Z Pflanzenkr Gallenkd d: Zeitschrift für Pflanzenkrankheiten und Gallenkunde
Z Pflanzenkr Pflanzenpathol Pflanzenschutz [Sonderh] d: Zeitschrift für Pflanzenkrankheiten, Pflanzenpathologie, und Pflanzenschutz. [Sonderheft]
Z Pflanzenkr Pflanzenschutz d: Zeitschrift für Pflanzenkrankheiten und Pflanzenschutz
Z Pflanzenp[hysiol] d: Zeitschrift für Pflanzenphysiologie
Z Pflanzenz[ücht] d: Zeitschrift für Pflanzenzüchtung
Z PflErnähr Bodenk d: Zeitschrift für Pflanzenernährung und Bodenkunde
Z PflErnähr Düng Bodenk d: Zeitschrift für Pflanzenernährung, Düngung, und Bodenkunde
Z PflKrankh PflSchutz d: Zeitschrift für Pflanzenkrankheiten und Pflanzenschutz
Z PflPhysiol d: Zeitschrift für Pflanzenphysiologie
Z Pflzücht d: Zeitschrift für Pflanzenzüchtung
ZPG d: Zentrale Projektgruppe
- e: Airship Group
- e: Zero Population Growth
ZPH e: Zephyrhills
ZPh d: Zeitschrift für Psychologie
ZPHF d: Zeitschrift für Philosophische Forschung
Z Phil Forsch d: Zeitschrift für Philosophische Forschung
ZPhilos F[orsch] d: Zeitschrift für Philosophische Forschung
ZPhon d: Zeitschrift für Phonetik und Allgemeine Sprachwissenschaft
Z Phonetik Sprachwiss Komm Forsch d: Zeitschrift für Phonetik, Sprachwissenschaft, und Kommunikationsforschung
Z Phon Sprachwiss Kommunikationsforsch d: Zeitschrift für Phonetik, Sprachwissenschaft, und Kommunikationsforschung
Z. Photogramm. Fernerkund. d: Zeitschrift für Photogrammetrie und Fernerkundung
Z Phys [A] d: Zeitschrift für Physik. [Sektion A. Atoms and Nuclei]
Z Phys B d: Zeitschrift für Physik. Sektion B. Condensed Matter and Quanta
Z Phys C d: Zeitschrift für Physik. Sektion C. Particles and Fields
Z Phys Chem Abt A d: Zeitschrift für Physikalische Chemie. Abteilung A. Chemische Thermodynamik, Kinetik, Elektrochemie, Eigenschaftslehre
Z Phys Chem Abt B d: Zeitschrift für Physikalische Chemie. Abteilung B. Chemie der Elementarprozesse, Aufbau der Materie

Z Phys Chem Frankf Ausg Neue Folge d: Zeitschrift für Physikalische Chemie. Frankfurter Ausgabe. Neue Folge
Z Phys Chem [Frankfurt/Main] d: Zeitschrift für Physikalische Chemie [Frankfurt/Main]
Z Phys Chemie Stoechiom Verwandschaftsl d: Zeitschrift für Physikalische Chemie, Stoechiometrie, und Verwandschaftslehre
Z Phys Chem [Leipzig] d: Zeitschrift für Physikalische Chemie [Leipzig]
Z Phys Chem Materialforsch d: Zeitschrift für Physikalisch-Chemische Materialforschung
Z Phys Chem Neue Folge d: Zeitschrift für Physikalische Chemie. Neue Folge
Z Phys Chem Neue Fo [Wiesbaden] d: Zeitschrift für Physikalische Chemie. Neue Folge [Wiesbaden]
Z Phys Chem [Wiesbaden] d: Zeitschrift für Physikalische Chemie [Wiesbaden]
Z Phys Ch [F] d: Zeitschrift für Physikalische Chemie [Frankfurt]
Z Phys Ch [L] d: Zeitschrift für Physikalische Chemie [Leipzig]
Z Phys Diät Ther d: Zeitschrift für Physikalische und Diätetische Therapie
Z Physik d: Zeitschrift für Physik
Z Physiol Chem d: Zeitschrift für Physiologische Chemie
Z Physiol Chem Hoppe-Seylers d: Zeitschrift für Physiologische Chemie. Hoppe-Seylers
Z Physiother d: Zeitschrift für Physiotherapie
ZPI e: New York State Psychiatric Institute, Medical Library Center of NewYork
ZPID d: Zentralstelle für Psychologische Information und Dokumentation, Trier
Z Pilzkd d: Zeitschrift für Pilzkunde
ZPJIAK U: Zbirnyk Prat Jewrejskiej Istorychno-Arkheologichnoj Komisji
ZPL e: Zero-Phonon Line
Z Plast Chir d: Zeitschrift für Plastische Chirurgie
ZPM e: State University of New York, College at Purchase
ZPMPA d: Zeitschrift für Psychotherapie und Medizinische Psychologie
ZPN e: Impedance Pneumograph
ZPO d: Zivilprozeßordnung
Z Polit[ik] d: Zeitschrift für Politik
Z POl N F d: Zeitschrift für Politik. Neue Folge
ZPP e: Zinc Protophorphyrin
ZPPA d: Zeitschrift für Wissenschaftliche Photographie, Photophysik, und Photochemie
ZPPAUP f: zone de protection du patrimonie urbain architectural et paysager
ZPPF e: Zero Power Reactor Facility
ZPPR e: Zero Power Plutonium Reactor
ZPPT e: Zero Power Physics Test
ZPR d: Zentrum für Paralleles Rechnen, Universität zu Köln
- e: Zero Power Reactor
- e: Zero Power Resistance

Z/PR s: Zona de Peligro Relativo
Z Präklin Geriatr d: Zeitschrift für Präklinische Geriatrie
Z Präklin Klin Geriatr d: Zeitschrift für Präklinische und Klinische Geriatrie
Z Prakt Anästh d: Zeitschrift für Praktische Anästhesie, Wiederbelebung, und- Intensivtherapie
Z Prakt Geol d: Zeitschrift für Praktische Geologie
Z Präventivmed d: Zeitschrift für Präventivmedizin
Zpravy c: Zpravy pro Cestinare
Zpr Cesk Keram Sklarske Spol c: Zpravy Ceskoslovenske Keramicke a Sklarske Spolecnosti
Z Prikl Meh i Tehn Fiz R: Zurnal Prikladnoi Mehaniki i Tehnicesknoj Fiziki
ZprMK c: Zpravodaj Mistopisne Komise CSAV
ZPS e: [William H.] Zimmer Nuclear Power Station
- e: ZEH Plotting Software
ZPSAN se: Zbornik Radova. Srpske Akademije Nauke
ZPSEA P: Zeszyty Naukowe Politechnika Slaska. Energetyka
ZPSIA d: Zeitschrift für Psychologie und Physiologie der Sinnesorgane
ZPSK d: Zentrale Personal- und Strukturkommission
- d: Zeitschrift für Phonetik, Sprachwissenschaft, und Kommunikationsforschung
ZPSMA d: Zeitschrift für Psychosomatische Medizin
ZPSS e: Zion Probabilistic Safety Study
- P: Z Polskich Studiow Slawistycznych
Z Psych Hyg d: Zeitschrift für Psychische Hygiene
Z Psychol d: Zeitschrift für Psychologie
Z Psycholog d: Zeitschrift für Psychologie
Z Psychol Physiol Sinnesorg d: Zeitschrift für Psychologie und Physiologie der Sinnesorgane
Z Psychol Z Angew Psychol d: Zeitschrift für Psychologie mit Zeitschrift für Angewandte Psychologie
Z Psychol Z Angew Psychol Charakterkd d: Zeitschrift für Psychologie mit Zeitschrift für Angewandte Psychologie und Charakterkunde
Z Psychos M d: Zeitschrift für Psychosomatische Medizin und Psychoanalyse
Z Psychosom Med d: Zeitschrift für Psychosomatische Medizin
Z Psychosom Med Psychoanal d: Zeitschrift für Psychosomatische Medizin und Psychoanalyse
Z Psychother Med Psychol d: Zeitschrift für Psychotherapie und Medizinische Psychologie
Z Psychot M d: Zeitschrift für Psychotherapie und Medizinische Psychologie
Z/PT s: Zona de Peligro Remoto
ZPT-D e: Zuclopenthixol Decanoate
ZPtK d: Zentrale Projekttutorienkommission
ZPV e: Zoomed Port Video
- f: Zone maritime Particulièrement Vulnérable

ZPW e: Poliklinik für Zahnärztliche Prothetik und Werkstoffkunde
ZPWCA P: Zeszyty Naukowe Politechniki Wroclawskiej. Chemia
ZPYFA d: Zeitschrift für Physikalische Chemie. Frankfurter Ausgabe. Neue Folge
ZQ e: Center in charge of a flight information region or an upper flight information region when the message is relevant to an IFR [Instrument Flight Rules]
ZQC e: Queensborough Community College of the City University of New York, Library, Bayside
ZQM e: State University of New York, College at Potsdam
ZQN e: Queenstown [New Zealand] [Airport]
ZQT e: Zero Quantum Transition
ZR d: Zentrum für Radiologie
- e: Area Control Center
- e: Zero Coupon Issue
- e: Zionist Record
- e: Zionist Review
- e: Zone of Responsibility
- e: Zoological Record [Online]
- f: Zone Régionale
- s: Zona de Resistencia
- s: Zona Reticularis
- se: Zadarska Revija
zr d: repräsentative z-Koordinate von Mp, Mp 2, Mp 3, Mp 4
Zr-95 e: Zirconium-95
ZRADA s: Zona Radi[o]activa
ZRC e: Zanzibar Revolutionary Council
- s: Zona de Reacción
ZRDI e: Zionic Research and Development Institute
ZREC e: Zoological Records
Z Rechtsmed d: Zeitschrift für Rechtsmedizin
Z Rechtspolit d: Zeitschrift für Rechtspolitik
Z Reich Geschmackstoffe d: Zeitschrift für Reich- und Geschmackstoffe
Z Rel Gg d: Zeitschrift für Religions- und Geistesgeschichte
Z Rel[ig-] [u] Geistesges[ch] d: Zeitschrift für Religions- und Geistesgeschichte
Z Reproduktionstech d: Zeitschrift für Reproduktionstechnik
ZREV s: Zona de Recepción de Evacuados
ZRFF e: Zero-Risk Fuel Fabrication
ZRFG d: Gesetz zur Förderung des Zonenrandgebiets [Zonenrandförderungsgesetz]
ZRG d: Zeitschrift der Savigny-Stiftung für Rechtsgeschichte
- d: Zeitschrift für Religions- und Geistesgeschichte
ZRGA d: Zeitschrift. Savigny-Stiftung für Rechtsgeschichte. Germanistische Abteilung
ZRGGB d: Zeitschrift für Religions- und Geistesgeschichte. Beihefte
ZRGGS d: Zeitschrift für Religions- und Geistesgeschichte. Sonderhefte
ZRH e: Zurich [Switzerland] [Airport]
Z Rheumaforsch d: Zeitschrift für Rheumaforschung

Z Rheumatol [Suppl] d: Zeitschrift für Rheumatologie. [Supplement]
ZRHMB d: Zeitschrift für Rheumatolgie
ZRI d: Zeitschrift für die Religiösen Interessen des Judentums
- e: Zosen
- in: Serui [Indonesia] [Airport]
ZRIO e: Zimbabwe Rhodesian Information Office
ZRL e: Zero Risk Level
- P: Zagadnienia Rodzajow Literackich
ZRM in: Sarmi [Indonesia] [Airport]
ZRN e: Zurn Industries, Inc.
- s: Zona de Reunión
ZRNI R: Zapiski Russkogo Naucnogo Instituta
ZRO e: Zero Corp.
- e: Zoological Record Online
- e: Zoological Record Outline
ZrO e: Zirconia
Z Roman Ph d: Zeitschrift für Romanische Philologie
Zroshuvane Zemlerob se: Zroshuvane Zemlerobstvo
Zrosh Zemlerob se: Zroshuvane Zemlerobstvo
ZRPBA se: Zbornik Radova. Poljoprivred-nog Fakulteta. Universitet u Beogradu
ZRS e: Russell Sage College
ZRSG e: Zoological Record Search Guide
ZRSNDI e: Zimbabwe Science News
ZRT e: Zero Reaction Tool
ZRTLS n: Zwolse Recks van Taal- en Leiterkundige Studies
ZRU d: Zeitschrift für den Russisch-Unterricht
ZRV e: Zero-Relative Velocity
ZRZ d: Zentrales Rechenzentrum
- S: Zbornik Radova. Svencılıste u Zagrebu
ZS d: Zeitschrift
- d: Zeitschrift für die Gesamte Staatswissenschaft
d: Zeitschrift für Semitistik und Verwandte Gebiete
- d: Zeitschrift für Slawistik
- e: Limit Switch
- e: SARSAT Centre [France]
- e: Union of South Africa
- e: Zero Sum
- f: Zone de Salaire
- f: Zone Suburbaine
- f: Zone Sud
- p: Zoosporangia
- s: Zona de Seguridad
Z/S e: Operational Display System
ZSA e: San Salvador [Bahamas] [Airport]
- e: Southern Tier Library System, Corning
ZSAK d: Zeitschrift für Schweizerische Archäologie und Kunstgeschichte
Zs Allg Erdk d: Zeitschrift für Allgemeine Erdkunde
Zs Anorg Chem d: Zeitschrift für Anorganische Chemie
Z Säugetierkd d: Zeitschrift für Säugetierkunde
ZSav d: Zeitschrift. Savigny-Stiftung für Rechtsgeschichte. Romanistische Abteilung
Z Savigny-Stift e: Rechtsgesch Kanon Abt
ZSavRG d: Zeitschrift. Savigny-Stiftung für Rechtsgeschichte. Romanistische Abteilung
Zs Berg- Hütten- u Salinen-Wesen d: Zeitschrift für das Berg-, Hütten-, und Salinenwesen
ZSC e: Stauffer Chemical Co., Information Services, Dobbs Ferry
- f: Zone Spéciale de Conservation
ZS Ch R: Zurnal Strukturnoj Chimii
Zschft f Ausl u Intl Privatr d: Zeitschrift für Ausländisches und Internationales Privatrecht
Zschft f Vergl Rechtswissenschaft d: Zeitschrift für Vergleichende Rechtswissenschaft
Zschft Luft- u Weltr-Recht d: Zeitschrift für Luftrecht- und Weltraumrechtsfragen
Zschft Rechtsvergl d: Zeitschrift für Rechtsvergleichung
Zschft Savigny-Germ d: Zeitschrift. Savigny-Stiftung für Rechtsgeschichte. Germanistische Abteilung
Zschft Savigny-Kanon d: Zeitschrift. Savigny-Stiftung für Rechtsgeschichte. Kanonistische Abteilung
Zschft Savigny-Rom d: Zeitschrift. Savigny-Stiftung für Rechtsgeschichte. Romanistische Abteilung
Z Sch G d: Zeitschrift für Schweizerische Geschichte
Z Sch Kg d: Zeitschrift für Schweizerische Kirchengeschichte
Z Schles Holst Gesch d: Zeitschrift. Gesellschaft für Schleswig-Holsteinische Geschichte
Z Sch R d: Zeitschrift für Schweizerisches Recht
Z Schw AKg d: Zeitschrift für Schweizerische Archäologie und Kunstgeschichte
Z Schweisstech d: Zeitschrift für Schweisstechnik
Z Schweiz Archäol Kunstgesch d: Zeit-schrift für Schweizerische Archäologie und Kunstgeschichte
Z Schweiz Arch Kunstgesch d: Zeitschrift für Schweizerische Archäologie und Kunstgeschichte
ZSchwG d: Zeitschrift für Schweizerische Geschichte
Z Schw KG d: Zeitschrift für Schweizerische Kirchengeschichte
ZSD s: Zona de Servicios de División
ZSDG d: Zeitschrift für Sudetendeutsche Geschichte
ZSDJ d: Zeitschrift für Statistik und Demographie der Juden
Zs D Morg Des d: Zeitschrift der Deutschen Morgenländischen Gesellschaft
ZSE e: Zimbabwe Stock Exchange
- f: Zone Sud-Est
- s: Zona de Servicios de Ejército
ZSEM d: Zeitschrift für Semitistik und Verwandte Gebiete
Z Semiotik d: Zeitschrift für Semiotik
ZSEV e: Z-Seven Fund
Z Sev-Dvin OIMK R: Zapiski Severo-Dvinskogo Obscestva Izucenija Mestnogo Kraja
ZSF f: Zone Sanitaire Franche
ZSG d: Zeitschrift für Schweizerische Geschichte
Zs Ges Naturw d: Zeitschrift für die Gesamten Naturwissenschaften
Zs Gletscherk d: Zeitschrift für Gletscherkunde
ZSHG d: Zeitschrift der Gesellschaft für Schleswig-Holsteinische Geschichte
ZSI d: Zentralstelle für Sicherheit in der Informationstechnik
- e: Z Solar Inertial
- e: Zytec Systems, Inc.
ZSIL e: Zytec Systems
Z Sinnephysiol d: Zeitschrift für Sinnephysiologie
ZSISA P: Zeszyty Naukowe Politechniki Slaskiej. Inzynieria Sanitarna
ZSJ e: St. John's University Library, Jamaica
- e: Zangri, S. J., Chicago
ZSK d: Zeitschrift für Schweizerische Kirchengeschichte
- se: Ze Skarbca Kultury
ZSKG d: Zeitschrift für schweizerische Kirchengeschichte
ZSKHA d: Zeitschrift für Kinderheilkunde. Referate
Zs Kryst d: Zeitschrift für Krystallographie und Mineralogie
ZSL d: Zeitschrift für Slawistik
- e: Zero Slot LAN
Z Slav Ph[il][ol] d: Zeitschrift für Slavische Philologie
Z Slawistik d: Zeitschrift für Slawistik
Zs Miner [Leonhard] d: Zeitschrift für Mineralogie [Leonhard]
ZSN e: Zurich Sunspot Number
ZSNED7 e: Zimbabwe Science News
ZSNeuOG d: Zivilschutzneuordnungs-gesetz
ZSNMAS l: Acta Rerum Naturalium. Musei Nationalis Slovaci Bratislava
ZSNUA e: Zeitschnft für Neurologie
ZSO f: Zone Sud-Ouest
ZSOB e: Zinc-Silver-Oxide Battery
Z Soz d: Zeitschrift für Sozialpsychologie
Z Sozialpsy d: Zeitschrift für Sozialpsychologie
Z Sozialreform d: Zeitschrift für Sozialreform
Z Soziol[og] d: Zeitschrift für Soziologie
Z Soz Psych[ol] d: Zeitschrift für Sozialpsychologie
Z Soz Ref d: Zeitschrift für Sozialreform
Z Sozreform d: Zeitschrift für Sozialreform
ZS Ph d: Zeitschrift für Slavische Philologie
Z Spiritusind d: Zeitschrift für Spiritusindustrie
Zs Prak G d: Zeitschrift für Praktische Geologie
ZSR d: Zeitschrift der Savigny-Stiftung für Rechtsgeschichte
- d: Zeitschrift für Schweizerisches Recht
- d: Zeitschrift für Sozialreform
ZS-RDS e: Zung Self-Rating Depression Scale

ZSRG d: Zeitschrift der Savigny-Stiftung für Rechtsgeschichte
ZSRK d: Zeitschrift. Savigny-Stiftung für Rechtsgeschichte. Kanonistische Abteilung
ZSRS e: Zung Self-Rating Scale
ZSS d: Zeitschrift der Savigny-Stiftung für Rechtsgeschichte
- d: Zentraler Service Switch
- e: Sassandra [Ivory Coast] [Airport]
- e: Zen Studies Society
ZSS f R d: Zeitschrift der Savigny-Stiftung für Rechtsgeschichte
ZSSGerm d: Zeitschrift. Savigny-Stiftung für Rechtsgeschichte. Germanistische Abteilung
ZSSKanon d: Zeitschrift. Savigny-Stiftung für Rechtsgeschichte. Kanonistische Abteilung
ZSSR d: Zeitschrift der Savigny-Stiftung für Rechtsgeschichte
ZSSRGGerm d: Zeitschrift. Savigny-Stiftung für Rechtsgeschichte. Germanistische Abteilung
ZSSRGKan d: Zeitschrift. Savigny-Stiftung für Rechtsgeschichte. Kanonistische Abteilung
ZSSRGROM d: Zeitschrift. Savigny-Stiftung für Rechtsgeschichte. Romanistische Abteilung
ZSSRO P: Zapsana Spolecnost s Rucenim Omezenym
ZSSROM d: Zeitschrift. Savigny-Stiftung für Rechtsgeschichte. Romanistische Abteilung
ZST d: Zentralabteilung Strahlenschutz
Z Staatsw d: Zeitschrift für die Gesamte Staatswissenschaft
ZST IF d: Zusatzstudium Informatik
Z Stomatol d: Zeitschrift für Stomatologie
Z Str R e: Schweizerische Zeitschrift für Strafrecht
Z Strukturn Him R: Zurnal Strukturnoi Himii. Akademija Nauk SSR. Sibirskoe Otdelenie
Z Str W d: Zeitschrift für die Gesamte Strafrechtswissenschaft
Z St W d: Zeitschrift für die Gesamte Staatswissenschaft
ZSV d: Schweizerische Zeitschrift für Volkswirtschaft und Statistik
Zs Vulkan d: Zeitschrift für Vulkanologie
Z Sw d: Zeitschrift für die Gesamte Staatswissenschaft
ZSW d: Zentrum für Studienberatung und Weiterbildung der Universität
- e: Prince Rupert [Canada] [Airport]
ZSWODE P: Zeszyty Naukowe Szkoly Glownej Gospodarstwa Wiejskiego. Akademii Rolniczej w Warszawie. Ogrodnictwo
ZSWRA P: Zeszyty Naukowe Szkoly Glownej Gospodarstwa Wiejskiego w Warszawie. Rolnictwo
ZSysTh d: Zeitschrift für Systematische Theologie
ZT d: Zeitschrift für Tierpsychologie
- d: Zürcher Taschenbuch
- e: Aerodrome Control Tower
- e: Union of South Africa
- e: Z-Axis of External Tank
- e: Zulu Time
- f: Zone Tempérée
ZTA e: Zeta Tau Alpha
- f: Zone Tempérée Australe
ZTAT e: Zero Turn-Around Time
ZTB d: Zürcher Taschenbuch
- f: Tete A La Baleine [Canada] [Airport]
- f: Zone Tempérée Boréale
Ztbl d: Zentralblatt
ZTC e: Zero-Temperature Coefficient
ZTCL e: Zylec Computers Limited
Z Tech Biol d: Zeitschrift für Technische Biologie
Z Techn Fiz R: Zurnal Techniceskoj Fiziki
Z Tech Phys d: Zeitschrift für Technische Physik
Z Tech Überwach d: Zeitschrift für die Technische Überwachung
Z Tech Univ [Berlin] d: Zeitschrift. Technische Universität [Berlin]
Z Tech Univ [Hannover] d: Zeitschrift. Technische Universität [Hannover]
ZTGAK d: Zeitschrift für Thüringische Geschichte und Altertumskunde
Ztg Gesunde d: Zeitung für Gesunde
ZTH g: Zakinthos [Greece] [Airport]
Z Theol Kir d: Zeitschrift für Theologie und Kirche
Z Th K d: Zeitschrift für Theologie und Kirche
Z Th Kirche d: Zeitschrift für Theologie und Kirche
Z Tierernähr Futtermittelkd d: Zeitschrift für Tierernährung und Futtermittelkunde
Z Tierphysiol [Tiernähr Futtermittelk[d]] d: Zeitschrift für Tierphysiologie, Tierernährung, und Futtermittelkunde
Z Tierpsychol [Beih] d: Zeitschrift für Tierpsychologie. [Beiheft]
Z Tierz Züchtungsbiol d: Zeitschrift für Tierzüchtung und Züchtungsbiologie
ZT-index e: Zircon-Tourmaline index
ZTJWG e: Zeus Target Joint Working Group
ZTK d: Zeitschrift für Theologie und Kirche
ZTL e: Touro Law Library, New York
ZTM e: Mid-York Library System
- e: Zone Temperature Monitor
ZTN e: Histria and Zamir Transnational Net
- e: Zinc Tannate of Naloxone
- f: Zone Territoriale du Nord
ZTO e: Zero-Time Outage
ZTOS P: Zydowskie Towarzystwo Ochrony Sierot
- P: Zydowskie Towarzystwo Opieki Spolecznej
ZTP e: Zero-Temperature Plasma
- P: Zydowskie Towarzystwo Prezeciwgruzliczego
ZTPHA d: Zeitschrift für Technische Physik
ZTPSA d: Zeitschrift für Psychologie
ZTR e: Zweig Total Return Fund Inc.
Ztrbl d: Zentralblatt
ZTR-index e: Zircon-Tourmaline-Rutile index
Z Tropenmed Parasitol d: Zeitschrift für Tropenmedizin und Parasitologie
Z Trop Med d: Zeitschrift für Tropenmedizin und Parasitologie
ZTRX e: Zytrex Corp.
ZTS f: Zone Territoriale du Sud
Ztsch f Angew Psychol d: Zeitschrift für Angewandte Psychologie und Psychologische Forschung
Ztsch f Angew Psychol Sammelforsch d: Zeitschrift für Angewandte Psychologie und Psychologische Sammelforschung
Ztsch Gesch Erzieh u Unterr d: Zeitschrift für Geschichte der Erziehung und des Unterrichts
Ztsch Mikr Fleischschau d: Zeitschrift für Mikroskopische Fleischschau und Populäre Mikroskopie
Ztsch Milltärärzte [Tokyo] d: Zeitschrift für Militärärzte [Tokyo]
Ztschr Ärztli Fortbild d: Zeitschrift für Ärztliche Fortbildung
Ztschr Augenh d: Zeitschrift für Augenheilkunde
Ztschr Fleisch u Milchhyg d: Zeitschrift für Fleisch- und Milchhygiene
Ztschr Genossensch Tierversich d: Zeitschrift für Genossenschaftliche Tierversicherung
Ztschr Gewerbe Hyg d: Zeitschrift für Gewerbe Hygiene
Ztschr Hyg d: Zeitschrift für Hygiene
Ztschr Hyg u Infektionskr d: Zeitschrift für Hygiene und Infektionskrankheiten
Ztschr Immunitätsforsch u Exper Therap d: Zeitschrift für Immunitätsforschung und Experimentelle Therapie
Ztschr Infektionskr Haustiere d: Zeitschrift für Infektionskrankheiten, Parasitäre Krankheiten, und Hygiene der Haustiere
Ztschr Klin Med [Berlin] d: Zeitschrift für Klinische Medizin [Berlin]
Ztschr Krebsforsch d: Zeitschrift für Krebsforschung
Ztschr Morphol u Ökol Tiere d: Zeitschrift für Morphologie und Ökologie der Tiere
Ztschr Ophth d: Zeitschrift für die Ophthalmologie
Ztschr Parasitenk [Berlin] d: Zeitschrift für Parasitenkunde [Berlin]
Ztschr Parasitenk [Jena] d: Zeitschrift für Parasitenkunde [Jena]
Ztschr Physiol Chem d: Zeitschrift für Physiologische Chemie
Ztschr Tokio Med Gesellsch d: Zeitschrift. Tokio Medizinische Gesellschaft
Ztschr Vergleich Physiol d: Zeitschrift für Vergleichende Physiologie
Ztschr Veterinärk d: Zeitschrift für Veterinärkunde
Ztschr Wissensch Mikr d: Zeitschrift für Wissenschaftliche Mikroskopie
Ztschr Wissensch Zool d: Zeitschrift für Wissenschaftliche Zoologie

Z Tuberk d: Zeitschrift für Tuberkulose
Z Tuberk[ulose] Erkr Thoraxogane d: Zeitschrift für Tuberkulose und Erkrankungen der Thoraxorgane
ZTUWA d: Zeitschrift Technische Überwachung
ZTZB d: Zentrale Technik Zentraler Betriebslenkung
ZT Zü d: Zeitschrift für Tierzüchtung und Züchtungsbiologie
ZU e: Union of South Africa
- e: Upper Area Control Center
- f: Zone Urbaine
ZUA e: Agana
- e: Central New York Union List of Serials, Syracuse
ZUB e: Allied Chemical Corp., Library, Solvay
ZuB d: Zürcher Bibelkommentare
ZUBEAQ f: Revue de Medecine des Accidents et des Maladies Professionnelles
Zubolek Pregl P: Zubolekarski Pregled
ZUC e: American Foundation for Management Research, Library, Hamilton, NY
ZU Ch d: Zeitschrift für Urologische Chirurgie
Zucker Beih e: Zucker Beihefte
Zucker Frucht Gemueseverwert d: Zucker- Frucht- und Gemüseverwertung
Zucker Sonderbeil e: Zucker Sonderbeilage
Zuck[er] [u] Süsswar[en] Wirt[sch] d: Zucker- und Süsswaren Wirtschaft
ZUD e: Bristol Laboratories, Library, Syracuse, NY
ZUE e: Carrier Corp.
- e: Zuni Energy
ZUFIAW e: Zambia Union of Financial Institutions and Allied Workers
ZUG e: Community-General Hospital, Staff Library, Syracuse
ZugabeVO d: Verordnung des Reichspräsidenten zum Schutze der Wirtschaft, Erster Teil: Zugabewesen [Zugabeverordnung]
ZugerNjb d: Zuger Neujahrsblatt
ZUH e: Education Opportunity Center of the State University of New York, Syracuse
ZUI e: General Electric Co., Electronics Park Library, Syracuse
ZUJ e: General Electric Co., Information Resources Library, Utica
ZUJCA P: Zeszyty Naukowe Uniwersytetu Jagiellonskiego. Prace Chemiczne
ZUK e: United Stales Veterans Administration, Hospital Library, Syracuse
ZUL e: Agway
Zulu e: Greenwich Mean Time
ZULU [time] e: Local time
ZUM d: Zeitschrift für Urheber und Medienrecht
- e: Churchill Falls [Canada] [Airport]
- e: Supreme Court, Fifth Judicial District, Law Library, Utica
ZUMA d: Zentrum für Umfragen, Methoden und Analysen in Mannheim
ZuMonAwV d: Verordnung zur Einreichung zusammengefaßter Monatsausweise nach dem Gesetz über das Kreditwesen
- d: Zusammengefaßte Monatsausweise-Verordnung
Z Umweltpolit d: Zeitschrift für Umweltpolitik
ZUN e: Saint Josephs Hospital, Health Center Library, Syracuse
ZUNBA R: Zhurnal Ushnykh Nosovykh i Gorlovykh Boleznei
Z Unfallmed Berufskr d: Zeitschrift für Unfallmedizin und Berufskrankheiten
Z Unter[s] Lebensm[ittel] d: Zeitschrift für Untersuchung der Lebensmittel
Z Unters Nahr Genussm Gebrauchsgegenstände d: Zeitschrift für Untersuchung der Nahrungs- und Genussmittel sowie der Gebrauchsgegenstände
Z Unters Nahr-u Genussmittel d: Zeitschrift für Untersuchung der Nahrungs- und Genussmittel
ZUO e: Utica Mutual Insurance Co., Library, New Hartford
ZUP e: Utica/Marcy Psychiatric Center, Ulica Campus Library, Utica
ZuP d: Zellstoff und Papier
ZUP f: Zone a Urbaniser en Priorite
ZUQ e: Saint Luke's Memorial Hospital Center, Medical Library, Utica
ZUR e: Hancock Airbase Library, Hancock Field
- e: Zurfund International Ltd.
Zürcher Beiträge d: Zürcher Beiträge zur Geschichtswissenschaft
d: Zürcher Beiträge zur Rechtswissenschaft
Zur Didak Phys Chem d: Zur Didaktik der Physik und Chemie
Zür. Geogr. Schr. d: Zürcher Geographische Schriften
Z.u.R.n.V. D: Zeit und Raum nach Vereinbarung
Z Urol d: Zeitschrift für Urologie
Z Urol Chir d: Zeitschrift für Urologische Chirurgie
Z Urol Nephrol d: Zeitschrift für Urologie und Nephrologie
Zür Univ Geol Inst-Eidgenoss Tech Hochsch Geol Inst Mitt d: Zürich Universität. Biologisches Institut - Eidgenössische Technische Hochschule. Geologisches Institut. Mitteilungen
Zür. Wald d: Zürcher Wald
ZUS e: Utica/Marcy Psychiatric Center, Marcy Campus Library, Utica
Zuschl d: Zuschlag
ZuständErgG d: Gesetz zur Ergänzung von Zuständigkeiten auf den Gebieten des Bürgerlichen Rechts, des Handelsrechts und des Strafrechts [Zuständigkeitsergänzungsgesetz]
ZUT e: Maria Regina College, Library, Syracuse
ZUU e: Masonic Medical Research Laboratory, Library, Utica
ZUV d: Zentrale Universitätsverwaltung
ZUW e: Munson-Williams-Proctor Institute, Library, Utica
ZUX e: Mohawk Valley Learning Resource Center, Utica Library, Utica
ZUY e: Special Metals Corp., Library, New Hartford
ZUZ e: Crouse-Irving Hospital, School of Nursing, Library, Syracuse
ZV d: Zeitung
- d: Zeitschrift für Versicherung
- d: Zeitschrift für Volkskunde
- s: Zona de Vigilancia
Zv e: Zvezda
Z&V d: Zeiten und Völker
ZVA e: Jervis Public Library
- e: Miandrivazo [Madagascar] [Airport]
- e: Zero order Variable Aperture nonredundant point transmitted
Zvaracsky Sb c: Zvaracsky Sbornik
ZVB e: Syracuse Research Corp., Library, Syracuse
ZVC e: Utica Public Library, Utica
ZVD e: Supreme Court of New York, Library, Syracuse
ZVE d: Zählervergleichseinrichtung
ZVEI d: Zentralverband der Elektrotechnik- und Elektronikindustrie
- e: Zentralverband der Elektrotechnischen Industrie
Z Verbraucherpol d: Zeitschrift für Verbraucherpolitik
Z Verbungsl d: Zeitschrift für Vererbungslehre
Z Ver Dtsch Ing d: Zeitschrift. Verein Deutscher Ingenieure
Z Ver Dtsch Zucker Ind d: Zeitschrift. Verein der Deutschen Zucker-Industrie
Z Ver Dtsch Zucker Ind Allg Teil d: Zeitschrift. Verein der Deutschen Zucker-Industrie. Allgemeiner Teil
Z Ver Dtsch Zucker Ind Tech Teil d: Zeitschrift. Verein der Deutschen Zucker-Industrie. Technischer Teil
Z Vererbungsl d: Zeitschrift für Vererbungslehre
Z Vergl Physiol d: Zeitschrift für Vergleichende Physiologie
Z Ver Hess Gesch Landesk d: Zeitschrift. Verein für Hessische Geschichte und Landeskunde
Z Ver Hessische Gesch d: Zeitschrift. Verein für Hessische Geschichte und Landeskunde
Z Verkehrssicherheit d: Zeitschrift für Verkehrssicherheit
Z Verkehrswiss d: Zeitschrift für Verkehrswissenschaft
Z Vermessungswes d: Zeitschrift für Vermessungswesen
Zverolek Obz c: Zverolekarsky Obzor
Z Vers Kund d: Zeitschrift für Versuchstierkunde
Z Versuchstierkd d: Zeitschrift für Versuchstierkunde
ZVF e: Zero-Velocity Fading
ZVfD d: Zionistische Vereinigung für Deutschland
ZVG d: Gesetz über die Zwangsversteigerung

ZVGAK d: Zeitschrift für Vaterländische Geschichte und Altertumskunde
Z Vgl Physiol d: Zeitschrift für Vergleichende Physiologie
ZVGMS d: Zeitschrift. Deutscher Verein für die Geschichte Mährens und Schlesiens
ZVHFA d: Zeitschrift für Vitamin-, Hormon-, und Fermentforschung
ZVHG d: Zeitschrift des Vereins für Hamburgische Geschichte
ZVHGLK d: Zeitschrift. Verein für Hessische Geschichte und Landeskunde
Z VitamForsch d: Zeitschrift für Vitaminforschung
Z Vitam-Horm- [u] Fermentforsch d: Zeitschrift für Vitamin-, Hormon-, und Fermentforschung
Z Vitaminforsch d: Zeitschrift für Vitaminforschung
ZVKOA R: Zhurnal Vsesoyuznogo Khimicheskogo Obshchestva Imeni D. I. Mendeleeva
ZVKPS d: Zeitschrift. Verein für Kirchengeschichte in der Provinz Sachsen und Anhalt
ZVLGA d: Zeitschrift des Vereins für Lübeckische Geschichte und Altertumskunde
ZVM e: Mohawk Valley Community College, Utica
ZVNDA R: Zhurnal Vysshei Nervnoi Deyatel'nosti Imeni I. P. Pavlova
ZVNL s: Zona de Vigilancia Naval Lejana
ZVNP s: Zona de Vigilancia Naval Próxima
ZVO R: Zapiski Vostochnovo Otdeleniia
Z Volksernähr d: Zeitschrift für Volksernährung
Z Volkskund[e] d: Zeitschrift für Volkskunde
ZVORAO R: Zapiski Vostochnovo Otdeleniia Imperatorskovo Ruskavo Arkheologicheskavo Obshchestva
ZVS d: Zentralstelle zur Vergabe von Studienplätzen, Dortmund
- e: Zero Voltage Switch[ing]
ZVT d: Zeitschrift für Verkehrswissenschaft
ZVTGA[K] d: Zeitschrift. Verein für Thüringische Geschichte und Altertumskunde
ZVUL s: Zona Vulnerable
ZVV d: Zeitschrift. Verein für Volkskunde
ZvV R: Zvezda Vostoka
ZVX e: Zygocactus Virus X
Z Vycisl Mat i Mat Fiz R: Zurnal Vycislitel'noi Malematiki i Matematicesknoj Fiziki
ZVZ e: Zavitz Technology, Inc.
ZV & ZV d: Zeitungsverlag und Zeitschriftenverlag
ZW d: Zeitwende Monatsschrift
- e: Air Wisconsin
- e: Zero Wait
ZWA e: Andapa [Madagascar] [Airport]
Z Wahrscheinlichkeitstheorie und Verw Gebiete d: Zeitschrift für Wahrscheinlichkeitstheorie und Verwandte Gebiete
Z Wahrsch V[erw Gebiete] d: Zeitschrift für Wahrscheinlichkeitstheorie und Verwandte Gebiete
Z Wasser Abwasser Forsch d: Zeitschrift für Wasser- und Abwasserforschung
Z Wasserrecht d: Zeitschrift für Wasserrecht
Z Wasser u Abwasserforsch d: Zeitschrift für Wasser- und Abwasserforschung
Z Wasser Versorg Abwasserkunde d: Zeitschrift für Wasser-Versorgung und Abwasserkunde
ZWBAA d: Zeitschrift für Wissenschaftliche Biologie. Abteilung A
ZWEI e: Zwei Was Eine Initially
Z Weltforstwirtsch d: Zeitschrift für Weltforstwirtschaft
Z Werkstofftech d: Zeitschrift für Werkstofftechnik
Z Werkstofftech J Mater Technol d: Zeitschrift für Werkstofftechnik/ Journal of Materials Technology
ZWF Z Wirtsch Fertigung e: ZWF. Zeitschrift für Wirtschaftliche Fertigung
ZWG d: Sudhoffs Archiv. Zeitschrift für Wissenschaftsgeschichte
ZWIBA d: Zeitschrift für Wissenschaftliche Insektenbiologie
Z Wien Ent[omol] Ges d: Zeitschrift. Wiener Entomologische Gesellschaft
Z Wien Entomol Ver d: Zeitschrift. Wiener Entomologe-Verein
Zwierzeta Lab P: Zwierzeta Laboratoryjne
Z Wirtschaftsgeog[raphie] d: Zeitschrift für Wirtschaftsgeographie
Z Wirtschaftsgruppe Zuckerind d: Zeitschrift. Wirtschaftsgruppe Zuckerindustrie
Z Wirtschaftsgruppe Zuckerind Allg Teil d: Zeitschrift. Wirtschaftsgruppe Zuckerindustrie. Allgemeiner Teil
Z Wirtschaftsgruppe Zuckerind Tech Teil d: Zeitschrift. Wirtschaftsgruppe Zuckerindustrie. Technischer Teil
Z Wirtschaftspol d: Zeitschrift für Wirtschaftspolitik
Z Wirtsch Fertigung d: Zeitschrift für Wirtschaftliche Fertigung
Z Wirtsch -u Soz-Wiss d: Zeitschrift für Wirtschafts- und Sozialwissenschaften
Z Wirt- und Sozialwiss d: Zeitschrift für Wirtschafts- und Sozialwissenschaften
Z Wiss Biol Abt A d: Zeitschrift für Wissenschaftliche Biologie. Abteilung A
Z Wiss InsektBiol d: Zeitschrift für Wissenschaftliche Insektenbiologie
Z Wiss Insektenbiol d: Zeitschrift für Wissenschaftliche Insektenbiologie
Z Wiss Mikrosk d: Zeitschrift für Wissenschaftliche Mikroskopie und für Mikroskopische Technik
Z Wiss Mikrosk Mikrosk Tech d: Zeitschrift für Wissenschaftliche Mikroskopie und für Mikroskopische Technik
Z Wiss Phot d: Zeitschrift für Wissenschaftliche Photographie, Photophysik, und Photochemie
Z Wiss Photogr Photophys Photochem d: Zeitschrift für Wissenschaftliche Photographie, Photophysik, und Photochemie
Z Wiss Zool d: Zeitschrift für Wissenschaftliche Zoologie
Z Wiss Zool Abt A d: Zeitschrift für Wissenschaftliche Zoologie. Abteilung A
ZWL d: Zeitschrift für Württembergische Landesgeschichte
- e: Wollaston Lake [Canada] [Airport]
- e: Zero Wavelength
ZWLG d: Zeitschrift für Württembergische Landesgeschichte
ZWMIA d: Zeitschrift für Wissenschaftliche Mikroskopie und für Mikroskopische Technik
ZWMZDP e: Chinese Journal of Microbiology and Immunology
ZWO n: Zuiver Wentenschappelijk Orderzock
ZWP d: Zeitschrift für Wissenschaftliche Photographie, Photophysik, und Photochemie
ZWPGV d: Zeitschrift. Westpreussischer Geschichtsverein
ZWR d: ZWR. Zahnärztliche Welt, Zahnärztliche Rundschau, Zahnärztliche Reform
ZWS d: Zellweger Syndrome
- d: Zentralwohlfahrtsstelle der Juden in Deutschland
- e: Zero Wait State
- e: Zonal Wind Stress
ZWST d: Zentralwohlfahrtsstelle der Juden in Deutschland
ZWTh d: Zeitschrift für Wissenschaftliche Theologie
ZWU e: Union College
ZWV d: Zentraler Wahlvorstand
ZWW e: Zonal Westerly Wind
ZXC e: City College of New York
ZXCFAR e: Zero Crossing Constant False Alarm Rate
ZXMP e: Zero Transmission Power
ZXX e: Exxon Corp., Information Center, Technical Service Coordinator, NewYork
ZY e: Aerodrome Security Services
- e: Zayre Corp.
Zy e: Zygon
ZYB e: Zionist Year Book
Zycie Weteryn P: Zycie Weterynaryjne
ZYDXDM e: Journal. Zhejiang Medical University
ZYFV e: Zucchini Yellow Fleck Virus
ZYG e: Zygote Resources
Zygote e: Zygote
ZYL e: Sylhet [Bangladesh] [Airport]
Zymot Chem Colloidi i: Zymologica e Chemica dei Colloidi
ZYMV e: Zuochini Yellow Mosaic Virus
ZYU e: New York University New York
ZYWE P: Zeszyty Wroclawskie
Zywienie Czlowieka Metab P: Zywienie Czlowieka i Metabolizm
ZZ d: Datum Position
- d: Zeitschrift für die Wissenschaft des Judentums
- e: Aircraft in Flight
- e: Zinziber

Zz e: Zinszahl
Z-Z e: Roll Axis
ZZA e: Zamak Zinc Alloy
ZZACA d: Zeitschrift für Zellforschung und Mikroskopische Anatomie
Z. Zellforsch. d: Zeitschrift für Zellforschung
Z Zellforsch Mikrosk Anat d: Zeitschrift für Zellforschung und Mikroskopische Anatomie
Z Zellforsch Mikrosk Anat Abt Histochem d: Zeitschrift für Zellforschung und Mikroskopische Anatomie. Abteilung Histochemie
ZZF d: Zentralamt zur Zulassung von Fernmeldeeinrichtungen
z1 ... zn d: Zielfunktionskoeffiziente
Z Zool Syst Evolutionsforsch d: Zeitschrift für Zoologische Systematik und Evolutionsforschung
ZZPInt d: Zeitschrift für Zivilprozeß International
ZZR e: Zigzag Riveting
ZzS c: Zbornik za Slavistiku
Z Zücht A d: Zeitschrift für Züchtung. A
Z Zücht B d: Zeitschrift für Züchtung. B
Z Zuckerind d: Zeitschrift für die Zuckerindustrie
Z Zuckerind Böhm d: Zeitschrift für die Zuckerindustre in Böhmen
Z Zuckerind Böhm Mähren d: Zeitschrift für die Zuckerindustrie in Böhmen-Mähren
Z Zuckerind Cech Repub d: Zeitschrift für die Zuckerindustrie der Tschechoslowakischen Republik
Z Zuckind d: Zeitschrift für die Zuckerindustrie
ZZW e: Zero-Zero Weather
ZZZZ e: Unknown Elements in Formatted Flight Plan